Separations Division

Core Programming Topic at the 2012 AIChE Annual Meeting

Pittsburgh, Pennsylvania, USA
28 October - 2 November 2012

Volume 1 of 2

ISBN: 978-1-62276-740-3

Printed from e-media with permission by:

Curran Associates, Inc.
57 Morehouse Lane
Red Hook, NY 12571

Some format issues inherent in the e-media version may also appear in this print version.

Printed by Curran Associates, Inc. (2013)

For permission requests, please contact AIChE
at the address below.

AIChE
3 Park Avenue
New York, NY 10016-5991

Phone: (203) 702-7660
Fax: (203) 775-5177

www.aiche.org

Additional copies of this publication are available from:

Curran Associates, Inc.
57 Morehouse Lane
Red Hook, NY 12571 USA
Phone: 845-758-0400
Fax: 845-758-2634
Email: curran@proceedings.com
Web: www.proceedings.com

TABLE OF CONTENTS

VOLUME 1

VOLUME 2

(15a) Chemical Grafting of Chromatographic Resins for the Selective Adsorption of Pegylated Proteins

Aguilar Oscar and Hernández-Martínez Agustín, Center of Biotechnology-FEMSA, Tec de Monterrey, Monterrey, Mexico

Chemical grafting of chromatographic resins for the selective adsorption of PEGylated proteins.

Hernandez-Martínez Agustín; Aguilar Oscar*

Centro de Biotecnología-FEMSA. Tecnológico de Monterrey, Campus Monterrey.

Av. Eugenio Garza Sada 2501 Sur, Col. Tecnológico Monterrey, México C.P. 64849

*corresponding author: alex.aguilar@itesm.mx

PEGylation reaction, or covalent attachment of a polyethylene glycol (PEG) molecule to a therapeutic protein is a recent strategy still in development aiming to increase the pharmacologic effect and decrease the degradation rate of the protein during its circulation in the body. One of the main challenges in the production of PEGylated proteins is related with the downstream processing, due to the complex mixture of reaction products, resulting in a population comprising mono- and di-PEGylated species as well as several positional isomers. The resolution and characterization of these positional isomers gains higher relevance due to the fact that in most of the cases only one of the isomers exhibits the desired therapeutic effect.

Recent reports of our group suggest that PEGyilated proteing exhibit a marked prefference for the PEG-rich phase of a PEG/salt aqueous two-phase system (ATPS)[1] suggesting a posible affinity partitioning that could be applied to a PEGylated surface such as the chromatografic resin.

Sepharose 6B was subjected to chemical modification according to the protocol reported by Sundberg and Porath[2] for the covalent attachment of PEG 550, PEG 2000 and PEG 5000. The amount of polymer covalently linked to the resin was determined to be from 138 to 180 mg PEG/g resin. An increase in the static binding capacity in HIC conditions was observed for the three modified resins (3.0, 3.6 and 2.6 mg BSA/mL resin respectively), as compared to the unmodified Sepharose 6B (1.2 mg BSA/mL), suggesting changes in hydrophobicity of the resins due to the PEG molecule.

Dynamic binding capacities were determined for the three modified supports in 1 mL Tricorn columns using a Äkta Prime system (GE Healthcare) under the conditions reported by Cisneros-Ruiz *et al.* (2009)[3] (Figure 1), showing a decrease in DBC fro BSA between the unmodified Sepharose compared to modified resins. Different capacities were observed between the molecular weight of the polymer used for grafting the chromatographic curface, mainly due to the difference in the chain length, having a direct impact on the hydrophobicity of the resin. Additionally a change in the conformation of the polymer attached to the resin creates a different environment for protein adsorption.

The potential application of these modified resins for the isolation and purification of PEGylated products is under study. Preliminaty results indicate the potential use of these resins for the isolation of PEGylated products from the reaction mixture, without the need for additional SEC or IEX previous steps (Figure 2). An optimization of the separation conditions is

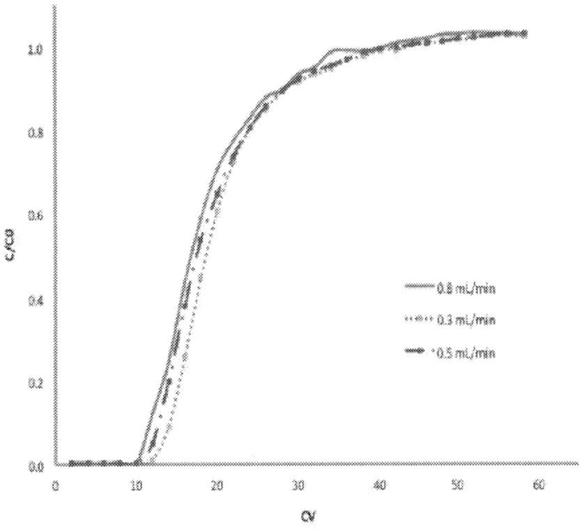

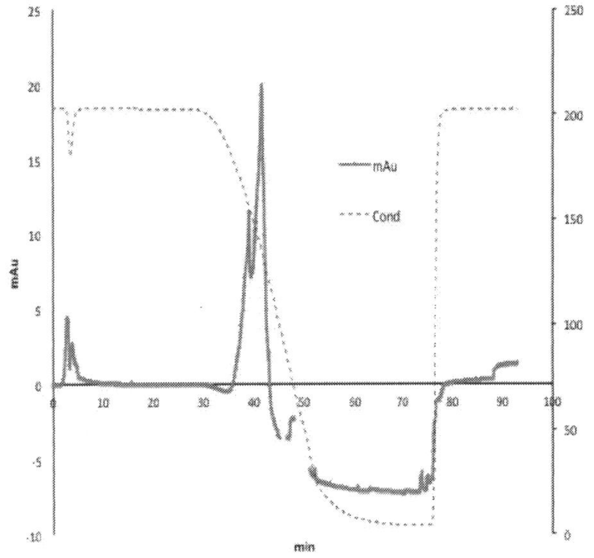

needed, but the potential for peak resolution would open the way for an integrated process for primary extraction and purification of PEGylated proteins from reaction mixtures.

1- González-Valdez J, Cueto L, Benavides J, Rito-Palomares M. Potential application of aqueous two-phase systems for the fractionation of RNase A and a-Lactalbumin from their PEGylated conjugates. *J Chem Technol Biotechnol.* 86 (2011) 26–33.

2- **Sundberg** L, Porath J. Preparation of adsorbents for biospecific affinity chromatography: I. Attachment of group-containing ligands to insoluble polymers by means of bifunctional oxiranes. *J. Chromatogr.* 90 (1974) 87-96

3- Cisneros-Ruiz M, Mayolo-Deloisa K, Przybycien TM, Rito-Palomares, M. Separation of PEGylated from unmodified ribonuclease A using sepharose media. Sep Pur Technol 65 (2009) 105–109

(15b) Control of Intein-Mediated Recombinant Protein Purification in a Chinese Hamster Ovary (CHO) Cell Expression System

Tzu-Chiang Han, Chemical and Biomolecular Engineering, The Ohio State University, COLUMBUS, OH and David W. Wood, Chemical and Biomolecular Engineering, The Ohio State University, Columbus, OH

Control of Intein-mediated Recombinant Protein Purification

in a Chinese Hamster Ovary (CHO) Cell Expression System

Tzu-Chiang Han

Dr. David W. Wood

Inteins are genetic elements that can self-excise from various host proteins in a post-translational protein splicing process. This activity can be modified by simple mutation to yield isolated, controllable cleaving activity at either the N- or C-terminus of the intein. By combining self-cleaving inteins with conventional affinity tags, self-cleaving affinity tags have been developed. Once the tagged target protein is purified, the affinity tag and intein components are removed through the controlled self-cleavage of the intein, in a single step without adding any protease. Although this efficient tag-removal method has been successfully applied in prokaryotic cell systems in our previous studies, premature cleaving of the tag *in vivo* has been a serious drawback for this method in eukaryotic cell systems. Here we present our most recent work in developing a next-generation intein for use in mammalian cells. In particular, we have engineered a new intein with enhanced control, where the control mechanism is compatible with CHO cell culture. This intein has been tested with several target proteins in CHO cell culture, and results of these experiments will be presented. By controlling *in vivo* intein cleavage in a CHO cell culture context, we have generated a potential platform technology for the manufacture of non-antibody glycoprotein therapeutics at large scale. Additionally, this intein has been modified for high-throughput applications in drug and target discovery and development.

(15c) Excipients Removal by Capto S Chromatography and UF/DF Process Development for High-Concentration Drug Substance Production

Gregory Waszak, Purification Process Development, Pfizer Inc, Chesterfield , MO, Larry Determan Jr., Bioprocess R&D, Pfizer Inc., Chesterfield and Joseph Martin Jr., Purification Process Development, Pfizer Inc, Chesterfield

The scope of this work is two-fold: 1) excipients removal from formulated mAb drug substance by Capto S chromatography and 2) UF/DF process development to make high-concentration drug substance (DS) for subcutaneous injection. During process scale-up, the project team decided to make a high-concentration mAb drug substance for subcutaneous injection and change the formulation. Since the pilot plants were not available for large-scale campaigns, a creative alternative was needed to produce 2 kg of antibody from formulated DS for these studies. Fortunately, a sufficient supply of formulated DS was available for reprocessing. A Capto S column was chosen over Protein A chromatography to remove excipients from formulated drug substance because of its higher binding capacity. In addition, Capto S has lower resin costs, takes less time to process, and uses milder elution conditions. Analysis showed that the Capto S run removed the excipients with yields of ≥ 96%. During the UF/DF process, the antibody was initially concentrated to 90 g/L , diafiltered, and concentrated to ≥ 180 g/L, then the retentate was collected. After the retentate was collected, a minimal volume of buffer was added for the UF rinse. Most of the rinse was combined with the retentate to hit the target protein concentration. UF/DF yields of 92% have been obtained. This method enabled > 90 % overall recovery of unformulated DS at ≥ 150 g/L.

(15d) Designing in Process Robustness: Minimizing Proteolysis Through Analytical and Chromatographic Tools

Janelle Konietzko[1], Andrew Englehart[2], Thomas Svab[2], Adam Kristopeit[2], Marc Wenger[3], Michael E. Laska[4] and Aaron R. Goerke[2], (1)Vaccine Process Development, Merck & Co., West Point, PA, (2)Vaccine Process Development, Merck & Co., (3)Bioprocess and Bioanalytical Research, Merck & Co., Inc., West Point, PA, (4)BioPurification Development, Merck & Co., West Point, PA

Host cell protein removal is a primary goal during purification process development for vaccines and therapeutic proteins, however the subset of protease impurities is a particular threat to product yield, quality, and product/process consistency. The presence of such enzymes in the process can lead to proteolytic clipping of the product, resulting in a heterogeneous mixture of degraded product fragments and isoforms. Resolution of the desired product from these closely-related degradates represents a significant separation challenge and can result in project delays and elevated developmental cost (and cost of goods) for vaccines and therapeutics. Due to the complexity and constraints of genetically removing proteases from the host cell system, the purification process is often tasked with solving these problems by minimizing degradation and eliminating the offending protease activity. The exact solution to the problem, however, is highly dependent on the particular host cell, product, and process conditions. In this work we outline a strategic methodology that has been consistently utilized to rapidly identify and minimize protease activity. Examples comprising unique experimental designs and statistics will be used to illustrate chromatographic and filtration development aimed at minimizing residual proteases while maintaining high product recovery. The resolving power of mixed-mode resins combined with tailored buffer/wash matrixes was instrumental in identifying conditions yielding the lower proteolytic activity. Rapid development methods, such as high-throughput screening, were exploited to shorten the development life cycle and map the chromatography design space for optimal removal of protease activity and product degradates. Novel fluorescent linkers enabling a low level of detection and real time analytics were used to detect protease and inform development decisions. By understanding the operating conditions that minimize protease activity, we have been able to design multiple tiers of robustness into several high value biological processes.

(15e) Proteomics Based Multivariate Random Forest Method for Prediction of Protein Separation Behavior During Downstream Purification

Ryan K. Swanson[1], Ruo Xu[2], Dan Nettleton[2] and Charles E. Glatz[1], (1)Chemical and Biological Engineering, Iowa State University, Ames, IA, (2)Statistics, Iowa State University, Ames, IA

Proteomics based multivariate random forest method for prediction of protein separation behavior during downstream purification

Ryan K. Swanson[1], Ruo Xu[2], Dan Nettleton[2] and Charles E. Glatz[1]

[1]Department of Chemical and Biological Engineering, Iowa State University, Ames, IA 50011

[2]Department of Statistics, Iowa State University, Ames, IA 50011

Optimizing the downstream purification process remains a significant challenge when using biological expression hosts for the production of recombinant proteins. An important reason why is due to the lack of knowledge of the host cell proteins' (HCP) separation behavior during the selected purification methods. The work to be presented addresses this issue by modeling the separation behavior of a HCP extract from an arbitrary expression host (corn) during commonly used chromatographic and non-chromatographic methods. Aqueous two phase system (ATPS) partitioning followed by two dimensional electrophoresis (2DE) provided data on the three physicochemical properties most commonly exploited during downstream purification for each HCP; isoelectric point, molecular weight and surface hydrophobicity. A multivariate random forest (MVRF) statistical methodology was then applied to this database of characterized proteins creating an accurate tool for predicting the separation behavior of a mixture of proteins for three commonly selected purification methods; cation exchange chromatography (CEX), hydrophobic interaction chromatography (HIC) and ammonium sulfate precipitation (ASP). The next logical goal will be to link these three methods in series (i.e. ASP followed by HIC followed by CEX) and test the prediction accuracy of the MVRF methodology during this simulated downstream purification process. Also discussed will be the future capabilities of this work including the ability to optimize a downstream purification process without requiring any initial product or time spent screening potential methods in the lab regardless of expression host.

(15f) Virus Purification by Aqueous Two Phase Extraction System

K. Saagar Vijayaragavan, Chemical Engineering, Michigan Technological University, Houghton, MI and Caryn L. Heldt, Department of Chemical Engineering, Michigan Technological University, Houghton, MI

AIChE 2012 Annual Meeting, Pittsburg, PA October 28-November 2, 2012

Division: Bioengineering

Group: Bioseparations and Downstream Processing

Virus Purification by Aqueous Two Phase Extraction System

K. SAAGAR VIJAYARAGAVAN (kskhrupa@mtu.edu) & CARYN L. HELDT (heldt@mtu.edu)

Department of Chemical Engineering, MTU

Rising social concerns over potential viral contamination and infection from blood plasma derivatives in the last few decades have raised the need of biotheraputic safety measures for downstream processes in the biopharmaceutical industry. One strong technique that is often applied to protein purification, but rarely to virus removal is aqueous two phase systems (ATPS). ATPS was first developed in Sweden by Huddleston in 1950's and consists of a liquid-liquid biphasic system widely recognized for separation and purification of biomolecules, like proteins and cell organelles. Due to its ecofriendly, economic, and easy scale up, it is possible to implement this separation system to different biotechnological applications. The partitioning of protein molecules into different aqueous phases is a highly complex phenomenon due to hydrogen bonding, charge interactions, Van der Waal's forces and surface hydrophobicity. Porcine Parvovirus (PPV), a non-enveloped DNA virus of 18-26 nm diameter, is a model virus used to test virus removal. It is a model of the human B19 parvovirus, which has been reported to contaminate biotheraputic products. PPV has been studied and recently validated to be more hydrophobic than typical blood and host cell proteins, by our laboratory. Based on this experimental data, we are proposing ATPS to effectively separate PPV to one of the liquid-liquid immiscible phases due largely on hydrophobicity. In our ATPS experiments we use both a polymer-polymer and polymer-salt system. We are currently focusing our efforts on a polymer-salt system comprising of PEG and three different salts, namely phosphate, sulphate and citrate. We have successfully identified the phase separation based on binodial phase curve lines and are using Alexa fluor-tagged PPV to determine the separation of the virus in individual phases of the biphasic system. We have successfully separated PPV with this system. Some of the factors affecting partition coefficient (K) are pH, neutral salts addition, salt concentration, PEG molecular weight and PEG concentration. The study will provide scientific understanding into viral clearance strategies and improve biotheraputic safety in manufacturing processes.

(15g) The Effect of Single-Use Retentate Recycle Bag Design On Diafiltration Efficiency

David M. Bohonak and Elizabeth Goodrich, Biomanufacturing Science Network, EMD Millipore Corporation, Billerica, MA

Performance of ultrafiltration/diafiltration (UF/DF) systems depends, among other factors, on highly effective mixing in the retentate recycle tank. Retentate mixing requirements are unique as compared to mixing requirements for other bioprocessing unit operations such as buffer preparation, because UF/DF processes must achieve a high level of tank homogeneity and buffer exchange (e.g. > 99.9%) under continuously changing level, protein concentration, and/or buffer concentration conditions. In ultrafiltration processes, poor retentate mixing may result in a higher protein concentration in the feed stream than intended, leading to reduced flux and issues with protein quality or solubility. In diafiltration processes, poor retentate mixing can lead to reduced exchange of buffer components and inability to meet a final excipient specification. Consideration of effective retentate mixing is of particular importance for processes employing single-use flow paths since component and system availability is more limited and less established for rigorous downstream separation capability than more mature stainless steel system designs.

The design of two different styles of commercially available, single-use retentate recycle bags was experimentally investigated, specifically to assess the effect on diafiltration performance. The use of active versus passive agitation, different inlet/outlet configurations, and a range of process parameters such as tank level, recirculation rate, and fluid viscosity were evaluated using a simulated diafiltration process to characterize the effect on solute clearance. A mechanistic model of solute clearance during diafiltration is introduced which explicitly accounts for unmixed regions or deadlegs. The model is used to explain the experimental results and illustrates shortcomings of common practices used to evaluate tank mixing in UF/DF processes.

(15h) Flux Enhancement in Recycling Cellulase From Enzymatic Hydrolyzate of Acid Treated Wheat Straw by Electroultrafiltration

Guoqiang Chen, National Key Laboratory of Biochemical Engineering, Institute of Process Engineering, Chinese Academy of Sciences, Beijing, China and Yinhua Wan, National Key Laboratory of Biochemical Engineering, Institute of Process Engineering, Chinese Academy Of Sciences, Beijing, China

Ultrafiltration has been applied extensively in concentration and separation of biomacromolecules. It also shows great potential in recovering both cellulase and β-glucosidase from lignocellulosic hydrolyzates by one step without addition of any chemical. In addition, ultrafiltration membrane reactor coupled with microfiltration or not, could recycle cellulase while removing end-products continuously during hydrolysis. In spite of these advantages, fouling and concentration polarization, which cause flux decline limiting application of membrane processes significantly, are two severe problems remaining to be resolved. In the present work, to enhance the membrane filtration performance in recycling cellulase from lignocellulosic hydrolyzate by ultrafiltration, an electric field was applied across the membrane. The feasibility of electroultrafiltration (EUF) to recycle cellulase was first determined by measuring the filter paper activity of cellulase after EUF of cellulase solution, and the effect of electric field under different operating conditions (molar concentration of buffer, acid treated wheat straw concentration, current and temperature) on permeate flux during EUF was examined indetail. The results showed that EUF was an effective method to recycle cellulase due to its high cellulase activity recovery and high flux. Under appropriate operating conditions, the fllux for 2% (substrate concentration, w/v) lignocellulosic hydrolyzate increased by a factor of 5.0 when an electric strength of 920 V/m was applied, compared to that without electric field. The improvement of flux was independent on substrate concentration. Therefore, EUF could be a useful approach to recycle cellulase for reducing hydrolysis cost.

(23a) Microstructural Characterization and Dissolution Behavior of Drug/Semicrystalline Polymer Systems On Substrates

Hsin-Yun Hsu, Chemical Engineering, Purdue University, West Lafayette, IN and Michael T. Harris, School of Chemical Engineering, Purdue University, West Lafayette, IN

Solid dispersions have been used to improve the bioavailability of poorly water-soluble drugs. However, controlling the stability of the products can still be challenging. Recently, solidification on certain substrates can result in the control of polymorph and crystallization kinetics. The presence of the substrate surface provides heterogeneous nucleation sites that enables the control and scale-up of solidification processes.

In this study, the Drop on Demand technique is used to dispense a solution of Naproxen (NAP) and polyvinylpyrrolidone (PVP) on Chitosan or HPMC films to investigate the role of the substrate surface on the crystallization of the solid dispersion. XRD curves show that NAP has higher crystallinity when crystallized on Chitosan film compared to the HPMC film. The SEM and SHG images also show that NAP forms larger crystal sizes on the Chitosan film. AFM was used to characterize the films and it shows that the Chitosan film has a rougher surface that may provide more sites for NAP to nucleate. The dissolution experiments of those samples on different substrates were also conducted and the results are correlated to their crystallinity and microstructure on the films.

(23b) Slurry Crystallization of Water-Insoluble Drug Substance – Overcoming Challenges in Solubility and Miscibility Requirements for Solvents and Anti-Solvents

Alexander Chu-Kung, Engineering, Abbott Laboratories, North Chicago, IN and Hsien-Hsin Tung, Global Pharmaceutical Research & Development, Abbott Laboratories, North Chicago, IL

In the crystallization development of water-insoluble drug substances, a common issue encountered is the availability of proper solvents and anti-solvents. Ideally, a proper solvent would possess a sufficient solubility at the beginning to dissolve the drug substance, and then its solubility can be sufficiently reduced by adding a solvent-miscible anti-solvent, to crystallize out the drug substance. The dual requirements of solubility and miscibility for both solvents and anti-solvents can be challenging from time to time.

In our case, we present a unique slurry crystallization approach as one solution to bypass such constraints. In the 1st part of the slurry crystallization process, the drug substance is slurried in the solvent which has only limited drug substance solubility. The slurry is seeded with the drug substance of the right crystal form, which subsequently converts all un-dissolved drug substance to the correct crystal form. By this approach, the solubility constraint to dissolve the entire drug substance is successfully bypassed.

In the 2nd part of the slurry crystallization process, the slurry is charged to an anti-solvent, which has a very low solubility, in order to minimize the loss in the mother liquor. As mentioned, the anti-solvent may not be miscible with the solvent used in the 1st part of the process. To overcome this, the solvent is distilled off during the transfer so that its level is kept at a level below the miscibility boundary of solvent and anti-solvent. As a result, the miscibility requirement of solvent and anti-solvent can be successfully bypassed.

In this presentation, we demonstrated a case of slurry crystallization to illustrate the this features, using ethyl-acetate as solvent and water as anti-solvent.

(23c) A Microfluidic Platform for *In Meso* Crystallization and *In Situ* Crystal X-Ray Diffraction of Membrane Proteins

Daria S. Khvostichenko, Jeremy M. Schieferstein, Ashtamurthy Pawate and Paul J.A. Kenis, Department of Chemical & Biomolecular Engineering, University of Illinois at Urbana-Champaign, Urbana, IL

Membrane proteins are key actors in a number of biological processes, which is highlighted by the fact that they account for ~60% of all available drug targets. To better understand functional mechanisms, high-resolution crystal structures of proteins are required. The success rate of membrane protein crystallization trials, however, is notoriously low, and growing diffraction-quality crystals of membrane proteins is a major bottleneck in structural biology.

In meso crystallization exploits properties of certain lipids that self-assemble into continuous bilayer systems, known as lipidic mesophases, when mixed with water or aqueous solutions. During *in meso* crystallization trials membrane protein molecules are incorporated into the bilayers of the mesophase and are thus maintained in a native-like environment. In spite of a number of recent impressive successes, the *in meso* method is not used widely in the structural biology community, presumably due to extra steps required in the crystallization protocol compared to traditional crystallization from solutions, and due to the difficulties of handling highly viscous lipidic mesophases. Similarly, harvesting fragile protein crystals from gel-like lipidic mesophases as necessary for crystal X-ray diffraction data collection is challenging and may compromise the quality of crystals.

We present a microfluidic platform for membrane protein crystallization in lipidic mesophases and *in situ* X-ray analysis. The multi-well platform requires 60 nL of the protein solution and 25 nL of lipid per well, with a potential for further reduction in the sample amount, and automates simultaneous formulation of multiple samples. The platform design is optimized for diffusional mixing of the protein solution with the lipid to formulate protein-loaded mesophases. For fluid metering and routing the platform relies on the multilayer polydimethylsiloxane (PDMS) architecture with monolithic pneumatic valves. For X-ray transparency thin layers of PDMS patterned with features for metering and routing are sandwiched between layers of cyclic olefin copolymer (COC), which provides rigidity and servers as an evaporation barrier. We validate our approach by crystallizing membrane protein photosynthetic reaction center (*R. Sphaeroides*) and collecting protein crystal X-ray diffraction data on-chip.

(23d) Modeling and Simulation of Coaxial Crystallizers by Dynamically Coupled Population Balance, Macromixing, and Micromixing Models

J. Carl Pirkle Jr. [1], Lucas Foguth [1], Steven J. Brenek [2], Kevin Girard [2] and Richard D. Braatz [3], (1)Massachusetts Institute of Technology, Cambridge, MA, (2)Chemical R&D, Pfizer, Inc, Groton, CT, (3)Department of Chemical Engineering, Massachusetts Institute of Technology, Cambridge, MA

A leading method for the crystallization of pharmaceutical compounds is to rapidly mix an antisolvent with a solvent saturated with the desired drug. Many crystallizer designs have been explored to generate high supersaturation in such mixtures as an approach for generating consistent crystal nuclei that are subsequently grown to a desired size [1-5]. Compared to cross-flow mixers, coaxial jet mixers have negligible buildup of crystalline material on their surfaces and are less likely to plug. Coaxial mixers can be designed to deliver rapid turbulent mixing using short sections of pipe. As the energy required for mixing is provided by the inlet streams, with no moving metal parts and no bearings, these devices have simple maintenance and operation. Some experimental and modeling studies of coaxial crystallizers have been published to gain deeper understanding and to facilitate more efficient development and optimization of the coaxial mixer crystallization process [6-9]. Important design parameters for crystallization in coaxial mixers are the length of pipe downstream of the injection point, the velocity and temperature of the inlet streams, and the inner and outer pipe diameters.

This presentation describes an effort whose goal is to speed up the design of the coaxial crystallizers to tailor the crystal size distribution according to the bioavailability and drug administration requirements. Dynamic simulations of a confirmed coaxial crystallizer were carried out that simultaneously solve partial differential equations for macromixing, micromixing, and a population balance for the crystals. The computational model [10-11] was used, which replaced a quadrature-method-of-moments model used to simulate the time evolution of the particle size distribution by Rodney Fox [12] with a full spatially varying population balance model implemented using a high resolution finite-volume method. This presentation employs an extension of the model [10-11] to include the temperature effects on the crystallization. Our simulations were used to perform a parameter sensitivity analysis (see [13] for background on such analyses) to identify the key model parameters and to simulate variations on their values on the full crystal size distribution (CSD) in the antisolvent crystallization of Lovastatin, using kinetics reported in the literature [14]. The effects of inlet concentrations and stream flow rates on CSD were numerically investigated and compared with CSDs obtained in a dual-impinging jet crystallizer [11]. As observed in simulations of dual impinging jets, the mean crystal size and the width of the distribution were found to decrease with an increase in inlet stream velocity. The simulation results showed different degrees of inhomogeneity in the supersaturation and the nucleation and growth rates for different inlet stream flow rates.

To the authors' knowledge, this is the most detailed simulation study on coaxial crystallizers reported to date. The simulation results show the feasibility of tailoring a specific crystal size distribution by adjusting the operating conditions (such as inlet stream velocities) of the coaxial crystallizer. Such computational tools can provide a technological advancement for the process development in the pharmaceutical industry by providing a more in-depth understanding of the crystallization process, and by reducing the number of experiments required to determine the optimum operating conditions, to reduce the quantify of active pharmaceutical ingredient (API) needed for the experiments. Consequently, the crystallizer process design can be performed much earlier during the drug development process, where a limited quantity of API is available.

References

1. Midler, M., E.L. Paul, E.F. Whittington, M. Futran, P.D. Liu, J. Hsu, and S.-H. Pan, U.S. Patent 5,314,506, 1994.
2. Dauer, R., J.E. Mokrauer, and W.J. McKeel, U.S. Patent 5,578,279, 1996.

3. Lindrud, M.D., S. Kim, and C. Wei, U.S. Patent 6,302,958, 2001.
4. am Ende, D.J., T.C. Crawford, and N.P. Weston, U.S. Patent 6,558,435, 2003.
5. Wang, X., J.M. Gillian, and D.J. Kirwan, Quasi-emulsion precipitation of pharmaceuticals. 1. Conditions for formation and crystal nucleation and growth behavior, Crystal Growth & Design, 2006, 6, 2214-2227.
6. Wei, H., and J. Garside, Application of CFD modelling to precipitation systems, Chem. Eng. Res. Des., 1995, 75(2), 219-227.
7. Baldyga, J., and W. Orciuch, Barium sulphate precipitation in a pipe: An experimental study and CFD modeling, Chem. Eng. Sci., 2001, 56, 2435-2444.
8. Marchisio, D.L., A.A. Barresi, and M. Garbero, Nucleation, growth, and agglomeration in barium sulfate turbulent precipitation. AIChE J., 2001, 48, 2039-2050.
9. Öncül, A.A., K. Sundmacher, A. Seidel-Morgenstern, and D. Thévenin, Numerical and analytical investigation of barium sulphate crystallization, Chem. Eng. Sci., 2006, 61, 652-664.
10. Woo, X.Y., R.B.H. Tan, P.S. Chow, and R.D. Braatz, Simulation of mixing effects in antisolvent crystallization using a coupled CFD-PDF-PBE approach, Crystal Growth & Design, 2006, 6,1291-1303.OWTH
11. Woo, X.Y., R.B.H. Tan, and R.D. Braatz, Modeling and computational fluid dynamics-population balance equation-micromixing simulation of impinging jet crystallizers, Crystal Growth & Design, 2009, 9(1), 156-164.
12. Fox, R.O., Computational Models for Turbulent Reactive Flows, 2003, Cambridge: Cambridge University Press.
13. Varma, A., M. Morbidelli, and H. Wu, Parametric Sensitivity in Chemical Systems, 1999, Cambridge: Cambridge University Press.
14. Mahajan, A.J., and D.J. Kirwan, Nucleation and growth-kinetics of biochemicals measured at high supersaturations. J. of Crystal Growth, 1994, 144, 281-290.

(23e) Impact of a Reversible Solid State Form Change On Particle Morphology and Bulk Powder Properties of a Pharmaceutical Compound

Joshua D. Engstrom[1], Qi Gao[2], Daniel Roberts[3], Chenkou Wei[2], Chiajen Lai[4] and Jonathan Brown[1], (1)Drug Product Science and Technology, Bristol-Myers Squibb Company, New Brunswick, NJ, (2)Drug Product Science & Technology, Bristol-Myers Squibb Co., New Brunswick, NJ, (3)Drug Product Science & Technology, Bristol-Myers Squibb Co., New Brunswick, NJ, (4)Drug Product Science and Technology, Research and Development, Bristol-Myers Squibb Company, New Brunswick, NJ

For efficient formulation processing of a pharmaceutical drug product, it is desirable to have an active pharmaceutical ingredient (API) that handles well in terms of flow and bulk density, especially for products with high drug loadings. To improve bulk API powder property behavior, particle engineering techniques can be employed during the crystallization step to alter the morphology of the API crystals from needles, which normally have poor flowing powders, to bricks or rods. Although particle engineering can have a significant impact on particle morphology, the susceptibility of the API to crystalline form change from environmental factors such as relative humidity must also be carefully studied and controlled.

A case study is presented for Compound A whose thermodynamically desired crystalline form has an undesired morphology but can be modified to bricks or rods through a particle engineering approach. Compound A is known to crystallize as iso-structural solvates (Form I) in all solvent systems that have been studied. Upon isolation and drying, Form I desolvates to Form II which can reversibly convert to Form III in the solid state at a critical relative humidity. In this presentation, the crystal structures of Forms I, II and III will be discussed and used to explain how the form changes impact the integrity and morphology of the API crystal as well as the bulk powder properties. The importance of this work is to gain a fundamental understanding of how the API crystal morphology improvement made during crystallization can be retained in the post crystallization steps of drying, storage and handling. The thorough understanding of the crystal structures can provide valuable insights for handling bulk powders to maintain desired API powder properties for drug product processing.

(23f) The Morphological-Population Balance Model (M-PBM) Generator. Application to Additives Controlled Crystallization of KAP

Meenesh R. Singh and Doraiswami Ramkrishna, School of Chemical Engineering, Purdue University, West Lafayette, IN

This talk will present a comprehensive framework to develop M-PBM for crystal morphology distributions evolving in a supersaturated solution. The model considers faces on seed crystals and energetically favorable F-faces to assess the polyhedral symmetry and identify internal coordinates. A set of all possible morphologies is generated using a combinatorial approach. These morphologies are further decomposed to yield polyhedral cones representing morphology domains in crystal state space. A bidirected graph is constructed where the nodes and edges correspond to morphology domains and their transitions, respectively. The population balance equations are identified for each node accounting for morphology transformations in dynamic environments. The solution of M-PBM can be readily obtained by the method of characteristics which takes considerably less computation time as compared to other numerical methods. The application of M-PBM to additives controlled crystallization of Potassium Acid Phthalate (KAP) is considered. The optimal profiles of additives concentration are computed for maximizing purity, productivity and sphericity of KAP crystals.

(28a) Metal-Organic Framework (MOF) Materials and Membranes for Molecular Separations

Sankar Nair, Chemical & Biomolecular Engineering, Georgia Institute of Technology, Atlanta, GA

This talk will describe our recent progress in the synthesis and processing of MOF/ZIF materials for molecular separation applications, and their subsequent use in the fabrication of membrane/thin film devices for such uses. New separations technology is key to increased energy efficiency in the chemical sector, as well as the efficient production of clean and renewable fuels. It is hypothesized that MOF materials may find successful applications in this area. Towards this goal, we will first discuss approaches to controlling the pore structure, particle size, and morphology of small-pore MOF materials such as Cu-hfipbb and ZIFs-7/8/90 via mixed-linker synthesis, low-temperature sonochemical synthesis, and the non-solvent-induced-crystallization (NSIC) approach. We then discuss the processing and use of these MOF materials in the fabrication of membranes for separations such as CO_2/CH_4 and organic/water mixtures. Experimental and computational developments that aid the efficient screening, selection, and evaluation of suitable MOF materials will be discussed. Some key scientific and engineering questions whose resolution may enable the successful application of MOF materials in energy-related separations, will be discussed.

(28b) The Mixed Matrix Membrane Story

Santi Kulprathipanja, UOP, LLC, Des Plaines, IL

The talk will highlight the evolution of Mixed Matrix Membranes (MMM), developed in the mid-1980s. The talk will first provide an introduction into the strategic thinking that led to the development of the water stable adsorbent ADS-200. ADS-200 is a cellulose acetate bound zeolite used in the commercial UOP Sorbex process for adsorptive fructose separation. This concept of combining a polymer and zeolite matrix for use in an adsorptive separation process was an inspiration in the creation of MMM. The talk will conclude by describing the connection between the adsorption technology and membrane technology that led to the development of MMM.

(28c) Scale-up of SAPO-34 Membranes for CO2/CH4 Separation

Richard D. Noble[1], John L. Falconer[2], Hans Funke[1], Eric Ping[3] and Rongfei Zhou[3], (1)Chemical and Biological Engineering, University of Colorado, Boulder, CO, (2)Chemical Engineering, University of Colorado at Boulder, Boulder, CO, (3)Chemical & Biological Engineering, University of Colorado, Boulder, CO

SAPO-34 membranes have pores with a diameter of approximately 0.38 nm and have the potential to be used for large scale removal of CO_2 impurities from natural gas. Tubular SAPO-34 membranes have been shown previously to have high permeances and selectivities at high pressures, so that CO_2 can be removed while maintaining CH_4 at high pressures. Here we show that the SAPO-34 membrane preparation can be scaled up and simplified. The membranes were prepared on 7-channel monolith alumina supports to increase the membrane surface area per volume.

The SAPO-34 separation layers were grown on both single-channel and monolith supports using seeded synthesis where the SAPO-34 seeds were deposited on the supports either by dip-coating, rub-coating, or a newly developed seeding technique. The membranes prepared on monolith supports had average CO_2 permeances of 6.3×10^{-7} +/- 0.9 mol/(m^2sPa) and average selectivities of 56 +/- 4 for 50/50 CO_2/CH_4 mixtures at 46 bar feed pressure. These permeances and selectivities were probably limited by concentration polarization because the available mass flow controllers limited the maximum feed rate and resulted in high stage cuts when the seven-channel monoliths were used. Analysis of the monolith membranes by SEM indicate that the SAPO-34 layers were a few μm thick, and they were slightly thinner than SAPO-34 membranes prepared on single channel tubes.

(28d) Novel Membranes, Membrane Separation Techniques and Technologies

Kamalesh K. Sirkar, Chemical, Biological and Pharmaceutical Engineering, New Jersey Institute of Technology, Newark, NJ

There are numerous ongoing investigations in supposed frontiers in membranes and membrane separation technologies. In our laboratories we have been pursuing a few of them. This talk will identify the problems, the approaches we are studying and the membrane separation results being achieved. The problems involve thermal desalination methods, treating strongly oxidizing aqueous solutions, processing powerful organic solvent systems, and removing and recovering CO_2 from post combustion flue gas as well as precombustion syngas. In thermal desalination methods we will illustrate first our results on direct contact membrane distillation (DCMD) for temperatures between 50 to 130^{0}C and operation with salt concentrations between 0.2- 20 wt%. In airgap membrane distillation (AGMD) we will next show how a two hollow-fiber-set based compact device can be a highly productive desalination device without any need for a cooling water stream to start with. We will next focus on the performances of a perfluoropolymer membrane having exceptional chemical and thermal resistance. This membrane can concentrate a very dilute aqueous solution of H_2O_2 to as much as 40% H_2O_2 by simple pervaporation under almost ambient conditions. Further this membrane can dehydrate powerful aprotic solvents such as DMF, DMSO, DMAc with selectivities in the range of 1000-12000 characteristic of zeolite membranes in dehydration of ethanol. We will conclude by describing two new techniques and the corresponding membranes/processes to remove CO_2 from flue gas or from hot precombustion syngas at a high pressure.

(28e) An Overview of the Global Desalination Situation

Isabel C. Escobar, Chemical and Environmental Engineering, The University of Toledo, Toledo, OH

According to the International Water Management Institute (IWMI), one in three people globally endures some form of water scarcity, one-quarter of the world's population lives in areas where water is physically scarce, and over a billion people live where water is economically scarce. Existing water supplies may be limited in quantity or quality for meeting the increasing demands from population growth and industry expansion. In many arid and semi-arid areas, providing the large volume of water required for industrial, agricultural, recreational and potable applications is especially difficult. The United Nations declared 2005-2015 the "Water for Life" Decade because water is crucial for sustainable development. The goals of the "Water for Life" Decade are to reduce by half the proportion of people without access to safe drinking water, to stop unsustainable exploitation of water resources, to aim to develop integrated water resource management and water efficiency plans, and to halve the proportion of people who do not have access to basic sanitation.

Membrane processes are now a proven and reliable method of providing high-quality, cost-effective water. Membrane technologies have immediate applications to treatment of fresh, brackish and seawaters, as well as wastewater reclamation. With innovative module design and engineering, micro- and ultra-filtrations have become effective and economical for drinking water production, particularly for removal of microorganisms. Desalination is becoming more and more a commodity to satisfy the growing demand for water.

(49a) Mathematical Modeling of Rtsa Using Polymeric/Supported Amine Hollow Fibers-Single Fiber Model

Fateme Rezaei[1], Ryan P. Lively[2], David S. Sholl[3], Christopher W. Jones[3], William Koros[2], Yoshiaki Kawajiri[3] and Matthew Realff[4], (1)School of Chemical & Biomolecular Engineering, Georgia Institude of Technology, Atlanta, GA, (2)Chemical & Biomolecular Engineering, Georgia Institute of Technology, Atlanta, GA, (3)School of Chemical & Biomolecular Engineering, Georgia Institute of Technology, Atlanta, GA, (4)School of Chemical and Biomolecular Engineering, Georgia Institute of Technology, Atlanta, GA

Abstract

The use of novel polymeric hollow fiber contactors loaded with CO_2 adsorbents has been recently demonstrated as a new and scalable process configuration for post-combustion CO_2 capture. The hollow fiber morphology allows coupling of efficient heat transfer with effective gas contacting, potentially giving lower parasitic loads on the power plant compared to traditional contacting strategies using solid sorbents[1]. In this study, a two dimensional mathematical model of a rapid pressure swing adsorption (RTSA) process is developed to predict the fiber-sorbent performance during post-combustion CO_2 capture from flue gas. In particular, the model has been developed in several stages moving from simple models with no mass transfer to more complex physical situations with coupled heat and mass transfer between the gas, fiber sorbent and water flow in the bore. The first stage of this project is focused on developing a single fiber model to simulate a three-step RTSA system accounting for adsorption, heating/desorption and cooling steps. The sensitivity of the model to parameter values such as water velocity, initial temperatures, fiber dimension and heat and mass transfer characteristics of polymer, sorbent and water are evaluated accordingly by considering the amount of utility and overall parasitic load on power plant usage for this process. These models are implemented using gPROMS ® version 3 for simulating a complete RTSA cycle. The outcome of the modeling effort is used to guide the experimental team in the development of the fiber and sorbent properties, and to provide input to the techno-economic modeling of the process.

Reference

[1] R.P. Lively et al. Hollow Fiber Adsorbents for CO_2 Removal from Flue Gas, *Ind. Eng. Chem. Res.* 2009, *48*, 7314–7324.

(49b) Improved Pressure Swing Adsorption Cycles for Ethanol PSA-Distillation Systems

Fan Wu[1], Armin D. Ebner[2] and James A. Ritter[2], (1)University of South Carolina, Columbia, SC, (2)Department of Chemical Engineering, University of South Carolina, Columbia, SC

Pressure swing adsorption (PSA) has been the customary approach for azeotrope breaking and ethanol drying in the corn-ethanol industry. This PSA process is typically carried at 4 atm and 440 K using zeolite 3A as the adsorbent and a relatively simple 2-bed 4-step pressure vacuum swing process (PVSA). The question addressed in this work was: Is it possible to improve the PSA process performacne by adding more complexity to the PVSA cycle schedule in terms of the number of beds and additional cycle steps? The work done to date tends to indicate that the answer to this question is, yes. In this presentation the role of four different, continuous feed, PSA cycle schedules on the performance of ethanol drying will be evaluated. The PSA variables to be discussed include the number of beds, the number of equalization steps and cycle time. The performance will be judged in terms of the ethanol purity, ethanol recovery and separations cost.

(49c) Hot-Gas Pressure Swing Adsorption for Isopropanol Dehydration

Yujun Liu[1], Shawn D. Feist[2], Christopher M Jones[3], Nathan P Francis[3], Walter R Cooney Jr.[3] and Daniel R Armstrong[3], (1)Engineering and Process Sciences, The Dow Chemical Company, Freeport, TX, (2)Enginerring and Process Sciences, The Dow Chemical Company, Midland, MI, (3)The Dow Chemical Company

Many processes can be used to dehydrate the azeotropic mixture of water and isopropanol. Membrane pervaporation, vapor permeation, and azeotrope and extractive distillations can all be used for this purpose. But both membrane separation and distillation are limited by their ability and/or cost to produce isopropanol with less than 1.5 wt-% water. Hot-gas pressure swing adsorption (HGPSA) is ultimately capable of delivering water concentrations less than 10 ppm if required and is generally less energy and less capital intensive compared to extractive distillation and membrane technology.

A HGPSA process was designed and implemented in a Dow process to dry isopropanol. Experimental studies were first performed to evaluate the feasibility and validate a HGPSA mathematical model. The dynamic model was then used to design the HGPSAS process. Aspen ADSORPTION® was used to improve the design of the process. In this presentation, both experimental and mathematical modeling results will be discussed. The implemented HGPSA process, which is able to remove water from isopropanol to non-detectable levels, will also be presented.

(49d) Novel TSA Process for Enhanced Hydrocarbon Recovery From Produced Water

Anand Venkatesan and Phillip C. Wankat, School of Chemical Engineering, Purdue University, West Lafayette, IN

"Produced Water" is the largest byproduct stream by volume associated with oil and gas (O&G) production (75 billion barrels (bbls) of produced water worldwide annually). The O&G industry in USA produces about 18 billion bbls of produced water (≈7 bbl of water per bbl of oil). Management of produced water is a challenge due to its complex composition which includes compounds ranging from simple hydrocarbons to complex polynuclear aromatics and a variety of salts. The cost of managing large volumes of produced water can be as high as 60% of cost of oil per bbl making it an important component in the overall cost of producing oil.

Produced water, often considered as waste is now being viewed as a potential profit stream. A process for high recovery and low cost desalination of the produced water is required. Such a process will reduce the water consumption in O&G extraction (75-225 billion bbls worldwide annually) by improving recyclability. In addition, a recovery of the hydrocarbons present in produced water is desired as it can be used as a feedstock for downstream operations.

Membrane technologies such as reverse osmosis (RO) are used to desalinate produced water. Hydrocarbon fouling and/or precipitation of salts on the membrane surface limits the water recovery. Thus, we propose an adsorption/ion exchange (IEX)/RO desalination process with an adsorption pretreatment to remove the hydrocarbons and other dissolved organic substances, an ion exchange system to remove divalent salts, and an optimized multi-stage RO system to produce both pure water and concentrated brine that is used to regenerate the ion exchanger. Since the IEX/RO system has been studied previously[1,2], the current focus of the research is on the adsorption system.

Since the feed is liquid, the energy requirements are minimized by regeneration with liquid. In order to remove heavy components, regeneration is done by water at high temperature with pressure high enough to prevent boiling[3]. Feasibility of this process for binary aqueous solutions containing benzene or pyrene has been established by simulation using Aspen Chromatography(Regeneration water: Feed Water ratio 0.3). Since high temperature adsorption data is scarce, the isotherms and mass transfer coefficients were extrapolated from lower temperature data. Simulation results for multicomponent interacting solutes will also be presented.

1. A. Venkatesan and P. C. Wankat, Desalination 271 (2011) 122-131
2. A. Venkatesan and P. C. Wankat, Desalination 286 (2012) 176-186
3. J. K. Kim, G. Natarajan, and P. C. Wankat, Adsorption 9 (2003), 117-123

(49e) Pressure Vacuum Swing Adsorption Cycles for Landfill Methane Purification

Lutfi Erden[1], Armin D. Ebner[2] and James A. Ritter[2], (1)Chemical Engineering, University of South Carolina, Columbia, SC, (2)Department of Chemical Engineering, University of South Carolina, Columbia, SC

Pressure vacuum swing adsorption cycles have been designed for the purification of methane out a stream previously enriched in methane from landfill gas at room temperature and 800 kPa and using BPL activated carbon as the adsorbent. The feed stream contains 86 vol% methane with the remainder being either 14 vol% N_2 or 12 vol% N_2, 1 vol% CO_2 and 1 vol% O_2. The product stream must contain at least 97 vol% CH_4 and less than 0.2 vol% O_2 and must recover more than 90% of the CH_4 fed. This study will reveal the role of equalization, heavy reflux, countercurrent depressurization and light reflux steps. In particular, the role of countercurrent heavy reflux with recycle to the feed will be discussed. This study will also analyze the role of low pressure in the cycle and the influence of carbon dioxide and oxygen in the feed. In addition to the main performance parameters of CH_4, purity, CH_4 recovery and feed throughput, the separation costs will also be discussed.

(49f) Analysis of Concentration Swing (Magnitude of Concentration Change during a Cycle) in Pressure Swing Adsorption

Tsutomu Hirose, Advanced Adsorption Technology, Fukuoka, Japan

Pressure swing adsorption is a time-dependent cyclic process. Although the average product concentration is primarily important information, the concentration swing, i.e. magnitude of concentration change during a cycle, is another important index of performance. The present author analyzed the latter topic by a "short cycle time approximation" in an old paper [1] and gave the result in a closed analytical form in case of a linear isotherm of adsorption. The same approach was extended to the case of a non-linear isotherm [2] and the result was given in terms of a first-degree linear differential equation with variable coefficients with respect to the average concentration. In this presentation, a concept of HETP (height equivalent to a theoretical plate) [3] is applied to evaluate the integrating factor involved in the original linear differential equation. The final result of the concentration swing in a non-linear isotherm case is expressed in terms of a summation of the corresponding result of linear isotherm weighted by the average gradient of the isotherm over each theoretical plate. The present closed form result is discussed in comparison with more rigorous numerical simulation and experimental results for dehumidification and CO2 removal.

[1] T.Hirose, T.Minoda, J. of Chem. Eng. Japan, 19, 300(1986)

[2] T.Hirose, J. of Chem. Eng. Japan, 20, 270(1987)

[3] T.Hirose, M.Yoshida, A.Kodama, M.Goto, Fundamentals of Adsorption, 7, 724(2002)

(49g) A Pilot Plant Study of a VSA Process for CO2 Capture From Power Plant Flue Gas

Vemula Rama Rao[1], Shreenath Krishnamurthy[1], Sathish Kumar Guntuka[2], Arvind Rajendran[3], Mohammad Aman Ullah[3], Paul Sharratt[2], Iftekar A. Karimi[1] and S. Farooq[1], (1)Department of Chemical and Biomolecular Engineering, National University of Singapore, Singapore, Singapore, (2)Process Science and Modelling, A*Star. Institute of Chemical and Engineering Sciences, Singapore, Singapore, (3)School of Chemical and Biomedical Engineering, Nanyang Technological University, Singapore, Singapore

Science and Engineering Research Council (SERC) of Singapore in 2009 launched a research programme on Carbon Capture and Utilization (CCU). One project under the programme is entitled, *Development and Demonstration of an Adsorption-based System for Post-combustion Carbon Capture*, which will be the basis of this presentation. It is a collaborative project among the adsorption and process systems research groups at National University of Singapore (NUS), Nanyang Technological University (NTU) and Institute of Chemical and Engineering Sciences (ICES).

A pilot plant, designed based on the results from a detailed simulation study, has been constructed and commissioned. It uses 1 m long columns with 0.3 m internal diameter and is rated to capture ~ 3 tonnes CO_2 per cubic meter of adsorbent per day. Operated from a PLC, the pilot plant is designed to track all the process variables as a function of time. Special attention has been given for tracking power consumption by the vacuum pumps so that a reliable estimate of energy penalty may be obtained.

In the first phase, CO_2 capture and concentration from dry flue, synthetic gas containing 15% CO_2 in N_2 on Zeochem 13X using a simple 4-step Vacuum Swing Adsorption (VSA) cycle has been studied. Detailed breakthrough results, effects of various process variables on purity, recovery and productivity, energy consumption data will be presented at the meeting.

HIGH EFFICIENCY OXYGEN SEPARATION FROM AIR USING PEROVSKITE-TYPE OXYGEN SORBENTS WITH HIGH TEMPERATURE PSA

Jun Izumi, ADSOTECH, Ltd.
Norio Miura, Kyushu University
Tomoya Fujimine, Tokyo Gas Co.

Abstract

From the perspectives of alternative energy and energy savings, it is very important to develop highly-efficient oxygen production processes. It is expected that the combination of perovskite-type oxygen sorbents and the heat-recovery PSA technique will result in a highly-efficient oxygen production method, having the lowest oxygen production cost among all other air separation processes.

Introduction

In this study, in order to better understand oxygen separation from air using Perovskite-type oxygen sorbents with high temperature PSA, sample pellets were formed and calcined, and evaluation of oxygen selectivity was conducted in an oxygen-nitrogen binary system under high-temperature adsorption conditions with a small column apparatus and a small experimental apparatus.

Experimental

Small Column Apparatus Evaluation of Oxygen Adsorption/Desorption on Perovskite-type Sample

A small column adsorption technique was used to obtain adsorption equilibrium and rate characteristics of a bi-component system of oxygen and nitrogen. Figure 1 shows a schematic diagram of the small column apparatus. Table-1 shows the experimental conditions of this evaluation.

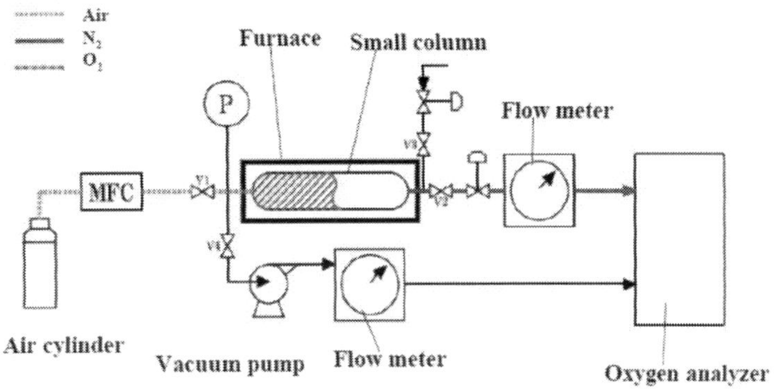

Figure-1 Schematic Diagram of HTPSA-O_2 Small Column Apparatus

Table-1 Experimental Conditions of HTPSA-O_2 Small Column Apparatus

Adsorbent			LSCF1991
Adsorption pressure		[kPa]	120
Desorption pressure		[kPa]	3
Adsorption temperature		[deg-C]	120
Loaded adsorbent weight		[g]	17
Shape of small column		[mm]	10mm ϕ ID × 300mmL
Sequence	Pressurization	[seconds]	5
	Adsorption	[seconds]	60
	Desorption	[seconds]	60
Experimental period		[months]	3

Table-2 presents the results of this small column apparatus evaluation.

Table-2 Results of HTPSA-O_2 Small Column Apparatus Experiment

Time		after 1 month	after 3months
Inlet O_2 conc. at adsorption stage	[vol%]	20.8	20.8
Outlet O_2 conc. at adsorption stage	[vol%]	15.6	15.9
Recovered O_2 conc. at desorption stage	[vol%]	39.8	40.2
Recovered O_2 volume at desorption stage	[mlN/g]	2.01	2.04
N_2 adsorbed amount/O_2 adsorbed amount S at 590 deg-C	[-]	≒0	≒0

Small Column Apparatus Evaluation of Oxygen Adsorption/Desorption on Perovskite-type Sample

Having obtained the adsorption equilibrium and rate characteristics of a bi-component system of oxygen and nitrogen, this HTPSA technique was scaled up to a small scale experimental apparatus. Figure 2 shows a schematic diagram of the small scale experimental apparatus, while Table-3 shows the operational conditions. Figure-3 shows the appearance of the HTPSA-O_2 small scale experimental apparatus.

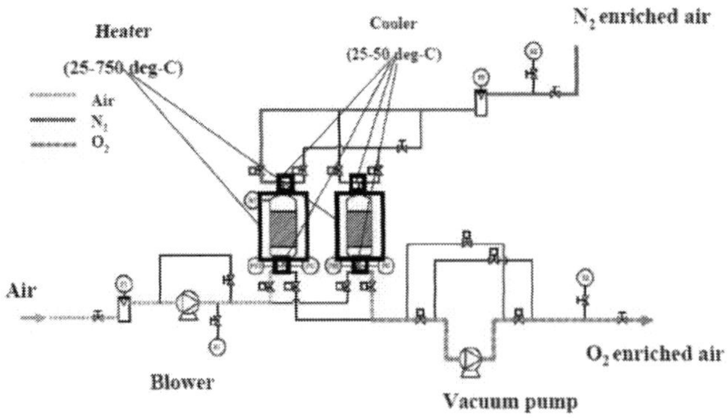

Figure-2 Schematic Diagram of HTPSA-O_2 Small Scale Experimental Apparatus

Table-3 Experimental Conditions of HTPSA-O_2 Small Scale Experimental Apparatus

Adsorbent		**LSCF1991**	
Separation system		**VPSA 2towers**	
Adsorption pressure		**[kPa]**	**120**
Desorption pressure		**[kPa]**	**3**
Adsorption temperature		**[deg-C]**	**600-800**
Loaded adsorbent weight		**[kg/tower]**	**2.5**
Shape of tower		**[mm]**	**50mm ϕ ID × 760mmL**
Tower quantity		**2**	
Sequence	**Pressurization**	**[seconds]**	**5**
	Adsorption	**[seconds]**	**60**
	Desorption	**[seconds]**	**60**
Inlet gas rate (maximum)		**[lN/h]**	**1,000**

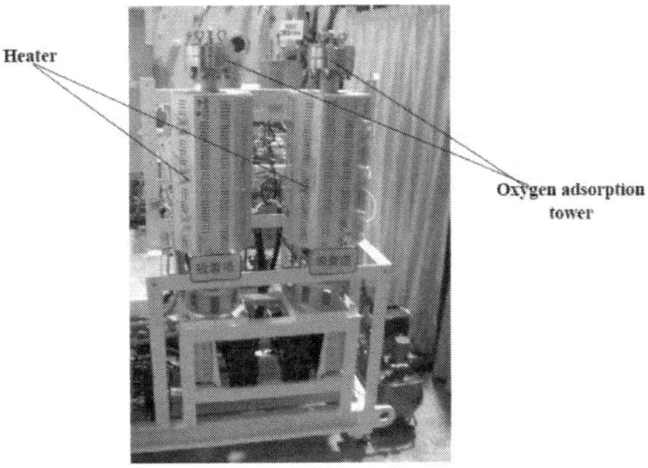

Figure-3 Appearance of HTPSA-O_2 Small Scale Experimental Apparatus

Figure-4 shows the mass balance of oxygen and nitrogen separation with this small scale experimental apparatus.

Table-4 presents a performance comparison between HTPSA-O_2 (the present study) and the conventional process.

.

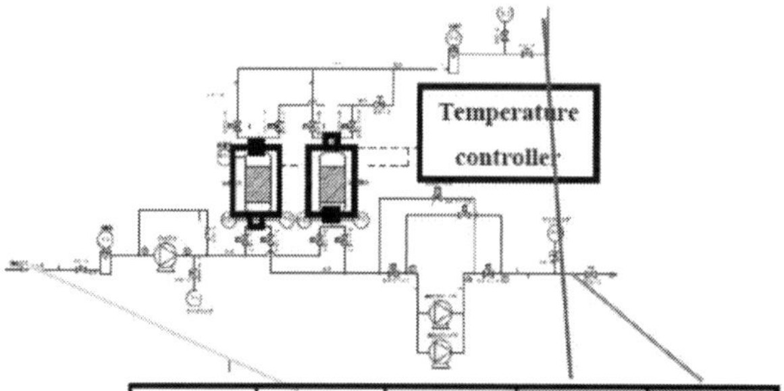

Line		0	1	2
N_2+Ar	(vol%)	79.2	94.5	22.5
O_2	(vol%)	20.8	5.5	77.5
total	(vol%)	1	1	1
N_2+Ar	(lN/h)	748	693	55
O_2	(lN/h)	192	3	189
total	(lN/h)	940	696	244

Figure-4 Mass Balance of Oxygen and Nitrogen Separation

Table-4 Performance Comparison between HTPSA-O_2 (Present Study) and Conventional Process.

Oxygen production process		This study (HTPSA-O_2)	Conventional (PSA-O_2 (Li-LSX))
Oxygen production capacity	[m^3N-O_2/h]	1000	
Produced oxygen conc.	[vol%]	80	93
Blower adiabatic efficiency	[%]	80	80
Vacuum pump adiabatic efficiency	[%]	65	65
Heat recovery ratio	[%]	92	92
Blower electric power consumption rate	[kWh/m^3N-O_2]	0.06	0.09
Vacuum pump electric power consumption rate	[kWh/m^3N-O_2]	0.13	0.29
Heater electric power consumption rate	[kWh/m^3N-O_2]	0.06	
Oxygen electric power consumption rate	[kWh/m^3N-O_2]	0.25	0.38

Summary

1. From the standpoints of alternative energy and energy savings, it is very important to develop highly-efficient oxygen production processes.
2. We studied a novel high-temperature PSA (pressure swing adsorption) process for oxygen production, using perovskite-type oxygen sorbents.
3. This adsorbent can selectively and reversibly adsorb a considerable amount of oxygen at high temperatures (>400℃).
4. The oxygen to nitrogen selectivity is infinitely high, and therefore the electrical power consumption rate of oxygen production will be much lower than that of cryogenic separation.

5. The formed perovskite-type oxygen sorbent was loaded into a small column, which was installed in a small-scale PSA instrument. Good air separation performance was observed (infinite oxygen to nitrogen selectivity, large oxygen adsorption capacity, fast adsorption-desorption kinetics).
6. Mass balance, heat balance, and oxygen production costs have been calculated, and a low electrical power consumption rate of about 0.2 kWh/ m^3N is estimated.
7. It is expected that the combination of perovskite-type oxygen sorbents and the heat-recovery PSA technique will result in a highly-efficient oxygen production method, having the lowest oxygen production cost among all other air separation processes.

Reference

1. Speaker, Izumi J. and Norio Miura (2006), Japan Patent **4721967**

Acknowledgement

A part of this work belongs to "Research and Development Program for Innovative Energy Efficiency Technology, Exploratory research phase (Preliminary research):Preliminary research on development of a highly-efficient oxygen production instrument using perovskite-type oxygen sorbents, (FY2010) " which Adsorption Technology Industries and Kyushu Universities contracted with New Energy and Industrial Technology Development Organization (NEDO).

(53a) Real-Time Operation and Control of Integrated Ultrafiltration-Reverse Osmosis Membrane Desalination System

Larry Gao[1], Han Gu[2], Anditya Rahardianto[2], Panagiotis D. Christofides[3] and Yoram Cohen[4], (1)Chemical & Biomolecular Engineering, UCLA, Los Angeles, CA, (2)Chemical & Biomolecular Engineering, University of California, Los Angeles, Los Angeles, CA, (3)Department of Chemical and Biomolecular Engineering, University of California, Los Angeles, Los Angeles, CA, (4)Chemical and Biomolecular Engineering, University of California, Los Angeles, Los Angeles, CA

Reverse osmosis (RO) membrane desalination has emerged as one of the leading methods for water desalination due to the low cost and energy efficiency of the process. Recent advances in model-based control and energy optimization for RO desalination [1-3] have made it possible to develop self-adaptive operational strategies for RO membrane desalination systems. The goal of self-adaptive operation is to enable the plant to autonomously respond to changes in feed water salinity and fouling propensity. In order to combat membrane fouling, RO feed water pre-treatment is crucial for effective RO desalination. In recent years, microfiltration (MF) and ultrafiltration (UF) systems have been increasingly explored for effective feed-filtration. While the concept of UF/RO integration is appealing, there are significant technical challenges in situations where plant footprint (e.g., for shipboard desalination) is at a premium, the use of cleaning chemicals must be kept to a minimum, and feed water quality often fluctuates. To demonstrate the feasibility of an integrated UF/RO system and the operational benefits of advanced process control, the UCLA team developed a compact second generation RO (CoM2RO) desalination system integrated with MF-UF feed filtration. The CoM2RO implements two unique controllers: a novel RO model-based controller capable of simultaneously self-adjusting the feed flow rate and trans-membrane pressure in order to maintain the desired water productivity and quality, and a self-adaptive UF operational strategy that varies backwashing intensity depending on detected fouling levels. These controllers enable integration of the two systems such that the UF can directly utilize the RO concentrate for backwashing and minimizes RO energy consumption, while the RO system adapts to any changes in pressure, flow rate, or salinity that result from the cyclic nature of the UF backwashing. Real-time sensors data along with embedded computing capability allow on-line membrane element characterization and overall system performance assessment (with appropriate data normalization) that are used for rapid control and optimization. A field evaluation of the present advanced RO/UF system control was carried out for seawater desalination. Experiments demonstrated that the pre-treatment operation significantly improved the RO feed water conditions (Turbidity, florescence reading, particle size) which the RO controller successfully adapted to while maintaining constant RO permeate production and quality. Real-time sensor data allowed for on-line system characterization of fouling levels in the UF system, thereby enabling adjustment of operating conditions as well as fault detection and significantly lengthening the operational duration of the system. Results of the study demonstrated that the real-time information regarding membrane characterization enabled fault detection and rapid evaluation of the impact of varying fouling and feed conditions on the performance of both UF and RO systems.

1. Zhu, A., P. D. Christofides and Y. Cohen, "Effect of Thermodynamic Restriction on Energy Cost Optimization of RO Membrane Water Desalination," Industrial and Engineering Chemistry Research, 48, 6010-6021 (2009).
2. Bartman, A.R., A.H. Zhu, P.D. Christofides, Y. Cohen, "Minimizing Energy Consumption in Reverse Osmosis Membrane Desalination using Optimization-Based Control," J. Process Control, 20, 1261-1269 (2010).
3. Zhu, A., P. D. Christofides, and Y. Cohen, "Energy Consumption Optimization of Reverse Osmosis Membrane Water Desalination Subject to Feed Salinity Fluctuation," Ind. Eng. Chem Res, 48, 9581–9589(2009).

(53b) Optimal Plant Operation of Industrial Reverse Osmosis Desalination

Mingheng Li, Department of Chemical and Materials Engineering, California State Polytechnic University, Pomona, CA and Brian Noh, Inland Empire Utility Agency, Chino, CA

A comprehensive first-principles based mathematical model, which explicitly accounts for membrane hydraulic permeability, pump characteristic curves and retentate pressure drop, is developed to describe the reverse osmosis desalination process. Parameters used in the model are derived from data collected in an industrial desalination plant. A model-based optimization framework is then developed to reduce pump energy consumption while maintaining the same permeate rate. The model predicted results match well with plant trials, and a 10% reduction in energy consumption is demonstrated.

References

[1] Li, M. "Optimization of Multitrain Brackish Water Reverse Osmosis (BWRO) Desalination," Industrial Engineering Chemistry Research, 51, 3732-3739, 2012.
[2] Li, M. "Optimal Plant Operation of Brackish Water Reverse Osmosis (BWRO) Desalination," Desalination, 293, 61-68, 2012.
[3] Li, M. "Reducing Specific Energy Consumption in Reverse Osmosis (RO) Water Desalination: An Analysis from First Principles," Desalination, 276, 128-135, 2011.
[4] Li, M. "Minimization of Energy in Reverse Osmosis Water Desalination using Constrained Nonlinear Optimization," Industrial and Engineering Chemistry Research, 49, 1822-1831, 2010.

(53c) Novel Cellulose Esters for Forward Osmosis Membranes

Rui Chin Ong[1], Jincai Su[1], Bradley J. Helmer[2], Jos S. de Wit[2] and Tai-Shung Chung[3], (1)National University of Singapore, Singapore, Singapore, (2)Eastman Chemical Company, Kingsport, TN, (3)Department of Chemical and Biomolecular Engineering, National University of Singapore, Singapore, Singapore

In view of the tremendous efforts for the invention of materials and techniques for forward osmosis (FO) membrane fabrication, in this work, we focused our study in the use of novel cellulose ester materials for FO membrane fabrication. A wide range of cellulose esters with different degree of distributions (DS) of different functional groups which include hydroxyl (OH), acetyl (Ac), propionyl (Pr) and butyryl (Bu) groups were specially synthesized for this study. The effects of DS of functional groups of the cellulose esters have on the resultant membranes' morphology and performance were carefully studied. The resultant membranes have distinguished morphology and these cellulose esters exhibited distinguished membrane performance under forward osmosis process due to their differences in the functional group distribution. Fundamental properties such as water permeability, water solubility, salt partition coefficient and salt permeability of each cellulose ester were characterized to determine the effects of functional group distributions on the membrane performance. This study showed that desired membrane performance can be achieved by altering the DS of various functional groups for cellulose esters.

(53d) Harvesting of Pure Water From Air and Combustion Exhaust Gas by Membrane Pervaporation

Wei Liu, Pacific Northwest National Laboratory, Richland, WA

When there is scarcity in natural water resources, supply of pure, potable water becomes a problem for sustainable living and industrial development. On the other hand, presence of water in air is ubiquitous and water is also produced from combustion of biomass and hydrocarbon fuels. This presentation presents a brief review of various water recovery technologies reported in the literature and introduces a new membrane technology for production of water from humid air and/or combustion flue gas via a pervaporation process. In the pervaporation process, as a water-containing gas stream flows over a membrane surface, water vapor is selectively pulled out of the mixture into the another side of the membrane and is compressed to above its dew point by a vacuum pump so that the water vapor is condensed at ambient temperature and collected as liquid water. Important performance requirements of the membrane are 1) high water permeance, 2) high selectivity of water vapor over air, 3) robustness of the membrane, and 4) low cost. The thin flat sheet molecular sieve membrane recently developed in our research group will be presented to address these critical requirements for a practical membrane water production system. The new membrane provides very high water vapor permeance and a large window of operation conditions, such as temperature and water vapor content. Due to the water molecular specificity of the zeolite membrane material, the membrane pervaporation produces pure water in one-step without need of further purification.

(53e) Design and Development of Hollow Fiber Membrane for Desalination and Salt Recovery Via Integrated Membrane Distillation-Crystallization

Felinia Edwie, Chemical and Biomolecular Engineering, National University of Singapore, Singapore, Singapore and Tai-Shung Chung, Department of Chemical & Biomolecular Engineering, National University of Singapore, Singapore, Singapore

Over the past few years, the water research worldwide has carried a similar theme of water crisis with the continuous increment of world population and limited availability of fresh water resources were identified as the major causes. In this regard, desalination of saline water has great potencies to suffice the world's increasing demand of fresh water considering the high percentage of saline water on earth. Among other membrane-based desalination process, reverse osmosis (RO) technology has been proven reliable and becomes the most preferred option for desalination. Nevertheless, the large scale application of RO is often questioned for the brine disposal issue and its unfavorable impacts on the environment. These issues have urged the development and innovation of more sustainable desalination processes with the ultimate aim to maximize the water recovery and the process efficiency. For these reasons, various hybrid desalination processes have been proposed, for instance membrane distillation (MD) based hybrid desalination technology. MD process is an integration of membrane-based and thermal-based desalination processes in which a porous hydrophobic membrane is employed as a thermal insulator and physical barrier between the hot feed where water evaporation occurs and the cold distillate where the diffused water vapor is condensed. Therefore, the permeation flux in MD process is less dependent on the solution concentration as the transport of water vapor is driven by the vapor pressure gradient across the membrane. This allows MD operation to concentrate the saline solution up to saturation level enabling the subsequent recovery of the salt by means of the crystallizer. Therefore, by integrating membrane distillation with crystallizer (MDC), not only the valuable compounds can be harvested but the issue of brine disposal can be also mitigated.

As compared to MD operation with a low concentration of saline solution, membranes operated with a high concentration are more prone towards salt scaling facilitated membrane wetting which results in flux decay and deterioration of separation performance. Therefore, in this study we investigated the performance of various hollow fiber membrane configurations under MDC operation with a highly concentrated sodium chloride (NaCl) aqueous solution. The membranes were developed in our laboratory which include single- and dual-layers membrane configurations along with combination of hydrophobic-hydrophilic characteristics. Heat and mass transfers along membrane module were modeled to determine the change of membrane permeability with increasing salt concentration. Upon MDC operation, the membrane performances were retested to define the recovery of membrane permeability. It was found that the morphology underneath membrane surface played important role in mitigating scaling facilitated membrane wetting and membrane with cellular structure exhibited a better wetting resistance as compared to globular structure. Moreover, the performance of membrane with cellular structure was maintained even at the widest trans-membrane temperature gap at which the scaling propensity is the highest. Lastly, NaCl was crystallized and recovered from the supersaturated MDC retentate solution. The effects of crystallizer cooling rate on the resultant salt crystal nucleation and growth were investigated.

(53f) Prevention of Environmental Pollution with Advanced High Capacity Water Storage

Lev Reznikov, Engineering, Global Fabrication, Inc., DuBois, PA and Luke Sicard II, Engineering, Global Fabrication Inc., DuBois, PA

Oil and gas companies throughout the world are doing their part to protect nature and the environment, and consumers are paying closer attention than ever to make sure they do.

Damage to local wildlife can bring negative publicity, regulatory fines, and other long-term expenses to your operation.

Global Fabrication, Inc. designed and introduced the most efficient on the market waste water storage tanks that preventing any considerable leakages to the surroundings.

Full assembly of the high capacity tanks takes much less time that the competitive storages.

Novel technological concept provides not only significant cost reduction technology, but also practically full prevention of leakages.

It is our promise to safe life of all kinds of birds and animals!

A system of weaved netting around your Big Holding water storage tanks can easily prevent small and large creatures from harming themselves or damaging your equipment. The result is a commonsense system that's as good for business as it is local wildlife.

A system of weaved netting around your Big Holding water storage tanks can easily prevent small and large creatures from harming themselves or damaging your equipment. The result is a commonsense system that's as good for business as it is local wildlife.

(83a) Robust Particle Size Control During API Crystallization

Nathan Domagalski[1], Brendan C. Mack[2], Amanda Rogers[1], Chenchi Wang[2], Jose E. Tabora[3] and Lindsay Hobson[2], (1)Chemical Development, Bristol-Myers Squibb Co., New Brunswick, NJ, (2)Bristol-Myers Squibb Company, New Brunswick, NJ, (3)Chemical Development, Bristol Myers Squibb, New Brunswick, NJ

A robust process offers flexibility to large scale manufacturing by permitting execution within broad parameter ranges without impact to product quality. In this paper the development of a controlled API batch crystallization will be described which delivers both a highly reproducible and predictable particle size specification. The process includes simultaneous cooling and wet milling that favors secondary nucleation over growth under low supersaturation conditions. Investigation includes a fractional factorial design of experiments (DoE) approach studied on a greater than 100 kg pilot plant scale. This DoE enabled identification of key process parameters using an equipment train nearly equivalent to commercial manufacturing. Collected data served as the basis for an empirical model used to estimate particle size and establish a demonstrable design space for the crystallization process. Comparison with similarly executed batches showed that the model may be used to accurately predict future performance. The described crystallization procedure was superior to conventional wet milling in producing unimodal particle size distribution in the isolated product.

(83b) Effects of an Impurity On the Morphology of an API: Using ATR-FTIR, FBRM and PVM to Confirm a Hypothesis

Christopher S. Polster[1], Michael A. Lovette[1], Lori R. Hilden[1] and Christopher L. Burcham[2], (1)Chemical Product Research and Development, Eli Lilly and Company, Indianapolis, IN, (2)Chemical Product Research and Development; Particle Design Laboratory, Eli Lilly and Company, Indianapolis, IN

During the crystallization of multiple 1-2 kg batches of API, a substantial change in morphology from needles (aspect ratio 10-20) to plates (aspect ratio 1-2) was observed in some of the lots. An unknown impurity found at trace levels was found to correlate with the change in morphology. This impurity was isolated chromatographically and identified as a dimeric species of the parent compound. Several milligrams of the impurity were isolated and used to spike into subsequent crystallizations in order to confirm the hypothesis that the impurity alone was responsible for the morphology change. These crystallizations were characterized by using ATR-FTIR to monitor solute concentration changes with time, FBRM to monitor particle chord length (along with aspect ratio) changes with time, and PVM was used as visual confirmation of the aspect ratio changes. Crystallizations (over a range of impurity levels) show an evolution from expected needle morphology to plate morphology. The results find that the aspect ratio is altered by slowing the growth of the expected axial face.

(83c) Microfluidic Platforms for Time Resolved Laue Crystallography

Sarah L. Perry[1], Sudipto Guha[2], Ashtamurthy Pawate[2], Zhong Ren[3] and Paul J.A. Kenis[4], (1)Institute for Molecular Engineering, University of Chicago, Chicago, IL, (2)Department of Chemical & Biomolecular Engineering, University of Illinois at Urbana-Champaign, Urbana, IL, (3)Department of Biochemistry and Molecular Biology and the Consortium for Advanced Radiation Sources, University of Chicago, Chicago, IL, (4)Department of Chemical & Biomolecular Engineering, University of Illinois at Urbana Champaign, Urbana, IL

While efforts in microfluidics for protein crystallization have developed dramatically in the past ten years [1-3], the idea of coupling microfluidic technology with advanced protein crystallography techniques for *in situ* analysis is an area where significant advances can be made. Microfluidic devices are particularly applicable for Laue methods [4], where polychromatic X-rays allow for significantly faster data collection that enables the analysis of tiny or fragile crystals that are susceptible to radiation damage as well as kinetic experiments [5,6]. This high rate of data collection enables time-resolved crystallography experiments by matching the X-ray exposure time to the timescale of chemical and structural changes in the protein. Thus Laue crystallography provides a very elegant platform for performing kinetic experiments that directly probe changes that occur during enzymatic function.

Microfluidic platforms have the benefit of not only enabling experiments at very small volumes, but also by creating an environment free of inertial or convective effects and allowing for exquisite control over local conditions and gradients. We have developed microfluidic platforms for protein crystallization which utilize this high degree of control for both the growth of high quality protein crystals, and also as a triggering mechanism for use in time resolved crystallography experiments. Here we report on microfluidic crystallization platforms that have been optimized for the growth of a large number of crystals for *in situ* time resolved Laue analysis. We used photoactive yellow protein (PYP) as a model system for the first time resolved crystallography experiments. The enzymatic reaction is first triggered by a laser light pulse and then a series of X-ray diffraction images are taken at varying time delays. This large set of diffraction data is then analyzed to yield a time resolved picture of the various structural changes within the crystal. We are working on extending the capabilities of our microfluidic systems to include triggering of enzyme reactions by flowing a solution over the crystals. This capability will enable the time resolved structural analysis of a large range of protein systems, which had previously been inaccessible for study.

1. Hansen, C. and S.R. Quake, *Microfluidics in structural biology: smaller, faster... better.* Current Opinion in Structural Biology, 2003. **13**(5): p. 538-544.

2. van der Woerd, M., D. Ferree, and M. Pusey, *The promise of macromolecular crystallization in microfluidic chips.* Journal of Structural Biology, 2003. **142**(1): p. 180-187.

3. Perry, S.L., et al., *Microfluidic Generation of Lipidic Mesophases for Membrane Protein Crystallization.* Crystal Growth & Design, 2009. **9**(6): p. 2566-2569.

4. Ren, Z., et al., *Laue crystallography: coming of age.* Journal of Synchrotron Radiation, 1999. **6**: p. 891-917.

5. Cornaby, S., et al., *Feasibility of one-shot-per-crystal structure determination using Laue diffraction.* Acta Crystallographica Section D, 2010. **66**(1): p. 2-11.

6. Bourgeois, D., et al., *Time-resolved methods in biophysics. 6. Time-resolved Laue crystallography as a tool to investigate photo-activated protein dynamics.* Photochemical & Photobiological Sciences, 2007. **6**(10): p. 1047-1056.

(83d) Crystalline Particle Formation and Polymorphism From Uniform Droplet Evaporation

Kelly M. Carver and **Ryan C. Snyder**, Chemical Engineering, Bucknell University, Lewisburg, PA

Particle size, shape and structure are all important product qualities with direct implications for the pharmaceutical, food product and dye industries. Crystal size directly impacts dissolution rate. Crystal shape affects both downstream processes such as filtering, washing and drying as well as material properties such as mechanical strength, bioavailability and surface functionality. Finally, different crystal structures, or polymorphs, impact the entire scope of the product from melting point and solubility to patent considerations. Thus, methods to manipulate and control size, shape and structure are continuously being sought. Droplet evaporation, commonly used industrially through spray drying, offers some advantages such as a reduction in production steps when compared to traditional crystallizations; however, it is notoriously difficult to control and model the product properties because of the varied droplet sizes and flow patterns of the droplets. Monodisperse droplet generation offers a useful tool for better controlling droplet evaporation by eliminating the challenges associated with droplet size distributions in spray dryers. Monodisperse droplet generators have previously been used to produce uniform particle sizes (Nandiyanto and Okuyama 2011, Vehring 2008); however, crystalline properties have not been the primary focus of previous work.

In this talk, we will highlight our recent work on producing crystalline particles through the evaporation of monodisperse droplets produced using a vibrating orifice aerosol generator (VOAG). In this study, we use succinic acid as a test case molecule. The experiments show that we not only can produce monodisperse particle sizes, but that our particles are also fully crystalline, containing no amorphous characteristics. As with other crystallization methods, solvent choice is a key variable in determining the morphology of the resulting particles, and we demonstrate the variability in morphology across several solvents (e.g. acetone, water, isopropanol). We will show that solvent choice indeed impacts the morphologies of these particles and demonstrate that the morphologies fall into a new class of particle shapes that previously have not been obtained from droplet evaporation. Additionally, we will show that the polymorphs obtained through this method are unique when compared to the crystallization of succinic acid from bulk solution. Finally, we will discuss the potential for this type of droplet evaporation as a tool for studying polymorphism of organic molecules in general and as a good system to develop fundamental models that could later be applied to industrial scale processes such as spray drying.

References

Nandiyanto, A. B. D.; Okuyama, K. *Adv. Powder Technol.* **2011**, *22*, 1-19.

Vehring, R. *Pharm. Res.* **2008**, *25*, 999-1022.

(83e) Influence of Additives On Crystal Morphology of Organic Molecular Crystals

Zubin B. Kuvadia, Department of Chemical Engineering, University of California, Santa Barbara, Santa Barbara, CA and Michael F. Doherty, Chemical Engineering, UC Santa Barbara, Santa Barbara, CA

Crystal morphology has a great impact on the end-use efficacy of solid products as well as on the downstream performance of the entire manufacturing and formulation process. Foreign molecules like additives or impurities influence the crystal morphology to a significant extent by inhibiting the growth mechanism of certain crystal faces. The additives can be byproducts of the preceding step of reaction or they could be specially designed to obtain specific crystal morphologies for focused applications, for example, using crystal growth inhibitors to reduce the size of kidney stones[1].

Because of the specific chemical structure of structurally similar additives, they can incorporate or lock into the original lattice only in certain configurations, hence only some of the crystal faces can recognize these additive molecules. Once attached, they interfere with the first turn of a rotating growth spiral and hence slow down the growth rate of the face [2]. In this work, we develop a generic probabilistic scheme for quantitatively estimating imposter recognition on each crystal face using a combination of mean field theory and energy minimization. Based on this recognition, the mechanistic effect of impurity on the first turn of the spiral and hence modified growth rates and modified crystal habits are computed. These concepts are generalized for all molecular crystals, including non-centrosymmetric molecules. Finally, we present work on specific systems of immense industrial and scientific interest such as paracetamol/PAA[3] and L-cystine/L-CDME[1] and compare our results with experiments from the literature and our laboratory.

References:

[1]: Rimer J . D, et al, Science **330**, 337(2010)

[2]: Sizemore J.P. and Doherty M.F, Cryst. Growth Des., **9**,6 (2009)

[3]: Chow et al, Intnl. Journal of Pharmaceutics, **24** (2-3) (1985)

(83f) Impact of Pre-Heating On Nucleation Kinetics and Polymorph

Srividya Ramakrishnan[1], Vasu Dev Rudraraju[2], Sundaralakshmi Kanniah[1] and Rashmi Agrawal[1], (1)Dr. Reddy's Laboratories Ltd., Hyderabad, India, (2)Aurigene Discovery Technologies Ltd., Hyderabad, India

Polymorphs play an important role in pharmaceutical development as they can impact the stability and bioavailability of the active pharmaceutical ingredient (API). However, isolating and stabilizing the desired form can often be a challenging task as several factors can affect the crystallization of a particular polymorph. In this study, the impact of thermal history of the under-saturated solution on the nucleation kinetics was studied for an API. It was found that pre-heating a solution of the API at a temperature higher than the dissolution temperature could significantly increase the induction time or suppress the nucleation temperature during cooling crystallization. In addition, pre-heating was also found to have an effect on the polymorph that nucleated. While typically the metastable form would nucleate during cooling, pre-heating resulted in concomitant nucleation of the metastable as well as stable form, with a correlation between the duration of pre-heating and the proportion of the stable form in the polymorphic mixture. This suppression of nucleation by pre-heating may due to a conformational transition at higher temperatures. Conversion to the wrong conformer(s) would not only reduce the effective supersaturation but might also inhibit nucleation. Spectroscopic studies are under way to study the conformers and pre-nucleation aggregates in solution.

Communication number: IPDO IPM-00269

(83g) Integration of Biocatalysis and Crystallization towards the Manufacture of Enantiomerically Pure Compounds

Luis G. Encarnación-Gómez, School of Chemical and Biomolecular Engineering, Georgia Institute of Technology, Atlanta, GA, Andreas S. Bommarius, Chemical & Biomolecular Engineering, Georgia Institute of Technology, Atlanta, GA and Ronald W. Rousseau, School of Chemical & Biomolecular Engineering, Georgia Institute of Technology, Atlanta, GA

Crystallization of enantiomerically pure compounds from racemic mixtures that form racemic compounds requires an enantiomeric excess greater than the eutectic point between the compound and the pure enantiomer. It is estimated that roughly 90% of racemic mixtures fall into this category; hence, development of a process that utilizes such an approach could be of significant value.

To achieve a necessary enantiomeric excess, we are investigating the chemoenzymatic reaction of racemic mixtures of amino acids. We have demonstrated the possibility of performing such a reaction at high concentrations, even at those that require substrate dissolution. Preliminary results show that even in supersaturated solutions, enantiomeric excesses greater than 99% can be achieved. The chemoenzymatic reaction of (RS)-phenylalanine to obtain (S)-phenylalanine has been performed in undersaturated, saturated, and supersaturated solutions. The reaction proceeds through the selective oxidation of the (R)-amino acid utilizing D-amino acid oxidase (D-AAO), which produces an achiral imino acid that is reduced into a racemic amino acid utilizing ammonia-borane. Moreover, extending such reaction to saturated or supersaturated solutions allows us to operate crystallization processes such that enantiomerically pure compounds are continuously recovered in a solid phase. Therefore, the implementation of this reactive separation process is beneficial because of the following factors: (1) this process allows higher productivity, (2) the rate of reaction is maximized, (3) the process has potential to be run continuously, and (4) organic solvents or low-MW chiral acid/bases (i.e., the formation of diastereomers) are not needed during separation.

(84a) Polymer Based Microfluidic Reactor for Liquid-Liquid Extraction Applications

Sachit Goyal[1], Amit V. Desai[1], Robert Lewis[1], David Ranganathan[2], David E. Reichert[2] and Paul J.A. Kenis[1], (1)Department of Chemical & Biomolecular Engineering, University of Illinois at Urbana Champaign, Urbana, IL, (2)Radiological Sciences Division at Mallinckrodt Institute of Radiology, Washington University School of Medicine at St. Louis, St. Louis, MO

Liquid-liquid extraction (LLX), which involves reactions occurring at the aqueous-organic interface in a two-phase system, finds application in diverse fields, ranging from purification and isolation of biomolecules to organic synthesis. Conventional techniques for LLX suffer from several limitations, including inefficient handling of small volumes, long extraction times due to long diffusion lengths, and long times required to form stable interfaces. In this context, working at the micro-scale enables effective handling of small volumes, short extraction times, and rapid formation of stable interfaces. We report on solvent-resistant polymer-based microfluidic reactors for LLX applications. These microreactors provide quantitative information on the extraction process compared to conventional techniques, due to the superior control over the spatial and temporal properties of the interface.

Microfluidic systems for LLX extractions involving droplets have been reported previously, where the droplet of one phase is encapsulated in the other phase. However, these systems are not suitable for applications where off-chip analysis or post processing of the solutions is required, as the separation of phases is challenging. In contrast, micro-reactors that enable co-flow of the aqueous and organic phases allow for more convenient phase separation. Microreactors based on hard materials (glass and silicon) have been developed for co-flowing aqueous-organic streams; however, these reactors require complex fabrication that is time and cost intensive. We have developed polymer based solvent resistant microreactors fabricated via a simple and inexpensive procedure. We systematically optimized several factors affecting the interface stability, including geometry of the micro-channel, surface functionalization, and the flow rates of the two immiscible streams.

In this work, we report the design, fabrication, and application of thiolene (NOA 81) and perfluoropolyether (SIFEL) based micro-reactors for liquid-liquid extraction applications, including purification of cyclotron-produced radioisotopes that are of interest in medical imaging, and studying distribution of pharmaceuticals between two phases to evaluate molecule's uptake in the human body.

(84b) Organophosphorus Solvents for Lactic Acid Extraction From Dilute Aqueous Solution

Susanne Lux and Matthäus Siebenhofer, Institute of Chemical Engineering and Environmental Technology, Graz University of Technology, Graz, Austria

Separation of lactic acid from dilute aqueous solution is complex. It still accounts for a major part in lactic acid production costs. Due to high polarity and the hydrophilic character of lactic acid, it is poorly extracted with conventional organic extractants. On the contrary, reactive extraction based on reversible adduct formation is a promising, sustainable alternative. Phosphorus-bonded oxygen-bearing extractants provide good partition of carboxylic acids due to their high Lewis basicity. Temperature dependence of the liquid-liquid partition equilibrium allows extraction at room temperature and re-extraction by stripping with water at elevated temperature.

Efficiency and selectivity of various organophosphorus compounds such as tri-*n*-butylphosphate, tri-*n*-octylphosphine oxide and Cyanex 923 were investigated. Shellsol T was used as a non reactive, unpolar diluent. The liquid phosphine oxide Cyanex 923 proved best with respect to both lactic acid partition and hydrodynamic properties.

Liquid-liquid phase equilibria were investigated and modeled based on the law of mass action. Experimental data and model data showed good agreement. The effect of water co-extraction and lactic acid selectivity from multi acid solution was considered too.

(84c) Recycling of Valuabe Metals From Battery Scrap Leachate by Supported Liquid Membrane Technology

Hannes Noll, Matthäus Siebenhofer and Andreas Pfennig, Institute of Chemical Engineering and Environmental Technology, Graz University of Technology, Graz, Austria

In recent years the usage of Lithium Ion Batteries (LIB) has notably increased. The demand for recycling processes for the valuable components of spent Lithium ion batteries, such as Cobalt, Copper, and Lithium etc. is highly desireable.

The aim of this work was to develop a complete recycling processes for valuable metals from aqueous solutions by implementation of Supported Liquid Membrane Technology (SLMT).

In the first process step the electrode materials were separated from the non-recyclable body of the Lithium Ion Battery. The target constituents were dissolved by acid leaching with HNO_3 and H_2O_2 at elevated temperature. Leaching solutions typically contain Co^{2+}, Al^{3+}, Cu^{2+} Li^+ Fe^{3+} and traces of Ni^{2+} and Mn^{2+} ions. The trivalent ions Fe^{3+} and Al^{3+} were removed by precipitation with LiOH leaving Cobalt, Copper and Lithium in solution.

The separation of Cobalt, Copper and Lithium was conducted with Supported Liquid Membrane Technology. Microporous Polyethylene membranes with a thickness of 2 mm, a pore size distribution of 7-12 μm and a porosity of 0,36 were used as membrane supports. The pertraction experiments were conducted in a lab scale unit, providing two chambers for the aqueous donor and acceptor solutions, separated by the porous membrane support with the extractants. Cobalt and Copper were separated with either DEHPA or Cyanex 272, leaving Li^+ finally in the raffinate phase for further processing. Cobalt and Copper were separated through extraction with LIX 84. Different flux of the metal ions into the acceptor solution was applied as selectivity criterion.

Based on the experimental results the complete recycling route was validated.

Separation of Phenols from Coal Tar and Coal Liquefaction Oils via Forming Eutectic Solvents

Yucui Hou[a], Kun Pang[b], Wujie Guo[b], Weize Wu[b,*]

[a]*Department of Chemistry, Taiyuan Normal University, 030031, Taiyuan, China.*

[b]*State Key Laboratory of Chemical Resource Engineering, Beijing University of Chemical Technology, 100029, Beijing, China.*

EXTENDED ABSTRACT

Phenolic compounds, which are basic materials of the organic chemical industry, are mainly products derived from direct coal liquefaction or coal tar. The present method to separate phenol compounds from oil mixtures is to chemically extract phenols using aqueous alkaline solutions (such as aqueous NaOH) and then to acidify the extract by mineral acids (such as aqueous H_2SO_4) to recover the phenols. However there are disadvantages: the use of large amounts of both strong alkalis and acids and the production of excessive amounts of waste water containing phenols.

In this work, ammonium salts was found to be successful as eutectic media for separating phenols from direct coal liquefaction oil and coal tar. First, ammonium salts have been used to efficiently separate phenols from modeling oils (hexane, toluene and p-xylene as model oil) via forming eutectic solvents. Ammonium salts, especially choline chloride (ChCl), were found to be successful as eutectic media for separating phenols (phenol and cresols) from oils (hexane, toluene and p-xylene). ChCl shows a short eutectic equilibrium time less than 3 min (see Figure 1) and an in-sensitivity to eutectic temperature. Significantly, no ChCl was found in the upper oil phase at equilibrium with the eutectic solvent. ChCl in the eutectic solvent can be recovered by diethyl ether and reused without mass loss and reduction in separation efficiency.

Then, ChCl were used to separate phenolic compounds in coal tar and coal liquefaction oil via forming eutectic solvents (see Figure 2). ChCl also shows a short eutectic equilibrium time for coal tar and coal liquefaction oil. The extraction efficiencies of phenols from coal tar (distillate at 130-260 °C) with a total phenol

* Corresponding author. E-mail: wzwu@mail.buct.edu.cn. Tel./Fax: +86 10 6442 7603.

content of 24.1% and direct coal liquefaction oil (distillate at 20-220 °C) with a total phenol content of 34.7% are 76% and 86%, respectively, with only ChCl to phenol mass ratio of 1:1. Significantly, no ChCl was also found in the upper oil phase. ChCl in the eutectic solvents can also be recovered by diethyl ether and reused without mass loss and reduction in separation efficiency.

Compared with the traditional methods, this proposed method involving the formation of eutectic solvent with ChCl avoids the use of alkali and acid and the production of phenol containing waste water and shows a promising foreground in coal chemical engineering technology.

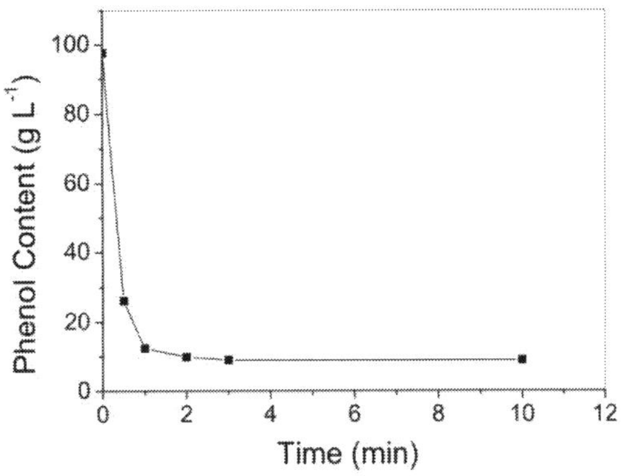

Figure 1. Phenol mass concentration in toluene versus time for eutectic to form. Separation conditions: 298.2 K; model oil, 20 cm^3 toluene; initial phenol content in toluene, 97.7 g·L^{-1}; ChCl, equimolar to phenol in oil.

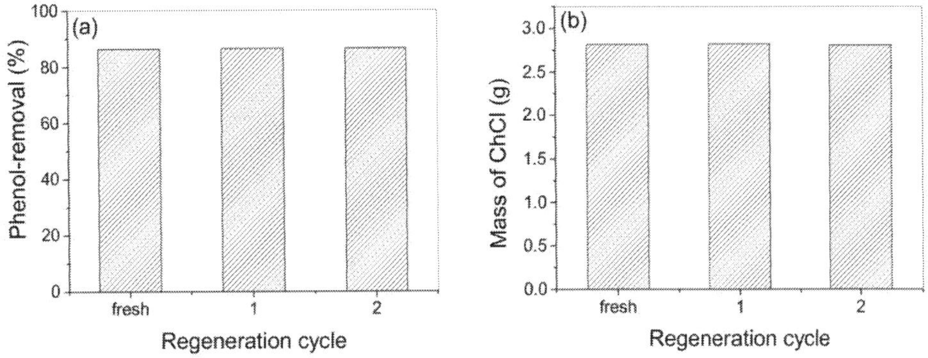

Figure 2. (a) Phenol-removal efficiency versus regeneration cycle; (b) Mass of ChCl versus regeneration cycle. Separation conditions: 313.2 K, 20cm^3 coal liquefaction oil.

KEYWORDS

Phenolic compounds, Separation, Choline Chloride, Eutectic solvent, Diethyl ether

ACKNOWLEDGEMENT

This work is financially supported by the National Basic Research Program of China (2011CB201300) and by Shanxi Scholarship Council of China (2011-086).

(84e) Deasphalting of Oil Sands Bitumen Using Supercritical Solvent Mixtures

Mustafa Al-Sabawi[1], Deepyaman Seth[2], Theo de Bruijn[1] and Yuming Xu[1], (1)CanmetENERGY, Natural Resources Canada, Devon, AB, Canada, (2)Alberta Innovates - Technology Futures, Edmonton, AB, Canada

The processing of heavy oil and bitumen is becoming increasingly important due to the decreasing supply of light crude. Conventional bitumen upgrading processes, such as vacuum distillation and coking, are energy-intensive and the liquid products are accompanied by significant amounts of coke and gases. In addition, the liquid products obtained are of poor quality and have high aromatic content. Using solvent deasphalting (SDA) can reduce energy consumption by as much as 30% compared to conventional upgrading processes.

Experiments involving the solvent deasphalting of bitumen using various solvent mixtures at supercritical conditions were carried out in a two-liter batch autoclave to determine the effect of modifier type and concentration on deasphalted oil (DAO) yield and quality. Athabasca bitumen was used as the feedstock and n-pentane was used as the primary solvent for extraction. Acetone, toluene, methanol and ethyl acetate were each added to n-pentane as modifiers in different concentrations to form the solvent mixtures. Extraction temperatures and pressures were maintained at around 200°C and 1100 psi to achieve supercritical conditions for solvents, and the solvent-to-feed volume ratio was kept around 6.5. Higher DAO yields were obtained with increasing modifier concentrations in n-pentane. DAO yields ranged from 79 to 92 wt% of the bitumen, with n-pentane/toluene mixtures providing the highest yields and n-pentane/methanol mixtures producing the lowest. Increases in DAO yield were accompanied by higher nickel, vanadium, nitrogen, sulfur and micro carbon residue content in DAO. Moreover, it was found that n-pentane/methanol and n-pentane/acetone solvent mixtures had lower selectivities for metals impurities than the other modifiers, while n-pentane/toluene had the highest selectivities.

(84f) Fractionation and Characterization of Pyrene Pitch Oligomers

David F. Esguerra and Mark C. Thies, Chemical and Biomolecular Engineering, Clemson University, Clemson, SC

Carbonaceous pitches can be used as precursors for a variety of high-performance carbon materials, including cathodes for lithium ion batteries, high thermal conductivity carbon fibers, and carbon-based composites. These pitches can be produced from various resources, such as coal tar, aromatic decant oil, and pure polycyclic aromatic hydrocarbons (PAHs) using either thermal or catalytic polymerization processes. Thies and co-workers have developed a method for fractionating thermally polymerized, petroleum-based pitches into narrow molecular weight (mol wt) cuts via dense-gas (also called supercritical) extraction (DGE or SCE). DGE can in principle be applied to all types of pitches, regardless of their method of production. Fractionating pitches in order to isolate the individual oligomers for study is a useful technique for helping us develop a better understanding of the molecular structure–bulk property relationships of pitches and their consituents.

In this study, pyrene pitch was fractionated into its constituent oligomers via DGE using N-Methyl-2-pyrrolidone (NMP)/toluene mixtures as the extractive solvent. The fractionation of synthetic, oligomeric pitches derived from PAHs such as pyrene is strongly limited by their lower solubilities in traditional aromatic solvents, such as toluene. The addition of NMP as a co-solvent to toluene was found to significantly increase the solubility of the pyrene pitch and allowed for the extraction of oligomers up to trimer.

The experimental setup consisted of a 17-mm i.d. x 2-m long packed column operated in a semi-continuous mode, in which solvent flows continuously through a pitch charge. A positive temperature gradient along the column from bottom to top (330, 350, and 380 °C) was used to enhance selectivity. NMP/Toluene mixtures of up to 30 mol % NMP and pressures of up to 80 bar were used to obtain fractions of different mol wt distributions and purities. The use of pure toluene under supercritical conditions (T_c = 321 °C, P_c = 42.2 bar) resulted in poor yields and the extraction of only monomer and low mol wt oligomers (i.e., dimer), whereas supercritical NMP/toluene mixtures resulted in higher yields and the extraction of oligomers up to trimer. The critical temperature and pressure of pure NMP have been estimated as T_c = 447 °C and P_c = 45.8 bar.

To our knowledge, this study represents the first time that NMP/toluene mixtures under supercritical conditions have been used for the fractionation of high mol wt PAHs such as pyrene pitch, and that pyrene oligomers up to and including trimer have been isolated. These results represent a significant step towards producing pyrene oligomeric standards (i.e., fractions of high purity) that can be used for establishing composition–property relationship in carbon materials.

(84g) The Purification of Polymeric Microspheres Via Soxhlet Extraction Using Liquid DME

Dong Woo Cho, School of Chemical and Biological Engineering and Institute of Chemical Processes, Seoul national univerrsity, Seoul, South Korea, Injung Jin, School of Chemical and Biological Engineering, Seoul National University, Seoul, South Korea, Kyoung Shil Oh, Petrochemicals and polymer R&D, LG Chem Research Park, Daejeon, South Korea, Won Bae, R&D Institute, Miwon Specialty Chemical Co., Ltd., Gyeonggi, South Korea and Hwayong Kim, Department of Chemical and Biological Engineering, Seoul National University, Seoul, South Korea

Polymer particles, used in cosmetic, pharmacy, and electric uses, require the purification for the removal of un-reacted monomer and residual surfactant used for the synthesis. For the purification of the synthesized polymer particle, the extraction using the organic solvent or supercritical fluid as a solvent was applied. In case of the organic solvent, the use of such common organic solvents, though, has been associated with several problems, including waste treatment and product separation disadvantage. Especially, the removal of amorphous fluorine-base polymer surfactants, the fluorinated or chlorinated solvents, such as CFCs and hexafluoro isopropyl alcohol, were used. Theses solvent cause not only waste treatment problem, but also ozone depletion. In case of supercritical fluid, especially $scCO_2$, high-pressure operating pressure, over 310 bar, causes the necessary apparatus to be quite costly. In an effort to circumvent these issues, the Soxhlet extraction using liquid DME was conducted. In this study, the Soxhlet extraction using liquid DME was conducted for on the removal of surfactants, siloxane-based surfactant and fluorine-based surfactant. After the extraction, the morphology was checked using FE-SEM and the residual concentration of surfactants was analyzed using EDS. Compared with n-hexane and liquid CO_2, liquid DME produced the more extracted microspheres than others produced. The residual surfactant concentration of the extracted microspheres was decreased dramatically.

(91a) Experimental and Theoretical Process Analysis of Membrane-Assisted Reactive Distillation

Johannes Holtbruegge[1], Sebastian Heile[1], Philip Lutze[2] and Andrzej Górak[1], (1)TU Dortmund University, Department of Biochemical and Chemical Engineering, Laboratory of Fluid Separations, D-44227 Dortmund, Germany, (2)Department of Biochemical and Chemical Engineering, Laboratory of Fluid Separations, TU Dortmund University, D-44227 Dortmund, Germany

Experimental and Theoretical Process Analysis of Membrane-Assisted Reactive Distillation

Johannes Holtbruegge, Sebastian Heile, Philip Lutze, Andrzej Górak

TU Dortmund University, Laboratory of Fluid Separations

Emil-Figge-Strasse 70, D-44227 Dortmund, Germany

Phone: +49 (0) 231 755-2342, E-Mail: johannes.holtbruegge@bci.tu-dortmund.de

Keywords: Process Intensification, Hybrid Process, Reactive Distillation, Membrane Separation, Modeling, Experimental Investigation

Due to decreasing fossil fuel reserves, the development of innovative process concepts that lead to economical and ecological benefits is absolutely necessary. Promising concepts to increase sustainability are integrated and intensified processes. Those processes integrate different phenomena or operations within one apparatus or allow for strong interactions between unit-operations with different separation characteristics in different apparatuses. Two examples are reactive distillation which superimposes chemical reaction and distillation in one apparatus as well as membrane-assisted hybrid separation processes that make use of a membrane to overcome thermodynamic limitations. As reactive distillation (RD) is nowadays the front-runner to overcome chemical equilibrium-limited reaction systems the combination of RD with a membrane separation is applicable for equilibrium-limited systems that exhibit strong thermodynamic non-ideality thus having a high potential for process improvement. To establish membrane-assisted reactive separation processes in industry, detailed experimental and model-based investigations of hybrid processes have to be performed. Thus it is possible to get on the one hand important process know-how and based on this on the other hand process design tools for membrane-assisted reactive separation processes. One of those tools is a process analysis tool which supports the user to the decision whether the process window of the considered phenomena for integration may happen at the same time and place as well as which process parameters are the sensitive ones to be intensified/improved to reach a defined process improvement.

In this work, both the process analysis tool as well as the experimental results in a pilot plant for the development of detailed models will be highlighted for a transesterification process.

Therefore, the transesterification of propylene carbonate (PC) with two molecules of methanol (MeOH) to form the target product dimethyl carbonate (DMC) and the byproduct 1,2-propanediol (PDO) in a chemical equilibrium-limited reaction is chosen as case-study.

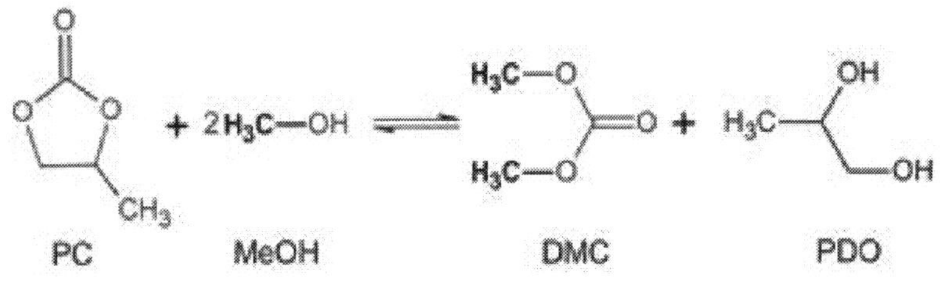

Reaction kinetics and chemical equilibrium for this chemical system were determined using sodium methoxide as homogeneous catalyst since no suitable heterogeneous catalyst has been found. The chemical equilibrium constant was found to be very low (K_{eq} = 0.2 at T = 343 K) yielding in low reactant conversions. By process analysis on the reaction, it was observed that by ensuring a molar excess of methanol or selective removal of one product species an almost complete propylene carbonate conversion can be obtained. To establish this while also having low recycle streams, a reactive distillation process is suggested for this system. With a molar excess of methanol in the feed stream it is possible to reach a complete propylene carbonate conversion while obtaining an azeotropic mixture consisting of dimethyl carbonate and large amounts of methanol in the upper part of the column. To overcome this azeotrope the integration with a membrane at the top of the column is identified. Hence, this mixture is fed to a vapor permeation membrane that separates unreacted methanol from the mixture so that dimethyl carbonate is enriched on the front side of the membrane. Unreacted methanol can be recycled afterwards to the RD column to guarantee a high methanol excess. A flowsheet of the membrane-assisted reactive distillation process is shown in Figure 1.

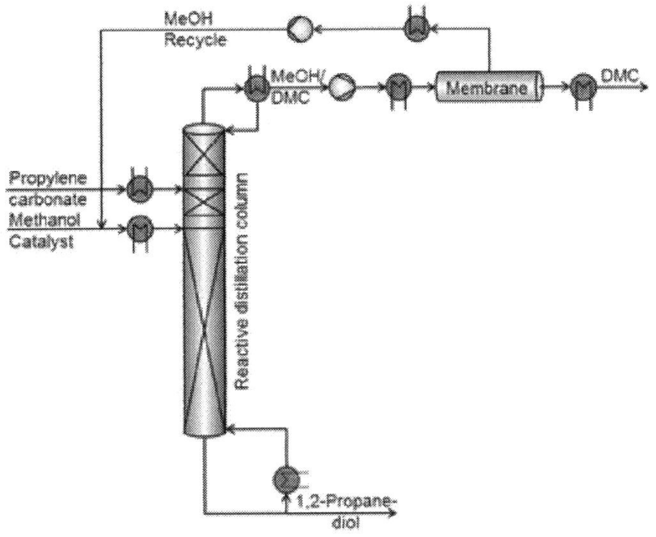

Figure 1: Simplified flowsheet of the membrane-assisted reactive distillation process

The integrated process is determined to be promising enabling the need for detailed models. Hence, experimental investigations of reactive distillation were performed in a pilot-scale RD-column with an inner diameter of 0.05 m that was equipped with Sulzer BX packing elements. The total packing height was 5.2 m and the total feed flow rate was 4 kg/h. The achieved composition and temperature profiles along the column were used successfully to validate the rate-based simulation model.

The vapor permeation was investigated in a lab-scale plant equipped with a flat-sheet membrane module. The separation characteristics of the dense, hydrophilic membrane Sulzer PERVAP^{TM} 1255 were determined by investigating the influence of feed composition, feed pressure and feed temperature on the transmembrane flux and membrane selectivity.

For the process analysis, a detailed simulation model comprising the rate-based model for reactive distillation and an empirical solution-diffusion model was set up. The model parameters for the membrane separation were determined from the lab-scale permeation experiments.

A process analysis on the detailed simulation model is used to show the influence of important operational parameters like reflux ratio and reactant ratio on the propylene carbonate conversion, the product purities and the necessary membrane area.

Acknowledgements:

The financial support of the German Federal Ministry of Education and Research for the project "Energy Efficiency Management and Benchmarking in the Process Industry" is gratefully acknowledged.

(91b) Hybrid Processes to Separate Enantiomers

Andreas Seidel-Morgenstern, Physical and Chemical Foundations of Process Engineering, Max-Planck Institute for Dynamics of Complex Technical System, Magdeburg, Germany and Heike Lorenz, Physical and Chemical Foundations of Process Engineering, Max Planck Institute for Dynamics of Complex Technical Systems, Magdeburg, Germany

Hybrid processes to separate enantiomers

Andreas Seidel-Morgenstern and Heike Lorenz

Max Planck Institute for Dynamics of Complex Technical Systems

Sandtorstr. 1, 39106 Magdeburg (Germany)

Enantiomers are pairwise occurring molecules, which are non-superimposable mirror images one of the other. Due to homochirality of life, there is a large interest and need to produce pure enantiomers in the pharmaceutical, fine chemical, food and agrochemical industries. Their provision is a challenging task since standard non-selective chemical synthesis always leads to racemic (50:50) mixtures and there is tremendous interest in the mentioned industries to develop innovative methods allowing for a faster access to pure enantiomers.

The first essential information for a rational selection of appropriate separation processes is the identification of the type of phase diagram for the specific chiral compound of interest. In the simplest but rare case that the chiral compound crystallizes as a conglomerate, it is most attractive to apply directly preferential crystallization [1]. However, more frequently racemic compounds are formed during crystallization from racemic feed mixtures. In these cases an initial enrichment is required prior to crystallizing a pure enantiomer. This enrichment might be provided by a partially selective synthesis or can be generated by an initial alternative separation process.

The presentation will summarize results of several case studies devoted to combine membrane separation [2] and preparative chromatography [3, 4] with subsequent enantioselective crystallization. The specific degree of enrichment required for successful crystallization was specified always based on preliminary measurements of ternary phase diagrams. Some of the examples were studied in the frame of the European project INTENANT (INTegrated synthesis and purification of single ENANTiomers), which attempted to combine the potential of the two rivaling general approaches, namely the development of a) enantioselective synthesis methods and b) physical methods aiming to separate efficiently mixtures of the two enantiomers.

The attractive incorporation of racemizing the counter-enantiomer into process schemes will be finally also discussed.

References

[1] G. Coquerel, *Preferential Crystallization in Novel Optical Resolution Technologies*, N. Sakai, R. Hirayama,

R. Tamura, Eds., Springer (Berlin, Heidelberg), 1-51, 2007.

[2] L. Gou, S. Robl, K. Leonhard, H. Lorenz, M. Sordo, A. Butka, S. Kesselheim, M. Wolff,

A. Seidel-Morgenstern, K. Schaber, *A Hybrid Process for Chiral Separation of Compound Forming Systems*,

Chirality 23 (2011) 118-127.

[3] H. Kaemmerer, Z. Horvath, J. W. Lee, M. Kaspereit, R. Arnell, M. Hedberg, B. Herschend, M. J. Jones,

K. Larson, H. Lorenz, A. Seidel-Morgensten, *Separation of Racemic Bicalutamide by an Optimized Combination of Continuous Chromatography and Selective Crystallization,* Org. Process Res. Dev. 16 (2012) 331–342.

[4] J. von Langermann, M. Kaspereit, M. Shakeri, H. Lorenz, M. Hedberg, M. J. Jones, K. Larson, B. Herschend, R. Arnell, E. Temmel, J.-E. Bäckvall, A. Kienle, A. Seidel-Morgenstern, *Design of an Integrated Process of Chromatography, Crystallization and Racemization for the Resolution of 2',6'-Pipecoloxylidide,* Org. ProcessRes. Dev. 16 (2012) 343-352.

(91c) A Simplified Conceptual Design Methodology for Double-Feed Extractive Distillation Processes

Libardo Estupinan Perez[1], Greg P. Dechaine[2] and Carlos A. Martinez Riascos[1], (1)Chemical and Environmental Engineering, Universidad Nacional de Colombia, Bogota, Colombia, (2)Chemical and Materials Engineering, University of Alberta, Edmonton, AB, Canada

Shortcut methods have an important role in the conceptual design of distillation processes because they rapidly provide numerical and qualitative constraints for important design parameters such as minimum reflux ratio, number of stages, process feasibility, possible splits, and, specifically for extractive distillation, minimum solvent feed ratio and maximum reflux ratio. The majority of past research in this area has dealt with single feed columns, with very little emphasis being placed on double feed extractive distillation processes.

In this work, a simplified method using reversible distillation models has been applied to the conceptual design of double feed extractive distillation columns. The proposed methodology uses the boundary value method (BVM) and the constant molar overflow (CMO) assumption for modeling the extractive column. The primary novel contribution of this work is the use of continuation methods for calculating the curves that determine the feasibility region and all the design parameters that can be derived from such a calculation. Combining the continuation methods with the column equations yields a simple method for calculating the pinch curves that border the feasibility region. We have found that these extractive distillation pinch curves, in addition to determining the feasibility region, are also useful for calculating an approximate minimum feed ratio independent of other parameters, as well as accurately pinpointing the minimum and maximum reflux ratios. Applications on extractive distillation systems allow us to conclude that the method is reliable, fast, and much easier to implement than other methods presented in the literature.

Simultaneous Esterification of Mixed Alcohols with Acetic Acid by Reactive Distillation

Karen A. Cañon-Rubio[1], **Jairo A. Duran**[1], Gerardo Rodriguez[1] and **Alvaro Orjuela**[1],

[(1)] Department of Chemical and Environmental Engineering, Universidad Nacional de Colombia, Bogota, Colombia, Grupo de Procesos Químicos y Bioquímicos. .

Abstract

Reactive distillation (RD) is a multifunctional operation that enhances selectivity and conversion through product removal in systems where conversion is limited by chemical equilibrium. Similarly, RD prevents the formation of undesired products, hot spots, avoiding azeotropic conditions, among other benefits. In this sense, RD become a potential processing alternative to overcome equilibrium limitations observed in esterification systems, primarily in those where high molecular weight and branched alcohols or acids are involved. Some of those branched; high molecular weight alcohols are present in bio-based products such as fusel oil, a well-known by-product in bioethanol production. In Colombia, Fusel oil is used as fuel in bioethanol plants, but its main components, isoamyl alcohol, ethanol and water can be separated and upgraded to produce fragrances and flavors additives, among other products with high value. In this work the mixed esterification of isoamyl alcohol and ethanol with acetic acid was studied by simulation using a RD process. Also, ethanol dehydration was considered as a feasible reaction. Design specifications were focused on obtaining high purity streams of ethyl acetate and isoamyl acetate, which are widely used in industry as paint solvent and artificial flavors, respectively.

Keywords: reactive distillation, fusel oil, mixed esterification, ion exchange resin, total annual cost.

1. Experimental

a. Phase Equilibrium

Reliability of the simulation results depends on the thermodynamic basis for all the systems involved. In this way, the present work considers 7 compounds and the consequent 21 binary systems listed in the Table 1.

Table 1 Binary systems considered.

	i	*j*		*i*	*j*
1	Ethanol	Water	11	Acetic Acid	Ethyl Acetate
2	Ethanol	Acetic Acid	12	Acetic Acid	Isoamyl Acetate
3	Ethanol	Isoamyl Alcohol	13	Isoamyl Alcohol	Ethyl Acetate
4	Ethanol	Ethyl Acetate	14	Isoamyl Alcohol	Isoamyl Acetate
5	Ethanol	Isoamyl Acetate	15	Ethyl Acetate	Isoamyl Acetate
6	Water	Acetic Acid	16	Ethanol	Diethyl Ethyl Ether
7	Water	Isoamyl Alcohol	17	Water	Diethyl Ethyl Ether
8	Water	Ethyl Acetate	18	Acetic Acid	Diethyl Ethyl Ether
9	Water	Isoamyl Acetate	19	Isoamyl Alcohol	Diethyl Ethyl Ether
10	Acetic Acid	Isoamyl Alcohol	20	Ethyl Acetate	Diethyl Ethyl Ether
			21	Isoamyl Acetate	Diethyl Ethyl Ether

Vapor-Liquid equilibrium data from literature and experimental work were fitted with NRTL-HOC activity model [1-3]. Experimental data for 2 systems were obtained in our laboratory using a Fischer Labodest VLE 602 apparatus. For Isoamyl Alcohol-Ethyl Acetate system, 80% of all the experimental data are thermodynamically consistent, passing Van Ness and Herington consistency tests. The difference between experimental and predicted data is less than 2% for this system.

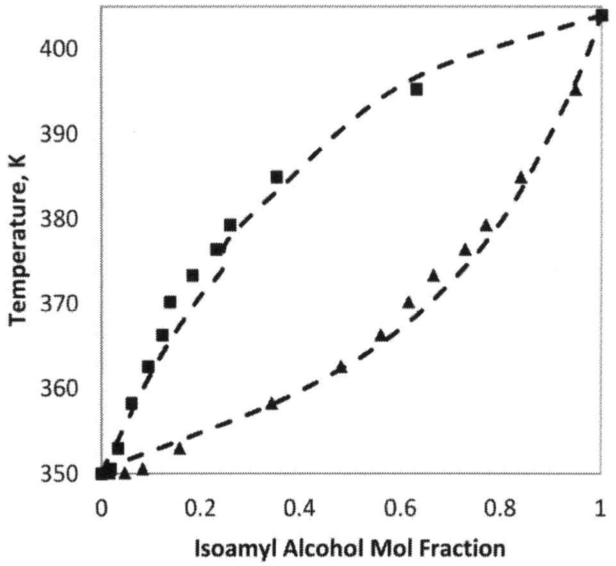

Figure 1 Measured vapor-liquid equilibrium data for Isoamyl Alcohol-Ethyl Acetate 101.3 kPa.

b. Kinetics

Three reactions were considered, for two of them, kinetics expressions from literature were found [4-6]. For the last one, a full factorial experimental design with three levels was used changing temperature, feed ratio and catalyst loading. The esterification was carried out using Amberlyst 70 cation exchange resin as catalyst, and runs were performed in a stainless steel batch reactor between 353–373 K. Kinetic data were fitted with Langmuir-Hinshelwood-Hougen-Watson (LHHW) and pseudo-homogeneous (PH) models; pre-exponential factor, adsorption constants and activation were obtained.

From the kinetic models, PH shows good fit for high temperatures, feed ratio and catalyst loading. However, for lower conditions, the predicted information is not close to reality. On the other hand, LHHW kinetic model shows better fitting in the range of experimental conditions, obtaining differences between experimental and predicted data is less than 5%. One of the runs at 370 K,1.25 Alcohol:Acid and 6 % of catalyst is show in **Error! Reference source not found.**.

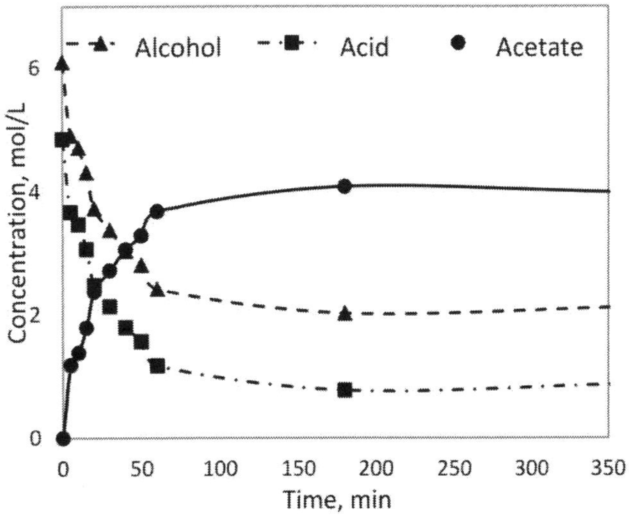

Figure 2 Isoamyl Alcohol esterification with Acetic Acid at 370 K, 1.25 alcohol/acid and 6% H+Eq.

2. Production Schemes and Simulation

Simulation was made using Aspen Plus® 7.2, and Radfrac model was used to simulate the catalytic column. Equilibrium stage model was assumed along the column, including reactive, stripping and rectifying stages. Considered schemes are show in Figure 3, concentration and temperature profiles for each one were calculated based on simulation results.

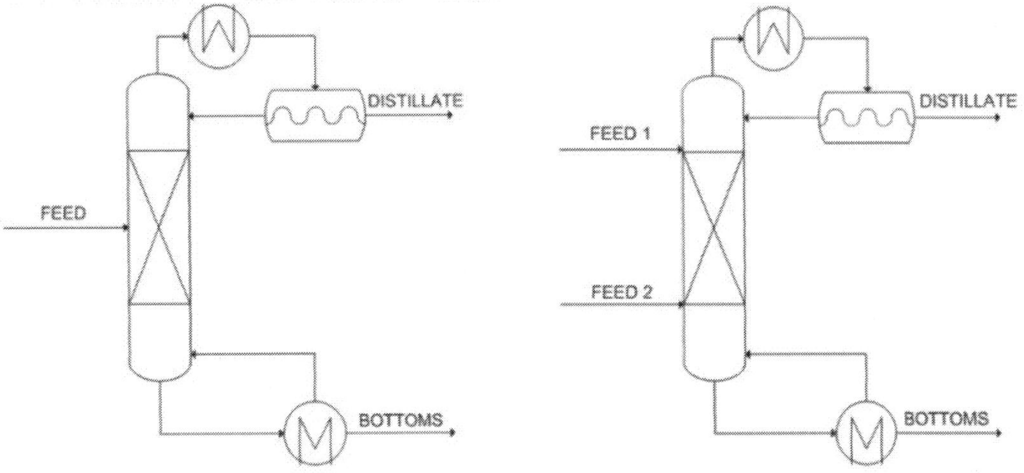

Figure 3 Production schemes tested. (left) RDC with one feed. (right) RDC with two feeds.

Sensibility analyses were made using minimization of Total Annual Cost (TAC) as objective function. The determination of number of reactive and non-reactive stages, reflux ratio, feed stage and feed ratio, etc. were found.

3. Conclusions

Vapor – Liquid equilibrium parameters regressed with NRTL-HOC for the 21 systems are in good agreement with the experimental data. Experimental kinetic data are correlated using LHHW and pseudo-homogeneus (PH) model. It was found that PH model do not describe the entire reactive behavior instead of LHHW. From both production schemes, DRC with one feed is the most suitable configuration in order to reduce the Total Annual Cost (TAC) and achieve the product purity. For the moment experimental validation of the presented configurations are out of the scope of this study.

References

1. Orjuela, A., et al., *Phase equilibria for reactive distillation of diethyl succinate: Part I.* . Fluid Phase Equilibria, 2010. **290**(1–2): p. 63-67.
2. Orjuela, A., et al., *Diethyl succinate synthesis by reactive distillation.* Separation and Purification Technology, 2012. **88**(0): p. 151-162.
3. Duran, J., *Esterificación del alcohol isoamílico con ácido butírico: Estudio del equilibrio químico y de fases.* 2012, Universidad Nacional de Colombia: Bogotá.
4. Orjuela, A., et al., *Kinetics of mixed succinic acid/acetic acid esterification with Amberlyst 70 ion exchange resin as catalyst.* Chemical Engineering Journal, 2012. **188**(0): p. 98-107.
5. Calvar, N., B. González, and A. Dominguez, *Esterification of acetic acid with ethanol: Reaction kinetics and operation in a packed bed reactive distillation column.* Chemical Engineering and Processing: Process Intensification, 2007. **46**(12): p. 1317-1323.
6. Teo, H.T.R. and B. Saha, *Heterogeneous catalysed esterification of acetic acid with isoamyl alcohol: kinetic studies.* Journal of Catalysis, 2004. **228**(1): p. 174-182.

(91e) Conceptual Design of Single-Feed Hybrid Multiproduct Reactive Distillation Column for Selectivity Engineering

Shabih Ul Hasan, Ranjan Malik and Sanjay Mahajani*, Department of Chemical Engineering, Indian Institute of Technology, Bombay, Powai, Mumbai 400076

***Abstract*—A conceptual design methodology for the synthesis of single feed hybrid multiproduct reactive distillation (HMPRD) column to obtain desired selectivities is presented, considering van de vusse reaction as a model reaction scheme. The methodology is based on the boundary value method already well established for non-reactive distillation columns and thus utilizes the concept of operation leaves of rectifying and stripping sections; the overlap of which indicates the potential feasible split. Based on the desired product composition, we determine the locus of reactive stage compositions in the RD column. Its intersections with the non-reactive stage composition lines (SCLs), allows rapid screening of feasible designs of HMPRD columns. A technique for the distribution of catalyst loading is also provided in case the catalyst loading is large for a single stage. The technique ensures that the desired selectivity remains unaffected after the distribution of catalyst loading. The applicability of this approach is restricted to a system with single reactant (i.e. single feed configurations). The results are useful for initialising more rigorous calculations.**

***Keywords*—Reactive Distillation, Design, Feasible Stage Composition Region, Selectivity**

I. Introduction

BECAUSE of the interactions between reaction and distillation in the combined unit, the synthesis and design of reactive distillation column is very complex. Simulations that are based on equilibrium-stage or non equilibrium stage modeling are widely used for designing an appropriate RD configuration [1], [2]. However, this approach may be at times tedious and undesirable in the early stage of process synthesis due to multiple feasible design alternatives. In some cases, the desired performance may not be feasible due to non-ideality or complexity of the reaction scheme. It is therefore necessary that one first identifies the maximum achievable performance (e.g. product composition, selectivity etc.) and the conceptual design associated with it. This design can be further fine-tuned with the help of rigorous simulation and optimization studies. The early studies on the conceptual design of reactive distillation mainly used transformed composition variables proposed by Doherty and co-workers [3], [4], [5] & [6]. These methods generate designs when a close approach to reaction equilibrium is assumed throughout the column i.e. either for high reaction rates or large residence times, which cannot always be attained in practice. Hence, subsequently, methods were also developed for the kinetically controlled reactions wherein Damkohler number is introduced to bring the effect of kinetics [7], [8] & [9]. The columns considered in all these studies were fully reactive wherein, kinetic fixed points may impose limitations on the performance. So far, conceptual design methods mostly consider simple reactions, and usually non-hybrid columns (i.e. fully reactive columns). Hence, there is a need in research in this area, for modifying the existing methods or finding new approaches which would be able to deal with more complex reactions, and Hybrid Configurations. To date, there is no systematic method for the identification and rapid screening of alternate configurations for a reactive distillation column giving best configurations in terms of selectivity for the complex reaction schemes. The present work contributes in this regard and presents a systematic conceptual design methodology based on BVMs to check whether the desired selectivity in the composition space is attainable with hybrid multiproduct reactive distillation (HMPRD) column or not. The approach allows rapid screening for the best HMPRD configuration and the corresponding operating parameters based on the desired selectivity. The methodology developed in this work offers a systematic and easy-to-use tool for the synthesis and design of reactive distillation columns for selectivity engineering.

II. Design Methodology for Single-Feed HMPRD Column

A. General Considerations

In this section, we discuss the prerequisites and key definitions required to understand the design methodology proposed in this work. We start with non-reactive distillation column and then introduce the reaction related attributes.

Boundary value method (BVM) — Reference [10] first proposed the boundary value design method to determine the minimum reflux ratio for the column involving ideal, non-ideal and homogeneous azeotropic distillation. In these methods, the composition trajectories are simultaneously calculated starting from fully specified top and bottom product compositions, and reflux (or reboil) ratio for the given feed composition and quality. If the two trajectories intersect, then a feasible design exists for the specified

Shabih Ul Hasan is with the Department of Chemical Engineering, IIT-Bombay, Powai, Mumbai 400076, India, (email: shabihamu@iitb.ac.in)
*Sanjay Mahajani is with the Department of Chemical Engineering, IIT-Bombay, Powai, Mumbai 400076, India, (Tel.: +91 22 2576 7246; 2578 2545; Fax: +91 22 2572 6895.email: sanjaym@iitb.ac.in)
Ranjan Malik is with the Department of Chemical Engineering, IIT-Bombay, Powai, Mumbai 400076, India,(email: rkmalik@iitb.ac.in)

reflux (or reboil) ratio; the number of stages in each section is given by the intersection point. The minimum reflux ratio is the value of the reflux ratio at which the two profiles just touch each other and this occurrence can be verified visually when the two profiles are plotted in a ternary diagram. A generic algorithm that uses the geometric properties of the multicomponent distillation systems is also proposed which does not require visualization of profiles in any form [14].

Operation leaves and column feasibility— Reference [11] introduced the concept of a product operation leaf, defined as the region of all possible liquid composition profiles at any reflux (or reboil) ratio leading to the specified product. This operation leaf is partly bounded by two extremes: total and minimum reflux. These two operating conditions are equally impractical, but they are useful in setting limits on the separation. A part of the boundary of the operation leaf comes from the pinch point curve on which, all the trajectories terminate at sufficiently large number of stages. More details can be found in [11]. The overlap section of the operation leaves indicates the region where the intersection of the potentially feasible split exists. There exists at least one continuous composition profile between the specified top and bottom products calculated at finite reflux and reboil ratios. The other necessary condition for the feasible design is the material balance around the column i.e. co-linearity of the feed and product compositions should be satisfied. Moreover, it should be kept in mind that the intersection of the composition profiles is not a sufficient condition for column feasibility. Since number of stages is a discrete variable, for a continuous composition profiles to exist in the column there must be an exact overlap of points from the rectifying and stripping trajectories at the intersection.

Stage Composition Lines (SCLs) — Reference [12] proposed stage composition line approach to column design and showed that the intersection of a pair of stage composition lines is a necessary and sufficient condition for staged distillation column feasibility. A stage composition line is defined as the locus of liquid compositions on a given stage at any reflux (boil-up) ratio. Unlike composition profiles, every point on a stage composition line can be reached by operating at the appropriate reflux or reboil ratio. Rectifying and stripping SCLs can be calculated from each product, and every intersection from a finite number of intersections represents a feasible combination of operating parameters. The advantage of using SCLs over composition profiles is that every point on a SCL is attainable by operating at the appropriate reflux or boil-up ratio. This is because SCLs are continuous loci, unlike composition profiles, which are sets of discrete points and hence intersection of continuous profiles is easier to obtain than finding points on two profiles that coincide.

B. Geometric Interpretation of CSTR & HMPRD Column

The earlier sections were dedicated to the analysis and interpretation of the stage compositions in the non-reactive distillation columns. The next step is to link these concepts to hybrid reactive distillation column wherein, the reaction takes place on one or more stages. Here, we view the column as a reactor and relate it to the conventional reactors for which geometric interpretation is well studied through the attainable region approach. Reference [13] presented geometrical interpretation of CSTR in terms of process vectors viz. reaction and mixing. In order to know whether a point in the composition space is attainable or not, they derived a condition that the rate vector at that point is collinear with the mixing vector of feed and product stream. Now we consider the reaction and the mixing vectors for any reaction scheme given by Eqn. 1.

$$\left.\begin{array}{l} r(c) = dc/dt \\ v(c,c^*) = c^* - c \end{array}\right\} \qquad (1)$$

The component material balance of CSTR gives the required condition. It must be noted that the reaction takes place at the product composition in the case of CSTR. The graphical interpretation is shown in Fig. 1a.

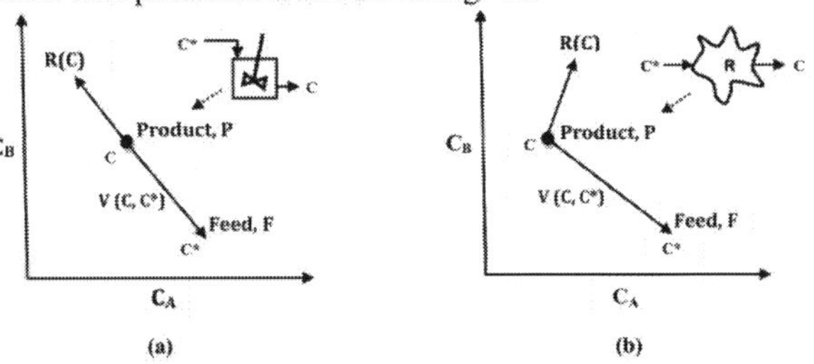

Fig. 1. Graphical representation of (a) CSTR (b) Arbitrary Reactor (R)

On the other hand, in an arbitrary reactor (R), if the composition at which reaction takes place is different than the product composition, then the collinearity condition is no longer valid (Fig. 1b). The composition at which reaction takes place is dependent of the type of reactor. One can control this composition by introducing separation attributes as is done the case of reactive distillation.

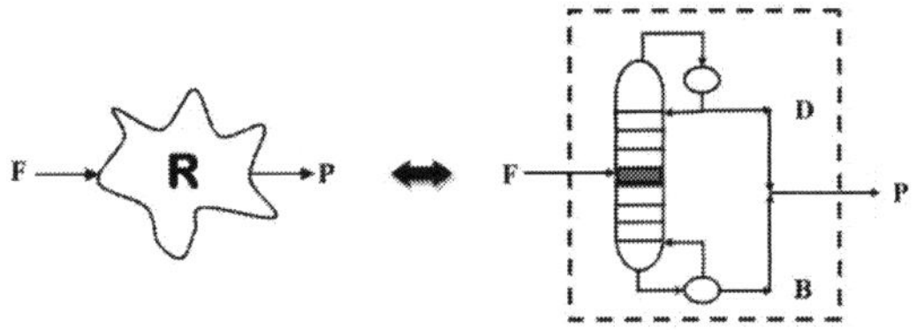

Fig. 2. Analogy between Arbitrary reactor (R) and HMPRD Column

Now, as shown in Fig 2, let us consider a hybrid distillation column with single reactive stage, wherein reaction does not take place at a composition corresponding to the product (P) given by mixing of the overhead and bottom streams. The composition of this reaction stage depends on the distillation attributes such as reflux ratio, feed location, number of stages etc. Now, the next exercise is to know- which are the points in the composition space that can be the potential reactive stage composition giving the desired selectivity corresponding to the point to be attained? The following section determines the locus of such feasible reactive stage compositions.

C. Key Definitions

In this section we define two new definitions viz. Locus of Reactive Stage Compositions (LRSCs) and Feasible Stage Composition Region (FSCR) which are used later to develop the design methodology for Single feed HMPRD column.

Locus of Reactive Stage Compositions (LRSCs)

Consider an equimolar van de Vusse type reaction scheme (eq. 2) with rate constants, [k1 k2 k3] = [1 1 3] and the feed as pure reactant A

$$\left.\begin{array}{l} A \xrightarrow{desired\ (k_1)} B \xrightarrow{undesired\ (k_2)} C \ \ and\ \ A \xrightarrow{undesired\ (k_3)} D, \\ \left[r_A\, r_B\, r_C\, r_D\right] = \left[-k_1 x_A^* - k_3 x_A^{*2},\ -k_2 x_B^* + k_1 x_A^*,\ k_2 x_B^*,\ k_3 x_A^{*2}\right] \end{array}\right\} \quad (2)$$

Material balances for components A and B on any arbitrary reactor are given by

$$x_{A,P} - x_{A0} = \tau.\sum_{k=1}^{nR} r_{k,A}(x^*) \quad (3)$$

$$x_{B,P} - x_{B0} = \tau.\sum_{k=1}^{nR} r_{k,B}(x^*) \quad (4)$$

Where, nR = total number of reactions; τ = residence time, $x_{i,P}$ is the product composition of component i , $x_{i,0}$ is the inlet composition of component i and x* denotes the compositions at which the reaction takes place in the reactor.

Divide (3) by (4) we get

$$\frac{x_{A,P} - x_{A,0}}{x_{B,P} - x_{B,0}} = \frac{-k_1 x_A^* - k_3 x_A^{*2}}{k_1 x_A^* - k_2 x_B^*} \quad (5)$$

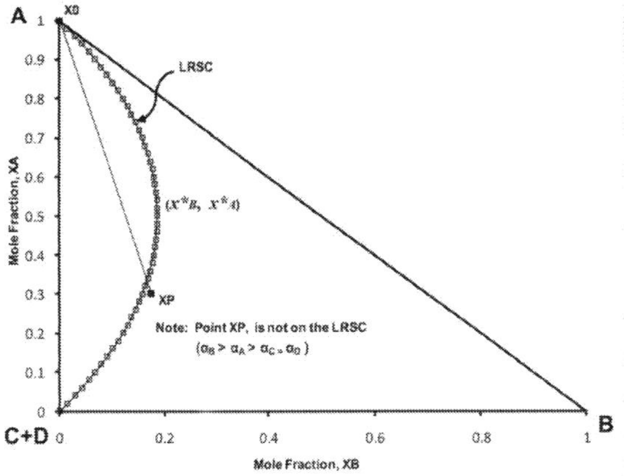

Fig. 3. Locus of reactive stage compositions (LRSC)

The L.H.S of Eqn. (5) is the slope of line joining points (x_{B0}, x_{A0}) and (x_{BP}, x_{AP}) as shown in Fig. 3. Hence, equation (5) can be rewritten as follows:

$$Slope\ AP = \frac{-k_1 x_A^* - k_3 x_A^{*2}}{k_1 x_A^* - k_2 x_B^*} \quad (6)$$

Hence, for any desired composition P, Eqn. (6) gives a relation between x_A* and x_B* to obtain the locus of reactive stage compositions (LRSC) as shown in Fig. 3. It should be noted here that by choosing a point P in the composition space implies fixed selectivity which we call the desired selectivity and hence if the reaction occurs at any of the compositions on the LRSC then the desired selectivity is ensured.

Feasible Stage Composition Region (FSCR)

As we know, it is not necessary that all the compositions on LRSC can be attained as the stage composition in the distillation column. For instance, it is probably intuitive that none of the stage composition is free from the component (A) with intermediate volatility, in a simple distillation column (i.e. single feed two product column). Therefore our intention is to find a systematic method to determine the feasible stage composition region (FSCR) for the distillation of ternary mixtures in a single feed distillation column for any reflux ratio and any number of stages. This region thus depends only on the feed composition and the vapor liquid equilibrium. Once we are able to find the FSCR for the given feed composition and volatalities then it is easy to visualize which of the reactive stage compositions on the LRSC are feasible in the composition space and can be attained on any one of the stages of simple distillation column. This new concept of FSCR is further used in this work to develop the design methodology for single feed HMPRD column. Before constructing FSCR we need to first construct the feasible distillate and bottom composition region [14].

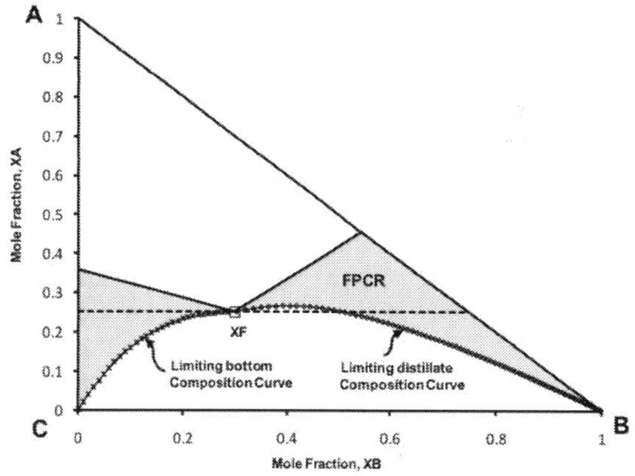

Fig. 4. Feasible Product Composition Region (FPCR)

Reference [15] & [16] used a pinch tracking method to calculate the feasible distillates and bottoms in a simple distillation column. The pinch-tracking methods are more accurate because they use the staged distillation column model equations and incorporate finite reflux ratios. Fig. 4 shows the Feasible Product Composition Region (FPCR) for a given feed composition (0.3, 0.25) and volatilities as [αA, αB, αC] = [1.9, 3.25, 1], obtained by using pinch tracking method of [16] The limiting distillate and bottom compositions are given by the locus of vapor pinch points and liquid pinch points, respectively. The product compositions below these curves cannot be accomplished by distillation in a single feed column [14].

Similarly by choosing the bottom compositions on the locus of liquid pinches we can obtain the locus of feed pinches for indirect split. And it turns out that this locus of feed pinches is same as the locus of liquid node pinches in rectifying and stripping sections which is evident in Fig. 5. Below this locus of feed pinches none of the stage composition is feasible in a simple distillation column. Hence this locus acts as the boundary between feasible and infeasible stage compositions i.e. below this locus all the stage compositions are infeasible.

Thus for a given feed composition and volatalities Fig. 6 represents the feasible stage composition region (FSCR) &

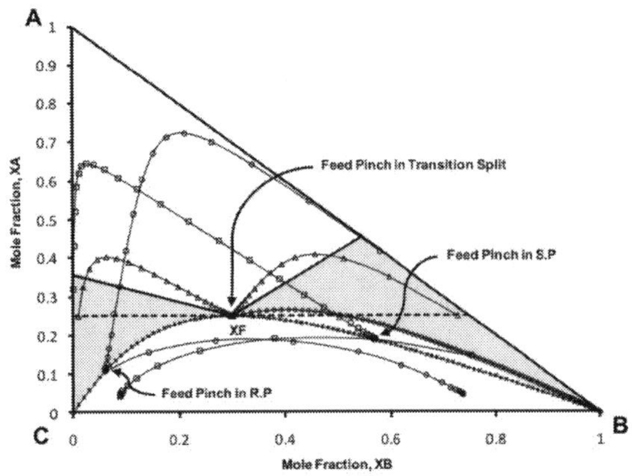

Fig. 5. Locus of liquid pinches in rectifying and stripping sections

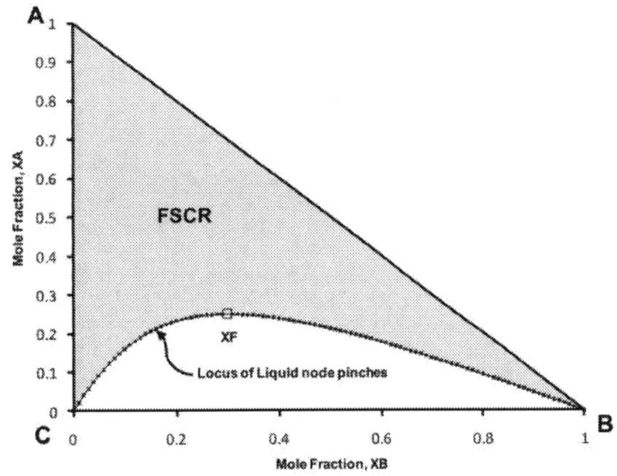

Fig. 6. Feasible Stage Composition region (FSCR)

the lower boundary of this FSCR can be mathematically obtained by solving the following set of three equations simultaneously.

$$\frac{x_{2,p}-x_{2,F}}{x_{1,p}-x_{1,F}}=\frac{y_{2,p}-x_{2,p}}{y_{1,p}-x_{1,p}}\ ,\ y_i=\frac{\alpha_i x_i}{\sum \alpha_j x_j}\ ,\ \sum x_i=\sum y_i=1 \qquad (7)$$

Where, $x_{i,p}$ = liquid mole fraction of component i at pinch point, $y_{i,p}$ = vapor mole fraction of component i at pinch point, $x_{i,F}$ = liquid mole fraction of component i in Feed, α = relative volatility.

III. Design procedure

Consider a reactor-separator system (Fig. 7a) and a single feed hybrid multiproduct reactive distillation column with one reactive stage (Fig. 7b). The approach presented in this work exploits the fact that for a given conversion and under the assumption of dominance of distillation over reaction, the column profiles of non-reactive (Non-RDC; Fig. 7a) and reactive distillation column (RDC; Fig. 7b) are almost identical. In other words, irrespective of whether we achieve the desired conversion inside the column or outside the column, the column profiles do not very significantly. To illustrate the above point we developed a conceptual design methodology for single feed HMPRD column by considering van de Vusse reaction as a model reaction scheme. It should be noted here that in this work when we say RDC, it means single feed HMPRD column with feed stage being a reactive stage.

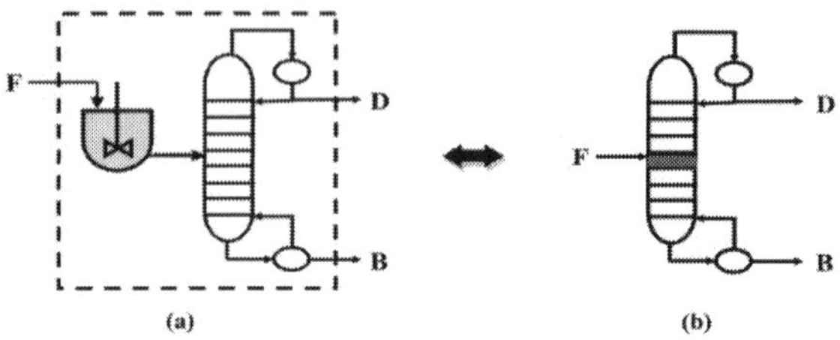

Fig. 7: (a) Reactor separator (Non-RDC) (b) HMPRD column (RDC)

Example: Equi-molar van de Vusse scheme and non-azeotropic mixtures

In this case, a simplified equimolar van de Vusse scheme (Eqn. 2) with rate constants, [k1 k2 k3] = [1 1 3] and volatilities as [αA αB αC αD] = [3 5 1 1] is chosen as an illustrative example. The feed is pure reactant A. Since the components C and D are characteristically similar, i.e. both are side products not undergoing any subsequent reaction and have same volatilities, we can group them together and the mixture can considered as a pseudo-ternary mixture for its graphical representation in 2-D space. Here onwards, we use the same reaction scheme, kinetics and volatility pattern for all the designs unless otherwise stated. In the following part of this section, we take you through each step of the developed design procedure and present the corresponding results.

Step-1: Select point P (Fig. 3) according to the desired selectivity in the composition space. This point represents the overall output composition of HMPRD column (Fig. 7b), obtained by mixing streams distillate (D) and bottoms (B). Hence, it can be considered as the feed to non-RDC (Fig 7a).

Step-2: Join point P (0.175, 0.3) with A (0, 1) to calculate the slope of line AP, and hence obtain the Locus of Reactive Stage Compositions (LRSC) using Eqn. 6.

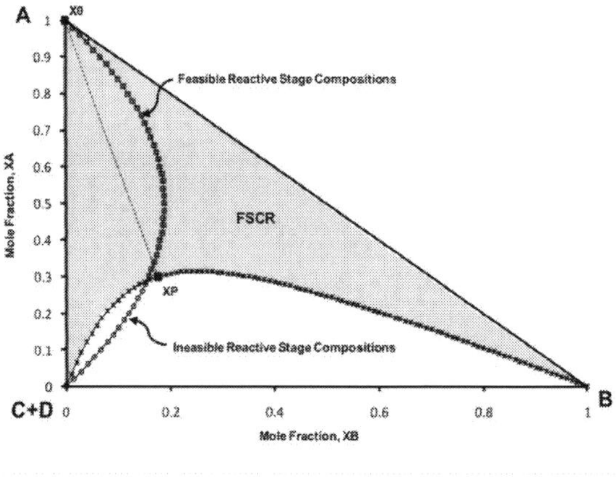

Fig. 8. Feasible and infeasible reactive stage compositions

Step-3: Obtain feasible stage composition region (FSCR) by solving the set of equations given by Eqn. (7) simultaneously and see whether any or some of the reactive stage compositions on LRSC lie in FSCR or not. If none of the reactive stage compositions lie in FSCR then it means that the point P is not attainable and it is not possible to get the desired selectivity with a single-feed HMPRD column. Fig.8 shows that many points on the LRSC are potential reactive stage compositions lying in FSCR.

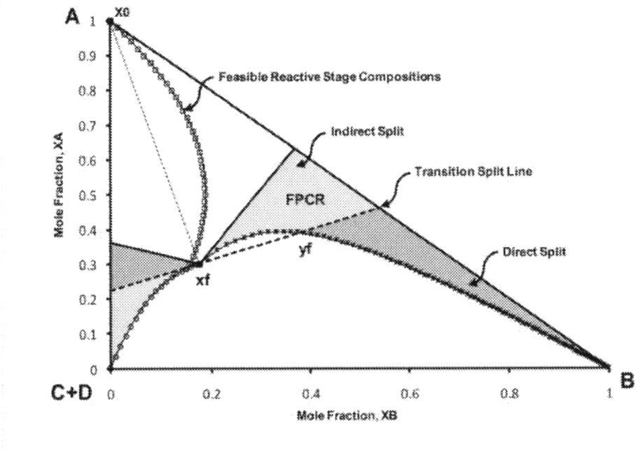

Fig.9. Feasible product Composition Region (FPCR)

Step-4: Draw the feasible product composition region (FPCR) by using pinch tracking method [16], as shown in Fig. 9 and choose the product specifications such that the overlap section of rectifying operation leaf (ROL) and stripping operation leaf (SOL) consist of feasible reactive stage compositions as shown in Fig.10.

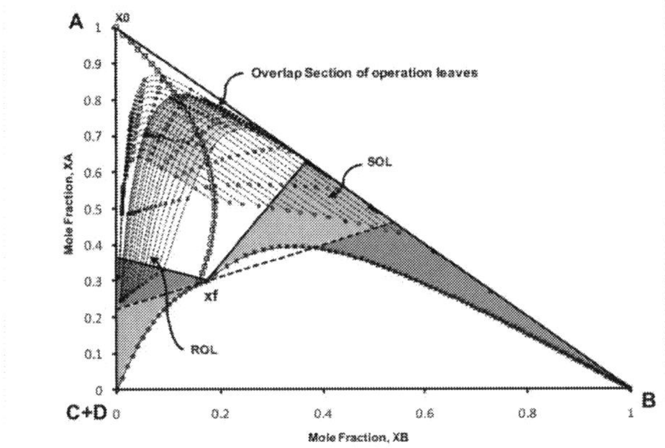

Fig. 10. Overlap of Rectifying Operation Leaf (ROL) and Stripping Operation Leave (SOL)

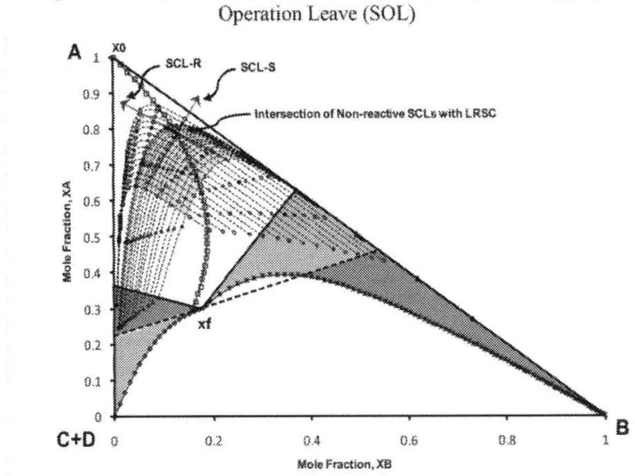

Fig.11. Common intersection point of non-reactive SCLs with LRSC

Step-5: Find the common point of intersection of non-reactive stage composition lines (i.e. stage composition line of rectifying (SCL-R) and stripping (SCL-S) sections) with the LRSC as shown in Fig.11 and obtain a feasible design meeting our requirements as shown in Fig.12. As discussed later, there are many such feasible designs possible.

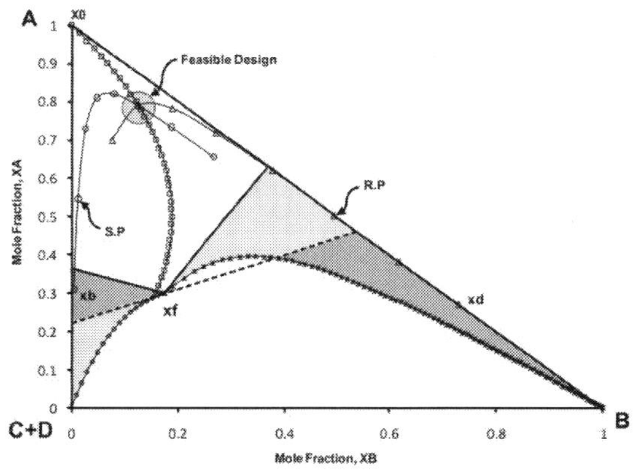

Fig. 12. Feasible design satisfying our requirements

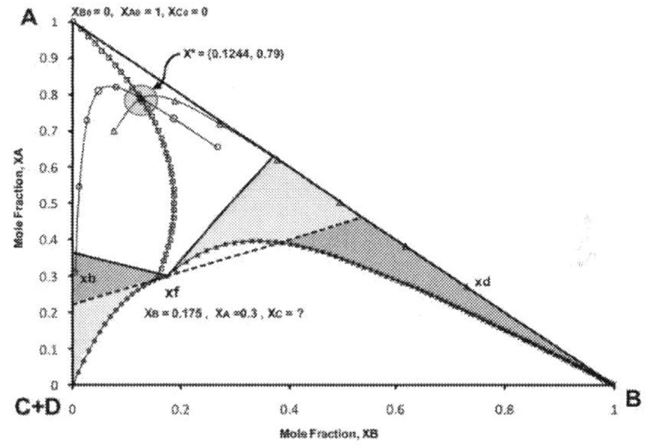

Fig. 13. Calculation of Damkohler Number and composition of 3rd component in feed

Step-6: At this stage, as regards to the side products, we only work with their total composition, i.e. ($X_{C,P} + X_{D,P}$). As we visualize RD as a combination of reactor followed by distillation column in series (Fig. 7), we need to know the composition of all the components including C and D in the feed that is sent to the distillation column. Hence, for further analysis, it is necessary that we know composition of the individual components C and D as a result of the reaction. The concentration of C is related to the extent to which individual reactions take place which in turn depends on the Damkohler number (Da) number i.e. the ratio of characteristic liquid residence time to the characteristic reaction time (Eqn. 8).

We determine Da (Eqn. 9) by applying material balance for component A for any arbitrary reactor in which reaction takes place at the composition X* as shown in Fig.13. Further, by using material balance for component C (Eqn. 10), one can find the composition of C in feed to the distillation column (Fig. 7).

$$Da=\left(\frac{W_{cat}*k_{ref}}{F}\right) \tag{8}$$

$$x_{A,0} - x_{A,P} + Da\left(\frac{-k_1 x_A^* - k_3 (x_A^*)^2}{k_{ref}}\right)=0 \tag{9}$$

$$x_{C,0} - x_{C,P} + Da\left(\frac{k_2 * x_B^*}{k_{ref}}\right)=0 \tag{10}$$

Step-7: Obtain the column profile for RDC, with the feed as pure A and the feed stage as the reactive stage, using same design and operating parameters as used for the Non-

RDC case. It should be noted here that to get identical column profiles one has to put catalyst on feed stage only, this is because feed compositions to non-RDC & RDC are different hence the column profile near the feed stage gets disturbed if we put catalyst other than feed stage as a result we couldn't get the identical profiles of Non-RDC & RDC. Hence, to overcome this feed effect, feed stage is chosen as reactive stage because we know that, for the same reflux ratio, R and number of stages, N the feed stage has the largest conversion as a single stage [17]. Fig. 14 and Fig. 15 show that the column profiles of Non-RDC & RDC exactly coincide. Hence, the desired selectivity for the given complex reaction scheme is attainable through HMPRD column with design specifications given in Table I.

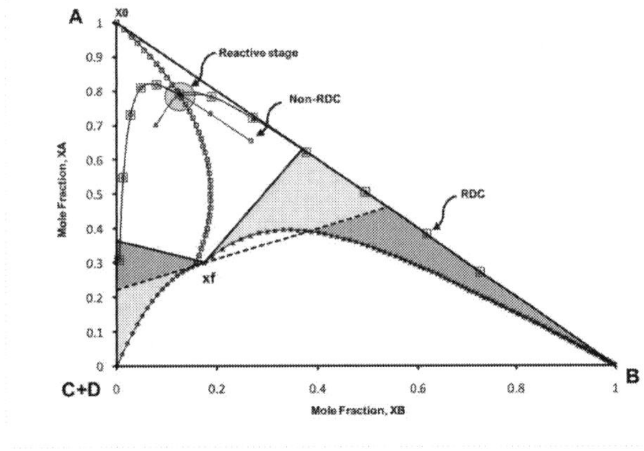

Fig. 14. Identical column profiles for RDC and Non-RDC in triangular diagram

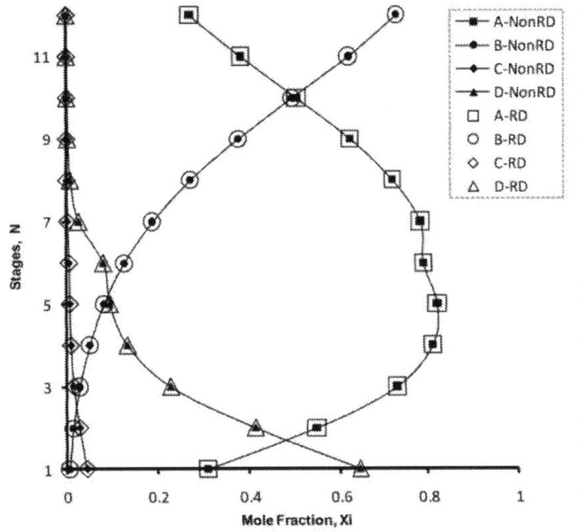

Fig. 15. Column profiles of RDC and Non-RDC.The coincidence indicates that the design is feasible

IV. Distribution of Catalyst Loading

In this section we provide a technique for the distribution of catalyst loading in case the catalyst loading is large for a single stage. Consider the two configurations in Fig. 16, the configuration having single reactive stage we call it SRC and that having multiple reactive stages with distributive feed we call it MRC. The approach exploits the fact that if we are able to maintain the extent of reaction in MRC same as in SRC then we will be able to distribute the catalyst loading without affecting the desired selectivity.

TABLE I
Design Specifications for single feed HMPRD column

Column Configuration	*Non-RDC*	*RDC*
Number of components, *NC*	4	4
Volatility, α	[3 5 1 1]	[3 5 1 1]
Number of stages, *N*	12	12
Feed location, *Nf*	6th stage	6th stage
Feed composition, *Xf*	(0.3, 0.175, 0.033)	(1,0,0)
Number of reactive stages, *Nr*	-	1
Location of reactive stage, *Nrl*	-	6th stage
Damkohler number, *Da*	-	0.263
Reflux ratio, *R*	36.96	36.96
Feed quality, *q*	Saturated liquid	-
Reboil ratio, *S*	*[DbyB*(R+q)]-1+q*	11.6931
Product specification (b=bottom, d=distillate) (A=comp 1, B=comp 2)	Xb_B=0.0043, Yd_B=0.729, Yd_A=0.27	-
End compositions after mixing the distillate & bottom streams	(0.3,0.175, 0.033)	(0.3,0.175, 0.033)

Where, Xb_B = liquid mole fraction of component B in bottoms, Yd_A= vapor mole fraction of component A in distillate, DbyB = ratio of distillate to bottom flowrate.

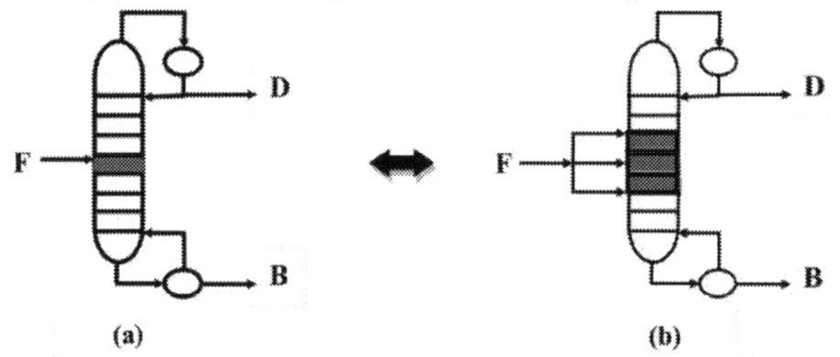

Fig. 16: (a) Single Reactive stage (SRC) (b) Multi-Reactive stage (MRC)

The general procedure for distributing the catalyst loading is as follows:

1. Introduce the number of reactive stages at the feed stage as per the requirement and distribute the catalyst equally on each reactive stage, keeping the number of stages in rectifying and stripping section same as in SRC.
2. Distribute the feed flow rate equally on each reactive stage as shown in Fig. 16 (b).
3. Start simulations for MRC with the operating parameters of SRC, further go on decreasing the value of reflux and reboil ratio until the reaction extent of MRC is same as SRC so that the desired selectivity will not be affected.

To illustrate the above technique we obtain another feasible design of HMPRD column (SRC) following the above developed algorithm, the design specification of which is given in Table II. The extent of reaction on each stage is a direct function of the stage composition, and Damkohler number. Since reaction and distillation goes simultaneously on the stage, therefore reaction extents also depend upon reflux and reboil ratios. It is clear that the extent of reaction for non-reactive stages is zero. The extent of reaction in the reactive section of MRC is calculated as the sum of reaction extents on individual reactive stages.

Example: Again for the equimolar van de Vusse scheme Eqn. (2) with rate constants, *[k1 k2 k3]* = [1 1 3], the extent of reaction for components A and B is given by

For Single- reactive stage column (SRC):

$$\xi_{A,SRC} = Da*\left(k_1 x_A^* + k_3 x_A^{*2}\right),\ \xi_{B,SRC} = Da*\left(k_1 x_A^* - k_2 x_B^*\right) \quad (11)$$

For Multi- reactive stage column (MRC):

$$\left.\begin{aligned}\xi_{A,MRC} &= \xi_{A,1} + \xi_{A,2} + - - - - - \xi_{A,Nr}\\ \xi_{A,MRC} &= \frac{Da}{Nr}\left(k_1 x_{A,1}^* + k_3 x_{A,1}^{*2}\right) + - - - + \frac{Da}{Nr}\left(k_1 x_{A,Nr}^* + k_3 x_{A,Nr}^{*2}\right)\end{aligned}\right\} \quad (12)$$

$$\left.\begin{aligned}\xi_{B,MRC} &= \xi_{B,1} + \xi_{B,2} + - - - - - \xi_{B,Nr}\\ \xi_{B,MRC} &= \frac{Da}{Nr}\left(k_1 x_{A,1}^* - k_2 x_{B,1}^*\right) + - - - - + \frac{Da}{Nr}\left(k_1 x_{A,Nr}^* - k_2 x_{B,Nr}^*\right)\end{aligned}\right\} \quad (13)$$

$$\xi_{A,MRC} = \xi_{A,SRC},\ \ \xi_{B,MRC} = \xi_{B,SRC} \quad (14)$$

Where, $\xi_{i,\ MRC}$ = extent of reaction of component i in MRC, *Da* = Damkohler number, *Nr* = total number of reactive stages.

It is clear that for a given reactive stage composition [X*= (0.38, 0.1767)] and Damkohler number (Da=0.86), the extent of reaction for SRC can be calculated from Eqn. (11). Now start simulations with the operating parameters of SRC and further go on decreasing the values of reflux and reboil ratios unless the constraint given by Eqn. (14) is satisfied as shown in Fig. 17.

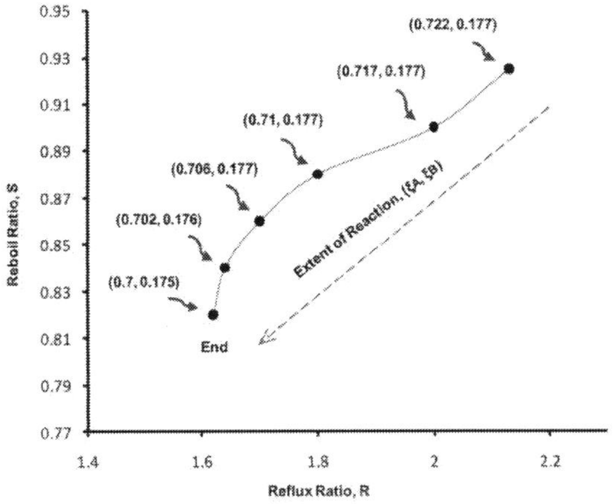

Fig. 17: Extent of reaction for different operating parameters

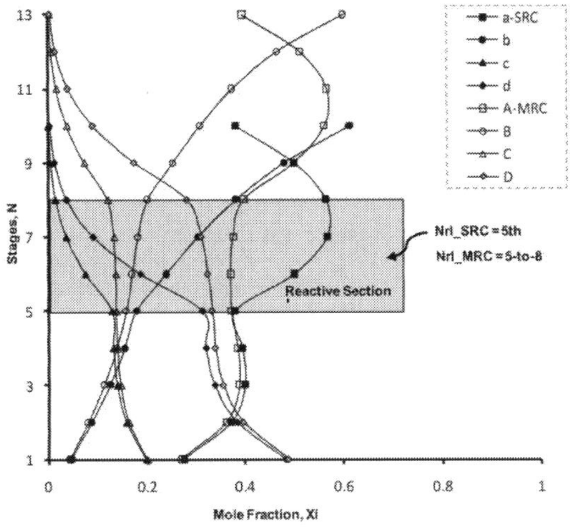

Fig. 18: Column profiles for SRC and MRC for the same extent of reaction

The column profiles for SRC and MRC for the same extent of reaction is shown in Fig. 18 and the design specifications are given in Table II.

TABLE II
DESIGN SPECIFICATIONS FOR SRC AND MRC

Column Configuration	*SRC*	*MRC*
Number of components, *NC*	4	4
Volatility, *α*	[3 5 1 1]	[3 5 1 1]
Number of stages, *N*	10	13
Feed location, *Nf*	5th stage	5-8th stage
Feed composition, *Xf*	(1,0,0)	(1,0,0)
Total feed flow rate, F	-	1
Feed flow rate of each stream, Fi	-	0.25
Number of reactive stages, *Nr*	1	4
Location of reactive stage, *Nrl*	5th stage	5-8th stage
Damkohler number, *Da*	0.86	0.86
Da per stage	0.86	(0.86/4)
Reflux ratio, *R*	2.13	1.62
Reboil ratio, *S*	0.925	0.82
End compositions after mixing the distillate & bottom streams	(0.3,0.175, 0.1521)	(0.3,0.175, 0.1521)

V. CONCLUSION

In this paper, a systematic conceptual design methodology, based on the BVM, is developed for single feed HMPRD column for complex reaction schemes. The approach allows rapid screening for the best HMPRD configuration in terms of desired selectivity, and the corresponding operating parameters, for a HMPRD column with a specified performance (reaction conversion and product specifications). Further, a technique for the distribution of catalyst loading by increasing the number of reactive stages is also provided in case the catalyst loading is large for a single stage. The technique ensures that the desired selectivity remains unaffected after the distribution of catalyst loading. The concepts developed in this work for the synthesis and design of single-feed HMPRD column can be used to analyse more complex configurations, such as double-feed HMPRD columns and can be extended to include azeotropic mixtures as well.

REFERENCES

[1] J. Slava, L. Jelemensky, J. Markos, "Numerical algorithm for modelling of reactive separation column with fast chemical reaction," Chemical Engineering Journal, 150, pp. 252-260, 2009.

[2] Z.H. Luo, X.Z. You, J. Zhong, "Design of a reactive distillation column for direct preparation of dichloropropanol from glycerol," Industrial and Engineering Chemistry Research, Vol-48, pp. 10779-10787, 2009.

[3] D. Barbosa, M.F. Doherty, "The simple distillation of homogeneous reactive mixtures," Chem. Eng. Sci. (43), pp. 541-550, 1988a.

[4] D. Barbosa, M.F. Doherty, "Design and minimum reflux calculations for single-feed multicomponent reactive distillation columns. Chemical Engineering Science (43), pp. 1523–1537, 1988b.

[5] S. Ung, M.F. Doherty, "Synthesis of Reactive Distillation Systems with Multiple Equilibrium Chemical Reactions," Industrial and Engineering Chemistry Research (34), pp. 2555-2565, 1995.

[6] J. Espinosa, P. Aguirre, G. Perez, "Some aspects on the design of multicomponent reactive distillation columns including non-reactive species," Chemical Engineering Science (50), pp. 469–484, 1995a.

[7] G. Buzad, M.F. Doherty, "Reactive Distillation by Design," Trans. IChemE, 70, Part A, pp. 448-458, 1992.

[8] S.M. Mahajani, A.K. Kolah, "Some design aspects of reactive distillation columns," Industrial and Engineering Chemistry Research (35), pp. 4587–4596, 1996.

[9] N. Chadda, M.F. Malone, M.F. Doherty, "Feasibility and Synthesis of Hybrid Reactive Distillation Systems," AIChE Journal 48 (12), pp. 2754–2768, 2002.

[10] S.G. Levy, D.B. Van Dongen, M.F. Doherty, "Design and synthesis of homogeneous azeotropic distillations. 2. Minimum reflux

calculations for nonideal and azeotropic columns," Industrial and Engineering Chemistry Fundamentals, Vol-24, pp. 463–474, 1985.
[11] F.J.L. Castillo, D.Y.C. Thong, G.P. Towler, "Homogeneous Azeotropic Distillation. 1. Design Procedure for Single-Feed Columns at Nontotal Reflux," Industrial and Engineering Chemistry Research, Vol-37, pp. 987–997, 1998.
[12] D.Y.C. Thong, F.J.L. Castillo, G.P. Towler, "Distillation design and retrofit using stage-composition lines," Chemical Engineering Science, Vol-55, pp. 625-640, 2000.
[13] D. Glasser, D. Hildebrandt, C. Crowe, "A geometric approach to steady flow reactors: the attainable region and optimization in concentration space," Industrial and Engineering Chemistry Research, Vol-26, Issue-9, pp. 1803-1810, 1987.
[14] M.F. Doherty, M. F., Malone, "Conceptual Design of Distillation Systems," McGraw Hill, 2001.
[15] M. Wahnschafft, J.W. Koehler, E. Blass, A.W. Westerberg, "The Product Composition Regions of Single-Feed Azeotropic Distillation columns," Industrial and Engineering Chemistry Research, Vol-31, pp. 2345-2362, 1992.
[16] Z.T. Fidkowski, M. F. Doherty, M. F. Malone, "Feasibility of Separations for Distillation for Non-ideal Mixtures," A.I.Ch.E. Journal, Vol-39, Issue-8, pp. 1303–1321, 1993.
[17] A.W. Westerberg and J.W. Lee, "Visualization of stage calculations in ternary reacting mixtures, "Computers and Chemical Engineering, Vol-24, pp. 639-644, 2000.

Study on Multi-scale of Composition in a Reactive Distillation column for Gasoline Desulfurization

Yonghong Li[1,2,3], *Benshuai Guo*[3]

(1) National Engineering Research Center of Distillation Technology, Tianjin, China
(2) Key Lab. for Green Chemical Technology of State Education Ministry, Tianjin, China
(3) School of Chemical Engineering and Technology, Tianjin University, 300072, China

Introduction

Reactive distillation (RD) is one of the important methods of process intensification in chemical engineering, as it has great potential to lower process costs and reduce environmental emissions. The RD residue curve maps are highly useful tools to visualize and elucidate conceptual designs of reactive distillation processes[1]. However, present computation of reactive residue curve assumes the homogeneous composition in the liquid phase, which would neglect the multi-scale structure of the reactive distillation system. As traditional mean filed (MF) models cannot always accurately capture the complex dynamics possibly resulting in lack of understanding of the interactions underpinning the system[2], a multi-scale model is required to analyze the reactive distillation system. The objective of this work is to study the muti-structure of a reactive distillation process with heterogeneous catalyst.

Muti-scale analysis

Macro-scale analysis

Considering a simple reactive batch distillation process, assuming that only one reaction occurs in the liquid system, the equations for computation of the reactive residue curves are written as [3]

$$\frac{dx_i}{d\tau} = x_i - y_i + \alpha\left\{\left[\nu_i - \left(\sum_{k=1}^{NC} \nu_k\right)x_i\right]\frac{r}{r_0}\right\} \qquad i=1,2,3,\ \ldots\ldots$$

The mole fractions of component i in the liquid phases and in vapor phases should be obtained with respect to "warped" time τ by solving the above Equations.

Micro-scale analysis

Considering an lighter reactant molecule in the solution of the reactive distillation system, it has two paths to undergo—be transferred to the gas phase or participate the reaction. The two control mechanisms in the micro-scale suggest that the reactive distillation couldn't be described thoroughly in the view of macro-scale, as it neglects the competition of the reaction and distillation in the micro-scale. Therefore, a meso-scale description is needed to correlate the micro-scale and macro-scale.

Meso-scale analysis

According to the different paths of micro-scale molecules, the homogenous liquid phase (x_i) in the macro-scale is divided into two phases (reaction phase x^{R}_{i} and distillation phase x^{D}_{i}) in the meso-scale. Here the concept phase is different from that in the thermodynamics, and it refers to the set of molecules with different behaviors.

Multi-scale model

Considering a batch distillation system with one heterogeneous reaction, We described the multi-scale model as follows:

① Equation of mass balance in the system

$$\frac{dx_i}{d\tau} = x_i - y_i + \alpha\left\{\left[v_i - \left(\sum_{k=1}^{NC} v_k\right)x_i\right]\frac{r}{r_0}\right\} \tag{1}$$

② Equation of phase equilibrium

$$y_i = K_i x_i^{D} \tag{2}$$

③ Equation of mass balance in the liquid phase

$$x_i = \varepsilon x_i^{D} + (1-\varepsilon)x_i^{R} \tag{3}$$

where ε is the mole ratio of the distillation phase.

④ Equation of reaction kinetics

$$r = f(x_i^{R}) \tag{4}$$

The distribution of the distillation phase and reaction phase is determined by their time and space. For the entire system, the reaction and distillation have equal time (t_D=t_R). The space of distillation and reaction (V_D and V_R) can be described by the multiple of the film thickness δ and surface area a respectively. The film thickness of the reaction space and distillation space can be assumed to be equal, as they are both decided by the flow conditions of the liquid. Therefore, the distribution of the reaction phase and distillation phase can be determined by the surface area of reaction space and distillation space.

$$\varepsilon = \frac{a_D}{a_D + a_R} \tag{5}$$

It is hard to derive this relation theoretically, so we use the adsorption of component i in the liquid phase on a heterogeneous catalyst for reference. As adsorption is a step of heterogeneous reaction, it is reasonable to use the relation of θ_i with C_i to describe the relation of concentration between the reaction phase and distillation phase.

$$x_i^{R'} = \frac{K_i x_i^{D}}{1 + \sum K_i x_i^{D}} \tag{6}$$

However, the sum up of $x^{R'}_{i}$ computed by equation (6) is not equal to 1, for the presence of unoccupied activity sites on the catalyst. Therefore, the mole fraction of reaction phase computed from equation (6) has to be normalized.

Based on the above analysis of reactive distillation with heterogeneous catalyst, a general conclusion could be derived: for a system with two control mechanisms, a multi-scale model is needed to give a detail description. Both the competition in the micro-scale and the compromising in the macro-scale should be included in the multi-scale model, and the bridge between the micro-scale and the macro-scale could be established by a meso-scale description.

Therefore, the key of multi-scale modeling for a system with two control mechanisms is to establish a meso-scale description, which on one hand needs to represent the competition between different mechanisms in the micro-scale, and on the other hand needs to be related with the composition in the macro-scale.

Analysis of particle-fluid system

Circulating fluidized beds have been used in a variety of industrial plants. As a typical example of complex system, extensive research has been carried out both experimentally and theoretically. Li[4] proposed an energy-minimization multi-scale (EMMS) model for CFB, based on the division of the flow system into dense phase and dilute phase.

Considering the heterogeneous structure of CFB, Li divided the flow system in to a gas-rich dilute phase and a solid-rich dense phase, and attributed the poor performance of predicting to the false assumption of the mean field of drag coefficient. He defined different drag coefficients for the dense phase, dilute phase and inter phase and established the stability conditions for particle-fluid system.Here we would like to illustrate the complexity of the particle-fluid system from the view of control mechanisms. In the macro-scale, the flows of particle and fluid are compromised. However, considering an element of one particle and little fluid in the micro-scale, there are two different control mechanisms-the behaviors of the element could be controlled by either the particle or the fluid, which would have a competition between each other, and this competition is not represented in the conventional macro-scale equations for describing the dynamics of CFB.

Therefore, meso-scale description is needed to bridge the gap of macro-scale and micro-scale. The competition of different control mechanisms leads to the two phases (dense phase and dilute phase) in the meso-scale, and the compromising of particle and fluid in the macro-scale could be derived from the equations of mass balance. Besides, the stability conditions are the expression of the control mechanisms, one dominated or compromising with each other.

Conclusion

There is a contradiction between macro-scale description of reaction distillation and the competition of the molecular motion in micro-scale. To establishe a multi-scale model for a system of gasoline desulfurization in a RD column, the liquid phase should be divided into reaction phase and distillation phase in the meso-scale according to the control mechanisms of the molecular motion in the micro-scale.

Acknowledgments

We aregratefultotheFundof National NaturalScience Foundation ofChina (No. 20976129) and the Program of Universities' InnovativeResearchTerms (No. IRT0936) for financial support.

References

1. C.Thiel, K. Sundmacher and U. Hoffmann (1997). Chem. Eng. J. 66, pp. 181-191.
2. H. Bostjan and T. Constantinos(2009). AIDIC Conference Series, pp. 167-176.

3. A.S. Granados-Aguilar, T. Viveros-García, E.S. Pérez-Cisnero(2008). Chem. Eng. J. 143, pp. 210-219.
4. J. Li, W. Ge, J. Zhang, M. Kwauk (2005). Multi-Scale Comromise and Multi-Level Correlation in Complex Systems--Challenges and opportunities for Chemical Engineering. Keynote speech at the 7th World Congress of Chemical Engineering, Glasgow, Scotland, pp. 10-14.

(109a) Recent Developments in Modeling the Dynamic Behavior of Confined Fluids

John Edison and **Peter A. Monson**, Chemical Engineering, University of Massachusetts, Amherst, MA

Recently we have developed a dynamic mean field theory (DMFT) that describes the dynamics of adsorption and desorption of fluids in mesoporous materials. The theory predicts the evolution of the density distribution of the fluid in the porous material after a step change in the bulk pressure or chemical potential. An important feature of DMFT is that it is fully consistent with the mean field density functional theory (DFT) description of the thermodynamics. In addition the nucleation mechanisms in phase transitions of confined fluids are emergent features of the calculations. For instance the theory describes the nucleation of capillary condenssation by liquid bridging. In this presentation we describe applications of DMFT to several systems. For pore networks we study the influence of pore connectivity upon uptake and desorption dynamics. Studies of adsorption/desorption dynamics of mixtures show the nature of the equilibration of the composition distribution in the system. We also show how DMFT can be used to study of nonequilibrium steady states encountered in membrane transport.

(109b) Molecular Modeling of Nanoporous Materials for Gas Storage and Separations

Youn-Sang Bae[1], Christopher E. Wilmer[2], Ki Chul Kim[2] and **Randall Snurr**[2], (1)Chemical & Biomolecular Engineering, Yonsei University, Seoul, South Korea, (2)Chemical and Biological Engineering, Northwestern University, Evanston, IL

Molecular modeling has become a powerful tool for predicting the adsorption of gases and liquids in nanoporous materials such as zeolites and metal-organic frameworks. This talk will give an overview of some recent examples. The thermodynamics of adsorption from the liquid phase tends to be complex and highly nonideal. In collaboration with Gino Baron and Joeri Denayer, we have investigated the adsorption of n-alkane mixtures in the zeolite LTA-5A under liquid-phase conditions. Advanced simulation techniques, such as parallel tempering, can capture the complex selectivity behavior observed in experiments such as selectivity inversion and azeotrope formation. We have also combined simulation and experiment to determine the adsorption selectivities for propene/propane mixtures in a series of isostructural M-MOF-74 materials (M=Co, Mn, and Mg) with high concentrations of open metal sites. Adsorption experiments, ideal adsorbed solution theory, breakthrough experiments, first-principles calculations, and grand canonical Monte Carlo (GCMC) simulations reveal that Co-MOF-74 exhibits a high thermodynamic propene/propane selectivity (ca. 46), due to strong pi-complexation between the open Co^{2+} sites and the propene molecules. These high selectivities occur at ambient pressure and temperature, conditions favorable for industrial adsorptive separations. An unusual increase of the propene/propane selectivity with increasing pressure is observed due to emergent behavior of the system that arises from the appropriate pore size relative to the size of the propene molecules. In the context of natural gas storage, we have recently developed a computational method for generating all possible MOF structures (subject to certain constraints) from a given library of MOF building blocks (nodes and linkers). From a library of 102 building blocks, for example, we created over 137,000 hypothetical MOFs on the computer and screened them for methane storage using short atomistic GCMC simulations. This method has the potential to greatly speed up the development of new MOFs for adsorption applications.

(109c) Some Examples of Cation and Framework Influences On Selective N2 Adsorption

Charles G. Coe[1], Thomas Gaffney[2], John Kirner[3], Herbert Klotz[4], James MacDougall[3], Scott Weigel[5] and Brian Toby[6], (1)Chemical Engineering, Villanova University, Villanova, PA, (2)Entegris, San Diego, (3)Air Products and Chemicals, Inc., (4)Air Products and Chemicals, (5)Exxon Mobil Research Corporation, (6)Argonne National Lab

Some Examples of Cation and Framework Influences on Selective N_2 Adsorption

Charles G. Coe[1*], Thomas. R. Gaffney, John. F. Kirner, Herbert. C. Klotz, James. E, MacDougall, Scott. J. Weigel ,and Brian. H. Toby

[1] Corporate Science and Technology Center, Air Products and Chemicals, Inc., 7201 Hamilton Blvd, Allentown, PA 18195 (*presently at Chemical Engineering Department, Villanova University, PA)

Selective N_2 adsorption in cation-rich zeolites has been known for over 40 years and provides the basis for the equilibrium-based production of O_2 from air. At the heart of this important process is a highly tailored adsorbent that combines the proper composition and pore structure to allow the economical production of 100 ton/day quantities of oxygen.

After giving an introduction to adsorptive separations, examples from our past research at Air Products will be reviewed and used to illustrate the remarkable importance of cation siting and structure in controlling selective N_2 adsorption. A multidisciplinary approach involving experimental synthesis, modeling, and extensive analytical characterization elucidated some new insights leading to improved air separation processes as well as other new applications. We found that both the N_2 capacity and selectivity are strongly influenced by not only the framework structure, but the type, size, location, and number of accessible and effective cations present in the zeolite . Comparisons of adsorptive properties on a series of Na, Ca, and Li exchanged X and Y type zeolites were made. In combination with known cation sitings for some of the compositions along with some additional structural findings, these studies showed that the site specific effectiveness of Li and Ca for selective N_2 adsorption was very different. Whereas Ca is effective in site II, Li is only effective in site III. Structural studies on CaLSX (Si/Al = 1.0) and ab initio calculations on six-ring clusters explain the observations and show there is a substantial difference in partial charges of Li and Ca due to local coordination geometries.

The importance of cation siting is not limited to X-type adsorbents. We also found an unexpected maximum in the N_2 capacity at ambient conditions for chabazite at a specific Si/Al ratio of 2.0 and developed a model to describe the strong variation in N_2 capacity with composition and aluminum distribution. The model also allowed us to define chromatographic chabazite-based adsorbents that can analyze trace oxygen in argon at ambient conditions, greatly simplifying the analytics for this important analysis. This review will hopefully reinforce the critical importance of using fundamental approaches to optimize adsorbent properties for a given application.

(109d) Theoretical and Experimental Studies of Scanning Adsorption-Desorption Isotherms On Mesoporous Materials

Alexander V. Neimark[1], Richard T. Cimino[1], Matthias Thommes[2] and Katie Cychosz[2], (1)Chemical and Biochemical Engineering, Rutgers, The State University of New Jersey, Piscataway, NJ, (2)Quantachrome Instruments, Boynton Beach, FL

Although the phenomenon of adsorption hysteresis has been attracting a lot of attention among both experimental and theoretical communities for several generations of researchers, its adequate description is still lacking. Indeed, the interplay between the thermodynamic and geometrical factors gives rise to distinct features of hysteresis in different materials. Experimentally, these distinct features are prominently displayed in the behavior of scanning isotherms, which provide additional information about the pore network geometry, including its connectivity and pore size distribution, which cannot be revealed from the main adsorption and desorption branches. Theoretically, it has been well understood since the seminal works of Everett [1] that the description of scanning isotherms cannot be achieved based on the models of adsorption in individual pores; it is necessary to take into account a cooperative nature of capillary condensation and desorption processes in three dimensional pore networks with distributed geometrical parameters of individual pores. In this work, we present high resolution adsorption and desorption scanning isotherms on porous materials of different yet well characterized structure: regular hexagonal arrays of pore channels in SBA-15 silica, cubically ordered three dimensional gyroid structure in KIT-6 silica, and ordered three dimensional network of spheroidal cages in 3DOm carbons. The classical system of Xenon adsorption on disordered pore network of Vycor porous glass [1] is served as a reference system. An original theoretical model is presented, which distinguish the materials, at which the scanning hysteresis can be described based on the assumption of individual pore domains, and the materials, at which the pore blocking effects are significant. For the latter case, a percolation model is suggested, which is capable of a quantitative description of the scanning isotherms and revealing the pore network connectivity and the size distribution of constrictions (necks) from the experimental data. At the same time, our research revealed more questions than answers that requires further studies.

1. Everett, D. H. In *The Solid-Gas Interface;* Flood, E. A.; Ed.; Decker: New York, 1967; Vol. 2, pp 1055-1113.

(109e) Hydrocarbon Separation in Nanopores: About Host-Guest Interactions and Molecular Freedom

Joeri Denayer, Department of Chemical Engineering, Vrije Universiteit Brussel, Brussels, Belgium

Many adsorptive separation processes are based on the use of nanoporous solids. Often, molecular separation is achieved as a result of differences in energetic interactions with the framework of the adsorbent. This allows separation of mixtures of polar and apolar molecules, separation of molecules with unsaturated bonds from saturated molecules or selective adsorption of molecules with acidic or basic character. For example, the interaction between double bonds in alkenes with cations in a zeolite allows the selective removal of alkanes from alkane/alkene mixtures.

The confinement of molecules inside pores of molecular dimensions also implies significant losses of freedom. Such entropic effects counterbalance the favorable energetic attractions, but more interestingly, in some cases allow to separate molecules with very similar properties. Typically, the freedom retained by the molecules strongly depends on the degree of pore filling. As a result, separation selectivity depends strongly on operating conditions. In this paper, several examples of such entropic based separation will be discussed.

(109f) Minimizing Axial Dispersion in Liquid Chromatography Using Microfabricated Pillar Array Columns

Gert Desmet[1], Jeff op De Beeck[1], J.G.E. (Han) Gardeniers[2] and Wim de Malsche[1], (1)Chemical Engineering, Vrije Universiteit Brussel, Brussels, Belgium, (2)Mesoscale Chemical Systems, MESA+ Institue for Nanotechnology, University of Twente, Enschede, Netherlands

The present contribution aims at illustrating and demonstrating how micro-machining technology can give a boost to High Performance Liquid Chromato-graphy (HPLC). Currently, HPLC is routinely used in nearly every chemical analysis lab. Despite its high degree of maturity, the technique however suffers from serious performance limitations when faced to the complex samples that need to be separated to solve the current state-of-the-art problems in the biological and pharmaceutical research (e.g., proteomics and metabolomics), the food and environmental analysis, etc…..

The currently used packed bed HPLC columns are clearly underachieving because of the packing disorder and the concomitant large degree of band broadening. To solve this packing disorder problem, the present contribution will focus on the possibilities of advanced photolithographic etching techniques such as the Bosch-process to produce perfectly ordered porous support columns with optimized hydrodynamic shape and optimized external porosity.

At the conference, rapid multi-component separations and the possibility to achieve very high separation efficiencies in pressure-driven LC will be demonstrated. The possibility to combine the pillar array technology with the latest sol-gel synthesis methods to produce monolithic silica columns with increased structural homogeneity will be discussed as well.

(109g) Recent Advances in Pressure Swing Adsorption Technology

James A. Ritter, Department of Chemical Engineering, University of South Carolina, Columbia, SC

Pressure swing adsorption (PSA) has been around for over half a century. Although it was commercialized for air drying in the late 1950's, the original Skarstrom patents mention many additional applications including air separation for the production of oxygen, air separation for the production of nitrogen, and even bulk vapor separation. These applications have all been commercialized.

PSA easily separates oxygen from air, nitrogen from air, and water vapor from ethanol vapor. One of the best success stories for PSA is its use in hydrogen purification from steam methane reforming. However, lesser known applications have also been commercialized, including gasoline vapor recovery, solvent vapor recovery, methane purification from natural gas wells, methane upgrading from landfill gas, and even carbon dioxide and water vapor removal from cabin air. The list seems to go on and on, with new separations or purifications being studied or even commercialized, like ethylene recovery from polyethylene production, as a more recent example.

Some of these PSA processes are based on equilibrium separation. Others are based on kinetic separation. Many of them utilize some type of aluminosilicate zeolite, while others utilize an activated carbon, and yet others utilize a carbon molecular sieve. Some of these PSA processes effect separation with a simple 2-bed 4-step cycle sequence, much like the original Skarstrom cycle. Others are utilizing a much more complex 16-bed 13-step cycle sequence. And, just about all combinations of bed numbers between 2 and 16 and cycle step numbers between 4 and 13 are being utilized in commercial and developmental PSA processes.

This presentation will provide an overview of some of the more recent advances in PSA technology. This overview will be based on recent journal literature, recent patent literature, and even unpublished results. This overview will also have a bias towards recent projects Professor Ritter and his team have worked on recently, and on-going projects currently being carried out by Professor Ritter and his group. A desirable outcome of this presentation will be a rudimentary understanding of why complexity in a PSA process is needed for some separations but not for others.

(119a) Effective Application of Established Treatment Technologies for Metals, Metalloids, and Oxyanions

James Patterson, Patterson Environmental Consultants, Chicago, IL

The underlying theories and fundamentals of application of treatment technologies for control of metals, metalloids and oxyanions are well established. However, the complexity of the wastewater matrix to be treated can introduce unanticipated responses to the applications of such technologies, including when discharge limits are unusually restrictive due for example to water quality-based effluent limitations, or in scale-up from treatability investigations to full-scale application. This Presentation will provide an overview of such theories and applications, and address pollutant speciation, wastewater matrix, and treatment technology aspects which can enhance or interfere with the effectiveness of treatability performance. Such aspects include target pollutant oxidation state and speciation, interfering (or enhancing) interactive species, wastewaters containing multiple (potentially non-compatible) target species, and other issues. Effective pretreatment can be requisite to the efficient application of the appropriate treatment technology. Pretreatment options may include extraction or chemical modification of complexing or other interfering species, electrochemical pre-oxidation or reduction of a target speci, and management of adsorptive surface area and electrical charge in applications of inherent primary, or applied secondary co-precipitation. Improved awareness and understanding of such issues support cost- and performance-effective application of treatment technology.

(119b) Impact of Water Quality and Assessment of Adsorption Technologies for Arsenic Removal

Kashinath Banerjee, Veolia Water Solution & Tech, Moon Township, PA, Gary L. Amy, Water Desalination and Reuse Research Center, 4700 King Abdullah University of Science and Technology, Thuwal, Saudi Arabia and Tapas K. Das, School of Engineering, Saint Martin's University, Lacey, WA

Abstract

The sources of arsenic in drinking water are both natural and anthropogenic. In order to protect public health, the U.S. EPA, World Health Organization (WHO), and the European Commission have decided to lower the MCL of arsenic in drinking water to < 10 μg/L. This stringent arsenic standard will inevitably require many utilities to upgrade their present systems or consider new treatment options. Extensive literature and vendor surveys revealed that a large number of commercial and experimental adsorbents are available for arsenic removal. Some of the commercially available materials are matured products (e.g., activated alumina) that have been widely tested, revealing both their attributes and limitations, while others are more recent products. Results from several investigators (DeMarco *et al.*, 2003; Greenleaf and SenGupta, 2005; Jekel et al., 1999; Pena *et al.*, 2005) indicate that some of these adsorbents are very promising.

Bench-scale studies were conducted to develop a comparative assessment of several commercially available adsorbents for arsenic removal, and to assess their performance in the presence of phosphate, vanadium, and silica. The primary objective of this work was to identify the impacts of water quality, and to define adsorbent properties that will minimize the effects of water quality and/or maximize arsenic adsorption capacity. Additionally, the effect of temperature on the adsorption kinetics and capacities were investigated. Iron modified activated alumina (AA-FS50), granular ferric hydroxide (GFH), granular ferric oxide (GFO), and a titanium-based media (MetSorb G) were used as the adsorbent. In order to better understand the adsorption mechanics of the selected adsorbents, each was characterized according to its important physical/chemical properties, including particle size, bulk density, pH of zero-point charge (pH_{zpc}), surface area, and mineralogy. The concentrations of dissolved arsenic and vanadium were analyzed using Inductively Coupled Plasma – Mass Spectroscopy (ICP-MS); Ion Chromatography (IC) was used to analyze phosphorus. The concentrations of molybdate-reactive (monomeric) silica in the samples were measured, and polymeric silica was estimated by the difference between the total and monomeric silica.

Results revealed that pH has a significant impact on the adsorption capacities. AA-FS50 showed a moderate decrease in adsorption capacity as pH increased, and phosphate had a small impact. GFO media was only slightly affected by increased pH, but was significantly impacted by phosphate, and monomeric silica had some minor effects. GFH performance decreased as pH increased, with phosphate and monomeric silica exhibiting some impact. The adsorption performance of MetSorb G also decreased as pH increased, and phosphate and monomeric silica had some impact, although the interfering effects were less at a lower pH for this media. The two iron oxide media and MetSorb G generally exhibited comparable capacities. For iron oxide based media, rate data are best described by a pseudo-first order kinetic model at each temperature and pH condition studied. At lower pH values, arsenic (V) exhibits greater removal rates than arsenic (III). An increase in temperature increases the overall adsorption reaction rate constant values for both arsenic (V) and arsenic (III). An examination of thermodynamic parameters reveals that adsorption of arsenic (III) as well as arsenic (V) by iron oxide media is an endothermic process and is spontaneous at the specific temperatures investigated.

References

DeMarco, M.J., SenGupta A.K., Greenleaf J.E., 2003. Arsenic removal using a polymeric/inorganic hybrid sorbent, *Water Res.* 37, 164-176.

Greenleaf, J. E. and SenGupta, A. K., 2005, *Investigating New Arsenic Removal Technologies*, Chapter 9.3.2, p. 524-

544; In Das, T. K., (ed.), *Toward Zero Discharge: Innovative Methodology and Technologies for Process Pollution Prevention,* John Wiley and Sons, Hoboken, NJ.

Jekel, M, and Seith, R., 1999. Comparison of conventional and new techniques for the removal of arsenic in a full-scale water treatment plant. Proceedings World Water Conference, Buenos Aires, Argentina.

Pena, M.E., Korfiatis, G.P., Patel, M., Lippincott, L., Meng, X., 2005. Adsorption of As (V) and As (III) by nanocrystalline titanium dioxide. *Water Res.* 39, 2327 – 2337.

(119c) Removal of Arsenic by a Nano-Particle Blend Hollow Fiber Membrane

Jinsong He, Civil and Env Engineering, National University of Singapore, Singapore, Singapore and J. Paul Chen, Department of Civil & Environmental Engineering, National University of Singapore, Singapore, Singapore

The objective of this research is to prepare a based nano-particle blend hollow fiber membrane to adsorb arsenate from groundwater. A new nano-sized metal oxide was first prepared, which was then used in fabrication of the membrane. The hydrophobicity, porosity, mechanical strength, water flux and arsenate adsorption capacity of the membrane were investigated. The additional zirconium based nano-particle increased the hydrophilicity, which resulted in a significant enhancement on the water flux of the blend membrane. The batch adsorption experiments indicated that the membrane could effectively remove arsenate in a wide pH ranging from 2 to 9. The adsorption equilibrium could be reached in 24 h, and the maximum adsorption capacity was above 120mg/g. Ionic strength did not obviously affect the adsorption. The co-anions such as F^-, HCO_3^-, $SiO_3^{2-}$and PO_3^{2-} had different effects on the adsorption. In addition, the membrane could remove arsenate efficiently though the continuous filtration. The mechanism for the removal of arsenate was proposed based on the FTIR and XPS studies.

COMPETITIVE ADSORPTION OF NICKEL, MANGANESE, CHROMIUM AND CADMIUM FROM AQUEOUS SOLUTIONS BY DIFFERENT ADSORBENTS

Malyuba Abu-Daabes [a,*], *Hani Abu Qdais* [b], *Hatem Alsyouri* [c]

[a] *Pharmaceutical-Chemical Engineering Department, School of Applied Medical Sciences, German-Jordanian University, Amman, Jordan*

[b] *Department of Civil Engineering, Faculty of Engineering, Jordan University of Science and Technology, Irbid, Jordan*

[c] *Department of Chemical Engineering, Faculty of Engineering and Technology, University of Jordan, Amman, Jordan*

**corresponding author:*

Telephone: +962-6-530-0666, Fax: +962-6-5300-672, Email address: malyuba.abudaabes@gju.edu.jo

Abstract

Landfilling is the primary solid waste disposal method in Jordan. Since Jordan is considered one of the poorest countries with water resources, it is crucial to protect its groundwater resources from any possible risk of contamination from landfill leachates. A survey on leachates from landfills of different ages and control systems in Jordan, performed by our group during the years 2009 and 2010, showed that heavy metals concentrations in leachates were high exceeding the allowable limits of both Jordan Industrial wastewater and US Environmental Protection Agency. The main heavy metals with high risk are identified to be: Cr (0.66-20.5 mg/L), Mn (0.41-10.37 mg/L), Cd (0.48-4.5 mg/L), and Ni (0.74-6.69 mg/L). In the present study, competitive adsorption of these four metals on different adsorbents including: natural untreated shrimp shells (SH), granular activated carbon (GC), acid-treated granular activated carbon (AGC) and acid-treated powder activated carbon (APC) were investigated under simulated conditions. All adsorption isotherms were generated using the bottle point method, following the ASTM standard procedure (ASTM D3860-89a). Untreated granular carbon (GC) was the best adsorbent for Ni removal from a simulated solution with 90% removal efficiency. AGC and APC both showed very high selectivity for Cr over other present metals with removal efficiency of 99%. For Mn, shrimp shells (SH) was the best adsorbent with 79% removal efficiency, while both GC and SH adsorbents have comparable efficiency for Cd removal from the simulated solution (92%). Order of metals selectivity for SH adsorbent was Cd > Mn > Ni > Cr, for GC the selectivity was Cd > Ni > Cr > Mn, for both AGC and APC the order was Cr > Cd > Ni > Mn.

Key words: competitive metal ions, Leachate, Nickel, Manganese, Chromium, Cadmium, activated carbon, shrimp shells, chitin.

Introduction

Landfilling remains the predominant management method for the disposal of Municipal Solid Waste which leads to the formation of leachates. Leachates may act as a vehicle for the introduction of heavy metals to groundwater resources. Contamination of water resources by heavy metals is a great environmental problem, mainly due to their high toxicity even at very low concentrations (Jaramillo et al., 2009). Heavy metals are not biodegradable and cannot be removed by natural processes from the ecosystems. A previous study by our group on leachates collected from three landfills with different ages and control systems showed that leachates from landfills possess high contamination potential to the limited water sources in Jordan. Heavy metals, mainly Cr, Mn, Cd, and Ni, were found to exceed the allowable limits set by Jordan Industrial Estates Corporation and US Environmental Protection Agency for industrial wastewater discharge. Pollution of heavy metals may arise from different sources. Cd, Cr, Mn and Ni are mainly introduced to the environment from spent rechargeable and household batteries, electroplating, (Xara et al. 2009; Karnchanawong and Limpiteeprakan 2009) and tannery industry (Gotvajn et al., 2009).

One of the promising methods for heavy metals removal from wastewater is the use of natural and surface-modified adsorbents for selective removal of metal ions. Activated carbon is widely used in many applications. In this study the effect of modifying activated carbon surface with acidic groups on competitive metal ions removal from aquatic solution is evaluated. Shrimp shells are also used as a low cost natural adsorbent without any surface modification.

Materials and Methods

1) Adsorbents

Activated carbon was used both in granular (GC) and powder form. All carbon was boiled in deionised water, dried in an oven at 110 °C for 24 hours. Further surface modification was performed by boiling part of them with 4M HNO_3 solution for 1 hour with total condensation. After cooling, the solution of the acid-treated granular activated carbon (AGC) and acid-treated powder activated carbon (APC) was decanted and the solids were washed with deionised water several times until a neutral, constant pH was achieved. Shrimp shells (SH) were washed thoroughly, grounded and sieved without further modification.

2) Characterization of Adsorbents

Surface area and pore size distribution of different adsorbents were measured following the Brunauer, Emmett, and Teller (BET) method (Quantachrome Instruments, Model Autosorb-1, USA).

Table 1. Adsorbents characterization

Adsorbent	BET	HK cumulative pore volume	HK pore wedth
	m^2/g	cc/g	nm
GC	635	2.99E-01	1.398
AGC	591	2.79E-01	1.388
APC	892	4.22E-01	1.388
SH	25.6	9.95E-03	1.058

3) Adsorption isotherms

Stock solutions of heavy metals of 100 mg/L concentration (Ni(II), Mn(II), Cr(VI)) and 50 mg/L Cd(II) were prepared separately by dissolving appropriate quantities of nickel (II) nitrate hexahydrate, cadmium nitrate in nitric acid 0.5 mol/L (Fluka, Switzerland), manganese (II) sulfate monohydrate and chromium (VI) oxide (Merck, Germany). Adsorption isotherms were generated using the bottle point method, following the ASTM standard procedure (ASTM D3860-89a). Adsorption isotherms were generated at 25 °C, using simulated solution of the main heavy metals found in the landfill leachates. The composition of the simulated solution was 100 mg Ni/L, 100 mg Cr/L, 100 mg Mn/L, and 50 mg Cd/L. The pH of the solution was adjusted to 6. Ten ml of the simulated solution was added to 15-ml bottles that contained preweighed amounts (25, 50, 100, 500 and 1000 mg) of the adsorbents. All samples were agitated for 3 days in a temperature controlled water bath to ensure that equilibrium was achieved. The samples were then centrifuged and filtered using 0.45 μm syringe filters before being analysed. The samples were analyzed for their heavy metals content using Inductive Coupled Plasma Technique (ICP, Perkin-Elmer, Optima DV2000, USA). Blank samples were used with every isotherm in order to account for any loss and adsorption on the walls of the tube.

The equilibrium adsorptive quantities of the metal ions on the adsorbents surfaces were calculated using the following relation:

$$q_e = \frac{C_0 - C_e}{M} \times V \qquad (1)$$

Where $q_e$the amount of metal ion adsorbed per unit mass of adsorbent (mg/g), C_0 and C_e are the original ion concentration and the equilibrium solution concentration (mg/L), respectively, M is the mass of adsorbent, V is the volume of solution (L).

The percent removal of the metal ion by the adsorbent was calculated using the following mass balance relation:

$$\%\ Removal = \frac{C_0 - C_e}{C_0} \times 100\% \qquad (2)$$

Results and Discussion

1) Properties of the adsorbents

The highest surface area is reported for the powdered activated carbon treated with acid (APC) (892 m^2/g) and the lowest one for shrimp shells (25.6 m^2/g), as shown in Table 1. All measured samples have microporous structure with pore sizes less than 2 nm based on Horvath-Kawazoe (HK) model for the determination of pore size distribution in microporous sorbents such as activated carbon and zeolites. Acid modification of granular activated carbon lowered its surface area from 635 m^2/g to 591 m^2/g for GC and AGC, respectively. Previous work showed that oxygenation of activated carbon with 4M HNO_3 significantly increased the concentration of acidic surface oxygen complexes on the surface (carboxylic and carbonyl groups), reduced the total number of basic groups on the surface and caused the surface to be more hydrophilic (Franz et al., 2000; Abu-Daabes and Pinto, 2004; Jaramillo et al., 2009). These acidic sites are bonded to the edges of the graphitic basal plane.

2) Competitive adsorption of heavy metals

Adsorption isotherms for all four metal ions, Ni^{+2}, Cr^{+6}, Mn^{+2}, and Cd^{+2}, are shown in Figure 1 (a-d). All isotherms don't show uniform curve which could be due to the competition between the ions for the adsorption sites as well as the repulsion and synergy between them. Similar behaviour for competitive adsorption of metal ions was observed by Covelo et al.,(2004). In general, the mechanism of heavy metal ions adsorption on the surface of activated carbon is by electrostatic attraction to the carbon surface, ion-exchange and surface complexation (Jaramillo et al., 2009; Radovic et al., 2001). Adsorption of heavy metals on the unmodified carbon surface was suggested to be through interactions of the metal ion with the π electrons of the carbon basal planes according to the reactions (Jaramillo et al., 2009):

$$-C\pi + H_2O \leftrightarrow -C\pi\text{-}H^+ + OH^- \qquad (3)$$

$$2(-C\pi\text{-}H^+) + M^{2+} \leftrightarrow (-C\pi)_2 - M^{2+} + 2H^+ \qquad (4)$$

In the presence of acidic functional groups on the surface of activated carbon the mechanism become surface complexation or ion-exchange depending on the pH. If the solution has low pH, the weakly acidic groups are more likely to be protonated and will undergo ion-exchange reactions with the metal ions, as described in reaction (5).

$$2(-COOH) + M^{2+} \leftrightarrow (-COO)_2Cu + 2H^+ \qquad (5)$$

At higher pH, the acidic groups will be nonprotonated and the metal ions will bind to them by surface complexation, releasing less than one hydrogen ion by each metal ion adsorb (Jaramillo et al., 2009).

The selectivity sequences for heavy metal ions depends on the characteristics of the ions such as ionic radius ($Cd^{2+} > Ni^{+2} = Mn^{+2} > Cr^{+6}$), ionic potential or charge-to-radius ratio ($Cr^{+6} > Ni^{+2} = Mn^{+2} > Cd^{+2}$) and electronegativity (Ni > Cd > Cr > Mn) (Sposito, 1989; McBride, 1994; Covelo et al., 2004). The selectivity of each adsorbent for heavy metals under simulated conditions is shown in Figure 2 for the 500 mg adsorbents isotherms. The results show that acid treatment of both granular carbon and powder carbon (AGC and APC, respectively) changed the order of metals selectivity compared to the basic granular carbon (GC). Furthermore, acid treatment of granular carbon (AGC) increased Cr removal efficiency by 30% while it decreased for Ni, Mn and Cd by 23%, 36% and 37%, respectively. This can be explained by the much higher ionic potential of Cr^{+6} compared to other ions which will lead to strong attraction to the acidic surface groups of activated carbon and less accessibility for the ions with lower ionic potential.

Shrimp shells compose of 20% chitin by mass of dry weight, 37% protein and 43% minerals (Einbu, 2007). Chitin is a natural aminopolysaccharide. The results of this study showed that untreated shrimp shells have very high selectivity for Cd^{2+} and Mn^{2+}. This could be due to the presence of amine groups on the chitin of SH. The amine functional groups will form metal complexation. The order of metals affinity for different adsorbents under simulated conditions is summarized in Table 2.

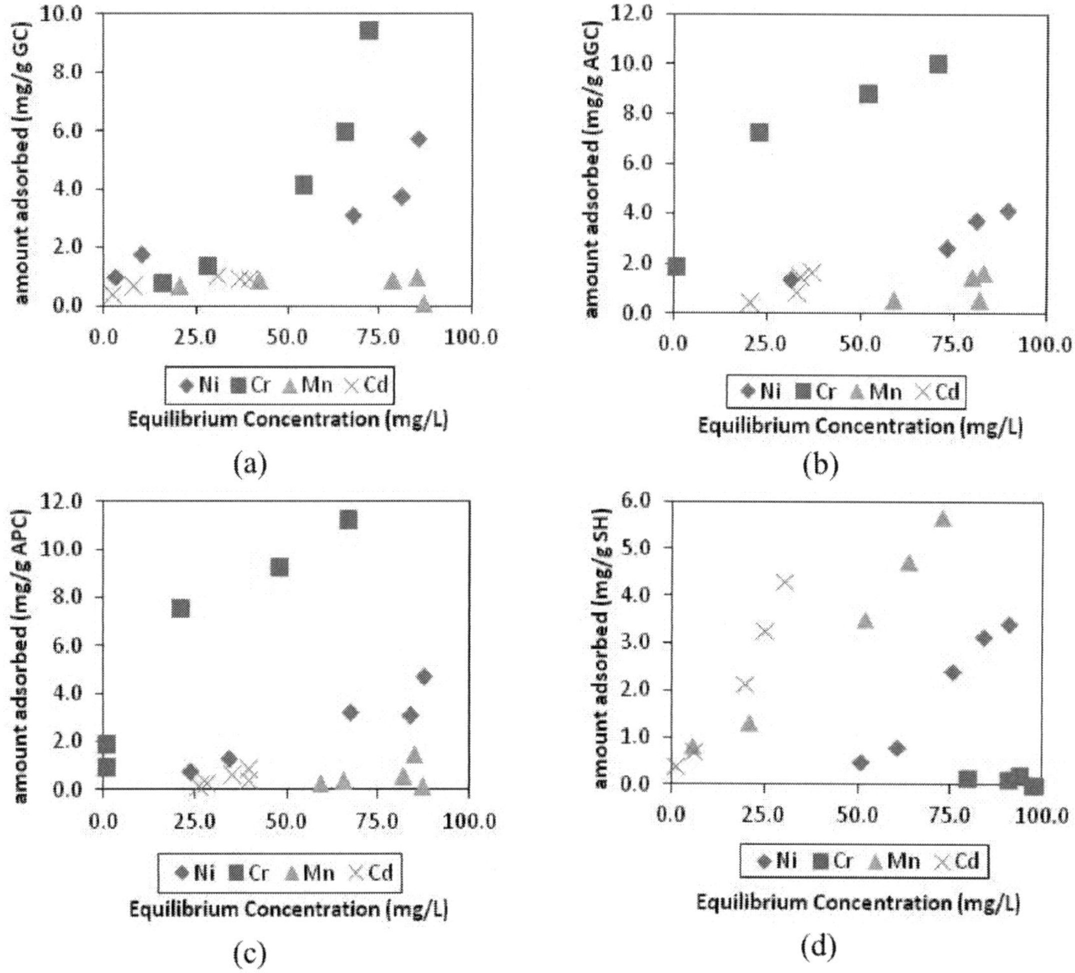

Figure 1. Adsorption Isotherms of Metal Ions on Different Adsorbents

Table 2. Order of metals selectivity for different adsorbents

Adsorbent	Metals order
GC	Cd > Ni > Cr > Mn
AGC	Cr > Cd > Ni > Mn
APC	Cr > Cd > Ni > Mn
SH	Cd > Mn > Ni > Cr

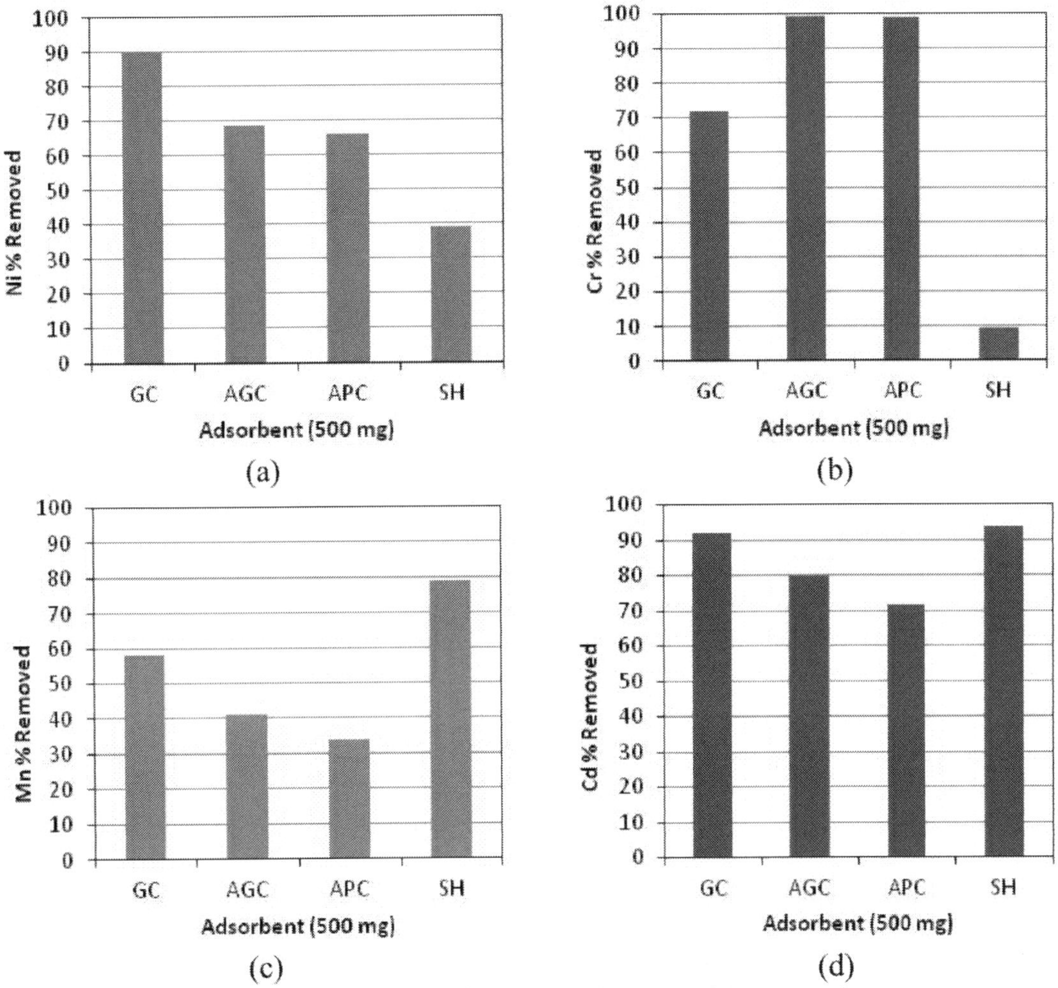

Figure 2. Percent removal of Metal Ions on Different Adsorbents

Conclusion

In this study activated carbon, surface modified activated carbon and unmodified natural shrimp shells were evaluated as potential adsorbents for competitive removal of Ni^{+2}, Cr^{+6}, Mn^{+2}, and Cd^{+2} from aqueous solutions. Untreated granular carbon was the best adsorbent for Ni^{+2} removal from a simulated solution with 90% removal. AGC and APC both showed very high selectivity for Cr^{+6} over other present metals with removal efficiency of 99%. This was due the high ionic potential of Cr^{+6} compared to other ions. For Mn^{+2}, shrimp shells (SH) was the best adsorbent with 79% removal efficiency, while both GC and SH adsorbents have comparable efficiency for Cd^{+2} removal from the simulated solution.

Acknowledgments

The authors wish to thank the German-Jordanian University, the Higher Council for Science and Technology in Jordan, and the European Union presented by Support to Research and Technological Development & Innovation Initiative & Strategies in Jordan (SRTD) Project for their financial support to this project.

References

Abu-Daabes, M., Pinto, N.G. (2004) "Effect of Surface Oxygen Complexes of Activated Carbon on Phenol Adsorption from Single and Mixed Non-Aqueous Solvents", *Separation Science and Technology*, 39, 2997-3009.

Covelo, E. F., Andrade Couce, M. L., Vega, F. A. (2004) "Competitive Adsorption and Desorption of Cadmium, Chromium, Copper, Nickel, Lead, and Zinc by Humic Umbrisols", *Communications in Soil Science and Plant Analysis*, 35, 2709-2729

Einbu, A. (2007), "Characterisation of Chitin and a Study of its Acid-Catalysed Hydrolysis", Ph.D. Thesis, Norwegian University ofScience and Technology.

Franz, M., Arafat, H.A., Pinto, N.G. (2000), "Effect of chemical surface heterogeneity on the adsorption mechanism of dissolved aromatics on activated carbon", *Carbon*, 38, 1807–1819

Gotvajn, Z., Tisler, T., Zagorc-Koncan, J. (2009), "Comparison of different treatment strategies for industrial landfill leachate". *Journal of Hazardous Materials*, 162, 1446-1456.

Jaramillo, J., Go´mez-Serrano, V., A´ lvarez P.M. (2009), "Enhanced adsorption of metal ions onto functionalized granular activated carbons prepared from cherry stones". *Journal of Hazardous Materials*, 161, 670–676

Karnchanawong, S., Limpiteeprakan, P. (2009), "Evaluation of heavy metal leaching from spent household batteries disposed in municipal solid waste". *Waste Management*, 29, 550–558.

McBride, M.B. Environmental Chemistry of Soils; Oxford University Press: New York, 1994.

Radovic, L.R., Moreno-Castilla, C., Rivera-Utrilla, J. (2001), "Carbon materials as adsorbents in aqueous solutions", in: L.R. Radovic (Ed.), Chemistry and Physics of Carbon, 27, Dekker, New York, pp. 227–405.

Sposito, G. (1989), "The Chemistry of Soils", Oxford University Press: New York,.

Strelko, V., Malik, D.J. (2002) "Characterization and metal sorptive properties of oxidized active carbon", *J. Colloid Interf. Sci.* 250 213–220

Xará, S., Delgado, J., Almeida, M., Costa, C. (2009), "Laboratory study on the leaching potential of spent alkaline batteries", *Waste Management*, 29, 2121–2131.

(119e) Arsenic Mobilization In Alluvial Soils of Punjab, North-West India Under Flooded Conditions

Kuldip Singh[1], Dhanwinder Singh[2], H. S. Hundal[1] and Raj Kumar[2], (1)Soils, Punjab Agricultural University, Ludhiana, India, (2)Soil Science, Punjab Agricultural University, Ludhiana, India

Elevated arsenic contaminant in drinking water above the WHO (1996) safe limits of 10 μgL^{-1} was reported at many locations in underground water of Punjab, North-West India. In rural areas, flooded irrigations and maintenance of standing water in paddy fields likely to create anoxic conditions in the soils over-lying above the shallow aquifers. The objective of the present investigation was to study the mechanism of arsenic mobilization in surface alluvial soils. Fifty gram of each of 18 soil samples were transferred to 200 ml polythene bottles, and 100 ml of deionized water was added to maintained flooded conditions. These soil-water systems were kept in BOD incubator at 25^0C for equilibration under anaerobic conditions of 24 and 240 hours period, respectively. After equilibration periods, pH, EC, soluble HCO_3 and Cl^- were determined for each soil-water equilibrated systems. Supernatant aliquots were withdrawn using a syringe, centrifuged, filtered and acidified with a drop of nitric acid prior to analysis for Fe, Mn, SO_4^{2-} and PO_4^{3-} on ICAP-AES. Arsenic in the filtrates solution was determined using hydride generation assembly on ICAP-AES.

Total dissolved As in 24 hour soil-water equilibrated systems varied from 3 to 16 μgL^{-1} with mean value of 9 μgL^{-1} (Figure 1), whereas its concentration range in different soil-water suspensions after 240 hour of equilibration period enhanced to 33 to 1761 μgL^{-1} with mean value of 392 μgL^{-1}. The median, 75th and 90th percentiles concentrations of As were 8, 12 and 14 μgL^{-1} after 24 hour and eventually elevated to 190, 490 and 1013 μgL^{-1}, respectively after 240 hours of anaerobic conditions. Arsenic concentration was enhanced with decreasing redox potential. The results showed antagonistic relationship of arsenic concentration with the magnitude of redox potential status of alluvial soil-water suspensions (Figure 2a). Dissolved Fe concentration versus redox status of soil-water systems determined after two equilibration periods are illustrated in Figure 2b. As the pe decreased, the Fe concentration declined in different alluvial soil water systems.

The effect of redox potential on soluble Mn concentrations determined at different flooding periods in alluvial soil-water suspensions is shown in Figure 2c. As the pe decreased, the soluble Mn concentration enhanced in soil water systems. The soluble concentrations of Mn was ranged between 0.034 to 0.407 mgL^{-1} with a mean value of 0.132 mgL^{-1} after 24 hour equilibration period, but were considerably enhanced to 0.449-44.44 with mean value of 22.398 mgL^{-1} after 240 hour of equilibration periods. The behavior of SO_4^{2-} in different alluvial soil-water systems under different redox status is shown in Figure 2d. With decrease in pe, SO_4^{2-} concentration declined in all soil water systems. Total dissolved concentration of SO_4^{2-} anion in soil-water suspensions also considerably lowered to range of 0.003-25.30 mgL^{-1} with mean value of 1.95 mgL^{-1} after 240 hour of equilibration periods from 4.56 to 81.6 mgL^{-1} with a mean value of 25 mgL^{-1} after 24 hours of equilibration periods. Under reducing conditions, sulfate concentrations decreased significantly.

Pourbaix (pe-pH) diagrams for arsenic and ferromagnesian oxides (Figure 3 and 4) were constructed by to identify the chemical species for each of these elements in different soil-water suspensions. Aqueous arsenic species, $HAsO_4^{2-}$ and $H_3AsO_3^0$ along with the possible solid phase, orpiment (amorphous As_2S_3) as were identified in soil-water systems (Figure 3). The mono valent arsenate oxy-anion ($HAsO_4$) specie was predominant in all the soil-water suspensions except soils 2, 4 and 8, where arsenite ($H_3AsO_3^0$) specie was identified after 24 hour of anaerobic condition. The predominance of neutral arsenite ($H_3AsO_3^0$) specie was observed after 240 hours of reduced incubated period in all soil water systems except soils 12, 15 and 16 in which orpiment (As_2S_3) as solid phase was identified. The neutral arsenite ($H_3AsO_3^0$) and arsenate oxy-anions ($H_2AsO_4^-$) has been found to be predominantly occurred in the alluvial aquifers of Punjab between the pH of 7.05 to 8.25 and under reduced pe value from 0.77 to -4.49, respectively. Thus, it is conclusively evident that two chemical aqueous species controlling As chemistry and its toxicity in soils are arsenate and arsenite. Arsenate is the thermodynamically stable form under aerobic conditions. Arsenite is the predominant species under anaerobic conditions and it is a neutral species at natural pH and is more soluble, mobile and phototoxic than arsenate. The orpiment (As_2S_3) solid phase domain predominates near the bottom of pe-pH diagram and lying over the H_2O / H_2g stability line in the Pourbaix diagram (Figure 3). After 240 hours of equilibrium period, arsenite ($H_3AsO_3^0$) specie was predominant in all alluvial soil-water suspensions. The increase in arsenic concentration was also primarily due to an increase in As (III) species in all the soil-water suspensions under more reduced conditions. In some soil water systems (soil number 12, 15 and 16) apparent precipitation of orpiment was identified on pe-pH diagram. It is evident that total arsenic concentration continuously enhanced in these soil-water systems with decreased magnitude of redox potential. The simultaneously drastic decline in concentration of both Fe and SO_4^{2-} in contrast to accumulation As in reduced soil-water suspensions of alluvial Holocene validated the formation of iron sulfide mineral in reduced alluvial soil-water systems of Punjab. Because of the differences in solubility of iron (DG= -117.11 Kcal $mole^{-1}$) versus arsenic (DG = -382.64 Kcal $mole^{-1}$) sulfides, precipitation of iron sulfide may remove sulfide from solution but not arsenic since the precipitation rates of iron sulfide was very rapid compared to amorphous arsenic sulfide.

The aqueous species of Fe^{2+}, Mn^{2+} and $MnHPO_4$ and solid phase of hematite (Fe_2O_3) were predominant in all the alluvial soil-water equilibrated solutions after 24 hours within the pe and pH ranges of -1.75 to 0.77 and 7.25-8.25, respectively (Figure 4). After 240 hours of incubation period, reduction potential status range in soil-water suspensions declined between -4.49 to -2.74 within pH range of 7.05 to 7.94. Aqueous species of Fe^{2+}, Mn^{2+} and $MnHPO_4$ and solid phase of hematite [Fe_2O_3] were predominant in soil-water suspensions of soils 1, 5, 6, 11 and 14. Pourbaix diagram (Figure 4) for iron illustrate that declined in iron was due to the predominantly precipitation of iron sulfide [FeS_2] and or as vivianite [$Fe_3\ (PO_4)_2\ 8H_2O$] minerals. The reduced solid phase vivianite [$Fe_3\ (PO_4)_2\ 8H_2O$] was found in soil 2, 3, 7 and 13, respectively. Pyrite [FeS_2] was found to be predominant in soil 8, 9, 10, 12, 16, 17 and 18, respectively. All the water composition falls in the stability field for aqueous Mn^{2+} ion (Figure 4) and its concentrations varied from 0.45 to 44.44 mgL^{-1}. Stability diagram for manganese showed that Mn^{2+} ion is thermodynamically more stable and predominant across a wide pe-pH range than for Fe^{2+} ion. Consequently, the Mn^{2+} ion remain in solution in aqueous environment in which Fe^{2+} ions is oxidized to Fe^{3+} and precipitated as Fe $(OH)_2.7Cl_{0.30}$ below pH 6.8 at pe > -1.8 and Fe_2O_3 above pH 6.8 at pe < -1.8 and thus eventually contributed to higher dissolved Mn^{2+} concentration under reduced environments in soils. Iron and manganese concentration in the anoxic soil-water systems were as high as 1.634 and 44.44 mgL^{-1}, respectively. The largest contributor of ferrous to the equilibrated soil-water interaction appears to be an iron oxide phase (probably maghemite, Gamma-Fe_3O_3, an alteration of magnetite). These results showed that dissolve Fe concentration were probably controlled by hematite (Fe_2O_3) under alkaline oxic condition and by iron sulfide/pyrite (FeS) or in some soils by vivianite [$Fe_3(PO_4)_2 8H_2O$] under reduced environment, where soluble phosphate concentration was high. Under reduced condition (developed after 240 hours of flooding), soil solution Fe and SO_4^{2-} are governed by newly formed precipitates of pyrite (FeS_2).

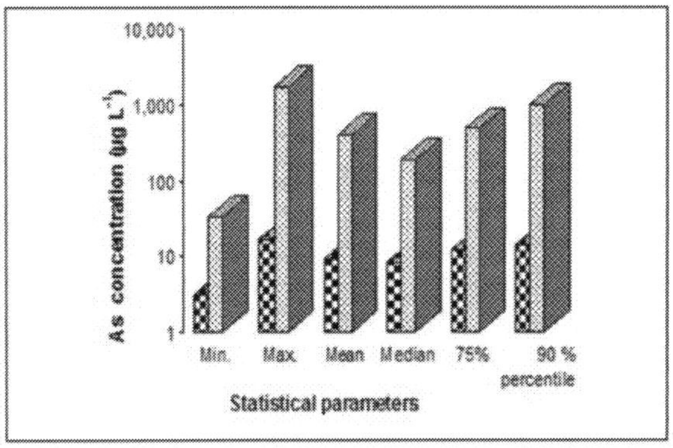

Fig.1 As concentrations measured in 18 alluvial soil-water suspensions of Punjab after equilibration of 24 (▩) and 240(▤) hours

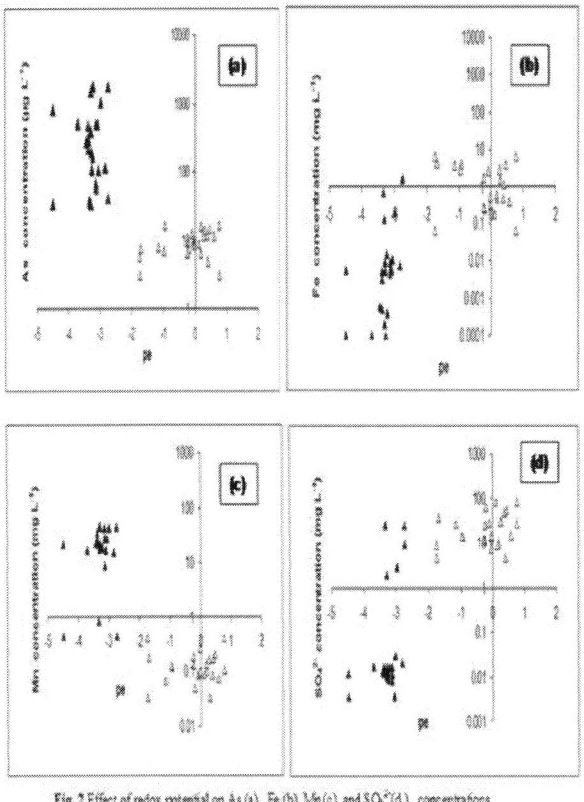

Fig. 2 Effect of redox potential on As (a), Fe (b), Mn (c) and SO_4^{2-} (d) concentrations measured in 18 alluvial soil-water suspensions after 24 (?) and 240 (?) hours of equilibration periods

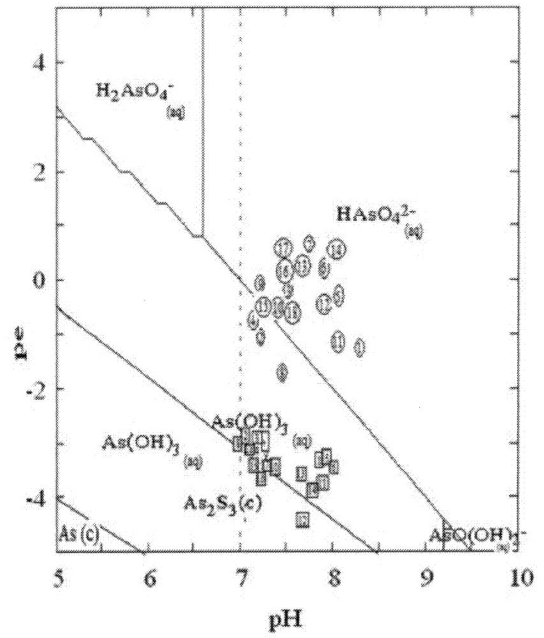

Fig. 3 Arsenic species identified in alluvial soil-water suspensions after equilibration of 24 (O) and 240 (?) hours at 25 °C. The concentration of $[AsO_4^{3-}]$ =6.70µM, $[PO_4^{3-}]$=0.37mM, $[SO_4^{2-}]$= 0.25mM$[CO_3^{2-}]$=0.20mM, $[Cl^-]$=0.13mMand ionic strength=0.020M were fixed to construct the Pourbaix diagram

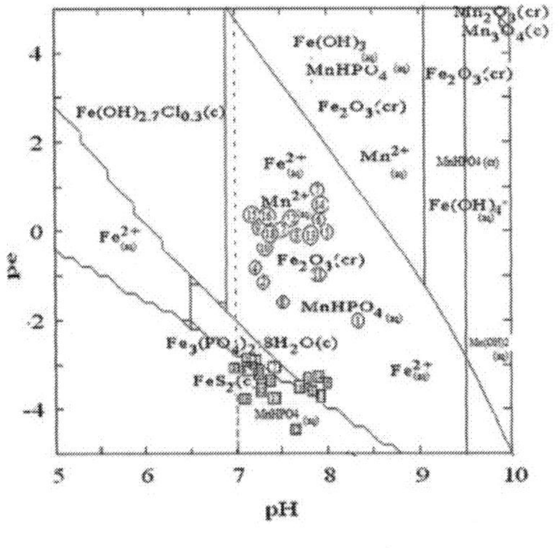

Fig.4 Iron and manganese species identified in alluvial soil-water suspensions after equilibration of 24 (O) and 240 (?) hours at 25 °C. The concentration of [Fe=15µM], [Mn=0.27mM], =6.70uM, [PO43-] =0.37mM, $[SO_4^{2-}]$= 0.25mM,$[CO_3^{2-}]$=0.20mM, $[Cl^-]$=0.13mM and ionic strength=0.020M were fixed to construct the Pourbaix diagram

ADSORPTION OF LEAD FROM AQUEOUS SOLUTIONS BY COTTON FABRICS MODIFIED WITH CITRIC ACID

Refugio Bernardo García Reyes, Saúl H. Martínez Treviño, Alcione García González and Ricardo Gómez González*

Universidad Autónoma de Nuevo León, Facultad de Ciencias Químicas, Av. Universidad s/n, Cd. Universitaria, C.P. 66400, San Nicolás de los Garza, N. L., México. FAX: +52(81)83294000 Ext. 6282, TEL: +52(81)83294000 Ext. 6288
*E-mail: bernardogarciareyes@yahoo.com.mx

ABSTRACT

The objective of this research was to modify cotton (i.e. cellulose) fabrics for the removal and recovery of lead (Pb(II)) from aqueous solutions. Cotton fabrics were modified with 0.6 M citric acid solutions and characterized by infrared spectroscopy (ATR-FTIR) and potentiometric titrations. Also, adsorption, desorption, and kinetic experiments in batch systems were conducted.

When modifying cotton fabrics with citric acid, density of functional groups increased up to 2.47 mmol/g in comparison with the non-modified fabric that had 0.61 mmol/g. In addition, estimated equilibrium constants for the modified fabrics can be associated in part to carboxylic groups of the citric acid. The appearance of carboxylic groups on modified fabrics was also confirmed by ATR-FTIR analyses. Maximum adsorption capacity of Pb(II) on modified fabrics, estimated with Langmuir isotherm, was 73.3 y 42.3 mg/g at pH 4 y 3, respectively, whereas non-modified cotton fabrics did not adsorb Pb(II) from water. Decrement in adsorption capacity suggests that Pb(II) and H^+ ions compete for the same adsorption sites. On the other hand, it was possible to desorb around 92% of Pb(II) previously adsorbed on the modified cotton with using 0.1 M HCl solutions since the chemical equilibrium is shifted. Finally, adsorption kinetics can be predicted with a pseudo-second order model.

In summary, cotton fabric treated with 0.6 M citric acid has an adsorption capacity of Pb(II) up to 3.5 times greater than the activated carbon, and the adsorbed metal can be recovered almost completely with 0.1 M HCl. These results suggest that modified cotton fabric is a highly competitive adsorbent for the treatment of metal-containing wastewater.

1. INTRODUCTION

Consumption of water polluted with lead (Pb(II)) is associated to human diseases such as anaemia, kidney damage, sterility, and it is suspected to be carcinogenic. For these reasons, several treatment technologies have been applied for the removal of this pollutant from water, for example: chemical precipitation, coagulation-flocculation, ion exchange, and adsorption with activated carbon, among others (Stumm y Morgan, 1996; Metcalf and Eddy, 2004). Adsorption process is one of the most commonly used technologies because it is useful as a polishing step in a water treatment plant. Nevertheless, it is important to explore the possible application of new adsorbents for removing lead from aqueous solutions. For instance, lignocelullosic materials can be used as new adsorbents because these are worldwide available and low-cost in comparison with polymeric resins and activated carbons. Moreover, it was reported that modification of lignocelullosic materials with citric acid enhanced the adsorption capacity of copper in comparison with non-treated adsorbents (Marshall *et al.*, 2007; Zhu *et al.*, 2008). In this way, the adsorption of metal cations in chemically modified lignocelullosic materials can be enhanced significantly. Nonetheless, to the best of our knowledge, at this time is not reported the density of functional groups added to the adsorbent and their equilibrium constants before and after chemical modification.

In this work, the effect of citric acid modification of cotton fabrics on lead adsorption from aqueous solutions is explored by means of ATR-FTIR analyses, potentiometric titrations, adsorption/desorption and kinetic experiments.

2. METHODOLOGY

2.1 Chemical modification. Thirty grams of cotton fabrics were added in a precipitated flask with 210 mL of 0.6 M citric acid solution. Suspension was heated at 60°C for 1 hour and, afterwards, water was evaporated at about 100°C until dryness. The remaining residues were dried at 120°C in an oven for 24 hours to carry out the esterification on cotton fabrics. Next, modified cotton fabrics were rinsed out with deionized water at 50°C and were dried at 100°C and stored in desiccators.

2.2 Adsorbent characterization. Non-modified and modified cotton fabrics were characterized by means of attenuated total reflection Fourier transform infrared spectroscopy (ATR-FTIR) and potentiometric titrations (Yun *et al.*, 2001; Garcia-Reyes *et al.*, 2009) to identify quantitative and qualitatively the functional groups incorporated to the adsorbent materials, respectively.

2.3 Adsorption experiments. Adsorption tests were conducted in batch systems at 30°C, 150 rpm and the selected pH (3 or 4) to keep lead in solution. A mass of 40 mg of cotton fabrics (non-modified and modified) were added to 40 mL of Pb(II) solutions with different concentration (5-100 mg/L). Solution pH was daily adjusted (with HNO_3 or NaOH) along the Pb(II) adsorption tests at the chosen value (pH 3 or 4) until the equilibrium was reached. Aliquots were taken at the beginning and the ending of adsorption experiments to determine the metal concentration by atomic absorption spectroscopy (ASS).

2.4 Desorption trials. Cotton fabrics, previously used in lead adsorption experiments, were regenerated with 40 mL of 0.1 N HCl solutions after being separated by filtration and rinsed with 30 mL of deionized water at pH 4. These suspensions were continuously stirred at 150 rpm during 24 hours. After that, an aliquot was taken to determine the Pb(II) concentration in solution by AAS.

2.6 Adsorption kinetics. To determine the time required to achieve the equilibrium and to estimate the parameters of kinetic models, adsorption rate experiments of Pb(II) on cotton fabrics were carried out in a packed column with recirculation at initial Pb(II) concentration of 0.256 mM, 30°C, initial pH of 4 and a feeding flow of 10 mL/min. Previous to the adsorption kinetic experiments, modified cotton fabrics were conditioned by stirring the adsorbent in a solution with pH 4 for 24 hours.

3. RESULTS AND DISCUSSION

Figure 1 shows the infrared spectra of the adsorbents used in this research. It is observed that chemical modification of cotton fabrics shown a peak about 1740 cm^{-1} (absent in non-modified cotton fabrics) that can be associated to carboxylic groups of the citric acid (see Figure 1) covalently bound to hydroxyl groups of cellulose structure. It was reported that carboxylic functional groups play a major in adsorption of metal cations from aqueous solutions (Tiemann *et al.*, 1999; Garcia-Reyes *et al.*, 2009).

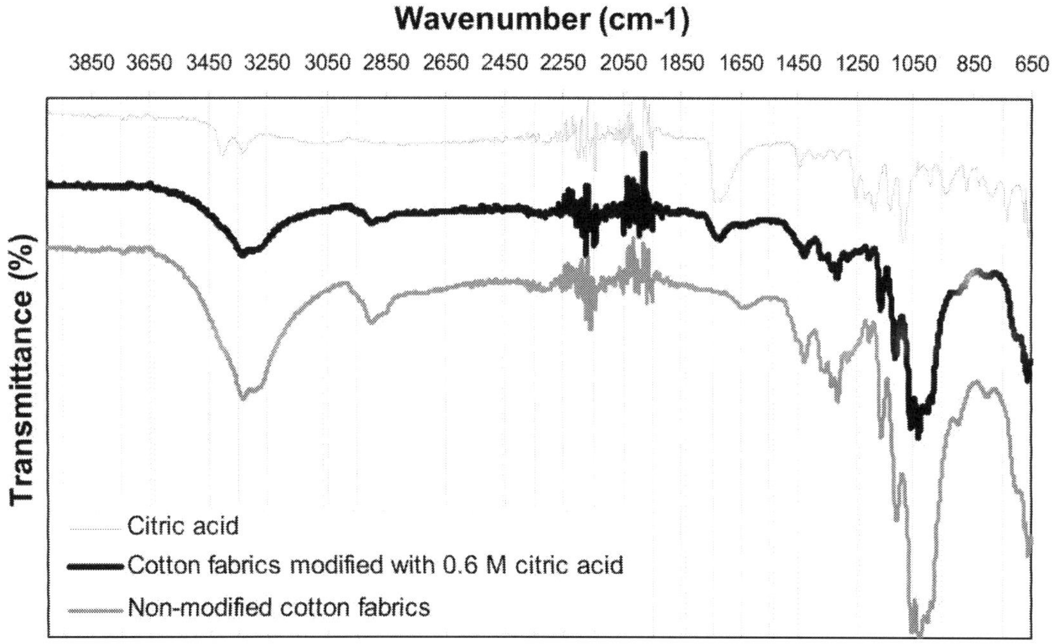

Figure 1. Infrared spectra (ATR-FTIR) of the studied adsorbents.

Table 1 shows the quantity of functional groups per unit mass of adsorbent (b_j) and their equilibrium constants (pk_{aj}) of non-modified and modified cotton fabrics. Non-modified cotton fabrics presented one functional group with an estimated quantity of 0.61 mmol/g and a pk_a value of 8.71 probably due to the hydroxyl groups of cellulose (peaks at 3350 and 1030 cm^{-1} in Figure 1) (Silverstein and Webster, 1998). In contrast, cotton fabrics modified with citric acid showed two functional groups (b_1=0.67 y b_2=1.80 mmol/g) with different pk_a values (4.06 and 8.08). These equilibrium constants are close to two of three pk_a values reported for citric acid (3.13, 4.8 y 6.40). It is important to mention that the functional group of modified cotton fabrics with pk_a=4.06 may correspond to a carboxylic group of citric acid and, therefore, these functional groups can be used as adsorption sites of positively-charged species in water such as Pb(II) species (Pb^{2+}, $PbNO_3^+$, $PbOH^+$, and Pb_2OH^{3+}) (Tiemann *et al.*, 1999; García-Reyes *et al.*, 2009) meanwhile pH of solution remains below 5.6 to avoid Pb(II) precipitation.

Table 1. Density of functional groups (b_j) and their equilibrium constants (pk_{ai}) of the studied adsorbents.

Adsorbent	b_1 (mmol/g)	pk_{a1}	b_2 (mmol/g)	pk_{a2}
Non-modified cotton fabrics	n.d.	n.d.	0.61	8.71
Modified cotton fabrics	0.67	4.06	1.80	8.08

n.d.: Not detected.

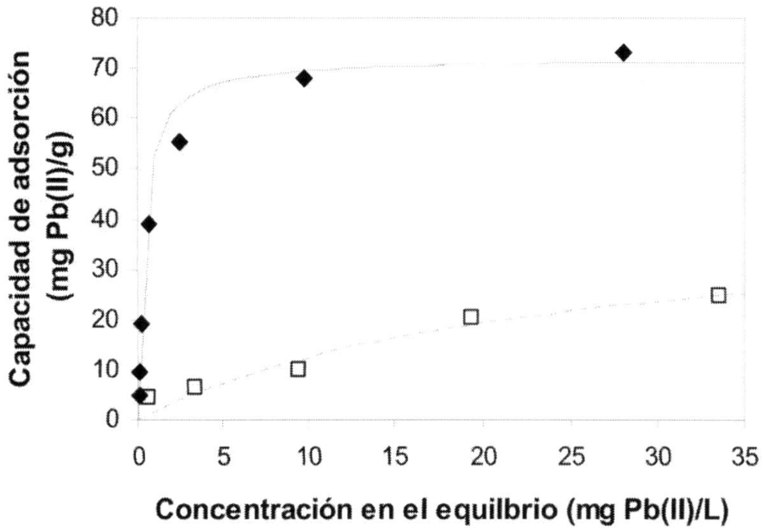

Figure 2. Adsorption isotherms of Pb(II) on cotton fabrics modified with 0.6 M citric acid at 30°C, pH 3 (□) y 4 (♦). Symbols represent experimental data and lines were predicted with the Langmuir isotherm.

Figure 3 shows the adsorption rate of Pb(II) on modified cotton fabrics without pH control. Kinetic experiments were conducted at initial pH 5 but final (at equilibrium) pH decreased to 3.36. This result seems to be reasonable because pH was not corrected along the kinetic experiments and Pb(II) species displaced H^+ ions from protonated carboxylic groups (covalently bound on the cotton fabrics). Adsorption kinetics was well predicted with the pseudo-second order model (Ho y

McKay, 1999). This result suggests that lead species interact with adsorption sites by physical forces such as electrostatic charges.

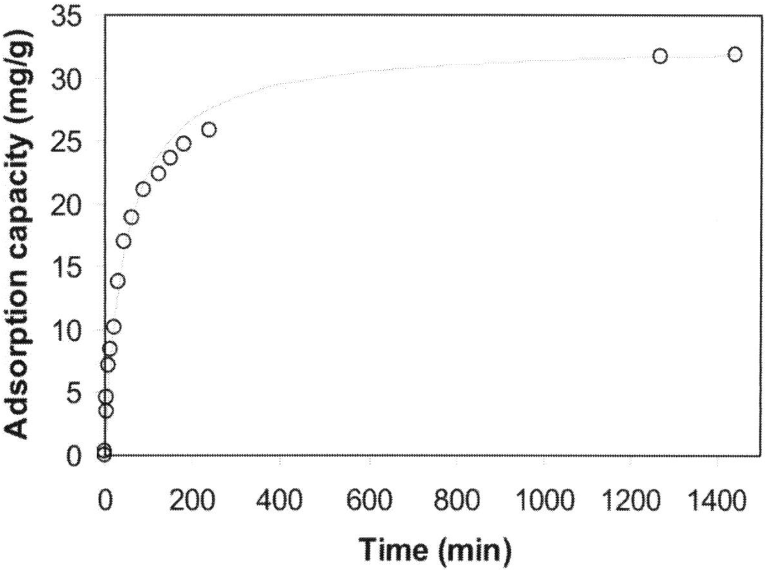

Figure 3. Adsorption kinetics of Pb(II) on modified cotton fabrics at 30°C, initial pH 5, initial concentration 53.18 mg/L, volumetric flow of 10 mL/min, and adsorbent dosage of 1 g/L. Experimental data are represented by symbols and line was predicted with the pseudo second order model.

4. CONCLUSIONS

- ✓ Chemical modification of cotton fabrics by citric acid incorporate functional groups (*i.e.* carboxyl groups) capable of binding metal cations such as Pb(II) from aqueous solutions.
- ✓ Adsorption capacity of Pb(II) on modified cotton fabrics at pH 4 was around 3.5 times greater than activated carbon.
- ✓ It is possible to recover up to 92% of the lead previously adsorbed on the modified cotton with using 0.1 M HCl solutions.
- ✓ Adsorption kinetics can be predicted by a pseudo-second order model and it suggests that lead species interact with adsorption sites by physical forces such as electrostatic charges.

5. ACKNOWLEDGMENTS

This research was financially supported in part by the Secretaría de Educación Pública (PROMEP/103.5/11/6627) y la Universidad Autónoma de Nuevo León (PAICYT IT577-10).

6. REFERENCES

Acharya, J., Sahu, J.N., Mohanty, C.R. and Meikap, B.C. (2009) Removal of lead(II) from wastewater by activated carbon developed from Tamarind wood by zinc chloride activation. Chemical Engineering Journal 149(1-3), 249-262.

Garcia-Reyes Refugio Bernardo, Rangel-Mendez Jose Rene, Alfaro-De la Torre Ma. Catalina (2009), Chromium (III) uptake by agro-waste biosorbents: chemical characterization, sorption-desorption studies, and mechanism, Journal of Hazardous Materials, 170(2-3) 845-854.

Ho, Y.S. and McKay, G. (1999) Pseudo-second order model for sorption processes. Process Biochemistry 34(5), 451-465.

Marshall, W.E., Akin, D.E., Wartelle, L.H. and Annis, P.A. (2007) Citric acid treatment of flax, cotton and blended nonwoven mats for copper ion absorption. Industrial Crops and Products 26(1), 8-13.

Metcalf and Eddy (2004), Waste Water Engineering Treatment and Reuse, 4th Ed., Ed. Mc Graw Hill, USA, 1819 + XXVIII.

Silverstein, R. M., and Webster, F. X. (1998), Spectrometric Identification of Organic Compounds. Wiley & Sons, New York.

Stumm, W. & J. J. Morgan, Aquatic Chemistry: Chemical Equilibria and Rates in Natural Waters, 3rd. ed., John Wiley & Sons, New York, 1996.

Tiemann, K. J., J. L. Gardea-Torresdey, G. Gamez, K. Dokken, S. Sias (1999), Use of X-ray absorption spectroscopy and esterification to investigate Cr(III) and Ni(II) ligands in alfalfa biomass. Environmetal Science and Technology, 33(1) 150-154.

Yun, Y. S., D. Park, J. M. Park, B. Volesky (2001), Biosorption of trivalent chromium on the brown seaweed biomass, Environmetal Science and Technology, 35(21) 4353-4358.

Zhu, B., T. X. Fan, D. Zhang (2008), Adsorption of copper ions from aqueous solution by citric acid modified soybean straw. Journal of Hazardous Materials, 153(1-2) 300-308.

PREPARATION AND CHARACTERIZATION OF VTMS-MODIFIED SILICALITE-1 FILLED PDMS COMPOSITE PERVAPORATION MEMBRANE AND APPLICATION IN BUTANOL SEPARATION FROM AQUEOUS SOLUTION

Haoli Zhou, National Key Laboratory of Biochemical Engineering, Institute of Process Engineering, Chinese Academy of Sciences, Beijing, China
Yi Su, National Key Laboratory of Biochemical Engineering, Institute of Process Engineering, Chinese Academy of Sciences, Beijing, China
Yinhua Wan, National Key Laboratory of Biochemical Engineering, *Institute of Process Engineering, Chinese Academy of Sciences, Beijing, China*

1. Introduction

Pervaporation (PV) is a membrane technology for liquid separation. The economics of the pervaporation process is strongly affected by the membrane performance. In order to make the pervaporation process economically competitive with conventional separation processes, membranes with high flux and high selectivity are needed [1]. So far, huge efforts have been devoted to the development of inorganic-organic membranes, such as silicalite-1 filled polydimethylsiloxane (PDMS) membrane, because of their high separation performance and very good film-forming properties [2-5]. It was reported that the selectivity of butanol/water obtained with silicalite-1 filled PDMS unsupported dense membranes could be up to about 160 in the separation of 1.5 wt% acetone, butanol and ethanol (ABE)-water solution [6]. However, though the unsupported dense membranes in pervaporation exhibit high selectivity, their permeation fluxes have been found to be very low, which limits their applications in industry [4, 6].

In the present study, in order to increase the viscosity of membrane solution and subsequently improve the stabilization of the casting solution, a new method of using "platinum cure" was used to assist in the pre-polymerization of coating suspension. This is based on the fact that most PDMS systems used for mixed matrix membranes consist of α, ω-vinyl-terminated siloxane chains and methyl-hydride silicone copolymer crosslingking agents. The hydrosilylation reaction (polymerization) between the vinyl groups and hydride groups is promoted by a platinum agent, often a Karstedt-type catalyst [1]. Such a material is commonly referred as an "addition cure" or "platinum cure" silicone rubber [1, 7]. When the "platinum cure" is added in the coating solution besides a two-component PDMS polymer, silicalite-1 and n-heptane, the hydrosilylation reaction could be enhanced, and it resulted in an increased viscosity of the coating solution. Moreover, a high viscous coating solution would stabilize the silicalite-1 suspense. Thus, a more stabilized suspension could be obtained. Furthermore, this method is more convenient in practice than the one with elevated temperature. Since only a small amount of "platinum cure" was added into the membrane solution, the effect of the addition on the total cost of membrane was negligible.

The objective of the present study is to develop a vinyltrimethoxysilane (VTMS)-modified silicalite-1/PDMS composite membrane with the help of "platinum cure". In order to choose a suitable support, different supports were examined for their effects on the separation performance of the composite membrane for separation of acetone-butanol-ethanol-water solution because the membrane will be used in the separation of acetone, butanol and ethanol from fermentation broth, and the interfacial views between top selective layer and support were detected by SEM and EDS. Then different concentrations of coating solution were studied. Then the home-made composite membrane was characterized in detail by pervaporation in butanol-water solution at different temperatures and feed concentrations in order to compare its properties with those in other reports.

2. Experimental

2.1 Materials

Polydimethylsiloxane (PDMS) consisting of two components (prepolymer and crosslinker) was purchased from GE Toshiba Silicones Co., Ltd, Japan. Silicalite-1 was prepared by the method reported by Meagher and Huang [26]. Surface modification of silicalite-1 by coupling agents was reported previously [27]. Porous ultrafiltration (UF) membranes of poly[acrylonitrile] (PAN) (20 kDa) and poly(ether sulfone) (PES) (5 kDa) were purchased from SEPRO Corporation, USA, and Megavision Membrane Technology & Engineering Co., Ltd, China, respectively. n-Heptane, acetone, butanol, and ethanol were supplied by Beijing Chemical Plant, Beijing, China. All the chemicals were of analytical grade and were used without further purification.

2.2 Preparation of silicalite-1-filled PDMS composite membrane

Composite membrane was prepared using a UF flat sheet membrane as support. Prior to coating operation, the UF support was immersed in distilled water at room temperature overnight. Residual water at the UF membrane surface was quickly wiped off with a filter paper when the UF support was removed from water. The suspension (coating solution) was prepared by dissolving PDMS (prepolymer and crosslinker in a 10:1 weight ratio) and VTMS-modified silicalite-1 in a 1:1.9 weight ratio to give a 10 wt% coating solution in n-heptane unless otherwise indicated. The mixture solution was stirred for about 30 min to assure a uniform suspension at 25 ℃. Then a 0.1 wt% concentration of "platinum cure" was added in the suspension and stirred for 5-6 h, so that the two components of PDMS were able to partially polymerize. This pre-polymerization was controlled by monitoring the increase of viscosity visually, until silicalite-1 suspended steadily in the solution when it was kept still for 20 min. When the membrane suspension reached the desired viscosity, it was poured and spread over the surface of the support (size is 20 cm × 15 cm) with a coating knife (200 μm). The coated membrane was dried by evaporating the n-heptane at room temperature overnight and then cured at 353 K in a vacuum oven for 9 h to obtain a completed cross-linking. With this technique, the intrusion of the PDMS polymer into the porous substrate during fabrication of the composite membrane could be reduced. By varying the concentration of membranes constituents (PDMS+VTMS-modified silicalite-1) solution, composite membranes with selective layers of variable thickness could be prepared.

2.3 Morphologic analysis

Scanning electron microscopy (SEM) was used to study the morphology of the composite membrane and to measure the selective layer thickness. The composite membranes were fractured in liquid nitrogen. The fractured section of membrane or membrane surface was coated with a conductive layer of sputtered gold to make it conductive. Scanning was done using a JSM-6700F Field Emission Scanning Electron Microscope (JEOL Corporation, Japan). To determine the intrusion degree of the membrane solution into the support visually, a silicium concentration profile along the section of the composite membranes was taken also using JSM-6301F Scanning Electron Microscope (JEOL Corporation, Japan), equipped with energy dispersive spectroscopy (EDS). In this case, the fractured section was coated with a conductive layer of sputtered carbon.

2.4 Contact angles analysis

Water contact angles were measured by a Dataphysics OCA 20 Optical Contact Angle Measuring Device (Dataphysics Corporation, Germany) at room temperature. Distilled water (8 μL) was dropped on the UF supports surface, and contact angles were quickly determined before the films commenced to swell. The average of four measured values for a sample was taken as its water contact angle.

2.4 Pervaporation experiments

The schematic diagram of pervaporation process was described elsewhere [28]. All measurements were carried out in duplicate; an average value obtained in the duplicate experiments was reported in this work. Pervaporation experiments were performed with a flat composite membrane in a circular flat pervaporation cell, with a radial flow of the fluid from centre to the outer region of the cell. The cell was made of stainless steel and could be fitted with a membrane giving an effective membrane area of 72.34 cm^2.

A Longer BT600 Peristaltic Pump (Longer Corporation, China) was used to recirculate the feed (butane-water or acetone-butanol-ethanol-water mixtures) from the feed tank to the pervaporation unit. The feed flow rate through the pervaporation unit was monitored using a ratemeter and kept at 2 L/min unless specially specified. During the experiments, the upstream of the membrane was maintained at atmospheric pressure and the downstream pressure at 350 Pa all the time unless specially specified. The temperature of the pervaporation set up was controlled within $\pm$ 0.1 °C by a Guohua Water Bath (Guohua Corporation, China). The initial experiment was allowed to run for two hours to make sure that a steady state was obtained for measurements. After each measurement, the operating condition changed, and it took half an hour to regain the steady state. The permeation flux (J) was obtained by weighing the permeate (W) collected in a cold trap filled with liquid nitrogen for a given time interval (t), and its composition was determined by an Agilent 7890 Gas Chromatography (Agilent Corporation, USA), via a heated flame ionization detector (Agilent Corporation, USA). The PV performance of the membrane was evaluated in terms of the permeation flux and selectivity.

The permeation flux is given by:

$$J = \frac{W}{At} \qquad (1)$$

where A is the effective membrane area.

The selectivity is defined as:

$$\alpha_{iw} = \frac{\omega_i^p / \omega_w^p}{\omega_i^f / \omega_w^f} \qquad (2)$$

where ω^p and ω^f are the weight fractions of solutes in the permeate and feed, respectively. The subindices i and w refer to butanol and water, respectively.

In most cases, the condensed solution in the permeate splits into two phases: the aqueous phase and organic phase due to the immiscibility of butanol and water. The concentration of butanol in the permeate was measured by diluting the condensed solution with water to generate a single phase.

3. Results and Discussion

3.1 Characterization of VTMS-modified silicalite-1

Fig. 1 shows the SEM picture of VTMS-modified silicalite-1 particles synthesized in our laboratory, with an average diameter of 1 μm. The adsorption capacities of the VTMS-modified silicalite-1 for butanol were reported previously and its adsorbed amount for butanol was high, up to 98.2mg n-butanol/g of silicalite-1 [7]. Previous work [8, 9] also showed that surface modification of the silicalite-1 by vinyltrimethoxysilane (VTMS) or vinyltriethoxysilane (VTES) could improve the affinity between the silicalite-1 and the polymer and consequently increase the selectivity of the membrane. In this work, the VTMS-modified silicalite-1 was used in the preparation of the composite membranes.

Fig.1 SEM photographs of VTMS modified silicalite-1 particles.

3.2 Preparation of VTMS-modified silicalite-1/PDMS composite membrane

3.2.1 Effect of coating solution on the composite membrane

In the preparation of a VTMS-modified silicalite-1-filled PDMS composite membrane, the selective layer thickness is determined by the coating solution concentration when the thickness of coating knife is kept constant. However, the inorganic filler VTMS-modified silicalite-1 is insoluble in any organic solvent and has negative affinity towards organic polymers and a higher density than most polymers. Thus when the viscosity of coating solution is very low, it is difficult to disperse VTMS-modified silicalite-1 particles homogeneously in a polymer solution. Fig. 2(a) shows SEM views of the composite membrane prepared with a low viscosity of coating solution. Precipitation of VTMS-modified silicalite-1 and pinholes during the preparation can be seen.

It is therefore very important to maintain the homogeneous and stable VTMS-modified silicalite-1-PDMS suspension. In order to prevent the well dispersed VTMS-modified silicalite-1 from precipitation, trial and error experiments taught us that the viscosity of VTMS-modified silicalite-1-PDMS suspension should be high enough. Thus to solve this problem, a "platinum cure" was added in the coating solution to enhance the cross-linking degree of PDMS polymer in the dilute membrane solution. After the coating solution containing "platinum cure" and the membrane constituents had been stirred intensely for about 6 h, a partial polymerization took place, which resulted in an increase in viscosity of the suspension. In this manner, a stabilized suspension was obtained. Then the solution was coated onto a UF supporting membrane with a 200 μm coating knife. The selective layer thickness and the dispersion of the VTMS-modified silicalite-1 in the polymer matrix was determined by SEM analysis of the cross-section of these composite membranes. It can be seen that VTMS-modified silicalite-1 distributes evenly in the selective layer [Fig. 2(b1)], and no pinholes can be seen [Fig. 2(b2)].

Fig 2(b) also shows that the active layer of composite membrane is uniform and its thickness distribution is good. Therefore, the separation performance is expected to increase accordingly.

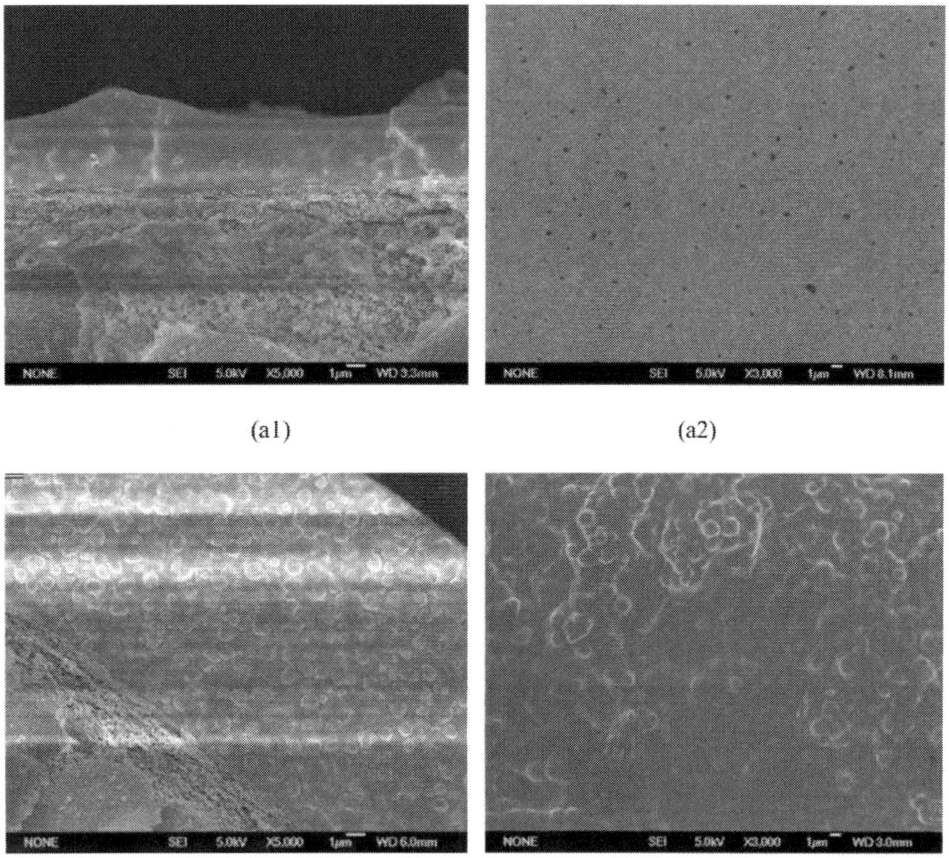

(a1) (a2)

(b1) (b2)

Fig. 2 SEM photographs of the PAN-siliclaite-1-PDMS membrane obtained in the low viscous coating solution: (a1) cross-sectional view, (a2) top view and PAN-siliclaite-1-PDMS membrane obtained in the high viscous coating solution: (b1) cross-sectional view, (b2) top view.

3.2.2 Membrane support effects

In the present work, two kinds of porous supports, PES 5 K and PAN 20 K were studied tentatively to find a suitable support for a composite membrane. In order to diminish the effect of possible pore collapse on the membrane performance, the two supports were saturated completely by water before they were cast with membrane solution. The PAN and PES support membranes were coated with an identical coating solution containing a 10 wt% n-heptane solution of membrane constituents (65 % VTMS-modified silicalte-1 contents). The comparison results are listed in Table 1. It can be seen that membrane selectivity and flux, which determine pervaporation performance of a composite membrane, varied with the supports used. It is believed that the hydrophobic/hydrophilic properties of the support significantly affected the separation performance of the composite membrane. The measurement of water contact angle of these two supports showed that the water contact angles of PAN and PES membranes were 28.1° and 48.3 °, respectively, suggesting that the PAN was more hydrophilic than the PES. By the pre-wetting method, the pores of the PAN support prior to coating were full of water due to the capillary effect. The intrusion of the hydrophobic PDMS polymer into the porous PAN support during the fabrication of the composite membrane could be greatly reduced. The SEM coupled EDS technique allows a spectroscopic analysis of the intrusion of PDMS in the support.

Table 1 Pervaporation performance of two composite membranes with different supports. Feed: 0.5 wt% acetone + 0.16 wt% ethanol + 1.0 wt% butanol; Temperature: 323 K.

Membrane	Separation factor (Butanol/water)	Total flux (g/m^2h)
PAN-siliclaite-1-PDMS	95	830
PES-siliclaite-1-PDMS	4	390

Fig. 3 shows the SEM pictures and EDS spectrum of the cross section of composite membranes. A red line was drawn in the picture, along which the Si signal was recorded with EDS. Fig. 3(a) shows the SEM picture and EDS spectra of PAN-VTMS-modified-silicalite-1-filled PDMS composite membrane. The high peak corresponded to the high Si concentration in the selective layer. In the support, the Si signal decreased quickly to zero, implying that no intrusion of PDMS in the PAN support occurred. Fig. 3(b) shows the SEM picture and EDS spectrum of PES-VTMS-modified-silicalite-1-filled PDMS composite membrane obtained in the same way. The Si signal coming from the support layer could indicate the intrusion of PDMS in the PES support, which would in turn increase the mass transfer resistance of butanol and water, and consequently leading to a lower flux. This could explain why the flux of the PES-silicalite-1-PDMS composite membrane was lower than that of the PAN-silicalite-1-PDMS composite membrane.

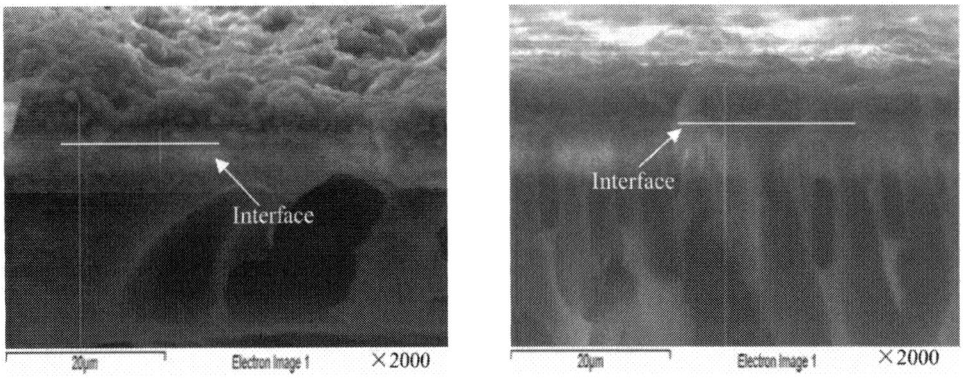

(a) (b)

Fig. 3 SEM photographs and Si concentration profiles (EDS, red-line) of the cross section of the composite membranes. (a) PAN-siliclaite-1-PDMS composite membrane, (b) PES-siliclaite-1-PDMS composite membrane

The continuity and interfacial stability between the top layer and support is another very important factor affecting membrane performance. Fig. 4 shows the SEM photographs of the

PAN-siliclaite-1-PDMS and PES-siliclaite-1-PDMS composite membranes. As can be seen from Figs. 4(a2) and 4(b2), no difference can be observed in the top view. However, the differences at the interface between the PDMS selective layer and the support in the two composite membranes are very clear as shown in Figs. 4(a1) and 4(b1). In Fig. 4(a1), a uniform VTMS-modified silicalite-1/PDMS layer, with a thickness of 5-6 μm on top of a PAN support and a good adhesion, can be observed between the dense PDMS top active layer and the PAN support, while in Fig. 4(b1), the air gap can be clearly seen between the top layer and PES support, which may be caused by the intrusion of PDMS into the pores of the PES membrane due to its more hydrophobic nature [10].

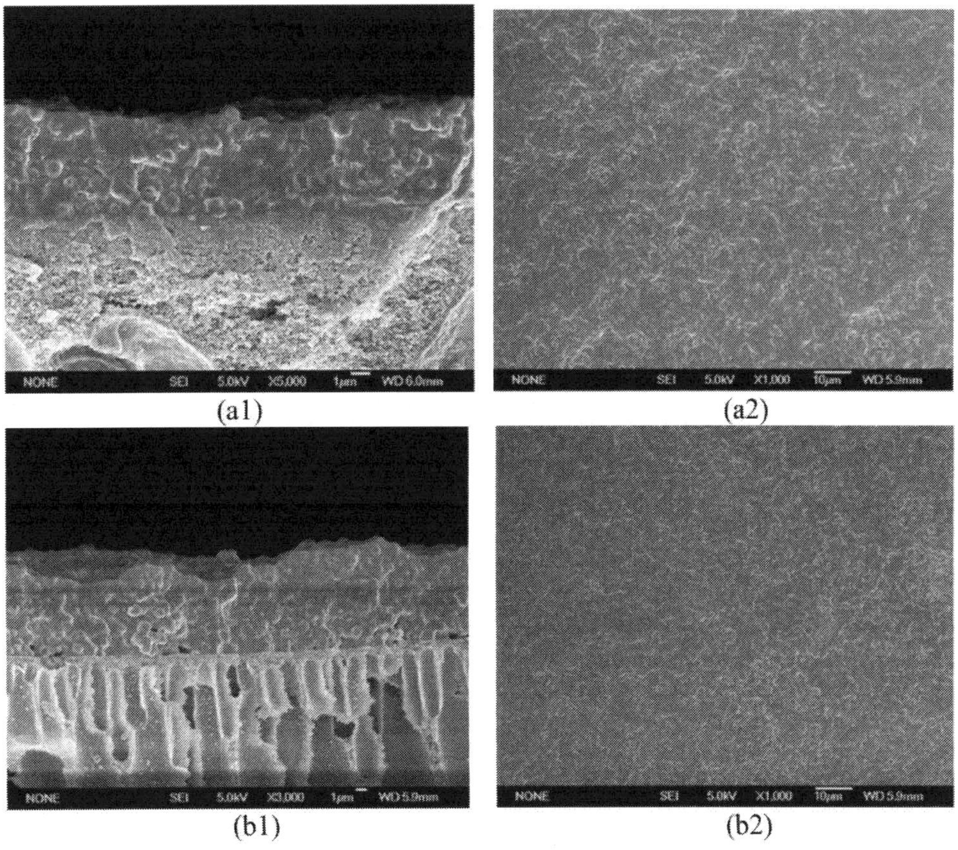

Fig. 4 SEM photographs of composite membranes. PAN-siliclaite-1-PDMS membrane: (a1) cross-sectional view, (a2) top view and PES-siliclaite-1-PDMS membrane: (b1) cross-sectional view, (b2) top view.

3.2.3 Effect of concentration of coating solution

The coating solution for the composite membrane was two constituents: membrane constituents (PDMS and silicalite-1) and the solvent n-heptane. When the coating solution was coated on the UF membrane support, the n-heptane evaporated and the residual membrane constituents formed the selective layer of the composite membrane. Therefore, in the preparation of the composite membrane, the thickness of the composite membrane normally decreases with decreasing the concentration of membrane constituents in the coating solution. However, when the selective layer of composite membrane is very thin, it would lead to defects or pinholes on the active layer surface due to the irregular packing of polymer chains and incomplete coalescence of polymer molecules in the selective layer of the composite membrane [11].

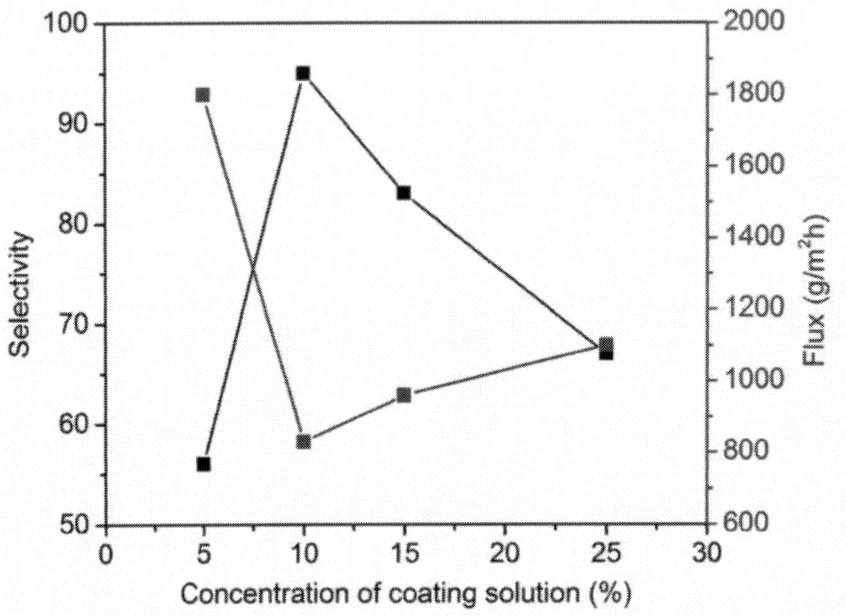

Fig. 5 Effect of concentration of coating solution on the separation performance of composite membrane in the pervaporation of acetone-butanol-ethanol-water solution. Feed: 0.5 wt% acetone + 1.0 wt% butanol + 0.16 wt% ethanol; Temperature: 323 K.

Given that fact that the VTMS-modified silicalite-1 has a particle size in micron range, there exists a minimal thickness where the prepared composite membrane is defect-free. It is worthy examining the effect of concentration of coating solution on the performance of the composite membrane with varied thickness. By varying the concentration of membrane constituents in the coating solution, composite membranes with different thicknesses were obtained. It was found that a membrane with a thickness of about 2 μm could be obtained with 5 % concentration of membrane constituents in the coating solution. Its flux was high up to about 1800 g/m^2h. However, the selectivity was only 55, as shown in Fig. 5. The low selectivity could be attributed to the scattered pinholes at the membrane surface (detected by SEM), which resulted from irregular packing of polymer chains and incomplete coalescence of polymer molecules. When the concentration of membrane constituents increased to 10 %, the thickness increased to 5-6 μm, and no pinholes could be found in the selective layer surface (also detected by SEM). Thus the selectivity of butanol/water was increased (up to about 95), while the flux decreased to 830 g/m^2h. When the concentration of membrane constituents increased further to 15 % and 25 %, the selectivity decreased. This is because the hydrosilylation reaction would increase at a higher concentration of coating solution, thus resulting in a very high viscosity of the coating solution. In this case, pinholes formed in the active layer (also detected by SEM) when the n-heptane evaporated, and the selectivity decreased accordingly, as shown in Fig. 5.

Table 2 shows comparisons of separation performance of the pervaporation membrane with that of previously reported membranes in pervaporative separation of butanol-water solution. It was found that the VTMS-modified silicalite-1/PDMS composite membrane exhibited reasonable selectivity towards 1-butanol, and the total flux was very high in comparison with that of previously reported membranes.

Table 2 Summary of silicone-based membranes for butanol separation in butanol-water binary mixtures

No.	Membrane	Feed concentration (wt%)	Feed temperature (K)	Total flux ($g/m^2 h$)	Separation factor	Ref.
1	PDMS homogeneous membrane	1	323	145.3	49.6	[3]
2	Thin-film silicalite-silicone membrane	1	343	607	93	[3]
3	Silicalite-silicone membrane	1	323	52	58	[9]
4	Poly(1-trimethylsiyl -1-propyne)(PTMSP)	1	343	999	70	[10]
5	Silicalite-silicone composite membrane	1	343	1353	70	This work
6	Silicalite-silicone composite membrane	7	310	954	29	This work
7	Silicalite-silicone homogeneous membrane	1	323	220	143	This work

3.3 Pervaporative separation of butanol from aqueous solution

3.3.1 Effect of temperature

Fig. 6 shows the effect of the operating temperature on separation performance. As can be seen, when the feed temperature increased from 303 K to 343 K, the total flux increased from 230 g/m^2h to 1353 g/m^2h. In pervaporation, permeating molecules diffuse through the free volumes of the membrane. Thermal motion of polymer chains in amorphous regions randomly produces free volumes. As the temperature increases, the frequency and amplitude of polymer jumping chains increase, and consequently the free volume of the membrane increases. Thus, at higher temperatures, the diffusion rate of individual permeating molecules increases, leading to high permeation fluxes [12].

As shown in Fig. 6, when the feed temperature increased from 303 K to 343 K, the selectivity increased from 40 to 70 in the separation of the butanol-water solution. However, when the temperature increased from 343 K to 348 K, the selectivity decreased to 65, whereas the flux increased further to 1700 g/m^2h. The decrease of selectivity at high temperature was probably caused by thermal motion of polymer chains in the membrane. As the temperature increased above 343 K, non-selective cavities were produced due to the too large free volume, which favored the diffusion of butanol and water but more strongly for water as the molecular size of water is smaller than that of butanol [13].

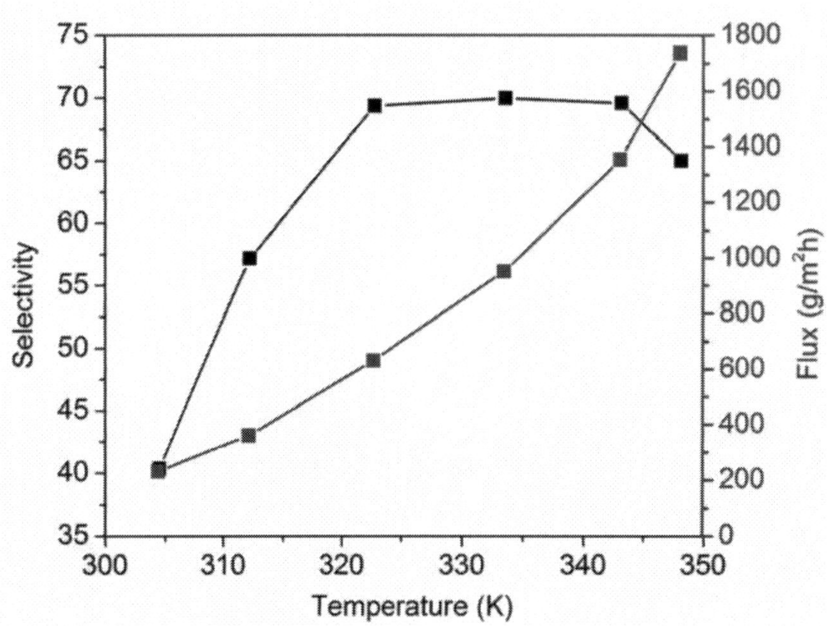

Fig. 6 Effect of feed temperatures on pervaporation performance of the composite membrane in the pervaporation of butanol-water solution. Feed: 1.0 wt% butanol; Permeation pressure: 350 Pa.

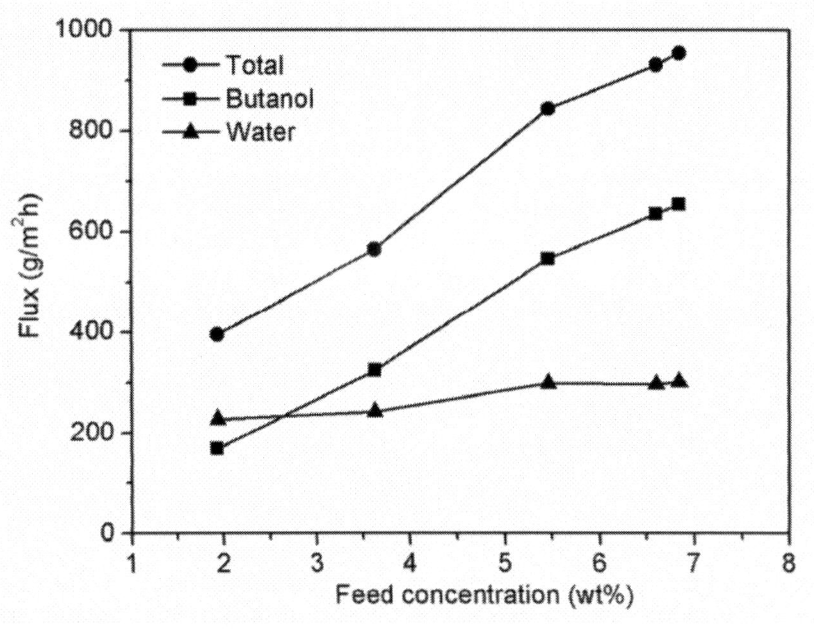

Fig. 7 Effect of feed concentrations on the fluxes in the pervaporation of butanol-water solution. Temperature: 310 K.

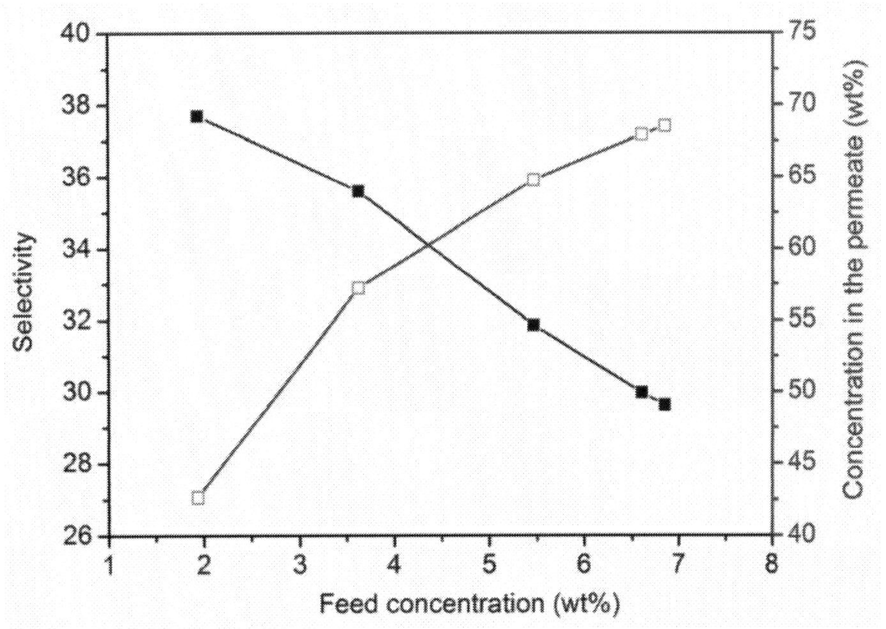

Fig. 8 Effect of feed concentrations on the separation factor and permeate concentration in the pervaporation of butanol-water solution. Temperature 310 K, flow rate 2.0 L/min.

3.3.2 Effect of feed concentration

Figs. 7 and 8 show the effects of butanol concentration on the flux and selectivity. As illustrated in Fig. 9, when the butanol concentration increased from 2.0 wt% to 7.0 wt%, the butanol flux increased linearly. It can be explained that an increase in feed concentration caused a rise in activity and partial pressure, hence enhancing the PV driving force and consequently leading to higher fluxes. The water flux was almost the same with increasing feed butanol concentration in the butanol-water solution due to little variation of water concentration [14]. Similar behavior (constant water flux, linear increasing organic flux) was also observed for butanol extraction using a PDMS membrane by Favre et al. [15], and they ascribed the relative constancy of the water flux to the constant water activity in the dilute aqueous solution.

4. Conclusions

High separation performance VTMS-modified silicalite-1/PDMS composite membrane with PAN as support was successfully prepared and applied in pervaporation to separate butanol from aqueous solution. The experimental results showed that hydrophilic support was favorable for the silicalite-1/PDMS composite membranes. Using "platinum cure" in the pre-polymerization of the coating solution, the VTMS-modified silicalite-1/PDMS composite membranes with an active layer thickness of 5-6 μm were prepared, resulting in a butanol/water selectivity of 70 and fluxes up to 1353 g/m^2h. Thus pre-polymerization of the coating solution using "platinum cure" could serve as a useful and promising method for the preparation of high performance composite membranes. In the separation of butanol-water solution, higher temperature brought out higher fluxes and higher selectivity. Butanol flux increased with the increase in the feed concentration, but not significantly for water flux.

Acknowledgements

This work is sponsored by the National Natural Science Foundation of China (Grant No. 21176239) and the National High-Tech R&D Program (863 Program) of China for the financial support (Grant No. 2012AA03A607).

References

1. L.M. Vane, V.V. Namboodiri, T.C. Bowen, Hydrophobic zeolite-silicone rubber mixed matrix membranes for ethanol-water separation: Effect of zeolite and silicone component selection on pervaporation performance, J. Membr. Sci. 308 (2008) 230-241.
2. H. Sun, L. Lu, X. Chen, Z. Jiang, Surface-modified zeolite-filled chitosan membranes for pervaporation dehydration of ethanol, Appl. Surf. Sci. 254 (2008) 5367-5374.
3. R. Mahajan, R. Burns, M. Schaeffer, W. Koros, Challenges in forming successful mixed matrix membranes with rigid polymeric materials, J. Appl. Polym. Sci. 86 (2002) 881-890.
4. J. Huang, M. Meagher, Pervaporative recovery of n-butanol from aqueous solutions and ABE fermentation broth using thin-film silicalite-filled silicone composite membranes, J. Membr. Sci. 192 (2001) 231-242.
5. B. Moermans, W. Beuckelaer, I. Vankelecom, R. Ravishankar, J. Martens, P. Jacobs, Incorporation of nano-sized zeolites in membranes, Chem. Commun. 2000 (2000) 2467-2468.
6. B. Smitha, D. Suhanya, S. Sridhar, M. Ramakrishna, Separation of organic-organic mixtures by pervaporation-a review, J. Membr. Sci. 241 (2004) 1-21.
7. Gelest Reactive Silicones catalog, 2004. Also available as "An Organosilicon Chemistry Primer" at http://www.gelest.com/Library/01Primer.pdf.
8. H. Zhou, Y. Su, X. Chen, S. Yi, Y. Wan, Modification of silicalite-1 by vinyltrimethoxysilane (VTMS) and preparation of silicalite-1 filled polydimethylsiloxane (PDMS) hybrid pervaporation membranes, Sep. Purif. Technol. 75 (2010) 286-294.
9. S. Yi, Y. Su, Y. Wan, Preparation and characterization of vinyltriethoxysilane (VTES) modified silicalite-1/PDMS hybrid pervaporation membrane and its application in ethanol separation from dilute aqueous solution, J. Membr. Sci. 360 (2010) 341-351.
10. S. Claes, P. Vandezande, S. Mullens, R. Leysen, K. De Sitter, A. Andersson, F.H.J. Maurer, H. Van den Rul, R. Peeters, M.K. Van Bael, High flux composite PTMSP-silica nanohybrid membranes for pervaporation of ethanol/water mixtures, J. Membr. Sci. 351 (2010) 160-167.
11. F. Xiangli, W. Wei, Y. Chen, W. Jin, N. Xu, Optimization of preparation conditions for polydimethylsiloxane (PDMS)/ceramic composite pervaporation membranes using response surface methodology, J. Membr. Sci. 311 (2008) 23-33.
12. P. Sampranpiboon, R. Jiraratananon, D. Uttapap, X.S. Feng, R.Y.M. Huang, Separation of aroma compounds from aqueous solutions by pervaporation using polyoctylmethyl siloxane (POMS) and polydimethyl siloxane (PDMS) membranes, J. Membr. Sci. 174 (2000) 55-65.
13. L. Aouinti, M. Belbachir, A maghnite-clay-H/polymer membrane for separation of ethanol-water azeotrope, Appl. Clay Sci. 39 (2008) 78-85.
14. E. Fouad, X. Feng, Use of pervaporation to separate butanol from dilute aqueous solutions: Effects of operating conditions and concentration polarization, J. Membr. Sci. 323 (2008) 428-435.
15. E. Favre, Q. Nguyen, S. Bruneau, Extraction of 1-butanol from aqueous solutions by pervaporation, J. Chem. Technol. Biotechnol. 65 (1996) 221-228.

(123b) Preparation of Mesoporous Silica Nanoparticles Enhanced Thin-Film Nanocomposite (TFN) Membrane for Water Purification

Jun Yin[1], Eun-Sik Kim[1], John Yang[2] and Baolin Deng[1], (1)Department of Civil & Environmental Engineering, University of Missouri-Columbia, Columbia, MO, (2)Department of Agriculture and Environmental Sciences, Lincoln University, Jefferson city, MO

A thin-film nanocomposite (TFN) membrane containing mesoporous silica nanoparticles (NPs) was prepared by the in situ interfacial polymerization (IP) process. Aqueous *m*-phenyl diamine (MPD) and organic trimesoyl chloride (TMC)-NPs mixture solutions were used in the IP process. Mesoporous silica NPs (~100 nm) and non-porous spherical silica NPs (~100 nm) were synthesized and used as the fillers to fabricate the TFN membrane at concentrations ranging from 0 to 0.1 wt.%. The membranes were characterized by scanning electron microscopy (SEM), transmission electron microscopy (TEM), atomic force microscopy (AFM) and attenuated total reflection Fourier transform infrared (ATR FT-IR) spectroscopy, and their performances were evaluated based on the pure water permeability and salt rejection. Results indicated that the mesoporous silica NPs dispersed well in polyamide (PA) thin-film layer and improved membrane performances under optimal concentrations. By increasing its concentration, hydrophilicity, roughness and zeta potential of TFN membranes all increased. Notably, the pure water flux increased around 64% with the incorporation of mesoporous NPs, while maintaining high rejections of NaCl and Na_2SO_4 (over 97% and 98%, respectively). When compared with membranes containing non-porous NPs, the membranes containing mesoporous NPs showed enhanced permeability, suggesting that the short flow paths through the hydrophilic porous structure of mesoporous NPs had played an important role in water permeation.

(123c) Development of Multi-Walled Carbon Nanotubes (MWNTs)/Polysulfone (PSF) Mixed Matrix Hollow Fiber Membranes for Enhanced Water Treatment

Jun Yin and Baolin Deng, Department of Civil & Environmental Engineering, University of Missouri-Columbia, Columbia, MO

Due to the carbon nanotubes' unique one-dimensional tubular structure and superior mechanical and chemical properties, we explored if the incorporation of oxidized multi-walled carbon nanotubes (MWNTs) into polysulfone (PSF) hollow fiber membranes could affect the membrane performance. Polyvinylpyrrolidone (PVP) and 1-Methyl-2-pyrrolidinone (NMP) were used as porogen and solvent, respectively, in the phase inversion spinning process, and deionized water was used as bore fluid and coagulant. MWNTs was first purified and oxidized in the mixed acid solution (H_2SO_4/HNO_3 = 3/1 in volume) and then used as the filler at concentrations ranging from 0 to 1 wt.% based on the amount of PSF. Results indicated that at three different PSF concentrations (15, 18, 20 wt.%), pure water flux of all membranes showed similar characteristics, with increasing MWNTs concentration, the flux first increased and then gradually decreased. The optimized blend membranes showed a significant increase of pure water flux (60% to 100%) while maintaining the same ability for solute rejection. The water contact angle of the mixed matrix membrane slightly decreased with increasing filler concentration, which indicated the improved membrane surface hydrophilicity, and the fouling resistance of the membrane to protein was also enhanced.

(123d) Mixed-Matrix Membranes with Zeolite-Like MOFs for CO2/CH4 Separation

Ayse Kiliç, Cigdem Atalay-Oral, Ahmet Sirkecioglu, S. Birgül Tantekin-Ersolmaz and M. Göktug Ahunbay, Department of Chemical Engineering, Istanbul Technical University, Istanbul, Turkey

The separation of CO_2 from industrially important gas mixtures like natural gas, syngas and flue gas is an enormously crucial issue in chemical industry. Zeolite-like metal organic frameworks (ZMOFs) are promising candidates for separation and purification applications. They have anionic frameworks and extra-framework ions in the pores of molecular dimensions increasing the interactions with guest molecules and improving their separation, storage, or ion-exchange capability. The charge-compensating extra-framework ions inside the cavities of ZMOFs can be ion-exchanged with alkali metals and this is expected to increase their affinity for CO_2 and improve natural gas (CO_2/CH_4) purification performance.

In the present study, *sod*-ZMOF crystals having a sodalite topology were synthesized under solvothermal conditions. Following characterization of the crystals (XRD, TGA, SEM analyses), they were incorporated into Matrimid® polyimide (PI) matrix to form dense mixed-matrix membranes (MMMs) by solvent-casting method. Although several gas adsorption studies with ZMOFs have been reported in the literature, they are used as fillers in mixed-matrix membranes and their gas separation properties are investigated for the first time in this work. SEM images of the MMMs, showed no apparent voids or defects at the filler/polymer interface and that ZMOF particles were dispersed homogenously in the polymer matrix. Physical and thermal properties of the prepared membranes (e.g. film density, T_g) were analyzed. The CO_2/CH_4 separation performances of the membranes were determined by measuring pure gas permeabilities. As-synthesized *sod*-ZMOF samples were also ion-exchanged with Na and K cations to examine the effect of ion-exchange on the material characteristics and gas separation properties of the MMMs. The contribution of ion-exchanged alkali cations is observed to change gas separation performance of *sod*-ZMOF significantly.

(123e) Aquaporin Based Biomimetic Membranes

Chuyang Y. Tang[1], Yang Zhao[1], Changquan Qiu[1], Xuesong Li[2], Ardcharaporn Vararattanavech[2], Wenmin Shen[2], Jaume Torres[2], Claus Hélix Nielsen[3], Xiao Hu[4], Rong Wang[5] and Anthony G. Fane[6], (1)Singapore Membrane Technology Centre and School of Civil and Environmental Engineering, Nanyang Technological University, Singapore, Singapore, (2)Nanyang Technological University, Singapore, Singapore, (3)Aquaporin A/S, Copenhagen, Denmark, (4)Nanyang Technological Universtiy, Singapore, Singapore, (5)Singapore Membrane Technology Centre, Nanyang Technological University, (6)Singapore Membrane Technology Centre, Nanyang Technological University, Singapore, Singapore

Nature has developed a most efficient way for water transport across an osmotic pressure gradient via aquaporin proteins. The aquaporins or water channel proteins, typically bound in phospholipid cellular membranes, are highly permeable to water but highly retentive to solutes. This makes the water delivery across a cell at a sufficiently low energy cost. This presentation will review the difference designs reported in the recent literature for fabricating aquaporin-based biomimetic membranes. Their advantages and potential issues will be discussed. The presentation will then focus on our latest developments in this topic. The permeability and rejection of aquaporin proteins in different vesicular structures have been characterized using stopped flow measurements. An artificial membrane has been developed to mimic the natural cellular membranes by incorporating aquaporins into an ultrathin rejection layer. The resulting aquaporin biomimetic membrane showed good mechanical stability, water permeability, and salt rejection, with strong potential for water purification, wastewater reuse and seawater desalination applications.

(123f) An Inorganic Nanotube/Polymer Composite Membrane Platform for Molecular Separations

Dun-Yen Kang[1], Sankar Nair[2] and Christopher W. Jones[1], (1)School of Chemical & Biomolecular Engineering, Georgia Institute of Technology, Atlanta, GA, (2)Chemical & Biomolecular Engineering, Georgia Institute of Technology, Atlanta, GA

Nanotubes possess intriguing mass transport properties, but have so far not been successfully integrated into an economical molecular separation device. This talk will describe the creation, characterization, and applications of a nanotube-based membrane separation platform by incorporating inorganic (e.g., single-walled metal oxide) nanotubes into polymeric membranes. In this talk, we will first describe the fabrication and characterization of defect-free nanotube/polymer composite membranes, containing aluminosilicate single-walled nanotubes embedded in a poly(vinyl alcohol) matrix. Next, we will describe the permeation behavior of molecules such as alcohols and water through these membranes by experimental and molecular simulation techniques. Finally, we introduce a mass transport model with specific applicability to composite membrane systems containing tubular fillers, and use it to analyze the membrane permeation properties. We show that our new model provides analytical guidance for the fabrication and understanding of membranes containing nanotubular fillers. This talk provides an overview of the state-of-the-art nanotube-based membranes for separation applications.

(123g) Carbon Nanotube-Polyamide Nanocomposite Membrane for Water Desalination

Hang-yan Chen[1], Wai Fong Chen[2], Anil Surapathi[3], Eva Marand[2], J. Karl Johnson[4], De-Li Chen[5] and Xiao-hong Shao[6], (1)Chemical & Petroleum Engineering, University of Pittsburgh, Pittsburgh, PA, (2)Chemical Engineering, Virginia Tech, Blacksburg, VA, (3)Virginia Tech, Blacksburg, VA, (4)Dept. of Chemical and Petroleum Engineering, University of Pittsburgh, Pittsburgh, PA, (5)Dept.of Chemical and Petroleum Engineering, University of Pittsburgh, Pittsburgh, PA, (6)Beijing University of Chemical Technology, Beijing, China

We have developed asymmetric carbon nanotube/polyamide membranes for water desalination applications. The carbon nanotubes are functionalized with zwitterionic groups to block the transport of ions while permitting a high flux of water. Both experiments and non-equilibrium molecular dynamics simulations were carried out to characterize the transport of fluids through functionalized carbon nanotubes, including entrance effects. The simulations utilized a model carbon nanotube membrane consisting of four carbon nanotubes embeded between two graphene sheets. The ends of each tube were functionalized with one or more zwitterion groups. The carbon nanotube membrane was embedded into a periodically repeated saltwater box in order to study the flux of water and salt ions through the memberane. A pressure drop of about 200 MPa across the memberane was maintained in order to generate a high water flux over a limited simulation time. The ion rejection ratio was found to be 100% when two zwitterions were affixed to the ends of (20,0) nanotubes. The simulations show that ions are rejected through a combination of electrostatic and steric effects. Experimental sorption studies also suggest that the zwitterionic groups are highly hydrolized and their presence selectively blocks the entry of molecules into the carbon nanotube pores. Water transport measurements show a salt rejection ratio of around 98% with water flux near 40 l/m^2hr. Furthermore, both the water flux and salt rejection increase in proportion with the area density of the carbon nanotubes in the membrane.

(131a) Highly Selective Faujasite-Type Zeolite Membrane for Isopropyl Alcohol/Water Vapor Mixtures

Masahiko Matsukata, Makoto Toyoda and Yasushi Sekine, Department of Applied Chemistry, Waseda University, Tokyo, Japan

In petrochemical industry, most of production processes involve distillation towers for dehydration of products such as various types of alcohol. Generally, distillation separation of water from products consumes a large amount of energy since vaporization of water requires a large amount of heat. Combination of a membrane separation unit with distillation can drastically reduce energy demand for the dehydration of alcohols. Faujasite-type zeolite is a kind of zeolite with both high stability in the presence of steam and strong hydrophilic nature. In this study, we investigated the preparation method of a compact faujasite-type of zeolite membrane, and separation properties were examined for mixtures of isopropyl alcohol(IPA)/water system. Faujasite-type membranes were prepared on the outer surface of tubular porous alumina support (o.d.= 10 mm, average pore diameter = 150 nm) by a secondary growth method including dip-coating of seed crystals and successive hydrothermal synthesis using aluminosilicate hydrogel. Vapor permeation and separation measurements were performed using a stainless steel tube module cell. By optimizing conditions for the dip-coating of seed crystals on the tubular support, seed crystals suitably dispersed in the pores of alumina tube. A compact faujasite-type membrane was formed by growing such seeds supported inside the pores of tubular support. The faujasite-type membrane developed in this study highly separated water from water/IPA mixtures. In the case of an mixture of water/IPA (20/80 wt/wt), the membrane showed a large flux of water of ca.$1x10^{-6}$ mol m^{-2} s^{-1} Pa^{-1} with a high separation factor of water/IPA exceeding 3000 at 373 K.

(131b) Insights Into Ion Transport Pathways in Ceramics by in-Situ Neutron Diffraction

Steven McIntosh[1], Rosemary Cox-Galhotra[2], Ashfia Huq[3] and Jason Hodges[3], (1)Chemical Engineering, Lehigh University, Bethlehem, PA, (2)Department of Chemical Engineering, University of Virginia, Charlottesville, VA, (3)Spallation Neutron Source, Oak Ridge National Laboratory, Oak Ridge, TN

The search for higher conductivity materials for solid oxide fuel cell (SOFC), and ceramic ion-transport membranes (ITMs) is increasingly leading towards more crystallographically complex materials. The traditional materials palette is based on a cubic perovskite structure, with randomly distributed oxygen anion vacancies. Ion transport, which occurs by ion hopping between these sites, is thus directionally homogenous. In contrast, the oxygen anion vacancies in higher order layered-perovskite materials are suggested to be localized within channels or planes within the crystal structure, leading to localization and heterogeneity in the ion transport pathway.

Powder neutron diffraction provides an ideal tool to examine the bulk structure of these materials. Unlike X-ray diffraction, where X-rays interact only weakly with oxygen atoms, neutron diffraction provides information on the position, occupancy, and thermal motion of both the cations and oxygen anions in the materials of interest. This enables the oxygen non-stoichiometry, lattice expansion, and preferred anion transport pathways to be determined in a single experiment. Since these parameters vary with temperature and gas atmosphere, it is necessary to perform these measurements at *in*-situ conditions.

We will present recent insights into the preferred ion transport pathways in layered perovskites as determined by neutron diffraction between 550 and 825°C at pO_2 from 10^{-1} to 10^{-4} atm.

(131c) Bilayer MFI Zeolite Membranes: Hydrogen Separation and Long Term Stability

Haibing Wang, School of Mechanical, Aerospace, Chemical and Materials Engineering, Arizona State University, Tempe, AZ and Jerry Y. S. Lin, Department of Chemical and Materials Engineering, Arizona State University, Tempe, AZ

Metallic or amorphous microporous inorganic membranes perm-selective for hydrogen separation suffer from thermal or chemical stability problems. MFI type zeolite membrane offers good stable and high hydrogen performance, making it attractive for application in membrane reactors for water gas shift (WGS) reactionfor hydrogen production. The hydrogen permselectivity of MFI zeolite membranes can be improved by intracrystalline pore modification. However, more than one order of magnitude reduction in H_2permeance usually accompanies with the significant improvement in H_2/CO_2 separation factor of MFI zeolite membranes modified by catalytic cracking deposition (CCD) of methyldiethoxysilane. In this work, a thin, high quality ZSM-5 top layer was synthesized on a thick silicalite bottom layer to form a ZSM-5/silicalite bilayer membrane. The ZSM-5/silicalite bilayer zeolite membrane was supported on a porous a-alumina support coated with an yttria stabilized zirconia (YSZ) intermediate barrier layer for membrane stability improvement. The CCD modification of the bilayer zeolite membrane resulted in an improvement of H_2/CO_2 separationfactor of the membrane by over five-fold from 5 to25 with only about 31% reduction in H_2permeance from $1.9x10^{-7}$ to $1.3x10^{-7}$mol/m^2.s.Pa at 450°C. Long-term hydrothermal and chemical stability of the modified bilayer zeolite membrane were studied under water gas shift (WGS) reaction conditions at high temperatures. The stability testing results showed that the H_2 recovery and the H_2permeance of the CCD modified membrane increased by 2.1% and 2.7%, respectively, and the H_2/CO_2 separation factor decreased from 22.4 to 21.7, afterthe membrane reactor was operated for WGS reaction for 28 days (500°C, $H_2O/CO=3$, GHSV=60,000 h^{-1}). This membrane was further tested under WGS reaction conditions (25% H_2: 25% CO: 25% H_2O: 25% CO_2, 500°C, no WGS catalyst was packed) in the presence of 400 ppm H_2S for 24 days, and no significantchange in H_2permeance and H_2/CO_2 separation factor was observed. The results demonstrated that the modified bilayer MFI zeolite membranes with an yttria stabilized zirconia (YSZ) intermediate barrier layer have excellent long-term hydrothermal and chemical stability under WGS reaction conditions.

DUAL-PHASE CERAMIC MEMBRANE FOR OXYGEN SEPARATION

Weishen Yang, Xuefeng Zhu, Dalian Institute of Chemical Physics, Chinese Academy of Sciences, Dalian, China

Abstract

Mixed conducting dual-phase membranes, made of ionic conducting oxides and electronic conducting oxides, have attracted considerable attention since the two phases can be separately tailored according to the specific requirements of a given application. Here, for the potential applications in membrane reactors and oxygen separation, the design of dual-phase membranes is discussed in detail from the point of view of solid state chemistry. Comparative investigation is focused on the dual-phase membranes made of an oxide ionic conducting oxide and a mixed conducting oxide and the dual-phase membranes made of an oxide ionic conducting oxide and a pure electronic conducting oxide. Microstructure effects of dual-phase membranes on oxygen exchange reactions and bulk oxide ionic transport were investigated by AC impedance spectroscopy, high resolution transmission electron microscopy and oxygen permeation.

1. Introduction

Oxygen permeation through the dense ceramic membrane is due to directional motion of oxygen ions under an oxygen partial pressure gradient. Many perovskite-type materials were developed to meet the requirements for practical applications such as the production of pure oxygen or oxygen-enriched air and partial oxidation of natural gas to syngas ($CO+H_2$). Perovskite oxides with cobalt doping of the B-site usually possess considerable permeability but poor stability under reducing environments [1]. Although these membrane materials can sustain hundreds or thousands of hours for syngas production without failing, their long-term stability under large oxygen partial pressure gradients is not guaranteed because of material decomposition to a depth up to tens of micrometers [2,3]. Cobalt-free perovskite-type membrane materials with low permeability are more durable, but still suffer from decomposition under syngas conditions though they can keep perovskite structure in hydrogen atmosphere [4-6].

Dual-phase composite membranes have been suggested as good substitutes with compositions that can be adjusted according to practical requirements. One phase is usually an oxide ionic conductor for oxide ionic transport, and the other is a noble metal or perovskite-type oxide for electron transport. Research about dual-phase membranes started in the 1990s for dual-phase membranes made of a noble metal and an oxide ionic conductor [7, 8]. However, the cermet membranes usually had a low permeation flux and high costs. Investigations into oxide ceramic dual-phase membranes made of an oxide ionic conductor and a pure electronic conducting perovskite oxide started in 2000 [9–11]. These membranes are less expensive than cermet membranes and are highly stable under reducing conditions, but they are limited by their low permeation fluxes.

Figure 1 shows our idea for the design of dual-phase membranes. Among the ionic conductors shown in figure 1, Y or Sc stabilized zirconia, doped ceria, doped bismuth oxides, $La_2Mo_2O_9$ and $La_{1-x}Sr_xGa_{1-y}Mg_yO_{3-\delta}$ ($0 \leqslant x \leqslant 0.2$, $0 \leqslant y \leqslant 0.2$) possess high ionic conductivity and are suitable as the ionic conducting phase for dual-phase membranes. However, when the stability under reducing environments is considered, only Y or Sc stabilized zirconia, doped ceria and $La_{1-x}Sr_xGa_{1-y}Mg_yO_{3-\delta}$ are suitable. Usually, perovskite oxides are used as the second phase for electronic transport and therefore, when the compatibility between the ionic conducting phase and the perovskite phase is considered, only doped ceria is left. Zirconia reacts easily with perovskite oxides to produce zirconates and pyrochlores when they are sintered together at elevated temperatures, and the resulting new compounds, with poor ionic and electronic conductivity, block the transport of ions and electrons. Elements in electronic conducting perovskite oxides easily diffuse into the $La_{1-x}Sr_xGa_{1-y}Mg_yO_{3-\delta}$ lattice and produce a new perovskite oxide, which usually has normal stability and permeability [12]. Therefore, these two ionic conductors are unsuitable unless a compatible electronic conductor is used.

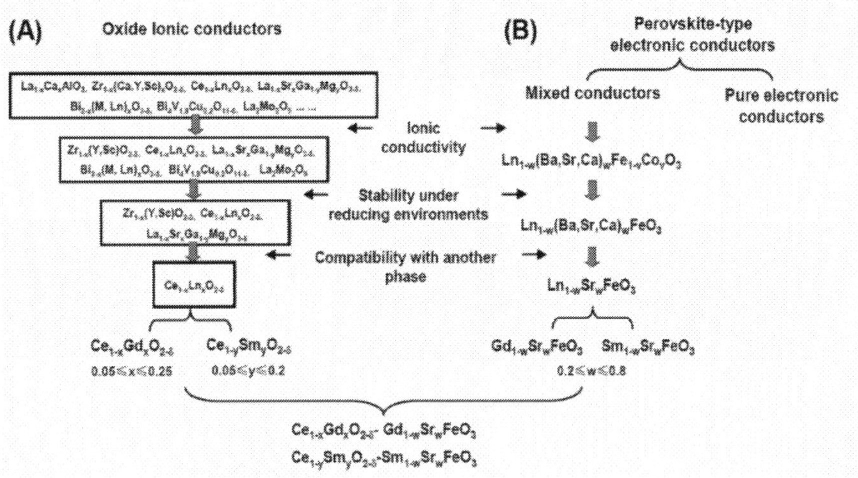

Figure 1 Selections of oxide ionic conductors and perovskite-type electronic conductors for dual-phase membranes. Ln: lanthanide elements.

The perovskite electronic conductors can be classified into two groups, i.e. a pure electronic conductor (or one where the ionic conductivity is extremely low, such as $Ln_{1-x}(Ba,Sr,Ca)_x(Mn,Cr)O_3$) and a mixed conductor (such as $Ln_x(Ba,Sr,Ca)_{1-x}Co_yFe_{1-y}O_{3-\delta}$. If the pure electronic conductor or mixed conductor with poor ionic conductivity is used as the second phase in dual-phase membranes, it blocks the transport of oxide ions in the membrane bulk between ionic conducting grains [13]. However, when the mixed conductor with high ionic conductivity is used as the second phase, it contributes to, rather than blocks, the transport of ions. If the stability under a reducing environment is considered, options with cobalt or other reducible metal ions doped into perovskite must be excluded, and only Fe-based perovskite oxides remain, as shown in figure 1. If the compatibility between doped ceria and $Ln_x(Ba,Sr,Ca)_{1-x}FeO_{3-\delta}$ perovskite is considered, barium and calcium should be excluded. Because barium doped perovskite oxides easily react with ceria to produce $BaCeO_3$, and calcium easily diffuses into the ceria lattice which reduce the ionic conductivity. Of course, strontium doped perovskite oxides can also react with ceria to produce $SrCeO_3$. However, the low tolerance factor of 0.87 gives Sr^{2+} a higher chemical potential in $SrCeO_3$ than that in $Ln_{1-x}Sr_xFeO_3$, so $SrCeO_3$ does not emerge in the dual-phase composite membranes; this has been verified in previous work [13-15]. Therefore, we can obtain dual-phase membranes with the composition of $Ce_{1-x}Ln_xO_{2-\delta}$-$Ln_{1-y}Sr_yFeO_{3-\delta}$. Among the doped ceria compounds, Gd^{3+} and Sm^{3+} doped ceria have the highest conductivities, so the matched perovskite oxides should be $Gd_{1-y}Sr_yFeO_{3-\delta}$ and $Sm_{1-y}Sr_yFeO_{3-\delta}$. The stability of the perovskite can be improved by doping Al^{3+} or Ga^{3+} into the B-site [16].

Here, dual-phase membranes consisting of $Ce_{0.85}Sm_{0.15}O_{1.925}$-$Sm_{0.6}Sr_{0.4}FeO_{3-\delta}$ (SDC-SSF) were investigated. In addition, $Ce_{0.85}Sm_{0.15}O_{1.925}$-$Sm_{0.6}Sr_{0.4}CrO_{3-\delta}$ (SDC-SSCr) which consists of an ionic conductor and pure electronic conductor, was investigated for comparison purposes to well illuminate our design idea of dual-phase materials.

2. Experimental

All the dual-phase membranes were prepared via the one-pot method. The dual-phase membranes made of SDC and SSF with weight ratio of 65:35, 75:25 and 85:15 were donated as SDC65-SSF35, SDC75-SSF25, and SDC85-SSF15, respectively. And the dual-phase membrane made of SDC and SSCr with a weight ratio of 75:25 was donated as SDC75-SSCr25. The detailed description of the preparation process can be found elsewhere [15]. The green disks were sintered at 1450 °C for 3 h. All the dual-phase membrane samples had relative density no less than 95%. After polished with SiC sandpaper and cleaned in ethanol, one or both sides of the membranes were coated with a $La_{0.6}Sr_{0.4}CoO_{3-\delta}$ (LSC) porous layer to improve the oxygen surface exchange rate [15]. Other perovskite oxides or fluorite oxides investigated in this paper were synthesized via the EDTA-citric acid method as described above.

Silver rings were used to seal the membranes at 961 °C for oxygen permeation experiments. The effective inner surface areas of the membrane discs were controlled at around 0.9 cm^2. Dried air (100 mL min^{-1}) was used as the feed and high purity helium was used as the purge gas on the other side of the membrane. Effluents were analyzed by a gas chromatograph (GC, Agilent 6890) equipped with a 3 meter 13 X column. Oxygen permeation fluxes through membranes were calculated based on the effluent flow rate and the concentration of oxygen in the effluents. The leakages were no more than 1% for all the oxygen permeation experiments and were subtracted when calculating the oxygen permeation fluxes.

The total conductivities (four-probe DC) were measured with bar-shaped samples about 20×5×2 mm^3 in air by four-probe methods for the measurement of SSF, SSCr and SDC65-SSF35. The total conductivities of other samples with much higher resistance than those of wires were measured by a two-probe method. The area specific resistances of the membranes were typically measured in ambient air at 600~800 °C by impedance spectroscopy using a Solartron 1287 electrochemical interface and a Solartron 1260 frequency response analyzer interfaced to a computer, with measurements in the frequency range 10 to 10^6 Hz. The thickness of the polished membranes was 1.5 mm. The Pt electrode areas were controlled at 0.33 cm^2. The phase structure of the membrane materials were determined by X-ray diffraction (XRD, Rigaku D/Max-RB, Cu Kα radiation) in the 2θ range 20-80° with a step width of 0.02°. Surface morphologies of the membranes and local EDX analyses were observed on a scanning electron microscope (SEM, Quanta 200 FEG, FEI Company) operating at 20 kV. The morphologies of the grain boundaries and selected area electron diffraction (SAED) of perovskite and fluorite grains were obtained on a high resolution transmission electron microscopy (HR-TEM, FEI Tecnai F30) operated at an accelerating voltage of 300 kV.

3. Results and discussion

The XRD patterns of the SDC-SSF dual-phase membranes with various ratios of the two phases is shown in figure 2a. All the membranes contain a fluorite phase and a perovskite phase, and no impurities were detected even after sintered at high temperatures. That is to say, the two phases have good structural compatibility with each other. However, it was found that a thick layer of dark green oxides was produced on the SDC75-SSCr25 membrane surface after it sintered under the same conditions as those for SDC-SSF. From the XRD pattern of the powder of SDC75-SSCr25, the two phases are shown to be compatible with each other. For the as-sintered SDC75-SSCr25 membrane, the phase composition is more complex than SDC-SSF because the $Cr^{3+/4+}$ ions in the perovskite oxide are easily oxidized to $Cr^{5+/6+}$ during the high temperature sintering in air, which results in the enrichment of oxides containing chromium on the surface of the membrane, as shown in figure 2c [17,18]. SEM pictures (figure 2d) also clearly show the enrichment of Cr-contained oxides on the membrane surface. Nevertheless, after the membranes had been polished by ~100 μm, the XRD pattern is very similar to that of the as-prepared powder. That is to say, the diffraction peaks of Cr-contained oxides disappear, and chromium enrichment takes place on the surface layer, no deeper than 100 μm from the XRD results.

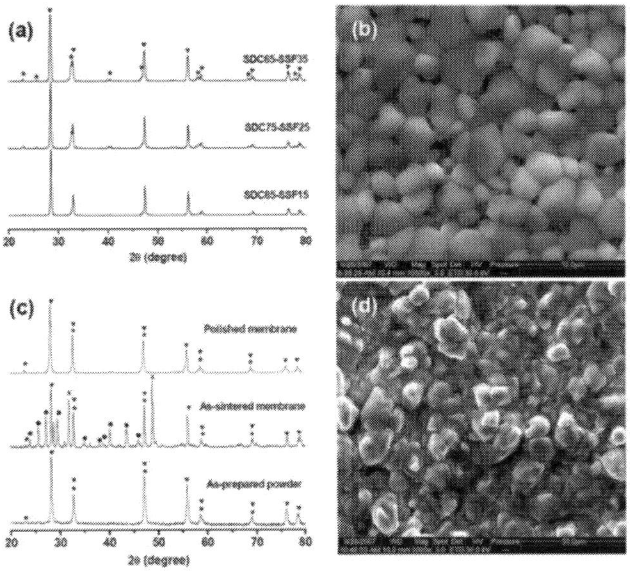

Figure 2 XRD patterns and SEM pictures of the as-prepared dual-phase materials. (a, b) SDC-SSF membranes; (c,d) SDC75-SSCr25 materials, (●) $SrCrO_4$; (▾) Fluorite; (★) perovskite; (×) $Sr_{2.67}Cr_2O_{8.33}$.

The total conductivities of the dual-phase materials and the single-phase compounds are plotted in figure 3. The total conductivity of SSF reaches its maximum value (82 S cm^{-1}) at about 400 °C, and then decreases with further increase of temperature. The phenomenon of conductivity depending on temperature is induced by the loss of oxygen that reduces the charge carriers generated from the acceptor (e.g. Sr^{2+}) dopants at high temperatures. The conductivity of the SDC increases with temperature, as displayed in figure 3. A useful indication as to whether the doping amount of SSF

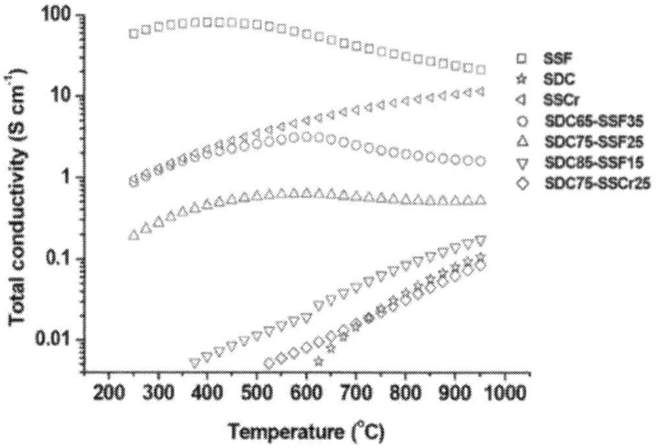

Figure 3 Total conductivities of dual-phase materials and the single-phase oxides.

is enough can be inferred based on the temperature dependence of the conductivity of the dual-phase materials. In brief, if the total conductivity of the SDC-SSF decreases with rising temperature in the range of 600~950 °C, the weight ratio between SDC and SSF is too low, such as SDC85-SSF15; in contrast, if the total conductivity increases with temperature, the ratio is too high such as SDC65-SSF35; if the total conductivity keeps almost constant, the ratio is just proper, such as SDC75-SSF25. The conductivity of SSCr reaches 11 S cm^{-1} at 950 °C. However, the total conductivity of SDC75-SSCr25 is even lower than that of SDC, and this falls outside of the three possibilities. One reasonable explanation is that the oxide containing Cr blocks the ionic transport through the materials and itself cannot reach the percolation limit even if 25 wt.% perovskite oxide is doped. That is to say, the microstructure of the dual-phase membranes is important, which has significant influence on both electronic and ionic transport.

HR-TEM and SAED analysis were employed to observe the grain boundaries. As shown in figure 4a, clear grain boundaries between the fluorite grains and perovskite grains were found for the SDC75-SSF25 dual-phase membrane, and clear grain boundaries between the adjacent fluorite grains are also observed (see figure 4b). However, there are some impurities appearing in the grain boundaries between the fluorite and perovskite grains, and the grain boundaries between fluorite grains for the SDC75-SSCr25 (see figure 4c, d). The appearance of the impurities blocks the ionic transport between fluorite grains, so as to degrade the oxygen permeation flux as displayed in the following section. The composition of the impurities is difficult to be determined by HR-TEM-EDX technology due to its width less than 2 nm. But compared with the composition of the two kinds of dual-phase membranes, the difference is only on the B-site element of the perovskite phase. It was reported that pure electronic conducting oxides contianing Mn or Cr are aways found to have high vaporization and easily deposit in grains boundaries of between adjacent fluorite grains. Therefore, one can safely inferred that the impurities are Cr-contained oxides which are produced during the densification of fluorite phase and the vaporization of chromium under high temperature sintering.

Figure 5 shows the temperature dependence of the oxygen permeation fluxes of SDC-SSF(Cr) dual-phase membranes with the air side and both sides coated with $La_{0.6}Sr_{0.4}CoO_3$ porous layers. To obtain accurate Arrhenius activation energies (Ea) for oxygen transport through those dual-phase membranes, the oxygen partial pressure gradient was fixed at 21 kPa/0.5 kPa by adjusting the helium flow rate during the change of temperature. The oxygen permeation Ea of membranes mainly depends on the ionic transport energy, if surface exchange limitation is

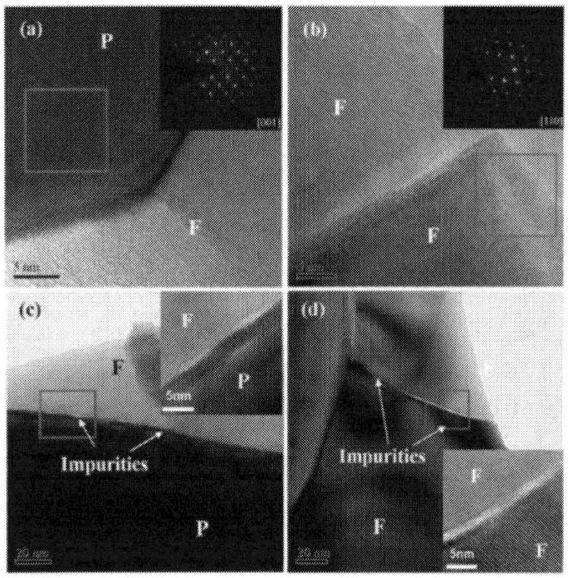

Figure 4 HR-TEM pictures and SAED of the grain boundaries of dual-phase membranes. (a,b) SDC75-SSF25, (a) the grain boundary between the perovskite grain and the fluorite grain along the [001] direction of perovskite grain; (d) the grain boundary between the fluorite grains along [110] directions; (c,d) SDC75-SSCr25, (c) the grain boundary between the perovskite grain and the fluorite grain; (d) the grain boundary between fluorite grains. "F" and "P" denote the fluorite grain and perovskite grain, respectively.

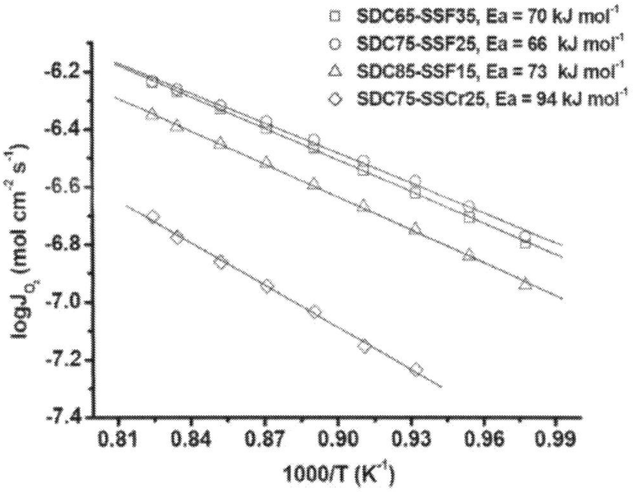

Figure 5 Temperature dependence of oxygen permeation fluxes of dual-phase membranes with both sides coated with $La_{0.6}Sr_{0.4}CoO_3$ porous layers. Thickness: 0.5 mm for SDC65-SSF35, SDC75-SSF25, SDC85-SSF15; and 0.38 mm for SDC75-SSCr25.

eliminated. In fact, the surface limitation was removed after both sides of membranes had been coated with $La_{0.6}Sr_{0.4}CoO_3$ porous layers. Since SDC is the main phase for the transport of oxide ions, the permeation Ea is largely determined by the ionic conduction Ea of SDC. However, the Ea of SDC in the temperature range 800~950 °C is 85.0±0.1 kJ mol^{-1}, based on the data shown in figure 3, which is larger than the oxygen permeation activation energy of SDC-SSF when both sides are coated with $La_{0.6}Sr_{0.4}CoO_3$. Two reasons might explain the decrease of activation energies. One is that some elements of the perovskite phase merge into the fluorite lattice and reduce the oxygen transport activation energy in the crystal. Another probable reason is that the resistance of the fluorite grain boundaries is reduced by the appearance of perovskite-fluorite grain boundaries in the SDC-SSF dual-phase system. However, the permeation activation energy of SDC75-SSCr25 is larger than the ionic conduction Ea of SDC. It can be confidently inferred that the Cr-based oxide with extremely low ionic conductivity blocks transport of oxide ions between fluorite grains, or the enrichment of impurities in the fluorite grain boundaries blocks the oxide ionic transport between SDC grains, as illustrated by the HR-TEM pictures. Furthermore, some ionic or pure electronic conducting phase is totally surrounded by another phase, which blocks these parts from transporting ions or electrons, and leads to low dipolar conductivity. Therefore, it is expected that the traditional dual-phase membranes made of ionic and pure electronic conducting oxides have poor permeability and high permeation Ea.

4. Conclusions

The design and experimental investigation of ceramic dual-phase membranes was reposted in this paper. All the selected dual-phase composite oxides were synthesized via a one-pot method, and show good compatibility of the two phases. Cr-enrichment on the membrane surface during sintering was found for the SDC75-SSCr25 dual-phase membrane. The total conductivity of SDC75-SSF25 remains constant at high temperature, which reveals a proper weight ratio between ionic conductor and mixed conductor. Microstructure investigation by HR-TEM shows impurities appear in the grains boundaries of the SDC75-SSCr25, which blocks ionic transport between fluorite grains and leads to the low oxygen permeation flux. However, the SDC-SSF membranes show clear grain boundaries whether between perovskite grains and fluorite grains or between fluorite grains. Oxygen permeation experiments show that its electronic conductivity is high enough for the transport of oxygen across the SDC75-SSF25 membrane, but which is not the situation for the SDC75-SSCr25 dual-phase membrane. To summarize, Fe-based mixed conducting perovskite oxides are good electronic conductors to ceria-base dual-phase membranes for electronic transport.

Acknowledgements

The authors gratefully acknowledge the financial support of National Natural Science Foundation of China (20801053, 21061130534/B0303) and Director Foundation of State Key Laboratory of Catalysis (R201010).

References

1. Yang W.S., Wang H.H., Zhu X.F., Lin L.W. (2005), "Development and application of oxygen permeable membrane in selective oxidation of light alkanes", *Top. Catal.*, vol. 35, pp.

155-167.
2. Jin W., Li S., Huang P., Xu N., Shi J., Lin Y.S. (2000), "Tubular lanthanum cobaltite perovskite-type membrane reactors for partial oxidation of methane to syngas", *J. Membr. Sci.*, vol. 166, pp. 13-22.
3. Bouwmeester H.J.M. (2003), "Dense ceramic membranes for methane conversion", *Catal. Today*, vol. 82, pp. 141-150.
4. Zhu X.F., Wang H.H., Cong Y., Yang W.S. (2006), "Partial oxidation of methane to syngas in $BaCe_{0.15}Fe_{0.85}O_{3-\delta}$ membrane reactors", *Catal. Lett.*, vol. 111, pp. 179-185.
5. Zhu X.F., Wang H.H., Cong Y., Yang W.S. (2006), "Structural stability and oxygen permeability of cerium lightly doped $BaFeO_{3-\delta}$ ceramic membranes", *Solid State Ionics*, vol. 177, pp. 2917-2921.
6. Kharton V.V., Shaula A.L., Snijkers F.M.M., Cooymans J.F.C., Luyten J.J., Yaremchenko A.A., Valente A.A., Tsipis E.V., Frade J.R., Marques F.M.B. and Rocha J. (2005), "Processing, stability and oxygen permeability of $Sr(Fe,Al)O_3$-based ceramic membranes", *J. Membr. Sci.*, vol. 252, pp. 215-225.
7. Chen C.S., Boukamp B.A., Bouwmeester H.J.M., Cao G.Z., Kruidhof H., Winnubst A.J.A., Burggraaf A.J. (1995), "Microstructural development, electrical properties and oxygen permeation of zirconia-palladium composites", *Solid State Ionics*, vol. 76, pp. 23-28.
8. Lee T.H., Yang Y.L., Jacobson A.J. (2000), "Electrical conductivity and oxygen permeation of $Ag/BaBi_8O_{13}$ composites", *Solid State Ionics*, vol. 134, pp. 331-339.
9. Kharton V.V., Kovalevsky A.V., Viskup A.P., Yaremchenko A.A., Naumovich E.N., Marques F.M.B. (2000), "Oxygen permeability of $Ce_{0.8}Gd_{0.2}O_{2-\delta}$- $La_{0.7}Sr_{0.3}MnO_{3-\delta}$ composite membranes", *J. Electrochem. Soc.*, vol.147, pp. 2814-2821.
10. Kharton V.V., Kovalevsky A.V., Viskup A.P., Shaula A.L., Figueiredo F.M., Naumovich E.N., Marques F.M.B. (2003), "Oxygen transport in $Ce_{0.8}Gd_{0.2}O_{2-\delta}$-based composite membranes", *Solid State Ionics*, vol. 160, pp. 247-258.
11. Yi J., Zuo Y., Liu W., Winnubst L., Chen C. (2006), "Oxygen permeation through a $Ce_{0.8}Sm_{0.2}O_{2-\delta}$–$La_{0.8}Sr_{0.2}CrO_{3-\delta}$ dual-phase composite membrane", *J. Membr. Sci.*, vol. 280, pp. 849-855.
12. Shaula A.L., Kharton V.V., Marques F.M.B. (2004), "Phase interaction and oxygen transport in $La_{0.8}Sr_{0.2}Fe_{0.8}Co_{0.2}O_3$-$(La_{0.9}Sr_{0.1})_{(0.98)}Ga_{0.8}Mg_{0.2}O_3$ composites", *J. Euro. Ceram. Soci.*, vol.24, pp. 2631-2639.
13. Zhu X.F., Yang W.S. (2008), "Composite membrane based on ionic conductor and mixed conductor for oxygen permeation", *AIChE J.*, vol. 54, pp. 665-672.
14. Li Q.M., Zhu X.F., Yang W.S. (2008), "Single-step fabrication of asymmetric dual-phase composite membranes for oxygen separation", *J. Membr. Sci.*, vol. 325, pp. 11-15.
15. Zhu X.F., Wang H.H., Yang W.S. (2008), "Relationship between homogeneity and oxygen permeability of composite membranes", *J. Membr. Sci.*, vol. 309, pp. 120-127.
16. Zhu X.F., Li Q.M., Cong Y., Yang W.S. (2008), "Syngas generation in a membrane reactor with a highly stable ceramic composite membrane", *Catal. Comm.*, vol. 10, pp. 309-312.
17. Fergus J.W. (2004), "Lanthanum chromite-based materials for solid oxide fuel cell interconnects", *Solid State Ionics*, vol. 171, pp. 1-15.
18. Mori M., Yamamoto T., Ichikawa T., Takeda Y. (2002), "Dense sintered conditions and sintering mechanisms for alkaline earth metal (Mg, Ca and Sr)-doped $LaCrO_3$ perovskites under reducing atmosphere", *Solid State Ionics*, vol. 148, pp. 93-101.

MODELING CHARGED-DEFECT TRANSPORT IN PROTON CONDUCTING CERAMIC MEMBRANES

Robert J. Kee, Huayang Zhu*, Einar Vøllestad**, Michael Sanders†, and Ryan O'Hayre†*
**Mechanical Engineering, Colorado School of Mines, Golden, CO, USA*
***Dept. of Chemistry, University of Oslo, Oslo, Norway*
†Metallurgical and Materials Engineering , Colorado School of Mines, Golden, CO, USA

Introduction

Certain doped perovskite ceramics, such as yttrium-doped barium zirconates or barium cerates (BCZY), are proton conductors. Thus, they are potentially valuable for application as separation membranes and in membrane reactors [1, 2]. The most natural thought is that these materials are hydrogen-selective permeable membranes. However, depending upon the gas compositions across the membrane, and possibly electrode polarization, the membranes can effectively transport H_2, H_2O, and O_2. It is important to recognize the distinction between hydrogen-permeable membranes (e.g., palladium) and proton-conducting membranes, which do not conduct H_2 directly. The apparent gas-phase H_2 flux depends upon the gas-phase composition at the membrane surfaces.

Figure 1 illustrates a shell and tube configuration in which a thin (order 10 microns) BCZY membrane is supported on a porous-ceramic tube. The subject of this paper is the development of a model that can predict the radial electrochemical defect transport through the membrane. The model also predicts the apparent gas-phase fluxes.

At elevated temperatures around 600 °C and above, the BCYZ materials are mixed ionic-electronic conductors (MIEC). As such, these materials enable multicomponent charged-defect transport of protons $\mathrm{OH}_\mathrm{O}^\bullet$, oxygen vacancies $\mathrm{V}_\mathrm{O}^{\bullet\bullet}$, electrons e', and electron holes $\mathrm{h}^\bullet$ [3, 4]. The model is based upon formulating and solving defect-conservation equations where the defect fluxes are represented via the Nernst-Planck equations. At the membrane surfaces the defect reactions are assumed to be in equilibrium with the gas phase, thus establishing boundary conditions for the transport equations. Internal electric-potential gradients are formed within the membrane to maintain electroneutrality. Thus the defect fluxes depend upon both concentration gradients and the local electric fields. The membrane performance depends upon physical parameters, including defect mobilities and equilibrium constants for the gas-surface defect reactions. The steady-state system of governing equations forms an ordinary-differential-equation boundary-value problem that is solved computationally using finite-volume discretization. The mathematical structure is somewhat unusual because of the requirement to determine the internal electric fields.

Defect transport model

The temporal and spatial variations of the defects can be represented in a mass-conservation equation as

$$\frac{\partial [X_k]}{\partial t} + \nabla \cdot \mathbf{J}_k = 0, \tag{1}$$

where $[X_k]$ are the charged-defect molar concentrations. The defect fluxes $\mathbf{J}_k$ are represented generally in terms of electrochemical-potential gradients as

$$\mathbf{J}_k = -u_k [X_k] \nabla \tilde{\mu}_k, \tag{2}$$

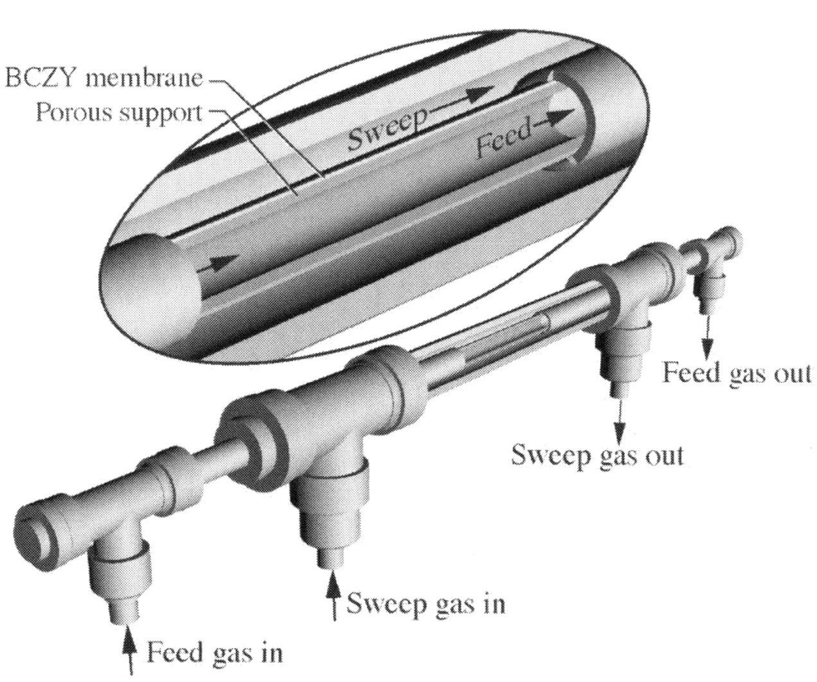

Figure 1: Shell and tube membrane configuration.

where u_k are defect mobilities and $\tilde{\mu}_k = \mu_k + z_k F\Phi$ are the electrochemical potentials (μ_k are the chemical potentials, z_k are the defect charges, Φ is the electrostatic potential, and F is the Faraday constant). For the ideal dilute mixtures, the defect diffusion fluxes $\mathbf{J}_k$ can be expressed by Nernst–Planck equation as

$$\mathbf{J}_k = -D_k\nabla[X_k] - z_k F u_k [X_k]\nabla\Phi. \tag{3}$$

The relationship between mobilities u_k and diffusion coefficients D_k is expressed by the Nernst-Einstein relationship as $u_k = D_k/RT$. Thus, in terms of the diffusion coefficients, the transport fluxes can be written as

$$\mathbf{J}_k = -D_k\left(\nabla[X_k] + \frac{z_k F}{RT}[X_k]\nabla\Phi\right). \tag{4}$$

In Eq. 4 the defect diffusion coefficients D_k are considered to be independent. In other words, it is being implicitly assumed that the defects are dilute and defect-defect interactions are neglected.

For the circumstances considered in this paper, the membrane is not polarized by electrodes and there is no external circuit. Thus, the electric-potential field is generated internally to preserve overall charge neutrality. With defect diffusion coefficients that are independent, charge imbalances would be produced by diffusion alone. Any charge separation by diffusion would lead to an unphysical violation of charge neutrality. The coulombic forces associated with charge separation produce an electric-potential gradient $\nabla\Phi$ that serves to affect the transport to maintain charge neutrality [5].

The charge-neutrality constraint may be expressed as

$$\sum_k z_k F\mathbf{J}_k = \sum_k z_k F\mathbf{J}_k^{\mathrm{c}} + \sum_k z_k F\mathbf{J}_k^{\mathrm{e}} = 0. \tag{5}$$

In this expression, the defect fluxes are represented as a sum of diffusion contributions,

$$\mathbf{J}_k^{\mathrm{c}} = -D_k\nabla[X_k], \tag{6}$$

and migration contributions,

$$\sum_k z_k F \mathbf{J}_k^{\mathrm{e}} = -\sigma \nabla \Phi, \tag{7}$$

where the total conductivity σ is the sum of individual defect conductivities,

$$\sigma = \sum_k z_k^2 F^2 [X_k] \frac{D_k}{RT}. \tag{8}$$

The electric-potential gradient can be expressed locally in terms of the diffusion fluxes as

$$\nabla \Phi = \frac{\sum_k z_k F \mathbf{J}_k^{\mathrm{c}}}{\sigma}. \tag{9}$$

Defect chemical equilibrium

The defect concentrations at the surface of the membrane are presumed to be in equilibrium with the gas phase at the gas-membrane interface. Specifically, the following four reactions (Kröger–Vink notation) are considered:

$$\mathrm{H_2O} + \mathrm{V_O^{\bullet\bullet}} + \mathrm{O_O^{\times}} \rightleftharpoons 2\mathrm{OH_O^{\bullet}}, \tag{10}$$

$$\frac{1}{2}\mathrm{O_2} + \mathrm{V_O^{\bullet\bullet}} \rightleftharpoons \mathrm{O_O^{\times}} + 2\mathrm{h^{\bullet}}, \tag{11}$$

$$\frac{1}{2}\mathrm{H_2} + \mathrm{O_O^{\times}} \rightleftharpoons \mathrm{OH_O^{\bullet}} + \mathrm{e'}, \tag{12}$$

$$\mathrm{h^{\bullet}} + \mathrm{e'} \rightleftharpoons \mathrm{null}. \tag{13}$$

Assuming that the surface defects are in equilibrium with each other and with the gas phase, equilibrium constants may be written as

$$K_{\mathrm{OH}} = \frac{[\mathrm{OH_O^{\bullet}}]_{\mathrm{L}}^2}{[\mathrm{O_O^{\times}}]_{\mathrm{L}} [\mathrm{V_O^{\bullet\bullet}}]_{\mathrm{L}}\, p_{\mathrm{H_2O}}}, \tag{14}$$

$$K_{\mathrm{h}} = \frac{[\mathrm{O_O^{\times}}]_{\mathrm{L}} [\mathrm{h^{\bullet}}]_{\mathrm{L}}^2}{[\mathrm{V_O^{\bullet\bullet}}]_{\mathrm{L}}\, p_{\mathrm{O_2}}^{1/2}}, \tag{15}$$

$$K_{\mathrm{e}} = \frac{[\mathrm{OH_O^{\bullet}}]_{\mathrm{L}} [\mathrm{e'}]_{\mathrm{L}}}{[\mathrm{O_O^{\times}}]_{\mathrm{L}}\, p_{\mathrm{H_2}}^{1/2}}, \tag{16}$$

$$K_{\mathrm{eh}} = [\mathrm{h^{\bullet}}]_{\mathrm{L}} [\mathrm{e'}]_{\mathrm{L}}. \tag{17}$$

It should be noted that the four constant-pressure equilibrium constants K_p are not independent, but are related through the gas-phase equilibrium as

$$K_{\mathrm{H_2O}} = \frac{K_{\mathrm{h}} K_{\mathrm{e}}^2}{K_{\mathrm{OH}} K_{\mathrm{eh}}^2} = \frac{p_{\mathrm{H_2O}}}{p_{\mathrm{H_2}} p_{\mathrm{O_2}}^{1/2}}. \tag{18}$$

Table 1: Charged-defect diffusion coefficients, $D_k = D_k^\circ \exp(E_k/RT)$.

	D_k^0 ($m^2 s^{-1}$)	E_k (kJ/mol)	D_k (650°C) ($m^2 s^{-1}$)	D_k (850°C) ($m^2 s^{-1}$)
$OH_O^\bullet$	1.55×10^{-8}	43.0	5.72×10^{-11}	1.55×10^{-10}
$V_O^{\bullet\bullet}$	1.90×10^{-9}	70.0	2.08×10^{-13}	1.05×10^{-12}
$h^\bullet$	1.05×10^{-6}	97.0	3.41×10^{-12}	3.23×10^{-11}
e'	1.70×10^{-7}	90.0	1.37×10^{-12}	1.11×10^{-11}

In these expressions, the molar concentrations $[X_k]$ are measured in mol m^{-3} and the gas-phase partial pressures p_k are measured in atm. The equilibrium constants for each of the reactions can be evaluated as

$$-RT \ln K_p = \Delta G = \Delta H - T\Delta S, \tag{19}$$

where ΔG, ΔH, and ΔS are the change of the free Gibbs energy, enthalpy, and entropy, respectively.

In addition to the defect chemistry at the surface, overall charge neutrality must be enforced as

$$2[V_O^{\bullet\bullet}]_L + [OH_O^\bullet]_L + [h^\bullet]_L - [e']_L - [Y'_{Ce,Zr}]_L = 0. \tag{20}$$

Moreover, the number of oxygen-ion sites in a pervoskite lattice must be maintained as 3.

$$[V_O^{\bullet\bullet}]_L + [OH_O^\bullet]_L + [O_O^\times]_L = 3. \tag{21}$$

In the foregoing expressions the nomenclature $[X_k]_L$ is used to indicate that the concentrations are evaluated at the lattice scale, not molar concentrations as in Eqs. 1-18. The two concentration representations are related by the molar volume V_m of the lattice as

$$V_m = \frac{[X_k]_L}{[X_k]}. \tag{22}$$

The doping concentration is typically on the order of 10%, in which case $\left[Y'_{Ce,Zr}\right]_L = 0.1$. Molar volumes for perovskites are typically on the order of 5×10^{-5} m^3 mol^{-1}.

Material properties

It is evident from the foregoing analysis that material properties are needed before solving any particular problem. However, for the materials of interest relatively little data has been reported. Table 1 reports the defect diffusion coefficients used in the present paper. Table 2 reports the values of ΔH and ΔS that are used to evaluate the defect-reaction equilibrium constants (Eq. 19). The doping concentration used in the present analysis is taken to be $\left[Y'_{Ce,Zr}\right]_L = 0.19$

Computational algorithm

Although the model is entirely capable of representing transient behavior, the results reported in the present paper assume steady-state radial transport through a membrane. The model is solved in MATLAB using the FSOLVE function to minimize the residual of a finite-volume representation of the conservation equations. That is, $\nabla \cdot \mathbf{J}_k = 0$, with $\mathbf{J}_k$ being represented in Nernst-Planck form (Eq. 4). The

Table 2: Thermodynamics for surface defect reactions

	ΔH (kJ/mol)	ΔS (J/mol K)	K (650°C)	K (850°C)
$H_2O + V_O^{\bullet\bullet} + O_O^{\times} = 2OH_O^{\bullet}$	-95.5	-113.2	3.10×10^{-01}	3.38×10^{-02}
$\frac{1}{2}O_2 + V_O^{\bullet\bullet} = O_O^{\times} + 2h^{\bullet}$	-135.0	-130.0	7.05×10^{-00}	3.08×10^{-01}
$h^{\bullet} + e' = \text{null}$	+266.2	0.0	8.65×10^{-16}	4.16×10^{-13}

local electric-potential gradients are evaluated to enforce charge neutrality (Eq. 9). In one-dimensional radial coordinates

$$\nabla \cdot \mathbf{J}_k = \frac{1}{r}\frac{d(rJ_k)}{dr} = 0, \tag{23}$$

where J_k are the radial components of the defect-flux vector. In the finite-volume discretization, the concentrations are represented at the cell centers and the fluxes J_k are evaluated at the cell faces.

Electrostatic potential

At each finite-volume cell face during the course of the iteration, the diffusion flux is evaluated in terms of the defect concentrations (i.e., Eq. 6). Then, using Eq. 9, $\nabla\Phi$ is evaluated at each cell face, enabling the net flux to be evaluated at the cell faces via the Nernst–Planck equation (Eq. 4).

Strictly speaking, the electrostatic potential Φ is not a dependent variable of the conservation equations. Nevertheless, the radial profile $\Phi(r)$ can be evaluated computationally from the local gradients at the cell faces $\nabla\Phi = d\Phi/dr$. That is,

$$\Phi(r) = \int_0^r \frac{d\Phi}{dr}dr. \tag{24}$$

For the purposes of evaluating the $\Phi(r)$, the inner membrane surface is taken to be at a reference potential as $\Phi(r_{\text{in}}) = 0$.

Boundary conditions

The defect concentrations at the surfaces of the membrane form the necessary boundary conditions. Assuming that the only active gas-phase species are H_2, O_2, and H_2O, the surface defect concentrations may be determined using a relatively straightforward iteration. Using the vacancy concentration $[V_O^{\bullet\bullet}]$ as the iteration variable, Eq. 14 can be rearranged as

$$[O_O^{\times}]_L = \frac{1}{K_{OH}p_{H_2O}}\frac{[OH_O^{\bullet}]_L^2}{[V_O^{\bullet\bullet}]_L}. \tag{25}$$

Based on the site constraint (Eq. 21), the following second-order quadratic equation can be solved analytically to represent $[OH_O^{\bullet}]_L$ as a function of $[V_O^{\bullet\bullet}]_L$,

$$\frac{1}{K_{OH}p_{H_2O}}\frac{[OH_O^{\bullet}]_L^2}{[V_O^{\bullet\bullet}]_L} + [OH_O^{\bullet}]_L + [V_O^{\bullet\bullet}]_L = 3. \tag{26}$$

Because both $[OH_O^{\bullet}]_L$ and $[O_O^{\times}]_L$ have been already expressed as functions of $[V_O^{\bullet\bullet}]_L$ (Eqs. 25 and 26), the hole and electron concentrations are represented indirectly as functions of $[V_O^{\bullet\bullet}]_L$ by

$$[e']_L = K_e p_{H_2}^{1/2}\frac{[O_O^{\times}]_L}{[OH_O^{\bullet}]_L}, \tag{27}$$

$$[\mathrm{h}^\bullet]_\mathrm{L} = \frac{K_\mathrm{eh}}{[\mathrm{e}']_\mathrm{L}}. \tag{28}$$

With all the other defects being represented as functions of $[\mathrm{V_O^{\bullet\bullet}}]_\mathrm{L}$, the overall charge neutrality constraint (Eq. 20) is a function of a single variable, $[\mathrm{V_O^{\bullet\bullet}}]_\mathrm{L}$. Thus, given the gas-phase composition, the equilibrium surface defect concentrations can be evaluated by solving Eq. 20 iteratively for $[\mathrm{V_O^{\bullet\bullet}}]_\mathrm{L}$. The boundary conditions for the conservation equations (Eq. 1) are converted to $[X_k] = [X_k]_\mathrm{L}/V_\mathrm{m}$.

Gas-phase fluxes

Although the membrane model itself is concerned primarily with defect fluxes, it is the effective gas-phase fluxes that are of primary concern to applications such as separations and membrane reactors. Thus, it is important to establish the relationships between gas-phase and defect fluxes at the gas-membrane interfaces. The present model assumes that in the immediate proximity of the membrane surface, the gas phase is equilibrated. The gas-phase equilibrium composition can be different from that in the bulk gas flow, such as in the tube or sweep channels (e.g., Fig. 1). Generally speaking, as long as the gas-phase thermodynamic properties are available, the gas-phase equilibrium composition, represented as partial pressures p_k^eq, can always be evaluated from the non-equilibrated gas-phase composition p_k. For the mixture of H_2, O_2, and H_2O considered here, the equilibrium composition is evaluated easily without software for a general free-energy minimization.

The present model evaluates the gas-phase H_2 and H_2O fluxes by splitting the proton $\mathrm{OH_O^\bullet}$ flux based on the gas-phase equilibrium partial pressures. Specifically, the H_2 and H_2O fluxes at the membrane interface are evaluated as

$$\mathbf{J}_{\mathrm{H_2}} = \frac{1}{2}\mathbf{J}_{\mathrm{OH_O^\bullet}}\left(\frac{p_{\mathrm{H_2}}^\mathrm{eq}}{p_{\mathrm{H_2}}^\mathrm{eq} + p_{\mathrm{H_2O}}^\mathrm{eq}}\right), \tag{29}$$

$$\mathbf{J}_{\mathrm{H_2O}} = \frac{1}{2}\mathbf{J}_{\mathrm{OH_O^\bullet}}\left(\frac{p_{\mathrm{H_2O}}^\mathrm{eq}}{p_{\mathrm{H_2}}^\mathrm{eq} + p_{\mathrm{H_2O}}^\mathrm{eq}}\right). \tag{30}$$

The gas-phase O_2 flux follows as

$$\mathbf{J}_{\mathrm{O_2}} = -\frac{1}{2}\left(\mathbf{J}_{\mathrm{V_O^{\bullet\bullet}}} + \mathbf{J}_{\mathrm{H_2O}}\right). \tag{31}$$

Results and discussion

Consider an example problem in which a tubular membrane is 50 μm thick, with an inner radius of 2 mm. The operating temperature and pressure are $T = 750$ °C and $p = 1$ atm. The gas-phase composition on the inside of the tube is 96% H_2, 3% H_2O, and 1% Ar. The gas-phase composition on the outside of the tube is 1% O_2, 1% H_2O, and 98% Ar. Figure 2 illustrates defect and electrostatic-potential profiles and Table 3 lists defect and gas-phase fluxes. A positive flux means that the flux is directed radially outward. So, for example, the proton flux and gas phase H_2 flux are proceeding from the interior of the membrane toward the exterior. The electron-hole flux is radially inward, from the exterior of the membrane toward the interior. For the circumstances of this example, the electrostatic potential variation is relatively small at around 20 mV. Nevertheless, as discussed below, the migration contribution can be significant. The very small electron concentrations (order of 10^{-8} mol m^{-3}) are not plotted in Fig. 2.

Table 3: Defect fluxes and gas-phase fluxes (mol s^{-1} m^{-2}). The membrane is 50 μm thick, with the inner gas-phase composition being 96% H_2, 3% H_2O, and 1% Ar, and the outer composition being 1% O_2, 1% H_2O, and 98% Ar.

	$V_O^{\bullet\bullet}$	$OH_O^{\bullet}$	e'	$h^{\bullet}$	H_2	O_2	H_2O
Inner	-3.49×10^{-06}	3.12×10^{-04}	6.87×10^{-15}	-3.05×10^{-04}	1.51×10^{-04}	-6.18×10^{-07}	4.73×10^{-06}
Outer	-3.41×10^{-06}	3.05×10^{-04}	6.71×10^{-15}	-2.98×10^{-04}	2.32×10^{-09}	-7.44×10^{-05}	1.52×10^{-04}

Table 4: Defect fluxes and gas-phase fluxes (mol s^{-1} m^{-2}). The membrane is 50 μm thick, with the inner gas-phase composition being 96% H_2, 3% H_2O, and 1% Ar. and the outer composition being 21% O_2, 3% H_2O, and 76% Ar.

	$V_O^{\bullet\bullet}$	$OH_O^{\bullet}$	e'	$h^{\bullet}$	H_2	O_2	H_2O
Inner	1.34×10^{-05}	4.06×10^{-04}	5.71×10^{-15}	-4.33×10^{-04}	1.97×10^{-04}	-9.77×10^{-06}	6.16×10^{-06}
Outer	1.31×10^{-05}	3.97×10^{-04}	5.58×10^{-15}	-4.23×10^{-04}	1.89×10^{-10}	-1.06×10^{-04}	1.98×10^{-04}

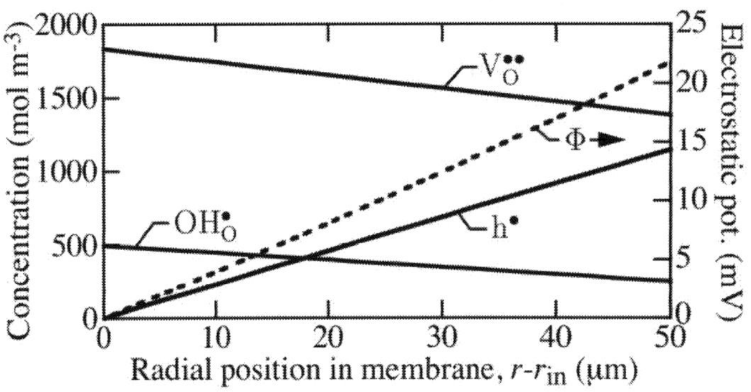

Figure 2: Concentration and electrostatic-potential profiles for a 50 μm membrane with the inner gas-phase composition being 96% H_2, 3% H_2O, and 1% Ar. and the outer composition being 1% O_2, 1% H_2O, and 98% Ar.

Despite the relatively high vacancy concentrations, because of the low vacancy diffusion coefficient, the vacancy fluxes are very small compared to the proton and hole fluxes. Notice that the vacancy flux (Table 3) is negative, meaning it is directed radially inward, but the vacancy concentration gradient is negative (Fig. 2). By diffusion alone, the negative concentration gradient would produce a radially outward (positive) flux. The so-called "uphill" is the result of migration using the internal electrostatic potential gradient that is needed to preserve charge neutrality [6]. The migration force acts to transport the positively charged vacancy toward regions of higher electrostatic potential, thus counteracting the diffusive transport. Although the net mass transport for each defect species must be the same at all radial locations, the fluxes are slightly different as a result of the areas changing as a function of radius. Thus, as indicated in Table 3 the fluxes are slightly different on the inner and outer membrane surfaces.

Note that the apparent gas-phase fluxes are greatly different at the inner and outer membrane surfaces. This is the result of gas-phase equilibrium interactions at the membrane surfaces (i.e., Eqs. 29-31). It is important to note that there is not a simple relationship between proton flux within the membrane

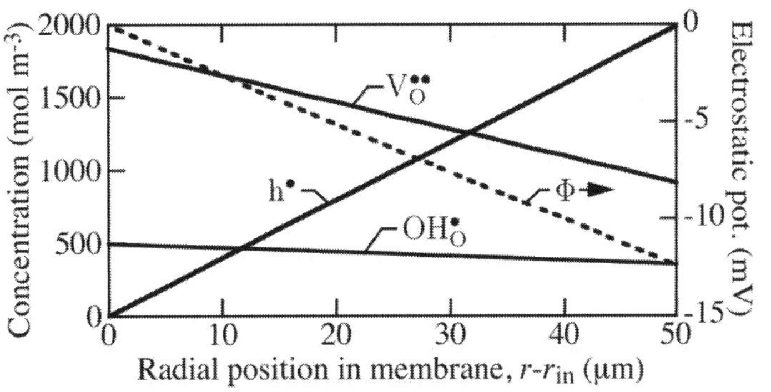

Figure 3: Concentration and electrostatic-potential profiles for a 50 μm membrane with the inner gas-phase composition being 96% H_2, 3% H_2O, and 1% Ar. and the outer composition being 21% O_2, 3% H_2O, and 76% Ar.

and gas-phase H_2 fluxes entering into and emerging from the membrane surface.

Figure 3 and Table 4 show results from a second example where the outer gas-phase composition is changed to 21% O_2, 3% H_2O, and 76% Ar, with everything else remaining the same as in the first example. In this case, because of the higher O_2 concentration, the hole concentration on the outer edge of the membrane has increased greatly. Notice also that the sign of the electric-potential gradient has changed.

Acknowledgements

We gratefully acknowledge numerous insightful discussions with Dr. Grover Coors (CoorsTek) and Prof. Neal Sullivan (Colorado School of Mines). This work was supported by the Office of Naval Research via grant N00014-12-1-0201.

References

[1] L. Li and E. Iglesia. Modeling and analysis of hydrogen permeation in mixed proton–electronic conductors. *Chem. Engr. Sci.*, 58:1977–1988, 2007.

[2] K. Li. *Ceramic membranes for separation and reaction*. Wiley, West Sussex, England, 2007.

[3] M. Oishi, S. Akoshima, K. Yashiro, K. Sato, J. Mizusaki, and T. Kawada. Defect structure analysis of B-site doped perovskite-type proton conducting oxide $BaCeO_3$. Part 1: The defect concentration of $BaCe_{0.9}M_{0.1}O_{3-\delta}$ (M=Y and Yb). *Solid State Ionics*, 180:127–131, 2009.

[4] M. Oishi, S. Akoshima, K. Yashiro, K. Sato, J. Mizusaki, and T. Kawada. Defect structure analysis of B-site doped perovskite-type proton conducting oxide $BaCeO_3$. Part 2: The electrical conductivity and diffusion coefficient of $BaCe_{0.9}Y_{0.1}O_{3-\delta}$. *Solid State Ionics*, 179:2240–2247, 2008.

[5] M.H.R. Lankhorst, H.J.M. Bouwmeester, and H. Verweij. Thermodynamics and transport of ionic and electronic defects in crystalline oxides. *J. Am. Ceram. Soc.*, 80:2175–2198, 1997.

[6] M.D. Sanders and R.P. OHayre. Coupled transport and uphill permeation of steam and oxygen in a dense ceramic membrane. *J. Membrane Sci.*, 376:96–101, 2011.

(131f) First Principles Assessment of Perovskite Dopants for Proton Conductors with Chemical Stability and High Conductivity

Sung Gu Kang, Chemical and Biomolecular Engineering, Georgia Institute of Technology, Atlanta, GA and David S. Sholl, School of Chemical & Biomolecular Engineering, Georgia Institute of Technology, Atlanta, GA

Perovskite-type oxides are one useful approach for proton conduction. A longstanding question in development of electrolytes has been to find the dopants that give high proton conductivity coupled with good chemical stability. We used first-principles calculations to predict the chemical stability with respect to carbonate formation and proton conductivity in doped $BaZrO_3$.

We use efficient methods to examine a wide range of possible dopants. In specific, we present results for BaZrO3 doped with Y, In, Ga, Sc, Nd, Al, Tl, Sm, Dy, La, Pm, Er, and Ho. These calculations identify the doped $BaZrO_3$ that are already known to have favorable properties, but also identify as promising a number of materials that have not been examined previously. We investigated the physical origins of the trends in chemical stability and proton mobility among different dopants. Our data allows us to consider several possible physical descriptors for characterizing doped perovskites as proton conductors.

(138a) Solventpro – the Solvent Selection and Design Framework

Igor Mitrofanov, CAPEC, Department of Chemical and Biochemical Engineering, Technical University of Denmark, Kgs. Lyngby, DK-2800, Denmark, Gürkan Sin, Department of Chemical and Biochemical Engineering, Technical University of Denmark, Kgs. Lyngby, Denmark and Rafiqul Gani, CAPEC, Department of Chemical and Biochemical Engineering, Technical University of Denmark, Kgs. Lyngby, Denmark

Solvents are widely used in a number of applications, for example, as separation agents, reaction mediums, cleaning agents and product carriers. The use of solvents may be classified as either required for processing (separation, reaction, cleaning) and/or as part of a chemical-based product (cosmetic products, delivery devices). While there are some solvent selection databases and special software for very specific applications, a comprehensive database and/or framework for the general solvent selection/design problem is lacking. Also, in many instances, solvents are selected from a list mainly by experience (previous experience or expert knowledge) and/or experiment based trial and error. Therefore, a systematic generic approach to solvent selection and design for all types of solvent use is highly desirable, even though it provides some challenges in terms of database management, inclusion of expert knowledge, model-based solvent design and integration of the various methods and tools within a generic framework. Such a systematic approach should ideally be model/data-based, encouraging innovation, and focused on experiment-based verification only as the last step before solvent application. An important additional aspect in solvent selection and design is the consideration of environmental impact, which is a crucial factor in the disposal (or waste) of industrial solvents. That is, the solvent selection and design framework must also incorporate the "Green Chemistry Principles".

In this paper, the computer-aided solvent selection and design framework, a comprehensive approach to solvent selection problem, called SolventPro, will be presented. The framework integrates different methods and tools needed to manage the complexity and solve a wide range of solvent related problems in an efficient, flexible and robust manner. In particular, new models for solubility predictions, multi-level property estimation as well as molecular and solvent blend design, solvents database, and, solution of interesting problems from industry will be highlighted. In particular, application of SolventPro to design and analysis of various solvent-based extraction processes will be illustrated.

An important feature to highlight of the SolventPro software is the development of problem specific templates for various solvent related problems such as, solvents for organic synthesis (including complex reaction systems), solvent-based separations (separations involving vapour-liquid, liquid-liquid and/or solid-liquid equilibrium systems), solvents for phase transfer catalysis (PTC) reactions, solvents for pharmaceutical industry (API solubility), solvents in formulations (as part of cosmetic and other solvent based formulations), and as cleaning agents. The goal of the template is to guide the average user through the essential and desirable steps in solvent selection and design. The expert may also use the general interface and create their own template for the types of solvent related problems they usually solve. The solvents database currently contains information about more than 1000 organic solvents and their properties, including environmental and transportation properties, and, there are about 1000 ionic liquids (IL), which may also be studied as potential solvents for extraction-based separation processes. There is an efficient search engine to find the solvents (organic or IL) and a corresponding method for computer aided molecular-mixture design. Property models library include state of the art group contribution^{plus}based models for pure component as well as mixture property predictions. These predictive models allow the selection and/or design of innovative solvent based processes and/or products. Another feature of the model library is models used for solvent performance analysis/verification. For example, driving force diagrams for solvent-based separations; saturation diagrams for solid solubility; controlled release predictions for product delivery; and many more.

The SolventPro software has been developed using the Net Framework and Visual C++6 programming tool/language. It is based on plug-in architecture which means that it possible to extend the framework by adding new modules without changing source code or the architecture of the framework. SolventPro has been tested on different types of solvent related problems in the area of process-product design. In this way, the SolventPro software belongs to the F3 category, that is, Fast, Flexible and Future.

(138b) Liquid-Liquid Chromatography: Application of a Predictive Thermodynamic Model for the Selection of the Stationary and Mobile Phase

Andreas Frey, Chair of Separation Science and Technology, Friedrich Alexander University Erlangen-Nuremberg Friedrich Alexander University Erlangen-Nuremberg , Erlangen, Germany and Mirjana Minceva, Chair of Separation Science and Technology, Friedrich Alexander University Erlangen-Nuremberg , Erlangen, Germany

Liquid-liquid chromatography: Application of a predictive thermodynamic model for the selection of the stationary and the mobile phase

A. Frey, M. Minceva

Chair of Separation Science and Technology, Friedrich Alexander University

Erlangen-Nuremberg

Liquid-liquid chromatography (LLC) is a unique separation technology, which combines the principles of extraction and chromatography. The separation is achieved as a result of different partitioning of the solutes between two liquid phases, while one of the phases is kept stationary by means of centrifugal forces. The limitless choice of solvents for the formation of biphasic liquid systems (the mobile and stationary phase) requires a systematic method for the selection of the appropriate solvent combination for a specific separation task. At the moment, this selection is performed by "trial and error" experimental screening procedures, using only a little number of solvents and solvent combinations.

A decisive parameter for the separation performance in LLC is the partition coefficient of the solutes present in the feed mixture. A good separation resolution, a reasonable separation time and not excessively diluted products can be obtained if the partition coefficient of the solute of interest is in the range from 0.4 to 2.5.

In this work a general approach for the selection of the mobile and the stationary phase for a specific LLC separation task is presented. The approach is strongly supported by the use of the a priori predictive thermodynamic method "Conductor-like Screening Model for Real Solvents" (COSMO-RS).[1]

In the first step, potential solvents are selected based on criteria defined by the user. The solubility of the target solutes in the selected solvents is evaluated using the solvent capacity, calculated with COSMO-RS. After that, biphasic liquid systems (obtained by mixing of at least three solvents) are chosen from available liquid-liquid phase diagrams. In a next step, a screening of the pre-selected biphasic liquid systems is performed using the partition coefficient calculated with COSMO-RS.[2]It is demonstrated that with the proposed approach the extensive experimental effort needed for the selection of the biphasic liquid system can be significantly reduced, if not completely eliminated.

---

[1] E. Hopmann, A. Frey, M. Minceva, J. Chromatogr. A 1238 (2012) 68

[2] E. Hopmann, W. Arlt, M. Minceva, J. Chromatogr. A 1218 (2011) 242

(138c) Molecular Dynamics Simulation of Interfacial Water Extraction by TBP/n-Dodecane

Xianggui Ye[1], Shengting Cui[1], Valmor F. de Almeida[2] and Bamin Khomami[1], (1)Material Research and Innovation Laboratory, Department of Chemical and Biomolecular Engineering, University of Tennessee, Knoxville, TN, (2)Energy and Environmental Sciences Directorate, Oak Ridge National Laboratory, Oak Ridge, TN

Solvent extraction is a commonly used process in industrial nuclear fuel reprocessing. Metal cations, acid, and water are typically co-extracted from the aqueous phase into the organic phase mixture of an extracting agent, such as tri-*n*-butyl phosphate (TBP), and a diluent, such as n-dodecane. Quantification of solute extraction by molecular dynamics (MD) simulations is an open area of research. Major challenges include development of molecular models and long-time simulations to determine relevant transport rates. To this end, we have been developing practical force fields and performing corresponding two-phase simulations to predict water solubility in an organic phase consisting of TBP/dodecane. Water extraction is important for model/simulation validation and for practical operation of extraction processes since it competes for available solvent agent molecules. Extensive molecular dynamic simulations of pure water in direct contact with a TBP/n-dodecane organic phase have been performed in the NPT ensemble at 25 °C and 1 bar using LAMMPS with the Verlet integrator and a Nosé-Hoover thermostat and barostat with a time step of 1 fs.

Our results indicate that the atomic charges on the TBP molecular model strongly affect water molecule extraction into the organic phase. The charges are systematically varied so to reproduce the experimental dipole moment of the TBP molecule. Higher dipole moments of TBP increase the roughness of the water/organic interface by reducing the interfacial tension. A "rougher" interface increases the interfacial area, and facilitates water hydrogen bonding breaking which is found to be the rate limiting process for water extraction. We have quantitatively characterized the aqueous/organic interface, including: the distribution profiles of water, TBP, and n-dodecane across the interface, the preferential orientation of TBP, and the interfacial coverage by TBP molecules. Moreover, we have conducted extensive MD simulations (~100 ns) on large systems (~360,000 atoms) to obtain the saturated water solubility in the organic phase, which is found to compare well with experiments. Our simulations indicate that water extraction is intricately affected by TBP amphiphilicity, interface topology, dynamics of hydrogen bond breaking, water cluster formation, and temperature.

(138d) Marangoni Convection At Deformable Single Rising Droplets – a Numerical Investigation of Fluid Dynamics and Mass Transfer

Roland F. Engberg, Fluid Process Engineering, University of Paderborn, Paderborn, Germany and Eugeny Y. Kenig, Fluid Process Engineering, University of Paderbron, Paderborn, Germany

Marangoni convection at deformable single rising droplets
– a numerical investigation of fluid dynamics and mass transfer

R.F. Engberg, E.Y. Kenig

University of Paderborn, Chair of Fluid Process Engineering,

D-33098 Paderborn, Germany

Exact knowledge on fluid dynamics and mass transfer characteristics of rising droplets is essential for the design of liquid-liquid extraction processes. In particular, the complex interaction of various transport phenomena has to be taken into account. Mass transfer, for instance, can lead to a tangential variation of interfacial tension which may cause Marangoni convection. This phenomenon can affect the fluid dynamics of rising droplets considerably; in general, the rise velocity is reduced [1].

Along with experimental investigations, numerical simulations contribute more and more to the understanding of complex transport phenomena at moving boundaries. In this work, the impact of Marangoni convection on mass transfer and fluid dynamics of single droplets in liquid-liquid systems was studied numerically. For this purpose, we developed a CFD code which employs the classical level set method to capture the moving interface. A well-known drawback of this method is unsatisfactory volume (or mass) conservation – usually, the droplet volume decreases in the course of a simulation. Therefore, we implemented two recently published methods [2,3] that improve the level set approach ensuring exact conservation of the initial droplet volume. The code was supplemented with a concentration-dependent interfacial tension and with a model for mass transfer taking the concentration jump at the interface into account. The model equations were implemented in the open-source CFD toolbox OpenFOAM®.

To the best of our knowledge, no *three-dimensional* simulations of mass transfer with Marangoni convection at *deformable* droplets in liquid-liquid systems have been published up to now. Wang et al. [4] performed axisymmetric two-dimensional simulations of deformable droplets which are unable to describe the inherent three-dimensional character of Marangoni convection. In contrast, Wegener et al. [1] investigated Marangoni convection with three-dimensional simulations, however their study was limited to rigid spherical droplets.

In this work, full three-dimensional simulations of deformable toluene droplets rising in water were performed. The concentration of the transferring component acetone was varied. While the mass transfer rates were found to be higher than the experimental and numerical results published by Wegener et al [1], the fluid dynamics and the reduction of the rise velocity is well reproduced.

[1] Wegener, M.; Eppinger, T.; Bäumler, K.; Kraume, M.; Paschedag, A.R.; Bänsch, E: Transient rise velocity and mass transfer of a single drop with interfacial instabilities-Numerical investigations. Chem. Eng. Sci. 64 (2009), 4835-4845.

[2] Groß, S.: Numerical methods for three-dimensional incompressible two-phase flow

problems. PhD thesis, RWTH Aachen, 2008.

[3] Hartmann, D.; Meinke, M.; Schröder, W.: The constrained reinitialization equation for level set methods. J. Comput. Phys. 229 (2010), 1514-1535.

[4] Wang, J.F.; Wang, Z.H.; Lu, P.; Yang, C.; Mao, Z.S.: Numerical simulation of the Marangoni effect on transient mass transfer from single moving deformable drops. AIChE Journal 57 (2011), 2670-2683.

(138e) A Simplified and Optimized Novel Design of RDC-Columns

Enes Aksamija, Rene Prieler, Kathrin Zoerweg, Andreas Pfennig and Matthäus Siebenhofer, Institute of Chemical Engineering and Environmental Technology, Graz University of Technology, Graz, Austria

The applicability of Computational Fluid Dynamics (CFD) to extraction column research is constantly growing due to the incessant increase of calculation power. So far, the interfering phenomena are still far too complex to be predicted without using a bunch of models and assumptions. At present, CFD-simulations can provide hydrodynamic parameters (axial dispersion coefficients, bubble hold-up, droplet size distributions etc.) which are crucial for apparatus design.

In the present work, the continuous phase flow fields of two configurations of the *Rotating Disc Contactor* (RDC150 & RDC100) were recorded with *Particle Image Velocimetry* (PIV). The mean velocity vector fields obtained from PIV were subsequently used to evaluate the simulation setup in *Ansys Fluent* 14.

The residence time distribution of the continuous phase was obtained by injecting saturated salt solution and by monitoring the electric conductivity at four different axial positions. The obtained values for the axial dispersion coefficient were compared with *Euler-Langrange* Simulations (stochastic tracking method) and showed excellent agreement.

Based on the validated simulation settings (models and parameters) continuous phase simulations combined with the Euler-Langrange framework were used to optimize the ratio of the column internals. The shaft diameter, the compartment height, the rotor diameter and the stator diameter were systematically altered in order to find the optimum compartment geometry that leads to lowest axial dispersion. In a next step the same tracking method was applied to quantify and optimize radial mixing within one compartment. With the target of minimal axial dispersion and maximal radial mixing an optimized RDC geometry was proposed.

Subsequent dual phase simulations (Euler-Euler model combined with population balances) were performed to evaluate an optimized and stabilized flow pattern in dual phase operation. Smaller droplets and narrow drop size distributions were observed for the optimized internals geometry, confirming optimization in terms of overall column performance.

(138f) RDC-Design and Operation Efficiency

Enes Aksamija, Department of Chemical Engineering and Environmental Technology, Graz University of Technology, Graz, Austria and Matthäus Siebenhofer, Institute of Chemical Engineering and Environmental Technology, Graz University of Technology, Graz, Austria

Although beeing classified as "out of fashion" RDC-extractors still do a great job in liquid-liquid extraction.

Advantageous application is recommended when the feed phases do have problematic properties such as high viscosities or the formation of crud layers in separation zones. Therefore design research for optimization of RDC operation is still on the agenda.

The dimension of compartment geometry and its effect on operation have been object of research for decades with different design recommendations. In the present work a lab size RDC column with two different compartment geometries was investigated. Operation characteristic and mass transfer efficiency were investigated on the basis of RTD- and mass transfer experiments. RDT was carried out with pulse tracer and monitoring of the conductivity along the column height. Mass tranfer was investigated with the test system SST-butanol-water. Although the test system is not in accordance with EFCE recommendation it has several operational and safety related advantages.

From basic investigations a significant effect of compartment geometry on dispersed phase distribution in the single compartment was derived. The outcome of these hydraulic optimizations shows very even holdup along the apparus height with symmetric eddies for large compartment height. Low compartment height induces uneven hold-up distribution of the dispersed phase along the apparatus height with eddies of different direction of rotation, even in neighbouring compartments.

But RTD as well as mass transfer experiments confirm that latter design is advantageous over symmetric compartment hydraulics, worth being considered in practical application.

(148a) Novel Materials for Carbon Capture, Utilization and Storage (CCUS)

Ah-Hyung Alissa Park, Earth and Environmental Engineering & Chemical Engineering, Columbia University, New York, NY

Historically, the atmospheric concentration of CO_2 fluctuated naturally on the timescales of ice ages. Concerns, however, stem from the recent dramatic increase in CO_2 concentration, which coincides with global industrial development. This rise is mainly due to the high use of fossil fuels. In order to meet the ever-increasing global energy demands while stabilizing the CO_2 level in the atmosphere, it is widely believed that current carbon emissions must be reduced by at least a factor of three. The containment of CO_2 involves three operations: separation, transportation, and storage. Until now, these technologies have been developed independent of one another, which has resulted in complex and economically challenging large-scale designs. The future direction of carbon management technologies now focuses on the integration of CO_2 capture and storage schemes as well as CO_2 utilization. In this seminar, two novel carbon capture, utilization and storage (CCUS) technologies will be introduced. CO_2 capture fluids based on the Nanoparticle Organic Hybrid Materials (NOHMs) are currently developed and their absorption isotherms are characterized as a function of CO_2 partial pressure and temperature (i.e., combustion and gasification conditions). NOHMs are a new class of organic-inorganic hybrids that consist of a hard nanoparticle core functionalized with a molecular organic (sometimes polymeric) corona that possesses high degree of tunability. NOHMs are non-volatile and stable over a very wide temperature range, which make them interesting materials for various energy and environmental applications. Once captured, CO_2 needs to be stored for permanent disposal. The geological storage of carbon dioxide has been considered to be the most economical method of carbon sequestration, while mineral carbonation is a relatively new and less explored method of sequestering CO_2. The advantage of carbon mineral sequestration is that it is the most permanent and safe method of carbon storage, since the gaseous CO_2 is fixed into a solid matrix of Mg-bearing minerals (e.g., serpentine) forming a thermodynamically stable solid product. These carbon sequestration technologies can be integrated into the existing or new energy conversion systems in order to achieve their overall sustainability.

(148b) On the Selection of Ionic Liquids for the Conversion of Cellulosic Biomass Feedstock

Sameer Parvathikar, Andrew R. Tadd and Johannes W. Schwank, Chemical Engineering, University of Michigan, Ann Arbor, MI

The use of ionic liquids (ILs) as solvents represents a promising route for the conversion of cellulosic biomass feedstock. 1,3-dialkylimidazolium ILs in particular, have become a popular and thoroughly-studied reaction media for this. Reactions such as the dissolution and hydrolysis of cellulose, the dehydration of glucose, etc. have been demonstrated. While these ILs have proven to be quite capable of processing these large biomolecules, their widespread adoption has been limited due to their toxicity and inhibitive costs.

It is clear that there is a need for less expensive, and environmentally benign alternatives. The objective of this work is to develop a deeper understanding of the interaction between the solute (biomolecules) and the solvent, which is necessary to effectively choose these alternative solvents. A number of solvent properties like hydrogen bond formation, acidity, etc., will be studied for both imidazolium-based ILs and lower-cost alternatives. Using probe reactions like the isomerization and dehydration of sugars, property-performance relationships will be established. The effect of using different catalysts to affect the reaction environment and thence product yields will also be examined. This work will provide insights into rationally selecting ILs to perform targeted chemistries.

(148c) Production of Hydroxymethylfurfural From Sugars or Renewable Biomass Feedstock As a Potential Platform Intermediate for Bio-Plastics

Wei Liu, Pacific Northwest National Laboratory, Richland, WA

There has been world-wide interest in making plastics out of renewable biomass feedstock for recent years. Hydroxymethylfurfural (HMF) is viewed as an attractive alternate to terephthalic acid (TPA) for production of polyesters (PET) and polyamides. An economic production process needs to be developed to make HMF at a cost competitive with current petroleum-derived feedstock. Conversion of sugars into HMF has been studied in numerous publications. In this presentation, a brief overview of those literature reports pertinent to a practical process development will be presented. A complete ionic liquid-based catalytic process recently developed at PNNL will be discussed. This new process comprises integration of ionic liquid catalytic conversion with adsorption separation. The ionic liquid catalytic reaction is conducted under moderate reaction conditions and provides both high reactivity and selectivity, which is practical for scale-up. The main challenge lied in separation of reaction mixtures. Thus, novel membrane-coated zeolite adsorbent particles are invented, and the adsorption process is demonstrated for isolation of 99% pure HMF product and complete recovery of the ionic liquid and un-converted species. The new process enables nearly 100% conversion of fresh fructose feed into HMF product in the recycled ionic liquid. A simplified process flow diagram is conceived based on these innovations, and the key equipment such as reactor and adsorbent bed is sized for a plant of 200,000 ton/year of fructose processing capacity. The proposed HMF production process is much simpler than the current paraxylene (PX) manufacturing process from petroleum oil, which suggests substantial reduction to the capital cost and energy consumption is possible. At the equivalent value to PX on the molar basis, there can be a large gross margin for HMF production from fructose and/or sugars.

(148d) Large-Scale Energy Storage: New Generation Vanadium Mixed Acid Redox Flow Battery

Liyu Li, UniEnergy Technologies, LLC, Mukilteo, WA

Electrical energy storage is considered as a key enabler to significant penetration of intermittent renewable power from solar and wind, as well as implementation of the Smart Grid. Among the most promising technologies is the all-vanadium redox flow battery (VRFB), which is capable of storing electricity on a MW/MWh scale in the form of two pairs of reduced and oxidized vanadium species (V^{2+}/V^{3+} and V^{4+}/V^{5+}) dissolved in two separate liquid electrolytes, for a discharge duration of hours or even longer. VRFBs offer the flexibility of separate design of power (kW) and energy (kWh): power depends on stack design, while energy depends on the volume and vanadium concentrations in electrolytes. With electrodes free of repetitive mechanical or structural stresses, VFRBs potentially deliver virtually unlimited cycle life. In addition, VRFBs are inherently safe as flowing electrolytes carry away the heat generated during operation. Other advantages include quick response, deep discharge ability, capability to withstand a fluctuating power supply, tolerance to over-charge and over-discharge, and fast charge and discharge. Substantial advancement has been made in VRFBs using the traditional vanadium sulfate electrolytes in the past years. However this technology cannot meet the economic and performance requirements for broad market penetration. While there is much room to improve the economy and performance by engineering and system integration, the current technologies are inherently limited by the current electrolyte chemistries. To prevent irreversible vanadium oxide precipitation, active control of operating temperatures in the range 10-35°C is essential, as is vanadium concentrations (i.e. energy capacity) $\leq$ 1.7M (1.5M in practical systems). Many impurities have to be controlled down to ppm levels to maintain the stability of the electrolytes.

Recently, new vanadium redox chemistry with SO_4^{2-}/Cl^- mixed supporting electrolytes was invented.[1] To stabilize the vanadium electrolytes, chloride was added into the aqueous sulfate electrolyte solution to form new complexes of vanadium species. The new complexes were proved to be much more stable over the current vanadium sulfate chemistries, leading to: i) an extension of operation temperatures to a range of -5 to 60°C, potentially eliminating the need for external heat management; ii) an increase in concentrations of vanadium species (V^{5+}, V^{4+}, V^{3+}, V^{2+}) to 2.5M, representing an energy capacity increase of 70% when compared to the all-vanadium sulfate chemistry; iii) achievement of nearly 90% energy efficiency at 50mA.cm^{-2}, with negligible degradation over a few hundred cycles. In this presentation, the structure of the vanadium species in the new electrolyte solutions and the performance and large-scale implementation potential of this advanced VRFB system will be discussed.

Acknowledgements

Part of the work was performed at Pacific Northwest National Laboratory, and supported by US Department of Energy.

References

1. Li*, Kim, Wang, Vijayakumar, Nie, Chen, Zhang, Xia, Hu, Graff, Liu, Yang*, *Advanced Energy Materials*, 1, 394~400, 2011.8th International Vanadium Symposium: Chemistry, Biological Chemistry, & Toxicology (vanadium8.org)

(148e) High Surface Area Transparent Conducting Oxide Electrodes and Their Application to Energy Technologies Based On Electrochemical Processes

Arnold J. Forman, Zhebo Chen and Thomas F. Jaramillo, Chemical Engineering, Stanford University, Stanford, CA

The ability to control the interfacial area and light trapping characteristics of structured electrodes enables new levels of performance in modern optoelectronic devices. Textured, high surface area electrodes (HSEs) combine large interfacial areas and increased utilization of active material with a light trapping morphology – key parameters for boosting device performance. Transparent conductive oxide (TCO) HSEs with tunable pore sizes on the nanometer to micron scale facilitate diffusion through the structure and serve as a broadly applicable platform for functionalization in many fields including electrochemical sensors, catalysis, (super)capacitors, batteries, photovoltaics and photoelectrochemistry. In this talk we describe the development of optically transparent, electrochemically stable, physically robust HSEs with tunable roughness factors (RF) from 1 through >100. We present a bottom-up approach to future device design and a low-cost, scalable synthetic route for production of transparent HSEs of several TCO compositions. Lastly, we demonstrate the utility of these HSEs by incorporating thin films and particles of electrochemically active materials for enhanced performance. Examples will include adopting the HSE platform for improved supercapacitors and solar photoelectrochemical water-splitting, among others.

(148f) Novel Wave-Shaped MnO2 Cathode Supported by a Partially Porous Nickel Layer for Fuel Cell/Battery

Mike Musil, Institute of Industrial Science, The University of Tokyo, Tokyo, 153-8505, Japan and Atsushi Tsutsumi, Institute of Industrial Science, The University of Tokyo, Tokyo, Japan

Increasing global demand for green energy has boosted the development of various means of energy generation and storage. Both fuel cells and batteries show big potential for an increasing amount of applications, however they still suffer from various drawbacks. Among others, fuel cells possess a low power density, which can mainly be ascribed to the reaction that occurs at a three-phase interface between electrode, electrolyte and cathodic and anodic reactants. Batteries, on the other, hand lack high energy density, whose reason lies in the limited amount of active material in the electrodes.

To eliminate each one's drawback, a so called Fuel Cell/Battery (FCB) has been suggested in previous works. As the name implies, it can work both as a fuel cell and battery. When working as latter, the best results were obtained when coating MnO_2 via electrochemical deposition on carbon fibers. However, oxygen charging – crucial when running in fuel cell mode – was fairly poor. Therefore, a novel cathode structure is presented in this work.

On the one hand, a high surface area to enhance the oxygen charging rate is obtained via a wave-shaped structure in the micrometer scale. Thanks to the obtained surface increase, a reaction rate improvement by up to one order of magnitude is expected compared to a flat electrode. To tackle the problem of having to charge electrically insulating MnO_2 via oxygen while enabling electron collection, a novel, tunnel-looking structure has been developed. Via template method, a several micrometer thick Nickel electrode with one micrometer long pillars has been obtained through electrochemical deposition. On top of the pillars, MnO_2 is deposited, creating a triple layer consisting of MnO_2, porous Nickel, and bulk Nickel.

(149a) Practical Membranes for CO2 Capture

Christina Myers[1], David Luebke[1], Wei Shi[2], Shan Wickramanayake[3], Brian Kail[3], Robert L. Thompson[3] and Hunaid Nulwala[4], (1)US DOE/NETL, Pittsburgh, PA, (2)URS , South Park, PA, (3)URS, South Park, PA, (4)Department of Chemistry, Carnegie Mellon University, Pittsburgh, PA

Membrane technology has advanced greatly in the last 10 years, and due to its potential cost advantages, membranes are now being applied to a wide variety of processes. One of the major focal points for research is in the utility of membranes in CO_2 separations, particularly in coal gasification plants.

Our research examines supported ionic liquid membranes (SILMs) and mixed matrix membranes (MMMs) in both flat and hollow fiber configurations. NETL's membrane development team includes researchers from a wide variety of backgrounds including computational modeling, materials synthesis, membrane fabrication, performance testing, and systems analysis.

SILMs were chosen for their improved transport properties. Gas diffusivities in the IL transport media are higher than those observed in polymers leading to greater permeabilities. More importantly, the materials are stable at elevated temperatures and have negligible vapor pressure, making them suitable for gasification applications. Earlier testing of the various SILMs demonstrated their stability for gas separations at 300°C, and other modifications increased permeability and selectivity. While the studies have demonstrated the potential for laboratory gas separation, the work is expanding to include performance studies under more realistic gas compositions. The presentation will detail evolution of the SILM technology from inception toward practicality.

(149b) Theoretical and Experimental Approach to Nitrogen Selective Membranes

Ekin Ozdogan, Panithita Rochana and Jennifer Wilcox, Department of Energy Resources Engineering, Stanford University, Stanford, CA

The main objective of the proposed research project is to develop N_2-selective catalytic membrane technology with potential sustainability benefits of indirect CO_2 capture and ammonia synthesis. The N_2-selective membrane technology benefits from the driving force of N_2 flue gas (73 wt.%) stream for indirect CO_2 capture as it provides atomic nitrogen on the permeate side of the membrane during separation. With these properties if the H_2 is provided as a sweep gas alongside this novel technology, additional benefits, such as formation of ammonia as a byproduct of the N_2 separation process would be observed.

Within this study, N_2 in particular, has been investigated to determine the mechanism of surface adsorption, dissociation and subsequent atomic diffusion into the bulk crystal structure. Theoretical calculations have been carried out using Vienna *ab initio* Simulation Package (VASP)[i],[ii]. The electron exchange correlation effects are described by a generalized-gradient approximation (GGA) using the Perdew-Burke-Ernzerhof functional[iii] with a plane-wave expansion with a cutoff of 600 eV. The surface Brillouin zone integration has been calculated using a 7x7x7 Monkhorst-Pack mesh[iv]. For absorption energy estimates, binding energy calculations of atomic N in the BCC structure of bulk V and V-Ru alloys have been simulated using a unit cell of 16 atoms with an optimized lattice constant of 2.98Å (3.03Å experiment). Atomic N was found to be stable in the octahedral, face-octahedral and tetrahedral interstitial crystal sites (Figure 1.). It was determined that the octahedral site is the most favorable binding site for N within bulk V, with a binding energy of -2.132 eV. N binding in V is nearly two orders of magnitude stronger compared to the well-known H binding[v] case (-0.076 eV). These strong N binding energies indicate that N may have difficulty diffusing through the material. We found through alloying V with Ru, that this binding energy could be tuned so that it approaches that of H. Alloying V with 6.25 at.% Ru yields a weaker N binding energy of -0.889. It is anticipated that increasing the content of Ru in the alloy will lead to a further reduction in the stability of N within the crystal lattice. Optimal N permeability includes a careful balance between N solubility and diffusivity within a given material.

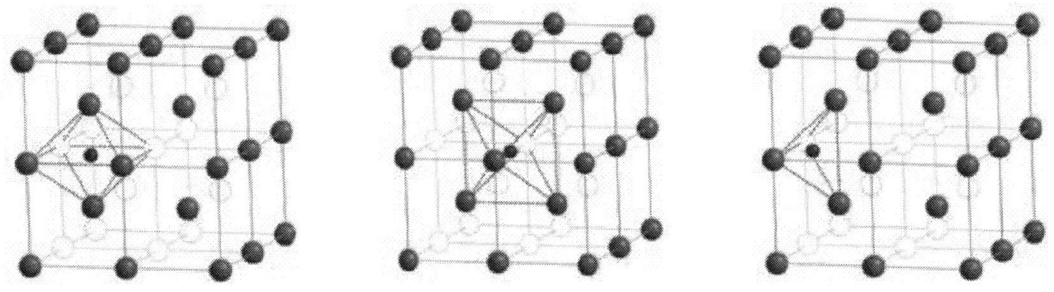

(a) O-site (b) Face-O site (c) T-site

Figure 1 Schematic representation of atomic nitrogen in the interstitial crystal sites

(a) octahedral, (b) face-octahedral and (c) tetrahedral.

Binding energy calculations are carried at different interstitial sites of Vanadium bulk structure. Nitrogen was found to be stable primarily at O-sites within the bulk V lattice. Bader charge analysis has been carried out to investigate the mechanism associated with bulk absorption phenomena. Additionally, alloys of ruthenium (Ru)-V have been studied indicating that Ru can be used as a dopant to tune the electronic structure of the bulk to enhance atomic diffusion. Surface calculations on the 3 low-index surfaces, V(110), V(100) and V(111) based on DFT indicated that the V(111) surface binds N_2 the strongest at the fcc site, with an adsorption energy of 1.4 eV.

Experimental efforts involve pure and mixed gas permeation to determine the N_2 and CO_2 permeabilities of membrane foils comprised of Group V metals and their different composition alloys with Ru over a range of temperatures from 500 to 1000 °C. The schematic of the permeation test set-up is given in Figure 2. In this study, linear regression analyses of the flux isotherms were performed to obtain Sieverts' Constant (n) and it was shown to be 0.5 for N_2 permeation through V and Nb membranes. Unlike the lighter gases such as H_2 and N_2 permeating through the metallic membranes by solution-diffusion mechanism, CO_2 was shown to diffuse only through the grain boundaries or defects on the dense metal surfaces. As the CO_2 fluxes were compared with N_2 fluxes for the pure foil membranes, N_2 fluxes were found to be 2 to 3 orders of magnitudes higher than the CO_2 fluxes. Characterization of the foils before and after the permeation tests was performed by XRD, XPS and SEM techniques to investigate the surface and subsurface structural changes in the membranes. In total, characterization and flux measurements were compared alongside electronic structure predictions of DFT to determine the performance of Group V metals and their alloys for N_2-selective membrane application.

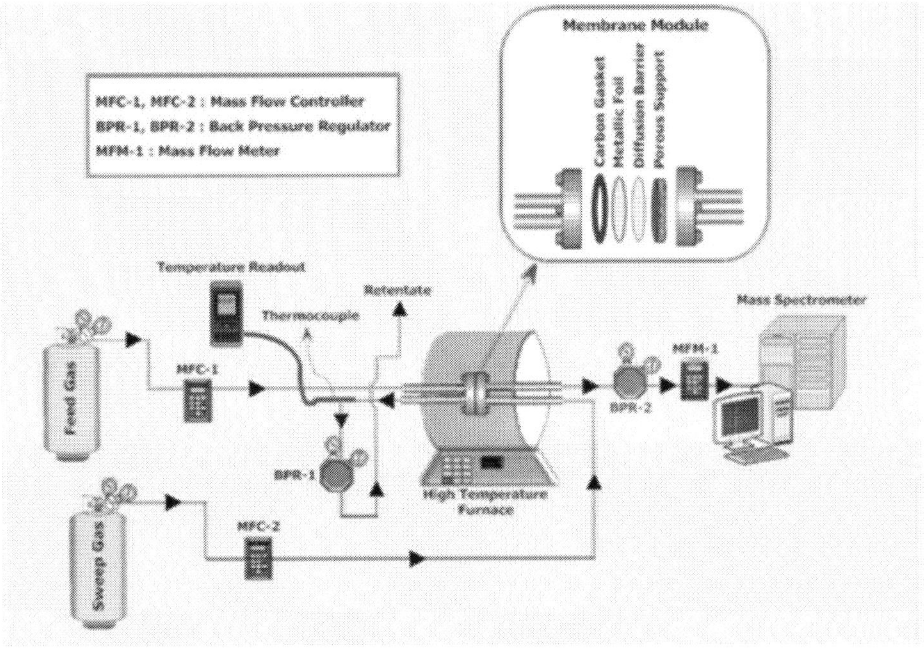

Figure 2.Schematic of Experimental Set-Up

References:

[i] G. Kresse and J. Hafner, *Phys. Rev. B* **48** 13115 (1933).

[ii] G. Kresse and J. Furthmuller, *Comput. Mater. Sci.* ***6*** *15 (1996).*

[iii] P. Perdew, K. Burke, and M. Ernzerhof, *Phys. Rev. Lett.* **77** 3865 (1996).

[iv] H.J. Monkhorst and J.D. Pack, *Phys. Rev. B* **13** 5188 (1976).

[v] S. Aboud and J. Wilcox, *J. Phys. Chem. C* **114** 10978-10985.

(149c) Poly(ionic liquid) Membranes for CO2/N2 Separation: Insight From Atomistic Simulation

Weijie Fang[1], Zhonglin Luo[2] and Jianwen Jiang[1], (1)Chemical and Biomolecular Engineering, National University of Singapore, Singapore, Singapore, (2)National University of Singapore, Singapore, Singapore

Polymeric ionic liquids (poly(IL)s) are a new class of membrane materials for gas separation. In this study, we conduct atomistic simulation to investigate the membrane structures of poly(IL)s and their capability for CO_2/N_2 separation. The poly(IL)s examined are poly[VBIM][TF_2N] and poly[VBIM][SCN]; for comparison, ionic liquids ([BMIM][TF_2N] and [BMIM][SCN]) are also considered. The densities, solubility parameters, ion mobilities and heats of vaporization for the poly(IL)s and ILs are predicted and compared with available experimental data. The fractional free volume increases in the order of poly[VBIM][TF_2N] < poly[VBIM][SCN] < [BMIM][TF_2N] < [BMIM][SCN]. The poly(IL)s and ILs with [TF_2N]$^-$ have a high percentage of large voids (> 3 Å). The predicted solubility and diffusion coefficients CO_2 and N_2 are consistent with experimental data. The analysis of radial distribution functions reveals that gas molecules in poly(IL)s have strong interactions with the end groups of poly-cation chains, but with both cations and anions in ILs. This study provides an atomistic insight into experimental observation, in which the anions of a poly(IL)s do not play an important role in CO_2 sorption while the anions of ILs do.

(149d) Multifold CO2 Transport Property Enhancement of Interfacially Constrained Nanocomposite Membranes

Lakshmi Suhasini Kocherlakota[1], Tiep Hoang Pham[2] and René M. Overney[1], (1)Chemical Engineering, University of Washington, Seattle, WA, (2)Chemical Engineering, University of Washington, Seattle

Multifold CO_2Transport Property Enhancement of Interfacially Constrained Nanocomposite Membranes

Lakshmi S. Kocherlakota, Tiep Hoang Pham, René M. Overney*

AIChE 2012-Abstract

Polymeric membranes are gaining wide attention in separation processes owing to their economic viability and operational flexibility. Poly(l-trimethylsilyl-1-propyne)(PTMSP), a high free volume glassy polymer, exhibits extraordinarily high gas permeability coefficients and high organic-vapor/permanent-gas selectivities. These unusual transport properties in PTMSP are attributed to its high fractional free volume. In this paper, we will address the impact on interfacial constraints on PTMSP's free volume and transport properties. In particular, we found that within the submicrometer regime towards silicon-oxide surfaces, PTMSP membranes reveal CO_2/helium selectivities that exceed bulk PTMSP transport properties by up to three-fold.[1] This finding was attributed to a further increase in free volume, as verified by molecular energetic mobility analysis[2] (intrinsic friction analysis, IFA) that showed an enhancement of the methyl side groups in ultrathin membranes compared to bulk membranes[1]. For scale-up purposes, "inverse" nanocomposites, made up of nanorods (100 to 200 nm diameter) of PTMSP within anodic alumina oxides sieve membranes were produced, yielding a two-fold increase of CO_2 permeability over virgin bulk membranes, and about a 20% improvement of the selectivity of CO_2extraction from helium and nitrogen. While PTMSP membrane are notorious for aging under ambient conditions, the embedded inverse nanocomposite systems show improved longevity.

[1] Kocherlakota LS, Daniel B. Knorr Jr. , Laura Foster, René M. Overney, Enhanced gas transport properties and molecular mobilities in nano-constrained poly[1-(trimethylsilyl)-1-propyne] membranes, Polymer 2012; doi:10.1016/j.polymer.2012.03.067

[2] Knorr DB, Kocherlakota LS, and Overney RM. Journal of Membrane Science 2010;346(2):302-309.

(149e) Nanocomposite Membrane of a Polymer of Intrinsic Microporosity and Zeolitic Imidazolate Frameworks for Gas Separation

Qilei Song[1], Shaheen A. Al-Muhtaseb[2] and Easan Sivaniah[1], (1)Cavendish Laboratory, Physics, University of Cambridge, Cambridge, United Kingdom, (2)Chemical Engineering, Qatar University, Doha, Qatar

Developing energy-efficient and environmentally-friendly separation processes has become an important research topic in dealing with global issues, such as carbon capture, natural gas production, and water purification. Membrane separation technology has great potential for solving these challenges as an alternative to conventional industrial processes. Nanocomposite membranes, also known as mixed matrix membranes, are promising materials for gas separation with potentially enhanced permeability and selectivity. In this study, we report the preparation and gas separation performance of nanocomposite membranes combined of zeolitic imidazolate frameworks (ZIF-8) nanocrystals in a polymer of intrinsic microporosity (PIMs).

The PIMs polymers present microporosity due to their unique contorted and rigid molecular structure, and have shown great potential for a wide range of applications, such as membrane for gas separation, gas storage, and catalysis. As membrane for gas separation, PIMs have shown high gas permeability but have selectivity that need to be enhanced to a satisfactory level for industrial applications. Zeolitic imidazolate frameworks (ZIFs), a sub-family of MOFs, have attracted significant attention with tuneable pore sizes, exceptional chemical stability, and versatile structures analogous to that of inorganic zeolites. The ZIFs nanoparticles with sub-nanometer channels could present synergetic effects as fillers for mixed matrix membranes: (1) disruption of the packing of polymer chains; and (2) molecular sieving effect. The PIMs/ZIFs composite membrane is a representative example of combination of *amorphous microporous* polymer and *crystalline microporous* MOFs. Therefore, understanding the physical and chemical properties of this composite, such as porosity, free volume, thermal stability, and gas sorption properties, is not only important for developing high performance composite membrane for gas separation, but also useful for many other applications, such as adsorbent and catalysis.

In this study, the as synthesised ZIF-8 nanoparticles, dispersed as colloids in solvent solution, were incorporated into the PIM-1 polymer matrix forming a nanocomposite membrane, which gives uniform dispersion of ZIF-8 in the polymer matrix. A series of composite membrane were fabricated from solution casting of the PIM-1/ZIF-8 solution with loading of ZIF-8 up to the significantly high values of ~40 wt%. The nanocomposite membranes were further treated with various techniques, such as methanol treatment, thermal treatment, and photochemical modification, aiming to improve the permeability or selectivity. The pure gas permeation properties were evaluated in a constant-volume variable-pressure apparatus (time lag method). Mixed gas permeation was tested in a similar membrane cell equipped with a gas chromatography.

For the nanocomposite membrane without methanol treatment, the gas permeability decreased with the loading of ZIF-8 nanoparticles, while the selectivity was maintained without significant loss. After methanol treatment, the permeability of nanocomposite membrane increased with the loading of ZIF-8 nanocrystals and surpassed the Robeson's upper bound. The effects of annealing temperature (100-300°C) on the gas permeation performance of pure polymer and nanocomposite membrane were monitored. For the pure polymer membranes annealed at high temperature (up to 300°C), the gas permeability slightly decreased while the selectivity was found to be stable. For the PIM-1/ZIF-8 nanocomposite membranes, the gas permeability decreased with the annealing temperature, interestingly, the selectivity could be significantly enhanced. Particularly, for the composite membrane with 20 wt% loading of ZIF-8 annealed at 300°C, the CO_2 permeability could reach approximately 1000 Barrer with selectivity of both CO_2/N_2 and CO_2/CH_4 as high as 30. The ideal selectivity of other gas pairs, such as O_2/N_2, H_2/N_2 and H_2/CH_4, were all enhanced surpassing the Robeson's upper bound.

Gas sorption properties of these nanocomposites were also examined with various types of gases, so their potential as adsorbents or gas storage materials were also evaluated.

Acknowledgements

We acknowledge the NPRP grant from the QNRF, the Engineering and Physical Sciences Research Council (EPSRC, UK), and the China Scholarship Council.

Notes: Corresponding author: es10009@cam.ac.uk (Dr. Easan Sivaniah)

References

1. N. B. McKeown and P. M. Budd, *Chem. Soc. Rev.*, 2006, **35**, 675-683.

2. T.-H. Bae, J. S. Lee, W. Qiu, W. J. Koros, C. W. Jones and S. Nair, *Angew. Chem., Int. Ed.*, 2010, 49, 9863-9866.

3. T. C. Merkel, B. D. Freeman, R. J. Spontak, Z. He, I. Pinnau, P. Meakin and A. J. Hill, *Science*, 2002, **296**, 519-522.

(149f) Steam Stability of Layered Zeolites Suitable for Hydrogen Transport

Bahman Elyassi, Xueyi Zhang and Michael Tsapatsis, Chemical Engineering and Materials Science, University of Minnesota, Minneapolis, MN

In this study, we demonstrate potential of zeolite nanosheets with 6-member ring (6MR) apertures for preparation of hydrogen permselective membranes. The preparation technique is based on exfoliation of layered zeolites into their constituent nanosheets, formation of closely packed monolayers on porous substrates, and then performing secondary growth. Steam stability of these zeolites in conditions relevant to steam methane reforming and water gas shift reactions over a period of about 3 months using X-ray diffraction (XRD), scanning electron microscopy (SEM), transmission electron microscopy (TEM), multinuclear solid state NMR, and Ar adsorption analyses was investigated. It was found that zeolites with lower concentrations of structural defects perform better in steam. In line with this observation, enhanced steam stability of the zeolites treated with silicon tetrachloride ($SiCl_4$) was seen. This work highlights the importance of zeolite synthesis or post-synthesis methods suitable to eliminate structural defects for applications where high temperature steam is present.

(149g) Physical Aging and Carbon Dioxide Plasticization of Thin Polyetherimide Films

Jianzhong Xia[1,2], Tai Shung Chung[2], Pei Li[2], Norman R. Horn[1] and Donald R. Paul[1,2], (1)Chemical Engineering, University of Texas at Austin, Austin, TX, (2)Chemical and Biomolecular Engineering, National University of Singapore, Singapore, Singapore

Industrial gas separation membranes have selective dense layers with thicknesses around 100 nm. It has long been assumed that these thin layers have the same properties as thick (bulk) films. However, recent research has shown that thin films with such thickness experience accelerated physical aging relative to bulk films and, thus, their permeation properties can differ significantly from the bulk. This paper explores the aging and carbon dioxide plasticization behavior of this new polyethermide and compares the results with those for PSF and Matrimid® 5218 from the previous work. Particular attention is devoted to the permeation response of thin films for a variety of CO_2 conditions due to the current strong interest in CO_2 related separations, such as the removal of CO_2 from natural gas and flue gases. The CO_2 plasticization of Extem thin films is proven to be a function of film thickness, aging time, exposure time, pressure and prior history.

(149h) The Design and Modeling of a Controlled Atmosphere Food Storage System Based On Membrane Technology

Hang Yin, Chemical and Biomolecular Engineering, National University of Singapore, singapore, Singapore, Na Peng, Chemical and Biomolecular Engineering, National University of Singapore, Singapore, Singapore, Kok Lin Chong, National University of Singapore, singapore, Singapore, Glenn Lipscomb, Department of Chemical and Environmental Engineering, University of Toledo, Toledo, OH and Tai-Shung Chung, Department of Chemical & Biomolecular Engineering, National University of Singapore, Singapore, Singapore

Fruits and vegetables continue to respire after harvesting. In addition to refrigeration, controlled atmosphere storage technology can prolong the shelf-life of fruits and vegetables by creating a specific atmosphere. In this study, an innovative storage system based on green membrane technology is evaluated for food preservation. This is a novel study of hollow fiber membrane modules in a controlled atmosphere storage (CAS) system to maintain an environment that prolongs the lifespan of fruits and vegetables during transportation and storage. Compared to conventional CAS systems that utilize nitrogen flushing, the proposed membrane controlled atmosphere storage (MCAS) system provides a nitrogen-enriched air stream to control the concentration of both oxygen and carbon dioxide in the storage container. Detailed mathematical models to describe mixed gas permeation in a hollow fiber membrane module and guide system design are presented. The performance of MCAS systems under on-off and continuous operation modes is evaluated and a case study of the newly developed MCAS system in food transportation presented.

(162a) Mechanism of Water Adsorption in Hydrophobic Nanospaces of Disordered Carbons

Suresh K. Bhatia and Thanh X. Nguyen, School of Chemical Engineering, The University of Queensland, Brisbane, Australia

The understanding of the mechanism of water adsorption in the hydrophobic nanospaces of carbons is critical to many industrial processes for gas separation and water purification, and to emerging nanotechnologies for desalination, CO_2 capture from flue gas, and separation by nanofluidic devices. While there have been numerous attempts at simulating water adsorption in hydrophobic carbons using idealized models of independent slit pores, only qualitative agreement with experiment has been achieved, and the answer to the difficult question of how water enters such spaces has remained elusive. Using grand canonical Monte Carlo (GCMC) simulations with a realistic model of a disordered hydrophobic carbon, we show that the key to the puzzle is the connectivity of the structure – overlooked by independent slit pore models. Our simulations and data confirm that significant amount of water adsorbs below the saturation pressure in purely hydrophobic nanopores, and it is demonstrated that this occurs only when pore entries are sufficiently large to allow the passage of stable hydrogen-bonded water clusters. We investigate the effect of pore connectivity through synthetic models of connected and unconnected slit pores, and show that the connectivity to narrow water-filled pores mediates the adsorption of water in large hydrophobic nanospaces. This unique feature is not observed for nonpolar or weakly polar gases (e.g. Ar or N_2) at subcritical conditions, and explains why the Kelvin equation fails to estimate the condensation pressure for water. The results open the door for the design and tailoring of efficient adsorbents for CO_2 capture, in which the co-adsorption of water vapor which saturates flue gas is inhibited.

(162b) NET Adsorption for Multicomponent GAS Equilibrium

Dr.Orhan Talu, Chemical and Biomedical Engineering, Cleveland State University, Cleveland, OH

The *net* adsorption thermodynamic framework was recently introduced as an alternative to *excess* and *absolute* adsorption (Gumma and Talu, *Langmuir* 2010, 26(22), 17013-17023). It is a fundamentally different thermodynamic description of adsorption since the placement of Gibbs dividing surface, necessary for any treatment of adsorption, is different than the known/used alternatives. Net adsorption is a natural choice for microporous solids. It removes any ambiguity in theory and simulations of adsorption and circumvents all questionable measurements/quantities in experiments.

Initial development of net adsorption was prompted by the interest in adsorptive gas storage and by the need to unambiguously compare experiments to molecular simulation results. Previously, our research group and others have shown that: 1) non-adsorbing helium assumption commonly used to place the Gibbs dividing surface in experiments does not hold (e.g. Gumma and Talu, *Adsorption J.* 2003, 9, 17-28), and 2) molecular simulation results can only be compared to excess data by performing more simulations with helium (e.g. Talu and Myers, *AIChE J.* 2001, 47(5), 1160-1168). As such, previous work has concentrated on pure component adsorption equilibrium and dynamics with net adsorption.

This contribution involves extension of net adsorption framework to gas mixtures. Equilibrium formulations will be presented, net and excess adsorption will be compared with a set of experimental data, and implications of net framework on column dynamics for adsorption processes will be discussed.

(162c) A Bird's Eye View of Porous Materials for Carbon Dioxide Separation and Capture

Christopher E. Wilmer[1], Omar K. Farha[2], Benjamin J. Sikora[1], Youn-Sang Bae[3], Joseph T. Hupp[4] and Randall Snurr[1], (1)Chemical and Biological Engineering, Northwestern University, Evanston, IL, (2)Department of Chemistry, Northwestern University, Evanston, IL, (3)Chemical & Biomolecular Engineering, Yonsei University, Seoul, South Korea, (4)Department of Chemistry, Northwestern university, Evanston, IL

Recently Bae et al. summarized research on over 40 metal-organic frameworks (MOFs) assessed for their use in either carbon capture and sequestration (CO2/N2 separations) or natural gas purification (CO2/methane separations). In addition to common performance metrics such as adsorption and selectivity at a given pressure, Bae et al. considered regenerability, deliverable capacity, and an aggregate metric called the "sorbent selection parameter." Here, we calculate the same performance metrics as Bae et al. for 137,000 hypothetical MOF structures using our recently developed hypothetical MOF-generation algorithms; single component grand canonical Monte Carlo (GCMC) simulations; and a new, computationally efficient method to estimate partial atomic charges in periodic systems. This analysis yields a useful "bird's eye" view of the porous material space, and we will present several new structure-property relationships that are relevant to carbon dioxide separation and capture. Additionally, we will compare structure-property relationships found for another industrially important application: separating xenon from krypton.

(162d) The Separation of Sulfur Dioxide From Air Using Magnetic Nanoparticles

T. Grant Glover, SAIC, Gunpowder, MD, Daniel Sabo, Chemistry and Biochemistry, Georgia Institute of Technology , Atlanta, GA, Lisa Vaughan, Georgia Institute of Technlogy , Atlanta, GA, Joseph Rossin, Guild Associates, Dublin, OH and Z. John Zhang, Chemistry and Biochemistry, Georgia Institute of Technology, Atlanta, GA

The Separation of Sulfur Dioxide from Air Using Magnetic Nanoparticles

T. Grant Glover*, Daniel Sabo[+], Lisa A. Vaughan[+], Joseph A. Rossin[#], Z. John Zhang[+]

*SAIC, Gunpowder Branch, P.O Box 68, APG, Maryland 21010

[+]Georgia Institute of Technology School of Chemistry and Biochemistry,

901 Atlantic Drive, Atlanta, Georgia, 30332

[#]Guild Associates, Incorporated, 5750 Shier-Rings Road, Dublin, Ohio 43016

The adsorption of sulfur dioxide on magnetic nanoparticles is examined. Microbreakthrough experiments were used to quantify the ability of magnetic nanoparticles to adsorb sulfur dioxide from air. The results show that the magnetic nanoparticles out perform traditional activated carbon. The adsorption capacity of the materials is noteworthy given the 150 m^2/g surface area of the magnetic nanoparticles. Additionally, the nanoparticles exhibit a 23% decrease in the remnant magnetization, a 20% decrease in saturation magnetization, and a 9% decrease in coercivity. Adsorbents that provide a magnetic signal upon adsorption could have broad implications on adsorption-based separations.

(162e) Total Reflux Dual Piston PSA

Wenli Dang[1], Daniel Friedrich[2], Maria-Chiara Ferrari[2] and **Stefano Brandani**[2], (1)SCCS - School of Engineering, Univerisity of Edinburgh, Edinburgh, United Kingdom, (2)SCCS - School of Engineering, University of Edinburgh, Edinburgh, United Kingdom

We present the results from the total reflux dual piston Vacuum and Pressure Swing Adsorption apparatus developed at the University of Edinburgh. The system has been recently extended to include the dynamic response of gas phase and solid phase temperature at two positions along an adsorption column, in addition to total pressure and column pressure drop. The temperature responses have allowed further development of the model of the system, which now includes also the heat generated by the piston movement. Representative experiments with helium and pure and binary mixtures of nitrogen and carbon dioxide on 13X pellets will be presented. The new apparatus is used to validate models of adsorption columns over a very wide range of conditions: vacuum to 25 bar; 20-200 °C; cycle times as fast as 1 s. The fact that the movement of the pistons can be controlled independently allows also to run the system under both linear and non-linear frequency response conditions.

The authors would like to acknowledge the support from the Engineering and Physical Sciences Research Council under grants EP/G062129/1 and EP/F034520/1

Minimum-Column-Volume Design of Capture Chromatography for Mo Recovery

Lei Ling*, Pei-Lun Chung*, Amanda Youker†, Dominique C. Stepinski†, George F. Vandegrift†, and Nien-Hwa Linda Wang*

*School of Chemical Engineering, Purdue University, West Lafayette, IN 47907

*Email: wangn@ecn.purdue.edu, Phone: 765-494-4081, Fax: 765-494-0805

†Argonne National Laboratory, Argonne, IL 60439

Abstract

Capture chromatography is widely used for many different applications. An adsorption column is used to recover a target component from a finite batch of feed solution. In designing this process, one needs to specify the column size and feasible operating velocities that can satisfy a specific batch size and loading time, a column pressure drop limit and a dimensionless breakthrough concentration, which is the ratio of breakthrough concentration to feed concentration at the end of loading. A low dimensionless breakthrough concentration (0.01) gives a high capture yield (>99%).

A systematic design method based on key dimensionless groups has been developed to determine the minimum column volume, the maximum operating velocity and the maximum column length for a given batch of feed. For a system with known solute, sorbent, solvent and temperature, when loading time is fixed, the dimensionless column length is inversely proportional to the loading factor. This hyperbolic curve is defined as the operating line. Along an operating line, the loading factor and dimensionless breakthrough concentration increase when dimensionless column length decreases. On the other hand, the minimum dimensionless column length needed for a fixed dimensionless breakthrough concentration was found to increase with increasing loading factor. The curve that shows the relation between the two dimensionless groups is defined as the constant breakthrough limit curve, which can be obtained either from experiments or rate model simulations based on intrinsic adsorption and mass transfer parameters. The minimum column volume that can satisfy both loading time and breakthrough concentration requirements was found from the intersection point of the operating line with the constant breakthrough limit curve. If intra-particle diffusion controls wave spreading, and feed concentration is within a linear isotherm region, the constant breakthrough limit curve is only a function of the isotherm. Additionally, the maximum linear velocity and column length that satisfy the desired pressure limit can also be found from this intersection point and the minimum column volume.

This design method was verified with the frontal data of Molybdenum (Mo) capture from an aqueous uranyl sulfate solution using a titania sorbent. Sensitivity analysis was carried out to take into account errors in estimated parameters and variations of material properties or operating conditions. The stripping yield of Mo was then verified using lab scale columns with a safety factor obtained from the sensitivity analysis. The results showed that Mo loaded during two hours can be fully captured and then fully recovered by stripping with 1 M ammonium hydroxide solution.

This design method requires a small number of bench-scale experiments for the design of a large-scale process. Scaled-up designs showed that a relatively small titania column (9 cm in diameter and 20 cm in length) is needed to capture 250 liters of feed of 0.002 mM Mo in two hours. If some of the system parameters and operating conditions are varied, the design can be readily modified without further simulations or experiments. This method has potential to be applied to design capture processes for many other types of separations based on high affinity sorbents.

(162g) Small-Angle X-Ray Scattering (SAXS) Investigation of Porous Monolithic Carbons From Phenolic Precursor

Dipendu Saha, Andrew E. Payzant and Amit K. Naskar, Materials Science and Technology Division, Oak Ridge National Laboratory, Oak Ridge, TN

In this work, we have synthesized monolithic carbonaceous materials from varieties of synthetic and natural phenolic precursors upon bulk-polymerization with and without the presence of surfactants. A series of carbon monoliths were synthesized with varying surfactant concentration from each of the precursors and it was observed that nonporous to nanoporous morphologies vary with the surfactant concentration and leave their fingerprints in small-angle X-ray scattering (SAXS) patterns. The understanding of evolution of pore structures in bulk carbon materials is important for its applications. The pore textural properties and surface fractality of the porous carbons were investigated by analyzing the SAXS patterns and compared with the results of nitrogen adsorption-desorption studies. The inherent advantage of SAXS analysis is the identification of closed or impenetrable pores in these carbons that cannot be characterized by any adsorption studies. Micropores and macropores leave very prominent and distinct fingerprint in the SAXS patterns that can be readily identified from their scattering patterns. All the SAXS patterns were modeled with a suitable equation that identifies the contribution from micropore, external surface or macropore and background and hence can be compared with adsorption studies. The distance distribution functions were also calculated from the SAXS patterns and a comparison is made with the pore widths obtained from the adsorption data. The effect of degree of interfacial tension varied by surfactant doping on the structure of carbons is also investigated by wide angle X-ray scattering (WAXS) analysis.

(170a) Identification, Characterization and Development of a Betaine Co-Crystal for a Pharmaceutical Compound

Bing-Shiou Yang[1], Zhibin Li[1], Ka Yip[1], Soojin Kim[1], Daniel Fandrick[2], Jeff Song[2], Nathan Yee[1] and Chris Senanayake[1], (1)Boehringer Ingelheim Pharmaceuticals, Inc., Ridgefield, CT, (2)Boehringer Ingelheim Pharmacuticals, Inc., Ridgefield, CT

Solid form screen and crystallization development are one of the key activities to move the compound from the research to development stages. In many cases, salt or co-crystal forms are required to provide better developability compared to the free form. In this case, a betaine co-crystal for a development compound for the treatment of HIV was identified after extensive salt screening. This novel co-crystal, due to its superior physiochemical properties compared to the free form and other salt forms, was selected as the development solid form. Single crystal X-ray confirmed the formation of co-crystal based on the interaction between betaine and the compound. The major challenge for the development and scale-up of the co-crystal was the inconsistent stoichiometry between the compound and betaine. A process was developed to address this challenge by proper solvent selection and crystallization design which was later successfully scaled up in the Kilo Lab to multiple hundred gram scale before the project was terminated. Identification of the stable operational region for the co-crystal formation will also be discussed.

(170b) Polymorph Screening Using Self-Assembled Monolayers and Scale-up of Metastable Forms

Jie Chen and Allan S. Myerson, Chemical Engineering, Massachusetts Institute of Technology, Cambridge, MA

Beginning in the 1940s with the work of Bigelow[1] self-assembled monolayers (SAMs) have been used to engineer the chemistry of surfaces and have been recently applied to template the nucleation and growth of organic, inorganic and macromolecular crystals. SAMs are ordered molecular assemblies that are formed spontaneously by the adsorption of a surfactant-like molecule with a specific affinity of its headgroup to a substrate. Our group developed a patterned bifunctional SAM with hydrophilic islands surrounded by hydrophobic regions. Arrays of small hemispherical droplets can form on those hydrophilic islands when wetted with polar solutions. Each of those droplets serves as an independent crystallization experiment and thus offers a platform for high throughput polymorph screening[2-5].

In this work we performed polymorph screening of a pharmaceutical active ingredient using patterned SAMs. Polymorphic forms of the crystals obtained on SAM islands were analyzed using Raman microscopy. Melting temperatures of potential novel forms were measured using hot-stage microscopy. The influence of droplet size and crystallization method (evaporation vs. anti-solvent diffusion) was also investigated. Three unreported as well as two reported forms of the compound under investigation were obtained on patterned SAM surfaces. Efforts were made to reproduce those three unreported forms in bulk using various crystallization methods with fast kinetics. At the end, we were able to make one of the three unreported forms using electrospray and further characterize it using powder xray diffraction.

[1] Bigelow, W. C.; Pickett, D. L.; Zisman, W. A., *J. Coll. Sci.* **1946**, *1*, 513-538.

[2] Lee, A. Y.; Lee, I. S.; Dettet, S. S.; Boerner, J.; Myerson, A. S., *J. Am. Chem. Soc.* **2005**, *127*, 14982-14983.

[3] Lee, A. Y.; Lee, I. S.; Myerson, A. S., *Chem. Eng. Technol.* **2006**, *29*, 281-285.

[4] Lee, I. S.; Lee, A. Y.; Myerson, A. S., *Pharm. Res.* **2008**, *25*, 960-968.

[5] Singh, A.; Lee, I. S.; Myerson, A. S., *Cryst. Growth Des.* **2009**, *9*, 1182-1185.

(170c) Transformation of an Amorphous API to Crystalline Hydrate and Co-Crystals

Fang Wang, Ernest Carra, Anna Chiu and Jesper Jernelius, Chemical Development, Gilead Sciences, Foster City, CA

This research relates to a novel crystalline hydrate form of a Gilead development compound, GS-A, and novel crystalline co-crystals of GS-A with different co-formers.

Prior to the discovery of the crystalline hydrate form, GS-A existed for many years as an amorphous solid. The first crystalline form was discovered through solvent mediated transformation of the amorphous material in water over an extended period of time, which produced GS-A hydrate. The interaction between water molecules and GS-A molecules, especially hydrogen bonding between the two, appears to support the crystalline structure of the hydrate form. The crystalline monohydrate is more stable and much less hygroscopic than the amorphous form, which make it a more suitable form for development.

Development of a crystallization process to isolate GS-A crystalline hydrate on large scale will be discussed. The parameters investigated include: solvent system, order of addition and seed loading. Solvent systems which produced GS-A as the crystalline hydrate include: methanol/H_2O, ethanol/H_2O, 2-propanol/H_2O, di-isopropyl ether/toluene/H_2O and isopropyl acetate/heptane/H_2O. One set of conditions were optimized to produce GS-A crystalline hydrate on multi-kilogram scale.

Additionally, GS-A+L-tartaric acid, GS A+D-tartaric acid, GS-A+malonic acid, GS-A+L-malic acid, GS-A+D-malic, and GS-A+benzoic acid co-crystals were discovered. In two cases, co-crystals were formed through slow crystallization from a resin at ambient temperature: L-Tartaric acid and benzoic acid. Crystallization from solution was effective in producing co-crystals of GS-A with L-tartaric acid, malonic acid, L-malic acid, and D-malic acid. This discovery is based on the expected significant degree of hydrogen bonding between GS-A and the co-former molecules. These co-crystals of GS-A show improved physical properties (such as their anhydrous nature and higher melting points) compared to the crystalline monohydrate.

(170d) Solvent Compatible Microfluidic Platforms for Solid Form Screening of Pharmaceutical Parent Compounds

Sachit Goyal[1], Yuchuan Gong[2], Geoff G.Z. Zhang[2] and Paul J.A. Kenis[1], (1)Department of Chemical & Biomolecular Engineering, University of Illinois at Urbana Champaign, Urbana, IL, (2)Solid State Sciences, Abbott Laboratories, North Chicago, IL

We present solvent compatible microfluidic platforms for solid form (salts, co-crystals, and polymorphs) screening of pharmaceutical parent compounds (PC). Success in finding a crystalline solid form of a PC with optimized CMC properties using conventional screening procedures is limited due to the limited experimental conditions that can be investigated while utilizing small quantities of PCs available (≈ 10 mg) in the early stages of drug development. Microfluidics enable screening with reduced sample volumes of PC solutions [1] and thereby can allow screening of multiple conditions using the limited amount of PC. This will help reduce the time and money invested in the solid form development of PCs by enabling early identification of all possible solid forms of PCs.

Previously, we had reported on hybrid polymer based microfluidic platforms that permitted combinatorial mixing of PC and precipitants (salt or cocrystal formers or antisolvents) solutions on-chip in arrays of 24/48, 50 to 200 nL wells, which is a drastic reduction in the volume of PC needed per condition compared to traditional approaches (~5 to 100 ul per condition [2]). The platforms were designed, fabricated, and validated to screen for solid forms employing different modes of crystallization – free interface diffusion mixing, antisolvent addition, and solvent evaporation. The developed platforms enabled on-chip solid form identification via Raman spectroscopy. These platforms were compatible with mild organic solvents such as alcohols, acetonitrile, and water. However, they were incompatible with a variety of strong organic solvents employed in pharmaceutical crystallization.

Here, we present microfluidic platforms with enhanced solvent compatibility thereby enabling investigation of a much wider range of conditions on-chip. These platforms enable enhanced control over the rates of solvent evaporation compared to previously reported platforms and thereby, high propensity to form crystals on-chip [2]. In this presentation, the design, fabrication, and application of the solvent compatible platforms to solid form screening will be covered. Model compounds such as theophylline and piroxicam have been used to validate the platform's ability to screen for salts, cocrystals, and polymorphs. On-chip Raman analysis was employed to identify the salts and cocrystals.

References

[1] C. Hansen, E. Skordalakes, J. Berger, S. Quake, Proc. Nat'l Acad. Sci., 2002, 99, 16531-6.

[2] S. Talreja, D.Y. Kim, A.Y. Mirarefi, C.F. Zukoski, P.J.A. Kenis, J. Appl. Crystallography, 2005, 38, 988-995.

(170e) Computational Crystal Structure Prediction of Organic Molecules

Erik E. Santiso, Andrei V. Kazantsev, Manolis Vasileiadis, Claire S. Adjiman and Constantinos C. Pantelides, Department of Chemical Engineering, Imperial College London,Center for Process Systems Engineering, London, United Kingdom

Crystallization is an extremely important unit operation the food, chemical, and pharmaceutical industries. The experimental determination of all possible crystal forms of a given material is, however, a very challenging problem due to the large experimental parameter space that needs to be explored. Predicting crystal structures *in silico*, however, is also challenging: for many materials there are multiple crystal configurations very close in energy to each other, and there are no general empirical potentials available that can predict their energies with the required accuracy. This means that accurate *ab initio* calculations are needed to predict crystal structures

In this work, we present our most recent work in a systematic methodology to identify crystal structures for small to intermediate size organic molecules. Our methodology involves an initial global search for candidate polymorphs with the CrystalPredictor[1] algorithm, followed by a more accurate minimization of the lowest-energy structures with the CrystalOptimizer[2] algorithm. Both algorithms use quantum mechanical calculations augmented with an empirical dispersion correction to estimate the energies of different crystalline conformations. In CrystalOptimizer, we have found that using Local Approximate Models (LAMs) is an effective way to reduce the computational cost associated with the *ab initio* calculations. In this work, we show how this methodology can be extended to the global search in CrystalPredictor to produce a set of candidate structures that is improved in terms of both the geometries and the energy rankings. We illustrate this methodology with several examples.

[1] P.G. Karamertzanis, C.C. Pantelides, "Ab Initio Crystal Structure Prediction - I. Rigid Molecules", Journal of Computational Chemistry, 2005, 26, 304-324.
[2] A.V. Kazantsev, P.G. Karamertzanis, C.S. Adjiman, C.C. Pantelides, "Efficient Handling of Molecular Flexibility in Lattice Energy Minimization of Organic Crystals", Journal of Chemical Theory and Computation, 2011, 7, 1998-2016.

(170f) Computational Study of Polymorph Selection in Glycine

Nathan Duff, Chemical Engineering, University of California, Santa Barbara, CA and Baron Peters, Chemical Engineering, University of California Santa Barbara, Santa Barbara, CA

Controlling polymorph selection is of great importance for pharmaceutical efficacy and mechanical properties. We recently developed RMSD local order parameters that can distinguish the local environments of individual molecules in the α, β, and γ-glycine polymorphs.[1] The solvated polar γ-glycine crystallite in aqueous solution forms a thicker surface melted layer than uncharged α-glycine. The addition of NaCl to the solvent screens the surface charge of the polar γ-glycine crystallite allowing a refreezing of its surface melted layer. In contrast, the surface melted layer of the non-polar α-glycine is unaffected by the NaCl. Experimental results have shown that α-glycine forms first out of aqueous solution, while γ-glycine forms first out of salt solution.[2] Our results suggest that double layer formation and surface melting are important for selection between polar and non-polar polymorphs.

References:

1. Duff N, Peters B. Polymorph specific RMSD local order parameters for molecular crystals and nuclei: α-, β-, and γ-glycine. J. Chem. Phys. 2011;135:134101.

2. Yang X, Lu J, Wang XJ, Ching CB. Effect of sodium chloride on the nucleation and polymorphic transformation of glycine. J. Cryst. Growth. 2008;310:604-611.

(170g) Influence of Polymeric Additives On the Nucleation Kinetics, Growth, and Polymorph Selectivity of Calcium Salts

Mitali China, Priyanka Rajaram and Marina Tsianou, Department of Chemical and Biological Engineering, University at Buffalo - The State University of New York (SUNY), Buffalo, NY

We investigate the factors that control the nucleation and growth of important calcium biominerals such as calcium carbonate, calcium oxalate, and calcium phosphate. We establish that macromolecules (polyelectrolytes, polypeptides, surfactants) can control the final morphology and polymorph by i) face-selective adsorption on crystals (thermodynamic regime) or by ii) temporarily stabilizing the primary nanoparticles formed thus leading to mesoscale crystallization (kinetic regime). A combination of techniques which include Scanning Electron Microscopy (SEM), X-ray Diffraction (XRD), Fourier Transform Infrared Spectroscopy (FTIR), Thermogravimetric Analysis (TGA), and Calcium Ion Selective Electrode have been used to evaluate the effect of macromolecules on the nucleation, growth, and the morphology of obtained crystals. For example, sodium polystyrene sulfonate (PSS) was found to affect the morphology of the produced crystals and to preferentially promote the crystallization of calcium oxalate dihydrate (COD) over the thermodynamically stable calcium oxalate monohydrate (COM). At sufficiently high PSS concentrations, all calcium ions participate in complex formation with PSS and thus, the crystallization of calcium oxalate was completely inhibited. The effects of PSS on the nucleation and growth of calcium oxalate were remarkably reduced or completely eliminated when a cationic surfactant molecule was introduced in the system.

(187a) Layer Melt Crystallization of Long-Chain Aldehydes From Hydroformylation Processes

Thorsten Beierling, Department of Biochemical and Chemical Engineering, Laboratory of Thermodynamics, Technische Universität Dortmund, Dortmund, Germany and Gabriele Sadowski, Department of Biochemical and Chemical Engineering, Laboratory of Thermodynamics, Technische Universität Dortmund, Dortmund, Germany

The hydroformylation of long-chain olefins, an industrially highly relevant reaction, produces a mixture of isomeric long-chain aldehydes. The purified substances are used as intermediates for further bulk and fine chemicals, such as diluent or for the flavour and fragrance industry. Conventional methods like stand-alone distillation tend to be costly or even infeasible, due to a decreasing difference in boiling points with increasing chain lengths. Layer melt crystallization in contrast is a technically proven and selective method for separating isomeric compounds and has generally the advantages of an easy scale up, facile solid-liquid separation and process handling. Despite these advantages, the promising method of separating isomeric long chain aldehydes with melt crystallization has not been investigated yet.

In this work the applicability of layer melt crystallization for various isomeric long chain aldehyde systems with different chain lengths was evaluated. Necessary parameters like related phase-equilibria were measured and modeled. Layer melt crystallization experiments were conducted, in which achievable growth rates and the corresponding purities were studied. The growth rates were modeled using the physical properties of the aldehyde systems as a function of process conditions like concentration, temperature and cooling rate. Additionally, the efficiency of sweating as a post-purification step was investigated. The sweating behavior can mathematically be described in high accordance to the experimental values as a function of temperature, crystal purity and layer thickness.

Summing up we show that melt crystallization should, due to its high efficiency, be considered as the method of choice for the separation of isomeric long-chain aldehydes. Furthermore our results reveal useful methods for the extrapolation of melt crystallization including the sweating step to systems with different chain-lengths, notably without conduction of further experiments.

(187b) Hybrid Filtration-Drying-Dissolution Device for Compact Pharmaceutical Manufacturing System

Shin Yee Wong, Jie Chen and Allan S. Myerson, Department of Chemical Engineering, Massachusetts Institute of Technology, Cambridge, MA

Pharmaceutical manufacturing is typically done in batch vessels. The recent interest in continuous pharmaceutical manufacturing had triggered a lot of interest in both academic and industrial communities, driven by the many benefits of the transformation. To bring the idea of continuous manufacturing further, we are developing a bench-top sized Pharmacy-on-Demand (PoD) system, capable of producing formulated drugs, corresponding to a daily production rate of few kilograms of Active Pharmaceutical Ingredients (API). The realization of the compact PoD system would require innovation in all unit operations (chemical synthesis, crystallization, and downstream processes), and high level of controls and automations. In this study, a hybrid device that integrated all post-processing operations after the crystallization process was developed.

This device consisted of five major components: 1) top cover consisted of inlets for slurry (from crystallizer) and liquid excipients; 2) steel mesh (0.5μm) holder; 3) lower barb end connected to vacuum; 4) wall heating component; 5) side drain valve. The filtration, drying and dissolution operations were conducted in semi-batch manner. Once the crystallization process was completed, the slurry was pumped into the hybrid device. Then, the refined API was separated from the mother liquor through the steel mesh under vacuum condition. After the filtration, the API was dried under vacuum condition at the temperature controlled by power supplied to the wall heating element. Once dried, the API was re-dispersed in the liquid excipients and drained from the side valve. Initial testing had shown high separation and drying efficiency. More than 98% of the crystallized API was successfully recovered after re-dissolution with the liquid excipients. The hybrid device had successfully transformed the slurry obtained from the crystallization process into formulated drug products in a single integrated device.

In short, the invention of this hybrid device had essentially eliminated the need for multiple stand-alone units for filtration, drying, blending and dissolution. On top of its integrated function, the compact (max dimension = 90mm) device had also significantly simplified the implementation of the downstream purification process of the PoD system.

Design and Characterization of a Continuous Stirred Tank Crystallizer

G. Morris, G. Hou, M. Barrett, B. Glennon

Solid State Pharmaceutical Cluster, The School of Chemical and Bioprocess Engineering, University College Dublin, Belfield, Dublin 4, Ireland

Keywords: Mixed Suspension Mixed Product Removal Crystallizer, Benzoic Acid, Cooling Crystallization, Process Analytical Technology

Paper ID#: 270744

Introduction

This study presents the development of a continuous crystallization platform consisting of a single stirred tank Mixed Suspension Mixed Product Removal (MSMPR) crystallizer. A photo and schematic drawing of the MSMPR setup appear in Figures 1 and 2 respectively. A recirculation loop is incorporated such that the withdrawn product is recycled to a dissolution vessel which serves as the feed tank to the crystallizer. Two OptiMax™ workstations (Mettler-Toledo) are used as both the feed/dissolution tank and the MSMPR crystallizer. Difficulties can often arise in the continuous removal of representative, non-classified product from lab-scale MSMPRs as a result of the relatively low volumes involved and consequently slow flowrates and removal velocities observed **[1]**. Intermittent withdrawal, whereby a certain percentage volume of the MSMPR is periodically removed at a high velocity so as to aid the potential for iso-kinetic removal and mitigate classification has been applied previously and is adopted in this work **[2, 3]**. This quasi-continuous transfer of the MSMPR product to the feed/dissolution vessel is achieved using vacuum with 7.1% of the suspension volume removed per transfer.

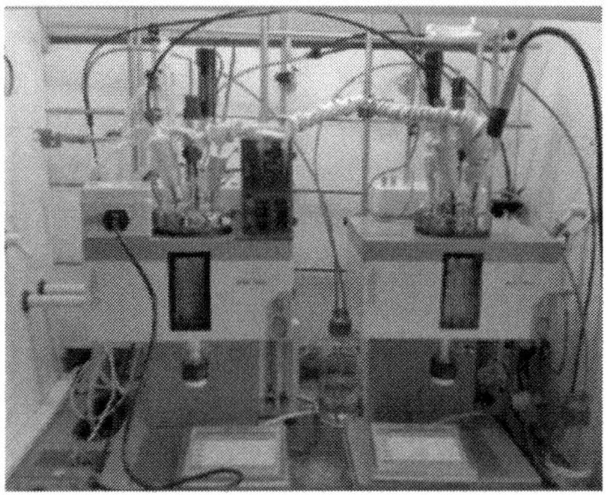

Figure 1: Image of the MSMPR setup showing feed/dissolution vessel on the left and MSMPR on the right with FBRM in-situ

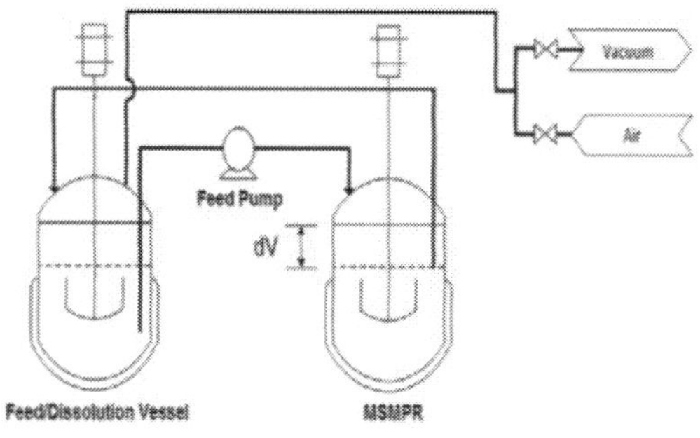

Figure 2: Schematic drawing of the MSMPR setup showing application of vacuum in recycling material from MSMPR to feed/dissolution vessel

In-situ process analytical technologies (FBRM, PVM), together with periodic sampling of the withdrawn product at steady-state and during the dynamic period of the process, have been applied in order to characterize the continuous cooling crystallization of benzoic acid from water and ethanol. Comparison is drawn between *high magma density* crystallizations (feed saturated at 40°C with the crystallizer at 18°C) and *low magma density* crystallizations (feed saturated at 15°C with the crystallizer at 0°C). For the low magma density runs the influence of residence time on affecting the steady-state particle size was investigated. For each of these studies, continuous operation was started with a suspension in the MSMPR at time zero by first batch cooling the saturated feed to the desired MSMPR temperature. In addition, a further set of experiments was carried out to investigate whether the mode of start-up bears any influence on the final steady-state product obtained and the route to steady-state.

Characterization with FBRM and Periodic Sampling

The crystallization is monitored in-situ using FBRM and PVM. This allows real time tracking of the changes in the particle size distribution and attainment of steady-state. Figure 3 shows the dynamic response in the FBRM total counts and mean chord length trends for a high-magma density crystallization proceeding to steady-state.

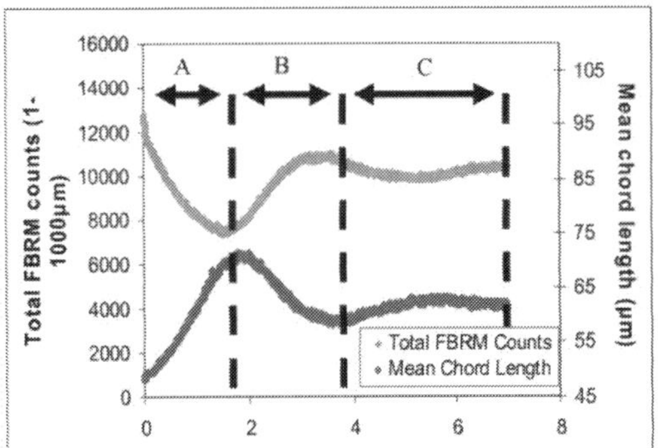

Figure 3: Tracking MSMPR to steady-state with FBRM for a high magma density run of 60 minute residence time. Showing regions: (A) wash-out period, (B) nucleation response and (C) attainment of steady-state

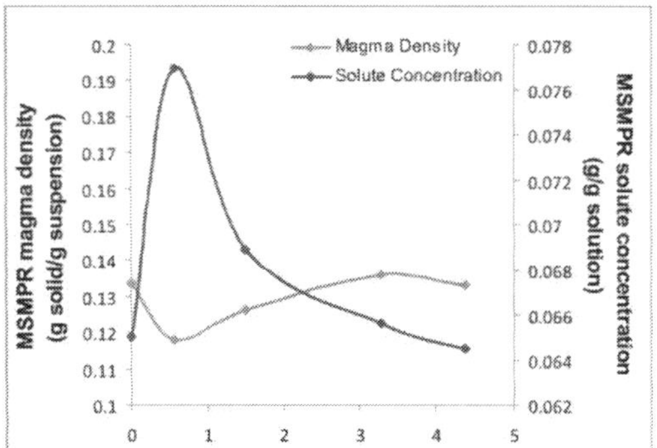

Figure 4: MSMPR solute concentration and magma density trends from start-up to steady-state for a 15 minute residence time high magma density run

As the crystallization proceeds initially there is a wash-out period (A) where particles are withdrawn in the absence of nucleation as the supersaturation builds. Over this period crystal growth is observed which is confirmed by an increase in the FBRM size trends and in-situ PVM images (Figure 5). Supersaturation increases to a maximum just before region (B), where the kinetics of the system force a period of constant nucleation which steadily consumes some of the supersaturation and slightly shifts the CSD back towards the fine end. Eventually the rate at which new crystals are formed by nucleation balances the rate at which they are withdrawn from the MSMPR and the system tends towards a steady-state (C).

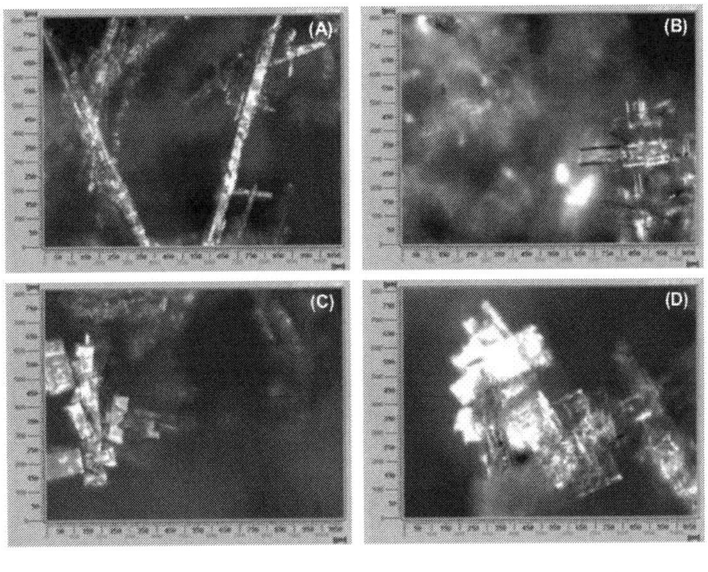

Figure 5: In-situ PVM images from a low magma density run of 15 minute residence time showing the change in particle size during MSMPR start-up. Taken at: (A) time zero after batch cooling at 1.5°C/min, (B) at 1 residence time, (C) at 2 residence times and (D) at 3 residence times

High & Low Magma Density Crystallizations

Comparing high and low magma density crystallizations it was observed that the level of the nucleation response in the dynamic route to steady-state is reduced when the slurry density in the crystallizer is relatively low (Figure 6), demonstrating the dependency of the nucleation rate on magma density. Consequently for the high magma density process the steady-state chord length distribution is shifted more towards the fine end (Figure 7).

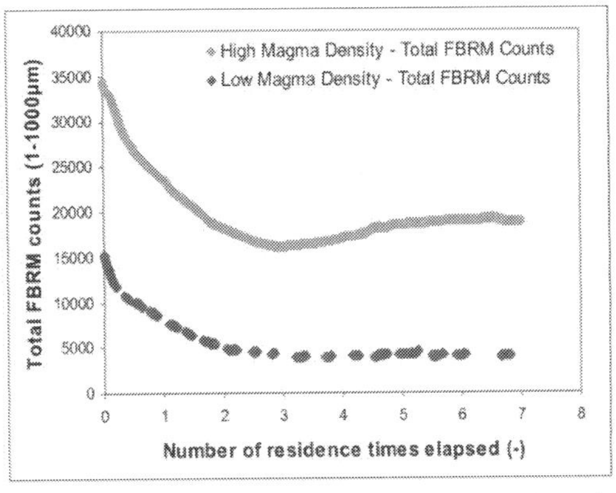

Figure 6: FBRM total counts trend for high and low magma density MSMPR runs of 15 minute residence time. For the low magma density run FBRM fouling was an issue so periodic cleaning of the probe was adopted

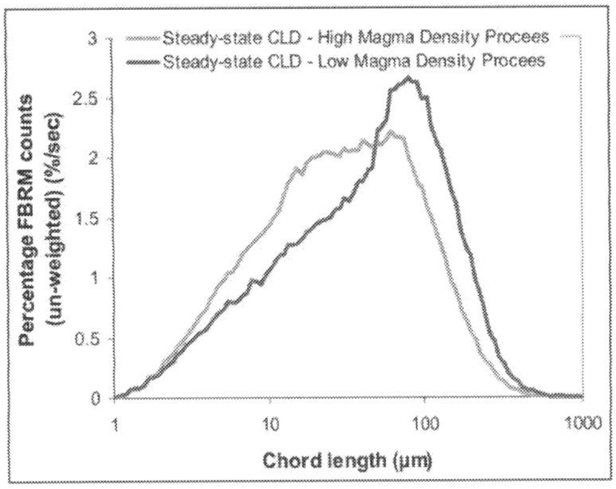

Figure 7: Final steady-state (un-weighted) chord length distributions (CLDs) for both high and low magma density processes

Evolution of Particle Size with Increasing Residence Time

In an MSMPR, for a given temperature drop between feed and crystallizer, it is possible to access different positions on the phase diagram by varying the residence time and as a result manipulate the steady-state supersaturation. The speed of the crystallization kinetics and the meta-stable zone width will ultimately dictate the supersaturation range accessible, and whether changing residence time will be favourable in terms of manipulating particle size advantageously. The benzoic acid system used in this study has a small meta-stable zone width and is more growth dominated. Consequently, the accessible range of supersaturation is limited but particle size has been shown to grow with residence time (Figure 8). Figure 9 demonstrates that steady-state particle size increases whilst MSMPR solute concentration and hence supersaturation deceases with increasing residence time.

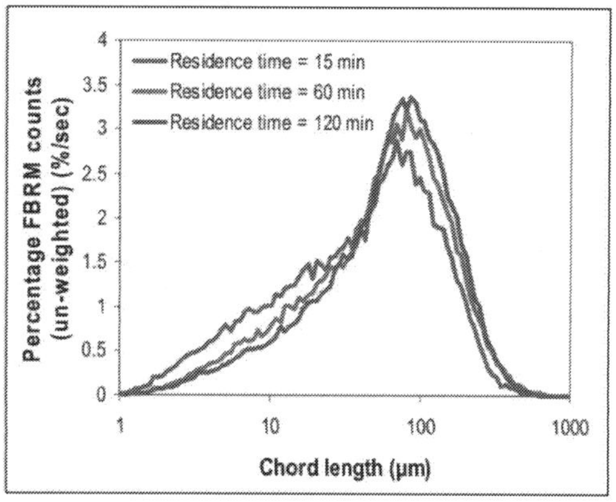

Figure 8: Steady-state CLDs (un-weighted) for low magma density runs of varying residence time

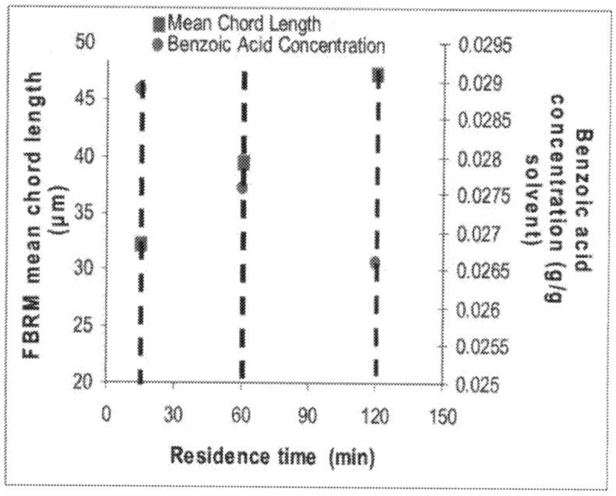

Figure 9: Demonstrating increase in particle size with residence time whilst solute concentration decreases

Influence of Start-Up Mode on Attainment of Steady-State

Several start-up regimes for running an MSMPR could potentially be employed. To examine whether particular choices in this area can have a significant bearing on the steady-state particle size obtained by the system, three methods of starting the MSMPR were investigated whilst keeping all other conditions the same:

(i) *Batch cooling the saturated feed to the desired MSMPR operating temperature, therefore allowing continuous operation to be started with an equilibrium suspension in the MSMPR at time zero*

(ii) *Starting with a clear solution which is saturated at the MSMPR operating temperature*

(iii) *Seeding a solution that is saturated at the MSMPR operating temperature with isolated steady-state product from a previous run, therefore creating a suspension of the final product at time zero*

All start-up methodologies were assessed through high magma density benzoic acid crystallizations, where the feed to the MSMPR is saturated at 40°C and the crystallizer is operated at 18°C. The resulting steady-state chord length distributions from each start-up regime are seen to be similar in all cases (Figure 10). This suggests that the final steady-state condition attained by the system is largely unaffected by the start-up methodology adopted.

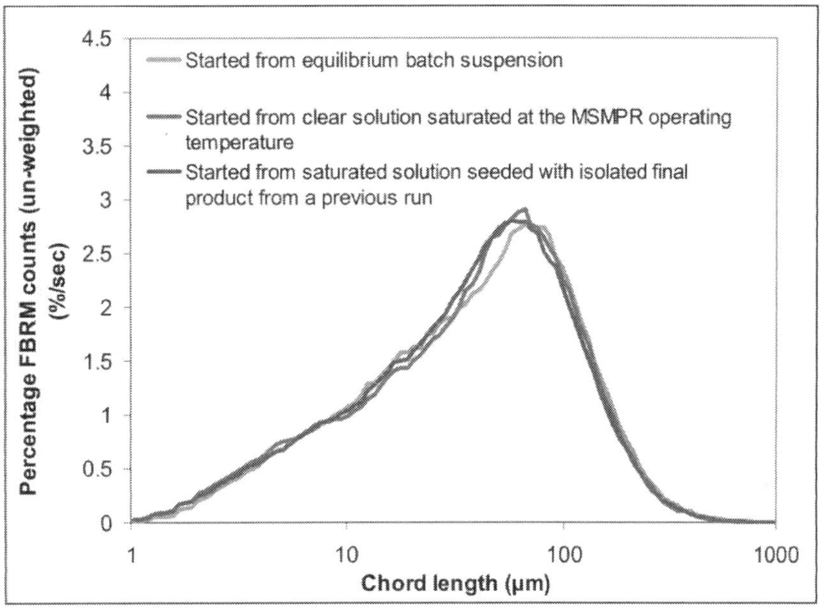

Figure 10: Steady-state CLDs (un-weighted) for high magma density crystallizations with different start-up regimes

Comparison of the FBRM total counts trends during the dynamic route to steady-state suggests that starting with saturated solution that is seeded with final product material may offer the quickest route to steady-state (Figure 11). Unlike the equilibrium batch start-up, in this case the initial particle size distribution is the same as the final and as a result the system does not require periods of significant oscillation in the size distribution to develop the steady-state CSD for the continuous process. This therefore may represent an optimum start-up strategy for MSMPRs, as the time required to reach steady-state is reduced and hence wastage of material whilst the final product is potentially out of spec is minimised.

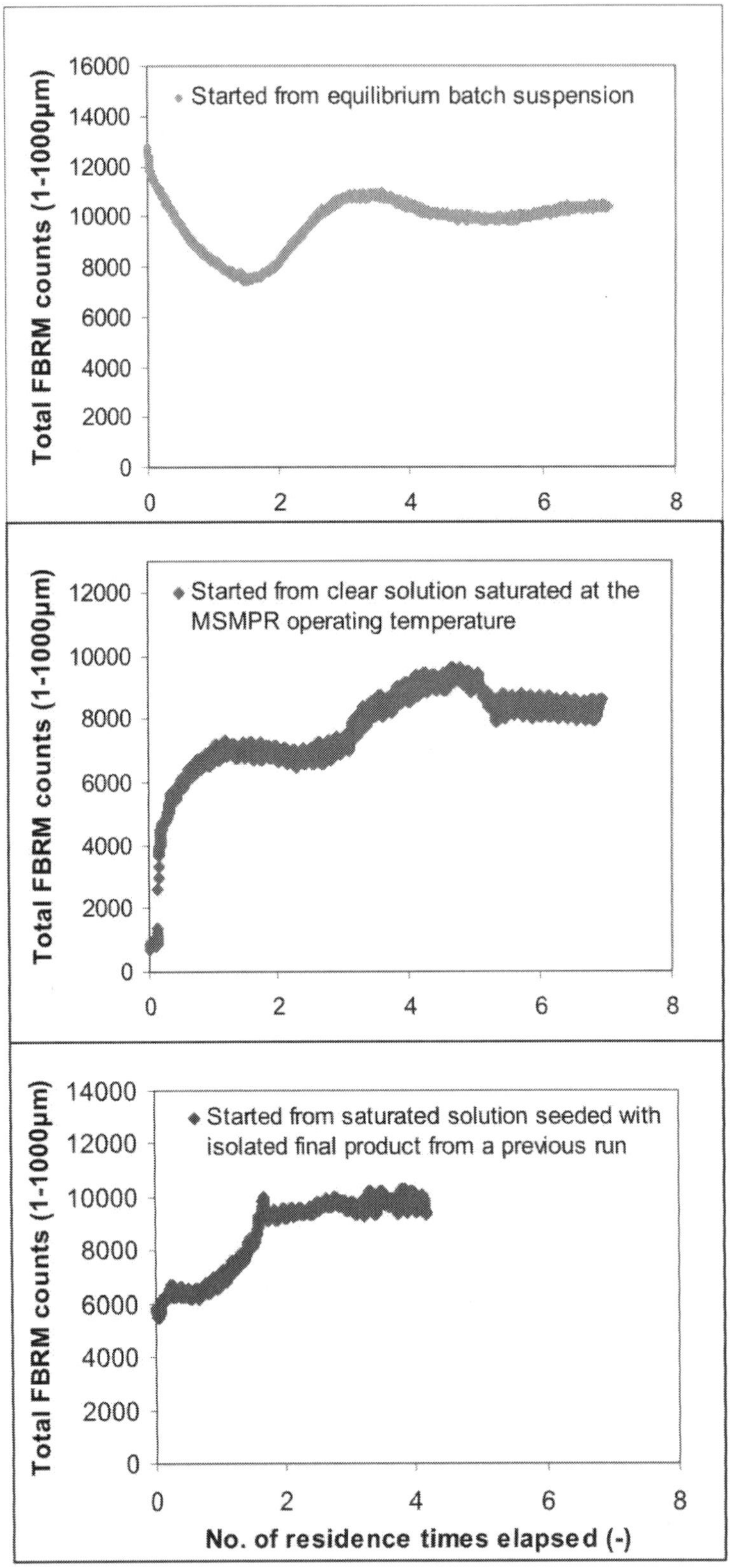

Figure 11: Comparison of the dynamic response in the FBRM total counts trends as the system proceeds to steady-state for all three start-up methodologies investigated

References

[1] A.D. Randolph, M.A. Larson, Theory of Particulate Processes, 2nd edition, Academic Press, New York, 1988.
[2] V.A. Chan, H.M. Ang, Journal of Crystal Growth 166 (1996) 1009-1014
[3] E. Kougoulos, A.G. Jones, K.H. Jennings, M.W. Wood-Kaczmar, Journal of Crystal Growth 273 (2005) 529–534

(187d) X-Ray Compatible Microfluidic Platforms for On-Chip Structure Determination of Membrane Proteins

Sudipto Guha, Ashtamurthy Pawate, Sarah L. Perry and Paul J.A. Kenis, Department of Chemical & Biomolecular Engineering, University of Illinois at Urbana-Champaign, Urbana, IL

Microfluidics has found extensive use in fields as diverse as energy conversion, medical and clinical diagnostics, point-of-care medical devices, proteomics and protein crystallization. Microfluidics allows for ease of sample preparation, small sample sizes, ease of scale-up, as well as integration of various sample analysis techniques on a single platform[1,2].

Here, we focus on the application of microfluidic platforms for membrane protein crystallization. Membrane proteins (MPs) play a crucial role in many important biological processes including energy and material transduction across cellular membranes, molecular recognition and immune response. They are amphiphilic, structurally diverse, and extremely responsive to the surrounding environment. Efforts to understand the structure-function aspects of these proteins through X-ray crystallography have been hampered by difficulty in obtaining high quality crystals[3]. Identifying the optimal crystallization condition(s) requires systematic screening of a huge chemical space (salts, buffers and precipitants), which requires a lot of protein sample that is generally not available. Additionally, current methods of structure determination are not optimized for handling small and fragile crystals that MPs tend to form.

We present an array-based, X-ray compatible microfluidic platform comprised of cyclic olefin copolymer (COC) and a thin polydimethylsiloxane (PDMS) membrane needed for valve actuation. The array-based design allows for screening for upto 96 conditions using $< 6\mu L$ of protein solution. After screening, a similar chip is used for structure determination of the protein by merging data from multiple crystals grown on-chip. The use of X-ray compatible materials allows *on-chip* analysis, thus preventing damage to fragile protein crystals that might occur during manual harvesting and mounting of crystals. As proof of concept, we have screened crystallization conditions for a soluble protein, phosphonoacetate hydrolase (PhnA). After obtaining the optimal condition and using anomalous data collected solely on-chip, we determined its structure to a resolution of 1.99 Å.

We are now in the process of validating our platform by screening and solving the structure of two MPs - photosynthetic reaction center (RC) and cytochrome ba_3-type oxidase. We have collected data on RC to 3.0 Å and are currently optimizing the crystallization conditions and chip design to get higher quality data.

(1) Hansen, C.; Quake, S. R. *Current Opinion in Structural Biology* **2003**, *13*, 538-544.

(2) Li, L.; Ismagilov, R. F. *Annual Review of Biophysics* **2010**, *39*, 139-158.

(3) Stevens, R. C. *Current Opinion in Structural Biology* **2000**, *10*, 558-563.

(187e) Investigation of a Counter-Current Crystallization Process for Separation of Systems Forming Solid Solutions

Erik Temmel[1], Sebastian Wloch[1], Uwe Müller[2], Detlef Grawe[3], Robert Eilers[4], Heike Lorenz[5] and Andreas Seidel-Morgenstern[6], (1)Max Planck Institute for Dynamics of Complex Technical Systems, Magdeburg, Germany, (2)Hapila Gera GmbH, Gera, Germany, (3)Jesalis Pharma GmbH, Jena, Germany, (4)Hapila Gera GmbH, Magdeburg, Germany, (5)Physical and Chemical Foundations of Process Engineering, Max Planck Institute for Dynamics of Complex Technical Systems, Magdeburg, Germany, (6) Physical and Chemical Foundations of Process Engineering, Max-Planck Institute for Dynamics of Complex Technical System, Magdeburg, Germany

Investigation of a counter-current crystallization process for separation of systems forming solid solutions

Erik Temmel[1], Sebastian Wloch[1], Uwe Müller[2], Detlef Grawe[3], Robert Eilers[2], Heike Lorenz[1], Andreas Seidel-Morgenstern[1,4]

[1]*Max Planck Institute for Dynamics of Complex Technical Systems, Magdeburg, D-39108, Germany*

[2]*HAPILA GmbH, Gera, D-07552, Germany*

[3]*Jesalis Pharma GmbH, Jena, D- 07745, Germany*

[4]*Otto von Guericke University Magdeburg, Chair of Chemical Process Engineering, Magdeburg, D-39108, Germany*

The occurrence of complete or partial solid solutions in solution crystallization is still the focus of several recent investigations. One reason is that the underlying mechanisms of the formation of mixed crystals are not clarified in every detail. On the other hand this special solid liquid equilibrium (SLE) limits the applicability of commonly used crystallization processes and complicates separation tasks. Especially for pharmaceuticals this can lead to a huge economical drawback.

Process design for crystallization from solution is mainly based on the solid-liquid equilibria (SLE) of the substances of interest [1]. For the, in case of ternary mixtures, commonly appearing simple eutectic and intermediate compound-forming systems several opportunities for crystallization based separation from a solvent are known and are well understood, e.g. processes based on applying preferential crystallization in case of two dissolved enantiomers. Besides, the tie lines of systems involving solid solutions do not lead to a corner of the pure substance in the ternary phase diagram (figure 1). Therefore it is not possible to separate these mixtures within one crystallization step and there is no option to apply the commonly used crystallization processes. For many examples such a miscibility behavior is published including inorganic systems like salts containing one common ion [1, 2] or simple organic substances like amino acids [3]. For more complex chiral molecules the occurrence of complete mixed crystal formation seems to be rare [4]. However, there are more cases reported where partial solid solutions are present at the pure enantiomer side of conglomerate and racemic compound-forming chiral systems.

Some theoretical studies suggest a separation opportunity based on fractional crystallization [5-7]. By means of repeated dissolution and crystallization it is possible to separate systems forming solid solutions. This principle can be realized by exploiting a cascade of batch crystallizers. Furthermore, it was shown that a counter-current between the solid and liquid phase in such a process would lead to optimal process performance in terms of yield and productivity. Nevertheless, realization and automation of a counter-current crystallization cascade is quite complicated. Between every crystallization stage a solid/liquid separation is required. Additionally, the automatic transport of the crystalline solid phase between the stages is difficult and would lead to huge instrumental effort.

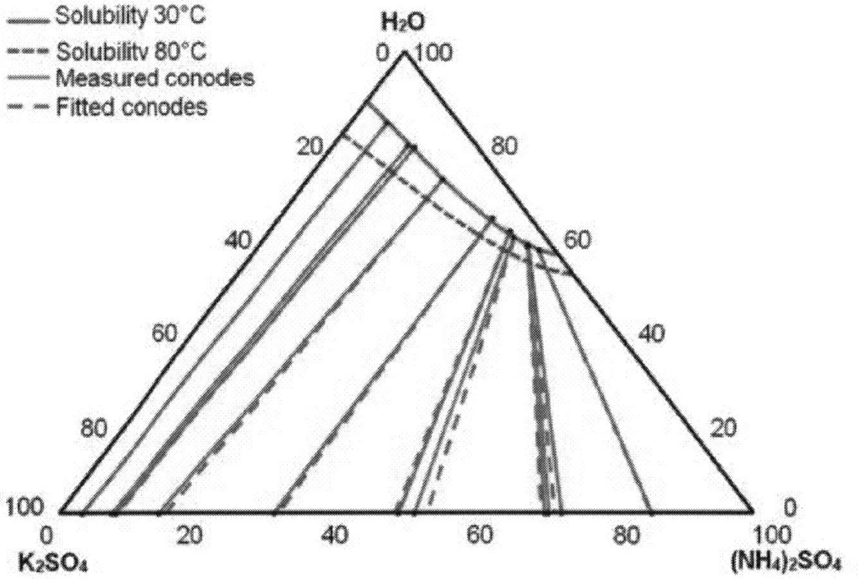

Figure 1: Ternary phase diagram of the system K_2SO_4 / $(NH_4)_2SO_4$ / H_2O for 30°C and 80°C with conodes indicated.

Recently an automatic counter-current crystallization cascade was developed by the HAPILA Gera GmbH [8]. In this plant exclusively liquid phases are transported. The solid phase is dissolved after the crystallization to avoid the expensive transport of the crystalline phases in the solid state. Hence, it is possible to proof this up to now only theoretically considered separation opportunity for systems involving solid solutions.

In the contribution the SLE of two different systems forming complete solid solutions will be shown. Furthermore, the mathematical description of the process of interest will be explained and verified by a manually operated counter-current crystallization in a rotary evaporator. The validated model was then used to run a semi-automatic pilot plant followed by a comparison of theoretical and experimental results. Finally a partial solid-solution forming system was theoretically investigated in a simulation study.

References

[1] Mullin, J. W., *Crystallization*, Boston, Butterworth-Heinemann (1997)

[2] Dejewska, D.; Szymański, T., *Cryst. Res. Technol.*, 1998, 33, 757-765

[3] Kurosawa, I. W.; Teja, A. S.; Rousseau, R. W., *Fluid Phase Equilib.*, 2004, 224, 245-249.

[4] Jaques, J.; Collet, André; Wilen, *Enantiomers, racemates, and resolutions*, S. H., Florida, Krieger (1994)

[5] Lin, S. W; Ng, K. M.; Wibowo, C., *Comput. & Chem. Eng.*, 2008, 32, 956–970

[6] Balawejder, M.; Galan, K. ; Elsner, M. P.; Seidel-Morgenstern, A.; Piatkowski, W.; Antos, D., *Chem. Eng. Sci.*, 2011, 66, 5638–5647

[7] Matz, G., *Chem. Ing. Tech.*, 1980, 52, 562-570

[8] HAPILA GmbH, Offenlegungsschrift DE 10 2008 023 833 A1, 2009

(187f) Designing a Fit-for-Purpose Agglomeration Method for Pharmaceutical Crystals

Chenchi Wang, Daniel Hsieh, Mario Hubert and Chiajen J. Lai, Research and Development, Bristol-Myers Squibb Company, New Brunswick, NJ

Designing a Fit-for-Purpose Agglomeration Method for Pharmaceutical Crystals

Chenchi Wang, Daniel Hsieh, Mario Hubert, Chiajen Lai.

Research and Development, Bristol-Myers Squibb Company, New Brunswick, New Jersey

In pharmaceutical development various particle engineering techniques are commonly employed to alter crystal morphology, improve crystal size, or increase purity of the active pharmaceutical ingredient (API). Crystal agglomeration is an alternative method to improve slurry and powder properties, especially for materials with a crystal habit or size that is difficult to change by classical crystallization techniques. In a crystal agglomeration process, the primary crystals can be directed to form agglomerates in a multi-solvent system. The formation of agglomerates can significantly improve slurry filterability or bulk powder properties compared to the primary crystals. A "fit-for-purpose" method will be presented that can be used to identify suitable conditions to encourage crystal agglomeration. Multiple case studies will be presented in detail to demonstrate how specific slurry or powder properties can be attained through the crystal agglomeration approach.

(187g) Utilization of Population Balance Models to Develop a Continuous Crystallization Process; Determination of Crystallization Kinetics

Christopher L. Burcham[1], Michael A. Lovette[2], Christopher S. Polster[2], Sean K. Bermingham[3] and Hassan Mumtaz[3], (1)Chemical Product Research and Development; Particle Design Laboratory, Eli Lilly and Company, Indianapolis, IN, (2)Chemical Product Research and Development, Eli Lilly and Company, Indianapolis, IN, (3)Process Systems Enterprise, London, United Kingdom

Due to material and time constraints, experimentation alone often cannot provide a means for design optimization of continuous crystallization processes within the pharmaceuticals industry. While several conventional process design software packages offer the ability to handle process optimization, such software packages often lack a robust means for predicting the effect of changing process variables on the corresponding particle size distributions (PSD). In the crystallization of active pharmaceutical ingredients, specific PSD requirements are often a crucial to the manufacturability and bioavailability of the final drug (solid dosage form). We use a combined population balance and process design software (e.g., gCRYSTAL, Process Systems Enterprises) as a means for process optimization for continuous crystallization. In performing this process optimization, we use rate constants for nucleation and growth that were obtained from regressed experimental data using the same software package. Furthermore, we are able to perform this process optimization with the assertion of constraints on the PSD for the optimal solution.

In this talk, experiments are conducted to provide the necessary data needed to regress kinetic parameters associated with crystallization; nucleation, growth and agglomeration. This talk will discuss both the experimental and computation aspects of the work and subsequent refinements as continuous crystallization results are collected later in development. This is the first of two discussions at this meeting, illustrating our proposed "universal" work flow for the design of continuous crystallizations.

(187h) Utilization of Population Balance Models to Develop a Continuous Crystallization Process; Process Design Using Optimization

Christopher L. Burcham[1], Michael A. Lovette[2], Christopher S. Polster[2], Sean K. Bermingham[3] and Hassan Mumtaz[3], (1)Chemical Product Research and Development; Particle Design Laboratory, Eli Lilly and Company, Indianapolis, IN, (2)Chemical Product Research and Development, Eli Lilly and Company, Indianapolis, IN, (3)Process Systems Enterprise, London, United Kingdom

Due to material and time constraints, experimentation alone often cannot provide a means for design optimization of continuous crystallization processes within the pharmaceuticals industry. While several conventional process design software packages offer the ability to handle process optimization, such software packages often lack a robust means for predicting the effect of changing process variables on the corresponding particle size distributions (PSD). In the crystallization of active pharmaceutical ingredients, specific PSD requirements are often a crucial to the manufacturability and bioavailability of the final drug (solid dosage form). In this talk, we use a combined population balance and process design software (e.g., gCRYSTAL, Process Systems Enterprises) as a means for process optimization for continuous crystallization. In performing this process optimization, we use rate constants for nucleation and growth that were obtained from regressed experimental data using the same software package (the details of this regression are contained in the talk "Utilization of Population Balance Models to Develop a Continuous Crystallization Process; Determination of Crystallization Kinetics," which is also being given at this conference). Furthermore, we are able to perform this process optimization with the assertion of constraints on the PSD for the optimal solution. This talk concludes with our insights regarding the implementation of process optimization as a part of a proposed "universal" work flow for the design of continuous crystallizations.

(204a) The Importance of Transport Processes in Silica-Supported, Polyethyleneimine-Impregnated CO2 Sorbents

David S Mebane[1], Curtis B Storlie[2], Leslie M. Moore[2], K. Sham Bhat[2], Joel D. Kress[2], Daniel J. Fauth[3] and McMahan L. Gray[3], (1)National Energy Technology Laboratory, Morgantown, WV, (2)Los Alamos National Laboratory, Los Alamos, NM, (3)National Energy Technology Laboratory, Pittsburgh, PA

THe U.S. Department of Energy's Carbon Capture Simulation Initiative (CCSI) is a DOE-wide project focused on the accelerated upscaling of new carbon capture technologies through science-based simulation. Polyethyleneimine (PEI)-based sorbent processes are one of the primary focuses of the project. Experimental evidence indicates that CO_2 transport through the bulk of the amine polymer is a significant limiting factor in the capacity of these sorbents. A microstructurally sensitive model for the uptake of CO_2 in mesoporous silica-supported PEI sorbents has been implemented. The model includes a two-step reaction mechanism for the adsoprtion of CO_2 in dry conditions, along with transport in the mesopores and in the bulk of PEI. This talk will present the results of a parametric study that demonstrates the importance of the stability of diffusive intermediates – with the zwitterions as a prime candidate – in the sorbent capacity. Statistically sensitive comparisons with experimental data will also be made.

(204b) Impact of Silane Agent Addition On the Stability of Amine Sorbents for Post-Combustion CO2 Capture

Sonia Hammache[1], Adefemi Egbebi[1], McMahan L. Gray[2] and Henry W. Pennline[2], (1)URS-NETL, Pittsburgh, PA, (2)U.S. Department of Energy, National Energy Technology Laboratory, Pittsburgh, PA

Impact of Silane Agent Addition on the Stability of Amine Sorbents for Post-Combustion CO_2 Capture

Sonia Hammache[1, 2], Adefemi Egbebi[1, 2], McMahan L. Gray[1], and Henry W. Pennline[1]

[1] US DOE-National Energy Technology Laboratory, Pittsburgh, PA

[2] URS, Pittsburgh, PA

Amine-based sorbents are being considered as an alternative to the energy intensive aqueous monoethanolamine (MEA) scrubbing process for post-combustion CO_2 capture from coal-fired power plants. A successful sorbent must be stable under moist conditions present during adsorption and regeneration steps. This communication evaluates the stability of amine sorbents under humid conditions and the impact of silane agent addition on the stability of these sorbents.

Amine sorbents were prepared by impregnation of polyethyleneimine (PEI) on silica cross-linked with different aminosilanes. One such aminosilane is 3-aminopropyltrimethoxy silane (APTES). Sorbents with four PEI/APTES weight ratios were prepared: 0, 0.4, 1, and 2.3. CO_2 adsorption-desorption cycling was conducted on these sorbents using a fixed bed reactor. The baseline performance of the sorbents was determined by testing them under dry cycles. CO_2 sorption was carried out at 60°C using a gas mixture of 10% CO_2/He at 100 sccm. Desorption was conducted under helium, starting at 60°C and then ramping up to 105°C. These sorbents were then tested using a gas composition of 10% CO_2-8% H_2O/He to determine both their stability when exposed to H_2O and the impact of silane addition on the moisture loading. The sorbents were regenerated under steam conditions at 105°C as well. The gas effluent was monitored with a mass spectrometer.

It has been found that as the amount of PEI in the sorbent increased, the CO_2 uptake increased as well; from 1.4 for a PEI/APTES ratio of 0 to 2.6 mmolCO_2/g sorbent for a PEI/APTES ratio of 2.3. CO_2 capture in the presence of 8% H_2O did not negatively impact the sorbents. Indeed, a positive impact was seen on the CO_2 uptake of the sorbents, and the increase is proportional to the increase in the PEI/APTES ratio. For a PEI/APTES ratio of 0 only a 4% increase in CO_2 uptake was seen in contrast to an 18% increase for a PEI/APTES ratio of 2.3. Regenerating the sorbents under steam conditions had a slight negative impact on the sorbents with PEI/APTES ratios of 0 and 0.4, whereas the remaining sorbents were fairly stable.

The impact of multiple steam cycles on the aforementioned sorbents and on sorbents cross-linked with various silane agents will be presented as well as their characterization prior to and following steam exposure.

(204c) Characterization and CO_2 Sorption Properties of Carbon-Based Materials for Carbon Capture Applications

Erik C. Rupp, Jennifer Wilcox and Jiajun He, Department of Energy Resources Engineering, Stanford University, Stanford, CA

Carbon capture and sequestration (CCS) has become an important option for mitigating CO_2 emission and global climate change by capturing CO_2 from large point sources and sequestering it underground. However, the feasibility of carbon capture is mainly limited by the high cost of carbon capture, and to be more specific, the significant energy requirements during the regeneration of aqueous amine solutions. Sorbent technologies for CO_2 capture have several advantages over the traditional amine-based solvent absorption approaches. For instance, within a sorbent-based approach, water is absent, which may decrease the energy requirement associated with regeneration, since heating water is the greatest energetic expense associated with CO_2 capture using solvent-based approaches. Another benefit of sorbents is the flexibility associated with the choice in pore size, surface functionality, and connectivity, in addition to the advantageous heat properties of materials such as carbon.

To overcome solvent-based regeneration expenses, micro and mesoporous sorbents have become promising candidates. A considerable variety of sorbent materials have been investigated for the application of CO_2 capture, including metal organic frameworks, zeolites, carbon nanotubes and mesoporous silicas, including SBA-15 and MCM-41. However, the CO_2 sorption capacity and kinetics under realistic flue gas conditions remain unclear. This research focuses on two sets of CO_2 sorbents: amine-functionalized multiwalled carbon nanotubes (aminated MWNTS) and biomimetic micro- and mesoporous carbon-based materials. Multi-walled carbon nanotubes were functionalized with the aminosilane compounds aminopropyltriethoxysilane (APTES) and N-dimethylaminopropyltrimethoxy-silane (DMAPS) and the resulting materials were characterized and tested for CO_2 capture under relevant conditions. As the first step in developing biomimetic sorbents, silica-based sorbents were functionalized with zinc in order to mimic CO_2 hydration that takes place with carbonic anhydrase, as a means to easily capture and release CO_2.

Breakthrough experiments have been performed in a temperature-controlled packed-bed reactor with small amounts of sorbent (i.e., < 250 mg). Experiments were performed with a resolution of 0.1 mmol CO_2 per gram of sorbent using an Extrel MAX300-LG quadrupole mass spectrometer downstream of the packed bed. The inlet gas was either a mixed gas (CO
with potential contaminants such as SO_2, HCl and NO_x doped in at part per million levels post-combustion. The breakthrough experiments provide an understanding of CO_2 uptake, sorbent repeatability and impact of flue gas contaminants on capture. In addition, a Quantachrome Autosorb iQ_2 automated gas sorption analyzer is used to determine surface area, pore volume and pore size distribution of the tested CO_2sorbents.

(204d) Large-Scale Computational Screening of Adsorbent Materials for Carbon Capture

Li-Chiang Lin[1], Adam H. Berger[2], Richard L. Martin[3], Jihan Kim[3], Joseph A. Swisher[1], Kuldeep Jariwala[3], Chris H. Rycroft[3,4], Abhoyjit S. Bhown[2], Michael W. Deem[5], Maciej Haranczyk[3] and Berend Smit[1], (1)Chemical and biomolecular Engineering, University of California-Berkeley, Berkeley, CA, (2)Electric Power Research Institute , CA, (3)Lawrence Berkeley National Laboratory, (4)University of California-Berkeley, (5)Rice University, Houston, TX

Carbon capture and sequestration (CCS) is an important strategy for reducing the negative environmental impact of energy derived from fossil fuels. The current technology to capture CO_2, amine scrubbing, is very energy intensive and can decrease the efficiency of a coal-fired power plant by as much as 30%. Recently, nano-porous adsorbent materials such as zeolites, ZIFs, and MOFs have been found to potentially provide a more energy-efficient way for CO2 separation. In principle, there are hundreds of thousands of possible candidates that might be used for carbon capture. Fully characterizing all those possible materials with traditional molecular simulation technique is computationally prohibitive. We have developed a new computational approach to enable us to evaluate such large number of materials, across different classes, for their performance in carbon capture. In this study, we don't rank the materials just based on a single adsorption property, such as selectivity or breakthrough time. Our metric, called parasitic energy, constitutes the minimized electric load which is imposed upon a power plant by using the material in a temperature-pressure swing capture process at it's own optimal operation condition, followed by compression. For the determination of the parasitic energy, our computational procedure allows us to rapidly calculate all the important thermodynamic and geometric properties of a material. This efficient approach exploits the parallel processing power of graphics processing units (GPUs) as well as material informatics methods. For instance, the time required to calculate a CO_2 adsorption isotherm of a material over a wide range of pressure can be decreased by as much as two orders of magnitude compared to the traditional grand canonical Monte Carlo (GCMC) simulation on CPUs. From the outcome of this large-scale screening, we have identified a number of materials with an energy penalty almost 35% lower than the near-term technology. More importantly, insights about the optimal structures for carbon capture were found that should be very helpful for the future design and synthesis.

Reference:

L.-C. Lin et al., "In Silico Screening of Carbon Capture Materials", *Nature Materials*, in press. (2012)

(204e) Functionalized Silica-Polymer Composite for CO2 Capture

Gifty Osei-Prempeh[1], Garth Thomas Jr.[1] and Asad Davari[2], (1)Department of Chemical Engineering, West Virginia University Institute of Technology, Montgomery, WV, (2)Department of Electrical and Computer Engineering, West Virginia University Institute of Technology, Montgomery, WV

Atmospheric CO_2 concentration has increased significantly in the last 50 years due to increased utilization of fossil fuels. Fossil fuel use is projected to peak in 2050 hence an effective means of controlling CO_2 emissions from coal burning power plants, vehicles and other emission sources are seriously in need.

Solid adsorbents for CO_2 capture presents an effective, less energy intensive method. Numerous solid adsorbents have been tailored and tested to determine an effective CO_2 capture material. This capture material must possess high CO_2 capture capacity, excellent adsorption-desorption kinetics and good stability.

This work involves the synthesis of polystyrene (core)-silica (shell) composites for use as carbon dioxide adsorbent. The polymer-silica composites may provide better column packing than nanoporous silica. Composites are known to enhance properties of polymers whilst utilizing the surface properties of the inorganic phase. The incorporation of functional groups such as amines in both the organic and inorganic phases has shown improved CO_2 capture capacity. Material properties such as surface area, pore size, functional group loading, particle size and CO_2 capture capacity of polymer-silica composites will be compared to those of nanoporous silica. CO_2 adsorption and desorption kinetics will also be compared.

(204f) Nanoclay-Based Solid-Amine Sorbents for Carbon Dioxide Capture

Elliot Roth[1], Sushant Agarwal[1] and Rakesh Gupta[2], (1)Chemical Engineering, West Virginia University, Morgantown, WV, (2)Department of Chemical Engineering, West Virginia University, Morgantown, WV

Nanoclay-Based Solid-Amine Sorbents for Carbon Dioxide Capture

The objective of the research project is to develop an efficient, low cost, regenerable solid sorbent for carbon dioxide adsorption from large point sources, which can be regenerated with more realistic recycling schemes. The main regeneration scheme that has been studied in the literature for solid adsorbents has been temperature swing desorption while using nitrogen as a sweep gas. This technique, while very useful in the lab to show regeneration is possible, is not realistic as one again ends up with a mixture of CO_2 and N_2 as an outlet stream. In this work we show that a low cost nanoclay-based solid sorbent can be regenerated with more industrially relevant regeneration methods. We show that it is possible to use pressure-swing vacuum desorption to regenerate the sorbent. We also show that CO_2 containing moisture itself can be used as a sweep gas with temperature swing to regenerate the sorbent.

The sorbent developed here is composed of a montmorillonite nanoclay, commonly used in the production of polymer nanocomposites, grafted with commercially available amines. Aminopropyltrimethoxysilane (APTMS) was chemically grafted on to the edge hydroxyl groups of the clay, and polyethylenimine (PEI) was then attached to the surface of the clay by electrostatic interactions. FTIR analysis was used to confirm the amine grafting. The amount of amine loaded onto the support was determined by TGA techniques. The treated clay was initially analyzed for CO_2 adsorption in a pure CO_2 stream in a TGA. At atmospheric pressure, the optimized adsorption temperature was determined to be between 75°C and 85°C. The maximum CO_2 adsorption capacity for clay treated with APTMS at 85°C in pure CO_2 at atmospheric pressure was 6.7 wt% CO_2, but with the additional treatment of attaching PEI to the surface of the clay the maximum CO_2 adsorption capacity increased to 9.7 wt% CO_2. In a more realistic simulated flue gas of 10% CO_2 and 90% N_2, the adsorbents had essentially the same overall maximum CO
treated with APTMS and PEI.

Adsorption studies conducted by us in pure CO_2 at room temperature and under pressures from 275-2070 kPa showed that the average adsorption capacity of the adsorbents did not change significantly under the pressures examined, indicating that the uptake of CO_2 was due mainly to chemical reaction and not due to the physical absorption of CO_2. The average CO
of APTMS and PEI treatment on the clay increased the average adsorption capacity to 11.39 wt% CO_2. While these results are comparable to those available in the literature for other solid sorbents, the cost of the sorbents developed here is expected to be, as much as an order of magnitude, lower than that of competing sorbents.

Initial desorption studies were conducted using pure N
completely regenerating the adsorbent. Multiple cycles of adsorption and desorption can be carried out with very little loss in CO_2 capture capacity. Other regeneration schemes that we have studied that are more commercially realistic are vacuum regeneration and regeneration using CO_2 with a small amount of water at 155°C. Vacuum regeneration was studied by first adsorbing CO_2 on the sample using a TGA and then placing the sample in a vacuum oven at 65°C-85°C at 93 kPa vacuum for 1 hour. The sample was then returned to the TGA were CO_2 was adsorbed onto the sample. The amount of CO_2 adsorbed was compared and showed that vacuum desorption is successful at regenerating the adsorbent. Regeneration using CO_2 and water was also successful in regenerating the adsorbent at the temperature of 155°C. The CO_2 and water studies were conducted in the TGA and water was incorporated into the reaction gas by

bubbling the CO_2 through water. Although the desorption temperature using CO_2 and water was higher than using nitrogen, this scheme is a much more realistic way to regenerate the adsorbent on a large scale since water can easily be condensed out of the exit stream forming an almost pure CO_2 stream for sequestration. This work shows that a nanoclay based solid adsorbent can be regenerated using more industrially relevant regenerations schemes that have not been studied much in the literature.

Kinetics and Modeling of CO_2 Adsorption on Amine Functionalized Mesoporous SBA-15 Sorbents

Arunkumar Samanta[1], An Zhao[1], Partha Sarkar[2], Rajender Gupta[1]
[1]Department of Chemical and Materials Engineering, University of Alberta, Canada
9107 - 116 St, ECERF, Edmonton, AB, T6G 2V4, Canada
Email:asamanta@ualberta.ca, Tel: 780-492-6861, Fax: 780-492-2881
[2]Environmental & Carbon Management Division, Alberta Innovates - Technology Futures,
250 Karl Clark Rd., Edmonton, AB, T6N 1E4, Canada

Introduction

Adsorption processes using porous solid sorbents, such as activated carbon[1], zeolites[2], micro/mesoporous silicas and their functionalized counterparts [3,4] capable of capturing CO_2 from flue gas streams reversibly have been extensively investigated and have shown great potential compared to other separation techniques for CO_2 capture[5-7]. Conventional physisorbents, such as activated carbons and zeolites were proved not to be satisfactory for post-combustion CO_2 capture, since it largely depends on the physical characteristics of solid adsorbents and show a poor CO_2 separation performance with regards to CO_2 adsorption capacity, selectivity and durability. Adsorption processes using functionalized ordered mesoprous silica sorbents have shown a great potential for post-combustion CO_2 capture from flue gas compared to the other separation techniques. Ordered mesoporous silicas generally provide excellent structural properties as supports, such as large pore volume, pore size and surface area [8-9]. In addition to this, numerous active sites disperse highly on the inner pore surface which benefit and promote the distribution of amine moieties on the surface. The immobilization of amino groups anchored to the surface of solid support by means of physical impregnation or covalent modification via silane chemistry offers stable interaction with acidic CO_2 molecules, meanwhile largely avoids the corrosion and toxic problems.

In this work an experimental and theoretical investigation on the adsorption of CO_2 onto amine-functionalized mesoporous SBA-15. The adsorption of CO_2 on the functionalized sorbents have been measured by thermogravimetric method over the CO_2 partial pressure range of 10-100 kPa and temperature range of 303 – 373 K under atmospheric pressure. The functionalization of SBA-15 silicas with tetraethylenepentamine (TEPA) has been achieved using conventional wet impregnation technique. The structural properties of the sorbents have been characterized by nitrogen adsorption/desorption, SAXS, SEM, TEM and FTIR techniques. Thermal swing adsorption cycles over a range of temperatures and time in a simulated flue gas were also explored. Different chemisorption kinetic models have been used to analyze the adsorption data. The model was validated with the experimental results of isothermal adsorption measurements of CO_2 on SBA-15/TEPA.

Experimental

Material Synthesis and Characterization

The mesoporous silica support used in this work was synthesized according to the method reported elsewhere [10]. It was synthesized using amphiphilic triblock copolymer Pluronic P123 (MW_{avg} = 5800, Aldrich) as the organic structure-directing agent, tetraethyl orthosilicate (TEOS, Aldrich) as silica source and HCl as pH controlling agent. In a typical synthesis procedure, 4g of Pluronic P123 was dissolved in 30 ml of deionized water and 120 ml 2M HCl solution while stirring for 2 hrs at room temperature. Then the resultant solution was transferred into a Teflon autoclave and solution temperature was raised to about 40°C. 8.5 g of TEOS was added dropwise to the homogenous solution under vigorous stirring for about 30 mins and a precipitated product appeared. The resulting gel was stirred at that temepature for about 24 h at low stirring rate and then aged at 100°C for 48 hrs without stirring. Then, the white precipitated solid was filtered, washed with deionized water and dried in air at room temperature for 12 hrs and in an air oven at 100°C for 12 hrs. The template was removed from the product by calcination in flowing air at 550°C for 5 hrs.

The resulting SBA-15 materials, after drying, calcination and subsequent impregnation, were characterized by N_2 adsorption/desorption, SAXS, SEM, TEM and FTIR.

CO_2 Adsorption and Regeneration Study

CO_2 adsorption/desorption measurement was performed on a thermogravimetric analyzer(Q 600, TA Instruments). Pure CO_2 (purity:99.99%) or CO_2/ N_2 at 1 atm was used for the adsorption measurements runs and N_2 was used as purging gas for CO_2 desorption. In a typical adsorption run, about 10 mg of the sorbent were placed in platinum pan. After the sorbent was heated to 105 °C in a N_2 stream with a flow rate of about 100 cm^3. min^{-1} for about 30 min to remove all moisture and CO_2 adsorbed from air, the temperature was decreased to desired adsorption temperature. The gas was then switched from N_2 to CO_2 and the temperature was kept constants at desired adsorption temperature for about 1 hr. The CO_2 capturing capacity of the solid sorbent was calculated in $mmol.g^{-1}$ dry sorbent from the weight gain of the sample in the adsorption process.

Results and Discussion

Sorbent Characterization

The N_2 adsorption isotherms of calcined SBA-15 shows typical type IV isotherm indicative of defined mesoporosity in the framework. As shown in Table 1, the BET surface area, mesopore area and mesopore volume are about 1099 $m^2.g^{-1}$, 923 $m^2.g^{-1}$ and 1.78 $cm^3.g^{-1}$, respectively for calcined SBA-15,. The pore size (in the range of 10-30 nm) was estimated using BJH method. Five distinct reflective peaks are identified from the SAXS diffraction spectra of SBA-15 and SBA-15/TEPA. The estimated d-spacing (d_{100}) is in the range of 9.7-10.3 nm. The morphology of the mesoporous SBA-15 support studied using SEM and TEM techniques (as shown in Figure 1). The SBA-15 is observed to contain rope like material arranged in bundle of about 4.3 μm diameter and length of 35 μm. Besides, FTIR spectroscopy of SBA-15/TEPA provided clear evidence for amine impregnation.

Table 1. Structural Properties of SBA-15 Support

Support	S_{BET} (m^2/g)	S_{meso} (m^2/g)	V_{tot} (cm^3/g)	V_{micro} (cm^3/g)
Calcined SBA-15	1099	922.9	1.844	0.065
As-synthesized SBA-15	222.6	222.6	0.597	~0

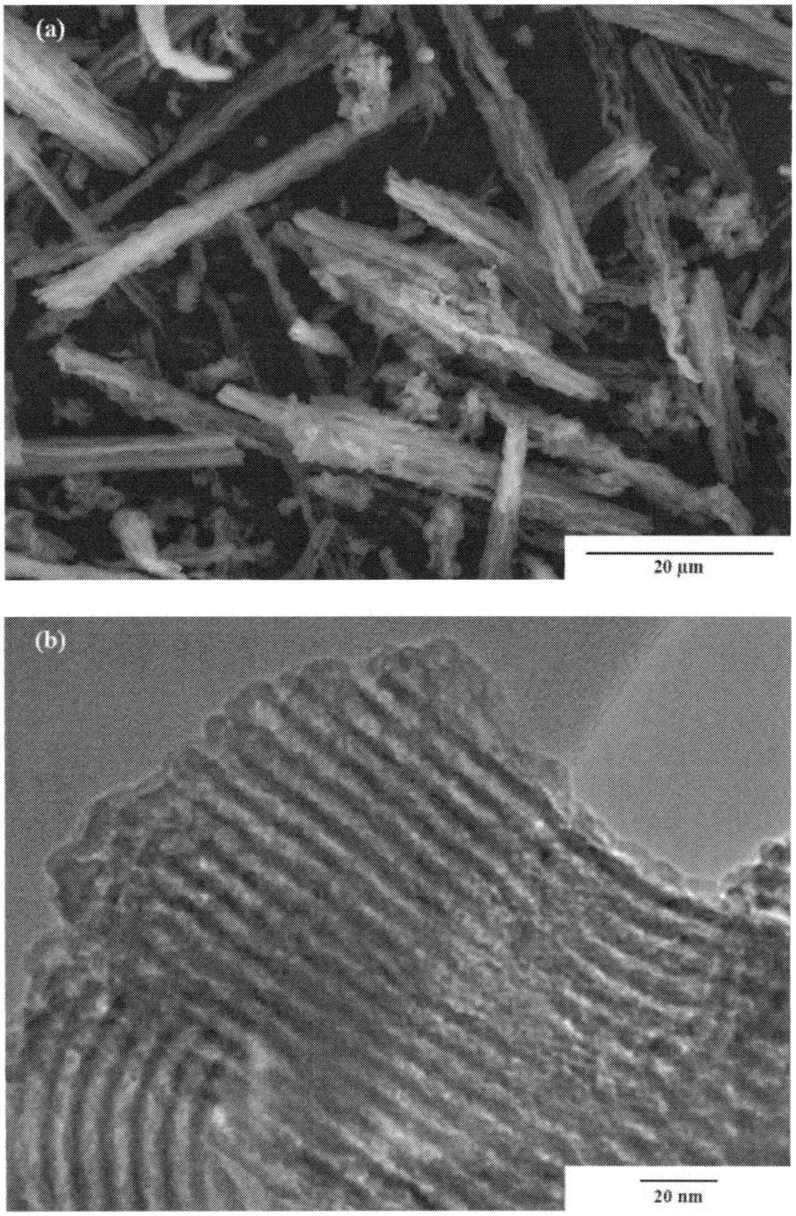

Fig. 1. (a) SEM and (b)TEM of Calcined SBA-15

CO_2 Adsorption Study

CO_2 adsorption capacities of the amine impregnated solid sorbent supports, such as Norit AC, Sigma AC (C5510), SBA-15, and MCM-41 has been determined at 303, 323, 348 and 373 K in pure CO_2 and various partial pressures of CO_2 (in N_2). Typical CO_2 adsorption capacity results are shown and compared in Figure 2. From the preliminary screening study for TEPA and PEI impregnated sorbents, it has been found that SBA-15/TEPA outperforms all other sorbents under the same conditions. In particular, SBA-15/TEPA sorbents shows better CO_2 adsorption capacity than the sorbent loaded with same amount of PEI. SBA-15/TEPA exhibits the highest CO_2 adsorption capacity and reached CO_2 uptake value as high as 4.51 mmol-CO_2/g-dry sorbent within few minutes of adsorption test at 348 K and pure CO_2 environment. Amine modified activated carbon does not show satisfactory performance on adsorption capacity because of the limitations such as relatively lower pore volume and surface area.

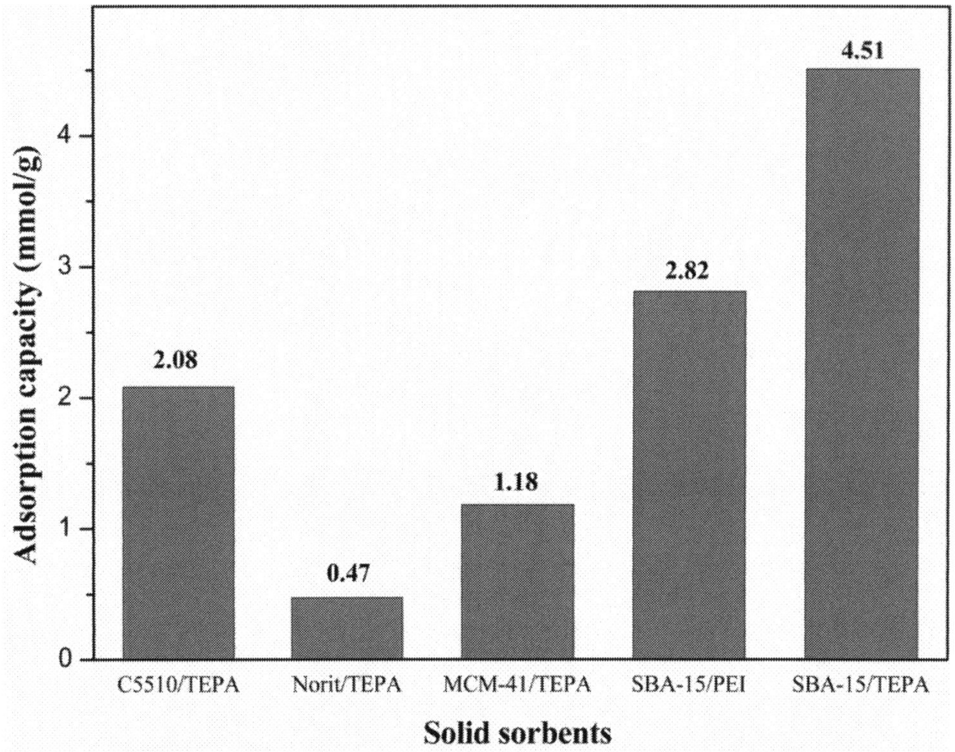

Fig. 2. Comparison of CO_2 Adsorption Capacity on Different Solid Supports Impregnated with 60 wt% TEPA at 348 K for 1 hr in Pure CO_2 Feed Gas.

To study the influence of adsorption temperature on TEPA impregnated SBA-15 sorbents, adsorption experiments were carried out at 303, 323, 348 and 373 K. When the temperature increases from 303 to 348 K, there is a large difference observed in adsorption capacity that varies from 2.73 to 4.51 $mmol.g^{-1}$. One possible reason is that higher temperature promotes the adsorption reaction and increases the accessibility of CO_2 to available active sites on the surface of TEPA. On the other hand, all the sorbents exhibit a lower adsorption capacity at low CO_2 partial pressure. Preliminary test with ~2.0 (vol) % moisture in pure CO_2 shows CO_2 adsorption capacity of about 5.05 $mmol.g^{-1}$ indicating moisture favors in CO_2 adsorption.

A series of impregnated sorbent using calcined SBA-15 containing 50- 70 wt% TEPA were also prepared to investigate the effect of amine loading on adsorption capacity. It is observed that the adsorption capacity increases from 3.57 $mmol.g^{-1}$ to 4.60 $mmol.g^{-1}$ with increasing the amine loading from 50 wt% to 70 wt%. However, the enhancement of adsorption capacity with respect to amine efficiency does not follow similar trend. For example, SBA-15/TEPA 60 exhibits the highest amine efficiency 0.290, while SBA-15/TEPA70, which presents the highest adsorption capacity, has relatively lower amine efficiency value of 0.248.

About ten cycles of adsorption-desorption were performed to study the cyclic adsorption-desorption performance of the SBA-15/TEPA60 sorbent using temperature swing process. From the cyclic adsorpion-desorpiton performance, it is observed that after ten cylces of adsorption-desorption, the sorbent shows a gradual decrease in capacity. This is probably due to the continuous degdation of TEPA during the repeated adsorption-desorption processes.

CO_2 Adsorption Kinetics

A fast adsorption kinetics have been found with more than 90% of the total CO_2 uptake on TEPA-SBA-15 sorbents occurring within the first few minutes of adsorption. All SBA-15/TEPA show a sharp increase of adsorption during the initial adsorption stage. Three different kinetic models expressed by Eq.(1), Eq. (2) and Eq.(3), have been used to analyze the experimental CO_2 uptake data:

$$\frac{\partial q_t}{\partial t} = k(q_e - q_t) \quad (1)$$

$$\frac{\partial q_t}{\partial t} = k(q_e - q_t)^2 \quad (2)$$

$$\frac{\partial q_t}{\partial t} = k_n t^{m-1} \, (q_e - q_t)^n \quad (3)$$

where q_e and q_t are the sorption capacity at equilibrium and at time t, respectively; k, k_n, n and m are the model constants. Figure 3 shows the experimental results of isothermal adsorption measurements of CO_2 on functionalized SBA-15 sorbents at 348 K and the corresponding values predicted by various kinetic models.

The significantly higher values of AAD% for pseudo first-order and pseudo second-order equations suggest the experimental CO_2 uptake for TEPA impregnated SBA-15 do not follow pseudo first-order and pseudo second-order models strictly. Both the models overestimate the CO_2 uptake at the initial phase of adsorption, when the significant fast adsorption occurs. While in case of fractional-order kinetic model, agreement was very good over the entire adsorption region with maximum AAD between experimental CO_2 uptake and that calculated from Eq.(3) is 2.4 %. The significantly lower value of AAD% for fractional-order kinetic model suggests its applicability for TEPA impregnated SBA-15 adsorbents. Hence, in this work, experimental CO_2 uptakes have been correlated by fractional-order kinetic model as expressed by Eq. (3).

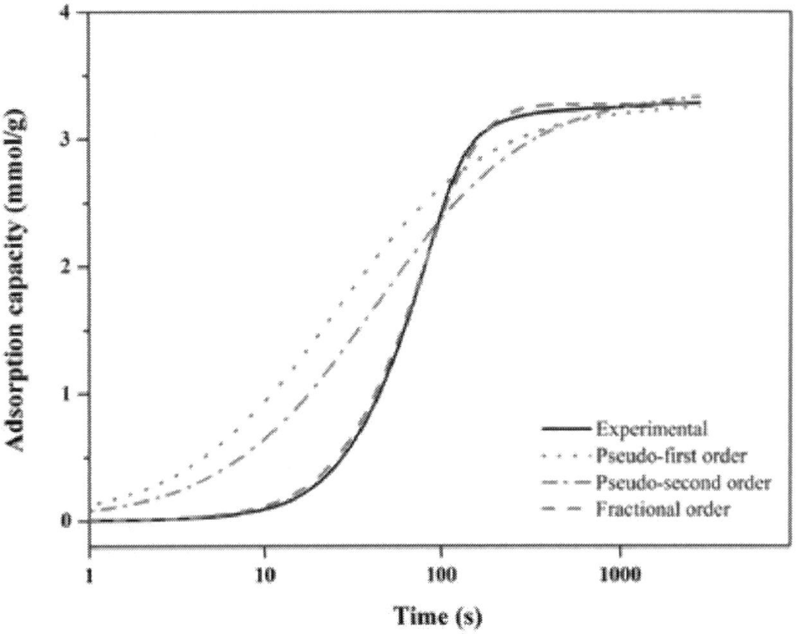

Fig. 3. Typical Comparison of Experimental and Model Predicted CO_2-uptake on SBA-15/TEPA60 in 10% CO_2/90% N_2 at 348 K.

Conclusions

This work presents a theoretical and experimental investigation on the adsorption of CO_2 onto amine- functionalized mesoporous SBA-15. The CO_2 sorption/desorption experiments performed by thermogravimetric method revealed that the well-dispersed TEPA inside SBA-15 exhibited a CO_2 sorption capacities as high as 4.51 mmol.g^{-1} for the sorbent SBA-15/TEPA60 at 348 K. The textural properties of the raw materials and in particular, the mesopore content seems to be the dominant factor for creating a good dispersion of the amines into the pore channels of the silica support. The CO_2 capture kinetics is found to be fast and reached 90% of the total capacities within the first ten minutes. Different chemisorption kinetic model has been developed to analyze the experimental data. The model is validated with the experimental results of isothermal adsorption measurements of CO_2 on functionalized SBA-15 sorbents by thermogravimetric method. Good agreement between the model results and experimental results indicates that a general kinetic model described by Eq. (3) can effectively represent the dynamics of CO_2 mass transfer rates on TEPA functionalized SBA-15 sorbents for wide range of operating conditions. The cyclic CO_2 adsorption-desorption revealed that the TEPA impregnated SBA-15 sorbents exhibited good cyclic stability. The results suggest that mesoporous SBA-15 impregnated with TEPA shows a great potential as adsorbents for post-combustion CO_2 capture.

Acknowledgements

The financial support of the Carbon Management Canada and Canadian Centre for Clean Coal/Carbon and Mineral Processing Technology is acknowledged.

References

1. Siriwardane, Ranjani V.; Shen, Ming -S.; Fisher, Edward P. and James A. Poston (2001), "Adsorption of CO_2 on Molecular Sieves and Activated Carbon," *Energy & Fuels*, 15(2), pp. 279-284.
2. Harlick, Peter J. E. and F. Handan Tezel (2004), "An Experimental Adsorbent Screening Study for CO_2 Removal from N_2," *Microporous and Mesoporous Materials*, 76(1–3), pp. 71-79.
3. Xu, X.; Song, C.; Miller, Bruce G. and Alan W. Scaroni (2005). "Influence of Moisture on CO_2 Separation from Gas Mixture by a Nanoporous Adsorbent Based on Polyethylenimine-Modified Molecular Sieve MCM-41," *Industrial & Engineering Chemistry Research*, 44(21), pp. 8113-8119.
4. Franchi, Robert S.; Harlick, Peter J. E. and Abdelhamid Sayari (2005), "Applications of Pore-Expanded Mesoporous Silica. 2. Development of a High-Capacity, Water-Tolerant Adsorbent for CO_2," *Industrial & Engineering Chemistry Research*, 44(21), pp. 8007-8013.
5. Choi, Sunho; Drese, Jeffrey H. and Christopher W. Jones (2009), "Adsorbent Materials for Carbon Dioxide Capture from Large Anthropogenic Point Sources," *ChemSusChem*, 2, pp. 796-854.
6. Sayari, Abdelhamid.; Belmabkhout, Y. and Rodrigo Serna-Guerrero (2011), "Flue Gas Treatment via CO_2 Adsorption," *Chemical Engineering Journal*, 171, pp. 760-774.
7. Samanta, Arunkumar.; Zhao, An.; Shimizu, George K. H.; Sarkar, Partha and Rajender Gupta (2012), "Post-Combustion CO_2 Capture Using Solid Sorbents: A Review," *Industrial & Engineering Chemistry Research*, 51, pp. 1438-1463.
8. Joo, Sang H.;Ryoo, Ryong; Kruk, Michal and Mietek Jaroniec (2002), "Evidence for General Nature of Pore Interconnectivity in 2-Dimensional Hexagonal Mesoporous Silicas Prepared Using Block Copolymer Templates," *The Journal of Physical Chemistry B*, 106(18), pp. 4640-4646.
9. Ryoo, Ryong; Ko, Chang H.; Kruk, Michal; Antochshuk, Valentyn and Mietek Jaroniec (2000), "Block-Copolymer-Templated Ordered Mesoporous Silica: Array of Uniform Mesopores or Mesopore-Micropore Network?" *The Journal of Physical Chemistry B*, 104(48), pp. 11465-11471.
10. Zhao, Dongyuan; Feng, Jianlin; Huo, Qisheng; Melosh, Nicholas; Fredrickson, Glenn H.; Chmelka, Bradley F. and Galen D. Stucky (1998), "Triblock Copolymer Syntheses of Mesoporous Silica with Periodic 50 to 300 Angstrom Pores," *Science*, 279(5350), pp. 548-552.

Chemisorption of Carbon Dioxide on Chemically Treated Fly Ash Surface: experimental and kinetics modeling

Abstract ID# 276815

*[1]Reyad A. Shawabkeh, [1]Ali Yaumi and [1]Ibnelwaleed Hussien

[1] Department of Chemical Engineering, King Fahd University of Petroleum & Minerals, Dhahran, Saudi Arabia

* AIChE members, corresponding Author, rshawabk@kfupm.edu.sa

The concentration of CO_2 in the atmosphere is elevated by the combustion of fossil fuels for the generation of electricity, heavy oil, coal, oil shale, and exhausts from automobiles [1]. CO_2 as a greenhouse gas has a major impact on the environment and climate change which causes global warming [2]. The ecological systems, sea levels and the atmosphere are generally being affected by the rising level of CO_2. Currently, humans are emitting around 29 billion tons of CO_2 into the atmosphere per year. Around 43% remains in the atmosphere while the rest is absorbed by vegetation and the oceans [3].

The consequences of CO_2 emissions include reduction in productivity of agricultural products, melting of ice caps due to rise in sea level, and reduction in ozone layer [4]. The stabilization of CO_2 in the atmosphere can be potentially achieved by capturing and sequestration from the main source of emission. Separation and recovery of CO_2 emitted by power plants can be achieved by processes such as liquid solvent absorption, membrane separation, pressure/temperature swing adsorption and cryogenic techniques [5,6]. Amine-based processes and wet scrubbing systems are currently heavily used in chemical plants to separate CO_2, however, due to the large amount of water needed in these systems and being energy intensive processes they may not provide an optimum choice for capture and sequestrate this gas [1]. The others have high regeneration energy and large equipment sizing, costly or having low separation efficiency [7].

In contrast, adsorption is another potential technique that can be used in the separation of CO_2 as a result of its low energy requirement, low cost, ease of applicability over a wide range of temperature and pressure [1,7-12]. Nevertheless, the success of using adsorption as an efficient technique to separate CO_2 depends on the using of durable and easily regenerable adsorbent with high selectivity and adsorption capacity [2]. Among conventional adsorbent that can be used for this purpose are activated carbons, zeolites, fly ash, metal-organic-framework (MOFs) and different metal oxides [13-19]. There are promising results from the use of zeolites and MOFs in separating CO_2 from gas stream. It has been generally found that at lower temperature, these materials have high adsorption capacity because of large internal pore volume, regularity in structure, molecular size of its pores and diverse framework in chemical composition [20]. However, with increasing temperature, their adsorption capacity decreases rapidly. Also, their adsorption capacities are severely affected by the presence of other counter adsorbates such as water. On the other hand, activated carbons do not require removal of moisture; they have large surface area, with easily modified pore structures and low cost. Their use as an adsorbent for CO_2 separation is costly [1].

The combustion of coal and fossil fuel results in a fly ash as a by-product [17]. This ash is generally dumped in landfill with little usage in agriculture[21] and recovery of metals [22]. A growing attention has been given to fly ash as an economical adsorbent for removal of metal ions from wastewater [23]. Indeed, this waste material could have a potential use as adsorbent after surface modification and treatment. This is because it has a large percentage of carbon and can be further treated to obtain higher surface area and pore volume for adsorption of CO_2 from gas stream.

In this study fly ash from power generation plant was chemically treated with 30% ammonium hydroxide and used for adsorption of carbon dioxide from CO_2/N_2 mixture. After treatment the activated ash was tested for its surface area, pore size distribution, morphology and attached functional groups.

The surface area and pore size were measured for both untreated and treated samples using N_2 adsorption at 77 K and showed an increase from 68 to 318 m^2/g and from 0.036 to 0.678 cm^3/g, respectively, with average pore diameter of 13.28. SEM morphology illustrated porous, spherical particles with an average particle size of 50 μm. FTIR analysis provided the existence of amine functional groups appeared at 3178, 2929 and 2770 cm^{-1}.

The treated fly ash was tested for adsorption of CO_2 using isothermal packed column and found that both chemical and physical adsorption of CO_2 takes place on the surface of the treated ash. The maximum adsorption capacity was 249 mg/g achieved when the initial concentration was 795 mg/L operated at 40 °C, 1 L/min, and 11% relative humidity. The isotherm data were fitted to Langmuir, Freundlich and BET model and found that BET best fit these data with R^2= 0.97 (Fig 1).

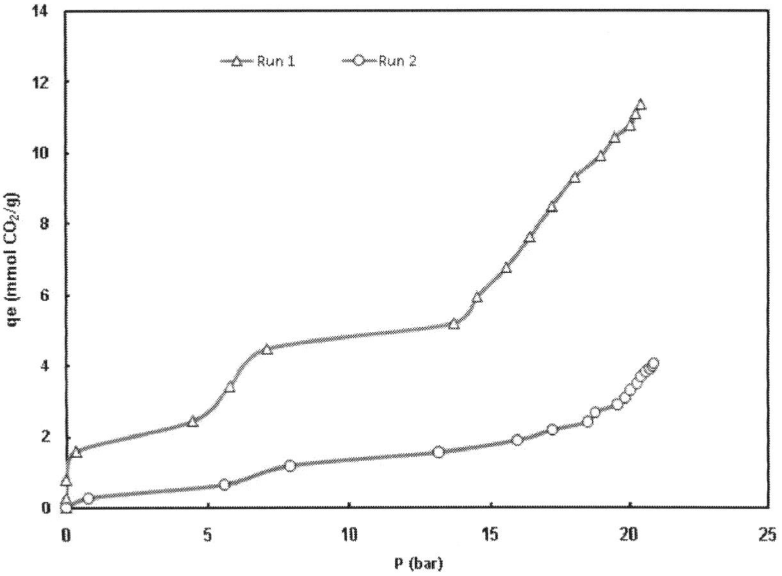

Fig 1: Adsorption isotherms of CO_2 by treated fly ash

Effect of kinetic parameters on the rate of chemisorption such as initial concentration of CO2 (350 -760 mg/L), flow rate (0.3-1 L/min) and humidity (10% -80% RH) of the gas, temperature (-20-40 °C) were studied. It is shown that the adsorption rate increases with the increase in flow rate, concentration and low humidity (Fig 2-5). Unlike physical adsorption, the chemisorption process resulted in increased adsorption with temperature in the range 0-40°C. Thermodynamic studies showed that the heat of adsorption was 80 and 178 kJ/mol at low and high relative humidity's suggesting that the process is endothermic.

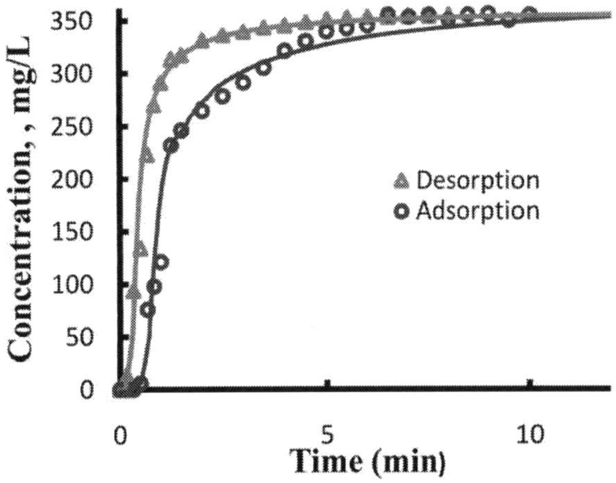

Fig 2: Adsorption-desorption curves for CO_2 by treated fly ash

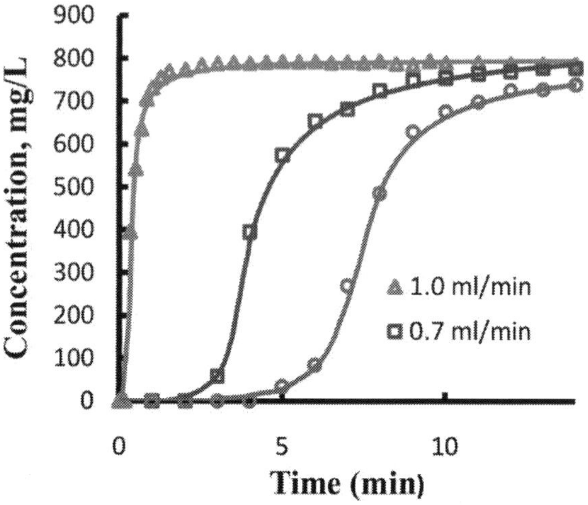

Fig 3: Effect of flow rate on adsorption of CO_2 by fly ash

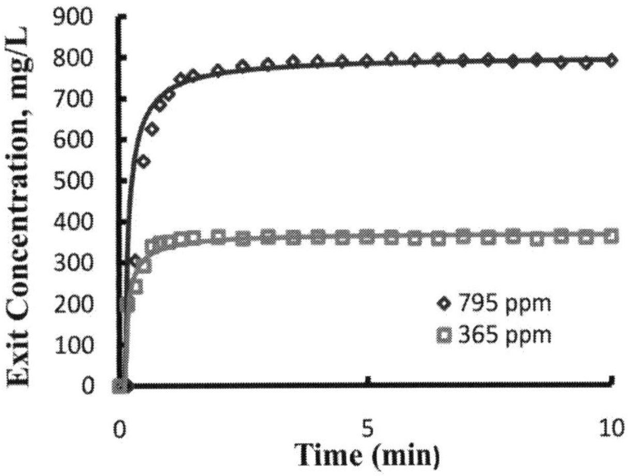

Fig 4: Effect of CO_2 concentration on adsorption

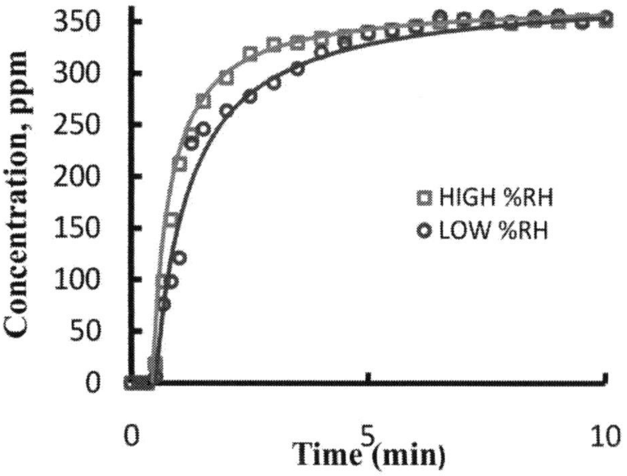

Fig 5: Effect of Humidity on adsorption of CO_2

Then a theoretical kinetic model was developed for the rate of chemisorption of carbon dioxide on chemically treated fly ash in packed bed column. It is developed based on assumption that surface reaction, adsorption and internal diffusion resistance are involved in the rate of chemisorptions according:

$$\frac{\partial C_A}{\partial t} = D_{eff}\left(\frac{\partial^2 C_A}{\partial r^2} + \frac{1}{r}\frac{\partial C_A}{\partial r}\right) - u\frac{\partial C_A}{\partial x} - kC_A^n$$

Where C_A is the concentration of carbon dioxide obtained at a distance r from the column entrance, D_{eff} is the axial diffusion and u is velocity of the gas of the gas inside the column, and k and n are the reaction constant and order, respectively.

The initial and boundary conditions for the above equation are

$$C_A = C_{Ao} \quad \text{at} \quad t = t, \quad r = R$$

$$C_A = 0 \quad \text{at} \quad t = 0, \quad r = r$$

$$\frac{\partial C_A}{\partial r} = 0 \quad \text{at} \quad t = t, \ r = 0$$

$$\frac{\partial C_A}{\partial x} = 0 \quad \text{at} \quad t = 0, \ x = L$$

The corresponding reaction constant and order, adsorption equilibrium constant and bed and particle diffusion coefficient were evaluated and found to be 2.4×10^{-4} (L.min/mol), 1.3, 5.0×10^{-3}(L/mol.g), 3.6×10^{-2}cm^2/g, respectively.

Effect of temperature and nature removal of CO_2 by the treated ash showed an endothermic behavior with heat of adsorption of 178.5 kJ/mol, which revealed a chemisorption process (Fig 6).

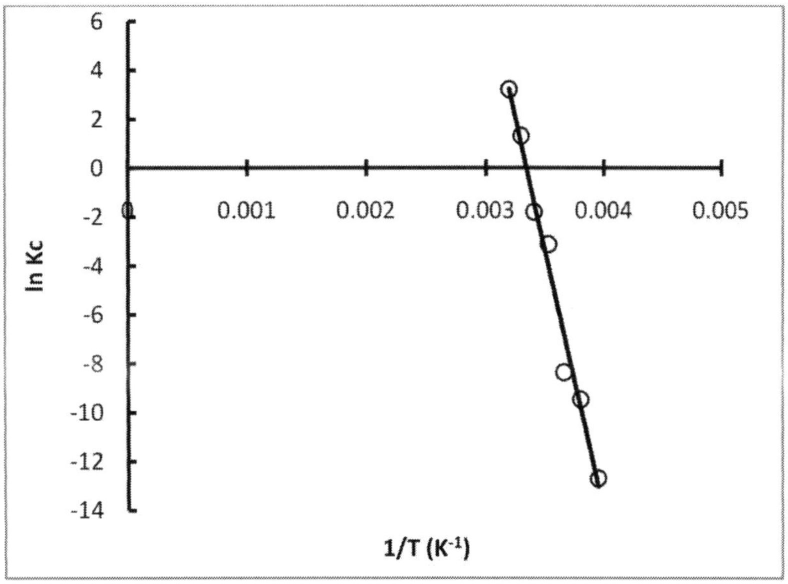

Fig. 6 Determination of heat of adsorption at flow rate of 0.3L/min, %RH = 80% and Concentration = of 330 mg/L

In conclusion, this study suggests that waste OFA which is mainly composed of alumino-silicate and unburned carbon can be regarded as a low cost adsorbent for adsorption of CO_2. The chemically treated waste OFA showed promising results for use in the adsorption of CO_2. Using ammonium hydroxide, chemical modification of fly ash was performed so as to improve the CO_2 adsorption capacity and removal efficiency.

1. Shafeeyan MS, Daud, Wan Mohd Ashri Wan,Houshmand, Amirhossein Shamiri, Ahmad. A review on surface modification of activated carbon for carbon dioxide adsorption. *Journal of Analytical and Applied Pyrolysis.* 2010;89(2):143-151.
2. Arenillas A, Smith, K. M., Drage, T. C. and Snape, C. E. CO2 capture using some fly ash-derived carbon materials. *Fuel.* 2005;84(17):2204-2210.
3. S. Kazama TT, K. Haraya. . Carbon dioxide and nitrogen transport properties of bis (phenyl) fluorine based card polymer membranes. *Journal of Membrane Science.* 2002;207(1):91–104.
4. Agency UEP. *Greenhouse Gases and Global Warming Potential Values*2002.
5. C.L. Leci. Financial implications on power generation costs resulting from the parasitic effect of CO2 capture using liquid scrubbing technology from power station flue gases. *Energy Conversion and Management.* 1996;37(6-8):915-921.
6. M.G. Plaza CP, B. Arias, M.D. Casal, C.F. Martín, J. Fermoso, F. Rubiera and J.J. Pis, . Different Approaches for the Development of Low-Cost CO2 Adsorbents. *J. Environ. Eng.* 2009;135:426–432.
7. Satyapal S FT, Trela J, Strange *J. Energy Fuels.* 2001;15:250.
8. Plaza MG, Pevida, C., Arenillas, A., Rubiera, F. and Pis, J. J. CO2 capture by adsorption with nitrogen enriched carbons. *Fuel.* 2007;86(14):2204-2212.
9. Shafeeyan MS, Daud, Wan Mohd Ashri Wan, Houshmand, and Amirhossein, Arami-Niya A. Ammonia modification of activated carbon to enhance carbon dioxide adsorption: Effect of pre-oxidation. *Applied Surface Science.* 2011;257(9):3936-3942.
10. Tan IAW, Ahmad, A. L. and Hameed, B. H. Enhancement of basic dye adsorption uptake from aqueous solutions using chemically modified oil palm shell activated carbon. *Colloids and Surfaces A: Physicochemical and Engineering Aspects.* 2008;318(1–3):88-96.
11. Tomeczek JP, H. . Kinetics of mineral matter transformation during coal combustion. *Fuel.* 2002;81(10):1251-1258.
12. Zou Yong VGMAAIER. Adsorption of Carbon Dioxide on Chemically Modified High Surface Area Carbon-Based Adsorbents at High Temperature. *Adsorption.* 2001;7:41-50.
13. Santos ET, AlfonsÃn C, Chambel AJS, et al. Investigation of a stable synthetic sol-gel CaO sorbent for CO 2 capture. *Fuel.* 2012;94:624-628.
14. Olivares-MarÃn M, Maroto-Valer MM. Preparation of a highly microporous carbon from a carpet material and its application as CO2 sorbent. *Fuel Processing Technology.* 2011;92(3):322-329.
15. Small CC, Hashisho Z, Ulrich AC. Preparation and characterization of activated carbon from oil sands coke. *Fuel.* 2012;92(1):69-76.
16. Rosas JM, Bedia J, RodrÃguez-Mirasol J, Cordero T. On the preparation and characterization of chars and activated carbons from orange skin. *Fuel Processing Technology.* 2010;91(10):1345-1354.
17. A.Boonpoke1 SC, N. Laosiripojana, S. Towprayoon and A. Chidthaisong. Synthesis of Activated Carbon and MCM-41 from Bagasse and Rice Husk and their Carbon Dioxide Adsorption Capacity. *Journal of Sustainable Energy & Environment* 2011;2:77-81.
18. Grande CAaR, Alírio E. Electric Swing Adsorption for CO2 removal from flue gases. *International Journal of Greenhouse Gas Control.* 2008;2(2):194-202.
19. Youssef Belmabkhout GDW, and Abdelhamid Sayari Amine-Bearing Mesoporous Silica for CO2 and H2S Removal from Natural Gas and Biogas. *Langmuir* 2009;25(23):13275–13278.
20. Weitkamp J. Zeolites and Catalysis. *Solid State Ionics.* 2000;131:175-188.
21. Tanigaki N, Manako, K., and Osada, M. Co-gasification of municipal solid waste and material recovery in a large-scale gasification and melting system. *Waste Management.* 2012;32(4):667-675.
22. Xu H, Shao, Q., Zhou, J., and Ma, X. Study on preparation of water treatment ceramic particles by using sludge and fly ash and the ceramic particles' application. *Advanced Materials Research.* 2012; 356-360,:1876-1881.

23. Widi Astuti AP, Endang Tri Wahyuni, and Made Bendiyasa. The Adsorption of Lead from Aqueous Solutions Using Coal Fly Ash: Effect of Crystallinity. *Engineering and Technology.* 2011;78.

(206a) Controlling the Composition of a Mixture of Polymorphs by Examining Seeded Smpt Kinetics

Ramya Kumar, Venu Nalivela and **Srividya Ramakrishnan**, Dr. Reddy's Laboratories Ltd., Hyderabad, India

Solvent mediated phase transformation (SMPT) occurs in a variety of organic and inorganic crystals. In this study SMPT kinetics were investigated in the context of generating polymorphic mixtures. It was required to crystallize an anti-epileptic drug as a binary mixture of forms with consistent composition for various reasons.

This API exhibited polymorphism with three forms, A, B, and C, which were found to exist as three pairs of enantiotropes. A mixture of Forms A and C was selected for development based on the thermodynamic order of stability which indicated that Form A was the most stable below 55°C, Form C at intermediate temperatures (between 56°C to 65°C) and Form B at high temperatures. Their enantiotropic relationship was exploited by heating a slurry of Form-A to a temperature slightly above the transition temperature, and seeding with Form C to trigger partial transformation to the marginally more stable Form-C.

Controlled implementation of SMPT to produce a mixture of the desired polymorphic composition required an investigation into the impact of various crystallization parameters such as transformation temperature, stirring rate (RPM), crystal size distribution of Form-A on the rate and extent of conversion to Form-C. This study was performed systematically through Design-of Experiments. Increasing the temperature was expected to provide a greater thermodynamic driving force and increase the rate of conversion of Form A to C. Though the DoE model confirmed that temperature was indeed the most significant variable, it was observed that kinetic parameters such as RPM and CSD of Form-A gained greater prominence close to the transition temperature. These were significant enough to even cause dissolution of Form-C seeds, thus preventing conversion of Form A to Form C. This interesting reversal of the thermodynamic script demonstrates that dissolution and growth kinetics in SMPT are determined by the dynamic equilibrium that exists between the solute in the liquid phase and the two solid phases, not just temperature and solubility. The ratio of the surface area between the metastable and stable crystals and the hydrodynamic environment in which this equilibrium operates were found to be as important as the thermodynamic driving force provided.

The second part of our work focuses on elucidation of the mechanism of nucleation and growth of the stable Form-C. Differences in conversion behavior were observed when the particle size of Form C seed used to trigger the conversion was changed. Using conversion-time profiles in combination with microscopic observations, two mechanisms of conversion were identified. With coarse Form-C seeds, cross-nucleation of Form-C out of the metastable Form-A crystal faces was observed which could be attributed to insufficient surface area for growth provided by the coarse seed. On the contrary, when micronized seed was used, the extent of surface cross-nucleation was considerably lessened as growth of the micronized seed was the primary route by which transformation to Form-C occurred. Thus the conversion mechanism followed was found to be dependent on the surface area of the seed.

Communication Number: IPDO IPM-00330

(206b) Amine-Functionalized Metal-Organic Framework Crystallization Process Design

Paul M. Schoenecker, Grace A. Belancik, Bogna Grabicka, William J. Koros and Krista S. Walton, Chemical and Biomolecular Engineering, Georgia Institute of Technology, Atlanta, GA

Metal-organic frameworks (MOFs) are porous crystalline structures consisting of metal centers connected via organic ligands. MOF crystals are typically formed via solvothermal methods, which consist of heating a mixture of metal salt, organic ligand, and solvent(s) in a sealed vessel. Most current MOF research is devoted to crystal structure identification and application specific experimentation. However, we examine the scalability of amine-functionalized UiO-66, a water-stable MOF with direct application potential, via batch crystallization vessels as well as a novel continuous-flow crystallization process. The flow-through crystallizer utilizes a single draft-tube in an attempt to preferentially produce larger MOF crystals and agglomerates. A recycle stream is also implemented to minimize solvent usage and provide seed crystals to the crystallizer feed. Characterization of the products is accomplished by determining the product yield as well as quality. In particular, crystal phase, size, and porosity are examined via pXRD, SEM, and BET modeling of N_2 adsorption at 77 K, respectively.

(206c) The Role of Meso-Mixing in Anti-Solvent Crystallization Processes

Des O'Grady, METTLER TOLEDO, Columbia, MD, Mark Barrett, School of Chemical and Bioprocess Engineering, University College Dublin, Dublin 4, Ireland, Brian Glennon, School of Chemical & Bioprocess Engineering, University College Dublin, Dublin 4, Ireland and Eoin Casey, School of Chemical and BioProcess Engineering, University College Dublin, Dublin, Ireland

This paper establishes links between the available turbulent mixing capability in a fed-batch reactor and important crystallization characteristics such as the Metastable Zone Width (MSZW) and product nucleation during the anti-solvent crystallization of benzoic acid. As demonstrated an increase in the locally dominant meso-mixing timescale, τ_D, facilitates the premature and subsequent sporadic nucleation of the solute. An alternative well mixed addition location, defined via Computational Fluid Dynamics (CFD) simulations and turbulent mixing theory, allows this dispersion limiting timescale to be reduced by over 50% across all anti-solvent addition rates assessed. Additionally, a more direct pathway from the feed point to the well mixed region of the impeller is also presented. These more favorable hydrodynamic conditions improve supersaturation dispersion and blending which in turn eliminate the premature and additional nucleation of the solute throughout the batch. Correlations between the population of fine particles and τ_D are also presented. These relationships demonstrate the sensitivity to the systems crystal size when hydrodynamic properties are improved.

(206d) Crystallization of Active Pharmaceutical Ingredients in Scrapped Surface Crystallizers

Jie Chen[1], Shin Yee Wong[1] and Allan S. Myerson[2], (1)Chemical Engineering, Massachusetts Institute of Technology, Cambridge, MA, (2)Department of Chemical Engineering, Massachusetts Institute of Technology, Cambridge, MA

Scraped-surface heat exchanger (SSHE) (Härröd, M. 1986) is used widely in the chemical and food industries, for example the separation of para-xylene from xylene mixtures, the separation of chlorinated benzenes, the production of ice cream and margarine and etc. The scrapping motion minimizes fouling and improves the heat-transfer rates by removing the solid layer on the wall and transferring solid to the bulk. Despite SSHE's proven efficiency, its application in pharmaceutical crystallization is very limited.

To explore the concept of SSHE, a laboratory scale scraped surface crystallizer (SSC) was designed and tested using several active pharmaceutical ingredients (API) of different crystal characteristics (needles, rods, fines, and aggregates). The crystallizer was designed using a MultiMax™ 50 mL reactor with a custom made PTFE anchor impeller. The close clearance (~ 1 mm) between the impeller and the reactor's interior wall prevented crystal encrustation on both the side and bottom wall. The temperatures of all crystallization experiments were controlled using the MultiMax™ station.

The performance of SSC was investigated for cooling and anti-solvent crystallizations. Crystallization kinetics, mixing efficiency, slurry loading tolerance of SSC were evaluated and compared to a conventional stirred tank crystallizer. The results show that SSC can operate at a significantly lower stirring speed for suspension of thick slurry. In addition the crystals produced from SSC are significantly larger, and higher yield was observed due to minimal encrustation on the crystallizer wall. This study showed that SSC is suitable for pharmaceutical crystallization applications, especially those processes with high solid loading of fines and needle-like crystals.

Härröd, M., "Scraped Surface Heat Exchangers", *J. Food Proc. Eng.*, **9**, 1, 1986.

(206e) Continuous Enantioseparation in a MSMPR Crystallizer – Experimental and Theoretical Investigation of Different Process Conditions

Kamila Galan, Physical and Chemical Foundations of Process Engineering, Max Planck Institute for Dynamics of Complex Technical Systems, Magdeburg, Germany, Martin Peter Elsner, Department of Applied Chemistry, Georg Simon Ohm University of Applied Sciences, Nuremberg, Germany and Andreas Seidel-Morgenstern, Physical and Chemical Foundations of Process Engineering, Max-Planck Institute for Dynamics of Complex Technical System, Magdeburg, Germany; Chemical Engineering, Otto von Guericke University, Magdeburg, Germany

Growing demands for products with constant high quality make continuous crystallization very attractive. Carrying out a process at a steady state can guarantee the continuity of product properties. This might be especially important, when one considers drugs produced in an enantiopure form.

Within this contribution the idea of a Mixed Suspension Mixed Product Removal (MSMPR) crystallizer for preferential crystallization is analyzed both experimentally and theoretically [1]. Preferential crystallization is applied to resolve one enantiomer from a racemic mixture of both. The separation process takes place in the metastable zone, where spontaneous nucleation is suppressed kinetically. Preferential crystallization is initiated by introducing seeds of only one enantiomer to the solution. The surface of these seeds drives resolution due to their growth. Executing preferential crystallization in a continuous mode requires a continuous supply of seeds together with racemic feed solution and a continuous withdrawal of the product suspension containing the seeded target enantiomer.

To gain insight into the process a mathematical model was developed assuming an ideally mixed vessel, uniform conditions inside the tank, identical crystals size distribution of crystals in the vessel and in the product stream. Population balance and mass balance equations as well as kinetics for crystal growth, primary and secondary nucleation were used to simulate the behavior of the solid and liquid phases. A simulation study using parameters of the enantiomer of threonine [2] was done to investigate the effects of different conditions on the process performance, including the saturation and crystallization temperature as well as mass flow rates of seeds and the feed solution. An optimum set of process conditions was chosen considering also the aspect of reducing the duration of the dynamic start-up phase.

The experimental part of the contribution will focus on the following aspects: first, observed resolution results of L- and D-threonine will be compared with model predictions. Effects of selected conditions will be discussed to identify the influence of key process parameters on the length of the dynamic start-up phase and the product quality (purity and mean characteristic size of the product). A comparison of the continuous process with the performance of a corresponding batch process will conclude this contribution.

[1] Jones A.G., *Crystallization Process Systems*, Butterworth-Heinemann, Oxford, 2002

[2] Qamar S., Elsner M.P., Hussain I., Seidel-Morgenstern A. *Seeding strategies and residence time characteristics of continuous preferential crystallization* Chemical Engineering Science, 71, 5-17, 2012

(206f) Continuous Crystallization Process Development

Haitao Zhang[1], Justin Quon[2], James M.B. Evans[3], Allan S. Myerson[4] and Bernhardt L. Trout[2], (1)Chemcial Engineering department, Massachusetts Institute of Technology, Cambridge, MA, (2)Massachusetts Institute of Technology, Cambridge, MA, (3)Department of Chemical Engineering, Massachusetts Institute of Technology, Cambridge, MA, (4)Chemical Engineering, Massachusetts Institute of Technology, Cambridge, MA

Batch-based manufacturing system currently employed by the pharmaceutical industry is costly and inefficient. A drug's active ingredients are synthesized in a chemical manufacturing plant and then shipped to a separate facility where they are converted into large batches of pills, liquids, or creams. With multiple interruptions, including transport to separate locations, the production of just one batch of drugs can take weeks. What's more, the manufacturing design and scale-up required to produce a new drug can be financially unsustainable and exceedingly time-consuming. Continuous Processes/Manufacturing offers a paradigm shift that will enable the industry to produce quality products more efficiently. Continuous crystallization has long been a standard method for the crystallization of bulk commodity chemicals and large-volume specialty chemicals, but it has only recently become a focus in the pharmaceutical and fine chemical industries. This presentation will:

• Highlight the current approaches to continuous crystallization

• Assess the advantages of a continuous crystallization process

• Analyze the key challenges and draw on real life case studies to present the key points

(207a) Gamma Scanning of a Kettle Vapor Line

Tony Cai[1], Simon Chambers[1], Lowell Pless[2] and Ron Carlson[2], (1)Fractionation Research, Inc., Stillwater, OK, (2)Tracerco, Pasadena, TX

As of May of 2011, FRI's Stillwater, Oklahoma, low pressure kettle reboiler contains two 6-inch windows. This affords an excellent view of the boiling action inside this industrial-size exchanger. This also allows heat exchange researchers to see entrained liquid entering the overhead vapor product nozzle, when rates become excessive. FRI's low pressure distillation column includes windows exactly opposite the nozzle that brings vapor from the kettle reboiler. Again, researchers can see the entrainment that issues from the reboiler. Tracerco engineers joined FRI engineers on-site at FRI's Stillwater facility to see if gamma scanning of the reboiler vapor return piping would allow quantification of the visually-observed entrainment rates. Radiation sources and detectors were positioned in two different places on the vapor piping. Detector data were used to calculate the amount of entrained liquid in the piping. The calculated values matched the visual observations very well. This presentation describes the qualitative and quantitative results. Video footage will be shown.

(207b) Tower Scanning As a Quantitative Tool

Lowell Pless, Tracerco, Pasadena, TX and Izak Nieuwoudt, Koch-Glitsch, Wichita, KS

Tower Scanning as a Quantitative Tool

Lowell Pless - Tracerco & Izak Nieuwoudt - Koch-Glitsch

Tower scanning is well established in the process industries as a qualitative tool to help troubleshoot separation towers. In this presentation it will be shown that through careful and innovative application of gamma scanning and neutron backscatter techniques, quantitative information about tray performance can be extracted. This will be illustrated by comparing data taken under very controlled conditions on a hydrocarbon tower equipped with windows, with the data obtained from scans of the tray downcomer and active area. Three sets of trays designed to flood under jet flood, downcomer backup or choke flood mechanisms were tested and studied. The experimental observations, tray model predictions, and scan data for each of these sets of trays will be compared. Based on this information it can be concluded that tower scanning can be used to generate quantitative information about tray performance.

(207c) Determination of Liquid Holdup in Structured Packing Using Neutron Radiography

Michael Basden[1], Esther Feng[1], John P. Farone[2], Daniel S. Hussey[3], David L. Jacobson[3] and R. Bruce Eldridge[1], (1)Chemical Engineering, The University of Texas at Austin, Austin, TX, (2)Dow Chemical (retired), Cedar Park, TX, (3)National Institute of Standards and Technology, Gaithersburg, MD

Similar to x-ray radiography, neutron radiography allows for non-intrusive and non-invasive observation of the inner-workings of an operating mass transfer device, providing insight into the phase distribution of the system. Neutron scans taken at the NIST Center for Neutron Research (NCNR) of an air-water contactor were used to determine local and average liquid holdup profiles in three stainless steel structured packings (Mellapak 250Y, Mellapak 500Y, and MellpakPlus 752Y). The tests were conducted at liquid loads ranging from 2.2-20GPM/ft^2 and f-factors ranging from 0.5-1.3 ft/s(lb/cuft)$^{0.5}$ in a 5.75" ID column. The results for Mellapak 250Y are validated against previous studies.

Results for local holdup profiles showed a strong dependence on vertical position in the column, generating a periodic profile in all packings at all conditions examined. The largest values for holdup were seen in the joint region between two adjacent packing elements. Holdup profiles for MellapakPlus 752Y exhibited lower holdup at the joint when compared to Mellapak 500Y.

(207d) Tray Capacity Limitations At Low Surface Tension

Daniel R. Summers, Sulzer Chemtech USA, Inc., Tulsa, OK

Tray vapor capacity correlations do not include a surface tension physical property parameter. Low liquid surface tensions on a distillation tray can lead to tiny droplet sizes, high liquid entrainment and limited vapor capacity. Several authors have examined surface tension in the past but today's leading vapor capacity correlations have omitted this important physical property parameter. This paper will examine very low surface tension data as well as a wide range of surface tension capacity data and then present a new tray vapor capacity correlation that uses surface tension as one of the primary correlating parameters. The intent of this paper is to show that tray designs of very high pressure trayed distillation towers, such as demethanizers and deethanizers, can be made without the need for "system factors" with this new correlation. It is the author's opinion that the system factor should be used strictly for those tray applications that are limited in capacity due to foaming and not be used as a "fudge factor" to account for capacity correlation deficiencies.

Study of gas-liquid two phase flow in multiphase contactor using low-intrusive measuring methods

Guanghua Zheng[1], Linda schlusemann[1],Markus Schubert[2], Uwe Hampel[2], Marcus Grünewald[1]

[1] *Ruhr-Universität Bochum, Institute of Thermo- und Fluid Dynamic, Bochum, Germany.*

[2] *Helmholtz-Zentrum Dresden-Rossendorf, Institute of Safety Research, Dresden, Germany.*

Abstract

Multiphase contactors are the most used apparatuses for reaction and separation in chemical engineering. Due to limitations imposed by the measuring methods, the analysis is made difficult as the hydrodynamic behavior under real process conditions is not completely understood. Several new measuring techniques have been developed for the investigation of multiphase flows in vessel cross sections in the last decades. Especially visualization techniques like tomography have been in the focus since these are noninvasive and thus non-intrusive methods, and enable the visualization of phase distributions.

However, currently developed nonintrusive methods have considerable drawbacks. The computer tomography methods can obtain high spatial resolution. In comparison, the temporal resolution is relative low. For the application of topographic measurement techniques in multiphase flow study, especially with fast changing flow patterns, a high temporal resolution is essential. Electrical tomography has a high temporal resolution. However, reconstruction algorithm is complex and the electrical field lines are not linear, therefore spatial resolution rate is relative low. Therefore, no exact mass balance could be established and the resulting phase fractions cannot be applied for model developments.

In this research project, Wire-Mesh Sensor (WMS) has relatively high temporal and spatial resolution is used for the study of phase distribution in multiphase contactors. The WMS comprises two planes, in each plane there are sensing wires. Wires from different planes are orthogonally arranged. Since water and air have different electrical permittivity values and thus, they produce different capacitance in the sensing points of the WMS. Images of phase distributions can be generated from the measurements of the local capacitance. Measurements of phase distribution with WMS can supply more accurate information of spatial phase distribution to update the theoretical models used in multiphase flow.

Keywords: Wire-Mesh Sensor, Packed Column, Bubble Column, Phase Distribution

Introduction

Multiphase contactors are widely used in chemical reaction and separation engineering. Due to the complex mathematical description and intrusive experimental measurement, the behavior of multiphase flow is not completely understood. Consequently design and scale-up are difficult. According to this lack of knowledge, further investigations to find a measurement technique, which can be used to identify multiphase flow, are in the focus of our research. Two types of apparatuses are studied: Packed columns as an example for liquid in gas dispersed flow and bubble column reactors as a representative for gas in liquid dispersed flow.

Packed columns employing gas-liquid counter-current flow are widely used across a wide range of industries, e.g., distillation, rectification, absorption and extraction processes [6,11,12]. In order to predict the process conditions accurately, knowledge of packing characteristics is a fundamental requirement for proper design of packed columns considering fluid dynamics and mass transfer. Gas pressure drop, liquid holdup, loading and flooding points are generally considered as important parameters in hydrodynamic studies. Lower pressure drop is required for many industrial applications, especially in vacuum distillation. Liquid holdup influences not only the pressure drop but also the mass transfer rates strongly [2]. Therefore, the phase distribution were widely studied [1,7,15,16,17].

Bubble column reactors are widely used for gas-liquid reactions continuously or in semi-batch [3, 4]. Since they are very successfully applied for technical important reactions, they represent an important type of reactor for the chemical industry. One of the most essential parameters for characterizing their behavior is the gas hold-up. Gas hold-up is defined as the fraction of gas in relation to the whole dispersed volume and determines gas residence time as well as the interface and consequently the mass transfer rate. Various measuring techniques [5,8,9,10] have been applied for the hydrodynamic investigation of bubble columns.

Experimental measurements

In this research, Wire-Mesh Sensor (WMS) is used for the study of phase distribution in multiphase contactors. The implemented wire-mesh sensor is a flow-imaging device and allows the investigation of multiphase flows with high spatial and temporal resolution. The wire mesh sensor method can be categorized as invasive, but low-intrusive method compared with other tomography methods. WMS has been applied in hydrodynamic study of trickle bed reactor [14] and bubble columns [13], however, the size of WMS was limited up to 100 mm.

Figure 1. Wire-Mesh Sensor

WMS illustrated in Figure 1 consists of 32×32 orthogonally arranged wires (stainless steel), which are arranged in two planes with a distance of 3 mm apart. The diameter of wires is 0.5 mm. This arrangement gives 1024 sensing points with 840 inside the circular cross-section of the column. Diameter of the column is 288 mm. Measurements frequency is set to 400 Hz (400 frames per second). Due to the fact that the gas and liquid phases have different permittivity, local permittivity can be measured with WMS, which reflects the local phase distribution. The WMS used in this work is based on the capacitance between the wires. Thus, liquid phase is deionized water (relative permittivity 5) and relative permittivity of gas phase is 1.

Wires from different planes are orthogonally arranged. Since water and air have different electrical permittivity values and thus, produce different capacitances in the sensing points of the WMS, images of phase distributions can be generated from the measurements of the local capacitance (Figure 2).

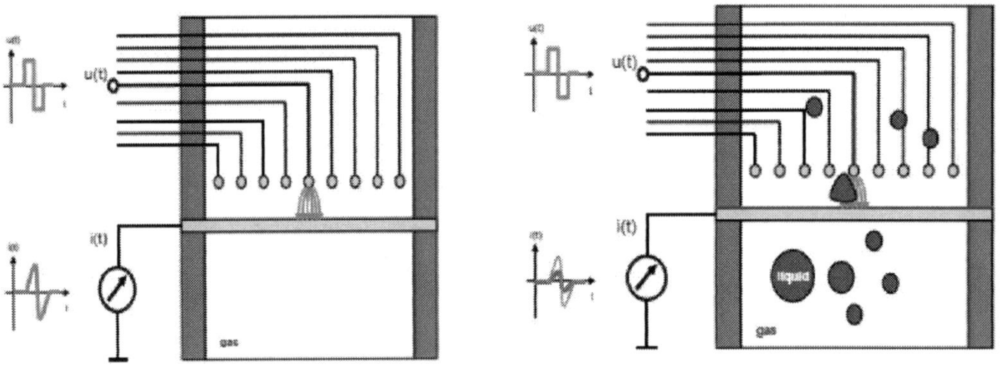

Figure 2. Measurement principle of Wire-Mesh Sensor

Measurements in packed column and bubble column were carried out in the column with the diameter of 288 mm. Gas phase is delivered using a compressor. Gas phase flows into the column upwards while liquid phase is in countercurrent directions. The process flow sheet is given in Figure 3.

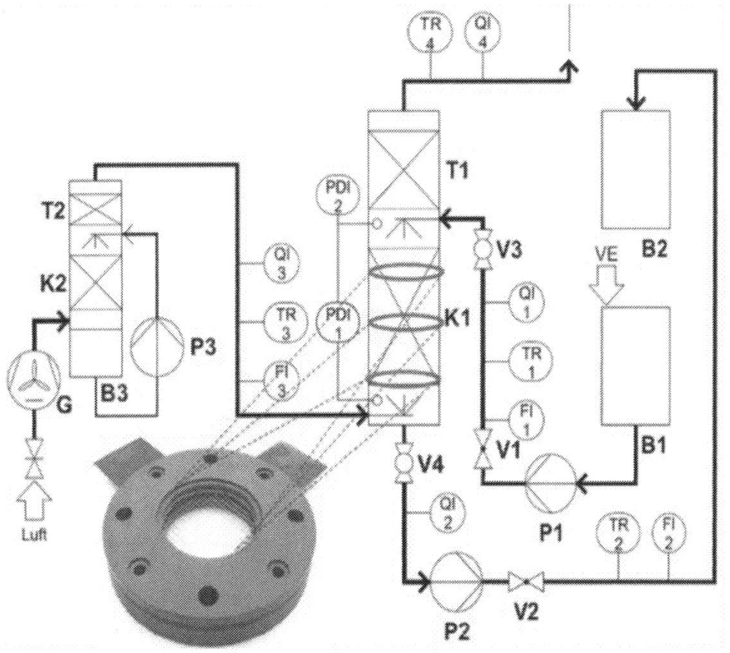

Figure 3. Process flow sheet for the experiments with WMS in packed column.

P1, P2 and P3: Pump	*B1: Liquid vessel (inlet)*	*B2: Liquid vessel (outlet)*
G: Radial ventilator	*T1,T2: Demister*	*V1,V2,V3 and V4: Valve*

Results and Discussion

Gas-liquid phase distribution of absorption process in packed column filled with structured packings was measured with WMS (shown in Figure 4).

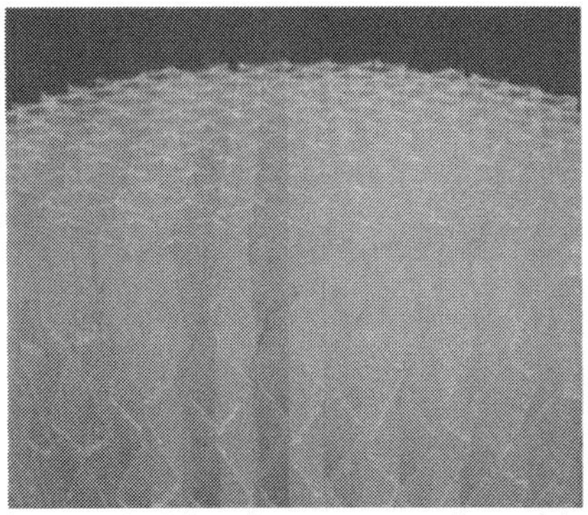

Figure 4. Plastic structured packing

Liquid load factor is 40 $m^3/m^2/h$, gas load factor is in this case 0.9 $Pa^{0.5}$ and packing height is 0.7 m (two packing elements). It is shown in Figure 5 that the liquid phase distributes properly in the cross section.

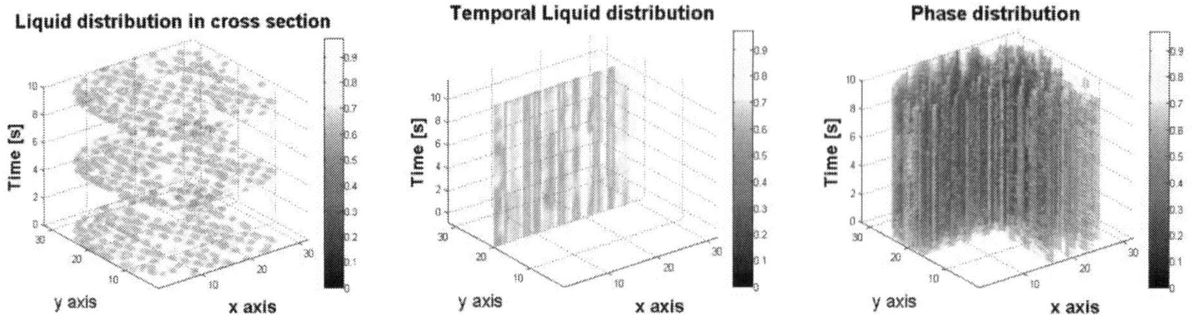

Figure 5. Visualization of phase distribution in structured packed columns. (Colour bar denotes the liquid phase fraction)

Gas load factor is increased to 3.9 $Pa^{0.5}$, the phase distribution measured with WMS is shown in Figure 6. Compared with the phase distribution at lower gas load factor (Figure 5), it is shown that the higher is the gas load factor, the stronger is the mixture of gas and liquid phase.

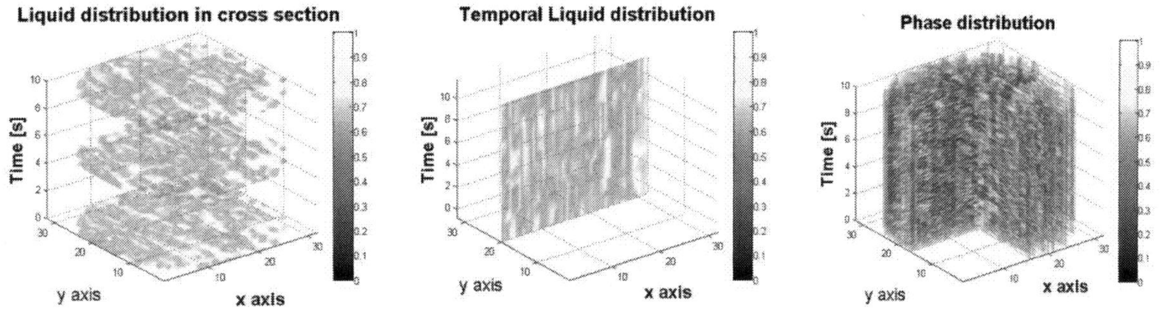

Figure 6. Visualization of phase distribution in random packed columns.

Furthermore, measurements of phase distribution in random packed column (25mm metal Pallring) have shown that serious maldistribution occurs near the column wall. Large amount of liquid flows along the column wall instead of flowing inside the column. Local liquid fraction in packed column with random and structured packing is illustrated in Figure 8. The results show that in random packed column the wall effect is serious while in structured packed column the maldistribution is mainly at the center region.

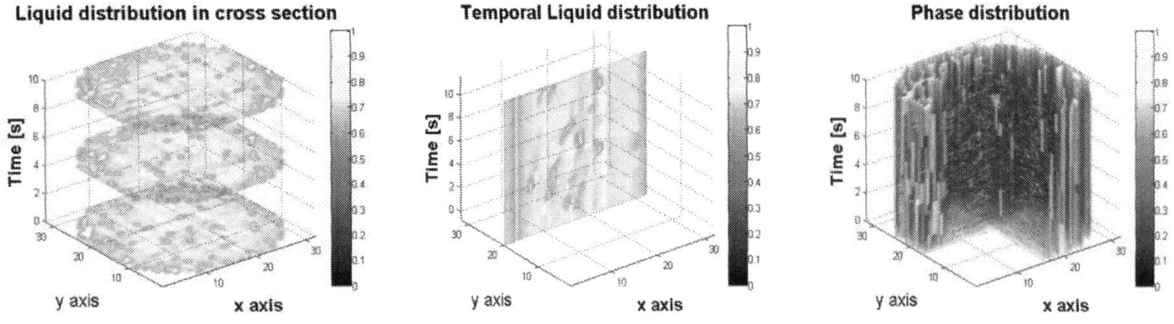

Figure 7. Visualization of phase distribution in random packed columns

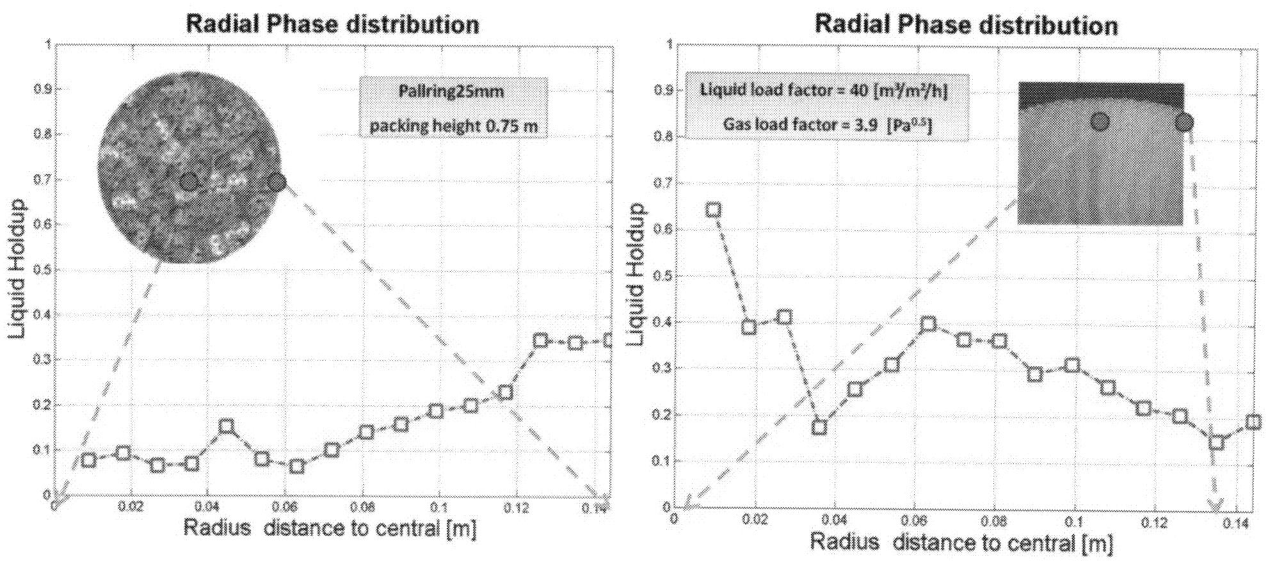

Figure 8. Local liquid phase fraction in radial direction

Measured liquid fraction in cross section with Wire-Mesh Sensor agrees well with the liquid Holdup measured with traditional method (valve-closing method). These results denote that the measurements of phase distribution with Wire-Mesh Sensor can be applied in hydrodynamic investigation of packed columns.

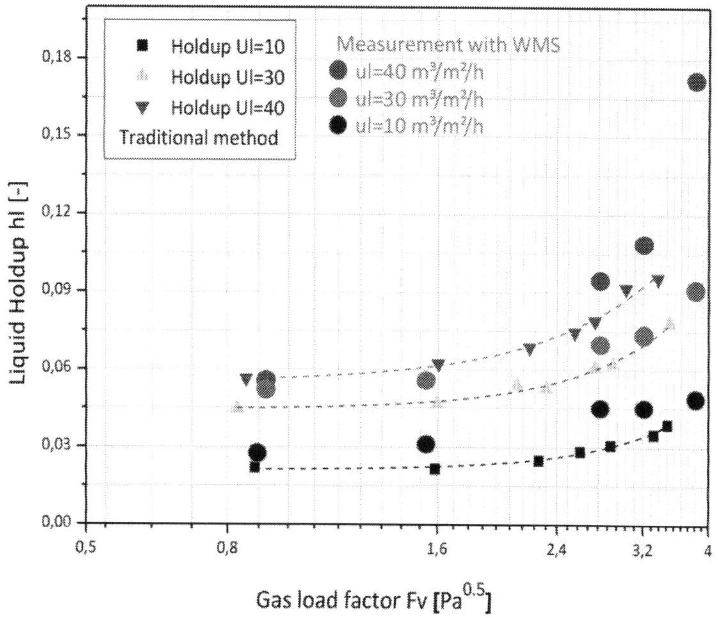

Figure 9. Comparison of liquid Holdup measured with WMS and traditional method

The second example of multiphase flow apparatuses - bubble columns - is studied. As in the first case the phase distribution in bubble column, was studied with a WMS implemented in the same column used for packed column experiments. It was shown that the results of WMS implemented in bubble column could be used to identify the flow regime. Comparing figures a, b and c in Figure 10 the difference in homogeneity and intensity of the results is evident. Whereas the homogeneous flow regime can be identified by small bubbles distributed uniformly in the column, the heterogeneous regime is characterized by large bubbles rise upwards in the middle of the column.

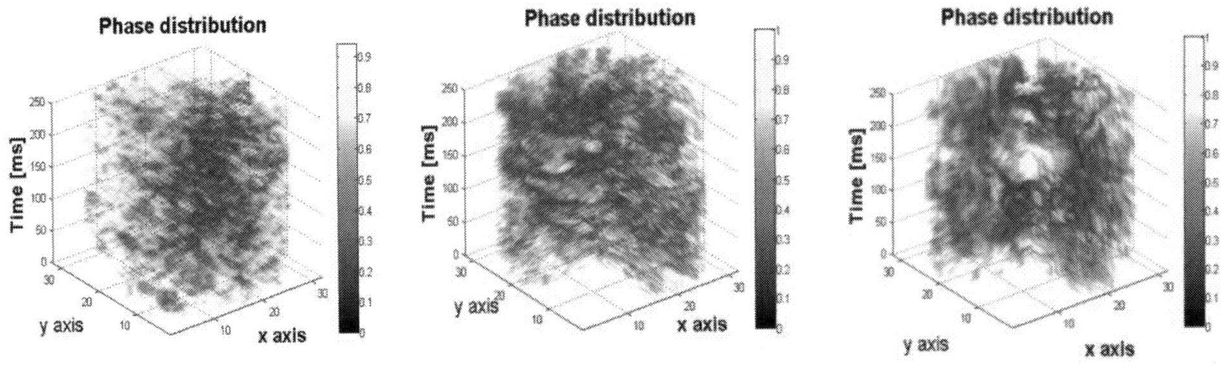

(a) (b) (c)

Figure 10. Visualization of flow regime at various gas flow velocities: (a) u_G=2 cm/s; (b) u_G=8.4 cm/s; (c) u_G=16.8 cm/s (Colour bar denotes the bubble phase fraction)

In addition orientating studies in our experimental facilities reveal that the measurement signals can also be used to determine bubble sizes and the bubble size distributions, respectively. Furthermore by integrating the results along concentric circles the radial gas holdup profile could be determined. Our experimental results show an excellent qualitative fit with results reported in literature and therefore could be used to set up a more detailed compartment based modeling approach. Nevertheless it is only right to stress that the summation of such bubble counting and bubble size detection methods slightly underestimate the overall gas holdup derived from level measurements. To improve the accuracy of such approaches the spatial resolution should be increased for further studies.

Concluding our experience with WMS implemented in bubble columns the measurement technique seems to be a reasonable method for the investigation of multiphase flow in bubble columns. Beside the reported applications WMS could also be used to discover back mixing effects in the liquid phase and the bubble velocities. We will report about such measurements within the near future.

To sum up, wire-mesh sensors are successfully applied in multiphase flows. The WMS have relative high spatial and temporal resolution and the intrusiveness is relatively low. Flow regime and bubble clusters can be spatially and temporally identified with WMS in bubble columns. Phase distribution in packed columns has been demonstrated to be applicable to randomly arranged packings as well as for structured packings. The successful application of the WMS in this study leads to the conclusion that the sensors can be applied in columns with technical relevant component systems under industrial conditions such as organic solutions under high pressure and temperature. Indeed, these are apparatuses which cannot be observed visually and the WMS is an instrument to explore phenomena inside.

References

1. Bemer, G. G. and F. J. Zuiderweg (1978), "Radial Liquid Spread and Maldistribution in Packed Columns under Different Wetting Conditions", *Chemical Engineering Science*, 33, pp. 1637-1643.
2. Billet, R. and M. Schultes (1993), "Predicting mass transfer in packed columns", *Chemical Engineering Technology*, 16(1), pp. 1-9.
3. Deckwer, W.-D. and R.W. Field (1992), "Bubble column reactors", *Wiley*.
4. Deckwer, W.D. and A. Schumpe (1993), "Improved tools for bubble column reactor design and scale-up", *Chemical Engineering Science*, 48(5), pp. 889-911.
5. Delnoij, E., et al. (2000), "Measurement of gas–liquid two-phase flow in bubble columns using ensemble correlation PIV", *Chemical Engineering Science*, 55(17), pp. 3385-3395.
6. Fair, J.R., et. Al. (2000), "Structured Packing Performance Experimental Evaluation of Two Predictive Models", *Industrial & Engineering Chemistry Research*, 39(6), pp. 1788-1796.
7. Hoek, P. J., J. A. Wesselingh, and F. J. Zuiderweg (1986), "Small scale and large scale liquid maldistribution in packed columns", *Chemical Engineering Research and Design*, 64, pp. 431.
8. Hubers, J. L., et. al. (2005), "X-ray computed tomography in large bubble columns", *Chemical Engineering Science*, 60(22), pp. 6124-6133.
9. Kemoun, A., et al. (2001), "Gas holdup in bubble columns at elevated pressure via computed tomography", *International Journal of Multiphase Flow*, 27(5), pp. 929-946.
10. Olmos, E., et al.(2003), "Description of flow regime transitions in bubble columns via laser Doppler anemometry signals processing", *Chemical Engineering Science*, 58(9), pp. 1731-1742.
11. Olujic, Z. and J. de Graauw (1990), "Experimental Studies on the Interaction Between the Initial Liquid Distribution and the Performance of Structured Packings", *Separation Science and Technology*, 25(13), pp. 1723-1735.
12. Onda, K., H. Takeuchi, and Y. Okumoto (1968), "Mass transfer coefficients between gas and liquid phases in packed columns", *Journal of Chemical Engineering Japan*, 1, pp. 56-62.
13. Prasser, H.-M., E. Krepper, and D. Lucas (2002), "Evolution of the two-phase flow in a vertical tube--decomposition of gas fraction profiles according to bubble size classes using wire-mesh sensors", *International Journal of Thermal Sciences*, 41(1), pp. 17-28.
14. Schubert, M., et al. (2010), "Spatially resolved inline measurement of liquid velocity in trickle bed reactors", *Chemical Engineering Journal*, 158(3), pp. 623-632.
15. Stikkelman, R., et al. (1989), "A study of gas and liquid distributions in structured packings", *Chemical Engineering Technology*, 12(1), pp. 445-449.
16. Stoter, C.F., Z. Olujic, and J. Graauw (1992), "Modeling of hydraulic and separation performance of large diameter columns containing structured packings", *IChemE Symposium Series No.* 128, pp. A201-210.
17. Sun, C.G., et al. (2000), "Modeling and Simulation of Flow Maldistribution in Random Packed Columns with Gas-Liquid Countercurrent Flow", *Chemical Engineering Research and Design*, 78(3), pp. 378-388.

(207f) Bibliography of Mr. Lowell Pless

Mike Resetarits, Simon Chambers and Tony Cai, Fractionation Research, Inc., Stillwater, OK

After earning a B.S. in chemical engineering from the University of Texas at Austin, Mr. Lowell Pless joined Celanese Chemical Company in Bay City, Texas, where he performed process and production engineering. He was the supervisor of the Oxo-Alcohols Unit for two years. Thereafter, he joined Calumet Lubricants in Princeton, Louisiana. In 1987, he joined, Tru-Tec Inc. and then Tracerco. With Tru-Tec and Tracerco, Mr. Pless became a well-known, and appreciated, figure in the world of separation columns. His expertise regarding gamma scanning and other radioisotope diagnostic tools is sought regularly, by engineers from across the globe. He has written many important papers, for the AIChE, for magazines and for the Tracerco newsletter. Some of those papers were coauthored by customer engineers who had sought and received Lowell's assistance. This Pittsburgh presentation briefly describes Mr. Lowell Pless' most important articles and papers. Copies of his full bibliography will be handed out to all attendees.

(220a) Multi-Objective Optimization Applied to Energy Efficient-Buildings

Ekaterina Antipova, Department of Chemical Engineering, University Rovira i Virgili, Tarragona, Spain

Multi-objective optimization applied to energy efficient-buildings.

ABSTRACT

E. Antipova[a] , D. Boer[b], L. Cabeza[c], G. Guillén-Gosálbez[a], L. Jiménez[a]

a. Department of Chemical Engineering, University Rovira i Virgili

b. Department of Mechanical Engineering, University Rovira i Virgili

c. Edifici CREA, Universitat de Lleida, Pere de Cabrera s/n, 25001-Lleida, Spain

The growth of population implies the growth in construction speed and housing cost. Affordable housing is one of the global problems in the world. Though the term "affordable" implies some compliance with the environmental regulations, the main target of the construction is to make the housing cheaper. However, cost reduction is not the unique criterion that should be taken into account when projecting a new building. Buildings are considered to be responsible for up to 50% of the total CO_2 emissions worldwide (if we consider both direct and indirect impacts). Hence, this is a sector with large potential for environmental improvements that demands adequate measures for mitigating global warming. The problem of energy efficient buildings is equally relevant for cold and hot climates. In the former, it is necessary to insulate the building properly to keep it warm in winter months, while in the latter, we need to cool it down in summer and to maintain a comfortable temperature in winters as the insulation of the buildings in these regions is usually poor because of the large-scale tendency to build low-cost housing.

When we talk about direct contribution of the construction to the environmental impact, we refer to the use of energy during operation of buildings (heating, cooling, ventilation, lightening and etc.), while indirect contribution accounts for the construction and maintenance of the buildings (embodied energy). It is important to make preliminary estimations to balance both constituents: operation and maintenance, as sometimes reduction in the operation impact causes the growth of the impact related to construction. This might happen, for instance, when we decrease the use of fossil fuels to heat the spaces by increasing the insulation of the building, thereby offsetting the savings in operation by the additional impact due to the production of isolating materials.

In this work we propose a systematic method for the retrofit of buildings considering economic and environmental aspects. The design is posed mathematically as a multiobjective mixed-integer nonlinear programming model (MINLP) that considers the simultaneous minimization of cost and environmental impact given a certain house to be retrofitted. Our model accounts for a set of retrofit measures to make a building more energy efficient, such as incorporation of alternative types of windows, variation of insulation of walls and roof, and employment of different types of solar collectors for hot water generation, among some others. The environmental performance is quantified following life cycle assessment (LCA) principles. A case study based on existing facilities of cubicles located in Lleida (Spain) is used to illustrate the capabilities of the approach. By applying the epsilon constraint method, a set of Pareto solutions representing the optimal trade-off between both conflicting objectives is generated. The systematic tool presented herein is intended to guide decision-makers towards the achievement of economic and environmental viability of the construction processes.

(220b) Thin Flat Sheet Molecular Sieve Membranes for Air Drying and Energy Recovery of Building Ventilation

Wei Liu, Pacific Northwest National Laboratory, Richland, WA

Dehumidification and/or humidity control of air, gas in general is commonly needed in industrial processes and in buildings. A great amount of energy is consumed world-wide with conventional approaches. In this presentation, a new, versatile membrane technology is introduced to reduce energy consumption and capital cost for these applications. The membrane is prepared by deposition of a water-selective molecular sieve (zeolite) membrane on a thin (50um), robust porous metal sheet support. Such a membrane provides a unique combination of performance attributes that could not be obtained with conventional membranes, such as mechanical strength, thermal and chemical stability, high thermal conductivity, and high water permeance. The membrane can be used to separate water vapor from a hot gas stream by pervaporation and result in more than 50% energy saving for air conditioning of buildings in hot humid climate. The membrane can also be used to conduct humidity and/or heat exchange for energy recovery of building ventilation. The exceptionally high water permeance (>1.0E-5 mol/m2/s/Pa) of this membrane dramatically reduces usage of membrane surface area and enables development of a compact, cost-competitive membrane device suitable for these applications involving large volume of gas flow.

(220c) Application of Multi-Element Structure Array On Activated Carbon Filter in HVAC Systems

Guomin Xu, Chemical Engineering, Auburn University, Auburn, AL and Bruce Tatarchuk, Department of Chemical Engineering, Auburn University, Auburn, AL

Activated carbon is widely used in industries to remove gas phase contaminants such as ozone and volatile organic compounds (VOC). Coarse particles are mostly used in commercial carbon filters mainly due to their low flow resistance despite the fact that they have a low mass transfer rate which leads low bed utilization and short breakthrough times. In contrast, fine particles have higher mass transfer rate but are limited by high pressure drop. This study aims to improve the carbon filter to obtain a high carbon loading while maintaining a low pressure drop. To reach this goal, a multi-element structure array (MESA), which combines multiple filters into single housing, was developed. A W-shape MESA bank can obtain a tenfold resistance reduction while introducing fourfold carbon load. A predictive pressure drop model for single honeycomb carbon filters, as well as MESA bank, was derived based on fundamental fluid dynamics equations and experimental data. The model can be used as a design tool to optimize carbon filter's parameter including filter depth, particle size, and filter count. Microfibrous entrapped sorbent (MFES) was prepared by entrapping fine particles (180-300μm) into 3D sinter-locked network of polymer or metal microfibers via a wet-lay process. These materials excel in mass transport and pressure drop due to their uniform structure with small particles and adjustable porosity. A polishing layer (a thin layer of MFES media placed in series with the packed bed) was used to improve the performance of bed by "polishing" the trace volume of gas after packed bed. Hexane adsorption breakthrough test was used to evaluate the performance of the composite bed based on the breakthrough time and percentage of adsorbent utilization. Experimental breakthrough curves showed that composite bed outperformed packed beds in single pass remove efficiency, breakthrough time and bed utilization due to a synergic effect between the high capacity of packed bed and the high contacting efficiency of MFES layer.

ENERGETIC AND EXERGETIC PERFORMANCES OF A SOLAR STORAGE COLLECTOR BASED ON CERAMICS MATRIXES

Leonardo Alonso[1], Mario R Bernal[2], Gabriel Camargo[2] and Fabio Sierra[3], (1)Agroindustrial Engineering, Corporación Universitaria del Meta, Villavicencio , Colombia, (2)Mechanical Engineering, Universidad Libre, Bogota, Colombia, (3)Mechanical Engineering, Universidad Nacional, Bogotá, Colombia

INTRODUCTION

In these times, the interest for materials and processes that integrate energy and economy are driven different studies in the solar collector fields. The commercial solar water heating systems (SDHW) have high investment cost which requires high initial investment and auxiliary energy used to compensate the solar energy during non sunny days. In Tunisia, Hazami et al, (1) designed and built a low cost new solar collector used for domestic water heating called integrated solar storage collector (ISSC) and characterized by an absorber matrix made up of a thin cement concrete slab which performs the function of both absorbing and storing of the solar thermal energy. There are other similar researches (2-6) looking for more economics systems of supply hot water for the population in different parts of the world. The (ISSC) technology has several advantages, if they are used in non developing countries:

a- Inexpensive, it does not require a storage tank (main cost factor).
b- Local production from available materials
c- Do not use auxiliary equipment for control purposes.

In the present study, a ISSC was built using two different ceramics materials as absorber matrices and the absorbers was tested under controlled laboratory experimental conditions. The data generated were statistically first and then in thermodynamics form.

Two solar collectors were built with 0.325 m^2 of area, and using as absorber matrix made up of a thin ceramics slab which performs the function of both absorbing and storing solar thermal energy. Two different commercial ceramics materials were used for each solar collector. The ceramics used are called red and white ceramic. Each solar collector has a glass, a container, polyurethane foam as insulating material and finally a metallic frame.

EXPERIMENTAL PROTOCOL

In order to achieve the objectives of the investigation, an experimental design was established. The ISSC were tested in laboratory environment and with the following conditions;

a- Incandescent lamps were using to simulate the solar radiation. Three levels of radiation were selected 400, 500 and 750 w/m^2

b- The water initial temperature was always the temperature at which the aqueduct water supplied in Bogota.(16 C).
c- The duration for each experiments were 4 hours
d- All measurements ended after 120 minutes of heating and turn off the lamps, the temperature sensors followed recording data until the initial condition was reached again

The incident radiation was measure with a DATALOGGER LI – 1400, and the acquisition a data collection systems were recorder with a NI FP 1000 equipment.

The following experimental procedure were used

1- Fill the containers (0.125 L, for each collector).
2- To put ISSC(with the thermocouples) parallel to radiation panel (0.5 m)
3- Turn on the radiation panel, with the incident radiation level chosen.
4- Recorder data.

The figures 1 and 2, shown the thermocouples position in the system and the solar collector structure.

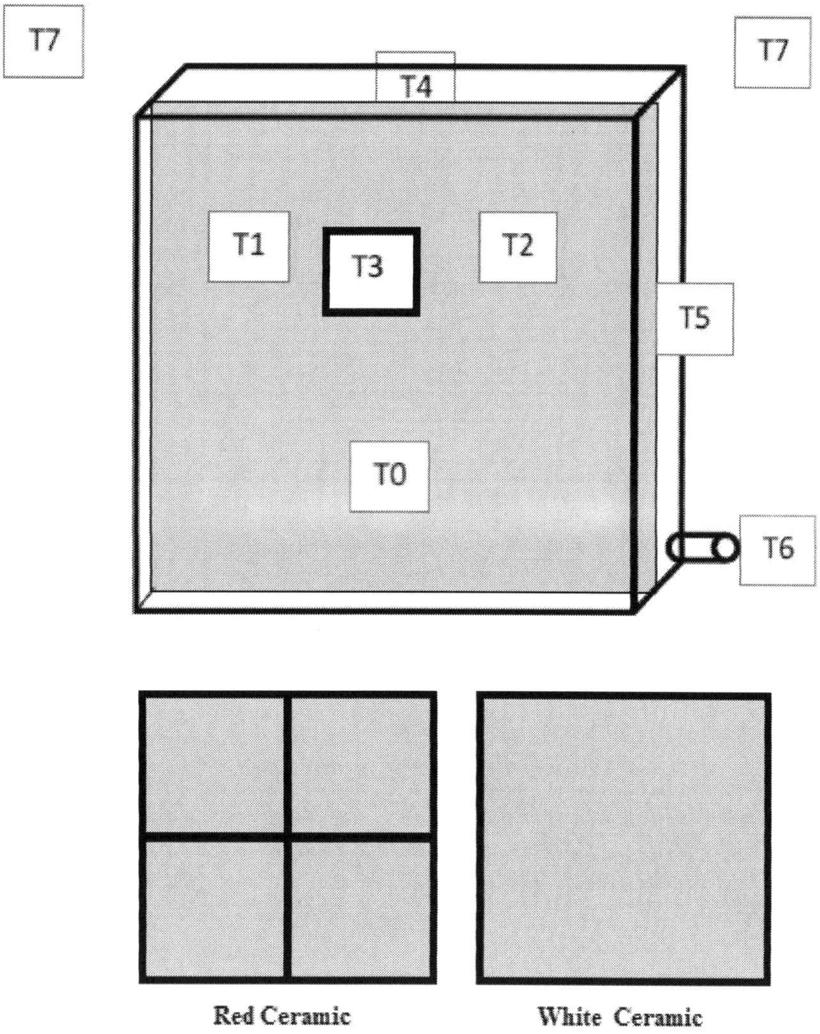

Figure 1. Thermocouple positions for the ISSC experiments

The temperature 4 was behind the collectors and the temperature 6 was for water inside the containers. The temperature 5 registered the temperature of a side of the ISSC and the thermocouple number 7 corresponded to the room temperature (environmental temperature).

The solar collectors' structure is shown in the figure 2. The ISSC in this study is a close system.

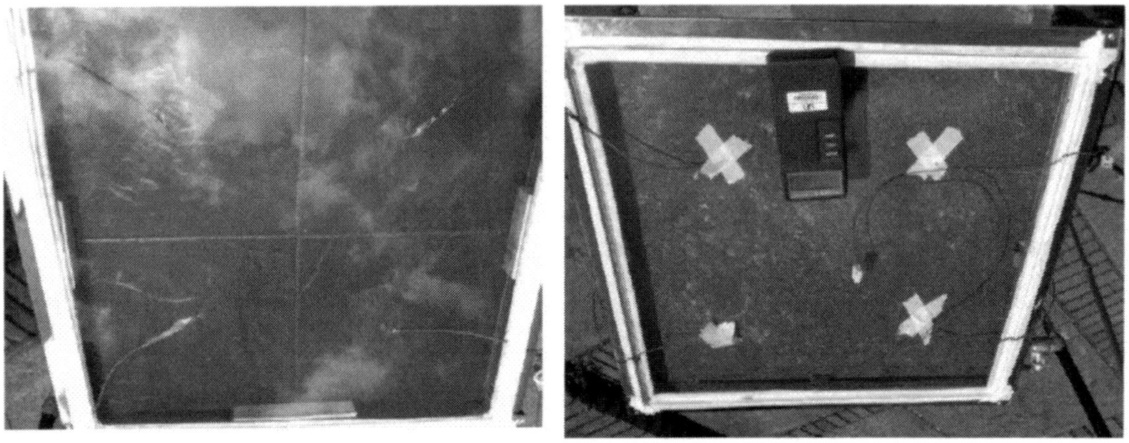

Figure 2. Solar Collector with ceramic materials as absorber

ENERGY AND EXERGY ANALYSIS

The energy analysis presented in this section is mainly based on the first law of thermodynamics and the main goal of this part is to know the energy taken by the water during the experiments for calculate the energy efficiency of the system.

The theoretical model used for the study of (ISSC) by Polo Bravo (7) is an energy balance on the collector as viewed in figure No. 3

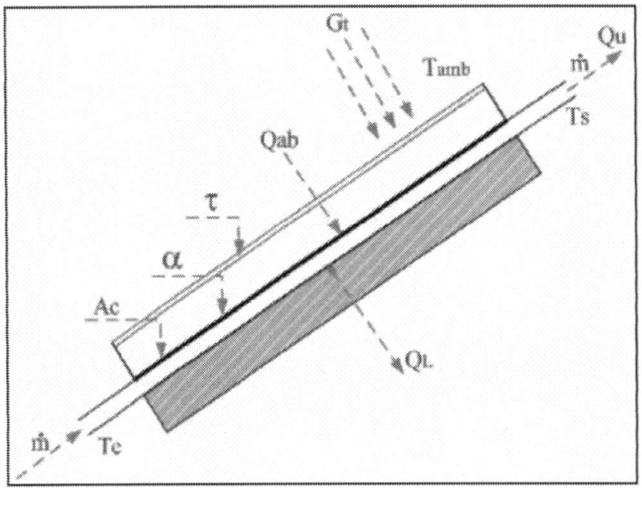

Figure No. 3: ISSC scheme with experiment energy balance (Polo Bravo)

Nomenclature:

Q_U: [W] useful power

Q_L: [W] power lost by radiation and convection

Qab :[W] power absorbed by the ceramic

Te,Ts :[°C] input temperature, output temperature

Tamb. [°C] environment temperature

M: [kg/s] mass flow of heat transfer fluid

Gt: [W/m^2] surface radiation

Ac: [m^2] absorber surface area

U_L: [W,/m^2 K] Global loss coefficient

$\propto\tau$: transmittance absorptance product

Energy flows mentioned are defined by:

$$QU = mc\,(Ts-Te), \quad [1]$$

$$QL = ULAc\,(Tm - Tamb), \quad [2]$$

$$Tm = (Te + Ts)/2 \quad [3]$$

$$Qab = \tau \propto Gt\,Ac \quad [4]$$

The definition of the instantaneous efficiency η is:

$\eta = QU / Gt \bullet Ac = m \bullet c \bullet (Ts - Te) / Gt \bullet Ac$ = Useful energy / incident energy [5]

The parameters, mass (m), specific heat (c) of the sensor must be known, while Ts, Te, AC and GT must be measured experimentally.

The useful energy (QU) can also be defined as the difference between the energy absorbed (QAB) and the losses of the collector (QL). If all the solar radiation absorbed and converted into heat is transferred to the fluid (heat transfer factor FR = 1), then:

$$QU = Qab - QL = \tau \bullet \propto \bullet Gt \bullet Ac - UL \bullet Ac - (Tm - Tamb) \quad [6]$$

Dividing by (Gt • Ac), we obtain the equation:

$$\eta = \tau \cdot \alpha - UL\,(Tm - Tamb)/\,Gt \qquad [7]$$

Is known as the characteristic equation of the instantaneous efficiency flat plate collector, and can be plotted as a graph on a straight $\eta = f\,[(Tm\text{-}Tamb) / Gt]$. In mathematical interpretation of this line, $\tau \bullet \alpha$: is the intersection with the y-η, and UL is the slope of the curve:

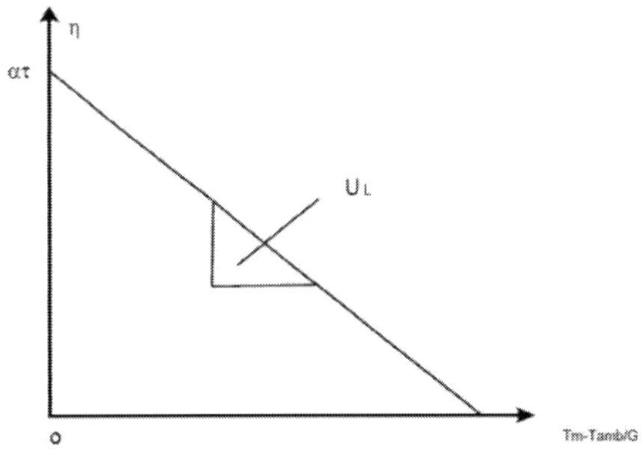

Figure 4 Characteristic of the instantaneous efficiency of a flat plate collector according to operating conditions Tm, Tamb and Gt.

RESULTS

The different results are shown in the figures No 5 to No 10, where the temperatures for each part of the solar collector vs. time are drawn. When these data are used in the energetic analysis the global efficiency and the instantaneous efficiency are evaluated. For all levels of radiation the white ceramics materials presents a higher efficiency than the red one material. The energy is stored in the absorber materials and then transferred to water when the lamps are turn off.

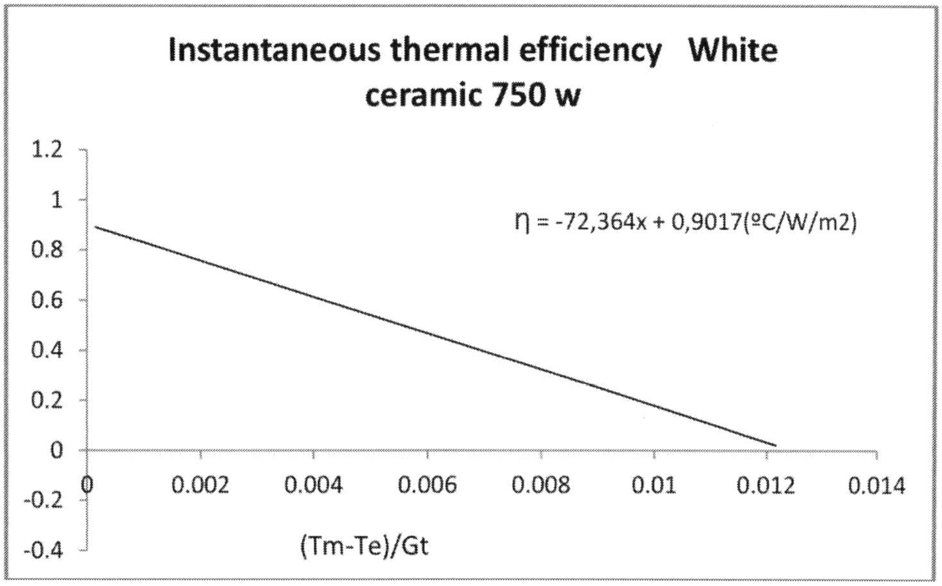

Figure 5 Instantaneous thermal efficiency curve 750W

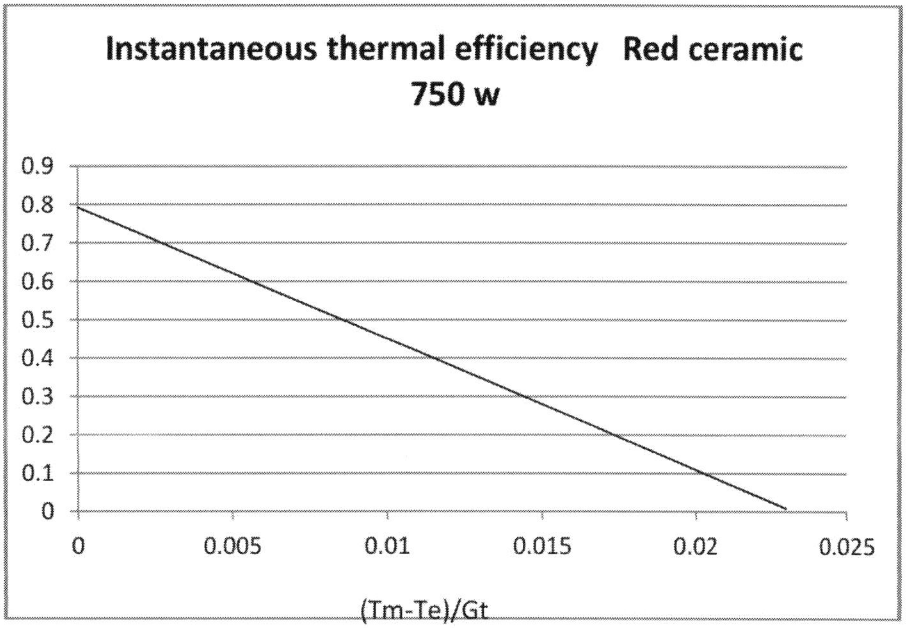

Figure 6 instantaneous thermal efficiency curve 750W

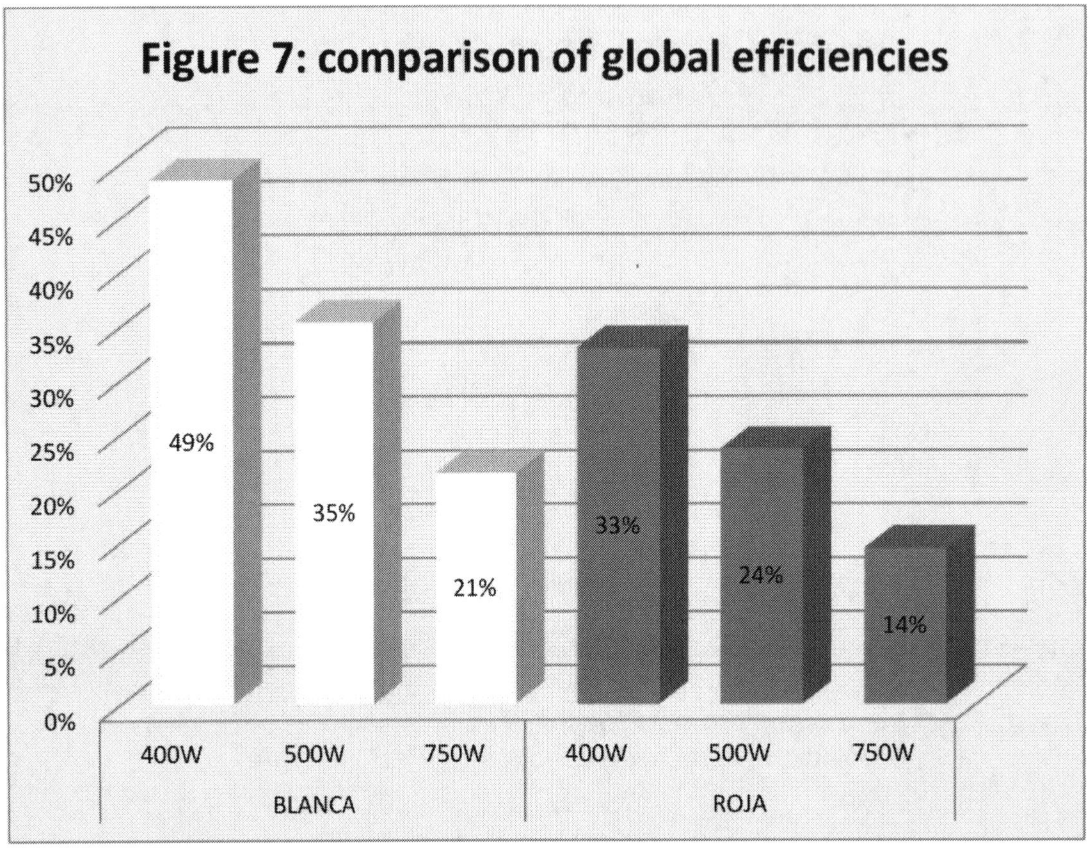

REFERENCES

1. Majdi Hazami, S. K. (29 de Enero de 2010). energetic and exergetic performances of an economical and available integrated solar storage collector based on concrete matrix. *Energy Conversion and Managment.*
2. C. Dharuman *, J. A. (2006). Performance evaluation of an integrated solarwater heater as an option for building energy conservation. *Energy and Buildings 38 (2006) 214–219*(38), 214-219.
3. Callister, W. D. (1996). *Introducción a la ciencia e ingeniería de los materiales, Volumen 2.*
4. Casanova Colás, J. (1993). *Curso de Energía Solar.* Valladolid: Universidad de valladolid.
5. Cengel, Y. A. (2006). *Thermodynamics: An Engineering Approach, 5th ed.* New York: McGraw-Hill:.
6. D. Luna a, *. J.-P. (2009). Solar timber kilns: State of the art and foreseeable developments. *Renewable and Sustainable Energy Reviews*, 1446-1455.
7. D.U. Tulyaganov a, b. S. (2009). Synthesis and properties of lithium disilicate glass-ceramics in the system SiO2–Al2O3–K2O–Li2O. *Ceramics international*, 3013-3019.
8. Division of Agricultural Engineering and Energy Management, Central Arid Zone Research Institute, Jodhpur. (1999). Solar water heaters based on concrete collectors. *Energy*, 703-716.

9. Ebru Kavak Akpinar *, F. K. (2010). Energy and exergy analysis of a new flat-plate solar air heater having. *ELSEVIER*, Article in press.
10. Energy Information Administration. (2010). *International Energy Outlook.*
11. Exiplast S.A. (2010). Cotizacion.
12. Galimberti, P. (2009). *diseño de laboratorio de ensayos fotovoltaicos.* Andalucia.
13. Heim*, D. (2010). Isothermal storage of solar energy in building construction. *Renewable energy*, 788-796.
14. Instituto de Tecnológia cerámica. (24 de Junio de 2011). *Tecnociencia*. Recuperado el 24 de Junio de 2011, de http://www.tecnociencia.es/entidades/itc/tcn_itc5.html
15. Iturrate, E. (1998). Curso básico de teleteccion con ENVI. Buenos Aires.
16. Jorge L. Alvarado a, *. W. (2008). Passive cooling systems for cement-based roofs. *Building and Environment*, 1869-1875.
17. JUAN CARLOS BURBANO J., Á. H. (Agosto de 2006). DISEÑO Y CONSTRUCCIÓN DE UN CALENTADOR SOLAR DE AGUA OPERANDO POR TERMOSIFON. *Scientia et Technica* (31).
18. JUAN CARLOS BURBANO J., Á. H. (2007). ENSAYOS Y PRUEBAS DE UN CALENTADOR SOLAR DE AGUA OPERANDO POR TERMOSIFON. *Scientia et Technica*(34).
19. L. Zhu a, R. H. (2010). Comprehensive energy and economic analyses on a zero energy house versus a conventional house. *ELSEVIER*, 1043-1053.
20. Llesa, L. B. (s.f.). Materiales no metálicos resistentes a la corrosión.
21. Marc Medrano, A. G. (2010). State of the art on high-temperature thermal energy storage for power generation. *Renewable and Sustainable Energy Reviews*, 56-72.
22. Marwa M. Hassan a, Y. B. (2006). Design, construction and performance prediction of integratedsolar roof collectors using finite element analysis. *Science Direct*, 1069-1078.
23. Murcia, H. R. (2008). Desarrollo de la energía solar en Colombia y . *revista de ingenier ía. Universidad de los Andes*, 83-89.
24. Na Zhu, Z. M. (2009). Dynamic characteristics and energy performance of buildings using phasechange materials: A review. *Energy Conversion and Management*, 3169-3181.
25. (ojo). Propiedades termicas de metales, ceramicas y polimeros.
26. Ontiveros, R. I. (s.f.). Energias .
27. Otiveros, R. I. (2009). Energias Alternativas ¿luz natural? (94).
28. T.N.Anderson a, b. M. (2009). The effectofcolouronthethermalperformanceofbuildingintegrated solar collectors. *Solar EnergyMaterials&SolarCells*, 350-354.
29. TALLER ECOLOGISTA. (2009). *Manual de construcción de colectores solares termicos.* ROSARIO ARGENTINA.
30. Universidad Libre. (2007). *Documento registro calificado Maestria en ingenieria con enfasis en energias alternativas.* Bogotá.
31. UPME,IDEAM. (2005). *Atlas de radiacion solar.* Bogotá.
32. William F. Pickard, A. Q. (2009). Renewable an sustainable energy revews. *ELSEVIER*, 1934-1945.
33. Xinping Zhou a, b. F. (2010). A review of solar chimney power technology. *ELSEVIER*, 2315-2338.

(220e) Boosting the Energy Efficiency of Ethylene Production

Martin Schmid[1], Bernhard Prettenhofer[1] and Thomas Wallek[2], (1)Institute of Chemical Engineering and Environmental Technology, Graz University of Technology, 8010 Graz, Austria, (2)Institute of Chemical Engineering and Environmental Technology, Graz University of Technology, Graz, Austria

Due to the large amount of different equipment and the complex set-up of heat exchanger-networks in order to increase the recovery of heat and cold, ethylene plants are among the most complex petrochemical processes. Changes with positive effects in one part of the process can lead to effects in other parts of the plant with negative impacts on the whole process that are difficult to detect and to quantify. Process simulation is a powerful tool to display these correlations. A detailed model of the crack-gas-separation part of an ethylene plant was generated with KBC-PetroSim® based on process flow sheets, equipment data sheets, laboratory tests of products and intermediates and process data obtained from the process control system during a three day test run under full load. The complete model contains 268 devices and 995 streams with 74 recycles. About 81% of the energy to run the gas separation is delivered by the crack-furnaces, the rest has to be imported from other sources within the refinery. The cryogenic processes for the separation of the light gases are counted among the major energy consumers and are composed of a 3-stage propylene refrigeration system and a 4-stage ethylene refrigeration system. Due to economic and technical issues the plant cannot always be operated at the design point. Especially partial load conditions are crucial for the compressors of the cryogenic systems. Recycle streams, which ensure a minimum flow rate to critical equipment cause a reduced energy efficiency of the whole process. The simulation of the plant shows how to adapt the control strategy in order to move energy demand from sections that are working under full capacity to these critical sections and obtain energy savings of 2% based on the imported energy. Using the model of the plant as a tool for pinch and capacity analysis it is possible to develop different scenarios for the modification of the heat exchanger network, which allows a higher interaction between the two refrigeration systems. The so obtained savings of energy are up to 7.6% based on the imported energy and 8938 t/a of CO_2 emissions.

(220f) Automating Molecular Modeling of Gasification with the Kinetic Modelers Toolbox

Scott R. Horton[1], Yu Zhang[2], Craig A. Bennett[1], Frank Petrocelli[2] and Michael T. Klein[1], (1)Chemical and Biomolecular Engineering, University of Delaware Energy Institute, Newark, DE, (2)Tonnage Gases, Equipment and Energy - Technology, Air Products and Chemicals, Inc., Allentown, PA

Automating Molecular Modeling of Gasification with the Kinetic Modelers Toolbox

Scott R. Horton*, Craig A. Bennett, Michael T. Klein

Department of Chemical & Biomolecular Engineering, University of Delaware

Francis P. Petrocelli

Air Products and Chemicals, Inc.

A molecule-based kinetics model was developed for gasification, a prevalent chemical process used in the production of syngas from a variety of feedstocks including coal, biomass, and municipal solid waste. It is envisioned that this modeling approach will have applicability across a wide range of gasification reactor types and operating conditions, and represents a more-rigorous treatment for modeling gasification than most previous efforts described in the literature, in that it is not reliant on the definition of arbitrary, "lumped" pseudo-components nor on the assumption that the reaction trajectory achieves chemical equilibrium.

The generally macromolecular feeds to a gasification unit were described, structurally, as independent molecules or a collection of various interconnected cores (following either a polymeric distribution or confined to a lattice). Gasification reaction processes fall into three major categories: pyrolysis, gasification, and combustion. The feed structures were subjected to a series of reaction and logistical rules via the Interactive Network Generator (INGen) software in order to obtain a reaction network comprising all three reaction classes.

Polymeric and lattice-type species were conceptually broken into "cores" and "substituents", reacted independently, and reformed into molecules by following the Attribute Reaction Model (ARM) credo. This methodology removed the permutative explosion of species from the balance equations, and placed it on the post reaction analysis. Isomeric detail was collapsed into representative species. Each species was broken into a series of composition components (such as ring, group, and element counts).

The balance equations for the reaction network were automatically generated and solved using the Kinetic Model Editor (KME) software. The specifics of the reactor design led to the modeling of interconnected ideal reactor zones, wherein the conditions of reaction could be independently varied. It was then possible to use process, experimental, and published data to optimize the kinetic parameters.

Self-assembly and Superhydrophobicity in Membrane Development for Water Application

Suzana Pereira Nunes
Water Desalination and Reuse Center (WDRC),
King Abdullah University of Science and Technology (KAUST),
23955-6900 Thuwal, Saudi Arabia,
suzana.nunes@kaust.edu.sa

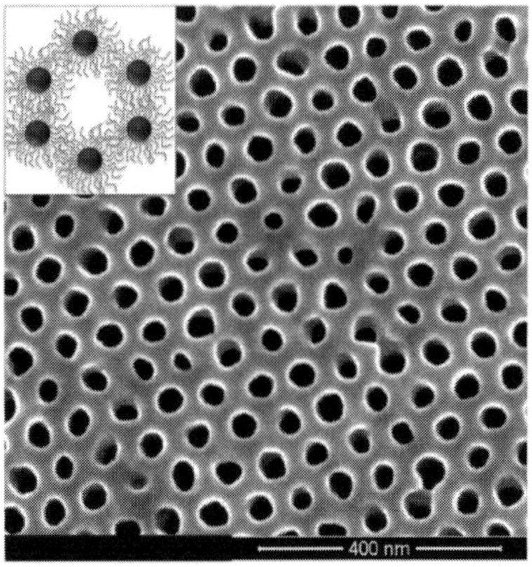

Figure 1. Isoporous membrane obtained by block copolymer micelle self-assembly

The development of sustainable technology to attend the demands of fresh water is one of the big challenges in engineering today. Water is essential to keep high health standards and to guarantee food supply. Membrane technology has been more and more applied for water desalination and reuse. Membranes are the key components in seawater desalination plants by reverse osmosis (RO) and water treatment by ultra- and nanofiltration. Even in countries like Saudi Arabia, where petroleum is abundant and thermal desalination is still a strong technology, the number and capacity of membrane driven plants is increasing. RO plants with capacity higher than 200,000 m^3/day are in operation. In many large-scale applications membrane technology is established and competitive and RO is one of the cases. However other technologies are gaining importance and finding niches where they could be advantageous from the point of view of energy consumption or hybridization with other technologies. Among these technologies are forward osmosis, membrane distillation and innovative membrane-biotechnology processes. These are topics under investigation by different groups at the Water Desalination and Reuse at KAUST. The implementation and optimization of emerging membrane technologies require development of new materials for membranes, exact morphology control and surface modification.

The development of new polymers and membranes for water application is the main activity of my group at KAUST.

When developing better membranes, the main challenges are the exact control of

- hydrophilicity-hydrophobicity
- pore size distribution and porosity
- chemical and thermal stability
- processability in large scale

We currently work on the synthesis of new polymers and copolymers to address the topics above, as well as new mechanisms of pore formation in membranes.

The talk will focus on 2 of our recent developments:

1) Isoporous membranes based on block copolymers self-assembly
2) Superhydrophobic porous membranes for membrane distillation

Isoporous membranes based on block copolymers self-assembly

Asymmetric porous membranes have been successful for decades in applications, which vary from ultrafiltration to reverse osmosis. High water flux and easy manufacture in large scale are large advantages. The main drawback is the broad distribution of pore sizes. In collaboration with the group of Klaus Peinemann at KAUST (Nunes, Sougrat et al. 2010; Nunes, Behzad et al. 2011; Nunes, Karunakaran et al. 2011) we have been preparing membranes with exceptionally sharp pore size distribution and much higher porosity than other available membranes for the range of ultrafiltration (Figure 1). By using different methods of morphological characterization, which include cryo transmission and scanning electron microscopy, atomic force microscopy, focused ion beam and synchrotron small-angle x-ray scattering, we demonstrated for the first time that the mechanism of pore formation is based on block copolymer micelle assembly and can be controlled, by choosing the right selective solvent mixture and or by adding complexing additives to the casting solution. In collaboration with the group of Wiesner (at Cornell) and Vainio (at DESY) (Dorin, Marques et al. 2012), we proved for different block copolymers that the final membrane morphology is already determined by the assembly in the casting solution. In this way, diblock copolymers with 2D hexagonal order in solution lead to 2D hexagonal pores, while body centered cubic (BCC) order in solution lead to BCC pore arrangement. The big advantage of this method of membrane manufacture, when compared to other block copolymer assembly methods, is that no additional etching or post solvent treatment is required and the membranes can be manufactured as flat sheet and hollow fibers in machines analogous to those used for industrial membrane preparation. Due to the high porosity and asymmetry, water fluxes higher than 800 L m^{-2} h^{-1} bar^{-1} were achieved. The pore sizes are sensitive to pH, what opens the possibility of using the membranes as chemical gates for biomedical devices.

Superhydrophobic porous membranes for membrane distillation (MD)

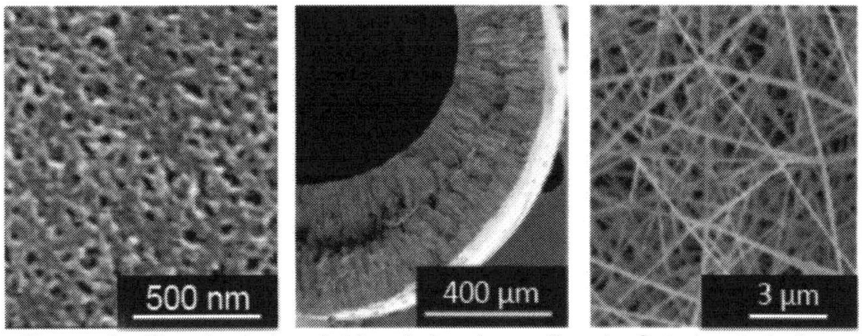

Figure 2. Fluorinated polytriazole/ polyoxadiazole flat-sheet, hollow fiber and electrospun membranes

Another very important challenge in membrane development for water application is the control of surface hydrophobicity-hydrophilicity. This is known to have a decisive role in the fouling control, but is also a key issue when development membranes for membrane distillation. In membrane distillation (MD) saline or wastewater are kept in contact with a porous hydrophobic membrane and heated to a temperature below the water boiling point. The other side is kept at a much lower temperature. The partial vapor pressure difference arising due to the temperature difference across the membrane is the driving force for water transport. Membranes with high porosity and temperature resistance are required, but the most important characteristic is exceptionally high hydrophobicity to avoid liquid water leakage through the pores. For this application we synthesized fluorinated polytriazoles and polyoxadiazoles. The polymers are soluble in polar organic solvents and could be manufactured into asymmetric flat sheet and hollow fiber membranes, as well as electrospun with very high porosity (Figure 2). Contact angles as high as 162° were obtained for water on electrospun membranes. In collaboration with the membrane distillation group at KAUST(Maab, Francis et al. 2012), the membranes were tested for seawater desalination achieving water fluxes above 80 L m^{-2} h^{-1} with feed temperature of 80°C.

Co-authors and acknowledgements: The self-assembly membranes in this work at KAUST have been prepared in collaboration with Klaus-Viktor Peinemann, Pradeep Neelakanda, Madhavan Karunakaran and Debora Marques. The membranes for MD have been manufactured by Husnul Maab and Lijo Francis and tested in collaboration with the KAUST MD group coordinated by Gary Amy and Noreddine Gaffour.

References

*Dorin, R. M., D. S. Marques, et al. (2012). "Solution Small-Angle X-ray Scattering as a Screening and Predictive Tool in the Fabrication of Asymmetric Block Copolymer Membranes." ACS Macro Letters **1**(5): 614-617.*

Maab, H., L. Francis, et al. (2012). "Synthesis and fabrication of nanostructured hydrophobic polyazole membranes for low-energy water recovery." Journal of Membrane Science.

*Nunes, S. P., A. R. Behzad, et al. (2011). "Switchable pH-Responsive Polymeric Membranes Prepared via Block Copolymer Micelle Assembly." ACS nano **5**(5): 3516-3522.*

*Nunes, S. P., M. Karunakaran, et al. (2011). "From micelle supramolecular assemblies in selective solvents to isoporous membranes." Langmuir **27**(16): 10184-10190.*

*Nunes, S. P., R. Sougrat, et al. (2010). "Ultraporous films with uniform nanochannels by block copolymer micelles assembly." Macromolecules **43**(19): 8079-8085.*

(222b) Water and Ion Transport in Polymer Membranes for Water Purification

Benny Freeman, Deptartment of Chemical Engineering , The University of Texas at Austin, Austin, TX

Polymer membranes are critically important in addressing urgent global needs in the 21st century for reliable, sustainable, efficient access to clean energy and clean water. Polymer membranes have emerged as a leading technology to desalination water (reverse osmosis and nanofiltration) and are being explored for energy generation in applications such as reverse electrodialysis and pressure retarded osmosis. Furthermore, efforts are under way to develop additional applications of membranes for water purification, such as forward osmosis and membrane-assisted capacitive deionization. In each of these applications, control of water and ion transport across polymer membranes is critically important for optimizing performance of such membranes. This presentation focuses on the fundamentals of ion and water transport in polymers. Structure/property correlations are shown for a variety of polymers, including uncharged and charged materials. The solution/diffusion model is described for water and ion transport. The role of free volume in governing diffusion of solutes through hydrated polymers is demonstrated. The existence of a water/salt permeability/selectivity tradeoff relation is shown for polymers being considered for such applications. Comparisons are made to similar physics governing gas transport in polymers used for gas separations applications.

(222c) Membrane Distillation Processes for Water Desalination

Kamalesh K. Sirkar and Dhananjay Singh, Chemical, Biological and Pharmaceutical Engineering, New Jersey Institute of Technology, Newark, NJ

Membrane distillation for desalination employs heated brine in contact with a porous hydrophobic membrane to generate water vapor which is recovered in a variety of ways on the other side of the membrane. Direct contact membrane distillation (DCMD) employs colder distilled water on the other side to condense the water vapor. Air gap membrane distillation (AGMD) devices utilize a cold surface to condense the water vapor and generate the distillate. This presentation will focus primarily on hollow fiber membrane-based devices for DCMD as well as AGMD. For AGMD the colder condensation surface is provided by a solid nonporous hollow fiber. Novel two hollow fiber-based AGMD devices will be illustrated and the relation of AGMD to DCMD will be explored. For DCMD an overview of the device performance, resistance to scaling, concentration cascades and energy requirements will be provided for a variety of brines. Results of studies with brine and produced water will be provided for feeds at temperatures in the range of 100 C-130 C as well. The relative performances of a variety of polymeric and ceramic membranes will be discussed.

(222d) Effect of Preparation Variables On Morphology and Water Flux of PVDF Hollow Fiber for Membrane Distillation

Lan Ying Jiang, School of Metallurgical Science and Engineering, Central South University, Changsha, China

Metallurgy is one of the pillar industries of China. The conventional pyrometallurgy suitable for high grade ore reserves is facing the significant emission problem. Hydrometallurgy can provide wide and varied techniques and their combinations to process the lean and complex ores. Nevertheless it is demanding in process materials and energy with great amounts of liquid and/or solid wastes as side effect.

Due to its mechanisms, membrane-based separation possesses the advantages of energy efficiency and bio-friendliness. Application of membrane-based separation in metallurgical industry does not have a long history, but seems to be a growing prospect as it has achieved obvious economic and environmental benefits against the aforementioned background.

This seminar will firstly present the current status and potential of applying membrane separation in the hydro- and pyro-metallurgical field with detailing the flowsheets, environmental influence and economic analysis. Our research and development work in the fabrication of micro-porous hydrophobic membranes for membrane distillation (MD) will also be highlighted.

Porous poly (vinylidenefluoride) (PVDF) hollow fiber membranes were fabricated in current work by the dry/wet spinning and phase inversion. Effects of PVDF concentration, glycerol content and external coagulant composition on membrane morphology and performance were investigated. Taguchi experimental design with a set of the nine experiments was used to elucidate the relationship between the above conditions and water flux in DCMD. The highest flux in desalinating 3.5wt.% NaCl aqueous solution was 23kg/m^2-hr at average feed temperature of 62.5°C with 99.9% rejection. Analysis of variance revealed that glycerol content was the more influential factor than the other two. Predictions regarding the performance of membranes involving other preparation conditions were in agreement with experimental results.

The FESEM study showed that finger-like macrovoids were developed underneath the inner surface of all the fibers, which indicated rapid phase demixing occurring at the lumen skin of the membrane. Increasing the ethanol content in external coagulant facilitated the macrovoids growth at inner skin but depressed those at the shell side; higher PVDF content in dopes inhibited the macrovoids expansion; the macrovoids in fibers prepared with 5% glycerol was bigger as compared with those in fibers containing no glycerol and 10% glycerol. The inner skin of the fibers prepared with glycerol in dopes was porous. Gas permeation test demonstrated that the pore size of the shell skin ranged from 2nm to 50nm. This range corresponds well with that revealed by FESEM. Mechanisms for the membrane morphologies evolved were discussed using XRD characterization and the characteristics of phase inversion.

The membrane surface morphology (i.e. pore size) correlates positively with both gas flux and DCMD flux, with the latter being influenced more obviously. It was also found that the surface morphology influenced membrane long-term performance in desalination of waste acid containing dissolved salts discharged from metallurgical process using both DCMD and vacuum membrane distillation (VMD).

(223a) Effect of Confinement On Ionic Liquid Properties for Carbon Capture Processes

Ruth E. Baltus, Chemical and Biomolecular Engineering, Clarkson University, Potsdam, NY and Laila Banu, Chemical & Biomolecular Engineering, Clarkson University, Potsdam, NY

Work in our laboratory has focused for a number of years on examining the potential of room temperature ionic liquids for post combustion carbon capture processes. Experimental work has involved measurements of carbon dioxide solubility and diffusivity in different ionic liquids as well as on gas permeation measurements through supported ionic liquid membranes (SILMs) prepared using anodic alumina films as the support membrane.

Results from these studies have raised questions about the impact of a solid interface on the properties of ionic liquids. Gas permeation through SILM systems with different pore size indicates both carbon dioxide and nitrogen permeation is larger when the ionic liquid is confined within a 20 nm pore size membrane when compared to a 100 nm pore size membrane. Ideal selectivity values are generally the same in the two membrane systems. A comparison of measured membrane permeation to values predicted from bulk solubility and diffusivity values shows higher carbon dioxide permeance in the membranes compared to values predicted from bulk properties of the ionic liquids.

Gas uptake into confined ionic liquids have also been experimentally measured, with solubility and diffusivity of carbon dioxide in the confined ionic liquids compared to bulk phase characteristics. To date, results have been mixed with results from some ionic liquids showing no impact of the solid interface whereas others indicate a significant difference in ionic liquid properties when confined. In this presentation. Results from these studies will be presented and discussed.

(223b) Metallic Based Aqueous Amine Membranes for the Separation of Carbon Dioxide From Fossil Fuel Flue Gas

Brian L. Bischoff and Lawrence E. Powell, Oak Ridge National Laboratory, Oak Ridge, TN

According to the US Department of Energy, there is scientific evidence that CO_2 emissions from fossil fuels have caused climate change and that a reduction of these emissions is essential for reducing the risk of future devastating effects. At present, the leading technology for the post combustion capture of CO_2 relies on the separation of CO_2 from the flue gas stream by employing chemical absorption with an aqueous monoethanolamine (MEA) solution. The separation is followed by compression of the gas to a liquid or supercritical state for transportation, and ultimately long term storage in geological formations. However, there is little experience with large scale amine separation systems especially in oxidizing environments typical of a fossil fuel flue gas stream. Furthermore, there are concerns about the degradation rate and evaporative losses of the amine solution and the amount of energy required for regeneration.

We propose using metallic supported liquid amine membranes to overcome the shortcomings of the large energy intensive amine process. Because the amine solution is immobilized in the pores of a thin nanoporous layer there is a lower vapor pressure thereby reducing the evaporation rate and the thin layer requires a small fraction of the amine compared to the absorption method. The membrane can be operated at a constant temperature, lower than the absorption process regeneration temperature, which reduces the degradation rate of the amine. We prepared a series of robust membranes based on a stainless steel support structure and a very thin nanoporous aluminum oxide layer that holds the aqueous amine solution. We will present laboratory results showing flux data and selectivity data for a range of amines under varying temperatures. We will also present stability data which demonstrates a strong promise for long-term application.

(223c) A Comprehensive Multi-Step Description of the Permeation of Hydrogen in a Dense Metal Membrane Over a Broad Temperature Range

Nicholas Deveau[1], **Yi Hua Ma**[2] and Ravindra Datta[1], (1)Chemical Engineering, Worcester Polytechnic Institute, Worcester, MA, (2)Worcester Polytechnique Institute, Worcester, MA

Development of cheaper and more durable dense metal membranes are being sought for generation of pure hydrogen suitable for fuel cells from coal. Toward this end, we have developed a comprehensive multi-step model for describing permeation of hydrogen in such membranes. The various steps include: s_1) dissociative adsorption on the surface, followed by: s_2) solution and then: s_3) diffusion of hydrogen atoms within the bulk metal matrix, and finally: s_4) evolution from the bulk to the surface, and: s_5) associative desorption on the permeate side.

The efficacy of the model is ascertained by first applying it to Pd and Pd-Cu membranes, for which much literature exists. Of course, for these membranes, typically Sieverts' law is assumed, which is limited to high temperatures (> 300 °C) and low hydrogen pressures, when diffusion (step, s_3) is controlling, and hydrogen concentration in bulk metal is low. Over a broader range of temperatures spanning the crystallographic phase-transition in the Pd-H and Pd-Cu-H systems, in fact, the slopes are very different at low and high temperatures, and also there is a pronounced peak in permeability in the intermediate temperature range. The latter behavior has not yet been theoretically explained.

Our approach considers the 5-step network in analogy to an electrical circuit and describes well the permeation behavior of hydrogen in Pd and Pd-Cu membranes, including the peak at intermediate temperatures. Further, it can adequately describe the absorption isotherms of the metal-hydrogen system. A comparison of step resistances also allows a systematic reduction to a three-step (adsorption, diffusion, and desorption control) model that is adequate over a broad range of temperatures and pressures.

(223d) High Temperature Metal Membranes for Hydrogen Permeation without Platinum Group Metals

J. Douglas Way, Colin A. Wolden, Kehinde Adeyemo and Mayur Ostwal, Chemical and Biological Engineering, Colorado School of Mines, Golden, CO

Group V metal membranes (tantalum, vanadium, niobium and their alloys) have extremely high theoretical permeability for dissociated hydrogen[1], but are unable to self-catalyze the hydrogen dissociation reaction. Such membranes need to be coated with an active catalyst layer, typically palladium or a Pd-alloy, increasing the materials cost of the membrane. Additionally, the use of a metallic surface catalyst layer limits the upper operating temperature of the membrane, as temperatures higher than 623 K tend to promote alloying, which causes loss of flux. As group V metals are also prone to low-temperature hydrogen embrittlement, this limits their applicability.

Transition metal carbides, nitrides, and sulfides have been studied as substitutes for platinum group catalysts in applications requiring hydrogen dissociation, such as the water-gas shift reaction. The goal of this study was to determine if these lower cost materials could be used as catalyst layers for group V metal membranes.

Molybdenum carbide (Mo_2C) was chosen as the test transition metal catalyst, and vanadium as the group V bulk metal. Magnetron sputtering was used to apply layers of Mo_2C to both sides of commercial cold-rolled vanadium foils with thicknesses ranging from 50 to 125 microns. Membranes sputtered at temperatures of 473 K or below had a cubic structure, and were the most permeable to hydrogen, while higher-temperature (873 K) sputtering produced an orthorhombic lattice[2].

These variations in catalyst structure produced corresponding variations in performance. All membranes were single-gas tested in hydrogen and helium at temperatures from 823 to 1073 K and feed gas pressures up to 690 kPa gauge (100 psig). No membranes had any detectable helium leakage through defects, and membranes without Mo_2C layers had no hydrogen flux, demonstrating that all hydrogen dissociation activity was caused by the catalyst.

At temperatures below 1023 K, the structure and thickness of the carbide catalyst layer had the most influence on the hydrogen permeability. As temperature increased, the permeability reached a maximum at 1023 K, and then decreased as predicted by Steward[1] due to decreasing hydrogen solubility in the vanadium foil. The highest H_2 permeability was 5.9×10^{-8} $mol.m/m^2.s.Pa^{0.5}$ at 1023 K, about 70% higher than pure Pd. We believe these are the first measurements of hydrogen permeability for vanadium at high temperature. The hydrogen flux for this 50 micron thick membrane at a differential feed pressure of 690 kPa was 0.7 $mol/m^2.s$. Membranes display long term stability (> 150 hrs) at high temperature with no loss of flux and TEM analysis after testing confirms that the Mo_2C did not reduce or otherwise alloy with the vanadium foil. Effects of process parameters such as V foil thickness, catalyst layer thickness, sputter conditions, and permeation temperature will be discussed and compared to Steward's predictions[1].

1. Steward, S.A., *Review of Hydrogen Isotope Permeability Through Materials*, 1983, Lawrence Livermore National Laboratory.

2. Gade, S.K., S.J. Chmelka, S. Parks, J.D. Way, and C.A. Wolden, *Dense Carbide/Metal Composite Membranes for Hydrogen Separations Without Platinum Group Metals.* Advanced Materials, 2011. **23**(31): p. 3585–3589.

(223e) Long Term H2 Permeance and Selectivity Stability of Pd and Pd-Au Large Scale Membranes in Actual Syngas Atmospheres

Federico Guazzone[1], Jacopo Catalano[2], Ivan Mardilovich[2], Nikolas Kazantzis[3] and Yi Hua Ma[4], (1)Chemical Engineering, Worcester Polytechnic Institute, Worcester, MA, (2)Worcester Polytechnic Institute, (3)Chemical Engineering, WPI, Worcester, MA, (4)Worcester Polytechnique Institute, Worcester, MA

The integration of composite Pd based membrane reactors (WGS or MSR reactions) into large scale H_2 producing units entails valuable cost savings for pre-combustion CO_2 sequestration. In addition, it was recently demonstrated that composite Pd and Pd-Au membranes were stable for more than 250 h in actual syngas atmosphere (450°C, 13 bara) and that H_2 permeate purity was higher than 99.9%. The Pd-Au membranes tested earlier were prepared by the deposition of a thin (0.15-0.2 μm) layer of Au on top of the dense Pd layer. The membranes showed some initial flux decline attributed to poisoning of the layer by unknown contaminants and a maximum recoverability of 77% after 300 h in pure H_2 at 450°C. The objective of the present work was to further investigate the stability of permeance and purity and the of Pd/Au membranes with a higher Au content on the Pd surface of 0.5 μm thickness.

Composite Pd and Pd/Au porous stainless steel membranes having an area of more than 200 cm^2 (2.54 cm OD, 25.4 cm L) were prepared by WPI's patented technology and tested at the National Carbon Capture Center (NCCC) in Wilsonville, Alabama. The Syngas composition was slightly different from the previous study since in this second run, the syngas was shifted before entering the membrane module. Hence, CO concentrations were lower than 1% and H_2O/CO ratio was higher than 1. The operation conditions and H_2S concentration levels were kept the same as in the previous study. A Pd-0.5 μm/Au membrane showed a remarkably stable hydrogen flux for over 475 h albeit an initial hydrogen flux decline of ~40% after 20 hours of exposure to the syngas stream. The hydrogen purity, as high as 99.9%, was also stable for over 475 hours. Upon test at NCCC, the permeance of the Pd-0.5 μm/Au membrane was totally recovered after 160 h at 450°C in a pure H_2 atmosphere. Hence, a thicker Au layer enhanced drastically the recoverability of membranes exposed to actual syngas atmospheres which represents a breakthrough accomplishment in the development of composite Pd membranes for WGS-CMR applications.

In an attempt to understand the causes of the initial permeance flux decline, a preliminary XPS study of the surface of Pd coupons exposed to actual syngas was performed and showed the presence of carbon contamination in the Pd bulk. Further analysis would be needed to confirm this finding.

(223f) Composite Catalytic Membranes: Enhancing H2 Recovery Membranes by Appropriate Coupling with Water-Gas-Shift Reaction

Elva Romero, Daejin Kim and **Benjamin Wilhite**, Artie McFerrin Department of Chemical Engineering, Texas A&M University, College Station, TX

Membranes for gas purification are a promising alternative to existing cryogenic, absorbtion and pressure-swing adsorption systems. However, permselective membrane materials are often limited by surface inhibition (e.g., CO inhibition of palladium surfaces) or low gas permselectivities (~50:1 for most polymeric membranes). To address these limitations, our research group has developed a composite catalytic-permselective (CCP) membrane design approach, in which a permselective film (e.g., polymeric, metallic) is integrated with a porous catalytic film such that reaction and diffusion within the catalytic layer directly modifies the gas composition at the surface of the underlying permselective barrier [1]. This modification of gas composition has been predicted to enhance permeation rates, membrane permselectivities and reduce surface corrosion/inhibition.

In this talk, we will review general design rules for implementing this strategy, and demonstrate their extension to the specific application of hydrogen purification from diesel reformate mixtures using water-gas-shift as the enhancing catalytic reaction. Detailed two-dimensional simulations will be presented comparing the resulting CCP design against reference cases of (i) an untreated membrane, and (ii) an untreated membrane employed in a packed-bed membrane reactor configuration. Experimental results for the case of an electroless-plated 30-micron Pd film coupled with water-gas-shift will be presented for comparison with model predictions. Lastly, a discussion of how this approach may be extended to polymeric systems in order to enhance permselectivity will be presented.

[1] B.A. Wilhite. Composite Catalytic-Permselective Membranes: A Novel Strategy for Enhancing Selectivity and Permeation Rates via Reaction and Diffusion. Ind. Eng. Chem. Res., 2011, 50(17), 10185-10193.

(227a) Understanding Membrane Separation of Gases in the Transition Flow Region Through Comsol Multiphysics Modeling

Kishori Deshpande, Core Research and Development, Dow Chemical Company, Freeport, TX, Jim Sturnfield, Dow Chemical Company, Freeport, TX, Alexia Finotello, The Dow Chemical Company, Freeport, TX and John Pendergast, Core R&D, The Dow Chemical Company, Freeport, TX

The pore sizes of many membranes presently being studied for separating the components in gas mixtures are on the scale of nanometers. Depending on the specific gases and pressures being used, this scale will put the flow either in the Knudsen flow regime or in the transition regime. The differential flow of the individual gas components gives the relative transport of the gases through the membrane. The solubility of the gases into and out of the membrane provides another mechanism of separation. The solubility can be enhanced through facilitated transport using carriers which react reversibly and selectively leading to improved permeate selectivity. However, before vesting resources in experimental programs, it is critical to develop a model to enable screening of viable solutions and avoid resource redundancy. These models need to consider the impact of the different mechanisms on separation and predict the membrane selectivity and yield for facilitated transport in different Knudsen number regimes. In this work, we investigate methods of combining the different mechanism with a multiphysics model to understand the resultant separation. In particular, we look at implementations of some different transition flow models in this framework.

(227b) Nanoparticle Permeation Induced Water Penetration and Ion Transport Through a Lipid Membrane

Bo Song, Huajun Yuan, Cynthia Jameson and Sohail Murad, Chemical Engineering, University of Illinois at Chicago, Chicago, IL

Nanoparticles are generally considered excellent candidates for targeted drug delivery. However, ion leakage and cytotoxicity induced by nanoparticle permeation is a potential problem in such drug delivery schemes because of the toxic effect of many ions. In this research, we have carried out a series of coarse-grained molecular dynamics simulations to investigate the water penetration, ion transport and lipid molecule flip-flop in a protein-free phosphor-lipid bilayer membrane during nanoparticle permeation. The effect of ion concentration, pressure differential across the membrane, nanoparticle size and permeation velocity have been examined in this work. Some conclusions from our studies include: 1) The number of water molecules in the interior of the membrane during the nanoparticle permeation increases with the nanoparticle size and the pressure differential across the membrane, but is unaffected by the nanoparticle permeation velocity or the ion concentration; 2) Ion transport is sensitive to the size of nanoparticle as well as the ion concentration difference between two sides of the membrane; We also observe no selectivity between anions and cations; 3) The lipid molecule flip-flop is directly related to the size of nanoparticle and ion concentration differential, while it is an inversely related to the pressure differential and the nanoparticle permeation velocity. The findings described in our work will lead to a better understanding of passive water and ion transport during the permeation of nanoparticle and help develop more efficient nanocarrier drug delivery systems while avoiding cell cytotoxicity for intracellular delivery of therapeutics.

(227c) Modeling Permeate Flux in Ultrafiltration of Non-Newtonian Polysaccharide Solutions

Michelle C. Almendrala, Philippine Sugar Research Institute Foundation, Makati City, Philippines

Modeling permeate flux in ultrafiltration of non-Newtonian polysaccharide solutions

Michelle C. Almendrala*, 1, Shang-Tian Yang[2], Jonathan L. Salvacion[3]

[1]School of Chemical Engineering and Chemistry, Intramuros, Manila, Philippines 1002

[2]Department of Chemical and Biomolecular Engineering, Columbus, Ohio, USA 43210

[3]Department of Chemical Engineering, University of the Philippines, Diliman, QC, Philippines *e-mail: michael.almendrala@yahoo.com

Abstract

The development of mathematical models for permeate flux behavior of non-Newtonian polysaccharide solutions in ultrafiltration has been studied using hollow fiber membrane modules. The gel polarization model was modified to express the permeate flux as function of the operating parameters. The modified gel model integrating the average wall shear stress per unit length of the membrane (g_W/L) has been proposed to predict permeate flux. Two empirical equations based on the modified gel polarization model were also developed to express the dependence of the resulting permeate flux on the properties of the solution and operational variables. These correlations were able to predict the permeate flux behavior on the assumptions made for the constant and variable gel layer concentrations. An empirical model was also developed incorporating the factors affecting flux. The developed model is capable of predicting the dependence of the permeate flux on operating conditions and solution properties. It was demonstrated in this model that the effects of temperature, concentration, transmembrane pressure, shear rate and pH on permeate rate are interdependent. For all these models, the resulting permeate fluxes have been shown to be controlled primarily by the wall shear rate or the feed flowrate. Achievement of high fluxes depends therefore, upon operating at flow conditions that maximize the rate of mass transfer from the membrane surface. In laminar flow systems, this is achieved by operating at high fluid velocities across the membrane surface.The predicted results based from the mathematical models were similar and found to be in good agreement with the experimental data.

Keyword*s*: ultrafiltration; permeate flux; gel polarization

(227d) CFD Analysis of the Convective Transport of Rod Shaped Particles in Cylindrical Pores

Basavaraju Nagendra Agasanapura, Chemical & Biomolecular Engineering, Clarkson University, Potsdam, NY, Ruth E. Baltus, Chemical and Biomolecular Engineering, Clarkson University, Potsdam, NY and Shankar Chellam, Department of Civil and Environmental Engineering, Department of Chemical Engineering, University of Houston, Houston, TX

Rod-shaped bacteria and non-spherical viruses are frequently encountered in membrane feed waters. Their removal from drinking water supplies is essential to providing biologically stable tap water, from wastewater before discharging to receiving water bodies and to produce water free of viruses, cells, and cell debris during downstream processing. The design of efficient membrane separation systems to separate these pathogens requires an understanding of the transport of non-spherical particles through porous membranes. Previous modeling efforts have focused almost exclusively on spherical particles.

In this research, computational fluid dynamics (CFD) has been used to model the migration of rod- shaped particles in a cylindrical pore. The analysis involves incorporation of both steric restrictions, which limit the positions and orientations available to a rod in a pore, and hydrodynamic resistances experienced by the particle because of the presence of the pore wall. Simulations involved solving the Navier Stokes equation defined in an Arbitrary Lagrangian Eulerian (moving mesh) frame to predict the translational motion of a rod-shaped particle at a fixed orientation where rotational motion of the particle was neglected. The analysis yields the particle lag coefficient, which is defined as the ratio of the particle velocity to the fluid velocity when no particle is present. Using a centerline approximation, particle reflection coefficients were determined by integrating the lag coefficient over all sterically allowed particle orientations.

In all cases, including hydrodynamic resistances increased the particle reflection coefficient relative to a model that only considered steric limitations for rod shaped particles. Results obtained for particles with the same rod diameter but different particle lengths (i.e., different aspect ratio) shows that including hydrodynamic wall interactions in predictions of the reflection coefficient has a larger impact for particles which are closer to spherical (smaller aspect ratio) when compared to particles with greater aspect ratios. Predictions of the reflection coefficient for different sized particles with the same aspect ratio indicate that including hydrodynamic resistances has a larger impact for smaller particles, with little difference between predictions made with and without hydrodynamic resistances for particles larger than ~ 50% of the pore size.

CFD simulations have also been performed with the inclusion of particle rotational motion (i.e., torque). Results from these simulations will also be discussed in this presentation.

(227e) Electrophoretic Annular Separator with Combined Poiseuille and Electroosmotic Flow

Jeffery W. Thompson[1], Holly A. Stretz[2] and Pedro E. Arce[1], (1)Chemical Engineering, Tennessee Technological University, Cookeville, TN, (2)Department of Chemical Engineering, Tennessee Technological University, Cookeville, TN

With recent advances in materials science such as two photon polymerization (Pfleging et al., 2008, Bhuian et al., 2007), new potential now exists for customizing the geometrical design of a microfluidic device structure or capillary electrophoresis geometry in addition to the pore structure in nanocomposite gels (Thompson et al., 2012). These devices have many applications including microreactors and electrokinetic separators, but the focus of this contribution is specifically capillary electrophoresis in an annular channel. The contribution focuses on assessing the role of five key parameters, namely ratio of inner to outer radii of the annulus, electrical double layer thickness, inner to outer wall charge ratio, ratio between EOF and hydrodynamic flow, and ratio of convection to diffusion from a modeling point of view. The model combines both pressure driven Poiseuille flow and electroosmotic flow and inputs this flow into the solute equation. By using area-averaging on the solute equation, optimal times of separation are computed and studied.. Three of the five parameters, mentioned above, could be used in order to reduce optimal separation time: wall charge ratio, ratio between EOF and hydrodynamic flow, and ratio between inner to outer radii. Rectangular channels tended to produce the smallest optimal time of separation, a result which can be explained by comparison of the relative magnitudes of the effective velocity and effective diffusivity in the annulus. The results are not only useful for the capillary model but, also, as guiding explanations to recent experimental findings in nanocomposite gel material (Thompson, Dissertation).

1. Bhuian, B., Winfield, R.J., O'Brien, S., Crean, G.M., 2007. Pattern Generation Using Axicon Lens Beam Shaping in Two-Photon Polymerization. Applied Surface Science 254, 841-844.

2. Pfleging, W., Lu, Y., Washio, K., Willem;, H., Amako, J., 2008. Beam Shaping Effects in two photon polymerization of hybrid materials systems. SPIE Proceedings 6880.

3. Thompson, J.W., Stretz, H.A., Arce, P.E., Ploehn, H.J., Gao, H., He, J., 2012. Effect of Magnetization on Gel Structure and Protein Electrophoresis in Polyacrylamide Hydrogel Nanocomposites. Journal of Applied Polymer Science In Press.

(227f) Salt Transport Structure/Property Relationships and Modeling in Polymer Membranes for Water Purification and Power Generation

Geoffrey M. Geise, Deptartment of Chemical Engineering & The Texas Materials Institute, The University of Texas at Austin, Austin, TX, Benny D. Freeman, Department of Chemical Engineering, The University of Texas at Austin, Austin, TX and Donald R. Paul, Chemical Engineering, University of Texas at Austin, Austin, TX

Providing sustainable supplies of water and energy is a critical global challenge. Polymer membranes dominate desalination and could be crucial to power generation applications. These desalination and power generation applications include reverse osmosis (RO), nanofiltration (NF), forward osmosis (FO), pressure-retarded osmosis (PRO), electrodialysis (ED), membrane capacitive deionization (CDI), and reverse electrodialysis (RED). Improved membranes with tailored water and salt transport properties are required to extend and optimize these technologies. Water and salt transport structure/property relationships must be used to provide the fundamental framework for optimizing polymer materials for water and salt transport property-critical applications.

Sodium chloride permeability, sorption, and diffusion properties of poly(ethylene glycol) (PEG) hydrogels, sulfonated poly(arylene ether sulfone) random copolymers, and sulfonated styrenic pentablock copolymers were studied to probe the influence of polymer charge, water content, polymer structure, and micro- as well as nano-scale structural heterogeneity on salt transport properties. Salt permeability, sorption, and diffusion properties are sensitive to the uncharged or charged (e.g., sulfonated) nature of the polymer. While salt permeability is generally sensitive to water content, charged polymers exhibit salt permeability properties that are fundamentally different from those properties observed for uncharged polymers, such as PEG. The sodium chloride permeability of charged polymers increases with increasing upstream sodium chloride concentration in the range of 0.01 – 1.0 mol/L; the salt permeability of uncharged polymers, in contrast, often decreases as salt concentration increases due to decreased water content and free volume as a result of osmotic de-swelling. The salt permeability behavior observed for charged polymers is a result of covalently bound fixed charge groups, such as sulfonate groups, that affect the salt sorption and diffusion properties of charged polymers. Salt sorption can be modeled using a combination of Donnan exclusion theory and salt partitioning to further describe the influence of fixed charge concentration, water content, and micro- as well as nano-scale polymer structure heterogeneity on salt sorption and transport properties.

MODELLING OF ELECTRODIALYTIC REMOVAL OF MULTIPLE IONS FROM SYNTHETIC SOLUTION

Jogi Ganesh Dattatreya[a], Shilpi Jain[a], Navneet Kumar[a], Sujay Chattopadhyay[a], P.K. Bhattacharya[b], Amiya Kumar Ray[a]

[a]*Department of Paper Technology, I.I.T.Roorkee, Saharanpur Campus, India*
[b]*Department of Chemical Engineering, I.I.T.Kanpur, India*

email: dattut@gmail.com, avneetkum@gmail.com, shilpi621@gmail.com, sujay1999@gmail.com, pkbhatta@iitk.ac.in, amiyakumarray@gmail.com

Abstract:

Drinking water is slowly becoming a very precious commodity with rapid urbanization of world. Uncontrolled pollution in every aspect of mother earth (water, soil, environment and air) is posing challenge to flora and fauna. Days are not very far when we will have no ground water available for drinking purposes. Therefore, means to recycle used water through low cost technology needs to be devised. Commercial waste water treatment reduces BOD/COD and provides water that can be drained in channels and rivers. To make this water drinkable we need to critically analyze its contents and pass it through adsorption bed, membrane module and UV treatment. Electrodialysis (ED) is claimed to be more energy economic process compared to conventional reverse osmosis and ultra filtration.

Factors governing electrodialysis process are (i) Membrane, (ii) type of ions, (iii) medium and (iv) potential applied. For a given ion, the efficiency of separation depends on membrane type (surface charge, porosity, resistance etc.). The rate of adsorption, diffusion, desorption of ions from one side to the other decides the resistance to ion transport.

Current density was estimated theoretically for a known concentration and flow and the behavior was noted for batch recirculation electrodialysis. The cell was operated below limiting current density (lcd). A linear concentration profile (with increase in current density) was assumed between the bulk to membrane surface while the concentration at membrane surface was obtained from 'limiting current density'. This value approaches zero while the current density approaches limiting value.

Process parameters (flow, voltage and concentration) influencing removal efficiency of Na^{+}, Mg^{2+}, Cl^{-1}, SO_4^{-2} ions were estimated. Limiting current density and current densities of solutions containing multi ions were estimated using the model. The limiting current density of the process was estimated theoretically using Nernst-Planck equation [1and2] and found to decrease with time which supports experimental findings. Diffusivity and mass transfer coefficients were calculated [4and5] and found to match well with literature [3]. The proposed model can be used (within the range specified here) for estimation and design of an electrodialysis stack to achieve a desired output.

Key words: *electrodialysis, limiting current density, current density, mass transfer coefficient.*

Introduction

Rapid urbanization and industrialization has taken a toll on almost all of the readily available natural resources, but in all of them one of the most affected are the water bodies. Water is an essential part of not only our daily life but also contributes to various luxuries of life. No process industry can function without the use of fresh water and neither the contaminated water is suitable for drinking/cooking purposes. Thus the focus has now been shifted to the purification of water by various processes. Many processes like osmosis, reverse osmosis, UV radiation, electrodialysis have been in existence but the feasibility of any process depends on various factors as: economic factor, efficiency of the process, the location of the plant etc.

Electrodialysis(ED) is based on the principle of a normal dialysis process but in the presence of an external electric voltage. Electrodialysis has been extensively applied in water softening, demineralisation of organic acids, recovery of salts in pharmaceutical industry etc. All these applications involve use of ion exchange membranes, and ions responsible for water hardness (e.g. Ca^{2+}, Mg^{2+} etc.) are exchanged with soft ions (Na^{+} and K^{+}). Ion exchange membranes(IEMs) (fluoro carbon based proton exchange) have also seen application in fuel cells and they have high ionic conductivity, low gas permeability, low swelling, high mechanical strength, high resistance to oxidation, reduction and hydrolysis [1].

The basic ED cell consists of alternate layers of cation exchange membranes (CEM) and anion exchange membranes (AEM). Feed solution is fed to one of the compartments formed between the two membranes and is then passed through all the compartments for the removal of ions. A potential is applied at the end of the stack which facilitates the removal of the ions. Some of the compartments become rich in ions and are called the Concentrate solutions whereas other compartments become scarce in ions and are the Diluate compartments.The electrodes are always rinsed with the same electrolyte solution to ensure proper conductivity in the cell and to reduce the resistance for the flow of current. Electro-neutrality will be always maintained so that equivalent number of cation/ anion will exchange between diluate and concentrate solutions.

The major variable that concerns the process is the Limiting Current Density (LCD). It is the maximum amount of current that can be passed through the given solution per unit area that facilitates the removal of ions from the feed solution. Any further increase in current above the limiting current density will only increase the cost of production.

Theory

Since the most important factor for the modelling of electrodialysis (ED) process is the limiting current density (LCD), thus the major focus is to find the current density for the given set of ions. For finding the LCD, the following assumptions were made to develop a model following the Nernst idealization for the process: [2, 3, 4]

- Osmotic pressure difference was equal to the difference in HCl concentration across the membrane.
- Boundary layers adjacent to the membranes were completely static.
- The influences of the flow profile of the fluid on the model were negligible for both dilute and concentrate channels. Thus, the solution in the interior of a solution compartment was thoroughly mixed so that the concentration of the electrolyte at any point in this zone was similar.
- There was no change in the thickness of the boundary layer or the gradient along the flow channel.
- The flow dynamics were similar in all compartments.
- The distribution of pressure and current was uniform.
- Trans-membrane pressure was zero.
- There was no solution leakage in the membrane.
- Transport due to convection is ignored.

The flux of ions passing through the membrane can be expressed by Nernst Plank equation as [5]

$$N_j = -D_j \frac{\partial C_j}{\partial x} - \frac{z_j C_j F D_j}{RT} \frac{\partial \psi}{\partial x} \quad (1)$$

Where, x is the distance from boundary layer edge in contact with the diluate towards the membrane, D_j is the diffusivity of ion, 'C_j' is concentration of ion, 'R' is the universal gas constant, 'T' is the temperature in K, 'z' is the charge of the diffusing ion, and $\frac{\partial \psi}{\partial x}$ is the voltage gradient applied

Expression for LCD was derived by using a linearized form of Nernst-Planck equation; with a reasonable approximation of ionic concentration distributions near the boundary layer are linear. [The expression for the linearized Nernst-Planck equation is [5]

$$\frac{t_j^m i}{zF} = \frac{D_j(C_{j,b} - C_{j,m})}{\delta} - \frac{z_j D_j F \xi C_{j,m}}{RT} \quad (2)$$

Where, δ is the boundary layer thickness, $C_{j,b}$and $C_{j,m}$ is concentrations of ions in bulk and at the membrane surface respectively and where, ξ is the potential gradient and is expressed as.

$$\xi = \frac{\Psi}{\delta} \quad (3)$$

Where δ is the boundary layer thickness.

In the above equation the first part, i.e., $\frac{D_j(C_{j,b} - C_{j,m})}{\delta}$ denotes the role played by diffusion while the second part i.e., $\frac{z_j D_j F C_{j,m} \xi}{RT}$ reflects the effect of voltage on ion-transfer through the membrane.

Boundary layer thickness δ is estimated from film theory from the salt mass-transfer coefficient, which is usually determined based on the salt diffusivity and a mass-transfer correlation which is a suitable one. [5, 6, 7]

$$\delta = \frac{D_j}{K} \quad (4)$$

Where D_j and K are diffusivity and mass transfer coefficient of the ion in the solution. The mass transfer coefficient can be calculated from the expression of Sherwood number [5, 8] as

$$Sh = \frac{K.l}{D_j} \quad (5)$$

Sherwood number is expressed as a function of Reynolds number and Schmidt number
[5, 8]. It is always necessary to select a suitable expression for the Sherwood number based on the cell geometry and spacer configuration [9, 10, 11]

$$Sh = 0.29.Re^{0.5}.Sc^{0.33} \quad (6)$$

Schmidt's number [7,8] $Sc = \mu/\rho D_j$ (7)

Reynolds's number [7,8] $Re = \frac{\rho v l}{\mu}$ (8)

Diffusivity of ion or an electrolyte in a solution at infinite dilution is estimated from Nernst-Haskell equation 9 [12] as

$$D_{AB}^{\circ} = \frac{RT[(1/z_+)+(1/z_-)]}{F^2[(1/\lambda_+)+(1/\lambda_-)]} \quad (9)$$

Where z_+ and z_- denotes the charges on a cation and anion respectively and λ_+ and λ_- denotes the limiting ionic conductance in the solvent,

D_{AB}° - Diffusion coefficient at infinite dilution, based on molecular concentration, cm^2/s

T - Temperature, K

R - Gas constant, 8.314 J / mol. K)

λ_+, λ_- - limiting (zero concentration) ionic conductance, (A/cm^2) (V/cm) (g-equiv/cm^3)

z_+, z_- - valences of cation and anion, respectively

F – faraday, 96,500 C/g-equiv

The bulk concentration $C_{j,b}$ can be determined during the electrodialysis experiment. If the process is operated well below the limiting current density situation, the concentrations of the ions at the membrane surface can be found out using the following relation (equation 10) [13]

$$C_{j,m} = C_{j,b}(1 - i/i_{lim}) \quad (10)$$

where, i and i_{lim} are current density and LCD respectively. Based on linearized Nernst diffusion equation, the LCD is expressed by equation (11) [6].

$$i_{lim} = \frac{C_{j,b}D_j z_j F}{\delta(t_j^m - t_j^s)} \quad (11)$$

Where, t_j^m and t_j^s are transport numbers inside membrane and electrolyte solution respectively.

Substituting equations (10) and (11) in equation (2), we get the equation for the current density as

$$i = \frac{C_{j,b} D_j z_j \xi \frac{F}{RT}}{\frac{t_j^m - t_j^s}{z_j F} + \frac{\xi \delta (t_j^m - t_j^s)}{RT} - \frac{t_j^m z_j}{F}} \qquad (12)$$

Experimental limiting current density

The diluate and concentrate solutions were recycled back to a single beaker (1 L) so that the concentration of the feed solution remains constant throughout the experiment. Distinct experiments were carried out for solutions of NaCl, NaCl + Na_2SO_4, Na_2SO_4 + $MgSO_4$, Na_2SO_4+$MgSO_4$ at flow rate of 5.8 l/h, corresponding to superficial velocity of 0.0025 m/s, and Reynolds number of 14. The limiting current intensity, I_{lim}, was determined according to the method of Cowan and Brown, by plotting the ED stack electrical resistance ($\Delta V/I$) versus the reciprocal of the current intensity, 1/I [16] and identifying the point where the electrical resistance started to increase. The limiting current intensity corresponded to the reciprocal of the fitted polynomial minimum and the limiting current density was finally obtained by dividing the limiting current intensity by the effective membrane area, A (6.4516*10^-4 m^2).

Table 1: List of the values of various experimental parameters used

Parameter	Value
Temperature, T	298 K
Distance between adjacent membranes, l	5×10^{-3} *m*
Area of the membrane, *A*	6.4516*10^-4 m^2
Velocity of feed stream, v	*0.0025 m/s*
Density of feed stream, ρ	*1000kg/m*^{3}

Table 2: Ionic diffusivities at infinite dilution and 20.0 ◦C

Ion	D_j ×10^9 (m^2/s)
Na^+	1.16
Mg^{2+}	0.613
Cl^-	1.77
SO_4^{2-}	0.923

Data at 25.0 ◦C [**R.C. Reid, J.M. Prausnitz, T.K. Sherwood, The Properties of Gases and Liquids, 3rd ed., McGraw-Hill Book Company, Inc., New York, 1977.**] corrected for 20.0 ◦C through the Stokes–Einstein equation.

Experiment no.: 1

Electrode Rinsing Solution- Na_2SO_4 (5mol/m^3)

Diluate: Na_2SO_4 (5mol/m^3) + NaCl (5mol/m^3)

Concentrate: Na_2SO_4 (5mol/m^3) + NaCl (5mol/m^3)

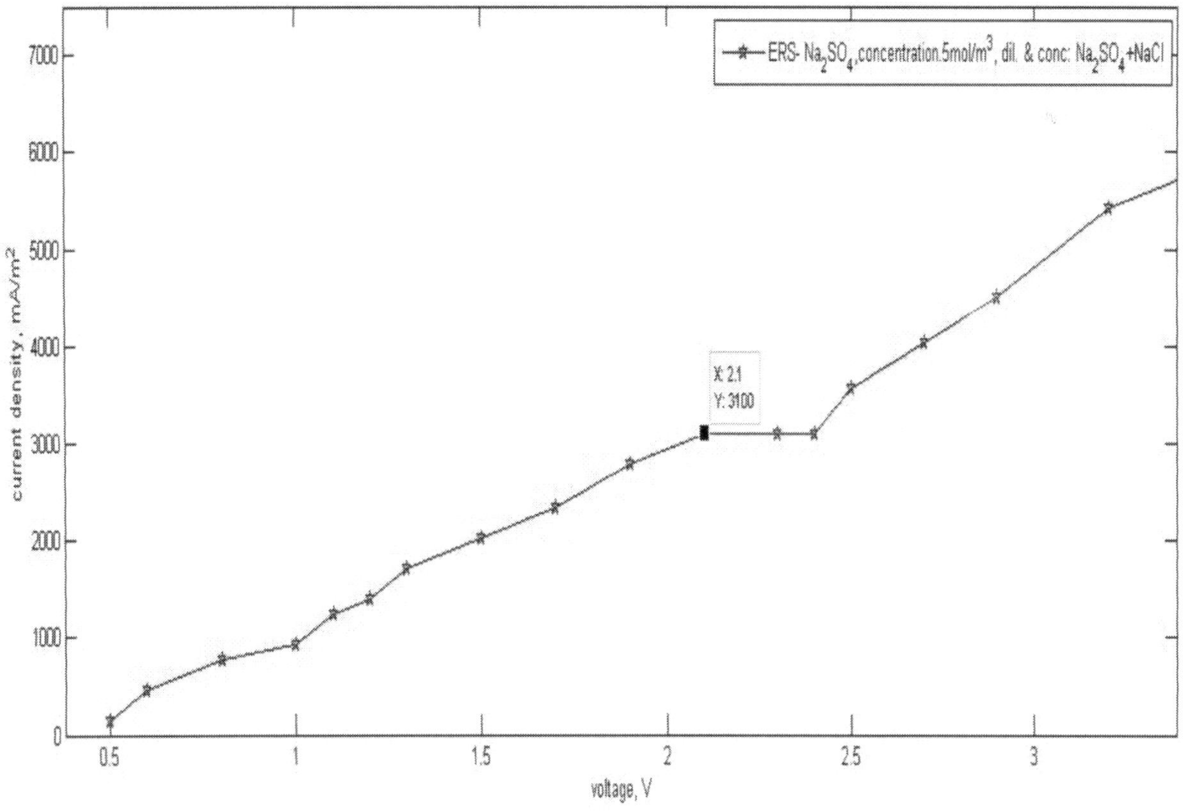

Figure 1: Plot of Current density Vs Applied Voltage for Na_2SO_4 (5mol/m^3) + NaCl (5mol/m^3)

For the above set of ions the $(LCD)_{theoretical}$ is found to be [equation 12] = 2.9267 A/m^2

Now from the above plot it is found to be around 3.1 A/m^2 which is in close approximation to the experimental value.

Experiment no: 2

Electrode Rinsing Solution: $MgSO_4$ (5mol/m^3)
Diluate: Na_2SO_4 (5mol/m^3) + $MgSO_4$ (5mol/m^3)
Concentrate: Na_2SO_4 (5mol/m^3) + $MgSO_4$ (5mol/m^3)

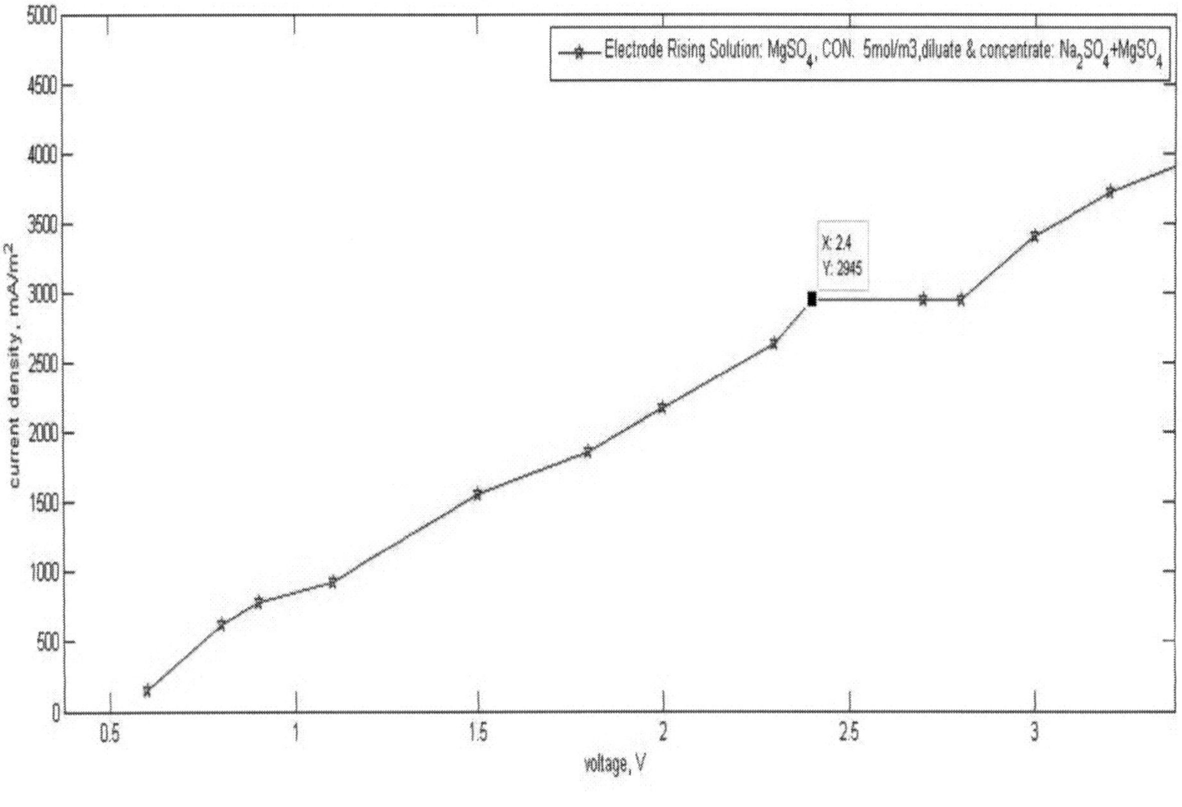

Figure 2: Plot of Current density Vs Applied Voltage forNa_2SO_4 (5mol/m^3)+$MgSO_4$(5mol/m^3)

For the above set of ions the $(LCD)_{theoretical}$ is found to be [equation 12] = 2.982 A/m^2

Now from the above plot it is found to be around 2.945 A/m^2 which is in close approximation to the experimental value.

Experiment no.: 3

Electrode Rinsing Solution: Na_2SO_4 (5mol/m^3),

Diluate: Na_2SO_4 (5mol/m^3) + $MgSO_4$ (5mol/m^3)

Concentrate: Na_2SO_4 (5mol/m^3) + $MgSO_4$ (5mol/m^3)

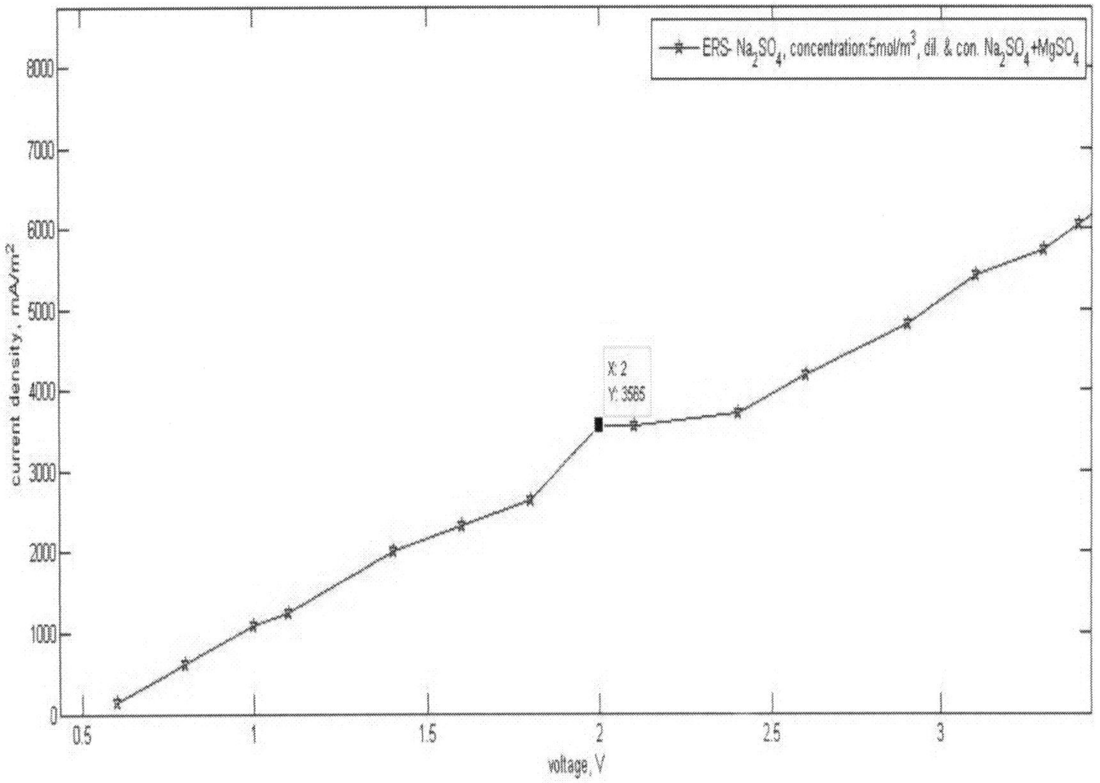

Figure 3: Plot of Current density Vs Applied Voltage forNa_2SO_4 (5mol/m^3)+$MgSO_4$(5mol/m^3)

For the above set of ions the (LCD) $_{theoretical}$ is found to be [equation 12] = 2.982 A/m^2

Now from the above plot it is found to be around 3.565 A/m^2 .

Results and Discussions:

Effect of voltage on current density

The current initially increases with the increase in applied potential (ohmic region), becomes constant momentarily and then increases abruptly with further increase in the applied potential. At LCD the concentration of ions along the surface of membrane remains unchanged i.e. as soon as ions are attracted towards the membrane, they pass through it. Thus, with increase in potential does not further increase the current. Further increase in potential leads to electro-convection currents in the solution and this would lead to failure of application of Film theory, the basis of the above mentioned theoretical analysis [14, 15]

Effect of ERS on current density:

The main purpose of using electrode rinse solution is to facilitate the diffusion of ions across the membranes. Usually the diluate solution resistance increases with time and to compensate this overall resistance a rinse solution is used. But as seen in experiment no. 2 and 3 (Fig. 2 and Fig. 3), when different ERS were used for the same set of ions, there was a noticeable difference in the experimental current density. This interesting phenomena can be explained as, higher the charge over ions higher will be the size of hydrated ions in electrode rinse solution and thus diffusion of ions will reduce across the membrane. Thus, LCD will be low as compared to those cases where ionic charge will be lower.

Prediction of Limiting Current density at varying flow rates from model:

At lower feed rate, the LCD was found to be quite low. This may be attributed to less amount of ions flowing towards the membranes per unit time. As the feed rate is increased, LCD starts increasing as both the turbulence and the amount of ions reaching the bulk/membranes per unit time has increased (hence the boundary layer thickness along the membrane has decreased). This phenomenon is clear from Fig. 4 in which LCD for a solution containing Na^+ ion is plotted against Reynolds number. In this case at high Reynolds number (> 125) the increase in LCD w.r.t. Reynolds number is almost a straight line. The linear variation of LCD depends on type of ions present in feed solution. Also it can be inferred from Fig. 4 that at a constant Reynolds number, limiting current density is more for the solution of higher concentration.

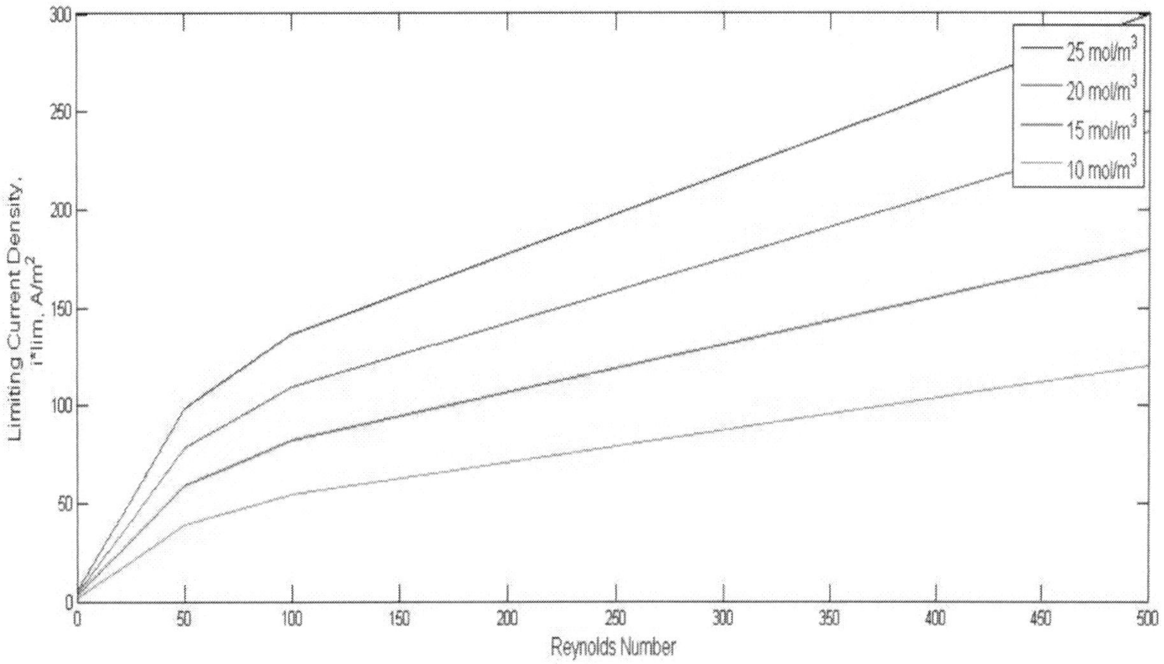

Figure 4: Plot of Limiting current density vs. Reynolds Number for four different concentrations (10mol.m^{-3}, 15mol.m^{-3}, 20mol.m^{-3}, and 25mol.m^{-3}).

Conclusion

1. Concentration of ions along the surface of membrane remains unchanged at LCD. With increase in electrode potential the current does not rise beyond the limiting value. Instead, this increased potential leads to electro-convection currents in the solution which disturbs the uniformity of the film leading to failure of film theory.
2. LCD increases with increase in concentration of ions and Reynolds number.
3. Ionic charge and the hydrodynamic radius of the ions in electrode rinse solution decide LCD.

References:

1. M.Y. Kariduraganavar, R. K. Nagarale, R. K. Kittur x, S. S. Kulkarni, Ion-exchange membranes preparative methods for electrodialysis and fuel cell applications, Desalination 197 (2006) 225.
2. V. Montiel, A. Aldaz, J.M. Ortiz, J.A. Sotoca, E. Exposito, G. Gallud, Brackish water desalination by electrodialysis: batch recirculation operation modeling, Journal of Membrane Science 252 (2005) 65–75.
3. N. Boniardi, R. Rota, G. Nano, B. Mazza, Analysis of the sodium lactate concentrationprocess by electrodialysis, Separations Technology 6 (1996) 43–54.
4. M. Law, T. Wen, G.S. Solt, Thickness and concentration profile of the boundary layer in electrodialysis, Desalination 109 (1997) 5–103.
5. V. Geraldes, M.D. Afonso, Limiting Current Density in the Electrodialysis of Multi-Ionic Solutions, J. Membr. Sci. 360 (2010) 499.
6. H. J. Lee, H. Strathmann, S.H. Moon, Determination Of The Limiting Current Density In Electro-Dialysis Desalination As An Empirical Function Of Linear Velocity, Desalination 190 (2006) 43.
7. R.B. Bird , W. E. Stewart, E. N. Lightfoot, Transport Phenomenon, 594.
8. R.E. Treybal, 'Mass-Transfer Operations', Third Edition, Molecular diffusion in fluids.
9. M.S. Isaacson, A. A. Soninlnd, Sherwood Number and Friction Factor correlations for clectrodialysis systems, with application to process optimization, Ind.Eng. Chem., Process Des. Dev., 15, No. 2 (1976) 313-321.
10. O.V. Grigorchuk, V.I. Vasil'eva, V.A. Shaposhnik, Local characteristics of mass transfer under electrodialysis demineralization, Desalination 184 (2005) 431–438.
11. V. A. Shaposhnik, O. V. Grigorchuk, Mathematical model of electrodialysis with ion exchange membranes and inert spacers, Russian Journal of Electrochemistry, 46, No. 10 (2010) 1264-1271.
12. B.E. Poling, J.M. Prausnitz, J.P. O'Connell, The Properties of Gases and Liquids, Fifth edition, Mc-Graw Hill publications.
13. V.M.Barragan and C.Ruiz-Bauza, Current–Voltage Curves for Ion-Exchange Membranes: A Method for Determining the Limiting Current Density, J. Coll. Int. Sci. 205 (1998) 365.
14. I. Rubinstein, B. Zaltzman, O. Kedem, Electric fields in and around ion-exchange membranes, J. Membr. Sci. 125 (1997) 17–21

15. B. Zaltzman, I. Rubinstein, Electro-osmotic slip and electroconvective instability, J. Fluid Mech. 579 (2007) 173–226.
16. D.A. Cowan, J.H. Brown, Effect of turbulence on limiting current in electrodialysis cells, Ind. Eng. Chem. 51 (12) (1959) 1445–1448

(241a) Reflections On 50 Years in Chemical Engineering!

William B. Krantz, Chemical and Biological Engineering, University of Colorado, Boulder, CO

Reflecting on 50 Years in Chemical Engineering!

William B. Krantz

This presentation will review the somewhat circuitous path traversed by the speaker during the 50 years since he received his B.S. in chemical engineering from the University of Illinois. Although this paper is being presented in a session sponsored by AIChE's Separations Division, the speaker spent the first 20 years of his professional life working in fluid mechanics, interfacial science, synfuels, and self-organization in nature. He initiated a research program in membrane science during the middle age of his professional life. This progression does not seem particularly logical and most certainly was in no way premeditated. As many of us have found in our life, we sometimes end up doing what we are doing by way of earlier decisions that we did not think were all that important at the time. This presentation will try to show how the threads that weave together the fabric of one's professional life somehow come together, sometimes almost accidentally.

This presentation will also serve as an introduction to some of the other speakers in these sessions, some who are former students and others who are professional colleagues with whom the speaker has interacted in some capacity. This reflection on 50 years in chemical engineering hopefully will acknowledge appropriately the wonderful influence my former students and colleagues have had on my professional life. Hopefully this presentation will be inspirational for young people in that it provides a datum point from one person's professional life that underscores the opportunities that a career in chemical engineering provides.

(241b) Nanostructured Polymer Membranes with Sub-1-Nm Pores for Molecular Size Separations Via the Cross-Linking of Lyotropic Liquid Crystal Assemblies

Richard D. Noble, Department of Chemical and Biological Engineering, University of Colorado, Boulder, CO, Douglas L. Gin, University of Colorado, Boulder, CO, Brian Weisenauser, Chemistry and Biochemistry, University of Colorado, Boulder, CO and Blaine Carter, Chemical and Biological Engineering, University of Colorado, Boulder, CO

The development of polymers with ordered, interconnected, sub-1-nm size pores for molecular size separation is presented. These nanostructured polymers are made by cross-linking lyotropic liquid crystal (LLC) assemblies formed by polymerizable surfactants in the presence of water. Supported LLC polymer membranes with a type I bicontinuous cubic (QI) architecture made with a 1st-generation gemini phosphonium LLC monomer possess a 3-D interconnected water pore system with a uniform pore size of ca. 0.75 nm. These LLC membranes are able to cleanly size-exclude hydrated salt ions and a variety of small organic solutes from water (i.e., water desalination) with good permeabilities. The water desalination/nanofiltration performance of this LLC polymer membrane is superior to that of conventional nanofiltration membranes, and close to that of commercial reverse-osmosis membranes. The relationship between the nanoscale architecture of these polymerized LLC assemblies and their separation will be discussed. The synthesis of a new, more modular QI-phase LLC monomer platform, and the separation performance of new LLC membranes made from this monomer, will also be presented.

(241c) High Temperature Dual-Phase Inorganic Membranes for Carbon Dioxide Separation

Jerry Y. S. Lin[1,2], Tyler Norton[3], Zebao Rui[4] and Jose Ortiz[3], (1)School for Enginering of Matter, Transport and Energy, Arizona State University, Tempe, AZ, (2)Department of Chemical and Materials Engineering, Arizona State University, Tempe, AZ, (3)Arizona State University, Tempe, AZ, (4)School for Engineering of Matter, Transport and Energy, Arizona State University, Tempe, AZ

Electrical generation from fossil fuels results in emission of carbon dioxide into atmosphere causing global warming. The challenge is how to effectively capture carbon dioxide before other alternative non-fossil energy resources are developed to replace the fossil fuels. There are three strategies to capture carbon dioxide from power plants burning fossil fuels: post-combustion carbon dioxide removal, pre-combustion carbon removal andoxyfuelcombustion processes. High temperature CO_2 perm-selective inorganic membranes will play a critical role in these processes. All microporous inorganic membranes showing high CO_2/N_2 selectivity at room temperature become non-selective for CO_2 separation at high temperatures due to low diffusion controlled selectivity for molecules of similar sizes. The talk will focus on new dual-phase membranes that permeate only carbon dioxide at high temperatures. The dual-phase membranes consist of a metal or oxygen ionic conducting oxide phase and a molten carbonate phase which conduct respectively electrons or oxygen ions and carbonate ions. When using metal as the solid phase, the membranes permeate both CO_2 and O_2 in 2:1 ratio. Dual-phase membranes consisting of a mixed electronic and ionic conducting or oxygen ionic conducting ceramic solid phase permeate CO_2 with CO_2permeace upto 10^{-7}mol/m^2.s.Pa with CO_2/N_2 selectivity up to 3000 in temperature range from 500-900°C. The activation energy for CO2 permeation through these dual-phase membranes is in the range of 80-110 kJ/mol, depending on the ceramic phase. CO_2 permeation flux is controlled by oxygen ionic conductivity in the ceramic phase. Results will be presented showing improvement of CO_2permeance by using a ceramic phase with higher oxygen ionic conductivity or preparing membranes with small thickness.

(241d) Enhancing Membrane Performance - IDEAS THAT FIT the Bill

Anthony G. Fane, Singapore Membrane Technology Center, Nanyang Technological University, Singapore, Singapore, Rong Wang, Singapore Membrane Technology Centre, Nanyang Technological University and Chuyang Y. Tang, Singapore Membrane Technology Centre and School of Civil and Environmental Engineering, Nanyang Technological University, Singapore, Singapore

The Singapore Membrane Technology Centre (SMTC) has a broad interest in the development of membrane processes for water and the environment. Advancing new ideas, scrutinizing old ones and their communication are keys to our activities. In these tasks Bill Krantz has been a guide and a guru. This presentation highlights some Krantz-inspired (K-i) developments at SMTC to enhance membrane performance. Characterization of membranes and process phenomena provides valuable insights. Techniques, such as evapoporometry to measure pore sizes and ultrasonic time domain reflectometry (UTDR) for fouling detection are powerful K-i techniques. At SMTC we have used UTDR to observe the metastability of colloidal fouling and to monitor biofilm development using a novel approach. Our membrane integrity sensor can be used to detect 1 broken fibre in 106 fibres, and is more sensitive than particle counting or turbidity, thanks to novel K-i design and performance metrics. In order to better control membrane fouling we are evaluating vibrations for both low pressure and high pressure membranes. To reduce energy in desalination we are examining reflux and recycle concepts. Our interests include membranes for special needs (acute and chronic water supply), and one K-i approach is to combine hydrogels and membranes. These projects illustrate how curiosity and enquiry can enhance the performance of membranes for mankind.

(241e) Study of Cell Proliferation in Porous Scaffold in a Taylor-Couette Bioreactor

Jian Qiao, Chemical and Biomolecular Engineering, National University of Singapore, Singapore 117576, Singapore and **Chi-Hwa Wang**, Department of Chemical and Biomolecular Engineering, National University of Singapore, Singapore, Singapore

Taylor vortices can be formed by subjecting a viscous liquid to shear stresses in an annular space between two rotating cylinders and a stationary bottom surface. In recent years, Taylor vortex flow is employed in many practical applications such as reaction, filtration and extraction. In this study, Taylor vortex flow is applied in a bioreactor to culture cells that are seeded in a degradable porous scaffold, taking its advantages of low shear stress, high mass transfer rate and easy implementation for scaling up. The biodegradable porous scaffold is fabricated from solvent-free supercritical gas foaming technique. The cell distribution in the scaffold is investigated by scanning electron microscope (SEM) and Confocal microscope. In contrast, the cell number in the scaffold is measured by PicoGreen dsDNA assay kit to analyze the cell seeding efficiency into the scaffold and proliferation rate inside the bioreactor. The cell proliferation rate is compared with the conventional bioreactor, and the optimal operation conditions are explored.

Using Extraction Columns for Recovery Of Valuable Products from Biomass

Authors:

Donald J. Glatz
Thomas D. Lightfoot

Koch Modular Process Systems, LLC.

Prepared for Presentation at

AIChE National Meeting

Pittsburgh, PA
October - November 2012

INTRODUCTION

In recent years there has been a large emphasis on the production of organic chemicals from biomass generated via processes such as fermentation and algae treatment. Most of these processes produce low concentrations of organic chemicals that are higher boiling than water, and as such distillation is not a viable first step for the recovery and purification process. Liquid-Liquid Extraction (LLE) is an ideal unit operation for the removal of many of these compounds either directly from the broth itself (whole broth) or from the broth after some pre-treatment step to separate the "bad acting" components such as cell mass, lignin, etc. (1). Selection of the type of column is critical in order to optimize the performance and cost for the extraction process. This paper will present case studies of several different biomass feed materials that were tested in pilot plant columns. For each case, results from the pilot plant testing are summarized, including both the raw data and the expert evaluation of the hydraulic behavior in the column. Both types of information are very important in selecting the best column for a given application and then using the pilot data to scale up to a demonstration plant or a full scale production plant. The conclusions from this paper will reinforce the need for pilot plant testing each individual process, especially for complex systems such as biomass.

PROCESS EVALUATION

1. Solvent Selection and Laboratory Testing

Solvent selection is the first step in the development of any LLE process. This topic has been covered in other references (2) and as such will not be covered here. Once the solvent has been selected, it is important to perform laboratory scale tests to generate the liquid-liquid equilibrium data and also to evaluate the hydraulic behavior for the system. This information can be generated by various methods, with the preference here to use a round-bottom flask, with half-moon impellor. Starting with the feed solution and fresh solvent, a series of mix-decant runs are made, using the raffinate from each preceding run and fresh solvent so as to generate equilibrium data from the feed concentration down to the required raffinate concentration. This equilibrium data is useful for determining the solvent to feed ratio and number of theoretical stages that will be required to achieve the desired recovery of the chemical product from the feed broth.

For biomass feeds, a qualitative evaluation of the hydraulic behavior between the phases is just as important as the raw equilibrium data for selecting the type of extraction column that will be used for testing and eventual scale-up to a production column. Systems that mix and separate quickly, without a tendency for the formation of an emulsion can be processed using a column with rotating internals. However, years of experience has shown that the reciprocating type agitation used by a KARR® Column is superior to rotating internals for systems that have a tendency to emulsify (3). Many biomass systems, especially those produced by fermentation processes or by algae ponds demonstrate this tendency towards emulsification upon contact with the extraction solvent. As will be shown in some of the examples presented herein, the KARR® Column can be effectively used for these systems.

2. Pilot Plant Testing

Following generation of the laboratory data and selection of the column type for scale-up, the next step is pilot plant testing of the extraction column. Here, the performance of the column is optimized with respect to capacity, height, S/F, agitation speed and temperature. This paper will cover pilot testing the extraction step for three different systems.

Once the extraction column performance has been optimized, the actual compositions for the extract and raffinate stream can be identified. These compositions are then used to study the design for the downstream distillation steps. The number and design for the distillation columns is dependent on the various components and compositions in the two streams. If the components are identified and VLE data is available, then the distillation columns can often be designed via computer simulations and good engineering understanding of the internals to be used. However, where this information is not well known, then testing of the distillation steps may be required.

3. Complete System Design

As mentioned above, in addition to the extraction column, downstream distillation columns are usually required to recover the solvent for recycle and also to selectively recover and purify the chemicals of interest. However, it is not the scope of this paper to evaluate these distillation steps. It is critical to have a good understanding of the composition of the LLE feed as well as the final compositions of both the extract and raffinate phases leaving the extraction column so that the distillation columns can be designed accordingly.

PILOT PLANT EQUIPMENT SET-UP AND PROCEDURE

The typical equipment set-up for testing the extraction step is shown in Figure 1. Typically, a set-up for extraction consists of the following:

- A glass shell column for visual observation of the phase behavior in the column. The actual column type and configuration depends upon the process requirement.
- A multi-stage manifold can also be used for the inlet location of the disperse phase, thus allowing for variation in the total height of the column.
- Typically the feed and solvent are charged directly from 55-gallon drums using dip tubes and variable speed pumps. Flow rates are monitored by mass flow meters and adjusted as required to maintain the desired rates.
- The bottoms flow rate (heavy phase outlet) and interface are manually controlled using a variable speed bottoms take-off pump. Light phase overhead is allowed to flow by gravity to a receiver.

- Inlet heat exchangers and electrical heating tape wrapped around the column shell are used to adjust and maintain the required temperature within the column.

Test Procedure

For all the feed materials used in this evaluation, the aqueous broth was the heavy phase and the continuous phase in the extraction process. The solvent was the light phase (lower density than the broth) and was the dispersed phase, thus the interface was at the top of the column. The following generalize procedure was used for these test programs:

- Initially, the column was filled with water (not broth) as the continuous phase. Once full, the flow of the broth and solvent were started and the flow rates of each were set.
- The agitator was turned on at a low speed (lower that expected for the process).
- When the interface appeared in the top disengaging chamber, the bottoms pump was turned on and adjusted as required to control the interface at a specified level.
- Heat exchangers and heating tape were adjusted as required to produce the targeted temperature within the column.
- The agitation speed was set as specified by the engineer; this was time 0.
- During start-up each day, a total of five (5) column turnovers were performed before the initial samples were taken. A column turnover is defined as the total column volume divided by the combined flow of all inlet streams. After the first run, and following adjustments of the variables, a total of 3-4 column turnovers were performed before additional sampling.

APPLICATION 1 – Fermentation Broth

This application involved the extraction of carboxylic acids from an aqueous fermentation broth generated in a cellulosic ethanol process. Ethyl acetate was the solvent of choice. Initial attempts by the end user to develop the LLE process step in a rotating disc contactor (RDC) proved to be unsatisfactory due to poor product recovery (< 90%) and a high solvent to feed ratio required ($\geq$ 1.7).

Although the authors did not perform the shake tests for this specific application, enough information was provided by the end user to demonstrate that the process had a high tendency towards emulsification. Thus, this process was tested in a 25 mm diameter KARR® Column. The KARR® Column demonstrated excellent operability, generating fine dispersion of the solvent phase at low agitation speed (as expected for systems which tend to emulsify). Results from the pilot runs are shown in Table 1. A review of this data shows:

- The solvent to feed ratio was improved to 1.5.
- At room temperature (20-25 °C), the acid recovery was optimized at approximately 96-97%.
- At slightly elevated temperature (40-45 °C) the acid recovery was increased to 98-99%.

- For scale-up purposes, the capacity was selected as 650 GPH/ft^2, although successful operation was observed at 700 GPH/ft^2.

Following this pilot plant testing, a production scale KARR® Column was designed and fabricated. This column has been installed as part of a complete modular system including a number of downstream distillation columns for recovery and purification of the specified carboxylic acid and recycle of purified solvent.

APPLICATION 2 – Woody Biomass

This process involved the recovery of organic compounds from woody-biomass liquor generated in a pulp and paper plant. The scope was to develop a process (LLE followed by several distillation step) to recover and purify several compounds. The aqueous feed stream was actually a condensate from an evaporator, and as such lignin and other cellulosic by-products were not present.

As a first step, shake tests were performed using an organic solvent. This solvent proved to be acceptable for the extraction of the key components, and liquid-liquid equilibrium data was generated. Also, it was observed that this system mixed and separated easily without any tendency to emulsify. As such, the SCHEIBEL® Column, which has rotating internals, was selected as the column of choice for pilot plant testing and design. The SCHEIBEL® Column demonstrated excellent operability, generating fine dispersion of the solvent phase at moderate to high agitation speed. Pilot plant testing verified the shake test observation that the system did not demonstrate any tendency towards emulsification. Therefore, the SCHEIBEL® Column was an excellent choice for this process. Results from the pilot runs are shown in Table 2. A review of these data shows:

- Excellent recovery of Acid A was achieved for nearly all runs. Since Acid B was more difficult to extract, Acid B extraction was the determining factor for evaluation of the column performance and subsequent scale up to production equipment.
- Run #9 was selected as the design basis for the production system. Thus, the capacity was 600 GPH/ft^2, the S/F = 1.5 and the expected recovery of Acid B was 93% (minimum).
- There was also some indication that once this process is in production mode, that the S/F could be reduced to 1.3. This has the obvious benefit of reducing the operating cost by minimizing the amount of solvent for distillation and recycle.

Following pilot testing, a complete extraction / distillation system was designed for recovery and purification of 3 components in the woody biomass feed. Implementation of this system is pending.

APPLICATION 3 – Algae Broth

This application involved the extraction of an organic product from an aqueous broth generated in an algae pond using an organic solvent selected by the end user. Initial attempts by the end user to develop the LLE process step in a batch mix-decant mode of operation demonstrated that the system was highly sensitive to emulsification and the phases were difficult to separate. Therefore a KARR® Column was recommended for testing and scale-up. Also, due the properties of the product, higher temperatures (> 60 °C) were required to optimize the mass transfer.

The results from testing in a 1" (25 mm) diameter column with 12' plate stack height are shown in Table 3. Based upon this data and observation of the column performance during testing:

- It was obvious that the KARR® Column was the best column type for this application. The reciprocating action of the perforated plates allowed for solvent dispersion in the broth, as well as phase separation within the column. As shown, low agitation speeds were necessary, as expected for these types of systems. Although some runs were made at 75 SPM, the column was observed to be near flooding at this speed (unstable operation). As such, the majority of the runs were made at 50 SPM.
- At 500 GPH/ft^2 and 12' plate stack height, the product recovery was 98% for S/F ratio of 0.5 – 1.0. The product recovery decreased when using 8' plate stack height.
- Pushing the capacity to 700 GPH/ft^2 did appear to reduce the product recovery. However, this was only one data point. At 900 GPH/ft^2 the column operation was unstable, and considered too close to flooding for safe operation.

At the time of this paper, this application was still in development mode with a demonstration size plant pending.

CONCLUSION

Liquid-liquid extraction is an excellent first step for recovery and purification of chemicals from biomass broths created by fermentation, algae, or chemical processing. This is especially true when the chemicals to be recovered are higher boiling than water. For systems that mix and separate easily, rotating columns such as the SCHEIBEL® Column should be considered. However, when the system has a strong tendency to emulsify, then the reciprocating-mixing action of the KARR® Column has been shown to be a better design. Pilot plant testing in the specified extraction column is strongly recommended for process development and generation of the data for accurate scale-up.

LITERATURE CITED

1. Cusack, R., Fremeaux, P., and Glatz, D., "A Fresh Look at Liquid-Liquid Extraction, Extractor Design and Specification", Chemical Engineering, February 1991.

2. Cusack, R., Glatz, D., "Apply Liquid-Liquid Extraction to Today's Problems", Chemical Engineering, July 1996.

3. Cusack, R., Fremeaux, P., and Glatz, D., "A Fresh Look at Liquid-Liquid Extraction", Chemical Engineering, March 1991.

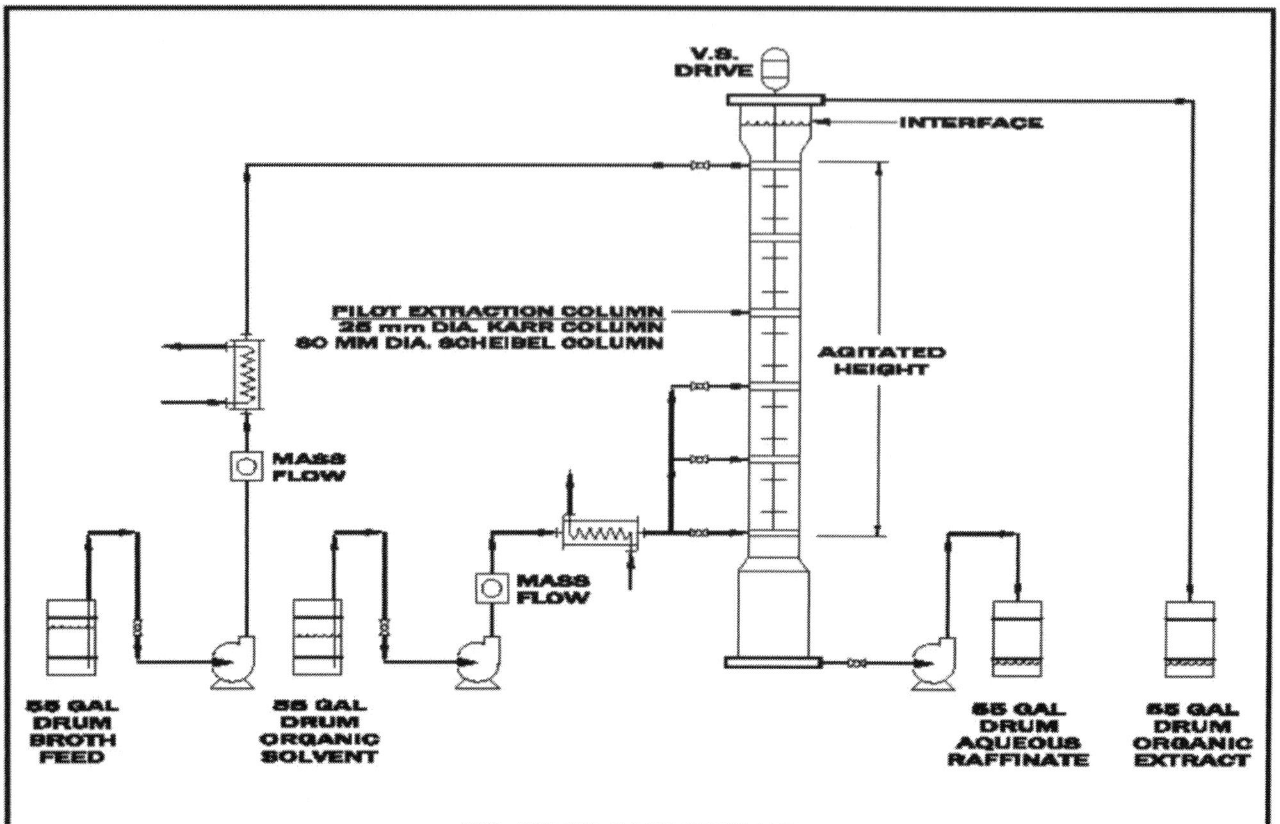

PILOT PLANT SET-UP

FIGURE - 1

KOCH MODULAR PROCESS

TABLE 1

Pilot Plant Data for Fermentation Broth Extraction

All runs performed in a 25 mm diameter, KARR® Column

RUN #	Plate Stack Ht. (feet)	Capacity GPH/ft^2	Temp. C	S/F Ratio	Agitation SPM	Acid Recovery %
1	12	700	23	1.7	30	95.0
2	12	700	23	1.7	40	97.5
3	12	700	23	1.7	50	Flooded
4	8	700	23	1.7	30	93.3
5	8	700	23	1.7	40	Flooded
6	10	700	23	1.7	30	96.3
7	12	650	23	1.5	30	96.5
8	12	500	23	1.5	40	97.6
9	12	700	23	1.5	30	94.3
10	12	650	45	1.5	30	98.3
11	12	650	45	1.5	50	98.7
12	12	650	45	1.5	60	Flooded

Table 2

Pilot Plant Data for Woody Biomass Extraction

All runs performed in a 80 mm diameter, SCHEIBEL® Column

RUN #	Column Height Number of Stages	Capacity GPH/ft^2	S/F Ratio	Agitation SPM	Product Recovery (%) Acid A	Acid B
1	60	550	1.5	400	99.6	92.8
2	60	550	1.5	500	99.6	92.5
3	40	550	1.5	500	99.2	90.4
4	50	550	1.5	500	99.4	91.9
5	60	550	1.3	500	99.4	92.4
6	40	550	1.3	500	99.4	89.1
7	40	550	1.8	500	99.6	92.4
8	60	550	1.8	500	99.7	94.8
9	60	600	1.5	500	99.7	93.4
10	60	600	1.0	500	96.0	76.3
11	40	650	1.3	500	98.3	88.0
12	60	650	1.3	500	99.6	91.4

Table 3

Pilot Plant Data for Algae Broth Extraction

All runs performed in a 25 mm diameter, KARR® Column

RUN #	Plate Stack Ht. (feet)	Capacity GPH/ft2	Temp. C	S/F Ratio	Agitation SPM	Oil Recovery %
1	12	500	65	1	30	88.3
2	12	500	65	1	50	97.8
3	12	500	65	1	75	99.0
4	12	500	65	2	30	96.3
5	12	500	65	2	75	89.3
6	12	500	65	1	50	98.4
7	8	500	65	2	50	93.1
8	8	500	65	1	50	90.8
9	12	700	65	1	50	85.1
10	12	900	65	1	40	Flooded
11	12	500	65	0.8	50	98.2
12	12	500	65	0.5	50	97.6

(242b) Purification of Arabitol From Fermentation Broth of Debaryomyces Hansenii Using Glycerol As Substrate

Abdullah Loman, Chemical and Biomolecular Engineering, The University of Akron, akron, OH and Lu-Kwang Ju, Department of Chemical Engineering, The University of Akron, Akron, OH

Abstract

Arabitol, a five carbon sugar alcohol, has many potential industrial applications. This work deals with the recovery of arabitol from fermentation broth containing arabitol, unconsumed glycerol and other minor impurities. Arabitol was produced by fermentation of *Debaryomyces hansenii* in a suitable medium containing glycerol as the carbon source. To collect and purify arabitol from the broths, studies were done and the following process developed. The supernatant was first treated with 8 g/l of activated carbon for decoloration and removal of some impurities. Then glycerol was removed by acetone extraction. To obtain the necessary phase equilibrium information, solubility and liquid-liquid equilibrium studies were done with simulative three-component acetone-glycerol-xylitol systems. Xylitol, an enantiomer of arabitol, was used because it was much less expensive and more readily available than pure arabitol. Selective extraction of glycerol by acetone from the broth was accordingly designed and optimized. The extraction remainder, containing the desirable arabitol, was dried to remove the residual acetone and then extracted for arabitol with hot butanol (900 ml per liter of supernatant at 90°C).White arabitol crystals were collected from the extract cooled to room temperature. The overall recovery and purity of arabitol crystals were 66% and 95%, respectively. Arabitol loss occurred primarily during the acetone extraction to remove glycerol. Lowering the unconsumed glycerol concentration in fermentation broth is critically important to the overall arabitol production cost.

(242c) Application of High Voltage for the Destabilization of Water-in-Oil Emulsion for the Extraction of Lipids From Microorganisms without Removal of Water

Emmanuel Revellame[1], William Holmes[1], Rafael Hernandez[1], L. Antonio Estévez[2] and W. Todd French[1], (1)Dave C. Swalm School of Chemical Engineering, Mississippi State University, Mississippi State, MS, (2)Department of Chemical Engineering, University of Puerto Rico, Mayaguez, PR

The conventional biofuel production process from microbial sources involves lipid extraction, fuel conversion (i.e. transesterification, catalytic cracking) and downstream processing. Other processes combine the first two steps into a single step (i.e. in situ transesterification, fluidized-bed catalytic cracking) to improve the process economics. However, all these processes are sensitive to water, which for microbial lipid sources constitutes up to 98% (weight) of the growth medium. Removal of water has proved to be the bottleneck for biofuel production from microbial sources (>50% of the biodiesel cost from activated sludge). Thus an alternative extraction strategy is needed. In this study, ternary systems containing alkane + lipids + water were established to design a water tolerant extraction scheme. Initial results indicated that for these ternary systems, emulsions could be formed in different regions of the phase diagram. These emulsions might complicate the phase separation. Thus, an electrostatic coalescence process was sought to break these emulsions. This process, which is important in the petroleum industry for removal of water from crude oil, involve application of electric field (~13-20 kV) to break emulsion drops and could be used to separate lipids in microbial consortia in the presence of water.

Reactive Extraction of Nicotinic Acid using Tri-*n*-Octylphosphine Oxide (TOPO) Dissolved in a Binary Diluent Mixture

Sushil Kumar[1*] **Dipaloy Datta**[1] **and B V Babu**[2]

[1]Department of Chemical Engineering
Birla Institute of Technology and Science (BITS), PILANI – 333031 (Rajasthan), INDIA

*Presenting and Corresponding Author:
E-mail: sushilk2006@gmail.com; skumar@bits-pilani.ac.in
Phone : +91-1596-245073 Ext 215; Fax: +91-1596-244183
Homepage: http://universe.bits-pilani.ac.in/pilani/skumar/profile

Abstract

The aim of present work is to intensify the recovery of nicotinic acid from fermentation broths using reactive extraction. Tri-*n*-octylphosphine oxide (TOPO) in a diluent mixture of an active (also called modifier, MIBK) and an inactive (inert, kerosene) diluent (1:1 v/v) is employed. The effect of acid concentration and TOPO concentration on degree of extraction is determined. TOPO/diluent system favors the formation of 'not overloaded' complexes of polar acid-TOPO structures with the Z factors restricted mainly between 0.032 and 0.684. The maximum value of K_D is found to be 4.168 with TOPO (0.365 mol/L) at an acid concentration of 0.02 mol/L. The mathematical model, based on mass action law, is proposed to estimate the values of partition coefficient (P) and dimerization constant (D) in physical extraction and equilibrium constants (K_E) and number of reacting extractant molecules per acid molecule in chemical extraction. The model predicted values of K_E are showing good correlation with $R^2 > 0.98$ and maximum value of $SD = 0.092$.

Introduction

Nicotinic acid (pyridine-3-carboxilic acid) is a water-soluble vitamin and precursor to coenzymes (NADH, NAD^+, $NADP^+$, and NADPH) which serves an important role in the redox reactions taking place inside the human living cells for the metabolism activity. Niacin helps in both DNA repair and formation of steroid hormones in the adrenal gland. A deficiency of niacin can cause pellagra, a serious disease that has paralyzed mankind for centuries. Also, a mild deficiency slows down the metabolism of the body, causing decreased tolerance to cold (Kumar and Babu, 2009). The production of organic acids by biochemical fermentation route is comparatively a clean and a green technology. The biosynthesis process produces nicotinic acid at a lower rate and also the concentration of the acid in the fermentation broth is found to be very low. Therefore, to make the fermentation route efficient and effective, there is a need to develop novel fermentation processes and efficient separation techniques. The reactive extraction with higher distribution coefficient is proposed to be an efficient and eco-friendly primary separation process (Kertes and King, 1986; Kumar and Babu, 2008).

Phosphorus-bonded, oxygen containing extractants have a phosphoryl group and a stronger Lewis basicity than those of carbon-bonded, oxygen-containing extractants. Phosphorus-bonded, oxygen-containing extractants can only co-extract small amounts of water and show low solubilities in water. When organophosphorus extractants are used, the solvation has a higher specificity. Tri-n-butyl phosphate (TBP) and tri-*n*-octylphosphine oxide (TOPO) are used effectively for the separation of carboxylic acids from aqueous solution (Mei et al., 2001; Keshav et al., 2008; Keshav et al., 2009; Kumar et al., 2008; Kumar and Babu 2009). These extractants are dissolved in organic diluents (ketones, alcohols, hydrocarbons) to provide appropriate physical properties (density, viscosity, etc.,) to the extractant-diluent system. The diluents are categorized in two groups based on their activity: (i) inactive (inert) diluents, and (ii) active diluents (modifiers). The presence of polar functional groups in the modifiers enables them to act as better solvation medium for the acid-extractant complex by the formation of hydrogen bond. Also, a modifier enhances the extracting power of organophosphoric extractant as compared to an inert diluent in the extraction of organic acids (Mariya et al., 2005; Yankov et al., 2004).

The present work is aimed to intensify the recovery of nicotinic acid using reactive extraction with tri-octyl phosphine oxide (TOPO) in a diluent mixture consisting of an active and an inactive diluent [MIBK + kerosene (1:1 v/v)].

Materials and Methods

The reagents used in this equilibrium study are the carboxylic acid [nicotinic acid (NA)], extractant [tri-*n*-octylphosphine oxide (TOPO)], and diluents [4-methylpentan-2-one (MIBK) and kerosene] and listed in Table 1 with their physical properties.

The aqueous solutions of nicotinic acid are prepared in the concentration range 0.02 to 0.12 mol/L using distilled water. Organic solutions are prepared by varying the concentration of TOPO (0.1 and 0.5 mol/L) in the mixture of MIBK + kerosene (1:1 v/v). MIBK is used as the active diluent (modifier). Kerosene is used as inactive diluent to control density and viscosity of the organic phase. The extraction equilibrium experiments are carried out with equal volumes (20 ml) of the aqueous and organic solutions in conical flasks of 100 ml and shaken at 100 rpm for 8 hours in a temperature controlled reciprocal shaker bath (HS 250, Remi Labs, India) at constant temperature (298 K). After attaining equilibrium, the mixture of aqueous and organic phases is kept for separation in separating funnel (125 ml) for 2 hrs at 298 K. After separation of both phases, the aqueous phase acid concentration is analyzed by titration using NaOH solution of 0.01 N with phenolphthalein as an indicator and also by UV-Vis Spectrophotometer (Evolution 201, Merck, India at 262 nm). The acid concentration in the organic phase is calculated by mass balance. The equilibrium p*H* values of aqueous solution are measured by a digital p*H*-meter (PCT 40, ArmField Instruments, UK).

The experimental data are analyzed by calculating distribution coefficient ($K_D = C_{org}/C_{aq}$), degrees of extraction [$E = K_D / (1 + K_D)$] and loading ratios ($Z = C_{org}/[\overline{T}]_o$). Where, C_{org} and C_{aq} are the total carboxylic acid concentration in the organic phase and in the aqueous phase, respectively, at equilibrium; and $[\overline{T}]_o$ is the initial extractant concentration in the organic phase.

Results and Discussion

The equilibrium results for physical and chemical extraction of NA from aqueous solution with TOPO as an extractant in diluent mixture [MIBK + kerosene (1:1 v/v)] are presented in Figure 1 and Table 2, respectively.

Physical Extraction

The understanding of diluent effect on the reactive extraction of acids can be determined by taking into account three important factors: (i) partial dissociation of the acid molecules in the aqueous phase, (ii) partition of the undissociated acid molecules between aqueous and organic phase, and (iii) dimerization of the undissociated acid in the extract phase.

To determine the physical extraction parameters (partition coefficient = P and dimerization constant = D), the following equation is fitted linearly in the Origin 8.0, a software package (Kertes and King, 1986).

$$K_D^{Diluent} = P + 2DP^2[HC] \quad (1)$$

The values of $K_D^{Diluent}$ of nicotinic acid with diluent mixture are found to be in the range of 0.074 to 0.149 which are less than one. The values of P and D are obtained as 0.0598 and 115.66, respectively (Figure 2). The use of non-polar solvents such as kerosene in the extraction of carboxylic acid at higher initial acid concentration may also lead to the formation of a stable emulsion and dimer in the organic phase. Therefore, a modifier (active diluent) is generally added with the inert diluent to avoid the formation of a stable emulsion and dimer in the organic phase. The MIBK (aprotic or proton acceptor, $\mu = 2.79$, $E_T = 39.4$) is an active diluent, which show higher extractability compared to inactive diluents (kerosene).

Chemical Extraction

The lower values of distribution coefficient in case of pure diluent mixture suggest the use of an extractant to improve the extraction efficiency of the recovery process. Therefore, the equilibrium chemical extraction experiments for the recovery of nicotinic acid are carried out using TOPO dissolved in MIBK + kerosene (1:1 vv). The extraction degree decreases with an increase in acid concentration in the aqueous phase. This may be due to the lower amount of TOPO used in the initial organic phase, and TOPO concentration is found to be a limiting parameter for the extraction. Generally, for a high initial acid concentration, the distribution coefficient (K_D) may decrease with an increase in the acid concentration in aqueous solution using TOPO with diluent mixture. The distribution coefficients (K_D) and degree of extraction (E) are found to increase with an increase in TOPO concentration (0.10 to 0.50 mol/L). TOPO/diluent system favors the formation of 'not overloaded' complexes of polar acid-TOPO structures with the Z factors restricted mainly between 0.032 and 0.684. With TOPO (0.365 mol/L) dissolved in MIBK + kerosene (1:1 v/v), the maximum value of K_D is to be 4.168 for an acid concentration of 0.02 mol/L.

The experimental results are also interpreted in terms of the distribution coefficient of acid by chemical extraction (K_D^{Chem}) with extractants (TBP and TOA) dissolved in diluent mixtures and given by Eq. (2).

$$K_D^{Chem} = \frac{\overline{C}_{HC} - \nu \overline{C}_{HC}^{Diluent}}{C_{HC}} \quad (2)$$

where ν is the volume fraction of diluent mixture.

The overall distribution coefficient (K_D^{Total}) by physical and chemical extraction is obtained from the following equation.

$$K_D^{Total} = \nu K_D^{Diluent} + K_D^{Chem} \quad (3)$$

The Z values less than 0.5 indicate a formation of (1:1) acid-TOPO complex in the organic phase. Therefore, with assumption of (1:1) acid-extractant complexes in the organic phase, the following model equations of reactive extraction mechanism are developed incorporating the effect of physical extraction.

$$K_E[\mathrm{HC}] = \frac{\overline{[(\mathrm{HC})(T)]}}{\overline{[\mathrm{T}]}_o - \overline{[(\mathrm{HC})(\mathrm{T})]}} = \frac{\overline{C}_{\mathrm{HC}} - \nu \overline{C}_{\mathrm{HC}}^{\mathrm{Diluent}}}{\overline{[\mathrm{T}]}_o - (\overline{C}_{\mathrm{HC}} - \nu \overline{C}_{\mathrm{HC}}^{\mathrm{Diluent}})} = \frac{(Z - \nu m)}{1-(Z-\nu m)} \tag{4}$$

The m is the loading of acid in organic phase by diluent mixture.

$$m = \frac{\overline{C}_{\mathrm{HC}}^{\mathrm{Diluent}}}{\overline{[\mathrm{T}]}_o} \tag{5}$$

The plots of $\dfrac{(Z-\nu m)}{1-(Z-\nu m)}$ versus [HC] yield a straight line with a slope representing the corresponding K_E value of the reactive extraction. The equilibrium complexation constants (K_E) predicted by the Eq. (4) are presented with the coefficient of determination (R^2) and standard deviation (*SD*). The model predicted values of K_E are showing good correlation with $R^2 > 0.98$ and maximum value of $SD = 0.092$. The highest value of K_E is found for lower acid and extractant concentration which shows a faster mass transfer of solute into the organic phase.

Conclusion

The equilibrium reactive extraction of nicotinic acid is studied considering a diluent mixture [MIBK + kerosene (1:1 v/v)] with and without extractant (TOPO) at isothermal condition (298 ± 1 K). The MIBK (aprotic or proton acceptor, $\mu = 2.79$, $E_T = 39.4$) is an active diluent, which show higher extractability compared to inactive diluents (kerosene). The degree of extraction decreases with an increase in acid concentration and increases with an increase in TOPO concentration. The maximum value of the K_D in the chemical extraction are found to be 4.168 with TOPO (0.365 mol/L) in MIBK + kerosene (1:1 v/v) for an acid concentration of 0.02 mol/L. The Z values less than 0.5 indicate a formation of (1:1) acid-TOPO complex in the organic phase.

References

(1) Kertes, A. S.; King, C. Extraction Chemistry of Fermentation Product Carboxylic Acids. *Biotechnol. Bioeng.* 1986, *28,* 269-282.

(2) Keshav, A.; Chand, S.; Wasewar, K.L. Equilibrium Studies for Extraction of Propionic Acid using Tri-*n*-Butyl Phosphate in Different Solvents. *J. Chem. Eng. Data* **2008,** *53,* 1424-1430.

(3) Keshav, A.; Chand, S.; Wasewar, K.L. Reactive Extraction of Acrylic Acid using Tri-*n*-Butyl Phosphate in Different Diluents. *J. Chem. Eng. Data* **2009,** *54,* 1782–1786.

(4) Kumar, S.; Wasewar, K.L.; Babu, B.V. (2008) Intensification of Nicotinic Acid Separation using Organophosphorous Solvating Extractants: Reactive Extraction. *Chem. Eng. Technol.* 31, 1584-1590.

(5) Kumar, S., Babu, B.V. (2008) Process intensification for separation of carboxylic acids from fermentation broths using reactive extraction, *J. Fut. Eng. Technol.* 3, 19.

(6) Kumar, S.; Babu, B. V., Extraction of Pyridine-3-carboxylic Acid Using 1-Dioctylphosphoryloctane (TOPO) with Different Diluents: Equilibrium Studies. *J. Chem. Eng. Data* **2009**, *54* (9), 2669-2677.

(7) Kumar, S.; Babu B. V. Process Intensification of Nicotinic Acid Production via Enzymatic Conversion using Reactive Extraction. *Chem. Biochem. Eng. Q.* 2009, *23*, 367-376.

(8) Mariya, M., Albet, J., Molinier, J., Kyuchoukov, G. Specific Influence of the Modifier (1-Decanol) on the Extraction of Tartaric Acid by Different Extractants. *Ind. Eng. Chem. Res.* 2005, *44*, 6534-6538.

(9) Mei, F.; Qin, W.; Dai, Y., Extraction Equilibria of Benzoic Acid with Tributyl Phosphate in Kerosene and 1-Octanol. *J. Chem. Eng. Data* **2002**, *47* (4), 941-943.

(10) Yankov, D., Molinier, J., Albet, J., Malmary, G., Kyuchoukov, G. Lactic Acid Extraction from Aqueous Solutions with Tri-n-Octylamine Dissolved in Decanol And Dodecane. *Biochem. Eng. J.* **2004**, *21* (1), 63-71.

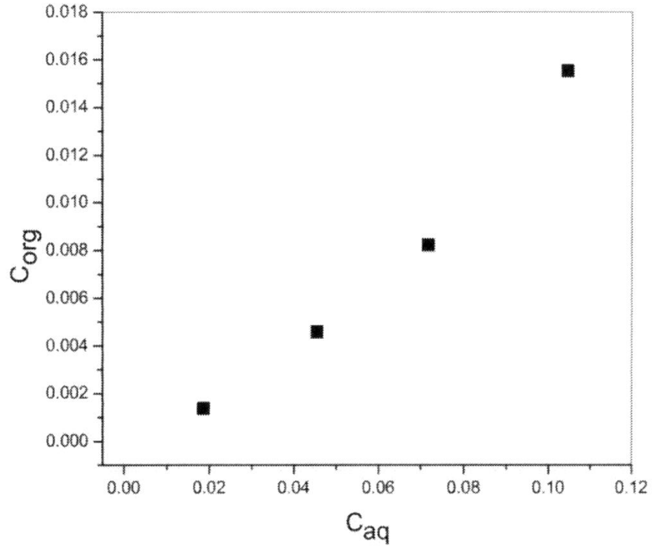

Figure 1. Physical equilibria for extraction of nicotinic acid at 298 K

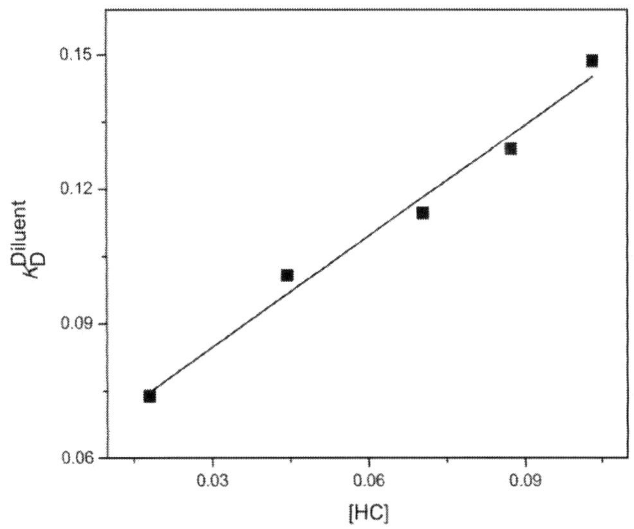

Figure 2. Determination of values of *P* and *D* for extraction of nicotinic acid at 298 K with MIBK + kerosene (1:1 v/v)

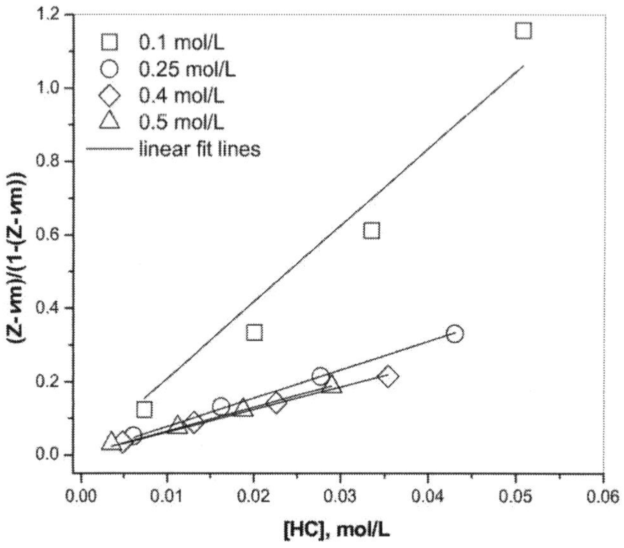

Figure 3. Equilibrium complexation constant (K_E) determination for extraction of nicotinic acid at 298 K with TOPO in MIBK + kerosene (1:1 v/v)

Table1. Reagents Used in this Equilibrium Study

S. No.	Reagents Name	Supplier	Purity / *w*	Mol. Wt. / kg·mol^{-1}	Specific gravity / kg·m^{-3}	Viscosity (mPa·s)
1.	nicotinic acid	HIMEDIA, India.	0.995	123.11	-	-
2.	Tri-*n*-octyl phosphine oxide (TOPO)	Fluka, USA	0.98	353.68	809	8.325 (25 ^{0}C)
3.	MIBK	Spectrochem. Pvt. Ltd., India	0.998	100.16	801	0.58 (20 °C)
4.	Kerosene	Commercial	-	170	780-810	2.1-2.2 (15.6 °C)

Table 1. Equilibrium data for the extraction of nicotinic acid using TOPO dissolved in MIBK + kerosene (1:1 v/v)

$[\overline{T}]_o$ mol·L^{-1}	Experimental results								Model predicted values					
	C_{HC} mol·L^{-1}	$\overline{C}_{HC}$ mol·L^{-1}	K_D^{total}	K_D^{chem}	E	Z	m	pH_{eq}	K_D^{total}	K_D^{chem}	Z	K_E	R^2	SD
0.10	0.00774	0.01226	1.584	1.415	61.30	0.123	0.014	3.44	1.898	1.828	0.147	20.94 ± 1.42	0.986	0.092
	0.02064	0.02936	1.422	1.212	58.72	0.294	0.046	3.22	1.643	1.547	0.339			
	0.03419	0.04581	1.339	1.111	57.26	0.458	0.082	3.11	1.433	1.324	0.490			
	0.05161	0.06839	1.325	1.039	56.99	0.684	0.155	3.02	1.283	1.142	0.662			
0.25	0.00645	0.01355	2.101	1.909	67.75	0.054	0.006	3.48	1.949	1.883	0.050	7.76 ± 0.083	0.999	0.005
	0.01677	0.03323	1.982	1.734	66.46	0.133	0.018	3.27	1.913	1.822	0.128			
	0.02839	0.05161	1.818	1.557	64.51	0.206	0.033	3.15	1.818	1.715	0.207			
	0.04387	0.07613	1.735	1.417	63.44	0.305	0.062	3.06	1.744	1.610	0.306			
0.40	0.00516	0.01484	2.876	2.649	74.20	0.037	0.003	3.53	2.499	2.436	0.032	6.21 ± 0.14	0.998	0.006
	0.01355	0.03645	2.690	2.403	72.90	0.091	0.011	3.32	2.502	2.416	0.085			
	0.02322	0.05678	2.445	2.144	70.98	0.142	0.021	3.20	2.419	2.322	0.140			
	0.03613	0.08387	2.321	1.956	69.89	0.210	0.039	3.10	2.357	2.231	0.213			
0.50	0.00387	0.01613	4.168	3.901	80.65	0.032	0.003	3.60	3.24	3.185	0.025	6.51 ± 0.13	0.999	0.005
	0.01161	0.03839	3.307	3.011	76.78	0.077	0.009	3.35	3.215	3.139	0.075			
	0.01935	0.06065	3.134	2.815	75.81	0.121	0.016	3.24	3.134	3.048	0.121			
	0.02968	0.09032	3.043	2.651	75.27	0.181	0.031	3.14	3.066	2.955	0.182			

(242e) Extraction and Back-Extraction Studies of Picolinic Acid Using Tri-n-Octylamine Dissolved in 1-Decanol

Dipaloy Datta, Chemical Engineering, Birla Institute of Technology & Science, Pilani, Pilani, India and Sushil Kumar, Chemical Engineering, Birla institute of Technology and Science (BITS), Pilani, India

Extraction and Back-Extraction Studies of Picolinic Acid Using Tri-*n*-Octylamine Dissolved in 1-Decanol

Dipaloy Datta[1] and Sushil Kumar[2]

Department of Chemical Engineering

Birla Institute of Technology and Science (BITS), PILANI – 333031 (Rajasthan), INDIA

[1] E-mail: dipaloy@bits-pilani.ac.in; dipaloy@gmail.com

Homepage: http://universe.bits-pilani.ac.in/pilani/dipaloy/profile

[2]E-mail: skumar@bits-pilani.ac.in; sushilk2006@gmail.com

Phone : +91-1596-245073 Ext 215; Fax: +91-1596-244183

Homepage: http://universe.bits-pilani.ac.in/pilani/skumar/profile

Abstract

The chemical industry has come under increasing pressure to make chemical production more eco-friendly due to its reliance on fossil resources, its environmentally damaging production processes and its toxic byproducts and waste. Within this framework, bio-based chemistry and biotechnologies offer great prospects. Microbial production of organic acids is a promising approach for obtaining building-block chemicals from renewable carbon sources (Hatti-Kaul et al., 2007). Picolinic acid acts as a bidentate ligand with two active groups such as a carboxyl group (-COOH) and a pyridinic nitrogen atom (-N). The acid is well known for its efficient chelating activity and can chelate metals like Cu, Fe, Ni, Zn, Cd, Pb, Mn, Cr, and Mo in the human body (Suzuki et al., 1957). 3-hydroxyanthranilic acid oxygenase enzyme can oxidize 3-hydroxyanthranilic acid by the catabolism of tryptophan through kynurenine to picolinic acid (Smith et al., 2007). A major drawback of fermentation is low acid product yield and concentration which leads to the difficulty in recovery of the product from the very dilute solution. To improve the biological production of picolinic acid, it is necessary to develop a cheap and environment friendly recovery method (Wang et al., 2005). Among the several separation processes, reactive extraction using a suitable extractant has been found to be a most promising alternative to recover carboxylic acids from the fermentation broth and dilute aqueous solution. Tertiary amines especially tri-*n*-octylamine (TOA) are found to be effective extractant to recover different carboxylic acids by reactive extraction (Kertes and King, 1986; Tamada et al., 1990; Datta and Kumar, 2011).

Literature is widely available on the reactive extraction of different carboxylic acids from aqueous streams, but work on reactive extraction and back-extraction of picolinic acid is limited. Therefore, the present work is aimed to intensify the recovery of picolinic acid (0.01 to 0.05 kmol×m^{-3}) using reactive extraction with TOA ($[\overline{T}]_{in}$ = 0 to 0.344 kmol×m^{-3}) using 1-decanol as active (polar) diluent. The experimental data are analyzed by calculating distribution coefficient ($K_D = C_{org}/C_{aq}$), degrees of extraction [$E = K_D / (1 + K_D)$] and loading ratios ($Z = C_{org}/[\overline{T}]_{in}$). In the experiment, equal volumes (20 ml) of organic and aqueous phases are equilibrated by shaking for 8 h at 298 K in a constant temperature water bath (Remi Labs, HS 250, India). After reaching equilibrium, the phases are allowed to settle for 2 h to have a clear separation of the phases. The aqueous phase is sampled by a pipette. Picolinic acid concentration in the aqueous phase (C_{aq}) is measured by titration using NaOH solution of 0.01 N. Acid concentration in the organic phase (C_{org}) is calculated by a mass balance. The initial and equilibrium pH of aqueous solution are measured using a digital pH-meter (ArmField Instruments, PCT-40, UK). The back-extraction of picolinic acid is carried out by pure water (temperature swing regeneration) at 353 K with 1:4 volume ratio of phases.

The equilibrium isotherms are drawn between organic and aqueous phase concentrations of picolinic acid at four initial acid concentrations and three concentrations of TOA in Figure 1. It is observed that the distribution coefficients ($K_D^{physical}$ << 1) of picolinic acid in 1-decanol alone are not sufficiently high. Also the degree of extraction is very low. Physical extraction of picolinic acid with 1-decanol has been found unsuitable. The hydrophilic nature of picolinic acid [dipole moment, μ = 4.42 D (Kulkarni et al., 1978); log P = -0.97 (Leo et al., 1971)] makes it poorly extractable by common organic solvents. Therefore, the equilibrium chemical extraction experiments for the recovery of picolinic acid are also carried out with TOA (0.115 - 0.344 kmol·m^{-3}) dissolved in 1-decanol and isotherms are shown in Figure 1. Improved extraction efficiency in terms of K_D is observed when an extractant is used with diluent as compared to that of diluent alone. The maximum value of K_D in the chemical extraction is found to be 7 with TOA (0.344 kmol·m^{-3}) at 0.01 kmol·m^{-3} of picolinic acid initial concentration. The distribution coefficients are found to be higher at lower concentration of acid (0.01 kmol·m^{-3}). It can be observed that at a fixed concentration of picolinic acid, the distribution coefficient increases with an increase in the TOA (0.115 to 0.344 kmol·m^{-3}) concentration, whereas upon varying the acid concentration for a fixed extractant concentration, K_D values decrease (Figure 2). The maximum removal of picolinic acid is 87.5% with TOA (0.344 kmol·m^{-3}) in 1-decanol at 0.01 kmol·m^{-3} of picolinic acid initial concentration. The distribution coefficient increases from 3 to 7 when the amount of TOA is increased from 0.115 to 0.344 kmol·m^{-3}. The effect of loading ratio (Z) on the extraction efficiency is shown in Figure 3. It can be seen that loading ratio decreases with increasing concentration of TOA at a fixed concentration (0.01 kmol·m^{-3}) of acid. The same trend is observed for other concentrations of acid. Also, at higher acid concentration, the organic phase is more loaded with the acid molecule compared to low acid concentration. The extraction mechanism of picolinic acid also depends on the pH and the pK_a of acid. In the present study, the values of equilibrium pH are found to be in the range of 3.55 to 4.11 which is in between the pK_a's of the acid (pK_{a1} = 1.01, pK_{a2} = 5.29; John, 1972).

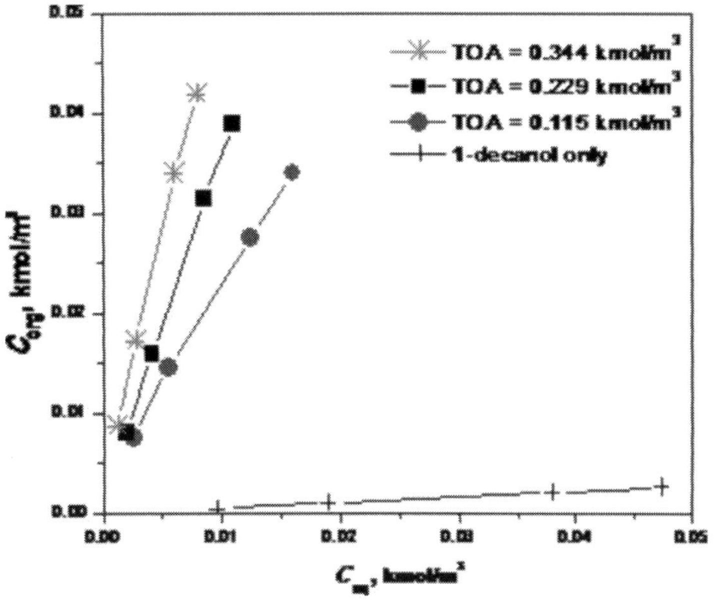

Figure 1. Equilibrium isotherms of picolinic acid (0.01 to 0.05 kmol·m^{-3}) for different concentrations of TOA (0.115 to 0.344 kmol·m^{-3}) dissolved in 1-decanol

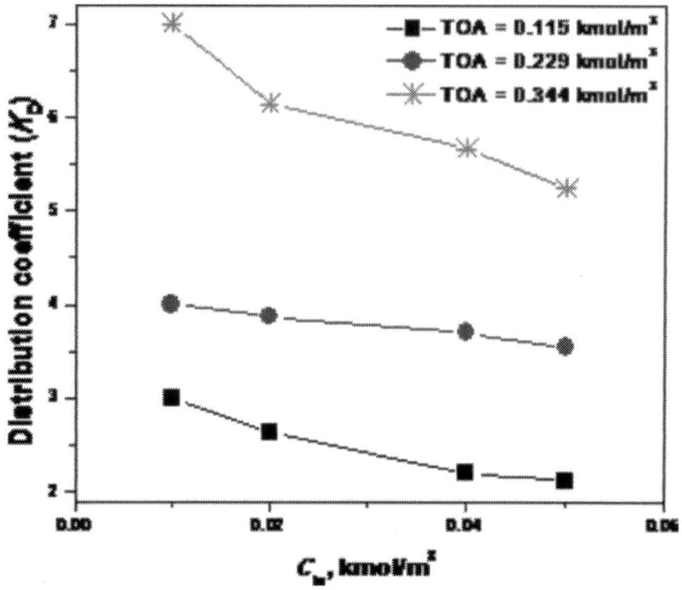

Figure 2. Effect of initial acid and extractant concentration on K_D

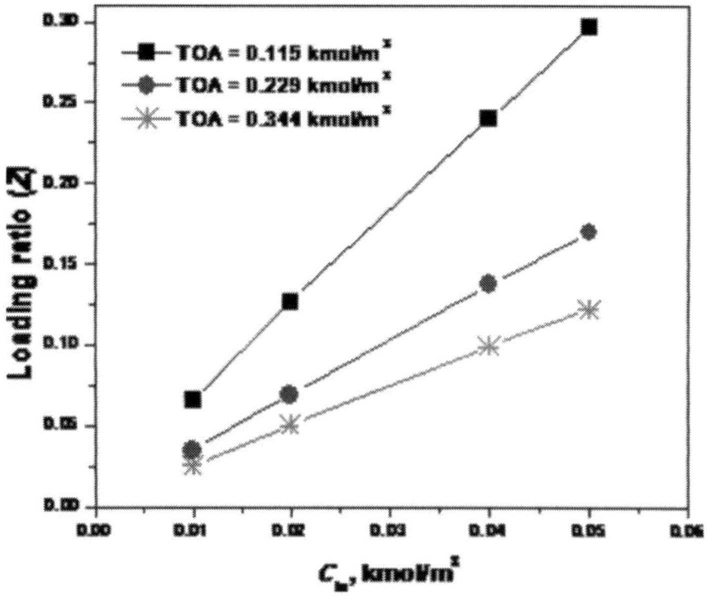

Figure 3. Effect of initial acid and extractant concentration on *Z*

An equation is derived using mass action law to relate K_D with *m, n* and equilibrium constant (K_E) of equilibrium reaction as:

$$K_D = mK_E\left([\overline{T}]_o - K_D n\frac{C_{aq}}{m}\right)^n \frac{C_{aq}^{\,m-1}}{(1+K_a/[H^+])^m} \quad (1)$$

The values of *m, n* and K_E are estimated by optimizing the error between the experimental and predicted values of K_D

using Eq. (1). The estimated values of m are found to be near about one with TOA imply that there are mainly formation of (1:1) acid-TOA complex in the organic phase (Table 1). The higher values of K_E are found at lower extractant concentration which shows faster mass transfer of the solute into the organic phase.

Table 1. Values of *m*, *n* and K_E

TOA ($kmol \cdot m^{-3}$)	m	n	K_E
0.115	0.94	1	21.51
0.229	1.0	1	19.37
0.344	0.89	1	11.75

In the back-extraction of picolinic acid (Figure 4), it can be seen that with an increase in the concentration of TOA, the slope of the isotherm decreases i.e. distribution coefficient of back-extraction ($K_{D'} = C_{aq}/C_{org}$) of picolinic acid is reduced. Though, at higher concentration of TOA may provide higher extraction of acid but would make the regeneration process difficult. The regeneration of the extractant loading with high concentration of acid (C_{in} = 0.05 $kmol \cdot m^{-3}$) will be easier (Z = 0.297 at $\overline{[TOA]}_0$ = 0.115 $kmol \cdot m^{-3}$) and higher distribution of acid ($K_{D'}$ = 4.037) can be achieved. Less loading of the extractant with the acid (Z = 0.122 at $\overline{[TOA]}_0$ = 0.344 $kmol \cdot m^{-3}$) results in lower distribution of acid ($K_{D'}$ = 0.714) and incomplete regeneration of the extracting agent.

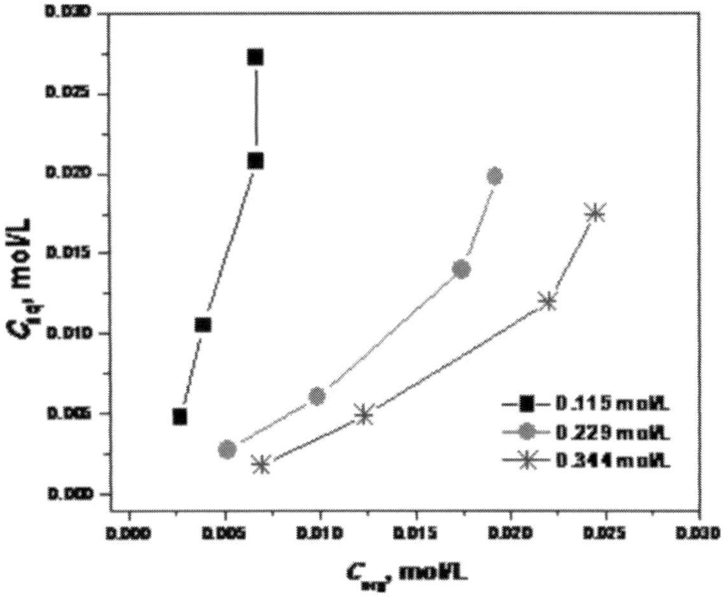

Figure 4. Back-extraction isothermal curve of picolinic acid using TOA in 1-decanol

REFERENCES

Datta, D.; Kumar, S. Reactive Extraction of Glycolic Acid Using Tri-n-Butyl Phosphate and Tri-n-Octylamine in Six Different Diluents: Experimental Data and Theoretical Predictions. *Ind. Eng. Chem. Res.* 2011, *50*, 3041-3048.

Hatti-Kaul, R.; Tornvall, U.; Gustafsson, L.; Borjesson, P. Industrial Biotechnology for the Production of Bio-Based Chemicals – A Cradle to Grave Perspective. *Trends in Biotechnol.* 2007, *26*, 119-124.

John, A.D. *Lange's Handbook of Chemistry* XXth Ed. McGraw-Hill Book Co. Inc., New York, USA. 1972.

Kertes, A. S.; King, C. J. Extraction Chemistry of Fermentation Product Carboxylic Acids. *Biotechnol. Bioeng.* 1986, *28*, 269-282.

Kulkarni, G. V.; Ray, A.; Patel, C. C. Molecular Orbital Studies on Nicotinic Acid, Nicotinamide and Related Compounds. *J. Molecular Structure.* 1978, *49*, 373-382.

Leo, A.; Hansch, C.; Elkins, D. Partition Coefficients and Their Uses. *Chem. Reviews.* 1971, *71*, 525-616.

Smith, A. J.; Stone, T. W.; Smith, R. A. Neurotoxicity of Tryptophan Metabolites. *Biochem. Soc. Transactions.* 2007, *35*, 1287-1289.

Suzuki, K.; Yasuda, M.; Yamasaki, K. Stability Constants of Picolinic and Quinaldic Acid Chelates of Bivalent Metals. *J Phys Chem.* 1957, *61*, 229-231.

Tamada, J. A.; Kertes, A. S.; King, C. J. Extraction of Carboxylic Acids with Amine Extractants. 1. Equilibria and Law of Mass Action Modeling. *Ind. Eng. Chem. Res.* 1990, *29*, 1319-1326.

Wang, L. R.; Fang, Y. UV-Raman Study and Theoretical Analogue of Picolinic Acid in Aqueous Solution. *J. Molecular Spectroscopy.* 2005, *234*, 137-142.

(242f) Modeling Partition Equilibria for Uranyl Nitrate, Nitric Acid and Water Extracted by Tributyl Phosphate in Dodecane

Stepan Hlushak[1], Jean-Pierre Simonin[2], Olivier Bernard[2], Christian Sorel[3] and Philippe Moisy[4], (1)Institute for Condensed Matter Physics, Lviv, Ukraine, (2)Laboratoire PECSA, Université Pierre et Marie Curie, France, (3)Commissariat al Energie Atomique (CEA), Bagnols–sur-Ceze, France, (4)CEA Marcoule, France

A new model for the extraction of uranyl nitrate, nitric acid and water from aqueous solution to organic phases containing tributyl phosphate (TBP) was recently developed. It is based on the model developed by Naganawa and Tachimori (H. Naganawa, S. Tachimori, Bull. Chem. Soc. Jap. 70, 809, (1997)), which describes the extraction of nitric acid and water by assuming the formation of 10 complexes involving TBP molecules, nitric acid and water in the organic phase. It is demonstrated that the introduction of 6 additional organic phase complexes allows to accurately describe extraction of water, nitric acid and uranyl nitrate into organic phase composed of 30% in volume of TBP (PUREX process) at 25°C. The thermodynamic activities of the components in the aqueous phase are modeled by the extended Pitzer expression, which gives a good representation of the osmotic coefficient for pure binary aqueous solutions of uranyl nitrate and nitric acid up to 4 mol L^{-1} and 10 mol L^{-1}, respectively. The stoichiometries of the 6 additional complexes were obtained one by one by the differential evolution meta heuristic optimization algorithm. The results suggest the formation of complexes containing 2 and 3 TBPs for 1 uranyl nitrate with additional water molecules.

(255a) Counterion Effects On Protein Adsorption Equilibrium and Kinetics in Polymer-Grafted Cation Exchangers

Ernie Perez Almodovar, Chemical Engineering, University of Virginia, Charlottesville, VA and **Giorgio Carta**, Department Chemical Engineering, University of Virginia, Charlottesville, VA

Protein binding equilibrium and mass transfer kinetics are studied for cation exchangers containing charged polymer grafts as well as for a macroporous matrix in pH 5 acetate buffers using sodium, tetrabutyl ammonium (TBAH), arginine, and calcium as counterions and a monoclonal antibody (mAb) as a model protein. Dynamic light scattering shows that there is no significant effect of the counterion type on the mAb aqueous diffusivity. The counterion type also does not affect substantially the structure of the polymer grafts, nor does it affect the stoichiometry of the protein ion exchange process. While no counterion effects are also observed on the protein mass transfer kinetics for the macroporous matrix, very large effects are seen for the polymer grafted matrices with protein adsorption rates increasing dramatically in the order $Ca^{++} > Arg^{+} > Na^{+} > TBAH^{+}$. This order is the same order in which the relative protein binding strength decreases. Accordingly, the counterions leading to weaker protein binding also lead to faster protein diffusion. Although the quantitative aspects are different, the same trends hold for different proteins (lysozyme, and lactoferrin) and for an agarose-based matrix also containing grafted polymers (Capto™ S) . The underlying mechanism is qualitatively consistent with protein transport occurring through a hopping process driven by the adsorbed protein concentration within the apparently flexible network structure formed by the grafted polymers. From a practical viewpoint, the results show that improved protein adsorption kinetics in polymer-grafted cation exchanger and, hence, improved performance, can be obtained by selecting particular counterions.

(255b) Purification of Monoclonal Antibodies by Affinity Precipitation Using Thermally Responsive Elastin-Like Polypeptides(ELPs) Fused to IgG Binding Domains: High Throughput Analysis and Scale up Considerations

Rahul D. Sheth[1], Bhawna Madan[2], Wilfred Chen[2] and Steven M. Cramer[1], (1)Department of Chemical and Biological Engineering, Rensselaer Polytechnic Institute, Troy, NY, (2)Chemical and Biomolecular Engineering, University of Delaware, Newark, DE

Elastin like polypeptides (ELPs) are smart biopolymers which exhibit reversible precipitation above a certain temperature called as inverse transition temperature (T_t). ELPs fused to IgG binding domains can be used as novel tools for antibody purification. Robotic high-throughput screening was employed to investigate the performance of an ELP-Z affinity precipitation system, and the effects of operating conditions on IgG precipitation, elution and product quality were examined. High levels of IgG recovery were obtained using a range of conditions with good results occurring even at low ELP:IgG molar ratios and relatively high elution pH (pH 4.2). Product quality issues such as aggregation were observed to be very sensitive to elution conditions and were evaluated using SEC, DLS and Dye binding studies. Conditions which exhibited both low aggregation and high product recovery were identified using a multidimensional high throughput screen and the robustness of the process was evaluated. Finally, key scale-up properties such as precipitation kinetics, precipitate morphology and particle size distributions were also examined at various stages of the process using DLS, confocal microscopy and turbidity measurements. This work sets the stage for the industrial implementation of this technology.

(255c) Purification of Monoclonal Antibody From Trangenic Nicotiana Benthamiana

Andrew Fulton[1], Huafang Lai[2], Qiang Chen[3] and **Chenming Zhang**[1], (1)Biological Systems Engineering, Virginia Tech, Blacksburg, VA, (2)Arizona State University, Tempe, AZ, (3)The Biodesign Institute, Arizona State University, Tempe, AZ

Monoclonal antibodies are the fast growing drug molecules targeting the treatment of diseases ranging from arthritis, immune, and infectious diseases to cancer. Due to its unique application principle, antibodies are commonly produced in large quantities. Plants, such as *Nicotiana benthamiana*, offer a unique production platform for bio-therapeutics due to their ability to produce large amounts of biomolecules in a relatively quick manner. However, purification of a target protein from plant is an arduous task due to the presence of toxic compounds in ground plant tissue. The current methods for antibody purification from plant extract include expensive affinity column steps (Protein A/G). Here, we report an alternative method of purifying a mAb from *N. bethamiana*. and compare the method to a process involving Protein A. The process involves an UF/DF step, a precipitation of native impurities step using a charged polymer, a hydrophobic interaction chromatography and a hydrophobic charge induction chromatography. This process compared favorably with Protein A based process based on the elimnation of host cell proteins.

(255d) Manufacturing-Scale Implementation of Enhanced Harvest Process Controls to Inhibit Colored Protein-Adduct Formation in Prokaryotic Host Cell Derived Recombinant Protein

Deepa Nadarajah[1], Kimberly Kaleas[2], Jane Gunson[3], Michael W Laird[3] and Rick St. John[1], (1)Genentech, Inc., South San Francisco, CA, (2)Purification Development, Genentech Inc, South San Francisco, CA, (3)Late Stage Cell Culture, Genentech, Inc, South San Francisco, CA

Manufacturing-Scale Implementation of Enhanced Harvest Process Controls to Inhibit Colored Protein-Adduct Formation in Prokaryotic Host Cell Derived Recombinant Protein

Deepa Nadarajah, Kimberly Kaleas, Jane Gunson, Michael Laird, Rick St. John

Prokaryotic cells are widely used as host cells for the production of recombinant therapeutic proteins. It was discovered that recombinant proteins produced in prokaryotic host cells are susceptible to modification by a colored adduct when exposed to reducing environment. The colored adduct was identified as 1-4 dihydroxy-2-napthoic acid (DHNA) and is produced inherently as part of the prokaryotic host cell's metabolic (menoquinone) pathway. DHNA covalently attaches to cysteine residues in the recombinant protein in reducing environment during the harvest operation. An oxidative environment eliminates protein modification by DHNA and was achieved by modifying the process controls that focused on delivering and maintaining positive dissolved oxygen levels during different stages of the harvest operations. This talk will focus on the successful implementation of enhanced harvest process controls at manufacturing-scale to produce a drug substance bulk that does not contain detectable DHNA-recombinant protein adduct.

(255e) Operating Window Determination for Chromatography of Biomolecules On the Basis of the Iso-Resolution Curve

Ryo Maeda, Noriko Yoshimoto and **Shuichi Yamamoto**, Bio-Process Engineering Laboratory, Yamaguchi University, Ube, Japan

We have proposed a concept "iso-resolution curve" for linear (salt concentration) gradient elution of proteins in electrostatic interaction chromatography (ion-exchange chromatography, IEC). The iso-resolution curve was found to be useful for understanding the relationship between separation time and buffer (solvent) consumption [1].

We further extended the iso-resolution curve concept to stepwise elution chromatography, in which the mobile phase for elution is changed discontinuously [2]. The following two model separation systems were chosen: polyphenol separations by polymer-based reversed phase chromatography (RPC) with ethanol-water mobile phase [2,3] and IEC of proteins.

The distribution coefficient values *K*were determined from linear gradient elution experiments by using our model. For both cases, the distribution coefficients depended strongly on the mobile phase modulator (ethanol for RPC and salt for IEC). HETP values as a function of mobile phase velocity were determined by linear gradient elution (LGE) experiments as well as isocratic elution experiments.

Based on the iso-resolution curve we can choose the mobile phase modifier concentration which can provide the desired separation time and buffer consumption. However, in the actual operations the acceptable operating windows must be considered as variations for the equipment such as pumps and mobile phase compositions are unavoidable. It was found that when the separation time is short and the buffer consumption is small, a small change in the mobile phase modifier results in significant variations. The operating window is small and the mobile phase modifier concentration must be carefully prepared.

References

1) Yamamoto, S., Kita, A., Theoretical background of short chromatographic layers. Optimization of gradient elution in short columns, Journal of Chromatography A, 1065, 45-50 (2005)

2) Hosono, M., Maeda, R., Yoshimoto, N., Yamamoto,S., Rational Method for Designing Efficient Chromatography Processes based on the Iso-Resolution Curve, Chem. Eng. Technol., vol.35, 198-203 (2012)

3) Yamamoto, S., Hakoda, M., Oda, T., Hosono, M., Rational method for designing efficient separations by chromatography on polystyrene-divinylbenzene resins eluted with aqueous ethanol, Journal of Chromatography A , 1162, 50-55(2007)

(255f) Separation and Characterization of Raft Associated Membrane Species Using a Patterned Supported Lipid Bilayer

Mark J. Richards[1], Ling Chao[2] and Susan Daniel[1], (1)Chemical and Biomolecular Engineering, Cornell University, Ithaca, NY, (2)Chemical Engineering, National Taiwan University, Taipei, Taiwan

The role of lipid rafts is important in understanding cell processes and disease states; however, identifying the key components of these lipid domains is difficult and impedes research progress. Here we present a new platform to sort, separate, and characterize membrane-bound molecules based on their affinity for raft phase domains using a heterogeneous supported lipid bilayer (SLB) consisting of patterned lipid raft and non-raft domains. Adjacent SLBs consisting of liquid-ordered and liquid-disordered phases are formed in predefined locations. Membrane-bound biomolecules loaded into the device are convected laterally in the two-dimensional plane along the heterogeneous supported bilayer by a shear force induced from hydrodynamic flow of the bulk aqueous phase. During axial convection, the membrane-bound biomolecules diffuse within the 2-D heterogeneous bilayer plane, partitioning into their preferred lipid phase. Patterns for the lipid phases are designed to facilitate the sorting and collection of separated species. The main advantages of this new method over existing methods used to identify and assay raft species include separating membrane species within a membrane environment, near physiological conditions, and without artifacts associated with detergent, high salt, or alkaline pH. These criteria are particularly crucial to the separation of lipid-linked proteins and transmembrane proteins to avoid denaturation. This new capability to separate membrane-bound species near native conditions based on affinity for certain lipid phases can be used to identify intrinsic membrane raft residents, characterize how post-translational modifications shift the affinity of analogs to a particular lipid phase and dynamically observe the effect of membrane composition on membrane species behavior.

(255g) The Role of Material Scale in Bioseparations

Jyothirmai J. Simhadri[1], Holly A. Stretz[2] and Pedro E. Arce[1], (1)Chemical Engineering, Tennessee Technological University, Cookeville, TN, (2)Department of Chemical Engineering, Tennessee Technological University, Cookeville, TN

Study of transport in porous or structurally-modified media such as nanocomposite gels i.e., gels having nanoparticles embedded into the polymer matrix plays an important role is separation of bio-solutes i.e., proteins, DNA, etc. under different driving forces such as 1D or 2D electric-fields, and pressure gradient. Porous or fibrous media with a heterogeneous structure can be modeled by considering the material as a network of small and large domains having idealized geometries connected to each other through which the solute particles transport. For the current analysis, models of a single domain having two idealized parallel wall sub-domains connected in series to form an expansion or constriction cross-section are studied. Two different geometries such as rectangular and cylindrical are considered. Analytical expressions are derived for the effective transport coefficients such as effective velocity and effective diffusivity as a function of parameters dictating the geometry of the pore, and also the bulk flow rate. The results show the strong dependence of geometry, and also the flow direction on the effective transport parameters which eventually offer insights into the behavior of the system for separation of biomolecules.

(256a) Consistent Measurements of Hydraulic and Mass Transfer Packing Characteristics

Micah Perry[1], Chao Wang[2] and Frank Seibert[1], (1)SRP, University of Texas at Austin, Austin, TX, (2)Dept. of Chemical Engineering, University of Texas at Austin, Austin, TX

Packed column design is strongly dependent on the accuracy of models predicting the mass transfer performance of packings. This research utilizes a consistent approach to measuring mass transfer and hydraulic characteristics of both random and structured packings in order to create new and improved models for pressure drop, liquid holdup, contact area, and liquid and gas mass transfer coefficients. Experimental methods have been created to measure the following characteristics of any packing:

-Pressure drop (dry and irrigated)

-Liquid Holdup

-Flood Point

-Effective Mass Transfer Area

-Liquid Film Mass Transfer Coefficients

-Gas Film Mass Transfer Coefficients

The purpose of this paper is to update the progress of this research by adding the results of characterizations completed on several new packings. This paper will also show progress made in modeling efforts utilizing this new data. Data will also be compared to existing models for contact area, liquid film, and gas film mass transfer coefficients. Updating the previous version of this presentation, 4-8 new packings have been studied and modeled, and this new data will be presented.

(256c) Globally Optimal Networks for Multi-Pressure Distillation of Homogeneous Azeotropic Mixtures: Application to Ternary Mixtures

Paul Ghougassian, Chemical Engineering, UCLA, Los Angeles, CA and Vasilios Manousiouthakis, Chemical & Biomolecular Engineering Department,, University of California Los Angeles, Los Angeles, Los Angeles, CA

In this paper, a methodology for the globally optimal synthesis of a network of vapor-liquid equilibrium flash separators that can operate at multiple pressures and separate an azeotropic mixture is presented. The objective function minimized is the total flow entering the network flashes. The proposed synthesis methodology employs the Infinite DimEnsionAl State-Space (IDEAS) conceptual framework, which is shown to be applicable to the problem under consideration. The resulting infinite linear programming (ILP) IDEAS formulation is shown to possess several properties that allow its simplification. The approximate solution of this IDEAS ILP is pursued through the solution of a number of finite dimensional linear programs (FLP) of ever increasing size, whose optimum values form a sequence that converges to the ILP's infimum. The proposed optimal design methodology is general in nature and can be used to separate any number of pressure-sensitive azeotropic mixtures, with or without using an entrainer. The method is demonstrated on a case study involving the dual pressure separation of a ternary mixture of Methanol-Acetone-Water which exhibits a minimum boiling azeotrope, without using an entrainer. Comparison with a traditional two-column design reveals the benefits of employing the IDEAS framework not only to realize the separation with minimum total flow but also to reach outlet concentrations and flow rates combinations that are unattainable with the traditional design.

A SERIES OF NOVEL HIGH-EFFECTIVE MASS TRANSFER UNIT ELEMENTS IN DISTILLATION COLUMNS BASED ON SIC CERAMIC FOAM MATERIAL

Xin Gao[a, b], Xingang Li[a, b], Hong Li[a, b *], Jinsong Zhang[c]

[a] *School of Chemical Engineering and Technology, Tianjin University, Tianjin 300072, China*

[b] *National Engineering Research Center for Distillation Technology, Tianjin 300072, China*

[c] *Institute of Metal Research, Chinese Academy of Sciences, Shenyang 110015, China*

Introduction

In recent years, there has been a growing interest in solid silicon carbide foam for applications such as heat exchangers, burners and catalysts or structured catalyst supports[1-4]. Stemmet et al.[5] initiated the study of countercurrent hydrodynamics in foams and observed a low pressure drop with a biphasic system. They also obtained relatively high mass transfer coefficient in terms of an overall volumetric gas–liquid mass transfer coefficient[6]. Lévêque and Rouzineau[7] investigated the hydrodynamic performance and mass transfer efficiency of the ceramic foam structured packing in gas–liquid countercurrent operation, and observed the terms of pressure drop and flooding were quite low compared with classical distillation packing metal materials (M250Y, CY, and Pall rings) and mass transfer performance was very good. Altogether ceramic foams could be considered as a good potential material for distillation.

A series of high-effective mass transfer unit elements in the form of SiC ceramic foam material were developed by Tianjin University used in distillation columns. The mass transfer unit elements include SiC foam valve tray, SiC foam tray and structured SiC foam packing. The foam preparation method was developed by Institute of Metal Research, Chinese Academy of Sciences, which was based on the impregnation of a polyurethane foam with a homogeneous mixture of silicon, charcoal and phenolic

* Corresponing author
E-mail : lihongtju@tju.edu.cn

resin. The mass transfer unit elements of SiC ceramic foam was manufactured by Liaoning ZhuoYi New Materials Co. Ltd. in China.

In this paper, The hydrodynamics characteristic and mass transfer performance of the mass transfer unit elements with foam structure material were evaluated experimentally using pilot-scale installations available at National Engineering Research Center for Distillation Technology in China. A comparison of experimental data was made with the performance of conventional mass transfer unit elements under same test conditions.

Experimental and Method

The experimental rig for hydraulic studies comprises a transparent organic glass column (600mm in diameter × 2500mm height). The inclined drip catcher and the screen packing above the test tray are used to collect the liquid entrainment. The column was operated in a crosscurrent mode with an air–water system. Experiments were conducted at room temperature (25□) under atmospheric pressure (101.325 kPa). The liquid flowed from a tank through a centrifugal pump and flow meter and was slowly supplied to the tray. The outflow is gathered in the tank for recycle. The gas flow was supplied by centrifugal fan at the bottom of the column and was measured using a pitot tube. The pressure drop was measured using a U-tube manometer filled with water. The quantity of liquid weeping and entrainment rate were obtained by gravimetric method. Each measurement was carried out after the system had reached the steady state. The opening ratio of sieve tray can't be as high as SiC foam trays due to the limitations of mechanical strength.

The mass transfer efficiency was characterized by the measurement of overall efficiency by total reflux experiments. Experiments are performed using a stainless steel distillation column (300 mm in diameter×3500 mm height). The experiments were carried out with a standard cyclohexane/n-heptane (molar ratio of 2:3) mixture. The mixture was heated in the reboiler and cooled by water at the top of the column, and total reflux was done at different heating voltages under atmospheric pressure.

Samples were drawn from the top and bottom of the column, respectively, and compositions determined by a WAY-2WAJ Abbe refractometer analysis at 40 □.

Results and Discussions

This work presented an investigation of the use of SiC ceramic foam as a distillation column internal. The hydrodynamics characteristic and mass transfer performance of the mass transfer unit elements with foam structure material were evaluated experimentally using pilot-scale installations. The results showed that the overall performance of the SiC foam valve tray was better than that of F1 float valve tray; the SiC foam tray was better than sieve tray; and the structured SiC foam packing was better than structured wire gauze packing. The hydrodynamics of these mass transfer unit elements were simulated using CFD techniques. The agreement between the experiments and CFD simulations was found to be very good, suggesting that CFD simulations could be used for design and scale-up purposes.

Keywords: distillation, SiC ceramic foam, column internal, process intensification

ACKNOWLEDGEMENT

The authors are grateful for the financial support from the National Basic Research Program of China (2009CB219907), Program for Changjiang Scholars and Innovative Research Team in University (No. IRT0936), National Natural Science Foundation of China (No. 21176172) and Nationial Key Technology R&D Progran (No. 2011BAE03B07).

Reference

[1]G. Indranil, Heat transfer correlation for high-porosity open-cell foam,Int J Heat Mass Tran, 52(2009)1488-1494.

[2] J.T. Richardson, D. Remue, J.K. Hung, Properties of ceramic foam catalyst supports: mass and heat transfer,Applied Catalysis A: General, 250(2003)319-329.

[3] J.T. Richardson, Y. Peng, D. Remue, Properties of ceramic foam catalyst supports: pressure drop,Applied Catalysis A: General, 204(2000)19-32.

[4] M.V. Twigg, J.T. Richardson, Theory and Applications of Ceramic Foam Catalysts,Chemical Engineering Research and Design, 80(2002)183-189.

[5] C.P. Stemmet, J.N. Jongmans, J. van der Schaaf, B.F.M. Kuster, J.C. Schouten, Hydrodynamics of gas–liquid counter-current flow in solid foam packings,Chem Eng Sci, 60(2005)6422-6429.

[6] C.P. Stemmet, M. Meeuwse, J. van der Schaaf, B.F.M. Kuster, J.C. Schouten, Gas–liquid mass transfer and axial dispersion in solid foam packings, Chem Eng Sci, 62(2007)5444-5450.

[7] J. Lévêque, D. Rouzineau, M. Prévost, M. Meyer, Hydrodynamic and mass transfer efficiency of ceramic foam packing applied to distillation, Chem Eng Sci, 64(2009)2607-2616.

(256e) Critical Assessment of Energy-Saving Potential of Extractive Divided-Wall Column

Paul Hen-Chia Hsu, Yi Chang Wu and **I-Lung Chien**, Department of Chemical Engineering, National Taiwan University, Taipei, Taiwan

One of a common way to separate azeotrope in industry is to add a heavy entrainer so that the relative volatility of the original two components can be greatly enhanced. Thus, one original component can go overhead in an extractive distillation column and the other component will go with the heavy entrainer to the column bottoms. A second entrainer recovery column is designed to separate this stream so that the entrainer can be recycled back to the extractive distillation column. By observing the composition profile of stripping section of the extractive distillation column, a prominent remixing effect is exhibited with one composition going through a maximum and then decreasing toward the column bottoms. A way to eliminate this remixing effect to save energy is to design a thermally coupled two-column system so that the vapor traffic of the extractive distillation column is provided by a sidedraw from the entrainer recovery column. A special internal wall can be designed so that the original two-column system becomes one extractive divided-wall column (EDWC).

However, one potential drawback on the energy-saving of the EDWC design is that two reboilers in the original design should be combined into one reboiler. Since the entrainer is the heaviest component in the system, there are cases with reducing total reboiler duty but adversely increasing of the total steam cost of the separation system. Three industrial separation systems of acetone-methanol-water, isopropanol-water-DMSO and dimethyl carbonate-methanol-aniline will be used as demonstrating examples for critical assessment of the energy-saving potential of EDWC design. Dynamic controllability and operability of the EDWC system are also investigated to properly reject common feed rate and feed composition disturbances with a suitable overall control strategy.

(256f) Dynamic Control of a Divided-Wall Column Used for High-Purity Separation of Aromatic Reformate

Hao Wu[1], Zhi Cai[2], Jun Wang[2] and Hao Ling[1], (1)Department of Petroleum Processing, East China University of Science and Technology, Shanghai, China, (2)Sinopec Jiujiang Company, Jiujiang, China

The separation of ternary mixtures by divided-wall column has been reported in many literatures, but the dynamic behavior of divided-wall column is rarely seen when separating industrial feed. This work aims to separate an industrial aromatic reformate by a divided-wall column. The aromatic reformate contains 12 aromatic components. Very high product purities of 99.97 wt.% and 99.95 wt.% were specified for distillate and side stream products, respectively. The optimum economic design of the divided-wall column was calculated based on minimizing total annual cost at steady state. The control structure includes four control loops besides regular flow and level control. A control loop at the top of the prefractionator was installed to avoid the remixing of toluene and minimize energy consumption. Three control loops were used to obtain distillate, sidstream and bottom specified product purities. Composition control, single temperature control and differential temperature control were all implemented in Aspen Dynamics. Results showed that the three control methods worked very well in the face of 10% throughput and feed composition disturbances.

(269a) Remarkable CO2 Uptakes in MOFs Whilst in the Presence of Water Vapor

Philip L. Llewellyn, Estelle Lenoir and Christelle Vagner, MADIREL (UMR7246), Aix-Marseille Univ. / CNRS, Marseille, France

The search for solutions for economic CO_2capture from flue gases is ever continuing. Capture using physical adsorption could be viable as long as the right adsorbent is found and indeed amine functionalized materials could be of interest.

Our group is interested in the initial evaluation of various adsorbent materials for a number of different processes. In this respect, Metal Organic Frameworks are a particularly interesting family of materials due to the possibility to vary the metal centre and/or organic linker to give a huge variety of physical and chemical properties. The problem here is that with such a wide variety of materials to hand, which are of interest for CO_2capture ?

Indeed, the capture of CO_2 from flue gases poses a number of problems in terms of low CO_2 concentration (below 20%), low total pressure (1 bar), and the eventual presence of water. Indeed, whilst many MOFs seem to be unstable in the presence of water, some studies suggest that CO_2capture in moderate humidity's enhance CO2 uptake. This is in contrast with zeolites and activated carbons where water has a detrimental effect.

Our work will show several cases where water plays a determining role during CO_2uptakes in various MOFs. Uptakes are measured using chromatography and simultaneously we are able to directly measure the heats using microcalorimetry.

Two separate cases will be considered : (1) mesoporous MOFs and (2) a bifunctional microporous MOF. These systems will be compared with a reference activated carbon and a zeolite. We show that in each case, considerable increases in CO_2 capture can be attained in the presence of water vapor and we will give tentative explanations of the phenonena in play.

(269b) CO2 Capture by VSA with SIM-1: A Success Story From Nanometers to Kilograms

David Farrusseng[1], Sonia Aguado[1], Jerome Canivet[1], Javier Perez-Pellitero[2], Gerhard D. Pirngruber[3], Nicolas Bats[2], Virginie Moizan[2], Francois Guilhamasse[4], Marc Clausse[5] and Bogdan Harbuzaru[2], (1)IRCELYON, CNRS, Villeurbanne, France, (2)IFP Energies nouvelles, Solaize, France, (3)Catalysis and Separation, IFPEN, Solaize, France, (4)CNAM, Paris, France, (5)Cnam - IFFI (EA 21), Paris, France

The discovery of MOFs with pore size in the mesopores is a breakthrough discovery in materials science. For H_2 storage, capacity is directly correlated with the pore volume. However, the effect of pore size on CO_2 adsorption capacity and heat adsorption has not been investigated so far. When the pore size (dp) is approaching the diameter of the adsorbates, the solid-fluid interactions are the greatest. Despite the incentive for developing porous materials that can exhibit confinement effect the development of MOFs in the ultramicroporous range (i.e dp $<$ 1 nm) has not paid much attention so far. In this work, we present the story of the development of a functionalized Zinc Imidazolate adsorbent, called SIM-1, for CO_2/N_2 separation. We will show:

- the effect of pore size and polarity of ZIFs on CO_2 adsorption isotherms. GCMC simulation was applied for the rational design of ZIF adsorbents,

- the pilot-scale synthesis of 3kg of SIM-1 powder,

- the shaping of SIM-1 as extrudates,

- CO_2/N_2 separation by VSA at semi-pilot scale,

- CAPEX and OPEW will be discussed

The moderate heat of adsorption of CO2 on SIM-1 allows fast VSA cycling. In addition, because of the hydrophobicity of SIM-1, humid stream can be fed with a minor impact of the CO2 capture.

We thank ACACIA 31 program (ANR, Axelera) and our partners IFPEN, CNAM.

(269c) Post-Combustion Carbon Dioxide Capture Using Solid Amines in a Pressure Swing Adsorption Process

Anahita Abdollahi[1], Armin D. Ebner[2] and James A. Ritter[2], (1)University of South Carolina, Columbia, SC, (2)Department of Chemical Engineering, University of South Carolina, Columbia, SC

Solid amines are comprised of an adsorbent material, like silica gel, that contains an amine based organic compound, either physically or chemically, attached within its pores. The presence of this amine makes the silica gel almost exclusively selective to CO_2. The specific solid amines being explored in this work also offer significant CO_2 working capacities at relatively high temperatures, e.g., at 80 to 100 °C; and unlike zeolites they are not negatively affected by the presence of water vapor, which is typically around 10 vol% in flue gas. Furthermore, these silica based solid amines exhibit a low working capacity for water vapor, which allows for significant water vapor to pass through the bed and exit in the light product along with most of the nitrogen. The current work has been focusing on the development of a novel solid amine based PSA cycle for CO_2 capture from feeds that necessarily contain water vapor. The results will be evaluated in terms of both CO_2 removal and water vapor destination.

(269d) Composite Metal Oxide Materials Synthesized by Ultrasonic Spray Pyrolysis for CO2 Capture in a Sorption-Enhanced Water-Gas-Shift Process

Maryam Sayyah, Department of Chemical and Biomolecular Engineering, University of Illinois at Urbana-Champaign, Urbana, IL, Yongqi Lu, Prairie Research Institute, University of Illinois at Urbana-Champaign, Champaign, IL, Massoud Rostam-Abadi, Prairie Research Institute , University of Illinois at Urbana-Champaign, Champaign, IL and Kenneth S. Suslick, Chemistry, University of Illinois at Urbana-Champaign, Urbana, IL

Combining high-temperature CO_2 adsorption with the water-gas-shift (WGS) reaction has several advantages over conventional integrated gasification combined cycle power plants, which presently uses solvent-based CO_2 capture (e.g. Selexol^{TM}): high-temperature sorption of CO_2 in a sorption-enhanced water-gas-shift (SEWGS) process would eliminate the need for multiple catalytic WGS reactors and for a separate low-temperature CO_2 capture unit and would avoid gas cooling and reheating steps prior to entering a gas turbine. Developing a cost-effective sorbent with a stable working capacity under WGS conditions, however, has still remained a challenge.

In this work, for the first time, we report a facile synthesis of composite CaO-based materials for SEWGS applications via ultrasonic spray pyrolysis (USP). USP is a continuous and industrially scalable method for synthesis of materials with different morphologies. Materials here consist of inert refractory metal oxides dispersed in a CaO matrix. Inert additive oxides can provide structural stability for the sorbents at high temperature. A variety of characterization methods such as SEM, (S) TEM, XRD, and TGA were employed in order to explain the performance of sorbents during multicycle testing of CO_2 adsorption and desorption. A few sorbents were identified to be promising and have undergone further testing under simulated WGS reaction conditions. Finally, the possibility of synthesis of hybrid sorbent/catalyst materials via USP will be discussed in this presentation.

(269e) On the Mechanism of Morphological Enhancement of CaO by Hydration

Zhenchao Sun, Hao Chi, Pengpeng Qi and Liang-Shih Fan, William G. Lowrie Department of Chemical and Biomolecular Engineering, The Ohio State University, Columbus, OH

As an effective and efficient approach for CO_2 separation from coal-combustion flue gas, CaO-based chemical looping processes attracted world-wide research attention in the past decades. Similar to many other chemical looping particles, the reactivity and recyclability of pure CaO steadily deteriorate as the cyclic reaction proceeds. One of the most effective sorbent reactivation strategies to date is to use hydration onto the deactivated CaO, which restores the surface area and pore volume. The hydration mechanism has been studied for decades, but most previous studies followed the "black-box" approach and tried to explain this phenomenon by varying single reaction parameter, such as temperature, pressure, water-to-CaO ratio, *etc*. However, a hydration process, especially a water hydration, is a multi-phase reaction, which involves several physical and chemical sub-steps. Hence, many previous theories on this mechanism are inconsistent and sometimes conflicting. This work explores the increases of surface area and pore volume of CaO by hydration. Firstly, a widely-believed mechanism: "physical attrition theory" is experimentally examined, which shows its limitation in explaining this phenomenon. To explain this increase of morphological properties by hydration, a typical water hydration process is examined by dividing it into four independent chemical and physical sub-steps. The morphological changes of $Ca(OH)_2$ and derived CaO by each sub-step are measured by BET. During the first step – the intrinsic chemical conversion from CaO to $Ca(OH)_2$, the formed $Ca(OH)_2$ disintegrates because of its low tensile strength and weak crack resistance, which explains the increases of surface area and pore volume by steam/moisture hydration as well as the rapid heat release during hydration. Physical interaction with water slightly decreases the surface area and pore volume, possibly by mobilizing micro-particles into the porous structure of bigger particles and inducing stronger particle agglomeration. The $Ca(OH)_2$ solid can further chemically bond water molecules, which significantly enlarges the solid volume during water-bonding and generates a more porous structure during dehydration. The final precipitation of dissolved $Ca(OH)_2$ decreases the solid's surface area and pore volume. This decrease is attributed to the formation of micro-particles from solution, which plug some of the surface pores on the bigger particles during the drying process.

(269f) Flame Synthesized CaO Based Sorbents for CO2 Capture – Influence of Metal Ions As Dopants

Rajesh Koirala, **Krishna Reddy Gunugunuri** and Panagiotis Smirniotis, Chemical Engineering Program, School of Energy, Environmental, Biological and Medicinal Engineering, University of Cincinnati, Cincinnati, OH

Growing energy demand and subsequent CO_2 production call for highly efficient CO_2 capture process, which is benign to the environment. Flame synthesized doped Calcium-based novel refractory sorbents have shown excellent performance in CO_2 adsorption. In our previous studies we incorporated several metals (Co, Si, Ti, Cr, Hf, W, Y, La, Ce, Zr) into calcium oxide lattice to improve stability of the sorbent during the CO_2 adsorption experiments. Among these dopants, Zirconia doped CaO with molar ratio of 10:3 showed enormous adsorption stability during 100 cycles of operation. We reported that formation of orthorhombic structured $CaZrO_3$ in CaO-Zr sorbent is responsible for the observed stability. However, Ca/Zr (10:3) sorbent exhibits lower adsorption capacity. Therefore the present study mainly aimed at finding a different metal dopant ion which can stabilize the CaO sorbent during multi-cyclic operation as well as exhibits high adsorption capacity. For this purpose various metal doped CaO based sorbents were synthesized using flame technique. The sorbents were characterized by BET, XRD, SEM and TEM techniques. CO_2 capture experiments were performed in a TGA at 700 ^oC. Among the various dopants Al doped CaO sorbent exhibit optimal performance. The optimal sorbent were tested under severe condition of high temperature. These interesting findings will be discussed in the presentation.

(269g) Carbon Dioxide Uptake On Sorbents with Narrow Pore Window Aperatures: Zeolite (NaKA), (Silico)Aluminum Phosphates (AlPO4:s and SAPOs) and Titanium Silicate (CTS-1)

Ocean Cheung, Qingling Liu, Zoltan Bacsik, Jing Li, Jie Su, Xiaodong Zou and Niklas Hedin, Department of Materials and Environmental Chemistry, The Berzelii Center EXSELENT on Porous Materials, Stockholm University, Stockholm, Sweden

Adsorption driven CO_2 capture needs adsorbents with a range of desirable properties; these properties include high CO_2 adsorption capacity, high CO_2 selectivity over other gases including N_2 and CH_4, satisfactory mechanical and thermal properties. Here, we summarize results on our recent studies on microporous materials with narrow window apertures. Such narrow apertures can enhance the CO_2-over-N_2 selectivity as N_2 has a larger kinetic dimension than CO2 in microporous solids. We showed that zeolite NaKA with an optimal 17% K^+ exchange was highly selective for CO_2 sorption (ideal selectivity 172, capacity 3.4 mmol/g, 298 K 101 kPa) and we synthesized nano-sized zeolite NaKA for improved adsorption kinetics. Our studies on aluminum phosphates ($AlPO_4$:s) and silico aluminum phosphates (SAPOs) showed that many of these materials had significant ideal CO_2-over-N_2 selectivity (AlPO-21, ideal selectivity = 32, 273 K 101 kPa) and high CO_2 adsorption capacities (SAPO-56, 5.4 mmol/g at 273 K, 101 kPa). The phosphates based materials were also less hydrophilic than traditional zeolite sorbents and had almost no lost in CO_2 capacities during cycle adsorption. Certain version of contracted titanium silicate (CTS-1) can be used for upgrading of natural gas and biogas. Our study on titanium silicates revealed that CTS-1, in particular its Mg^{2+} form (Mg-CTS-1), offered high CO_2 capacity (2.1 mmol/g, 273 K 101 kPa) and high ideal selectivity over N_2 (ideal selectivity ~ 14). We studied these adsorbents with a wide range of techniques, including volumetric adsorption, in situ infrared spectroscopy and in situ powder X-ray diffraction. These studies demonstrated the strengths of these adsorbents, and explored the potential of using different materials in adsorption driven carbon capture and storage (CCS).

Sorption of Carbon Dioxide on Binderless Beads of 13X Zeolite

José A. C. Silva[1*], Kristin Schumann[2] Alírio E. Rodrigues[3]

[1] Instituto Politécnico de Bragança, ESTiG, Apartado 134, 5301-857 Bragança, Portugal

[2] Chemiewerk Bad Köstritz GmbH, Heinrichshall 2, 07586 Bad Köstritz, Germany

[3] LSRE, Faculdade de Engenharia, Universidade do Porto, Rua do Dr. Roberto Frias, S/N, 4200-465 Porto, Portugal

*Corresponding author, email: jsilva@ipb.pt

Keywords: Binderless 13X zeolite; Sorption of CO_2/CH_4; ZLC technique; Adsorption Equilibrium

Abstract

The sorption equilibrium of CO_2 and CH_4 in binderless beads of 13X zeolite has been investigated between 313 and 373 K and pressure up to 4 atm. The amount adsorbed of CO_2 and CH_4 is around 5.2 mmol/g_{ads} and 1.2 mmol/g_{ads}, respectively, at 313 K and 4 atm. Comparing these values with the ones in literature the value of CO_2 is 20% higher than in CECA 13X binder pellets.

The crystal diffusivity of CO_2 measured experimentally by the ZLC technique is 5.8×10^{-15} m^2/s and 1.3×10^{-15} m^2/s at 373 and 313 K, respectively. These values are in the same order magnitude of the ones measured by a frequency response and pulse chromatography techniques reported in literature. The ZLC desorption curves for CH_4 were measured under an equilibrium regime and no kinetic data is obtained.

Introduction

To be used as adsorbents zeolite powder need to pelletized and this reduces it´s working capacity in 20% or more which is the amount of inert clay binder generally used to give the necessary mechanical strength to the pellets in order be used in packed-columns and at the same time reduce pressure drop. Recently, a new technology was applied where the binder matter is converted to zeolite matter during the manufacturing procedure [1]. The resulting binderless beads can increase in this way the working capacities of existing zeolite adsorbent technologies.

The objectives of this work is to obtain data of equilibrium and kinetics of sorption of CO_2 and CH_4 on a new type of binderless beads of 13X zeolite. The equilibrium data is measured in a breakthrough apparatus and the kinetic data by the ZLC technique. Data obtained is compared to published one on binder pellets of the same type. Another goal is to determinate thermodynamic and kinetic parameters that are useful for the development of adsorption separation processes such as the ones calculated from modelling of equilibrium and kinetic of sorption: heats of sorption, Henry´s constants, equilibrium constants, working capacities, inter and intracrystalline diffusivities. Through this work some ideas about the use ZLC technique for the measurement the diffusvity in porous adsorbents are revised by establishing a procedure to analyse conveniently such results introducing a simple criteria to evaluate which kind of systems can be measured macroscopically by ZLC.

Results and Discussion

The powder of 13X from which the binderless beads were formed is from Chemiewerk Bas Kostritz GmbH (Germany) with a Si/Al ratio of 2.35. Metakaolin is used to manufacture the pellets. The synthesis and characterization procedure is described in detail elsewhere [1].

Adsorption equilibrium

The experimental single component sorption isotherms measured for CO_2 and CH_4 in 13X are shown in Figure 1 a) and b, respectively. Data were collected at three temperatures: 313, 343 and 373 K and for partial pressures (helium as inert) up to 4 atm. We can observe in Figure 1 that CH_4 isotherms are practically linear and marked Type I for CO_2 in the range of temperature and pressure studied. Figure 1 shows that CO_2 is the more strong adsorbed component with an amount adsorbed that reaches more than 5 mmol/g_{ads} at 313 K and partial pressure around 4 atm. For the same pressure and temperature the amount adsorbed of CH_4 is much smaller being almost 1.2 mmol/g_{ads}. This means also a selectivity CO_2/CH_4 around 4.2.
The sorption equilibrium of CO_2 was modeled with the Fowler isotherm that accounts for lateral interactions between molecules adsorbed molecules in adjacent sites,

$$\frac{1}{p}\frac{\theta}{(1-\theta)} = b\exp(-2w\theta / RT) \quad (1)$$

where $\theta = q/q_m$ is the degree of filling of sites, b is an equilibrium constant, p the pressure, q the amount adsorbed and q_m is the amount adsorbed at the saturation of the adsorbent, w is the extra energy when sorbate molecules occupy adjacent sites, R the ideal gas constant and T the temperature.

In the case o CH_4 isotherms they were fitted with the Langmuir model but are practically linear until 4 atm of pressure.

$$\frac{1}{p}\frac{\theta}{(1-\theta)} = b \qquad (2)$$

The parameters of the model isotherms are shown in Table 1.

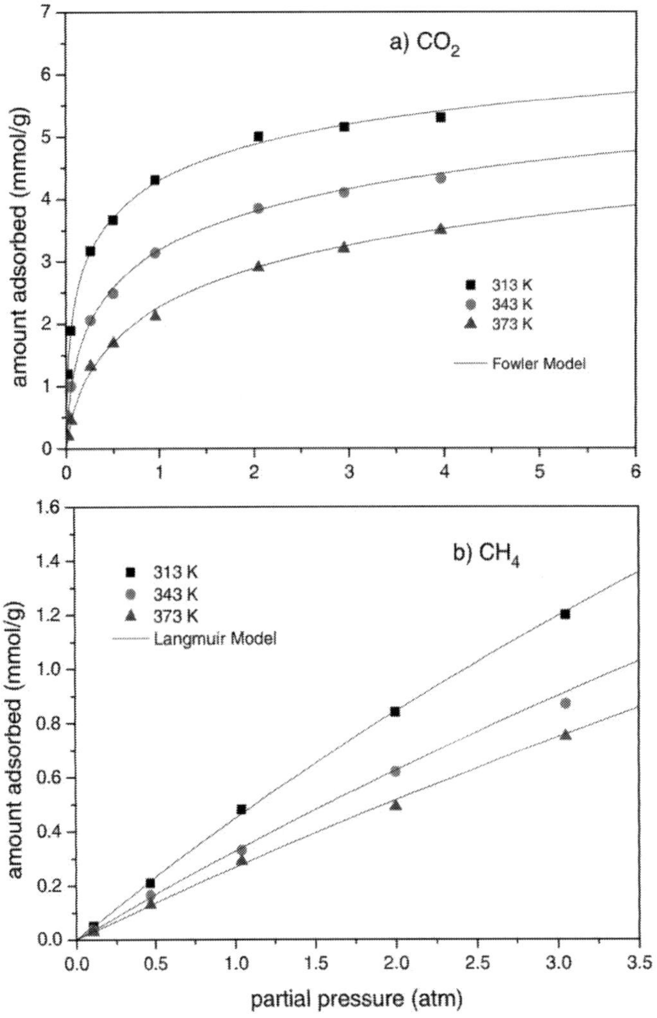

Figure 1. Adsorption equilibrium isotherms of CO_2 (a) and CH_4 (b) and on binderless beads of 13X zeolite.

Table 1. Isotherm model parameters for sorption of CO_2 and CH_4 in binderless beads of 13X zeolite.

Isotherm Models		CO_2 Fowler	CH_4 Langmuir
q_m	(mmol/g_{ads})	7.4	7.4
$-\Delta H$	(kJ/mol)	43.1	8.9
w	(kJ/mol)	6.1	-
		313 K	
H^2	(mmol/g_{ads}.atm)	143	0.45
b	(atm^{-1})	21.3	0.0643
		343 K	
H^2	(mmol/g_{ads}.atm)	38.5	0.34
b	(atm^{-1})	4.76	0.0462
		373 K	
H^2	(mmol/g_{ads}.atm)	11.1	0.27
b	(atm^{-1})	1.49	0.0374

[2] Henry's constant calculated from Virial plots

Kinetics of sorption

The ZLC technique was used to measure kinetic data. When the operational parameter of the ZLC model [2] L is higher than 10 kinetic data can be obtained in an easy way using Equations (3) and (4),

$$\ln\left(\frac{c}{c_0}\right) = \ln\left(\frac{2}{L}\right) - \pi^2 D_c / r_c^2 t \tag{3}$$

$$L = \frac{1}{3}\frac{F}{KV_s}\frac{r_c^2}{D_c} \tag{4}$$

where c is the outlet concentration of the ZLC, c_0 is the saturation concentration, t is the time, D_c is the crystal diffusivity, r_c the crystal radius, L a model parameter, F the purge flowrate, V_s the volume of adsorbent and K a dimensionless adsorption equilibrium constant (Henry´s law constant). Since ZLC technique is a macroscopic technique the time scale of the experiments should be preferably above 10 s. This time scale will depend of diffusion time constants r_c^2/D_c. If Equation 3 is represented in terms $t = f(D_c/R_c^2)$ it assumes the form,

$$t = \frac{\ln\left(\frac{(2/L)}{(c/c_0)}\right)}{\pi^2} \frac{r_c^2}{D_c} \qquad (5)$$

For L =10 and c/c_0 =0.1 which means 90% of the desorption curve we get,

$$t_{0.1} = 7.02x10^{-2} \frac{r_c^2}{D_c} \qquad (6)$$

Equation (6) equation can be used for a criterion to evaluate which type of systems can be measured by ZLC experiments in a macroscopic time.

For the present system since the diffusion of CO_2 reported in literature [3] is $1.8x10^{-15}$ m^2/s and the crystal size is 2 μm the value of $t_{0.1}$ calculated from Eq (6) is 555 s which means that it is possible to carry out a macroscopic ZLC experiment for CO_2 in zeolite 13X. For CH_4 values in literature are around $1x10^{-9}$ m^2/s and for 2 μm crystals we get $t_{0.1}$ = $7.02x10^{-5}$ s a value that turns the measurement of crystal diffusivity of CH_4 by ZLC impossible unless we have crystals of at least 1000 μm (1 mm) to get $t_{0.1}$ = 17.5 s.

Taking the previous remarks into consideration the crystal diffusivity of CO_2 on 13X was measured by ZLC. Figure 2 shows the effect of temperature in the ZLC desorption curves where it can be seen a strong temperature dependence of desorption time. The diffusion coefficient of CO_2 in 13X at 373 K is around $5.4x10^{-15}$ m^2/s which is comparable with the one reported in literature [3]. Since the batch of beads studied range from 1.2 mm to 2.0 mm we isolate the smaller beads from the larger ones and perform two different experiments to confirm is ZLC desorption curves are sensitive or not to the size of the beads. If they are not sensitive it means that the sorption rate is dominated crystal diffusion. Figure 3 show the experiments performed at 373 K under a similar flowrate of 50 mL/min being clear that the size of the beads does not change the shape of the desorption curves. Accordingly, we believe that ZLC experiments for CO_2 were performed under a kinetic regime controlled by intracrystalline diffusivity being the crystal diffusivities (2 μm crystal size) for CO_2 in 13X at 373 K of $5.64x10^{-15}$ m^2/s, at 343 K of $4.13x10^{-15}$ m^2/s and at 313 K $1.09x10^{-15}$ m^2/s.

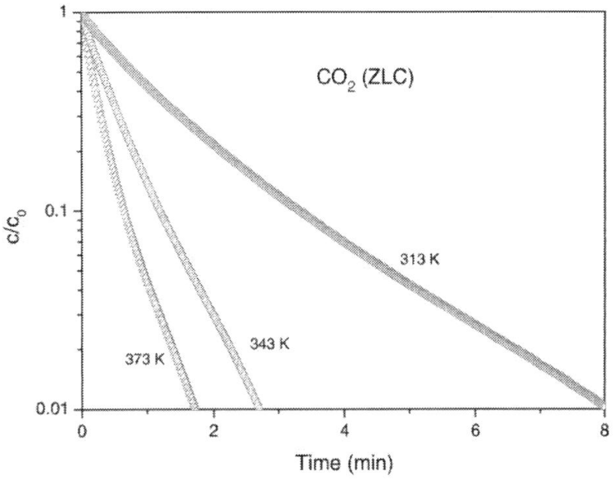

Figure 2. Effect of temperature on ZLC desorption curves of CO_2. Semilog plot of c/c_0 *vs* time t.

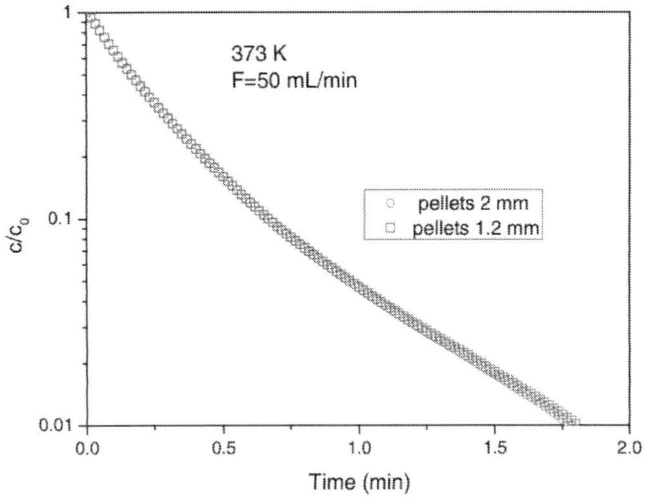

Figure 3. Effect of beads size on ZLC desorption curves of CO_2. Semilog plot of c/c_0 *vs* time t.

Figure 4 shows the desorption curve performed for CH_4 at 313 K at a purge flowrate of 15 mL/min. As expected the experiment is very fast (lower than 10 s for 99% of the desorption curve) and it is clearly seen that the intercept of the long time asymptote is 1 in the semi-log plot corresponding physically to the expected equilibrium regime. To validate this experiment we plot the predicted ZLC experiment using the isotherm data represented in Table 1 for CH_4 at 313 K. Figure 4 shows that the experiment is validated. This means that for CH_4 diffusion data in macropores and micropores could not be measured by ZLC unless we have crystals or beads with a much bigger size. For the other temperatures of 343 and 373 K experiments are too fast that data is unreliable.

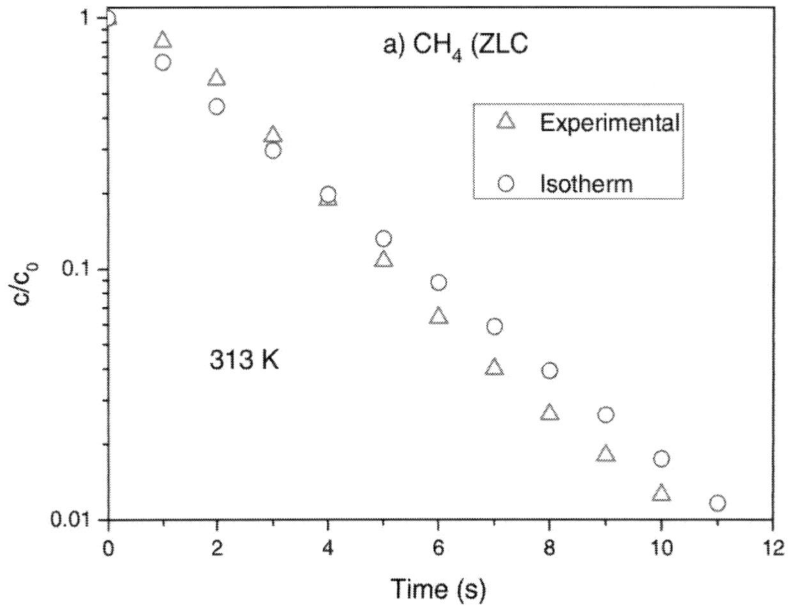

Figure 4. ZLC desorption curves of CH_4 at 313 K and purge flowrate of 15 mL/min. Semilog plot of c/c_0 *vs* time *t*.

Conclusions

We performed a study of the sorption and kinetics of CO_2 and CH_4 in binderless beads of 13X zeolite between 313 and 373 K at partial pressure up to 4 bar.

The sorption equilibrium isotherms of CO_2 are pronounced Type I and show an amount adsorbed of CO_2 around 5.2 mmol/g_{ads} at 4 bar which is 20% higher than values reported in literature for CECA 13X pellets with binder. The sorption equilibrium of CO_2 was successfully modeled with the Fowler isotherm that accounts for lateral interactions between molecules adsorbed in adjacent sites. The CH_4 isotherms were fitted with the Langmuir model but are practically linear until 4 atm of pressure.

It is demonstrated that CO_2 diffusion in crystals of 13X can be measured by ZLC in a macroscopic time using pellets with crystals of 2 μm of size. However, for CH_4 that is impossible even for macropore diffusion. The diffusivity of CO_2 in 13X at 373 K measured is around 5.4×10^{-15} m^2/s which is in the same order of a value reported in literature measured by a frequency response technique.

A simple equation that can be used to predict if a ZLC experiment can be performed in a macroscopic time is derived and applied for the present system proving that sorption of CO_2 in 13X zeolite can be measured macroscopically. The equation developed is also useful to check if diffusivity data calculated are consistent with a macroscopic experiment. The time of the experiments (that should be above a few seconds) is directly related with the diffusion time constant by the following expression: $t_{0.1} \geq 7.02x10^{-2} \frac{r_c^2}{D_c}$.

Acknowlegements

We acknowledge Chemiewerkt Bas Kostritz GmbH (Germany) for kindly provide the samples of binderless 13X zeolite studied in this work.

References

[1] K. Schumann, B. Unger, A. Brandt, F. Scheffler , Micropor. Mesopor. Mater. 154 (2011) 119-123.

[2] M. Eic, D. M. Ruthven, Zeolites 8 (1988) 40-45.

[3] G. Onyestyák, L. V. C. Rees, J. Phys. Chem. B 103 (1999) 7469- 7479.

(287a) The Importance of Mass Transfer Limitations in the Application of Forward Osmosis (Keynote Talk)

Robert W. Field, Department of Engineering Science, University of Oxford, Oxford, United Kingdom and Jun Jie Wu, School of Engineering and Computing Sciences, Durham University, Durham, United Kingdom

Abstract

The growth in world population, an increase in demand for resources, a changing global climate, and pollution of available water resources are all exerting unprecedented demands on fresh water supplies around the world. Furthermore, rising energy costs and global warming, both linked to use of fossil-fuels, are driving the need for low energy, more sustainable forms of water reclamation. Forward Osmosis (FO) has attracted growing interest recently as a highly promising technology for desalination and water reuse. The basis of FO is osmosis itself, a natural and spontaneously occurring process. It has attracted wide interest for its potential applications in a number of areas [1] including desalination [2], wastewater treatment [3-4], and food processing [5]. A related membrane process, known as pressure retarded osmosis (PRO), has also been used for osmotic power generation [6].

The FO process offers some advantages over conventional pressure-driven membrane processes such as (i) potentially lower energy consumption and (ii) minimal disturbance to sensitive components/species in the feed solution and it is the former that we address herein. It is widely recognized that the water flux in an FO process is severely limited by internal concentration polarization (ICP) which is a term for describing the phenomenon by which there is either dilution of the high-osmotic-pressure draw solution (DS) or undesirable concentration of the feed solution (FS) inside the FO support structure [see ref 1 for more details]. The former occurs if the support faces the draw solution; the latter if the support faces the feed.

The overall aim of the analysis is to ask whether even an ideal FO membrane of very large A and K will be superior to existing RO systems, and if so under what circumstances. Initially we have ourselves with three specific questions:

(1) What is the relationship between the flux, ratio of osmotic pressures and mass transfer coefficients?

(2) Can external mass transfer coefficients be neglected?

(3) Can explicit equations be obtained if $J/K < 1.0$?

Two answers to the third question will be covered in the presentation. Regarding the first question a sample result is presented in Figure 1. This is based upon the widely used equations found in [1] and elsewhere. These do not include a full allowance for mass transfer in the feed and draw solution boundary layer mass transfer coefficients, k_f and k_d. The importance of both K and the osmotic pressure ratio is apparent.

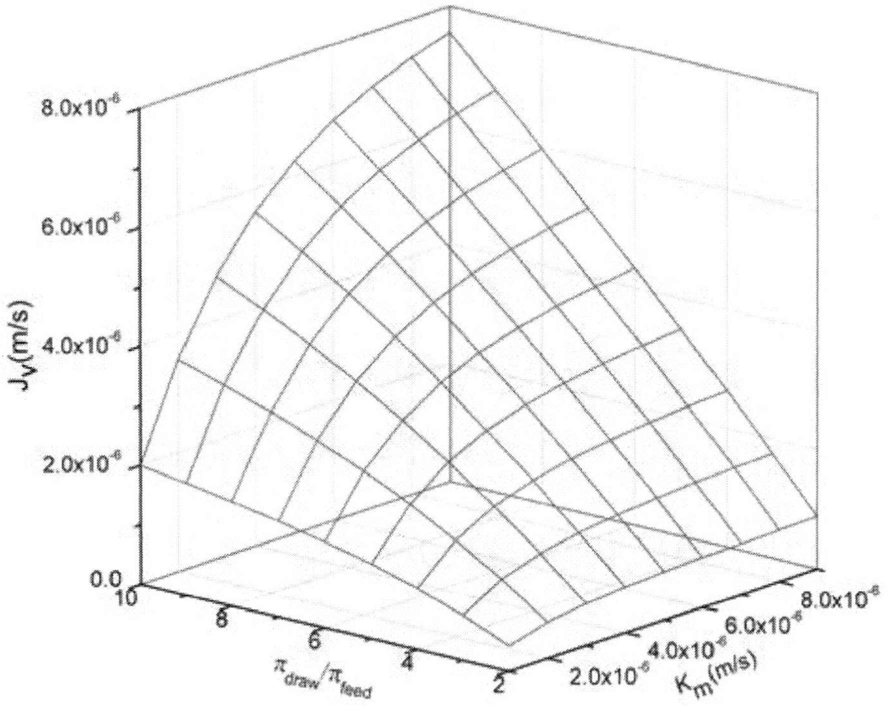

Figure 1 Predicted fluxes for various values of (π_{draw}/π_{feed}) and mass transfer coefficient K (as defined in Ref 1) for active layer facing draw solution. (Draw by Gaoqi Zhang)

As the mass transfer coefficient of support has improved in practice (made thinner and with great porosity) allowance for k_f and k_d becomes essential. Under circumstances in which the membrane parameters A and K become large (i.e. a tendency towards an ideal FO membrane), it can be shown that the flux tends to

$$J = \frac{Kk_f k_d}{Kk_d + k_f K + k_f k_d} \ln\left(\frac{\pi_d}{\pi_f}\right) \quad (1)$$

which for very large K would simplify to:

$$J = \frac{k_f k_d}{k_d + k_f} \ln\left(\frac{\pi_d}{\pi_f}\right) \quad (2)$$

Considerations of limitations due to scaling coupled with equation (2) suggest that even if membranes with high A, were available FO would be unattractive for high salinity feeds. This is certainly not contradicted by recent experimental results [7]. The main conclusion is that in the area of water treatment application of FO should be directed at low salinity feed waters.

Less Common Symbols Used

A Overall water permeability coefficient (m/s Pa)

K Membrane support mass transfer coefficient (m/s)

pi ratio = π_d/π_f where d refers to draw and f to feed

[1] Tang CY, She Q , Lay WCL, Wang R,Field RW and Fane AG Modeling double-skinned FO embranes Desalination 283 (2011) 178-186

[2] J.R. McCutcheon, R.L. McGinnis, M. Elimelech, Desalination by ammonia-carbon dioxide forward osmosis: Influence of draw and feed solution concentrations on process performance, Journal of Membrane Science 278 (2006) 114–123.

[3] A. Achilli, T.Y. Cath, E.A. Marchand, A.E. Childress, The forward osmosis membrane bioreactor: a low fouling alternative to MBR processes, Desalination 238 (2009) 10–21.

[4] W.C.L. Lay, Y. Liu, A.G. Fane, Impacts of salinity on the performance of high retention membrane bioreactors for water reclamation: a review, Water Research 44 (2010) 21–40.

[5] M.I. Dova, K.B. Petrotos, H.N. Lazarides, On the direct osmotic concentration of liquid foods. Part I: Impact of process parameters on process performance, Journal of Food Engineering 78 (2007) 422–430.

[6] S. Loeb, Large-scale power production by pressure-retarded osmosis, using river water and sea water passing through spiral modules, Desalination 143 (2002) 115–122.

[7] Y. Xu, X. Peng, C.Y. Tang, Q.S. Fu, S. Nie, Effect of draw solution concentration and operating conditions on forward osmosis and pressure retarded osmosis performance in a spiral wound module, Journal of Membrane Science 348 (2010) 298–309.

(287b) An Advanced Design of Membrane Structure for Wastewater Treatment Through Forward Osmosis

Jincai Su[1], Rui Chin Ong[1], Bradley J. Helmer[2], Jos S. de Wit[2] and Tai-Shung Chung[3], (1)National University of Singapore, Singapore, Singapore, (2)Eastman Chemical Company, Kingsport, TN, (3)Department of Chemical & Biomolecular Engineering, National University of Singapore, Singapore, Singapore

An advanced design of membrane structure for wastewater treatment through forward osmosis

Jincai Su[1], Rui Chin Ong[1], Bradley J. Helmer[2], Jos S. de Wit[2], Tai-Shung Chung[1,*]

[1]*Department of Chemical & Biomolecular Engineering, National University of Singapore, 4 Engineering Drive 4, Singapore 117576*

[2]*Eastman Chemical Company, P.O. Box 1972, Kingsport, Tennessee 37662*

**Corresponding author, Email: chencts@nus.edu.sg, Fax: (65)-6779 1936*

Key words: Forward osmosis; Hollow fiber; Double-skin; Wastewater treatment

Abstracts

Forward osmosis (FO), utilizing the natural phenomenon of osmosis, is an emerging membrane process driven by the osmotic pressure gradient created across a membrane by two flowing streams of varying concentrations. Considered as an energy-effective technology, FO has attracted much attention in water-related R&D activities. The 1st key component of the FO process is the membrane material which should be semipermeable, i.e., allowing water to permeate through while blocking all the solutes in the draw and feed solutions. To maximize the water permeation flux, FO membranes are designed with asymmetric structure, i.e., an ultra thin dense layer on the top of a highly porous sublayer. With this structure, either running the draw solution or the feed along the porous sublayer would result in severe internal concentration polarization (ICP), which would greatly reduce the effective osmotic pressure gradient across the membrane. Nonetheless, running the draw solution along the porous sublayer would be more practical since the feed may contain foulants which would be easily stuck in porous matrix. Once the membrane sublayer is fouled, it is very difficult to clean. To date, how to effectively reduce the effect of ICP is still a big challenge.

We have designed and fabricated very interesting hollow fiber FO membranes with two apparently dense skins by manipulating the phase inversion rates during the spinning and the degrees of subsequent thermal annealing at the inner and outer layers. The surface pores within the inner skin layer show a very narrow size distribution with a mean radius of 0.34 nm. Being also relatively dense, the outer skin layer is able to keep certain solutes such as macromolecules or multi-valence ions from entering the sublayer and avoid their accumulation within the sublayer. The double-skinned hollow fiber membranes have been used for Sucrose concentration and treatment of wastewater containing mixed metal ions through a forward osmosis-nanofiltration (FO-NF) hybrid system (Fig. 1). The FO system draws water from the wastewater while the NF system is used to regenerate the diluted draw solution. The experimental results have revealed great potential of the newly developed double-skinned CA hollow fiber membranes for wastewater water reclamation and macromolecule recycle.

Fig. 1. The FO–NF hybrid system for wastewater reclamation

(287c) Charged and Hydrophilized Polybenzimidazole (PBI) Membranes for Forward Osmosis

Michael Flanagan, Department of Chemical and Environmental Engineering, University of Toledo, Toledo, OH and **Isabel C. Escobar**, Chemical and Environmental Engineering, The University of Toledo, Toledo, OH

Forward osmosis is the movement of water across a selectively permeable membrane. The driving force for water permeation through the membrane is the difference in osmotic pressure between the feed and draw solutions. Polybenzimidazole (PBI) is a material with excellent chemical resistance and high mechanical and thermal stability that is a promising material for forward osmosis separations. Drawbacks associated with the use of PBI as a membrane material include low hydrophilicity and surface charge neutrality at neutral pH values.

In this study, PBI membranes were cast using the phase-inversion technique in the form of asymmetric flat sheets, and membrane surfaces were functionalized using different modifying agents with the goal of increasing hydrophilicity, increasing surface charge, and reducing membrane pore sizes. An increase on the negative charge of the membrane surface was expected to yield an increased rejection of ions and of charged species in the feed solution. An increase in hydrophilicity was expected to reduce fouling propensity and enhance wettability of the membrane surface. Lastly, a reduction in pore sizes was expected to allow for greater steric effects. In order to modify the membranes, the surfaces of the membranes were first activated with 4-(chloromethyl) benzoic acid (CMBA). The modifying agents selected for membrane functionalization included: taurine, para-phenylene diamine, ethylene diamine, and poly(acrylamide-co-acrylic acid).

Membranes were characterized using Fourier transform infrared spectroscopy in attenuated reflectance mode (FTIR-ATR), zeta potential, environmental scanning electron microscopy (ESEM), contact angle measurements, and total organic carbon (TOC). Functionalization, surface charge, increased hydrophilicity, and reduced pore size were all verified. Pure water permeability and monovalent salt rejection were tested in a pressure driven mode for comparison between both virgin and modified membranes. Salt rejection was investigated using various sodium chloride feed concentrations, and a range of pH values.

Flat sheet membranes were also tested in forward osmosis (FO) applications. The FO process involved use of an ammonium bicarbonate draw solution and a sodium chloride feed solution. All modified membranes showed enhanced water permeability and increased salt rejection over the unmodified membrane surface.

(287d) Development of High-Flux and Fouling-Resistant Reverse Osmosis Membranes for Brackish Water Desalination

Lin Zhao, Department of Chemical & Biomolecular Engineering, The Ohio State University, Columbus, OH and W. S. Winston Ho, The Ohio State University, Columbus, OH

Development of High-Flux and Fouling-Resistant Reverse Osmosis Membranes for Brackish Water Desalination

Lin Zhao, C.-Y. Chang, and W.S. Winston Ho, The Ohio State University

A novel high-flux and fouling-resistant reverse osmosis membrane was synthesized and characterized under brackish water desalination conditions. The *o*-aminobenzoic acid-triethylamine salt was added into *m*-phenylenediamine (MPD) solution to react with trimesoyl chloride (TMC) during the interfacial polymerization. The membrane preparation conditions including MPD concentration, TMC concentration, and interfacial polymerization time were optimized using brackish water desalination tests with 2000 ppm NaCl solution at 225 psi and 25°C. The membrane synthesized under the optimal conditions was post-treated with aqueous solutions containing glycerol, sodium lauryl sulfate, and camphorsulfonic acid-triethylamine salt to further increase the water flux. The resulting membrane showed a flux of 54.4 gallons/ft^2/day and a salt rejection of 98.6%. This membrane outperformed the other membranes including BW30, LE, XLE, and ESPA3, evaluated at their favorable desalination conditions, respectively. The fouling-resistant property of the synthesized membrane was obtained by physically coating a cross-linked polyethylene glycol (PEG-200) layer on top of the thin film. The effects of cross-linked PEG-200 solution concentration were investigated under brackish water desalination conditions. Dodecyltrimethylammonium bromide, a cationic foulant, caused a 52.9% flux reduction of the uncoated membrane and a 30.8% flux reduction of the coated membrane. Tannic acid, an organic foulant, caused a 15.6% flux reduction of the uncoated membrane and a 4.0% flux reduction of the coated membrane. The modified membrane surface morphology was analyzed using atomic force microscopy and scanning electron microscopy. The results showed a smoother membrane surface which confirmed the successful modification.

(288a) High-Throughput Characterization of the H_2-D_2 Exchange Activity of Pd_xCu_{1-x} Alloys

Gamze Gumuslu[1,2], Petro Kondratyuk[1,2], James B. Miller[1,2] and Andrew J. Gellman[1,2], (1)Chemical Engineering, Carnegie Mellon University, Pittsburgh, PA, (2)NETL, US Department of Energy, Pittsburgh, PA

Pd alloys have been widely studied as they are used as membrane materials to separate H_2 from mixed gas streams. In the mechanism of hydrogen permeation through the membrane the first step is dissociation of molecular H_2 on the alloy surface. In this study, we prepare Pd_xCu_{1-x} alloys and characterize the H_2 dissociation activities as a function of composition, x. The conventional approach in alloy research is to prepare and characterize a series of single composition samples, which is time consuming. To address this challenge, we have developed a high-throughput methodology that enables characterization of a range of alloy compositions in a single experiment. The approach is based on use of Composition Spread Alloy Films (CSAFs), high-throughput libraries which include all possible alloy compositions of an alloy in the form of a thin film deposited on a substrate. In this work, Pd_xCu_{1-x} CSAFs were prepared by evaporative co-deposition of the components onto Mo substrates, followed by annealing at 700 K and H_2-D_2 exchange ($H_2 + D_2 \rightarrow 2HD$) was used to probe the activity of film surface across composition and structure space.

H_2-D_2 exchange experiments were performed using a multichannel microreactor array, at atmospheric pressure and over a temperature range from 60 to 320°C. The array distributes 100 inlet-outlet pairs across the surface of the CSAF, enabling measurement of H_2 conversion at 100 discrete compositions on the CSAF in a single experiment. Binary Pd-Cu CSAFs, spanning the composition range 5-95% Pd, possess the features of the bulk phase diagram at the annealing temperature (700 K) and reveal increasing activity with increasing Pd content and increasing temperature. The maximum exchange activity was observed at ~70% Pd on the CSAF surface. The CSAF activity results are consistent with those obtained in a series of conventional experiments conducted over fixed beds of single composition bulk Pd-Cu alloys. Hence, this study shows that high-throughput methods provide reliable data and can be further used to characterize binary or higher order alloys.

(288b) Long Term Thermal Stability of Pd and Pd-Alloy Composite Membranes

Hani Abuelhawa[1], Oyvind Hatlevik[1], Stephen N. Paglieri[2], Aadesh X. Harale[3] and J. Douglas Way[4], (1)Chemical and Biological Engineering, Colorado School of Mines, Golden, CO, (2)TDA Research, Inc., Wheat Ridge, CO, (3)Saudi Arabian Oil Company, Dhahran, Saudi Arabia, (4)Chemical Engineering, Colorado School of Mines, Golden, CO

Long Term Thermal Stability of Pd and Pd-alloy Composite Membranes

Hani Abuelhawa[1], Oyvind Hatlevik[1], Stephen N. Paglieri[2], Aadesh Harale[3], and J. Douglas Way[1]

[1]Department of Chemical & Biological Engineering, Colorado School of Mines,

Golden, CO 80401, USA

[2]TDA Research Inc., Wheat Ridge, CO 80033, USA

[3] Saudi Arabian Oil Company, Dhahran, 31311, KSA

The integration of a Pd-based membrane within a catalytic reforming reactor to produce hydrogen has received much more attention during the last few decades **[1]**. In addition to being permselective to hydrogen, Pd-based membranes are capable of maximizing the conversion of thermodynamically-limited reactions, such as the steam reforming of methane, beyond the equilibrium conversion based on the feed composition, by simply withdrawing the hydrogen from the reaction mixture as it forms. However, there are still some issues that prevent this technology from being fully commercialized, such as the long term thermal stability and the chemical tolerance of these membranes in real-world mixed gas environments.

Guazzone, et al. **[2]** has investigated the long term thermal stability of electrolessly plated pure Pd membranes by identifying the mechanism that leads to the leak evolution in hydrogen atmosphere. They concluded the leak evolution is a thermally-activated process induced by the self-diffusion of Pd at temperatures above 400°C. In this work, we will investigate the thermal stability of Pd and Pd-alloy membranes such as PdRu, PdAu, and/or PdAg. Previous literature **[1]** has investigated the performance and properties of Pd-alloy membranes, however, the long term thermal stability of these alloys did not receive so much attention.

In this paper, long term high temperature stability data for a 5.1 micron thick sequentially plated PdAu10wt% membrane prepared by the electroless plating method on a porous-zirconia/porous-stainless-steel support will be discussed. The membrane has been tested for approximately 1000 hours and showed superior thermal stability at 500°C and below, in comparison to similar pure Pd membranes. The membrane also maintained an ideal H_2/N_2 pure gas flux ratio of at least 1000 at 100 psig for more than 1000 hours of testing. As shown in Figure 1, nitrogen leak did not grow significantly in the temperature range (450-500°C) during several hundred hours of testing. Figure 2 compares the performance of PdAu membrane to a pure Pd of similar thickness at 500°C. Addition of Au to Pd enhanced the thermal stability of the membrane by reducing the leak rate by at least one order of magnitude. Additionally, by comparing the leak rate of this PdAu membrane with the results of Guazzone, et al. **[2]** for a pure Pd membrane of comparable thickness, the leak growth rate in the PdAu membrane is also lower by at least one order of magnitude. The calculation of the activation energy of the leak growth rate in Pd-alloy membranes and comparing it to the Pd self-diffusion activation energy could give some insights in revealing the leak growth mechanism in these membranes.

References

[1] Ø. Hatlevik et al., Separation and Purification Technology 73 (2010) 59–64

[2] F. Guazzone et al., AIChE Journal, (2008) Vol. 54, No. 2, 487-494.

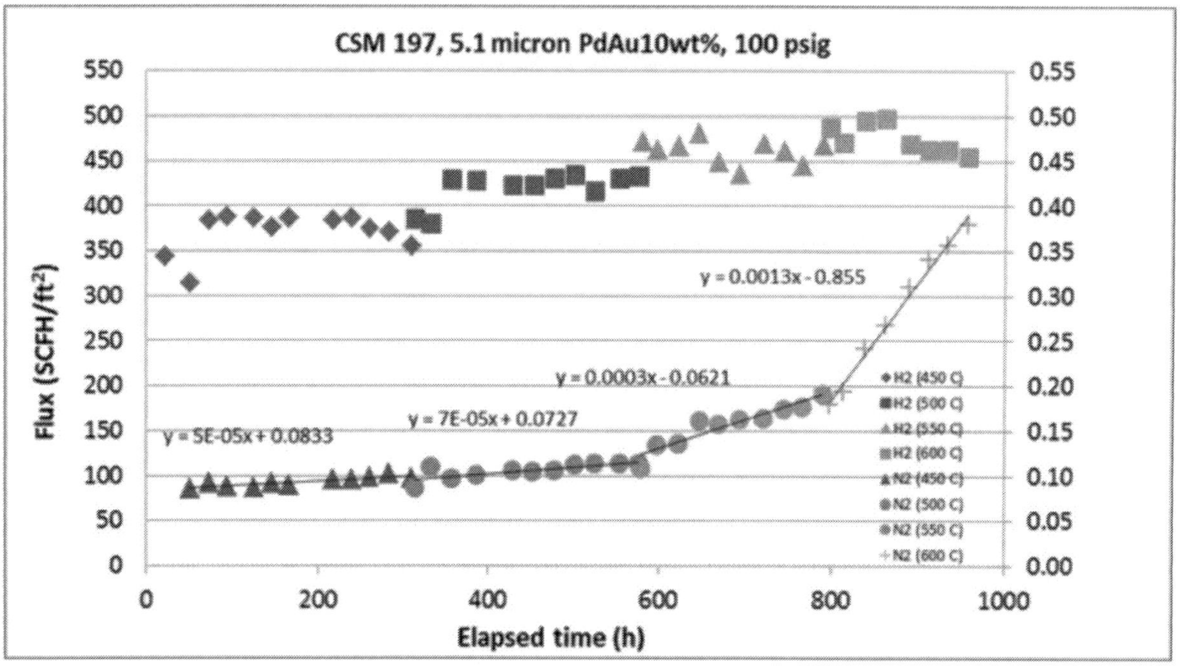

Figure (1): The high temperature thermal stability data for the 5.1 micron PdAu10wt% membrane deposited on zirconia/SS porous support.

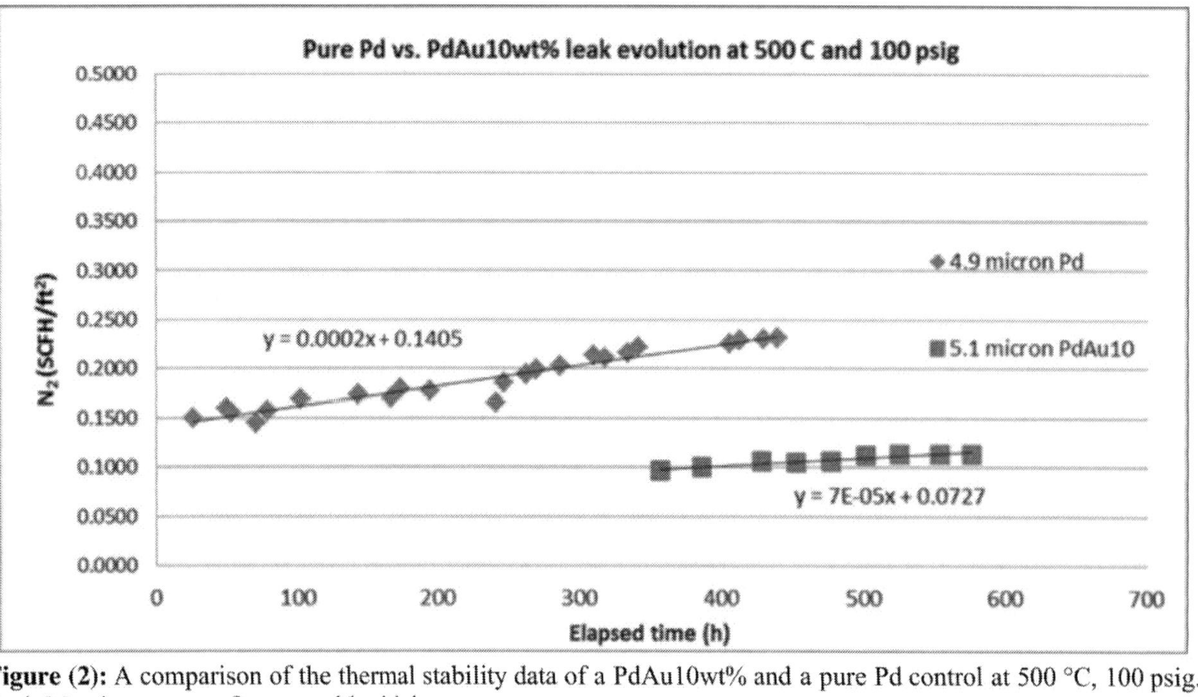

Figure (2): A comparison of the thermal stability data of a PdAu10wt% and a pure Pd control at 500 °C, 100 psig. Both Membranes are of comparable thickness.

(288c) Sulfur Resistant Pd-Ag-Au Alloys for Hydrogen Separation

James B. Miller[1,2], Fernando Braun[3], Laura Cornaglia[3], Petro Kondratyuk[1,2], Benoit Fleutot[1], Bret H. Howard[2] and Andrew J. Gellman[1,2], (1)Chemical Engineering, Carnegie Mellon University, Pittsburgh, PA, (2)NETL, US Department of Energy, Pittsburgh, PA, (3)INCAPE, Santa Fe, Argentina

Separation of hydrogen from mixed gas streams is a key unit operation in the generation of carbon-neutral fuels and electricity from fossil- and bio-derived feedstocks. Pd-based membranes have received significant attention for the separation application in advanced coal gasification processes because of their near-perfect selectivity for hydrogen. In practice, pure Pd suffers from several limitations, including high cost, poor mechanical strength, and deactivation by minor components, most notably H_2S. H_2S can compromise membrane performance by reacting with Pd to form a low-permeability Pd_4S scale on the membrane surface. Pd has been alloyed with minor components to improve structural properties, improve permeability (i.e., Ag) and impart resistance to H_2S (i.e., Au). To achieve both high permeability and H_2S tolerance, we are investigating PdAgAu ternary alloys.

In this work, we synthesized $Pd_{83}Ag_2Au_{15}$ and $Pd_{74}Ag_{14}Au_{12}$ alloys by sequential (Pd/Ag/Au) electroless plating onto stainless steel. After deposition, samples were annealed in H_2 at 600°C for 48 hr to obtain homogeneous composition and achieve complete formation of the FCC solid solution. Both alloys and a 25 μm thick Pd foil reference sample were characterized by XRD, SEM-EDS and XPS before and after treatment in 1000 ppm H_2S and H_2 at 350°C for 3 and then 30 hours. As expected, the Pd reference sample displayed clear XRD evidence of bulk sulfide formation upon H_2S exposure. A Pd_4S phase started to form at 3 hr and grew with increasing exposure time. The alloys, in contrast, displayed no evidence of bulk sulfide formation. EDS measurements of bulk composition were consistent with the structural characterization results: after 30 hours of H_2S exposure, the S content of the Pd foil was 20 at% (i.e., complete conversion to Pd_4S), and no S was detected in either alloy. XPS depth profiling experiments revealed very weak S signals at the terminal surfaces of the H_2S-exposed alloys, but those disappeared before a depth of 10 nm. The depth profiles also revealed co-segregation of Au and Ag to the surface of the alloy; preferential location of Au at the alloy surface may be related to its resistance to bulk sulfidation.

In a preliminary permeation test, a $Pd_{75}Ag_{12.5}Au_{12.5}$ alloy delivered H_2 fluxes comparable to those of pure Pd. This alloy foil was fabricated by flame melting pure metals and then rolling to a thickness of 100 μm. The permeance was measured over the range of 350 to 750°C using pure H_2. Near future experiments will measure the alloy permeance in 1000 ppm H_2S / H_2 at several temperatures.

(288d) Microstructured Palladium-Based Membranes for Hydrogen Purification

Zachary W. Dunbar and Deryn Chu, Army Research Laboratory, Adelphi, MD

Hydrogen Purification with Microstructured Palladium Alloy Composite Membranes

Zachary W. Dunbar and Deryn Chu

Abstract:

Portable fuel cells have potential to fill a unique power niche in the US Army for small to intermediate power demands, such as battery charging, drone power, small gensets, and auxillary power units. Mobile fuel reformation is sought by the Army to provide mobile hydrogen-on-demand for fuel cell applications, using logistic fuels as a hydrogen carrier. Unfortunately, running a fuel cell on pure logistic fuels reformate is troublesome, due to its high concentrations of carbon monoxide, hydrogen sulfide, and coking precursors.

To alleviate this issue, a purification step is necessary to clean up the gas stream being processed by the fuel cell. Palladium based membrane technology is an ideal candidate for this operation.

In this work, microfabrication processes are used to engineer a microstructured metallic support for both pure palladium and palladium-copper alloy membranes. The resulting membranes are entirely metallic, and have planar geometry. The active layer for hydrogen purification is varied between 250 and 500 nanometers, while the nickel mechanical support is approximately 10 microns in thickness. The membranes are tested under pure hydrogen and synthetic water gas shift atmospheres. Hydrogen flux, selectivity, hydrogen sulfide sensitivity and long term durability are evaluated.

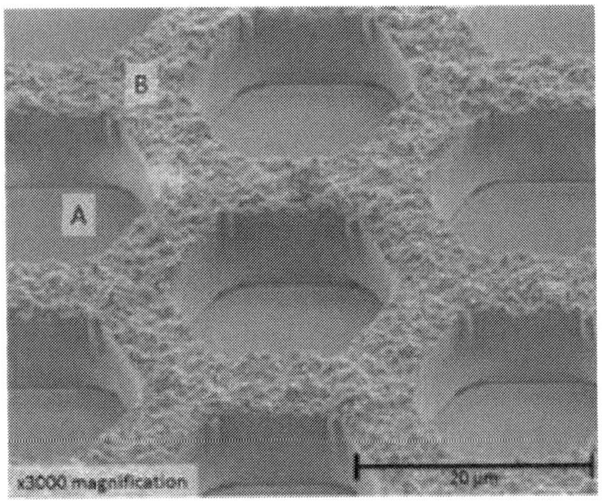

Figure 1. Composite membrane (permeate side) A – palladium thin film; B – electroplated nickel metal for mechanical support

Hydrogen permeation through Pd membranes with/without deposited Cu catalyst

Abdulrahman Y. Alraeesi, UAE University, Al Ain, UAE

Tracy Q. Gardner, Colorado School of Mines, Golden, CO, USA

Introduction

Palladium-based membranes have been considered recently for use in catalytic membrane reactors [1-3]. Palladium membranes can be used to separate or purify hydrogen in separation processes and can also be used in conjunction with hydrogenation reactions to produce valuable products. An incredible effort has been made in the field of hydrogen purification using palladium and palladium alloy membranes, (e.g. Pd60%-Cu40%, Pd75%-Ag25%, etc.) including many studies aimed at understanding the fundamentals of hydrogen transport through such metal membranes [4-5].

This study presents the fundamental aspects of hydrogen transport through palladium membranes, with and without deposited catalyst films. Understanding the transport through such systems is important for properly analyzing and predicting the performance of membranes and associated catalysts in catalytic membrane reactors. Hydrogen permeation is investigated through palladium-based membranes under conditions that would be used in applying them as catalytic membrane reactors. The potential for using palladium-based membranes with deposited catalyst films as catalytic membrane reactors is to improve reaction yield and selectivity for hydrogenation reactions.

In this study, copper layers were deposited on palladium-based foils by electroplating (electrodeposition). A new approach involving masking the foil to control the deposition geometry was established. Deposited copper layers were oxidized in some cases prior to permeation measurements, and the effects of catalyst on permeation and vice versa were studied

Sieverts' law (see equation 1), an equation relating the flux of hydrogen through a metal membrane to the hydrogen partial pressures on the feed and permeate sides [3, 6]:

$$J_{H2} = \frac{Q_{H2}}{L}\left(P_{H2,f}^{0.5} - P_{H2,p}^{0.5}\right) \qquad (1)$$

Above equation is often used to model and analyze hydrogen flux through palladium-based membranes whether or not the flux depends on the square root of the partial pressures.

Experimental Procedures

In this section the plating bath (as existed in NREL) as well as the sample preparation and deposition conditions are discussed.

Plating Bath

The electroplating experiments were carried out in the conventional three-electrode glass electrochemical cell shown in Figure 1. Use of the apparatus and all chemicals were provided by Mark Reimann at the National Renewable Energy Laboratory (NREL).

The palladium foil was mounted on the left, the platinum electrode is in the middle, and the reference electrode is on the right. The data acquisition system was used to control the deposition parameters. The potentiostat was used (along with software) to set the voltage, the current, and the time for deposition and the variation of these parameters during deposition is displayed. A pre-prepared 0.05M $CuSO_4.5\ H_2O$ solution was used for all depositions.

Figure 1 Conventional three-electrode glass electrochemical cell (at NREL)

<u>Sample Preparation and Deposition Conditions</u>

The palladium foils on which films were electrodeposited were ~47 mm in diameter. They were heated in air to 573 K for ~1 h, rinsed in acetone, washed in distilled water, dried using nitrogen gas, and in some cases were masked with ParaFilm. The foils were then mounted in the bath. After deposition, the foils were washed in distilled water and dried with nitrogen gas. Since the goal was to deposit the copper on only one side of the palladium foils, ParaFilm was used to mask the backside of most of the foils. The ParaFilm was cut to fit the foil and was heated slightly on a hot plate while in contact with the foil until the film stuck to the foil. Then the masked foil was mounted in the solution for deposition. To deposit copper islands on one side of some palladium foils, one side of the foil was masked with ParaFilm as mentioned and the other side was covered with a ParaFilm mask with holes made using a paper punch. Paper punches making ~3 mm and ~6 mm diameter holes were used to make small and large catalyst islands.

A constant voltage of ~0.8 V was applied (assuming constant current of 70 mA) for some foils, and a constant current of 70mA was applied (assuming constant voltage of 0.8 V) for other foils. The sample naming convention included the following:

- Substrate or foil material: Pd
- On which side deposition made, permeate as "1", both sides as "2", half of permeate side as "1/2"
- Film type: continuous "C" or islands "Is"
- Constant voltage "V" or constant current "C" applied
- Deposition time in seconds.

For example; Pd2CV240 means that the substrate or the membrane material is palladium (Pd) and copper was deposited as continuous film on both sides of the foil for 240 seconds at a constant voltage of ~0.8V.

Copper Annealing and Copper Oxide Reduction by Hydrogen

The deposited material was mainly metallic copper, partially oxidized to cuprous oxide (Cu_2O). However, the catalyst we desired was cupric oxide (CuO) as it is the catalyst for the reverse water gas shift reaction.

Figure 2 (a) shows a Pd/Cu foil that was annealed at 473 K for ~4 h. The surface is a combination of orange, green, and purple colors indicating partial oxidation of some of the copper to cuprous oxide (Cu_2O) and copper hydroxide ($Cu(OH)_2$). In Figure 2 (b) the surface is black after annealing at 673 K for ~4 h, indicating that the film oxidized to cupric oxide (CuO).

Figure 2 Palladium foils with deposited copper films (a) annealed in-situ in flowing air at 473 K for 4 h (b) annealed in stagnant air at 673 K for 4 h.

Four samples of Pd/Cu were annealed for 18 hours at 673 K to ensure that CuO is formed. Then for reduction study, an annealed sample was placed in a 100 mm long quartz tube (6 mm ID, 13 mm OD) from Technical Glass Products, Inc. Results showed that at lower temperatures (473 K) the reduction was slower and had not completed after ~15 minutes hydrogen exposure, suggesting it is possible to measure short-time hydrogen permeation through CuO deposited films on Pd foils at or below 473 K.

Transient and Long-Time Hydrogen Permeation through Pd-Based Foils

For different samples of interest, short-time transition and long-time hydrogen fluxes through Pd and Pd/CuO foils were addressed. Particularly, uncoated Pd, Pd with continuous film of Cu on one side, and islands of Cu on one side, were considered for permeation. Measurements were taken at 473 K to minimize CuO reduction. The membrane module used is shown in Figure 3.

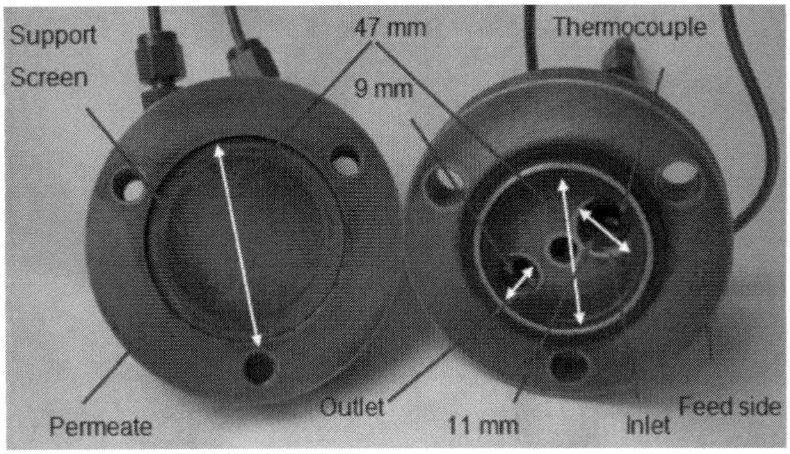

Figure 3 Membrane module

Results and Discussion

Transient Hydrogen Permeation through Undeposited Palladium (Pd-3 Foil)

The Pd-3 foil was rinsed with acetone and activated ex-situ in stagnant air at 673 K and ambient pressure for one hour. Transient hydrogen permeation was measured for the first 30 minutes, He purged both sides for at least 10 minutes and then the transient hydrogen permeation was repeated (i.e. ΔP_{H2} ~ 59 kPa). The transient hydrogen fluxes for five runs show that all decreased with time (Figure 4). The fluxes were similar for four runs and declined by about 33% in each case over the first 30 minutes. For the fifth run, after 13 hours hydrogen exposure followed by helium purge, the flux was higher and declined by only ~8% during the first 30 minutes. As suggested, the membrane structure may have changed over time during the continuous hydrogen flow. Results show that the fluxes followed Sieverts' Law and were reproducible during the period of hydrogen exposure following last run.

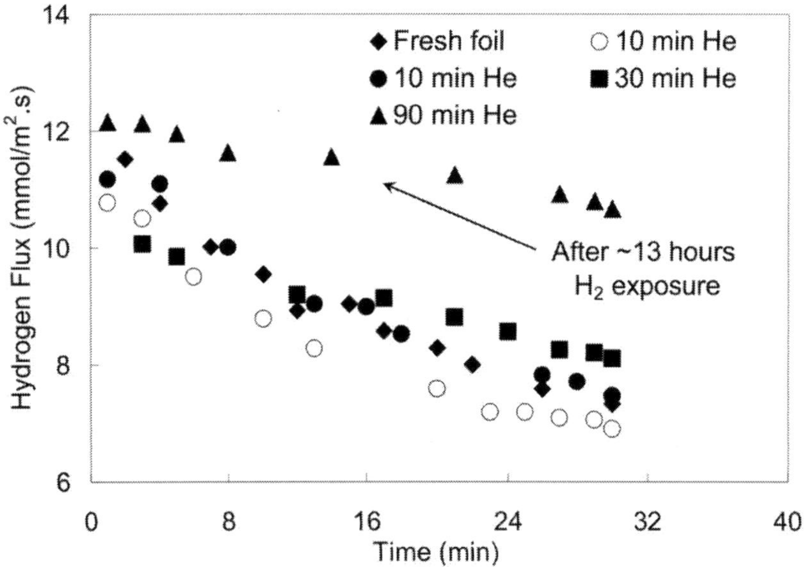

Figure 4 Short-time transient hydrogen permeation fluxes through 25 μm thick Pd-3 foil (area ~13 cm^2) at 473 K run sequentially (in order listed in legend) after various timed He purges between runs. $P_{H2,f}$ = 152 kPa and $P_{H2,p}$ = 93 kPa

H_2 Permeation through Pd Membrane with CuO Film on Permeate Side

The Pd1CC60 foil (estimated 0.064 μm Cu thick) was used for permeation. The area exposed for deposition was ~24 cm^2, but the area for permeation was ~13 cm^2, as it is for all membranes tested in the large module. The membrane was annealed at 673 K in ambient air for ~24 hours to form the cupric oxide (CuO) layer. The module was heated to 473 K under helium flow on both sides.

Figure 5 shows the changes in fluxes within 30 minutes from H_2 conduction to the feed zone for the last run at each pressure. Early fluxes (measured around 7 minutes after hydrogen exposure) were ~0.62 mmol/m^2.s for all feed pressures, but at longer times (~20 minutes) the fluxes for higher feed pressures were higher. This could be a result of the reduction process and/or diffusion of Cu into the Pd metal occurring at different rates for the different conditions. For example, copper oxide reduction to copper, and subsequent copper diffusion into the membrane, may occur faster at higher feed pressures because hydrogen transports through the membrane faster due to the higher driving force, resulting in faster hydrogen permeation (see Figure 6 for SEM image).

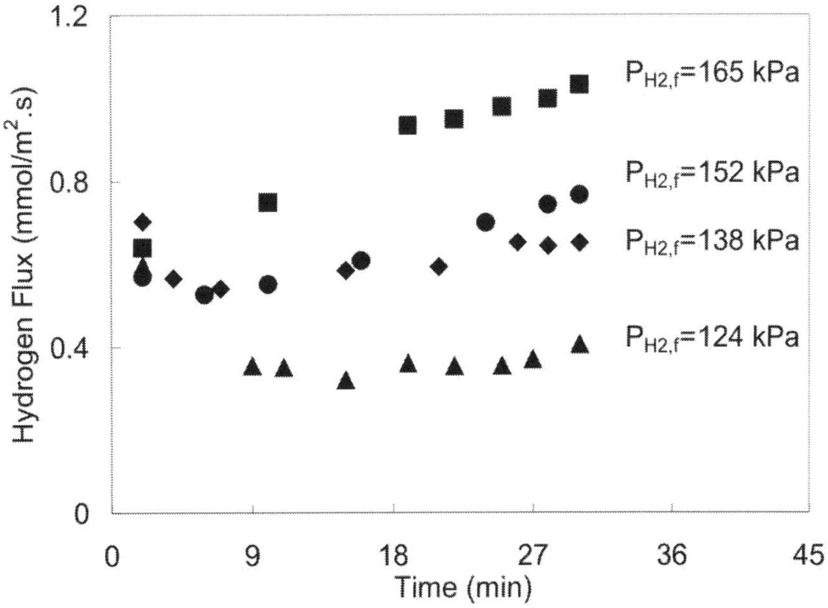

Figure 5 Short-time transient H_2 fluxes through 25.064 μm thick Pd1CC60 foil at 473 K and feed pressures of 124 kPa (after 15 min air), 138 kPa (after 20 min air), 152 kPa (after 12.5 h air), and 165 kPa (after 13 h air) (permeate at ~93 kPa)

Figure 6 SEM image of Pd1CC60 foil after annealing followed by hydrogen permeation studies

Long-Time H_2 Permeation through Coated and Uncoated Palladium Foils

Long-time fluxes after thermal cycling and long-time exposure to hydrogen through the uncoated Pd-3, Pd1CC60, Pd1CC30, Pd1CC10, Pd1/2CV10, Pd1IsV60, and Pd1IsC7 foils at 473 K were measured at feed pressures of 124, 138, 152, and 165 kPa (and $P_{H2,p}$ ~93 kPa).

Figure 7 shows the plot of fluxes versus ($P_{H2,f}^{0.5} - P_{H2,p}^{0.5}$) (deemed the "Sieverts' law plot") for the above samples.

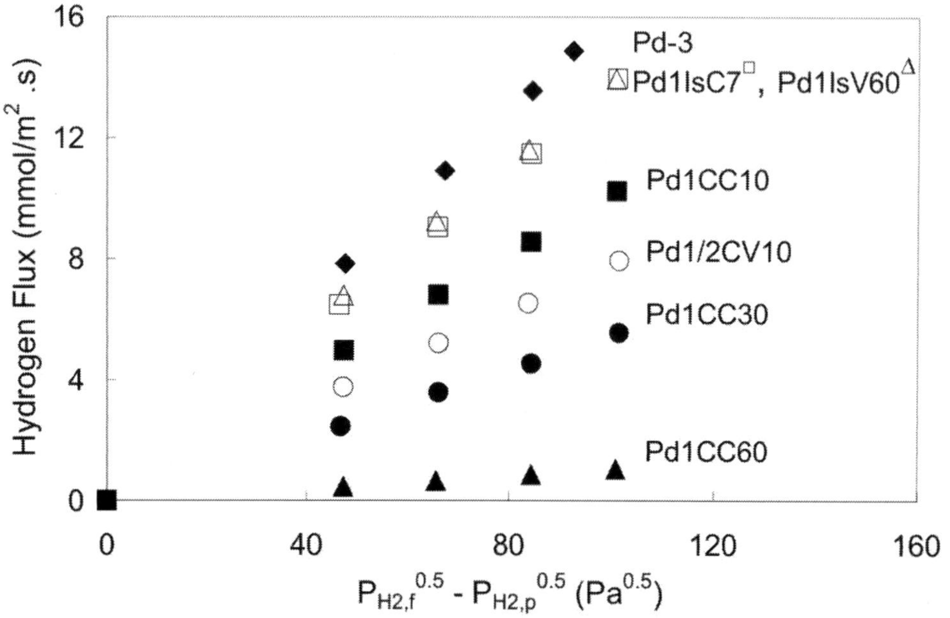

Figure 7 Long-time H_2 fluxes through uncoated Pd-3 and Pd/Cu foils at 473 K and feed pressures of 124, 138, 152, and 165 kPa with permeate at 93 kPa (Include straight lines to make it easy to see if Sieverts' law is followed)

Fluxes through the uncoated Pd-3 were the highest, ranging from ~8 to 15 mmol/m^2.s. At the other extreme, fluxes of ~0.5 to 1 mmol/m^2.s were measured for the Pd1CC60; the thickest continuous film studied. The two Pd/Cu foils with islands (Pd1IsV60 and Pd1IsC7) had nearly identical fluxes even though the deposition times were 60 and 7 seconds, respectively. Compare fluxes through the foils with continuous films, it is clear that the flux decreases with Cu film thickness.

Conclusions

Copper catalyst was uniformly deposited on Pd membranes with different geometries and thicknesses via electrodeposition using a novel masking technique developed in this work. Deposition was as continuous films or as islands on one side of the foils (permeates side). These films (or islands) were annealed to cupric oxide that reduced to Cu, which subsequently diffused into the membrane during long-time H_2 permeation studies.

Short-time and long-time transient studies after exposing the membranes to various gas atmospheres and temperatures elucidated the effects of the membrane "history" on performance. The transient studies indicated that the membrane structure changes during permeation, resulting in faster hydrogen permeation. It was also concluded that diffusivity in the membrane material likely increases with hydrogen coverage for the conditions studied. Finally, fluxes through all of these membranes followed Sieverts' law.

Acknowledgements

Alraeesi would like to thank Mark Reimann at the National Renewable Energy Laboratory (NREL) for his help in preparing the samples and for his advices and support. The authors also gratefully acknowledge the financial support provided by United Arab Emirates University (UAEU) and the Petroleum Institute (PI) in Abu Dhabi. Special thanks to the Center for Space Resources (CSR) at Colorado School of Mines to use the lab in order to accomplish this work.

References

1. H.P. Hsieh, Inorganic membrane reactors - a review, AIChE Symp. Ser. 268, 85, (1989), pp. 53-67.
2. G. Saracco and V. Specchia, Catalytic inorganic membrane reactors: Present experience and future opportunities, Catal. Rev.-Sci. Eng., 36, (1994), pp. 305.
3. S.N. Paglieri, J.D. Way, Innovations in Palladium Membrane Research, Separation and Purification Methods, 31(1), (2002), pp. 1-169.
4. F. Roa, M.J. Block, J.D. Way, The influence of alloy composition on the H_2 flux of composite Pd-Cu membranes, Desalination, 147(1-3),(2002), pp. 411-416.
5. J.N. Keuler, L. Lorenzen, Developing a heating procedure to optimise hydrogen permeance through Pd-Ag membranes of thickness less than 2.2 μm, Journal of Membrane Science,195(2), (2002), pp. 203-213.
6. J. Shu, B.P.A Grandjean, A. Van Neste, S. Kaliaguine, Catalytic palladium-based membrane reactors: a review, The Canadian Journal of Chemical Engineering, 69, (1991), pp.1036-1060.

(300b) Crystallization Kinetics Identification within a Generic Modeling Framework

Kresten Troelstrup Meisler[1], Noor Asma Fazli Abdul Samad[1], Nicolas von Solms[2], Krist V. Gernaey[3] and Rafiqul Gani[4], (1)CAPEC, Department of Chemical and Biochemical Engineering, Technical University of Denmark (DTU), Kgs. Lyngby, Denmark, (2)CERE, Deparment of Chemical and Biochemical Engineering, Technical University of Denmark (DTU), Kgs. Lyngby, Denmark, (3)Department of Chemical and Biochemical Engineering, Technical University of Denmark (DTU), Kgs. Lyngby, Denmark, (4)CAPEC, Department of Chemical and Biochemical Engineering, Technical University of Denmark, Kgs. Lyngby, Denmark

Crystallization kinetics identification within a generic modeling framework

Kresten Troelstrup Meisler[1] (kretm@kt.dtu.dk), *Noor Asma Fazli bin Abdul Samad, Krist V. Gernaey*[1] (kvg@kt.dtu.dk), *Nicolas von Solms*[1] (nvs@kt.dtu.dk) and *Rafiqul Gani*[1] (rag@kt.dtu.dk)

(1) Department of Chemical and Biochemical Engineering, Søltofts Plads, Building 229, Technical University of Denmark, DK-2800 Lyngby, Denmark

Crystallization serves as an efficient separation process for compounds, which are solid in their pure form at the given separation conditions. Crystallization is an essential step in the production of many pharmaceutical products as the active pharmaceutical ingredients (APIs) are often separated efficiently through this operation. The monitoring and analysis of crystallization operations has recently received increased attention due to the growing need to control the final crystal size distribution (CSD) in a relatively narrow range as well as to measure and monitor the final product characteristics CITATION EAa10 \l 1030 (1). This monitoring and analysis requires models that describe the multiple phenomena and their interplay encountered in crystallization operations in order to have the ability to design effective control strategies and the possibility of planning crystallization operations. To this end, models of the phenomena (kinetics, solubility, etc.) are needed within a generic framework that allows the generation of the needed models to achieve a more complete description of specific crystallization operations. For example, certain size distributions can be desired in order to obtain specific properties for the product such as rate of dissolution. Growth of the crystals occurs in multiple dimensions and the growth rates of the facets determine the shape of the crystals. The size distribution is obtained because of different competing phenomena such as nucleation and growth, each with different kinetics. A full representation of a crystallizer requires models for all these kinetic phenomena and saturation descriptions (constitutive equations) coupled with descriptions of the equipment and an operational policy. Such a description is possible within a generic framework where the models can be combined, reused and identified. This framework has been established CITATION Noo11 \l 1030 (2) and tested for simulation of selected crystallization processes.

The objective of this work is to provide the ability to establish the kinetics of a crystallization operation systematically and efficiently. Initially an operational scenario is defined for which the specific balance equations are set up. The constitutive models are chosen based on the system description and the objectives of the model. Once the model is ready, that is, all model parameters are available, it can be used for simulation of the corresponding crystallization operations. If an established model for the desired kinetics is found within the model library, then this model is used. If not available, or in case a new model is desired, then identification of model parameters is performed. This identification step requires measured data, which may be available in different forms such as single crystal growth data or chord length measurements (for example, data from Focused Beam Reflectance Measurements, FBRM). Use of such data requires appropriate translational policies to convert measurements into one or more variables that are described by the model. A preliminary version has been developed and tested to obtain information about the development of crystal size distribution (CSD) in a given operation. The measurement types that can be handled through the modeling framework are continuously being expanded to accept more data types with information relevant for the crystallization operations. The measurements can be used for offline analysis and parameter regression. With an established kinetic model it is thus possible to translate the predicted model results into a form, which is directly comparable (and visualized) to the data available for evaluation of the model. Furthermore, if there is model – data mismatch, the measured data can be used for online parameter estimation.

The expanded model framework combined with the systematic approach to establish the kinetic models for use in general crystallization operations in combination with monitoring tools will be presented using case studies involving different scenarios for crystallization operations.

<>Works Cited

BIBLIOGRAPHY 1. *Optimal seed recipe design for crystal size distributionn control for batch cooling crystallisation processes.* **E. Aamir, Z.K. Nagy, C.D.Rielly.** 2010, Chemical Engineering Science, Vol. 65, pp. 3602-3614.

2. *A generic multi-dimensional model-based system for batch cooling.* **Noor Asma Fazli Abdul Samad, Ravendra Singh, Gürkan Sin, Krist V. Gernaey, Rafiqul Gani.** 2011, Computers and Chemical Engineering, Vol. 35, pp. 828–843.

(300c) Development of an Empirical Framework for Monitoring CSD Using FBRM in Batch Crystallization

Huayu Li, Martha Grover, Yoshiaki Kawajiri and Ronald W. Rousseau, School of Chemical & Biomolecular Engineering, Georgia Institute of Technology, Atlanta, GA

Focus Beam Reflectance Measurement (FBRM) allows in-situ monitoring of crystallization processes. In order to utilize FBRM for control of crystallizers, accurate and robust measurement of crystal size distribution must be realized. To fulfill this goal, models for crystallization kinetics and the sensor are needed; the former is a population balance equation, while the latter converts the measured chord-length distribution (CLD) to the crystal size distribution (CSD) [1-3]. In our previous study, we proposed an empirical model for suspensions of paracetamol crystals in a non-solvent toluene [4]. The CLD obtained for a population of crystals of a specific size is referred to as a fingerprint. By obtaining fingerprints for different crystal sizes in toluene, the model was validated by investigating the linearity and additivity of fingerprint vectors.

In the present work, we apply the empirical model to a crystallization process. As paracetamol crystals of a certain size range are added into the saturated solution, the increase in total counts and changes in the chord-length histogram are recorded and analyzed to build the empirical model. We use a microscope and ATR-FTIR to confirm there is no significant transformation of crystal shape and no dissolution. Our investigations include the following: (1) a regime where the linearity of number of crystals and chord counts holds is identified, (2) linear additivity among different sizes of crystals is verified, (3) detectability of different crystal sizes by the FBRM is investigated, and (4) the most appropriate inversion technique is identified.

With the crystal size distribution estimated by the empirical model, liquid concentration and temperature recording, the full scope of the crystallization process is obtained for simulation purposes, such as parameter estimation and control scheme development. Such a framework is directly generated from practical FBRM data and requires no complicated physical modeling. Therefore, the small-scaled model can be applied in real-time conveniently to monitor and control crystallization processes.

References

1. Yu, Z.Q., P.S. Chow, and R.B.H. Tan, "Interpretation of focused beam reflectance measurement (FBRM) data via simulated crystallization," *Organic Process Research & Development*, 2008. 12(4): p. 646-654.
2. Ruf, A., J. Worlitschek, and M. Mazzotti, "Modeling and experimental analysis of PSD measurements through FBRM," *Particle & Particle Systems Characterization*, 2000. 17(4): p. 167-179.
3. Kail, N., W. Marquardt, and H. Briesen, "Estimation of particle size distributions from focused beam reflectance measurements based on an optical model," *Chemical Engineering Science*, 2009. 64(5): p. 984-1000.
4. Li, H., Y. Kawajiri, M. Grover, and R. W. Rousseau, "Estimation of Crystal Size Distribution with an empirical model of FBRM," AIChE Annual Meeting, Minneapolis, MN, October 2011

(300d) Apply PAT-Enabled Platform to Realize the QbD Control Strategy for Crystallization Processes

George Zhou[1], Aaron Moment[2], Shane T. Grosser[3], Paul Fernandez[2], Luke Schenck[4] and John Higgins[5], (1)Analytical Development and Commercialization, Merck & Co., Inc., Rahway, NJ, (2)Chemical Process Development and Commerciliazation, Merck & Co., Inc., Rahway, NJ, (3)Chemical Process Development and Commercialization, Merck & Co., Inc., Rahway, NJ, (4)Chemical Process Development and Commercialization, Merck & Co, Inc., Rahway, NJ, (5)Process Analytical Technologies, Merck and Company, Inc., West Point, PA

It is critical to achieve the desired crystal form for an active pharmaceutical ingredient (API) because polymorphic behavior may affect its chemical stability, solubility, morphology, density and bioavailability. The automated API platform enabled by process analytical technologies (PAT) provides the flexibility for crystallization process development. It can significantly enhance our capability to explore the design space of crystallization processes and improve the ability in maintaining the crystallization performance under different or unexpected scenarios of disturbances and scale-up variations. In this presentation, this enhancement of capability will be illustrated through several case studies. For example, risk assessment and corresponding DoE studies on crystallization and PSD indicate that the final PSD of a high volume API strongly depends on the seeding-point temperature which in turn relies on the solution composition. Therefore, by successfully utilizing this PAT-enabled automated API platform to map out the solubility for a through-process stream across the multi-dimensional solvent composition space as a function of temperature, control space of the crystallization process can be defined. It has enabled a QbD control strategy which incorporates the accurate measurement of batch composition in real-time for a robust process that reduces cycle time, and increases yield and productivity as well as realizes QbD benefits - greater process stability, process portability, and process flexibility.

(300f) Theoretical and Numerical Analysis of Critical Seed Loading in Batch Crystallization

Yu-Ti Tseng, Chemical Engineering, National Taiwan University, Taipei, Taiwan and Jeffrey D. Ward, Department of Chemical Engineering, National Taiwan University, Taipei, Taiwan

Doki et al. (2002) introduced the idea of critical seed loading for batch crystallization. The idea is that for a given crystallization process there is a certain seed mass that will effectively suppress nucleation and lead to a product with a large averages crystal size and narrow size distribution. Furthermore, the critical seed loading ratio depends primarily on the average seed size and much less on other process variables. Doki et al. determined the critical seed loading chart experimentally and developed the following empirical expression for critical seed loading:

$$C_s^* = 2.17 \times 10^{-6} L_s^2$$

where is the critical seed loading (initial seed mass divided by theoretical crystal yield) and L_s is the seed volume mean size in microns. Later authors (Hojjati et al. (2005) and Kubota et al. (2009)) determined critical seed loading ratios experimentally for other solute-solvent systems.

In this work, we show that critical seed loading charts can also be produced by numerical simulation based on kinetic models of crystallization processes. Results based on processes for the crystallization of potassium sulphate and potassium nitrate show good agreement with the correlation suggested by Kubota and coworkers. This further supports the utility of the critical seed loading concept for rapid process development and suggests that the critical seed loading can be determined for other solute-solvent systems using kinetic models already published in the literature.

Furthermore, we show that with certain reasonable assumptions it is possible to develop an implicit analytical expression for the critical seed loading as a function of seed mean size. These results are helpful for understanding the concept of critical seed loading intuitively and using it in process development.

References

Doki N, Kubota N, Yokota M, Chianese A. Determination of critical seed loading ratio for the production of crystals of uni-modal size distribution in batch cooling crystallization of potassium alum. *J Chem Eng Jpn.* Jul 2002;35(7):670-676.

Hojjati H, Rohani S. Cooling and seeding effect on supersaturation and final crystal size distribution (CSD) of ammonium sulphate in a batch crystallizer. Chemical Engineering and Processing 2005;44(9): 949-57.

Kubota N, Onosawa M. Seeded batch crystallization of ammonium aluminum sulfate from aqueous solution. Journal of Crystal Growth 2009;311(20): 4525-29.

(311a) Wave Formation On Nonparallel Axisymmetric Flows or the Unstable Career of William B. Krantz

Richard L. Zollars, Voiland School of Chemical Engineering and Bioengineering, Washington State University, Pullman, WA

When one thinks of the career of Bill Krantz, "unstable" does not come immediately to mind. Perhaps "cosmopolitan" as he has worked at universities in Turkey, Germany, England, India, the Netherlands, Australia, and Singapore in addition to Colorado, Cincinnati and Notre Dame in the US. "Multitalented" would be appropriate as he has published seminal papers in the areas of fluid mechanics, in-situ combustion of coal, the formation of patterned ground in the Arctic, and membrane separations. But at the heart of almost all of his research, since his days as a graduate student, is stability analysis. Whether this is applied to a free surface flow, a combustion front during an in-situ gasification process, or the formation of an asymmetric membrane Bill Krantz has been able to see the similarities and make significant contributions in these seemingly disparate areas. And so, "unstable" is an apt description.

My work with Bill involved a stability analysis of a free surface flow over a cone. The reason for studying such a seemingly esoteric phenomena lies in the effect that waves have on mass transfer in such devices as wetted-wall columns and gas-liquid absorption towers. While Bill had studied wave formation on a planar flow, the flow over packings in an absorption tower are obviously more complex. Flow over a conical surface offered the advantage of a known geometry but with a more complex underlying flow pattern (2-D rather than 1-D) than previously studied.

To study the stability of such a system, the two dimensional Navier-Stokes equations for an axisymmetric flow were non-dimensionalized using an unspecified characteristic length. The two equations were cross differentiated and subtracted to eliminate the pressure term. When a stream function was used to replace the velocity terms, a single equation for this system was derived. This equation is a partial differential equation, containing fourth-order derivatives in both the streamwise (x) and cross-stream (y) directions, with non-constant coefficients. In nondimensionalizing the Navier-Stokes equations, a parameter (δ) representing the ratio of the characteristic length for the diffusion of vorticity in the cross-stream direction to the characteristic length for the diffusion of vorticity in the streamwise direction arises. For low Reynolds number flows, this parameter will be small, thus suggesting the use of a regular perturbation analysis.

The resulting equation then can be solved sequentially for each increasing order of magnitude of the perturbation parameter (δ). The zeroth-order solution turns out to be the solution to Problem 3.W_4 in Bird, Stewart and Lightfoot (*Transport Phenomena*, John Wiley & Sons, Inc., 1960). This solution assumes that the flow profile at any point along the conical surface is semi-parabolic, i.e., it neglects the influence of the cross-stream velocity on the streamwise velocity profile. When the equation representing the next higher order of the perturbation parameter is solved, the effect of the cross-stream velocity is introduced. The effect of this cross-stream velocity is to make the basic flow thinner thus increasing the average streamwise velocity.

With the stream function of the basic flow now solved, the stream function for the oscillating flow is expressed as a sum of a non-time dependent function (basic flow) and a time-varying function (dependent on both the streamwise and cross-stream directions as well as time). Since a solution corresponding to small amplitude oscillations is sought, the resulting equation can be linearized to yield the 2-D analog of the Orr-Sommerfeld equation. Again, the presence of the small parameter (δ) suggests the use of a perturbation analysis to solve this equation.

Because of the thinning nature of the basic flow in the streamwise direction, the stability of the flow can be expressed in either absolute or relative terms. A flow which is unstable in an absolute sense is one in which the amplitude of the oscillations increase. A flow which is unstable in a relative sense is one in which the amplitude of the oscillations is not increasing in size but where the ratio of the amplitude of the oscillation to the depth of the unperturbed flow is increasing. Thus a flow that is unstable in the absolute sense is also unstable in the relative sense but a flow can be unstable in the relative sense while simultaneously being stable in the absolute sense.

From this analysis it is determined that there exist conditions under which the flow of a Newtonian fluid down a conical surface will be unstable in an absolute sense. However, as these instabilities move down the cone they will eventually become stable (in the absolute sense but still unstable in the relative sense) regardless of the magnitude of the instability near the apex of the cone. Under other conditions, no unstable oscillations (in the absolute sense) are observed but oscillations in the relative sense do exist. As these oscillations (unstable in the relative sense) move further down the cone, they will eventually become stable in both absolute and relative terms. Increasing the flow rate or decreasing the apex angle of the cone both lead to more unstable conditions, in both an absolute and relative sense. This solution also predicts that as oscillations move down the cone their wave velocity will decrease, thus leading to shorter wave lengths as oscillations move down the conical surface. While no experiments were performed to verify these results, certain geologic features do provide qualitative support for these predictions.

This same approach can be used to also predict the stability of flows over other axisymmetric flows, such as flows on cylindrical surfaces. Unlike the flows over conical surfaces, where no temporally growing oscillations are predicted, flows on cylindrical surfaces do allow temporally growing solutions. In the case of flows over cylindrical surfaces the effect of the surface tension of the liquid is to make the flow more unstable in all situations.

These solutions, for both the basic flow and the perturbed flow on either cylinders or cones, were the first of their kind to be reported in the literature. But more than that, they represent the faith that Bill Krantz had in a young graduate student. As is typical for Bill the mentoring does not stop on graduation. Through Bill's inspiration and guidance I have followed his lead into academia and into government service at the National Science Foundation. This is not the only time that Bill Krantz has worked to bring out the best in his students. In fact, this is common for Bill Krantz. Hopefully many more students in the future will benefit from his guidance and friendship.

(311b) Effect of Residual Solvent On the Gas Separation Properties of Mixed Matrix Membranes

Aylin Kertik, Graduate School of Science, Engineering, and Technology, Istanbul Technical University, Istanbul, Turkey and S. Birgül Tantekin-Ersolmaz, Department of Chemical Engineering, Istanbul Technical University, Istanbul, Turkey

Mixed matrix membranes (MMMs), typically composed of inorganic particles, such as zeolites, distributed in a polymeric matrix, have been studied extensively over the last two decades. Many factors including the physical properties of the polymer forming the matrix, and the zeolite particles (such as particle size and zeolite pore size), as well as the interaction between the zeolite and the polymer (resulting in the formation of a zeolite/polymer interphase with distinct properties) have a tremendous effect on the MMM performance, which is hard to predict prior to experimentation. It is a well-known phenomenon that smaller particles provide more zeolite/polymer contact area, thus possibly favor a more pronounced effect of the interphase on the final membrane performance [1]. Additionally, the solvent type used in MMM fabrication and thermal treatment conditions are among the parameters which may affect the gas permeation properties. Therefore, a suitable annealing process must be applied to the membranes, in order to control the effect of residual solvent on the overall properties of MMMs. In this study, we investigated the effect of thermal treatment conditions and hence the residual solvent on the single and binary gas transport properties of MMMs.

MMMs are prepared by embedding micron- and submicron-sized zeolites (SAPO-34) into Ultem® (T_g, 215°C) polymeric matrix. NMP (n-methyl pyrrolidone) (b.p. 205°C) was used as solvent. SAPO-34 is a silicaaluminophosphate with a pore size of 3.8 Å, favoring the passage of CO_2 over CH_4 and N_2, and that of O_2 over N_2. MMMs were prepared with 20% zeolite content through optimisation and application of a priming procedure to enhance interaction between the zeolite and the polymer phases. The MMMs prepared were annealed at 150°C (below T_g and b.p. of NMP), 200°C (below T_g, close to b.p. of NMP), and 220°C (above T_g and b.p. of NMP) for 48 hours.

SEM analysis showed good adhesion, TGA analysis showed decreased amounts of solvent, and DSC measurements showed increased T_g values, all increasing with higher annealing temperature. Permeabilities (measured at 35°C and 4 bar upstream pressure) of all gases in MMMs generally decreased whereas selectivities increased, which became more pronounced with decreasing zeolite particle size. CO_2 permeability decreased for membranes annealed below T_g, whereas a slight increase was observed for membranes annealed above T_g. CO_2/CH_4 selectivity significantly increased with increasing annealing temperature reaching to an ideal selectivity of 120 for the membrane containing submicron-sized particles and annealed at 220°C. This was attributed to the plasticization effect of residual solvent favoring the smaller penetrant, as well as the preferential sorption of CO_2 in NMP compared to CH_4, decreasing with the decreasing amount of residual solvent. Despite the competitive sorption phenomena occurring in permeation of gas mixtures, mixed gas measurements (at 35°C and 4 bar) of MMM with submicron-sized particles and annealed above T_g showed a CO_2/CH_4 selectivity of 52 at 50/50 molar composition which is higher than that of pure polymer (39 at 50/50) but lower than ideal selectivity for the same membrane. These results lead us to the conclusion that residual solvent affects the gas transport properties in MMMs significantly; the residual solvent trapped in the zeolite/polymer interphase is relatively more difficult to remove than that in the bulk of the polymer and therefore smaller size zeolite particles enhance the influence of residual solvent on the MMM transport properties.

[1] Tantekin-Ersolmaz, S.B., Atalay-Oral, C., Tatlıer, M., Erdem-Senatalar, A., Schoeman, B., Sterte, J., "Effect of Zeolite Particle Size on the Performance of Polymer-Zeolite Mixed- Matrix Membranes", Journal of Membrane Science, 175(2), 285-288 (2000).

(311c) Links Between Synthetic Membrane Formation and Geologic Ice Lens Formation

Rorik Peterson, Mechanical Engineering, Univ. Alaska, Fairbanks, Fairbanks, AK

There is an intriguing similarity between the physics that occurs during some synthetic membrane formation processes, and the geological formation of discrete lenses of ices during freezing of a wet, natural soil. Relatively weak but long-range inter-molecular forces can lead to different structures depending on the chemical and/or thermal gradients present during formation. Because these two processes usually occur in widely different environments, laboratories and chemical process facilities versus alpine mountain ranges and arctic tundra, the connection remained elusive for quite some time. Professor William Krantz, whose wide-ranging interests reached into both of these areas, was one of the pioneers who helped recognize this similarity, and helped develop some important analytic tools for describing the ice-lens formation process. The more comprehensive descriptions of this "separation" process, soil particles from pure ice, involve processes that occur on size scales spanning 7 orders of magnitude, from nanometer to centimeter. Effective scaling of the equations that describe these processes allows for some strikingly compact yet accurate predictive models. The continuing advancements in numerical computing power has recently allowed for further exploration of the lens-formation process under a wider range of circumstances. These simulations have not only confirmed many of the assumptions underlying the analytic scaling arguments, but also allowed for exploration of higher-dimensional (2D and 3D) systems. The perhaps even more interesting occurrence of a particular type of arctic patterned ground emerges naturally from some of these numerical simulations, and may also have an analog in synthetic membrane formation.

(311d) Beyond TIPS (Thermally Induced Phase Separation)

Dongmei Li, Department of Chemical and Petroleum Engineering, University of Wyoming, Laramie, WY

Thermally induced phase separation (TIPS) is a membrane formation process in which phase separation is caused by large temperature gradients between a polymer - solvent solution at an elevated temperature (here the solvent is called a diluent), and a casting surface at a much lower temperature. Sequentially, a second solvent will be used to extract the diluent trapped during the phase separation process to generate various microporous structures. Among the advantages of TIPS process are: 1) enabling membrane fabrication from of a wide spectrum of polymers that can not be dissolved by any solvent at room temperature; 2) the capability of creating final membrane microstructures by tuning the casting surface temperature alone.

As one of three of my Ph.D. advisors, there is no doubt that Prof. Bill Krantz is one of the most influential people in my academic career. Working with Bill has taught me not only how to model a physical system and how to use the material science and engineering background and knowledge to various fields, but how not to limit oneself to one field or one narrow area. In this presentation I will share the positions I held in industry as well as a couple of my ongoing research projects at University of Wyoming that is based on materials and membrane technology. I hope that my career path, enriched by working with Prof. Bill Krantz, will help current graduate students in the membrane science and technology field envision the diverse opportunities that lie ahead of.

(311e) Comprehensive Integrity Verification System for Low Pressure Membrane Systems

Adrian Yeo[1], Corliss Lin[1], William B. Krantz[2] and Anthony G. Fane[3], (1)Membrane Instruments and Technology Pte Ltd, Singapore, Singapore, (2)Chemical and Biological Engineering, University of Colorado, Boulder, CO, (3)Singapore Membrane Technology Center, Nanyang Technological University, Singapore, Singapore

Low pressure membranes are used for water treatment as well as pre-treatment in water reclamation plants and desalination.

The main advantage of using MF or UF in the drinking water treatment is their high level of microbial pathogen removal to meet stringent water quality regulations. However, pathogens will pass through the membrane system and contaminate the product if the membrane integrity is compromised. The Membrane Filtration Guidance Manual published by the USEPA details both direct and continuous integrity testing for water treatment plants.

Any process in a water treatment system is assigned a log removal value (LRV). Depending on the quality of the source water, different LRV targets are assigned to various plants. The maximum removal credit that a membrane filtration process may receive is the lower value of either:

- The removal efficiency demonstrated during challenge testing; OR
- The maximum log removal value that can be verified by the direct integrity test used to monitor the membrane filtration process

Hence, the ability of a plant to accurately monitor the integrity of the plant directly affects the design and cost of the plant.

Wastewater reclamation via nanofiltration (NF) or reverse osmosis (RO) requires pre-treatment of the feed water using microfiltration (MF) or ultrafiltration (UF) membrane modules. The performance of these pre-filtration modules is critical to maintaining efficient operation of the NF or RO membrane modules. Any compromised pre-filtration modules can cause fouling of the NF or RO membrane modules via the deposition of particulate materials that would otherwise have been removed by the upstream modules. This fouling of the NF or RO membrane modules causes increased energy usage per unit product which can cause the plant to shut down for cleaning or replacement. Higher module replacement rates will also lead to high operating, maintenance cost and unit cost of water.

This paper describes a comprehensive integrity management system that helps plant operators determine the health of their membrane systems as well as the LRV credit. The system has three components

- LRV Determination
- Integrity Monitoring
- Diagnostic Testing

The backbone of this system is the Membrane Integrity Sensor (MIS) developed at the Nanyang Technological University.

(317a) Summary Results of Epri's Engineering and Economic Case Studies of Post Combustion Capture Retrofit Applied to Various North American Host Sites

Des Dillon, Advanced Generation, EPRI, Palo Alto, Ca, CA

Summary Results of EPRI's Engineering and Economic Case Studies of Post Combustion Capture Retrofit Applied to Various North American Host Sites.

Dr. Desmond Dillon[1], Robert Chu[2] and Dr. Gerald Choi[2]

[1]The Electric Power Research Institute, EPRI, Palo Alto, Ca, USA

[2] Nexant Inc, San Francisco, Ca, USA

Abstract

With more than 300 GW of installed coal-fired plants in the US, accounting for about half of all electricity being generated domestically, it is anticipated that the existing coal fleet would need to be retrofitted for CO_2 capture in order to significantly reduce greenhouse gas emissions to acceptable levels. Amongst the available and developing capture technologies of pre-combustion, oxy-combustion and post-combustion capture (PCC), the latter is viewed as a viable and commercially available technology for existing plant retrofit application.

However, retrofitting PCC to an existing coal fired plant presents significant challenges in itself. Key issues arise such as:

· Limited space for new plant equipment

· Limited heat available for process integration

· Limitations of the existing steam turbine

· Cooling water limitations

· Replacement power considerations

· Complicated pipe routings.

With an objective to better understand these factors and quantify their effect, the Electric Power Research Company (EPRI) of California USA, recently completed a series of detailed economic and engineering studies examining the feasibility of retrofitting post combustion capture (PCC) to existing pulverised coal (PC) and/or circulating fluidized-bed (CFB) power plants, for five different North American "host" sites.

To effectively undertake this study EPRI assembled a Retrofit Team with engineering consultants of Nexant Inc. and architect engineers of Bechtel Power Corporation of Frederick, Maryland. Nexant's expertise in process plant economic optimization was complimented with Bechtels practical knowledge and experience in power plant retrofit costing, design and build.

The sites selected for the study included:

· Edison Mission Groups 1500 MW Powerton Station, operated by Mid West Generation in Pekin, Illinoi, USA

· Intermountain Power Agencies 950 MW Intermountain Generating Station in Delta, Utah USA

· Nova Scotia Powers 320MW (2 x 160MW units) Lingan Generating Station in Nova Scotia, Canada

· Great River Energys 1100MW Coal Creek Station in Underwood, North Dakota, USA

· First Energys 176MW CFB Unit at Bayshore Generation Station in Oregon, Ohio, USA

Each of the host sites offered a unique combination of unit sizes and ages, planned emissions control systems, fuel types, steam conditions, boilers, turbines, and cooling systems. Collectively, they represented a critical sample of five different types of existing coal-fired plants within the North American fleet. The data from these "real world" sites provided a broad insight into generalized cost and performance impact estimates for a variety of power plant configurations.

Specifically the studies:

- Assessed the most practical CO_2 capture efficiency configuration based on the existing site constraints
- Determined the space required for the CO_2 capture technology and the interfaces with the existing systems
- Estimated performance and costs for the post combustion CO_2 capture and compression plants
- Assessed the features of each plant that materially affect the cost and feasibility of the retrofit

The paper will give a summary overview of the studies.

Electrochemical Membrane for Carbon Dioxide Separation and Power Generation

Stephen Jolly, Hossein Ghezel-Ayagh, Jennifer Hunt, Dilip Patel
FuelCell Energy, Inc., 3 Great Pasture Road, Danbury, CT 06813

William A Steen, Carl F Richardson
URS Corporation, 9400 Amberglen Blvd, Austin, TX 78729

Olga A Marina
Pacific Northwest National Laboratory, 902 Battelle Blvd., Richland, WA 99352

FuelCell Energy, Inc. (FCE) has developed a novel system concept for separation of carbon dioxide (CO_2) from greenhouse gas (GHG) emission sources using an electrochemical membrane (ECM). The salient feature of the ECM is its capability to produce electric power while capturing CO_2 from flue gas, such as the flue gas from an existing pulverized coal (PC) plant. Laboratory scale testing of the ECM has verified the feasibility of the technology for CO_2 separation from simulated flue gases of PC plants as well as combined cycle power plants and other industrial facilities. Recently, FCE was awarded a contract (DE-FE0007634) from the U.S. Department of Energy to evaluate the use of ECM to efficiently and cost effectively separate CO_2 from the emissions of existing coal fired power plants. The overarching objective of the project is to verify that the ECM can achieve at least 90% CO_2 capture from flue gas of an existing PC plant with no more than 35% increase in the cost of electricity (COE) produced by the plant. The specific objectives and related activities planned for the project include: 1) conduct bench scale tests of a planar membrane assembly consisting of ten or more cells of about 0.9 m^2 area each, 2) develop a detailed design for an ECM-based CO_2 capture system applied to an existing PC plant, and 3) evaluate the effects of impurities (pollutants such as SO_2, NO_x, Hg) present in the coal plant flue gas by conducting laboratory scale performance tests of the membrane. The results of this project are anticipated to demonstrate that the ECM is an advanced technology, fabricated from inexpensive materials, based on proven operational track records, modular, scalable to large sizes, and a viable candidate for >90% carbon capture from existing PC plants. In this paper, the fundamentals of ECM technology including: material of construction, principal mechanism of operation, carbon capture test results and the benefits of applications to PC plants will be presented.

SYSTEM CONCEPT

To address the concerns about climate change resulting from emission of CO_2 by coal-fueled power plants and other industrial plants, FCE has developed **C**ombined **E**lectric **P**ower **a**nd **C**arbon-dioxide **S**eparation (**CEPACS**) system concept, as a novel solution (US Patent 7,396,603 B2) for greenhouse gas emission reduction. The ECM-based CEPACS system utilizes the Company's well-established Direct FuelCell® (DFC®) technology. The system concept works as two devices in one: it separates the CO_2 from the exhaust of other plants such as an existing coal fired plant and simultaneously, using a supplementary fuel, produces clean and environmentally benign electric power at a very high efficiency.

A simplified diagram of the CEPACS system concept is shown in Figure 1. CO_2-containing flue gas from a coal-fired (combustion-based) power plant, such as the exhaust from a PC power plant or other industrial sources, is utilized as the oxidant for the ECM cathode. The key feature is that the ECM utilizes the CO_2 of the flue gas as a reactant for the electrochemical reaction to produce power, while synergistically transferring CO_2 from the flue gas to the anode exhaust stream. A supplementary fuel such as natural gas or biogas (from a digester) is internally reformed in the fuel cell to (convert methane as shown in Equation 1 and) provide the hydrogen needed to complete the electrochemical power generation cycle.

$$CH_4 + 2H_2O \rightarrow CO_2 + 4H_2 \qquad (1)$$

Hydrogen is used as a reactant at the ECM anode.

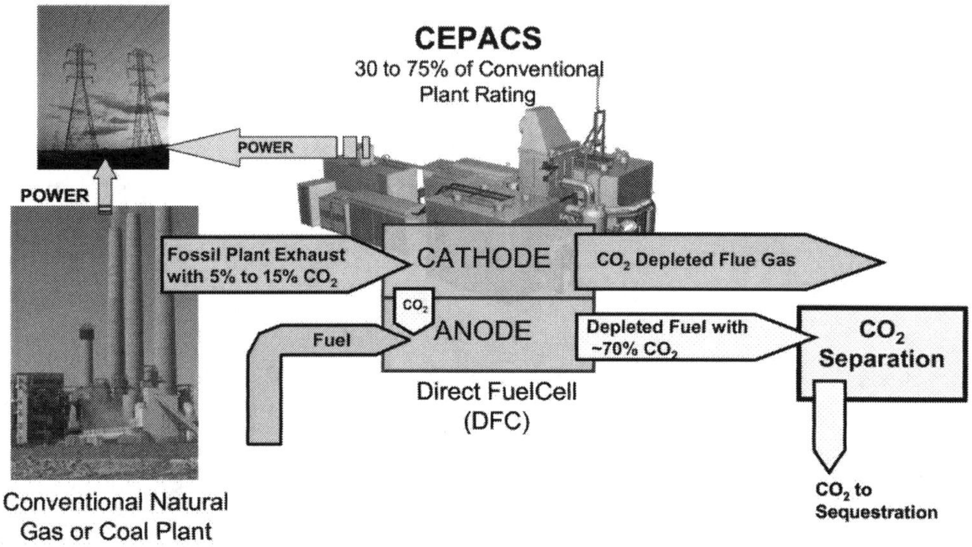

Figure 1. CEPACS CO_2 Separation and Power System Concept:
The system can be used with a variety of CO_2-containing greenhouse gases (GHG)

CEPACS separates the CO_2 from the flue gas because of the unique underlying mechanism of ECM operation. The operating principle of the ECM is shown in Figure 2, along with the electrochemical reactions involved. Carbon dioxide and oxygen present in the flue gas are used as reactants at the cathode. Carbonate ions produced at the cathode migrate to the anode side via the electrolyte in the cell. Carbon dioxide is liberated at the anode as carbonate ions react with H_2. In addition to CO_2, H_2O is produced at the anode as H_2 is consumed by the electrochemical reaction. The transport of carbonate ions internally in an ECM cell and the flow of electrons in the external circuit results in electric power generation as a byproduct of the electrochemical CO_2 separation process. Overall, the operating mechanism of the ECM cell results in the separation and transfer of CO_2 into the anode exhaust stream which has a much reduced volumetric flow rate (resulting in a CO_2-rich stream) compared to the original flue gas stream.

The CO_2-rich anode exhaust gas is processed (post-processing) in the ECM-based CEPACS system to further concentrate and compress the CO_2 for sequestration. The CO_2 rich anode exhaust contains water vapor which can be condensed out. It also contains a small amount of unused fuel (H_2) which can be easily separated. Water condensed during post-processing of the CO_2-rich anode exhaust stream is used to provide water (steam) needed for internal reforming of CH_4 in supplementary fuel, eliminating the need for external process water. H_2 separated during post-processing of CO_2-rich anode exhaust stream is recycled to provide additional preheat in the system and as part of the supplementary fuel (thereby reducing NG fuel needed). The captured CO_2 can then be processed for sequestration.

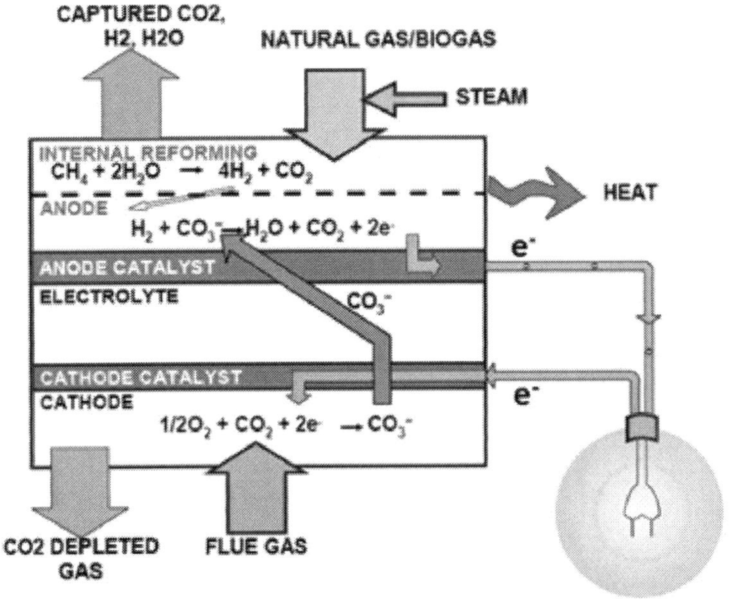

Figure 2. Separation of CO_2 in a Carbonate Fuel Cell:
Carbon dioxide is used at the cathode as an oxidant and transferred to the anode via the carbonate electrolyte

The CO_2-lean cathode exhaust (flue gas after CO_2 removal) is vented to atmosphere after recovering the heat for process use (e. g. preheating of feed streams).

The ECM cell operates at 550 - 650°C and atmospheric pressure. Unlike conventional membrane technologies that rely on pressure (partial pressure) differentials and permeability properties, ECM separates CO_2 at a rate dependent on the electrical current drawn. The flue gas does not need to be pressurized and vacuum operation is not required on the permeate side. Because of fast electrode kinetics, the ECM cell does not require high CO_2 concentration in the flue gas feed. Planar geometry and large gas flow channels of the ECM cell enable processing of large volumetric flow of flue gas feed without significant back pressure.

TECHNOLOGY STATUS

The fuel cells used in the CEPACS system are built from the same materials and by the same fabrication processes as those used in the Company's DFC-based power plant products. DFC (ECM) technology has reached commercial maturity and is readily available as a manufactured product used in building fuel cell stacks. Currently, the mass produced cells of about 9000 cm^2 area, suitable for large-scale applications, are commercially available. These cells have already been engineered for multi-layer stacking. The stacks of ECMs are incorporated in modules, which make up the MW scale building blocks for large power plants and CEPACS

Systems. Figure 3 shows a diagram of the ECM production and modularization process, which is performed at the Company's manufacturing plant in Torrington, Connecticut.

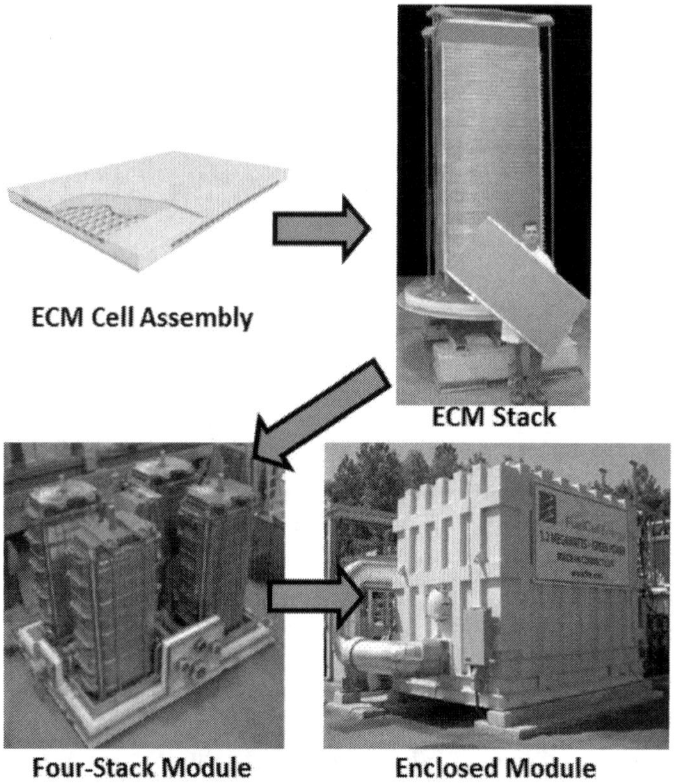

Figure 3. ECM Module Assembly Process: Cells fabricated at FCE's manufacturing plant in Torrington, CT are stacked and then modularized to form the building blocks usable in a CEPACS system

The CEPACS enabling technology has been deployed in MW-scale power plants for stationary power applications in three sizes: 300 kW, 1.4 MW, and 2.8 MW units. A 1 MW (nominal) unit (DFC1500) installed at a customer site is shown in Figure 4. FCE services over 50 DFC power plant sites around the globe.

Figure 4. 1MW DFC1500 Direct FuelCell Power Plant:
The DFC products are suitable for large multi-MW systems

CEPACS SYSTEM TECHNO-ECONOMIC EVALUATION

Monoethanolamine (MEA) based scrubbing technology is considered to be the current benchmark technology for separating CO_2. However, the energy and efficiency penalties of using amines for CO_2 capture in PC plants are substantial. About 22-30% of plant gross power is used up by the amine system, dropping the plant efficiency to <30%. Operation of the CEPACS system is set apart from other technologies by generation of electric power rather than consuming it, resulting in an increase in the power output of the retrofitted power plant.

A preliminary Technical and Economic Feasibility Study has been performed. The study included assessments of cost and performance of the ECM-based CO_2 capture and compression system for a 550 MW (net AC) PC plant. While the ECM-based CEPACS system captures 90% of CO_2 from the PC plant flue gas, it generates additional (net AC) power (~441.15 MW) after compensating for the auxiliary power requirements of CO_2 capture and compression. Figure 5 presents the plant performance results showing the impact on PC plant efficiency. The net electrical efficiency of the PC plant equipped with CEPACS-based CO_2 capture was estimated to be 39.8% (based on higher heating values - HHV of coal and natural gas fuels used by PC plant and CEPACS system, respectively). The efficiency of the PC plant with no CO_2 capture is estimated to be 36.8% (coal HHV). The later number is based on U.S. DOE/NETL system study[1]. The DOE study also reports efficiency of the PC plant with amine-based CO_2 capture to be 26.2% (coal HHV). For comparison, the DOE study assumes fixed (nominal) net plant output of 550 MW. To compensate for the auxiliary power requirements of CO_2 capture and compression, the PC plant with amine-based CO_2 capture is bigger (672.7 MW gross) than the PC plant with no CO_2 capture (582.6 MW gross). The CEPACS-based CO_2 capture also has other attractive features/benefits. The PC plant equipped with CEPACS system has lower emissions of NOx, SOx and Hg compared to a PC plant equipped with amine-based scrubber for CO_2 capture. The CEPACS system produces excess process water thereby reducing the total water usage of the PC plant equipped with CEPACS-based CO_2 capture.

[1] Cost and Performance Baseline for Fossil Energy Plants, Volume 1: Bituminous Coal and Natural Gas to Electricity, Revision 2, DOE/NETL-2010/1397, November 2010.

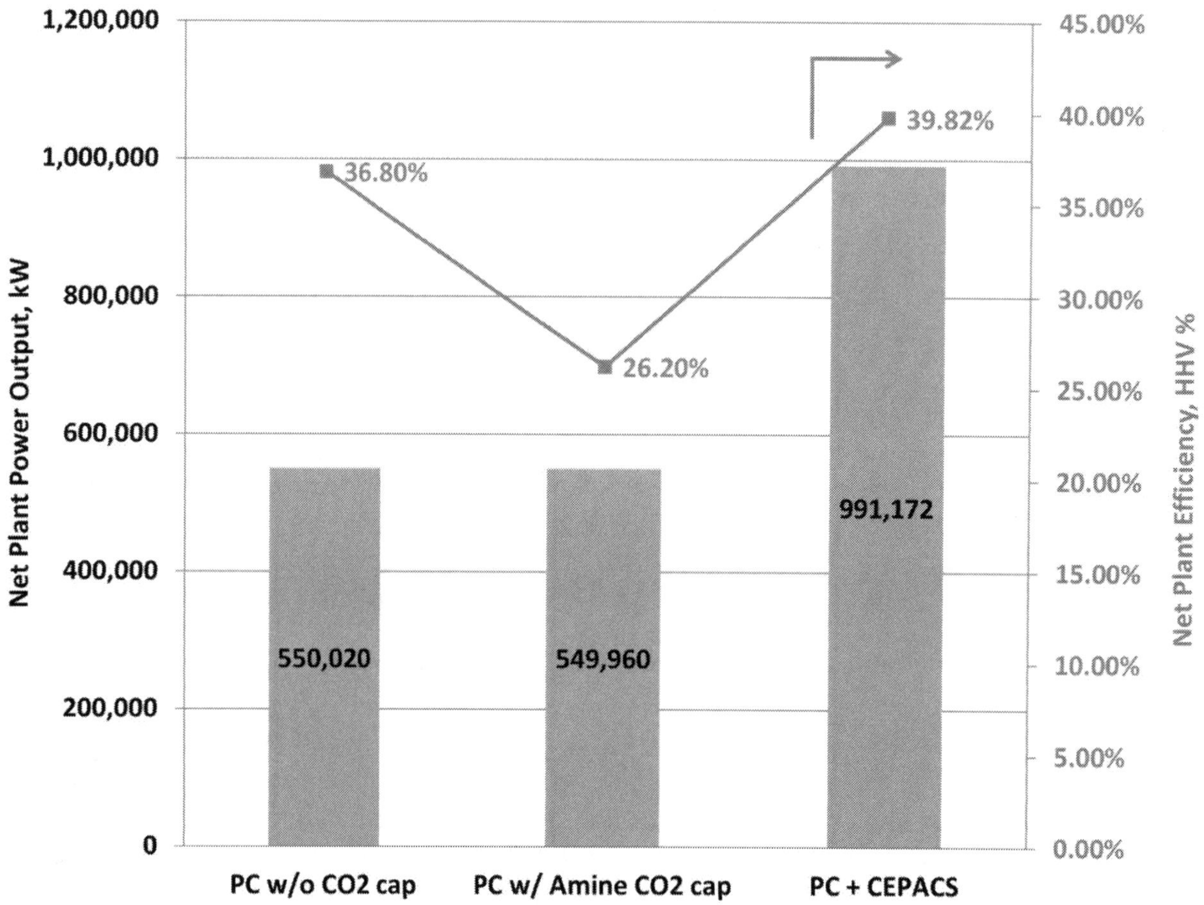

Figure 5. Impact of CO_2 capture on PC Plant Performance

The economic analysis included estimation of CEPACS plant cost and subsequently the incremental cost of electricity (COE) for ECM-based CO_2 capture. Figure 6 presents the results showing the impact on PC plant COE and provides a comparison with the competing amine-based CO_2 capture technology (PC w/o CO_2 capture is referred as Case 1). The fuel costs were based on the coal price of $38.18/ton and the natural gas price of $6.55/MMBtu. The cost estimation methodology outlines in DOE's guideline document was followed for the analysis[2]. The CEPACS technology has the potential to meet/exceed DOE's goal of <35% increase in COE for post-combustion CO_2 capture. The cost of CO_2 capture for a PC plant equipped with CEPACS system was estimated to be $14/ton which is significantly lower than $60/ton for a PC plant equipped with an amine-based scrubbing system.

[2] Quality Guidelines for Energy System Studies: Cost Estimation Methodology for NETL Assessment of Power Plant Performance, April 2011, DOE/NETL-2011/1455.

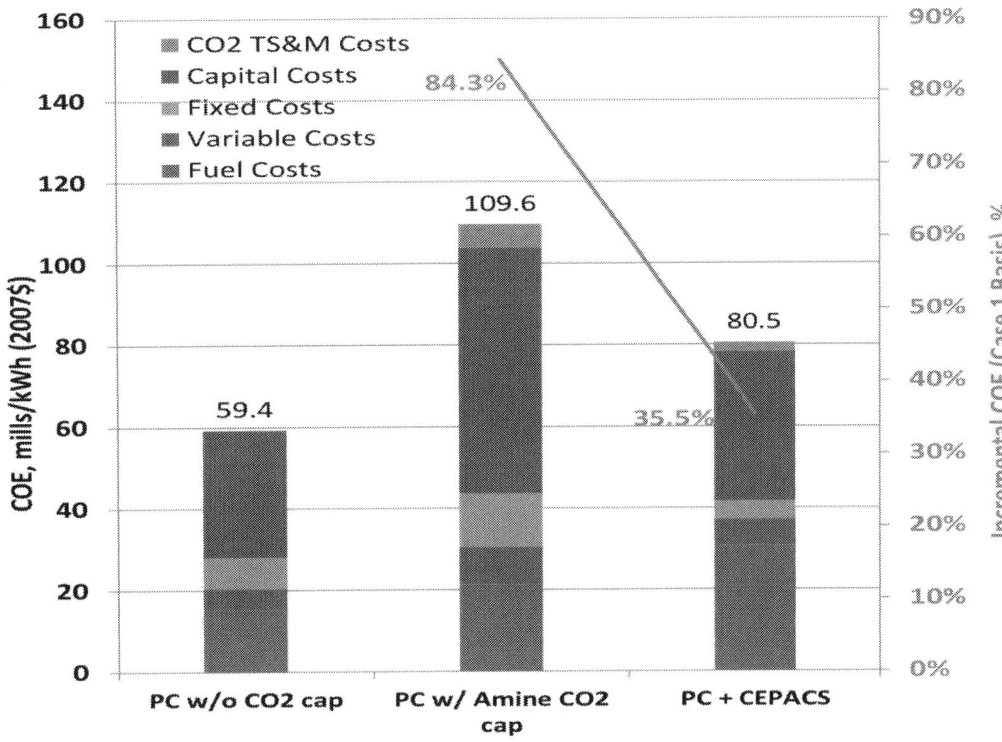

Figure 6. Impact of CO_2 capture on PC Plant Cost of Electricity

(317c) Modeling, Simulation and Optimization of Absorption, Membrane and Adsorption-Based CO2 Capture for Variable Feed CO2 Concentration and Feed Flow

M. M. Faruque Hasan, Richard C. Baliban, Josephine A. Elia and Christodoulos A. Floudas, Chemical and Biological Engineering, Princeton University, Princeton, NJ

Studies on four leading technologies for industrial CO_2 capture are performed. The technologies are MEA-based chemical absorption, multi-stage membrane, pressure swing adsorption (PSA) and vacuum swing adsorption (VSA). Each process is rigorously modeled, optimized, and energy integrated to capture and compress at least 90% of CO_2 from the feed for sequestration, over a range of feed CO_2 compositions and total feed flow rates. Emphasis is given on understanding and quantifying how the cost of CO capture changes with CO feed composition and flow rate of the gas mixtures coming from various chemical, power and heat generation plants. The novel features of this study include:

(i) variation in feed CO_2 composition (ranging from 1 to 90%) and total feed flow rate (ranging from 0.1 to 10 kmol/s);

(ii) CO_2 capture coupled with compression to 150 bar;

(iii) rate-based modeling of the MEA absorption-based CO_2 capture coupled with heat integration;

(iv) multi-stage superstructure model for CO_2 capture using membranes;

(v) detailed and grey-box modeling of PSA and VSA alternatives for the adsorption-based CO_2 capture;

(vi) cost-based comparison between the absorption, membrane, PSA and VSA-based CO_2 capture technologies for different emission scenarios; and

(vii) explicit expressions for the investment and operating costs as function of feed CO_2 composition and total feed flow rate.

In this talk, we will discuss our approach towards the modeling, simulation and optimization of each of the four technologies. A rigorous simulation-based optimization model is proposed to determine the minimum annualized cost of the MEA-absorption process. The MEA-absorption process is further heat integrated through heat exchanger network optimization. A novel mathematical model is developed for the optimization of multi-stage and multi-component separation of CO_2 using membranes, which can be also used for a range of membrane-based gas separation applications. A 4-step cyclic process with multiple adsorption columns, purging and product recovery at moderate to low vacuum is configured for the PSA and VSA processes. A PDE-based non-isothermal adsorption model is used, which are fully discretized and optimized via a Kriging model.

We will also present a cost-based technology comparison over a range of feed gas compositions and flow rates. This may assist the decision-makers to select the cost-appropriate technology for CO_2 mitigation by taking the diverse emission scenarios into consideration. Furthermore, we will discuss an explicit input-output model for the investment and operating costs for the first time. This would provide a basis towards scaling up a CO capture process, and lay the foundation for (i) process synthesis studies, and (ii) CO supply chain network studies in the carbon management domain.

(317d) Pressure Swing Adsoprtion: Design and Optimization for Pre-Combustion Carbon Capture

Alexander W. Dowling[1], Lorenz T. Biegler[1] and David C. Miller[2], (1)Chemical Engineering, Carnegie Mellon University, Pittsburgh, PA, (2)U.S. Department Of Energy, National Energy Technology Laboratory, Morgantown, WV

Pressure Swing Adsorption (PSA) is an efficient method for gas separation, and is a potential candidate for carbon dioxide (CO_2) capture from power plants. However, few PSA cycles have been designed for this purpose and the optimization of PSA systems for CO_2 capture remains a very challenging task. In this study, we present a systematic optimization-based formulation for the synthesis and design of novel PSA cycles for CO_2 capture in an IGCC flowsheet, which can simultaneously produce hydrogen (H_2) and CO_2 at a high purity. Here, we apply a superstructure-based approach to simultaneously determine optimal cycle configurations and design parameters for PSA units. This approach integrates automatic differentiation, efficient ODE solvers for the state and sensitivity equations of the PSA model and state of the art NLP solvers. Three optimization formulations are proposed and two PSA case studies are considered. The first case study considers a binary separation of H_2 and CO_2 at high purity, where specific energy is minimized, while the second considers a larger five component separation.

(317e) Assessment and Analysis of Post-Combustion Carbon Capture with Membrane Separation Technology

Xiangping Zhang, Norwegian University of Science and Technology, Trondheim, Norway, Xuezhong He, Norwegian University of Science and , Trondheim, Norway and Truls Gundersen, Energy and Process Engineering, Norwegian University of Science and Technology, Trondheim, Norway

Large scale CO_2 Capture and Storage has been considered to be one of the most promising ways of mitigating anthropogenic CO_2 emissions, but energy penalty, high economic costs and possible environmental impacts are important barriers for its deployment. One of the effective solutions is to develop lower cost and industrially feasible technologies, especially for the post-combustion capture technologies that have a major challenge due to the large gas volumes, low CO_2 concentration and low partial pressure. Among the different technologies, MEA-based systems have been the most likely candidate from an industrial perspective, but some inherent drawbacks, like solvent degradation and large energy consumption, limit its applications. Due to the rapid development of gas separation membrane technology and its inherent benefits for carbon capture[1, 2], it has attracted considerable attention. It is expected to have the potential to overcome the drawbacks of MEA-based systems, but most of the research is still done in lab level or pilot scale. From the viewpoints of industrial application and process design, understanding deeply the technical and economic benefits or drawbacks of the new technology along with fundamental experimental research is of significant importance for further development, and can also provide possible research directions to accelerate the industrial application of the technology.

However, there are only a few reports related to systematic analysis of membrane systems compared to the large amount of work focusing on material synthesis and experimental research[2, 3]. Bounaceur et al. performed relatively early a systematic analysis of the separation performance and energy cost of a single stage module and optimized the pressure ratio[4], but they did not perform any economic analysis. Zhao et al. described a detailed parameter study of mass and energy balances for a single membrane[5] and also the energetic and economic analyses of a multi-stage process[6]. Merkel et al. investigated the influences of membrane parameters and process configurations on the energy consumption and capture cost, and some important suggestions were proposed for future studies[7], however, more detailed economic models should be used. Hussain et al. investigated the feasibility of carbon capture from flue gas by facilitated transport membranes, and a real gas composition was considered[8]. However, some results were inconsistent and only a few of these studies focused on thermodynamic analysis for this energy intensive process. Britan et al. studied the feasibility of membrane technology of mixed gas separation with pure thermodynamic principles[9], but did not consider the real process. Thus, further systematic analysis and assessment is important and necessary.

In this work, a systems methodology integrated with process simulation, technological and economic analysis as well as exergy analysis is applied to assess a membrane capture technology and compare it with MEA-based systems. An Advanced Super Critical (ASC) Boiler and Turbine delivering 819 MWe (gross) being part of a coal based power plant is chosen as a base case for the study. The flue gas after ESP and FGD to remove ash and sulfur is fed into the membrane separation system. The aim of the work is to try to clarify the advantages and disadvantages with using membranes for post-combustion carbon, and to indicate what the main obstacles are and how far from commercialization the technology is.

At first, the current research on polymeric membranes for post-combustion CO_2capture is reviewed, including the type of membranes, application conditions, permeability and selectivity, and some emerging materials. Then the membrane separation process is simulated with a cross flow model in PRO/II. Under the different process configurations, membrane parameters and operational parameters, the mass and energy balances, the power consumption as well as the membrane area are estimated. Afterwards, the influence of key parameters on membrane area and energy consumption is studied. A cost model is established to estimate the capture cost, and used to optimize the key parameters. Both the energy consumption and the capture cost of the membrane technology are compared with MEA-based systems. Finally, the exergy analysis method is used to identify the bottlenecks of the energy consumption in a multi-stage membrane process.

The simulation results show that the energy consumption and membrane area of a membrane process are influenced greatly by CO_2 permeability and CO_2/N_2 selectivity, which are the two key parameters in carbon capture from flue gas. Unfortunately, there is frequently a trade-off between permeability and selectivity, which is also a big challenge in material development. Higher CO_2 permeance can reduce the specific energy consumption and membrane area considerably. For a given CO_2permeance, the specific energy consumption decreases with increasing selectivity, however, more membrane area is needed. But in general, continuously increasing selectivity will not influence the separation performance so much, which means that permeance could still be the dominating parameter.

Besides the membrane parameters, the configuration of the separation process also plays a key role for energy requirement and capture cost. Due to the limitation of the membrane performance for post-combustion capture at present, it is hard to reach both 90% capture ratio and 90% CO_2purify with 1 stage membrane separators, otherwise a very high pressure ratio across the membrane is required and tremendous energy is needed to compress the gas and create vacuum, which is unpractical. Multi-stage processes represent a good option to resolve this problem, but more than 2 stages will be very complex from an operational point of view. The simulation and optimization results in this work also show how the capture load in each stage influences the total energy consumption and capture cost.

Depending on the influence of membrane parameters and flowsheet structure, referring to the concept of Robeson Plot[10], a 3-dimensional profile which reflects the relationship between permeability, selectivity and capture cost is drawn, which provides a roadmap to develop the suitable membrane material which has the possibility to compete with an MEA-based system.

Different from the MEA-based system which is a typical chemisorption process, the membrane separation of $CO_2$from a gas mixture is a physical process and the driving force is the pressure differential of the gases between the feed and permeate side. The energy demand to produce the high pressure ratio between upstream and downstream is noticeable. Thus, understanding the energy consumption level is a key point for process design, and exergy analysis is a basic method for this purpose. It is helpful to indicate where the exergy losses occur, the magnitude of these losses, and what the bottleneck parts are. Taking a two-stage process as a case study, the total exergy loss due to irreversibilities, waste heat and waste mass in each unit is calculated and ranked. The results show that the largest exergy loss occurs in the 1st membrane stage, followed by the 2nd stage. The two separators contribute to about 80% of the total exergy loss due to irreversibilities.

Compared with an MEA-based system, the gas separation membrane technology has shown some advantages for post combustion carbon capture, such as less environmental footprint, ease for scaling up and operability, without auxiliary retrofitting measures or costs in the power plant. In this work, some important aspects of this new technology were investigated using an integrated assessment method and making a comparison with an MEA-based system. This study will be important both for membrane development and future industrial applications.

Acknowledgements

This publication has been produced with support from the BIGCCS Centre, performed under the Norwegian research program *Centers for Environment-friendly Energy Research (FME)*. The authors acknowledge the following partners for their contributions: Aker Solutions, ConocoPhilips, Det Norske Veritas, Gassco, Hydro, Shell, Statoil, TOTAL, GDF SUEZ and the Research Council of Norway (193816/S60). We are grateful to the anonymous referees for their constructive comments on an earlier version of this paper.

Reference:

1. E. Favre, *Chem Eng J*, 2011, **171**, 782-793.

2. C. E. Powell and G. G. Qiao, *J Membrane Sci*, 2006, **279**, 1-49.

3. C. A. Scholes, S. E. Kentish and G. W. Stevens, *Recent Patents on Chemical Engineering*, 2008, **1**, 52-66.

4. R. Bounaceur, N. Lape, D. Roizard, C. Vallieres and E. Favre, *Energy*, 2006, **31**, 2556-2570.

5. L. Zhao, E. Riensche, R. Menzer, L. Blum and D. Stolten, *J Membrane Sci*, 2008, **325**, 284-294.

6. L. Zhao, E. Riensche, L. Blum and D. Stolten, *J Membrane Sci*, 2010, **359**, 160-172.

7. T. C. Merkel, H. Q. Lin, X. T. Wei and R. Baker, *J Membrane Sci*, 2010, **359**, 126-139.

8. A. Hussain and M. B. Hagg, *J Membrane Sci*, 2010, **359**, 140-148.

9. I. M. Britan, I. L. Leites and T. N. Vasilkovskaya, *J Membrane Sci*, 1991, **55**, 349-352.

10. L. M. Robeson, *J Membrane Sci*, 1991, **62**, 165-185.

(317f) Dynamic Modelling and Analysis of Post-Combustion CO2 Capture Processes

Niall Mac Dowell, Chemical Engineering, Imperial College London, London, United Kingdom and Nilay Shah, Centre for Process Systems Engineering, Department of Chemical Engineering, Imperial College London, London, United Kingdom

There have been several design and optimisation studies in the context of amine-based post-combustion CO_2 capture (PC-CC) from coal-fired power stations, and it is well accepted that the principal contributor to the capital-cost (CAPEX) of amine-based PC-CC is the absorber, with the majority of operating cost (OPEX) being accounted for by solvent regeneration[1]. However, the majority of these studies consider only the maximal stable generation (MSG) point of the power station. In reality, many power-plants do not operate consistently at the MSG. Thus, the cost optimal design of a PC-CC process must account for the dynamic operation of many fossil-fuel power-plants. The economic operation of a decarbonised fossil-fired power-plant is a complex trade-off between the time varying cost-of-electricity and cost of fuel. These variables will in turn affect the cost-optimal degree of CO_2 capture, and thus the design and operation of the CO_2 capture plant.

In this work, we present a new approach for the multi-period design of an integrated coal-fired power-plant and an alkanolamine-based CO_2 capture process. A validated[2] rate-based model of an absorber/stripper system for the chemisorption of acid gas in aqueous solvent solutions is used to represent the CO_2 capture process. To account for the non-idealities that are typical of amines and water, the statistical associating fluid theory for potentials of variable range (SAFT-VR)[3],[4] is used. This is a molecular approach, specifically suited to associating fluids. The SAFT formalism is used to represent some of the equilibrium reactions characterising the system, thereby simplifying the description of the chemical reactions[5]. The dynamic operation of the power-plant is described using data obtained from a real power station describing the variation in fuel burn rates in response to changing power demand. All models are implemented in the gPROMS[6] software package.

---

[1] Mac Dowell, N. et al, Energy. Environ. Sci., 3, 2010, 1645-1669

[2] Mac Dowell, N., Galindo, A., Jackson, G and Adjiman, C. S., Comp. Aided Chem. Eng., 28, 2010, 1231-1236

[3] Chapman, W.G., Gubbins, K.E., Jackson, G. & Radosz, M., Ind. Eng. Chem. Res. 29, 1709-1721., 1990

[4] Gil-Villegas, A., Galindo, A., Whitehead, P. J., Mills, S. J. & Jackson, G., J. Chem. Phys. 106 (10), 1997

[5] Mac Dowell, N., Llovell, F., Adjiman, C. S., Jackson, G and Galindo, A., Ind. Eng. Chem. Res., 49(4), 1883-1899, 2010

[6] Process Systems Enterprise (PSE) Ltd. http://www.psenterprise.com/index.html

(317g) Comparisons of Solvents for Post-Combustion CO2 Capture Using a Multi-Objective Analysis Approach

Anita S. Lee and John R. Kitchin, Department of Chemical Engineering, Carnegie Mellon University, Pittsburgh, PA

Gas scrubbing using amine solvents is one of the most promising first generation technologies for post-combustion CO_2 capture. However, the high parasitic energy demand associated with the leading commercially available amine solvent, monoethanolamine, has motivated the development of new amine solvents. [1-2] As new solvents are developed, the need arises to be able to quantify and compare improvements in energy and utility demands of CO_2 capture at the process level. In this work we uses ASPEN Plus to model the CO_2 capture and compression process for three different amines, monoethanolamine (MEA), diethanolamine (DEA), and Amino-2-methyl-1-propanol (AMP). Utilizing ModeFrontier and the modeling framework of Eslik and Miller[3] we identify amine specific optimal operating and process conditions, considering energy demands, water consumption, and capital costs as design objectives. These solutions provide the basis of comparison of performance between amines. Additionally, we analyze trade-offs between design objectives for each amine. This work illustrates the complexities associated with the CO_2 capture process and identifies an approach to comparing the performance of different solvents.

(1) Singh, D.; Croiset, E.; Douglas, P. L.; Douglas, M. A., Techno-economic study of CO2 capture from an existing coal-fired power plant: MEA scrubbing vs. O-2/CO2 recycle combustion. *Energ Convers Manage* **2003,** *44* (19), 3073.

(2) NETL, Cost and Performance Baseline for Fossil Energy Plants. **2010**.

(3) Eslick, J. C.; Miller, D. C., A multi-objective analysis for the retrofit of a pulverized coal power plant with a CO(2) capture and compression process. *Computers & Chemical Engineering* **2011,** *35* (8), 1488.

(317h) Post-Combustion Gas Permeation Carbon Capture System Models

Juan E. Morinelly, National Energy Technology Laboratory, Morgantown, WV and David C. Miller, U.S. Department Of Energy, National Energy Technology Laboratory, Morgantown, WV

Post-Combustion Gas Permeation Carbon Capture System Models

Juan E. Morinelly, David C. Miller

National Energy Technology Laboratory (NETL)

Gas permeation membranes have been proposed as an alternative to amine absorption for post-combustion carbon capture [1,2]. Industrial-scale membrane systems have been successfully applied in natural gas sweetening, and their use in other separations is expected to grow in the near future [3]. Under the constraints imposed by the performance limits of state-of-the-art gas permeation membranes multi-stage processes are required to obtain a high purity CO_2 stream suitable for sequestration. An effective design for such a process entails balancing of numerous factors, including the pressure ratio across the membrane stages, which affects the parasitic power load, and the size of the membrane modules, which affects the capital cost. Merkel et al. (2010) introduced a multi-stage process that incorporates permeate recirculation, the use of the boiler's air feed as sweep, compression of the feed, and pulling vacuum on the permeate side. The main challenge from a modeling perspective is that, unlike conventional CO_2 capture processes, this is not a purely 'end of pipe' process. The proposed process is integrated with other parts of the power plant. A modeling strategy that captures undesired effects such as higher flows due to the recirculation of gases is required in order to accurately evaluate its performance.

Two steady-state process models were built in Aspen Custom Modeler® (ACM) based on the work by Merkel et al. (2010) and Toy et al. (2011). These models include modules for the boiler, air pre-heater, and FGD units that can be tuned to a given pulverized coal (PC) power plant system. A one-dimensional, multi-component, distributed model was developed for the membrane stages within the process models. This device-scale gas permeation model incorporates pressure drop effects and can be adjusted to varying dimensions for the module as well as membrane properties. The process models are linked to a simulation interface that allows the optimization of key process parameters with respect to system-level functions such as the cost of electricity (COE). In addition to a description of the process models and their features, this presentation will describe the results of varying levels of membrane properties. The base power plant in this analysis is a supercritical PC with a net output of 650 MWe before the capture and compression systems are integrated. Constraints on the capture level (>90%), the purity of the sequestration stream (>95%), and the increase in flow of the flue gas stream out of the boiler (<15%) were included in the optimization routine.

References

1. Merkel, TC, Lin, H, Wei, X, Baker, R. Power Plant Post-Combustion Carbon Dioxide Capture: An Opportunity for Membranes. *J. Membr. Sci.*, 2010;359:126-139

2. Favre, E. Carbon Dioxide Recovery from Post-Combustion Processes: Can Gas Permeation Membranes Compete with Absorption? *J. Membr. Sci.*, 2007;294:50-59

3. Bernardo, P, Drioli, E, Golemme, G. Membrane Gas Separation: A Review/State of the Art. *Ind. Eng. Chem. Res.*, 2009;48:4638-4663

4. Toy, L. CO2 Capture Membrane Process for Power Plant Flue Gas. Presented at the 2011 NETL Capture Technology Meeting. Pittsburgh, PA, August 22, 2011.

(321a) Microchannel Rapid Temperature Swing Adsorption of CO2 for Methane Production

Kriston Brooks[1], Scot Rassat[2] and Paul Humble[2], (1)Energy and Environmental Directorate, Pacific Northwest National Laboratory, Richland, WA, (2)Pacific Northwest National Laboratory, Richland, WA

A rapid thermal swing adsorption system combined with Sabatier reactor was developed to demonstrate the applicability of a NASA microchannel Mars *In Situ* Propellant Production (ISPP) system. A thermal swing adsorption pump was designed to collect and compress CO_2 by rapidly temperature cycling eight cells filled with Molesieve 13X. Recuperation of energy between cells was used to minimize heat and cooling requirements. The collected CO_2 can be combined hydrogen from electrolyzed water to produce methane fuel using a microchannel Sabatier reactor.

Tests were performed to compress CO_2 by 10 fold using a feed stream of 10% CO_2 and 90% N_2 at 1, 2, and 4 minute cycle times. With the initial testing, a compression ratio of 5.3 was achieved with 71 to 79% thermal recuperation efficiency. Subsequent modeling work has refined the design and the number of cells required to improve the overall system efficiency and simplify the process. These results indicate similar thermal efficiencies can be achieved using a system with only four cells rather than eight. Modeling results further demonstrate design changes required to obtain improved efficiencies and compression ratios. Results of both the original experimental work and the more recent modeling efforts will be presented.

(321b) Solubilities of CO2 in Aqueous Solutions of Ionic Liquid and Monoethanolamine

Fan Xu[1], Haifeng Dong[2], Xiangping Zhang[2], Hongshuai Gao[3], Zhanli Wang[1], Suojiang Zhang[2] and Baozeng Ren[4], (1)Institute of Process Engineering, Beijing, China, (2)Institute of Process Engineering, Chinese Academy of Sciences, Beijing, China, (3)State Key Laboratory of Biochemical Engineering, Key Laboratory of Green Process and Engineering, Institute of Process Engineering, Chinese Academy of Sciences, Beijing, China, (4)Zhengzhou University, Zhengzhou, China

With the deterioration of climate change, CO_2 capture and sequestration have been paid much attention to reduce greenhouse gas emissions. Currently aqueous solutions of alkanolamines or their mixtures are commonly used in many industrial processes to capture CO_2. However, drawbacks such as solvent loss and high regeneration costs due to the high water content drive researchers for new superior technologies [1]. Ionic liquids (ILs), due to their wide liquid range, low point, tunable properties, negligible vapor pressure, high CO_2 solubility and reasonable thermal stability, are emerging as promising candidates to capture CO_2, but it is difficult to realize industrialization owing to its high viscous and high cost. Recently, the idea of mixing ILs and alkanolamines has been received great attention from the industries, since its advantages and capabilities in reducing the problem caused by the usage of conventional alkanolamines solutions are promising [2-4]. In this work, a low viscous IL ($[C_2OHmim][N(CN)_2]$) has been synthesized and the solubilities of CO_2 in aqueous mixture of IL-MEA were measured by instruments with high accuracy at temperature 313.15K and 333.15K, over CO_2 partial pressure ranging from 100 to 1000kPa and IL concentration varying from 5% to 30%. Then the corrected PVT equation has been used to calculate the absorption of CO_2. The results indicate that 0.583 mol CO_2/mol MEA could be reached under the condition of 40°C and 100.66 kPa when appropriate amount of IL has been added, which only dropped by 1.7% when compared with 30% MEA aqueous at the same condition. The results sufficiently demonstrated that the addition of the IL can reduce regeneration energy consumption to 27% with little affecting their absorption performance.

References

[1] Chinn, D.; Vu, D. Q.; Driver, M. S.; Boudreau, L. C. CO_2 removal from gas using ionic liquid absorbents. *US Patent*, 7,527,775 B2, 2009

[2] Camper, D.; Bara, J. E.; Gin, D. L.; Noble, R. D. Room-temperature ionic liquid – amine solutions: Tunable solvents for efficient and reversible capture of CO_2. *Ind. Eng. Chem. Res*., 2008, 47, 8496-8498.

[3] Zhao, Y.; Zhang, X.; Zeng, S.; Zhou, Q.; Dong, H.; Tian, X.; Zhang, S., Density,

Viscosity, and Performances of Carbon Dioxide Capture in 16 Absorbents of Amine+ Ionic Liquid+ H_2O, Ionic Liquid+ H_2O, and Amine+ H_2O Systems. *J. Chem. Eng., Data*, 2010, 55, 3513-3519.

[4] Ahmady, A.; Hashim, M. A.; Aroua, M. K., Absorption of carbon dioxide in the aqueous mixtures of methyldiethanolamine with three types of imidazolium-based ionic liquids. *Fluid Phase Equilibria*, 2011, 309, 76-82.

(321c) A Nove Amino-Functionalized Ionic Liquid for CO2 Capture

Jinzhi Zhang, Haifeng Dong and Xiangping Zhang, Institute of Process Engineering, Chinese Academy of Sciences, Beijing, China

Ionic liquids (ILs), especially amino-functionalized ionic liquids, have been found to be effective for CO_2 capture because of their unique properties such as negligible vapor pressure, high thermal stability and good absorption capacity[1-2]. As reported [3-5], ILs of more amine groups could afford higher CO_2 solubility. To explore deeply into this viewpoint, a new ionic liquid is synthesized and used in lab scale of CO_2capture experiments to identify the relationship between the structure characteristics and capture performance in terms of the absorption rate and capacity.

In present work, a novel amino-functionalized ionic liquid was synthesized and characterized by ^{1}H NMR, ^{13}C NMR, IR and MS. The physicochemical properties including glass transition temperature and thermal decomposition temperature were tested. Experiment result shows that this ionic liquid is capable of rapid capture 0.174g CO_2 / g IL at 303.15K and atmospheric pressure, which is the theoretical maximum for absorption of CO_2 chemically. The interactions of the ionic liquid with CO_2 were further investigated by comparison of both the FTIR and ^{13}C NMR spectra of the gas-untreated and gas-treated materials to support the experimental and the theoretical results.

Moreover, the synthesized ionic liquid could be recycled under vacuum at 100°C.

References

[1] J. L. Anthony, E. J. Maginn, J. F. Brennecke, *J. Phys. Chem. B* **2002**, *106*, 7315-7320.

[2] L. A. Blanchard, D. Hancu, E. J. Beckman, J. F. Brennecke, *Nature* **1999**, *399*, 28-29.

[3] Z. Xue, Z. Zhang, J. Han, Y. Chen, T. Mu, *International Journal of Greenhouse Gas Control* **2011**, *5*, 628-633.

[4] J. M. Zhang, S. J. Zhang, K. Dong, Y. Q. Zhang, Y. Q. Shen, X. M. Lv, *Chem.-Eur. J.* **2006**, *12*, 4021-4026.

[5] Y. Zhang, S. Zhang, X. Lu, Q. Zhou, W. Fan, X. Zhang, *Chemistry - A European Journal* **2009**, *15*, 3003-3011.

(321d) Assessing Anhydrous Tertiary Alkanolamines for Gas Purifications

David J. Heldebrant[1], Paul M. Mathias[2], Kash Afshar[2], Satish Reddy[2], Arnold R. Smith[3], Rubin J. McDougal[4], Louis V. Jasperson[4], Phillip K. Koech[5], Mark Bearden[6] and Charles J. Freeman[1], (1)PNNL, Richland, WA, (2)Fluor Corporation, Aliso Viejo, CA, (3)Fluor Corporation, Aliso Viejo, WA, (4)Wiltec Research Company, Provo, UT, (5)Pacific Northwest National Laboratory, Richland, WA, (6)Energy & Environmental, Pacific Northwest National Lab, Richalnd, WA

Assessing Anhydrous Tertiary Alkanolamines for Gas Purifications

Paul M. Mathias, Kash Afshar, Satish Reddy, Arnold R. Smith, Rubin J. McDougal
Louis V. Jasperson, Phillip K. Koech, Mark D. Bearden, Charles J. Freeman, David J. Heldebrant

Anhydrous tertiary alkanolamines chemically react with CO_2 and H_2S, with greater selectivity for the latter. This is in direct contrast to aqueous amine-based solvent systems with higher selectivity for CO_2 due to stronger chemical fixation than for H_2S. Anhydrous tertiary alkanolamines exhibit pressure-induced chemical fixation of CO_2 to form zwitterionic ammonium alkylcarbonate ionic liquids, while the same tertiary alkanolamines react with H_2S at atmospheric pressures to form hydrosulfide ionic liquids. This difference in capture pressure implies that anhydrous alkanolamines could be chemically selective for H_2S over CO_2 under specific conditions. We present here the first vapor-liquid equilibrium (VLE) data of anhydrous ethyldiethanolamine (EDEA) with CH_4, C_3H_8, H_2S and CO_2 at 10 and 50 °C measured by the PTx and PTxy methods. The data are modeled in Aspen Plus® using an electrolyte NRTL model. Data trends and phenomena are discussed for each gas. We also compare anhydrous EDEA's simulated performance for CO_2 and H_2S separations to other solvents such as Fluor Solvent, Econosolv, and aqueous methyldiethanolamine (MDEA) for representative gas-processing applications.

(321e) Development of a Novel Carbonate Absorption Process with Crystallization-Enabled High Pressure Stripping for Post-Combustion CO2 Capture: Kinetic Study of Bicarbonate Salt Crystallization

Qing Ye, University of Illinois at Urbana-Champaign, Champaign, IL

Development of a Novel Carbonate Absorption Process with Crystallization-Enabled High Pressure Stripping for Post-Combustion CO_2 Capture: Kinetic Study of Bicarbonate Salt Crystallization

Qing Ye[1,2], Manoranjan Sahu[2], Yongqi Lu[2] and Xinlei Wang[1]

1 Agricultural and Biological Engineering, University of Illinois at Urbana-Champaign

2 Illinois State Geological Survey, University of Illinois at Urbana-Champaign

A novel absorption process with high pressure stripping is being developed at our group for the post-combustion CO_2 capture application. The process employs a carbonate salt aqueous solution, such as potassium carbonate (PC), as a solvent. The CO_2 absorption operates at elevated temperature and the bicarbonate in the CO_2 rich solution from the absorber is crystalized in a separate cooling-crystallizer. Slurry of bicarbonate crystals is used in the stripper for CO_2 stripping at high pressure. Compared to conventional mono-ethanol-amine (MEA) processes, the carbonate-based process consumes less stripping heat and CO_2 compression work. A preliminary techno-economic analysis reveals that the energy use of the novel process is 40-50% less than that of its MEA counterpart.

Crystallization of the bicarbonate from the CO_2-rich PC solution is a key step in the carbonate-based process. Kinetic studies of the bicarbonate crystallization were performed under continuous mix suspension, mixed product removal (MSMPR) mode using an automated calorimetric reactor system. Parametric tests were conducted to investigate the effect of the PC concentration, CO_2 loading, addition of absorption promoter, temperature, residence time, stirring speed, seeding, etc. on the kinetics of the crystallization and morphology of crystal products. Crystal solids were characterized by particle size distribution (PSD), turbidity, SEM, and XRD analysis techniques. Kinetic data, such as the nucleation rate constant, size-dependent growth rate constant, and agglomeration/breakage factor, were obtained by fitting the differential PSD data under various operating conditions.

This presentation will provide a summary of the results from the crystallization kinetic study and discuss their implications for the crystallizer design.

Doped lithium orthosilicate for absorption of CO_2 at steam atmosphere with low CO_2 partial pressure

Abstract: Calcium and lithium containing oxides are regarded as options for the capture of CO_2 in sorption-enhanced steam reforming of methane (SEMSR) system. Among the latter, lithium orthosilicate (Li4SiO4) is considered as one of the most potential materials due to its large capacity, high selectivity and good cycle performance. Improved properties can be obtained by appropriately doping different elements into Li_4SiO_4. In this study, the best amount of doping elements (K, Fe) into Li_4SiO_4 were calculated by Factsage 5.5. Therefore, different lithium orthosilicate based absorbents, potassium doped and iron doped lithium orthosilicate according to the calculated results from Factsage 5.5 were prepared and characterized by XRD, SEM. For the study of practical applicability in SESMR, the absorption ability was investigated by TGA under low CO_2 partial pressure (5%~20%) influenced by different steam atmosphere (5%~30%). In addition, the cyclic absorption/desorption performance of different absorbents were also investigated. The results show that potassium and iron can enhance the absorption capacity at low CO_2 partial pressure and cyclic ability respectively and the best absorbent with suitable amount of potassium and iron was acquired.

Key words: CO_2 acceptors; Lithium orthosilicate; doping method

1 Introduction

CO_2 acceptors are widely used in industrial process such as sorption enhanced methane steam reforming, flue gas decarburization, energy storage system and etc. Thus, it is important to obtain a solid CO_2 acceptor which has high CO_2 absorption capacity at moderate temperature with good selectivity for CO_2 and good regeneration ability.

Thus far, several materials, such as hydrotalcite-like materials, CaO-based sorbents, and lithium-containing materials[1], have been proposed for CO_2 capture. Among these materials, lithium orthosilicate (Li_4SiO_4) is considered to be one of the most potential materials. Li_4SiO_4 has been reported to adsorb CO_2 more than 30 times faster than Li_2ZrO_3 and is lighter and cheaper than Li_2ZrO_3.

$$Li_4SiO_4 + CO_2(g) = Li_2CO_3 + Li_2SiO_3$$

Li_4SiO_4 has high CO_2 absorption capacity. However, the temperature in the real SESRM process is between 500□~600□. Moreover, the concentration of CO_2 is as low as about 10%. Under such condition, the CO_2 absorption ability of Li_4SiO_4 will decrease dramatically. In order to improve the defect, exotic elements can be doped into Li_4SiO_4. C. Gauer [2]etc. added elements such as Al, Fe and K into Li_4SiO_4, concluding that for the practical use of Li_4SiO_4, doping vacancies through doping Fe can not only increase the sorption ability but also the cycle performance. M. Seggiani etc[3]. doped K element using simple addition method and eutectic method into Li4SiO4 and compared the two methods drawing the conclusion that simple addition method can well promote the sorption ability while has a bad cycle performance. Mayra Y. Veliz-Enriquez etc[4]. prepared $Li_{2-x}K_xZrO_3$ with different quantities of K_2CO_3 and then concluded that the best x is 0.2 using solid solution theory and experiment.

In the real SEMSR process, not only can the temperature reach the best absorption temperature(700□) and can the concentration of CO_2 be high enough for sorption, but also the steam exist to influence the sorption ability of Li_4SiO_4. Hughes, Manovic, Reijers, Sun, Yan, Zeman etc[5-10]. investigated the effect of steam on absorption of CaO for CO_2 holding the opinion that steam can facilitate the sorption ability of CaO for CO_2. Esther Ochoa-Fernández[11] studied the effect of steam on pure Li_4SiO_4 for CO_2 absorption but the reason was not indicated.

In this article, Li_4SiO_4 with different quantities of K were prepared through solid synthesis using eutectic method. The best amount of K_2CO_3 was calculated by Factsage software and the results were proved by characterization method such as XRD, SEM and TG. The sorption ability of different samples for CO_2 was investigated under the same condition as in the real SEMSR process. Finally, the effect of sorption condition on sorption ability such as temperature, pressure of CO_2 and the amount of steam were analyzed.

2 Experimental part

2.1 Characterization of samples

Solid-state synthesis of analytical grade lithium carbonate and highly dispersed silicon dioxide (SiO_2) was performed by adjusting a molar ratio of 2.05:1 to obtain Li_4SiO_4 powder. After homogenization in a vibrating disc mill for a minute the initial mixture was calcined in air at 750□ for 6h.

For simple addition doping, the molar ratio of starting materials was adjusted to $Li_2CO_3:SiO_2:K_2CO_3=(2.05+A):1:(2A)$.

2.2 Characterization of samples

The sample was identified by X-ray diffraction (XRD). A diffractometer (RIGAKU D/MAX 2550 VB/PC, Japan) coupled to a copper-anode X-ray tube was used. The Kα wavelength (1.54056nm) was selected with a diffracted beam monochromator. Li4SiO4 was identified by the corresponding Joint Committee on Powder Diffraction Standards (JCPDS) in virtue of MDI Jade 5.0 software. The surface structure and particle size of the sample was characterized by scanning electron microscopy (SEM) (JEOL JSM-6360LV, Japan).

A WRT-3P scale was used to analyze the temperature ranges and kinetic behaviors of the CO2 sorption process.

3 results and discussions

3.1 Theoretic effects of doped K element on Li_4SiO_4

Doping substance which is able to form eutectic compound with Li_2CO_3 can improve the CO_2 sorption ability of Li_4SiO_4. There are two common alkali metal salt including K_2CO_3 and Na_2CO_3 that can form eutectic compound with Li_2CO3. Fig.1 and Fig.2 show the binary phase diagram of Li_2CO_3/K_2CO_3 and Li_2CO_3/Na_2CO_3. When the amount of K_2CO_3 is 0.4, the melting point of the eutectic compound is 500⍰. While the amount of Na_2CO_3 is up to 0.5, the melting point is 500⍰. K_2CO_3 and Na_2CO_3 can both decrease the melting point dramatically but K_2CO_3 is better for its smaller amount needed.

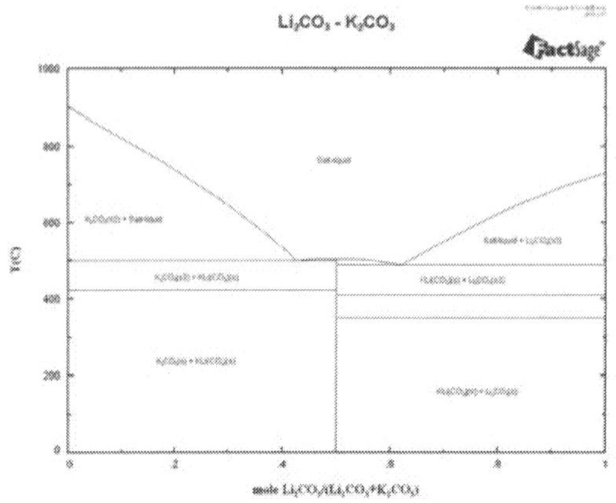

Fig.1 Li_2CO_3-K_2CO_3 binary phase diagram

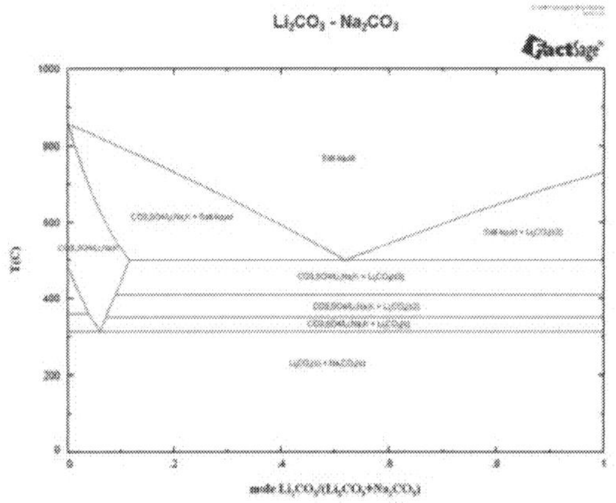

Fig.2 Na_2CO_3-K_2CO_3 binary phase diagram

Fig.3 shows the calculated equilibrium composition using Factsage software of starting materials with different amount of K_2CO_3

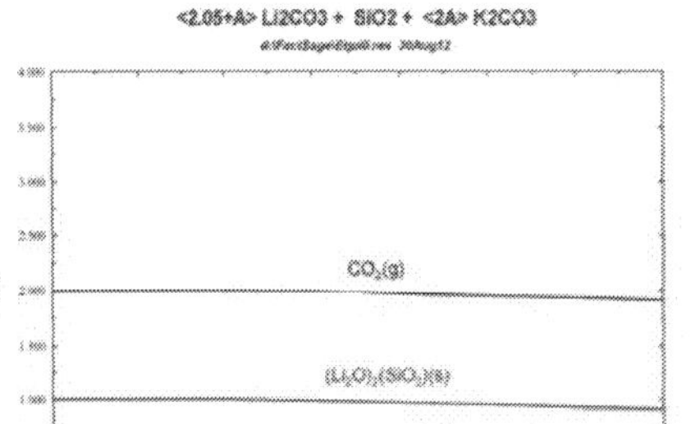

Fig.3 Calculated equilibrium composition of Li_2CO_3+SiO_2+K_2CO_3 calcined at 750℃ for 6h

The value of A in the experiment of C. GAUER is 0.1. With the increase of A, which means the increase of amount of K_2CO_3, Li_2CO_3 that remains in the product will disappear gradually. While A increases continually, Li_2SiO_3 appears as a by-product, which has no promotion for the sorption of CO_2. A new substance, KLi(CO_3), which is the eutectic compound formed by K_2CO_3 and Li_2CO_3, can be found with the increase of K_2CO_3. The melting point decrease more as the amount of K_2CO_3 increase thus the sorption ability for CO_2 is improved more. In summary, the value of A should be maintained between 0.04~0.06 so the absorbent can have more KLi(CO_3)

while less impurity.

3.2 influence of amount of K_2CO_3 to sorption ability of K-Li_4SiO_4

Table.1 shows the value of A in different samples. The molar ratio of starting materials is as follows: Li_2CO_3:SiO_2:K_2CO_3=(2.05+A):1:(2A)

Table.1 value of A in different samples

sample	1	2	3	4	5	6
A	0	0.06	0.08	0.10	0.12	0.14

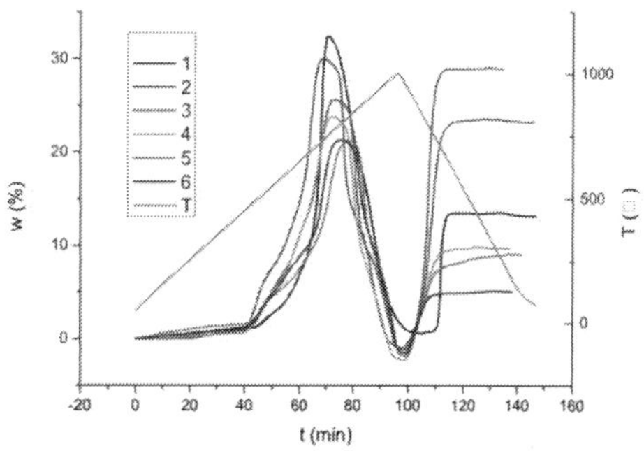

Fig.4 thermo analytical curve of different samples

The thermo analytical curve in Figure 4 was obtained by heating the samples in the pure CO_2 atmosphere with a heating rate of 10□/min from 0°C to 1000℃ and then with a cooling rate of 20℃/min from 1000℃ to room temperature. The samples which have K_2CO_3(No.2~No.6) has a clearly improvement of CO_2 sorption ability at 50min when the temperature is about 500□ compared with pure Li_4SiO_4(No.1), which indicates that the eutectic compound can indeed improve the CO_2 sorption ability. No.2 sample has the most significant increase at 500□. Nevertheless, the sorption ability decreases as the amount of K_2CO_3 increases. This can be attributed to the increase of the Li_2SiO_3, which has no positive effect for CO_2 sorption. The results are well consistent with the results calculated by Factsage before.

During the temperature decrease process, all samples absorb CO_2 for the second time. This process is rather short for the cooling rate is as twice as much as the heating rate. Even though, No.2 sample can have a weight gain of 30%, which is obviously better than other samples, within such a short time.

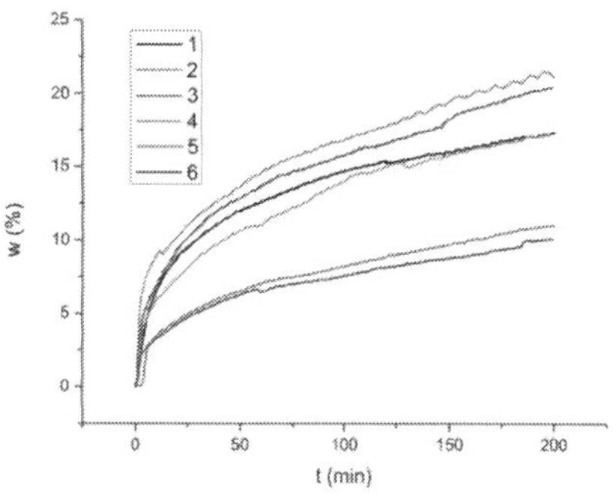

Fig.5 Isothermal curves of different samples at 575□ and 100% CO_2 atmosphere

In the real SESRM process, the low the temperature is, the less energy the reaction consumes. Fig. 5 shows the isothermal analysis curves of different samples at 575□ and 100% CO_2 atmosphere. No.2 and No.3 samples have better sorption performance than No.1 sample and No.2 has a bigger increase. Compare No.2 sample with No.1 sample, the rate of rapid reaction stage increases. This can be attributed to the formation of eutectic compound, which restrains the formation of the double shell because liquid phase eutectic compound can facilitate the diffusion of Li^+ and O^{2-} and thus higher conversion is needed to form the solid double shell. With the increase of K_2CO_3, the sorption ability, on the contrary, decreases. The reason is similar with the reason for the results of the thermal analytical curves.

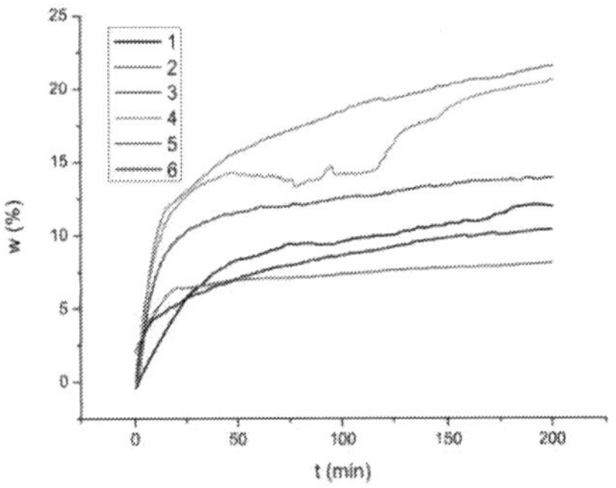

Fig.6 Isothermal curves of different samples at 575□ and 10% CO_2 atmosphere

Fig.6 shows the isothermal curves of different samples at 575□ and 10% CO_2 atmosphere. The rate of the rapid reaction stage of No.1 sample is significantly low than any other samples because of the low CO_2 concentration. While for the doped with K samples, No.2~No.6 samples, the rate of the rapid reaction stage increases apparently. This can be attributed to the formation of eutectic compound which can increase the amount of Li^+ and O^{2-} on the surface and thus facilitating the following reaction.

$$2Li^+ + O^{2-} + CO_2 \underset{k_4}{\overset{k_3}{\rightleftharpoons}} Li_2CO_3 \qquad (3)$$

For No.2 sample, the weight gain can easily reach to 13% within 20min and to 22% within 200min, which is clearly higher than that of No.1 sample. As the amount of K increases, the sorption ability decreases. The reason is also the same as the reason for the results of thermal analytical curves.

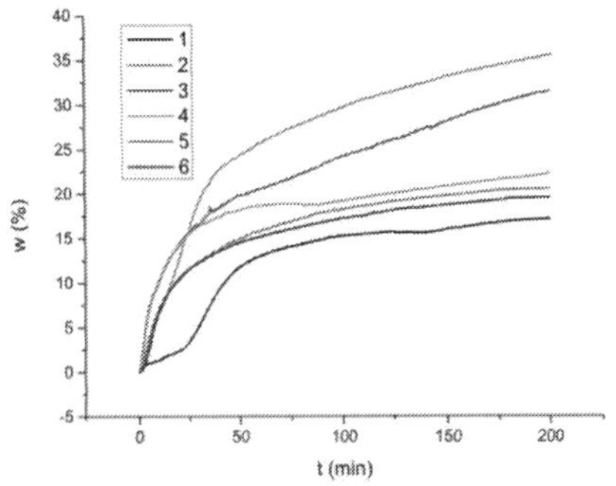

Fig.7 Isothermal curves of different samples at 575□, 10% CO_2 and 20% steam atmosphere

Fig.7 shows the isothermal curves of different samples at 575□, 10% CO_2 and 20% steam atmosphere. For No.1 sample, the weight gain within 200min increases from 13% to 15%, while fro No.2 sample, the weight gain increases from 22% to 35%. For the other samples, the sorption capacity also increases more than No.1 sample.

3.3 Characterization of samples

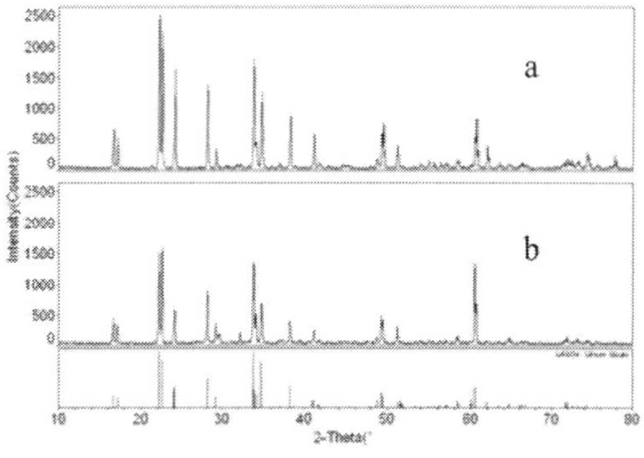

a-pure Li_4SiO_4; b-No.2 sample

Fig.8 XRD patterns of different samples

Fig.8 shows the XRD patterns of pure Li4SiO4 and No.2 sample. They were all fitted to JCPDS file 37-1472, and only the highly crystalline structure of Li_4SiO_4 was presented. This can be explained by the solid solution theory. Small quantities of K_2CO_3 can be dissolved into the Li_4SiO_4 solvent as a solid solute, so no substances including K can be found. This also explains why the sorption ability decreases while the amount of K increases.

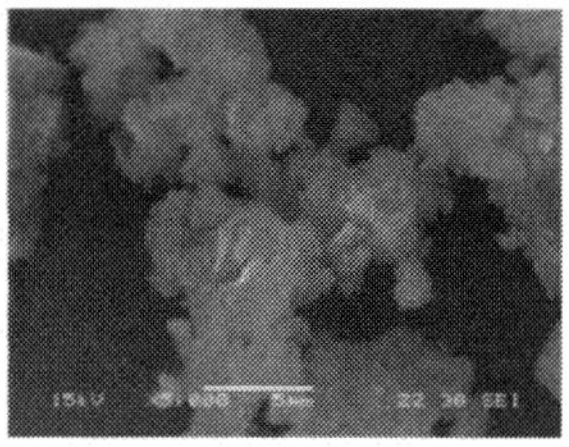

a-pure Li_4SiO_4

b-No.2 sample

Fig.9 SEM images of different samples

Fig.9 shows the SEM images of different samples. Pure Li_4SiO_4 and No.2 sample both have smooth surface and are fin powders. Pure Li_4SiO_4 has a smaller diameter of 10μm, while No.2 sample has an agglomeration phenomenon with a diameter of 30μm and has small particle which is Li_4SiO_4 with K on its surface. Thus, even though No.2 sample has bigger particle which is not good for chemical absorption, the sorption ability is much better than that of pure Li_4SiO_4. The results show that the formation of eutectic substances can improve the sorption ability significantly.

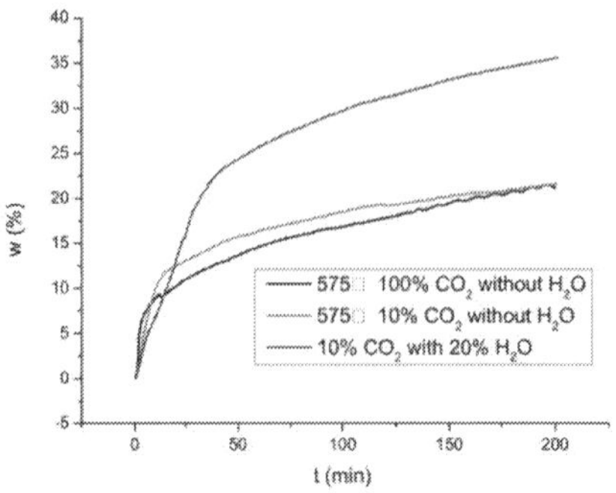

Fig.10 Isothermal curves of No.2 sample under different conditions

Fig.10 shows the Isothermal curves of No.2 sample under different conditions. For No.2 sample, the sorption ability maintains while the concentration of CO_2 decreases. While the water steam occurs, the sorption ability even improves with a weight gain of 20% with 30min. The improvement mainly displays in the extension of rapid reaction stage. When there is no steam, the rapid reaction stage ends when the weight gain is about 13% while 23% when there is steam. This can be attributed to the inhibition of the formation of double shell and the fact that steam can facilitate the mobility of Li^+ and O^{2-}.

4 Conclusion

(1) Doping K element into Li_4SiO_4 can form eutectic substance with the Li_2CO_3 produced by Li_4SiO_4 after CO_2 sorption thus facilitating the CO_2 sorption ability of Li_4SiO_4.

(2)It was comfirmed by calculation using Factsage and experiments results that the optimum doping ratio was Li_2CO_3:SiO_2:K_2CO_3=2.11:1:0.12. Then the rapid reaction stage of the adsorbent will be extended, and the maximum absorption capacity will be improved obviously.

(3)Under the same condition as in real SESRM process, the ability of CO_2 acceptors obtained maintains when the concentration of CO_2 decreases and the sorption ability even improves when steam occurs. The weight gain is more than 20% within 30min at 575℃, 10%CO_2 and 20% steam atmosphere.

5 References

[1] Li Li, Yuan Wenhui, Wei Chaohai. Adsorbents and adsorptive process for carbon dioxide at high temperature[J]. Chemical Industry and Engineering Progress, 2006, 25(8): 918-922

[2] Gauer C, Heschel W. Doped lithium orthosilicate for absorption of carbon dioxide[J]. Journal of Materials Science, 2006,41(8): 2405-2409

[3] M. Seggiani, M. Puccini, S. Vitolo. High-temperature and low concentration CO_2 sorption on Li_4SiO_4 based
sorbents: Study of the used silica and doping method effects[J]. International Journal of Greenhouse Gas Control 5 (2011): 741–748

[4] Mayra Y, Veliz-Enriquez, Gonzalo Gonzalez, Heriberto Pfeiffer. Synthesis and CO_2 capture evaluation of $Li_{2-x}K_xZrO_3$ solid solutions and crystal structure of a new lithium-potassium zirconate phase[J]. Journal of Solid State Chemistry,2007, 180: 2485–2492

[5] Robin W. Hughes, Dennis Lu, Edward J. Anthony, and Yinghai Wu. Improved long-term conversion of limestone-derived sorbents for in situ capture of CO_2 in a fuidized bed combustor[J]. Ind. Eng. Chem. Res, 2004, 43, 5529-5539

[6] Vasilije Manovic, Dennis Lu, Edward J. Anthony. Steam hydration of sorbents from a dual fluidized bed CO_2 looping cycle reactor[J]. Fuel, 87 (2008) 3344-3352

[7] Hendricus Th. J. Reijers, Sacha E. A. Valster-Schiermeier, Paul D. Cobden, and Ruud W. van den Brink. Hydrotalcite as CO_2 sorbent for sorption-enhanced steam reforming of methane[J]. Ind. Eng. Chem. Res. 2006, 45, 2522-2530

[8] Ping Sun, John R. Grace, C. Jim Lim, and Edward J. Anthony. Investigation of attempts to improve cyclic CO_2 capture by sorbent hydration and modification[J]. Ind. Eng. Chem. Res. 2008, 47, 2024-2032

[9] Aiyu Yan, Bin Liu, Yonglai Dong, Zhijian Tian, Dezheng Wang, Mojie Cheng. A temperature programmed desorption investigation on the interaction of $Ba_{0.5}Sr_{0.5}Co_{0.8}Fe_{0.2}O_{3\_\delta}$ perovskite oxides with CO2 in the absence and presence of H_2O and O_2[J]. Applied Catalysis B: Environmental 80 (2008) 24–31

[10] Frank Zeman. Effect of steam hydration on performance of limesorbent for CO2 capture[J]. International Journal of Greenhoue Gas control, 2, (2008) 20 –209

[11] Esther Ochoa-Fernández, Tiejun Zhao, Ph.D., Magnus Rønning, De Chen. Effects of steam addition on the properties of high temperature ceramic CO_2 acceptors[J]. Journal of Environmental Engineering, 2009, 6: 397-403

(324a) Pressure-Swing Distillation for Minimum and Maximum Boiling Homogeneous Azeotropes

William L. Luyben, Department of Chemical Engineering, Lehigh University, Bethlehem, PA

Pressure-swing azeotropic distillation uses two columns operating at two different pressures to separate azeotropic mixtures by taking high-purity product streams from one end of the columns and recycling the streams from the other end with compositions near the two azeotropes. This configuration can be economically used when changes in pressure significantly shift the composition of the azeotrope. The larger the shift, the smaller the required recycle flowrates, so the smaller the energy requirements in the two reboilers.

Pressure-swing can be applied to both minimum-boiling and maximum-boiling homogeneous azeotropic mixtures. With minimum-boiling systems, the distillate streams are recycled. With maximum-boiling systems, the bottoms streams are recycled. Intuition would lead us to expect that recycling distillate streams would be more energy intensive than recycling bottoms streams. A distillate recycle must be boiled up in the column. A bottoms recycle stream is not boiled up.

Therefore we would expect that less pressure sensitivity of the azeotropic mixture would be required to make the pressure-swing configuration in a maximum-boiling system economical than in a minimum-boiling system. The purpose of this paper is to explore this question. Results show that there is very little difference between the two systems in terms of the effect of pressure sensitivity and energy consumption.

(324b) Design of Efficient Systems for Multicomponent Distillation

Joshua Huff[1], Rakesh Agrawal[1] and Mohit Tawarmalani[2], (1)School of Chemical Engineering, Purdue University, West Lafayette, IN, (2)Krannert School of Management, Purdue University, West Lafayette, IN

In the traditional industrial applications of multicomponent distillation, a relatively limited number of distillation column arrangements are repeatedly applied to a wide variety of separations. However, there are many separation process solutions differing significantly in cost and energy consumption that carry out the same overall process. Therefore a systematic method to identify and design optimal multicomponent separation sequences is needed that does not rely on the inventive activity of a few experienced engineers. Even for a commonly perceived "mature" technology such as distillation, a method to elucidate all possible multicomponent separation schemes has only been developed recently.

This talk will focus on two aspects of multicomponent continuous distillation. First, a method to draw multicomponent distillation configurations will be introduced. It is known that the number of possible distillation configurations increases rapidly in hundreds and thousands as the number of components in an n-component mixture increase beyond three ($n > 3$). Therefore, the first challenge for a process engineer is to be able to draw all feasible configurations and then narrow down the search to a set of suitable candidates. A systematic procedure to draw distillation column configurations to separate an ideal to near ideal n-component mixture into product streams each enriched in one of the components will be presented. The method is simple and easy to use.

The second aspect of a system for multicomponent distillation design is a screening tool that allows rank listing of distillation configurations generated by the above procedure. While it is possible to compare systems based on rigorous modeling and simulation, it becomes impractical for the large number of possible configurations generated when n becomes large. Instead, the configuration's vapor duty as calculated by the Underwood equations is minimized. The total vapor flow in a column is closely related to both the reboiler heat requirement (a principal component of operating cost for distillation) and column width (a major component of capital cost). Thus, a nonlinear program written to minimize total vapor duty is able to produce a ranked list of energy-efficient distillation columns that is very similar to a ranked list generated by rigorous simulation, and takes much less computation. This program can be extended to design of configurations with less than n-1 columns; in addition, a study will be presented on whether the Underwood equations can be used as an accurate predictor of energy efficiency in real systems.

(324c) New Class of Thermal Coupling Links in Multicomponent Distillation

Gautham Madenoor Ramapriya, Anirudh A. Shenvi and Rakesh Agrawal, School of Chemical Engineering, Purdue University, West Lafayette, IN

Thermal coupling links between distillation columns have been known to provide significant savings in heat duty requirements. Thermal coupling links are two way liquid-vapor communications between distillation columns. Historically, thermal coupling links have been between reboilers and condensers associated with streams of similar compositions. The Petlyuk column is a classic example of such a thermal coupling link. In this work, we introduce a new class of thermal coupling links that involves the elimination of reboilers and condensers associated with streams of *dissimilar* compositions. The well-known Brugma configuration for separating a four-component feed mixture is one such configuration that demonstrates this class of thermal coupling links between distillation columns. However, our generalized class includes many possible novel thermal coupling links between distillation columns which result in several hitherto unknown configurations. This new class of thermal coupling links not only has the potential to save operating costs but also to provide capital cost benefits through the reduction of number of distillation columns and elimination of reboiler and condenser heat exchangers. Additionally, side-streams can be strategically withdrawn from the resulting thermally linked sections which have the potential to further provide significant energy savings. We present a systematic framework to identify all possibilities to thermally couple two distillation columns producing streams of dissimilar compositions. This framework can be used as an underlying principle to synthesize configurations resulting from such a class of thermal coupling links.

(324d) Production of Dialkly Carbonates Via Reactive-Extractive and Pressure-Swing Distillations Using Unifac-CI VLE Model Predictions

Emmanuel A. Dada, ChemProcess Technologies (CPT), LLC, League City, TX, Azizul Azri Mustaffa, Department of Chemical and Biochemical Engineering (CAPEC), Technical University of Denmark, Kgs. Lyngby, Denmark and Rafiqul Gani, CAPEC, Department of Chemical and Biochemical Engineering, Technical University of Denmark, Kgs. Lyngby, Denmark

Dialkyl carbonates like dimethyl carbonate (DMC) and diphenyl carbonate (DPC) are environmentally benign and biodegradable products and have found wide applications in polycarbonates plastics and as important solvents in the emerging battery markets. In particular, DMC is the solvent of choice for lithium ion battery solutions. The synthesis of dialkyl carbonates are generally carried out by continuous reactive distillation that produce azeotropic mixtures of the dialkyl carbonates and linear alcohols followed by the recovery of the dialkyl carbonates. The recovery of a typical dialkyl carbonate from the azeotropic mixture is achieved by either using pressure swing distillation process where the pressure sensitive azeotropic mixture is separated or by using extractive distillation process where the azeotropic mixture is contacted with an entrainer like phenol, aniline and diethyl oxylate to produce high purity dialkyl carbonate. In all of these distillation processes (reactive, pressure-swing, and extractive distillations), accurate predictions of the binary VLE systems involving the carbonate group (-OCOO-) and other components especially the linear alcohols in the production of high purity dialkyl carbonates are very important. The UNIFAC group interaction parameters (GIPs) involving the carbonate group are very limited in the literature. We have used the UNIFAC-CI model where all the group interaction parameters (GIPs) are predicted and/or fine tuned using the GC^{Plus}/UNIFAC-CI method in the latest ICAS regressed atom interaction parameters (AIPs). Typical GIPs generated will be presented. With these parameters, we were able to accurately predict the VLE of the azeotropic mixtures. The results were used for the design of the reactive-extractive and reactive-pressure swing distillations to produce high purity dialky carbonates. The energy consumption of the extractive distillation is less than the pressure-swing distillation. Overall, a heat integrated reactive and extractive distillation system provides the best option for energy savings than those without the heat integration.

(324e) A Novel Systematic Approach to Pressure-Swing Distillation Process Design

Andreas Scheithauer[1], Jakob Burger[1], Norbert Asprion[2], Christoph Grossmann[2] and Hans Hasse[1], (1)Laboratory of Engineering Thermodynamics, University of Kaiserslautern, Kaiserslautern, Germany, (2)Chemical and Process Engineering, BASF SE, Ludwigshafen, Germany

Pressure-swing distillation (PSD) is a well-known and widely used process for the separation of azeotropic mixtures. It is based on the pressure dependence of the azeotropic composition. Designing a PSD process is a multi-parameter optimization problem. Even in the most simple two column configuration and on the simplest reasonable level of description by an equilibrium stage model, there is a high number of degrees of freedom which have to be chosen (column pressures, reflux ratios, recycle ratio, overall and feed stage numbers). Finding the optimal numbers for these parameters, for given feed and product specifications, is far from being trivial. Astonishingly, only little work on that widespread and important design problem has been published in the literature so far. None of the existing approaches allows a systematic optimization of all relevant parameters.

Therefore, a new method was developed in the present work that only needs the educt and product specifications as input. The corresponding algorithm couples two modeling levels. On the first level, separation factors which are determined from the knowledge of the fluid property data are used for characterizing the effort for producing desired qualities in the columns. The choice of these factors allows to determine the top and the bottom composition of the columns. For the optimization, they are systematically varied. Furthermore, a range for the column-pressures is defined considering boundary conditions like maximum bottom or minimal top temperatures. Within that range, the column pressures are discretized. For each pair of known column pressures all streams can then be calculated from simple mass-balances. Therefore, on the second level, the two columns can be designed decoupled from each other. The column design itself is based on the equilibrium stage concept using a detailed vapor-liquid equilibrium model and the well-established concept of the N,Q-curve [1] and yields all information for estimating and minimizing the OPEX and CAPEX. The entire algorithm was automated. Due to the decoupling of the column designs the new method is fast and robust. The global economic optimum can be determined swiftly and reliably. This is demonstrated using different industrial relevant examples.

[1] S. Zeck: *Einfluß von thermophysikalischen Stoffdaten auf die Auslegung und den Betrieb von Destillationskolonnen,* Chemie Ingenieur Technik 62 (1990) 707-717

(324f) Practical Application of Column Profile Maps: Distributed Feed Crude Distillation Column

Emmanuel Kasese[1], Mark Peters[2], Diane Hildebrandt[3] and David Glasser[3], (1)School of Chemical and Metallurgical Engineering, University of the Witwatersrand (COMPS), Johannesburg, South Africa, (2)Wits University, Johannesburg, South Africa, (3)Centre of Material and Process Synthesis (COMPS), University of the Witwatersrand, Johannesburg, South Africa

Distillation columns with multiple feeds are often used in the process industry for the separation of multicomponent mixtures, though the energy-saving potential for different column variations like reflux reduction and stage reduction have not been fully addressed for complex columns (Bandyopadhyay et al., 2004). This paper will be founded on the premise that there are economic benefits in having feeds entering the column at different locations.

The CPM method, a novel graphical distillation design technique developed by Tapp et al. (2004), has already been established to be effective in distillation systems analysis and can be generalised for any number of components. This paper proposes the use of distributed feed on an Atmospheric Crude Distillation column (ACD) for a crude oil refinery with the aid of CPMs and a software simulation package (ASPEN®). Separation is not perfect in the commercial ACD, and that makes it an ideal candidate for distributed feed because of the non-sharp split product specifications (Holland, 2005). Since plotting CPM profiles for a system with more than 4 components makes it very hard to get a clear picture of the compositional profile plots, this issue can be circumvented by taking a representative three component mixture that fairly mimics the real multicomponent system. For the purposes of this study a constant volatility n-butane/n-decane/n-hexadecane mixture was chosen on the basis that a paraffinic crude oil would be processed. The CPM method would then be used to analyse the distributed feed scheme and the results can be generalised for the crude oil system by using the CPM results to initialise the ASPEN® simulation.

At the end of the study, the paper would have further validated the authenticity of the CPM method by having applied it on a well-established process thus proving that it has real practical applications rather than just scientific. Finally the question of whether the alternative distributed feed configuration proposed is worthy will be answered by looking at the approximate potential savings in capital, energy as well as operating costs.

References

Bandyopadhyay, S., Mishra, M. & Shenoy, U.V., 2004. Energy-Based Targets for Multiple-Feed Distillation Columns. *AIChE*, 50(8), pp.1837-53.

Holland, S.T., 2005. *Column Profile Maps: A Tool for the Design and Analysis of Complex Distillation Systems*. Thesis. Johannesburg: COMPS University of the Witwatersrand.

Tapp, M., Holland, S.T., Hildebrandt, D. & Glasser, D., 2004. Column Profile Maps. 1. Derivation and Interpretation. *Ind Eng Chem Res*, 43, pp.364-74.

(331a) C2U Performance Modeling for CO_2 Adsorption

Ronald W. Breault, Computational Sciences Division, National Energy Technology Laboratory, Morgantown, WV, E. David Huckaby, Computational Science Division, National Energy technology Laboratory, Morgantown, WV, Lawrence J. Shadle, National Energy Technology Laboratory, U.S. Department of Energy, Morgantown, WV, James Spenik, REM, Corporation, Morgantown, WV and Esmail R. Monazam, REM Engineering Services, U.S. DOE NETL, Morgantown, WV

The National Energy Technology Laboratory as well as other institutions are in the process of conceptualizing new technology as an alternative to amine based absorption processes for CO_2 capture from large sources such as Utility power generation facilities. Many, but not all of these advanced dry processes are based upon sorbents composed of supported polyamines. This particular investigation gives a first time comparison between the predicted performance and the experimental performance of one of these new class of sorbents. In the continuous process CO_2 from is absorbed from flue gas in the absorber, a cyclone separates the carbonated particles from the de-carbonated flue gas and then regenerates the sorbent, creating a concentrated stream of pure CO_2 for sequestration. In the current work, the reactors are run in a batch mode, that is without circulating solids. In this work, experimental data from adsorption and regeneration in a fluidized bed operated in batch mode are compared to simulation predictions from a 3-dimentional multiphase flow model complete with chemistry and thermal effects. The effects of gas flow rate and reactor temperature are explored and compared with experimental measurements. The specifics for the experimental facility and the model as well as the comparative analysis between the data and the simulation results will be discussed.

Development of PSA System for the Recovery of Carbon Dioxide and Carbon Monoxide from Blast Furnace Gas in Steel Works

H.Saima, Y. Mogi, T. Haraoka,
Environmental Process Research Department, Steel Research Laboratory, JFE Steel Corp,

1.Introduction

Climate exchange by carbon dioxide became the serious and global problem. In Japan, steel works discharge carbon dioxide about 15% of total discharging amount. From this point of view, The Japan Iron and Steel Federation (JISF) starts a new project, called "COURSE 50" (CO_2 Ultimate Reduction in Steelmaking process by innovative technology for cool Earth 50). The authors are trying to develop separation and recovery technology of carbon dioxide and carbon monoxide from Blast furnace gas by utilizing PSA technology.

This trial seemed to be easy by binding PSA for carbon dioxide and one for carbon monoxide. However, the amount of Blast furnace gas is enormous, such as more than several million m^3/hr and the concentrations of both carbon dioxide and monoxide are only around 20%. Furthermore, it is necessary to built up low cost process for the commercialization.

2.Laboratory PSA Examination

13 kinds of adsorbent from the market were compared by adsorption isotherm curve and choose 2 kinds of adsorbent. These adsorbents are tested by PSA testing apparatus in the laboratory. The inner diameter of adsorption tower in testing apparatus is 43mm and the height of it is 500mm long. We found Zeolum F-9 was suitable for carbon dioxide separation because of high adsorption capacity and adsorption selectivity.

From the importance of recovery cost for the commercialization, the authors estimated the recovery it at laboratory PSA apparatus, which was shown in photo 1, in the various separation conditions. Higher adsorption pressure leads higher recovery ratio although it also leads higher power consumption at blower. Simultaneously, lower desorption pressure leads higher recovery ratio and higher power consumption at vacuum pump. As the results of experiment and power estimation at blower and vacuum pump, it is suitable that adsorption pressure is around 150-200kPa (absolute pressure) and desorption pressure is around 10kPa (absolute pressure). Furthermore, cycle time is key factor of cost for construction. The authors thought the standard cycle time was 630 seconds. As the results of laboratory PSA apparatus, which was shown in Figure 1, any change in separation ability was recognized among the results of cycle time at 450, 630 and 720 seconds. And slight change was observed in the results of cycle time at 300 seconds.

Photo 1 PSA Laboratory Apparatus

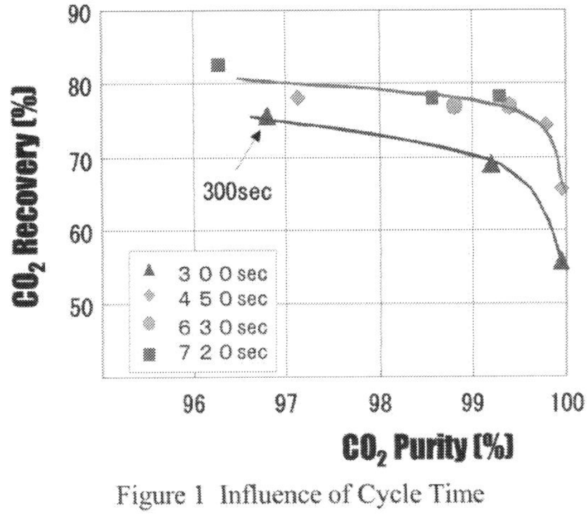

Figure 1 Influence of Cycle Time

The effects of CO_2 concentration in raw gas were also tested. Higher CO_2 concentration leads higher CO_2 recovery ratio. As shown in Figure 2, when CO_2 concentration is 32% in raw gas, CO_2 recovery ratio rises up to 1.6 times higher than one in original CO_2 concentration even the ratio in CO_2 concentration in raw gas is only around 1.4. This change is important because CO_2 concentration in blast furnace gas is thought to be increased in the future development. Injection of hydrogen to blast furnace is studied in the same project. If hydrogen is injected, CO_2 in the blast furnace gas will be increased because of operation conditions of blast furnace. By these results, recovery cost estimated to be 63% of original cost suggested.

3.Bench Scale Plant Examination

Base on laboratory PSA examination results, bench scale PSA plant called "ASCOA-3" (Advanced Separation system by Carbon Oxides Adsorption) was constructed as shown in Photo 2. The planned capacity of ASCOA-3 is 3 tons of CO_2 per day. The inner diameter and height of adsorption tower are 0.75m and 1.2m, respectively. Figure 3 shows the flow diagram of ASCOA-3. The actual blast furnace gas is compressed up to 300kPa, cools to 10 degree centigrade and removes water and sulfur compounds by adsorption with silica-alumina gel and activated carbon before feeding to PSA unit. At the first operation called RUN100, ASCOA-3 shows its enough ability. CO_2 recovery rises up to 4.2tons per day and CO_2 recovery ratio and purity exceeds the target as shown in Figure 4. At the second and third operation called RUN 200 and 300, various operation conditions were tested and actual power consumption was measured in these operations. Especially in RUN 300, CO_2 recovery rises up to 6.3 tons per day with shorter cycle time (225 seconds) and high CO_2 concentration in blest furnace gas (34%) even at low adsorption pressure (150kPa-absolute) as shown in Figure 5. From these results, recovery cost (63% of original one), which is suggested in laboratory PSA apparatus, is confirmed.

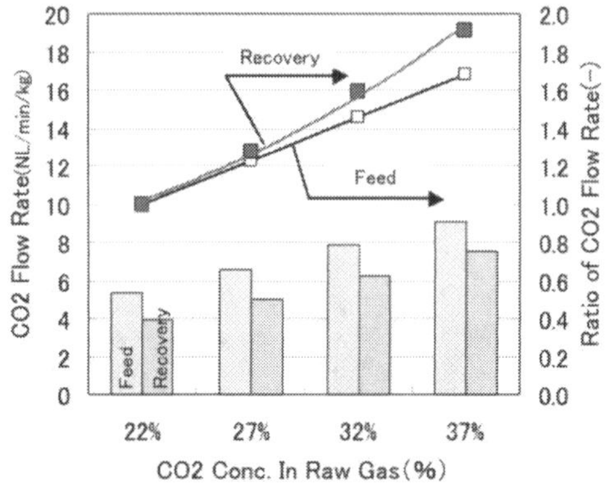

Figure 2 Influence of CO_2 Concentration in Raw Gas

Photo 2 Bird View of Bench Scale Plant (ASCOA-3)

One of the best feature of ASCOA-3 is that 9 resistance thermometers were installed in a adsorption tower as shown in Figure 6. They can measure the temperature of center, half radius and wall side positions of 3 heights. At adsorption step and rinsing step, rise of temperature which is caused by adsorption of carbon dioxide is observed from bottom to top as shown in Figure 7. On the other hands, they go down simultaneously at desorption step. It is easily known that the center position and the half radius position show the same temperature at any time. This phenomena are occurred by the following reasons. (i) a large heat capacity of the adsorbent, (ii) a small heat capacity of the gas, (iii) homogeneous gas flow

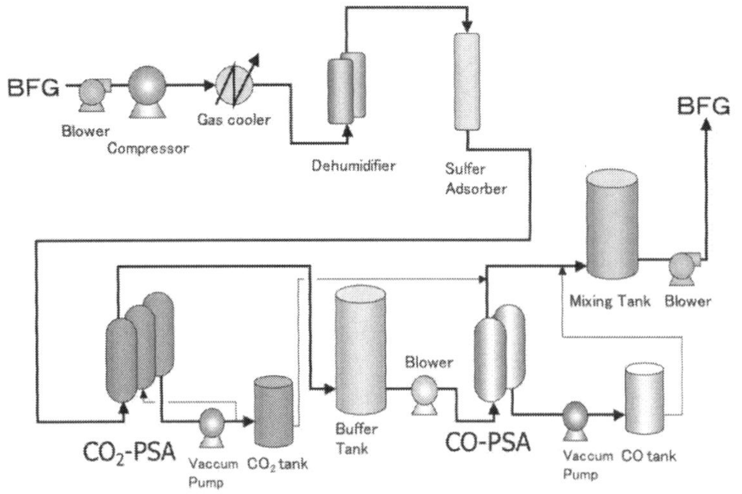

Figure 3 Flow Diagram of Bench Scale Plant (ASCOA-3)

caused by homogeneous packing of adsorbent. This result is very important. Because, if the adsorbent is packed homogeneously and gas flow is homogeneous, scale-up for the horizontal direction is easy and has no problem.

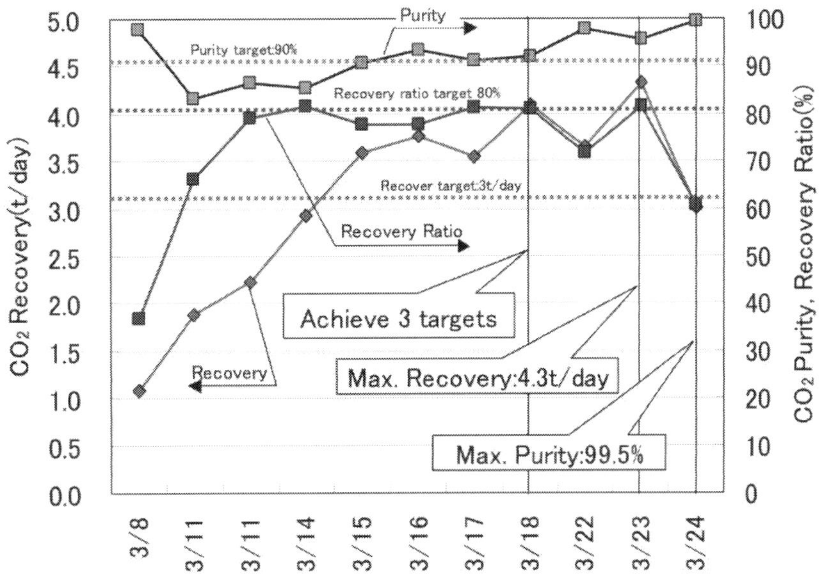

Figure 4 Operational Results of Bench Scale Plant (ASCOA-3)

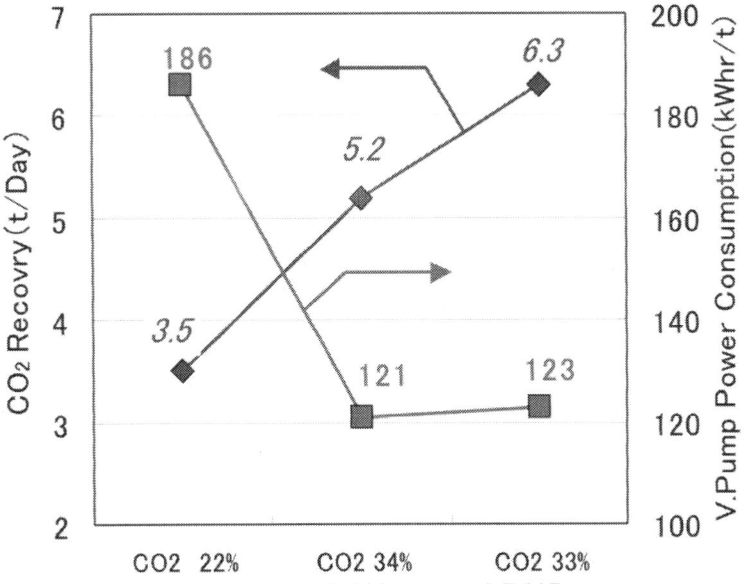

Figure 5 Influence of Cycle Time (C.T.) and CO_2 Concentration of Raw gas in Bench Scale Plant (ASCOA-3)

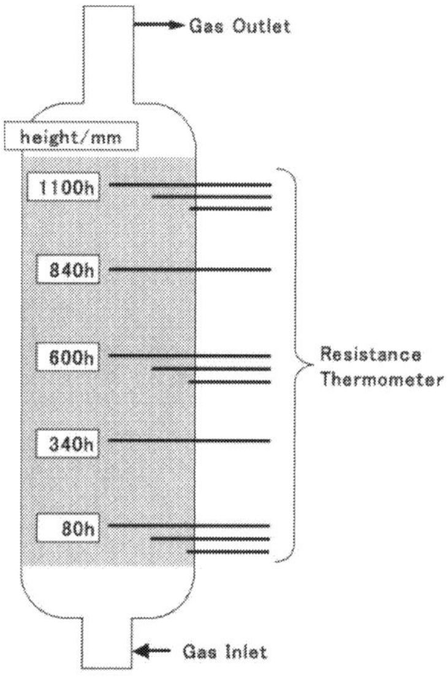

Figure 6 Resistance Thermometer Arrangement

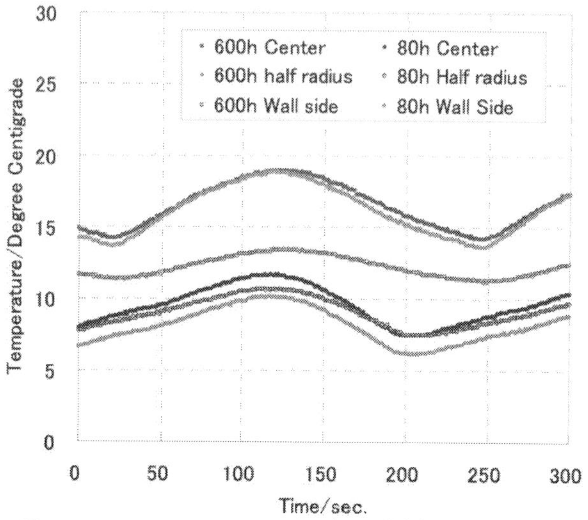

Figure 7 Temperature Distribution of Adsorption Tower

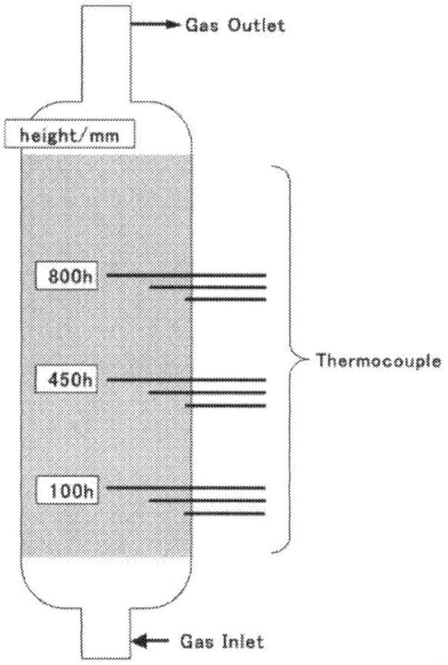

Figure 8 Thermocouple Arrangement

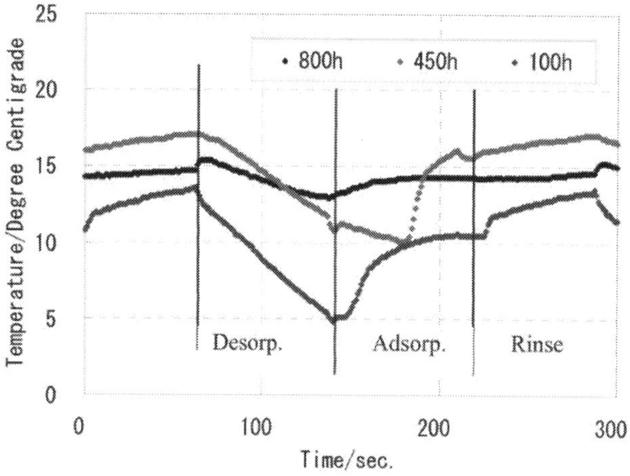

Figure 9 Temperature Distribution of Adsorption Tower with Thermocouples

Temperatures in Figure 7 were measured by resistance thermometers. A resistance thermometer is not easily broken but is lower in responsibility. Then the 9 thermocouples, which had high responsibility, were additionally installed in the adsorption tower as shown in Figure 8. Figure 9 shows the operational results of additional thermocouples. Adsorption, rinsing and desorption steps are clearly observed from temperature behavior. From this observation, adequate step time is easily known.

Those results were obtained at small PSA testing apparatus compared with commercial sized plant. The diameter and length of adsorption tower in commercial plant is presumed to be 6.5m and 20m, respectively. The volume of adsorption tower of commercial plant is near 1 thousand times larger than that of ASCOA-3. Especially, there are possibilities that the gas pressure and the velocity in the adsorption tower will not be uniform because the adsorption tower is enormously large as mentioned. The gas pressure distribution of lower face of adsorbent is calculated with "Fluent". Unexpectedly the results of calculation clearly show that pressure distribution is homogeneous and the maximum difference is only 0.03kPa as shown in Figure 10. This means there needs no special design to the adsorption tower from the viewpoint of pressure distribution.

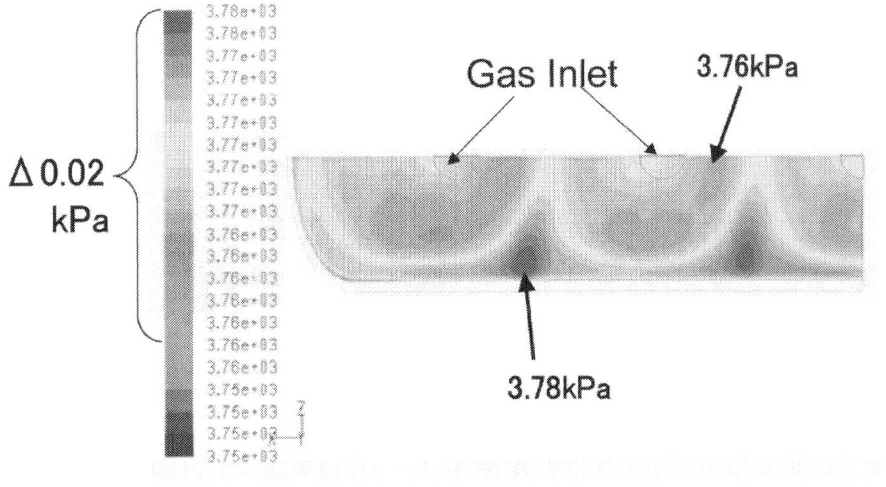

Figure 10 Calculation Results of Pressure Difference in Commercial Adsorption Tower

4.Conclusions

(1) The influence of cycle time is small and it can be reduced to 300 seconds.

(2) By increasing CO_2 concentration to 32%, CO_2 recovery rises up to 1.6 times higher than the original gas composition.

(3) Upper 2 results are also demonstrated in the bench scale plant called ASCOA-3.
(6.3 tons of CO_2 was recovered with the following conditions; Cycle Time = 225 seconds, CO_2 concentration of raw gas = 33%)

(4) At the same height, temperature moves at the homogeneous temperature. It shows the behavior of CO_2 adsorption and desorption are homogeneous and scale-up of horizontal direction is easy.

(5) Thermocouple clearly shows the adsorption, rinsing and desorption step.

(6) Simulation of pressure difference in commercial adsorption tower shows the pressure difference of the bottom of the adsorbent is homogeneous.

5. Acknowledgement

We are gratefully appreciated to NEDO (New Energy and Industrial Technology Development Organization), who is sponsor of the COURSE 50 project including this study.

6.References

(1) H.Saima, Y. Mogi, T. Haraoka, preprint of the 10th Conference of Fundamentals On Adsorption (2010)

(2) H.Saima, Y. Mogi, T. Haraoka, preprint of 6th Tokyo Conference on Advanced Catalytic Science and Technology (TOCAT 6,2010)

(331c) Post-Combustion CO2 Capture by Adsorption - a Theoretical Comparison of TSA, VSA and CSA

Gerhard D. Pirngruber[1], Damien Leinekugel-le-Cocq[2], Florent Guillou[3], Adrien Gomez[2] and Alain Favre[2], (1)Catalysis and Separation, IFPEN, Solaize, France, (2)IFPEN, Solaize, France, (3)IFP Energies Nouvelles, Solaize, France

Post-Combustion CO_2 Capture by Adsorption - a Theoretical Comparison of TSA, VSA and CSA

Gerhard D. Pirngruber, D. Leinekugel-le-Cocq, F. Guillou, A. Gomez, A. Favre

IFP Energies nouvelles, Rond point échangeur de Solaize, 69360 Solaize, France

The current state-of-the-art post-combustion capture technology is absorption of CO_2 by an aqueous monoethanolamine (MEA) solution. MEA has a very strong affinity to CO_2 and captures CO_2 efficiently even at low partial pressure, but the regeneration of the solvent by heating requires a large amount of energy. Moreover, degradation of the amine by oxygen in the flue gas causes corrosion problems and leads to a high net consumption of solvent (> 1.4 kg MEA/t CO_2 captured, i.e. > 4700 t per year for a 600 MW power plant). Therefore an intensive search for alternative solutions is going on [1].

It has been claimed that solid sorbents have the potential to significantly reduce the power consumption of CO_2 capture compared to MEA. The argument in favor of solid sorbents is that the combination of a lower heat of sorption, a higher CO_2 loading and a lower heat capacity of the solid compared to the aqueous solution of MEA lowers the energy consumption for regeneration. This reasoning is oversimplified because it does not take heat integration into account. A large fraction of the heat that is injected into the regenerator column of an amine absorption process can be recovered via a heat exchanger. Recovering heat from a solid is much more difficult than recovering heat from a liquid, which tilts the net energy balance in favor of absorption in the liquid phase. Berger and Bhown have recently carried out a purely theoretical analysis of the energy consumption of a TSA process [2]. Their calculations are only based on thermodynamic equilibrium properties of the solid. Two terms contribute to the energy consumption of a TSA: the sensible heat, required for raising the temperature of the solid from adsorption to regeneration temperature, and the latent heat, i.e. the heat of adsorption. There is a trade-off between the two terms. As a consequence, an absolute lower limit of the minimum energy consumption of a TSA process exists, which depends on the boundary conditions, like the partial pressure of CO_2 in the feed, the available temperature window, etc. In our contribution, detailed calculations of this minimum energy consumption have been carried out for two different TSA technologies: fixed bed and fluidized bed operation. Fixed bed operation is more favorable from a thermodynamic point of view, but heat transfer is a critical issue. Only the best case scenario (optimal sorbent, optimal operating conditions, optimistic assumptions) brings the energy consumption of the TSA process with solid sorbents down to the range achieved with demixing solvents (2.1 GJ/t [3]), i.e. the potential for breakthrough improvements is limited.

Another alternative is to use Pressure (or Vacuum) Swing Adsorption (PSA or VSA) technology in combination with solid sorbents. PSA exploits the difference of the partial pressure of CO_2 between the adsorption and the desorption step. In conventional PSAs desorption is carried out at near ambient pressure. Since the objective is to recover a stream of pure CO_2 (> 95%), i.e. the partial pressure of CO_2 under desorption conditions is close to 1 bar. The partial pressure of CO_2 in the feed must be higher than that. The molar concentration of CO_2 in flue gases is typically 15%; the pressure of the feed must, therefore, be at least 7 bar in order to have a positive partial pressure difference between adsorption and desorption. This strongly penalizes the cost and energy consumption of the capture process.[4] Moreover, a weak partial pressure difference between adsorption and desorption will lead to a low recovery of CO_2, i.e. an inefficient separation. An alternative option is to desorb CO_2 under vacuum, i.e. to run a VSA process. This also makes more sense from a point of view of adsorption thermodynamics because most CO_2 adsorbents that are considered for postcombustion CO_2 capture have very steep isotherms that reach a plateau at very low partial pressures. Lowering the desorption pressure brings about a larger gain in the delta loading of the adsorbent than increasing the adsorption pressure. Theoretically, the energy consumption of a VSA process is much lower than that of a TSA, but VSAs face serious technological and economic challenges. The desorption pressure in a VSA must be very low to achieve a high recovery of CO_2. At high vacuum, the volume flow strongly increases and reaches values that largely exceed the capacity of conventional vacuum pumps. As a consequence, several vacuum blower trains need to be installed in parallel. Moreover, several compression stages are needed to raise the pressure back to one atmosphere [4]. This strongly increases the capital cost of a VSA process and raises questions about its technical and economic feasibility.

We therefore propose a third alternative that avoids the shortcomings of TSA (energy wasted for heating and cooling the solid) and VSA (technological and economic handicap of regeneration under high vacuum). The idea is to perform regeneration isothermally, by using a purge gas. The purge gas strips CO_2 from the bed by an effect of dilution. We, therefore, call the process a CSA (Concentration Swing Adsorption). The purge gas must be easily separable from CO_2. Hence, water vapor is the most obvious choice. Water vapor at a moderate temperature levels is often abundantly available, sometimes as a waste product. In a CSA it can be effectively valorized, at minimum energy penalty for the power plant. A drawback of this mode of regeneration is, however, that it excludes the use of physisorbents, like zeolites and MOFs, because they will not able to adsorb CO_2 after regeneration in H_2O (adsorption of H_2O is too strong compared to adsorption of CO_2). Only basic sorbents, which are not inhibited by the presence of water in feed, can be chosen: immobilized amines, alkali carbonates, etc. The choice of the optimal operating conditions, the problems of pressure drop and the properties of the ideal adsorbents for a CSA process (adsorption thermodynamics and mass transfer properties) will be discussed in detail.

References

[1] Figueroa, J. D., Fout, T., Plasynski, S., McIlvried, H. and others, Advances in CO2 capture technology—The U.S. Department of Energy's Carbon Sequestration Program, Int.J.Greenhouse Gas Control, 2 (2008) 9-20.

[2] Berger, A. H. and Bhown, A. S., Comparing Physisorption and Chemisorption Solid Sorbents for use Separating CO2 from Flue Gas using Temperature Swing Adsorption, Energy Procedia, 4 (2011) 562-567.

[3] Raynal, L., Alix, P., Bouillon, P. A., Gomez, A. and others, The DMX™ process : an original solution for lowering the cost of post-combustion carbon capture, Energy Procedia, 4 (2011) 779-786.

[4] Ho, M. T., Allinson, G. W., and Wiley, D. E., Reducing the cost of CO2 capture from flue gases using pressure swing adsorption, Industrial & Engineering Chemistry Research, 47 (2008) 4883-4890.

(331d) Solid Sorbents As a Retrofit CO2 Capture Technology: 1 MW Pilot Update

Holly Krutka, Sharon Sjostrom, Travis Starns and Martin Dillon, ADA Environmental Solutions, Highlands Ranch, CO

ADA Environmental Solutions (ADA) is conducting a multi-year project to develop a post-combustion CO2 capture retrofit technology that utilizes solid sorbents in a temperature-swing adsorption process. This project is funded through the Department of Energy's National Energy Technology Laboratory's Innovations for Existing Plants (IEP) Program, EPRI, and several large power generation companies. The first step of this project was to identify and fully characterize a sorbent. The sorbent specific information was used as the basis to create a commercial-scale conceptual design. Multiple commercial designs were compared using techno-economic analysis. Once a commercial-scale design was selected it was reduced in scale to design a 1 MWe pilot. The key sorbent properties, processes considered, description of the commercial design, and the progress on the 1 MW pilot will be discussed. A majority of the project funds are being provided by the DOE National Energy Technology Laboratory's Innovations for Existing Plants (IEP) Program.

(331e) Techno-Economic Study of Ca-Looping Processes for Carbon Capture From Cement Plants

Dursun Can Ozcan, Hyungwoong Ahn and Stefano Brandani, SCCS - School of Engineering, University of Edinburgh, Edinburgh, United Kingdom

Cement is a key construction material, and gains on demand of cement are expected with a trend around six percent in 2012, reaching 3.78 billion tons [1]. The cement industry is one of the largest CO_2 emissions source, accounting for 2 Gt CO_2/year in 2007 [2]. The excessive CO_2 emissions of a cement plant are not only related to high energy demand but also the emission due to raw material calcination. A more efficient thermal management can reduce the carbon emissions from cement plants but the extent of capture is very limited since the heat required for its endothermic reaction has to be provided. Therefore, it is essential to deploy carbon capture technology on cement processes in order to achieve the target to reduce its CO_2emissions by up to more than 90%.

We have investigated process integration of a Ca-looping process with cement plants aiming to select optimal feed stream, determine operating conditions of the circulating fluidised bed process, and estimate CAPEX and OPEX. A model for a carbonator, where hydrodynamics in fast fluidisation, reaction kinetics as well as active fraction of the sorbent are taken into account, has been implemented in Matlab and incorporated into a cement process flowsheet using Honeywell's Unisim process simulator. The entire process simulation also includes a steam cycle supplementary to the Ca-looping process for heat recovery and an air separation unit for 95% oxygen supply to the calciner. It enables us to evaluate the incremental cost for carbon capture precisely. Different post-capture processes such as amine and membrane have also been investigated for the carbon capture for a cement plant, and the comparison of these processes with Ca-looping process can be conducted.

We would like to acknowledge the financial support from the Turkish Ministry of Education and the Engineering and Physical Sciences Research Council under grant EP/F034520/1 (Science & Innovation Award).

[1] CW Group *Global Cement Volume Forecast Report*, 2012.

[2] IEA, *Energy Technology Perspectives 2010*; International Energy Agency: Paris, 2010.

[3] European Cement Research Academy GmbH, *Carbon Capture Technology - Options and Potentials for the Cement Industry*, Technical Report, TR 044/2007, July 2007.

(331f) Carbon Dioxide Capture From Flue Gas by High Feed Throughput Pressure Swing Adsorption Processes

James A. Ritter[1], Atikur Rahman[2] and Armin D. Ebner[1], (1)Department of Chemical Engineering, University of South Carolina, Columbia, SC, (2)Chemical Engineering, University of South Carolina, Columbia, SC

A number of research groups from around the world have been developing PSA cycles for capturing CO_2 from flue gas by pressure swing adsorption (PSA). All of the published results so far suffer from low CO_2 purity, low CO_2 recovery, or low feed throughput. In other words, the PSA cycles either do not meet the US DOE CO_2 purity and recovery constraints, or the beds are so large that the capitol costs become prohibitive. It is submitted here that this was largely due to inefficient PSA cycle designs.

The objective of this presentation is to show that it is indeed possible to develop a high feed throughput PSA cycle that meets the US DOE constraints on CO_2 purity and recovery. In fact, a number of attractive PSA cycles have been developed that appear to be economical from both operating and capital costs. These PSA cycles utilize a commercial adsorbent and a unique process flow sheet that obviates the need to process water vapor through the PSA columns. An important, unanswered question is: what is the maximum feed throughput that can be achieved while meeting the US DOE constraints? In an attempt to answer this question, the latest results from Ritter and his group will be presented.

(331h) Optimization of a 4 Step PSA Cycle for CO2 Capture From Post-Combustion Flue Gas

Reza Haghpanah[1], Aniruddha Majumder[1], Ricky Nilam[1], Arvind Rajendran[1], Mohammad Aman Ullah[1], Iftekar A. Karimi[2] and S. Farooq[2], (1)School of Chemical and Biomedical Engineering, Nanyang Technological University, Singapore, Singapore, (2)Department of Chemical and Biomolecular Engineering, National University of Singapore, Singapore, Singapore

Carbon dioxide capture and concentration from post combustion flue gas is a widely studied contemporary topic. Naturally, the effectiveness of cyclic adsorption processes, such as pressure swing adsorption (PSA), for this important application has been investigated in many recent publications. In these studies, the lower operating pressure varied from 0.5 – 0.005 bar. Similarly, the higher operating pressure varied in the range 1 – 10 bar.

Attaining high purity and recovery (typically 90% and above for both) has the focus of these studies and a variety of cycles differing in number of steps and operations have been used. Energy penalty is another important performance indicator for evaluating a carbon capture process. Consequently, choice of feed pressure and evacuation level are important decisions that affect energy penalty, which have not been critically analyzed. Moreover, unlike a steady state process, productivity is another performance variable that poses a design challenge. Proper design of a cyclic adsorption process for carbon capture is, therefore, a multi-objective optimization problem that requires finding the operating conditions to minimize energy penalty and maximize productivity while meeting the stringent purity-recovery constraints.

In the proposed talk, we will report the development of a comprehensive non-isothermal, non-isobaric PSA process simulator that is very fast and robust to allow full stochastic optimization without resorting to any model reduction. Using a simple 4-step cycle, 13X zeolite and dry flue gas, the advantages and limitations of pressure swing and vacuum swing cycles will be discussed in detail.

(336a) Nanostructured Block Copolymer Membranes for Biofuel Production

A. Evren Ozcam, University California, Berkeley, Berkeley, CA, Ashish Jha, Chemical and Biomolecular Engineering, University California, Berkeley, Berkeley, CA and **Nitash P Balsara**, Chemical and Biomolecular Engineering, University of California, Berkeley, Berkeley, CA

We report on the effect of changing nanoscale morphology on pervaporation of alcohol/water mixtures through block copolymer membranes. Experiments were conducted using polystyrene-*b*-polybutadiene-*b*-polystyrene (SBS) and polystyrene-*b*-polydimethylsiloxane-*b*-polystyrene copolymers with polybutadiene (PB) and polydimethylsiloxane (PDMS) as the alcohol transporting blocks. The morphology of the membranes is governed by the composition of the block copolymer while the size of the domains is governed by the molecular weight of the block copolymer. The volume fraction of the transporting PB and PDMS microphases was varied from 0.63 to 0.93 and 0.59 to 0.83 for PB (f_{PB}) and PDMS (f_{PDMS}) microphases, respectively. The overall molecular weight of the copolymers, M_n, was varied from 34 to 207 kg mol^{-1}. Normalized ethanol permeability through the membrane, P_E/f_{PB}, and the ethanol selectivity, a_{EW}, increase with increasing M_n for SBS, but P_E/f_{PDMS} and a_{EW} are not a strong function of M_n for SDS based membranes. Master curves are obtained when all of the permeation data are plotted in the $P_E/f_{transporting}$, $P_W/f_{transporting}$, versus a_{EW} format.

(336b) Diffusive Water Transport: Relating Hydraulic Permeability to the Apparent Water Diffusion Coefficient in Water-Swollen Polymers

Geoffrey M. Geise, Deptartment of Chemical Engineering & The Texas Materials Institute, The University of Texas at Austin, Austin, TX, Benny D. Freeman, Department of Chemical Engineering, The University of Texas at Austin, Austin, TX and Donald R. Paul, Chemical Engineering, University of Texas at Austin, Austin, TX

A dense, nonporous polymer film's hydraulic water permeability, which is the water flux through the film normalized by the film's thickness and the pressure difference applied across the film, is a material property that can be measured experimentally to describe water transport. Permeation of water through dense, non-porous polymer membranes, such as those used in desalination processes (e.g., reverse osmosis (RO)), occurs via a solution-diffusion process where water first sorbs into the polymer at the membrane's upstream face and then diffuses down a chemical potential and water concentration gradient before desorbing at the downstream face of the membrane, and the diffusion process is the rate-determining step. The chemical potential and water concentration gradients in these nonporous polymers are generated thermodynamically, and they are related to the pressure difference applied across the polymer film. In order to relate the measured hydraulic water permeability to the apparent diffusion coefficient of water in the polymer, a constitutive equation is required. The traditional water transport equation, developed for RO membranes, represents one such constitutive equation, but this equation is only strictly applicable to polymers that sorb relatively little water. When this limited form of the constitutive equation, which is commonly reported in the literature, is applied to polymers that sorb considerable amounts of water (e.g., hydrogels), the calculated apparent water diffusion coefficients can exceed the self-diffusion coefficient of water. In several cases in the literature, this observation has been used to incorrectly conclude that water transport occurs via a pore-flow mechanism in dense polymer films.

The traditional and simplified water transport equation does not account for the convective frame of reference terms in Fick's law and thermodynamic non-ideality of the swollen polymer system. When these effects are appropriately accounted for, the apparent water diffusion coefficients in highly water swollen polymers, calculated from hydraulic water permeability measurements, are satisfactorily below the self-diffusion coefficient of water indicating diffusive, not pore flow, transport of water. A review of the full constitutive equation for water transport will be presented, and its use will be demonstrated by analyzing several case studies including data from the literature.

(336c) Thermodynamic Analysis of the Permeability of Gases in Glassy Polymers

Matteo Minelli, University of Bologna, Bologna, Italy, Aweke E. Gemeda, Dipartimento di Ingegneria Chimica, Mineraria e delle Tecnologie Ambientali (DICMA), University of Bologna, Bologna, Italy and Giulio C. Sarti, Chemical Engineering, Mining and Environmental Technology, University of Bologna, Bologna, Italy

Thermodynamic analysis of the permeability of gases in glassy polymers

Matteo Minelli, Aweke E. Gemeda, Giulio C. Sarti

Dipartimento di Ingegneria Chimica, Mineraria e delle Tecnologie Ambientali (DICMA)

Alma Mater Studiorum - Università di Bologna, via Terracini 28, 40131, Bologna, Italy.

Permeability and average diffusivity of gases and vapors in glassy membranes are known to vary with upstream pressure. For that dependence the well known Dual Mode Sorption (DMS) model offers valuable and effective expressions which have been successfully used for that purpose over the years. It is known that in several cases the two parameters entering such expressions, i.e. the diffusivities of the dissolved species, D_D, and of the adsorbed species, D_H, can be retrieved from steady state permeation data and then can be used also for transient transport problems and for time-lag calculations. The DMS expressions, however, are known to hold only in the low pressure range, until the so called plasticization phenomena are not effective; to that aim further parameters would be required to modify the general expressions. In addition, the ability of the DMS model to describe well both steady state and transient behaviors is often not as good as desirable and is sometimes questionable.

Revisiting the permeation behavior of gases and vapors in glassy polymers we have seen that an alternative procedure is possible and preferable, to our opinion. Indeed, by considering another notorious expression of the diffusive flux as the product of the negative mobility (or thermodynamic diffusion coefficient) and chemical potential gradient we obtain a rather simple and general expression containing at most two adjustable parameters: the gas mobility at infinite dilution and the coefficient characterizing the concentration dependence of mobility itself. The two parameters can be retrieved form steady state permeation data at different pressure, as for the DMS model. The new expression with at most two adjustable parameters allows us to represent rather well both the permeability behavior at all pressures, also in the high pressure range where plasticization is present, as well as the transient behavior.

The effectiveness of the new expression proposed has been checked by using a variety of permeation and sorption data including CO_2 transport in polycarbonate, in PPO, in PMMA as well as the permeability of different common gases in glassy polymers frequently encountered in different applications.

(336d) Non-Equilibrium Sorption and Anomalous Diffusion of Water in Glassy Polymers

Eric M. Davis and Yossef A. Elabd, Chemical and Biological Engineering, Drexel University, Philadelphia, PA

For the development of advanced barrier plastics, a more thorough understanding of the impact of moisture on glassy polymers is of great interest. In this study, the sorption and diffusion of water in two glassy polymers, poly(methyl methacrylate) (PMMA) and poly(styrene) (PS), was measured over a wide range of water vapor activities and temperatures using several experimental techniques, including quartz spring microbalance (QSM) and *in situ* time-resolved Fourier transform infrared-attenuated total reflectance (FTIR-ATR) spectroscopy. At temperatures below the glass transition, the same type of anomalous diffusion was observed in both polymers. This phenomenon is the result of diffusion and polymer relaxation occurring on a similar time scale, where the polymer relaxes in response to a sorption-induced stress. With FTIR-ATR spectroscopy, both water diffusion and polymer relaxation were measured simultaneously through unique and separate molecular signatures for the polymer backbone and water. These results will be presented along with the results from a diffusion-relaxation model that captures both the diffusion coefficient of water and the stress-induced relaxation time of the polymer. Additionally, the non-equilibrium or pseudo-equilibrium sorption of water in both polymers was compared, where the uptake of water in PS is an order of magnitude lower than PMMA. Interestingly, FTIR-ATR spectroscopy reveals that the states of water in PS are also significantly different than PMMA. This data was compared along with temperature-dependent diffusivities to understand the impact of the states of water on sorption and diffusion.

(336e) Transport Fundamentals of Micro-Encapsulated Small Molecules Embedded in a Polymer Matrix: Model Predictions and Experiment Findings

Ian J. Drake, Core Research, The Dow Chemical Company, Spring House, PA

Micro-encapsulation technology is recognized as a sizable market opportunity for a number of specialty chemical businesses supporting agricultural, adhesives/epoxy, protective coatings, food, and personal care markets. A synthetic approach that leads to the encapsulation of actives/functional molecules by establishing core-shell morphology is often speculated to addresses the needs of these market opportunities. Particles of a core-shell morphology could potentially enable the controlled delivery of a number of actives/functional molecules including bioactive small molecules (biocides, insecticides, herbicides, etc.), fragrances, drugs, dyes/pigments, coalescing agents, reactive intermediates (hardeners, accelerators and catalysts for epoxy and other 2K reactive systems) and photoactive agents.

While each application is different, common underlying physics connect the fundamental performance drivers. In the broadest sense, these diverse applications can be described as a micro-encapsulated small molecule distributed in some type of matrix continuum (air, water/solvent, or polymer composite). The nature of this work is to demonstrate a detailed Fickian diffusion model that describes the case for simultaneous diffusion of a small molecule out of an embedded encapsulated particle through a polymer film to an exterior free surface. Case studies were examined and the model was validated by simulating experimental data. Using ATR-IR and a uniquely designed microfluidic weathering cell, the diffusion coefficient and surface loss factor of the small molecule were determined for various polymer properties (composition and T_g). It will be shown that the model can be extended to extract performance information which is challenging to observe experimentally. The simulation output includes (1) the distribution of active small molecule through the film at any given time, (2) the surface concentration of small molecule at any time point and (3) the "replenishment" rate (or surface flux) of the small molecule to the surface. This type of simulation is particularly useful to understanding the migration of bioactives, fragrances, drugs and coalescing agents that might be distributed in a polymer composite film where the initial concentration of small molecule in the film is much smaller than 1%. Since many of the applications of interest require formulations where the particle size cannot be larger than 50 um, it will be shown that manipulating the diffusion coefficient of the active through choice of shell wall chemistry is critical to enabling a successful application.

(336f) Tube Diameter of Oriented and Stretched Chains Studied with Isoconfigurational Averaging

Jian Qin, Jungseob So and Scott Milner, Chemical Engineering, Pennsylvania State University, State College, PA

Tube is the key concept for understanding the flow properties of polymers. We investigate how the tube diameter varies with deformation in polymer melts using the isoconfigurational averaging method [1]. To conduct isoconfigurational averaging, we start with a well equilibrated (deformed) configuration, run molecular dynamics (MD) simulation multiple times, each with different initial velocities, and record bead positions at different times and along different MD trajectories. The bead positions form a cloud showing the shape, range and strength of the tube confinement, which enables us to study the tube diameter and other statistical properties of tube. We applied the isoconfigurational averaging to deformed ring and linear polymers simulated with the bead-spring model. Ring polymers are equilibrated by invoking various molecular rebridging moves in Monte Carlo simulations, and are subjected to a sequence of uniaxial compressions and stretches. The deformed configurations are allowed to re-equilibrate with the rebridging moves turned off, and then are analyzed using isoconfigurational averaging. We investigate the variation of the tube primitive path and the tube diameter with deformation, and compared the results to the predictions of affine deformation theory. For linear polymers, the force-extension law and the dependence of the tube diameter on stretching are obtained.

[1] Windsor Bisbee, Jian Qin, and Scott T. Milner, "Finding the Tube with Isoconfigurational Averaging", Macromolecules, 2011, 44(22):8972-898

(336g) Nanoconfined Self-Diffusion of Poly(isobutyl methacrylate)

Joshua M. Katzenstein, Dustin W. Janes, Haley E. Hocker, Justin K. Chandler and Christopher J. Ellison, Department of Chemical Engineering, University of Texas at Austin, Austin, TX

Even though the physics of nanoconfined polymers have been extensively studied for years with respect to properties such as glass transition temperature (T_g), viscosity, elastic modulus and others, diffusion of polymer chains along confining interfaces has not been widely studied. This is likely due to the lack of experimental techniques available for these measurements. Here we present a study of self-diffusion of poly(isobutyl methacrylate) (PiBMA) nanoconfined to thin films via a technique based on fluorescence recovery after patterned photobleaching (FRAPP). PiBMA is an ideal polymer for this study because it exhibits a film thickness-independent T_g down to $7R_g$, as measured by spectroscopic ellipsometry for both fluorophore and non-fluorophore containing polymers. Since the diffusion coefficient of polymers depends strongly on the proximity of diffusion temperature to T_g, this attribute allows a straightforward measure of nanoconfined diffusion without superimposed influence from T_g related nanoconfinement effects.

This FRAPP approach uses an epifluorescence microscope that allows for direct, *in situ*, visualization of polymer diffusion over several periods of a photobleached array. This visualization approach is more robust compared to measuring fluorescence intensity alone and also significantly increases the experimental throughput. The self-diffusion of PiBMA parallel to the confining interfaces was found to be film thickness independent down to ~$14R_g$, where R_g is the radius of gyration. The diffusion coefficient agreed well with the bulk diffusion coefficient predicted by the Rouse model for unentangled polymer diffusion. Furthermore, the diffusion coefficient is independent of the location of the fluorophore on the polymer backbone.

(355a) Carbon Molecular Sieve Membranes for N_2/CH_4 Separation

Xue Ning, School of Chemical & Biomolecular Engineering, Georgia Institute of Technology, Atlanta, GA and William J. Koros, School of Chemical & Biomolecular Engineering, Georgia Institute of Technology, Atlanta, GA

Many natural gas reserves are contaminated by a high content of nitrogen which has to be removed to meet pipeline specifications. Currently available nitrogen rejection technologies include cryogenic distillation, pressure swing adsorption (PSA), and lean oil absorption. Polymer-based membrane separation for N_2/CH_4 has not been pursued widely because of the small size difference between N_2 and CH_4 and the competing effects of diffusion and sorption. These challenges lead to unattractive permselectivity using traditional polymeric materials due to the broad distribution of segmental motions of polymer chains. Carbon Molecular Sieve (CMS) membranes overcome this problem by providing a hyper-rigid structure and distinctive bimodal pore size distribution.

Different pyrolysis conditions for Matrimid® polyimide and resultant separation performance of CMS membranes were studied in this work. It was shown that N_2/CH_4 selectivity was highly improved with increased pyrolysis temperature. Sorption tests were performed to deconvolute permeability into diffusion coefficient and sorption coefficient. The temperature effects on permeability, sorption coefficient and diffusion coefficient were also studied.

(355b) High Performance Carbon Molecular Sieves (CMS) Hollow Fiber Membranes for Natural Gas Separation

Nitesh Bhuwania, School of Chemical and Biomolecular Engineering, Georgia Institute of Technology, Atlanta, GA and William J. Koros, School of Chemical & Biomolecular Engineering, Georgia Institute of Technology, Atlanta, GA

High Performance Carbon Molecular Sieves (CMS) Hollow Fiber Membranes for Natural Gas Separation

Nitesh Bhuwania, William J. Koros

Nitesh.Bhuwania@chbe.gatech.edu

bill.koros@chbe.gatech.edu

School of Chemical and Biomolecular Engineering

Georgia Institute of Technology, Atlanta, GA

CMS membranes show the enhanced separation capability for natural gas purification (processing of CO_2 from CH_4 stream). For industrial application, it is important to achieve this high performance in hollow fiber membrane configuration. Previous studies in CMS hollow fibers have shown the ability to achieve the good separation efficiency (i.e. selectivity) for CO_2/CH_4 but with a drop in transport flux in comparison to polymer precursor fibers. This is caused by the densification of porous morphology in CMS fibers, termed as"sub-structure collapse".

CMS membranes are formed by heating the polymer precursors at higher temperatures above its decomposition point. The phenomenon of sub-structure collapse occurs during the thermal transition of polymer, which begins at its glass transition temperature (T_g). Therefore in our study, we look at stabilizing the precursors before they undergo the thermal decomposition. Commercial available "Matrimid®" and custom-synthesized 6FDA-BPDA/DAM polyimides were used as the precursor platform for our research.

The impact of pre-pyrolysis stabilization on fiber morphology and separation performance for final CMS fiber membrane is significant. Our technique not only show attractive separation performance but also overcomes some key challenges during the processing of CMS membranes. We will go in detail of our method during the talk.

Evaluation of Facilitated Transport Membrane for Olefin/Paraffin Separation

Majid Keyvani, LyondellBasell

Ali Hamza, and Glenn Towe, Imtex Membranes Corporation

Abstract:

Currently, the separation of olefins (ethylene, propylene) from paraffins (ethane, propane) on a commercial scale is accomplished almost exclusively by cryogenic distillation in petrochemical industries and is considered highly energy intensive. Consequently, there is an enormous economic incentive to explore alternative separation technologies with lower energy consumption and an opportunity to achieve related reductions in environmental impact from air pollutant and greenhouse gas (GHG) emissions.

Several attempts have been made in the past to develop alternative separation technologies for this application, including significant work on facilitated transport membranes. One of the main unresolved challenges was membrane instability over time. Imtex Membranes Corporation has developed a membrane that has shown performance stability over thousands of hours of operation on both coupons and spiral wound test units. Imtex membranes are based on chitosan, a polysaccharide material, and silver nitrate as a facilitation agent.

LyondellBasell and Imtex Membrane Corporation have set up a collaborative effort to evaluate the performance of Imtex membrane in the olefin/paraffin separation, namely C2 and C3 splitter applications.

The performance results were very encouraging for C2 and C3 splitters as greater than 99.5% purity and 90% recovery were achieved in both cases. Based on the membrane performance results, we believe that the technology presents potential opportunities in upgrading refinery grade propylene to polymer grade propylene as the membrane was evaluated under operating conditions comparable to those of commercial plants.

There is also a promising potential for butene/butane separations where distillation separations require large numbers of trays and extremely high reflux ratios, or are difficult to achieve at all. Applying Imtex membranes in C4 separation can have a very positive impact on the economics of olefin metathesis, an increasingly important source of propylene.

Introduction and Background:

A recent Department of Energy (DOE) study [1] indicated distillation operations account for 95 percent of the total separation energy used in the refinery and chemical processing industries with 40,000 distillation columns operating in more than 200 different processes. Another study stated that olefin production is the single most energy consuming process in the chemical industry. The breakdown of the specific energy consumption shows that 30 percent of the total yearly energy consumption is associated with separation processes and 80 percent of that energy is utilized for olefin/paraffin separation [2]. This large energy requirement provides incentives to improve and enhance the energy efficiency in olefin/paraffin separation.

In light of the importance of olefin/paraffin separations, and because of the energy cost associated with distillation, alternative technologies continue to be the focus of industry studies. Some review literature [3, 4] describes the current status of research. It cites reactive absorption where silver or copper salts are used to extract the olefin from bulk gas in an absorber/stripper type process as promising new techniques. Union Carbide and BP/Stone and Webster took this type of process to the pilot plant stage. Other processes that were evaluated include extractive distillation and molecular sieves, such as UOP SORBUTENE technology. Despite all the reported studies, however, a viable alternative to conventional distillation has yet to be developed.

One of the most comprehensive research efforts in this subject was carried out by a DOE sponsored program led by Membrane Technology Research, Inc. (MTR) partner SRI International, ABB Lummus Global [5]. The study concluded that silver salt-based facilitated transport membranes are not stable even in the presence of ideal olefin/paraffin mixtures. Silver tetrafluoroborate (AgBF4) was used with two polymer matrices: polyether-polyamide block copolymer, poly ethylene oxide (PEO), poly vinyl pyrrolidone (PVP), and poly1-trimethylsilyl-1-propyne (PTMSP). However, regardless of the polymer matrix used, decline in membrane performance occurred. They attributed the issue with an unrecognized phenomenon labeled as " olefin conditioning" implying that the very species targeted for separation by the carrier membranes (olefins) are a source of membrane instability.

Imtex Membranes Corporation successfully developed a synthetic membrane with a viable alternative to distillation in separating olefins from their respective mixtures with paraffin. The patented [6] chitosan based Imtex membrane addresses the shortcomings of the aforementioned alternative approaches, including the avoidance of the "olefin conditioning" phenomenon experienced by MTR study.

Due to the presence of hydroxyl (-OH) and the amino ($-NH_2$) groups, Chitosan is a very hydrophilic and chelating polymer with excellent film-forming characteristics and has an excellent adsorption capacity for a number of metal ions such as the Silver cation. As presented here, excellent olefin selectivities and permeabilities were obtained when using chitosan in olefin/paraffin separation tests. The facilitation mechanism of the Imtex membrane is presumed to be through the olefin molecules hopping from one silver cation site to another as well as through the expected mobility of the olefin-silver ion complex within the aqueous phase of the membrane. The membrane structure can be balanced by the degree of deacetylation as well as by the silver loading level utilized. These aspects make chitosan an excellent base membrane for facilitated transport membranes

LyondellBasell and Imtex Membrane Corporation have set up a collaborative effort to evaluate the performance of Imtex membrane in C2 and C3 splitter applications.

In this paper we report performance data for propylene/propane and ethylene/ethane separation. Promising preliminary results for butenes/butanes are also presented.

Experimental:

The Imtex membrane uses a flat sheet composite structure based on Chitosan material and AgNO3 as a facilitation agent. Chitosan is a linear polysaccharide produced commercially by deacetylation of chitin, which is the structural element in the exoskeleton of crustaceans (such as crabs and shrimp) and other arthropods. The membrane units used for these studies included both small coupons as well as spiral wound elements. Each test coupon had a surface area of approximately 12.6 cm^2 while the spiral wound elements used in this study had each a nominal surface area of 600 cm^2. The chitosan layer thickness is estimated to be 7 micron. The tests were conducted with various ethylene/ethane, propylene/propane and butane/butane feed compositions. The feed pressure ranged from 40 psig to 360 psig.

Below is a picture of the front panel (flow lines and valves) of the Imtex test Apparatus used at LyondellBasell also showing two housings of spiral wound elements.

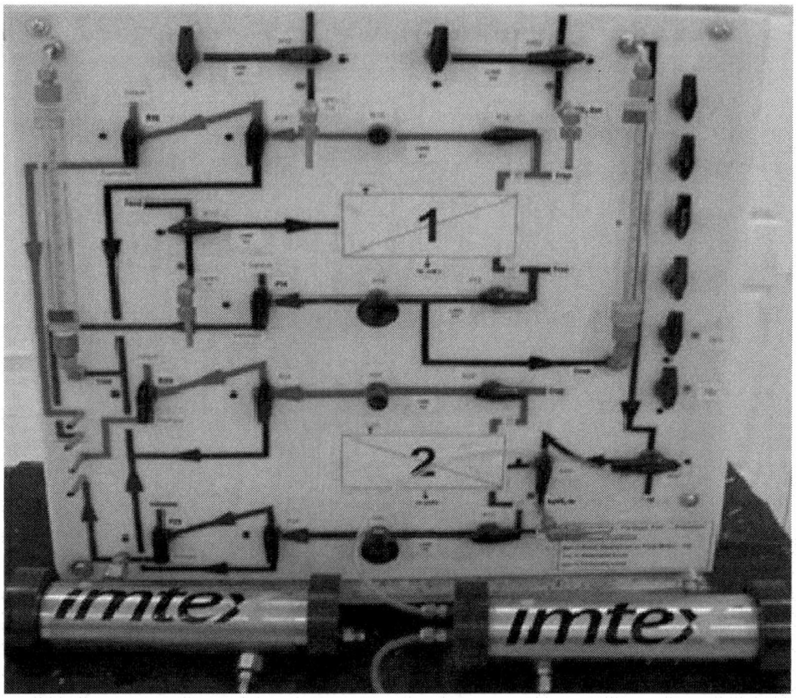

Results and Discussions:

Typical feed compositions to C2 and C3 splitters are 80/20 ethylene/ethane and propylene/propane. Using synthetic gas mixtures and operating at 60 psig feed pressure, selectivities greater than 100 were achieved using a spiral wound module made from first generation Imtex membrane LF1. This selectivity is higher than reported values in the literature for polymeric membranes (Figure 1) but not as good when compared with the reported data [7] for facilitated transport type membranes (Figure 2).

Figure 1. Membrane Performance Curve for Ethylene/Ethane

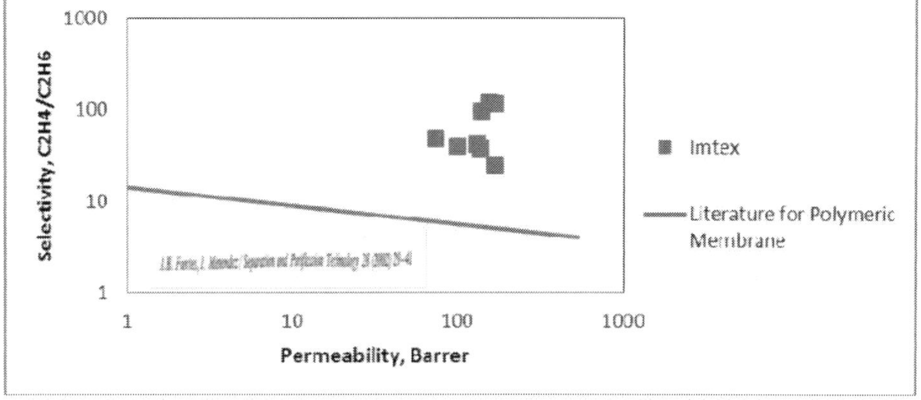

Figure 2-Imtex Membrane Performance against Literature (Facilitated Transport) for C2H4/C2H6

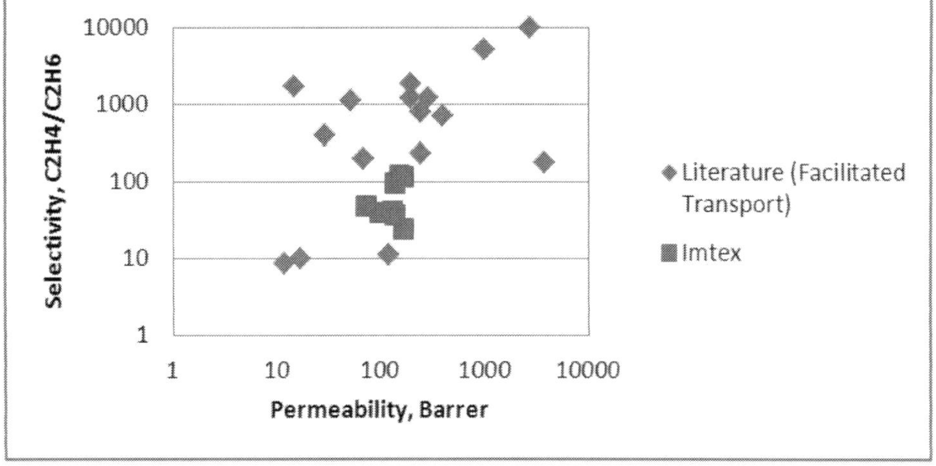

The ideal selectivity was calculated based on feed concentration instead of averaging feed and retentate concentrations:

Selectivity $\gamma = (x_{C2=} / x_{C2})_{perm} / (x_{C2=} / x_{C2})_{feed}$

Where x = mole fraction

The permeability was calculated based on the classic Fick's diffusion law. The STP (standard pressure, temperature) conditions defined as atmospheric and 68 °F and measured permeate flow rate was used as the flux (not total feed rate):

$$J = \frac{Perm(p_0 - p_l)}{l}$$

Where c= concentration, p = partial pressure, l= membrane thickness, J is flux, and Perm is permeability.

The excellent separation was seen over a wide range of feed composition. Figure 3 shows the performance of Imtex membrane LF1 as a function of ethylene content in the ethylene/ethane feed mixture.

Figure 3. Membrane Performance versus Feed Composition
Feed: Ethylene/Ethane Mixture at 66 psig and room temperature
Small spiral wound element with ~ 0.05 m² Imtex LF1 membrane

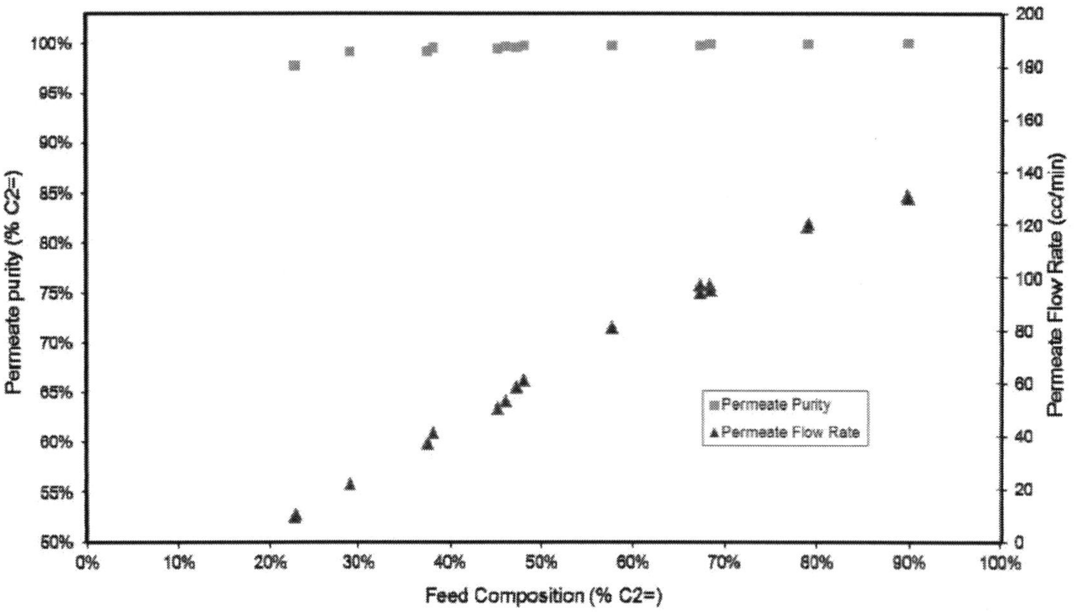

By increasing the feed pressure from 60 psig to 120 psig, the flux through the membrane almost doubled at the same product purity. Figure 4 shows performance (concentration of ethane in permeate stream) at varying pressures over time:

Figure 4- Ethylene/Ethane 80/20 Feed at Different Pressures:
Doubling Permeate Rate by increasing pressure while maintaining purity

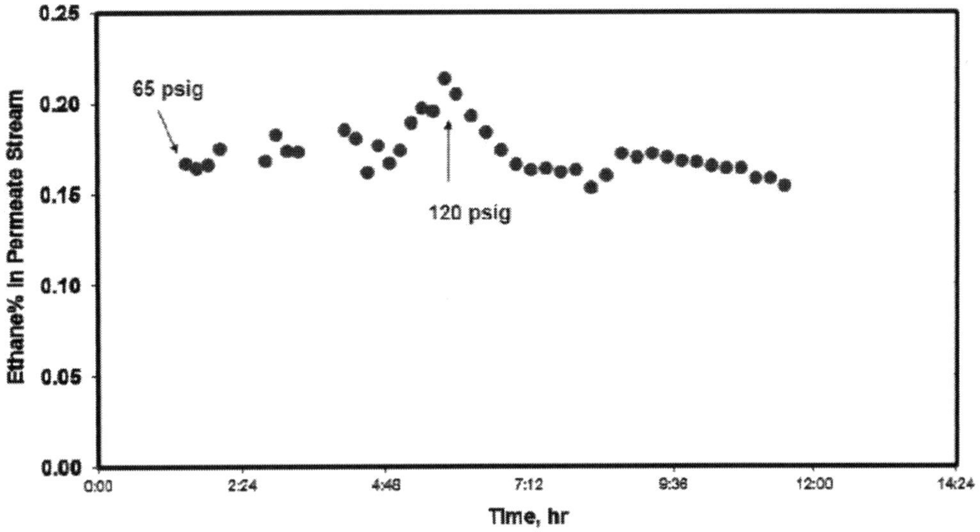

However, in spite of promising results for the C2 separation, we believe that the application for the C2 splitter faces many technical challenges, namely:

- C2 splitter is operated at cryogenic conditions and Imtex membrane has to be operated at ambient temperature (up to 50 $^{\circ}$ C).
- Permeate stream picks up water as it goes through the membrane matrix (has to be dried prior to returning back to column).
- There are acetylenic compounds present in the feed stream (e.g. acetylene, methyl acetylene (MA) and propadiene (PD), also known as MAPD). Ag complex strength in the membrane can be ranked strongest to weakest as followed PD>acetylene, methyl acetylene> ethylene> propylene. Thus any MAPD in the feed could end up with the olefin product on the permeate side.

On the other hand, the C3 splitter application does not have to deal with many of the above issues. Particularly, the membrane is best suited in upgrading the refinery grade propylene to a polymer grade.

The performance of Imtex membrane looks more favorable as compared against the published data on the application of facilitated membrane for propylene-propane separation. Figure 5 presents this comparison where a study was performed with 80/20 gas mixture of propylene/propane. Figure 6 shows the effect of feed pressure on the permeate stream purity and the flux:

Figure 5- Imtex membrane Performance against Literature (Facilitated Transport) for C3H6/C3H8

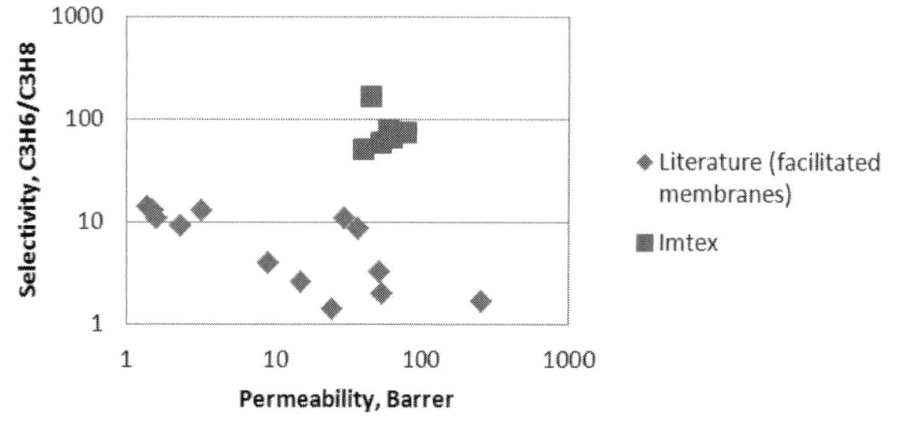

Figure 6- Propylene/Propane 80/20 Feed at Different Pressures:
Doubling permeate rate by increasing pessure while maintaing purity

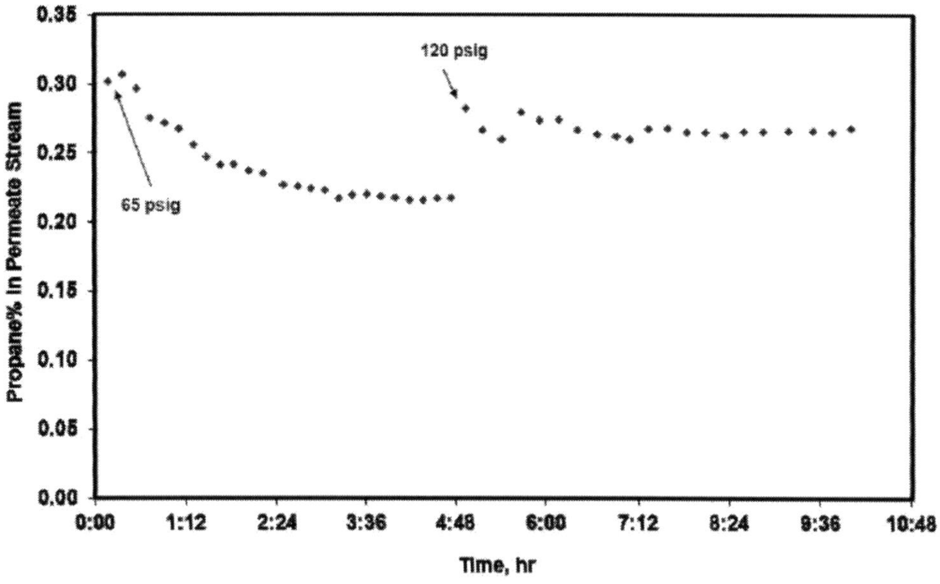

While Imtex membrane LF1 steadily showed high permeate purities (in both C2= and C3= cases) and higher permeabilities than those in the literature, for industrial applications, further improvement in flux was still needed.

Further development work led to making changes in the chemical and physical structure of the membrane. This resulted in the second generation of membranes referred to as HF3. This membrane still showed olefin product purities in the same range as LF1 but with an over three fold increase in the permeation rate.

Figure 7 below shows the performance of Imtex HF3 membrane for both C2s and C3s. The olefin mobility rate within the chitosan (along with a bonded Ag+ cation or as a hopping mechanism from one Ag+ site to another) seems to be similar for both ethylene and propylene. The molar or volumetric rates observed were at nearly the same level or at times slightly lower in the case of propylene. The difference in molecular weight between ethylene and propylene, however, reveals the more established mass flow rate difference as shown in Figure 7.

Figure 7. Imtex HF3 Membrane Performance
C2s and C3s Separation at 100 psig and room temperature

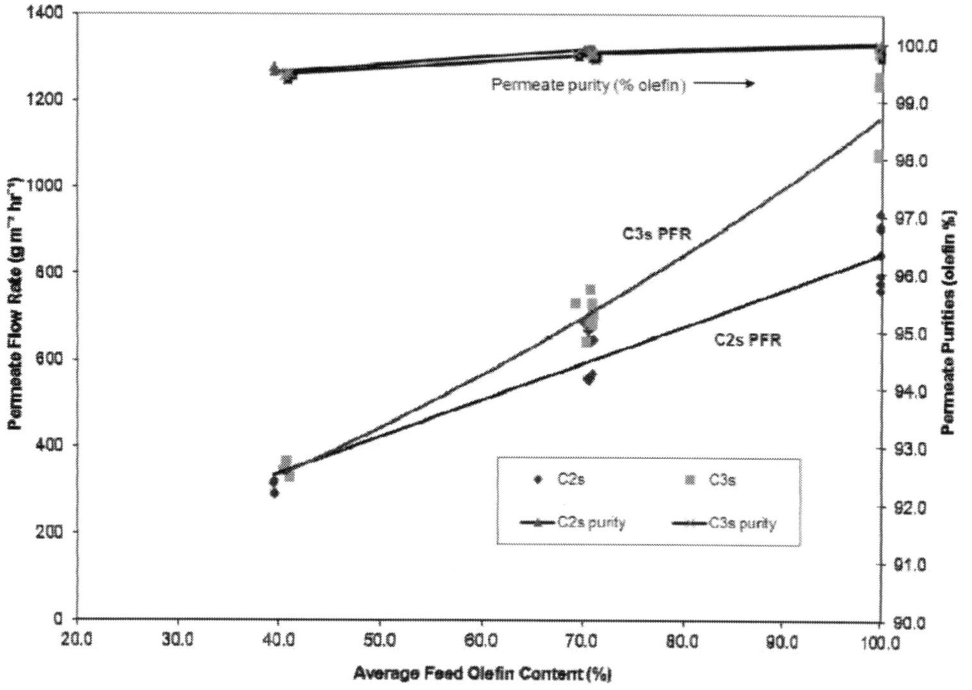

With HF3 being a promising candidate membrane for product scale-up and pilot demonstration purposes, R&D at Imtex is shifting more to product development and testing of spiral wound elements of various sizes. The picture below shows three sizes of rolled test elements.

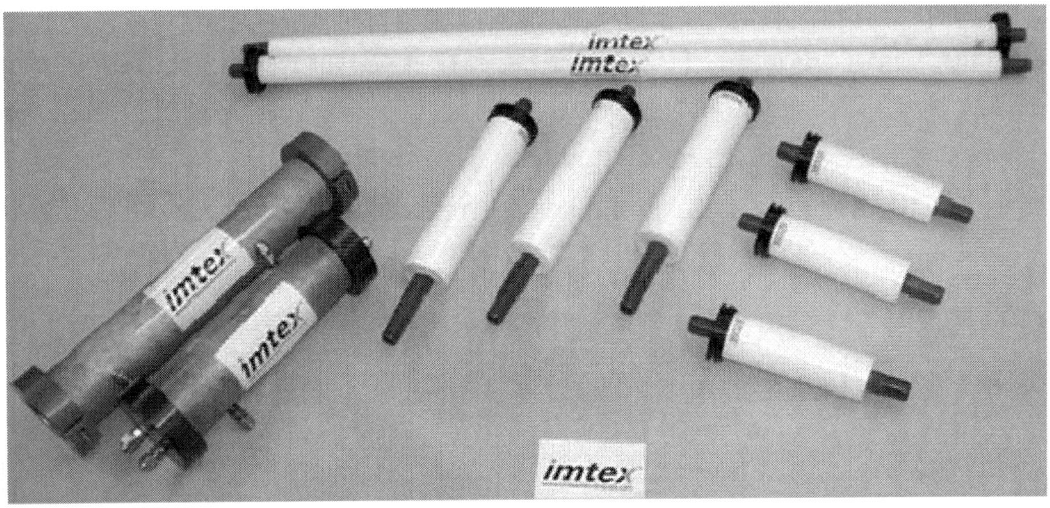

Using only one membrane module, we demonstrated reasonable propylene recovery producing polymer grade propylene from a 70/30 propylene/propane gas mixture as shown in Figure 8.

Figure 8- Effect of Stage Cut on Membrane Performance (60 Psig and room Temperature)

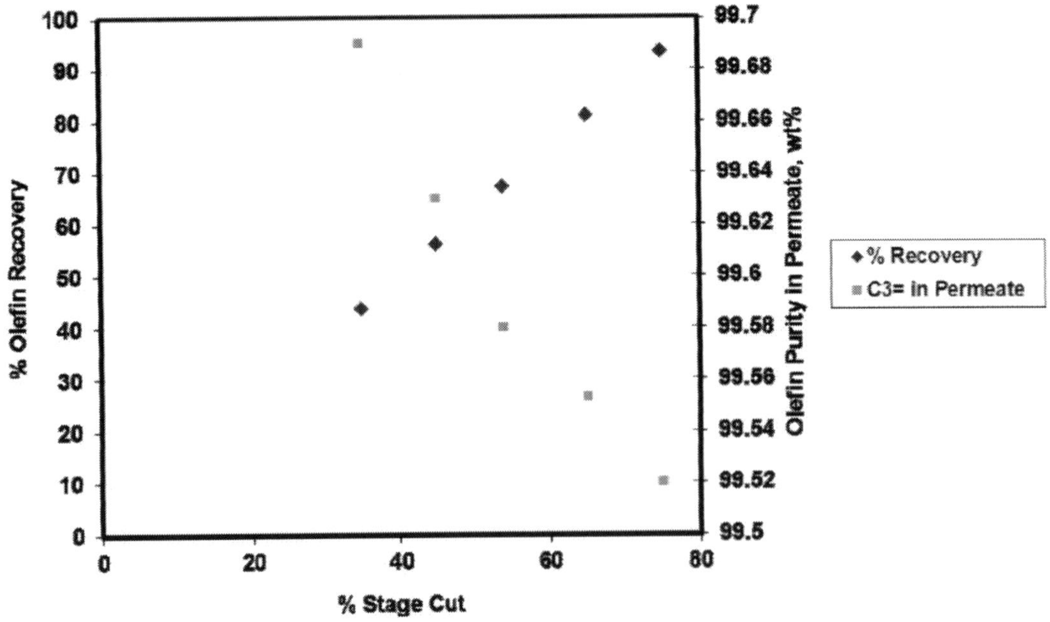

Multiple Membrane Modules in Series:

Currently the upgrading of the refinery grade to polymer grade is carried out in distillation columns; thus, the distillation bottom will produce 90%+ propane as second "product." Using the membrane technology to replace the distillation has the drawbacks of not being capable of producing propane in such high concentration. The implementation of a hybrid system to achieve debottlenecking may be an attractive option. A hybrid system will be technically easier and more economical than replacing the entire distillation system with a membrane unit. We presented various scenarios and performed an extensive simulation work for C2 splitter to quantify the energy saving through the retrofit of existing distillation columns with a membrane unit [8].

Another alternative approach is to use multiple membrane modules to get the propane to a reasonable concentration. We performed experiments with two (2) and four (4) modules in series.

The experimental results are presented below for 2 modules in series. As shown, while the polymer grade propylene is produced at the permeate side, the propane rich stream only was concentrated to 80 percent on the retentate side.

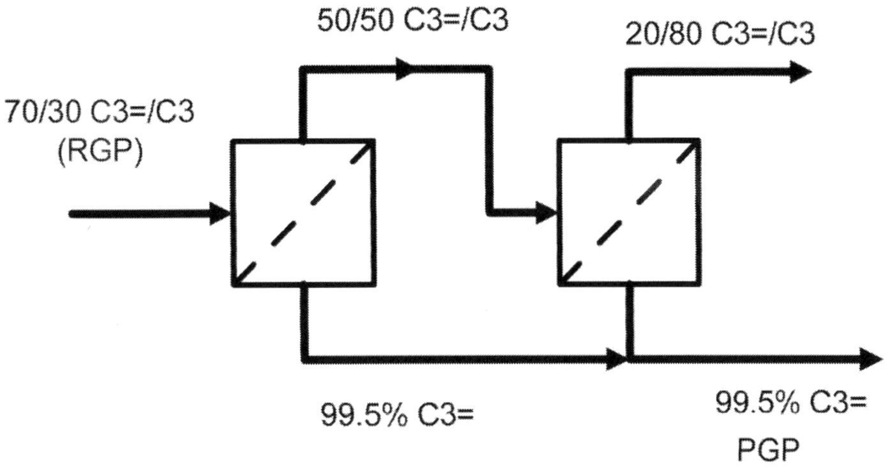

The experiments were performed at 66 psi. At this feed pressure and 20 percent olefin content, olefin feed side partial pressure would be approximately 16 psia. Then there is hardly any driving force against the nearly 14.7 psia on the permeate side. In theory, feed side pressure would be raised above 150 psig to get any permeation flow if olefin content in retentate is needed to be at 10 percent. Figure 9 shows the results.

**Figure 9- Simulating RGP Feed to Memebrane:
Propane/ Propylene 30/70 wt% @ 120 psig Feed Pressure**

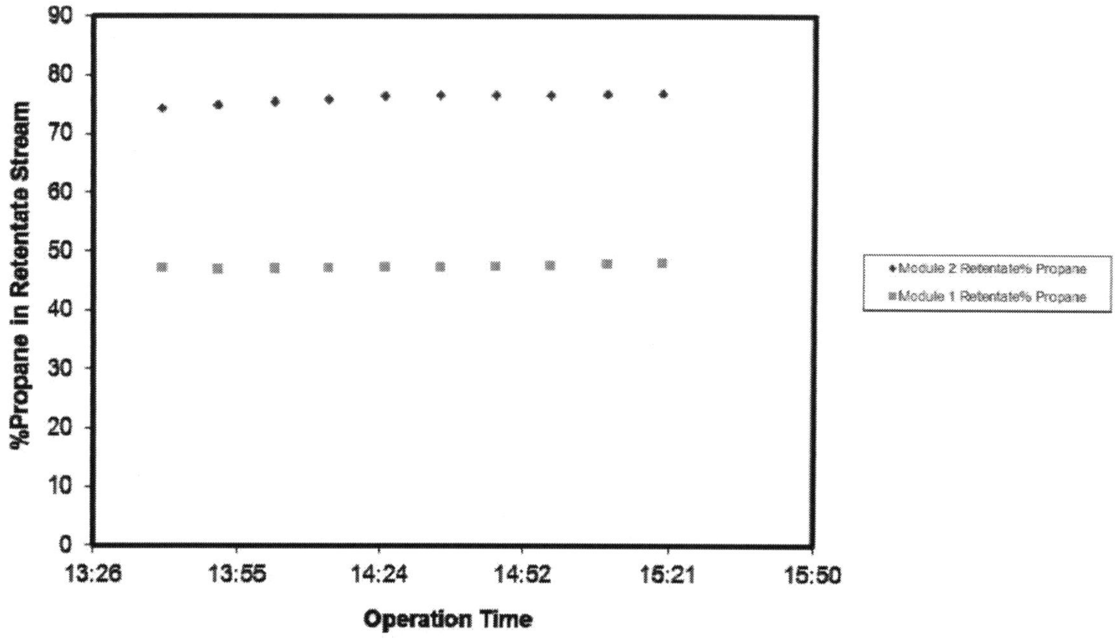

Potential Opportunities:

Energy savings of the membrane use in the olefin-paraffin separation is considered very attractive. Assuming that the membrane is in a hybrid distillation/membrane arrangement, energy savings were estimated to be on the order of $2 MM/yr for a commercial size C2 splitter. A retrofit of an existing splitter may include compressors for both permeate and retentate streams, and a dryer for the permeate stream in addition to the membrane skid. The $2MM energy savings alone does not justify the capital estimated for the retrofit. However, if the membrane were to be used for debottleneck purpose, and then the economics for C2 splitter are very favorable.

Another promising application requiring olefin-paraffin separation is in olefin metathesis where polymer-grade propylene is produced by combining the n-butene with ethylene. Butene/butane separation is desired for the operation and is not commercially attractive to use distillation. If successful, a membrane process would increase reactant utilization of metathesis plants and offers attractive energy savings by producing more propylene from the same volume of feedstock.

Imtex HF3 membranes were also tested in separating C4s. Testing was conducted with gas mixtures of 1-Butene/n-Butane at 40 psig and a temperature of 40 $^{\circ}$ C. As shown in Figure 10, n-butane content in the permeate stream was between 0.1 to 0.3%. The molar permeability of 1-butene was higher than either of ethylene and propylene. Accordingly, and in addition to the molar mass difference, the mass permeate rate for C4= would be significantly higher than for C2= or C3= if testing were performed at the same pressure.

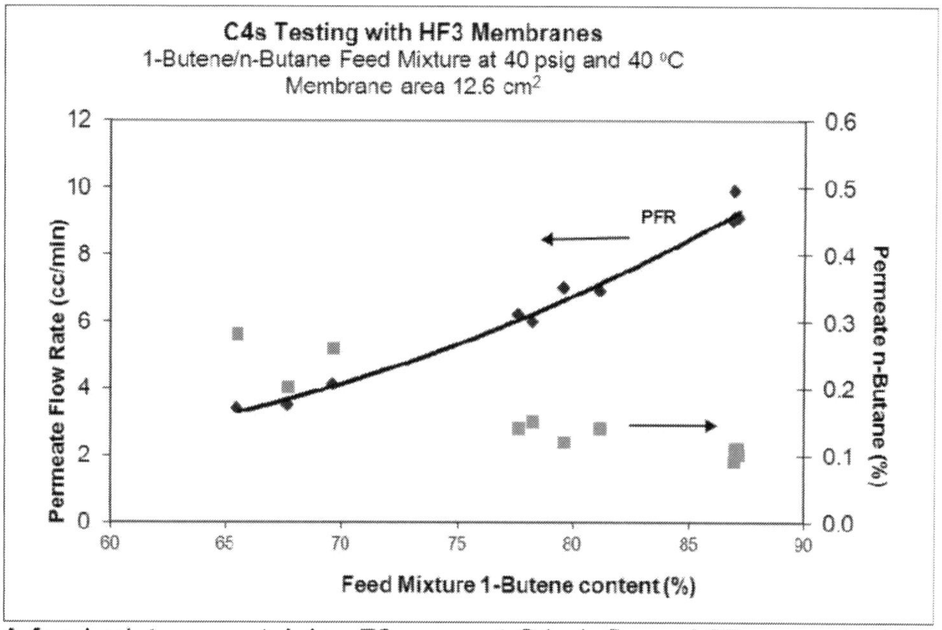

A feed mixture containing 73 percent C4 olefin and 27 percent of C4 paraffin was fed to the 500 cm^2 membrane unit. A total paraffin concentration of 0.57 weight percentage was measured in the permeate stream as shown in Figure 11.

Figure 11. Feed Mixture Containing 73% C4 Olefin and 27% C4 Paraffin fed to 600 Cm² Membrane Unit (65 psig pressure and room temperature)

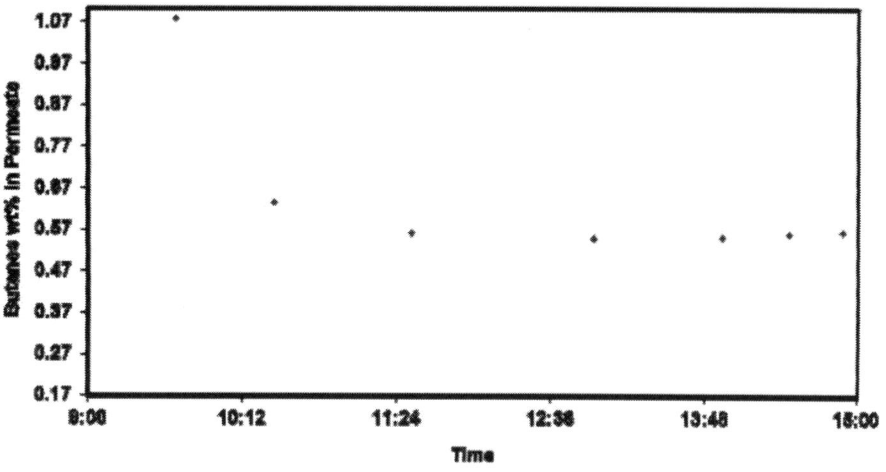

We believe that more investigations are needed to explore wider operating conditions (temperature and pressure) for C4s separations

Conclusions:

The preliminary work on various olefin-paraffin applications produced very encouraging results. In addition to the excellent selectivity and permeability, Imtex membranes demonstrated performance stability over time, which was one of the key shortcomings in prior attempts by others to use facilitated transport membranes in this application.

Currently Imtex is collaborating with The Shaw Group Inc. for the eventual commercial marketing and deployment of their technology. Work is underway at "proof of concept" stage. Prototype membrane elements have been made at commercial scale and a test program is underway at a Shaw pilot plant facility to test and refine the technology. Elements of this size will be the modular building blocks for commercial installations.

The technology has great potential to offer significant benefits in various applications beneficial to olefin manufacturing operations, such as debottlenecking C2 splitters and upgrading refinery grade propylene to chemical/polymer grade.

There is also a very promising potential for butene/butane separations where there are technical challenges facing distillation. A successful application of Imtex membranes can have a very positive impact on the economics of olefin metathesis which is used to produce polymer-grade propylene by combining n-butene and ethylene. Also, efficiencies can be realized in the production of n-butene co-monomers for linear low density polyethylene production.

References:

1- "Hybrid Separation/Distillation Technology: Research Opportunities for Energy and Emission Reduction", US DOE, Energy Efficiency and Renewable Energy April 2005.

2- "Energy Efficiency and Innovative Emerging Technologies for Olefin Production", Tao Ren, Martin Patel, and Kornelis Blok, presented in European Conference on Energy Efficiency, October 21-22, 2004.

3- R.B. Eldridge, "Olefin/Paraffin Separation Technology: A Review," Industrial & Engineering Chemistry Research, 32, 2208-12 (1993).

4- D.J. Safarik and R.B. Eldridge, "Olefin/Paraffin Separations by Reactive Absorption: A Review," Industrial & Engineering Chemistry Research, 37, 2571-81 (1998).

5- Separation of Olefin/Paraffin Mixtures with Carrier Facilitated Membranes; Final report by MTR for DOE; Contract # DE-FC36-04GO14151, March 2007.

6- P.S. Herrera, X. Feng, D. Payzant and J.H. Kim, "Process for the Separation of Olefins from Paraffins using Membranes," U.S. Patent 7,361,800 (April 22, 2008).

7- M. Azhin et al. "A Review on Olefin/Paraffin Separation Using Reversible Chemical Complexation Technology", Journal of Industrial and Engineering Chemistry, 14 (2008) 623-638.

8- Jose A. Caballero, et. al. "Design of Hybrid Distillation-Vapor Membrane Separation Systems"; Ind. Eng. Chem. Res. 2009, 48, 9151-9162.

(355d) Gas Transport of Paraffins and Olefins in Liquid Crystal Membranes

Feras Rabie, Chemical Engineering, Virginia Polytechnic Institute and State University, Blacksburg, VA and Stephen M. Martin, Chemical Engineering, Virginia Tech, Blacksburg, VA

Gas Transport of Paraffins and Olefins in Liquid Crystal Membranes

Abstract

A solubility induced phase separation (SIPS) method was used to form 8CB and MBBA liquid crystal domains in a polysulfone matrix. The membranes were characterized for their temperature dependent phase behavior and structural characteristics using DSC, POM, and FE-SEM. Paraffin and olefin separations were investigated at different temperatures and different liquid crystal phases using a mixed gas permeation apparatus. Propane and propylene were chosen to minimize size effects of diffusion and sorption.

We studied the effects of permeation and selectivity on intermolecular ordering of the small molecular liquid crystal. The liquid crystal phase plays an important role for diffusion and solubility on the solute. The crystalline state exhibited the lowest permeability with an increase in both propylene transport and propylene selectivity over propane as the crystalline domain sizes increased. Transport through the liquid crystals at their mesophase resulted in both an increase in permeability and selectivity. The contributing factor to this phenomenon was concluded to be solubility selectivity of propylene over propane due to enhanced intermolecular interactions between the ordered domains and the more polar propylene solute. A further increase in temperature resulted in a two fold magnitude increase in permeability while maintaining a selectivity of 4.

A cyanobiphenyl liquid crystal attached to a butadiene backbone with a flexible alkane spacer was synthesized for the investigation of its gas transport properties. Room temperature transport studies showed a helium selectivity of 60 over methane. This was primarily due to size selective diffusion across the membrane with very little solubility or sorption. The effect of propane and propylene transport on pressure and phase dependence was tested. Transport for temperatures under the LCPs glass transition was dominated by diffusivity. A transition to a smectic liquid crystal phase resulted in an increase in permeability and selectivity. The contributing factor was the increase of propylene solubility compared to that of propane. A feed pressure increase resulted in an increase in permeability selectivity, which was also due to an increase in solubility selectivity. Mixed gas transport experiments correlated well with single gas transport.

A series of methacrylate liquid crystal polymers with nitro azobenzene and cyano biphenyl mesogens were synthesized and characterized. These liquid crystal monomers were copolymerized with 2-ethylhexyl acrylate (2-EHA) using a free radical polymerization method. Liquid crystal phases were present for LC content of over 50 mol% for the cyanobiphenyl and 60 mol% for the nitro azobenzene. Mesophase stability increased with increasing LC content. A method to form acrylate networks for the fabrication of stable membranes was developed and partially characterized. We will test its transport properties in the future.

(355e) Synthesis and Characterization of ZIF-8 Membrane with Exceptionally High Propylene / Propane Separation Performance

Hyuk Taek Kwon, chemical engineering, texas A & M university, college station, TX, Sagar Lonkar, chemical engineering, Texas A & M university and Hae-Kwon Jeong, Chemical Engineering, Texas A & M University, College Station, TX

Synthesis and characterization of ZIF-8 Membrane with exceptionally high propylene / propane separation performance

Hyuk Taek Kwon[1], Sagar Lonkar[1], and Hae-Kwon Jeong*[1,2]

[1]Artie McFerrin Department of Chemical Engineering and [2]Materials Science and Engineering Program

Texas A&M University, College Station, TX 77843-3122

* Corresponding author: hjeong7@mail.che.tamu.edu

Metal-organic-frameworks (MOFs) have attracted intense research interest due to their potential application such as small-molecule separation[1], storage[2], catalysis[3], and others. Membrane-based gas separation is one of the most intriguing applications where MOFs can play a role due to their regular and sub-molecular size of pore structure, recognition and binding ability of specific molecules[4].

Several MOF membranes have been synthesized and tested for gas separation[5]. Despite their potentials, most of MOF membranes reported showed rather unimpressive separation performances mainly due to the flexible nature of their frameworks[6] as well as the likely poor microstructure. Recently Lai and co-worker have synthesized ZIF-8 membrane with exceptional hydrocarbon separation capability. The superior performance is likely due to improved microstructure[7], confirming the importance of controlling microstructure. Besides, there is also a strong need to develop a new synthesis method to prepare MOF membranes in a commercially-viable manner.

Here, we would like to present one-step in-situ route to prepare ZIF-8 membrane, which is highly reproducible. The formation of defect-free ZIF-8 membrane was completed within a couple of hours with exceptional mechanical stability. Most importantly, the membranes showed the exceptionally high separation performance of propylene / propane. We will also present synthesis of other MOF membranes such as HKUST-1 and ZIF-7.

References and Notes

1. J.-R. Li, R. J. Kuppler and H.-C. Zhou, Chemical Society Reviews **38** (5), 1477-1504 (2009).

2. L. J. Murray, M. Dinca and J. R. Long, Chemical Society Reviews **38** (5), 1294-1314 (2009).

3. J. Lee, O. K. Farha, J. Roberts, K. A. Scheidt, S. T. Nguyen and J. T. Hupp, Chemical Society Reviews **38** (5), 1450-1459 (2009).

4. M. Eddaoudi, J. Kim, N. Rosi, D. Vodak, J. Wachter, M. O'Keeffe and O. M. Yaghi, Science **295** (5554), 469-472 (2002).

5. M. Shah, M. C. McCarthy, S. Sachdeva, A. K. Lee and H.-K. Jeong, Industrial & Engineering Chemistry Research

51 (5), 2179-2199 (2011).

6. H. Bux, F. Liang, Y. Li, J. Cravillon, M. Wiebcke and J. r. Caro, Journal of the American Chemical Society **131** (44), 16000-16001 (2009).

7. Y. Pan and Z. Lai, Chemical Communications **47** (37), 10275-10277 (2011).

(355f) Design of ZIF/Polyimide Mixed Matrix Membrane for Olefin/Paraffin Separations

Chen Zhang, School of Chemical and Biomolecular Engineering, Georgia Institute of Technology, Atlanta, Georgia and William J. Koros, School of Chemical & Biomolecular Engineering, Georgia Institute of Technology, Atlanta, GA

<>C_2~C_4 olefins are consumed as feedstocks to produce a variety of significant chemicals. Membrane separations have been emerging as more energy-efficient alternatives to fractional distillation to enrich olefin from olefin/paraffin mixtures. The development of high-performance membrane materials has been the key to promote membrane separations as practical approaches for large-scale industrial gas separations.

Mixed matrix membranes, formed by dispersing molecular sieve particles/platelets in polymer matrices, are promising approaches for gas/vapor separations that combine the ease of processing polymers with the superior separation performance of molecular sieves. The selection of molecular sieving phase and polymer phase is crucial to fabrication of mixed matrix membranes with intact polymer-sieve adhesion and favorable gas transport properties. Zeolitic imidazolate frameworks (ZIFs) are microporous crystalline solids with zeolite or zeolite-like topologies. The favorable adsorption and molecular sieving properties of ZIFs enable the possibility of being used as selective adsorbents as well as membrane materials.

Our study suggests that with appropriate choice of ZIF and polymer, enhanced and industrially attractive permeability and permselectivity can be simultaneously obtained in the ZIF/polymer mixed matrix membrane. ZIF-8/6FDA-DAM mixed matrix dense films were fabricated with hydrophobic ZIF-8 as the molecular sieving phase and a high performance polyimide 6FDA-DAM as the polymer phase. Permeation properties of the neat 6FDA-DAM dense film and ZIF-8/6FDA-DAM mixed matrix dense films were studied with pure propylene (C_3H_6) and propane (C_3H_8) feeds as well as a 50/50 (vol %) C_3H_6/C_3H_8 mixture feed. Pure gas permeation showed significantly enhanced mixed matrix ZIF-8/6FDA-DAM membrane C_3H_6/C_3H_8 separation performance compared to the pure 6FDA-DAM membrane. A C_3H_6 permeability of 56 Barrers and C_3H_6/C_3H_8 ideal selectivity of 31 was found in ZIF-8/6FDA-DAM mixed matrix membrane with 48 wt % ZIF-8 loading. These results are 258 % and 150 % higher than the pure 6FDA-DAM membrane, respectively for permeability and selectivity. Mixed gas permeation also verified that selectivity enhancements were achievable in mixed gas environment by ZIF-8.

(355g) TR Polymers for Olefin/Paraffin Separation

Zachary P. Smith[1], David F. Sanders[1], Ruilan Guo[2], Benny Freeman[1], Donald R. Paul[1] and James E. McGrath[3], (1)Department of Chemical Engineering, The University of Texas at Austin, Austin, TX, (2)The University of Notre Dame, Notre Dame, IN, (3)Macromolecules and Interfaces Institute, Virginia Tech, Blacksburg, VA

Ethylene and propylene are currently the two largest, by volume, organic chemical feedstocks produced in the U.S [1]. These olefins must be purified for polyolefin synthesis by separating ethylene from ethane and propylene from propane, a thermally intensive process that requires approximately 0.12 Quads of energy per year and distillation columns containing nearly 200 trays [2]. This study focuses on determining the performance and plasticization resistance for a thermally rearranged (TR) polymer for olefin/paraffin separation.

Polyimides containing reactive functional groups *ortho*-position to the diamine can be chemically transformed into TR polymers at elevated temperatures [3]. For this study, a polyimide was prepared from 3,3'-dihydroxy-4,4'-diamino-biphenyl (HAB) and 2,2'-bis-(3,4-dicarboxyphenyl) hexafluoropropane dianhydride (6FDA) via chemical, thermal, and solid-state imidization [4, 5]. Chemical imidization produces polyimides with *ortho*-functional acetate groups, while thermal and solid-state imidization produce polyimides with *ortho*-functional hydroxyl groups. Furthermore, each imidization technique alters the non-equilibrium nature of the polymer film, thus producing polyimides with unique transport properties. All films were characterized by ^{1}HNMR, DSC, FT-IR, TGA-MS, and pure gas permeation.

For ethylene, ethane, propylene, and propane, increasing polyimide conversion dramatically increased gas permeation. For example, partial conversion of the chemically imidized HAB-6FDA polyimide to its corresponding TR polymer increased ethylene permeability by over two orders of magnitude. Furthermore, ethylene/ethane selectivity decreased by less than 10%. For ethylene/ethane and propylene/propane separation, HAB-6FDA TR polymers showed combinations of permeability and selectivity near the polymer upper bound [6, 7]. Propylene plasticization pressure curves were determined for TR polymers derived from all three synthesis routes, and the thermally imidized TR polymer had the highest plasticization pressure point of approximately 4 bar at 35°C.

1. Facts & Figures: Output Declines in U.S., Europe, in Chemical & Engineering News. 2010. 54-62.

2. Eldrige, R.B., Olefin/paraffin separation technology: A review. *Industrial & Engineering Chemistry Research*, **1993**. 32(10), 2208-2212.

3. Park, H.B., C.H. Jung, Y.M. Lee, A.J. Hill, S.J. Pas, S.T. Mudie, E. Van Wagner, B.D. Freeman, and D.J. Cookson, Polymers with cavities tuned for fast selective transport of small molecules and ions. *Science*, **2007**. 318(5848), 254-258.

4. Ghosh, M.K. and K.L. Mital, Polyimides: Fundamentals and applications. 1996, New York: Marcel.

5. Ohya, H., V.V. Kudryavtsev, and S.I. Semenova, Polyimide membranes: Applications, fabrications, and properties. 1996, Amsterdam: Gordon and Breach Publishers.

6. Burns, R.L. and W.J. Koros, Defining the challenges for C_3H_6/C_3H_8 separation using polymeric membranes. *Journal of Membrane Science*, **2003**. 211(2), 299-309.

7. Staudt-Bickel, C. and W.J. Koros, Olefin/paraffin gas separations with 6FDA-based polyimide membranes. *Journal of Membrane Science*, **2000**. 170(2), 205-214.

(355h) Asymmetric Hollow Fiber Membranes for Aggressive Sour Gas Purification

Brian E. Kraftschik and William J. Koros, School of Chemical and Biomolecular Engineering, Georgia Institute of Technology, Atlanta, GA

Carbon dioxide (CO_2) and hydrogen sulfide (H_2S) are present in high concentration in a large fraction of natural gas reserves. These species must be removed prior to transport through natural gas pipelines in order to prevent corrosion, increase the heating value of natural gas and because of the dangers associated with H_2S. A series of novel crosslinkable materials based on the glassy polyimide 6FDA-DAM:DABA, which has been used in the past for CO_2/CH_4 separations, has been developed for use in the simultaneous removal of H_2S and CO_2 from aggressive natural gas feeds. These materials exhibit significantly improved H_2S/CH_4 selectivity and plasticization resistance over the base polymer without sacrificing the outstanding CO_2/CH_4 permselectivity of 6FDA-DAM:DABA. Following dense film permeation and sorption studies, one of these crosslinkable 6FDA-DAM:DABA derivatives was identified as a promising candidate for asymmetric hollow fiber formation and testing. In dense film form, this material has an H_2S/CH_4 selectivity as high as 23.5 and a CO_2/CH_4 selectivity just below the intrinsic value when using a 20% H_2S/20% CO_2/60% CH_4 mixture at feed pressures up to 900 psi. The dense film characterization results as well as those for asymmetric hollow fiber spinning and preliminary permeation testing will be discussed.

(356a) Hydrogen Sorption in Polymers

Zachary P. Smith[1], Rajkiran Tiwari[1], David F. Sanders[1], Ruilan Guo[2], Benny D. Freeman[1], Donald R. Paul[1] and James E. McGrath[3], (1)Department of Chemical Engineering, The University of Texas at Austin, Austin, TX, (2)The University of Notre Dame, Notre Dame, IN, (3)Macromolecules and Interfaces Institute, Virginia Tech, Blacksburg, VA

For over 30 years, dense polymer membranes have been used for industrial applications in hydrogen separation, including syngas ratio adjustment, refinery off-gas purification, and ammonia purge gas recovery [1]. These membranes operate based on the solution-diffusion model, where permeability is defined as the product of an effective diffusion coefficient and an effective sorption coefficient. Determining sorption of H_2, however, is difficult owing to its highly non-condensable nature and, therefore, low sorption in polymer films. This study seeks to quantify how H_2 sorption changes during conversion of a thermally rearranged (TR) polymer and assess differences between H_2 sorption in TR polymers and in other polymers reported in the membrane literature.

TR polymers are synthesized from reactive polyimide precursors, which contain hydroxyl groups *ortho*-position to the diamine [2-4]. These polyimide precursors are synthesized through traditional polyimide chemistry [5, 6], cast as films, and converted to TR polymers in the solid state via a thermally induced reaction. For this study, we focus on a polyimide precursor prepared from 3,3'-dihydroxy-4,4'-diamino-biphenyl (HAB) and 2,2'-bis-(3,4-dicarboxyphenyl) hexafluoropropane dianhydride (6FDA). This polyimide, along with three partially converted HAB-6FDA TR polymers were tested in this study. The partially converted TR polymers were converted 39%, 60%, and 76% by heating the polyimide precursor under N_2 for different amounts of time and at different temperatures. To avoid polymer degradation, which occurs at sufficiently high temperatures, samples were not converted above 76%. Samples were characterized using ^{1}HNMR, TGA-MS, DSC, FT-IR, and XRD, and sorption was determined using a Rubotherm Magnetic Suspension Balance.

H_2 sorption in the HAB-6FDA polyimide is similar to that of Matrimid™, which is a polyimide used industrially for gas separations. Interestingly, thermal rearrangement induces significant changes in sorption capacity for HAB-6FDA. Compared to the polyimide, H_2 sorption increases by a factor of approximately 2.6 for the 76% converted HAB-6FDA TR polymer. This dramatic increase in sorption (i.e., a factor of 2.6) is largely driven by non-equilibrium free volume in the polymer and matches the increase in sorption observed for N_2, O_2, and CH_4 in the same material. Surprisingly, over the same extent of conversion, CO_2 sorption only increases by a factor of approximately 1.7. This difference is likely attributed to a quadrupolar moment that exists in CO_2 but not in H_2, N_2, O_2, and CH_4 [7]. This quadrupolar moment would encourage favorable interactions with polymer carbonyl groups [8] that are found in the HAB-6FDA polyimide, but are lost during the thermal rearrangement process. When compared to Matrimid™, polysulfone, AF2400, and the other partially converted TR polymers, H_2 sorption was highest in the 76% converted HAB-6FDA TR polymer.

The enthalpy of sorption was also determined as a function of conversion for HAB-6FDA. The HAB-6FDA polyimide had a similar enthalpy of sorption to Matrimid™. However, the enthalpy of sorption decreased in magnitude as the polyimide was converted to its TR polymer. This change in the enthalpy of sorption likely results from increasing free volume within the polymer film [9]. For the 76% converted HAB-6FDA TR polymer, the enthalpy of sorption decreased in magnitude to values similar to AF2400, which is another high free volume polymer.

1. Baker, R.W., Membrane technology and applications. 2004, Chichester: John Wiley & Sons, Ltd.

2. Park, H.B., C.H. Jung, Y.M. Lee, A.J. Hill, S.J. Pas, S.T. Mudie, E. Van Wagner, B.D. Freeman, and D.J. Cookson, Polymers with cavities tuned for fast selective transport of small molecules and ions. *Science*, **2007**. 318(5848), 254-258.

3. Tullos, G. and L. Mathias, Unexpected thermal conversion of hydroxy-containing polyimides to polybenzoxazoles. *Polymer*, **1999**. 40(12), 3463-3468.

4. Tullos, G.L., J.M. Powers, S.J. Jeskey, and L.J. Mathias, Thermal conversion of hydroxy-containing imides to benzoxazoles: Polymer and model compound study. *Macromolecules*, **1999**. 32(11), 3598-3612.

5. Ghosh, M.K. and K.L. Mital, Polyimides: Fundamentals and applications. 1996, New York: Marcel.

6. Ohya, H., V.V. Kudryavtsev, and S.I. Semenova, Polyimide membranes: Applications, fabrications, and properties. 1996, Amsterdam: Gordon and Breach Publishers.

7. Lin, H., E. Van Wagner, J. Swinnea, B.D. Freeman, S.J. Pas, A.J. Hill, S. Kalakkunnath, and D. Kalika, Transport and structural characteristics of crosslinked poly(ethylene oxide) rubbers. *Journal of Membrane Science*, **2006**. 276(1-2), 145-161.

8. Koros, W.J., Simplified analysis of gas/polymer selective solubility behavior. *Journal of Polymer Science: Polymer Physics Edition*, **1985**. 23(8), 1611-1628.

9. Broom, D.P., Hydrogen storage materials. 2011, London: Springer.

(356b) Symmetric and Asymmetric Zeolitic Imidazolate Frameworks (ZIFs)/Polybenzimidazole (PBI) Nano-Composite Membranes for Hydrogen Purification At High Temperatures

Tingxu Yang[1], Gui Min Shi[2] and Tai-Shung Chung[2], (1)NUS Graduate School for Integrative Sciences and Engineering, National University of Singapore, Singapore, Singapore, (2)Department of Chemical & Biomolecular Engineering, National University of Singapore, Singapore, Singapore

Hydrogen is one of the most strategically important sustainable fuel sources among many energy alternatives. In addition to be viable and versatile in widely applications such as fuel for transportation and feedstock for chemical industries, a hydrogen-based energy system has other advantages including low emissions and environmentally friendly, clean, and efficient use. Currently, the favored route of hydrogen production is from the steam reforming of methane followed by the water-gas shift reaction. The main by-product in the leaving stream from the operation is carbon dioxide. The removal of CO_2 from H_2 or its sequestration is a critical requirement for hydrogen to be a sustainable energy system as well as for the minimization of environmental impact.

We have molecularly designed high performance zeolitic imidazolate frameworks (ZIFs)/ polybenzimidazole (PBI) nano-composites for hydrogen separation at high temperatures, and demonstrated it in a useful configuration as dual-layer hollow fibers for the first time. By incorporating as-synthesized nanoporous ZIF-8 nano-particles into the high temperature but extremely low permeability polybenzimidazole (PBI), the resultant mixed matrix membranes show an impressive enhancement in H_2 permeability as high as hundred times without any significant deduction in H_2/CO_2 selectivity. The 30/70 ZIF-8/PBI dense membrane has a H_2 permeability of 105.4 Barrer and a H_2/CO_2 selectivity of 12.2. This performance is far superior to ZIF-7/PBI membranes and is the best ever reported data for H_2-selective polymeric materials in the literature. Meanwhile, defect-free ZIF-8-PBI/Matrimid dual-layer hollow fibers have been successfully fabricated without post annealing and coating by optimizing ZIF-8 nano-particle loadings, spinning conditions and solvent-exchange procedures. Two types of hollow fibers targeted at either high H_2/CO_2 selectivity or high H_2 permeance have been developed; namely, (1) PZM10-I B fibers with a medium H_2 permeance of 64.5 GPU at 180 $^{\circ}$C and a high H_2/CO_2 selectivity of 12.3, and (2) PZM33-I B fibers with a high H_2 permeance of 202 GPU at 180 $^{\circ}$C and a medium H_2/CO_2 selectivity of 7.7. This work not only molecularly designs novel nano-composite materials for harsh industrial applications such as syngas and hydrogen production, but also, for the first time, synergistically combines the strengths of both ZIF-8 and PBI for energy-related applications.

(356c) Hydrogen Permeability, Thermal Stability and Mechanical Strength of Melt-Spun (Ni1-xNbx)80Zr20 (for x = 0.3 to 0.6) Amorphous Alloy Membranes

Dhanesh Chandra[1], Sang-Mun Kim[2], Wen-Ming Chien[1], Narendra Kumar Pal[2], Stephen N. Paglieri[3], Michael D. Dolan[4] and Ted Flanagan[5], (1)Chemical and Materials Engineering, University of Nevada, Reno, Reno, NV, (2)University of Nevada, Reno, Reno, NV, (3)TDA Research, Inc., Wheat Ridge, CO, (4)Energy Technology, CSIRO, Pullenvale, Australia, (5)The University of Vermont

Amorphous Ni-based alloys are inexpensive hydrogen-selective membrane materials. One of the main barriers to the greater commercial use of these materials is their tendency to absorb hydrogen and embrittle, a process which decreases strength and causes membrane failure. The material also has a tendency to crystallize during a long-term operation at elevated temperatures,

a process which decreases both strength and hydrogen permeability. In this study, the hydrogen permeability and solubility of a series of $(Ni_{1-x}Nb_x)_{80}Zr_{20}$(for x = 0.3 to 0.6) amorphous alloy membranes was studied. Amorphous membrane ribbons with thicknesses of 40-60 μm were fabricated by melt spinning and coated with a thin (500 nm) layer of Pd surface catalyst by magnetron sputtering.

The hydrogen permeation properties of the membranes were investigated over the 350-450°C temperature range at different hydrogen partial pressures. The effect of Ni concentration on the hydrogen permeability and the thermal stability was also investigated; the hydrogen permeability decreases with the addition of Ni.

The activation energy of crystallization and the crystallization mechanism for binary Ni-Zr alloys were determined using the Johnson-Mehl-Avrami (JMA) equation. The Kissinger and Ozawa methods were applied to the non-isothermal kinetics derived from the heating rate dependence of the crystallization temperature.

The glass transition and crystallization kinetics of melt-spun $Ni_{60}Nb_{20}Zr_{20}$ amorphous alloy ribbons have been studied under non-isothermal and isothermal conditions using differential scanning calorimetry (DSC). The activation energies of crystallization, Ex, were determined to be 499.5 kJ/mol and 488.6 kJ/mol using the Kissinger and Ozawa equations, respectively. The Johnson–Mehl–Avrami equation has also been applied to the isothermal kinetics indicating a diffusion-controlled three-dimensional growth mechanism. In addition, the inter-metallic phases and morphology after membrane testing have been identified by X-ray diffraction (XRD) and scanning electron microscope (SEM).

(356d) Opportunities for CO2-Selective Membranes in Hydrogen Purification

Haiqing Lin, Jimmy Vu, Jenny He, Lloyd S. White and Tim Merkel, Membrane Technology and Research, Inc., Menlo Park, CA

Syngas operations require separation of CO_2 from hydrogen for various reasons, not only for CO_2 capture from IGCC operations, but for a number of existing refinery and petrochemical applications. Membrane technology is an attractive approach to CO_2 capture because of inherent advantages such as modularity, high energy efficiency and good reliability. MTR has been developing CO_2-selective Polaris^{TM} membranes, which are highly permeable to CO_2 and reject the other major components of syngas, including H_2, CO and CH_4. Polaris membranes have CO_2 permeances approximately ten times higher than other commercial CO_2-selective membranes. This combination of permeance and selectivity meets the target range necessary to yield a competitive membrane-based process for hydrogen purification and/or CO_2 capture and sequestration.

This presentation will describe how multi-stage and multi-step membrane process designs can be used to minimize membrane area, compressor power requirements, and hydrogen loss during hydrogen recovery from syngas. A preliminary process simulation study shows that a Polaris-based membrane process can achieve 99% hydrogen recovery, and capture 97% of CO_2 as a sequestration-ready liquid stream. Currently, MTR is working with the U.S. DOE National Carbon Capture Center (NCCC, operated by Southern Company) to conduct field tests of Polaris membranes at a coal-fired power plant. Results of the tests will be summarized, and critical issues such as long-term stability in practical applications and the impact of contaminants on system performance will be addressed. Future plans for running a 0.1 MW_e (500 lb/hr syngas) CO_2 capture/hydrogen purification membrane system and issues associated with eventual scale-up to full commercial operation will also be discussed.

(356e) Long-Term Thermal Stability and Chemical Tolerance of Supported Thin Film Pd-Au-Ru Membranes

Stephen N. Paglieri[1], Oyvind Hatlevik[2], Hani Abuelhawa[2] and J. Douglas Way[2], (1)TDA Research, Inc., Wheat Ridge, CO, (2)Chemical and Biological Engineering, Colorado School of Mines, Golden, CO

Long-term thermal stability and chemical tolerance of supported thin film Pd–Au–Ru membranes

S.N. Paglieri*,[1], Ø. Hatlevik[2], H. Abuelhawa[2], and J.D. Way[2]

[1]*TDA Research, Inc., Wheat Ridge, Colorado, USA*

[2]*Department of Chemical Engineering, Colorado School of Mines, Golden, Colorado, USA*

spaglieri@tda.com

Long-term thermal stability and chemical tolerance are essential for hydrogen membranes and membrane reactors. A 3.5 μm thick Pd–11Au–5Ru (wt.%) membrane was tested for >3000 h at 400-500°C in pure hydrogen and helium, and for 120 h in simulated water-gas shift (WGS: 50% H_2, 19% H_2O, 1% CO and 30% CO_2) mixtures at 400 and 450°C. The Pd–Au–Ru alloy membranes were thermally stable at 400°C and tolerant of WGS gases, but were attacked by WGS gases containing 20 ppm H_2S.

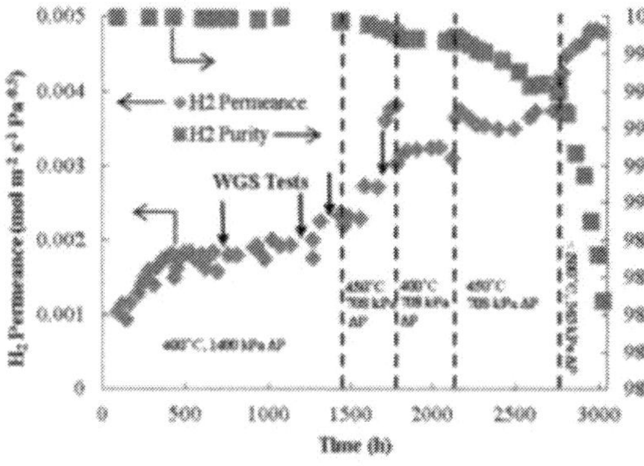

Figure 1. Hydrogen permeance and purity vs. time (and temperature) for a 3.5 μm thick Pd–11Au–5Ru membrane.

Pd-11Au-5Ru (3.5 μm thick) and Pd–10Au–5Ru (5.3 μm thick) membranes were deposited onto zirconia (ZrO_2)/porous stainless steel supports using electroless plating techniques. Simultaneous co-deposition of Pd–Ru was followed by displacement plating (a variant of electroless plating) of Au.[1] This sequence was repeated until a dense film was obtained. Preparation details and the testing apparatus have been described previously.[2] SEM/EDX, XRD, AA and DSC were used to characterize morphology, microstructure, chemical composition, and thermal stability.

The initial H_2/He pure gas permeation ratio for the Pd–11Au–5Ru membrane was ~20,000 at 400°C and 1.4 MPa pressure differential across the membrane. A minimal increase in helium leak rate through the membrane occurred at 400°C for >1000 h as shown in Figure 1, which indicated that the membrane was very stable, however, hydrogen flux was still increasing even after 1400 h. Subsequent annealing at 450°C for 300 h appeared to result in further alloying and higher flux, especially after exposure to simulated WGS gases for >120 h. At 450 and 500°C, the helium leak rate increased linearly over time, but it more than doubled when the temperature was increased to 500°C. A hydrogen permeance of 4.8×10^{-3} mol m^{-2} s^{-1} $Pa^{-0.5}$ was measured at 500°C.

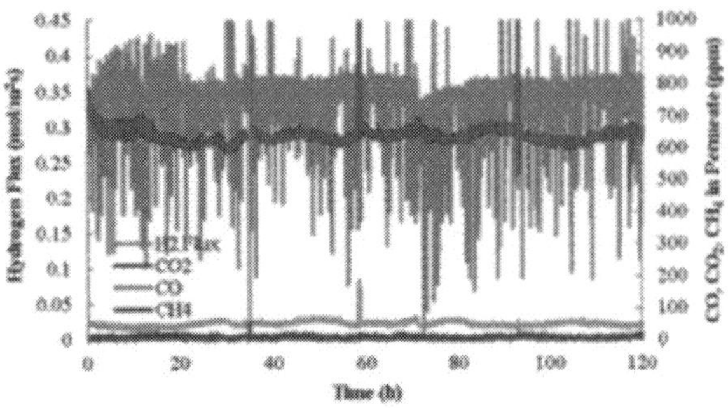

Figure 2. Hydrogen flux and purity vs. time for a 5.3 μm thick Pd-10Au-5Ru membrane under simulated WGS conditions: 400°C, 1.19 MPa ΔP, 50% H_2, 19% H_2O, 1% CO and 30% CO_2.

Exposing the Pd–11Au–5Ru and Pd–10Au–5Ru membranes to WGS conditions for >120 hours at either 400 or 450°C resulted in minimal increase in helium leak rate through the membrane (Figure 2). However, loss of feed CO_2 flow or exposure to excess CO for only several hours caused a permanent increase in the impurity leak rate. Next, the Pd–10Au–5Ru membrane was annealed at 400°C and exposed to WGS gases containing 20 ppm H_2S as shown in Figure 3, which resulted in an immediate 60% loss in hydrogen flux and a rapidly increasing concentration of impurities (CO, CO_2) in the hydrogen permeate.

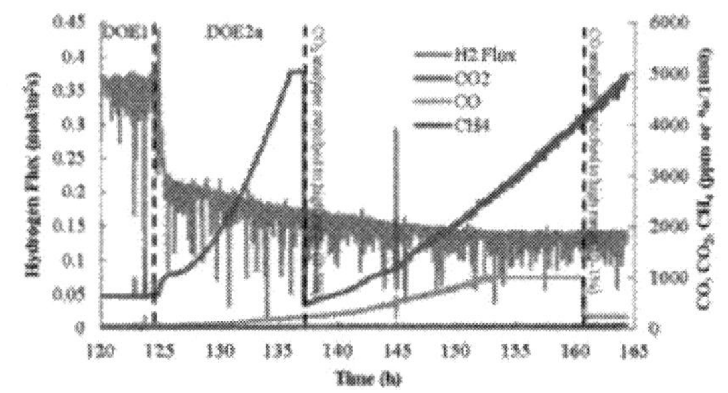

Figure 3. Hydrogen flux and purity vs. time for a 5.3 μm thick Pd-10Au-5Ru membrane under simulated WGS gases containing 20 ppm H_2S.

Pd–Au–Ru membranes were thermally stable at 400 and 450°C in pure hydrogen, and exposure to WGS conditions resulted in a minimal increase in the helium leak rate at both 400 and 450°C. Exposure to WGS gases apparently increased the hydrogen permeability by accelerating atomic rearrangement of the film (without increasing the impurity leak rate), as was previously observed for thicker Pd–Ru alloy membranes.[3] Exposure to WGS gases demonstrated the carbon tolerance, however, addition of 20 ppm H_2S to the WGS mixture rapidly degraded both flux and purity. Therefore, it appears that Pd–Au–Ru alloys may be less H_2S tolerant than binary Pd–Au or ternary Pd–Au–Pt alloys due to accelerated rearrangement in the presence of H_2S.[2b, 4] Potential mechanisms affecting pore growth rate in the membrane film include alloying, recrystallization, grain boundary diffusion and coalescence of microvoids. Development of alloys with greater durability is essential for the long-term operation of membranes and membrane reactors. The results for these Pd–Au–Ru alloy membranes will be compared to the performance of other binary (Pd–Au, Pd–Ru) and ternary (Pd–Au–Pt) alloy membranes and presented at the meeting.

References

1. Gade, S. K.; Keeling, M. K.; Davidson, A. P.; Hatlevik, Ø.; Way, J. D., Palladium–ruthenium membranes for hydrogen separation fabricated by electroless co-deposition. *Int. J. Hydrogen Energy* **2009,** *34*, 6484-6491.

2. (a) Hatlevik, Ø.; Gade, S. K.; Keeling, M. K.; Thoen, P. M.; Davidson, A. P.; Way, J. D., Palladium and palladium alloy membranes for hydrogen separation and production: History, fabrication strategies, and current performance. *Sep. Purif. Technol.* **2010,** *73* (1), 59-64; (b) Gade, S. K.; DeVoss, S. J.; Coulter, K. E.; Paglieri, S. N.; Alptekin, G. O.; Way, J. D., Palladium-gold membranes in mixed gas streams with hydrogen sulfide: Effect of alloy content and fabrication technique. *J. Membr. Sci.* **2011,** *378* (1-2), 35-41.

3. Ermilova, M. M.; Orekhova, N. V.; Skakunova, E. V.; Gryaznov, V. M., Changes in the catalytic activity and

hydrogen permeability of a palladium–ruthenium alloy membrane catalyst under the influence of reagents. *Bull. Acad. Sci. USSR* **1988,** *37* (4), 637-640.

4. Coulter, K. E.; Way, J. D.; Gade, S. K.; Chaudhari, S.; Alptekin, G. O.; DeVoss, S. J.; Paglieri, S. N.; Pledger, B., Sulfur tolerant PdAu and PdAuPt alloy hydrogen separation membranes. *J. Membr. Sci.* **2012,** *405–406*, 11-19.

(366a) Coupling of Crystallizers for Efficient Enantioseparation - Comparison of Two Different Process Strategies

Matthias Johannes Eicke, Physical and Chemical Foundations of Process Engineering, Max Planck Institute for Dynamics of Complex Technical Systems, Magdeburg, Germany, Guillaume Levilain, Max Planck Institute for Dynamics of Complex Technical Systems, Magdeburg, Germany, Martin Peter Elsner, Georg-Simon-Ohm University of Applied Sciences, Nuremberg, Germany and Andreas Seidel-Morgenstern, Physical and Chemical Foundations of Process Engineering, Max-Planck Institute for Dynamics of Complex Technical Systems, Magdeburg, Germany

Chiral molecules can be a particular challenge when used in drugs because of the existence of their stereoisomers. Although both enantiomers are merely mirror images, they can show drastically different effects in living organisms, which are chiral themselves and thus able to distinguish between the two molecules. Owing to this, it is essential to obtain enantiomers in their pure form. A wide array of methods has been studied and brought to application which on the one hand is concerned with the direct synthesis and on the other hand focuses on a more engineering biased approach that is, the separation of the racemate using well-known as well as novel techniques.

This contribution takes on the engineering route and puts the application of preferential crystallization (PC) as a separation technique at its center. It is well known that a racemic solution containing both enantiomers at equal proportions can be subcooled and still remain in a state where only the liquid phase exists. This phenomenon, called metastability, applies to other solutions as well but it is the key for applying PC. Within this region of kinetic inhibition it is possible to grow seed crystals of only one enantiomer. At some point however, the unseeded species will undergo nucleation and contaminates the product. In order to prevent this, the process must be stopped and the product has to be removed from the mother liquor. Certainly, this leads to great limitations with respect to yield because a considerable amount of product is still left in the dissolved state.

An improvement is possible for systems that crystallize as conglomerates such as DL-threonine/H_2O. It has been shown that two crystallizers coupled via a constant exchange of crystal free solution and seeded with opposite enantiomers can overcome these limitations [1, 2].

Recently more work has been done by the authors to further improve coupled PC. Different polythermal modes of operation have been investigated theoretically indicating the possibility for significant enhancements with respect to yield and productivity. Furthermore, a variant of coupled PC originally described in [3] has been investigated in detail both theoretically and experimentally for the model system DL-threonine/H_2O. The scheme of the process is depicted in Figure 1.

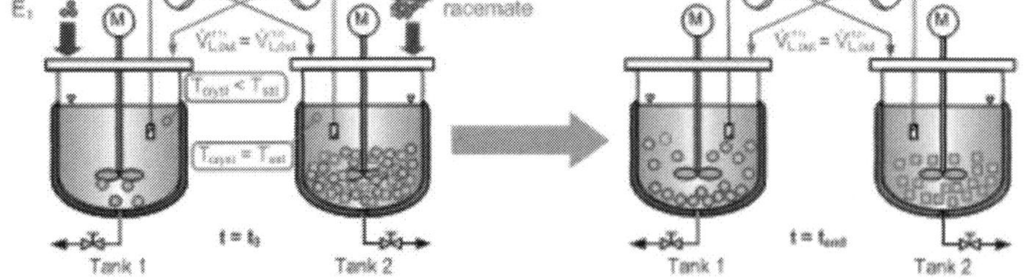

Figure 1: Operating principle for the connection of selective crystallization and dissolution.

Contrary to the first strategy, where both tanks are seeded with the respective pure enantiomers, this variant requires only one type of homochiral seeds. One crystallizer (Tank 1) holds a supersaturated racemic solution to which the preferred enantiomer E_1 is added in solid form. The other crystallizer (Tank 2) contains a saturated racemic solution with the same concentration to which solid racemate is added. Since the temperature in Tank 1 is below saturation level the seeds will grow reducing the level of supersaturation. Due to the exchange of clear solution a transient undersaturation with respect to E_1 will appear in Tank 2, leading to a selective dissolution of this enantiomer from the solid racemate. In this case preferential crystallization is happening in one tank while a selective dissolution in the other leads to a purification of the supplied racemate with respect to the second enantiomer E_2. The process thus yields two pure molecules while only one is needed as an investment.

This contribution will focus on how coupling crystallizers can help to make enantioseparation more efficient. The core will be a comparison of both process strategies with respect to their productivity but also their strengths and weaknesses. Further insight will be given by complementing experimental results with simulation studies directed at the influence of different process parameters.

1. Elsner, M.P., G. Ziomek, and A. Seidel-Morgenstern, *Efficient Separation of Enantiomers by Preferential Crystallization in Two Coupled Vessels*. AIChE Journal, 2009. **55**(3): p. 640-649.

2. Elsner, M.P., G. Ziomek, and A. Seidel-Morgenstern, *Simultaneous preferential crystallization in a coupled batch operation mode. Part II: Experimental study and model refinement*. Chemical Engineering Science, 2011. **66**(6): p. 1269-1284.

3. Merck & Co. Inc., *Resolution of Racemic Mixtures of Optically Active Enantiomorphs*. Patent DE1543238, 1965.

(366b) PBE-Aided Design of Slurry Milling Process to Reduce Particle Size of Intermediate and Active Pharmaceutical Ingredients

Carla Luciani, CPR&D, Eli Lilly & Co, Indianapolis, IN, Kevin D. Seibert, Chermical Product Research and Development, Eli Lilly & Co., Indianapolis, IN, Daniel Jarmer, CPRD, Eli Lilly & Co., Indianapolis, IN and Eric Moher, CPR&D, Eli Lilly & Co., Indianapolis, IN

A variety of milling processes are used in pharmaceutical industry to reduce the particle size of intermediate and active pharmaceutical ingredients. Slurry milling processes are of particular interest because they significantly reduce the transfer of heat to the product compared to dry mill equipment and they can eliminate a unit operation when crystallization-milling sequence are combined prior to the isolation. In spite of its extensive use in the pharmaceutical industry, the design of slurry milling processes are not trivial due to the multivariate nature of the process. One of the most intuitive ways to describe milling processes is the use of population balances equations (PBEs). However, relating PBE parameters (i.e., breakage rate and breakage kernel) to actual milling conditions (i.e., rotor tip speed or shear rate, slurry density, mechanical, morphological, and rheological properties of the milled material milled and the suspension medium, etc.) is challenging.

In this work, a PBE approach is applied to describe the slurry milling process of a pharmaceutical intermediate. Correlations between milling conditions and model parameters are proposed. Finally, the model is inverted to determine set of operating conditions that produce desired final particle sizes.

(366c) Development of a High Shear Crystallization Process for a Pharmaceutical Intermediate

Brenda Remy, Amit Joshi, Junying Fan, Jason Sweeney, Kenneth Natalie, Aghogho Pedro and Shawn Pack, Chemical Development, Bristol-Myers Squibb Co., New Brunswick, NJ

Batch crystallization processes for pharmaceutical intermediates are generally design to maximize the removal of impurities and to minimize yield losses. Crystallization performance is dependent on the ability to control several mass transport mechanisms such as growth, nucleation, impurity deposition etc. These mechanisms, in turn, can be impacted by controlling the amount of shear during the crystallization. Wet milling via rotor-stator devices is commonly employed in batch crystallization to control shear. While wet milling is mostly used as a particle size reduction technique, it can be used to control the purity of the isolated solids and to improve yield.

This talk will highlight the use of wet milling in a pharmaceutical intermediate crystallization to improve quality, purity and cycle time. The crystallization involves an initial mass nucleation followed by a solid-solid form conversion. Upon initial scale-up the process suffered from inconsistent yields and impurity rejection. These inconsistencies appeared to correlate with final product particle size. Wet milling achieved consistently smaller crystals and enabled faster desaturation leading to consistently higher yields. Additionally, wet milling improved the purity of the crystallized material by mitigating impurity inclusion. The use of a DoE study to determine the main parameters affecting desaturation and impurity removal will be highlighted. Scale up results from a pliot plant campaign will also be presented.

(366d) Developing a Selective Crystallization Process for a Complex Reaction System

Chenchi Wang, Truc Vu and Lucius Rossano, Research and Development, Bristol-Myers Squibb Company, New Brunswick, NJ

Developing a Selective Crystallization Process for a Complex Reaction System

Authors: Chenchi Wang, Truc Vu, Lucius Rossano.

Research and Development, Bristol-Myers Squibb Company, New Brunswick, New Jersey

A reactive crystallization process often involves multiple complex steps, such as chemical reaction, solid precipitation/crystallization, and particle size control if required. In this case study, an active pharmaceutical ingredient (API) is prepared in a telescoped process integrating deprotection, free basing and salt formation. Apart from the chemistry issues encountered that are not the focus of this presentation, the challenges of the crystallization lie in (1) maintaining chemical stability of the free base prior to crystallization, (2) quantitating the free base available for salt formation, (3) maximizing yield of the free base under process conditions, (4) enhancing purity by trace metal removal, and (v) achieving desired particle size for drug product development. In order to address these important process issues and improve process robustness, a two-step, selective crystallization process was developed. As a result of this development, a robust API synthesis step, incorporating this well-controlled reactive crystallization, was demonstrated at various scales and consistently generated material with desired purity, yield, and particle size. The designing and outcomes of this process will be discussed in details.

(366e) Crystal Nucleation Control Using Microscopic Online Imaging

Tod Canty, J.M. Canty Inc., Buffalo, NY

An innovative method of inline microscopic imaging has been developed to allow crystal viewing and measurement of size, shape, color and concentration in batch or continuous lab reactors ,pilot plant crystallizers and full production vessel at pressure(10,000psi) and temperature (-450 to 800 deg F). The high resolution images presented will show how this technology succesfully over comes the limitations of other systems where aprticle resoltions where limit to 30 micron and above due to front lighting even though the optical lenses where capable of higher pixel resoltuion. The particle imaging system captures high resolution 2 megapixel images down to .7 micron at speeds up to 1/100,000 of a second using back lighting. The system has a fiber optic lighting system that allows viewing through a wide variety of fluids (crude oil, polymer ,inks ,paint etc.). It does this at speeds through 10ft/sec in liquids. The software then analyzes the size and shape of particles compiling the data to determine particle distribution by length, width or volume along with ppm concentration. The Lab scale reactors then allow the user to optimize the seeding process to provide control of Nucleation to optimize yield and filterability.

(366f) Precipitation Kinetics of Aluminum Solids Formation During the Caustic Side Solvent Extraction (CSSX) Process

Rebecca K. Toghiani[1], Punith P. Naik[2], Jeffrey S. Lindner[2], Laura T. Smith[2] and Larry Pearson[2], (1)Dave C. Swalm School of Chemical Engineering, Mississippi State University, MS State, MS, (2)Institute for Clean Energy Technology, Mississippi State University, Starkville, MS

The Caustic Side Solvent Extraction Process (CSSX) process is used for cesium separation at the Savannah River Site (SRS). The CSSX process is comprised of 4 steps: 1) extracting cesium into the solvent; 2) scrubbed of the loaded solvent to remove co-extracted sodium and potassium; 3) stripping to transfer the cesium out of the solvent; and 4) washing of the stripped solvent to remove solvent degradation products if present. During the scrubbing process, aluminum precipitation has been observed. This is a concern as solids formation may have undesirable consequences during operation. The work reported was undertaken to identify conditions under which solids formation may occur and to identify an operating region where solids formation is avoided or minimized.

Experiments were conducted at room temperature using simulants based on different dissolution fractions of SRS Tank 25F contents. Combination of a given amount of simulant with a 0.05 M nitric acid solution mimicked the carryover during the CSSX process. Aluminum concentration and solution pH were measured over time for the prepared samples. Simulation of these experiments were also completed using the Environmental Simulation Program (ESP, OLI Systems, Inc.) and comparisons made. Solids formation was observed for a select subset of samples both experimentally and through ESP predictions. For these samples, a second set of samples was prepared. The temporal variation of aluminum concentration was measured, until solids were visually observed in each sample. The resulting data were analyzed using a first order reversible rate expression, and forward and backward reaction rate constants determined. Solids formed were isolated, dried and identified using polarized light microscopy.

(371a) Application of Split Intein in an ELP-Intein System

Changhua Shi, Chemical engineering, Ohio State University, Columbus, OH; Institute of bilological science and biotechnology, Donghua University, Shanghai, China, Qing Meng, Institute of biological science and biotechnology, Donghua University, Shanghai, China and David W. Wood, Department of Chemical and Biomolecular Engineering, the Ohio State University, Columbus, OH

Application of Split intein in an ELP-intein system

Changhua Shi, Qing Meng, David Wood

Inteins are internal protein sequences that can auto-excise from precursor proteins and splice together the flanking sequences. By substitution of amino acids, inteins can be transformed into self-cleavage elements, where cleavage is controlled by pH or thio-reagents. In our previous work, we have successfully combined a pH sensitive cleaving intein with a non-chromatographic purification tag (ELP, elastin like protein) to create a self-cleaving aggregation tag. The target protein can be easily purified by several steps of centrifugation and without expensive chromatography steps. However, the engineered intein usually has some level of pre-cleavage in vivo, depending on the target protein and expression temperature. This affects the final yield of the protein and the cost of production, especially for some highly expensive pharmaceutical proteins. This is also a problem that hampers the use of this technology at large scale. Here we introduce a split intein into our ELP-intein system to tackle this problem. The N terminal fragment of intein is removed from the ELP-intein-target protein precursor, and fused to a second ELP tag to be used as a switch for the cleavage activity of the tagged target precursor. The activity of truncated intein is tightly controlled by addition of the complementary N terminal fragment. The trans-cleaving efficiency of our split intein is similar to the full-sized intein. Our study also partially showed that the recognition and association of the intein fragments is strong even though there is a large tag, ELP, between them. Further, mutation of the first of amino acid of the target protein into the native conserved residues in this site, such as serine and cysteine can increase the cleavage efficiency. This dual ELP-intein technology can also be applied to do cost-effective protein purification in many other systems, such as yeast and mammalian cells.

(371b) Use of Flocculation to Improve Clarification Efficiency in a Microbial Feedstream

Jessica R. Molek, Jason M. Reck and Kent E. Goklen, Downstream Process Development, GlaxoSmithKline, King of Prussia, PA

E. coli is one of the fastest growing, least expensive, and highest product-to-volume systems available for the production of therapeutic proteins. Despite these advantages, efficient clarification of the product stream can be challenging due to its high solids content, propensity for cell lysis, as well as high concentrations of impurities such as DNA and host cell protein. While centrifugation can be used to remove larger particles, such as whole cells, many small particles, such as cell debris and protein DNA complexes, are difficult to sediment. Depth filtration can be used to remove fine particulates but capacities of the depth filters can be limited, especially by high DNA concentration. In this presentation, polycation and divalent cation precipitation are explored as methods to enhance the removal of cells, cell debris and DNA during centrifugation processes. Screening is performed using a combination of high throughput methods and bench scale experimentation. We examined the impact of several polycations and divalent salts, as well as changes to the pH and conductivity of the fermentation broth on the overall clarification process. Subsequent scale-up work has demonstrated a minimum four-fold increase in depth filter capacity using optimized flocculation conditions.

(371c) Fumaric Acid Recovery From Fermentation Broth by Activated Carbon Adsorption Followed by Desorption and Crystallization with Acetone

Kun Zhang, Chemical and Biomolecular Engineering, The Ohio State University, Columbus, OH, Lijie Zhang, Dept. of Pharmacy and Bioengineering, Chongqing University of Technology, Chongqing, China and Shang-Tian Yang, Department of Chemical and Biomolecular Engineering, Ohio State University, Columbus, OH

Fumaric acid, a dicarboxylic acid extensively used as a food acidulant and in the manufacturing of synthetic resins, can be produced from glucose in aerobic fermentation by the filamentous fungus *Rhizopus oryzae*. For fumaric acid recovery, acidification at a high temperature is usually required because of the low solubility of fumaric acid. Then, fumaric acid crystals can be obtained by cooling. However, the produced crystals may contain various impurities and need re-crystallization. In order to enhance the purity of fumaric acid, adsorption with activated carbon was investigated as a more economical separation method. The results from model fermentation broth (15 g/L fumaric acid, 5 g/L glucose, 5 g/L malic acid) showed that activated carbon had a high selectivity and high adsorption capacity (~250 mg/g) for fumaric acid, which generally increased with increasing the fumaric acid concentration in the solution. Desorption of fumaric acid from the activated carbon was then studied using acetone, which can effectively strip fumaric acid from activated carbon. After removing acetone by evaporation at 50 ^{o}C, fumaric acid crystals were obtained with a near 100% purity. The furmaric acid recovery yield from this process was affected by the amount of acetone and temperature used in desorption. After optimization, over 95% fumaric acid was recovered from activated carbon. Both activated carbon and acetone can be recovered and reused in the adsorption process. This process is effective and economical for fumaric acid recovery and purification, and should have great potential in industrial production of fumaric acid.

(371d) Membrane Fouling Mechanism in Ultrafiltration of Succinic Acid Fermentation Broth

Jianmin Xing, National Key Lab of Biochemical Engineering, Institute of Process Engineering, Chinese Academy of Sciences, Beijing, China, Caixia Wang, Institute of Process Engineering, Chinese Academy of Sciences, Beijing, China and Yinhua Wan, National Key Laboratory of Biochemical Engineering, Institute of Process Engineering, Chinese Academy Of Sciences, Beijing, China

Succinic acid (1, 4-butanedioic acid) is identified as an important building-block chemical which can be used for the synthesis of high value-added derivatives, such as 1, 4-butanediol (BDO), g-butyrolactone (GBL), tetrahydrofuran (THF), N-methyl-2-pyrrolidone (NMP), 2-pyrrolidone (2-Pyrr), succinimide, succinic esters, etc. Great efforts have been made to develop biotechnology methods to produce succinic acid from renewable resources since last decade.

However, fermentation process is still not economically feasible, mainly due to the low succinic acid concentration in the fermentation broth, and the high cost of succinic acid separation and purification. Considerable efforts have been made to improve the succinic acid concentration in the fermentation broth and several possible operation units such as adsorption, extraction and electrolysis have been investigated for the recovery of succinic acid.

In this study, fouling mechanisms of different membranes were fist studied in treating succinic acid fermentation broth during ultrafiltration process. The membrane fouling mechanism was studied during dead-end ultrafiltration. Different membranes were used and two models were applied to analyze the fouling mechanism. Resistance-in-series model was applied to determine the main factor that caused the operation resistance. Results indicated that most membranes tended to be fouled by cake layer or concentration polarization. Hermia's model, which is composed of four individual sub-models, was used to analyze the predominant fouling mechanism. Results showed that the fouling of RC 10kDa and PES 30kDa was controlled by the complete blocking mechanism, while PES 100kDa was controlled by the intermediate blocking and PES 10kDa was controlled by cake layer. This conclusion was also proved by SEM photos. Membrane characteristics were monitored before and after ultrafiltration by AFM and goniometer. Both contact angle and roughness of most membranes increased after ultrafiltration.

(371e) Separation of n-Butanol From Dilute Solution by Pervaporation Using Composite PDMS Membrane

Fangfang Liu, The Ohio State University, Columbus, OH, Kaijun Xiao, College of Light Industry and Food Science, South China University of Technology, Guangzhou, China and Shang-Tian Yang, Department of Chemical and Biomolecular Engineering, Ohio State University, Columbus, OH

Butanol has attracted increasing attention as an alternative to the fossil fuel. Before we can use butanol as a liquid fuel, it is necessary to refine the low-concentration butanol produced in fermentation. Pervaporation, as one of the membrane separation technologies, has high selectivity and low energy consumption compared to other separation techniques. Pervaporation can be used to separate azeotropic mixtures and other kind of mixtures, which are usually difficult to separate by conventional techniques like distillation. Poly(dimethyl siloxane) (PDMS) membrane is the most effective and widely used membrane for separating butanol from an aqueous solution. However, butanol flux and selectivity of PDMS membranes vary with fabrication methods and process conditions, and need further improvements for economical applications. In this work, different membrane fabrication techniques were explored, including plasma treatment, dipping, and electrospinning. The effects of membrane supporting materials, including poly(vinylidene fluoride) (PVDF), polysulfone (PS), and poly(tetrafluoroethylene) (PTFE), on the pervaporation performance of PDMS membrane were also studied. Further improvements in butanol selectivity and flux by the addition of hydrophobic fillers to the PDMS membrane were also investigated. Among the membranes studied, the PDMS membrane with a thickness of ~70 μm, supported by plasma treated PTFE and filled with 20 wt% CBV28014, gave the best separation performance, with a separation factor of 27 and a butanol flux of 48 $g/m^2 \cdot h$. The PTFE-supported PDMS membrane can concentrate a 15 g/L butanol solution to a 290 g/L butanol solution, and is promising for use in pervaporation for butanol recovery from dilute fermentation broth.

A TECHNIQUE FOR MASS BALANCE OF PROCESS DATA USING LINEAR COMBINATION OF LSQ AND LRSQ

A. I. A. Salama, W. Friesen, and T. Dabros
CanmetENERGY-Devon, Natural Resources Canada, Devon, AB, Canada

Introduction

In many oil sands, mineral, and chemical processing plants, plant audits are conducted, routinely, to evaluate the plant/process performance using mass recoveries and other separation criteria. The plant audit starts with sampling of plant streams and samples are collected using well-designed sampling scheme (Smith and Frew, 1983; Frew, 1983) which was improved, later, by Salama (1999a). The stream samples are subjected to analytical techniques to determine stream components (SCs). Furthermore, stream mass splits (SMS) / rates may be measured using online instrumentation. In most cases, the collected samples and SMS contain errors (e.g., sampling, measurement, and analytical) and must be balanced (or adjusted). The MB problem can be stated as: given a plant flowsheet and an associated set of observed (raw) data, find the optimal balanced set of SCs and SMSs such that an objective function is optimized. Simultaneously, a set of mass conservation constraints is satisfied.

However, in real time on-line optimization provides set points to the processes distributed control system (DCS) and therefore maintains the process near its optimum operating conditions (Ozyrut and Pike, 2004). Therefore, the process flowsheet mass balance and the real time on-line optimization are addressing two different objectives.

In almost all MB techniques, the stream consistency conditions (i.e., the sum of SCs in each stream adds up to 100%) are imposed and the SCs and SMSs are balanced at each node of the flowsheet. The only difference between the different techniques is the objective function of the optimization problem (which depends on the observed data error distributions or error model). In the:

- Conventional least error squares (LSQ) case, the objective function is defined as the sum of error (difference between the raw and estimate SCs) squares referenced to the standard deviations of raw SC errors. In this case, the stream data error distributions are assumed normal (Gaussian) (Kaput et al., 1993, Hodouin and Vaz Coelho, 1987, and Klimpel, 1979).
- Maximum likelihood technique, the data errors are assumed uniform or exponential to cover cases where many outliers are present in the raw data. For each error distribution a corresponding objective function is derived (Klimpel, 1979).
- Maximum entropy technique, the objective function is derived based on multinomial error distribution (Klimpel, 1979 and Kullback, 1959). The property of this technique is that no bias towards any particular data point (Jefferson and Scott, 1986, Kullback, 1959, Jaynes, 1957).
- Least relative error squares (LRSQ) case which is the subject of the present paper, the objective function is defined as the sum of, data error referenced to the raw data, squared.

Note that the focus of the present paper is to investigate the mass balance problem rather than modeling of the process. Using an observed data set try to search for a corresponding optimal data set such that a linear combination of the error and relative error squares is minimized. The advantages of the technique include:

- Cases where raw data error model is not available.
- Economic constraints may prevent the use of complete sampling campaign, especially, in early stages of new or proposed flowsheet developments.

- Conducting several preliminary tests to evaluate the feasibility and optimization of new or proposed or expanded flowsheet.

The process

One empirical method for obtaining data required in the engineering design of oil sands deposition process is to perform a simulation using a flume. A known amount of slurry is discharged continuously at one end of an open channel and flows toward the exit at the other end. A picture of the CanmetENERGY-Devon flume unit is shown in Figure 1. The mechanical dimensions of the unit are length (7.25 m), depth (0.6 m), and width (0.6 m). It should be emphasized that while the following developments are focused on a flume process, the results can be modified to handle any other general process (see the results of Salama, 2001 and 1999b).

Mathematical formulation

Let the raw data of feed, stages, and *ro* SCs be designated: p_{ij}, i=1,2,…, N_c, j=1,2,…, N_{st}+2. Where N_c, N_{st}, and N_{st}+2 are the number of stream components (SC)s, stages, and flume streams, respectively. Furthermore, the observed SMSs: (feed) $\boldsymbol{P}_1=\boldsymbol{P}^f$, (stages) $\boldsymbol{P}_j$, j=2,3,…,N_{st}+1, and (*ro*) $\boldsymbol{P}_{Ns+2}=\boldsymbol{P}^{ro}$, respectively. In most oil sands applications, the SCs are bitumen (hydrocarbons), minerals (solids), and water contents and are labeled B, M, and W, respectively. This order will be kept during the presentation. In this case the number of SCs N_c =3. The raw data may be arranged in an array $\boldsymbol{P}$

$$\boldsymbol{P}=\begin{bmatrix} p_{11} & p_{12} & \cdot & \cdot & \cdot & \cdot & p_{1N_{St}+1} & p_{1N_{St}+2} \\ p_{21} & p_{22} & \cdot & \cdot & \cdot & \cdot & p_{2N_{St}+1} & p_{2N_{St}+2} \\ p_{31} & p_{32} & \cdot & \cdot & \cdot & \cdot & p_{3N_{St}+1} & p_{3N_{St}+2} \\ \boldsymbol{P}_1 & \boldsymbol{P}_2 & \cdot & \cdot & \cdot & \cdot & \boldsymbol{P}_{N_{St}+1} & \boldsymbol{P}_{N_{St}+2} \end{bmatrix} \quad (1)$$

Let the optimal estimates or mass balanced data of SCs be designated as: x_{ij}, i=1,2,…,N_c, j=1,2,…, N_{st}+2. And the associated estimated SMSs: (feed) $\boldsymbol{X}_1=\boldsymbol{X}^f$, (stages) $\boldsymbol{X}_j$, j=2,3,…,N_{st}+1, (*ro*) and $\boldsymbol{X}_{Nst+2}=\boldsymbol{X}^{ro}$. The mass balanced data may be arranged in another array $\boldsymbol{X}$

$$\boldsymbol{X}=\begin{bmatrix} x_{11} & x_{12} & \cdot & \cdot & \cdot & \cdot & x_{1N_{St}+1} & x_{1N_{St}+2} \\ x_{21} & x_{22} & \cdot & \cdot & \cdot & \cdot & x_{2N_{St}+1} & x_{2N_{St}+2} \\ x_{31} & x_{32} & \cdot & \cdot & \cdot & \cdot & x_{3N_{St}+1} & x_{3N_{St}+2} \\ \boldsymbol{X}_1 & \boldsymbol{X}_2 & \cdot & \cdot & \cdot & \cdot & \boldsymbol{X}_{N_{St}+1} & \boldsymbol{X}_{N_{St}+2} \end{bmatrix} \quad (2)$$

Note that any other format different from the array representation is also applicable. In the present development the flowsheet estimated SMSs must satisfy the condition

$$\boldsymbol{X}^f=\boldsymbol{X}_1=\sum_{j=2}^{N_{st}+2}\boldsymbol{X}_j \quad (3)$$

i.e., the SMS of the feed equals the sum of stages and *ro* SMSs. Commonly, the SMS of the feed, $\boldsymbol{X}^f=\boldsymbol{X}_1$, is used as a reference and set equals 100% or unity; however, the label $\boldsymbol{X}_1$ will be kept in the subsequent equations. Using Eq. (3), the $\boldsymbol{X}_{Nst+2}$ (*ro* SMS, $\boldsymbol{X}^{ro}$) can be expressed as

$$\boldsymbol{X}^{ro}=\boldsymbol{X}_1-\sum_{j=2}^{N_{st}}\boldsymbol{X}_j \quad (4)$$

In Eq. (3), the circuit SMSs can be normalized relative to $\boldsymbol{X}_1$ as

$$1=\sum_{j=2}^{N_{st}+2}\frac{X_j}{X_1} \tag{5}$$

Furthermore, the flume feed, stages, and *ro* SCs must satisfy the SC consistency conditions or

$$\sum_{i=1}^{N_c} x_{ij}=1, \quad j=1,2,....,N_{st}+2 \tag{6}$$

Moreover, the SCs mass balance at the flume node of the flume circuit must satisfy the conditions

$$X_1 x_{i1}-\sum_{j=2}^{N_{st}+2} X_j x_{ij}=0, \quad i=1,2,....,N_c-1 \tag{7}$$

It should be emphasized that we considered only N_c-1 component mass balances (see Salama, 1999b). The reason is that the conditions in Eqs. (3-6) can be shown to imply the condition in Eq. (7) at $i=N_c$. This is important because the use of more constraints than what is needed may complicate the search for the optimal solution. One more note, if N_c=1, only one component is considered in Eq. (7).

Mass balance

Optimization Problem

Given the observed SCs: p_{ij}, i=1,2,…,N_c, j=1,2, …,N_{st}+2 and the associated SMSs: P_j, j=1,2,…, N_{st}+1, and $P_{Nst+2}=P^{ro}$, find the optimal SCs: x_{ij}, i=1,2,…, N_c, j=1,2,…,N_{st}+2 and the associated SMS: X_j, j=1,2, …,N_{st}+1, and $X_{Nst+2}=X^{ro}$, such that the objective function

$$\phi=\sum_{i=1}^{N_c}\sum_{j=1}^{N_{st}+2}\left(\frac{x_{ij}-p_{ij}}{\left[\alpha p_{ij}+(1-\alpha)\sigma_{ij}\right]}\right)^2+\sum_{j=1}^{N_{st}+2}\left(\frac{X_j-P_j}{\left[\alpha P_j+(1-\alpha)\sigma_j\right]}\right)^2 \tag{8}$$

is minimized and simultaneously the constraints in Eqs. (3-7) are satisfied, where σ_{ij}'s and σ_j's are the standard deviations of SC ij and SMS j, respectively. Note that in Eq. (8), selecting $\alpha=1$ corresponds to LRSQ case, selecting $\alpha=0$ corresponds to the conventional LSQ case, and for any general α, $\alpha\in[0,1]$, corresponds to linear combination of LSQ and LRSQ cases.

Necessary Conditions

The solution of the present optimization problem can be determined using the necessary conditions of the stationary point of the optimization problem Lagrangian defined as (Fletcher, 1987)

$$\begin{aligned}\mathscr{L}=&\sum_{i=1}^{N_c}\sum_{j=1}^{N_{st}+2}\left(\frac{x_{ij}-p_{ij}}{\left[\alpha p_{ij}+(1-\alpha)\sigma_{ij}\right]}\right)^2+\sum_{j=1}^{N_{st}+2}\left(\frac{X_j-P_j}{\left[\alpha P_j+(1-\alpha)\sigma_j\right]}\right)^2+\\&+\sum_{j=1}^{N_{st}+2}\lambda_j\left(1-\sum\nolimits_{i=1}^{N_c}x_{ij}\right)+\sum_{i=1}^{N_c-1}\psi_i\left(X_1x_{i1}-\sum_{j=2}^{N_{st}+2}X_jx_{ij}\right)\end{aligned} \tag{9}$$

where λ_j, j=1,2,…,N_{st}+1, $\lambda_{Nst+2}=\lambda^{ro}$, and ψ_i, i=1,2,…,N_c-1 are constant multipliers. The parameter $\alpha\in[0,1]$ is introduced to determine general optimization results. The cases of available raw data error model and relative error squares case are special cases of the general results.

The necessary conditions are:

for i=1,2,…,N_c-1

$$\frac{\partial\mathscr{L}}{\partial x_{i1}}=2\left(\frac{x_{i1}-p_{i1}}{\left[\alpha p_{i1}+(1-\alpha)\sigma_{i1}\right]}\right)\left(\frac{1}{\left[\alpha p_{i1}+(1-\alpha)\sigma_{i1}\right]}\right)-\lambda_1+\psi_iX_1=0 \tag{10}$$

$$\frac{\partial \mathscr{L}}{\partial x_{ij}} = 2\left(\frac{x_{ij} - p_{ij}}{\left[\alpha p_{ij} + (1-\alpha)\sigma_{ij}\right]}\right)\left(\frac{1}{\left[\alpha p_{ij} + (1-\alpha)\sigma_{ij}\right]}\right) - \lambda_j - \psi_i \boldsymbol{X}_j = 0, \quad j=2,\ldots,N_{st}+2 \tag{11}$$

and for $i=N_c$

$$\frac{\partial \mathscr{L}}{\partial x_{N_c j}} = 2\left(\frac{x_{N_c j} - p_{N_c j}}{\left[\alpha p_{N_c j} + (1-\alpha)\sigma_{N_c j}\right]}\right)\left(\frac{1}{\left[\alpha p_{N_c j} + (1-\alpha)\sigma_{N_c j}\right]}\right) - \lambda_j = 0, \quad j=1,2,\ldots,N_{st}+2 \tag{12}$$

and for $j=2,\ldots,N_{st}+1$

$$\begin{aligned}\frac{\partial \mathscr{L}}{\partial \boldsymbol{X}_j} &= 2\left(\frac{\boldsymbol{X}_j - \boldsymbol{P}_j}{\left[\alpha \boldsymbol{P}_j + (1-\alpha)\sigma_{ij}\right]}\right)\left(\frac{1}{\left[\alpha \boldsymbol{P}_j + (1-\alpha)\sigma_{ij}\right]}\right) + \\ &+ 2\left(\frac{\boldsymbol{X}^{ro} - \boldsymbol{P}^{ro}}{\left[\alpha \boldsymbol{P}^{ro} + (1-\alpha)\sigma^{ro}\right]}\right)\left(\frac{1}{\left[\alpha \boldsymbol{P}^{ro} + (1-\alpha)\sigma^{ro}\right]}\right)(-1) - \sum_{i=1}^{N_c-1} \psi_i \left(x_{ij} - x_i^{ro}\right) = 0\end{aligned} \tag{13}$$

and Eqs. (6 and 7). Using Eqs. (10-12) and after some manipulations we get

$$x_{i1} = p_{i1} + 0.5\left[\alpha p_{i1} + (1-\alpha)\sigma_{i1}\right]^2 \left[\lambda_1 - \psi_i \boldsymbol{X}_1\right] \tag{14}$$

for $i=1,2,\ldots, N_c-1$

$$x_{ij} = p_{ij} + 0.5\left[\alpha p_{ij} + (1-\alpha)\sigma_{ij}\right]^2 \left[\lambda_j + \psi_i \boldsymbol{X}_j\right], \quad j=2,3,\ldots,N_{st}+1 \tag{15}$$

where $\lambda_{Nst+2} = \lambda^{ro}$, for $i=N_c$

$$x_{N_c j} = p_{N_c j} + 0.5\left[\alpha p_{N_c j} + (1-\alpha)\sigma_{N_c j}\right]^2 \left[\lambda_j\right], \quad j=1,2,\ldots,N_{st}+2 \tag{16}$$

To facilitate the presentation, the following condition is assumed;

$$\sum_{i=1}^{N_c} p_{ij} = 1, \quad j=1,2,\ldots,N_{st} \tag{17}$$

and using Eqs. (14-16) and after some manipulations we get

$$\lambda_1 = \boldsymbol{X}_1 \frac{\sum_{i=1}^{N_c-1} \psi_i \left[\alpha p_{i1} + (1-\alpha)\sigma_1\right]^2}{\sum_{i=1}^{N_c} \left[\alpha p_{i1} + (1-\alpha)\sigma_1\right]^2} \tag{18}$$

$$\lambda_j = -\boldsymbol{X}_j \frac{\sum_{i=1}^{N_c-1} \psi_i \left[\alpha p_{ij} + (1-\alpha)\sigma_{ij}\right]^2}{\sum_{i=1}^{N_c} \left[\alpha p_{ij} + (1-\alpha)\sigma_{ij}\right]^2}, \quad j=2,\ldots,N_{st}+1 \tag{19}$$

where $\lambda_{Nst+2} = \lambda^{ro}$. It should be noted that, if the conditions in Eq. (17) are not satisfied, they don't cause any loss of generality because Eqs. (18 and 19) can be easily modified.

Demonstration

To visualize how the LSQ and LRSQ adjust raw data let us consider a two-dimensional case. The raw SCs are p_1 and p_2 and the corresponding estimates are x_1 and x_2 as shown in Figure 2. The constraint, $x_1+x_2=1$ is imposed. In the LSQ case the objective function is defined as $\phi=(x_1-p_1)^2+(x_2-p_2)^2$. The optimal solution is located when a concentric circle becomes tangent to the $x_1+x_2=1$ (X_L*). Close observation of the LSQ solution indicates that: a) for all raw points on the lines parallel above and below the line $x_1+x_2=1$ generates the same adjustments and b) if either one of the SCs p_1 or p_2 becomes greater than one the optimal solutions have negative values. Both results are disadvantages. The LRSQ solution is a directed line from the raw data point (p_1, p_2) to the $x_1+x_2=1$ is labeled X_R* (see Figure 2). The X_R* solution indicates that: a) the corrections between X_R* and P are proportional to the ratio of p_1 and p_2 squared, and b) if one of SCs, p_1 or p_2, becomes zero the corresponding optimal SCs, x_1* or x_2*, are close to zero. In this case both results are advantageous. It should be emphasized that Figure 2 is only given as demonstration but no further conclusions can be drawn.

A raw data set collected around the CanmetENERGY-Devon flume unit is used to generate the data presented in Table 1. The normalization of the raw data of Table 1 is given in Table 2. The proposed technique is implemented in spread sheet format using Microsoft Excel™. Furthermore, the solver module embedded within Excel is used to solve the optimization problem stated in the previous subsection. Using the data in Table 2 and implementing the proposed technique for $\alpha = 1$, i.e., LRSQ case, the results obtained are given in Table 3. The corresponding SC's distributions along the flume are shown in Figure 3. Close examination of Figure 3 indicates that; a) the estimated and the observed data are very close, which reflects the reliability of the raw data, b) naturally, most of the hydrocarbons and solids in the feed slurry are deposited longitudinally along the flume sections. The case $\alpha = 0.8$, i.e., linear combination of LRSQ and LSQ cases. In this case a set of σ_{ij}s is considered and is given in Table 4. The σ_{ij}s of *f* and *ro* streams are two's and other σ_{ij}s are ones. The reason for this selection is that the *f* and *ro* streams produce good representative samples and their analyses are more reliable. Implementation of the proposed technique produced the results given in Table 5. The corresponding SC's are shown in Figure 4. Using the SCs given in Tables 3 and 5, the corresponding cumulative SC recoveries are given in Figures 5 and 6. For $\alpha = 1$ and close examination of Figure 5 shows that ~ 96% of B and M are deposited longitudinally along the flume bed and the remaining ~ 4% are carried over in the *ro* stream. And ~ 20% of water is retained over the flume bed and ~ 80% are discharged from the flume bed. For $\alpha = 0.8$ and close examination of Figure 6 indicates that ~ 13% M and 2% B are discharged and the remaining material are deposited along the flume bed.

Conclusions

Based on the results presented, the following conclusions may be drawn:

- The proposed technique is recommended to handle cases where the raw data statistics (i.e., errors model) are not available.
- The dimension of the optimization problem is reduced to $N_{st}+N_c$ manipulating variables (N_{st} SMSs and N_c constant multipliers) and the $N_{st}+N_c$ equality constraints.
- The resulting balanced data are used to calculate performance measures such as performance (separation) measures.
- Deposition of solids and liquid on inclined surfaces can be modeled using the proposed technique results.
- Note that the mechanical dimensions of CanmetENERGY-Devon flume unit along with mass rate and SCs data can be used in the scale up to industrial oil sand flumes.
- Data can assist the process engineer to analyze/modify his process flowsheet.
- It should be emphasized that the development in this paper is given to demonstrate how the estimated data are related to the observed data in the mass balance of LSQ, LRSQ, and linear combination of LSQ and LRSQ cases.

Finally, it should be emphasized that the present technique can be extended to handle any other general process (using the details given in Salama, 2001 and 1999b).

Acknowledgement

The author would like to acknowledge support from the Government of Canada's interdepartmental Program of Energy Research and Development, PERD POL.

References

1. Draper, N.R. and Smith, H. (1966), "Applied Regression Analysis," John Wiley and Sons Ltd., New York.
2. Fletcher, R. (1987), "*Practical Methods of Optimization*,". John Wiley and Sons Ltd., New York.
3. Frew, J.A. (1983), "Computer-aided design of sampling schemes," *Int. J. Mineral Process.*, 11: pp. 255-265.
4. Hodouin, D. and Vaz Coelho, S. (1987), "Mass balance calculations around mineral processing unit using composition analyses within particle-size classes," *Int. J. Mineral Process.*, 21: pp. 65-82.
5. Jaynes, E.T. (1957), "Information theory and statistical mechanics," *Phys. Review*, 106: pp. 620-630.
6. Jefferson, T.R. and Scott, C.H. (1986), "The entropy approach to material balancing in mineral processing," *Int. J. Mineral Process.*, 18: pp. 251-261.
7. Kapur, P.C., Agrawal, M., and Fuerstenau, D.W. (1993), A two-tier approach to the mass balance of multi-level mineral processing data. Par 1: performance index and local balance around a node. *Int. J. Mineral Process.*, 39: pp. 209-223.
6. Kapur, P.C., Agrawal, M., and Fuerstenau, D.W. (1993), A two-tier approach to the mass balance of multi-level Mineral processing data. Par 2: splicing of stream flows and global balance over a circuit. *Int. J. Mineral Process.*, 39: pp. 225-237.
7. Klimpel, R. (1979), "Estimation of weight ratios given component make-up analyses of streams," *SME/ AIME Trans.*, 266: pp. 1882-1886.
8. Kullback, S. (1959), "Information Theory and Statistics," Dover Publications, New York.
9. Ozyrut, D.B. and R.W. Pike (2004), "Theory and Practice of simultaneous Data reconciliation and Gross Error detection for chemical processes," *Computers and Chemical Engineering.*, 28, pp. 381–402.
10. Salama, A.I.A. (1999a), "Sampling schemes for mass balance of flowsheets with multi-component streams." *Int. J. Mineral Process.*, 55: pp. 219-229.
11. Salama, A.I.A. (1999b), "The maximum entropy approach and comparison with the least-squares approach for flowsheet mass balance," *Int. J. Mineral Process.*, 57: pp. 219-241.
12. Salama, A.I.A. (2001), "Flowsheet Mass Balance with Partial Observation of Stream Components, *Int. J. Mineral Process.*, 63, pp. 59–77.
13. Smith, H.W. and Frew, J.A., 1983. Design and analysis of sampling experiment - a sensitivity approach. *Int. J. Mineral Process.*, 11: pp. 267-284.

Table 1 – A data set generated based on a CanmetENERGY–Devon observed data set

OBSERVED DATA (%)									
C \ S	*f*	1	2	3	4	5	6	7	*ro*
Hydrocarbon, B	4.6000	19.6000	19.4000	16.5000	13.4000	11.4000	12.6000	11.6000	0.2800
Minerals, M	6.5000	17.1000	14.1000	17.3000	16.2000	15.6000	15.5000	14.5000	0.3200
Water, W	88.9000	63.3000	66.5000	66.2000	70.4000	73.0000	71.9000	73.9000	99.4000
Sum	100.00	100.00	100.00	100.00	100.00	100.00	100.00	100.00	100.00

Table 2 – Normalization of the data set of Table 1

OBSERVED DATA (FRACTIONAL)									
C \ S	*f*	1	2	3	4	5	6	7	*ro*
Hydrocarbon, B	0.046	0.196	0.194	0.165	0.134	0.114	0.126	0.116	0.003
Minerals, M	0.065	0.171	0.141	0.173	0.162	0.156	0.155	0.145	0.003
Water, W	0.889	0.633	0.665	0.662	0.704	0.730	0.719	0.739	0.994
Sum	1.000	1.000	1.000	1.000	1.000	1.000	1.000	1.000	1.000
Mass rate	1200.000	97.660	78.791	66.354	46.934	29.975	14.221	3.079	862.986
Pi's	1.000	0.081	0.066	0.055	0.039	0.025	0.012	0.003	0.719

Table 3– Proposed technique results using the data in Table 2, α=1

ESTIMATED DATA - PROPOSED MASS BALANCE APPROACH										
Y_i's	1.000	0.0835	0.0665	0.0564	0.0397	0.0252	0.0119	0.0026	0.714	-7.201	
λ's	-7.172	0.586	0.476	0.392	0.276	0.176	0.083	0.018	5.144	-10.539	ψ's
Hydrocarbon, B	0.0496	0.190	0.190	0.162	0.133	0.113	0.126	0.116	0.003	0.000	
Minerals, M	0.0498	0.180	0.146	0.179	0.166	0.158	0.156	0.145	0.003	0.000	
Water, W	0.9006	0.630	0.664	0.659	0.702	0.728	0.718	0.739	0.994		
Sum	1.0000	1.000	1.000	1.000	1.000	1.000	1.000	1.000	1.000		
$\partial\mathscr{L}/\partial X$=0		0.000	0.000	0.000	0.000	0.000	0.000	0.000		5.1E-29	ϕ
Stream #	*f*	1	2	3	4	5	6	7	*ro*	Bal 1^2 + Bal 2^2	

Table 4 – Proposed σ_{ij} using LSQ

σ_{ij}									
C \ S	*f*	1	2	3	4	5	6	7	*ro*
Hydrocarbon, B	1.000	1.000	1.000	1.000	1.000	1.000	1.000	1.000	1.000
Minerals, M	1.000	1.000	1.000	1.000	1.000	1.000	1.000	1.000	1.000
Water, W	1.000	1.000	1.000	1.000	1.000	1.000	1.000	1.000	1.000
Pi's	1.000	1.000	1.000	1.000	1.000	1.000	1.000	1.000	1.000

Table 5 – Proposed technique results using the data in Tables 2 and Table 4, α=0.8

ESTIMATED DATA - PROPOSED MASS BALANCE APPROACH										
Y_i's	1.000	0.0819	0.0660	0.0557	0.0395	0.0253	0.0121	0.0028	0.717	-0.132	
λ's	-0.117	0.009	0.008	0.006	0.005	0.003	0.001	0.000	0.087	-0.148	ψ's
Hydrocarbon, B	0.0489	0.196	0.194	0.165	0.134	0.114	0.126	0.116	0.001	0.000	
Minerals, M	0.0530	0.172	0.141	0.173	0.162	0.156	0.155	0.145	0.010	0.000	
Water, W	0.8981	0.633	0.665	0.662	0.704	0.730	0.719	0.739	0.988		
Sum	1.0000	1.000	1.000	1.000	1.000	1.000	1.000	1.000	1.000		
$\partial\mathscr{L}/\partial X$=0		0.000	0.000	0.000	0.000	0.000	0.000	0.000		9.7E-31	ϕ
Stream #	*f*	1	2	3	4	5	6	7	*ro*	Bal 1^2 + Bal 2^2	

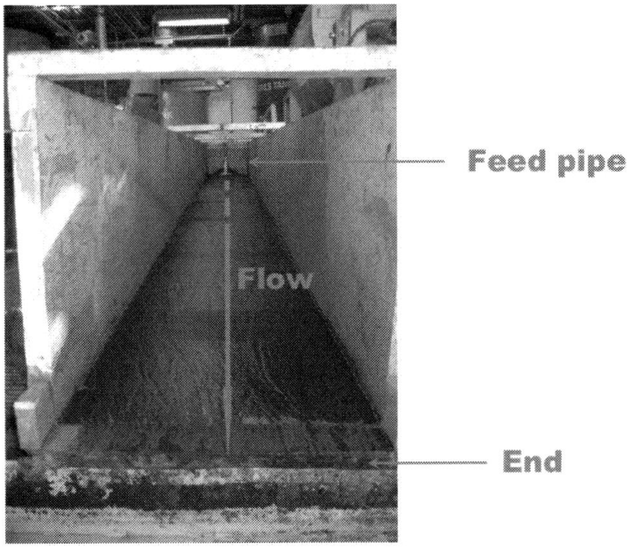

Figure 1 – CanmetENERGY –Devon Flume

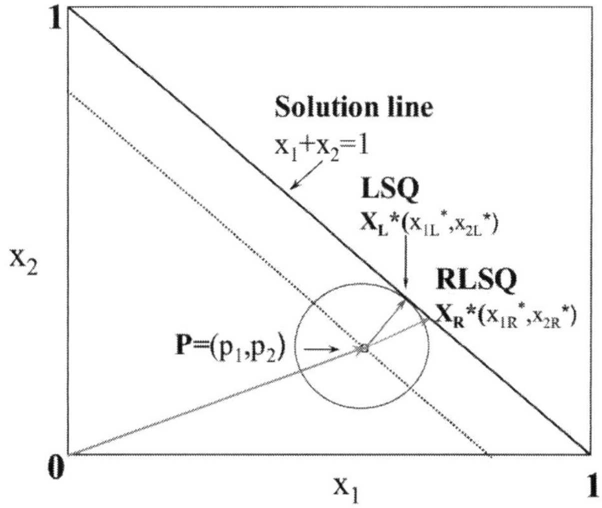

Figure 2 – A two-dimensional demonstration of LSQ and LRSQ formulations

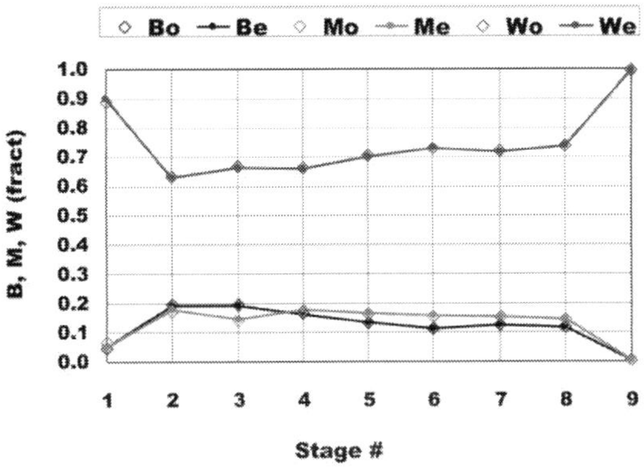

Figure 3 – Estimated SC distribution using the data of Table 2, α=1.0

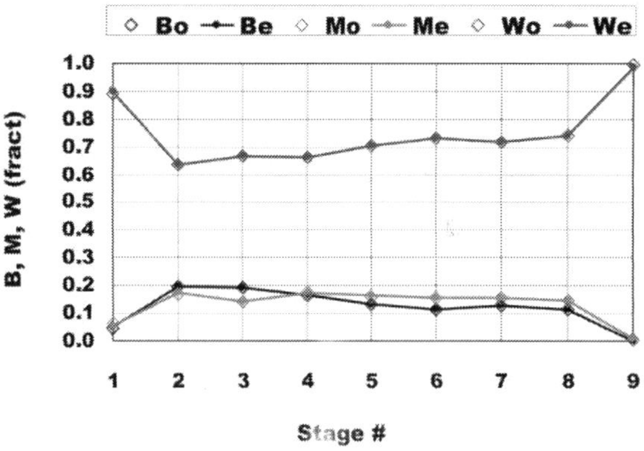

Figure 4 – Estimated SC distribution using the data of Table 2, α=0.8

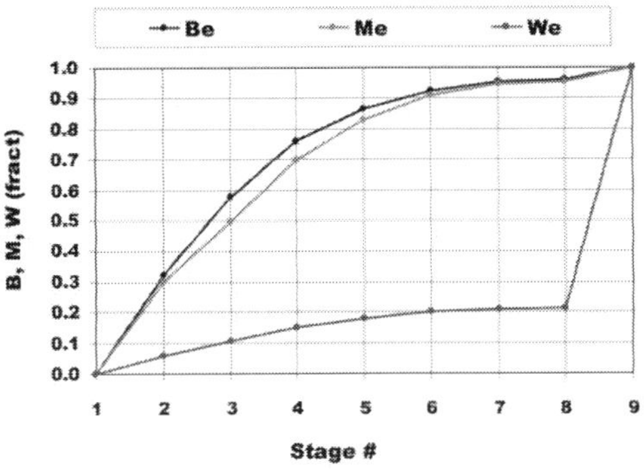

Figure 5 – Estimated SC cumulative recovery distributions α=1

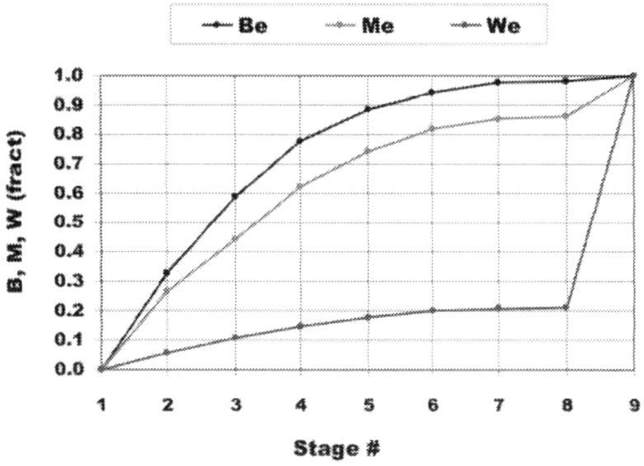

Figure 6 – Estimated SC cumulative recovery distributions α=0.8

(379a) Magnetically Responsive Membranes As Micromixers and Nanoheaters

Qian Yang, Ralph E Martin Department of Chemical Engineering, University of Arkansas, Fayetteville, AR, Heath Himstedt, Chemical and Biological Engineering, Colorado State Univeristy, Fort Collins, CO, Xianghong Qian, Ralph E Martin Department fo Chemical Engineering, University of Arkansas, Fayetteville, AR; Ralph E Martin Department of Chemical Engineering, University of Arkansas, Fayetteville, AR and S. Ranil Wickramasinghe, Department of Chemical Engineering, University of Arkansas, Fayetteville, AR

Magnetically Responsive Membranes as Micromixers and Nanoheaters

Qian YANG[1], Heath H. HIMSTEDT[2], Xianghong QIAN[1], S. Ranil WICKRAMASINGHE[1]

[1] Ralph E Martin Department of Chemical Engineering, University of Arkansas, Fayetteville 72701, USA;
[2]Department OF Chemical and Biological Engineering, Colorado State University, Fort Collins, CO 80523.

Two of the most important factors limiting the use of membrane technology are concentration polarization and membrane fouling.[1] Efforts have been made to solve these problems by means of optimizing module design or inducing mixing during operation, pretreatment of the feed and surface modification of the membrane. Feed pretreatment can only postpone the decline of membrane performance while surface modification is not able to suppress concentration polarization.[2] Therefore, there is still a need to find an efficient way to improve anti-fouling property of membranes and to reduce concentration polarization.

Herein, we report a novel approach to solve these problems by attaching magnetically responsive nanoparticles to the membrane surface. As can be seen from **Figure 1**, polymers (polyHEMA and poly NIPAAm) were grafted to membrane surfaces by surface initiated ATRP. After that, the bromine groups at chain ends were converted primary amine by Gabriel synthesis and then used for attaching carboxyl groups covered superparamagnetic Fe_3O_4 nanoparticles. The chain/nanoparticle density can be controlled by varying the ratio of active to inactive ATRP initiator in the initiator immobilization step. The modification and nanoparticle immobilization were monitored by XPS and SEM. XPS confirmed each step of modification. SEM images showed that the nanoparticles were successfully attached to the membrane surface and the density can be tuned in a broad range (see **Figure 2**).

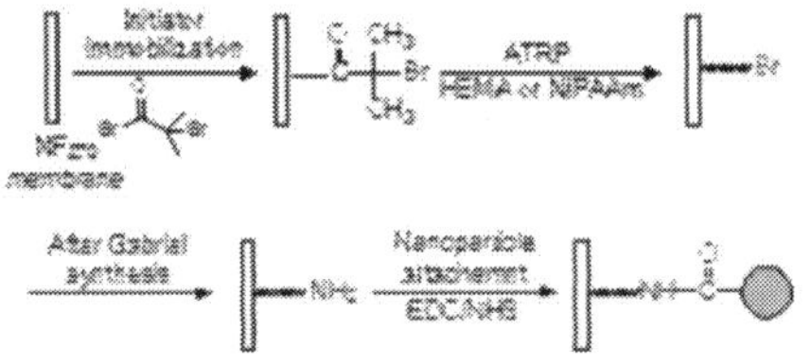

Figure 1. Schematic representation for the grafting of polyHEMA and immobilization of magnetic nanoparticles.

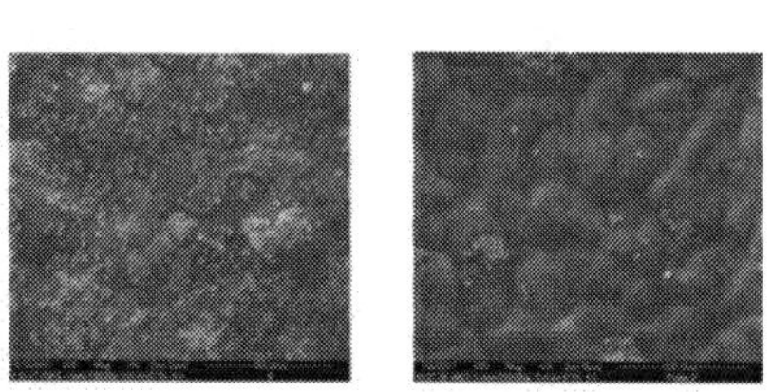

Figure 2. SEM images showing magnetic nanoparticles immobilized on membrane surface with high (left) and low (right) density.

With nanoparticles attached to the end of polyHEMA, the polymer chains acted as micromixers under oscillating magnetic field and mixing above the membrane surface was observed. This mixing resulted in a significantly improved membrane performance (flux and salt rejection) which can be ascribed to reduced concentration polarization. Tuning the space between these micromixers by diluting nanoparticle density leaded to optimization of membrane performance. On the other hand, with polyNIPAAm on the surface, nanoparticles were used as nanoheater in the presence of magnetic field to exert temperature change below and above LCST of this thermo-responsive polymer. Subsequently conformational changes of polyNIPAAm leaded to a dense/loose layer structure transformation on the membrane surface and changed membrane performance. This result demonstrated the possibility of using external magnetic field to control the membrane performance which is promising for industrial applications to realize instantaneous product control without interrupting the production line.

References

1. R.W. Baker, *Membrane Technology and Applications, 2nd ed.*; John Wiley: Chichester, (2004).

2. D. Rana and T. Matsuura, *Chem. Rev.*, **110**, 2448 (2010).

(379b) Membrane Facilitated Stripping of VOCs From Latex

Bridget Ulrich[1], Chaitanya Rachabattuni[1], Teng Zeng[2], Timothy C. Frank[3], Edward L. Cussler[1], William A. Arnold[2], Kent Carpenter[4] and Alon V. McCormick[1], (1)Chemical Engineering and Materials Science, University of Minnesota, Minneapolis, MN, (2)Civil/Environmental Engineering, University of Minnesota, Minneapolis, MN, (3)Engineering & Process Science, The Dow Chemical Company, Midland, MI, (4)Coatings Technology Center, The Dow Chemical Company, Spring House, PA

Membranes promise improved efficiency of distillation and gas-liquid contacting operations by increasing the interfacial area per unit volume and by breaking the link between hydrodynamics and flow rates. In particular, membranes allow independent control of flow rates to take best advantage of diffusion and thermodynamic behavior to effect separation without deleterious backmixing or flooding. Although this concept dates from 2003, few commercial developments of this technology have since been reported, apparently because of the challenge involved in displacing existing, more familiar, and already commercially-viable distillation technologies.

To be commercially successful, membranes must overcome a particularly vexing limitation of current technology without adding significant mass transfer resistance or fouling. With this in mind, we introduce a new membrane-based process involving stripping of volatile organic compounds (VOCs) from emulsion polymer latex used in water-borne paint formulations, an application driven by the market's demand for high-performance paints having little or no odor. Conventional distillation and stripping technology involves direct contact of steam, nitrogen, or air with the surfactant-laden latex. This typically causes severe foaming which necessitates a reduction in throughput and thus compromises productivity of conventional process equipment. By placing a membrane between the liquid and the vapor or gas, foaming is completely eliminated. Moreover, the membrane resistance frequently is negligible.

Thus the use of a membrane offers significant improvement in the stripping of latex products, allowing efficient vapor-liquid or gas-liquid contacting without foaming and permitting a more compact and cost-effective operation. This presentation describes process concepts and presents data demonstrating the utility of the membrane process for stripping latex (Dow patents pending).

Background: G. Zhang and E. L. Cussler, "Hollow Fibers as Structured Distillation Packing," *J. Membrane Sci.* **2003,** *215*, 185-193; and "Distillation in Hollow Fibers," *AIChE J.* **2003**, *49*, 2344-2351.

(379c) Tunable Membranes for Water Applications – an Overview

D.B. Bhattacharyya[1], Vasile Smuleac[2], Minghui Gui[2], Li Xiao[1], Austin Isner[1], Scott R. Lewis[3] and Andrew Tomaino[4], (1)Chemical and Materials Engineering, University of Kentucky, Lexington, KY, (2)Department of Chemical and Materials Engineering, University of Kentucky, Lexington, KY, (3)Chemical & Materials Engineering, University of Kentucky, Lexington, KY, (4)Chemical Engineering, University of Kentucky, Lexington, KY

The engineering advancement of separation processes with reduced energy consumption and minimal environmental impact is critical for sustainable operation. Membrane processes are finding wide applications ranging from water treatment to catalysis to advanced bio-separations. The development of tunable, nanocomposite membranes provide added opportunities in residence time control (through permeate flux) and responsive characteristics. Traditional MF membrane pores can be functionalized with single covalently-attached polypeptides and for the creation of LbL (layer-by-layer) assembly for NF type water related separations. Example membranes are: PVDF, PVP coated polycarbonate (PC), nylon-based anhydride activated membranes, Polyethylene-silica (PE-silica) composite membranes, alumina membranes, and cellulosic membranes. The dependence of conformation properties of polyelectrolytes/polypeptides on pH also provides tunable separations and membrane flux control. Layer-by-layer (LBL) assembly technique, most commonly conducted by intercalation of positive and negative polyelectrolytes or polypeptides, is a powerful, versatile and simple method for assembling nano-structures in pores. Non-stoichiometric immobilization of charged polyelectrolyte assemblies within confined pore geometries leads to an enhanced volume density of ionizable groups in the membrane pores. The presentation will include: (1) pore functionalization approaches, (2) tunable NF type separations, and (3) toxic pollutant separations from water. The authors would like to acknowledge the financial support of NIEHS-SBRP and NSF-IGERT program for this work.

A study of hydrophobic micro-porous membrane for membrane distillation

E. Drioli[1, 2, 3], A. Figoli[1], F. Macedonio[1, 2], S. Simone[1], A. Ali[2]

[1] National Research Council - Institute on Membrane Technology (ITM–CNR), Via Pietro BUCCI, c/o The University of Calabria,cubo 17C, 87036 Rende CS, Italy

[2] The University of Calabria - Department of Chemical Engineering and Materials, cubo 44A, Via Pietro BUCCI, 87036 Rende CS, Italy

[3] Hanyang University, WCU Energy Engineering Department, Room 917 9th Floor FTC Bldg., 17 Haengdang-dong, Seongdong-gu, Seoul 133-791 S. Korea

Extended Abstract

Membrane distillation (MD) is an emerging technology with promising applications in various fields involving the separation and purification of the effluent streams and crystallization from various solutions [1, 2]. MD is conventionally a temperature gradient based process in which two phases with different partial pressure are separated by a microporous membrane. The vapors from the side with higher partial pressure diffuse through the membrane and are collected on the distillate or permeate side. Different configurations can be used in this process. Four well known configurations implied in MD include direct contact membrane distillation (DCMD), vacuum membrane distillation (VMD), sweeping gas membrane distillation (SGMD) and air gap membrane distillation (AGMD). DCMD is the most studied type of membrane distillation due to its feasibility to separate the mixtures comprising of aqueous solutions and the simplicity of the set-up used [3]. The process possesses some unique advantages over the conventional processes, the worth mentioning being the theoretically complete rejection of the solute, the potential to use the waste grade heat or energy, operations carried out at low pressures and less strict requirements in terms of mechanical strength of the membranes used.

MD finds potential applications in various areas such as desalination, nuclear industry, textile industry, chemical industry, pharmaceutical and biomedical industries, food industry, etc. The unavailability of the membranes specifically designed for MD is the main hurdle in industrialization of the process. MD requires improved membranes, with specific morphology, high overall porosity and narrow pore size distribution with mean pore size ranging from 0.2 to 0.5μm, high liquid entry pressure, optimum thickness, high thermal stability, less susceptibility to fouling and long term permeance stability [4, 5]. Additionally and importantly, the membranes for MD must possess excellent hydrophobic character and minimum thermal conductivity to ensure an efficient and economical separation [6].

PVDF is widely utilized as polymeric material for preparing flat sheet and hollow fiber membranes to be used in membrane contactors and, in particular, in membrane distillation due to its intrinsic characteristics: high hydrophobicity coupled to better solubility in common organic solvents with respect to other hydrophobic polymers, such as poly(tetrafluoro ethylene) (PTFE), polypropylene (PP) and polyethylene (PE).

In this work, microporous hydrophobic poly(vinylidene fluoride) (PVDF) hollow fiber membranes have been prepared, characterized and tested in a membrane distillation set-up. The effect of different membrane preparation conditions and other parameters on MD performance has been investigated. It has been confirmed that the trans-membrane flux is a strong function of the membrane morphology which is influenced by the preparation conditions and dope solution composition.

The effect of processing conditions has been scrutinized by changing the flow rate of the inner bore fluid, composition of the bore fluid and that of the coagulant. The effect of composition of the polymeric dope has been analyzed by incorporating different additives into the dope and by changing the molecular weight and the composition of polymer into the dope solution. The membranes have been characterized in terms of morphology, mechanical properties, bubble point, maximum pore size, dimensions and trans-membrane flux in vacuum membrane distillation and direct contact membrane distillation configurations. The flux obtained has been related with the membrane characteristics which are function of the operating conditions and the polymeric dope composition. In this way, a correlation between the trans-membrane flux and the processing conditions and polymer dope composition has been sought.

Table1. Compositions and operating conditions applied for the membranes preparation

Fiber	*Polymer PVDF*	*Additives*	*Solvent*	*T (°C)*	*Air gap (cm)*	*Spinning rate (g/min)*	*P (bar)*	*Bore fluid*
M1	6012-18%	PVP(k17) 14%, H2O 6%	NMP 62%	85	14	11,5	1	NMP 30%, 10 ml/min
M2	6012-18%	PVP(k17) 14%, H2O 6%	NMP 62%	85	14	11,5	1	NMP 30% , 14 ml/min
M3	6012-18%	PVP(k17) 14%, H2O 6%	NMP 62%	85	25,5	10,4	1	H2O 100%, 10 ml/min
M4	1015-15%	PVP(k17) 15%, AMAL 10%	DMF 60%	60	25,5	11	1,1	IPA 30%, 17 ml/min

The morphology of the fibers has been noticed a function of dope composition and the operating conditions applied. The composition of the bore fluid, coagulant and air gap are considered as the important variants of operating conditions. The composition of the bore fluid has been varied to make it a soft or harsh coagulant. Less difference in solubility parameters of the solvent and bore fluid used attributes a soft character to the bore fluid and causes a delayed demixing between polymer rich and polymer lean phases. Consequently, a homogeneous, sponge-like structure is obtained. On the other hand, the application of harsh solvent like water as bore fluid favors the structure with finger-like morphology. The composition of bore fluid is stronger parameter influencing the fiber morphology as compared to that of the external coagulant. The effect of pore forming additive used in the study depends upon its concentration. At high concentrations, macro-voids free, sponge-like structure is obtained. The incorporation of secondary pore forming additive and reduced polymer contents in the dope generate highly porous structure.

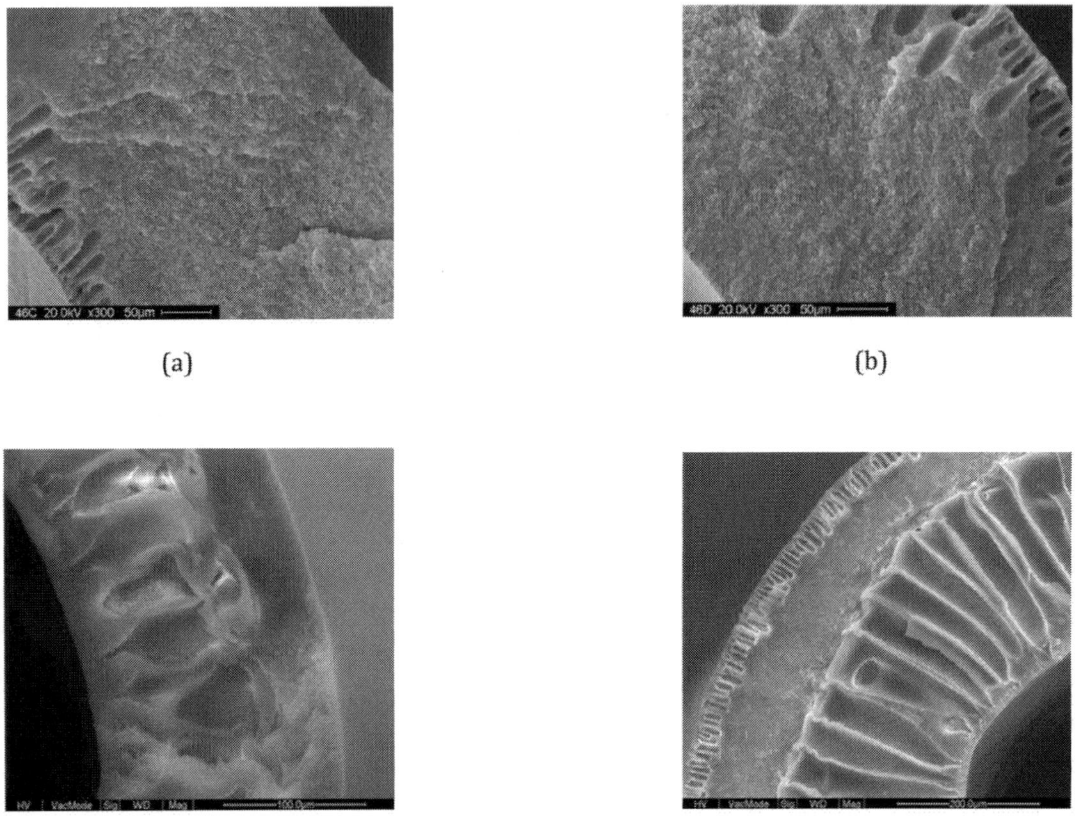

(a) (b)

(c) (d)

Figure1. Scanning electron microscopy images of fibers (a) M1 (b) M2 (c) M3 (d) M4

The mechanical properties of the fibers are closely associated with the microstructure of the fiber developed during the formation. The sponge-like structure appears due to the application of relatively weak solvent as the bore fluid imparts superior mechanical properties to the fibers. On the other hand, strong solvents generate finger-like or tear drop-like morphology and give rise to relatively inferior mechanical strength. The compositions and operating conditions leading to high porosity also depreciates the mechanical strength of the fiber.

The trans-membrane flux of the membranes prepared has been investigated in vacuum membrane and direct contact membrane distillation configurations and is strongly associated with the membranes porosity and the thickness. The amount of vapors transported through the membrane increases by increasing the membrane porosity. Similarly, reduced membrane thickness offers less resistance to the vapor transport through the membrane. The sponge-like structure offers more resistance to mass transfer and consequently, less trans-membrane flux is achieved. On contrary, the fibers with finger-like or tear drop-like morphology show less resistance to mass transfer and thus exhibit high mass transfer flux. The mass transfer flux increases by decreasing the membrane thickness.

Table2. Mechanical properties of the membranes prepared in the current study

Fiber type	*O.D. (mm)*	*I.D. (mm)*	*Thickness (mm)*	*Emod N/mm²*	*Rm N/mm²*	*▯break %*	*W Nm*	*PMI bubble point (bar)*	*PMI Pore size (μm)*	*Porosity (%)*
M1	1.80	1.04	0.38	70.34	3.43	251.97	0.60	0.40	0.47	80.90
M2	1.75	0.94	0.40	65.76	3.86	259.95	0.71	0.37	0.43	80.77
M3	1.59	0.70	0.45	57.15	2.84	192.88	0.30	0.20	0.14	79.11
M4	1.60	1.15	0.23	63.71	2.62	163.82	0.19	0.31	0.52	83.39

The fibers were tested in direct contact configuration by using double distilled water as feed and permeate. The experimentation was carried out at three different feed temperatures ranging from 50°C to 70°C with an interval of 10°C each. For each temperature condition, four different flow rates were applied varying from 100ml/min to 190ml/min. Trans-membrane flux increases by increasing feed temperature for each type of fiber due to the temperature dependence of vapor pressure. The flux was observed to be reliant on overall porosity, fiber morphology and membrane thickness. The maximum flux obtained reaches as high as 22L/m².h for temperature gradient of 45°C in direct contact membrane distillation.

Table3. Trans-membrane flux of the membranes in vacuum membrane configuration

Fiber	*Module active surface*	*Pvacuum (20-40 mbar)*	*VMD flux (H2O)*
	(mm²)	*Feed temperature (°C)*	*(Kg/h·m²·bar)*
M1	1611.57	50°C	~13.4
M2	1714.44	50°C	~13.5
M3	1120.98	50°C	~22
M4	1928.27	50°C	~50

The trans-membrane flux obtained in vacuum configuration was higher as compared to the direct contact one. The lower values of the flux obtained in DCMD configuration as compared to VMD configuration can be associated with high thermal losses and the additional resistance due to boundary layers present in this configuration. It is well recognized that the temperature polarization is quite active in DCMD configuration due to the direct contact of hot and cold phases with the opposite sides of thin membrane. It has been shown in [7] that the evaporation efficiency in VMD can reach as high as 90% which boosts the mass transfer flux in such configuration.

Table4. Flux obtained by using the fibers under various hydrodynamic and thermal conditions

Temperature (°C)	*Flow rate (ml/min)*	*Flux for different fiber types(L/m².h)*			
		M1	*M2*	*M3*	*M4*
50°C	100	5.746573	5.581777	6.696429	7.519694
	130	6.870902	6.512074	6.696429	8.556893
	160	6.895887	6.884192	7.366071	8.686543
	190	7.433067	7.44237	7.767857	9.334792
60°C	100	9.619263	10.07821	10.71429	12.44639
	130	9.794159	9.935564	10.49107	13.22429
	160	10.0565	9.395992	11.25	13.56138
	190	10.49374	9.147913	11.78571	14.00219
70°C	100	13.11718	14.08158	15.40179	18.66958
	130	13.55442	14.59635	16.07143	19.70678
	160	13.84591	14.57464	16.60714	20.66619
	190	14.42889	15.81504	17.00893	21.26258

Conclusions

Microporous hydrophobic PVDF hollow fibers were produced by the phase inversion method, varying polymer M_w and concentration, type and concentration of different additives (H_2O, PVP and AMAL) in the polymeric dope, composition of inner and outer coagulants.

The effect of all these parameters on the morphology of the produced hollow fiber membranes has been analyzed. A correlation between the developed morphology, fibers characteristics, in terms of thickness and porosity, and the trans-membrane water vapor flux in MD was observed. In general, fibers showing macro-voids in the structure possess higher porosity, reduced mechanical strength and enhanced flux. The best mechanical properties are obtained when a weak non-solvent, like NMP or EtOH solution, is used as the bore fluid and external coagulant.

It has been shown that the introduction of further pore forming additive (AMAL), combined with the relatively lower composition of polymer into the dope solution (Fiber type M4), gives rise to the structures exhibiting maximum trans-membrane flux working under both MD configuration (JVMD = 50 (L/m².h),

JDCMD = 22 (L/m^2.h). The fluxes obtained are upto 96% higher than the commercial membranes (PP Accurel). This is due to the fiber dual layer structure.

Comparison between results obtained working under Vacuum and Direct Contact mode confirms that VMD allows to reach higher partial pressure gradients and, hence, higher fluxes and plant productivity.

Relevant bibliography

[1]. Enrico Drioli, Francesca Macedonio, Aamer Ali , Membrane Distillation: Basic aspects and applications, A Journal of Membrane Science Virtual Special Issue

[2] M. Tomaszewska, M. Gryta, A.W. Morawski, The influence of salt in solutions on hydrochloric acid recovery by membrane distillation

[3]. E. Drioli, A. Criscuoli, E. Curcio, Membrane Contactors: Fundamentals, Applications and Potentialities, Membrane science and technology series, 11, Amsterdam; Boston: Elsevier, 2006

[4] M.S. El-Bourawi, Z. Ding, R.Ma, M. Khayet, A framework for better understanding membrane distillation separation process, Journal of Membrane Science 285 (2006) 4–29

[5] Mohamed Khayet, Membranes and theoretical modeling of membrane distillation: A review, Advances in Colloid and Interface Science 164 (2011) 56–88

[6] S. Simone, A. Figoli, A. Criscuoli, M.C. Carnevale, A. Rosselli, E. Drioli, "Preparation of hollow fibre membranes from PVDF/PVP blends and their application in VMD", Journal of Membrane Science 364 (2010) 219-232, doi:10.1016/j.memsci.2010.08.013

[7] Alessandra Criscuoli, Maria Concetta Carnevale, Enrico Drioli,Evaluation of energy requirements in membrane distillation, Chemical Engineering and Processing 47 (2008) 1098–1105

(379e) Membranes for Sorting Cells

Girish Kumar, Bo Chen, Carlos C. Co and **Chia-Chi Ho**, Chemical and Materials Engineering, University of Cincinnati, Cincinnati, OH

A new method to sort cells based on cell motility on micropatterned membranes or culture dishes is reported. The method permits real-time visualization of the entire cell population and requires less than 10^3 cells. The asymmetric micropatterns generated by microfabrication guide the path and direction of cell migration to allow mammalian cells autonomously sort on membranes based on their relative motility. The method has been demonstrated by sorting mixtures of same kind of cells expressing different signaling molecules, two different cell types, and cancerous and normal epithelial cells. A three dimensional version based on microchannels within the membranes has also been developed. The method requires no pre-labelling, external gradients or fields provides a useful tool in fractioning or concentrating fast/slow moving cells for cell separation and migration assays. Advances in the manufacture of membranes with designed microchannels for directing cell migration would enable rapid commercial-scale implementation of this cell motility assay.

(379f) Innovative Techniques for Membrane Fabrication and Characterization

Alan R. Greenberg, Department of Mechanical Engineering, University of Colorado Boulder, Boulder, CO

The polymeric membrane fabrication literature deals primarily with well-known processes such as phase inversion and interfacial polymerization that are commonly employed on a commercial scale. However, much can often be learned from novel fabrication processes that are best defined as hybrid techniques or utilize truly unusual casting conditions to create unique microstructures. Moreover, irrespective of the fabrication methodology utilized, there is a continuing and critical need for improved methods for overall characterization of membranes and membrane processes. Such processes are pivotal in a wide range of commercial applications including water desalination, industrial gas separation, and bioprocessing. Information obtained via characterization enables unique insights regarding fundamental membrane behavior and also provides a firm basis for improvements in the optimization of membrane-based processes. This presentation provides examples of developments in these areas by highlighting innovative membrane fabrication processes such as thermally assisted evaporative phase separation (TAEPS) and microgravity casting as well as novel membrane characterization techniques including evapoporometry (EP), pendant drop mechanical analysis (PDMA) and ultrasonic time-domain reflectometry (UTDR). The common connection among these examples is that they all were a part of the ongoing 30-year collaboration between Prof. William Krantz and me, a collaboration that has immeasurably enriched my career.

(385a) Assessing the Potential for Large-Scale Utilization of Fossil-Fuel Derived CO2

Jared Ciferno, U.S. Department of Energy, National Energy Technology Laboratory, Pittsburgh, PA and John J. Marano, JM Energy Consulting, Inc., Gibsonia, PA

Barring the imposition of a carbon tax (or equivalent cap-and-trade program) and corresponding international agreements, it is highly unlikely that carbon capture and storage/sequestration (CCS) will be implemented in any meaningful way in the near future. This is due to inherent inefficiency and higher costs associated with this option for mitigating climate change. Unfortunately, the science of climate change suggests that the world community must begin to act sooner rather than later. From an Industrial Ecology viewpoint, a preferable approach to carbon storage in perpetuity would be to re-use the carbon associated with the fossil-fuel derived CO_2 already being omitted. However, options for directly utilizing CO_2 are quite limited, with enhanced oil recovery (EOR) being the only physical use of CO_2 of any appreciable size. The concept of converting CO_2 to useful products has been explored in the past, but has fallen from favor in many policy circles due to the lack of a clear vision of how re-utilization of fossil carbon could be implemented. Given current in-action, it is worth re-visiting this concept.

Energetically, it is clear that the large-scale conversion of CO_2 to useful products requires a source of external energy that is not fossil-fuel derived; otherwise, the conversion process becomes itself a source of CO_2 emissions. Sources for the required energy could be direct solar or nuclear energy, or electricity generated from these primary energy resources. Obviously, a source of CO_2 is also required; however, all sources are of equal value. Atmospheric and mobile emissions are of little value due to the problems associated with collecting and concentrating them. Whereas, CO_2 from stationary point sources, and in particular large point sources such as coal and natural gas-fired power plants, are the "low hanging fruit" for implementing carbon re-use. Finally, useful products from CO_2 must be identified. It is suggested here that instead of pursuing a piecemeal approach consisting of an assortment of finished chemicals and material goods, a holistic approach be taken. Currently, the only existing commodity market that is big enough to absorb the large scale re-utilization of CO_2 lies in the petroleum sector. Thus, the basis for any carbon re-utilization platform should be the production of a petroleum substitute, a "synthetic" oil that may be refined and converted into a multitude of finished products.

This paper first compares and contrasts carbon storage with carbon utilization, and then moves on to quantify the scale of any carbon re-utilization effort; as well as, CO_2 life-cycle limitations of re-use. After all, in this approach much of the synthetic oil, like petroleum, will end up being (re-) combusted as an end-use transportation fuel. Thus, carbon re-use can only be considered as an interim strategy for cost effectively transitioning away from fossil fuels. We also briefly touch on the implications of large scale carbon re-use for CCS R&D, since the requirements for CO_2 capture, transport and short-term storage will be quite different than what is anticipated for long-term CO_2 sequestration. Finally, we briefly discuss the possible economic implications of this strategy. A follow-on paper at this conference identifies and analyzes what the present and best future options for producing a CO_2-derived synthetic oil might be.

(385b) Some Perspectives On the Use of Solar and Other Renewable Energy to Facilitate Carbon Management

Robert Wegeng, Pacific Northwest National Lab, Richland, WA

The recent and ongoing development of economical solar and wind energy systems presents a number of new options for carbon management. The purpose of this presentation will be to discuss how engineered solar thermochemical processes might be used to reduce carbon emissions from fossil-fuel powerplants and gas-to-liquids plants and how concentrated solar energy might be used to capture carbon from industrial emissions or the atmosphere for sequestration or use. Perspectives will include economics and Second Law efficiencies. Particular attention will be applied to technologies and strategies that might be applied in the U.S. Pacific Northwest.

(385c) Carbon Capture and Storage Directly From Flue Gas: The Effects of Chemical and Biological Catalysts On Magnesium Carbonate Formation

Edward J. Swanson, Chemical Engineering, Columbia University, New York, NY, Patrick V. Brady, Sandia National Laboratories, Albuquerque, NM and Ah-Hyung Alissa Park, Earth and Environmental Engineering & Chemical Engineering, Columbia University, New York, NY

The direct adsorption of CO_2 from flue gas and precipitation of mineral carbonates presents a transformative alternative to carbon capture and storage (CCS) schemes involving conventional post-combustion capture of CO_2 and its geologic storage. CO_2 capture processes generally involve energy intensive solvent regeneration and CO2 compression steps. Absorbing the carbon into an aqueous system and precipitating it as a thermodynamically stable solid carbonate in a single technology avoids both of these energy intensive steps in CO2 capture, while providing inherent storage of CO2. However, in order for this novel CCS technology to be viable, the kinetics of both CO2 absorption and mineral dissolution need to be sufficiently fast, particularly at the low partial pressure of CO2 commonly found in flue gas. In this study, both chemical (i.e., catechol and oxalate) and biological (i.e., carbonic anhydrase) catalysts were used to enhance the rates of CO2 hydration and serpentine dissolution. While these catalysts showed significant improvement over natural carbon mineralization kinetics, it was also important to investigate the effect of this catalytic system on the final formation of magnesium carbonate in terms of its chemistry and morphology. Experiments were performed in a lab-scale three-phase fluidized bed reactor, and optimum reaction conditions for carbonate precipitation were investigated.

(385d) Environmentally Benign Processes for Absorption and Chemical Fixation of CO2

An-Hua Liu, Ran Ma and Liang-Nian He Sr., State Key Laboratory of Elemento-Organic Chemistry, Nankai University, Tianjin, China

Carbon dioxide as the main green house gas is widely blamed to pay most of the responsibility for global warming. Hence, controlling CO_2 emission and further reducing its accumulation emerges to be an urgent and challenging research topic. CO_2 capture and storage/sequestration (CCS) to control its emission as well as chemical utilization of the captured CO_2 (CCU) has been considered as the rational technique to reduce or mitigate the so-called green-house effect. However, being the sustainable carbon source, CO_2 as a safe, cheap and abundant C_1 building block in organic synthesis also comes into light [1-3]. So chemical fixation of CO_2 into industrially useful materials and high value-added products turns to be one of the most attractive methods in recent years.

Amine-functionalized CO_2 absorbents have been intensively investigated in numerous studies due to the easy preparation, high CO_2 affinity and robust reversibility, whereas the intrinsic nature of 2:1 absorption stoichiometry could restrain their further development for pursuing of superior capture efficiency. In present work, such inherent disadvantage was circumvented by employing modified amino acid salts in polyethylene glycol, affording approximative 1:1 ratio CO_2 absorption. The used absorbent can be readily reverted and recycled without loss of its capture capacity at least 5 turns. The structure of captured absorbent species was identified by means of NMR and in situ IR spectroscopy, as well as computational calculations.

Meanwhile, using commercial available and recyclable basic ion-exchange resins, one kind of polystyryl supported tertiary amines, an efficient fixation of CO_2 by aziridines under mild conditions were achieved, leading to formation of 5-aryl-2-oxazolidinone with excellent regio-selectivities. Notably, neither solvent nor any additive was required, and the catalyst can be recovered by simply filtration and directly reused at least five times without significant loss of catalytic activity and selectivity [4]. The present protocol was successfully applied to cycloaddition reactions of epoxides/propargyl amines with CO_2/CS_2. In general, this solvent-free process thus represents environmentally friendly catalytic conversion of CO_2 into value-added chemicals and has great potential to be applied in various continuous flow reactors in industry.

References

[1] Aresta, M.; Dibenedetto, A. *Dalton Trans*. 2007, 28, 2975.

[2] Yang, Z.-Z.; He, L.-N.; Zhao, Y.-N. *Energy Environ. Sci*., 2011, 4, 3971-3975.

[3] He, L.-N.; Wang, J.-Q.; Wang, J.-L. Carbon dioxide chemistry: examples and challenges in chemical utilization of carbon dioxide, *Pure Appl. Chem*. 2009, 81(11), 2069-2080.

[4] Liu, A.-H.; He, L.-N.; Peng, S.-Y.; et al. *Sci. China Chem.*, 2010, 53(7), 1578-1585.

(385e) Process Ecoefficiency of Methanol and Dmc Production From CO2: A Comparative Analysis

Camila Ribeiro[1], José L. Medeiros[2], **Ofélia Araújo**[2], Paulo Coutinho[3] and Rita M.B. Alves[4], (1)Escola de Química (H2CIN), UFRJ - Universidade Federal do Rio de Janeiro, Rio de Janeiro, Brazil, (2)Escola de Química (H2CIN), UFRJ - Universidade Federal do Rio de Janeiro, Rio de Janeiro, Brazil, (3)management of innovation, Braskem, São Paulo, Brazil, (4)Braskem, São Paulo, Brazil

Process Ecoefficiency of Methanol and DMC Production from CO_2: A comparative analysis

Camila F. Ribeiro[1], Ofélia de Q. F. Araújo[1], José Luiz de Medeiros[1], Rita[2], Paulo Coutinho[2]

[1]Departament of Chemical Engineering, Escola de Química da Universidade Federal do Rio de Janeiro. Sala E-201, Bloco E, Centro de Tecnologia, Ilha do Fundão, CEP: 21949-900 Rio de Janeiro, Brazil.

E-mail adress: camila@eq.ufrj.br

Environmental performance of products and processes has gained importance as the industry is trying to minimize its environmental impacts. Of special concern to society, are climate effects associated to the rise of CO_2 atmospheric concentration. To mitigate emissions effects, processes are searched that use CO_2 as raw material. It is worth noting that an emission mitigating process should necessarily be a commodity as large production scale are required. With this orientation, methanol and DMC are promising alternative products: the first is a very attractive additive to diesel, and it blends with gasoline to yield a high octane fuel with low Reid vapor pressure. Methanol, on the other hand, is considered as a promising energy alternative besides displaying a wide range of industrial applications. However, in order to claim that a given technology is clean, its ecoefficiency should be addressed with quantitative sustainability indicators. In this sense, the present work selects a set of environmental metrics along with safety and economical (CAPEX and OPEX) indicators. A comparative analysis is presented for two potentially green processes: (i) methanol synthesis; and (ii) DMC synthesis. Furthermore, the later synthesis is approached in two alternatives, a direct and an indirect reaction path. In the methodology used, process simulation and optimization are employed to compare process performances at maximum yields.

(385f) High Efficiency Direct Electrochemical Conversion of CO2 to Liquid Fuel

Jingjie Wu[1], Frank Risalvato[2], Fusheng Ke[2], Perry Pellechia[3] and Xiao-Dong Zhou[2], (1)Chemical Engineering, University of South Carolina, Columbia, SC, (2)Department of Chemical Engineering, University of South Carolina, Columbia, SC, (3)Department of Chemistry and Biochemistry, University of South Carolina, Columbia, SC

In this paper, we will discuss a design of an electrochemical cell for CO_2 reduction implementing a proton exchange membrane (PEM) by adding a liquid buffer layer between the membrane and the cathode electrode. This configuration enables production of liquid fuels from CO_2 on Sn gas diffusion electrode at high efficiencies. Specifically, the formate was produced from the electrochemical reduction of CO_2 with an overpotential lower than 0.2 V. At such a low overpotential, the Faradaic efficiency and energetic efficiency of formate production with a buffer layer of 0.1 M $KHCO_3$ reaches 78.5% and 66.4%, respectively. Moreover, a maximum Faradaic efficiency of ~95% can be achieved. The production rate of formate increases with increasing cell potential to approximately 10 μmol min^{-1} cm^{-2} at a cell potential of -3.0 V. In contrast, hydrogen evolution reaction dominates at the cathode side in the PEM electrochemical cell without a buffer layer.

The enhancement of CO_2 reduction in the PEM electrochemical cell with a buffer layer is attributed to the optimization of H^+ and CO_2 concentration at the reactions sites in the cathode. The effects of H^+ concentration on the Faradaic efficiency were confirmed by the results of CO_2 reduction with a liquid buffer layer with different pH values. As a result, the Faradaic efficiency towards formation of formic acid can be further improved by optimizing the buffer layer thickness and subsequently tuning the H^+ concentration at the cathode surface. The optimized thickness of the buffer layer was found ~ 3 mm. The buffer layer thickness dependency of H^+ concentration at the cathode electrode will be reported.

(385g) Electrochemical Conversion of Carbon Dioxide to Useful Products

Arun Agarwal[1], Shan Guan[2], Davion Hill[2], Edward Rode[2] and Narasi Sridhar[2], (1)Research & Innovation, Det Norske Veritas (DNV), Dublin, OH, (2)Det Norske Veritas (DNV)

Utilization of captured CO_2 provides an advantageous alternative especially in locations where CO_2 sequestration in geologic media is not feasible. Making high value chemical products from CO_2, although not sufficient to convert a significant portion of the GHG emissions, provides an opportunity of extensive production of green chemicals when combined with renewable sources to feed the energy requirements. At DNV, we are developing the technology to economically convert CO_2 to formate salts/ formic acid using an electrochemical process. Tin is the preferred electrocatalyst due to high selectivity of CO_2 reduction to formate ion and high overpotential for cathodic hydrogen evolution. Gaseous CO_2 is continuously purged into the catholyte (2M KCl) through a porous tin electrode on which it is cathodically reduced. This cathode chamber is separated from the anodic chamber using a selective ion exchange membrane for allowing positive ion transport but not allowing the product (formate) ions to get oxidized on the anode. Oxygen evolution reaction occurs on a mixed metal oxide anode in either acidic or alkaline anolyte solutions to complete the electrochemical reaction. Electrode development in terms of high surface area of catalyst to match existing commercial electrochemical processes and reduce capital costs related to reactors will be presented. A process was developed to electroplate tin on high surface area carbon fiber substrate. Electroplating solution and substrate properties were evaluated for achieving maximum current densities at optimal product selectivity. Another focus area is understanding and optimizing the process chemistry. Selection of anolyte and catholytes, along with the applied potential regime, affects product selectivity, electrode degradation, expenditure due to chemicals consumption and the ability to reach high enough product concentrations before feasibly employing separation processes. In this presentation, a technical and economic analysis of the aforementioned factors affecting the commercial feasibility will be presented and evaluated as a guide for performance criteria of the proposed technology.

Progress of Purification of Industrial Exhaust Gas Using Novel High Gravity Technology

Liang-Liang Zhang[1], Jie-Xin Wang[1], Guang-Wen Chu[2], Jian-Feng Chen[1,2,*],

[1]*State Key Laboratory of Organic-Inorganic Composites, Beijing University of Chemical Technology, Beijing 100029, PR China*

[2]*Research Center of the Ministry of Education for High Gravity Engineering and Technology, Beijing University of Chemical Technology, Beijing, 100029, PR China*

Corresponding author. Tel: +86-10-64446466; fax: +86-10-64434784.
E-mail address: chenjf@mail.buct.edu.cn (J.-F. Chen)

Abstract

Recently, industrial emissions of acid gas pollutions, such as carbon dioxide, hydrogen sulfide, sulfur dioxide and nitrogen oxides, increase gradually with the global economy development, which causes the environmental deterioration of our world. Hence, investigation on new exhaust gas purification technology attracts much attention from industrial and academic circles. Here we show the research progress of our group on removal of carbon dioxide, hydrogen sulfide, sulfur dioxide and nitrogen oxides from industrial emissions and resourceful utilization of them. The theoretical principles and the typical large-scale industrial application instances are emphatically introduced. For example, the industrial desulfurization plant with the capacity to treat 80000 Nm3/h exhaust gas is built for sulfuric acid factory. The sulfur dioxide content in the exhaust gas decreases dramatically from 4000-5000 to 200 mg/m^3. An industrial trial plant with the capacity to treat 18000 Nm^3/h flue gas (30000 ton per year) has been built to capture carbon dioxide from the flue gas and the captured carbon dioxide is used for tertiary oil recovery. Our achievements by 20 years' research show that high gravity technology exhibits broad prospect for industrial applications on removal and capture acid gas pollutions from industrial emissions. High gravity technology could provide technical support for energy conservation and emission reduction.

Keywords: high gravity technology, carbon dioxide, sulfur dioxide, hydrogen sulfide, nitrogen oxides

Introduction

Acid gas pollutions from industrial emissions, such as carbon dioxide, hydrogen sulfide, sulfur dioxide and nitrogen oxides, increase gradually with the increase of the world's fossil energy consumption over the past century, which causes the environmental deterioration of our world.[1-4] There is little doubt that petroleum, natural gas, and coal will continue to be the primary global fuel and chemical feedstock sources for the next few years. Thus acid gas emissions from fossil fuel combustion are going on to be a major contributor to global climate change. The environmental problem, such as acid rain, photo chemical smog and global warming, will be more outstanding if there is no effective purification for the exhaust gas. Gas absorption is one of the most common methods for exhaust gas purification process, such as carbon dioxide capture, desulphurization and denitrification.[5-7] The gas-liquid reactors employed in these process, including packed tower, spray column, and bubble column, have significant limitations in mass transfer, leading to low efficiency and high cost.[8] Hence, the development of exhaust gas purification technology with high mass transfer efficiency is desired.

High gravity technology (also called Higee), carried out in a rotating packed bed (RPB), was originally invented by Colin Ramshow and his co-workers in 1979 for the separation process.[9-10] In previous studies, the mass transfer height of the RPB is only 1-2 cm and the volumetric gas-liquid mass transfer coefficients achieved in RPB is 1-3 orders magnitude larger than that in the conventional reactor or contactor, exhibiting excellent mass transfer characteristics.[11,12] The key point of the Higee technology relies on the simulated high-gravity environment created by the centrifugal force of the RPB. In the RPB, the fluids going through the packing are spread or split into very fine droplets, threads, and thin films by the strong shear force, resulting in a significant intensification of micromixing and mass transfer between the fluid elements and gas.[13-15] Moreover, a structured packing in RPB is specifically designed to have both a large gas-liquid interfacial area and a low pressure drop. Both of these characteristics reduce the amount of packing required and thus minimize the size of the equipment. So far, a number of high gravity technologies are already being practiced on laboratory scale or industrially demonstrated. Here we show the research progress of our group on removal of carbon dioxide, hydrogen sulfide, sulfur dioxide and nitrogen oxides from industrial emissions and resourceful utilization of them using high gravity technology.

The structure of a typical countercurrent RPB is illustrated in Fig 1. The key part consists of a packed rotator (6). The rotator is installed inside a stationary casing and it rotates at a speed of several hundred to several thousand rpm. Liquid, or slurry is introduced into the eye space of the rotator from the liquid inlet pipe (1) to spray onto the inside edge of the rotator through the slotted liquid distributor (5) at the center of RPB. The liquid entering the bed flows in the radial direction under centrifugal force, passes through the packing and outside space between the rotator and the casing, finally collects and leaves the RPB via the liquid outlet (4). The gas stream introduced from gas source through the gas inlet (3) flows inward from the outer edge of the RPB by a pressure-driving force, and finally leaves from the gas outlet (2). Gas and liquid streams have a countercurrent contact in the RPB.

Carbon dioxide capture by higee technology

Carbon dioxide capture by RPB has been widely studied, and the absorbants include NaOH,[16] K_2CO_3,[17] monoethanolamine (MEA),[18] 2-amino-2-methyl-1-propanol and the mixture, etc.[19] Recently, we invented a RPB of new structure based on the mass transfer mechanism research of the end effect zone for carbon dioxide capture and purification. The results show that the carbon dioxide absorption efficiency in the RPB of new structure can increase 10% compared to the conventional RPB. Based on these studies, an industrial trial plant has been built to capture carbon dioxide from the flue gas and the captured carbon dioxide is used for tertiary oil recovery. Figure 2 presents our newly setup trial plant for industrial CO_2 capture using RPB as absorber with 30% MEA solution as absorbent in Shandong province, China. This industrial trial plant has successfully captured CO_2 from the flue gas of a power plant and has been steadily running for several months. The plant has the capacity to treat 18000 m^3/h flue gas, corresponding to capturing up to 30000 ton per year of CO_2. It could be seen CO_2 capture by RPB is industrial feasible for CO_2 emission reduction.

Nowadays, using ionic liquid for CO_2 capture is of great interest due to their unique characteristics, i.e., thermal stability, negligible vapor pressure up to their thermal decomposition point, tunable physicochemical characteristic and high CO_2 solubility.[20-24] However, low gas-liquid mass transfer rate in the conventional gas–liquid contactors due to the high viscosity of ionic liquid leads to the significant limitation for large-scale application of CO_2 capture using ionic liquid.[25] Therefore, there is an urgent demand to develop a novel gas–liquid reactor or contactor for the intensification of the mass transfer efficiency in such a system. CO_2 absorption with ionic liquid using RPB gas-liquid contactor was firstly reported by our group.[26] The ionic liquid 1-n-butyl-3-methylimidazolium hexafluorophosphate ([bmim][PF_6]) was adopted as a experimental system. It was found that the RPB markedly enhanced the physical absorption of CO_2 in ionic liquid in a very short contact time within seconds. Only one cycle gas-liquid contact in the RPB could reach 60% of the saturation ratio of CO_2 in ionic liquid. The gas-liquid contact time to reach the same saturation ratio was shortened by hundreds of times in the RPB compared to the reported time-dependent absorption data. The effects of various operation conditions on liquid side volumetric mass transfer coefficient ($k_L\alpha$) were elucidated (Figure 3). Increasing the rotating speed from 1100 rpm to 3100 rpm doubled the $k_L\alpha$. The increases of the liquid flow rate and the absorption temperature were also beneficial to the enhancement of the $k_L\alpha$.

A correlation of $k_L\alpha$ for CO_2 absorption in ionic liquid was achieved in eq 1 from the experimental results shown in Figure 3.

$$\frac{k_L \alpha d_p}{D\alpha_t} = 8.183 Sc^{0.5} Gr^{0.37} We^{0.335} \tag{1}$$

In the equation, d_p is the spherical equivalent diameter of the packing, Sc is the Schmidt number, Gr is the Grashof number, and We is the Weber number. Because of the small liquid mass flux and high viscosity, the Reynolds number was very low (lower than 0.032) and had only a small change. So the correlation did not include

Reynolds number. According to eq 1, it was found that $k_L\alpha$ was proportional to the liquid viscosity to the power of -0.24. As compared to the results reported by Delaloye et al. for a packed tower,[27] liquid viscosity had less influence on mass transfer in RPB than in a packed tower. The $k_L\alpha$ was proportional to the centrifugal acceleration to the power of 0.37, indicating that the centrifugal force still revealed effective in enhancing mass transfer in highly viscous media.

Table 1 gives a comparison of the liquid side volumetric mass transfer coefficient in RPB and packed tower at the similar operating conditions. Although the packing in packed tower had much larger packing volume and surface area than those in RPB, $k_L\alpha$ for RPB was measured to be $0.95\text{-}3.9\times10^{-2}$ s^{-1}, while $0.63\text{-}1.9\times10^{-3}$ s^{-1} for packed tower. The former is at least one order of magnitude higher than the latter. This completely indicated that RPB still owned an excellent gas-liquid mass transfer performance for a high viscosity media. Thus it could be envisioned that RPB would exhibit a great potential for various industrial applications on ionic liquid, especially for CO_2 capture using ionic liquid.

Desulfuration by higee technology

We experimentally investigated desulfuration by amine, alkali and organic amine by higee technology and applied the higee to sulfur dioxide absorption in industrial experiment.[28] Based on these works, the industrial desulfurization plant with the capacity to treat 80000 Nm^3/h exhaust gas (the external diameter of the equipment is about 3 meter) is built for sulfuric acid factory in Zhejiang province of China (see

Figure 4). The sulfur dioxide content in the exhaust gas decreases dramatically from 4000-5000 to 200 mg/m^3, which is much lower than the new national discharge standard of China. Moreover, the pressure drop of the desulfuration equipment is about 600-800 Pa, which could effectively reduces the load of the fan for gas transportation. Compared to a conventional tower, the investment, space and the energy consumption of the higee technology dramatically decreased. So far, this technology has bee used in several sulfuric acid production line.

Removal of acid gas components from gas mixtures containing CO_2 and H_2S is very important in natural gas processing and treating refinery off gases and synthetic gases from coal and heavy oil gasification. In industrial gas processing, there is an increasing in gas absorption process for the selective removal of H_2S from gas streams with a high ratio of CO_2 to H_2S because H_2S must be removed for pipage and environmental reasons but CO_2 removal is not essential. Selective absorption of H_2S from gas mixture with CO_2 into N-methyldiethanolamine (MDEA) is investigated experimentally and theoretically in RPB by our group.[29] In the RPB, various rotating speed, gas flow rates, liquid flow rates and concentrations of MDEA aqueous solutions were studied by means of the evaluation of removal efficiency, selectivity and overall volumetric mass transfer coefficient (see Figure 5). The reaction-diffusion mass transfer model based on penetration theory for the selective absorption process is developed, accordingly. The results of experiment and model show that the uppermost function of RPB in selective absorption of H_2S is to restrain the CO_2 removal efficiency but sharply intensify the absorption of H_2S. Because the penetration of H_2S into the liquid film, it is just 2.0×10^{-9}s that H_2S need to be exhausted at 10^{-8}m into the liquid film. And the lifetime in RPB is 7 orders of magnitude bigger than this penetration time of H_2S. So the reaction and mass transfer of H_2S still can be completed even at the very short gas-liquid contact time within RPB and a high H_2S removal efficiency of around 99.76% also can be achieved. But for CO_2, the gas-liquid contact time in RPB is too short to reach a high mass transfer efficiency and the CO_2 removal efficiency is merely around 9.50%. Besides, the quantitative analysis based on model suggests that the existence of CO_2 has little effect on the absorption of H_2S while the H_2S has an apparent negative impact on the absorption of CO_2.

Denetrafication by higee technology

As a major pollutant flue gas, nitric oxide (NO) emission has received great attention of the world. Various technologies have been developed to reduce NO emission. One promising approach is wet scrubbing technology. The NO gas is absorbed in an aqueous solution with a suitable reactant, such as hydrogen peroxide (H_2O_2),[5] sodium chlorite ($NaClO_2$),[3] potassium permanganate ($KMnO_4$)[30] and metal chelate, etc.[31] Among these absorbents, ferrous chelate solution seems to be more attractive. Ferrous chelate solutions can rapidly react with the absorbed NO gas to form stable metal-nitrosyl complexes. Reaction mechanism is simply illustrated as follow:[32]

$$Fe^{II}(EDTA) + NO \leftrightarrow Fe^{II}(EDTA)(NO)$$

Some technologies like electrochemical and biochemical approaches have been developed to regenerate the loaded ferrous chelate solution, which make this absorbent economically viable. Various research groups have studied the reaction of NO with Fe^{II}(EDTA) using different reactor configurations and experimental conditions. The studies indicated that the absorption rate is strongly affected by mass transfer limitation of NO and Fe^{II}(EDTA) due to the high rate of the reaction in these traditional reactors, such as stirred cell and bubble column.[33] Thus, more efficient reactors are required to intensify the mass transfer of the reaction, thereby improving the absorption rate and NO removal efficiency. RPB is firstly employed to enhance the gas-liquid mass transfer between NO and Fe^{II}(EDTA).[34] The objective of this study is to systematically investigate the absorption process of NO by utilizing ferrous chelate solution as absorbent in the RPB. The effects of the operation parameters such as pH value of the absorbent, gravity level, gas-liquid ratio, liquid temperature and absorbent concentration are explored to provide guidance and optimization of this system. The pH value of 7 is suitable for the absorbent to achieve high removal efficiency. Increasing the gravity level from 30 to 150 g is beneficial to the enhancement of removal efficiency. When the gravity level of the RPB is higher than 150 g, the removal efficiency could reach 87%.

Conclusion

Over 20 years' exploration and application has demonstrated that higee is a promising and powerful process intensification technology, featuring miniaturization, high efficiency and ease of scale-up. Higee exhibits unique advantage in mass transfer limited reactions. The possibility of applying the higee technology to exhaust gas purification has been widely investigate and several processes including carbon dioxide capture and desulfuration have enjoyed commercial success.

Acknowledgement

This work was financially supported by National Natural Science Foundation of China (Nos. 21121064, 20990221 and 21176013), National “973” Program of China (No. 2009CB219903) and National "863" Program of China (No. 2009AA033301).

Notation

k_L = liquid side mass transfer coefficient, m/s

$k_L\alpha$ = liquid side volumetric mass transfer coefficient, s^{-1}

$k_G\alpha$ = gas side volumetric mass transfer coefficient, s^{-1}

$K_L\alpha$ = overall volumetric mass transfer coefficient, s^{-1}

L = liquid flow rate, mL/min

D = diffusivity, m^2/s

T = temperature, K

d = diameter of the liquid droplet, m

d_p=spherical equivalent diameter of the packing = $6(1-\varepsilon)/\alpha_t$, m

Greek letters

α = gas-liquid interfacial area, m^2/m^3

α_t = surface area of the packing, m^2/m^3

Literature Cited

1. Keith DW. Why capture CO_2 from the atmosphere. *Science*. 2009; 325: 1654-1655.
2. Hu GX, Sun ZG, Gao HY. Novel process of simultaneous removal of SO_2 and NO_2 by sodium humate solution. *Environ Sci Technol.* 2010; 44: 6712-6717.
3. Chien T, Chu H. Removal of SO_2 and NO from flue gas by wet scrubbing using an aqueous $NaClO_2$ solution. *J Hazard Mater.* 2000; B80: 43–57.
4. Ouwerkerk C. Design for selective hydrogen sulfide absorption. *Hydrocarbon Processing*. 1978; 57:89-94.
5. Liu YX, Zhang J. Photochemical oxidation removal of NO and SO_2 from simulated flue Gas of coal-fired Power plants by wet scrubbing using UV/H_2O_2 advanced oxidation process. *Ind Eng Chem Res*. 2011; 50: 3836-3841.
6. Chu H. Simultaneous absorption of SO_2 and NO in a stirred tank reactor with $NaClO_2$/NaOH solutions. *Water Air Soil Pollut.* 2003; 143: 337-350.
7. Boucif N, Favre E, Roizard D. CO_2 capture in HFMM contactor with typical amine solutions: a numerical analysis. *Chem Eng Sci*. 2008; 63: 5375-5385.
8. Stankiewicz A, Moulijn JA. Process intensification. *Ind Eng Chem Res.* 2002; 41: 1920-1924.
9. Ramshaw C. Mass transfer process. *European patent,* No.0002568. 1979.
10. Ramshaw C. Mallinson R. Mass transfer process. *US Patent,* No.4263255. 1981.
11. Munjal S, Dudukovic MP, Ramachandran P. Mass transfer in rotating packed beds: I. Development of gas-liquid and liquid-solid mass-transfer coefficients.

Chem Eng Sci. 1989, 44: 2245-2256.

12. Munjal S, Dudukovic MP, Ramachandran P. Mass transfer in rotating packed beds: II. Development of gas-liquid and liquid-solid mass-transfer coefficients. *Chem Eng Sci.*1989, 44: 2257-2268.
13. Basic, A.; Dudukovic, M. P. Liquid holdup in rotating packed bed: examination of the film flow assumption. *AIChE J.* 1995, 41: 301-316.
14. Guo K, Guo F, Feng Y, Chen J, Zheng C, Gardner N C. Synchronous visual and RTD study on liquid flow in rotating packed-bed contactor. *Chem En. Sci.* 2000, 55: 1699–1706.
15. Burns JR, Ramshaw C. Process intensification: visual study of liquid maldistribution in rotating packed beds. *Chem Eng Sci.* 1996, 51: 1347-1352.
16. Lin CC, Liu WT, Tan CS. Removal of carbon dioxide by absorption in a rotating packed bed. *Ind Eng Chem Res.*2003; 42: 2381-2386.
17. Yi F, Zou HK, Chu GW, Shao L, Chen JF. Modeling and experimental studies on absorption of CO_2 by Benfield solution in rotating packed bed. *Chem Eng J.* 2009; 145: 377-384.
18. Jassion MS. Carbon dioxide absorption and desorption in aqueous monoethanolamine solutions in a rotating packed bed. *Ind Eng Chem Res.*2007; 46: 2823-2833.
19. Tan CS, Chen JE. Absorption of carbon dioxide with piperazine and its mixtures in a rotating packed bed. *Sep Purif Technol.* 2006; 49: 174-180.
20. Zhang SJ, Chen YH, Li FW, Lu XM, Dai WB, Mori R. Fixation and conversion

of CO_2 using ionic liquids. *Catal Today*. 2006; 115: 61-69.

21. Zhang XC, Liu ZP, Wang WC. Screening of ionic liquids to capture CO_2 by COSMO-RS and experiments. *AIChE J*. 2008, 54: 2717-2728.

22. Cadena C, Anthony JL, Shah JK, Morrow TI, Brennecke JF, Maginn EJ. Why is CO_2 so soluble in imidazolium-based ionic liquids? *J Am Chem Soc*. 2004; 126: 5300-5308.

23. Freemantle M. New frontiers for ionic liquids. *Chem Eng News*. 2007; 85: 23-26.

24. Muldoon MJ, Aki S, Anderson JL, Dixon JK, Brennecke JF. Improving carbon dioxide solubility in ionic liquids. *J Phys Chem B*. 2007; 111: 9001-9009.

25. Betas ED, Mayton RD, Ntai I, Davis JH. CO_2 capture by a task-specific ionic liquid. *JACS*. 2002; 124: 926-927.

26. Zhang LL, Wang JX, Xiang Y, Zeng XF, Chen JF. Absorption of carbon dioxide with ionic liquid in a rotating packed bed contactor: mass transfer study. *Ind Eng Chem Res*.2011; 50: 6957-6964.

27. Delaloye MM, Stockar UV, Lu XP. The influence of viscosity on the liquid-phase mass transfer resistance in packed columns. *Chem Eng J*. 1991, 47: 51-61.

28. Li ZH. A study on modeling of mass transfer in RPB, Ph.D. Thesis, 2000.

29. Qian Z, Xu LB, Li ZH, Li H, Guo K. Selective absoption of H_2S from gas mixture with CO_2 by N-methyldiethanolamine in rotating packed bed. *Ind Eng Chem Res*. 2010; 49: 6196-6203.

30. Chu H, Chien TW, Li SY. Simultaneous absorption of SO_2 and NO from flue gas with $KMnO_4$/NaOH solution. *Sci Total Environ*. 2001; 275: 127-135.

31. Sada E, Kumazawa H, Takada Y. Chemical reactions accompanying absorption of NO into aqueous mixed solutions of Fe^{II}(EDTA) and Na_2SO_3. *Ind Eng Chem Fundam.* 1984; 23: 60-64.

32. Sada E, Kumazawa H, Hlkosaka H. A kinetic study of absorption of NO into aqueous solutions of Na_2SO_3 with added Fe^{II}(EDTA) chelate. *Ind Eng Chem Fundam.* 1986; 25: 386-390.

33. Demmink JF, Van Gils ICF, Beenackers AACM. Absorption of nitric oxide into aqueous solutions of ferrous chelates accompanied by instaneous reaction. *Ind Eng Chem Res.* 1997; 36: 4914-4927.

34. Zhang LL, Wang JX, Sun Q, Zeng XF, Chen JF. Removal of nitric oxide in rotating packed bed by ferrous chelate solution. *Chem Eng J.* 2012; 1: 624-629.

Table Captions:

Table 1. Comparison of liquid side volumetric mass transfer coefficient values between packed tower and RPB.

Table 1. Comparison of liquid side volumetric mass transfer coefficient values between packed tower and RPB

	packed tower	RPB
temperature, K	293	293
pressure, atm	1	1
packing volume, cm^3	98.17	50.26
packing surface area, m^2/m^3	2000	850
CO_2 concentration, mol %	10	10
gas flow rate, L/min	0.7-1.2	0.6-1
liquid flow rate, mL/min	36.5-73.0	29.2-102.2
gas-liquid ratio	11-33	6-34
$k_L \alpha$, 10^{-2} s^{-1}	0.063-0.19	0.95-3.9

Figure Captions:

Figure 1. The structure of the countercurrent RPB.

Figure 2. Industrial-scale RPB plant for CO_2 capture in power plant.

Figure 3. Effect of operation parameters on liquid side volumetric mass transfer coefficient in RPB.

Figure 4. Industrial-scale RPB plant for desulfuration in a sulfur acid factory.

Figure 5. Effect of operation parameters on Gas side volumetric mass transfer coefficient, H_2S removal efficiency and selective factor in RPB.

Figure 6. Effect of operation parameters on NO removal efficiency in RPB.

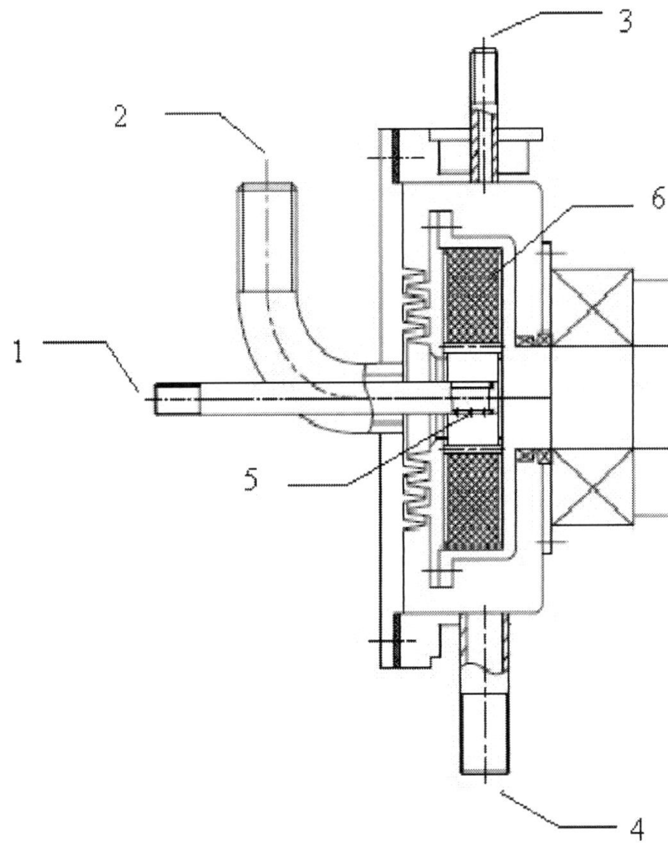

Figure 1. The structure of the countercurrent RPB.

(1) Liquid inlet. (2) Gas outlet. (3) Gas inlet. (4) Liquid outlet. (5) Liquid distributor. (6) Mesh packing.

Figure 2. Industrial-scale RPB plant for CO_2 capture in power plant.

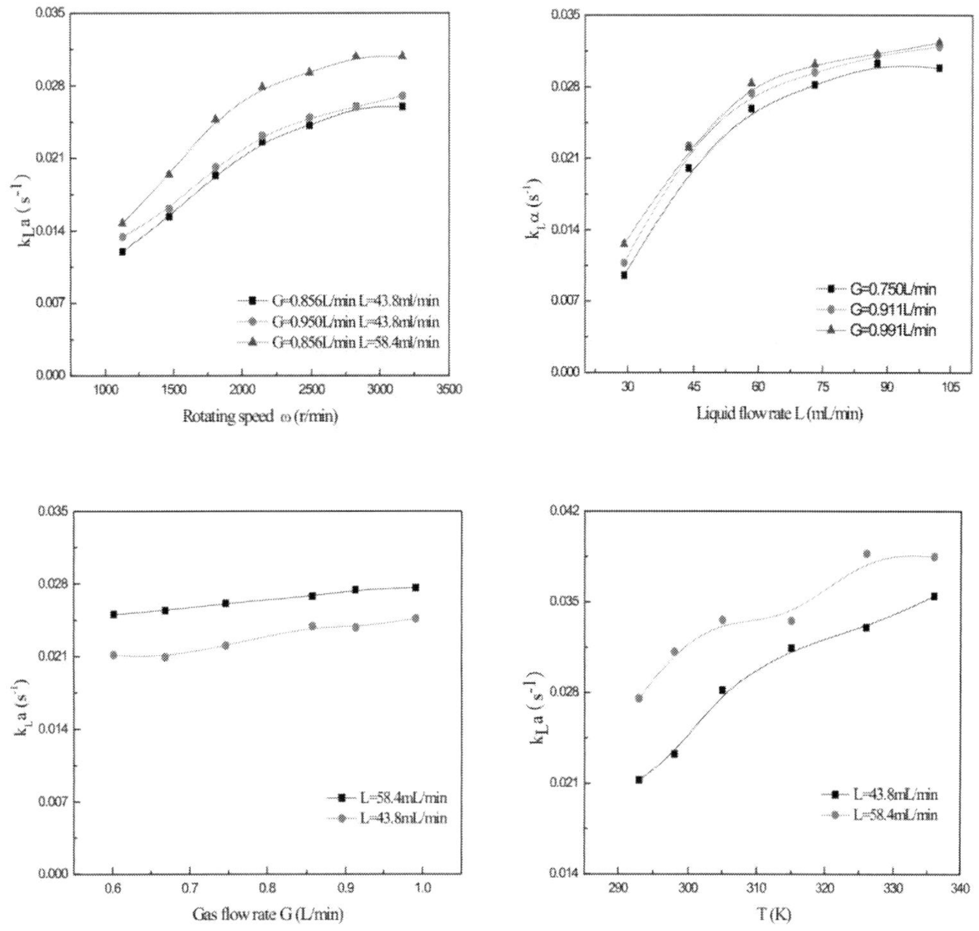

Figure 3. Effect of operation parameters on liquid side volumetric mass transfer coefficient in RPB.

Figure 4. Industrial-scale RPB plant for desulfuration in a sulfur acid factory.

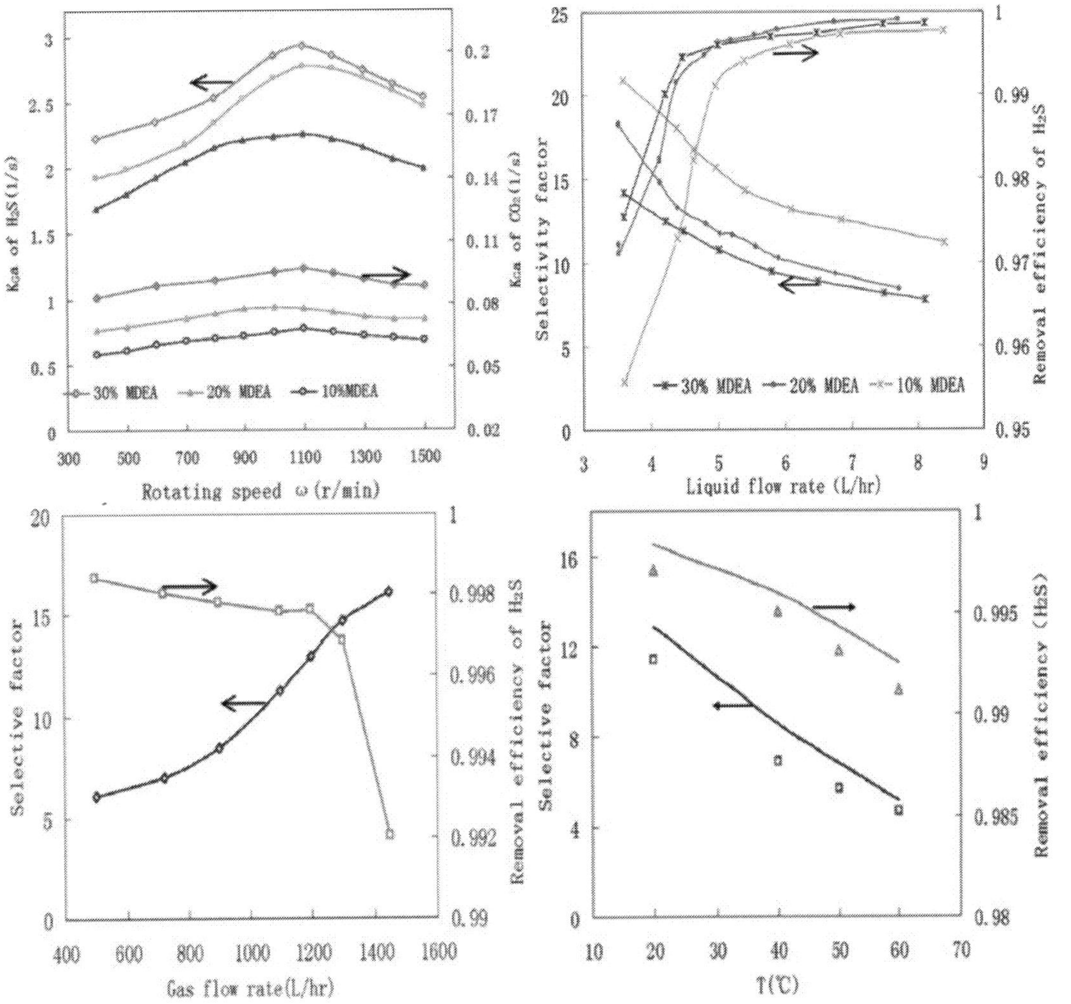

Figure 5. Effect of operation parameters on Gas side volumetric mass transfer coefficient, H_2S removal efficiency and selective factor in RPB.

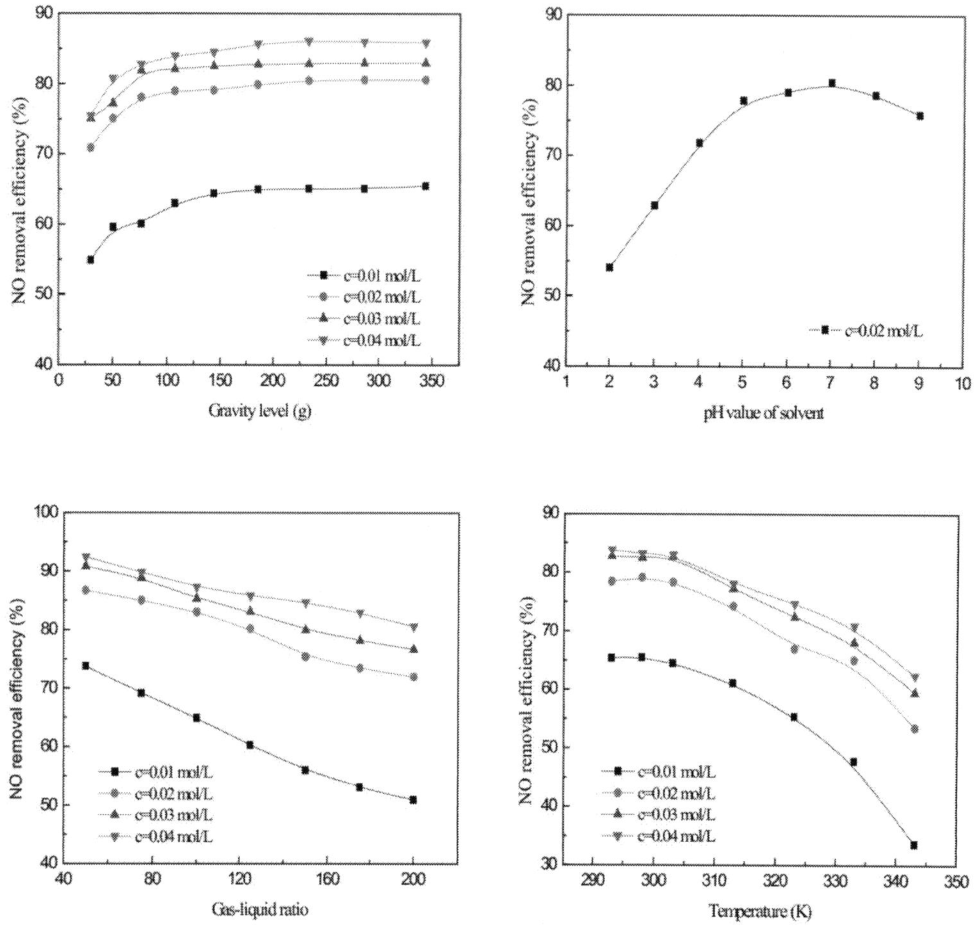

Figure 6. Effect of operation parameters on NO removal efficiency in RPB.

Pervaporation of High Boiling Point Organic Compounds with composite PDMS membrane

Paper number: 254903

Chunyan Chen, Zeyi Xiao, Kewang Deng, Weijia Li, Haidi Cui, Junqing Zhang

Introduction

The organic compounds with higher boiling points than water, such as propargyl alcohol, butanol, pyridine and acetic acid, are defined as high boiling point organic compounds (HBOCs). HBOCs are generally important industrial chemicals and widely used in industry. For example, propargyl alcohol is extensively used in oil industry (1, 2), iron and steel industry, and electroplating industry (3). Butanol has been regarded as an ideal fuel to relieve the depletion of petroleum fuel, due to its many superior properties over other fuels (4). Besides, pyridine is also an important solvent present in effluents from rubber and plastics, petrochemicals and some other organic chemical industries (5).

In HBOCs industries, the process streams for product extraction were generally aqueous mixtures with tiny content of HBOCs. It was obviously not a good method to remove the bulk of water to separate or concentrate HBOCs by usual evaporation technology, although we had to do so in current practical industries.

To avoid tedious distillation process and annoying bulky water evaporation, several separation technologies have been developed to recover HBOCs from aqueous mixtures, such as adsorption, extraction, gas stripping and pervaporation(6, 7). Pervaporation has been regarded as a promising technology owing to its energy saving and efficiency, as well as no product contamination.

As reported, polydimethylsiloxane (PDMS) membrane has shown good behaviors on separation of organic compounds from aqueous solutions, such as alcohols, acids, aldehydes, esters, ketones, and aroma compounds (8-14). In this work, experiments on pervaporation separation of three HBOCs, propargyl alcohol, butanol and pyridine, from their water mixtures were carried out by means of a composite PDMS membrane, to explore some common features on pervaporation separation of HBOCs-water mixture system.

Materials and methods

Experimental materials

The pervaporation membrane used here was a self-prepared composite PDMS membrane, with the module providing an effective membrane area of 0.024 m^2. The preparation procedure of the membrane and the structure of the module had been fully described in the previous work (9, 15).

Propargyl alcohol - butynediol - water mixture was a liquid mixture from the practical industry, offered by Tianyu Chemical Industry Corporation (Shandong, China). Aqueous butanol mixture was prepared directly using n-butanol of reagent grade and distilled water. Similarly, aqueous pyridine mixture was self prepared.

Experimental apparatus and process

A schematic diagram of the apparatus for pervaporation experiments is shown in Fig. 1. Mixture liquid of 2.5 L was circulating between the feed reservoir and upstream of the membrane module by a pump. The feed flow rate was regulated by a rotameter and the temperature was controlled by a thermostat. The absolute pressure at the downstream of the membrane was maintained at 2KPa by a vacuum pump. The permeate vapor through the membrane was condensed and collected in the cold trap immersed in a -30□ refrigerator. Samples were withdrawn every one hour to measure the concentrations of the concerned components in the feed reservoir and cold trap, respectively.

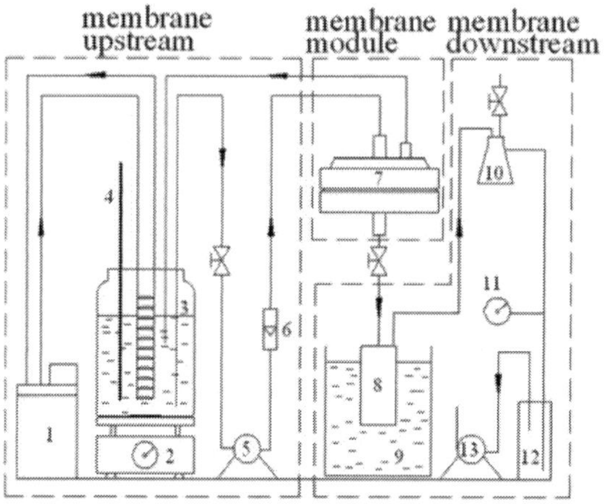

Fig.1. Schematic diagram of apparatus for pervaporation experiment

1. thermostat; 2. magnetic stirrer; 3 feed reservoir; 4 thermometer; 5. circulation pump; 6. rotameter; 7. membrane module; 8. cold trap; 9.refrigerator; 10. vacuum reservoir; 11. vacuum gauge; 12. desiccator; 13. vacuum pump

Analysis

The concerned organic components in this experiment were propargyl alcohol, butanol and pyridine. The detailed analysis methods were described as follows:

1) Propargyl alcohol: propargyl alcohol reacts with silver nitrate, and forms a soluble compound, releasing equivalent H^+ simultaneously. The formed compound solution is titrated with sodium hydroxide standard solution and then the concentration of propargyl alcohol could be determined by the following formula (1):

$$W = \frac{(C \cdot V) \times \frac{M}{1000}}{m} \times 100 \qquad (1)$$

Where W is the mass fraction of propargyl alcohol (wt%), C is the concentration of sodium hydroxide standard solution ($mol \cdot L^{-1}$), V is the volume of consumed sodium hydroxide solution (ml), M is the molar mass of propargyl alcohol ($g \cdot mol^{-1}$) and m is the mass of the sample (g).

2) Butanol was directly determined with a density meter DMA 4500 (Anton Paar, Austria). If the permeate separated into two phases, the permeate sample was diluted with distilled water prior to injection.

3) Pyridine was determined by a gas chromatography GC112A (Shanghai, China) equipped with a flame ionization detector and a FFAP capillary column (30mx0.32mm). The carrier gas was nitrogen and isobutanol was used as an internal standard. The temperature conditions were as following: Injector temperature 200□, detector temperature 200□, column temperature linearly ramping from 60□ to 180□ at 20□ per min.

Membrane behaviors

Membrane behaviors were evaluated with two parameters, selectivity and permeation flux.

1) Selectivity:

$$\alpha_{o/w} = \frac{Y_o / Y_w}{X_o / X_w} \tag{2}$$

Where X and Y is the mass fraction of components in the feed and permeate, respectively. Subscript "o" denotes organic compound and "w" water.

2) Permeation flux:

$$J = M_p / At \tag{3}$$

Where M_p is the weight of the collected permeate (g), t is the sampling time interval for the pervaporation (h) and A is the effective membrane surface area (m^2).

Results and discussion

Effects of feed flow rate on the pervaporation performance of HBOCs

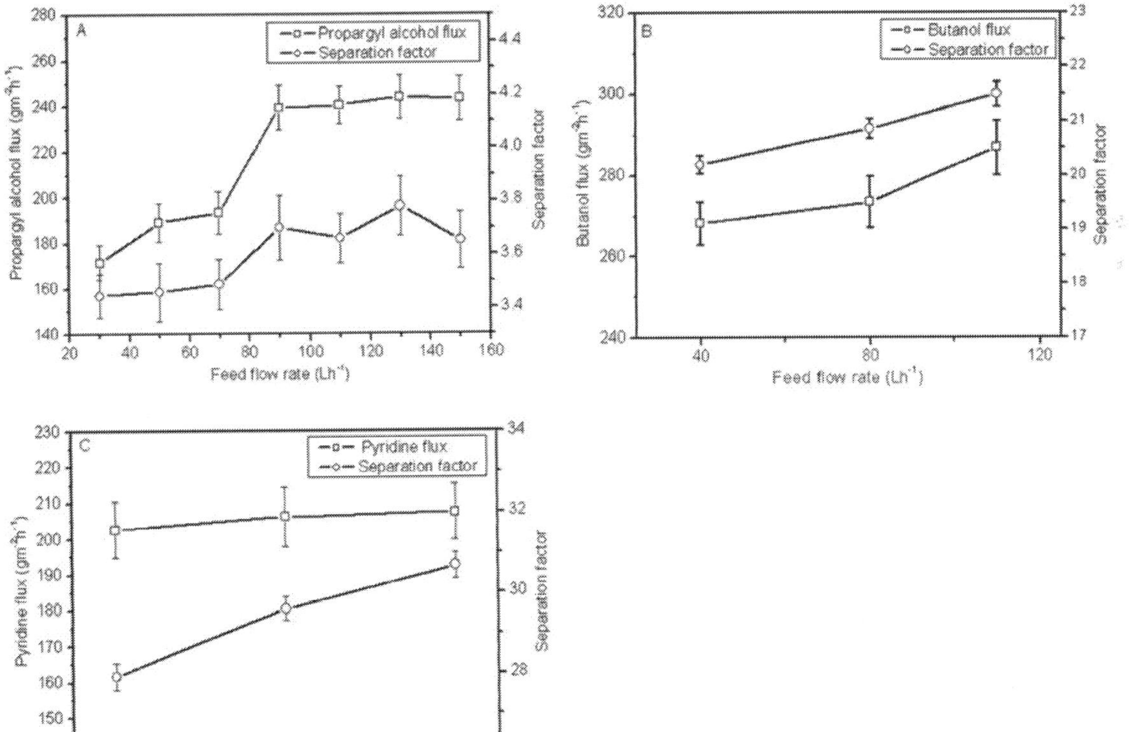

Fig.2. Effects of feed flow rate on HBOCs flux and selectivity (A)Propargyl alcohol (B)Butanol (C)Pyridine

The influence of feed flow rate on flux and selectivity of propargyl alcohol, butanol and pyridine was evaluated. As seen in Fig. 2, the HBOCs fluxes all increased with feed flow rate. According to the serial resistance model of membrane transport (16), higher feed flow rates produced higher flow velocities on the membrane surface, and then caused smaller boundary layer resistance for mass transfer through the membrane, finally resulted in increased fluxes. However, the increased degrees of HBOCs fluxes were different. With the feed flow rate increasing from 40 to 120 Lh^{-1}, the flux of propargyl alcohol, butanol and pyridine was varying in the range of 170-240, 260-290, 200-210 $gm^{-2}h^{-1}$, respectively. The difference may be attributed to the different adsorption characteristics of propargyl alcohol, butanol and pyridine.

Meanwhile, selectivities of the three HBOCs also increased with feed flow rate. Nevertheless, with the feed flow rate increasing, the ranges of selectivity for propargyl alcohol, butanol and pyridine were 3.4-3.8, 20-22, 27-31, respectively. Obviously, the increased degrees of selectivity for the three HBOCs were opposite to those of HBOCs fluxes, which demonstrated a trade-off between the variation degrees of flux and selectivity.

Effects of feed concentration on the pervaporation performance of HBOCs

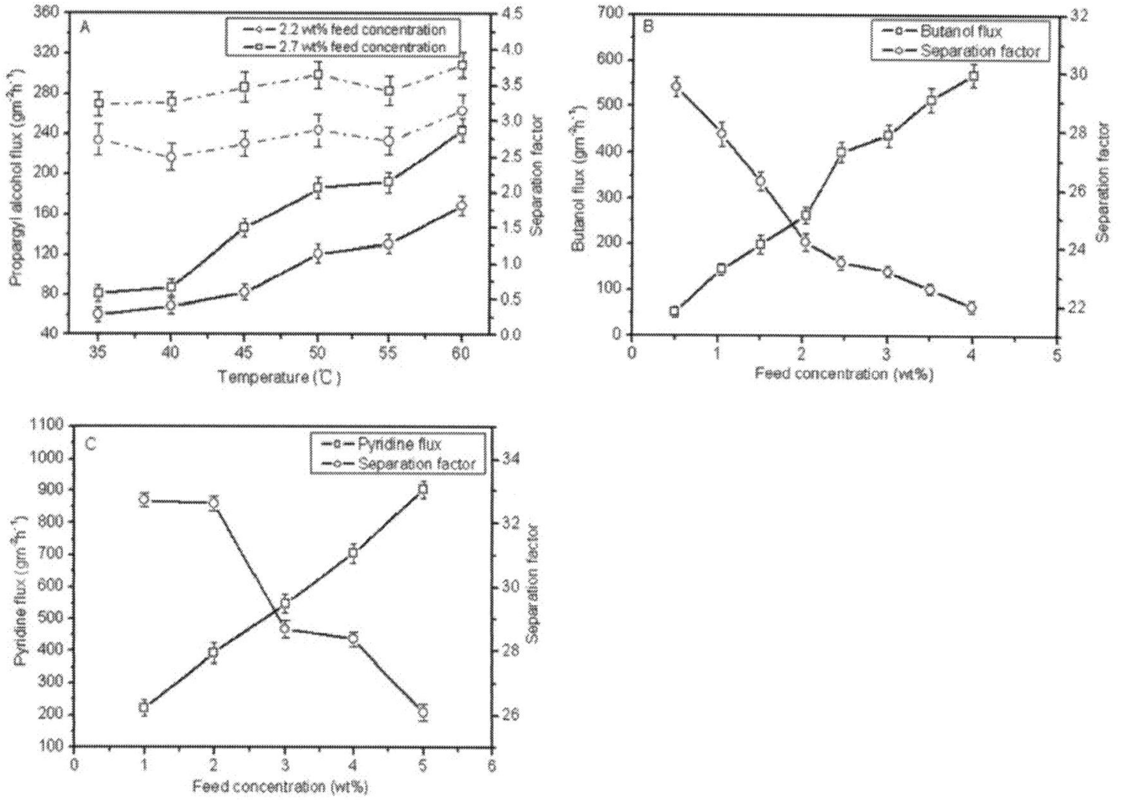

Fig.3. Effects of feed concentration on HBOCs flux and selectivity (A)Propargyl alcohol, solid lines for flux, dashed lines for selectivity (B)Butanol (C)Pyridine

The pervaporation behaviors of HBOCs at different feed concentrations were investigated. Fig. 3A shows propargyl alcohol flux and selectivity in the course of temperature at two different feed concentrations, while Fig. 3B and 3C show the pervaporation behaviors of butanol and pyridine in the course of feed concentration respectively. As the feed concentration increased so did the three HBOCs Fluxes. Higher feed concentration means stronger diffusion driving force, which could reduce the resistance of mass transfer and then result in higher flux. Therefore, increasing the feed concentration properly is also an alternative method for improving the pervaporation performance.

On the other hand, as observed in many membrane pervaporation researches, selectivities of butanol and pyridine decreased with the increase of feed concentration, since at high feed concentrations the denominator term in the selectivity relationship becomes large (17). But selectivity of propargyl alcohol at 2.7 wt% feed concentration was higher than that at 2.2 wt%. The unexpected result may be explained by the minor increase of the feed concentration, which needs to be confirmed in the further work.

Effects of temperature on the pervaporation performance of HBOCs

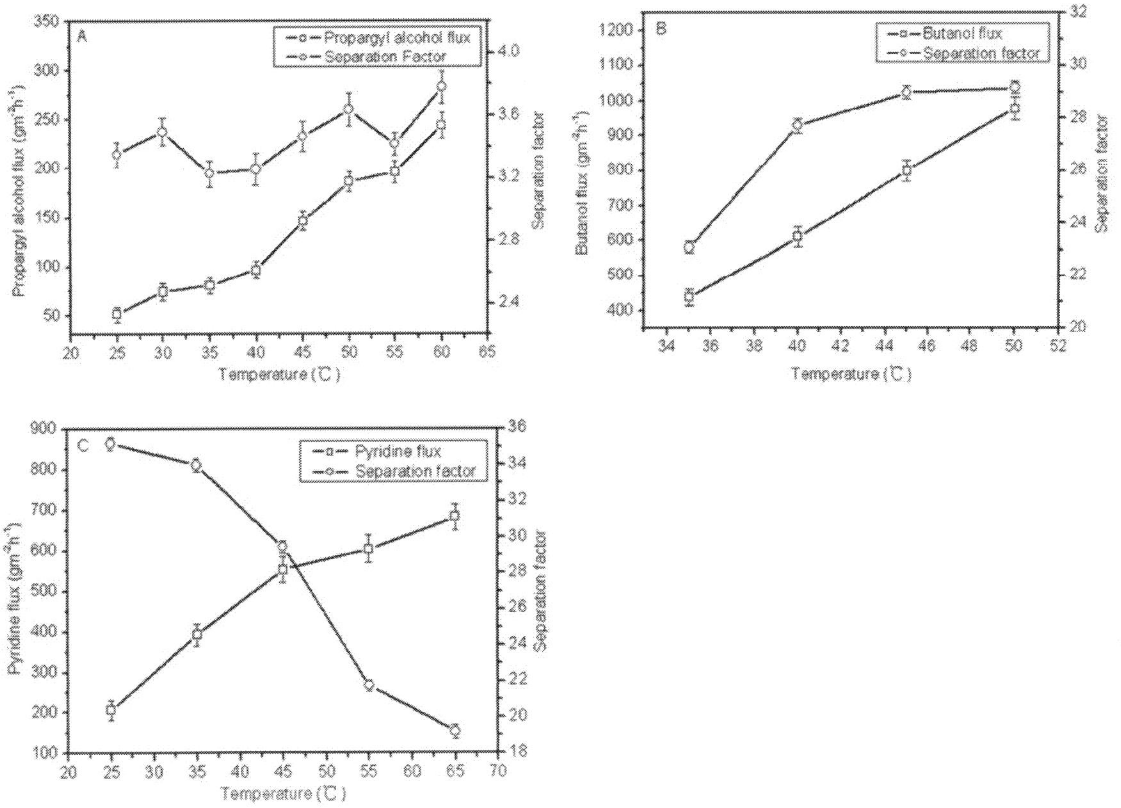

Fig.4. Effects of temperature on HBOCs flux and selectivity (A)Propargyl alcohol (B)Butanol (C)Pyridine

The effects of temperature on the pervaporation performance of propargyl alcohol, butanol and pyridine were presented in Fig. 4. As expected, the three HBOCs fluxes increased almost linearly with temperature. As the feed temperature rises, the vapor pressure in the feed compartment also increases, but the vapor pressure at the permeate side is not affected, which caused an increase of driving force, and then enhanced HBOCs fluxes with temperature. For example, propargyl alcohol flux was just 51.42 $gm^{-2}h^{-1}$ at the temperature of 25℃, the flux jumped to 243.94 $gm^{-2}h^{-1}$ when temperature went up to 60℃. It is reasonable to assume that, with further temperature increasing, a much higher flux of propargyl alcohol could be obtained by using PDMS membrane pervaporation.

On the other side, as seen in Fig. 4, selectivities of the three HBOCs behaved differently. For propargyl alcohol, the selectivity was just fluctuating in the range of 3.2-3.8 as the temperature increases, which may be considered to be temperature independent. For butanol, the selectivity increased with temperature, which was contrary to the common trade-off phenomenon existing between flux and selectivity in pervaporation process. However, a similar trend was observed by N.Qureshi et al (17). Their results showed that higher temperatures were required to desorb butanol on the basis of adsorption characteristic, which explains increased butanol selectivity at higher temperatures. For pyridine, the selectivity decreased with temperature, which was in accordance with the trade-off between flux and selectivity. It is known that pyridine has larger molecular size (0.60nm) than water (0.26nm), therefore the permeability of pyridine increases slower than water with temperature, resulting in decrease of selectivity. The differences among the three HBOCs selectivity tendencies could be attributed to the different affinities of the PDMS membrane for propargyl alcohol, butanol and pyridine.

Pervaporation characteristics of HBOCs

With regard to the separation of the HBOCs from water mixtures, the difference between pervaporation and distillation is significant. In distillation system, the separation mechanism is based on the boiling point of the components, which means the light component (water in most cases) is separated preferentially. While in pervaporation system, the separation mechanism is based on the diffusivity and solubility of the components, thus the HBOCs could be separated preferentially through hydrophobic organic membrane. For this reason, distillation could only obtain the low boiling point component first, but pervaporation is able to behave a 'reverse direction' selective separation for the mixtures by obtaining high boiling point component preferentially. Besides, distillation needs to be accomplished under high temperature with bulky facility and then causes intensive energy cost, while the operation condition of pervaporation is relatively moderate, which saves not only the cost of heating, but also the cost of the subsequent cooling process.

Especially, the solubility of butanol in water is 7.7 wt% at the temperature of 20□. In the butanol-water pervaporation experiment, the concentration of butanol in the permeate was in the range of 22.3-48.4wt%, which far exceeded its solubility limit, and thus the permeate was phase separated: the upper phase was substantially enriched in butanol (about 80 wt%), and the lower aqueous phase contained about 92 wt% water. The specific phenomenon is especially beneficial to butanol recovery. After the phase separation, the aqueous phase could be recycled to the feed stream to further increase the butanol concentration.

Conclusions

Pervaporation separations of three typical high boiling point organic compounds (HBOCs) mixtures were carried out with a composite PDMS membrane. These HBOCs could preferentially permeate the PDMS membrane and be selectively extracted from the mixtures through the membrane. To the authors' knowledge, this is the first report to present the conception of HBOCs and focus on the recovery of HBOCs by membrane pervaporation. The effects of feed flow rate, feed concentration and temperature on the pervaporation performance of the membrane were examined. The highest permeation flux of propargyl alcohol, butanol and pyridine was 243.24, 976.5 and 904.70 $gm^{-2}h^{-1}$ respectively, with the corresponding selectivity of 3.78, 29.65 and 26.09. Compared with the traditional distillation that separates different components in a mixture on the basis of boiling point, membrane pervaporation behaves a 'reverse direction' selective separation for HBOCs, which should be considered as the most principal feature on HBOCs pervaporation.

Reference

1 .Goncalves, R. S.;Azambuja, D. S.;Lucho, A. M. S.(2002) Electrochemical studies of propargyl alcohol as corrosion inhibitor for nickel, copper, and copper/nickel (55/45) alloy, *Corrosion Science*, 44 (3): 467-479.

2 .Avdeev, Y. G.;Podobaev, N. I.(2005) The role of acrolein in the inhibition of the acid corrosion of iron with propargyl alcohol, *Protection of Metals*, 41 (6): 592-596.

3 .Matsuda, I.;Asa, F.;Osaka, T.(2003) Simulations on the optimum conditions for propargyl alcohol to function as a shape control agent in the process of nickel electrodeposition onto a micropatterned substrate, *Electrochemistry*, 71 (11): 912-919.

4 .Tong, C.;Bai, Y.;Wu, J.;Zhang, L.;Yang, L.;Qian, J.(2010) Pervaporation Recovery of Acetone-Butanol from Aqueous Solution and Fermentation Broth Using HTPB-Based Polyurethaneurea Membranes, *Sep. Sci. Technol.*, 45 (6): 751-761.

5 .Mandal, M. K.;Bhattacharya, P. K.(2006) Poly(ether-block-amide) membrane for pervaporative separation of pyridine present in low concentration in aqueous solution, *J. Membr. Sci.*, 286 (1-2): 115-124.

6 .Sun, W. G.;Wang, X. W.;Yang, J. H.;Lu, J. M.;Han, H. L.;Zhang, Y.;Wang, J. Q.(2009) Pervaporation separation of acetic acid-water mixtures through Sn-substituted ZSM-5 zeolite membranes, *J. Membr. Sci.*, 335 (1-2): 83-88.

7 .Vane, L. M.(2008) Separation technologies for the recovery and dehydration of alcohols from fermentation broths, *Biofuels Bioproducts & Biorefining-Biofpr*, 2 (6): 553-588.

8 .Xiao, Z. Y.;Dahlan, M. H.;Xing, X. H.;Yoshikawa, Y.;Matsumoto, K.(2000) A membrane bioreactor with novel modules for effective biodegradation of toluene, *Biochemical Engineering Journal*, 5 (1): 83-88.

9 .Li, L.;Xiao, Z. Y.;Tan, S. J.;Liang, P.;Zhang, Z. B.(2004) Composite PDMS membrane with high flux for the separation of organics from water by pervaporation, *Journal of Membrane Science*, 243 (1-2): 177-187.

10 .Mohammadi, T.;Aroujalian, A.;Bakhshi, A.(2005) Pervaporation of dilute alcoholic mixtures using PDMS membrane, *Chemical Engineering Science*, 60 (7): 1875-1880.

11 .Tan, S. J.;Li, L.;Xiao, Z. Y.;Wu, Y. T.;Zhang, Z. B.(2005) Pervaporation of alcoholic beverages - the coupling effects between ethanol and aroma compounds, *Journal of Membrane Science*, 264 (1-2): 129-136.

12 .Xiao, Z. Y.;Li, L.;Zhang, Z. B.;Tan, S. J.(2004) Pervaporation of acetic acid/water mixtures through carbon molecular sieve-filled PDMS membranes, *Chem. Eng. J.*, 97 (1): 83-86.

13 .Tang, X. Y., Xiao,Z.Y, Shi, E.(2007) Fermentation of apple juice in a PDMS membrane bioreactor, *Proceedings of the 4th Joint China/Japan Chemical Engineering Symposium*.

14 .Shi, E.;Huang, W. X.;Xiao, Z. Y.;Tang, X. Y.;Xu, R. Q.;Zeng, F. J.(2007) Study on improvement of the quality in Chinese new-type liquor by pervaporation with polydimethylsiloxane membrane, *J. Food Process Eng*, 30 (1): 38-50.

15 .Tang, X. Y.;Wang, R.;Xiao, Z. Y.;Shi, E.;Yang, J.(2007) Preparation and pervaporation performances of fumed-silica-filled polydimethylsiloxane-polyamide (PA) composite membranes, *J. Appl. Polym. Sci.*, 105 (5): 3132-3137.

16 .Parnas, R. S.;Li, S. Y.;Srivastava, R.(2010) Separation of 1-butanol by pervaporation using a novel tri-layer PDMS composite membrane, *J. Membr. Sci.*, 363 (1-2): 287-294.

17 .Qureshi, N.;Meagher, M. M.;Hutkins, R. W.(1999) Recovery of butanol from model solutions and fermentation broth using a silicalite silicone membrane, *J. Membr. Sci.*, 158 (1-2): 115-125.

ANALYSIS AND MODELLING OF FROTH TREATMENT CASCADED COUNTER-CURRENT GRAVITY SETTLERS

A. I. A. Salama and T. Dabros
CanmetENERGY-Devon, Natural Resources Canada, Devon, AB, Canada

Introduction

The Canadian oil sand industry has adopted napthanic and paraffinic solvents (or diluents) technologies for froth treatment. The companies who have selected the paraffinic technology implemented single- and two-stage counter current (CC) gravity settler arrangements or flowheets. Furthermore, different technologies on tailing-solvent-recovery unit (TSRU) have been developed to strip solvent from the tailings of gravity settlers. The single- and two-stage settlers are integrated with different TSRU designs.

The objective of the present paper is to present a unified approach for the analysis and evaluation of CC flowsheets. To facilitate the development, it is proposed that the flowsheet stream mass rates and stream-components may be arranged as shown in the following arrays.

$$\mathbf{X} = \begin{bmatrix} \mathbf{Y}_1 & B_1 & M_1 & W_1 & S_1 \\ \mathbf{Y}_2 & B_2 & M_2 & W_2 & S_2 \\ . & . & . & . & . \\ . & . & . & . & . \\ \mathbf{Y}_7 & B_7 & M_7 & W_7 & S_7 \end{bmatrix}, \quad \mathbf{YX} = \begin{bmatrix} YX_{11} & YX_{12} & YX_{13} & YX_{14} \\ YX_{21} & YX_{22} & YX_{23} & YX_{24} \\ . & . & . & . \\ . & . & . & . \\ YX_{71} & YX_{72} & YX_{73} & YX_{74} \end{bmatrix}$$

where $\mathbf{Y}_i$'s, B_i's, M_i's, W_i's, and S_i's designate stream mass rates in kg/min or tones/hr, stream bitumen contents, mineral (solids) contents, water contents, and solvent contents, respectively. Stream-component contents are expressed in fractions or per cent. In the above arrays the number of rows (streams) N_s=7 and number of columns (stream-components+1) of array **X** is N_c+1=5 and number of columns (stream-components) of array **YX** is N_c=4. Note that the sequence (Y, B, M, W, and S) will be kept throughout the paper.

Single-stage flowsheet

Consider the signal flowgraph of single-stage flowsheet shown in Figure 1 (Shinners.1978 and Dunn & Chan, 1968)). Furthermore, the stream numbering shown in Table 1 is adopted to simplify the developments. It is obvious that I_j is expressed as

$$I_j = YX_{4j} + YX_{5j} \tag{1}$$

where I_j is component j mass rate at the inputs of node I. The other streams (i.e., node overflow and underflow) component mass rates can be expressed as

$$YX_{1j} = I_j = YX_{4j} + YX_{5j} \tag{2}$$

$$YX_{2j} = \left[P_{Ij}\right]YX_{4j} + \left[P_{Ij}\right]YX_{5j} \tag{3}$$

$$YX_{3j} = \left[\left(1 - P_{Ij}\right)\right]YX_{4j} + \left[\left(1 - P_{Ij}\right)\right]YX_{5j} \tag{4}$$

$$YX_{4j} = F_j \tag{5}$$

$$YX_{5j} = SI_j \tag{6}$$

where P_{Ij} is the fraction of component j mass rate at node I reporting to the node overflow . Moreover, from Eqs. (3 and 4), it can be shown that

$$YX_{2j} + YX_{3j} = IO_j + IU_j = F_j + SI_j \tag{7}$$

which verifies the relationships in Eqs. (3 and 4) and confirms the mass and stream-component mass (flux) balances. In this case, the relationships between the inputs, F_j and SI_j and the outputs, IO_j and IU_j, are simple and involve the stage partition (or separation) function (or operator) P_{Ij}. However, these relationships between inputs and outputs become complicated when we consider feedback (recirculation) loops as in the two-stage configurations which is considered next.

Two-stage flowsheet

Consider the cascaded two-stage CC configuration shown in Figure 2. Let Stage I and stage II are labeled nodes I and II. Furthermore, the stream numbering shown in Table 2 is adopted to simplify the developments. Using the signal flowgraph (Shinners.1978 and Dunn & Chan, 1968), it can be shown that I_j and II_j can be expressed in terms of P_{Ij}, P_{IIj}, YX_{7j}, YX_{8j}, and YX_{9j} as

$$I_j = \left[\frac{1}{\Delta_j}\right] YX_{7j} + \left[\frac{1}{\Delta_j}\right] YX_{8j} + \left[\frac{P_{IIj}}{\Delta_j}\right] YX_{9j} \tag{8}$$

$$II_j = \left[\frac{(1-P_{Ij})}{\Delta_j}\right] YX_{7j} + \left[\frac{(1-P_{Ij})}{\Delta_j}\right] YX_{8j} + \left[\frac{1}{\Delta_j}\right] YX_{9j} \tag{9}$$

where I_j and II_j are component j mass rate at the inputs of nodes I and II, respectively, and P_{Ij}, and P_{IIj} are the fractions of component j mass rate at nodes I and II inputs reporting to the node overflows, respectively. Furthermore, the subscript j designates component j. The other streams, i.e., node overflows and underflows component mass rates can be expressed as

$$YX_{1j} = I_{Ij} = \left[\frac{1}{\Delta_j}\right] YX_{7j} + \left[\frac{1}{\Delta_j}\right] YX_{8j} + \left[\frac{P_{IIj}}{\Delta_j}\right] YX_{9j} \tag{10}$$

$$YX_{2j} = \left[\frac{P_{Ij}}{\Delta_j}\right] YX_{7j} + \left[\frac{P_{Ij}}{\Delta_j}\right] YX_{8j} + \left[\frac{P_{Ij}P_{IIj}}{\Delta_j}\right] YX_{9j} \tag{11}$$

$$YX_{3j} = \left[\frac{(1-P_{Ij})}{\Delta_j}\right] YX_{7j} + \left[\frac{(1-P_{Ij})}{\Delta_j}\right] YX_{8j} + \left[\frac{(1-P_{Ij})P_{IIj}}{\Delta_j}\right] YX_{9j} \tag{12}$$

$$YX_{4j} = II_j = \left[\frac{(1-P_{Ij})}{\Delta_j}\right] YX_{7j} + \left[\frac{(1-P_{Ij})}{\Delta_j}\right] YX_{8j} + \left[\frac{1}{\Delta_j}\right] YX_{9j} \tag{13}$$

$$YX_{5j} = \left[\frac{(1-P_{Ij})P_{IIj}}{\Delta_j}\right] YX_{7j} + \left[\frac{(1-P_{Ij})P_{IIj}}{\Delta_j}\right] YX_{8j} + \left[\frac{P_{IIj}}{\Delta_j}\right] YX_{9j} \tag{14}$$

$$YX_{6j} = \left[\frac{(1-P_{Ij})(1-P_{IIj})}{\Delta_j}\right] YX_{7j} + \left[\frac{(1-P_{Ij})(1-P_{IIj})}{\Delta_j}\right] YX_{8j} + \left[\frac{(1-P_{IIj})}{\Delta_j}\right] YX_{9j} \tag{15}$$

$$YX_{7j} = F_j \tag{16}$$

$$YX_{8j} = SI_j \tag{17}$$

$$YX_{9j} = SII_j \tag{18}$$

The flowsheet Δ_j is expressed as

$$\Delta_j = 1 - \left(1 - P_{Ij}\right) P_{IIj} \tag{19}$$

Moreover, from Eqs. (11 and 15), it can be shown that

$$YX_{2j} + YX_{6j} = IO_j + IIU_j = F_j + SI_j + SII_j \tag{20}$$

which verifies Eqs. (11 and 15) and confirms the mass and stream-component mass balances. If YX_{8j} is set to zero, then Eqs. (10-15) simplify to

$$I_j = \left[\frac{1}{\Delta_j}\right] YX_{7j} + \left[\frac{1}{\Delta_j}\right]\left[\frac{P_{IIj}}{\Delta_j}\right] YX_{9j} \tag{21}$$

$$II_j = \left[\frac{\left(1 - P_{Ij}\right)}{\Delta_j}\right] YX_{7j} + \left[\frac{1}{\Delta_j}\right] YX_{9j} \tag{22}$$

$$YX_{1j} = I_{Ij} = \left[\frac{1}{\Delta_j}\right] YX_{7j} + \left[\frac{P_{IIj}}{\Delta_j}\right] YX_{9j} \tag{23}$$

$$YX_{2j} = \left[\frac{P_{Ij}}{\Delta_j}\right] YX_{7j} + \left[\frac{P_{Ij} P_{IIj}}{\Delta_j}\right] YX_{9j} \tag{24}$$

$$YX_{3j} = \left[\frac{\left(1 - P_{Ij}\right)}{\Delta_j}\right] YX_{7j} + \left[\frac{\left(1 - P_{Ij}\right) P_{IIj}}{\Delta_j}\right] YX_{9j} \tag{25}$$

$$YX_{4j} = II_j = \left[\frac{\left(1 - P_{Ij}\right)}{\Delta_j}\right] YX_{7j} + \left[\frac{1}{\Delta_j}\right] YX_{9j} \tag{26}$$

$$YX_{5j} = \left[\frac{\left(1 - P_{Ij}\right) P_{IIj}}{\Delta_j}\right] YX_{7j} + \left[\frac{P_{IIj}}{\Delta_j}\right] YX_{9j} \tag{27}$$

$$YX_{6j} = \left[\frac{\left(1 - P_{Ij}\right)\left(1 - P_{IIj}\right)}{\Delta_j}\right] YX_{7j} + \left[\frac{\left(1 - P_{IIj}\right)}{\Delta_j}\right] YX_{9j} \tag{28}$$

and

$$YX_{7j} = F_j \tag{29}$$

$$YX_{8j} = 0 \tag{30}$$

$$YX_{9j} = SII_j \tag{31}$$

The case where YX_{8j} is set to zero is the most common configuration which is adopted by the industry and is implemented at CanmetENERGY-Devon pilot plant.

Interrelationships

Relationship between $[S/B]_{CF}$ and $[S/B]_{IO}$

Using the circuit mass balance and using the bitumen and solvent flux balances,

$$[Q \cdot B]_{CF} = [Q \cdot B]_{IO} + [Q \cdot B]_{CU} \tag{32}$$

$$[Q \cdot Solv]_{CF} = [Q \cdot Solv]_{IO} + [Q \cdot Solv]_{CU} \tag{33}$$

where Q.B designates bitumen mass rate (kg/m) and the subscripts "CF", "IO", and "CU" designates the circuit feed, stage I overflow, and circuit underflow, respectively. It can be shown that

$$\left[\frac{S}{B}\right]_{CF} = \left[\frac{1+\delta_1^{CF-IO}}{1+\delta_2^{CF-IO}}\right]\frac{[Q.Solv]_{IO}}{[Q.B]_{IO}} \tag{34}$$

or

$$\left[\frac{S}{B}\right]_{CF} = \left[\frac{1+\delta_1^{CF-IO}}{1+\delta_2^{CF-IO}}\right]\left[\frac{S}{B}\right]_{IO} \tag{35}$$

where

$$\delta_1^{CF-IO} = \frac{\text{Solv Flux}_{CU}}{\text{Solv Flux}_{IO}} = \frac{[Q\cdot Solv]_{CU}}{[Q\cdot Solv]_{IO}} \tag{36}$$

$$\delta_2^{CF-IO} = \frac{\text{Bit Flux}_{CU}}{\text{Bit Flux}_{IO}} = \frac{[Q\cdot B]_{CU}}{[Q\cdot B]_{IO}} \tag{37}$$

For most practical cases, it can be shown that δ_1 and δ_2 are positive and very small quantities. Consider the special case of zero solvent and bitumen loss to the circuit underflow, then Eq. (35) reduces to

$$\left[\frac{S}{B}\right]_{CF} = \left[\frac{S}{B}\right]_{IO} \tag{38}$$

which is correct. Moreover, the special case of zero solvent loss to the underflow, then Eq. (35) reduces to

$$\left[\frac{S}{B}\right]_{CF} = \left[\frac{1}{1+\delta_2^{CF-IO}}\right]\left[\frac{S}{B}\right]_{IO} \tag{39}$$

Since δ_2^{CF-IO} is positive and small, then $[S/B]_{CF}$ is less than $[S/B]_{IO}$. A profile of $[S/B]_{CF}/[S/B]_{IO}$ versus incremental variations δ_1 and δ_2 is shown in Figure 3. Note that the region of interest is highlighted in red.

Relationship between $[S/B]_{IF}$ and $[S/B]_{IO}$

Using the Ith Stage bitumen and solvent flux balances

$$[Q\cdot B]_{CF} + [Q\cdot B]_{IIO} = [Q\cdot B]_{IO} + [Q\cdot B]_{IU} \tag{40}$$

$$[Q\cdot Solv]_{IIO} = [Q\cdot Solv]_{IO} + [Q\cdot Solv]_{IU} \tag{41}$$

it can be shown that

$$\left[\frac{S}{B}\right]_{IF} = \frac{[Q\cdot Solv]_{IIO}}{[Q\cdot B]_{CF} + [Q\cdot B]_{IIO}} = \frac{[Q\cdot Solv]_{IO} + [Q\cdot Solv]_{IU}}{[Q\cdot B]_{IO} + [Q\cdot B]_{IU}} = \left[\frac{1+\delta_1^{IF-IO}}{1+\delta_2^{IF-IO}}\right]\frac{[Q\cdot Solv]_{IO}}{[Q\cdot B]_{IO}} \tag{42}$$

or

$$\left[\frac{S}{B}\right]_{IF} = \left[\frac{1+\delta_1^{IF-IO}}{1+\delta_2^{IF-IO}}\right]\left[\frac{S}{B}\right]_{IO} \tag{43}$$

where

$$\delta_1^{IF-IO} = \frac{[Q\cdot Solv]_{IU}}{[Q\cdot Solv]_{IO}} \tag{44}$$

$$\delta_2^{IF-IO} = \frac{[Q\cdot B]_{IU}}{[Q\cdot B]_{IO}} \tag{45}$$

For most practical cases, it can be shown that δ_1^{IF-IO} and δ_2^{IF-IO} are positive and very small quantities. Consider the special case of zero solvent and bitumen loss to the underflow, then Eq. (43) reduces to

$$\left[\frac{S}{B}\right]_{IF} = \left[\frac{S}{B}\right]_{IO} \tag{46}$$

which is correct. Moreover, the special case of zero solvent loss to the underflow, then Eq. (43) reduces to

$$\left[\frac{S}{B}\right]_{IF} = \left[\frac{1}{1+\delta_2^{IF-IO}}\right]\left[\frac{S}{B}\right]_{IO} \tag{47}$$

Since δ_2 is positive and small, then $[S/B]_{IF}$ is less than $[S/B]_{IO}$. A profile of $[S/B]_{IF}/[S/B]_{IO}$ versus incremental variations δ_1 and δ_2 similar to Figure 3 can be constructed.

Relationship between $[S/B]_{IF}$ and $[S/B]_{CF}$

Using the developed it can be approximated to

$$\frac{\left[\frac{S}{B}\right]_{IF}}{\left[\frac{S}{B}\right]_{CF}} = \frac{\left[\frac{1+\delta_1^{IF-IO}}{1+\delta_2^{IF-IO}}\right]\left[\frac{S}{B}\right]_{IO}}{\left[\frac{1+\delta_1^{CF-IO}}{1+\delta_2^{CF-IO}}\right]\left[\frac{S}{B}\right]_{IO}} \tag{48}$$

$$\frac{\left[\frac{S}{B}\right]_{IF}}{\left[\frac{S}{B}\right]_{CF}} \simeq \left[\frac{1+\delta_1^{IF-IO}-\delta_2^{IF-IO}}{1+\delta_1^{CF-IO}-\delta_2^{CF-IO}}\right] \simeq 1+\left(\delta_1^{IF-IO}-\delta_2^{IF-IO}\right)-\left(\delta_1^{CF-IO}-\delta_2^{CF-IO}\right) \tag{49}$$

or

$$\left[\frac{S}{B}\right]_{CF} \simeq \left[1+\delta_1^{CF-IO}-\delta_2^{CF-IO}\right]\left[\frac{S}{B}\right]_{IO} \tag{50}$$

$$\left[\frac{S}{B}\right]_{IF} = \left(1+\delta_1^{IF-IO}-\delta_2^{IF-IO}\right)\left[\frac{S}{B}\right]_{IO} \tag{51}$$

From practice, we get

$$\left(1+\delta_1^{CF-IO}-\delta_2^{CF-IO}\right) \le \left(1+\delta_1^{IF-IO}-\delta_2^{IF-IO}\right) \tag{52}$$

Which indicates that $[S/B]_{IF} > [S/B]_{CF}$.

Two-stage flowsheet [s/b]'s

$$\left[\frac{S}{B}\right]_{IIF} = \frac{[Q\cdot Solv]_{IU}+[Q\cdot Solv]_{CF}}{[Q\cdot B]_{IU}} = \frac{[Q\cdot Solv]_{IU}+[Q\cdot Solv]_{CF}}{[Q\cdot Solv]_{IU}}\cdot\frac{[Q\cdot Solv]_{IU}}{[Q\cdot B]_{IU}} = \left(1+\delta_1^{IIF-IU}\right)\cdot\left[\frac{S}{B}\right]_{IU} \tag{53}$$

where

$$\delta_1^{IIF-IU} = \frac{[Q\cdot Solv]_{CF}}{[Q\cdot Solv]_{IU}} \tag{54}$$

Demonstration

It should be emphasized that the approach adopted in this paper for the analysis of any general flowheet uses the P_{xxj}'s of the different stages XX's of the flowheet. Using the P_{xxj}'s, the relationships between different streams component mass rate can be derived using the signal flowgraph approach.

This approach has been implemented for the derivation of the relationships presented for single- and two-stage CC flowsheets. Note that relationships are useful to visualize the relationships between different streams. Using practical experience, this can assist in examining the sensitivity between different stream-component mass rates. Or a mass balance data can used to derive the P_{xxj}'s of the flowheet.

The demonstration is restricted to single- and two-stage flowsheets. Each case will start by a given P_{xxj} of the flowsheet. It will be shown that how the stream-component mass rates are derived using the flowsheet input streams (froth and solvent) mass rates and composition. Furthermore, verification of the derived relationships is demonstrated by the mass balanced results.

Considering a two-stage CC flowheet and an associated P_{xxj}'s of the flowheet as shown in Table 3. The corresponding source data is given in Table 4. The relationships in Eqs. (8-31) are used to calculate the data given in Table 5. Note that the calculations are based on the input streams (froth and solvent) mass rates and composition and the P_{xxj}'s of Table 4. The results in Table 5 are balanced which verifies the relationships in Eqs. (8-31).

Conclusions

Cascaded CC single- and two-stage forth treatment flowsheets are analyzed using system analysis approach. Specifically, the signal flowgraph technique is adopted to express the relationships between input and output process streams of the flowsheet. Furthermore, an attempt is made to relate mathematically the solvent-to-bitumen ratio S/B in most common streams to each other. These S/B relationships in conjunction with some practical approximation produce results closely match actual measurement.

Acknowledgement

The author would like to acknowledge support from the Government of Canada's interdepartmental Program of Energy Research and Development, PERD POL.

References

1. Shinners S.M. (1978), "*Modern Control Systems Theory and Applications*," (2nd edition), Don Mills, Canada, Addison-Wesley.
2. Dunn W.R., Jr. and Chan S.P. (1968), "*An Algorithm for Calculating the Gain of a Signal Flowgraph*," NASA-CR-95760, N68-29177.

Table 1 Single-stage stream labeling

Stream label	#	Mass split	Component j mass rate
stage I input	1	Y_1	YX_{1j}
stage I OF	2	Y_2	YX_{2j}
stage I UF	3	Y_3	YX_{3j}
froth	4	Y_4	YX_{4j}
S_I	5	Y_5	YX_{5j}

Table 2 Two-stage stream labeling

Stream label	#	Mass split	Component j mass rate
stage I input	1	Y_1	YX_{1j}
stage I OF	2	Y_2	YX_{2j}
stage I UF	3	Y_3	YX_{3j}
stage II input	4	Y_4	YX_{4j}
stage II OF	5	Y_5	YX_{5j}
stage II UF	6	Y_6	YX_{6j}
froth	7	Y_7	YX_{7j}
S_I	8	Y_8	YX_{8j}
S_{II}	9	Y_9	YX_{9j}

Table 3 P_{XXj}'s and Δ_j of a two-stage CC flowsheet CC flowsheet

Two stage CC flowsheet P_j'S, and Δ_j's					
	Q (kg/min)	Bitumen (wt%)	Solids (wt%)	Water (wt%)	Solvent (wt%)
P_{Ij}	0.729	0.8558	0.000	0.000	0.900
P_{IIj}	0.729	0.5411	0.000	0.000	1.000
Δ_j	0.802	0.922	1.000	1.000	0.900

Table 4 A source data of a two-stage CC flowsheet

SOURCE DATA						
Sample Point	Q (kg/min)	Bitumen (wt%)	Solids (wt%)	Water (wt%)	Solvent (wt%)	Normal
IO	2.06	37.88	0.00	0.00	62.12	100.00
IU	0.77	17.14	18.80	45.50	18.56	100.00
IIO	1.50	4.76	0.00	0.00	95.23	100.00
IIU	0.55	10.89	26.03	62.99	0.09	100.00
F	1.34	63.01	10.82	26.17	0.00	100.00
S_{II}	1.26	0.00	0.00	0.00	100.00	100.00

Table 5 Calculated data of a two-stage CC flowheet and using Table 4

FEED DATA FROTH, $SOLVENT_{III}$, P_j'S, and Δ_j's							CALCULATED DATA					
Sample Point	Q (kg/min)	Bitumen (wt%)	Solids (wt%)	Water (wt%)	Solvent (wt%)	Normalize	Q (kg/min)	Bitumen (wt%)	Solids (wt%)	Water (wt%)	Solvent (wt%)	Normalize
IO							2.06	37.88	0.00	0.00	62.12	100.00
IU							0.77	17.14	18.80	45.50	18.56	100.00
IIO							1.50	4.76	0.00	0.00	95.23	100.00
IIU							0.55	10.89	26.03	62.99	0.09	100.00
F	1.34	63.01	10.82	26.17	0.00	100.00	1.95	65.53	8.58	25.89	0.00	100.00
S_{II}	1.26	0.00	0.00	0.00	100.00	100.00	1.98	0.00	0.00	0.00	100.00	100.00
IF							2.83	32.25	5.10	12.35	50.30	100.00
IIF							2.05	6.42	7.05	17.05	69.48	100.00
P_{Ij}	0.729	0.8558	0.000	0.000	0.900							
P_{IIj}	0.729	0.5411	0.000	0.000	1.000							
Δ_j	0.802	0.922	1.000	1.000	0.900							

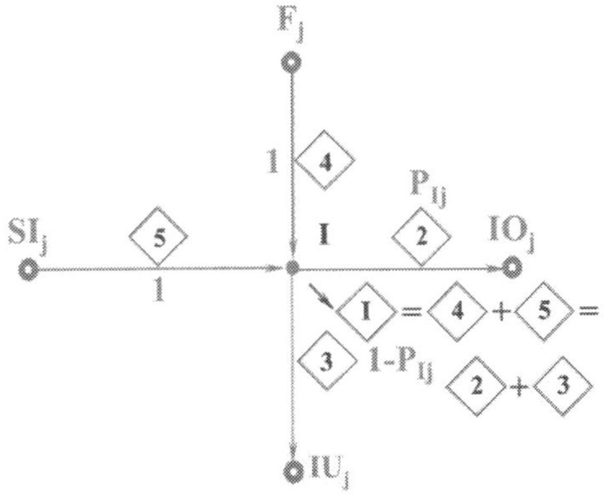

Figure 1 – Single-stage signal flowgraph

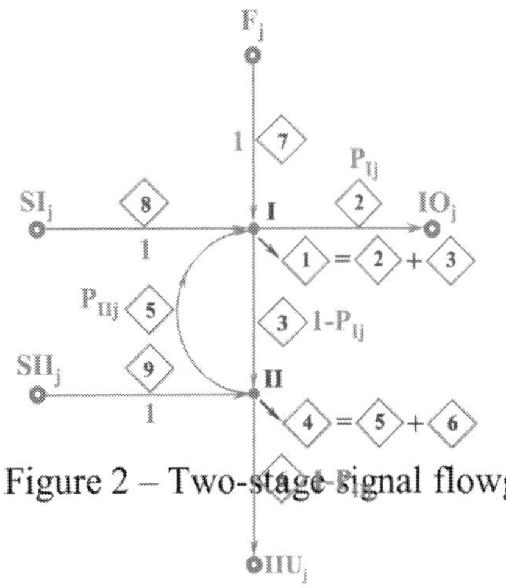

Figure 2 – Two-stage signal flowgraph

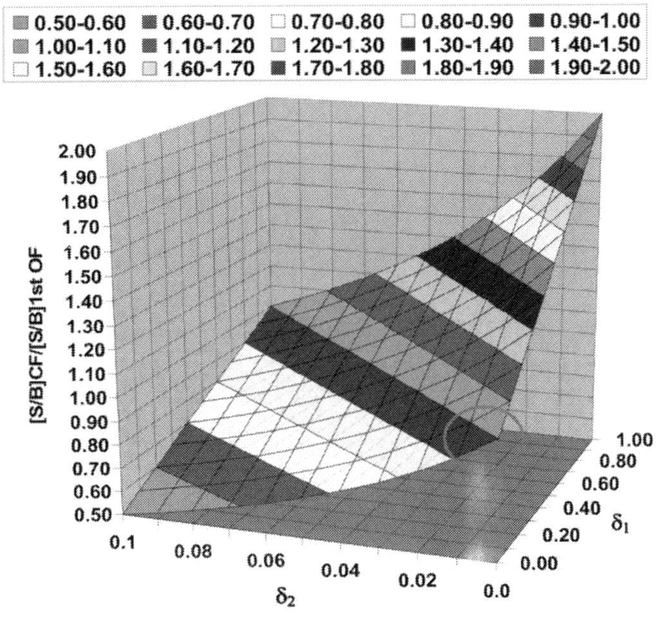

Figure 3 – A profile of $[S/B]_{CF}/[S/B]_{IO}$ versus incremental variations

(387d) A Comparative Study of Chemical and Physical Treatment of Greywater

Jaime Lopez[1], Marvin Ramos[1], Abdullah Al Qahtani[1] and Sedigheh Faraji[2], (1)CSULB, Long Beach, CA, (2)Chemical Engineering Department, California State University, Long Beach, Long Beach, CA

Water recycling is important for the purpose of water conservation in the state of California. Greywater is defined as water effluent that originates from kitchen, laundry rooms, bathrooms/washrooms, sinks, and showers and is composed of mostly chemicals and biological organisms. The treatment of greywater allows for the greywater to be recycled and reused for non-potable uses such as irrigation and washing in a house or residential complexes. Both physical and chemical methods can be used to treat greywater. In this study, the photocatalytic treatment of greywater by titanium dioxide nanoparticles under UV light is investigated and compared with single batch distillation and filtration. The parameters such as dissolved oxygen content, chemical oxygen demand, pH, and turbidity have been studied. The results show that filtration and distillation techniques are not viable because of their high energy requirements. The titanium dioxide nanocatalyst under UV light shows a good activity and stability for greywater treatment.

(387e) Using Structure-Property Relationships to Design Reversible Ionic Liquids for CO2 Capture

Jackson R. Switzer[1], Amy L. Rohan[1], Kyle M. Flack[2], Emily C. Nixon[2], Amber C. Rumple[2], Elizabeth J. Biddinger[1], Pamela Pollet[2], Charles L. Liotta[2] and Charles A. Eckert[1], (1)School of Chemical and Biomolecular Engineering, Georgia Institute of Technology, Atlanta, GA, (2)School of Chemistry and Biochemistry, Georgia Institute of Technology, Atlanta, GA

Reversible ionic liquids (RevILs) have been developed as alternatives to traditionally cited CO_2 capture solvents such as monoethanolamine (MEA). RevILs capture CO_2 through chemical reaction of CO_2 with an amine functionality to form an ionic liquid. The resulting ionic liquid is then capable of absorbing additional CO_2 through physical absorption. The design of reversible ionic liquids has been an important focal point in our research group. Using structure-property relationships we have shown an ability to tune the relevant CO_2 capture properties by altering the structure of the starting RevIL. Properties targeted include CO_2 capture capacity, ionic liquid viscosity, temperature of CO_2 release, temperature of RevIL evaporation, and the enthalpy of CO_2 absorption. We have examined several additional modifications to the structure of our RevILs and investigated their effects on CO_2 capture properties. Our progress on developing these structure-property relationships to further the design of optimized solvents for CO_2 capture will be discussed.

(387f) Effect of Imperfections in Microvoids On Optimal Time of Separations

Jyothirmai J. Simhadri[1], Holly A. Stretz[2] and Pedro E. Arce[1], (1)Chemical Engineering, Tennessee Technological University, Cookeville, TN, (2)Department of Chemical Engineering, Tennessee Technological University, Cookeville, TN

Research involving polymer gels with embedded nanoparticles of varying properties is quite attractive because of the multitude of potential applications, including separation of biomacromolecules for environmental proteomics in the assessment of contaminant impact on health issues; more effective and less time consuming clinical diagnostics, growth of tissue in synthetic or natural scaffold etc. Previously, analysis of the effects of the pore length and diameters on the transport of biomolecules was studied either analytically or computationally. In this research, we focus on computationally analyzing different pore models (associated with gel materials) and study the effect of geometry on the transport properties. In particular the role of the convective-diffusive transport via different values for the Peclet number will be assessed. Results will illustrate, for example, the role of pore-scale in conjunction with the effect of electrical fields in assessing different convective-diffusive regimes. Also, optimal time for the separations to occur will also be discussed.

Purification and Extraction of active components in *Polygonum cuspidatum* by Using Supercritical Fluid

Ruey Chi Hsu Ye Gen Chang

Department of Chemical and Materials Engineering, Chang Gung University

259 Wen-hwa 1st Road, Kwei-Shan, Tao-Yuan, Taiwan 333, R.O.C.

E-mail: rchsu@mail.cgu.edu.tw

Abstract

The using of natural products is one of principal topics of biotechnology development in our country. The important thing is how to purify the effect components from the product of extraction. Restveratrol, a polyphenol found in grape and other natural products(like *Polygonum cuspidatum*), was shown to have cancer chemopreventive activity and was found to act as an antioxidant and antimutagen and anti-initiation activity. The Polygonum cuspidatum is a cheap and abundant natural plant. Except resveratol, the chemical ingredients of *Polygonum cuspidatum* such as emodin, physcion and chrysophanol also have special medical activities.. The purification active compounds from *Polygonum cuspidatum* extraction that will be a economical ideal.

The purpose of this study is to separate resveratrol and emodin from *Polygonum cuspidatum* extracted by supercritical fluid chromatography (SFC). Several operating parameters were investigated in this study, including stationary phases, pressure, temperatures and modifier ratio of mobile phase.

Stationary Phases

In this study, the ODS column(150 mm long; 4.6mm dia.; 5 micron particle) as reverse phase column and the silica gel column (150 mm long; 4.6mm dia.; 5 micron particle) as normal reverse phase column. The typical SFC chromatograms are shown in Fig.1. The results shown that the reverse phase column is not suit to separate the emodin and resveratol, but the silica gel column can separate these two components very well under supercritical state.

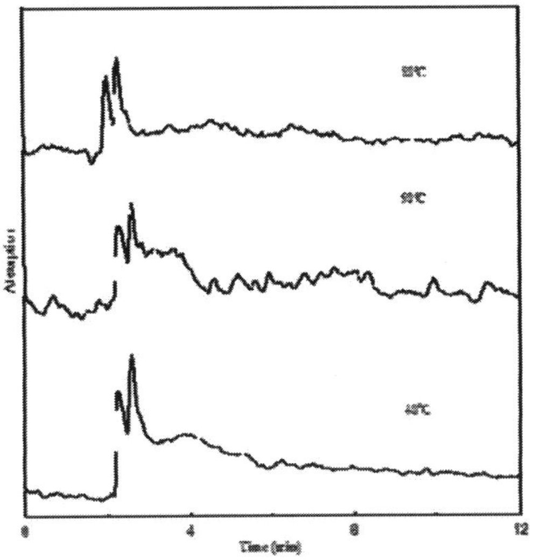

Fig1a. SFC chromatograms of ODS column

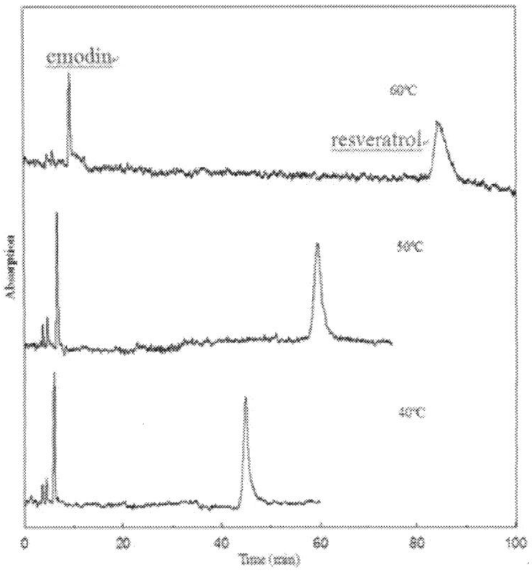

Fig1b. SFC chromatograms of silica gel column

Temperature Effect

The supercritical fluid can adjust solubility by tune the temperature. The temperature effect for the results of chromatography are shown in Fig.1b. It was shown that the elution time of the resveratrol were reduced from 85 to 45mines, under the temperature decreses from 60°C to 45 °C. Further decreasing the temperature ,the elution time of resveratrol seems to keeping constant. The pressure also effect the elute of emodin,but not signifiance as emodin.

Modifier

The modifier is play an important role in SFC, the methanol was chosen as the modifier in this study. The effects of the amount of modifier for emodin and resveratol are shown in Fig.2..It was shown that there has not appeared any peak in the both chromatograms under no addition of modifier. The more modifiers, the peak is sharper and the run time is reduced for the two components. Further, it can be obtained that the elution time of emodin is shorter than those for resveratrol. These were meaning that these two components can be separated by SFC. The elution time for different modifier concentration were shown in Fig.3. It cab can be get that the elution time reduce from 40min to 15 mines.

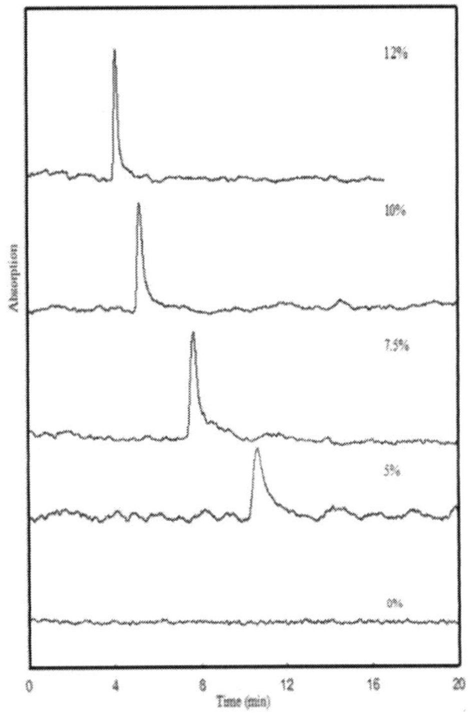

Fig2a. The chromatograms of emodin under various concentration of modifier.

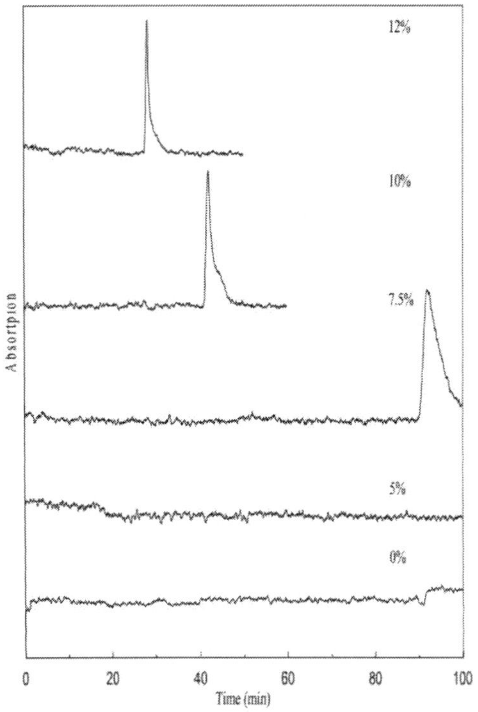

Fig2b. The chromatograms of resveratrol under various concentration of modifier.

Pressure Effect

The supercritical fluid can adjust solubility by tune the pressure. The pressure effect for the results of chromatography are shown in Fig.4. It was shown that the elution time of the resveratrol were reduced from 45 to 30 mines, under the pressure increases from 2000psi to 2500psi. Further decreasing the pressure ,the elution time of resveratrol seems to keeping constant. The pressure also effect the elute of emodin,but not signifiance as emodin.

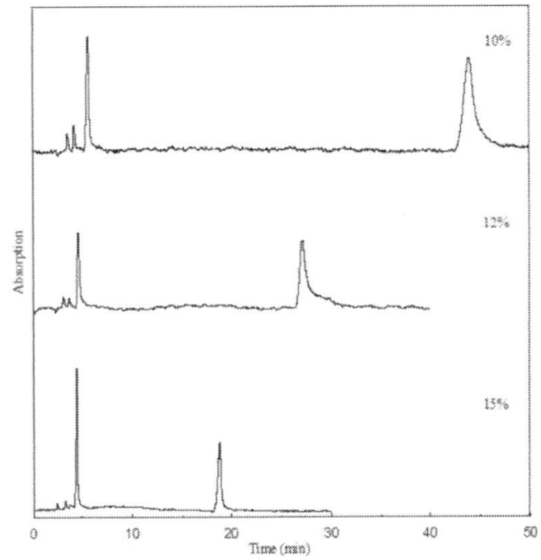

Fig.3. The elution time for different modifier concentration

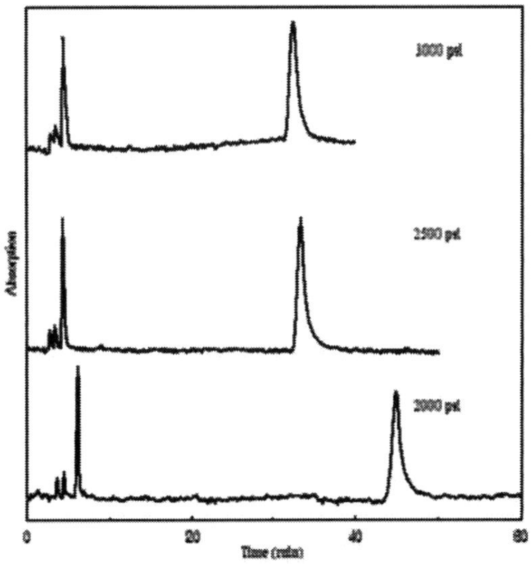

Fig.4. The elution time under different pressure

The best operating condition for chromatograph the extracted *Polygonum cuspidatum* was 2000psi and 40°C supercritical CO_2 with 10% (v/v) ethanol as modifier flowing through silica column for 50 min.

(387h) Liquid-Liquid Equilibria of Systems Containing Cottonseed Biodiesel + Glycerol + Ethanol At Different Temperatures

Francisca Mesquita[1], Aline Bessa[1], Daniel Lima[1], Hosiberto B. Sant'ana Sr.[2] and **Rílvia Santiago-Aguiar**[3], (1)Chemical Engineering, Federal University of Ceara, Fortaleza, Brazil, (2)Chemical Engineering, Federal University of Ceará, Fortaleza, Brazil, (3)Chemical Engineering Department, Universidade Federal do Ceará, Fortaleza, Brazil

Biodiesel presents several characteristics that can be considered as an appropriate alternative fuel in substitution of fossil fuels, e.g., renewable, biodegradable, and relatively clean-burning. According to the last Bulletin of the Brazilian Regulatory Agency (Agência Nacional do Petróleo, Gás Natural e Biocombustíveis - ANP), the most important raw material sources used for the production of biodiesel in Brazil are: soybean seeds (83.26%), followed by bovine fat (12.34%) and cottonseeds (2.17%). Biodiesel consists of a mixture of alkyl esters of fatty acids, generally produced from homogeneous transesterification reactions of vegetable oils and animal fats. Purification process of these transesterification reactions are extremely important in order to provide a biofuel with quality levels required by the international standards specifications. The purification step can be improved by using a liquid–liquid extraction process, generally by using aliphatic alcohols (methanol or more recently ethanol). This study reports LLE experimental data for ternary systems of cottonseed biodiesel + glycerol + ethanol at T = 293.15, 313.15 and 333.15 and atmospheric pressureK. The solubility (binodal) curves and tie-line compositions were obtained by means of density measurements. The reliability of the data has been tested by using Othmer–Tobias equation, with a squared correlation coefficient better than 0.99 for all systems studied. Selectivity and distribution coefficients have been determined from equilibrium data. The Nonrandom Two-Liquid Activity Coefficient Model (NRTL) model was used to predict binary interaction parameters for the biodiesel + glycerol + ethanol system. Root mean square deviations (rmsd) between experimental and calculated compositions in both equilibrium phases were about 0.89% for the system studied.

(387i) Design of Ionic Liquids for Aliphatic/Aromatic Separations

Amirhossein Mehrkesh and Arunprakash T. Karunanithi, Department of Civil Engineering, University of Colorado Denver, Denver, CO

In a supported liquid membrane the given solvent is immobilized inside the porous structure of the membrane which separates the feed phase from the receiving phase. The obvious advantage of such a system is that, it requires minimal amount of solvent and the solvent is continuously regenerated as the solute moves to the receiving phase. Ionic liquids are particularly suited for this application as they have very low vapor pressure and hence the solvent loss is negligible. The fact that the solubility of ionic liquids in the surrounding phases can be tailored, allows us to develop very stable supported liquid membranes.

Ionic liquids are molten salts that exist as liquids at a relatively low temperature. Ionic liquids consist of a large organic cation and a charge-delocalized inorganic or organic anion. These liquids are entirely composed of ions and are characterized by weak interactions between the cation and anion through hydrogen bonds and van der Walls forces. ILs are considered as 'green' solvents because of their extremely low vapor pressure due to the columbic attraction of the ions (< 10^{-9} bar). In addition ionic liquids have high thermal stability (thermal decomposition temperature of ~ 400 K).

The ionic liquids considered for this application needs to have the following characteristics a) should have high selectivity for aromatics; b) should have high aromatic distribution coefficient; c) moderate viscosity. In addition the ionic liquid should have relatively high long term thermal stability at higher temperatures so that the feed stream temperature/pressure need not be decreased significantly.

In this work we present a computer-aided ionic liquid design (CAILD) approach for designing task specific ionic liquids for separation of aromatics from aliphatics. This algorithm helps design optimal ionic liquid structure by appropriate selection/modification of the cation, the anion and/or the alky chains attached to the cation. The ionic liquids are tailored to have high selectivity, distribution coefficient and solubility for aromatics, wide liquid range, higher thermal stability and optimal physico-chemical properties (e.g. viscosity). The key criterion for ionic liquid selection for this application is aromatic selectivity.

(387j) Increased Productivity of Industrial Plant of Benzene Reforming

Dasha Smirnova, Resource-Saving Technologies, Saint-Petersburg State Institute of Technology (Technical University), St Petersburg, Russia

The study examined industrial plant of benzene reforming, which is part of the existing refinery. The device is intended for the production of aromatic hydrocarbons - benzene and toluene - from virgin gasoline fraction 62-105 ° C.

Purpose - Implementation of resource-saving technologies by improving the performance of the separation complex of catalysate of reforming, which is part of the existing facility, and increase the release of benzene when used with existing manufacturing equipment and energy resources.

The complex consists of three separate processing units - the unit of stabilization of catalysate, the unit of liquid extraction and the unit of secondary rectification. The raw materials other than gasoline fraction from the installation of the primary distillation of crude oil can be fed fractions containing benzene and toluene from the plant of summary xylenes.

To analyze the process and optimization calculations the model of the separation complex was developed in the program ASPEN HYSYS. We calculated the optimal structural and operational parameters of the separation system. It was established that in order to increase the load capabilities of the complex is necessary to optimize the separation of the two processing units - the unit of stabilization of catalysate and the unit of secondary rectification.

We shown The possibility and ways of extracting from the catalysate reforming of additional C6 + hydrocarbons in the unit of catalysate stabilization, as well as the ability of existing facilities to provide increased load on the block on 7 tone per hour. It will increase production of marketable products, in particular benzene.

The compliance with the quality of products of the block of liquid extraction after changes in stabilization process was confirmed in the model. The research results are confirmed in the course of industrial tests on the current industrial plant. The block of stabilisation was transferred up to the proposed scheme of operation on the basis of the experimental test.

To optimize the block of secondary rectification we identified factors affecting the quality of products, which are divided into two groups - the parameters of control of column of benzene separation(heat flow rate in cubic meters of the column, flow rate and temperature of irrigation), and raw material parameters (flow rate of benzene in the feed stream, the ratio of benzene / toluene in the feed stream). We established the main reason for the violation of separation in the column, namely overload of the unit by vapor in the case of increasing the proportion of benzene in the feed stream.

The analysis of the effectiveness of the operation of the process of benzene separation was done. To select the most effective position of sensor the term of "width of working range" is introduced. It refers to a range of values of the controlled parameter (in this case, temperature) at which the following restrictions on the production quality.

We developed An algorithm for optimization of technological modes of the block, as well as a prediction algorithm which allows to calculate the composition of raw secondary distillation unit and to develop appropriate technological parameters of the regime accordingly under change the flow of stable catalysate after the block of stabilization.

Thus, the proposed system of measures has increased the volume of benzene by 15% in compliance with quality requirements and restrictions on the work of production equipment. Results of the study introduced in the industrial process.

(387k) Measurement and Modelling of Vapour-Liquid Equilibria for Reactive Systems

Bruce P. Francois, School of Chemical Engineering, UKZN, Durban, South Africa

Measurement and Modelling of Vapour-Liquid Equilibria for Reactive Systems

B. P. Francois*[1], P. Naidoo[1], D. Ramjugernath[1]

[1]*Thermodynamics Research Unit, School of Engineering*

University of Kwa-Zulu Natal, King George V Avenue, Glenwood, Durban, 4041

* *Corresponding Author:*

Email: brucephilip.francois@gmail.com; Cell: +2784 811 0490

In many industrial applications, particular product conversions and system efficiencies cannot be attained with ordinary reaction vessels[1]. Heat sources and sinks are also required for flash purposes with the implementation of distillation columns following reactors being common practice to separate products. Reactive distillation is a fairly new process whereby the chemical reactor also serves as the still enabling separation of the products from the reaction mixture in a single step[1]. With this accomplished, equilibrium is disturbed and due to the continuous removal of products, conversions can be increased beyond what would normally be expected. Equipment costs are evidently reduced, heterogeneous catalyst may act as packing for distillation and energy costs substantially lowered by the utilisation of the heat of reaction in the separation of products[1]. Implementation of this energy efficient process is however restricted to systems whereby reactions take place in the liquid phase, reactions that are equilibrium limited with fast reaction rates and cases whereby the volatilities of components result in easy removal of the products from solution[2]. Some of the most notable applications include: esterification, etherification, hydrolysis and alkylation[2]. Due to the availability of numerous literature sources, the system involving the esterification of 1-pentanol with acetic acid was investigated in this study as a test system. The desired product; n-Amyl acetate, is a widely used solvent, extractant and polishing agent[3]. This system may however, form a ternary azeotrope in the mixture of amyl alcohol/ amyl acetate/ water. Implementation of reaction distillation may serve as a viable method to overcome this problem. Ethyl acetate is also a widely used solvent in industrial processes although few recent publications for VLE data for this quaternary system are available[4]. The esterification of ethyl alcohol and acetic acid was therefore investigated as a novel system. The quaternary VLE measurements were performed on a low pressure dynamic recirculating VLE apparatus. Chemical and phase equilibria were achieved in an equilibrium chamber packed with a heterogeneous solid ion-exchange catalyst, Amberlyst 15. Experiments were carried out at isobaric and isothermal conditions. The resultant VLE data were regressed for parameters of Gibbs excess model equations. The vapour phase was corrected using the Hayden-O' Connell equation of state. A distillation simulation model was developed in Aspen Plus for the system studied.

References:

1. Taylor. R, R. Krishna, 2000, Chemical Engineering Science, Vol. 55, pp 5183-5229

2. Luyben. W. L, C. C. Yu, 2008, John Wiley and Sons, USA

3. Lee. L. S, S. J. Lang, 1998, Fluid Phase Equilibria, Vol. 149, pp 57-74

4. Calvar. N, A. Dominguez, J. Togo, 2005, Fluid Phase Equilibria, Vol. 235, pp 215-222

(390a) Determination of Metastable Zone Width for Anti-Solvent Crystallization of Glycine

Debasis Sarkar and K Krishna Chaitanya, Chemical Engineering Department, Indian Institute of Technology Kharagpur, Kharagpur, India

The Metastable Zone Width (MSZW) is defined as the area between the saturation curve and the labile zone where nucleation occurs spontaneously. The metastability limit can thus be defined as the supersaturation at which the onset of nucleation can be detected when the supersaturation steadily increases under isothermal conditions. It is extremely important to determine the MSZW for a crystallization process as it reflects the nucleation characteristics of a system and determines the operating window for the process. The MSZW unlike solubility is not a thermodynamic property and must be determined experimentally for a given crystallization process. The MSZW can be determined by looking at changes in solution or solid phase properties upon onset of nucleation and several techniques have been proposed that exploit these properties for determination of the MSZW.

This work investigates the suitability of a simple optical probe consisting of a laser source, a light-dependent resistance, and a simple processing circuit for determination of MSZW of an anti-solvent crystallization process. The isothermal crystallization of glycine from glycine-water solution using ethanol as anti-solvent is used to measure the performance of the proposed measurement system.

(390b) Effect of PVP On Nucleation and Crystal Growth of RDX

Jun-Woo Kim, Department of Chemical and Biomolecular Engineering, Sogang university, Seoul 121-742, South Korea, **Jung-Soo Park**, Agency for Defense Development, Deajeon 305-600, South Korea, Hyoun-Soo Kim, Agency for Defense Development, Daejeon 305-600, South Korea and Kee-Kahb Koo, Department of Chemical & Biomolecular Engineering, Sogang University, Seoul 121-742, South Korea

Polyvinylpyrrolidone (PVP) was found to remarkably affect nucleation and crystal growth in the cooing crystallization of 1,3,5-trinitro-1,3,5-triazinane (RDX) from acetone: firstly, it was found that PVP acts an effective nucleation retarder. In the cooling crystallization of RDX from 30 ¢ªC saturated acetone solution at constant cooling rates of 0.25, 0.5, and 1.0 ¢ªC/min, values of maximum allowable undercooling(ΔT_{max}) were 7.5, 9.7, and 10.9 ¢ªC, respectively. On the other hand, the values of ΔT_{max} with PVP K30 of 100 ppm were found to increase to 18.4, 19.6, and 20.7 ¢ªC, respectively. According to well-known empirical relationship of nucleation rate (Nývlt et al.), nucleation order was estimated as 3.5 in the crystallization without additive, but that was 12.0 with PVP, which indicates that the nucleation mechanism is changed. Polar aprotic lactam ring in PVP has strong affinity with polar RDX molecules and therefore PVP is expected to interact with growing RDX cluster. In addition, RDX molecules can be easily adsorbed to any direction and part of PVP, because many branched lactam rings along the backbone of a chain in PVP molecule. As a result, PVP may penetrate into the emerging cluster, and thus disrupt the packing and lattice ordering of the RDX cluster. Secondly, PVP was confirmed as useful tailor-made additive for growth habit modification of RDX crystal. Morphology of RDX crystals obtained from acetone is prism-like with four trapezoid broad faces of {210} planes and eight isosceles triangle side faces of {111} planes. However, when PVP K30 of 100 ppm was added, morphology of RDX crystals was tetragonal bipyramid with eight faces of {111} planes. It indicates that the binding energy of PVP with (111) face of RDX is higher than that with (210) face, and thus the growth of (111) face of RDX was noticeably obstructed by PVP.

(390c) Application of Process Analytical Technologies (PAT) to Monitor a Continuous Crystallization

Eleftherios Kougoulos, Novan Inc, Durham, NC

A Mixed Suspension Mixed Product Removal Cooling crystallizer with Product Recycle using a heat exchanger was used on a laboratory scale to study the effects of agitation speed, solute concentration and mean residence time on the physical properties of an organic compound. A typical batch cooling crystallization operates at higher levels of supersaturation compared to continuous crystallization systems. Most continuous crystallization studies focus on achieving steady state operation with regards to particle size and size distribution. The monitoring of supersaturation levels during the start-up of a continuous crystallization process and its evolution towards achieving a constant supersaturation level at steady state operation has not received much attention. In many cases, it is possible to achieve steady state operation with regards to particle size but not with regards to supersaturation. On-line Focused Beam Reflectance Monitoring (FBRM) was introduced to monitor how many residence times elapsed in order to achieve steady state chord size and size distribution within the continuous crystallization system. Furthermore, Attenuated Total Reflectance- Ultraviolet (ATR-UV) Spectroscopy in combination with the use of chemometrics was used for the first time to monitor supersaturation on transitioning from a batch cooling to continuous cooling crystallization system including the operational establishment of steady state and constant supersaturation. During the start-up of the continuous crystallization system, an acicular crystal habit was produced, whereas at steady state operation in a continuous crystallization mode, a parallelepiped habit was evident when introducing Process Video Imaging (PVM) and light microscopy. The continuous cooling crystallization system allows for operation at a lower level of supersaturation and can therefore have a profound impact on growth rate on different crystal faces resulting in a change in crystal habit.

(391b) Study of the Mechanisms of Antibody Aggregation in Solution

Frida Ojala[1], Marcus Degerman[1], Thomas Budde Hansen[2], Ernst Broberg Hansen[2] and Bernt Nilsson[1],
(1)Department of Chemical Engineering, Lund University, Lund, Sweden, (2)Novo Nordisk A/S, Bagsværd, Denmark

In order to avoid aggregation during preparative separation of antibodies, it is desirable to have thorough knowledge of the mechanistic behavior of the reaction. A study of antibody aggregation occurring in a solution has been conducted with the main aim to find a methodology to determine the mechanisms of the reaction.

The aggregation reaction is sensitive to antibody and salt concentration as well as pH and temperature. A time series of the monomer, dimer and high molecular weight product (HMWP) under varying conditions were studied to produce a time dependent model of the system. The experiments showed a significant increase in dimer concentration and a low formation rate of HMPW at low salt concentrations. At high salt concentrations however, the situation was the opposite, with higher formation rate of HMWP and low formation rate of dimers. Calibrations to different reaction models were then conducted, estimating the kinetic parameters and searching for a likely mechanism.

(391c) Using a Platform Purification Process to Streamline Early Downstream Development

Adam Myrold[1], Yan Yao[1], Edward Glowacki[1], Hong Li[1] and Thomas Linden[2], (1)Protein Purification Development, Merck & Co., Union, NJ, (2)Protein Purification Development, Merck & Co., Inc., Rahway, NJ

The development of downstream processes for therapeutic proteins is extremely fast-paced and guided by short timelines. In order to meet these project timelines, a platform process can be used, and modified wherever necessary. This platform process approach allows for the development work to focus on fine-tuning the purification steps for optimal recovery and separation. Once the process is modified to meet the quality and recovery criteria, process characterization and robustness experiments are performed to define a working space for the large scale process. An example is presented, in which a platform monoclonal antibody purification scheme is applied to a mAb and slightly modified to meet the necessary quality criteria. Once the recovery and quality criteria were met, key studies were carried out to determine a robust operating space. The process was demonstrated at the pilot scale, and then further scaled-up in support of Safety Assessment/TOX studies. The utilization of a platform purification scheme allowed for the development of a TOX process in less than four months time.

(391d) Investigation of Biased Brownian Motion of Polyelectrolyte Under an Applied Electrical Field

Parvin Golbayani, Chemical Engineering, Tennessee Technological University, Cookeville, TN, Pedro Arce, chemical engineering, Tennessee Technological University, cookeville, TN and Robert Sanders, chemical engineering, Tennessee Technological University, Cookeville, TN

Key to this research is the determination of the effective diffusion (or biased diffusion under an electrical field) and effective mobility (also under an electrical field) whose competition determine the quality of the separation for a given type of protein. This competition can be efficiently assessed by the Optimal Time of Separation[1-2] that incorporates the two effects in a convenient manner to study the impact of the two transport based parameters, i.e. when the effective mobility wins over the dispersion (or biased effects due to the electrical field) the device leads to an efficient separation. Otherwise, we will achieve an efficient mixing.

The focus of this research is on the use of probabilistic- stochastic- based process to obtain such coefficients. This, to the best of our knowledge, would be the first effort to determine the effective coefficients such as biased diffusion and effective mobility for proteins (Polyelectrolyte) under an applied electrical field, i.e. "Biased Brownian Motion of Polyelectrolyte under an Applied Electrical Field". This research project is mostly focused on modeling and complemented by computational work. Results will be useful pieces of information at the device level and illustrated with parameters beneficial for practitioners. In summary, this project offers a very efficient path to obtain vital information to guide both experiments and new research relevant to both environmental proteomics and clinical diagnostics.

[1] Aris, R. "On the Dispersion of a Solute in a Fluid Flowing through a Tube ." Proceedings of the *Royal Society of London. Series A, Mathematical and Physical Sciences*, 235, 67,**1956**.

[2] Taylor, GI. "Dispersion of Soluble Matter in Solvent Flowing Slowly through a Tube." Proceedings of the *Royal Society of London. Series A, Mathematical and Physical Sciences* 1137,186 ,**1953**.

(391e) Optimal Resolutions for Microvoids with Imperfections in Gel- Electrophoresis

Jyothirmai J. Simhadri[1], Holly A. Stretz[2] and Pedro E. Arce[1], (1)Chemical Engineering, Tennessee Technological University, Cookeville, TN, (2)Department of Chemical Engineering, Tennessee Technological University, Cookeville, TN

Gel electrophoresis is the most widely used technique for the separation and purification of biomolecules (DNA and proteins). The nanometer-scale pore structure of the gel tuned either by templating or by adding nanoparticles has been shown to alter the electrophoretic mobility of the biomolecules thus improving separation efficiency. In this research, we propose to computationally analyze different pore models (associated with gel materials) and study the effect of geometry on the transport properties in this anisotropic-like media. This research also introduces a novel approach to investigate the morphological effects of nanocomposite gel electrophoresis by integrating the numerical simulations based on finite element method and population-based search algorithm such as differential evolution. Simulations are performed to study the solute transport by electromigration-diffusion in a microchannel with axially varying cross-section. Morphological parameters such as channel shape and size, as well as operational parameters such as electric field in axial and orthogonal directions were considered and found to have considerable effects on the electrophoresis separation resolution.

Please Check Your Paper One More Time.

- Then scroll all the way down to the bottom of this page and "Conclude Submission".

Synthetic Dye Decolorization by Three Sources of Fungal Laccase

Hamid Forootanfar, Pharmacuetical Biotechnology, Kerman University of Medical Sciences, Kerman, Iran

Abstract Text:

Decolorization of six synthetic dyes using three sources of fungal laccase with the origin of *Aspergillus oryzae*, *Trametes versicolor*, and *Paraconiothyrium variabile* was investigated. Among them, the enzyme from *P. variabile* was the most efficient which decolorized bromophenol blue (100%), commassie brilliant blue (91%), panseu-S (56%), Rimazol brilliant blue R (RBBR; 47%), Congo red (18.5%), and methylene blue (21.3%) after 3 h incubation in presence of hydroxybenzotriazole (HBT; 5 mM) as the laccase mediator. It was also observed that decolorization percent of all dyes was enhanced by increasing of HBT concentration from 0.1 mM to 5 mM. Laccase from *A. oryzae* was able to remove 53% of methylene blue and 26% of RBBR after 30 min incubation in absence of HBT but the enzyme could not efficiently decolorize other dyes even in presence of 5 mM of HBT. In the case of laccase from *T. versicolor*, only RBBR was decolorized (93%) in absence of HBT after 3 h incubation.

Title: Synthetic Dye Decolorization by Three Sources of Fungal Laccase
Submitter's E-mail Address: forootanfar@yahoo.com
Preferred Presentation Format: Oral
Keywords: Decolorization, Laccase, Organic Chemicals and Synthetic dyes

Session Selection: Novel Biological Technologies for Industrial Wastewater Treatment

First Author

Presenting Author
Hamid Forootanfar
Email: forootanfar@yahoo.com

FINAL STEPS

1. **Check spelling and contact information.**
2. **Make necessary corrections:**
   - Click any value in the Abstract Control Panel you want to change (e.g., Title, Author names)
   - Edit the information and click the submit button.
3. **Click** here **to print this page now.**

Conclude Submission

Bioremediation of Cu(II) Metal from Aqueous Solution by Acclimated Mixed Culture

Subhajit Majumder[1], Gayathri Gangadhar[2], Smita Raghuvanshi[3], Suresh Gupta[4,*]

[1]*Lecturer, Department of Chemical Engineering*
Birla Institute of Technology and Science (BITS), Pilani- 333031, Rajasthan, India
Email: subhajit@bits-pilani.ac.in; subh.jit@gmail.com

[2]*Post-Graduate Student, Department of Chemical Engineering*
Birla Institute of Technology and Science (BITS), Pilani- 333031, Rajasthan, India
Email: gayathri.star222@gmail.com

[3]*Assistant Professor, Department of Chemical Engineering*
Birla Institute of Technology and Science (BITS), Pilani- 333031, Rajasthan, India
Email: smita@bits-pilani.ac.in

[4]*Assistant Professor, Department of Chemical Engineering,*
Birla Institute of Technology and Science (BITS) Pilani, -333031, Rajasthan, India
Email: sureshg@bits-pilani.ac.in,sureshg25@gmail.com

Abstract

Bioremediation of metal pollutants from industrial wastewater using acclimated mixed culture is a very important aspect of environmental biotechnology. In the present study, the removal of divalent copper [Cu(II)] compound from aqueous solution using bioremediation has been investigated. Acclimated mixed culture developed from activated sludge is used for the bioremediation of Cu(II). At the end of the acclimation period, the biomass concentration and CFU obtained are 0.0187 g/L of solution and 3.2×10^8/mL (for 10^2 serial dilution) of sample, respectively. The effect of different operating conditions such as contact time, initial pH and amount of nutrient dosage on bioremediation of Cu(II) by acclimated mixed culture are studied in a batch mode. The effect of contact time is studied for an initial Cu(II) concentration ranging from 20-60 mg/L. With increasing initial Cu(II) concentration from 20 mg/L to 60 mg/L shows increase in Cu(II) uptake. The effect of initial pH is also studied by varying the solution pH in the range of 2 – 12. The maximum biomass growth obtained is 0.0115 g/L at pH = 7. The maximum percentage removal of Cu(II) is obtained for the initial Cu(II) concentration of 20 mg/L at the solution pH value of 7. Percentage removal of Cu(II) value is also compared with reported value of percentage removal of Cu(II). This study also shows that the microbial growth for Cu(II) uptake is significantly affected by nutrient supply. The optimum biomass growth obtained is 0.0248 g/L at half the standard composition of MSM.

Keywords: Bioremediation; Acclimated mixed culture; Divalent copper; Activated sludge; CFU; Contact time, Nutrient dosage; Percent removal; Standard composition; MSM.

---

* Corresponding author. Tel.: +91-1596-515454; Fax: +91-01596-244183
E-mail address: sureshg@bits-pilani.ac.in,sureshg25@gmail.com

INTRODUCTION

The typical problem of the effluents generated in the mining and processing of metal ores is the large volume of wastewater which contains high contents of various heavy metals. The removal of these toxic metal contaminants from wastewaters is one of the most important environmental concerns being faced by many countries these days. The most commonly applied technology to remove the contained metals is precipitation with lime or other agents. This leads to the precipitation of metals as hydroxides, which are then typically landfilled. Other treatment options are Chemical precipitation, ion exchange, reverse osmosis and solvent extraction from dilute solutions containing metal wastes (Rich and Cherry, 1987). These treatment techniques have significant disadvantages, such as incomplete metal removal, generation of toxic sludge and other wastes, high energy and reagent requirements. Moreover, these conventional methods are becoming inadequate to address current stringent regulatory effluent limits. Most of these methods are too costly to be implemented and adverse to environment. As a result, alternative, cost effective technologies are in high demand. The use of biological materials, including living and non-living micro-organisms, to remove and recover toxic or precious metals from industrial waste waters has focused attention over the years due to good performance, availability and low cost of raw materials (Ahluwali and Goyal, 2007; Benaissa and Elouchdi 2007; Bunluesin *et al.*, 2007). The ability of microbial culture to remove metal ions in solution was demonstrated in several studies by different researchers (Casiot *et al.*, 2003; Deng *et al.*, 2003; Loukidou *et al.*, 2003, Preetha and Viruthagiri, 2007; Lee et al., 2008).

Heavy metals known for their toxicity and non-biodegradability can lead to different ecological and health problems (Gadd, 1992). Copper [Cu(II)] in high doses causes anaemia, liver and kidney damages and stomach and intestinal irritation. Patients suffering from Wilson's disease are at greater risk for health effects from exposure to copper (ATSDR, 2002). Cu(II) forms highly stable complexes while interacting with the species containing oxygen and nitrogen donor compounds. Therefore, the harmful effects of Cu(II) can be remediated by removing them from water bodies using enriched microbial culture. The present study was undertaken to probe the potential of the mixed microbial culture to remove Cu(II) by bioremediation process from aqueous Cu(II) system.

MATERIALS AND METHODS

Cu(II) solutions

Anhydrous copper sulphate salt [$CuSO_4$, $5H_2O$] used was of analytical grade Merck (India) brand. Stock solutions of copper ions at 1000 mg/L concentration were prepared by dissolving copper sulphate in double distilled water. From the stock solution, the solutions of desired concentrations of Cu(II) (such as 20, 40 and 60mg/L) were prepared by diluting with appropriate volume of double distilled water.

Preparation of Media

The nutrient media, Minimal Salt Medium (MSM) was prepared with the following standard composition (in g/l): K_2HPO_4 – 0.8, KH_2PO_4 – 0.2, $CaSO_4.2H_2O$ – 0.05, $MgSO_4.7H_2O$ – 0.5, $(NH_4)_2SO_4$ – 1.0, $FeSO_4$ – 0.01 in distilled water. 100 mL of MSM was taken in 250 ml Erlenmeyer flask and was autoclaved at 121°C and 15 lb pressure for 20 min. The pH of MSM obtained after autoclaving was 6.7. Stock glucose solution of 10,000 mg/L was prepared by dissolving 10 g of glucose in 100 mL distilled water.

Microorganism Culture Conditions

An aerobic mixed microbial culture in the form of sludge was obtained from the Municipal Sewage Treatment Plant of Birla Institute of Technology & Science (BITS) Pilani. The sludge was then mixed thoroughly with water. It was allowed to settle for 3 h at room temperature (away from sunlight) in order to separate the supernatant and the dissolved impurities. The supernatant which include the dissolved impurities was discarded off and sludge was retained which retains microbial culture. 10 g of sludge obtained from first settling was taken and again thoroughly mixed with 100 mL of distilled water in a beaker. The shaking was carried out gently and then sludge was allowed to settle for short time (1 min) in order to screen out the stones or other particles. 50 mL of supernatant was then taken in a 50 mL centrifuge tube. The centrifugation was carried out for 2 min at 10,000 rpm at 4°C in a Centrifuge (Model: CPR-24, Remi Cooling Centrifuge, India). After centrifugation for 2 min, a clear pellet was obtained in the liquid. The portion of the upper liquid was removed carefully from the top of the centrifuge tube without disturbing the pellet. Then the pellet was taken out with the help of loop and transferred in the 250 mL flask in laminar airflow chamber (Model: MSW, Macro Scientific Works, India).

Enrichment Procedure

The enrichment of culture was carried out in an aseptic environment. A loop full of sludge obtained after centrifugation was added to 100 ml of MSM. The solution, was then kept in a BOD Incubator and Shaker (Model: MSW 132, Macro Scientific Works, India) at 37°C and at 100 rpm for 24 h. The growth of microbial culture was measured by optical density value in a UV–VIS Spectrophotometer (Model: Evolution 201, Thermo Scientific, India) at wavelength 260 nm. The value of optical density obtained for microbial culture was 0.1 which was an indicative of sufficient microbial culture. The enrichment of culture was carried out for a period of 6 days by decreasing the amount of glucose from 1,000 to 0 mg/L with a decrement of 200 mg/L in each day and increasing the Cu(II) concentration from 2 mg/L to 20 mg/L. The final acclimated culture was obtained only with the presence 20 mg/L of Cu(II) which was then used for the bioremediation study. The biomass concentration and CFU for the final acclimated culture were obtained as 0.0187 g/L of solution and 3.2×10^8/mL (for 10^2 serial dilutions) of sample respectively. This shows the extensive growth of microbial culture in the presence of Cu(II) so this culture can be further used for bioremediation study of Cu(II).

Bioremediation Study

The bioremediation of Cu(II) was studied for an initial Cu(II) concentration ranging from 20 – 60 mg/L in 250 mL Erlenmeyer flasks. This study was carried out to determine the effect of time on the bioremediation of Cu(II) at different initial concentrations of Cu(II). In these experiments, 100 mL of MSM was autoclaved and added with 2 mL of acclimated mixed culture obtained from the enrichment procedure and known amount of stock Cu(II) solution to maintain the required concentration. In order to maintain aerobic condition, the flasks were sealed with cotton and wrapped with soft paper. Then they were kept in BOD Incubator and Shaker which was maintained at 37°C and at 100 rpm throughout the bioremediation process. The samples were collected at different intervals for different concentrations ranging from 20 to 200 mg/L based on visual observation (turbidity). The effect of initial pH on the percent removal of Cu(II) was also studied. This study was performed in the pH value ranging from 2 – 12 and keeping the initial Cu(II) concentration as 50 mg/L. The pH of the solution was maintained by 1N NaOH and 1 N HCl solutions. Experiments were also conducted to find optimum concentration of minimal salt medium (MSM) for the bioremediation of Cu(II). This study was performed by varying the composition of standard MSM. The composition of various salts used in preparing the standard MSM solution (used in other bioremediation studies) was halved, doubled, tripled and quadrupled and bioremediation studies were carried out with these MSM solutions separately. In all the above studies, final samples were analyzed to find out biomass concentration, CFU and remaining amounts of Cu(II) present in the solutions.

Kinetic Study

The kinetic study was carried out for 32 h with 2 h time interval. 500 mL of 20 ppm Cu(II) solution was made in MSM. Ten samples of 30 mL MSM were taken in Erlenmeyer flask and was autoclaved at 121^0C and 15 lb pressure for 20 min. In the laminar hood 1 mL of the final culture is added to all the 10 samples and then kept in the BOD Incubator and Shaker which was maintained at 37°C and at 100 rpm. Samples were at an interval 2 h up to 32 h. The next step is the calculation of biomass and colony forming units (CFU).

ANALYTICAL METHODS

Isolation of Microbial Culture

1 ml of the aqueous Cu(II) sample was spread on a medium prepared from peptone, yeast and beef extract and agar on petri-dishes. Aqueous sample containing 100 μg of Cu(II)/ml was prepared from $CuSO_4$. Agar medium were prepared by dissolving 5 g NaCl, 5 g Peptone, 3 g Yeast extract, 17 g of Agar and 3 g Beef extract in 1000 ml distilled water. The pH of the agar medium was adjusted at 7.2 - 7.5. The medium was autoclaved. The CFU value obtained was 20

$\times 10^6$/ml (for 10^2 serial dilution) after 24 h of incubation at 37^0C. Isolated colonies were picked up with sterilized wire loop and streaked on agar medium plate. Plate was again incubated at 37^0C for 24 h.

Measurement of CFU

Agar medium was prepared. Petri-dishes and the solution were then taken into a laminar hood chamber where the agar solution was poured on to the petri-dishes carefully. It was allowed to dry for few minutes and the petri-dishes were kept in incubator for 24 h. The microbial sample was diluted 10 times and 0.1 ml of this sample was spread on the agar medium. Significant growth of the bacterial colonies was observed after 24 h of incubation at 37^0C.

The colonies were counted and then the number of cells per ml of the sample was calculated using Eq. 1. Single cell was streaked using a loop and tested for their Cu(II) tolerance in solid agar medium.

$$CFU/\text{ml of sample} = \frac{\text{no. of colonies}}{\text{ml of sample taken} \times \text{dilution factor}} \qquad (1)$$

Growth of Biomass

Microbial culture samples obtained in the bioremediation of Cu(II) were used to determine the biomass concentration. These samples were centrifuged to separate the biomass. Then, the biomass was dried and the concentration was calculated in g/L.

Concentration of Cu(II)

The supernatant of the different samples obtained in bioremediation studies were also analysed for the determination of remaining Cu(II) concentration in the solution. The concentration of free Cu(II) ions in the supernatant is determined using the cupric ion meter (HANNA Instruments).

RESULTS AND DISCUSSIONS

Growth of Biomass with Time and MSM Concentration

Biomass is biological material derived from living, or recently living organisms. Biomass is carbon based and is composed of a mixture of organic molecules containing hydrogen, usually including atoms of oxygen, often nitrogen and also small quantities of other atoms, including alkali, alkaline earth and heavy metals. These metals are often found in functional molecules such as the porphyrins which include chlorophyll which contains magnesium. MSM study aimed at the effect of nutrient supply to the microbes that later plays a significant role in the removal of the heavy metal Cu(II). From the above graph it is observed that the as the constituents of MSM is increased there is an increase in the biomass formed. Fig. 1 shows that the biomass doesn't increase continuously but it shows a slight decrease when the MSM constituents is tripled. Higher concentration of MSM inhibits the growth of microbial culture which leads to lower consumption of Cu(II). MSM solution contains various metal ions which hinders the Cu(II) absorption by the microorganisms. The CFU values obtained for 10^4 dilution were 4.8, 5 and 4.9 for halved, doubled and tripled concentration of MSM respectively.

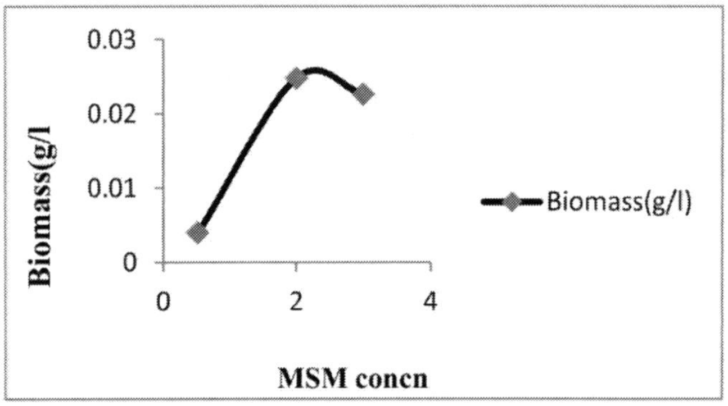

Figure 1 Biomass vs. MSM concentration for MSM study

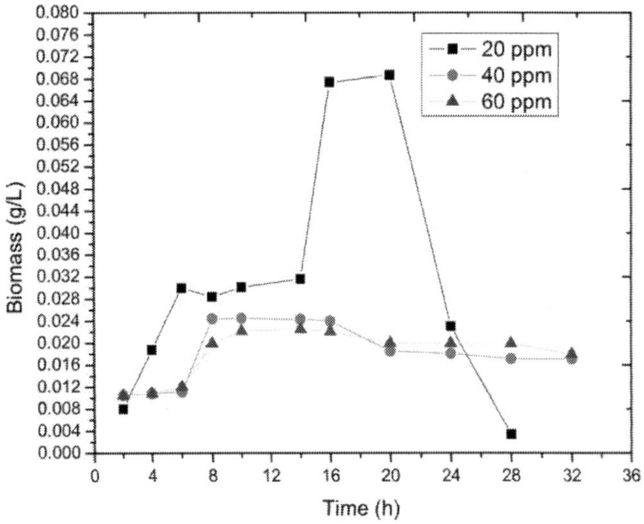

Figure 2 Biomass vs. time (h) for 20, 40 and 60 ppm

Fig. 2 shows that during the initial time the growth rate is slow. This phase is called the lag phase. Then there is a sudden increase in the growth called the log phase. The growth becomes stable after some time .This phase is the stationary phase. The microorganisms become saturated, so they cannot digest much copper ions. The next phase is the death phase which shows a significant decrease in the biomass indicating a decrease in microbial growth.

pH Study

From the pH study, biomass obtained shows that when the pH is acidic the microorganism's growth is less. Fig. 3 shows that as the pH is increased the growth rate also increases. Cu(II) reduction is enzyme-mediated, variation in pH will affect the degree of ionization of the enzyme, changing the protein's conformation and affecting the enzymatic activity of microbial cells.

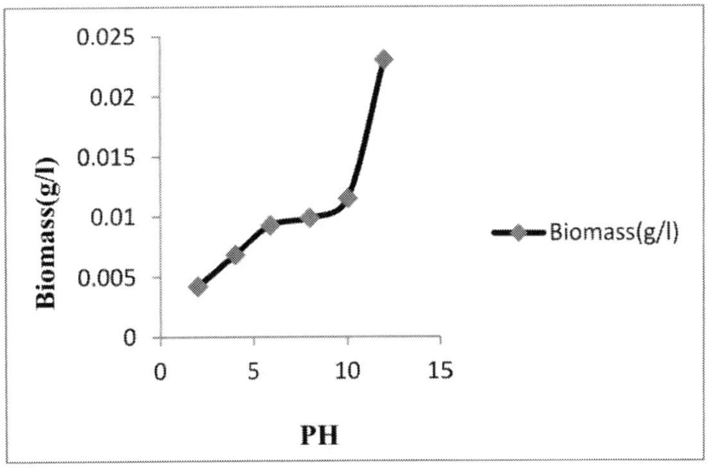

Figure 3 Biomass vs. time (h) for 40 ppm

CFU

The CFU/ml for 20, 40 and 60ppm concentration are shown in Figs. 4 and 5. The microbial growth for these concentrations is almost same during the beginning. But as the time of mixing is increased a sudden increase in growth is seen. This shows that as the time of mixing increased the microorganisms are more acclimated to survive in the new environment. The utilization of Cu(II) is ceased at stationary phase. A further increase in time of mixing shows a slight decrease in CFU showing a decline in microbial growth.

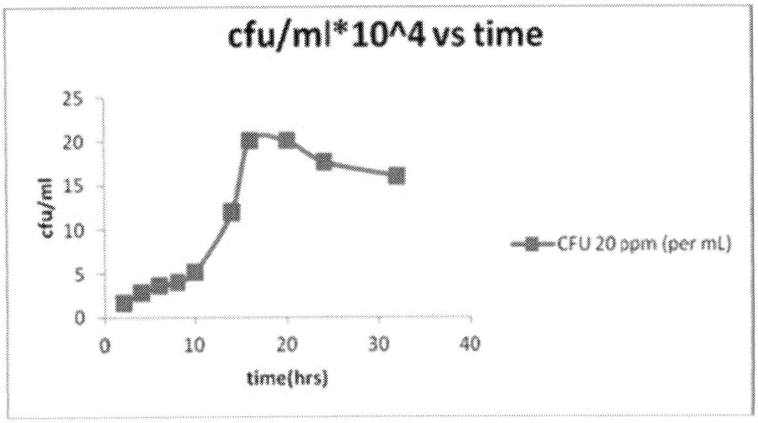

Figure 4 CFU/ml vs time (h) for 20 ppm

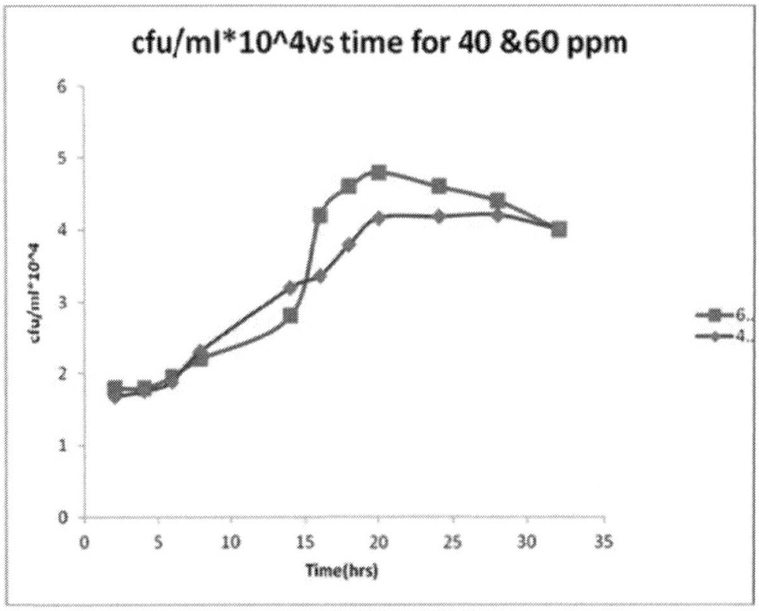

Figure 5 CFU/ml vs time (h) for 40 and 60 ppm

Final Cu(II) Concentration

Final copper concentration obtained after the kinetic study for initial Cu(II) concentrations of 20, 40 and 60 ppm is shown in Fig. 6. It is evident from the plot that Cu(II) was effectively removed by the acclimated mixed culture.

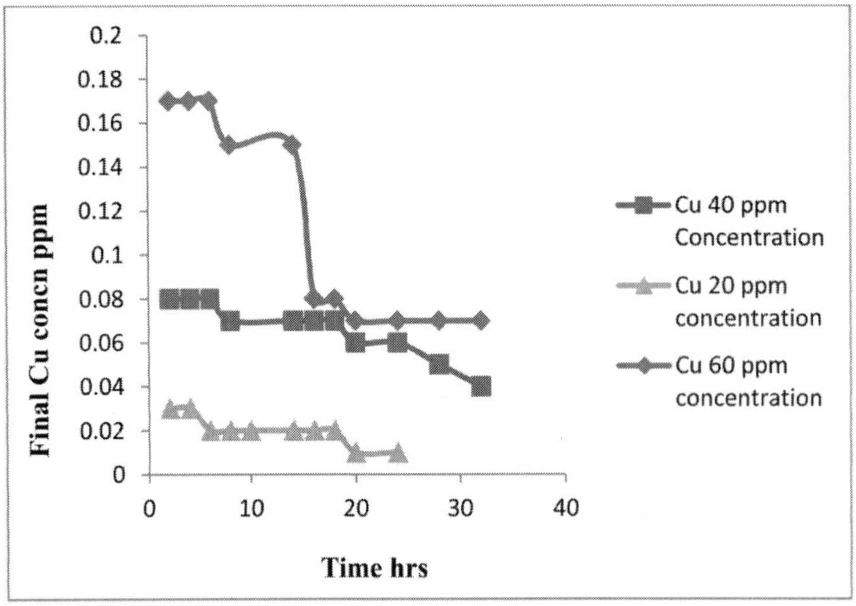

Fig 6 Final Cu concentration vs. time (h)

CONCLUSIONS

In the current study the emphasis was laid on the removal of Cu(II) ion from aqueous solution. Present investigation has examined the presence of indigenous organisms collected from secondary clarifier unit of sewage treatment plant at BITS, Pilani which was further used to decontaminate Cu(II) from aqueous solution. Bioremediation is found to be a better technology for the removal of cupric compounds [Cu(II)]. Final copper concentration vs. time shows that there is an effective removal of Cu(II) using acclimated mixed culture. The MSM study shows that the microbial growth gets enhanced with increasing concentration of MSM. But very high concentration does not favour growth. The pH study indicates that microbial growth is not favoured in high acidic medium.

REFERENCES

Ahluwalia, S. S., Goyal, D., Microbial and plant derived biomass for removal of heavy metals from wastewater , *Bioresource Technology*, 2007, 98(12) : 2243.

Benaissa, H., and Elouchdi, M. A., removal of copper ions from aqeous solution by dried sunflower leaves, *Chemical engineering and processing*, 2007, 46(7): 614.

Bunluesin, S., Kruatrachue, M., Pokethitiyook, P., Upatham, S. and Lanza, G. R. Batch and Continuous Packed Column Studies of Cadmium Biosorption by Hydrillaverticillata. *J. Bioscience Bioeng*. 2007, 103: 509.

Casiot, C., Morin, G., Bruneel, O., Personne, J. C., leblan, C. M., Dusquesne, C., Bonnefoy, V., and Elbazpoulichet, F., Bacterial immobilization and oxidation of arsenic in acid mine drainage (Carnoules creek, France), *Water research*, 2003, 37: 2929.

Deng X., Li Q. B., Lu Y. H., Sun D. H., Huang Y. L. and Chen X. R., Bioaccumulation of Nickel from Aqueous Solutions by Genetically Engineered Escherichia coli. *Water Research*. 2003, 37: 2505.

Lee E. Y., Lim J. S., Oh K. H., Jae Lee J. Y., Kim S. K., Lee Y. K. and Kim K., Removal of Heavy Metals by an Enriched Consortium. *The Journal of Microbiology*. 2008, 46: 23.

Loukidou, M. X., Matis, K. A., Zouboulis, A. I., Liakopoulou, K. M., Removal of As(V)from waste water by chemically modified fungal biomass, *Water research*, 2003, 37: 4544.

Preetha, B., Viruthagiri, T., Bioaccumulation of chromium(VI), copper(II) and nickel(II)ions growing by Rhizopus arrhizus, *Biochemical Engineering Journal*, 2007, 34: 131.

Rich, G., Cherry, K., Hazardous waste Treatment Technologies, *Pudram Publ. Co*., 1987: 169.

(391h) Role of Intermolecular Interactions On Ultrafiltration of Pegylated Proteins

Krisada Ruanjaikaen, Chemical Engineering, Pennsylvania State University, University Park, PA and Andrew L. Zydney, Chemical Engineering, The Pennsylvania State University, University Park, PA

PEGylated proteins are attractive as biotherapeutics due to the increased in vivo efficacy associated with the covalent attachment of polyethylene glycol (PEG) to the base protein. Previous studies have demonstrated the feasibility of using ultrafiltration for removal of residual unreacted proteins and PEG (Molek and Zydney, 2006) as well as more highly PEGylated species (Ruanjaikaen and Zydney, 2011) from the desired singly PEGylated product. In addition, ultrafiltration/ diafiltration (UF/DF) can be used for the final formulation of PEGylated proteins to achieve the desired final product concentration and buffer conditions. The objective of this study was to evaluate the effects of intermolecular interactions, mediated primarily by the PEG, on the performance of these ultrafiltration systems.

Experimental studies were performed with α-lactalbumin PEGylated with a 20 kDa activated PEG. Ultrafiltration data were obtained over a range of PEG concentration, solution conditions, and filtrate flux using Ultracel composite regenerated cellulose membranes (both as is and chemically-modified to provide a negatively-charged membrane). Data were analyzed using a recently developed theoretical model that accounts for the increase in effective size associated with the attached PEG along with ion exclusion from the PEG layer surrounding the charged protein (Molek et al., 2010).

The transmission of the PEGylated protein was significantly greater in the presence of high PEG concentrations due to the unfavorable intermolecular interactions between the free PEG and the PEGylated protein. The magnitude of this effect was consistent with previous thermodynamic models of PEG-PEG interactions. These intermolecular interactions also reduced the selectivity in ultrafiltration processes used for purification of the singly-PEGylated product. These results provide new insights into the effects of intermolecular interactions on the design and performance of ultrafiltration processes for the concentration and purification of PEGylated proteins.

References

Molek JR, Ruanjaikaen K, Zydney AL. 2010. Effect of electrostatic interactions on transmission of PEGylated proteins through charged ultrafiltration membranes. J Membr Sci 353:60–69.

Molek JR, Zydney AL. 2007. Separation of PEGylated α-lactalbumin from unreacted precursors and byproducts using ultrafiltration. Biotechnol Prog 23:1417–1424.

Ruanjaikaen K, Zydney AL. 2011. Purification of singly PEGylated α-lactalbumin using charged ultrafiltration membranes. Biotechnol Bioeng 108:822–829.

(392a) Theoretical and Experimental Investigations of N_2-Separation and Ammonia Formation Via Group V Metal Membranes

Kyoungjin Lee, Ekin Ozdogan, Panithita Rochana and Jennifer Wilcox, Department of Energy Resources Engineering, Stanford University, Stanford, CA

Nitrogen(N_2)-selective membrane technology is promising for post-combustion capture of CO_2 from coal- or natural gas-fired power plants, where CO_2 concentrations are not great enough for traditional CO_2-selective polymer membrane processes. A co-benefit of a N_2-selective membrane is the potential for ammonia (NH_3) synthesis by sweeping H_2 on the permeate side of the membrane, provided a H_2 source is available. The produced NH_3 can be used as a viable fertilizer or a clean energy source and may partially offset the costs of capture.

Previous investigations have shown that Group V metals strongly bind diatomic molecules such as N_2, O_2, and H_2, and that these metals also exhibit remarkably high atomic N diffusivities. In this study, Group V-based metallic membranes and their alloys with ruthenium (Ru) have been investigated for catalytic selective N_2 separation and subsequent NH_3 formation. Investigations include both theoretical modeling based upon electronic structure calculations and experimental testing using a high-temperature bench-scale membrane reactor. As the modeling component of this work enables screening the bulk and surface properties of all the possible Group V-based metallic alloys, the experimental efforts facilitate testing the activity of the most promising N_2-selective membrane candidates proposed from the theoretical calculations. This work will provide fundamental knowledge toward the design of a metallic membrane with optimal selectivity and permeability of N_2 for post-combustion CO_2 capture with simultaneous NH_3 production.

Density functional theory (DFT) calculations have been carried out on the investigation of the molecular N_2 adsorption at different low-index surfaces of vanadium (V) and niobium (Nb). DFT-based calculations have also been carried out for N_2 binding at different interstitial sites within the bulk of the V and Nb crystal. The binding energy of N at O-sites in bulk V is approximately an order of magnitude stronger compared to H binding in Pd. The strong-binding nature of N in these metal systems indicates that N may have difficulties in diffusing through the bulk material. The binding energy can be modified by alloying Ru of various concentrations. For instance, alloying V with 6.25 at.% Ru yields a substantially weaker N binding energy of -0.889 eV than that of -2.13 eV before alloying. This binding energy is approximately within an order of magnitude of H binding in Pd, which is a commercial example of a metallic membrane application. Bader charge analysis was carried out to investigate charge-transfer mechanisms associated with bulk nitrogen absorption.

In permeation experiments, we measured the flux of pure N_2 and CO_2, and their mixtures through foils comprised of V, Nb and their alloys with Ru. The concentration of the feed N_2-CO_2 gas mixtures has been adjusted to represent the flue gas out of coal- and natural gas-fired power plants. Due to the high temperatures required for the dissociation of N_2 and diffusion of atomic N through the metals, the membranes have been tested over a range of temperatures from 500 to 1000 °C. Permeability and selectivity of each metallic membrane were determined and the tested membranes have been characterized by different techniques such as x-ray photoelectron spectroscopy (XPS), x-ray diffraction (XRD), and scanning electron microscopy (SEM).

(392b) Gas Solubility and Selectivity Correlation to Free Volume in Ionic Liquids

Matthew S. Shannon[1], Jason M. Tedstone[2], Scott P.O. Danielsen[2] and Jason E. Bara[1], (1)Chemical & Biological Engineering, University of Alabama, Tuscaloosa, AL, (2)Chemical & Biological Engineering, University of Alabama, NSF-REU Site: Engineering Solutions for Clean Energy Generation, Storage, and Consumption, Tuscaloosa, AL

Introduction

Research using ionic liquids (ILs) in gas separation membranes is now entering its second decade. The versatility and tunability of ILs has resulted in a number of unique materials being produced for membrane-based CO_2 separations. Approaches to IL-based membrane materials include supported liquid membranes, polymerized ILs or poly(ILs), block co-polymers interfaced with ILs, poly(IL)-IL composites and gels. Given the advancements over the past several years in the state of IL materials development, we now see a need to re-visit some fundamental aspects of ILs in order to better outline a path forward for IL-based membranes.

Experimental and Results

While fractional free volume (FFV) is known to play a significant role in the properties and performances of polymer gas separation membranes, little focus has been given to developing relationships between FFV and IL-based materials. We recently calculated FFV values for 165 combinations of 1-*n*-alkyl-3-methylimidazolium ILs ([$C_n$mim][X]) using the COSMOtherm package. FFV was found to be a function of IL molar volume, and as such, earlier models for gas solubility based on Regular Solution Theory can now be reinterpreted in terms of free volume. In concert with experimental data, trends and relationships of gas solubility for various cation/anion pairs were directly linked to computational results. For gases such as CH_4 and N_2, it was determined that as FFV increases, solubility increases in a manner that is directly proportional to free volume. However, within a given IL family (e.g. same anion with various cations), solubility of CO_2 was observed to decrease with increasing FFV. Using experimental data, the influence of free volume on gas solubilities has been determined, and suggests that CO_2 selectivity is a function of the free volume to the (-3/2) power.

Conclusions

Furthermore, we are now exploring new ways to control IL FFV as a means of improving CO_2 solubility, selectivity and permeability in IL-based membranes, using COSMOtherm as a means of guiding materials design. Also, modeling of various functionalized positions of the imidazolium cation can aid in synthesizing high FFV polymer membranes. Initial experimental and computational results obtained for these new generations of IL-based membranes will also be presented.

(392c) Carbon Molecular Sieve Membranes for N_2/CH_4 Gas Separation

Xue Ning, School of Chemical & Biomolecular Engineering, Georgia Institute of Technology, Atlanta, GA and William J. Koros, School of Chemical & Biomolecular Engineering, Georgia Institute of Technology, Atlanta, GA

Many natural gas reserves are contaminated by a high content of nitrogen which has to be removed to meet pipeline specifications. Currently available nitrogen rejection technologies include cryogenic distillation, pressure swing adsorption (PSA), and lean oil absorption. Polymer-based membrane separation for N_2/CH_4 has not been pursued widely because of the small size difference between N_2 and CH_4 and the competing effects of diffusion and sorption. These challenges lead to unattractive permselectivity using traditional polymeric materials due to the broad distribution of segmental motions of polymer chains. Carbon Molecular Sieve (CMS) membranes overcome this problem by providing a hyper-rigid structure and distinctive bimodal pore size distribution.

Different pyrolysis conditions for Matrimid® polyimide and resultant separation performance of CMS membranes were studied in this work. It was shown that N_2/CH_4 selectivity was highly improved with increased pyrolysis temperature. Sorption tests were performed to deconvolute permeability into diffusion coefficient and sorption coefficient. The temperature effects on permeability, sorption coefficient and diffusion coefficient were also studied.

(392d) Ethylene Oxide Boosting in Oxygen-Assisted Membrane Reactor in Ethylene Epoxidation Process

Marzieh Shokrollahi Yancheshmeh, Mohammad R. Gholipour, Mohammad R. Rahimpour, Mitra Jafari, Davood Iranshahi and Sina Seifzadeh Haghighi, Department of Chemical Engineering, School of Chemical and Petroleum Engineering, Shiraz University, Shiraz, Iran

Abstract

In this study, a novel configuration is proposed to boost the ethylene oxide production rate in ethylene epoxidation process. Regarding this, a packed bed reactor with Ag-csr-Al_2O_3 catalyst is studied as a one dimensional, homogenous model under steady state condition. A variety of rate equations is applied into the model and the optimization results of kinetic parameters show that this kinetic $r_i = k_i P_E P_O^n / (1 + K_i P_E)^2$ is the most suitable one. The optimized kinetics' rates can predict industrial data with the maximum errors of 5.5%. In addition, initial concentrations of reactants decrease with the application of oxygen membrane where it can control reactants ratio which is exceedingly explosive. Furthermore, an appropriate distribution of oxygen along the reactor enhances the efficiency of final production so that the ethylene oxide increases significantly to 103%. Oxygen, transferring from air side to the reaction region, causes the remaining ethylene to react completely and produce much more ethylene oxide. Moreover, this oxygen transmission results in purifying air section from oxygen and reaching nitrogen mole fraction nearly 96%. Its temperature also rises dramatically from 300 K to 510 K. Furthermore, energy saving via this novel configuration is appealing from the economical viewpoint.

(392e) Segregation At the Surfaces of Cu8Pd7M Hydrogen Separation Alloys in the Presence of Adsorbed O and S

De Nyago Tafen[1,2], Omer Dogan[3], James B. Miller[4,5] and J. Baltrus[5], (1)National Energy Technology Laboratory , Albany, OR, (2)URS, Albany, OR, (3)National Energy Technology Laboratory, Albany, OR, (4)Chemical Engineering, Carnegie Mellon University, Pittsburgh, PA, (5)National Energy Technology Laboratory, Pittsburgh, PA

The CuPd B2 phase is stable only at relatively low temperatures; at high temperatures, it undergoes a phase transformation that reduces its hydrogen permeability. We have recently identified a set of alloying elements (Y, Al, and Mg) that can stabilize the B2 phase of the copper-palladium alloy by expanding its composition-temperature phase field. In this work, we examine surface segregation of these elements in the CuPd B2 lattice using both density functional theory calculations and experimental measurements. Trends in segregation, adsorption, and surface free energies are characterized. Theoretical results show that in vacuum, Y will substitute first for Cu and then for Pd at the sub-surface lattice site before segregating to the surface where it substitutes for Cu. However, in air, Y preferentially occupies surface sites due to its stronger oxygen affinity compared to Cu and Pd. XPS experiments reveal that surface segregation of Y is induced by formation of Y-oxides at the top-surface of the alloy. Preliminary XPS results for a Cu-Pd-Al alloy that had been exposed to H_2S show that, while adsorbed S does not penetrate far into the alloy, it induces significant segregation of Cu to the alloy surface at the expense of Pd. In the case of Cu-Pd-Mg, adsorbed S leads to strong segregation of Mg to the alloy surface at the expense of both Pd and Cu.

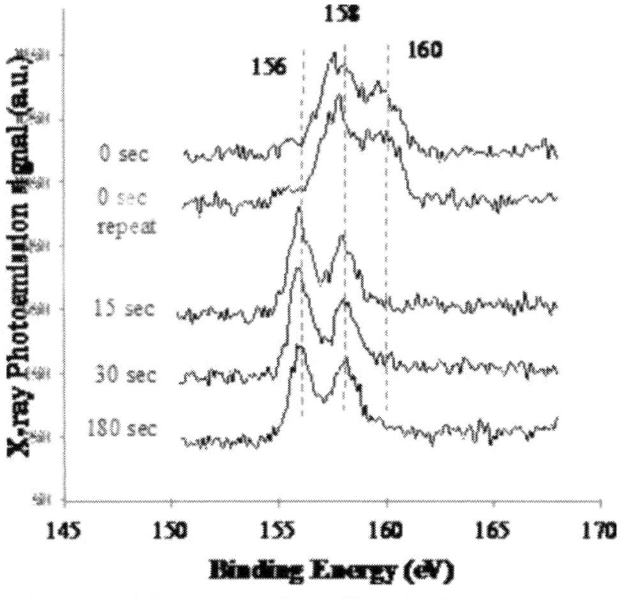

Figure 1: Y*3d* region of the XPS depth profile. The feature at ~160 eV is characteristic of Y-oxides ($3d_{3/2}$); the feature at 156 is characteristic of reduced Y ($3d_{5/2}$); the feature at ~158 eV is likely a combination of Y-oxide ($3d_{5/2}$) and reduced Y ($3d_{3/2}$). Y exists as oxide only at top-surface, before sputtering begins.

(392f) Physical Aging and CO2 Sorption in "High Free-Volume" Glassy Polymers

Rajkiran R. Tiwari[1], Zachary P. Smith[1], Haiqing Lin[2], Benny D. Freeman[1] and Donald R. Paul[1], (1)Department of Chemical Engineering, The University of Texas at Austin, Austin, TX, (2)Membrane Technology and Research, Inc., Menlo Park, CA

Thin film glassy polymers are mostly used as gas separation membranes due to their excellent permeability-selectivity properties; however, accelerated physical aging in free-standing thin films reduces permeability over time and limits membrane performance [1, 2]. Recently, "high free-volume" glassy polymers have gained significant interest in membrane applications mainly due to their high gas permeability [3]. Poly(1-trimethylsilyl-1-propyne) (PTMSP) is a "high free-volume" polymer with very high gas permeability; however, its high physical aging rate and solubility in many of the organic compounds in process streams of interest limit its practical utility. On the other hand, amorphous Teflon AF 2400 from DuPont has high gas permeability and good chemical stability, making it an interesting candidate as a gas separation membrane for some applications. The physical aging behavior and CO_2 sorption in thin films < 1 μm have been studied. These thin films show lower permeability than thick films but they do not seem to show the accelerated physical aging rates observed for other glassy polymers such as PSF, PI, and PPO. The selectivity did not change with physical aging. Results on physical aging rate and CO_2 sorption in thin Teflon AF 2400 films are compared with other commercially available amorphous Teflon materials having relatively lower free volume. The CO_2 sorption in thin films as measured by ellipsometry is compared with bulk films. The results are also compared with a polymer with intrinsic microporosity (PIM-1), which showed a faster physical aging compared to amorphous Teflon.

MODELING MASS TRANSPORT PROPERTIES IN BINARY COMPOSITE SYSTEMS

Dr. Matteo Minelli, Advanced Applications in Mechanical Engineering and Materials Technology Interdepartmental Center for Industrial Research, CIRI-MAM, University of Bologna, Italy
Prof. Ferruccio Doghieri, Department of Chemical Engineering, University of Bologna, Italy
Prof. John H. Petropoulos, National Center for Scientific Research Demokritos, Institute of Physical Chemistry, Athens, Greece

Introduction

The description of mass transport properties of low molecular weight species in heterogeneous polymeric systems is of great importance. The theoretical analysis of composite materials behaviors is indeed fundamental for proper material design of such systems for a variety of different applications. Among the others, immiscible polymer blends as well as block copolymers are often of interest for their ability to combine properties of two or more materials. On the other hand, semicrystalline polymers are a significant portion of polymers employed for many purposes. In recent years, the development of nanocomposite materials was employed for barrier applications by incorporating impermeable platelets in the polymer phase. This technology is also suitable for the design of membranes for gas separation, and to this aim, particles selective to one or more gases, such as for instance zeolites, are dispersed in the polymer phase.

All these examples are constituted by two, or more, well defined and well distinguished regions (of different length scale) characterized by their value of gas permeability P_i. The resulting permeability of the complete heterogeneous medium is then related to these P_i values but also to the relative weight of the two phases as well as the system morphology, i.e. the shape of the dispersed phase.

The aim of this work is to develop a valuable tool for the description of transport properties in composite media, able to evaluate the permeability as function of system morphology. Numerical calculations will be employed to model the two-phases system in a wide range of relative fraction of the two components and varying their characteristics. Results will be then compared with existent model equations in order to investigate their predictive ability and ranges of validity.

Review of principal models:

The investigation of overall permeability P of a heterogeneous medium with a dispersed phase A, in a continuous phase B has been investigated by many authors, and carefully reviewed by Petropoulos [1]. The main parameters assumed for such modeling effort are the permeability coefficients of the two moiety, P_A and P_B, (assumed to be constant for a given permeating species, independent of the concentration of the permeant), the composition of the system, as volume fraction of A and B (v_A and v_B). The shape of the dispersed particles is also relevant as well as their arrangement in the composite.

The simplest arrangements of a two phase media are those related to a unidimensional permeation flux with laminate domains parallel or normal to the direction of the penetrant permeation, namely:

$$P = v_A P_A + v_B P_B \quad (1)$$

$$\frac{1}{P} = \frac{v_A}{P_A} + \frac{v_B}{P_B} \quad (2)$$

The equations are readily derived from simple treatment of series resistance and parallel resistance, respectively. However, the structure of real materials is often more complicated and requires a more sophisticated analysis. Most of the approaches are developed by considering a certain dispersion of particles of one component in a continuous matrix, so that the two moieties can be distinguished as dispersed and the continuous phase.

In this concern, a pioneer work has been carried out by Maxwell, who derived the expression to describe a dilute dispersion of spheres [2,3]:

$$\frac{P}{P_B} = 1 + \frac{3\alpha}{\left(\frac{\alpha+2}{\alpha-1} - v_A\right)} \quad (3)$$

where α represents the permeability ratio between dispersed phase A and continuous one B:

$$\alpha = \frac{P_A}{P_B} \quad (4)$$

The same treatment could also be applied in two dimensions, corresponding to the dispersion of long cylinders oriented normally to the direction of permeation. This is described by the general Wiener equation, which enclosed also the three previous cases [4]:

$$\frac{P}{P_B} = 1 + \frac{(1+A)\alpha}{\left(\frac{\alpha+A}{\alpha-1} - v_A\right)} \quad (5)$$

The parameter A describes the arrangement and the geometry of the dispersed phase with respect to the continuous one. Indeed, A $\rightarrow \infty$ or A = 0 leads to the first two cases (Eq. (1) and (2)), whereas A = 2 or A = 1 yields to Maxwell model for spheres or for long transverse cylinders, respectively.

More concentrated dispersions have been derived by other authors, following different approaches [5]. Interestingly, Meredith and Tobias performed a rigorous treatment of packed congruent spheres arranged in a simple regular lattice and obtained the solution in the form as a truncated series:

$$\frac{P}{P_B} = 1 + 3v_A \left(\frac{\alpha+2}{\alpha-1} - v_A - \frac{k_1 v_A^{10/3}}{\left(\alpha+\frac{4}{3}\right) \Big/ (\alpha-1) - k_2 v_A^{7/3}} - \frac{k_2(\alpha-1) v_A^{14/3}}{\alpha+\frac{6}{5}} + \dots \right)^{-1} \quad (6)$$

where $k_1 = 1.315$, $k_2 = 0.409$ and $k_3 = 0.016$.

Other approaches address the problem of impermeable dispersions of flakes of different sizes and shapes to describe the transport properties of nanocomposites. Among the others, the first mathematical treatment was derived by Aris [6] for a ordered arrangement of flakes, later recalled by Cussler [7]. Minelli et al. also reviewed the fundamental work by Aris [8] and derived expressions for random arrangement of platelets of different sizes and shapes [9].

On the other hand, effective medium theory [10] can also be applied to model the small molecules transport in a heterogeneous system. Indeed, expressions were interestingly derived by Sax and Ottino to the description of permeability in polymer blends, accounting for different configurations and arrangements of the two moieties [11].

Numerical calculations description

In this work, the description of the transport properties in heterogeneous systems is performed by means of a continuous model, which accounts for the presence of phases with different characteristics. This technique has been already applied to the case of composite systems with a dispersion of impermeable flakes [8,9]

Briefly, steady state diffusion process of a single penetrant species in this two-phase medium is governed by Fick's law, which holds in the continuous permeable isotropic phase with uniform diffusivity:

$$\underline{J} = -D_0 \cdot (\underline{\nabla} c) \qquad (7)$$

$$0 = \underline{\nabla} \cdot \underline{J} \qquad (8)$$

The interphases between two domains are then treated as continuous boundaries; no partition function for the diffusing species between the two phases is then considered, but only a different value of diffusion coefficient. For the case of impermeable dispersed inclusion, this condition becomes the imposition of a null value of mass flux across the interface between the two phases.

The description of the problem is then completed by assigning the values of concentration at inlet and outlet boundaries of the system.

The numerical solution of the problem was finally approached by discretizing the computational domain and solving Fick's law with the control volume technique [12].

Numerical calculations of mass transport properties in these heterogeneous systems were performed on the smallest elementary unit in the idea of a simple lattice arrangement of cubes or parallelepipeds. In order to prevent numerical errors, which could affect the precision of the calculations, an attentive analysis of the mesh size was initially performed varying its values in order to produce results independent on mesh characteristics.

Results

The present model has been applied to the well-known case of isotropic dispersion of spherical inclusion in a continuous medium, which, in the past, has been widely investigated by many different points of view. These conditions indeed fulfill the hypothesis for Maxwell model (Eq. (3)) when the particle loading is sufficiently low or other more empirical treatments for higher values, as described by Petropoulos [1] (Eq.(5)).

Therefore, a simple lattice of spherical objects is considered within the range of inclusion loading between 0 and 0.50 (being π/6 the maximum loading that can be accessed with congruent packing of spheres), and having permeabilities ratios α (P_A / P_B) equal to 0, 0.01, 0.1, 10 and 100.
In Figure 1, results of these numerical calculations are reported together with the predictions given by Maxwell (Eq. (3)) and Meredith (Eq. (6)) equations.

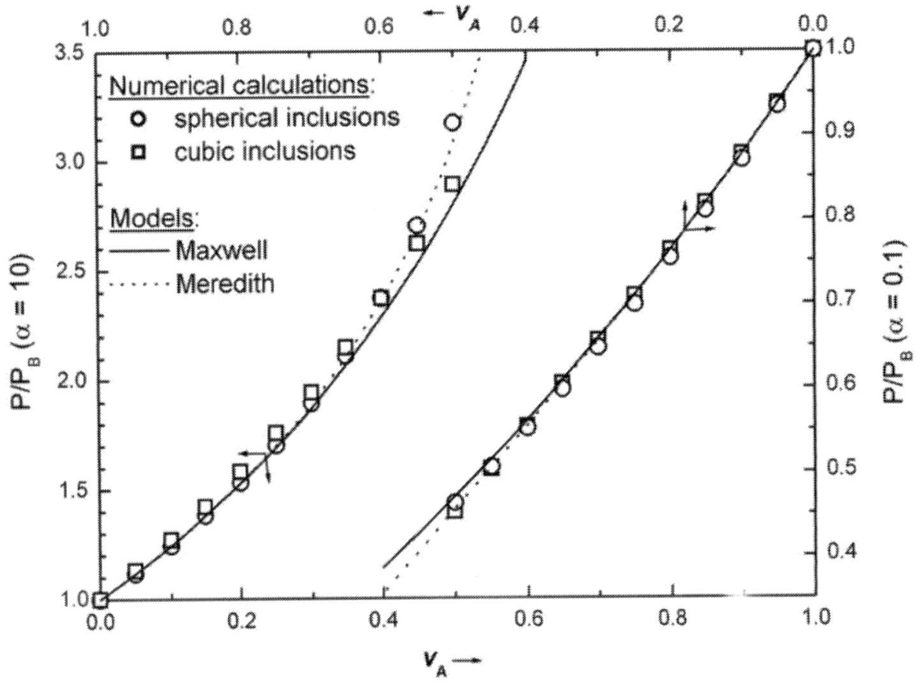

Figure 1, Relative permeability (P / P_B or P / P_A) in 3-D systems with spherical or cubic inclusions with different permeability ratio (P_A / P_B = 0.1 and 10): results from numerical calculations and predictions from Maxwell (Eq. (3)) and Meredith (Eq. (6)) models.

As one can see, the results from numerical calculations are in excellent agreement with the Maxwell model, especially at the lowest volume fractions (up to approximately 0.30). Therefore, this result can be seen as a proof of the accuracy of the present numerical tool for the description of the mass transport properties in heterogeneous systems. On the other hand, a deviation between Maxwell model and calculated permeability ratio P / P_B is apparent at higher inclusion loadings, higher than approximately 0.30 - 0.40. In those conditions, indeed, the hypothesis of diluted dispersion is not fulfilled and the interparticle distances are not sufficiently large to ensure that the permeant pathway around one sphere is undisturbed by the presence of the others. Meredith equation (Eq.(6)), instead, is able to describe the whole range of loadings that can be explored with a simple lattice of spheres.

The physical limit of about 0.50 for the loading of spherical inclusions arranged in a simple lattice can be overcame by considering cubical elements, which can be packed in configuration with volume concentrations up to unit value. This approximation allows for the study from a simulative point of view of the performance of a two-phase system with a dispersed moiety with even very high volume concentration.

Furthermore, differences among the two particles shapes can be observed at the highest v_A values inspected. In that range, indeed the distribution of the interparticle region is rather diverse and therefore the deviation in terms of overall permeability is more significant.

3-D cubic congruent isometric particles

As already mentioned, the arrangement of simple cubes on a regular lattice allows the investigations of permeability of composite systems in the full range of loading fractions from 0 to 1. Numerical calculations were indeed performed on such heterogeneous media exploring a wide range of different relative permeabilities (0, 0.01, 0.1, 10, 100). Results are reported in Figure 2.

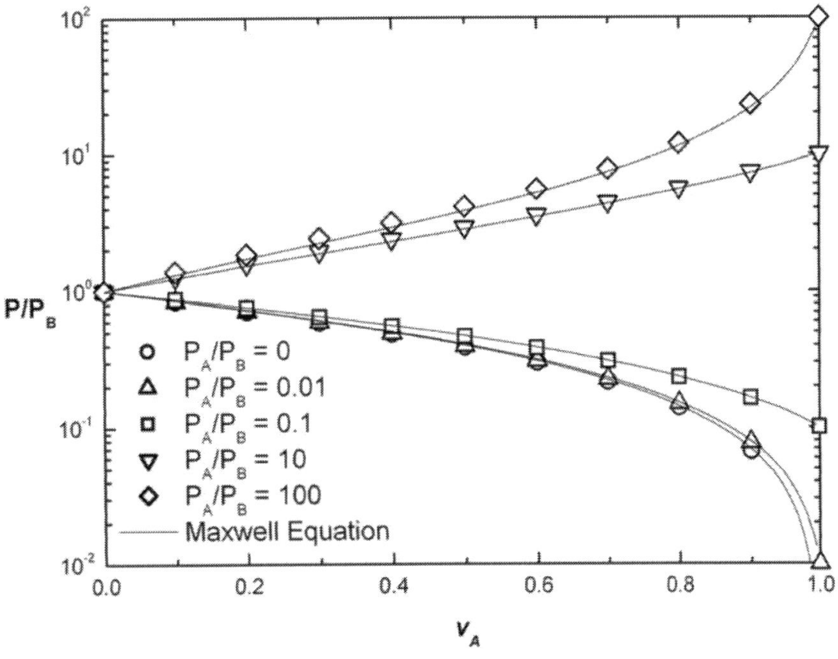

Figure 2, Relative permeability (P / P_B) in 3-D systems with cubic inclusions with different permeability ratio ($P_A / P_B = 0, 0.01, 0.1, 10, 100$): results from numerical calculations and predictions from Maxwell's model (Eq.(3)) for spherical particles.

Interestingly, Maxwell equation (Eq.(3)) is able to describe the permeability behavior of composites in the complete range of inclusions volume fractions. The agreement is remarkable especially at the highest values of v_A, whereas small deviations are observable between 0 and 0.50. In this range, it is also possible to compare the two different particle shapes, which can play a slight effect in determining the overall permeability of the composite, as illustrated in Figure 1.

1-D square rod particles

Another interesting family of fillers, from the applicative point of view, is that of long objects such as fibers or nano-rods, which are characterized by very high lengths compared to their thicknesses. To this aim, the present numerical tool has been applied to simulated permeability of composites obtained by the dispersion of monodimensional objects with square section (square rods) of different sizes. The case of indefinitely elongated rods has been first considered, and compared with the modeling work after Weiner above described (Eq.(5)).

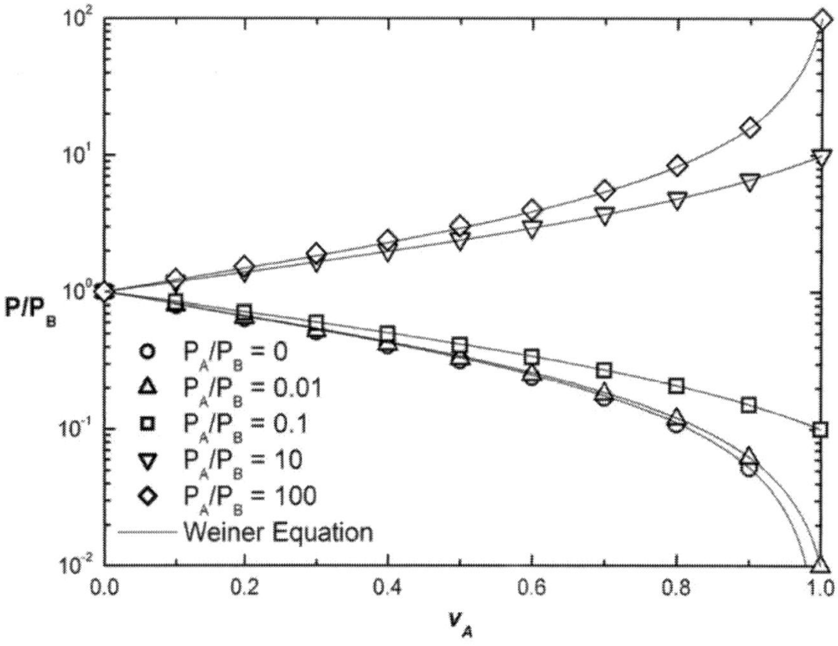

Figure 3, Relative permeability (P / P_B) in systems with indefinitely elongated square rods inclusions with different permeability ratio ($P_A / P_B = 0, 0.01, 0.1, 10, 100$): results from numerical calculations and predictions from Weiner's model (Eq.(4)).

In Figure 3, results from numerical calculation of heterogeneous systems with indefinitely elongated square rod inclusions with different permeability ratio ($P_A / P_B = 0, 0.01, 0.1, 10, 100$) are reported together with Weiner's equation model predictions (A = 1). As one can see, the results obtained from numerical calculations indicate that, although this model has been derived for long cylinders, the shape of their section does not affect significantly the ability of the equation to predict the permeability behavior of such composite systems.

The case of 1-D square rod particles has been investigated more in detail by exploring different sizes of particle inclusions. Hence, rods with different relative and finite lengths (namely twofold, fivefold and tenfold the section dimension) have been considered in order to complete the range between cubic particles and indefinitely elongated rods. Therefore, in the direction of diffusive flux, the same normal section of the inclusion is kept, whereas the particle is differently elongated (1, 2, 5, 10 and ∞) among that direction.

The increased particle lengths induces more significant variations on effective permeability, as expected, and the observed behaviors converge, at highest particle lengths, towards the values for infinitely elongated rod. This is illustrated in Figure 4, in which relative permeabilities P / P_A and P / P_B are reported for the case $\alpha = 10$ and $\alpha = 0.10$.

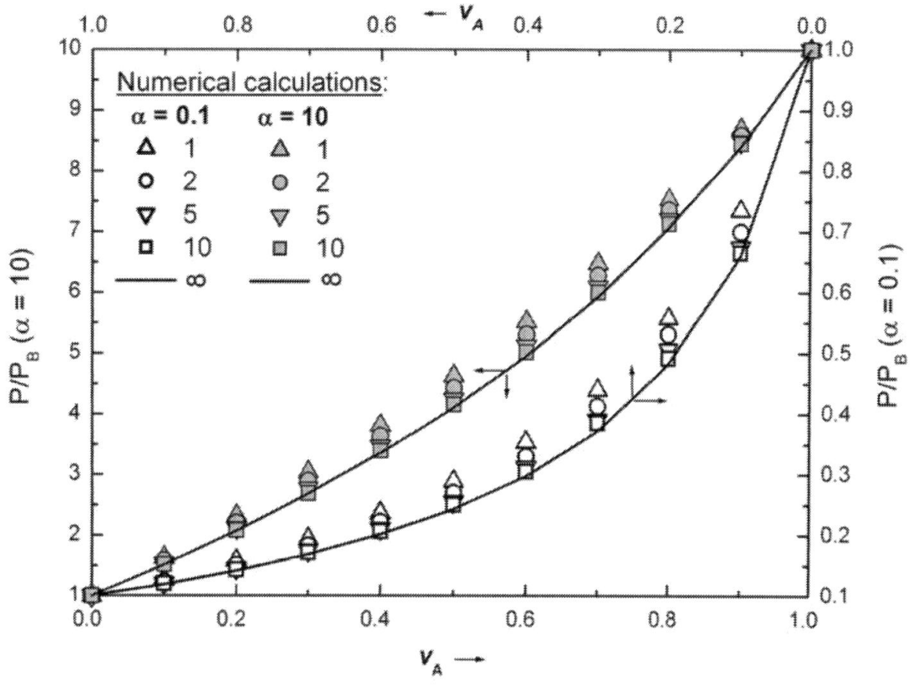

Figure 4, Relative permeability (P / P_B) in systems with elongated square rods (shape factor 1, 2, 5, 10 and ∞) inclusions with different permeability ratio ($P_A / P_B = 0.1$ and 10).

As expected, results for different particles elongation are comprised between the two curves given by cubical inclusions and indefinitely elongated rods. Hence, this suggest that Weiner equation (Eq.(5)) can be applied to describe such systems with an appropriate value of the parameter A.

Conclusions

The diffusion of small molecules in binary composite systems has been investigated from a modeling point of view. A finite volume technique is employed to perform numerical calculations on heterogeneous systems with two phases having different permeabilities. The arrangement of cubes in a simple lattice allowed for the investigation of the properties in the complete range of volume fraction of the dispersed phase (from 0 to 1). Results from numerical calculations are then compared with modeling equations already present in the technical literature. This showed that Maxwell equation, although it was derived in the limit of infinite dilution of dispersed spheres, is able to describe the permeability of the heterogeneous systems here investigated, in the whole range of volume fraction and permeability ratios inspected. Another configuration was then explored, having a uniform dispersion of rods with square section transverse to the main diffusive path. The effect of the length of such objects (from cubes to infinitely elongated particles) on the transport properties has also been explored and compared with the results from Weiner equation.

References

1. Petropoulos, John H. (1985), "A Comparative Study of Approaches Applied to the Permeability of Binary Composite Polymeric Materials", *J. Polym. Sci: Polym. Phys. Ed.* 23 pp 1309-1324.
2. Barrer, Richard M. (1968), "Diffusion and permeation in heterogeneous media" in "Diffusion in Polymers", John Crank and Geofrey S. Park, Eds. Academic, New York, Chap. 6.
3. Maxwell, James C. (1904) "A Treatise on Electricity and Magnetism", Clarendon Press, Oxford.
4. Nielsen, Lawrence E. (1974) " The Thermal and Electrical Conductivity of Two-Phase Systems" Ind. Eng. Chem. Fundam. 13 pp. 17-20.
5. Meredith, Robert E. and Tobias, Charles W. (1960), "Resistance to Potential Flow through a Cubical Array of Spheres", *J. Appl. Phys.* 31 pp. 1270-1273.
6. Aris, Rutherford (1986), "On a problem in hindered diffusion", *Arch. Ration. Mech. Anal.* 18 pp. 83–91.
7. Cussler, Edward L., Hughes, Stephanie E., Ward III, William J. and Aris, Rutherford (1988) "Barrier membranes", *J. Membr. Sci.* 38 pp. 161–174.
8. Minelli, Matteo, Giacinti Baschetti, Marco and Doghieri Ferruccio (2009), "Analysis of modeling results for barrier properties in ordered nanocomposite systems", *J. Membr. Sci.* 327 pp. 208–215.
9. Minelli, Matteo, Giacinti Baschetti, Marco and Doghieri Ferruccio (2011), " A comprehensive model for mass transport properties in nanocomposites", *J. Membr. Sci.* 381 pp. 10–20.
10. Landauer, Rolf "The Electrical Resistance of Binary Metallic Mixtures", *J. Appl. Phys.* 23 pp. 779-784
11. Sax, James and Ottino, Julio M. (1983) "Modeling of Transport of Small Molecules in Polymer Blends: Application of Effective Medium Theory", *Polym. Eng. Sci.* 23 pp. 165-176.
12. Hersteeg, Henk K. and Malalasekera, Weeratunge (2007) "An Introduction to Computational Fluid Dynamics: The Finite Volume Method", 2nd Ed., Prentice Hall, Harlow.

(392h) Effects of Polysulfone Support Layer On the Performance of Thin Film Composite (TFC) Membranes

Chuan Ding[1], Jun Yin[2] and Baolin Deng[1,2], (1)Department of Chemical Engineering, University of Missouri-Columbia, Columbia, MO, (2)Department of Civil & Environmental Engineering, University of Missouri-Columbia, Columbia, MO

Thin film composite (TFC) membranes are widely used in nanofiltration (NF) or reverse osmosis (RO) for desalination and waste water treatment. In this study, the phase inversion method was used to fabricate polysulfone (PSF) ultrafiltration membrane, and using it as a support layer, the interfacial polymerization (IP) process was applied to create a dense polyamide (PA) thin-film layer for salt rejection. The goal was to assess how the PSF concentration and the addition of polyvinylpyrrolidone (PVP) into the dimethylformamide (DMF) solvent could affect the PSF membrane characteristics and subsequently the performance of the thin film composite (TFC) membrane. The study demonstrated that the water permeability of PSF substrate membrane decreased when the polymer concentration increased. The TFC membrane had a good salt rejection but poor permeability by using the substrate membrane prepared at a higher polymer concentration. In the presence of PVP as an additive to the DMF solvent, PVP at a low concentration increased the water permeability but further increase of PVP concentration to 5% or 10% decreased the permeability. When the substrate membranes with increasing PVP concentration was used in the solvent, the water permeability of the prepared TFC membrane decreased but little change was observed in salt rejection. As a general conclusion, the performance of the TFC membranes depended highly on the casting parameters of the substrate layer.

Response of Zeta Potential to Fouling Status in MF/UF Processes

Kazuho Nakamura
Yokohama National University, Dept. of Chemical Engineering, Yokohama, Japan

Abstract

The responses of zeta potential of membrane to the pore blocking and cake formation during microfiltration of latex particle suspension were studied by monitoring both the filtration resistance and streaming potential during the filtration. The pattern of the change in the zeta potential during the filtration was quite different between the cake filtration and the pore blocking filtration. In the cake filtration the zeta potential changed from the intrinsic zeta potential of membrane to the zeta potential of cake layer depending on the proportion of the hydraulic resistance of the cake layer to the total filtration resistance. In the pore blocking filtration followed by the cake filtration jumped from the intrinsic zeta potential of membrane to a certain value between and just after the start of the filtration and became almost stable. The zeta potential at the blocked pore depended on the gap between surfaces of the blocked pore and the blocking particle and indicated the extent of pore blocking. It was confirmed that the change in the zeta potential during filtration will reflect both the location of pressure drop and the local zeta potential.

Introduction

The fouling phenomenon in microfiltration(MF) membrane can be qualitatively evaluated by analyzing the flux decline with filtration models. The flux decline in dead-end filtration can be explained by the pore constriction, pore blocking, and cake formation models. Pore constriction model is used to explain for the fouling caused by smaller foulant compared to pore size for instance the protein adsorption onto MF membranes[1]. The pore blocking(pore plugging) filtration depends on mainly the pore morphology. The membranes with cylindrical pore structure such as track-etched polycarbonate(PC) MF membranes will be highly affected by the pore blocking caused by particles while the membranes with a highly inter connected pore structure will be weakly affected by the pore blocking[2]. In the cake filtration model the flux decline can be explained by the building-up of cake layer which provides hydraulic resistance. The analyzing of the flux decline with these classical filtration models are useful as a simple method for prediction of the fouling mechanism although the actual fouling mechanism will be more complex than the hypothesises of these models.

The surface charge condition of membrane is a monitoring property of interest because the electrostatic interaction between membrane and foulant is considered as a major interaction causing fouling. The zeta potential estimated using Helmholtz-Smoluchowski (H-S) equation from streaming potential has been used as an index of the surface charge and applied for the characterization of membrane material, surface modification of membrane, effect of cleaning[1-3].

The relationship between fouling status and the zeta potential is not fully understood[4]. The change in zeta potential in the standard blocking filtration will reflect the adsorption status of pore surface during the filtration and the extent of the change in zeta potential depended on the surface coverage of the pore surface adsorbed by foulant[1]. In the cake filtration the zeta potential will reflect the charge condition of both membrane and cake layer[2-3]. In the pore blocking filtration the response of the zeta potential to the pore blocking status have not been clear.

In this study we focused on the response of zeta potential during the pore blocking filtration and the cake filtration in the dead-end filtration of latex particles with MF membranes. The purposes of this study was 1) to confirm the relationship among the zeta potential, the location of pressure drop, and local charge condition, 2) to study the response of zeta potential to the cake filtration, and 3) to study the response of zeta potential to the pore blocking filtration.

Theory

The hydraulic resistance is usually used as an index of hydraulic permeability of membrane and fouling layer.

$$J_v = \frac{1}{R_{total}} \frac{P}{\mu} \quad , \qquad R_{total} = R_m + R_b + R_c \tag{1}$$

where J_v is permeate flux, R_{total} is filtration resistance, P is filtration pressure, μ is viscosity of filtrate. The filtration resistance can be divided into three local resistances in series, which are hydraulic resistance of clean membrane R_m, hydraulic resistance at the pore entrance blocked by particles R_b, and hydraulic resistance of cake layer R_c. R_{total} is the sum of these local filtration resistances.
Figure 1 shows the location of these resistances. The local pressure drop can be written as following,

$$P_m = \frac{R_m}{R_{total}} P, \quad P_b = \frac{R_b}{R_{total}} P, \quad P_c = \frac{R_c}{R_{total}} P \ , \qquad P = P_m + P_b + P_c \tag{2}$$

where P_m is pressure drop in membrane, P_b is pressure drop at blocked pore, and P_c is pressure drop in cake layer. The pore blocking filtration will only occur at the early stage of filtration and be followed by the cake filtration.

In order to elucidate the dependence of the filtration resistances on filtration volume a simple model was developed based on the combined pore blocking and cake filtration models. The resistance provided by the pore blocking R_b is proportional to the ratio of $A_{blocked}$ to A_0,

$$R_b = \frac{A_{blocked}}{A_0} R_{b,\infty} = \left(1 - e^{-kv}\right) R_{b,\infty} \tag{3}$$

where $R_{b,\infty}$ is the hydraulic resistance at the end of the pore blocking filtration, which means $A_{blocked} = A_0$, and it can be an index of the extent of the pore blocking.

In the cake filtration model, the filtration resistance provided by the cake layer is proportional to the cake mass per unit filtration area, $C_0\, v$,

$$R_c = \left(1 - e^{-kv}\right) \alpha\, C_0\, v \tag{4}$$

where α is specific cake resistance.

The total filtration resistance R_{total} can be expressed as a function of v using Eqs.(2),(8), and (9).

$$R_{total} = R_m + R_b + R_c = R_m + \left(1 - e^{-kv}\right) R_{b,\infty} + \left(1 - e^{-kv}\right) \alpha\, C_0\, v \tag{5}$$

The zeta potential is usually used for the characterization of surface charge condition of porous membrane or particles. The zeta potential of membrane can be measured by the SP method. The SP is defined as the slope of linear regression of the relation between electrical potential difference E and filtration pressure P as $\Delta E / \Delta P$.

$$SP = \frac{\Delta E}{\Delta P} \tag{6}$$

The zeta potential, ζ_{obs}, can be determined by the H-S equation,

$$\zeta_{obs} = \frac{\mu\lambda}{\varepsilon_0 \varepsilon_r} SP \tag{7}$$

where ε_0 is the permittivity of free space, ε_r is relative dielectric constant, and λ is solution conductivity. The H-S equation underestimates the zeta potential because the H-S equation neglects the effect of surface conductance[28]. Hence, the zeta potential estimated using H-S equation is "apparent" or "observed" one. However, it is useful for practical application because of its simple formulation.

The electrical potential difference will be also generated in fouling layer. The overall SP can be written using the local values of SP[24,25].

$$\left(\frac{\Delta E}{\Delta P}\right)_{total} = \frac{R_m}{R_{total}}\left(\frac{\Delta E}{\Delta P}\right)_{membrane} + \frac{R_b}{R_{total}}\left(\frac{\Delta E}{\Delta P}\right)_{pore\ blocking} + \frac{R_c}{R_{total}}\left(\frac{\Delta E}{\Delta P}\right)_{cake\ layer} \tag{8}$$

Hence, the overall zeta potential, i.e. ζ_{calc}, can be estimated using the local hydraulic resistances and the local zeta potentials,

$$\zeta_{calc} = \frac{R_m}{R_{total}}\zeta_m + \frac{R_b}{R_{total}}\zeta_b + \frac{R_c}{R_{total}}\zeta_c \tag{9}$$

where ζ_m is the zeta potential of membrane, ζ_b is the zeta potential at blocked pore, and ζ_c is the zeta potential of cake layer.

Experimental

Membrane and particle

Polycarbonate(PC) membrane was used in pore blocking filtration experiments because this membrane has cylindrical pore structure which will be easily blocked by particles.

Polymethyl methacrylate(PMMA) latex particle(Soken Chemical & Engineering) having particle sizes of 0.40 μm and positive charge surface(PP40) was used. This PMMA particle is hard sphere having narrow particle size distribution and the cake layer formed by the particles is incompressible. In all experimental conditions the permeation of particles through membrane was not observed.

Feed suspension was prepared by dispersing the PMMA particles in 1mM KCl solution, which was filtered with a membrane filter having pore size of 0.5μm before use.

Experimental apparatus and filtration procedure

A dead-end membrane housing (25mm diameter, effective filtration area 3.0cm^2) with platinum black electrodes for streaming potential measurement was used. A pulse pattern of applied pressure was used for flux and SP measurements during filtration.

Flux and SP of clean membrane were measured with 1mM KCl solution. The filtration of PMMA particles suspension was conducted with the pulse pattern pressure. The hydraulic resistance and SP were monitored during the filtration.

Results and discussion

In the pore blocking filtration the filtration resistance will rapidly increase in the early stage of the filtration and it is expected that ζ_{obs} will be affected by the hydraulic resistance produced by pore blocking, R_b.

Figure 1 shows the changes in the normalized filtration resistance and ζ_{obs}. The normalized filtration resistance showed rapid increase in the early stage of the filtration and then increased gradually and linearly. This change suggests that the pore blocking occurred in the early stage of the filtration and then the cake layer was built up. The lines in the upper panel in Figure 1 shows the calculated lines using Eq.(5) for explanation of the proportion of the each hydraulic resistance to the total filtration resistance. The calculated line of Eq.(5) could be well fitted with the experimental result, which shows the validity of the filtration model. k, $R_{b,\infty}$, and α could be determined as fitting parameters. From the observation and analysis it is shown that the total filtration resistance would be governed by the hydraulic resistance produced by the pore blocking.

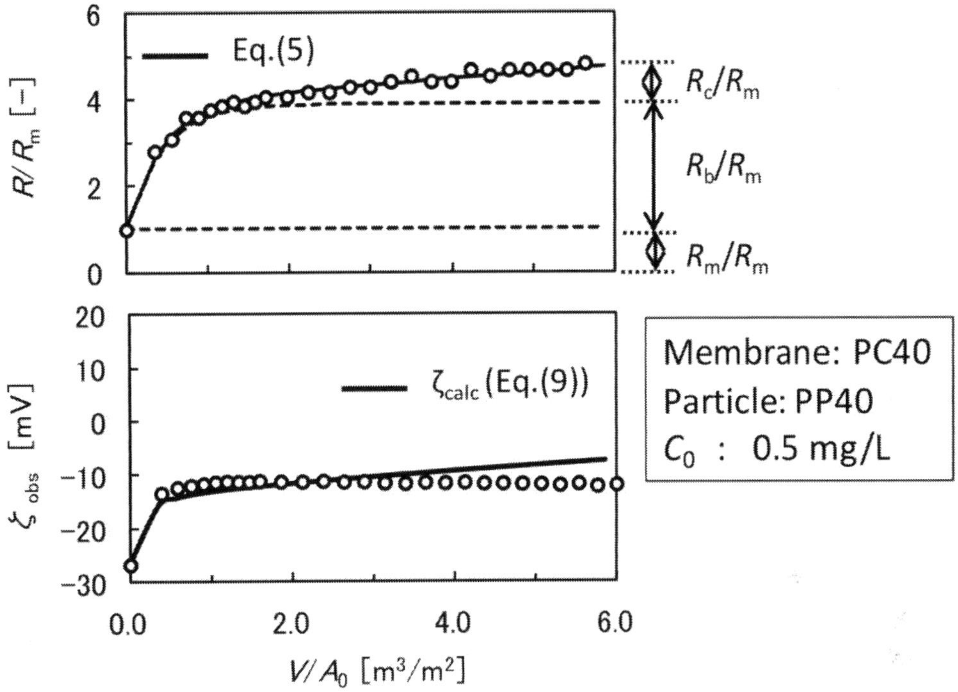

Figure 1 The change in the filtration resistance and ζ_{obs} during the pore blocking filtration.

In this filtration ζ_{obs} jumped from ζ_m to about -11mV just after the start of the filtration and became almost stable at about -11mV during the filtration in spite of the building up of the cake layer of particles with positive charge. This change was quite different from that observed in the cake filtration.

The solid line in Figure 1 (lower panel) shows ζ_{calc} estimated using Eq.(9). In the calculation, R_m and ζ_m was measured before the filtration, R_{total}, R_b, and R_c were determined from the analysis of the filtration resistance, ζ_c was fixed at -24mV, which was a typical value of the cake layer of PP40 observed in the cake filtration experiments, and ζ_b was determined as the fitting parameter. The ζ_{calc} fit to the early stage of the filtration, which shows the validity of Eq.(9) for the pore blocking filtration region, and ζ_b can be determined as –10.0mV.

The effect of feed concentration was also conducted with same membrane and particle. The fitting parameters obtained in the analysis of with Eqs.(5) and (9) are listed in Table 1. The values of ζ_b were between ζ_m and ζ_c and increased with the increase in the feed concentration. $R_{b,\infty}$, which is an index of the extent of pore blocking, decreased with the increase in feed concentration.

Table 1 Parameters determined by the analysis with Eqs.(5) and (9)

C_0 (kg/m^3)	α (m/kg)	k (m^{-1})	$R_{b,\infty}$ (m^{-1})	ζ_m (mV)	ζ_c (mV)	ζ_b (mV)
0.5	3.1×10^{12}	2.3	3.1×10^{10}	-26.5	24.0	-10.0
1.0	2.3×10^{12}	3.0	2.6×10^{10}	-23.7	24.0	-8.5
5.0	1.2×10^{12}	5.5	1.7×10^{10}	-27.7	24.0	0.5

These trends suggest that ζ_b increased with the decreased in the extent of pore blocking. Figure 2 shows the images of the pore blocking and the location of pressure drop. It is expected that in the lower concentration a pore is completely blocked by a single particle while in the higher concentration particles form bridging structure as a result of the rushing of particles to same pore simultaneously and the bridge structure cannot block the pore heavily. ζ_b will reflect the shift of the location at the dominant pressure drop from the blocked pore to the gap between particles with the increase in feed concentration.

From these observation it is shown that ζ_{obs} would reflect the extent of the pore blocking and be a useful tool to detect the cause of pressure drop.

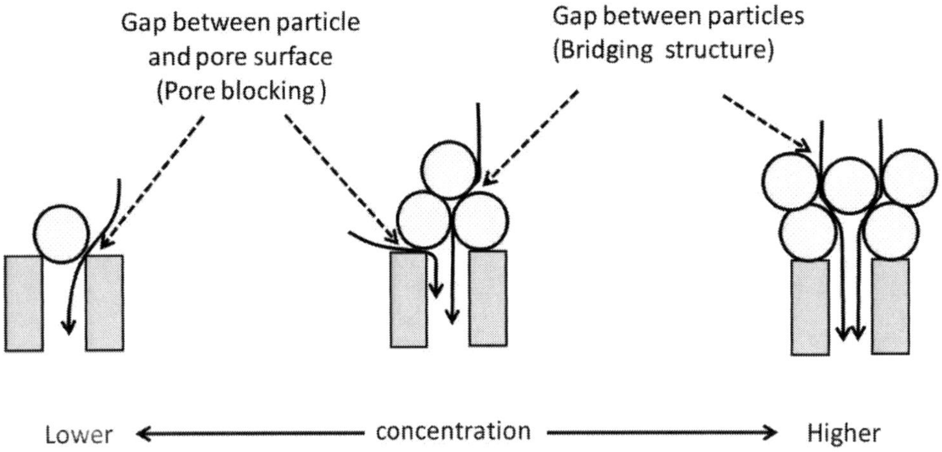

Figure 12 Effect of feed concentration on the location of dominant pressure drop in the pore blocking filtration.

References

1. K. Nakamura, K. Matsumoto, Properties of protein adsorption on a microfiltration membrane; The comparison between static and dynamic adsorption methods, Journal of Membrane Science, 285, 126-136(2006)
2. Kazuho Nakamura, Takako Orime, Kanji Matsumoto, Response of zeta potential to cake formation and pore blocking during the microfiltration of latex particles, J. Membr. Sci. 401-402, 274-281(2012)
3. Kazuho Nakamura, Takako Orime, Kanji Matsumoto, Zeta potential monitoring during microfiltration of humic acid, J. Chem. Eng. Jpn, 45(8), 583-589 (2012)
4. Hiroyuki Ohshima Ed., "Electrical Phenomena at Interfaces and Biointerfaces: Fundamentals and Application in Nano- Bio-, and Environmental Sciences, First Edition", John Wiley & Sons, Inc, Hoboken, NJ, 2012, Chapter 36

AUTHOR INDEX

AUTHOR INDEX

AUTHOR INDEX

AUTHOR INDEX

AUTHOR INDEX

AUTHOR INDEX

AUTHOR INDEX

AUTHOR INDEX

AUTHOR INDEX

AUTHOR INDEX

AUTHOR INDEX

AUTHOR INDEX

AUTHOR INDEX

AUTHOR INDEX

AUTHOR INDEX

AUTHOR INDEX

AUTHOR INDEX

Separations Division

Core Programming Topic at the 2012 AIChE Annual Meeting

Pittsburgh, Pennsylvania, USA
28 October - 2 November 2012

Volume 2 of 2

ISBN: 978-1-62276-740-3

Printed from e-media with permission by:

Curran Associates, Inc.
57 Morehouse Lane
Red Hook, NY 12571

Some format issues inherent in the e-media version may also appear in this print version.

Printed by Curran Associates, Inc. (2013)

AIChE
3 Park Avenue
New York, NY 10016-5991

Phone: (203) 702-7660
Fax: (203) 775-5177

www.aiche.org

Additional copies of this publication are available from:

Curran Associates, Inc.
57 Morehouse Lane
Red Hook, NY 12571 USA
Phone: 845-758-0400
Fax: 845-758-2634
Email: curran@proceedings.com
Web: www.proceedings.com

TABLE OF CONTENTS

VOLUME 1

VOLUME 2

(392k) Modified Correction-Factor Analysis for Recycled Membrane Extraction in Multipass Rectangular Modules

Ho-Ming Yeh, Chemical and Materials Engineering, Tamkang University, Tamsui, Taipei County, Taiwan

ABSTRACT:

The membrane extraction techniques are in principle variants of conventional liquid-liquid extraction but various shortcomings of the later are overcome by the former as membrane extraction use none or very little organic solvents, high enrichment facts can be obtained and there are no problems with emulsions. Also, they can provide unsurpassed cleanup efficiency and high extraction selectivity. The modified correction-factor analysis for recycled membrane extraction in parallel-flow multipass rectangular devices was introduced in present study, where the expressions of mass-transfer rate and modified correction-factor charts thus derived, are explicit. Unlike those in the conventional correction-factor analysis, the results can be readily calculated without using try-and-error method. The correction factors for recycled multipass operations are plotted in the modified correction-factor charts, they are function of flow rate, mass-transfer-area, distribution coefficient and overall mass-transfer coefficient. Experimental results confirm the predicted values for the extraction of acetic acid from aqueous solution by methyl isobutyl ketone in the double-pass device.

(392l) Ex-Situ Tests Based Correlations to Predict Chemical and Mechanical Durability of Polymer Electrolyte Membranes

Rameshwar Yadav, Gregory J. DiLeo, Nilesh Dale and Kev Adjemian, Zero Emission - Research, Nissan Technical Center North America, Farmington Hills, MI

Ex-situ Tests Based Correlations to Predict Chemical and Mechanical Durability of Polymer Electrolyte Membranes

R. Yadav, G. DiLeo, N. Dale, and K. Adjemian

Zero Emission – Research

Nissan Technical Center North America, Farmington Hills, MI-48331

Polymer electrolyte membrane (PEM) fuel cells are a high-potential technology to allow for zero-emission transportation. The durability of the PEM is a vital aspect that must be understood for the successful commercialization of PEM fuel cell vehicles. The ability to forecast the chemical and mechanical durability or lifetime of a membrane material in a fuel cell environment would be a valuable screening or development tool. This work proposes a chemical stability factor (CSF) and expands on a previously reported humidity stability factor (HSF), based on ex-situ testing, to predict the in-situ lifetime of PEM materials.

Chemical degradation of PEMs occurs through free-radical attacks (OH·, OOH·) on weak and vulnerable sites (-COOH, side group) in the PEM[1-3]. These free-radicals are generated from H_2O_2 that can be formed through a 2-electron O_2 reduction process at multiple locations in the membrane electrode assembly [4-5]. Open Circuit conditions (high potential, low RH) are favorable for the formation of free-radicals leading to severe chemical degradation of PEM. Variables such as temperature, O_2 concentration, water activity (relative humidity), membrane thickness, polymer type (PFSA, hydrocarbon), and processing techniques (reinforcement or chemical stabilizers) affect the chemical durability of a membrane [6-7]. In automotive applications, PEMs are also exposed to numerous cycles of wet and dry conditions in which swelling and shrinking, stress, creep, and fatigue of the membrane lead to membrane cracking and performance loss [8-9].

In-situ evaluation of PEM chemical durability and mechanical durability is lengthy in time even with accelerated stress tests *(i.e. OCV hold test: up to 4 weeks and dry/wet cycling: up to 10 weeks)*. Correlations based on quicker ex-situ tests to predict chemical durability and mechanical durability can help save significant in-situ testing times and resources. This work is aimed at developing such correlations defined as a chemical stability factor (CSF) for chemical durability and a humidity stability factor (HSF) for mechanical durability of PEMs. The CSF is a correlation based on oxygen permeation through the membrane and the fluoride emission from exposing the membrane to a solution of Fenton reagents. We have observed that the relative trend of PEM chemical durability obtained from OCV hold tests matches with the relative trend predicted by CSF. The correlation for HSF is based on the tensile strength of the membrane along with its swelling in liquid water. This factor has been previously reported by MacKinnon et al. [10] and in this work is applied to a wide variety of membranes. The relative trend of PEM mechanical durability from in-situ testing follows the relative trend predicted by HSF.

This work will discuss the fundamentals of these ex-situ-test-based correlations to predict in-situ chemical durability and mechanical durability. Results from OCV hold test hours, CSF, dry/wet cycling test, and HSF will be presented to show the reliability of these correlations.

Acknowledgement:

We thank the Composite Materials and Structure Centers, Michigan State University for measurement of oxygen permeation rate across PEMs.

References

1. N. E. Cipollini, "Chemical aspects of membrane degradation," *ECS Transactions*, 11(1), 1071-1082 (2007).

2. F. D. Coms, "The chemistry of fuel cell membrane chemical degradation," ECS Transactions, 16(2), 235-255 (2008).

3. A. Kabasawa, J. Saito, K. Miyatake, H. Uchida, and M. Watanabe," Effects of the decomposition products of sulfonated polyamide and Nafion membranes on the degradation and recovery of electrode," *Electrochem. Acta*, 54, 2754-2760 (2009).

4. A. Ohma, S. Suga, S. Yamamoto, and K. Shinohara, "Membrane degradation behavior during open-circuit voltage hold test," *J. Electrochem. Soc.*, 154(8), B757-B760 (2007).

5. V. O. Mittal, H. R. Kunz, and J. M. Fenton," Membrane degradation mechanisms in PEMFCs," *J. Electrochem. Soc.*, 154(7), B652-B656 (2007).

6. C. Chen and T. F. Fuller," The effect of humidity on the degradation of Nafion membrane," Polym. Degrad. Stab., 94, 1436-1447 (2009).

7. V. A. Sethuraman, J. W. Weidner, A. T. Haug, and L. V. Protsailo, "Durability of perfluorosulfonic acid and hydrocarbon membranes: effect of humidity and temperature," 155(2), B119-B124 (2008).

8. S. Vengatesan, M. W. Fowler, X. Z. Yuan, and H. Wang, "Diagnosis of MEA degradation under accelerated relative humidity cycling," 196, 5045-5052 (2011).

9. J. Kang and J. Kim, "Membrane electrode assembly degradation by dry/wet gas on a PEM fuel cell," 35, 13125-13130, (2010).

10. SM MacKinnon, TJ Fuller, FD Coms, MR Schoeneweiss, CS Gittleman, Y.-H Lai, R Jiang, and AM Brenner, "Membranes: Design and Characterization," *Fuel Cells – Proton Exchange Membrane Fuel Cells*, Elsevier, 741-754 (2009)

(392m) Computational Fluid Dynamics Simulation of a Carbon Hollow Fiber Membrane Module for IPA/Water Vapor Permeation

Nobuo Hara, Alessio Caravella, Miki Yoshimune, Kenji Haraya and Shigeki Hara, Research Institute for Innovation in Sustainable Chemistry, National Institute of Advanced Industrial Science and Technology (AIST), Tsukuba, Japan

We have developed a flexible carbon hollow fiber membranes from sulfonated poly(phenylene oxide) as a carbon precursor. These carbon hollow fiber membranes show excellent performances for gas and vapor separation, and thus efforts have been made to design and develop a membrane module with several hundreds of carbon hollow fibers. In doing that, it has been recognized that module geometries such as number of fibers and packing density of fibers are critical parameters, which influence has to be investigated.

For this purpose, 3D Computational Fluid Dynamics (CFD) simulations of a full-scale carbon hollow fiber membrane module are setup in the present work, analyzing its performance for separation of water and isopropyl alcohol (IPA) mixture in vapor phase. The dimension of membrane module corresponds to the real membrane module (diameter of 15mm, length of 200 mm). Hollow fibers are uniformly distributed along the radial direction with considering several inter-fiber distances (100 to 500 micrometers). The highest packing degree of hollow fibers considered in the simulation corresponds to 1867 fibers at a distance of 100 micrometers. The flow configuration is co-current with the feed on the shell-side. Simulations are performed exploiting the geometrical and physical symmetry. In this configuration, the vapor mixture of IPA and water is progressively enriched of IPA in force of the high permeability of the hollow fiber to water (water permeance of $1x10^{-6}$ mol m^{-2} s^{-1} Pa^{-1}, ideal water-IPA separation factor of 5000). In the simulation, the performances of the hollow fiber module are analyzed in terms of purity and recovery of IPA at the retentate. The present investigation allows the separation properties of a hollow fiber module to be evaluated in relation with the radial fiber distribution, which is discussed in detail.

(392n) Supercritical CO_2 Extraction of 1-Butanol From Aqueous Solutions Using a Hollow Fibre Membrane Contactor

Teresa Moreno, Steve J. Tallon and Owen J. Catchpole, Industrial Research Ltd, Lower Hutt, New Zealand

Hollow fibre membrane contactors have recently received much attention in a number of separation or purification processes due to their large specific areas, enhancing mass transfer and therefore extraction efficiencies. The so called porocritical extraction process couples the porosity of a hollow fibre membrane contactor (HFM) with the high diffusivity of a supercritical fluid (generally carbon dioxide) to create a non-dispersive contact between solvent and aqueous feed, allowing a high throughput capacity without flooding or foaming issues independently of density difference between phases. This study presents the application of this technique to the separation of 1-butanol from aqueous solutions. 1-butanol, or "biobutanol" is a second generation biofuel that can be produced by anaerobic fermentation; however, its yield is currently limited to less than 5 % by mass of the fermentation broth and is normally recovered by distillation, which is highly energy intensive. In this work, the efficiency of a lab-scale HFM contactor in the supercritical CO_2 extraction of 1-butanol from a model water/1-butanol mixture and from an actual fermentation broth is studied under a range of different flow rates, solvent-to-feed ratios and pressure conditions. Packed column fractionation was also carried out to enable comparison of extraction efficiency under a limited range of conditions.

(392o) Design of a Self-Cleaning Nanostructured Reverse Osmosis Membrane

Oishi Sanyal, Dept of Chemical Engineering and Materials Science, Michigan State University, East Lansing, MI and Ilsoon Lee, Department of Chemical Engineering and Materials Science, Michigan State University, East Lansing, MI

Reverse osmosis (RO) is a pressure driven membrane technology process that finds wide applications in water treatment processes. One of the major challenges faced by this technology is the tendency of the membrane to get fouled in presence of various foulants present in wastewater. This results in significant flux decline and lower rejection values that necessitate frequent replacement of the membrane. The current project thereby aims at preparation of an anti-fouling RO membrane having better longevity without compromising it's permeate flux and rejection properties and also without further increment in the operational cost of the process. Surface modification of commercial membranes using hydrophilic polymer brushes containing poly (ethylene glycol) (PEG) units has been implemented by several research groups. The polymer brushes can be grafted on the substrates via covalent linkages and these brushes provide resistance against fouling. Although addition of the brush layer lowers the initial flux value, the relative flux decline of the modified membrane is lesser compared to the unmodified membrane. If the underlying polyamide dense layer of commercial RO membranes be replaced by a more permeable membrane skin the overall permeability can be improved. Alternate deposition of oppositely charged polyelectrolytes to form nanostructured membrane skins has been reported as a promising technique in several papers. The layer-by-layer (LbL) technique of depositing polyelectrolytes allows control on deposition of ultrathin bilayers on a support both in terms of numbers as well as the thickness of the multilayer assembly. Appropriate choices of the polyelectrolytes and various deposition parameters lead to improvement in permeate flux and rejection properties of the resulting membrane. The combination of these two functionalities for membrane preparation has not yet been explored yet. We propose the LbL deposition of polyelectrolyte multilayers on a porous support and attachment of polymer brush units to the topmost layer by polymer grafting techniques. The attachment of PEG-based polymer brush layer can be achieved through polymer-on-polymer stamping method wherein a thin layer of polymer can be transferred onto the multilayer surface thereby yielding a multifunctional self-cleaning RO membrane.

(392p) Reverse Osmosis Membranes for Desalination with High Water Flux

Lin Zhao, Department of Chemical & Biomolecular Engineering, The Ohio State University, Columbus, OH and W. S. Winston Ho, The Ohio State University, Columbus, OH

Reverse Osmosis Membranes for Desalination with High Water Flux

Lin Zhao, C.-Y. Chang, and W.S. Winston Ho, The Ohio State University

Novel high-flux reverse osmosis membranes with hydrophilic additives were synthesized and characterized under brackish water desalination conditions. Four selected hydrophilic additives, *o*-aminobenzoic acid-triethylamine salt, *m*-aminobenzoic acid-triethylamine salt, 2-(2-hydroxyethyl) pyridine, and 4-(2-hydroxyethyl) morpholine, were added into *m*-phenylenediamine (MPD) solution to react with trimesoyl chloride (TMC) during the interfacial polymerization between MPD and TMC. The effects of different concentrations for each hydrophilic additive were evaluated using brackish water desalination tests with 2000 ppm NaCl solution at 225 psi and 25°C. After the additive concentration was optimized, the synthesized membranes showed 36.7 gallons/ft^2/day water flux and 98.2% salt rejection or greater. After the post-treatment using aqueous solutions containing glycerol, sodium lauryl sulfate, and camphorsulfonic acid-triethylamine salt, the water flux of the membrane was further increased. The resulting membrane showed a flux of 52.6 gallons/ft^2/day and a salt rejection of about 98.8% or greater. This membrane outperformed the other membranes evaluated, and it also exhibited a good stability in terms of water flux (average 50 gallons/ft^2/day) and salt rejection (average 99%) during the entire period of a 30-day test with 2000 ppm NaCl solution at 225 psi and 25°C. The good stability was verified by comparing FTIR spectra of the same membrane sample before and after the 30-day test.

(392q) Reverse Osmosis Membranes for Brackish Water Desalination with High Water Flux and Fouling Resistance

Lin Zhao, Department of Chemical & Biomolecular Engineering, The Ohio State University, Columbus, OH and W. S. Winston Ho, The Ohio State University, Columbus, OH

Reverse Osmosis Membranes for Brackish Water Desalination with High Water Flux and Fouling Resistance

Lin Zhao, C.-Y. Chang, and W.S. Winston Ho, The Ohio State University

A novel high-flux and fouling-resistant reverse osmosis membrane was synthesized and characterized under brackish water desalination conditions. The *o*-aminobenzoic acid-triethylamine salt was added into *m*-phenylenediamine (MPD) solution to react with trimesoyl chloride (TMC) during the interfacial polymerization. The membrane preparation conditions including MPD concentration, TMC concentration, and interfacial polymerization time were optimized using brackish water desalination tests with 2000 ppm NaCl solution at 225 psi and 25°C. The membrane synthesized under the optimal conditions was post-treated with aqueous solutions containing glycerol, sodium lauryl sulfate, and camphorsulfonic acid-triethylamine salt to further increase the water flux. The resulting membrane showed a flux of 54.4 gallons/ft^2/day and a salt rejection of 98.6%. This membrane outperformed the other membranes including BW30, LE, XLE, and ESPA3, evaluated at their favorable desalination conditions, respectively. The fouling-resistant property of the synthesized membrane was obtained by physically coating a cross-linked polyethylene glycol (PEG-200) layer on top of the thin film. The effects of cross-linked PEG-200 solution concentration were investigated under brackish water desalination conditions. Dodecyltrimethylammonium bromide, a cationic foulant, caused a 52.9% flux reduction of the uncoated membrane and a 30.8% flux reduction of the coated membrane. Tannic acid, an organic foulant, caused a 15.6% flux reduction of the uncoated membrane and a 4.0% flux reduction of the coated membrane. The modified membrane surface morphology was analyzed using atomic force microscopy and scanning electron microscopy. The results showed a smoother membrane surface which confirmed the successful modification.

(39aj) DFT Study of Coverage-Depended Adsorption of NH3 On TiO2-B (100) Surface

Xiaohua Lu, Xiaojing Guo, Lu Cai, Weijia Liu, Yudan Zhu and Linghong Lu, chemical engineering, Nanjing university of technology, Nanjing, China

The selective catalytic reduction (SCR) of NO_x with NH_3 to N_2 and H_2O is an efficient way to reduce the amount of NO_x from industrial plants and vehicles. V_2O_5/TiO_2 catalyst is the basis for the industrial reaction. Ammonia adsorption is critical for SCR of NO on the catalyst. It was reported that the SCR activity increases with the amount of NH_3 chemisorption on TiO_2 surfaces increasing [1]. However, the saturation coverage of NH_3 on TiO_2 surface such as rutile (110) surface was reported to be half a monolayer [2]. TiO_2 surface with high NH_3 chemisorption capacity is expected for SCR. Our previous study showed that TiO_2-B (100) surface is very unique. It is with high activity and a loose structure. The fully dissociative adsorption of water at 1 ML coverage on TiO_2-B (100) surface is still favored due to its unique open structure [3].

In order to investigate the NH_3 chemisorption capacity of TiO_2-B (100) surface, we studied the adsorption of ammonia on TiO_2-B (100) surface at coverages ranging from 1/6 to 1 ML using ab initio density functional calculations. The results show that the decrease rate of average molecular adsorption energy of NH_3 with coverage on TiO_2-B (100) surface is substantially lower than that on rutile (011) surface above 1/2 ML coverage due to its open structure. The variation pattern of average molecular adsorption energy with coverage on TiO_2-B (100) surface is different from that on other TiO_2 surfaces. It shows a linear dependence on the coverage of on TiO_2-B (100) surface, but accelerated degradation on rutile (011). The possibility of NH_3 molecule onto the Ti_{5c} site is nearly equal to forming dimer with adsorbed NH_3 on TiO_2-B (100) surface at 5/6 ML coverage.

[1] Yamazoe S., et al., Development of the efficient TiO_2 photocatalyst in photoassisted selective catalytic reduction of NO with NH_3. Catalysis Today, 2006. 111(3-4): p. 266-270.

[2] Siu W.K., Bartynski R.A., and Hulbert S.L., The role of defects at low concentrations in the NH_3/TiO_2(110) adsorption system: An Auger-photoelectron coincidence spectroscopy study. The Journal of Chemical Physics, 2000. 113(23): p. 10697-10702.

[3] Liu W.J., et al., A shortcut for evaluating activities of TiO_2 facets: water dissociative chemisorption on TiO_2-B (100) and (001). Physical Chemistry Chemical Physics, 2010. 12(31): p. 8721-8727.

(395a) Adsorption Equilibrium Studies of Bio-Based Fermentation Broth (Acetone, Butanol, and Ethanol) Onto Immobilization of Starch Sorbents

Dina Wahyu Indriani II, Chemical Engineering , Chung Yuan Christian University, Zhong Li, Taiwan; Agriculture Engineering, Agricultural Technology Faculty , Brawijaya University, Malang , Indonesia and Tsair-Wang Chung, Department of Chemical Engineering, Chung Yuan University, Taoyuan 320, Taiwan

Adsorption Equilibrium Studies of Bio-Based Fermentation Broth (Acetone, Butanol, and Ethanol) onto Immobilization of Potato Starch Sorbents and ZSM-5

Bio-based butanol has superior properties when compared to ethanol to be the gasohol and is gradually considered to be an important biofuel from the biomass fermentation. These properties include higher calorific value, low vapor pressure, low freezing point and miscibility with gasoline and diesel. In this context, in 2007, DuPont (US) and BP (UK) announced their plans to produce bio-butanol to be used as a new gasoline extender. [1] 1-Butanol can be produced by non-food products of fiber fermentation by *Clostridium acetobutylicum* or *C. berijerinckii* with acetone and ethanol as by-products. This acetone-butanol-ethanol (ABE) fermentation process used to be the main process for industrial production of butanol. The ABE solvent concentration is 20 g/L of which 1-butanol is only 8-13 g. As expected, the energy cost for traditional distillation to recover 1-butanol will be very high. The adsorption using silicalite (Al-free or high Si/Al zeolites) is the most energy efficient method compared to pervaporation, gas stripping and liquid-liquid extraction.[2]

Therefore, in this study, the high Si/Al ratio zeolites, the self-made silica immobilized starch sorbents and their mixtures will be conducted in our laboratory by using the gravimetric adsorption equilibrium apparatus and the dynamic adsorption system to evaluate the competitive adsorption behavior of 1-butanol with other compounds in the ABE solution. The results of this study can be applied to develop a high effectiveness industrial process of butanol recovery from ABE fermentation solution.

In the first year of this study, the selected sorbents will be used to adsorb 1-butanol, acetone, ethanol and water vapor for gas-phase single-component adsorption equilibrium study and analysis by our gravimetric adsorption equilibrium apparatus. The isotherm data were obtained in this study as well as the development of regression models for the isotherm data.

In the second year of this study, our dynamic adsorption system will be modified to allow the control of the required operating parameters. The breakthrough curves of the adsorption of each compound in ABE solution by the selected sorbent in a packed column will be obtained in this study and the effect of each operating parameter on the product concentration of 1-butanol will be discussed. It is important for the design of an industrial butanol recovery adsorption system form ABE solution.

Low-temperature preparation of titania-pillared montmorillonite with complex layered structure and the application in arsenic deep removal

Jingwei Guo, Ping Na, Yuan Li, Shimin Zhou, Xiaojiao Cai

School of Chemical Engineering and Technology, Tianjin University, Tianjin, China

1. Introduction

The Arsenic (As) contamination in environment which is notoriously toxic to man and other living organisms has draw worldwide concern [1-4] and consequently the world health organization (WHO) has recommended that the value of arsenic remain below 10μg/L in drinking water. In the present work, titania pillared montmorillonite (Ti-MMT) was prepared by hydrothermal method. Freeze-drying method instead of drying method was used to fabricate Ti-MMT with complex layered structure and homogeneous pore distribution. The removal of As (V) and As (III) in aqueous solutions was investigated.

2. Materials and methods

Titanium-pillared montmorillonite was prepared by a sol–gel method. The product was freeze-dried: the slurry was cooled at -50℃ for 4 h, and directly sublimated at room temperature under reduced pressure of 30 Pa. The product was sieved using a sieve of 100-180 mesh. Adsorption experiments were carried out by adding 0.10g of Ti-MMT to 500 mL of solutions containing 1 mg L^{-1} of arsenite or arsenate. Under UV irradiation or in dark, As(III) or As(V) adsorption experiments were carried out in a 1L quartz glass vessel under vigorous stirring, which contained 500 mL of As(III) or As(V) solution. The pH of solution was adjusted with 0.1mol L^{-1} HCl or 0.1mol L^{-1} NaOH. The reactor was located above the magnetic stirrer with a constant-temperature water bath set to 298 K. A 500 W high pressure mercury lamp (365nm, Beijing Lighting Research Institute, China) was placed outside the reactor as the UV light source. The samples were filtered using a 0.45 μm membrane syringe and analyzed by atomic fluorescence spectrometry (AFS, Beijing Rayleigh, China). The detection limit was determined to be 0.02 μg L^{-1}.Transmission electron microscopy (TEM) images were taken with a Japan Electronics JEM-2100F microscope (Japan). Nitrogen adsorption–desorption isotherms of Ti-MMT were obtained at 77 K (Quantachrome Corporation, NOVA-2000, USA). The adsorption and desorption branches of the isotherms were determined. The BET surface area and pore size distribution curve were calculated from the isotherms. The samples were pretreated in a vacuum (ca. 0.10 Torr) at 423 K for 2 h.

3. Results and discussion

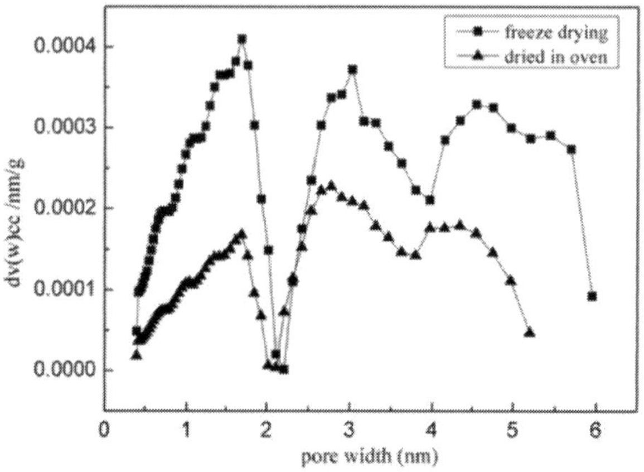

Figure 1. pore size distribution of Ti-MMT obtained by different dry methods: freeze-dried and dried in oven at 80℃.

The BET surface area of Ti-MMT prepared by drying at 80□ and freeze-drying are 89.7 and 201.6 m^2 g^{-1} respectively. The pore size distribution was shown in figure 1 and calculated using the density functional theory (DFT) method. The samples freeze-dried had trimodal pore-size distributions. One pore-size peak was above 4.5 nm, and the second was around 3.2 nm, the third was about 1.76 nm (the former two pores and the third are denoted as mesopores and micropores, respectively, according to International Union of Pure and Applied Chemistry (IUPAC)). However, for the samples dried at 80°C in an oven, the size of micropores and mesopores decreased. Materials containing macropores and micropores are expected to be used as support of catalysts and absorbents and for other applications. Freeze-drying method is effective in preparing materials with high porosity (90%). It was proposed that higher surface area benefit from small sintering shrinkage.

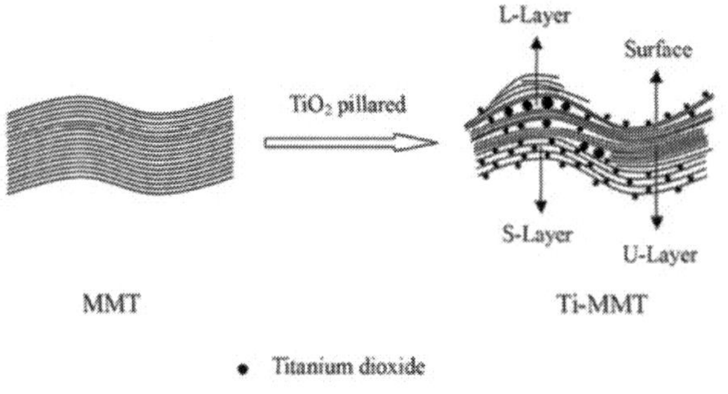

Figure 2. Schematic representation of the preparation of Ti-MMT: L-Layer (titania pillared in large spacing of layer); S-Layer (pillared in small spacing of layer); U-Layer (the un-pillared layer); Surface (titania on the outer surface of layer)

Four types of layers exist in Ti-MMT: S-Layer, L-Layer and U-Layer and surface (exfoliated layer) as described in schematic representation (figure 2) and TEM photographs (figure 3), especially to be mentioned, they are one-to-one relationship. Titania (anatase) was pillared in S-Layer, L-Layer and also exfoliated layer; as for U-Layer, no lattice of TiO_2 was observed (figure 3a), the spacing of layers changed to less than 1.42 nm (measured with software of DigitalMicrograph (Gatan, America)). The lattice fringes with a spacing of 0.35 nm are due to the spacing of the (101) planes of the anatase (figure 3b). Some of anatase particles were pillared in the space of multilayer (L-Layer), it was consistent with energy dispersive X-ray (EDX) analysis (Fig. 3c, inset). Figure 3d showed the photograph of S-Layer, the layer space is in the range 1.42-6nm. These kinds of layers have good stability and small particles of anatase exist in the interlayer, which created large number of

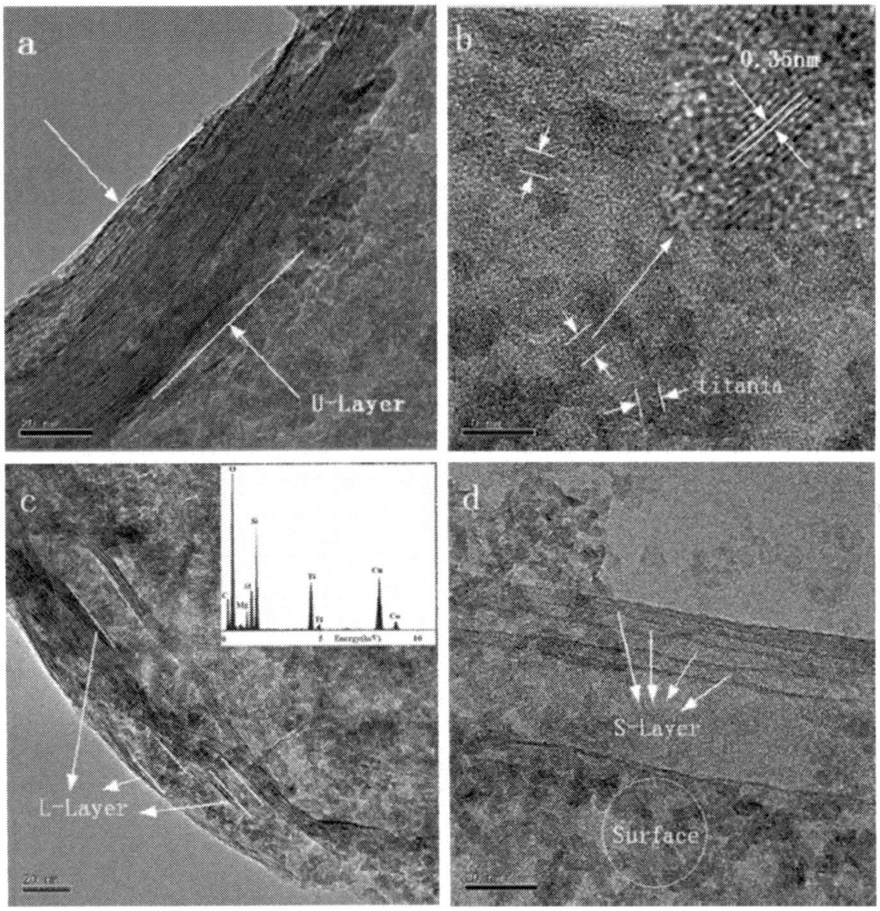

Figure 3 TEM photographs and X-ray (EDX) spectroscopic analysis of as prepared Ti-MMT

mesopores and micropores, they are highly insistent with the result in figure 3. During the preparation process, some of the ordered layers exfoliated, titania deposited on the outer surface. From the TEM photographs, it was concluded that titania formed at low temperature in the layers of MMT, and if we could turn all the U-layer and L-layer to S-Layer, the surface area would increase much more.

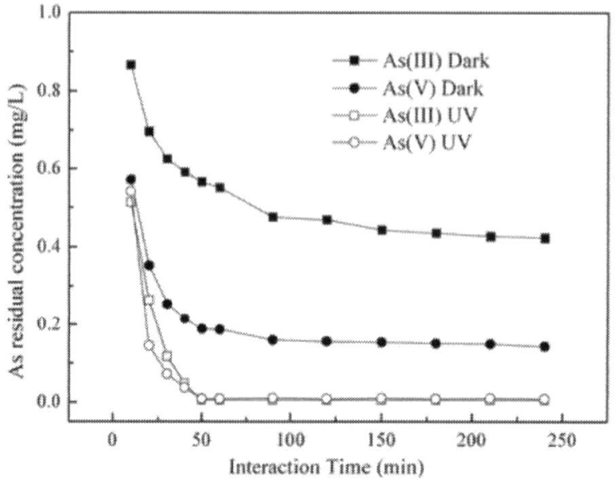

Figure 4 the adsorption of As(III) and As(V) under UV and in dark

In order to evaluate the efficiency of as-obtained product for arsenic removal in water intuitively, arsenic residual concentration was set as y in figure 4. We carried out the experiments under UV and dark conditions respectively. The uptake of As(V) and As(III) in dark system reached equilibrium approximately within 3h. The equilibrium concentration of As(III) is much higher than As(V), indicating that the strong affinity of As(V) with Ti-MMT. As to the removal of arsenic in UV system, excellent results can be observed that the residual concentration is 0.009 and 0.006 mg $L^{-1}$of As(V) and As(III) systems respectively, meanwhile, the adsorption capacity reached 4.96 and 4.97mg g^{-1}. It reached adsorption equilibrium within 1h. UV light greatly promoted the adsorption ability of As(V) and As(III). We attribute the good performance of Ti-MMT to three features: more active sites due to the higher specific surface area; facile diffusion of arsenic during the reaction because of the highly porous structure; and the pillared titania (anatase) prepared at low temperature. Under UV irradiation, the oxidation of As(III) to As(V) enhanced the removal of As(III), this principle is similar to arsenic removal with TiO_2 .

4. Conclusions

Titania (anatase) pillared MMT with homogeneous pore distribution and a surface area (201.6 m^2 g^{-1}) was obtained using hydrothermal method. According to the analysis of TEM photographs, it is the first time to confirm that the as prepared Ti-MMT has four types of layers, U-layer, S-Layer, L-layer and the exfoliated layer; titania was pillared in S-Layer, L-layer and deposited on the surface of the exfoliated layer. Under UV irradiation, the as prepared Ti-MMT has an adsorption capacity of 4.96 mg g^{-1} and 4.97 mg g^{-1} for As(V) and As(III), the residual concentration was 0.006 mg L^{-1} (As(III)) and 0.009 mg L^{-1} (As(V)). The high efficiency in arsenic removal resulted from the higher specific surface area, porous structure and the active anatase phase of TiO_2 prepared at low temperature.

Acknowledgments

Financial support from the Social Development of Science and Technology Projects of Yunnan Province (No. 2009CA038) is gratefully acknowledged.

References

[1] JQ Jiang (2001), "Removing arsenic from groundwater for the developing world—a review." Water Sci. Technol., pp. 89–98.
[2] WR Cullen, KJ Reimer (1989). Arsenic speciation in he environment, Chem Rev, pp. 713-764.
[3] JS Wang, CM Wai (2004), "Arsenic in drinking water—a global environmental problem." J Chem Educ, pp.207-213.
[4] JC Saha, AK Dikshit, M Bandopadhyay, KC Saha (1999), "A review of arsenic poisoning and its effects on human health." Crit Rev Env Sci Technol, pp.281-313.

(395ab) Predicting CO2 Adsorption in Polymers of Intrinsic Microporosity

Kyle E. Hart and Coray M. Colina, Department of Materials Science and Engineering, Pennsylvania State University, University Park, PA

Polymers of Intrinsic Microporosity (PIMs) are a novel class of ultra-rigid, glassy polymers that due to their contorted structure pack inefficiently, which creates accessible free volume. The size scale of this inherent free volume (< 0.7 nm) in combination with the ability of chemical functionalization of the polymer backbone gives this class of materials great potential for carbon dioxide storage and separation. The quality of any material at adsorbing a gas relies heavily on, and is dictated by, the amount of pore volume present and the energy of interactions with the pore wall. Both of these characteristics need to be understood and maximized if PIMs are to be realized as a carbon dioxide sequestration material. Atomistic molecular simulations offer unique insight into understanding the adsorption phenomena in a way that complements experiments, which facilitates a full understanding of the material's adsorbing capacity. The work presented here show predictive molecular simulations of both low and high pressure carbon dioxide adsorption of PIM-1, which compare well with available experimental data. In addition, several chemical and structural functionalizations were made to PIM-1 to determine the resulting effect on carbon dioxide adsorption as a means to increase both the enthalpy of adsorption and the total adsorbing capacity of linear polymers of intrinsic microporosity.

(395ad) Experiment and Simulation Study of Carbon Dioxide/Mixed Gas Adsorption in Elastic Layered Metal-Organic Frameworks

Trinh Tran, Francisco Sotomayor and Christian Lastoskie, Civil & Environmental Engineering, University of Michigan, Ann Arbor, MI

Elastic Layered Metal-organic framework (ELM) adsorbents are flexible two-dimensional coordination polymers that undergo abrupt reversible gated adsorption from an empty collapsed structure to a filled expanded state by cooperative sorption of guest molecules between square grid layers. The ELM gating transition is distinct from condensation transition commonly observed for other porous carbons as it is induced by ELM structural rearrangement rather than by pore size effects.

We report experimental measurement of CO_2 sorption isotherms on ELM11 [Cu(BF_4)2(bpy)2], ELM-12 [Cu(OTf)2(bpy)2] and ELM-22 [Co(OTf)2(bpy)2] (bpy=bipyridine; OTf = trifluoromethanesulfonate) at supercritical temperatures. We also present the hybrid Monte Carlo – molecular dynamics simulation results of CO_2 sorption selectivity for separation of CO_2 from post-combustion (CO_2/N_2) flue gas mixture and CO_2/H2 syngas mixture. Simulated CO_2 capacities above the gating pressure and CO2 isosteric heat of adsorption agree well with ELM-11 and ELM-12 experimental data. The spatial density distribution of pure CO_2 and gas mixtures adsorbed within the framework structure is mapped and transition in the preferred orientations of CO_2 within ELM-12 is noted. The possibility for tuning the structure for higher adsorption capacity as well as selectivity will also be discussed.

(395ae) Heat Integrated Moving Bed Adsorption Process for Carbon Dioxide Capture

Yongho Son, **Kiwoong Kim**, Daewook Kim and **Kwang Soon Lee**, Chemical and Biomolecular Engineering, Sogang Univ., Seoul, South Korea

Heat integrated moving bed adsorption process for carbon dioxide capture

Yongho Son, Kiwoong Kim, Daewook Kim, and Kwang Soon Lee*

Sogang university, South korea

(kslee@sogang.ac.kr)

A heat integrated moving bed adsorption (MBA) process is proposed as a potentially viable CO_2 capture process from large scale CO_2 emitting plants. The proposed process depicted in Figure 1 consists of an adsorption bed and two desorption beds through which the adsorbent particles are continuously circulated. The adsorption bed is run under a low temperature and 1 atm while the adsorbent and gas flow in opposite directions. The adsorbent particles discharged from the adsorption bed are carried to a desorption bed under a high temperature and 1 atm, and transferred to another desorption bed under a high temperature and a vacuum, and returned to the adsorption bed. Heat integration is designed to maximally recover the heat of adsorption from the adsorption bed to supply the heat of desorption required in the desorption beds.

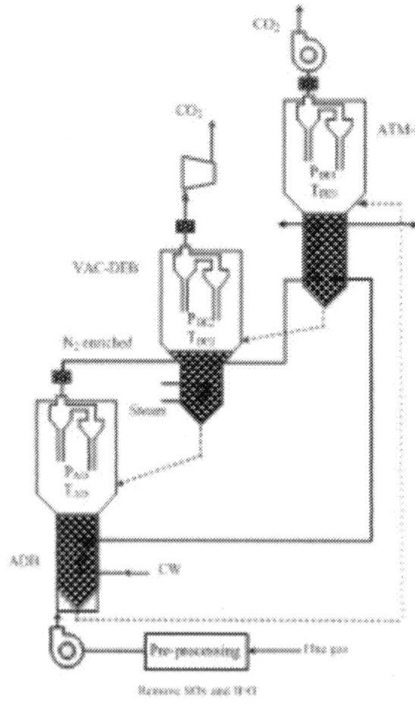

Fig. 1. The schematic diagram of MBA process.

The unsteady-state numerical model of the MBA process has been developed. The governing partial differential equation (PDE) model was set up from the mass and energy balances for the gas and solid phases and the associated constitutive equations. Then, an ordinary differential equation (ODE) model in time was derived by applying a collocation technique along the spatial coordinate. The process behavior with the variation of manipulable variables has been shown.

Afterward, the economy as a CO2 capture process from a large CO2 source was assessed. Economy evaluation using a numerical model was carried out in comparison with other existing techniques. Process optimization and economy evaluation was performed for the case that zeolite 13X is used as the adsorbent.

The lower bounds of the optimization were set as 65% recovery and 99.5% purity of CO2. The MBA process at optimum operating conditions exhibits the cost reduction of about 52% compared to the conventional MEA-absorption process by Fluor Daniel Co.

The MBA process is proposed only conceptually but reveals a great potential as a novel process that can replace the existing CO2 capture processes. In addition, the process idea may be transferred to solid-phase absorption processes and enhance the economic feasibility greatly.

Table 1. *Economy analysis of the MBA process and the MEA absorption process.*

	Steam ($4.40/ton)	Electric power ($0.07/kWh)	CW, 11° ($0.0045/$m^3$)	Etc.	Subtotal w/o SO_X, H_2O removal ($/ton)
MEA process [1]	7.90	2.77	0.34	2.66 (Added chemicals)	13.67
MBA process	3.88	2.13	0.05	0.22 (Adsorbents transport)	6.28

[1] D.G. Chapel, C.L. Mariz, and J. Ernest. *Recovery of Co2 from flue gases: commercial trends*, (1990), Aliso Viejo

(395af) QM and QM/MM Study On the Interaction of CO2 with CPL2

Paul Meza, chemical engineering, University of Puerto Rico at Mayaguez Camous, Mayagüez, PR, Maria Curet-Arana, Chemical Engineering, University of Puerto Rico, Mayagüez, PR and Rafael A. Soler-Crespo, Chemical Engineering, University of Puerto Rico, Mayaguez, PR

Metal-organic coordination networks (MOCNs) have high internal surface area, and the flexibility in their synthesis allows modifying the adsorptive properties of the pore very easily [1–6]. CPL2 (Cu2(pyrazine-2,3-dicarboxylate)2(4,4′-bipyridine)) is a MOCN that has a higher adsorption capacity for CO_2 than for other light molecules, such as O_2 and N_2 [7]. The objective of this project is to study CO_2 - CPL2 interactions using quantum mechanical (QM) calculations and ONIOM-EE, which is an integrated quantum mechanical molecular mechanical (QM/MM) method linked via electronic embedding. We have used two representative models for the CPL2 pore to reproduce binding sites in order to quantify energies of adsorption and to map possible adsorption mechanisms. For the QM calculations, we have compared two exchange/correlation functionals B3LYP and wB97XD, which has a specific treatment for Van der Waals forces. We have used a mix of 6-311G+ and LANL2DZ for the basis sets. The latter basis set was used solely on the transition metal atom. For the ONIOM-EE calculations, we have used UFF to describe the low level part of the system. Our results demonstrate that there is an unpaired electron in the Cu atom. Electronic energies of reaction to form a CPL2 - CO_2 complex range from -2 kJ/mol to -28 kJ/mol, and the distance between the CO_2 and the closest atom of the framework is around of 2.7 Å, indicating that there is a weak interaction on the CO_2-CPL2 complex.

Reference

[1] J.-R. Li et al., "Carbon dioxide capture-related gas adsorption and separation in metal-organic frameworks," *Coordination Chemistry Reviews*, vol. 255, no. 15–16, pp. 1791-1823, Aug. 2011.

[2] R. J. Kuppler et al., "Potential applications of metal-organic frameworks," *Coordination Chemistry Reviews*, vol. 253, no. 23–24, pp. 3042-3066, Dec. 2009.

[3] K. Biradha, A. Ramanan, and J. J. Vittal, "Coordination Polymers Versus Metal-Organic Frameworks," *Crystal Growth & Design*, vol. 9, no. 7, pp. 8-9, 2009.

[4] S. Bureekaew, S. Shimomura, and S. Kitagawa, "Chemistry and application of flexible porous coordination polymers," *Science and Technology of Advanced Materials*, vol. 9, no. 1, p. 014108, Apr. 2008.

[5] J. L. C. Rowsell and O. M. Yaghi, "Metal–organic frameworks: a new class of porous materials," *Microporous and Mesoporous Materials*, vol. 73, no. 1–2, pp. 3-14, Aug. 2004.

[6] C. Janiak, "Engineering coordination polymers towards applications," *Dalton Transactions*, no. 14, p. 2781, 2003.

[7] O. J. García-Ricard and A. J. Hernández-Maldonado, "Cu2(pyrazine-2,3-dicarboxylate)2(4,4′-bipyridine) Porous Coordination Sorbents: Activation Temperature, Textural Properties, and CO2 Adsorption at Low Pressure Range," *The Journal of Physical Chemistry C*, vol. 114, pp. 1827-1834, 2010.

(395ag) Large Screening of 1 Million Triazolium-Based Ionic Liquids for CO2 Capture

Fangyong Yan, Lawrence Berkeley National Laboratory, Berkeley , CA

Large screening of 1 million triazolium-based ionic liquids for CO_2 capture

Fangyong Yan[1,2**], Maciej Haranczyk[2**], Hunaid Nulwala[3,4**], Jihan Kim[1,2], Kuldeep Jariwala[2], Michael Lartey[3], Berend Smit[1,2*], David Luebke[3*]

[1] Department of Chemical and Biomolecular Engineering, UC Berkeley, Berkeley, CA

[2] Lawrence Berkeley National Laboratory, Berkeley, CA

[3]National Energy Technology Laboratory, Pittsburgh, PA

[4] Department of Chemistry, Carnegie Mellon University, Pittsburgh, PA.

** contribute equally to the work

* To whom correspondence should be addressed

Ionic liquids are organic salts existing as liquids near room temperature due to their asymmetric shape which prohibits close interaction between cations and anions. Recently, ionic liquids have attracted quite a bit of attention due to their excellent CO_2 solubility[1] and low vapor pressure[2]. The sheer number of possible ionic liquids, 10^{14} accessible structures[3], presents both an opportunity and a challenge. The task of down selecting to the best ionic liquid is daunting and experimentally impossible to perform.

To identify the best candidate for CO_2 capture application we have employed a combinatorial screening methodology to examine one million 1,2,3-triazolium cations paired with the TF_2N anion. The main purpose of this study is to understand the effect of the side groups on the overall properties of these ionic liquids. In our study, we have focused on effects of aliphatic chains attached to the triazolium ring and employed combinatorial approach to generate over 1 million cations and corresponding ionic liquids. We have then sampled the resulting chemical space using diversity selection based on geometrical parameters[4]. Further, we have employed density calculations to investigate trends and structure-property relationships in these materials. We have found interesting trends in self-diffusivity, solubility, and selectivity which can be used to screen our entire library of ILs. We will be synthesizing these ionic liquids using Cu(I) catalyzed click chemistry, and they are currently being evaluated. Our aim is to develop a rational design methodology and apply it to across all families of ionic liquids.

References

1. Cesar Cadena, Jennifer L. Anthony, Jindal K. Shah, Timothy I. Morrow, Joan F. Brennecke, and Edward J. Maginn, *J. Am. Chem. Soc.*, 2004, 126, pp 5300–5308.
2. Lynnette A. Blanchard, Dan Hancu, Eric J. Beckman, and Joan F. Brennecke, *Nature*, 1999, 399, pp 28-29.
3. Alexandre Varnek, Natalia Kireeva, Igor V. Tetko, Igor I. Baskin, Vitaly P. Solov'ev, *J. Chem. Inf. Model.*, 2007, 47, pp 1111–1122.
4. Maciej Haranczyk and Maciej Gutowski, *J. Comput Aided Mol Des.*, 2010, 24, pp 627–638.
5. Hunaid B. Nulwala , Chau N. Tang , Brian W. Kail , Krishnan Damodaran , Palwinder Kaur , Shan

Wickramanayake , Wei Shi and David R. Luebke, *Green Chem.*, 2011, 13, pp 3345-3349.

(395ah) Simple Prediction of Adsorption Henry Constants for Supercritical n Alkanes On 5A Zeolite

Kevin F. Loughlin, Chemical Engineering, American University of Sharjah, Sharjah, United Arab Emirates and Dana Abouelnasr, Department of Chemical Engineering, American University of Shatjah, Sharjah, United Arab Emirates

Simple Prediction of adsorption Henry Constants for Supercritical n alkanes on 5A zeolite

By Kevin F. Loughlin and Dana Abouelnasr, Department of Chemical Engineering , American University of Sharjah, PO Box 26666, University City, Sharjah UAE. email: kloughlin@aus.edu & dabouelnasr@aus.edu ¿Tel: (971) (6) 515-2986; FAX: (971) (6) 515-2979

ABSTRACT

A simple model is proposed to calculate Henry constants for supercritical n alkanes on 5A zeolite using mid isotherm data. The Gaussian isotherm contains three model parameters, q_{max} the saturation loading, s_e the standard deviation and P_{50} the log mean pressure for the entire isotherm. Since the slope of the Gaussian isotherm approaches zero as the pressure approaches zero, the equation *dlnq/dlnp* = 1 is used instead for the derivation of the Henry constants. Using this equation, the dimensionless temperature-independent Henry constant may be derived as

$$K_H^* = \frac{e^{-\left(z_H^2/2+\sigma_e z_H\right)}}{\sqrt{2\pi}\ \sigma_e}$$

where K_H^* is expressed as K_H*P_{50}/q_{max}. Z_H is the Z value for the Gaussian model at the point where *dlnq/dlnp* = 1, and σ_e is the standard deviation of the isotherm. For supercritical n alkanes on 5A zeolite, σ_e is equal to *ln 10* and Z_H is equal to -1.92. Upon substitution K_H^* becomes:

$$K_H^* = 2.28$$

The temperature-independence of K_H^* occurs because the temperature dependence of K_H is cancelled by the temperature dependence of P_{50} the log mean pressure for the entire isotherm. It is important to note that P_{50} is calculated at the middle of the isotherm. Therefore Henry's constants can be estimated for isotherm data far from the Henry's Law region. Using the equation, $K_H^* = 2.28$, the model is observed to satisfactorily predict the Henry's law data for all supercritical n alkane isotherms on 5A zeolite.

To completely predict Henry's constants for this system, a predictive model is incorporated based on a correlative approach for 63 isotherms for carbon numbers varying from 1 to 10. The calculated values from this approach are also in remarkable agreement with reported experimental values. A summary of the equations needed to predict the properties of supercritical isotherms of n alkanes on 5A zeolite is provided.

KEYWORDS: separations, Henry constants, 5A zeolite, n alkanes, supercritical isotherms

(395ai) Adsorption Isotherms At Different Temperatures May Intersect

Dana Abouelnasr, Department of Chemical Engineering, American University of Shatjah, Sharjah, United Arab Emirates and Kevin F. Loughlin, Chemical Engineering, American University of Sharjah, Sharjah, United Arab Emirates

Supercritical and Subcritical Adsorption Isotherms may Intersect

By Dana Abouelnasr and Kevin F. Loughlin, Department of Chemical Engineering , American University of Sharjah, PO Box 26666, University City, Sharjah UAE. email: dabouelnasr@aus.edu & kloughlin@aus.edu: (971) (6) 515-2982; FAX: (971) (6) 515-2979

ABSTRACT

Common sense dictates that adsorption isotherms for a given adsorbent-adsorbate system and at different temperatures should never intersect. Although this must be true for isotherms represented graphically as the loading, q, vs. pressure or vs. ln P, this might not be true when they are presented as θ vs. ln P. For example, isotherms for n butane on 5A zeolite clearly demonstrate this phenomenon. This work uses the Gaussian isotherm model to illustrate why this crossing may occur.

The Gaussian model transforms a type I adsorption isotherm into a straight line with slope equal to the standard deviation of the isotherm, σ_e. For two isotherms to never intersect, their linear representations must be parallel (i.e., σ_e must be the same for both isotherms). A comparison of σ_e for several isotherms indicates that this is not always true.

The isotherm standard deviation can be found through a linear regression fit using the Gaussian model. It can also be estimated in a relative fashion by eye, by comparing the range of ln P covered by each isotherm; the larger the range, the larger the standard deviation. This is illustrated in Figure 1, in which θ vs. ln P is displayed for two different standard deviations.

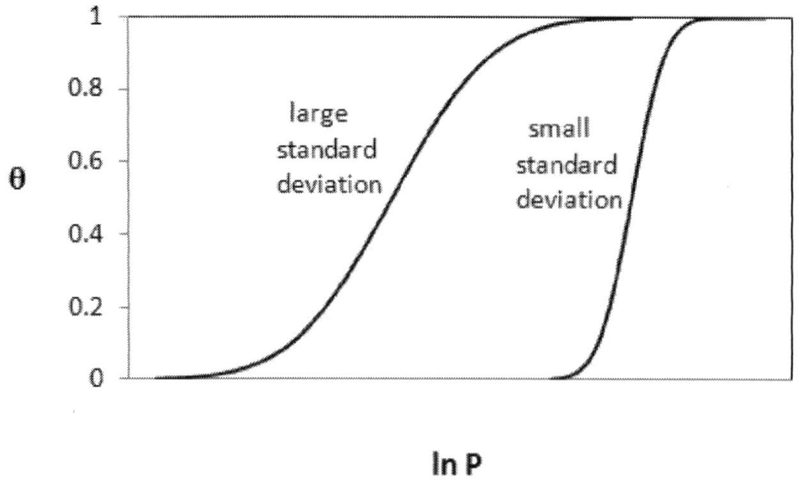

Figure 1: The effect of increasing standard deviation on isotherm width.

For example, the isotherms for the adsorption of several n alkanes onto 5A zeolite were collected from literature and assessed for consistency between studies. The resulting isotherms come from several different studies, and are remarkably consistent. These isotherms are shown in Figure 2, taken from our earlier work.

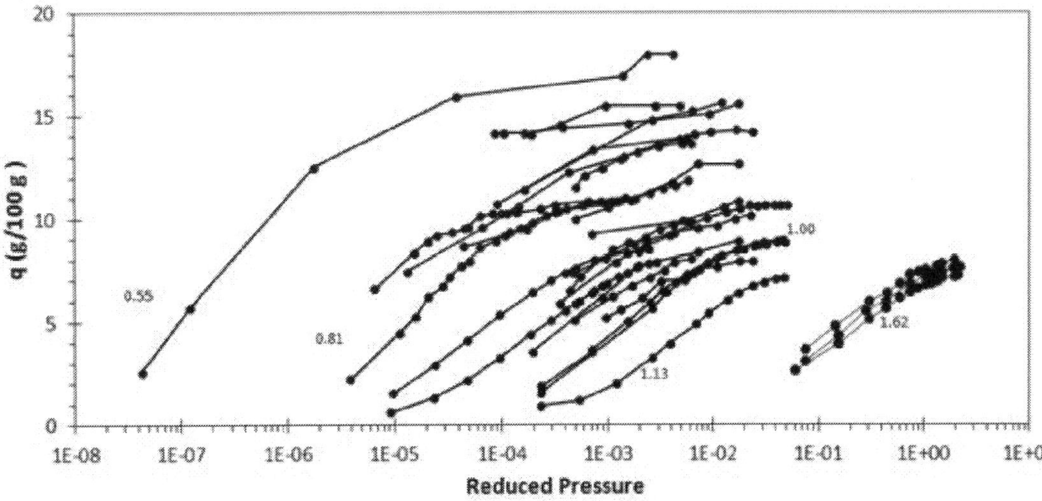

Figure 2: Isotherms for several n alkanes on 5A zeolite. Reduced temperatures for a few isotherms are displayed as values on the graph. Isotherms are taken from literature data.

The isotherms at low subcritical temperatures cover a much wider range over ln P than isotherms at higher subcritical temperatures or those at supercritical temperatures. This indicates that σ_e is large at low temperatures, and decreases as the temperature increases in the subcritical range. Since σ_e varies, then the slopes of the linearized isotherms must also vary, dictating that the lines must intersect. This also appears in the more conventional representation of the isotherms as θ vs. ln P, as is displayed in Figure 3 for n butane on 5A zeolite.

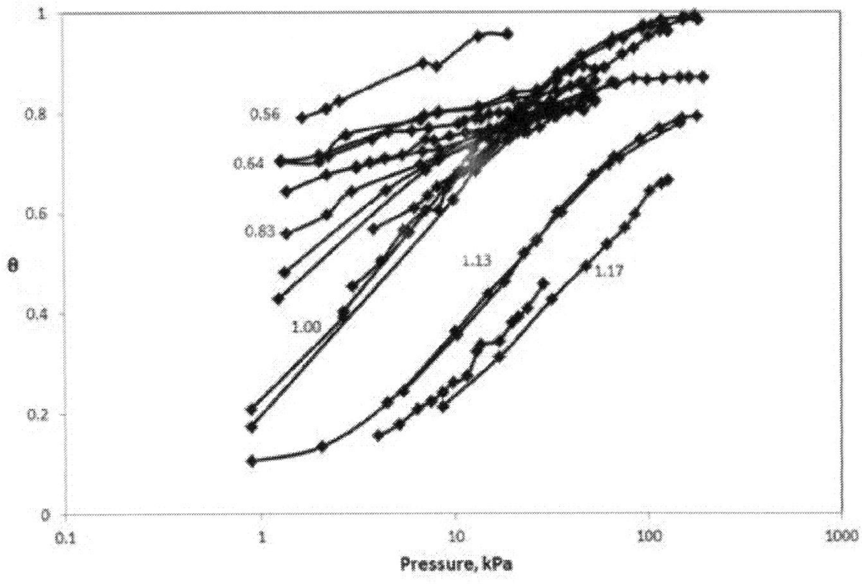

Figure 3: Intersecting isotherms for n butane. Reduced temperatures for a few isotherms are displayed as values on the graph. Isotherms are taken from literature data.

The intersection between subcritical and supercritical isotherms is readily apparent in Figure 3. The reason for the intersection can be observed to be the difference in the width of the isotherms (or the standard deviation). Therefore, adsorption isotherms can intersect when presented as θ vs. P or ln P.

KEYWORDS: separations, supercritical, subcritical, 5A zeolite, n alkanes, isotherms

(395ak) Gas Adsorption Simulation in Porous Material Made up with Silsesquioxane Units and Metal Catalytic Sites

Nethika S. Suraweera[1], Michael E. Peretich[2], Joshua Abbott[2], James Humble[2], Austin Albert[3], Craig E. Barnes[2] and David J. Keffer[4], (1)Chemical & Biomolecular Engineering, University of Tennessee, Knoxville, TN, (2)Chemistry, University of Tennessee, Knoxville, TN, (3)University of Tennesee, Knoxville, TN, (4)Chemical and Biomolecular Engineering, University of Tennessee, Knoxville, Knoxville, TN

Porous structures made up with spherosilicate building blocks and consist of metal catalytic sites exhibit favorable properties for the ability of adsorption of gases. Inspired by the experimental work, we modeled these amorphous structures and performed adsorption simulations for hydrogen, carbon dioxide and methane. This porous material has cubic silicate building blocks (silsesquioxane units: Si_8O_{20}), which are cross-linked by $SiCl_2O_2$, $SiClO_3$ and SiO_4 bridges and decorated with $OTiCl_3$ catalytic sites and $SiMe_3$ ends.

The structures for the simulations was developed first using a coarse grain model, followed by an energy optimization and replacing the coarse grains with atoms. The physical properties of the structure such as density, surface area, accessible volume and pore size distribution were compared with experimental vales and they are in good agreement.

Adsorption of gases was simulated by Grand Canonical Monte Carlo (GCMC) method. For CO_2 and methane United atom model and TraPPE fore fields were used to model the adsorbate molecules. For adsorption of hydrogen, UFF force fields were used and Feynman-Hibbs effective potential (FH) method was used to account for the quantum effects. Adsorption isotherms at 300K and 77 K were generated for a pressures range up to 100 bars. Energy of adsorption was calculated at each case. Analyzing the pair correlation functions was useful in identifying gas adsorption sites.

(395al) On the Adsorption of Thiol Molecules On a Nickel Catalyst: A Theoretical Approach

Damilola A. Daramola, Brian L. Hassler and Gerardine G. Botte, Chemical and Biomolecular Engineering Department, Center for Electrochemical Engineering Research, Ohio University, Athens, OH

Self-assembled molecules (SAMs) have been studied in literature and shown to modify the surface structure of the substrate on which they are adsorbed [1]. This surface modification results in a modification of the substrate's properties. Specifically, adsorbed sulfur and thiols have been shown to rearrange the structure of metals including Ni [2-4]. It is expected that this would result in a modification of the electronic and geometric properties of the Ni surface and subsequently enhance (or hinder) its catalytic properties.
Density Functional Theory (DFT) calculations have been performed using Crystal 09 [5] to study the effect of adsorbed thiols molecules on the properties of Ni. Two nickel surfaces were modeled: Ni (1 1 1) and Ni (1 0 0) because it has been reported that adsorption of sulfur causes a reconstruction of (1 1 1) to (1 0 0)[6]. Geometry optimizations and frequency calculations were performed to elucidate the adsorption geometry, adsorbate -- adsorbate interactions, and the electronic properties of the catalyst and these results will be presented at the meeting.

References

[1] S. T. Marshall, M. O'Brien, B. Oetter, A. Corpuz, R. M. Richards, D. K. Schwartz, and J. W. Medlin, "Controlled selectivity for palladium catalysts using self-assembled monolayers," *Nature Materials,* vol. 9, pp. 853-858, 2010.
[2] G. Somorjai, "Surface Reconstruction and Catalysis," *Annual Review of Physical Chemistry,* vol. 45, pp. 721-751, 1994.
[3] D. Mullins, D. Huntley, and S. Overbury, "The nature of the sulfur induced surface reconstruction on Ni(1 1 1)," *Surface Science,* vol. 323, pp. L287-L292, 1995.
[4] D. Alfonso, "Computational Studies of Experimentally Observed Structures of Sulfur on Metal Surfaces," *Journal of Physical Chemistry C,* vol. 115, pp. 17077-17091, 2011.
[5] R. Dovesi, V. R. Saunders, C. Roetti, R. Orlando, C. M. Zicovich-Wilson, F. Pascale, B. Civalleri, K. Doll, N. M. Harrison, I. J. Bush, P. D'Arco, and M. Llunell, *CRYSTAL09 User's Manual.* University of Torino, Torino, 2010.
[6] V. Maurice, N. Kitakatsu, M. Siegers, and P. Marcus, "Low-coverage sulfur induced reconstruction of Ni(111)," *Surface Science,* vol. 373, pp. 307-317, 1997.

(395am) Why DFT Fails in Describing Adsorption of Open-Shell Systems? A Combined Multiconfigurational and Model Hamiltonian Study of Cobalt On Graphene

Frerich J. Keil, Chemical Engineering, Hamburg University of Technology, Hamburg, Germany

Why DFT fails in describing adsorption of open-shell systems? A combined multiconfigurational and model Hamiltonian study of cobalt adsorption on graphene

In contrast to molecular closed-shell adsorbates or light monovalent impurities, theoretical investigation of adsorption of transition metal (TM) compounds is more challenging due to the presence of strong electron correlations. The results of the commonly used density functional therory (DFT) are usually controversial and strongly depend on the parametrization of the exchange-correlation functional. Motivated by the deficiency of the DFT in the description of TM adsorption on semimetallic surfaces, we present a scheme that can be used for their systematic investigation. As a test case we investigate the adsorption of Co adatom on graphene.

As a first step in this study, we apply the multiconfigurational complete active space self-consistent field formalism (CASSCF) in order to take into account the effects of electron correlations in a systematic way. By using this method it is possible to obtain accurate potential energy (PE) curves for different electronic states of a TM atom as well as binding energies. In Fig.1 three most important PE curves for the Co/graphene system are shown. The total adsorption energy curve of the TM atom is not typical for the adsorption of closed-shell adsorbates. Upon approaching the surface, the ground-state electronic structure undergoes several transitions, giving rise to two stable states: $3d^74s^2$ (high-spin) and $3d^94s^0$ (low-spin). In both cases the adsorption is weak (E_{ads} < 0.35 eV) in contrast to the strong binding, predicted by DFT studies (1-2 eV) [1].

In order to more deeply understand the adsorption of Co on graphene, we map the CASSCF results onto a mean-field model. To this end, we consider a simple Hamiltonian:

$$H = \sum_{i,\sigma} \varepsilon_i^\sigma n_i^\sigma + \frac{1}{2}\sum_i U_i n_i^\sigma n_i^{\bar\sigma} - \frac{1}{2}\sum_{\substack{ij,\sigma \\ i\neq j}} J_{ij} n_i^\sigma n_j^\sigma + \frac{1}{2}\sum_{\substack{ij,\sigma\sigma' \\ i\neq j}} U'_{ij} n_i^\sigma n_j^{\sigma'}$$

where the first term describes the one-electron contribution, while the rest part corresponds to electron-electron interactions. The parameters U_i, U'_{ij}, and J_{ij} determine intraorbital and interorbital Coulomb interactions as well as interorbital exchange interaction (Hund's rule coupling), respectively. The energies obtained at the CASSCF level for different electronic configurations allow us to extract information on the magnitude of these parameters.

In Fig.2 we show the calculated electron-electron interaction parameters as functions of the Co--graphene distance. The exchange coupling parameters J_{sd} and J_{dd} are almost constant and do not change over the entire range of distances. In contrast, the Coulomb repulsion U_{dd} significantly decreases as the atom comes closer to the surface. From the correspondence between the U_{dd} parameter and the particular electronic configurations of Co, it can be concluded that the interconfigurational transitions in the adsorption of Co on graphene are attributed to the reduction of the Coulomb repulsion in the Co 3*d* shell.

In the regime of small *U* our results, with respect to the Co state ($3d^9s^0$), are similar to that obtained by DFT using the standard LDA (GGA) functional [1,2]. On the other hand, the results of intermediate *U* ($3d^8s^1$) are comparable with the GGA+*U* [3] or hybrid functional calculations [4]. Being significantly dependent of *U*, the DFT results are applicable only to a certain range of adsorption distances, whereas the total adsorption energy curve cannot be described without external parameters. Inability to describe the total potential energy curves for TM atoms on graphene within the DFT scheme leads to difficulties in determination of binding energies. Another point that prevents DFT from giving accurate binding energies is the lack of non-local dispersion interactions in standard exchange-correlation functionals. With the present approach these problems can be widely overcome.

References [1] C. Cao *et al.*, Phys. Rev. B **81** (2010) 205424. [2] K.T. Chan *et al.*, Phys. Rev. B **83** (2011) 035405. [3] T.O. Wehling *et al.*, Phys. Rev. B **81** (2010) 115427. [4] D. Jacob *et al.*, Phys. Rev. B **82** (2010) 085423.

Figures

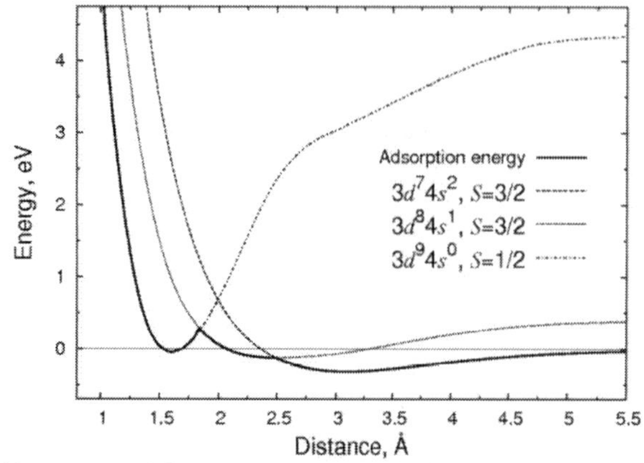

Fig. 1. Potential energy curves obtained for different electronic configurations of Co on graphene.

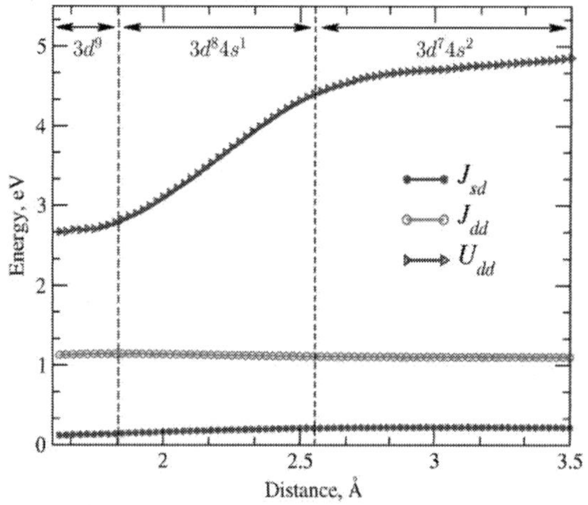

Fig. 2. Electron-electron interaction parameters for Co on graphene as a function of distance.

(395an) Adsorption in Zeolites of Components Encountered in Hexose Dehydration

Marta Leon Garcia, T. Dallas Swift, Vladimiros Nikolakis and Dionisios G. Vlachos, Catalysis Center for Energy Innovation, Department of Chemical and Biomolecular Engineering, University of Delaware, Newark, DE

The synthesis of chemical feedstocks from renewable sources, like biomass, has gained a lot of interest during the recent years. A promising reaction in this field is the acid catalyzed dehydration of carbohydrates, such as fructose and glucose, towards 5-hydroxymethylfurfural (HMF). HMF may serve as starting compound for several useful chemicals. Despite the fact that the dehydration of hexoses to HMF has been studied for many years, researchers are still searching for both the most suitable catalytic system and the best reaction medium. Zeolites have been proposed as advantageous heterogeneous catalysts compared to other materials, due to certain characteristics such as the possibility of tuning acidic and basic properties, hydrophilic and hydrophobic character, and adsorption and shape-selectivity. In fact, several interesting results on the dehydration of fructose and glucose using zeolites have been reported [1, 2]. However, HMF can undergo further rehydration to form levulinic acid and formic acid. Other side reactions are the oligomerizations of fructose and/or HMF leading to the so-called humins. In this regard, the study of the adsorption on the catalyst of the reactants, products and byproducts involved in the dehydration of sugars becomes crucial to understand the reactivity of such systems. Furthermore, zeolites without Bronsted acidity can also be considered as adsorbents that can selectively remove HMF from reaction medium before further degradation. Such a reactive adsorption process can be envisioned as an alternative to the bi-phasic reactive extraction configuration suggested in the literature [3]. In the present work, we studied the adsorption of fructose, glucose, mannose, 5-hydroxymethylfurfural, levulinic acid and formic acid in different zeolites as a function of temperature and zeolite SiO_2/Al_2O_3. The highest HMF loadings were observed for zeolite FAU. However, All zeolites types tested adsorbed higher amounts of HMF than any other component and for each zeolite type HMF adsorption increased with the hydrophobicity of the zeolites (higher SiO_2/Al_2O_3 ratios). The enthalpies of adsorption were also estimated from the adsorption isotherms measured at different temperatures. The strongest adsorption was determined for HMF, followed by levulinic acid. Finally, the competition between fructose and HMF for the actives sites on H-BEA-18 is studied by performing the adsorption experiments with binary mixtures. The adsorption of fructose decreases with increasing the amount of HMF, and it becomes negligible at high HMF concentrations. The results will be analyzed taking into account the effect of adsorbate-adsorbent interactions. Simulations using an adsorptive reactor model showed that using an adsorbent improves the HMF yield. The results will be analyzed taking into account the single and multicomponent adsorption isotherms and will be compared to that of a reactor using an organic phase for extraction.

[1] C. Moreau, R. Durand, S. Razigade, J. Duhamet, P. Faugeras, P. Rivalier, P. Ros, G. Avignon. Applied Catalysis A: General, 145, 211-224 (1996)

[2] K. Lourvanij, G. L. Rorrer. Industrial & Engineering Chemistry Research, 32, 11-19 (1993)

[3] X. Tong, Y. Ma, Y. Li. Applied Catalysis A: General, 385, 1–13 (2010)

(395ao) Innovative Air Cleaning For Fuel Cells Using Microfibrous Entrapped Sorbent and Composite Bed Design

Guomin Xu, Chemical Engineering, Auburn University, Auburn, AL and Bruce Tatarchuk, Department of Chemical Engineering, Auburn University, Auburn, AL

Innovative Air Cleaning For Fuel Cells Using Microfibrous
Entrapped Sorbent and Composite Bed Design

Guomin Xu and Bruce Tatarchuk

Department of Chemical Engineering, Auburn University, Auburn, AL

Fuels cell can be poisoned by some molecular contaminants such as H_2S, NH_3, COS, CO, VOCs at quite low concentration ranging from ppb to ppm levels. Current packed bed sorbent/catalyst bed has a tradeoff between removal efficiency (stack tolerance to the poison) and pressure drop (stack net output). In order to improve the performance of the air cleaning systems, a set of new technology consisting of microfibrous entrapped sorbent (MFES), multi-element structured array (MESA) and polish layer sorbent was developed. Various variables including removal efficiency, pressure drop, protection time, weight and volume, and form flexibility were considered to estimate the performance of filters. MFES has much higher heterogeneous contacting efficiency than packed bed, which make it has 5-log removal efficiency with a thin (1mm) layer media. The MESA can retain over 10-fold more sorbent than packed bed. The large capacity and large media area of MESA can significantly increase gas protection time while reducing filter's pressure drop. A polish layer is a very thin MFES media, just put following the packed bed to polish the trace volume of gas. A polish layer can improve the removal efficiency of packed bed and increase the utilization of bed.

Some molecular contaminants such as H2S, NH3, COS, CO, VOC are poison to fuel cells. Poison tolerance is ranging from ppb to ppm levels depending on the type of stack, the contaminant, and the operating temperature. To clean the anode air more effectively, an innovative microfibrous entrapped sorbent (MFES) and polishing layer design were developed.

MFES was prepared by entrapping fine activated carbon particles (180-300μm) into 3D sinter-locked network of polymer or metal microfibers via a wet-lay process. These materials excel in mass transport and pressure drop due to their uniform structure with small particles and adjustable porosity. A polishing layer (a thin layer of MFES media placed in series with the packed bed) was used to improve the performance of sorbent bed by "polishing" the trace volume of gas after packed bed. Hexane adsorption breakthrough test was used to evaluate the performance of the composite bed based on the breakthrough time and percentage of adsorbent utilization. Experimental breakthrough curves showed that composite bed outperformed packed beds in single pass remove efficiency, breakthrough time and bed utilization due to a synergic effect between the high capacity of packed bed and the high contacting efficiency of MFES layer.

V=-B+B2+4ADP2A

(395b) Dynamic Breakthrough Studies of Scale Inhibitor Used in Petroleum Sea Production

Carolina B. Veloso, Chemical Engineering Department, Universidade Federal do Ceará, Fortaleza, Brazil, Alvaro N. A. Silva, Dept. Engenharia Quimica,, Universidade Federal do Ceara-UFC, Fortaleza, Brazil, F. Murilo Luna, Dept. Engenharia Quimica, Universidade Federal do Ceara-UFC, Fortaleza, Brazil and Celio L. Cavalcante Jr., Chemical Engineering, Universidade Federal do Ceará, Fortaleza, CE, Brazil

Sea water contains significant sulfate concentrations ions and when mixed with the water produced during petroleum exploration, which may contain strontium, barium and magnesium ions, sulfate deposition may occur. Other possible scale that may develop in oil sea wells is originated from calcium carbonate, normally formed by high pressure differences and high temperatures in the petroleum wells. Some factors may affect the scale formation, like temperature, salinity and carbon dioxide partial pressure. The scales may deposit at the wellhead and in production lines, resulting in loss of productivity, and may require removal and prevention techniques, like the squeeze treatment. The squeeze technique involves injecting a scale inhibitor in high concentrations under pressure within the reservoir. Experiments using coreflood tests are normally performed to simulate the squeeze treatment in laboratory scale in order to estimate the interaction between the scale inhibitor and the petroleum rock. In this study we evaluate the use of a small column technique under constant temperature and flow to determine breakthrough curves of the inhibitors in these systems. Our experiments were performed at 303.15 *K* and flowrate of 0.1 *mL/min*. The effluent of the column at given times was dosed using a inductively coupled plasma optic emission spectrometer (ICP-OES) to estimate the phosphorous (scale inhibitor active phase) concentration, yielding amounts adsorbed of 0,032; 0,054; and 0,064 (*mg phosphorous)/(g rock)* for a rock bed of 57% porosity, with inhibitor concentrations of 1000, 2000 and 3000 *ppm*, respectively.

BATCH AND COLUMN BIOSORPTION OF Cu (II) ONTO SPENT COFFEE GROUND: EQUILIBRIUM AND DYNAMIC MODELING

N.E. Dávila-Guzmán[a], F.J. Cerino-Córdova[a], J. R. Rangel-Méndez[b], P.E. Díaz-Flores[c]
[a]Universidad Autónoma de Nuevo León, 66451, S.N. de los G., N.L., México.
[b]Instituto Potosino de Investigación Científica y Tecnológica, S.L.P., S.L.P, México.
[c]Academic Coordination Altiplano Region, Universidad Autónoma de San Luis Potosí, Matehuala, S.L.P., México

Introduction

Biosorption process conducted in a continuous system is one of the most important processes in wastewater treatment, as well as in the hydrocarbon separation, ion exchange processes, among others [1]. The column design requires knowledge biosorption concentration-time profile (breakthrough curves), which can be predicted by mathematical modeling. For this reason, it is essential to have a mathematical model that describes the operation, in order to design and optimize the performance of the process.

Different model has been used for modeling breakthrough curves, such as Bohart-Adams model, Thomas model and BSDT model. Nevertheless, the main disadvantage of these models lies in their simplicity, as they may present deviations raised high when the model assumptions do not apply to the process under study.

Therefore, in the present study, we evaluated the effectiveness of a mass transfer model based on material balance and characteristics of copper biosorption onto spent coffee ground (SCG) through an error function.

Materials and Methods

Biosorbent and Metal Solution.

Spent coffee ground (SCG) was collected locally and sieved through a #35 ASTM sieve. SCG was pretreated with 0.1 M NaOH solution at 40°C and 100 min^{-1} during 4h .The solid/liquid ratio used in the pretreatment was 100 g/L. After, the SCG was washed repeatedly with deionized water until pH close to 6.0 was achieved. Finally, the SCG was dried at 100°C for 6h and cooled to room temperature. All reagents used in this study were analytical-reagent grade. The Cu (II) solution was prepared by dissolving appropriate amounts of $Cu(NO_3)_2 \cdot 2.5H_20$.

Adsorption Batch Experiments

An adsorber system in a standard batch stirred tank configuration was used in the experiments. The absorber system was a 1.2 L glass vessel, 10 cm in diameter and 15 cm high, which was immersed into a bath temperature at 25°C. Six evenly spaced vertical baffles (1 cm wide) were fitted on the wall of the adsorber in order to prevent the formation of a vortex. When the stirrer rotates, the centrifugal action produces a rapid circulating flow of solution entering the cage at the bottom and leaving through the radial holes of the caging. A fixed mass of SCG (0.75g) was loaded within the cell of the stirrer for conditioning with 0.75 L acidified water (pH 4.5) at 400 min^{-1} for 90 min. After, the stirrer was turned off and the acidified water was replacement for new acidified water (0.5 L) at pH 4.5, then 0.25 L of concentrated metal solution was adding to obtain a desirable concentration value after dilution to 0.75

L, rapidly the stirrer was turned on at fixed speed. The experiments were performed during 60 min. The pH of the solutions was measured and kept at pH 4.5 during the biosorption test using 0.1 M NaOH solution. Samples were collected at different times, and the metal concentrations were analyzed by atomic absorption spectrometry (Thermo Scientific®). Biosorption isotherm study was carried out with different initial concentration of metal ion from 0.1-1.0 mM at 400 min^{-1}, 25°C, pH 4.5 and 0.750 g of SCG. The amount of Cu (II) adsorbed at equilibrium, q_e, was calculated as follows:

$$q_e = V(C_o - C_e)/m \qquad (1)$$

Where C_o and C_e are the Cu (II) initial and at equilibrium concentrations (mM), respectively; V is the solution volume in L; m is the dry weight of SCG in g; q is the uptake capacity at the time t in mmol/g.

Biosorption Column Experiments

The column internal diameter was 1 cm and length was 30 cm. After weighing the dry spent coffee ground (2 g), the biomass was washed with acidic water (pH 4.5) during 14 h in conical flask of 50 mL at 25°C y 200 min^{-1} in an orbital shaker. After filtering the sorbent, the column was packed. Then acidified water (pH = 4.5) was passed at a flow rate of 4 ml / min until the effluent pH was 4.5 (approximately 90 min.). The excess of water was drained from the bottom of the column without allowing the liquid level to drop below the bed of biomass. Not allowed the formation of air bubbles. The feed solution containing the total metal concentration of 0.1 mM was pumped downward through the column. The effect of flow rate on Cu (II) removal was studied at different volumetric flow rates (5.5, 7.7, 11 y 22 ml/min) fo a fixed bed depth of 14 cm and Cu (II) concentration of 0.1 mM.

The total quantity of metal mass biosorbed in the column (m_{ads}) was calculated from the area above the breakthrough curve multiplied by the flow rate:

$$m_{ads} = \frac{C_o Q}{1000m} \int_{t=0}^{t=t_{sat}} \left(1 - \frac{C}{C_o}\right) dt \qquad (2)$$

Dividing the metal mass biosorbed (m_{ads}) by the biosorbent mass (m) leads to the uptake capacity (q*) of the biomass. The total amount of metal ions sent to the column can be calculated form the following equation:

$$m_{total} = \frac{C_o Q t_{sat}}{1000} \qquad (3)$$

where C_o is the inlet metal ion concentration (mM), Q the volumetric flow rate (ml/min), t_{sat} is the saturation time (min), m_{ads} and m are the total mass adsorbed (g) and the biosorbent mass (g), respectively, and q* is the uptake capacity of the biomass (mmol/g). Total metal removal (R_{metal}) with respect to flow volume can be calculated form the ratio of metal mass biosorbed to the total amount of metal ions sent to the column (mtotal) as follows:

$$R_{metal} = \frac{m_{ads}}{m_{total}} \times 100\% \qquad (4)$$

Ion-exchange multicomponent model

The biosorption isotherm was simulated as an ion-exchange reaction type between cations (Ca^{2+}) and protons (H^+) already bound to the biomass, and the others present in the aqueous phase (Cu^{2+}). For a ternary mixture containing calcium, copper and hydrogen ions, the ion exchange reactions are represented by:

$$M^{2+} + 2HB \leftrightarrow MB_2 + 2H^+ \quad (5)$$

$$M^{2+} + CaB_2 \leftrightarrow MB_2 + Ca^{2+} \quad (6)$$

where M^{2+} is the metal ion (Cu^{2+}), HB are the protonated biosorption sites, B are the biosorption sites (i.e. hydroxyl, carboxyl) y CaB_2 are the neutralized biosorption sites with calcium ions. The ion-exchange multicomponent model is obtained from the ion-exchange reactions, Eq. (5) y (6), and by using the separation factors:

$$\alpha_H^M = \frac{y_M x_H}{x_M y_H} \quad (7)$$

$$\alpha_{Ca}^M = \frac{y_M x_{Ca}}{x_M y_{Ca}} \quad (8)$$

$$y_M = \frac{\alpha_{Ca}^M \alpha_H^M x_M}{x_M(\alpha_{Ca}^M \alpha_H^M - \alpha_H^M) + x_H(\alpha_H^M + \alpha_{Ca}^M) + \alpha_H^M} \quad (9)$$

where α_H^M y α_{Ca}^M are the separations factor between metal ion and the proton, and between the metal and the calcium ions, respectively. x_M , x_{Ca} and x_H are the dimensionless concentration of metal, calcium and hydrogen ions in the liquid phase, respectively. y_M, y_{Ca} and y_H are the dimensionless concentration of metal, calcium and hydrogen ions in the solid phase respectively.

Mass transfer column model

Assuming isothermal conditions, constant physical properties for the feed solution and axial dispersion, the mass balance in the liquid and solid phases assuming a linear driving force for the solid phase rate-controlling step, gives the following respective equations:

$$\frac{\partial x_M}{\partial \tau} + \frac{\partial x_M}{\partial \xi} + \frac{\rho Q}{\varepsilon_b C^\circ}\frac{\partial y_M}{\partial \tau} = \frac{1}{Pe}\frac{\partial^2 x_M}{\partial \xi^2} \quad (10)$$

$$\frac{\partial y_M}{\partial \tau} = \frac{K_T L S C^\circ}{Q \upsilon \rho}(x_M - x_M^*) \quad (11)$$

with initial and boundary conditions:

$$x_M = 1 \text{ a } \xi = 0; \quad \frac{\partial x_M}{\partial \xi} = 0 \text{ a } \xi = 1; \quad x_M = 0 \text{ a } y_M = 0 \quad (12)$$

The following dimensionless variables are also introduced:

$$\tau = \frac{t\upsilon}{L\varepsilon_b}; \qquad \xi = \frac{z}{L}; \qquad x_M = \frac{C_M}{C^\circ}; \qquad y_M = \frac{y_M}{Q} \tag{13}$$

where K_T is the mass transfer coefficient in the solid (cm/min), S is the superficial particle area per volume (cm^{-1}), L is the lenth of column (cm), Q is the flow rate (cm^3/min), ρ is the particle density (cm^3/g), υ is the interstitical velocity (cm/min), ε_b is the bed porosity, Pe is the Peclet number, x_M^* is the batch equilibrium relation from the ternary system (mmol/L), which is obtained by the following equation:

$$y_M^* = \frac{x_M^*}{\alpha_{Ca}^M x_M^* + 1} \tag{14}$$

The partial differential equations were solved numerically using FORTRAN software.

Nonlinear Regression Analysis

All the equilibrium model parameters were evaluated by nonlinear regression minimizing the error function of sum of the squared errors (SSE) by the Solver tool of Microsoft Excel ®. The error function was calculated from the Eq. (15):

$$\%\varepsilon = \frac{\sum_{i=1}^{n} \left| \frac{q_{exp_i} - q_{sim_i}}{q_{exp_i}} \right|}{n} \times 100\% \tag{15}$$

Results and Discussion

Batch biosorption equilibrium.

The results of the equilibrium studies are presented in Figure 1. The isotherm is concave with respect to the axis of the equilibrium concentration (C_e). The equilibrium uptake was increasing rapidly at low values of C_e, nevertheless on further increase in concentration the equilibrium uptake achieves a well-defined plateau, indicating the saturation capacity of biosorbent, with a value of 0.21 mmol/g. This type of biosorption isotherm indicates a very high adsorptive affinity and strong adsorbent-adsorbate interaction.

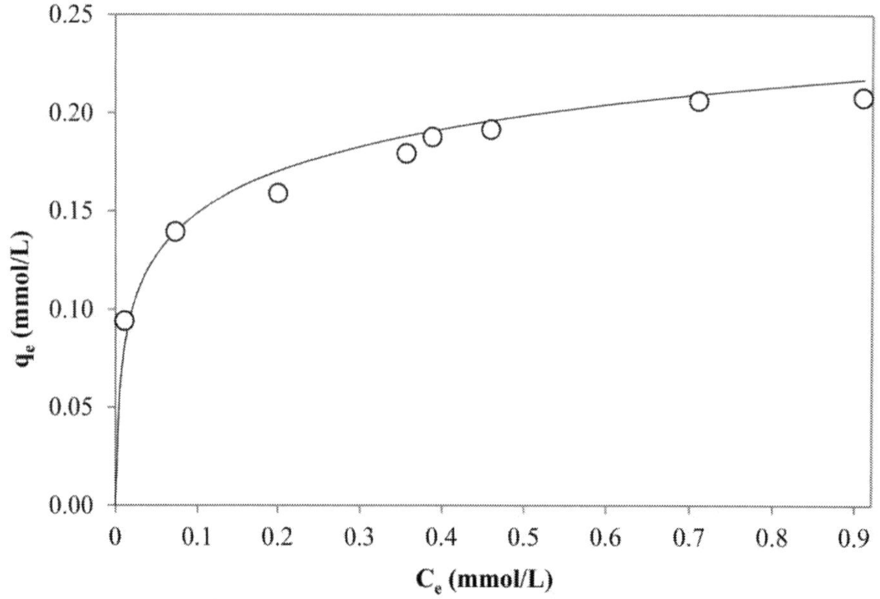

Figure 1. Copper biosoprtion isotherm.

Moreover, the analysis of equilibrium data was performed due to its importance in the development of an equation that can be used for design purposes [2]. The line in Figure 1, shows the simulation by using an ion-exchange multicomponent model. The model was able to predict the entire equilibrium curve, and the parameters of this model are showed in Table 1. The separation factor values were higher than unity, indicating that biomass has a higher affinity for metal ions than for calcium and hydrogen ions, which favors the adsorption of copper.

Table 1. Ion-exchange multicomponent model parameters..

Parameter	Value
Q (mmol/g)	0.46
α_H^M	4.78
α_{Ca}^M	4.52
SSE ($\times 10^4$)	85.1

Effect of flow rate on column biosorption

The effect of flow rate was evaluated on the dynamic column biosorption. The breakthrough curves at different flow rates (5.5, 7.3, 11 and 22 mL/min) are showed in Figure 2. When the flow rate decreased from 22 to 7.7 mL/min, the column efficiency and uptake capacity increase from 35.1 to 42.3%, and 0.0731 to 0.089 mmol/g, respectively (Table 1). The increase on efficiency and uptake capacity with decreasing the flow rate is due to the enhancement of the residence time of the copper ions, resulting in a higher contact time between these ions and the active sites on the biomass. Moreover, a further reduction in flow rate (5.5 mL/min) results in a decrease in the efficiency of the column. This can be explained considering that at very low flow rates (i.e. 5.5 mL/min), the axial dispersion phenomenon becomes more significant, generating a reduced contact between the adsorbate and adsorbent.

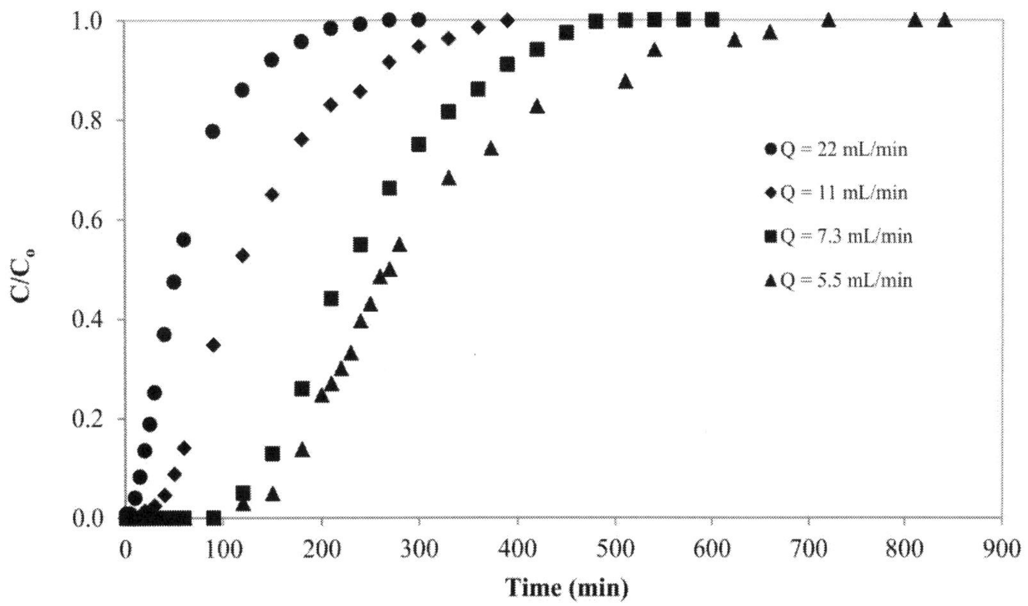

Figure 2. Effect of flow rate on breakthrough curves.

Table 1. Column efficiency and uptake capacity at different flow rates.

Q (mL/min)	q* (mmol/g)	m_{ads} (mmol)	m_t (mmol)	R_{metal} (%)
5.5	0.0838	0.167	0.396	42.3
7.3	0.0894	0.179	0.416	43.0
11	0.0754	0.151	0.429	35.1
22	0.0731	0.146	0.594	24.6

Column data modeling

In a packed column, diffusion and mixing of adsorbate in the bulk, is caused by concentration gradients and the non-uniformity of the fluid [3], generating the dispersion of the adsorbate in the axial and radial direction. The radial dispersion is generally neglected because the internal diameter of the column is several orders of magnitude greater than the diameter of the particles of the adsorbate, while the axial dispersion is considered when there are hydrodynamic conditions with low Reynolds numbers. In this study, a mass transfer model including axial dispersion and external diffusion was used to simulate the breakthrough curve of copper ions at 7.3 mL/min, the results are showed in Figure 3.

Mathematical model enables the estimation of external mass transfer coefficient (K_T) value, axial dispersion coefficient and the prediction of dimensionless concentration of copper in the effluent as the function of time, according with the low SSE value (0.36). The parameters values obtained were 3.53×10^{-6} m2/s and 1.08×10^{-5}m/s, for axial dispersion and external mass transfer coefficients, respectively.

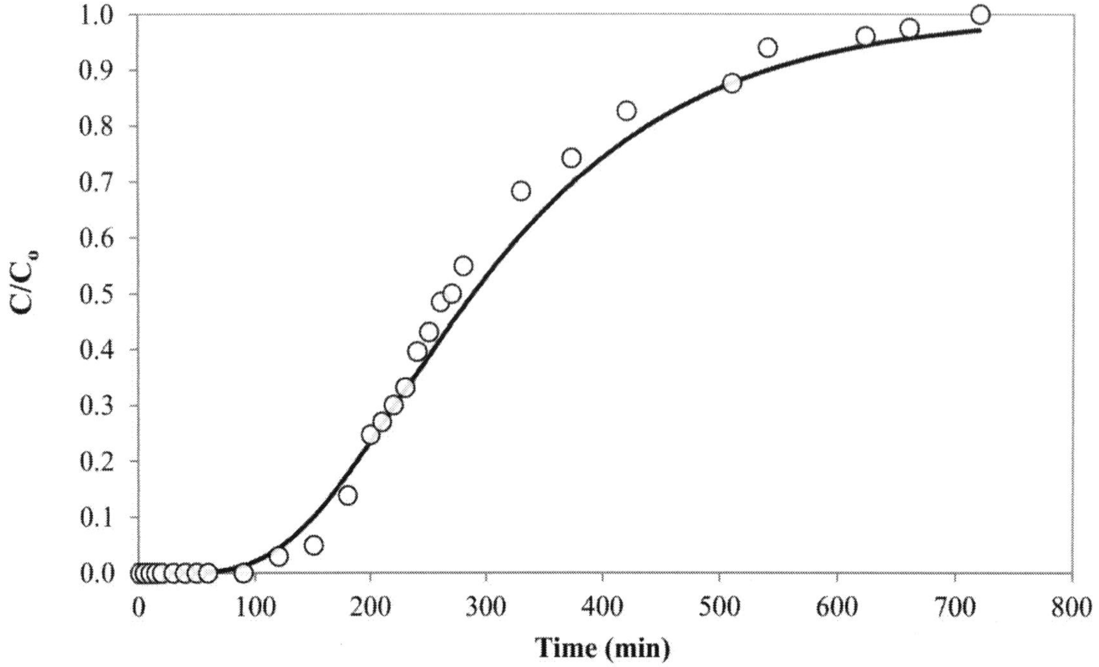

Figure 3. Mass transfer column model for copper biosorption.

REFERENCES

1. Baral, S.S., Das, S.N., Chaudhury, G.R., Rath, P. (2008), "Adsorption of Cr(VI) by treated weed Salvinia cucullata: kinetics and mechanism". Adsorption, pp. 111-121.
2. Volesky, B., Holan, Z.R. (1995), "Biosorption of heavy metals". Biotechnol Progress, pp. 235-250.
3. Mansour, A., Von Rosenberg, D.U., Sylvester, N.D. (1982), "Numerical solution of liquid-phase multicomponent adsorption in fixed beds". AIChE J., pp. 765-771.

ADSORPTION OF ORGANICS ON MSC5A IN SUPERCRITICAL CO2, COMPARING EXPERIMENT AND MOLECULAR SIMULATION

Kazuyuki Chihara, Masashi Shinsako, Shingo Ito, Hideaki Nagashima, Mai Hikita,
Meiji University, Kawasaki, Japan

Introduction

Adsorption of organics on adsorbents, such as activated carbon, in supercritical CO_2 could be mild [1, 2, 3] compared with adsorption in gas phase [4]. This is because of co-adsorption of CO_2 [5]. Amount adsorbed of organics can be changed by changing temperature and pressure of supercritical CO_2. Adsorption and desorption of organics could be easy and mild in supercritical CO_2 condition. New process of adsorptive separation using supercritical fluid might be possible [6]. However equilibria and dynamics for co-adsorption of organics and CO_2 have not been studied so much.

In this study, chromatographic measurements were made for the adsorption of benzene and toluene on molecular sieving carbon (MSC) in supercritical fluid CO_2. Supercritical fluids chromatograph (SFC) packed with MSC was used to detect pulse responses of organics. Adsorption equilibria for organics were obtained by moment analysis of the response peaks. Additionally, molecular simulation was performed by the Grand Caronical Monte Carlo method. The purpose of performing simulation is to elucidate an adsorption mechanism on the molecule level. Finally, experimental and simulated adsorption isotherm was compared.

Experimental

Chromatographic measurement

The experimental apparatus was shown in Figure 1. The carrier fluid of the chromatograph was supercritical CO_2 (critical temperature 304 K, critical pressure 73 atm) and it's mixture with the above-stated organics (benzene or toluene) respectively. The adsorbates used in the form of pulse were the same of organics mixed with supercritical CO_2.

For example, in the case of CO_2 mixed with benzene, the organic used in the form of pulse was benzene. The volumes of the pulse were fixed to be $4.0 \times 10^{-9} m^3$ as liquid. MSC 5A (HGK882, Takeda chemicals Co.) was crushed and screened to obtain particle size between 1.49×10^{-4} - 1.77×10^{-4}m (an average particle radius of 8.12×10^{-5} m). Properties of MSC 5A was shown in Table 1 [7]. 4.82×10^{-4}kg of these particles were packed into the chromatographic column of 15×10^{-2}m long and 4.6×10^{-3}m in diameter. The void fraction, ε, of the bed was determined to be 0.3548. Flow rate of supercritical CO_2 was $1.33 \times 10^{-7} m^3/s$ at 268 K and at 150, 200 and 250 atm respectively and flow rate of adsorbate (benzene or toluene) was $1.67 \times 10^{-11} m^3/s$, $5.00 \times 10^{-11} m^3/s$, $8.33 \times 10^{-11} m^3/s$, $1.67 \times 10^{-10} m^3/s$, $5.00 \times 10^{-10} m^3/s$ and $1.00 \times 10^{-9} m^3/s$ as liquid at room temperature (298 K).

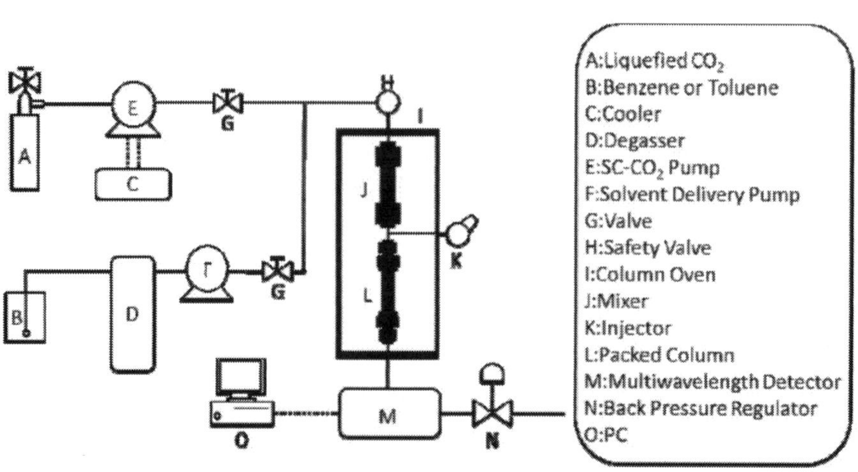

Figure 1. Experimental Apparatus (schematic)

Table 1. Properties of MSC5A

True density [g/cm^3]	1.8
Particle density [g/cm^3]	0.90
Macropore volume [cm^3/g carbon]	0.38
Micropore volume [cm^3/g carbon]	0.18
Porosity of macropore [-]	0.342
Macropore radius [μ]	2.0
Micropore opening [Å]	5.0

The column pressure was kept at 150, 200 and 250 atm respectively using back pressure regulator (880-81, Japan Spectroscopic Co., LTD.). The pressure drop across the adsorbent bed was estimated to be about 1.0 atm and was assumed to be negligible. The experimental column temperature was kept at 313, 333 and 353 K respectively using intelligent column oven (co-2065 Plus, Japan Spectroscopic Co., LTD.). Before experimental runs started, the adsorbent particles were regenerated and stabilized by feeding pure CO_2 for 2 hours at the experimental pressure and temperature. Pulse responses were detected using a multi wavelength UV detector (MD-2010 Plus, Japan Spectroscopic Co., LTD.) (195-650 nm). Response data were processed by a personal computer. Moment analysis of supercritical fluid chromatogram was tried, and the apparent adsorption equilibrium constant, K* obtained from first moment of response peak, as in references [8]. First moment can be expressed by Eq. 1.

$$\mu_1 = \frac{z}{u}\left[1 + \frac{(1-\varepsilon)\varepsilon_a}{\varepsilon}\left\{1 + \frac{\rho_p}{\varepsilon_a}(1-\gamma)K^*\right\}\right] \tag{1}$$

Molecular simulation condition

Cerius2 (Version4.6, MSI Inc.) was used throughout the simulations. DREIDING and Universal force field parameters [9, 10] in the Cerius^2 library were used. The Grand Caronical Monte Carlo method (under constant chemical potential, volume (V), and temperature (T)) was used to get the equilibrium amount adsorbed. The purpose of performing simulation is to elucidate an adsorption mechanism on the molecule level. The simulation was performed on the same conditions as an experiment in order to compare with experiments. Pseudo activated carbon sub-micropore model (MSC68-test04 model) was used as adsorbent.

Figure 2 shows the model had 6.8 A of distance between the centers of two graphitic carbon layers. The model consisted of 3504 carbons per unit cell and had 5.71 x 10^{-26} m^3 cell volume (9.00 x 10^{-9} m x 4.67 x 10^{-9} m x 1.36 x 10^{-9} m).

Figure 2. Pseudo activated carbon sub-micropore model (MSC68-test04 model)

Result and Discussion

Experimental Data

Figure 3 shows dependency of apparent adsorption equilibrium constants, K*, for toluene on the supercritical CO_2 density, which were determined by column temperature and pressure. According to Figure 3, the supercritical CO_2 density increased with decreases of equilibrium constants, K*.

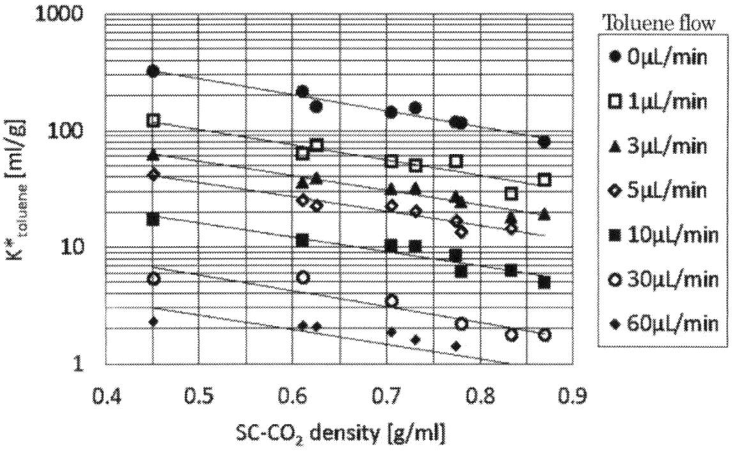

Figure 3. Dependency of apparent adsorption equilibrium constants, K*, for toluene on the supercritical CO_2 density

Figure 4 shows adsorption isotherm of benzene and toluene at four different experimental conditions. Because of obtaining amount of saturated sorption, q_∞, the adsorption isotherm were correlated by the Markham-Benton equation (Eq. 2). Basically, the amounts adsorbed became larger with decreases column pressure and increases temperature (decrease supercritical CO_2 density). But it`s not always be necessarily the same tendency benzene and toluene. It was considered that the differences in the physical properties of benzene and toluene.

$$\frac{q_i^*}{q_\infty} = \frac{K_i P_i}{1 + K_i P_i + K_{CO2} P_{CO2}} \quad (2)$$

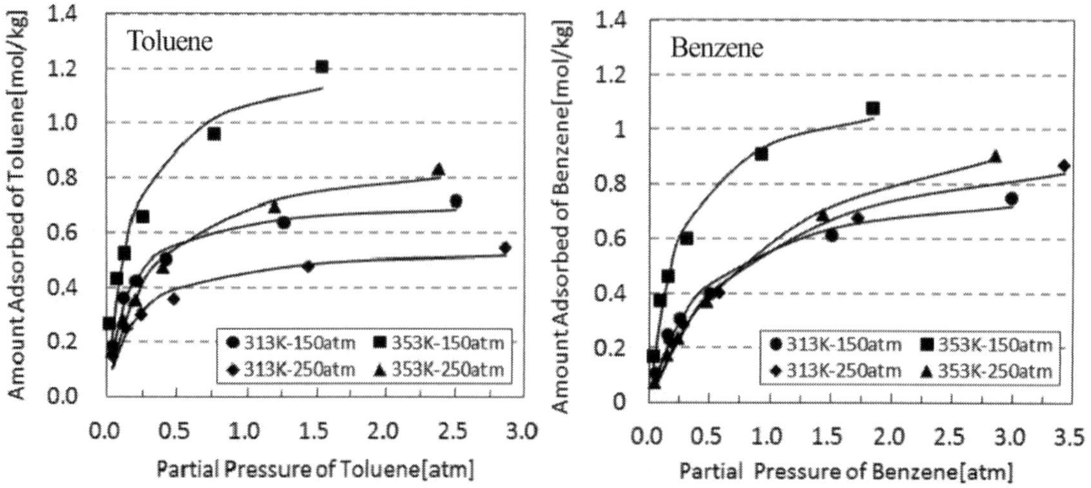

Figure 4. Adsorption isotherm for toluene and benzene

Figure 5 shows dependency of amount of saturated sorption, q_∞ for toluene on the supercritical CO_2 density. There is a good correlation between supercritical CO_2 density and amount of saturated sorption, q_∞. The supercritical CO_2 density may be related to the amount adsorbed of organics in the case of binary system of supercritical CO_2 and organics.

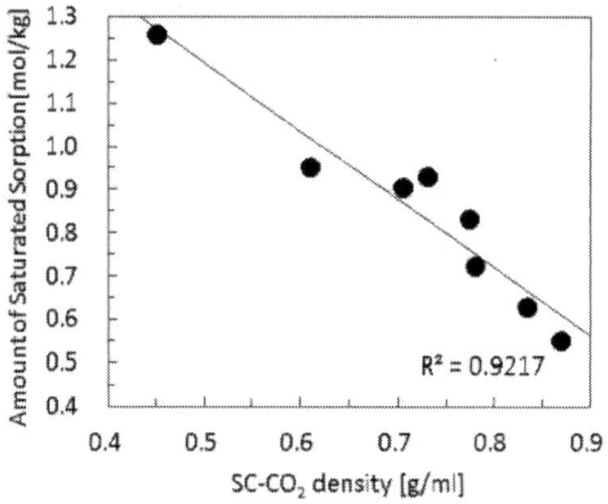

Figure 5. Dependency of amount of saturated sorption, for toluene on the supercritical CO_2 density

Molecular simulation result

First, single component adsorption simulation of benzene or cyclohexane were performed at 150 atm and 313 K respectively. According to Figure 6, MSC68-test04 model has molecular sieving mechanism. Whereas benzene average loading were 93.8 molecular/cell, cyclohexane were 1.7 molecular/cell at same simulation conditions.

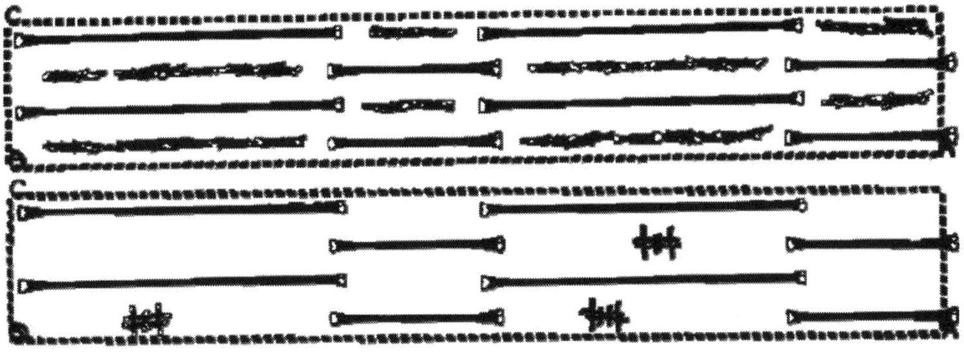

Figure 6. Adsorption state (top:benzene, bottom:cyclohexane)

How molecules of adsorbate are located were examined. The binary components benzene and CO_2 were used for the adsorbate here. The results are shown in Figure 7. Benzene and CO_2 were adsorbed into the adsorption space, which simulates micropore. Here, these adsorbed in parallel to layers in MSC68-test04 model.

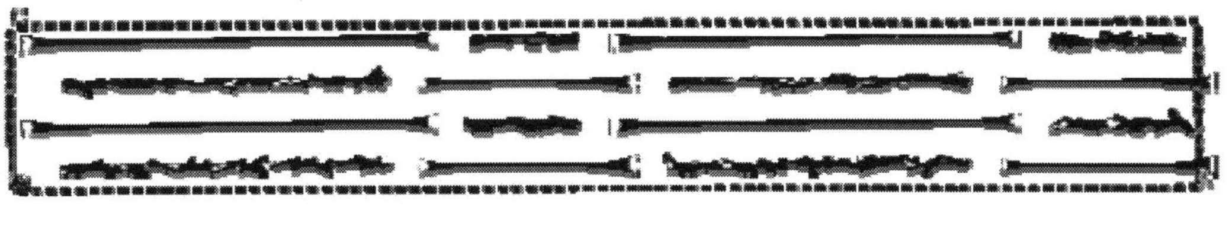

Figure 7. Adsorption state of benzene and CO_2

It is possible to calculate not only amount adsorbed of organics but CO_2 on a system of binary mixture by molecular simulation. This is an advantage compared to chromatographic measurements. Figure 8 shows adsorption isotherm for toluene, benzene and corresponding CO_2 respectively. Figure 8 are estimated that amount adsorbed of CO_2 is reduced under the influence of mixed a small amount of organics.

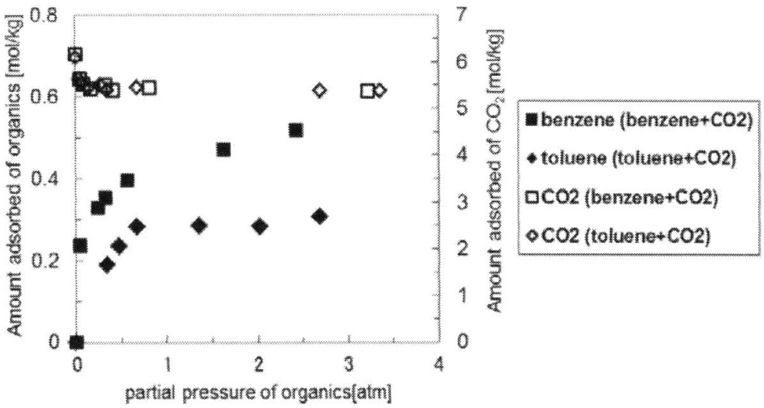

Figure 8. Adsorption isotherm for toluene, benzene and corresponding CO_2 respectively (simulation conditions : 250 atm, 313 K, Universal force field)

Figure 9 shows comparison of adsorption isotherm for a molecular simulation the force field and an experiment. Toluene and CO_2 were used for adsorbate. Here, simulation conditions are 353 K and 150 atm or 313 K and 250 atm. The amount of adsorption increased with increases of partial pressure of toluene.

Universal was used for the force field. As a consequence, the experimental amount adsorbed could not be simulated yet. In order to improve amount adsorbed using molecular simulation, we need to find more appropriate force field.

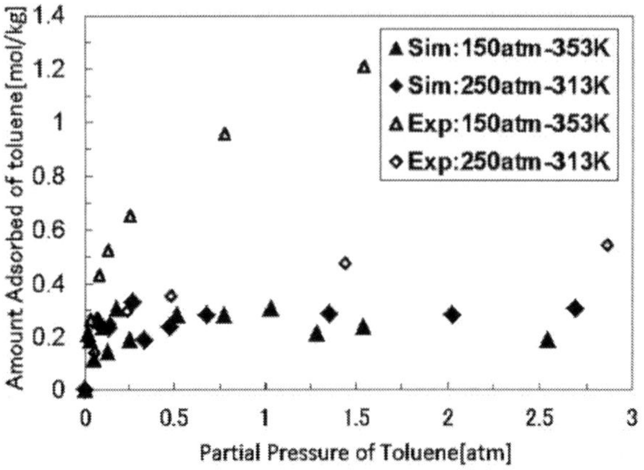

Figure 9. Comparison with experiment and molecular simulation : Adsorption isotherm for toluene

Conclusion

Chromatographic measurements were made for the adsorption of benzene and toluene on molecular sieving carbon (MSC) in supercritical fluid CO_2 mixed with organics. Supercritical fluids chromatograph (SFC) packed with MSC was used to detect pulse responses of organics. Adsorption equilibria parameters for organics were obtained by moment analysis of the response peaks. Dependences of adsorption equilibrium constants, K*, for toluene on the supercritical CO_2 density were obtained. In addition, molecular simulation of multicomponent adsorption was performed by Grand Caronical Monte Carlo method. The adsorption states between parallel graphite layers were observed. However, there are rooms for improvement in order to approach experimental amount adsorbed.

Nomenclature

ε	= void fraction in the bed	[-]
ε_a	= macropore fraction in the particle	[-]
K^*	= adsorption equilibrium constants	[ml/g]
K_i	= markham benton constants	[1/atm]
μ_1	= first absolute moment of the chromatographic peak	[s]
ρ_P	= particle density	[g/cm^3]
P_i	= partial pressure of i (i : benzene or toluene)	[atm]
q_i^*	= amount adsorbed of i (i : benzene or toluene)	[mol/kg]
q_∞	= amount of saturated sorption	[mol/kg]
u	= interstitial velocity of fluid	[cm/s]
y	= fraction of benzene of toluene in carrier	[-]
Z	= longitudinal position in the bed	[cm]

References

1. H. Shojibara, Y. Sato, S. Takishima and H. Masuoka (1995), "Adsorption Equilibria of Benzene on Activated Carbon in the Presence of Supercritical Carbon Dioxide," *Journal of Chemical Engineering of Japan*, 28, 3, pp. 245-249.
2. K. Chihara, K. Oomori, T. Oono and Y. Mochizuki (1997), "Supercritical CO_2 Regeneration of Activated Carbon Loaded with Organic Adsorbates," *Water Science and Technology*, 35, 7, pp. 261–268.
3. K. Chihara, S. Ito, H. Nagashima, M. Hikita, and R. Suzuki (2011), "Adsorption of Organics on MSC5A in Supercritical CO_2: Chromatographic Measurements & Stop & Go Simulation," *diffusion-fundamentals. org*, 16, 83, pp. 1-6.
4. J. Benkhedda, J. Jaubert, and D. Barth (2000), "Experimental and Modeled Results Describing the Adsorption of Toluene onto Activated Carbon, *Journal of Chemical and Engineering Data*, 45, 4, pp. 650-653.
5. K. Tanida, Y. Sato, H. Masuoka (1996),"Measurement and Correlation of Supercritical Carbon Dioxide Adsorption on Activated Carbon," *Kagaku Kogaku Ronbunshu*, 22, 2, pp. 385-391.
6. M. Sato, M. Goto, A. Kodama and T. Hirose (1998), "New Fractionation Process for Citrus Oil by Pressure Swing Adsorption in Supercritical Carbon Dioxide," *Chemical Engineering Science*," 53, 24, pp. 4095-4104.
7. K. Chihara, M. Suzuki, K. Kawazoe (1978), "Adsorption Rate on Molecular Sieving Carbon by Chromatography," *AIChE Journal*, 24, 2, pp. 237-246.
8. R Kumar, R. C. Duncan and D. M. Ruthven (1982), "A Chromatographic Study of Diffusion of Single Components and Binary Mixtures of Gases in 4A and 5A Zeolites," *The Canadian Journal of Chemical Engineering*, 60, 4, pp. 493-499.
9. S. L. Mayo, B. D. Olafson and W. A. Goddard III (1990), "DREIDING: a Generic Force Field for Molecular Simulations," *Journal of Physical Chemistry*, 94, 26, pp. 8897-8909.
10. A. K. Rappe, C. J. Casawit, K. S. Colwell, W. A. Goddard III and W. M. Skiff (1992), "UFF, a Full Periodic Table Force Field for Molecular Mechanics and Molecular Dynamics Simulations," *Journal of the American Chemical Society*, 114, 25, pp. 10024-10039.

(395e) MOF Syntheses Scale up and Crystallization Process Development

Paul M. Schoenecker, Bogna Grabicka, Grace A. Belancik and Krista S. Walton, Chemical and Biomolecular Engineering, Georgia Institute of Technology, Atlanta, GA

Metal-organic framework (MOF) synthesis and characterization has increased exponentially in recent years. Due in part to the nearly infinite set of structure possibilities, MOFs exhibit the potential for a plethora of applications including gas separation and storage, catalysis, drug delivery, as well as thin films. However, only a few MOFs are commercially available, and they are much less cost effective than many traditional porous materials such as activated carbons and zeolites. We examine the scalability of specific MOF syntheses including Mg MOF-74, UiO-66-NH_2, ZIF-65, and Mil-125 within sealed vessels as well as the potential to implement a continuous-flow reactor to reduce down time and increase MOF throughput. Flow-through crystallization processes are well developed for other materials and may play a vital role in the transition of MOFs to the applied level. Particular attention is paid to examining not only product output or quantity but also quality. Samples are characterized via yield, pXRD, and single component adsorption to allow direct comparison to smaller scale syntheses.

(395f) Development of Polymer-Templated Phenolic Resin-Based Carbons with Ultra Large Mesopores for Adsorption of Biomolecules

Nilantha Wickramaratne and Mietek Jaroniec, Chemistry & Biochemistry, Kent State University, Kent, OH

Mesoporous carbons (MCs) with high surface area, large pore volume, and high thermal, chemical and mechanical stability have received considerable attention because of their potential applications in separations, catalysis, gas and energy storage. For instance, carbons with large mesopores and high surface area are attractive for various applications including adsorption of large molecules such as vitamins, dyes, humic acids, etc. Although the hard-templating strategy is often used for the synthesis of carbons with large pores, it is less suitable for scaling up due to multi-step process and cost. An alternative and feasible strategy is the soft templating, which is reported here. This strategy has been successfully employed to obtain ultra-large mesoporous carbons (UMC) by using the evaporation-induced self assembly (EISA) in the presence of poly(ethylene oxide)-poly(butylene oxide)-poly(ethylene oxide) triblock copolymer (EO_{38}-BO_{46}-EO_{38}; Vorasurf 504) as a soft template. Resorcinol and formaldehyde were used as carbon precursors, whereas trimethyl benzene (TMB) was used as a micelle expander. N_2 adsorption isotherms reveal that the TMB-assisted synthesis of UMC resulted in carbons with large mesopores and relatively narrow pore size distribution (PSD) centered at about 27 nm. Carbons synthesized without TMB had also fairly large mesopores at the mean value of ~18 nm. Adsorption of lysozyme on the UMC materials was measured from buffered solution at pH = 10.8 and analyzed by using Langmuir adsorption isotherm equation. The maximum adsorption of lysozyme (= 31.0 μmol/g) was observed for the carbon sample obtained in the presence of TMB at 600°C.

SIMULATION STUDY OF MULTI-COMPONENT GAS ADSORPTION ON MSC5A BY CHROMATOGRAPHIC METHODS

*Kazuyuki Chihara, *Masashi Nomoto, Yuzo Amari, Yuki Teramura, Hidenori Nakamura,*
Meiji University, Kawasaki, Japan

Introduction

The combination of chromatographic method and moment analysis of the response peaks is one of the useful techniques to study adsorption equilibrium and adsorption rate (Chihara et al. 1978) . Perturbation chromatography with the mixed multi component adsorbate gas carrier (two adsorbates) has been applied to several studies on adsorption (Ruthven and Kumar 1979, Kumar et al. 1982). In this work, perturbation chromatography with multi component gas carrier (two adsorbates with inert gas) and non-equilibrium thermodynamics liner law was applied for discussion of the interference effect and the displacement effect (those are cross effects) on mass transfer in multi component gas adsorption as previous study (Chihara et al. 2006, 2008, 2009) for different gas mixture (He, N_2, CH_4) , (He, CO_2, CH_4). Moment analysis method and stop & go simulation method were utilized to obtain each mass transfer parameters of adsorbate gases. Dependency of micropore diffusion on amount adsorbed and correlation of micropore diffusion with chemical potential driving force for microporous adsorbent were discussed. Also, cross effect were discussed. Ruthven already pointed the dependency of micropore diffusivity on amount adsorbed in single component adsorption (Ruthven 1984). Tondeur et al gave general background on multi component perturbation chromatography for the first moment only (Tondeur 1996).

Experimental Method

Experimental Procedure and Conditions

The experimental apparatus was shown in figure1. The apparatus was similar to a conventional gas chromatograph. Adsorbent particles (molecular sieving carbon 5A, 20/30 mesh, Japan Enviro Chemical Ltd.,) were packed in a column (100cm x 3mm i.d.). Carrier gas was a mixture of two or three components among He, CO_2, C_2H_4. Perturbation pulse was introduced into the carrier gas stream. Introduction of pulses was performed by 6-way valve. The pulse size was 1cc, which meant injection period was 1.4 sec. Then pulse response was detected by TCD cell. Output signal of TCD was transmitted to a personal computer through RS232C. This signal was also transmitted to the personal computer. Simulated chromatogram by a personal computer can be overlapped on experimental chromatogram shown in the monitor screen. Further, moment of pulse response, which is shown in the monitor screen, can be automatically calculated by the personal computer. The first absolute moment and the second central moment were evaluated from the effluent peak C_e (t) as follows:

$$\mu_1 = \int_0^\infty C_e(t)\,t\,dt \Big/ \int_0^\infty C_e(t)\,dt \qquad (1) \qquad\qquad \mu_2{}' = \int_0^\infty C_e(t)(t-\mu_1)^2\,dt \Big/ \int_0^\infty C_e(t)\,dt \qquad (2)$$

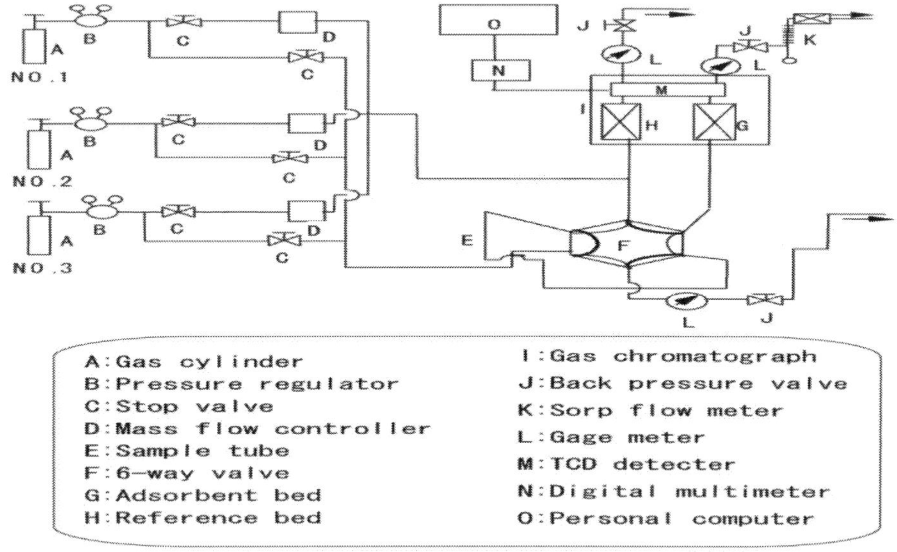

Figure I. Experimental apparatus

Moment Analysis

Basic equations are in ref. (Kumar et al. 1982) The resultant moment equations of the impulse response are

$$\mu_1 = \frac{z}{u}\left\{1 + \frac{(1-\varepsilon)\rho_p(1-y)K^*}{\varepsilon}\right\} \qquad (3)$$

$$\mu_2 = \frac{2z}{u}\left[Ez\left\{\frac{1+(1-\varepsilon)\rho_p K^*}{\varepsilon u}\right\}^2 + \frac{(1-\varepsilon)\rho_p(1-y)K^*}{\varepsilon Ksav}\right] = \frac{2z}{u}(\delta_d + \delta_f + \delta_a + \delta_i) \qquad (4)$$

where μ_1=first absolute moment of the chromatographic peak[s] μ_2=second central moment[s^2].

$$\delta_d = \frac{Ez}{u^2}\left\{1 + \frac{(1-\varepsilon)\rho_p(1-y)K^*}{\varepsilon}\right\}^2 \qquad (5) \qquad\qquad \delta_f = \frac{1-\varepsilon}{\varepsilon}\frac{R}{3k_f}\rho_p{}^2(1-y)K^{*2} \qquad (6)$$

$$\delta_a = \frac{1-\varepsilon}{\varepsilon}\frac{R^2}{15D_a}\rho_p{}^2(1-y)K^{*2} \qquad (7) \qquad\qquad \delta_i = \frac{1-\varepsilon}{\varepsilon}\frac{\rho_p(1-y)K^*}{15D}\bar{a}^2\exp(\sigma^2) \qquad (8)$$

where K^*=apparent adsorption equilibrium constant, Ez=axial dispersion coefficient based on void spaces in the bed, k_f=external mass transfer coefficient, D_a=diffusivity in macropores and D_i=diffusivity in micropores based on amount adsorbed gradient driving force. The arithmetic average radius $\bar{a}$ is 8.2 μm (Chihara, Suzuki and Kawazoe 1978). Equation(3) was used along with the experimental μ_1 to obtain the adsorption constant K^*. Equation (4) to (8) were used along with the experimental μ_2' to obtain the diffusivity in micropores D.

Stop & Go Simulation

Numerical solution for multi component chromatogram in time domain could be obtained by appropriate model equations with experimental conditions. This simulated chromatogram can be compared with experimental chromatogram to determine the equilibrium and the adsorption kinetic parameters. Here Markham-Benton equation as for adsorption equilibrium and linear driving force (LDF) approximation as for adsorption kinetics were adapted for numerical calculation, which was based on stop & go method (Chihara et al. 1986, Chihara and Kondo 1986). In particular, LDF model of adsorption kinetics was based on non-equilibrium thermodynamics.

For binary adsorbates, adsorption rate equations are

$$\gamma \frac{\partial q_1}{\partial t} = Ksav_{1,1}(q_1 * - q_1) + Ksav_{1,2}(q_2 * - q_2) \quad (9) \qquad \gamma \frac{\partial q_2}{\partial t} = Ksav_{2,1}(q_1 * - q_1) + Ksav_{2,2}(q_2 * - q_2) \quad (10)$$

where $Ksav$= Overall mass transfer coefficients.

Overall mass transfer coefficients ($Ksav$) for LDF model were determined. Then, micropore diffusivities were obtained by subtracting other mass transfer effects from overall resistance ($\gamma/Ksav$). Thus obtained micropore diffusivities were correlated with chemical potential driving force by consideration of Fick's diffusion equation, non-equilibrium thermodynamics and extended Langmuir equation (Karger and Bulows 1975).

$$\frac{D_{11}}{a^2} = \frac{D_1'}{a^2}\frac{\partial \ln p_1}{\partial \ln q_1} = \frac{D_1'}{a^2}\frac{1-\theta_2}{1-\theta_1-\theta_2} \quad (11) \qquad \frac{D_{12}}{a^2} = \frac{D_1'}{a^2}\frac{q_1}{q_2}\frac{\partial \ln p_1}{\partial \ln q_2} = \frac{D_1'}{a^2}\frac{\theta_1}{1-\theta_1-\theta_2} \quad (12)$$

$$\frac{D_{21}}{a^2} = \frac{D_2'}{a^2}\frac{q_2}{q_1}\frac{\partial \ln p_2}{\partial \ln q_1} = \frac{D_2'}{a^2}\frac{\theta_2}{1-\theta_1-\theta_2} \quad (13) \qquad \frac{D_{22}}{a^2} = \frac{D_2'}{a^2}\frac{\partial \ln p_2}{\partial \ln q_2} = \frac{D_2'}{a^2}\frac{1-\theta_1}{1-\theta_1-\theta_2} \quad (14)$$

For instance, $Ksav$ is related to D in single adsorbate case as

$$\frac{1}{Ksav} = \frac{a^2}{15D} + K^*\left(\frac{R}{3k_f} + \frac{R^2}{15\varepsilon D_a}\right) \quad (15)$$

For binary case, when using Equation (15), apparent K^* could be determined by appropriate slope of adsorption isothermal plane at perturbation point.

Result and Discussion

Single Adsorbate Carrier Mixed with He and The Same Adosorbate Pulse

Figure 2. shows an example of comparison of experimental chromatogram with simulated chromatogram for MSC5A to obtain *Ksav* for LDF model. Experimental conditions were 313 K, column pressure 5atm, flow rate 25cm/sec and He + CO_2 mixed gas carrier with CO_2 pulse. Here CO_2 concentration in the carrier gas was changed 20,40 , 60 and 80 %.

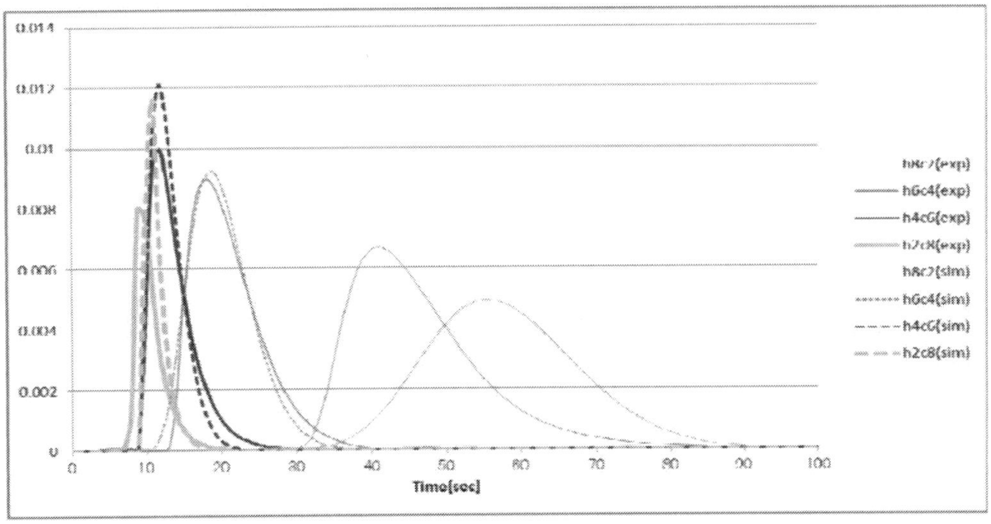

Figure2. Comparison of experimental peaks with simulation

Figure 3 shows the dependency of micropore diffusivity coefficient of CO_2 and C_2H_4 with amount adsorbed at 303, 313 and 323 K for MSC5A.

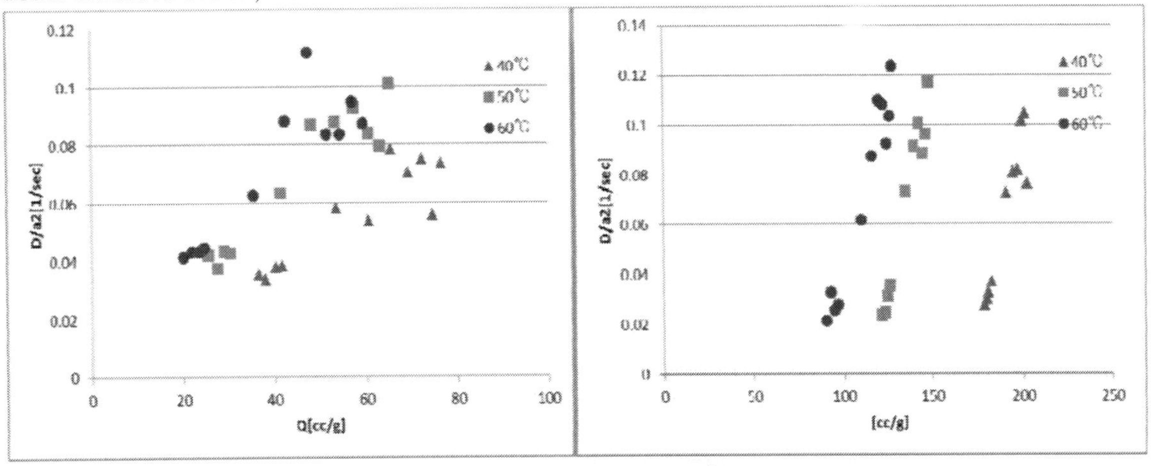

(a) (b)

Figure 3. (a)Dependency of micropore diffusivity coefficient on the amount adsorbed (CO_2)
(b) Dependency of micropore diffusivity coefficient on the amount adsorbed (C_2H_4)

Figure 4. shows the correlation of micropore diffusivity of CO_2 with chemical potential driving force at 303, 313 and 323K for MSC5A. Proportional relation was obtained, which means that micropore diffusion is based on chemical potential driving force.

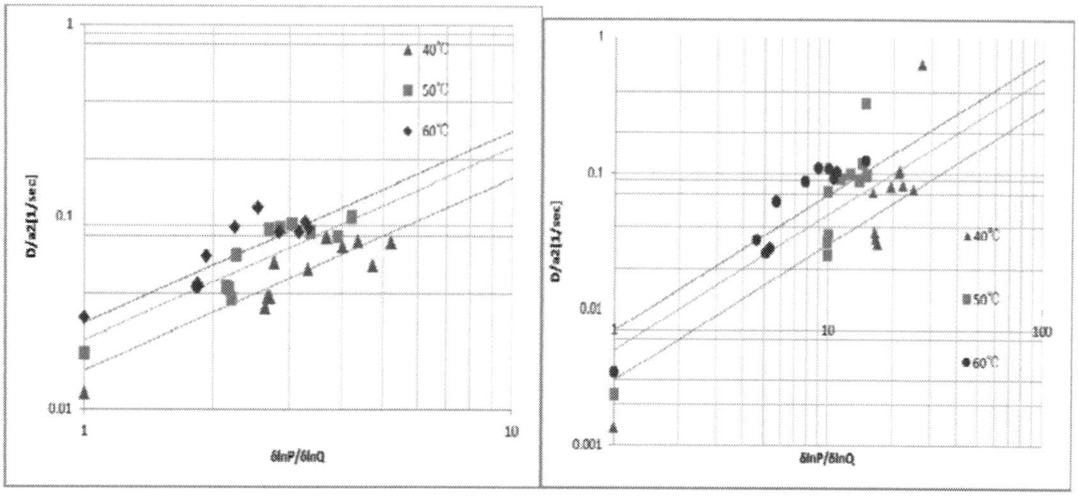

(a) (b)

Figure 4. (a) Correlation based on chemical potential driving force (CO_2)

(b) Correlation based on chemical potential driving force (C_2H_4)

Figure 5. shows Activation energy and the heat of adorption (CO_2 and C_2H_4).

		CO_2		C_2H_4	
		[kcal/mol]	[kJ/mol]	[kcal/mol]	[kJ/mol]
heat of adsorption	Adsorption equilibrium constant(10:0) used	8.0	33.5	8.8	37
	Langmuir constant used	6.9	29.0	11	44.5
activatation energy	D_0/a^2 used	6.2	25.8	9.7	40.4
	D'/a^2 used	9.4	39.4	8.8	36.8

Binary Adsorbate Carrier Mixed with He and Adsobate Pulse

Figure 5. show experimental and simulation results in an example case of binary adsorbate carrier mixed with He and an adsorbate pulse for MSC5A. Experimental conditions were 313 K, column pressure 5 atm, flow rate 25 cm/sec and He (10%) + CO_2 (30%) + C_2H_4 (60%) mixed gas carrier with C_2H_4 pulse. Figure6. is the comparison between experimental and simulated TCD peaks. Two simulated TCD peaks are those for accounted and not accounted for cross effect.

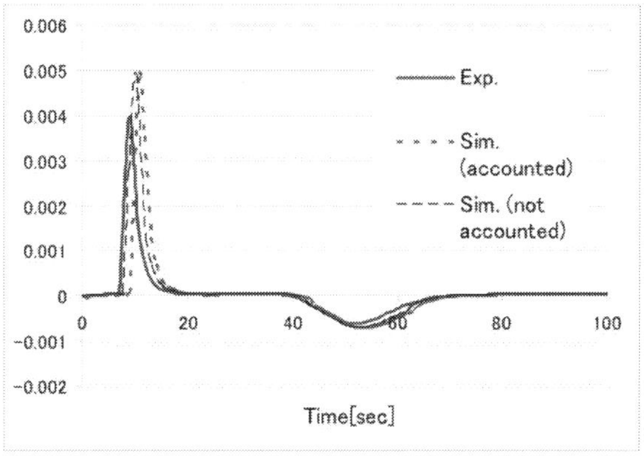

Figure 5. He (10%) + CO_2 (30%) + C_2H_4 (60%) mixed gas carrier with C_2H_4 pulse

Cross effect of micropore diffusion in multicomponent adsorption

Figure6. shows comparison with experiment and simulation about width of 1st and 2nd peak. In this study we couldn't get the significant between accounted and not accounted, but we could get the difference in all conditions.

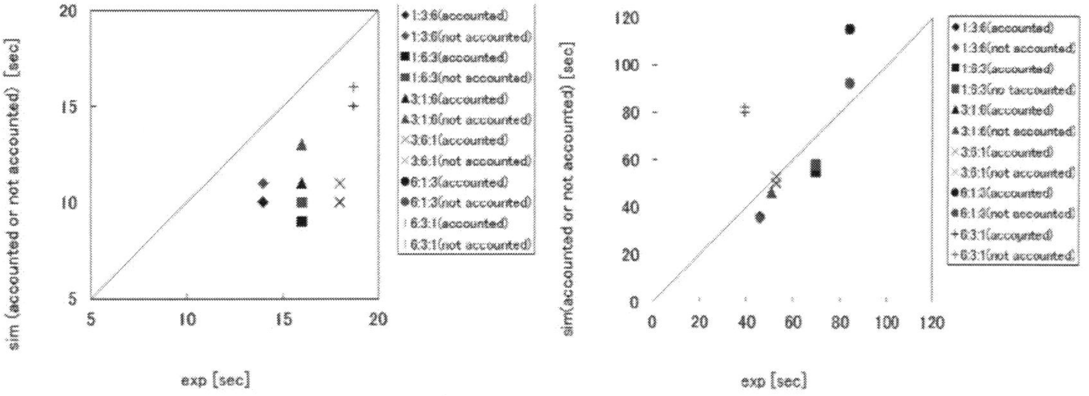

(a)

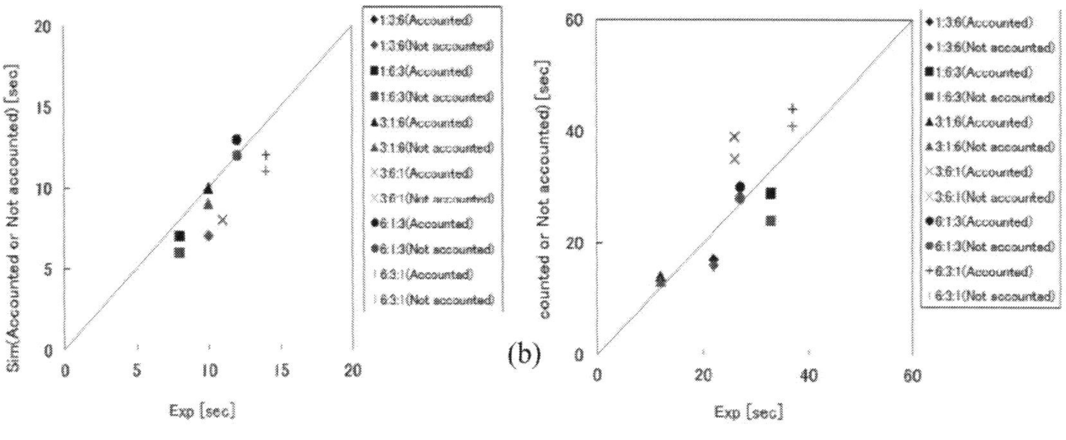

(b)

Figure 6. Comparison with experiment and simulation about width of 1st (Left)

(a) He + CO_2 + C_2H_4 - C_2H_4, (b) He + CO_2 + CH_4 - CH_4

Conclusion

Good agreements between experimental chromatogram and simulated chromatogram, which were based on the modeling of Stop & Go method, were observed in case of peturbation chromatography with mixed adsorbate gas carrier. And micropore diffusivities obtained were interpreted by chemical potential driving force consideration based on non-equilibrium thermodynamics law. Some possibility was shown for cross effect of micropore diffusion in multicomponent adsorption.

References

Chihara, K., Suzuki M. and Kawazoe K (1978) ; "Adsorption Rate on Molecular Sieving Carbon by Chromatography" *AIChE Journal Vol.24, 237*

Ruthven D.M. and Kumar R. (1979) ; "A Chromatographic Study of the Diffusion of N_2, CH_4 and Binary CH_4-N_2 Mixtures in 4A Molecular Sieve" *The Canadian Journal of Chemical Engineering, 57, 342*

Kazuyuki Chihara, Tatsuro Kinoda, Hidenori Nakamura and Yosuke Kaneko (2006); "Chromatographic Study of Multicomponent Gas Adsorption on MSC5A"*AIChE Annual Meeting*, San Francisco

Kazuyuki Chihara, Yuki Teramura, Hidenori Nakamura and Yosuke Kaneko(2008); "Multicomponent Gas Adsorption on MSC5A by Chromatographic Method and Stop&Go Simulation" *AIChE Annual Meeting*, Philadelphia

Kazuyuki Chihara, Yuk Teramura, Hidenori Nakamura and Yosuke Kaneko(2009);"Simulation Study of Multi-Component Gas Adsorption by Chromatographic Method and Stop&Go Simulation" *AIChE Annual Meeting*, Nashville

Ruthven D.M, (1984); "Principle of Adsorption and Adsorption Process" Wiley, 146 f

Tondeur D., Kabir H., Luo L., and Granger J. (1996); "Multicomponent adsorption equilibria from impulse response chromatography" *Chem. Eng. Science, 51(N15)* 3781-3799

Kumar R., Duncan R.C. and Ruthven D.M (1982).; "A Chromatographic Study of Diffusion of Single Components and Binary Mixtures of Gases in 4A and 5A Zeolites" *The Canadian Journal of Chemical Engineering vol.60, 342*

Chihara K. et al; "Simulation of Pressure Swing Adsorption for Air Separation" *Proc. 7th Int. zeolite Conf.*, 563(1986)

Chihara K. and Kondo A(1986).; "Simulation of Pressure Swing Adsorption –Three Gas Components and Three Adsorption Columns-" *2nd FOA,* 165

Karger J. and Bulows M(1975).; *Chem. Eng. Sci., 30, 893*

SIMULATION STUDY OF MULTI-COMPONENT GAS ADSORPTION ON MSC3A BY VOLUMETRIC TECHNIQUE METHOD

Kazuyuki Chihara, Masashi Nomoto, Yuzo Amari, Yuki Teramura and Yosuke Kaneko*
Meiji University Kawasaki, Japan

Introduction

MSC (Molecular sieving carbon) has micropores. In this study, the adsorption experiment was conducted by the volumetric technique. MSC 3A (KURARAY CHEMICAL CO., LTD) was used as adsorbent at 303K, 313K and 323K. Experimental adsorption of binary mixtures was compared with the simulation using the data obtained in equilibria and rates of adsorption of the single component. The characteristic value of MSC was shown in Table 1.

Table 1. MSC characteristic

Particle density [g/cm^3]	0.9
True density [g/cm^3]	1.9
Macropore volume [cm^3/g]	0.21
Micropore volume [cm^3/g]	0.13
Micropore opening [nm]	0.4

Experimental

The experimental apparatus was shown in Figure 1. The column was regenerated in a vacuum at 300℃ for 2 hours. The column was stabilized to experiment temperature with the PID control device. Gases were introduced after reaching the experimental temperature. At first, adsorbate gases was introduced to the gas mixing loop up to 1atm. The four-way valve was opened to let gas flow in the adsorbent column. Circulate pump was always operated. Change of the pressure in a column was recorded by the pressure sensor and chart recorder until the gases equilibrated. After equilibration, the pressure of the gases in the mixing loop was raised to 2atm and 3atm, and the same operation was repeated until equilibrated pressure exceeded 5atm. For the binary gas mixture experiment, while recording pressure change, composition of the gas mixture in a column was obtained by gas chromatography.

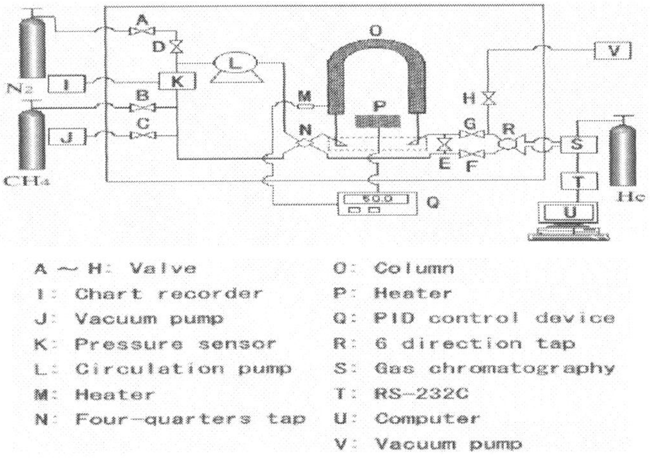

Figure 1. Experimantal apparatus

Analysis

In case of pure conponent pressure decay, theoretical curves which count for both intracrystalline diffusion and masstransfer theory pore month can be applied for curve fitting as Fig. 4. Numerical solution for multi component pressure decay in time domain could be obtained by appropriate model equations with experimental conditions. This simulated decay can be compared with experimental decay to determine the equilibrium and the adsorption kinetic parameters. Here Markham-Benton equation as for adsorption equilibrium and linear driving force (LDF) approximation as for adsorption kinetics were adapted for numerical calculation, which was based on stop & go method (Chihara et al. 1986, Chihara and Kondo 1986). In particular, LDF model of adsorption kinetics was based on non-equilibrium thermodynamics.

For binary adsorbates, adsorption rate equations are

$$\gamma \frac{\partial q_1}{\partial t} = Ksav_{1,1}(q_1 * - q_1) + Ksav_{1,2}(q_2 * - q_2) \qquad (1)$$

$$\gamma \frac{\partial q_2}{\partial t} = Ksav_{2,1}(q_1 * - q_1) + Ksav_{2,2}(q_2 * - q_2) \qquad (2)$$

where *Ksav*= Overall mass transfer coefficients.

Overall mass transfer coefficients (*Ksav*) for LDF model were determined. Then, micropore diffusivities were obtained by subtracting other mass transfer effects from overall resistance (γ/Ksav). Thus obtained micropore diffusivities were correlated with chemical potential driving force by consideration of Fick's diffusion equation and non-equilibrium thermodynamics (Karger and Bulows 1975).

$$\frac{D_{11}}{a^2} = \frac{D_1'}{a^2}\frac{\delta \ln p_1}{\delta \ln q_1} = \frac{D_1'}{a^2}\frac{1-\theta_2}{1-\theta_1-\theta_2} \qquad (3)$$

$$\frac{D_{12}}{a^2} = \frac{D_1'}{a^2}\frac{q_2}{q_1}\frac{\delta \ln p_1}{\delta \ln q_2} = \frac{D_1'}{a^2}\frac{\theta_1}{1-\theta_1-\theta_2} \qquad (4)$$

$$\frac{D_{21}}{a^2} = \frac{D_2'}{a^2}\frac{q_2}{q_1}\frac{\delta \ln p_2}{\delta \ln q_1} = \frac{D_2'}{a^2}\frac{\theta_2}{1-\theta_1-\theta_2} \qquad (5)$$

$$\frac{D_{22}}{a^2} = \frac{D_2'}{a^2}\frac{\delta \ln p_2}{\delta \ln q_1} = \frac{D_2'}{a^2}\frac{1-\theta_1}{1-\theta_1-\theta_2} \qquad (6)$$

For instance, *Ksav* is related to D in single adsorbate case as

$$\frac{1}{Ksav} = \frac{a^2}{15D} + K^*\left(\frac{R}{3k_f} + \frac{R^2}{15\varepsilon D_a}\right) \qquad (7)$$

For binary case, when using Equation (7), apparent K^* could be determined by appropriate slope of adsorption isothermal plane at perturbation point. D_{11} and D_{22} could be obtained from pure component experiment. Then D_{12} and D_{21} could be obtained by Eq. (4) and (5) with D_1' and D_2'.

Results & discussion

The adsorption isotherm

The adsorption isotherm was shown in Figure 2-(a), (b), (c). It was confirm that CH_4 has amounts adsorbed twice compared with N_2. And it was confirmed that CO2 has amounts adsorbed compared with CH_4. Experiment data was compared with the Langmuir and Toth equation. CH_4, CO_2 and N_2 were well correlated with both theoretical equations.

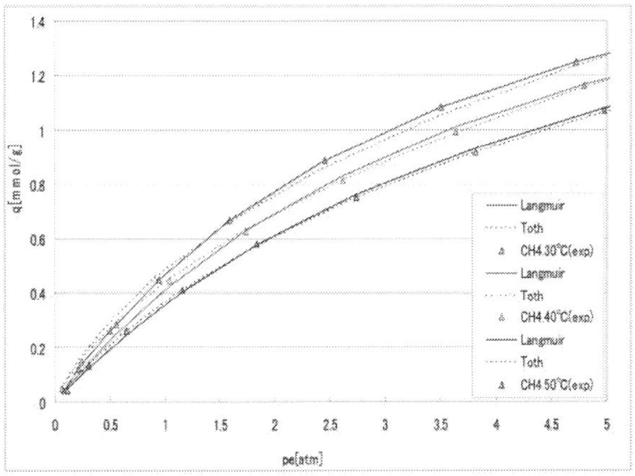

(a)

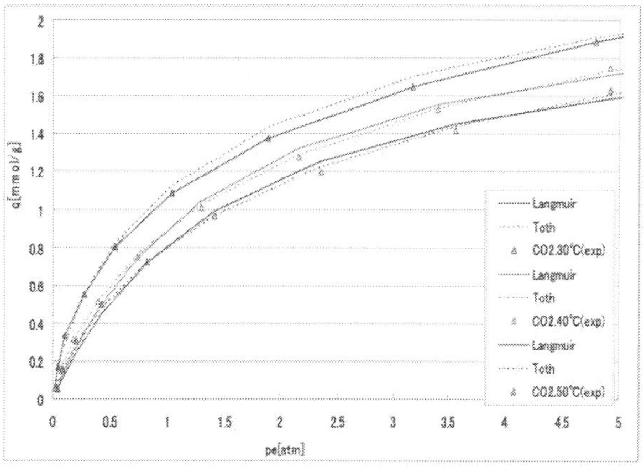

(b)

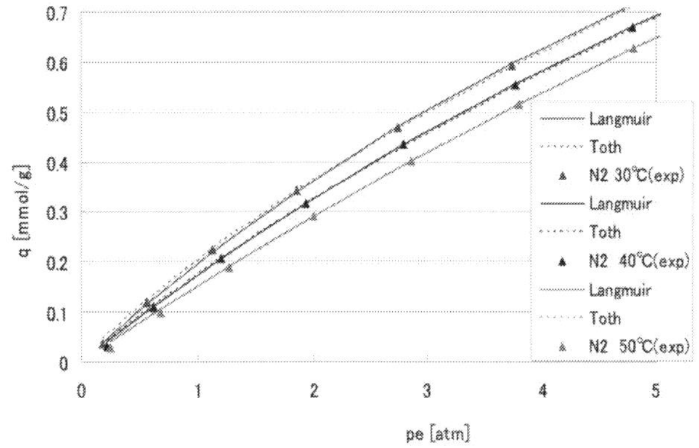

(c)

Figure 2. (a) Adsorption isotherm (CH_4), (b) Adsorption isotherm (CO_2) (c) Adsorption isotherm (N_2)

The value of the parameter of Langmuir and Toth equation was shown in Table 2.

Table 2. Parameter

	Gas	CO_2			CH_4			N_2		
	カラム温度 [K]	303	313	323	303	313	323	303	313	323
Langmuir	K [1/atm]	0.2711	0.2381	0.0567	0.2651	0.2227	0.1938	0.0771	0.0648	0.0567
	Q_∞ [mmol/g]	2.209	2.099	3.148	2.248	2.258	2.196	3.2216	3.2144	3.1476
Toth	K [1/atm]	0.4659	0.4606	0.0477	0.4659	0.4721	0.2895	0.105983	0.054544	0.0477
	Q_∞ [mmol/g]	6.026	9.933	3.601	6.026	9.933	3.291	4.89394	3.674903	3.6014
	t [-]	0.455	0.365	1.006	0.455	0.365	0.681	0.767523	0.998434	1.0064

Kinetics for single component

The pressure decay of the column was shown in Figure 3-(a), (b), (c). It took time until it reaches adsorption equilibria, so that pressure was high. Time spent about 5 times for CH_4 to adsorption rather than N_2. This is considered to be based on the characteristic of MSC. As for CH_4 with the large diameter of a molecule, adsorption speed becomes slow.

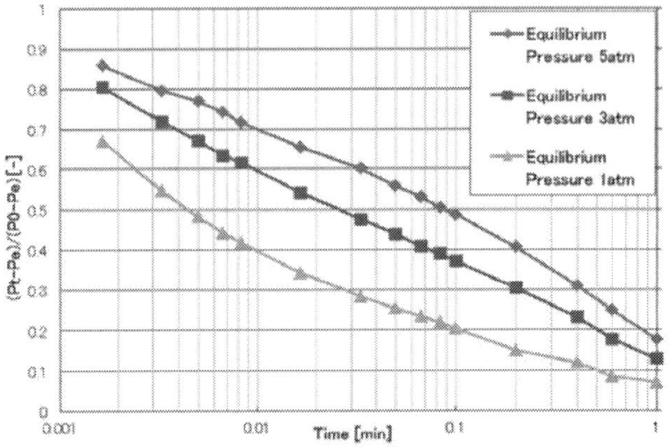

(a)

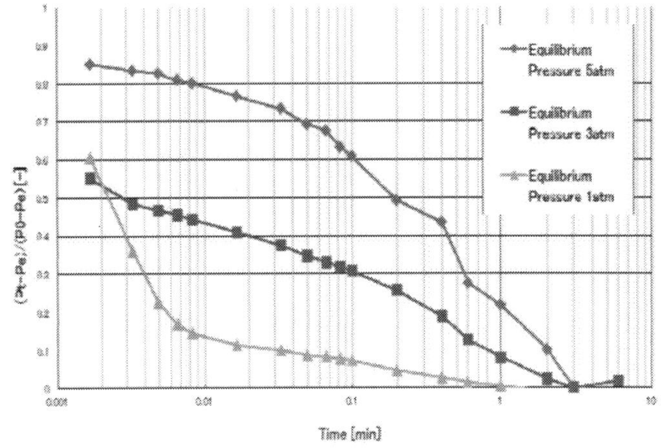

(b)

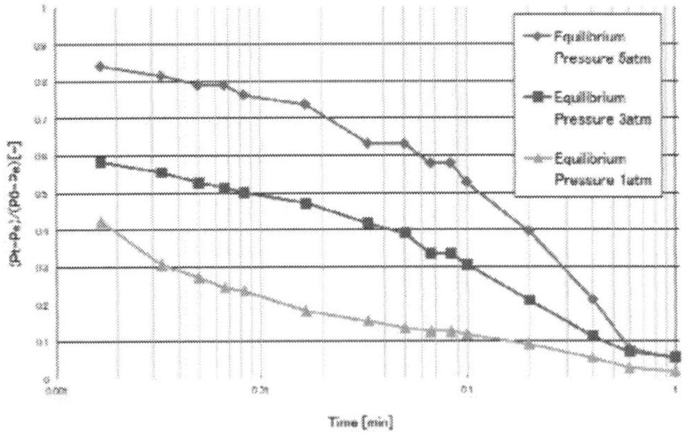

(c)

Figure 3. (a) Pressure decay (CH_4), (b) Pressure decay (CO_2), (c) Pressure decay (N_2) on the course of adsorption

Kinetics of single component was analyzed by fitting on the course of adsorption the pressure decay with the theoretical curves, and then D/a^2, $K_f a_p$ and $K_s a_p$ were calculated. The graph of fitting with theoretical curves was shown in Figure 4. The dependency on the amount of adsorbed of $K_s a_p$ was shown in Figure 5. The chemical potential was shown in Fig.6-(a), (b).

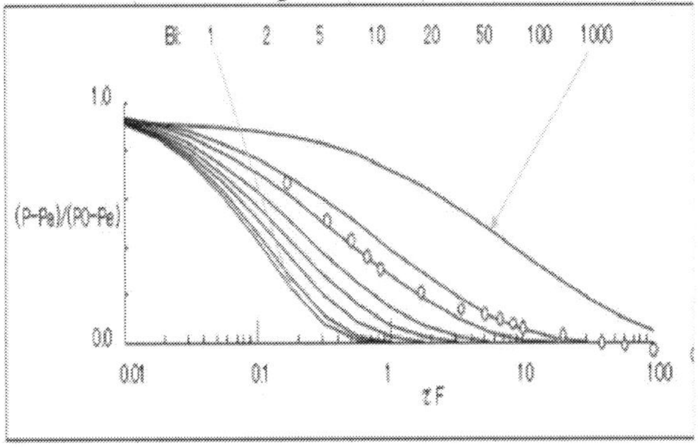

Figure 4. Curve fitting (CH_4 313K P_0=1 atm)

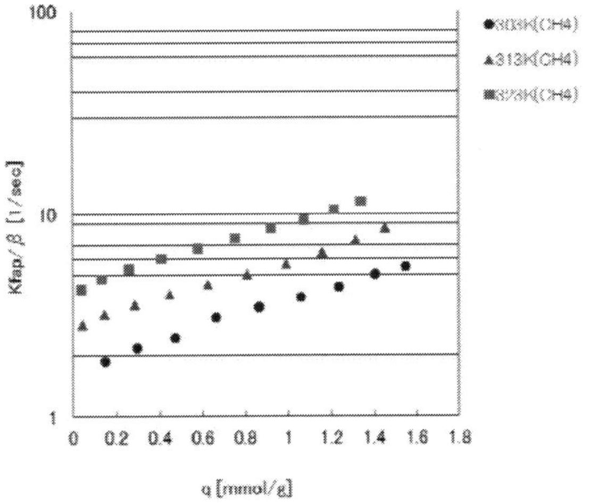

(a)

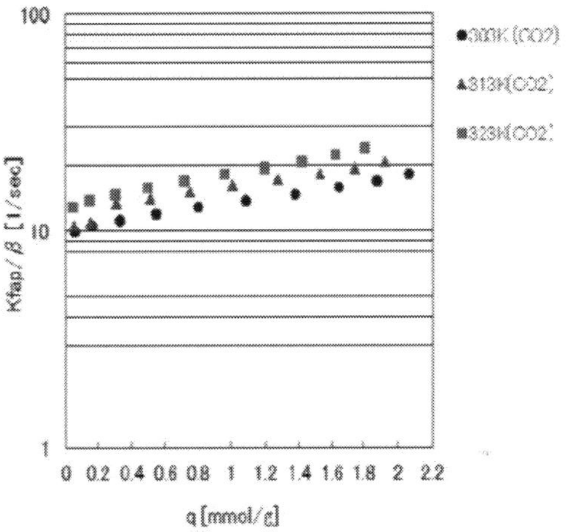

(b)

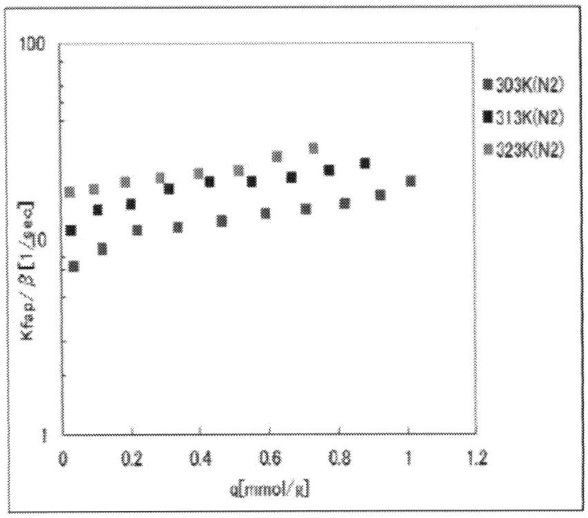

(c)

Fig.5 (a)Dependency $K_s a_p$ on amount adsorbed of CH_4

(b)Dependency $K_s a_p$ on amount adsorbed of CO_2

(c)Dependency $K_s a_p$ on amount adsorbed of N_2

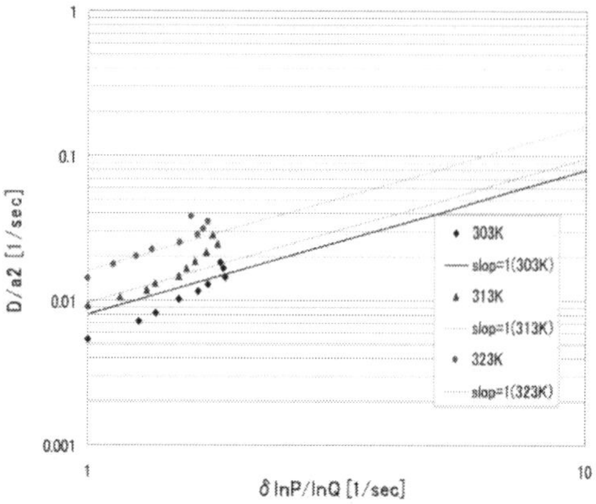

(a)Chemical potential(CH_4)

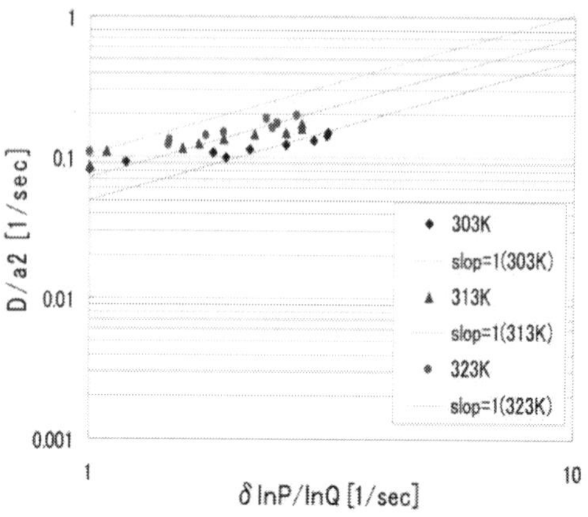

(b) Chemical potential(CO_2)

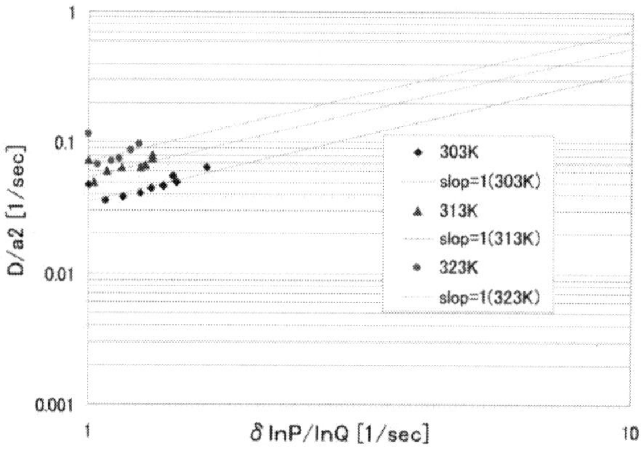

(c) Chemical potential(N_2)

Equilibria for binary components

Example of experimental binary adsorption data was shown in Figure 7 and 8. The data were compared with three kinds of theoretical calculation. The distinguishment of these three curves are difficult here.

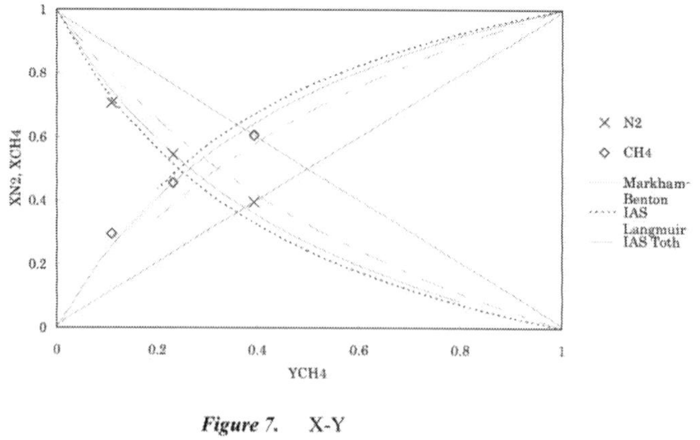

Figure 7. X-Y

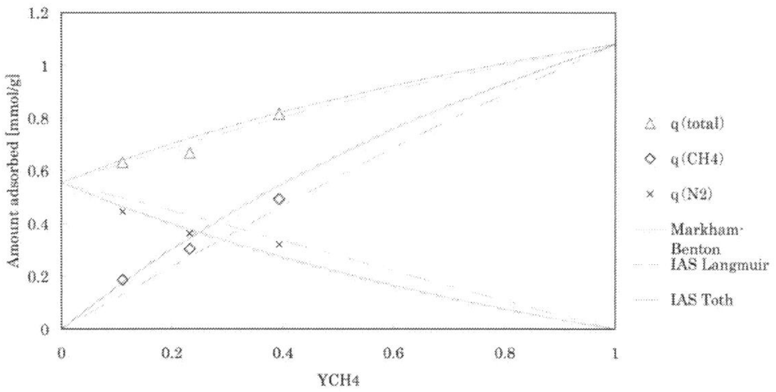

Figure 8. Adsorption equilibrium of N_2+CH_4

Kinetics for binary components

In Figure 9, the comparison of pressure decay between experimental data of volumetric method and simulated data. In conclusion of comparison of experiment and simulation, when inserted pressure was low, both datas were fitted. But when inserted pressure was high, experimental data lead to a minor difference in simulation.

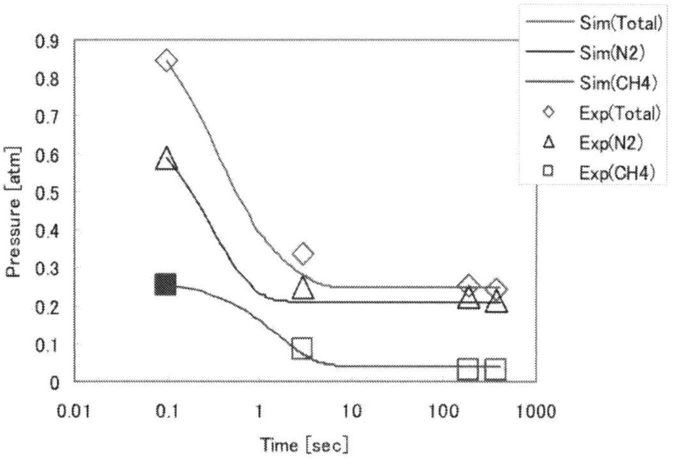

Figure 9. the comparison of pressure decay between experimental date of volumetric method and simulated data.

Conclusion

With the single gas experiment data, binary equilibrium was estimated and the pressure decay and the increase of the amount adsorbed of the binary gas mixture adsorption were simulated.

(395i) Metal-Organic Framework – Metal Oxide Nanoparticle Composites for Gas Adsorption

Katrina A Stults, Chemical & Biomolecular Engineering, Georgia Institute of Technology, Atlanta, GA and Krista S. Walton, Chemical and Biomolecular Engineering, Georgia Institute of Technology, Atlanta, GA

Metal-organic frameworks, highly porous materials consisting of metal nodes connected via organic linkers, are promising materials for a wide range of adsorption applications. Composite materials consisting of metal oxide nanoparticles included within metal-organic framework (MOF) particles were formed to increase the type and density of active sites. These composites were formed by synthesizing Cu-BTC around pre-formed metal oxide nanoparticles much larger than the pores of the MOF, including titanium dioxide and magnesium oxide. The Cu-BTC+MO composites were characterized to determine nanoparticle inclusion and changes to the MOF's pore system. Weight-normalized properties, such as surface area and pore volume, were reduced with the inclusion of the heavier nanoparticles but the contribution to the total pore volume from the smallest pores, attributed to the Cu-BTC pore system, is comparable for samples synthesized under similar conditions with and without nanoparticle incorporation. Static and dynamic adsorption experiments were conducted for environmentally and industrially relevant gases under dry and humid conditions.

(395j) Molecular Simulation of the CO_2 Deactivation Volume in Natural Gas Storage Vessels

Sebastiao M. P. Lucena, Dept. Eng. Quimica, Universidade Federal do Ceará, Fortaleza, CE, Brazil, Daniel Gonçalves, UFC, Fortaleza, Brazil and **Célio L. Cavalcante Jr.**, Chemical Engineering, Universidade Federal do Ceará, Fortaleza, Brazil

Molecular simulations were performed to evaluate the influence of CO_2 in the deactivated volume of storage vessels used in the adsorbed natural gas (ANG) technology. We compare the performance of metal-organic framework Cu-BTC and a hypothetical carbon material. The adsorption of carbon dioxide in the discharge pressure and room temperature (1 atm, 25 ° C) impairs the efficiency of charge and discharge cycles of the natural gas. Data for the adsorbing mixture CH_4/CO_2 were obtained for the charge and discharge pressures. Methane /CO_2 selectivity calculations were performed to compare the adsorbents Cu-BTC and the activated carbon. The metal-organic framework Cu-BTC presented similar CH_4 loading to the hypothetical activated carbon. Although Cu-BTC presented a higher CO_2 loading, this quantity is unable to effectively interfere with the charge/ discharge cycle efficiency. Considering that our activated carbon is a hypothetical material, we concluded that metal-organic framework Cu-BTC is a very promising adsorbent for ANG technology.

(395k) Standing Wave Analysis for 4-Zone SMB with Thermal Gradients

Nicholas Soepriatna, Phillip C. Wankat and Nien-Hwa Linda Wang, School of Chemical Engineering, Purdue University, West Lafayette, IN

The isothermal simulated moving bed (SMB) for binary separations has been shown to produce high throughput of high purity product. An inherent limitation of this process is its high desorbent consumption, which leads to high energy consumption in the separation device that recovers desorbent. In order to overcome this limitation, temperature swing adsorption (TSA) or thermal gradient have been integrated with the SMB process (thermal SMB)[1-3]. The thermal gradient process is operated at high temperature for adsorbent regeneration, and at cold temperature for desorbent regeneration. This results in a lower desorbent to feed (D/F) flow rate ratio and subsequently reduces energy consumption in the entire system. The optimal design of thermal SMB processes is considered a major challenge because the current initial design method (local equilibrium theory) only provides a region of acceptable design parameters, thus, requiring many simulation trials. In this study, Standing Wave Analysis (SWA), developed originally for 4-zone isothermal SMB fractionators, is extended to SMB systems with thermal gradients, and then is used to locate the optimal operating conditions for a 4-zone thermal SMB fractionation. The accuracy of SWA was validated by Aspen Chromatography simulations for toluene/p-xylene separation with silica gel as the adsorbent and n-heptane as the desorbent with two columns per zone. The system was operated in the temperature range of 0–80°C and it was shown that SWA provides very good prediction of optimal operating conditions.

References

1. Migliorini C, Wendlinger M, Mazzotti M. Temperature gradient operation of a simulated moving bed unit. *Ind Eng Chem Res*. 2001;40:2606–17.
2. Kim JK, Abunasser N, Wankat PC. Thermally Assisted Simulated Moving Bed Systems. *Adsorption*. 2005;11:579–584.
3. Jin W, Wankat PC. Thermal Operation of Four-Zone Simulated Moving Beds. *Ind Eng Chem Res*. 2007;46:7208–7220.

(395l) Development of Cathode Air Filters for PEM Fuel Cells Using Microfibrous Entrapped Sorbents

Abhijeet Phalle and Bruce Tatarchuk, Department of Chemical Engineering, Auburn University, Auburn, AL

The performance of Proton Exchange Membrane Fuel Cell (PEMFC) is degraded significantly as a result of poisoning of cathode catalyst by airborne contaminants, in particular SO_2 and NO_x in trace levels. Removal of air contaminants by adsorptive filtration is economical and easier way to mitigate cathode air contamination .This work thus aims at designing of adsorptive air filters having high adsorption capacity and high contacting efficiency for removal of SO_2 and NO_2 along with lower pressure drop across a filter. MnO_x/Al_2O_3 prepared by deposition precipitation route with $MnSO_4$ as a precursor showed the highest saturation (124.1 mg of/cc) and breakthrough capacity (33.7 mg/cc) for SO_2. Particularly, it showed 60% higher saturation capacity than commercial $KMnO_4/Al_2O_3$ sorbent. Granular PICA activated carbon was found to have the highest breakthrough capacity of 62.3 mg/cc and saturation capacity of 139.5 mg/cc for NO_2. Thus, two layers of filter media are required to efficiently remove both SO_2 and NOx.

For packed beds, use of smaller size particles give high single pass removal efficiency but generally at the expense of higher pressure drop, which increases parasitic power loss. Microfibrous media (MFM) prepared by entrapping micron size particles in sinter-locked network of bicomponent polymer fibers have been shown to give higher contacting efficiency with the lower pressure drop. Different filter configurations such as composite bed, pleated filters can be prepared with the use of MFM and their comparison with the conventional packed bed will be discussed in detail in order to find the optimal design of a filter. In one such study, the composite bed consisting of packed bed of MnO_x/Al_2O_3 (1.6cm thick) followed by thin polisher layer of microfibrous media (4mm) yielded ~200% higher breakthrough capacity for SO_2 than that for the packed bed of same sorbent.

(395m) Evaluation of Dry Sorbents On Sorption-Enhanced Water Gas Shift Reaction (SEWGS) in Coal Gasification Systems

Hong Lu[1], Yongqi Lu[1] and Massoud Rostam-Abadi[2], (1)Prairie Research Institute, University of Illinois at Urbana-Champaign, Champaign, IL, (2)Prairie Research Institute , University of Illinois at Urbana-Champaign, Champaign, IL

The University of Illinois at Urbana-Champaign and URS Group, Inc. are developing a sorption-enhanced water-gas shift (SEWGS) technology that combines water-gas-shift (WGS) reaction with CO_2 capture using dry sorbents for coal gasification systems. The key to the success of the SEWGS technology is to identify and develop highly efficient, highly stable sorbents with properties suitable for CO_2 capture at high pressure and high temperature syngas conditions. An ideal SEWGS sorbent will improve WGS efficiency, achieve high capacity of CO_2adsorption at high temperature, be regenerated at high pressure, and maintain high multi-cycle stability.

Our thermodynamic studies identified several sorbent candidates that possess properties desirable for the SEWGS application. Selected candidate sorbents were synthesized using a flame spray pyrolysis method and characterized using different techniques. Multiple cyclic CO_2 adsorption and regeneration performance were evaluated for the sorbents using a TGA at atmosphere pressure. Several sorbents demonstrated high CO_2 adsorption capacity, fast kinetics, and multi-cyclic stability at temperatures ranging between 400 and 700 °C. The performance of selected sorbents is further being evaluated for the SEWGS reaction using a high temperature, high pressure (HTHPR) reactor. This paper will present the recent results of the sorbent synthesis and performance evaluation using the TGA and HTHPR.

(395n) Intracrystalline Diffusion of Linear Alkanes and Linear Alcohols in SAPO-34

Célio L. Cavalcante Jr., Chemical Engineering, Universidade Federal do Ceará, Fortaleza, Brazil, Julien Cousin saint Remi, Chemical Engineering, VUB (Vrije Universiteit Brussel), Brussels, Belgium, Joeri Denayer, Dept. Chemical Engineering, Vrije Universiteit Brussel, Brussel, Belgium and Gino Baron, Department of Chemical Engineering, Vrije Universiteit Brussel, Brussels, Belgium

SAPO-34 is a silicon-, aluminum-, and phosphorus-based molecular sieve with CHA topology. It has a three-dimensional pore system with ellipsoidal shaped cages of 6.7 x 10 $Å^2$ interconnected via 8-membered ring windows with pore apertures of 3.8 x 3.8 $Å^2$. As a result of its small pore size, SAPO-34 is a highly selective catalyst and has been applied for methanol-to-olefins and methanol-to-gasoline catalytic processes. It can also be applied for CO2 – N2– CH4 separations. Adsorption of linear alkanes and linear alcohols in liquid phase conditions have been reported by Denayer et al. (J.Phys. Chem. C, 2008, 2011), showing chain length dependency of adsorption properties in this type of material. In this study, we present experimental measurements for intracrystalline diffusivities at different temperatures of linear alkanes (C5-C10) and linear C1-C8 alcohols in SAPO-34 samples using the ZLC (Zero-Length Column) method originally proposed by Eic and Ruthven (Zeolites, 1988). Activation energies are also reported for those systems.

(395o) Removal of Nickel Ions From Industrial Wastewater Using Sericite

Choong Jeon and Taik-Nam Kwon, Gangneung-Wonju National University, Gangneung, South Korea

The applicability of sericite for Ni(II) removal and recovery process in actual industrial wastewater which has Na(I) was studied. The removal efficiency for NI(II) could not be affected by Na(I) which generate ionic strength. The highest removal efficiency of sericite for Ni(II) was approached about 93% at pH 7.5 and silanol (SiO_2) and aluminol (Al_2O_3) groups are likely to play important role in adsorption process. Sericite had a high Ni (II) adsorption capacity of about 44mg/g at pH 7.5. All adsorption process was completed in 120min and removal efficiency of sericite was higher than that of Amberlite IR 120 plus resin. The effect of temperature on the removal efficiency of Ni(II) was negligible. In addition, previously adsorbed Ni(II) onto sericite could be easily and completely eluted by HNO_3. Therefore, it could be concluded that adsorption process using sericite can be applied to the recovery system for Ni(II) in actual wastewater treatment system and furthermore, the technique could sufficiently replace conventional treatment process such as solvent extraction and ion exchange resin.

SOLVENT RECOVERY BY PRESSURE SWING ADSORPTION(PSA)METHOD —INFLUENCE OF COEXISTENCE HUMIDITY—

Kazuyuki Chihara, Jun Kobayashi, Koki Aoshima, Takuya Chiyoda
Meiji University, Kawasaki, Japan

ABSTRACT

An organic solvent exhausted by the factory negatively affects the human body and the environment. The recovery and the removal are indispensable operations in a chemical industrial process. As a method of removing the organic solvent, adsorption in the chemical engineering field is used. There is pressure swing adsorption (PSA) method using the technique; An operation pressure is alternately changed with the high pressure and the low pressure. It is a method using the difference of the amount of adsorption **[1]**. Water vapor obstructs the adsorption of an adsorbate and results in gradual deterioration of an absorbent. It is necessary to examine how water vapor affects the organic solvent recovery. In the present study, water vapor content in the mixed gas was changed and the behavior of the concentration of the solvent in the product gas was examined with stripping reflux PSA unit like figure 1. CH_2Cl_2 and water vapor were chosen as an adsorbate. Silica gel (CARiACT Q-6) was chosen as an adsorbent. The recovery experiment of the organic solvent was conducted by PSA method. As a result, the concentration of the solvent in product gas increased when the ratio of water vapor increased as in figure 2. It is clear that water vapor has influences on the adsorption of CH_2Cl_2.

INTRODUCTION

Industrial case of the operation in PSA is air separation, CO separation, helium purification, carbon dioxide removal and so on. Recently, PSA method is used as the method of low-concentration organic-solvent recovery or removal. Water vapor obstructs the adsorption of an adsorbate and results in gradual deterioration of an absorbent. It is necessary to examine how water vapor affects the organic solvent recovery. In this sturdy, we investigate an influence which the water vapor has on PSA method. Please use this example paper as your primary reference in completing your paper. Your assistance in helping follow the styles and guidelines outlined for the following parts of your paper will help your association publish a more "professional looking" publication in the most cost-effective and timely manner possible.

EXPERIMENTAL

Stripping reflux PSA unit is shown in figure 1. Each column filled with adsorbent and attached to the unit. Air passed through silica gel column and dried. After CH_2Cl_2 or water flowed stream is obtained by passage of the part of the air through each bubbling unit which filled up with CH_2Cl_2 or water, the stream are mixed with the remaining dry air. A flow of the mixture gas was set as 1000 ml/min using mass flow controller, and then the gas passed through the column after controlling organic-solvent concentration (8000ppm).Two cylindrical Pyrex glass pipes were used as each adsorption column (diameter:2.7cm, height:30 cm, and capacity:172 cm^3). Furthermore, a pressure of the adsorption process was adjusted to 2.0 atm by check valve. Operating pathway of the gas was regulated by switching electromagnetic valve which were set for automatic operation, and also a pressure of the desorption and purge process was adjusted to 0.4 atm by vacuum pump.On the above conditions, Stripping reflux PSA unit was worked. First of all, we measured each flow of each gas with a soap film flow instrument. Secondly, we measured the solvent concentration in each gas brought together in the gas pack with gas chromatography equipment until it became in a steady state. Table 1 showed conditions of a PSA experiment.

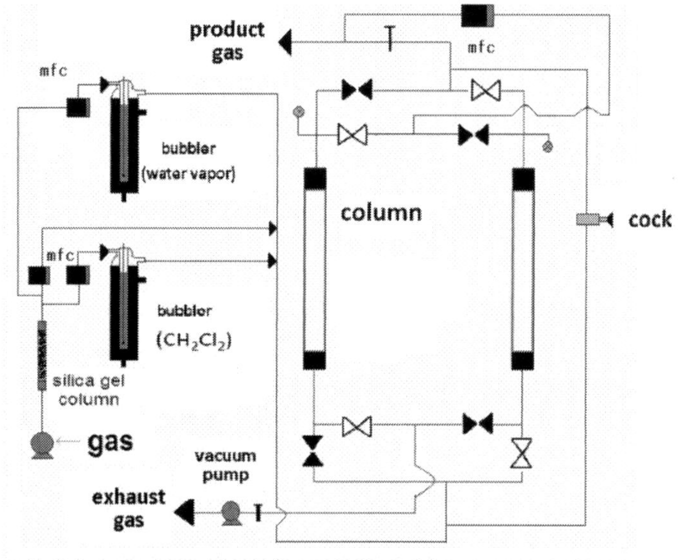

Fig. 1 Stripping reflux PSA unit

Table 1. Conditions of PSA experiment

Adsorbent	CARiACT Q-6(2.40-4.00mm)
Volume(column)	27.0mmΦ×300mm
Porosity	0.54
Solvent	CH2Cl2 and water vapor
Organic-solvent concentration	8000[ppm]
Supply gas mass flow	1000[ml/min]
Pressure of adsorption	2.0[atm]
Pressure of desorption	0.4[atm]
Ratio of purge	1
Relative humidity	0,50,100[%]
Cycle time	200

SIMULATION

Simulation was tried based on Stripping reflux PSA unit. Operation sequence is shown in Figure 2. It was consist of four sequence, pressurization step, adsorption step, evacuation step and purge step. We tried to reduce purge flow as much as possible by lowering the desorption pressure for maintenance cost reduction. Column conditions were shown in Table 2. Adsorbent conditions were shown in Table 3. Operation sequence times were shown Table 4.

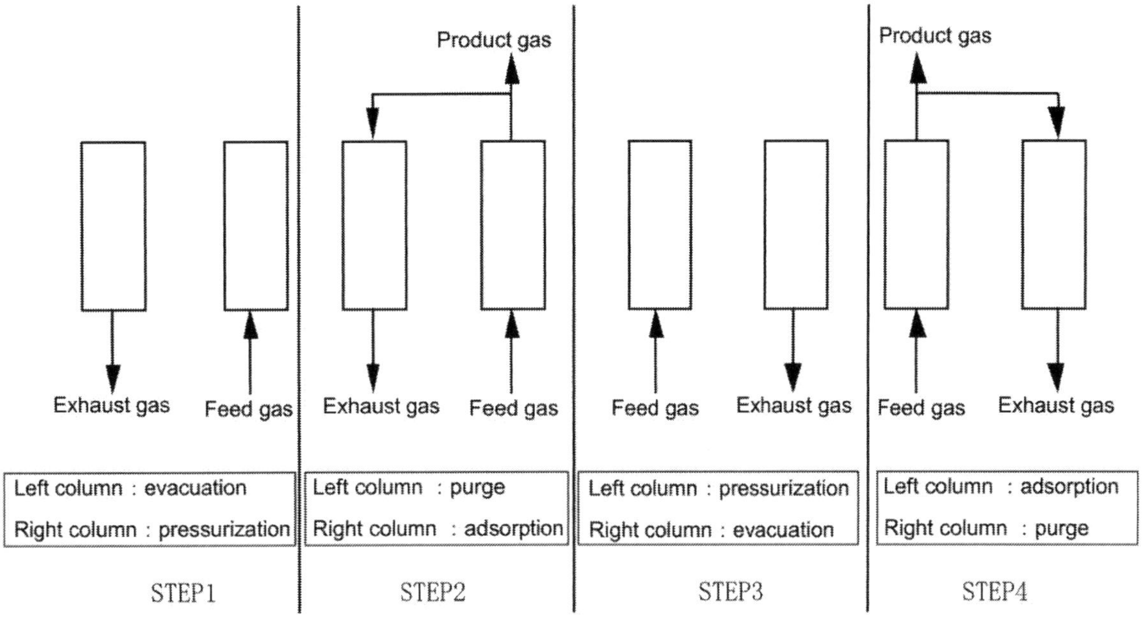

Fig. 2 Operation Sequence

Table 2. Column condition

radius [mm]	13.5
length [mm]	300
amount of adsorbent [g]	90
porosity [-]	0.54

Table 3. Adsorbate condition

Adsorbate	CH2Cl2	water vapor
q^{∞} [g/g]	0.06546	0.008170
γ/ksav [sec]	100	800

Table 4. Operation sequence time

Left column	Right column	Sequence time [sec]
pressurization	evacuation	20
adsorption	purge	160
evacuation	pressurization	20
purge	adsorption	160

Stop&Go Simulation

Stop & Go method was used as simulation method of Stripping reflux PSA. Fig. 3 showed Stop&Go method basic concept. The Cullum was split N piece toward the axial direction, and the column has cells with N. Stopping the flow at a certain time t, solvent has been adsorbing in each cell in minute time⊿t. The pressure within each cell was determined by a balance equation on gas phase, inlet flow and out flow during minute time⊿t. In pressurization process, solvent was adsorbed slightly at Stop. At Go, some of the product gas that has flowed is used most boost. In adsorption process, solvent was adsorbed at Stop. At Go, feed gas flowed and product gas discharged into next cell. In evacuation process, solvent was desorbed at Stop. At Go, desorbed solvent discharged as exhaust gas desorption. In purge process, solvent was desorbed at Stop. At Go, part of the product gas flowed and discharged solvent desorbed as exhaust gas desorption. Here Mass balance equation (Eq. 1) and Markham-Benton equation (Eq. 2) as for adsorption equilibrium and linear driving force (LDF) approximation (Eq. 3) as for adsorption rate were adapted for numerical calculation, which was based on Stop & Go method.

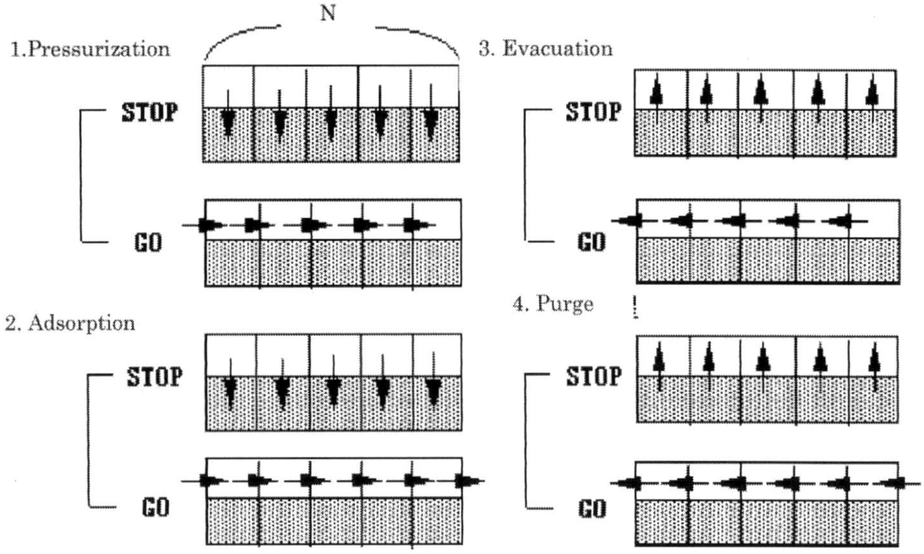

Fig. 3 Stop&Go method basic concept

Mass balance equation

$$\frac{m}{N}\frac{dq_{n,j}}{dt} = \frac{\varepsilon V}{N}\frac{dC_{n,j}}{dt} = Q_n C_{n-1,j} - Q_{n+1} C_{n,j} \qquad (1)$$

Markham-Benton equation

$$q_i* = \frac{K_i q_{\infty,i} P_i}{1+\sum K_i P_i} \qquad (2)$$

Linear driving force approximation

$$\gamma \frac{dq_{n,i}}{dt} = K_{si} a_v (q_{n,i}^* - q_{n,i}) \qquad (3)$$

RESULTS AND DISCUSSION

Adsorption Breakthrough Experiment

Figure 4 and 5 show that results of adsorption breakthrough experiment on each solvent. Regarding CH_2Cl_2, it takes more than 80 minutes to reach steady state. Regarding water vapor, it takes more than 200 minutes to reach steady state. The more solvent concentration in feed gas was increased, the faster solvent concentration in product gas reached to solvent concentration in feed gas.

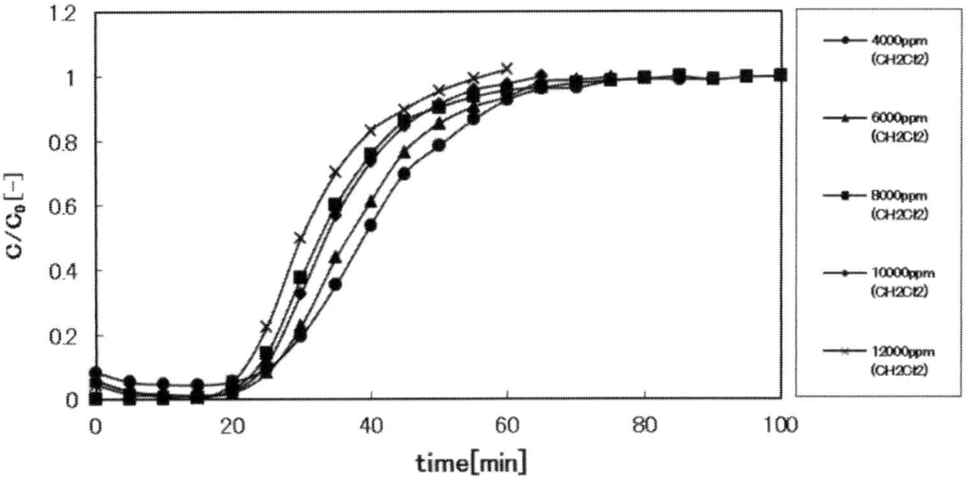

Fig. 4 Adsorption breakthrough experiment (CH_2Cl_2)

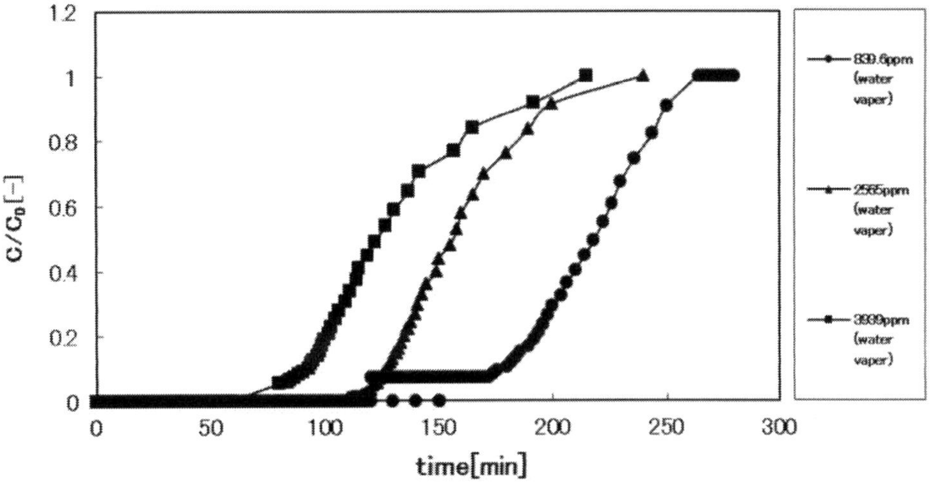

Fig. 5 Adsorption breakthrough experiment (water vapor)

Figure 6-9 show that results of Langmuir Plot and Adsorption isotherm on each solvent. We determined equilibrium adsorption capacity and Langmuir constant of each solvent by linear getting from Langmuir Plot (Fig. 6 and 7). Equilibrium adsorption capacity q^{∞} and Langmuir constant K of each solvent were shown in Table 5.

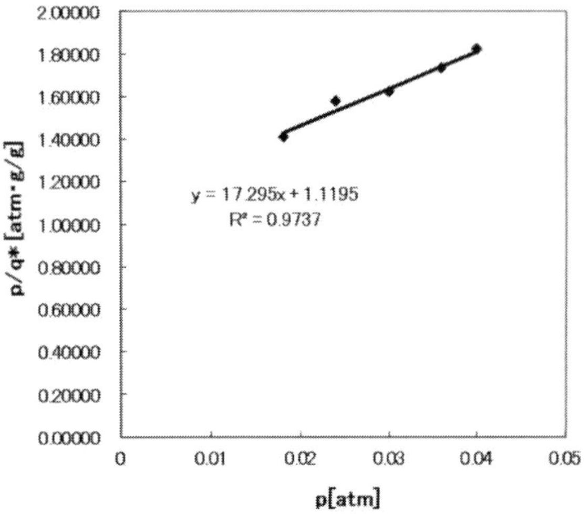

Fig. 6 Langmuir Plot (CH_2Cl_2)

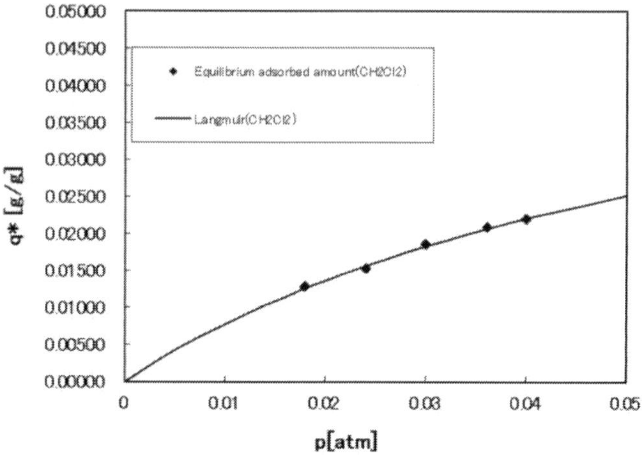

Fig. 7 Adsorption isotherm (CH_2Cl_2)

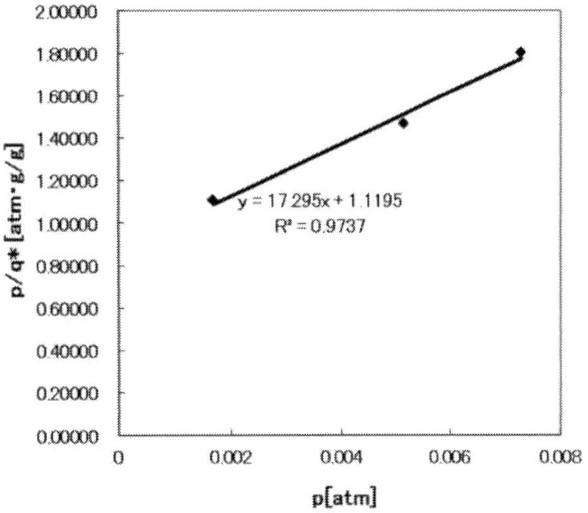

Fig. 8 Langmuir Plot (water vapor)

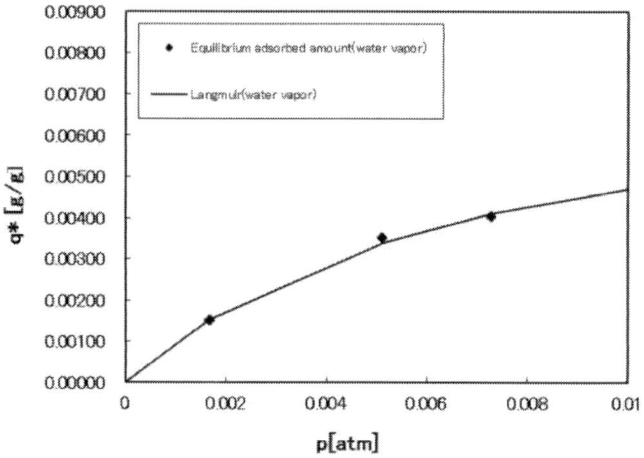

Fig. 9 Adsorption isotherm (water vapor)

Table 5. q^{∞} and K of each solvent

Adsorbent	CH2Cl2	water vapor
q^{∞} [g/g]	0.06546	0.008170
K [1/atm]	12.56	138.70

PSA Experiment and Simulation

Figure 10 shows that each solvent concentration changes in product gas when the water vapor in feed gas was changed. The more water vapor in feed gas was increased, the higher concentration of CH_2Cl_2 in product gas was. Regardless of water vapor content, each concentration of CH_2Cl_2 became steady state in 120 or more cycles. Meanwhile, the water vapor in product gas has started to emit in 200 cycles. Each concentration of the water vapor became steady state in 350 or more cycles. We found that

CH_2Cl_2 is adsorbed in the first adsorption is then water vapor.

Figure 11 shows that comparison of experiment and simulation about concentration of CH_2Cl_2 in product gas. Figure 12 shows that comparison of experiment and simulation about concentration of CH_2Cl_2 in exhaust gas. In product gas, simulated values and experimental values were close. On the other hand, in exhaust gas, simulated values and experimental values were away.

Figure 13 shows that comparison of experiment and simulation about concentration of water vapor in product gas. In product gas, simulated values and experimental values were away.

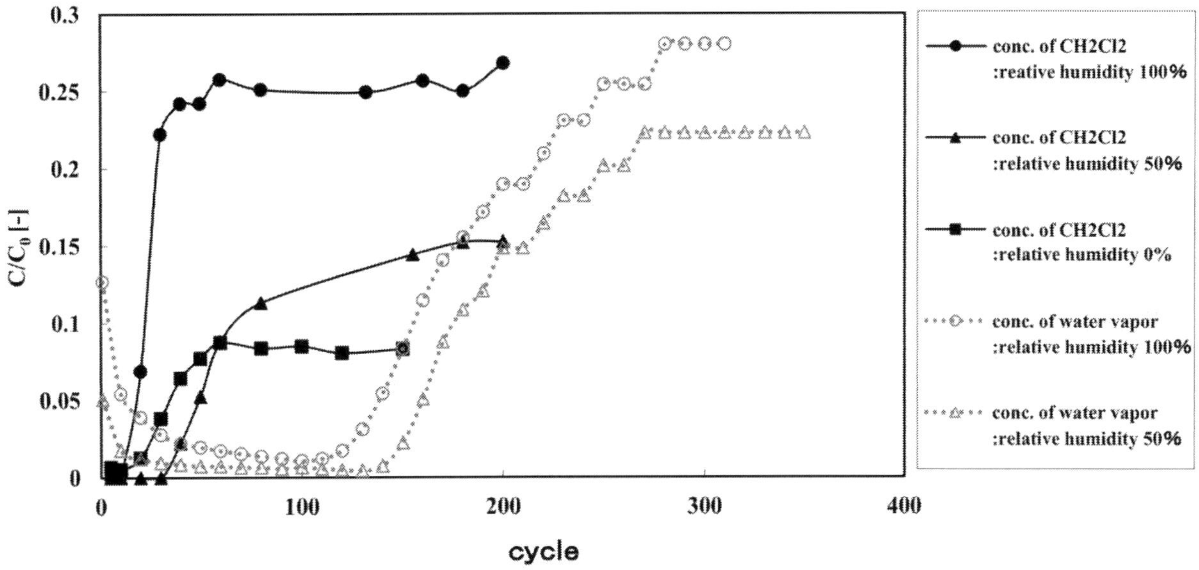

Fig. 10 Concentration of CH_2Cl_2 and water vapor in product gas

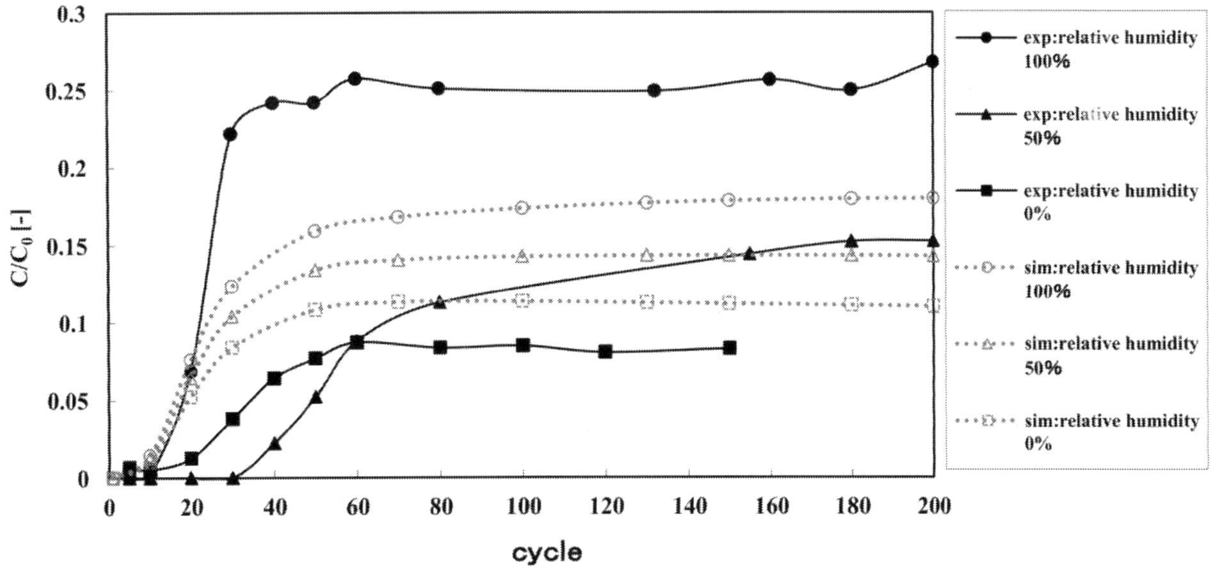

Fig. 11 Concentration of CH_2Cl_2 in product gas

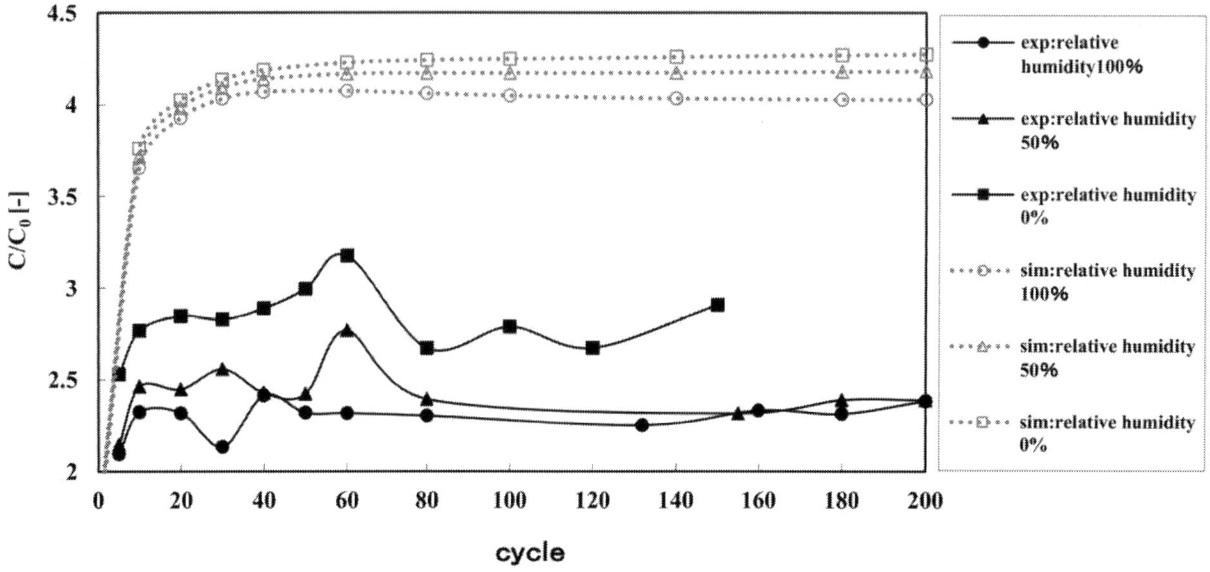

Fig. 12 Concentration of CH_2Cl_2 in exhaust gas

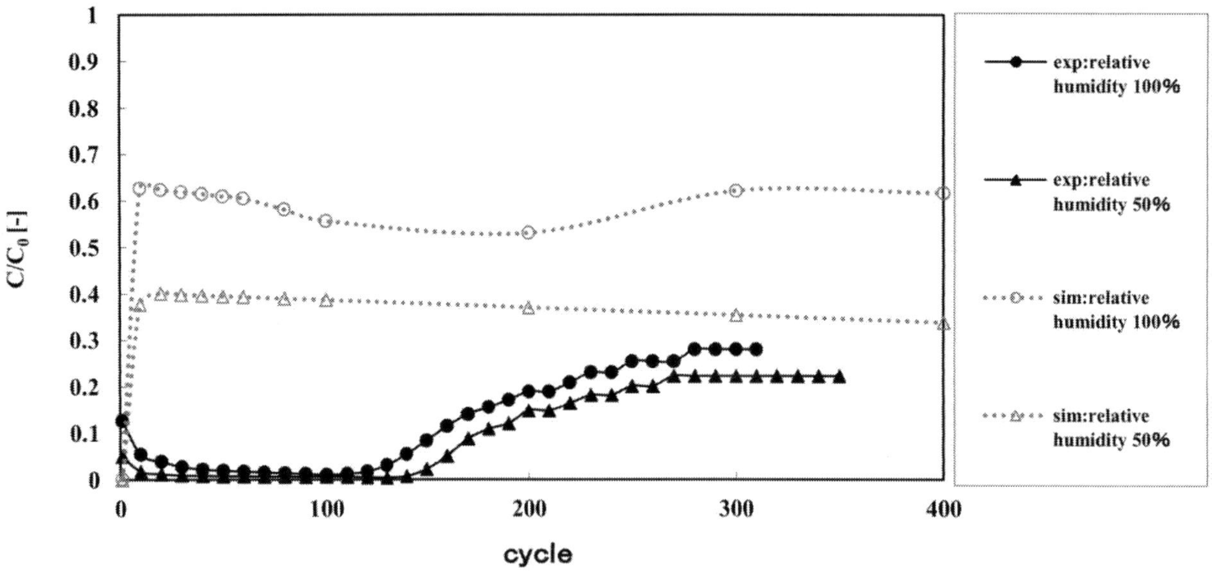

Fig. 13 Concentration of water vapor in product gas

CONCLUSIONS

The solvent concentration in product gas became higher when the water vapor content in feed gas increased (Fig.11). The results suggest that the water vapor obstructs the adsorption of CH_2Cl_2 in PSA method. Therefore, the water vapor in feed gas degraded the performance of CH_2Cl_2 separation in spite of using hydrophobic silica gel as adsorbent. We found that CH_2Cl_2 is adsorbed in the first adsorption and then, water vapor is adsorbed (Fig.10). In contrast, change of the water vapor content did not affect concentration of CH_2Cl_2 in exhaust gas (Fig.12).

Numerical calculation for PSA was tried using Stop & Go method. We found that experimental values of concentration of CH_2Cl_2 in the product gas are able to adapt to the simulation values by using Stop & Go method (Fig.11). However, experimental values of concentration of CH_2Cl_2 in the exhaust gas and the water vapor in product gas are not able to adapt to the simulation values (Fig.12, 13).

References

[1] K. Chihara and M. Suzuki, (1983)"Simulation of Nonisothermal Pressure Swing Adsorption," Journal of Chemical Engineering of Japan, Vol. 16, No. 1, pp. 53-61

ADSORPTION OF ORGANICS ON MSC5A IN SUPERCRITICAL CO_2, CHROMATOGRAPHIC MEASUREMENTS & STOP & GO SIMULATION

Kazuyuki Chihara, Shingo Ito, Masashi Shinsako, Hideaki Nagashima and Mai Hikita
Meiji University, Kawasaki, Japan

Introduction

Adsorption of organics on adsorbents, such as activated carbon, in supercritical CO_2 could be mild [1] [2] [3] compared with adsorption in gas phase [4]. This is because of co-adsorption of CO_2 [5]. Amount adsorbed of organics can be changed by changing temperature and pressure of supercritical CO_2. Adsorption and desorption of organics could be easy and mild in supercritical CO_2 condition. New process of adsorptive separation using supercritical fluid might be possible [6]. However equilibria and dynamics for co-adsorption of organics and CO_2 have not been studied so much [7].

In this study, chromatographic measurements were made for the adsorption of benzene, toluene and m-xylene on molecular sieving carbon (MSC) in supercritical fluid CO_2. Supercritical chromatograph packed with MSC was used to detect pulse responses of organics. Adsorption equilibria and adsorption dynamics for organics were obtained by moment analysis of the response peaks. Dependences of adsorption equilibrium constants, K*, and micropore diffusivity, D, on the amount adsorbed were examined. And, stop & go simulation of multicomponent adsorption equilibria was performed to get reasonable equilibrium parameters and rate parameters.

Experimental

The experimental apparatus was shown in Fig.1. The carrier fluid of the chromatograph was supercritical CO_2 (critical temperature 304 K, critical pressure 73 atm) and it's mixture with the above-stated organics (benzene, toluene or m-xylene) respectively. The adsorbates used in the form of pulse were the same of or the different from organics mixed with supercritical CO_2.

For example, in the case of CO_2 mixed with toluene, the organic used in the form of pulse was benzene, toluene or m-xylene. The volumes of the pulse were fixed to be 4×10^{-9} m^3 as liquid. MSC 5A (Takeda chemicals Co., HGK882.) was crushed and screened to obtain particle size between 1.49×10^{-4} - 1.77×10^{-4} m (an average particle radius of 8.12×10^{-5} m). 4.82×10^{-4} kg of these particles were packed into the chromatographic column of 15×10^{-2} m long and 4.6×10^{-3} m in diameter. The void fraction, ε, of the bed was determined to be 0.3548. Flow rate of supercritical CO_2 was 1.33×10^{-7} m^3/s at 268 K and at 150, 200 and 250 atm respectively and flow rate of adsorbate (benzene, toluene or m-xylene) was 1.67×10^{-10} m^3/s, 5.00×10^{-10} m^3/s and 1.00×10^{-9} m^3/s as liquid at room temperature (298 K).

The column pressure was kept at 150, 200 and 250 atm respectively (Back Pressure Regulator 880-81; Japan Spectroscopic Co., LTD.). The pressure drop across the adsorbent bed was estimated to be about 0.1MPa and was assumed to be negligible. The experimental column temperature was kept at 313, 333 and 353 K respectively (Intelligent Column Oven CO-2065 Plus; Japan Spectroscopic Co., LTD.). Before experimental runs started, the adsorbent particles were regenerated and stabilized by feeding pure CO_2 for 2 hours at the experimental pressure and temperature. Pulse responses were detected using a multi-wave length UV detector (MD-2010 Plus; Japan Spectroscopic Co., LTD.) (199-651 nm). Response data were processed by a personal computer. Moment analysis of supercritical fluid chromatogram was tried, and the apparent adsorption equilibrium constant, K* and time constant of micropore diffusivity, D/a^2 obtained from first and second moment of response peak, as in references [8]. Table 1 showed properties of MSC 5A [8].

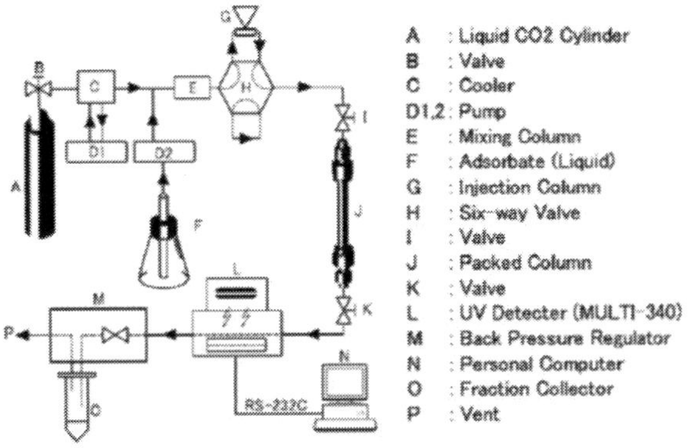

Fig. 1. Experimental setup

Table 1. Properties of MSC5A

True density [g/cm^3]	1.8
Particle density [g/cm^3]	0.90
Macropore volume [cm^3/g carbon]	0.38
Micropore volume [cm^3/g carbon]	0.18
Porosity of macropore [-]	0.342
Macropore radius [μ]	2.0
Micropore opening [Å]	5.0

Moment Analysis

Basic equations are in ref. (Kumar et al. 1982) [9]

The resultant moment equations of the impulse response are

$$\mu_1 = \frac{z}{u}\left\{1 + \frac{(1-\varepsilon)\rho_p(1-y)K^*}{\varepsilon}\right\} \quad (1)$$

$$\mu_2 = \frac{2z}{u}\left[Ez\left\{\frac{1+(1-\varepsilon)\rho_p K^*}{\varepsilon u}\right\}^2 + \frac{(1-\varepsilon)\rho_p(1-y)K^*}{\varepsilon Ksav}\right] = \frac{2z}{u}(\delta_d + \delta_f + \delta_a + \delta_i) \quad (2)$$

where μ_1=first absolute moment of the chromatographic peak [s] μ_2=second central moment [s^2].

$$\delta_d = \frac{Ez}{u^2}\left\{1 + \frac{(1-\varepsilon)\rho_p(1-y)K^*}{\varepsilon}\right\}^2 \quad (3)$$

$$\delta_f = \frac{1-\varepsilon}{\varepsilon}\frac{R}{3k_f}\rho_p^2(1-y)K^{*2} \quad (4)$$

$$\delta_a = \frac{1-\varepsilon}{\varepsilon}\frac{R^2}{15D_a}\rho_p^2(1-y)K^{*2} \quad (5)$$

$$\delta_i = \frac{1-\varepsilon}{\varepsilon}\frac{\rho_p(1-y)K^*}{15D}\bar{a}^2\exp(\sigma^2) \quad (6)$$

where K^*=apparent adsorption equilibrium constant, Ez=axial dispersion coefficient based on void spaces in the bed, k_f=external mass transfer coefficient, D_a=diffusivity in macropores and D=diffusivity in micropores based on amount adsorbed gradient driving force.

Equation (1) was used along with the experimental μ_1 to obtain the adsorption constant K^*. Equation (2) to (6) were used along with the experimental μ_2' to obtain the diffusivity in micropores D.

Stop&Go Simulation

Stop & Go method was used as simulation method of perturbation chromatography for investigating adsorption equilibrium and rate. Fig. 2 showed Stop&Go method basic concept.

Numerical solution for multicomponent chromatogram in time domain could be obtained by appropriate model equations with experimental conditions. This simulated chromatogram can be compared with experimental chromatogram to determine the adsorption equilibrium and rate parameters. Here Mass balance equation (Eq. 7) and Markham-Benton equation (Eq. 8) as for adsorption equilibrium and linear driving force (LDF) approximation (Eq. 9) as for adsorption rate were adapted for numerical calculation, which was based on stop & go method.

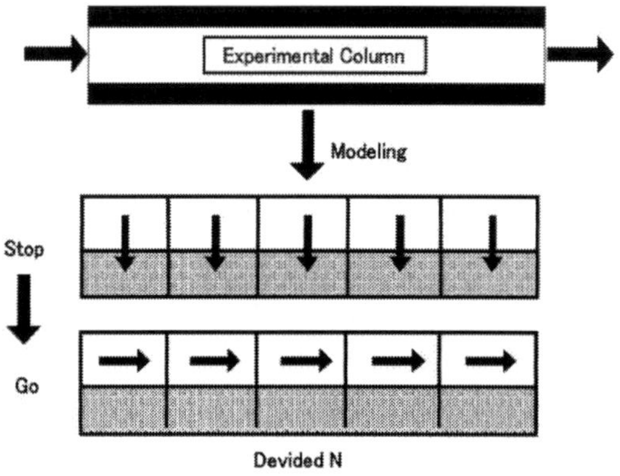

Fig. 2. Stop&Go method basic concept

Mass balance equation

$$\frac{m}{N}\frac{dq_{n,j}}{dt} = \frac{\varepsilon V}{N}\frac{dC_{n,j}}{dt} = Q_n C_{n-1,j} - Q_{n+1} C_{n,j} \tag{7}$$

Markham-Benton equation

$$q_i^* = \frac{K_i q_{\infty,i} P_i}{1 + \sum K_i P_i} \tag{8}$$

Linear driving force approximation

$$\gamma \frac{dq_{n,i}}{dt} = K_{si} a_v (q_{n,i}^* - q_{n,i}) \tag{9}$$

Results and Discussion

Single pulse data were first analyzed for toluene adsorption by moment method. Here, CO_2 adsorption was supposed to be as back ground. Later in Stop & Go simulation, pulse responses were simulated as binary or ternary components.

Experimental

Fig.3, 4 shows contribution of axial dispersion, external mass transfer, macropore diffusion and micropore diffusion in pulse response (μ_2) at 250 atm, 353 K. Fig. 3 illustrated that mass transfer is controlled by micropore diffusion, and according to Fig. 4, the four mass transfer resistances decrease with increases of adsorbate flow rate in the carrier (toluene).

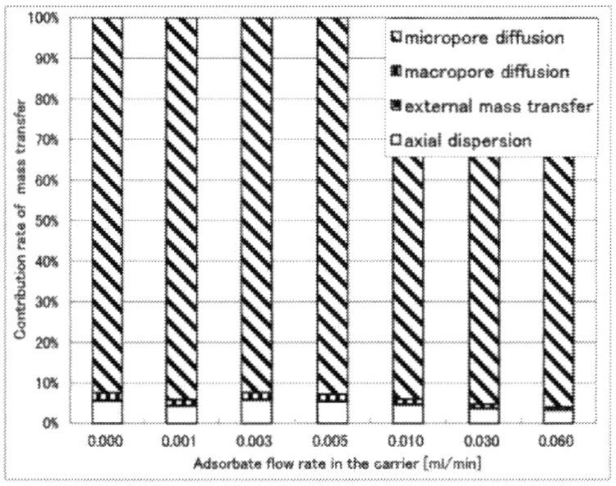

Fig. 3. Contribution rate of mass transfer of toluene (250 atm, 353 K)

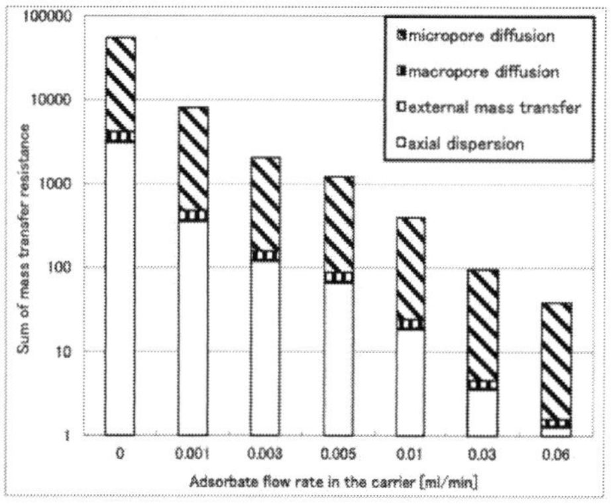

Fig. 4. Sum of mass transfer resistance of toluene (250 atm, 353 K)

Fig. 5 shows the dependency of apparent adsorption equilibrium constants, K*, for toluene on the supercritical CO_2 density, which were obtained from first moments of the pulses. The K* of benzene reasonably decreased, with increases of supercritical CO_2 density.

Fig. 6 shows the dependency of micropore diffusivity, D/a^2, for toluene on the supercritical CO_2 density as obtained from the second moments of the pulses. The increase of D/a^2 for toluene could be reasonably explained by chemical potential driving force. However, the explanation for dependency of D/a^2 values on the supercritical CO_2 density requires further consideration.

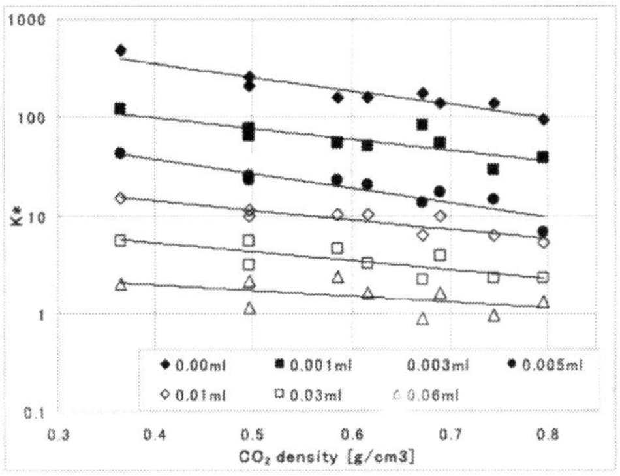

Fig. 5. Dependencies of K* on the SC-CO_2 density

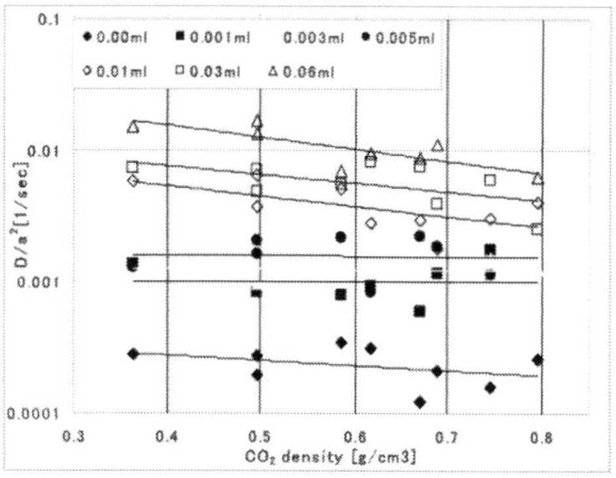

Fig. 6. Dependencies of D/a^2 on the SC-CO_2 density

Fig. 7 shows relationship of K* on the molarity of toluene. Amount adsorbed of toluene can be obtained by taking the integral of K* with respect to the molarity of toluene

Fig. 8 shows the apparent adsorption isotherm of toluene at 313 K, 250 atm and 353 K, 150 atm. Adsorption of organics in supercritical CO_2 tends to be larger as high-temperature and low-pressure. It is considered that the situation is competitive adsorption and the amount adsorbed of toluene decreases as CO_2 adsorption increase with increases of column pressure.

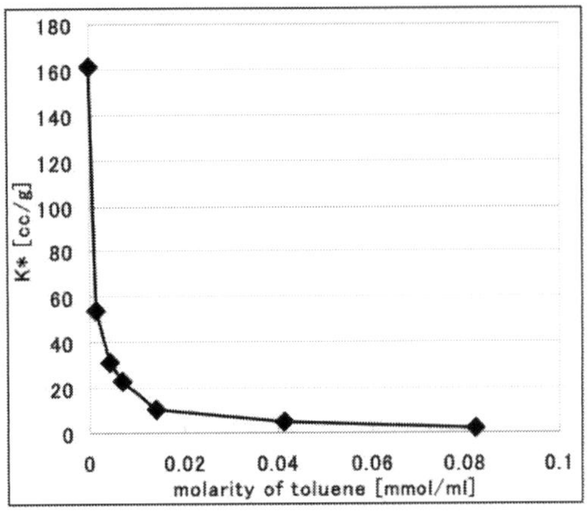

Fig. 7. Relationship of K* on the molarity of toluene

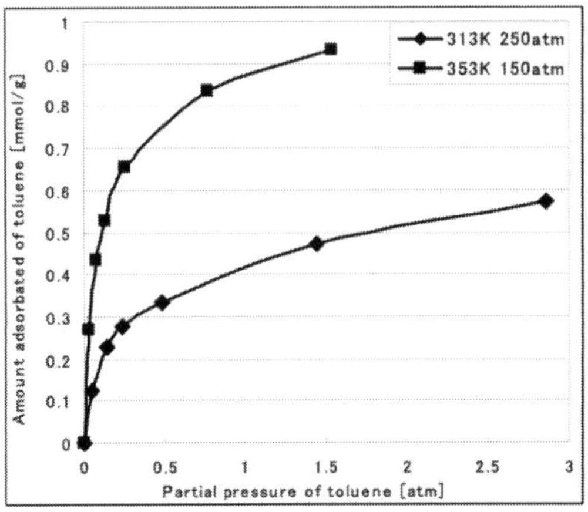

Fig. 8. Apparent adsorption isotherm

Stop & Go Simulation

A simulated chromatogram can then be compared with the experimental chromatogram to determine the adsorption equilibrium and rate parameters. A simulated system (cell model) has the inherent dispersion (mixing between cells), however we can neglect since it has too small effect by Fig.3.

Fig. 9, 10 shows comparison between experimental and simulated pulse response at constant temperature and pressure. Toluene and CO_2 were used for adsorbates. Conditions (Fig. 11) were 250atm, toluene flow 0.01 ml/min and toluene + CO_2 mixed carrier with toluene pulse. Here temperature was changed 313, 333 and 353 K. Conditions (Fig. 10) were 353 K, toluene flow 0.01ml/min and toluene + CO_2 mixed carrier with toluene pulse. Here pressure was changed 150, 200 and 250 atm. Good coincidence of peaks between experiment and simulation were observed as for peak retention and width. The simulated widths were adjusted. Conditions (Fig. 9) were 250atm, toluene flow 0.01 ml/min and toluene + CO_2 mixed carrier with toluene pulse.

Fig. 11 shows comparison between experimental and simulated pulse response. Conditions (Fig. 11) were 150atm, 313K, toluene flow 0.01 ml/min and benzene + CO_2 mixed carrier with toluene pulse. According to Fig. 11, 1st peak is toluene, 2nd peak is benzene. The width of the simulated pulse could be adjusted by rate parameter adjustment, but only minor adjustments were required.

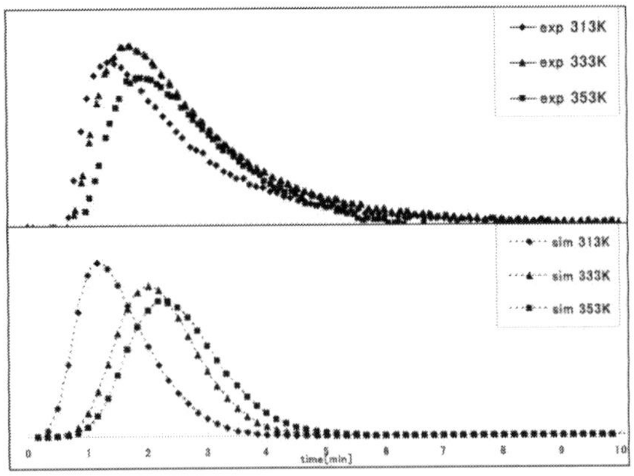

Fig. 9. Comparison between experimental and simulated pulse response at constant pressure

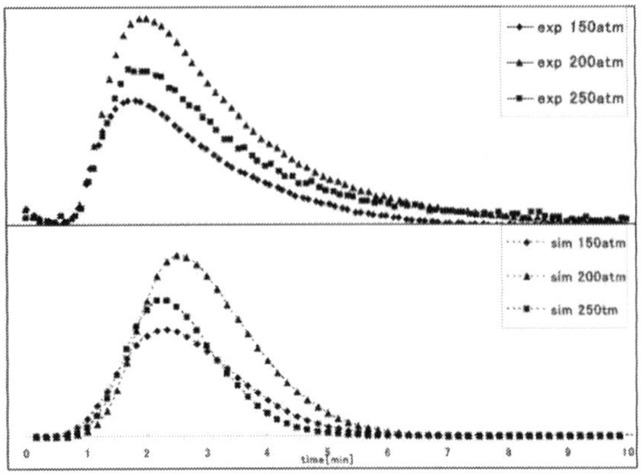

Fig. 10. Comparison between experimental and simulated pulse response at constant tempreture

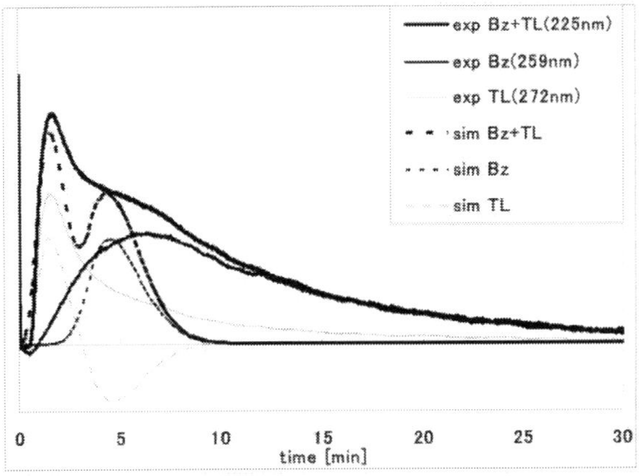

Fig. 11. Comparison between experimental and simulated pulse response (carrier: CO_2 + benzene, pulse: toluene, 150atm, 313 K, 0.01 ml/min)

Conclusion

Adsorption equilibrium and adsorption dynamics on MSC were evaluated for each organics in supercritical CO_2 fluid mixed with adsorbate by chromatographic measurement.

Adsorption of organics on MSC was adsorbed to micropore. The dependencies of apparent adsorption equilibrium constants, K*, and micropore diffusivity, D, of toluene on supercritical CO_2 density were obtained.

In the stop & go simulation, the experimental rate could be almost simulated by stop & go simulation.

References

1. H. Shojibara, Y. Sato, S. Takishima and H. Masuoka (1995), "Adsorption Equilibria of Benzene on Activated Carbon in the Presence of Supercritical Carbon Dioxide," *Journal of Chemical and Engineering of japan*, 28, 3 pp. 245-249.
2. K. Chihara, K. Oomori, T. Oono and Y. Mochizuki (1997), "Supercritical CO_2 Regeneration of Activated Carbon Loaded with Organic Adsorbates," *Water Science and Technology*, 35, 7 pp. 261-268.
3. K. Chihara, S. Ito, H. Nagashima, M. Hikita, and R. Suzuki (2011), "Adosorption of Organics on MSC5A in Supercritical CO2 : Chromatographic Measurements & Stop & Go Simulation," *diffusion-fundamentals. org*, 16, 83 pp. 1-6.
4. J. Benkhedda, J. Jaubert, and D. Barth (2000), "Experimental and Modeled Results Describing the Adsorption of Toluene onto Activated Carbon," *Journal of Chemical and Engineering Data*, 45, 4 pp. 650-653.
5. K. Tanida, Y. Sato, H. Masuoka (1996), "Measurement and Correlation of Supercritical Carbon Dioxide Adsorption on Activated Carbon," *Kagaku Kogaku Ronbunshu*, 22, 2 pp. 385-91.
6. M. Sato, M. Goto, A. Kodama and T. Hirose (1998), "New Fractionation Process of Citrus Oil by Pressure Swing Adsorption in Supercritical Carbon Dioxide," *Chemical Engineering Science*, 53, 24 pp. 4095-4104.
7. K. Chihara, S. Ito, H. Nagashima, M. Hikita, and R. Suzuki (2012), "Adsorption of Organics on MSC5A in Supercritical CO_2, Chromatographic Measurements and Stop and Go Simulation," Accepted for adsorption.

8. K. Chihara, M. Suzuki, K. Kawazoe (1978), "Adsorption Rate on Molecular Sieving Carbon by Chromatography," *AIChE Journal*, 24, 2 pp. 237-246.
9. R. Kumar, R.C. Duncan and D.M. Ruthven (1982), "A Chromatographic Study of Diffusion of Single Components and Binary Mixtures of Gases in 4A and 5A Zeolites," *The Canadian Journal of Chemical Engineering*, 60, 4, pp. 493-499.

Acknowledgment: Special research fund of 2011 fiscal year by Meiji univ. is appreciated.

AZEOTROPIC ADSORPTION EQUILIBRIUM OF MIXED ORGANIC SOLVENT ON FAU TYPE HIGH SILICA ZEOLITE -INTERPRETATION BY MOLECULAR SIMULATION AND DO'S METHOD-

Kazuyuki Chihara, Tatsuharu Taira, Natsuki Yamada, Yusuke Suzuki, Duong D. Do*

Department of Applied Chemistry, Meiji University, Kawasaki, Kanagawa, Japan

School of Chemical Engineering, University of Queensland, Brisbane, Australia

Introduction

In recent years, instead of the chlorofluorocarbon specified as the ozone-depleting substance, a chlorinated organic compound, hydrocarbon, alcohol, etc. are used as an alternative washing solvent, and they are used as a mixed solvent in many cases. Therefore, adsorption operation of binary systems is needed. Furthermore, since azeotropic adsorption may happen, in the case of the design of a solvent recovery system and operation, those adsorption equilibria are required as basic data. IAS theory can't treat the azeotropic adsorption. We applied the method of treating the azeotropic adsorption, that is, to treat the adsorbed phase as a combination of two independent adsorbed phases, and the IAS theory is applied for each adsorbed phase (Do, 2000) [1].

In this study, the adsorption equilibrium of the FAU type high silica zeolite was obtained by analyzing the adsorption phenomenon, using a Gravimetric adsorption experiment device and a molecular simulation. The adsorbate used was ethanol and trichloroethylene, which are the system of azeotropic mixture. The experiments were carried out using a batch system with circulation flow of the adsorbate mixture around the adsorbent to prevent the concentration distribution in gas phase, changing the total concentration and composition of ethanol and trichloroethylene, by stepwise dosing of mixed solvent vapor. For this system, one azeotropic point appeared. Simulation module in Materials Studio (Accelrys Inc.) was used throughout MC simulations. It calculated by the Sorption module, and the adsorption isotherm was obtained.

Two kinds of adsorption site or adsorption phase were observed in the cell (25.08 x 25.08 $Å^2$) by the Sorption simulation of binary adsorption equilibria for the system of EtOH-TCE(Fig.2). One phase containes EtOH and TCE (phase 1), and the other phase containes only EtOH (phase 2). The coordinates of two sites were determined. The ratio of EtOH amount adsorbed on both sites were counted from the simulated adsorbed cell. The ratio was found to be constant through the simulation. Using this constant ratio, the simulated or experimentally obtained amount adsorbed were divided for phase 1 and 2. Then adsorption isotherms were determined for both phases. Phase 1 was expressed as extended Langmuir equation. Phase 2 was expressed as Langmuir equation. Total adsorption isotherm was observed as total of both phase.

Experimenal

Gravimetric Method

Gravimetric method of laboratory-scale was carried out to remove organic solvent vapors by Y type high silica zeolite. The adsorbate used was ethanol and trichloroethylene, which are the system of azeotropic mixture. Experimental results for binary component system could be correlated satisfactorily.

Figure 1 showed experimental equipment for gravimetric analysis. The zeolite sample (about 0.3g) was placed in a quartz basket (E). Then the adsorbate in flask (B) was fed to adsorption tube . The whole equipments were in a constant temperature air bath. The temperature range was 293-298 K. The amount adsorbed was measured corresponding to the pressure of the vapor in the tube. The pressure was measured by pressure sensor (L) at higher pressure range (> 0.013 atm) and baratron (K) at lower pressure range (<0.013 atm). The concentration in the tube was determined by gas chromatography with FID, just after sampling from the sampling port.

The experiments were carried out using a batch system with circulation flow of the adsorbate mixture around the adsorbent to prevent the concentration distribution in gas phase, changing the total concentration and composition of ethanol and trichloroethylene, by stepwise dosing of mixed solvent vapor.

$$qt = \frac{P_0 - P_t}{RT} \times V \times M \qquad (1)$$

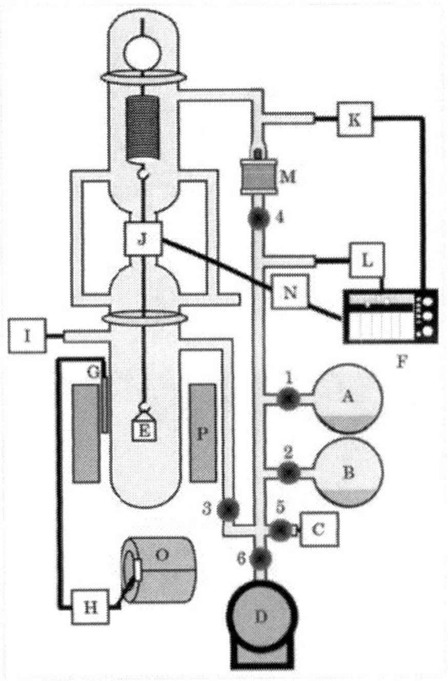

1~6	needle valve
A	adsorbate-flask(unused)
B	adsorbate-flask
C	hydraulic pump
D	Vacuum pump
E	Quartz basket
F	Data logger
G	Thermo detecter
H	Thermo controller
I	sampling port
J	Dtretch detector
K	Baratoron
L	Pressure sensor
M	Piston pump
N	Displacement meter
O	Nichrome heter
P	Mantle heater

Fig.1 Experimental equipment for Gravimetric Method

Simulation

Simulation module in Materials Studio (Accelrys Inc.) was used throughout MC simulations. The set of potential parameters was compass that was the corrected pcff. The Electrostatic parameters were used Ewald & Group. Ratio of the silica alumina (SiO_2/Al_2O_3) of the FAU type zeolite used in the experiment was 460, so that, siliceous model is used in the library of MS. It calculated by the Sorption module, and the adsorption isotherm was obtained. In MS, density field of ethanol and trichloroethylene that adsorbs zeolite was shown in Fig.2.

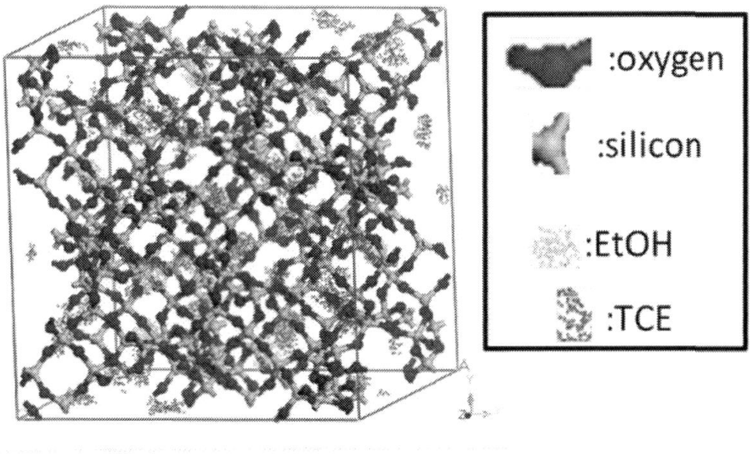

Fig.2 Density field of ethanol and trichloroethylene that adsorbs

Langmuir Plot

We made Langmuir Plot by use of results of simulation. Isothrem and Langmuir Plot were shown below. Langmuir constants obtained were shown below.

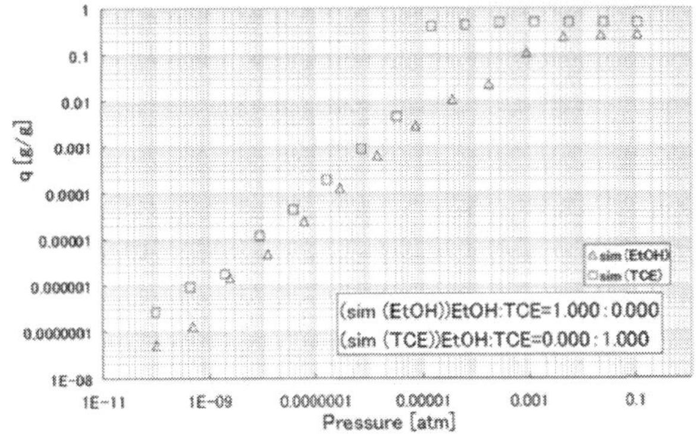

Fig.3 Adsorption isotherm

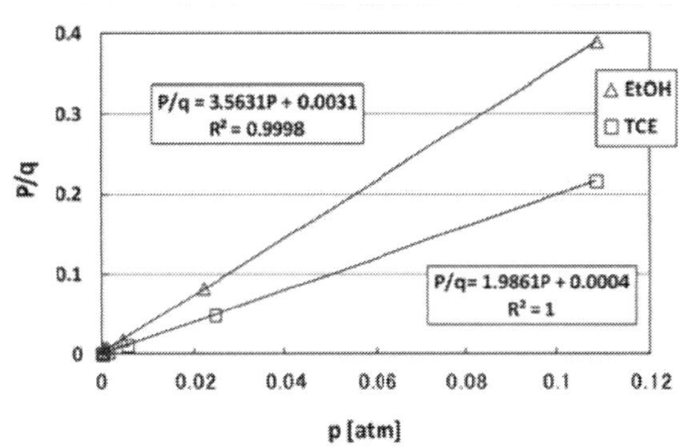

Fig.4 Langmuir Plot

Table 1 Langmuir constant

	q_∞ [g/g]	K [1/atm]
EtOH	0.28066	1131.9
TCE	0.50349	4589.8

Do's Method

We consider an adsorption of two components on zeolite. Two kinds of adsorption site or adsorption phase were observed in the cell (25.08 x 25.08 Å^2) by the Sorption simulation of binary adsorption equilibria for the system of EtOH-TCE (Fig.2). One phase containes EtOH and TCE (phase 1), and the other phase containes only EtOH (phase 2). The coordinates of two sites were determined. The ratio of EtOH amount adsorbed on both sites were counted from the simulated adsorbed cell with Materials Studio Scripting function by which field value could get.The ratio was found to be constant through the simulation. Using this constant ratio, the simulated or experimentally obtained amount adsorbed were divided for phase 1 and 2. Then adsorption isotherms were determined for both phases. Phase 1 was expressed as extended Langmuir equation. Phase 2 was expressed as Langmuir equation. Total adsorption isotherm was observed as total of both phase.These equations were shown below.

Phase1

$$\frac{q_{\mathrm{TCE}}}{q_{\infty \mathrm{TCE}}} = \frac{4589.8 P_{\mathrm{TCE}}}{1 + 4589.8 P_{\mathrm{TCE}} + 1131.9 P_{\mathrm{EtOH}phase1}} \quad (2)$$

$$\frac{q_{\mathrm{EtOH}phase1}}{0.10838 q_{\infty \mathrm{EtOH}}} = \frac{1131.9 P_{\mathrm{EtOH}phase1}}{1 + 4589.8 P_{\mathrm{TCE}} + 1131.9 P_{\mathrm{EtOH}phase1}} \quad (3)$$

Phase2

$$\frac{q_{\mathrm{EtOH}phase2}}{0.89162 q_{\infty EtOH}} = \frac{1131.9 P_{\mathrm{EtOH}phase2}}{1 + 1131.9 P_{\mathrm{EtOH}phase2}} \quad (4)$$

Total amount of adsorption

$$q_{\mathrm{TCE}} + q_{\mathrm{EtOH}} = \frac{q_{\infty \mathrm{TCE}} \times 4589.8 P_{\mathrm{TCE}} + 0.10838 \times q_{\infty \mathrm{EtOH}} \times 1131.9 P_{\mathrm{EtOH}phase1}}{1 + 4589.8 P_{\mathrm{TCE}} + 1131.9 P_{\mathrm{EtOH}phase1}} + \frac{0.89162 \times q_{\infty \mathrm{EtOH}} \times 1131.9 P_{\mathrm{EtOH}phase2}}{1 + 1131.9 P_{\mathrm{EtOH}phase2}} \quad (5)$$

The ratio of EtOH amount adsorbed on phase1 and phase2 is 0.89162:0.10838.

We made isotherms, q-X diagrams, X-Y diagrams by using these equations.

Result and Discussion

Comparing all of Binary Adsorption Isotherm

In the gravimetric method, the adsorption isotherm of binary systems was obtained. Figure 5 and 6 showed adsoption isotherm obtained for EtOH-TCE -Y-type zeolite system [2].Additionally in the MC simulation and simulation of Do's method ,the adsorption of binary systems were obtained. Figure 5,6 and 7 showed adsoption isotherm obtained for EtOH-TCE -Y-type zeolite system.In Figure 5,6 and 7,a ratio of TCE and EtOH is 0.815:0.185 on experiment and 0.800:0.200 on simulation and simulation of Do's method.The result of the binary systems experiment and simulation were able to reproduce the result of the binary breakthrough.Although we couldn't exactly match the results of simulation and the results of simulation of Do's method.The result of simulation of Do's method was shown decrease of adsorbed amount at lower pressure range ($<$ 0.001atm).

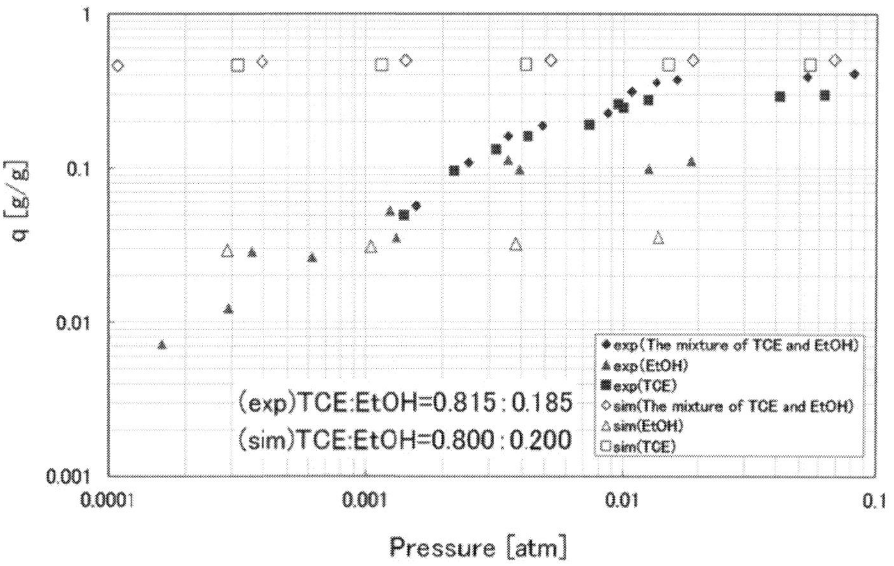

Fig.5 Adsorption isotherm observed and simulation

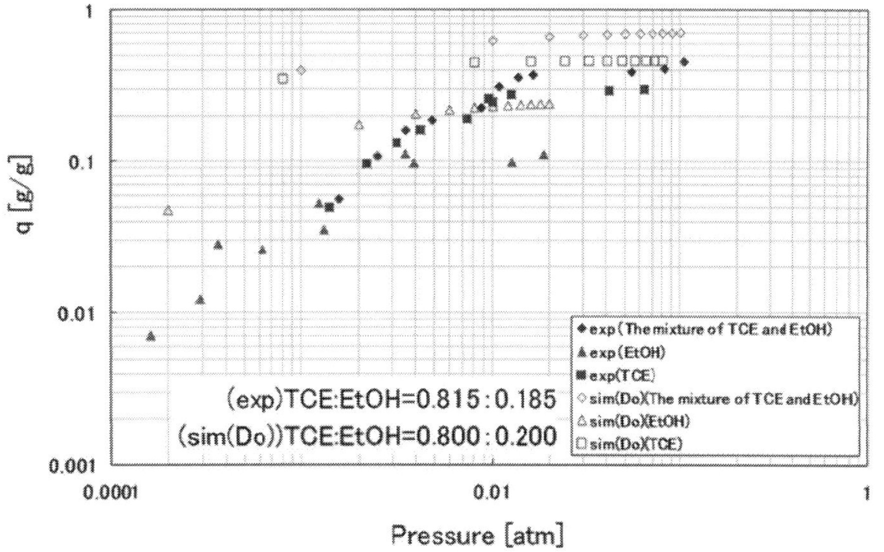

Fig.6 Adsorption isotherm observed and simulation of Do's method

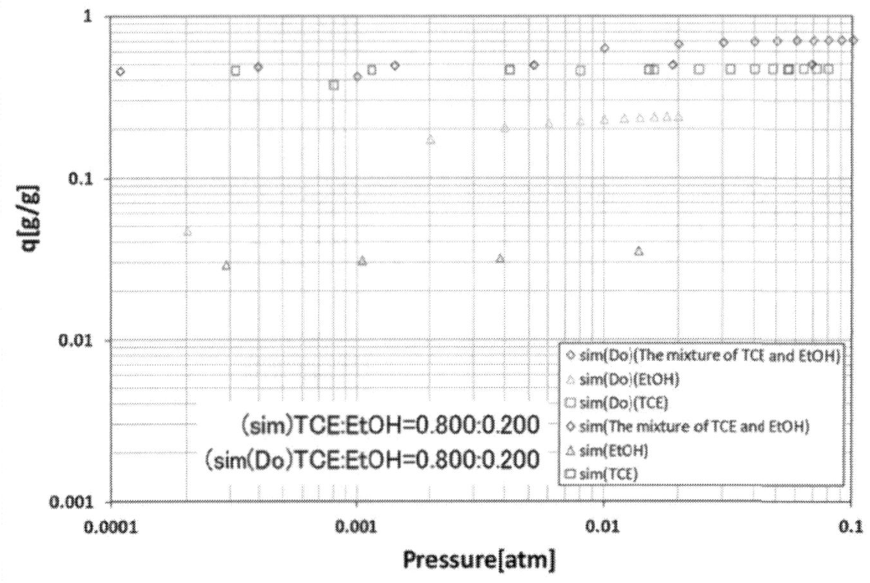

Fig.7 Adsorption isotherm simulation and simulation of Do's method

Adsorption q-X diagram

Binary adsorption equilibria for the system of EtOH-TCE of total concentration of 10000ppm and 70000ppmwere shown in Fig.8,9,10,11,12 and 13 in form of q-X diagram (q and X are the adsorbed amount and the mole fractions of gas phase concentration,respectively).Fig.8 and Fig.9 showed the experimental results.Fig.10 and Fig.11 showed the simulation results.Fig.12 and Fig.13 showed the simulation of Do's method results.

As for the gravimetric method and simulation of Do's method, q-X diagram showed almost the same tendency.

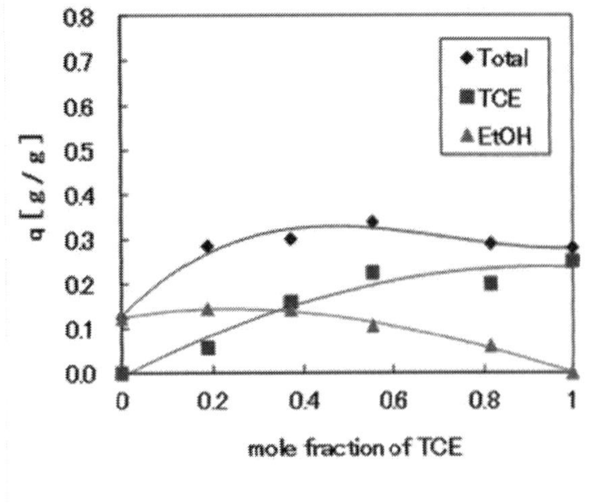

Fig.8 q-X diagram of experimental(10000ppm)

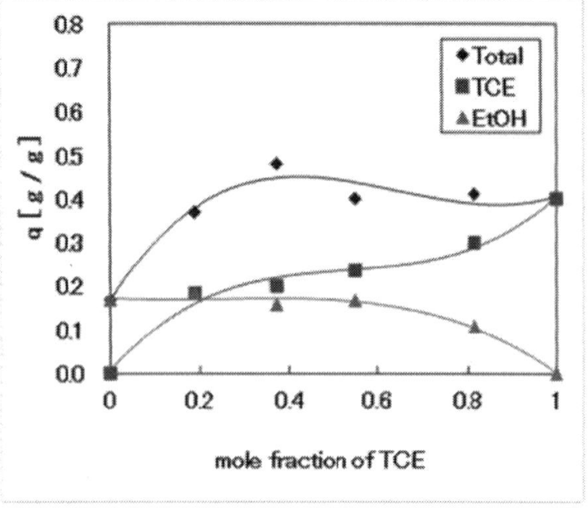

Fig.9 q-X diagram of experimental(70000ppm)

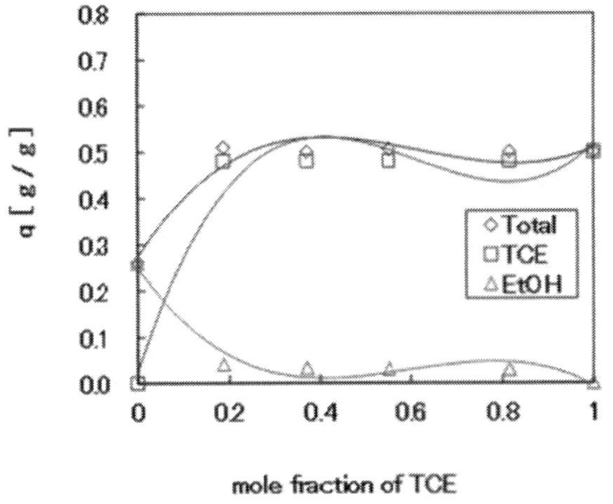

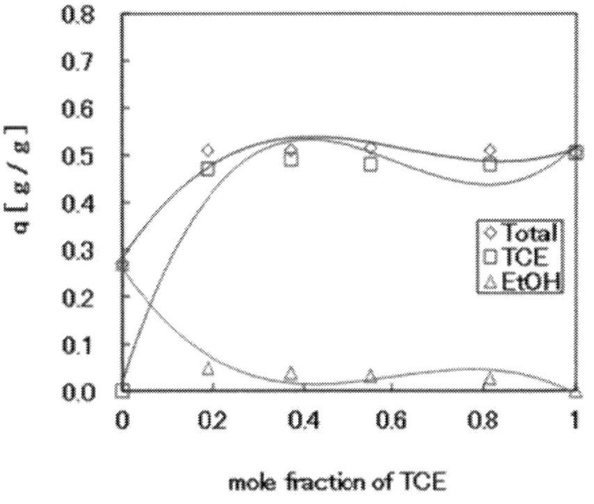

Fig.10 q-X diagram of simulation(10000ppm)

Fig.11 q-X diagram of simulation(70000ppm)

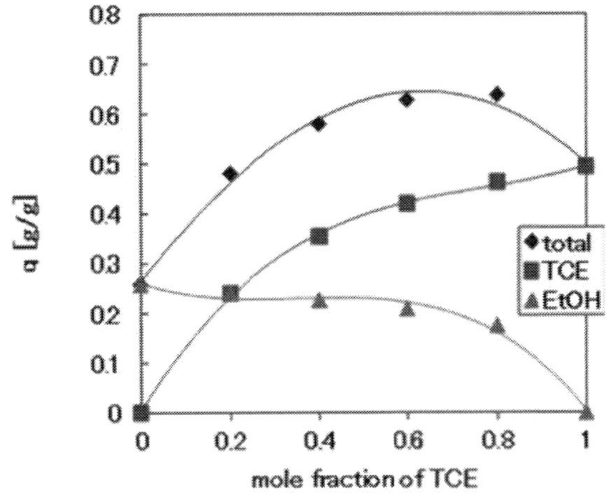

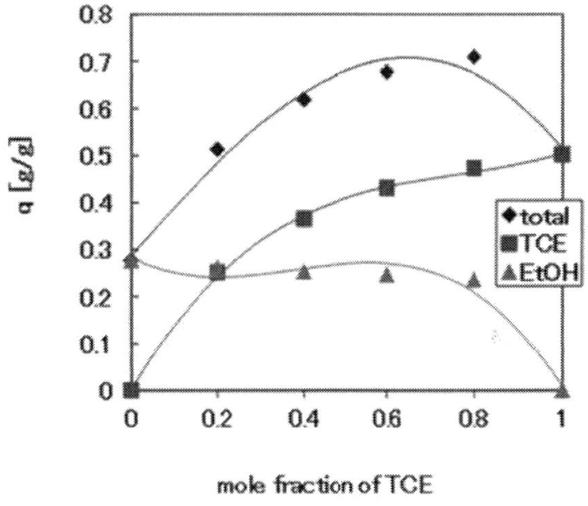

Fig.12 q-X diagram of simulation of Do's method(10000ppm)

Fig.13 q-X diagram of simulation of Do's method(70000ppm)

Adsorption X-Y diagram

Binary adsorption equilibria for the system of EtOH-TCE of total concentration of 10000ppm and 70000ppm were shown in Fig.14 and Fig.15 in form of X-Y diagram(X and Y are the mole fractions of gas phase concentration and the amount adsorbed at equiribrium,respectively).

As for the gravimetoric method and simulation of Do's method, azeotropic phenomenon was seen in the concentration(10000ppm,70000ppm).For this system, one azeotropic point appeared. Even if the concentration changes, the X-Y diagrams created experimental and simulation of Do's method showed almost the same tendency.

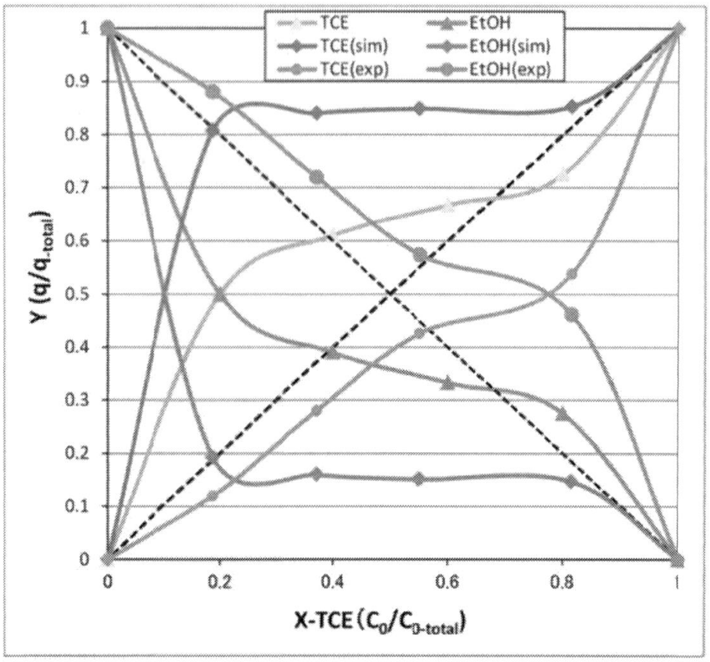

Fig.14 Comparison of adsorption equilibrium for experimental,simulation and simulation of Do's method(10000ppm)

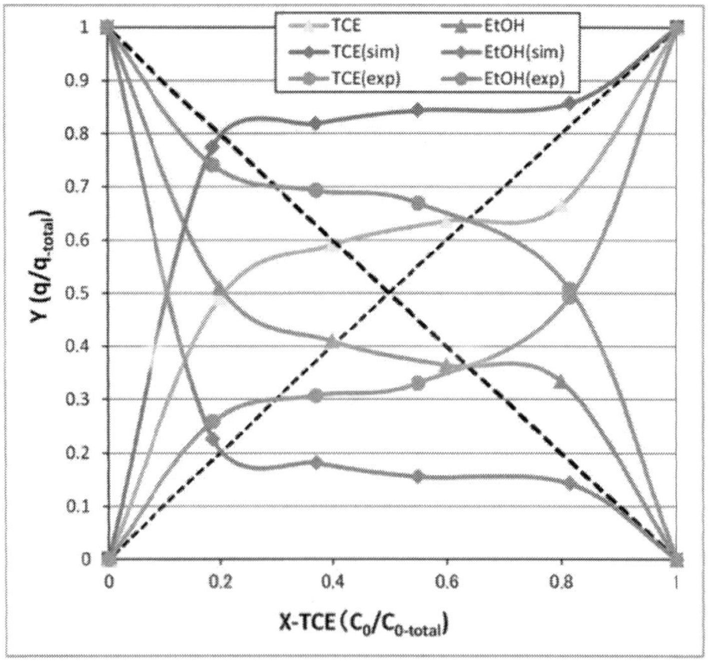

Fig.15 Comparison of adsorption equilibrium for experimental,simulation and simulation of Do's method(70000ppm)

Conclusion

The equilibrium amount of adsorption was able to be calculated as a tendency in MS and Do's method (Fig.5, 6, 7). In accuracy, the amount adsorbed of MS and Do's method are slightly different compared with the gravimetric method experiment. Azeotropic phenomenon was obtained in the result of experimental and simulation of Do's method.

References

1. Duong D Do (2000), "Adsorption Analysis:Equilibria and Kinetics," London , *Imperial College Press*, pp.219-221.
2. Chihara. K, Suzuki.Y, Tomita.S , Kabe.J (2011), "Adsorption equilibrium of the mixture of EtOH and TCE on FAU type high silica zeolite,"Adsorption 17, pp.201-203.

Acknowledgment: Special research fund of 2011 fiscal year by Meiji univ. is appreciated.

(395s) Adsorption Equilibria of Major Components (H2, CO, CO2, N2, CH4) From Coal Gas On Zeolite 13X, Zeolite LiX and Activated Carbon

Ki-Yong Yoon, **Dong-Kyu Moon** and Chang-Ha Lee, Chemical and Biomolecular Engineering, Yonsei University, Seoul, South Korea

Global warming, resulting from greenhouse gas emissions, is an important issue. The current situation has motivated the recovery of hydrogen from off-gases as a clean energy carrier even though certain off-gases contain relatively low fractions of hydrogen. Coal gas contains H_2 with CO, CO_2, N_2 and CH_4. Due to the large amounts of coal gas, a pressure swing adsorption (PSA) process is appropriate to achieve the proper purity level at low operating cost and high efficiency. More than two different adsorbents are applied to many PSA processes due to the adsorption affinity differences among the adsorbents. Therefore, the adsorption equilibrium data could be served as important data for selecting an optimal adsorbent of the hydrogen adsorption process of a first-aimed mixture, adsorption bed design and process simulation.

In this study, the adsorption equilibrium of the major components (H_2, CO, CO_2, N_2, CH_4) emitted from coal off-gases was measured on zeolite 13X, zeolite LiX, and activated carbon by a volumetric method. The adsorption equilibrium were measured at 298K, 308K, and 323.15K in the pressure range of 1 ~ 10bar. The adsorption isotherm and heat of adsorption among the adsorbents were compared.

(395t) Layered-Bed PSA Process for H2 Recovery From Coal Gas

Ki-Yong Yoon, Dong-Kyu Moon, Philip Jeon and **Chang-Ha Lee**, Chemical and Biomolecular Engineering, Yonsei University, Seoul, South Korea

H_2 syn-gas production by coal gasification has attracted much attention due to a lack of crude oil and global warming. In addition, H_2 mixutres from various sources, which were burned in the processes due to their low quality, are needed to recover H_2 by using an efficient separation process because of a great demand for H_2 as a clean fuel at gas turbine in IGCC and at a furnace in an iron/steel industry. A pressure swing adsorption (PSA) process can suggest a promising process to recover H_2 from a mixture because most of such mixtures were emitted at a high pressure condition.

In the present study, a layered two-bed PSA process was studied for H_2 recovery from coal gas with relatively low H_2 concentrations ($H_2/CO_2/CH_4/CO/N_2$; 38/50/1/1/10 vol.%). To design an energy-efficient PSA process, adsorption dynamics were compared among single adsorbent beds and activated carbon/zelite layered beds. For the layered bed, zeolite 13X and zeolite LiX were applied to the adsorption bed. In addition, the breakthrough experiments of various layered beds with different activated carbon/zeolite ratio were performed in the pressure range from 6.5 to 9.5 atm. Based on the dynamic results, the layered two-bed PSA process was performend and the PSA dynamic behaviors were evaluated by a non-isothermal dynamic model.

(395u) Solubility of CO2 in Saline Water and Adsorption of CO2 On Clay Minerals From Subcritical to Supercritical Conditions

Jiwon Choi[1], Philip Jeon[2], Hyeon Hui Lee[3], **Doo Wook Kim**[1] and Chang-Ha Lee[2], (1)Chemical and Biomolecular Engineering, Yonsei University, Seoul, South Korea, (2)Chemical and Biomolecular Engineering, Yonsei University, Seoul, South Korea, (3) Chemical and Biomolecular Engineering, Yonsei University, Seoul, South Korea

As carbon dioxide is well-known as a major contributor to greenhouse effect, capturing and sequestering CO_2 emission has emerged as one of the most important global issues. CO_2 storage in geological formations has been focused as a promising method to reduce anthropogenic CO_2 emission and especially, saline aquifers have the highest capacity for CO_2 storage. For this reason, the evaluation of the storage capacity and interaction was required between saline water and CO_2 and between constituent rocks and CO_2. Clay mineral is one of the major constituents of various rocks such as cap rock and coal mineral matter.

In this study, CO_2 sorption experiments were conducted to confirm the sealing integrity, and the storage capacity of CO_2 in the saline water was measured. Several kinds of clay minerals were selected as adsorbents, and CO_2 adsorption experiments were carried out by a gravimetric method. According to the practical reservoir conditions, the experiments were performed from CO_2 subcritical and supercritical conditions. The structural and textural changes of clay minerals due to CO_2 sorption were observed by comparing the BET analysis between the raw material and post-experiment samples. In addition, the solubility experiments of CO_2 in the standard sea water were conducted and the change of pH with pressure was observed.

(395v) NET Adsorption Thermodynamic Framework for Microporous Adsorbents

Dr.Orhan Talu and Qianqian Zhou, Chemical and Biomedical Engineering, Cleveland State University, Cleveland, OH

Adsorption thermodynamics is based on Gibbs dividing surface. Fixing the "location" of dividing surface has been a challenge both experimentally and theoretically, especially for microporous adsorbents which are used in all industrial processes. Often helium has been used in the past as a probe gas with the assumption that it does not adsorb. There is ample evidence in literature that helium, like all other fluids, do adsorb. Therefore, the entire thermodynamic treatment of adsorption has been based on this questionable assumption.

Recently, we introduced a novel approach to define the Gibbs dividing surface called "*Net Adsorption.*" Net adsorption completely eliminates the need for a probe fluid to locate the dividing surface, thus circumvents all questionable assumptions. In addition, using net adsorption framework greatly simplifies experimental and simulation efforts.

This poster will explain net adsorption, provide the thermodynamic basis, and show benefits of net adsorption over "*excess*" and "*absolute*" adsorption including some examples.

(395x) Use of a Three-Zone Simulated Moving Bed Process Based On the Amberchrom-CG161C Adsorbent for Continuous Removal of Acetic Acid From Biomass Hydrolyzate

Hee-Geun Nam[1], Chanhun Park[2], Heejeong Hwang[2], Se-Hee Jo[1] and Sungyong Mun[3], (1)Chemical Engineering, Hanyang University, Seoul, South Korea, (2)Department of Chemical Engineering, Hanyang university, (3)Hanyang University, Department of Chemical Engineering, Seoul, South Korea

In the production process of bio-ethanol from biomass, acetic acid is recognized as the key impurity to be removed from the sugar components that are generated by hydrolyzing biomass. To address this issue, the three-zone simulated moving bed (SMB) process based on the Amberchrom CG161C adsorbent was applied to remove acetic acid from the sugar components. First, the intrinsic parameters of all the relevant components were estimated through a series of single column experiments. Based on the estimated parameters, the optimal operating conditions of the three-zone SMB of interest were determined. To validate the optimized three-zone SMB process, the experimental unit of the SMB process was assembled and the relevant three-zone SMB experiment was conducted. The assay of all the resultant product samples verified that the SMB separation of interest was performed successfully as designed. The experimental data were also found to agree closely with the model predictions. Finally, a partial-discard strategy was applied to maintain the sugar concentration as high as possible.

(395y) Contracted Titanium Silicate (CTS-1): Aspects of Structure and CO2 Sorption

Ocean Cheung, Jing Li, Jie Su, Zoltan Bacsik, Xiaodong Zou and Niklas Hedin, Department of Materials and Environmental Chemistry, The Berzelii Center EXSELENT on Porous Materials, Stockholm University, Stockholm, Sweden

Engelhard Titanium Silicates (ETS) is a family of microporous titanium silicate materials first reported in 1989. Since then some variants were studied or implemented in applications. The Molecular Gate^{TM} technology can be used to upgrade CH_4 from mixtures of CH_4, N_2, and CO_2. It operates with a contracted titanium silicate (CTS-1) sorbent that relates to ETS-4. Surprisingly little is known for the structure and properties of such CTS-1 sorbents. We examined the potential of alkaline earth exchanged CTS-1 as sorbents for CO_2. All tested materials showed excellent levels of CO_2 uptake at 273 K. In particular, Mg^{2+} exchanged CTS-1 had high preferential CO_2 uptake (2.16 mmol/g, 273 K) with an ideal CO_2-over-N_2 selectivity of 14. CO_2 physisorbed rapidly with rates relevant for swing adsorption processes (80% of maximum capacity was reached in 45 seconds). We followed CO_2 sorption on Mg-CTS-1 by in situ infrared spectroscopy and characterized the physis- and chemisorbed CO_2 species. Vacuum and heat induced transformation of ETS-4 into CTS-1. In situ X-ray powder diffraction and electron diffraction were used to establish unit cell parameters of Mg-CTS-1. In summary, we showed that Mg-CTS-1 could potentially be a highly selective sorbent for carbon capture.

ADSORPTION RATE OF ORGANIC SOLVENT ON Y TYPE ZEOLITE

Kazuyuki Chihara, Natsuki Yamada,Tatsuharu Taira ,Yusuke Suzuki, Shinji Tomita, Kenta Saito*

Department of Applied Chemistry, Meiji University, Kawasaki, Kanagawa 214-8571, Japan

INTRODUCTION

In this study, adsorption measurement of various chlorinated hydrocarbons on Y-type zeolites were performed using the gravimetric method and the chromatographic method to examine the adsorption phenomena.Moreover, the relation between the calculation frequency and the amount of adsorption and the diffusivity was examined by the MD simulation.Molecular simulation has now become powerful tool for the study of adsorbed molecules in zeolites, and the Molecular Dynamics (MD) method is especially useful for predicting diffusion. However, information on forcefield parameters and charges are often inadequate, even in systems where the structure is well known. In this study, their forcefield was used to simulate adsorption isotherms and isosteric heats of adsorption for chloroform, trichloroethylene and tetrachloroethylene in Y-type zeolite. The results were compared with gravimetric and chromatographic experiment

EXPERIMENTAL

Gravimetric Method

Figure 1 showed experimental equipment for gravimetric method. The zeolite sample (about 0.3g) was placed in a quartz basket (J) Then the adsorbate in flask (B) was fed to adsorption tube (N). The whole equipment was in a constant temperature air bath kept at 303.15 K. The amount adsorbed was calculated corresponding to the pressure change of the vapor in the tube. The pressure was measured by pressure sensor (P) at higher pressure range (> 0.013 atm) and baratron (O) at lower pressure range (<0.013 atm). Adsorption curve obtained from this way was used for curve fitting program. In this way diffusion coefficients were obtained. Figure 2 shows the curve fitting equation and the theoretical curve was used at that time.

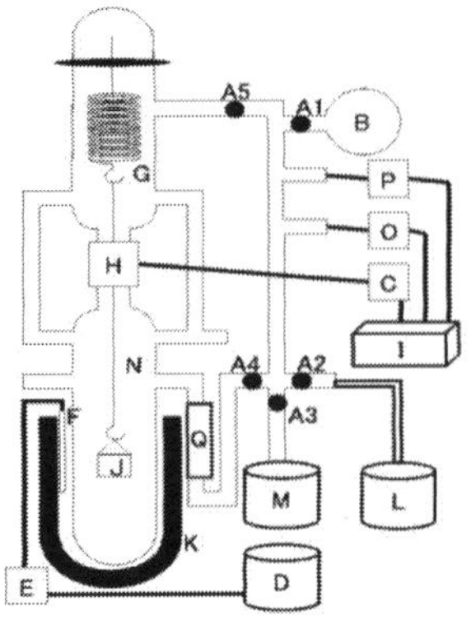

A1～A5	Needle valve
B	Adsorbate-flask
C	Displacement meter
D	Nichrome heter
E	Thermo controller
F	Thermo detector
G	Quartz spring
H	Dtretch detector
I	Data logger
J	Quartz basket
K	Mantle heater
L	Hydraulic pump
M	Vacuum pump
N	Adsorption tube
O	Baratoron
P	Pressure sensor
Q	Piston pump

Fig.1 Experimental equipment for gravimetric method

$$\frac{M(t)}{M(\infty)} = 1 - \sum_{n=1}^{\infty} \frac{6\alpha(1+\alpha)e^{-g_n^2 \tau_S}}{\left(9+9\alpha+\alpha^2 g_n^2\right) - \frac{(6+\alpha)\alpha g_n^2}{Bi} + \frac{\alpha^2 g_n^4}{Bi^2}} \quad (1)$$

$$\frac{\tan g_n}{g_n} = \frac{3Bi - \alpha g_n^2}{(Bi-1)\alpha g_n^2 + 3Bi} \quad (2)$$

$$Bi = \frac{a k_f}{\rho_P K^* D_S} = \frac{k_f a}{D_e} \quad (3)$$

$$\tau_S = \frac{Dt}{a^2} \quad (4)$$

Bi, τ s =dimensionless parameters. a = particle radius. ρ_P = the density of the particles.

K *= adsorption coefficient. Ds = effective diffusion coefficient in the particle size of the standard amount of adsorption. kf = film mass transfer coefficient. De = diffusion coefficient in the particle concentration standard. I also put out a percentage that represents the amount of adsorbate how much for the amount of adsorbed adsorbate poured into the path of the experiment(Fractional uptake(1/1+α). We have shown in the following equation was used to determine the Fractional uptake.

$$\frac{M_\infty}{VC_0} = \frac{1}{1+\alpha} \quad (5)$$

M(t) represents the amount of adsorbate moved inside the particles through the entire surface of the particles up to time t. M(∞)= equilibrium adsorption capacity. V= Volume in the experimental apparatus. C0 Introduction concentration. And creating performing curve fitting theoretical curve from these equations. Fig. 2 shows an example of dichloromethane.

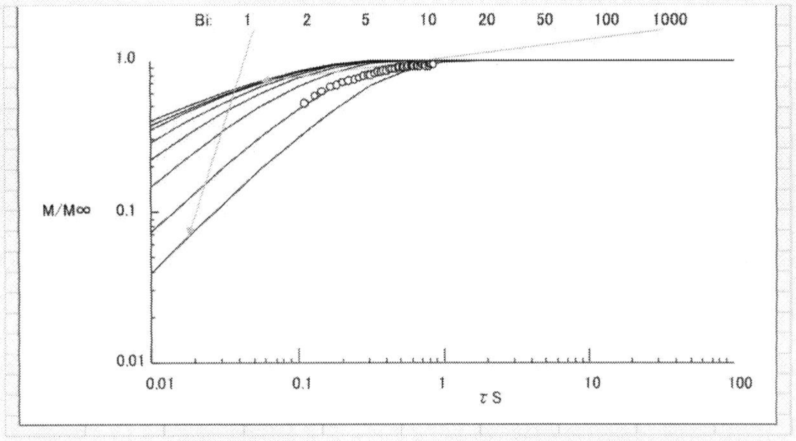

Fig.2 Theoretical curve(DCM 293K)

Chromatographic Method

In order to study the isotherm at lower pressure range, we used a pulse response method with a gas chromatograph. Figure 3 showed experimental equipment for chromatographic method.

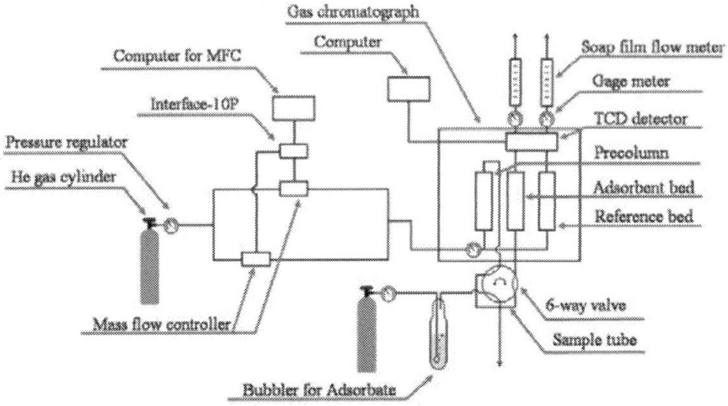

Fig.3 Experimental equipment for chromatographic method

Adsorption measurement of various chlorinated hydrocarbons on FAU-type zeolite was performed using the chromatographic method to examine the adsorption phenomena. Dichloromethane (DCM), chloroform (TCM), and trichloroethylene (TCE), were used as adsorbate, and HSZ-390HUD (Toso Co.) was used as adsorbent. By introducing the concentration pulse of adsorbate to the carrier gas stream at the inlet of the adsorbent column, the concentration elution curve at the outlet of the column was measured. Chromatographic peaks were analyzed by moment analysis equation. Van't Hoff plot and Arrhenius' plot were drawn with the use of adsorption equilibrium constant K* and micropore diffusivity D.

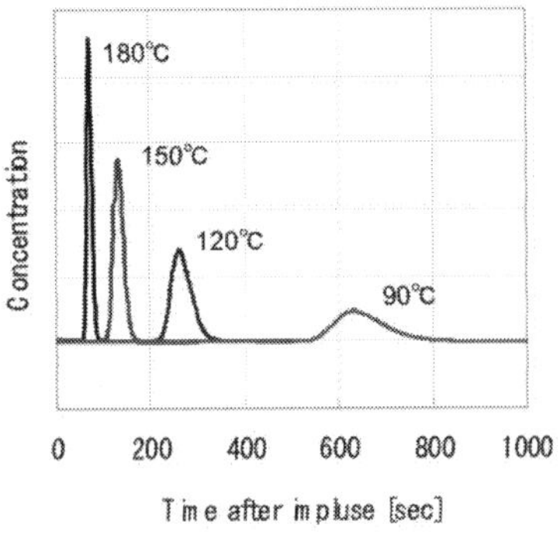

Fig.4 Typical chromatographic peaks

Adsorbate:chloroform[DCM]

Adsorbent:Y-type zeolite[HSZ-390HUD]

Moment analysis

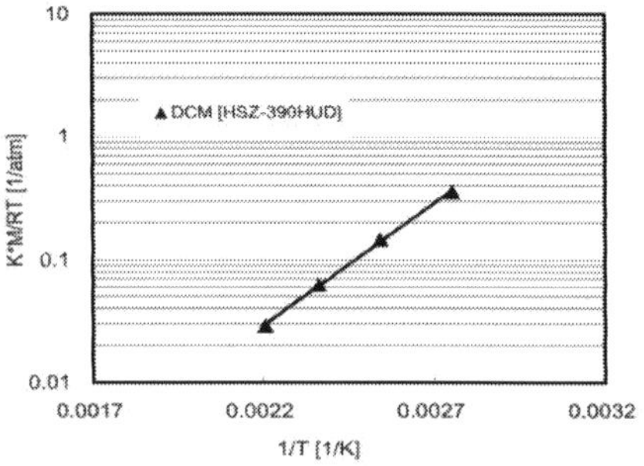

Fig.5 Typical Van't Hoff Plot

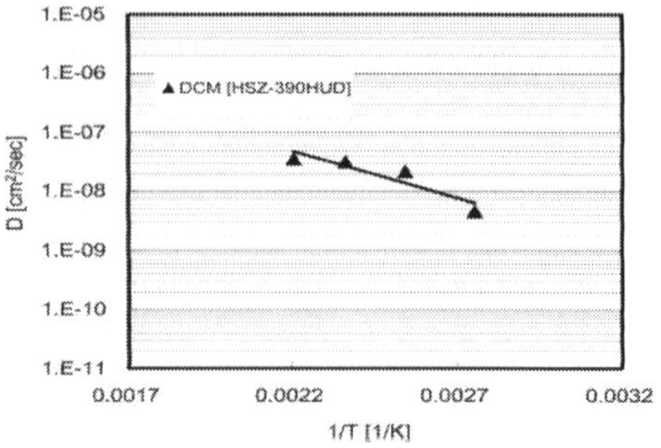

Fig.6 Typical Arrhenius' Plot

The obtained K^* divided by T was plotted against $1/T$ for various gases in Figure 5 to get the K^* value at 303K.

SIMULATION

Molecular Dynamics (MD) simulations for diffusion of the chlorinated hydrocarbons in the zeolites were performed using the Accelrys® MATERIALS STUDIO 5.5 simulation package. Self-diffusion coefficients were calculated from Mean Square Displacement (MSD) using the Einstein equations. The results were also compared with the experimental data. Figure7-9 shows a model of the MD simulation. Figure 10 shows an example of the mean square displacement.

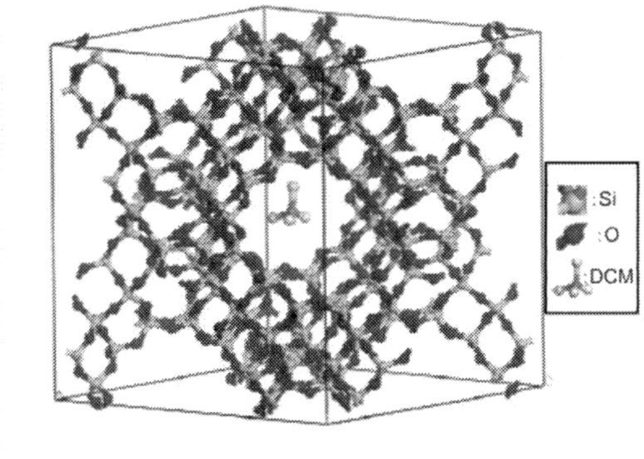

Fig.7 MD simulation model (DCM)

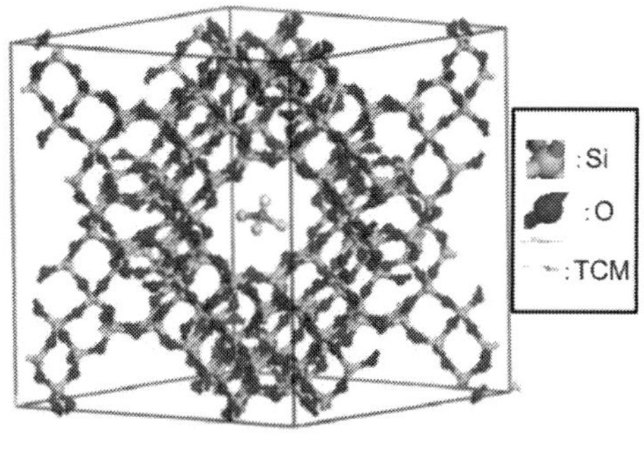

Fig.8 MD simulation model (TCM)

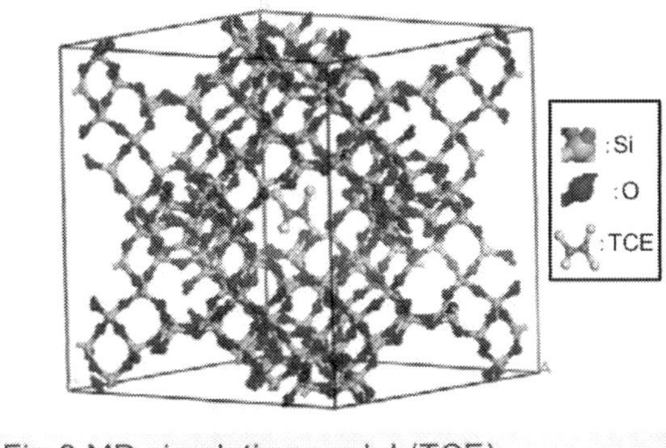

Fig.9 MD simulation model (TCE)

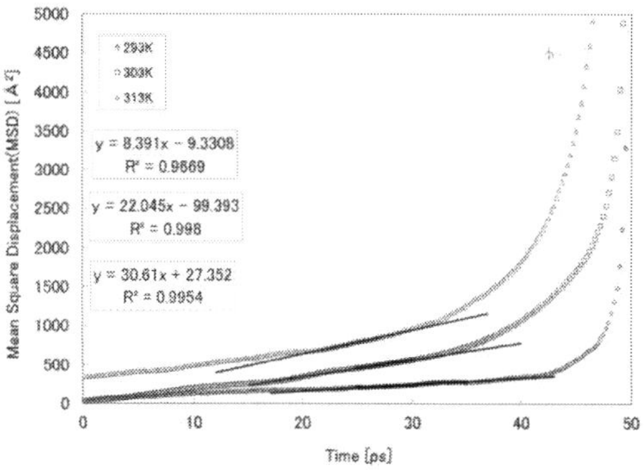

Fig.10 Mean square displacement(DCM)

RESULTS AND DISCUSSION

Fig. 11-13 shows the relationship between adsorption amount and adsorption speed parameters obtained by the gravimetric method experiment.

Fig. 14-16 shows the relationship between the mass transfer coefficient of the outer surface of the crystalline microporous diffusion coefficient.

The calculation of the diffusion coefficient D_0 adsorption amount zero reference, we have also carried out a calculation of activation energy.

Shown in Fig. 17 and Table 1.-3

I shows the time course of the amount of adsorption at 293K, 303K, 313K in dichloromethane system as an example.

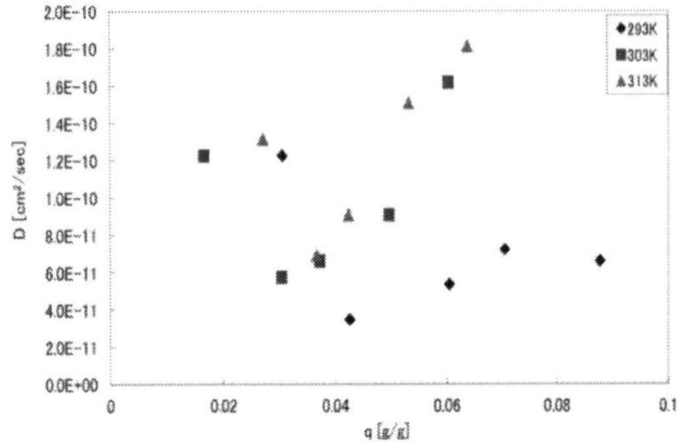

Fig.11 Relationship between the diffusion coefficient and the amount of adsorption(DCM)

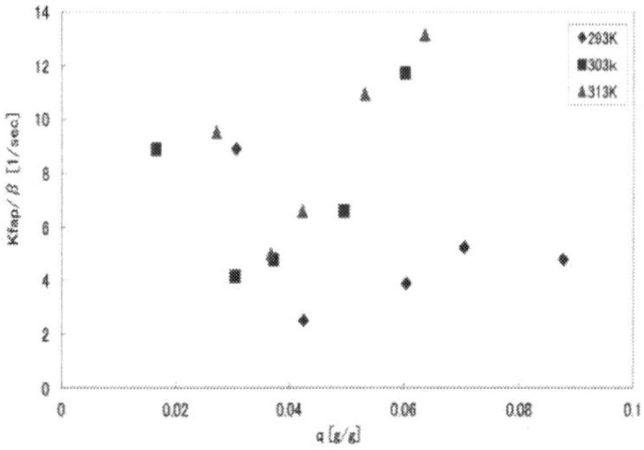

Fig.12 Relationship between mass transfer coefficient and the crystal outer surface adsorbed amount(DCM)

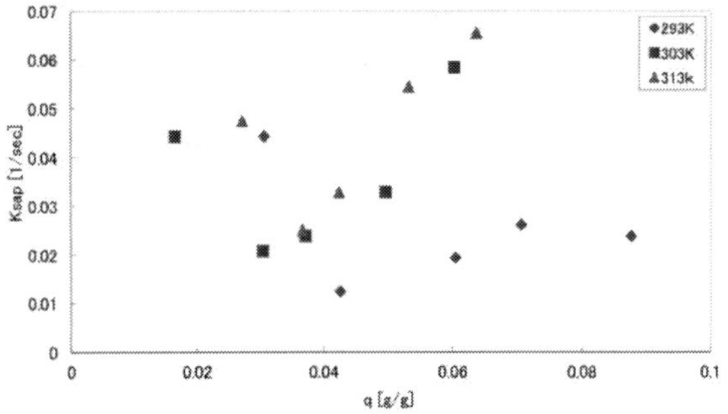

Fig.13 Relationship of overall mass transfer coefficient and the amount of adsorption(DCM)

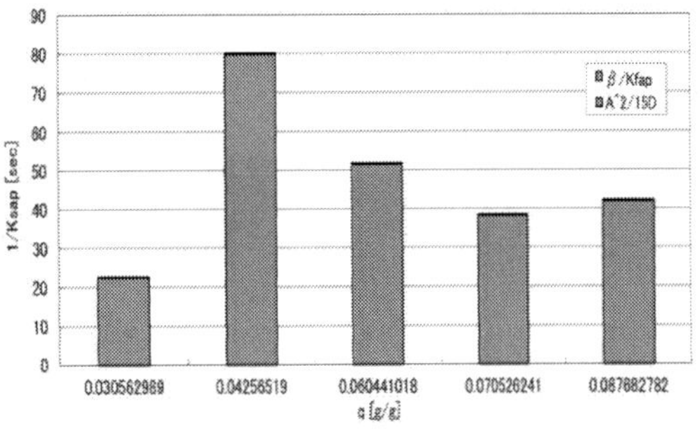

Fig.14 The sum of each mass transfer resistance (DCM 293K)

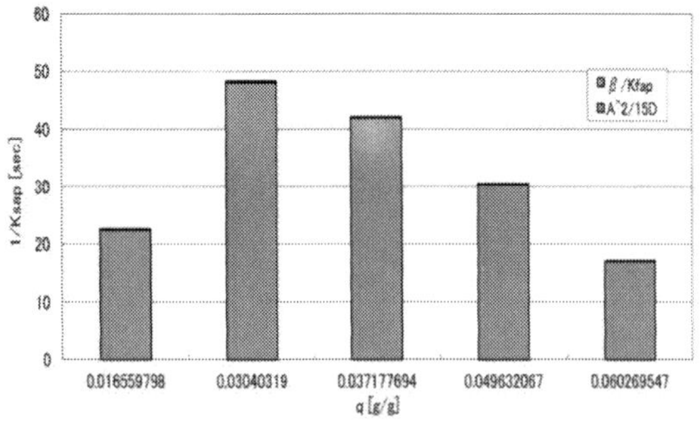

Fig.15 The sum of each mass transfer resistance (DCM 303K)

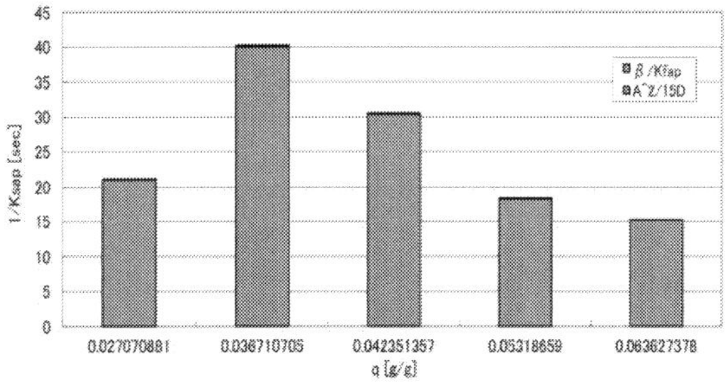

Fig.16 The sum of each mass transfer resistance (313K)

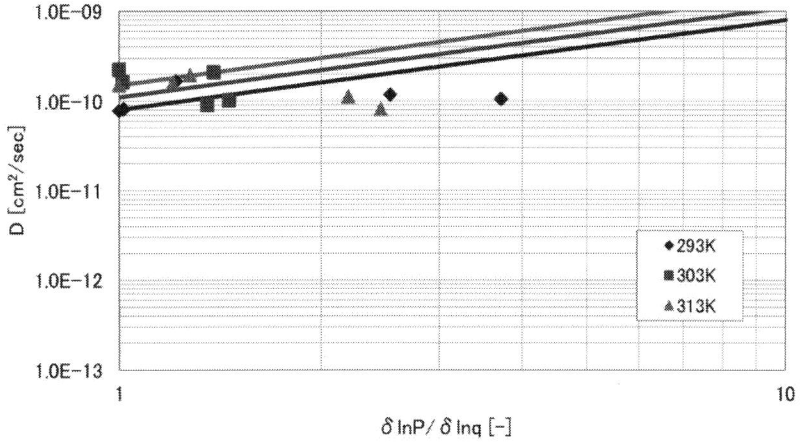

Fig.17 Calculation of the diffusion coefficient of the zero reference amount of adsorption

Figure 18-20 showed Diffusion coefficients obtained from experiments and simulations.

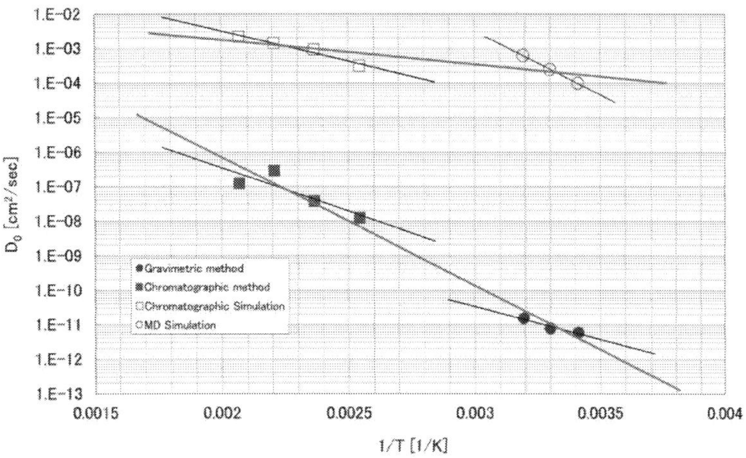

Fig.18 Diffusion coefficients obtained from experiment and simulation (TCE)

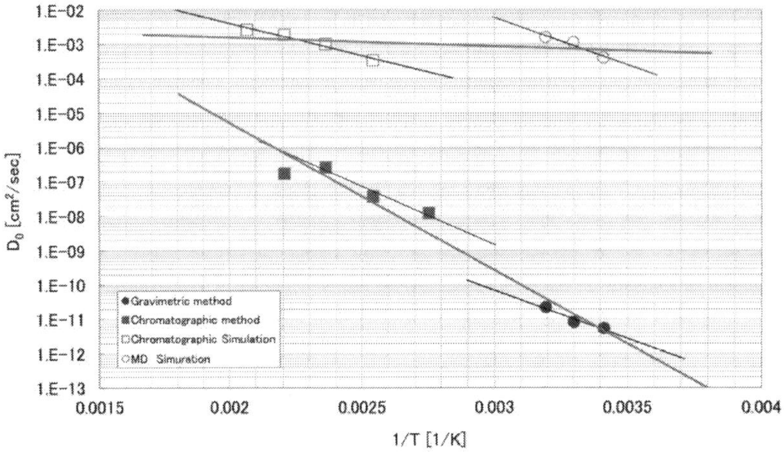

Fig.19 Diffusion coefficients obtained from experiment and simulation (TCM)

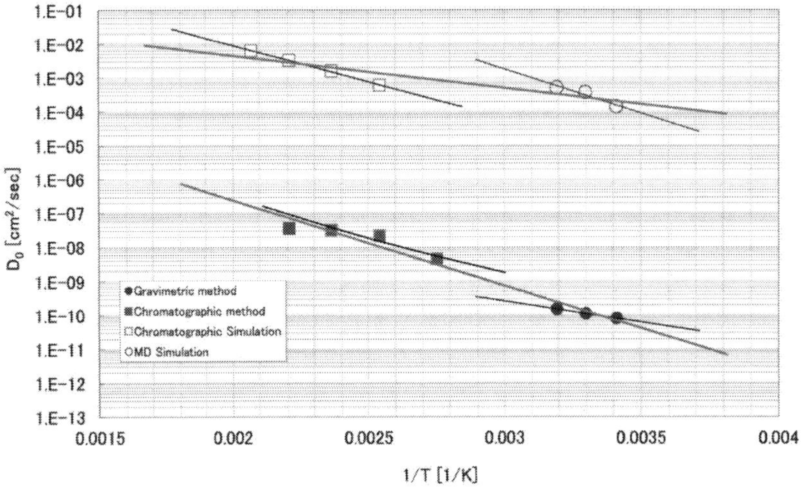

Fig.20 Diffusion coefficients obtained from experiment and simulation (DCM)

Table 1-3 showed activation energy obtained from experiments and simulations.

Table 1 . Activation energy obtained from experiments and simulations (TCE)

	Gravimetric method	MD sim.	Chromatographic method	Chromatographic sim.
Activation energy E[kJ/mol]	36.68	70.45	48.35	33.64

Table 2 . Activation energy obtained from experiments and simulations (TCM)

	Gravimetric method	MD sim.	Chromatographic method	Chromatographic sim.
Activation energy E[kJ/mol]	53.97	52.86	46.11	36.19

Table 3. Activation energy obtained from experiments and simulations (DCM)

	Gravimetric method	MD sim.	Chromatographic method	Chromatographic sim.
Activation energy E[kJ/mol]	51.59	49.61	46.11	36.19

CONCLUSIONS

In this study, we obtained diffusion coefficients from gravimetric and chromatographic method. I was not able to see a trend in the relationship between the amount of resistance in the outer surface and the adsorption percentage of resistance in the micropores and crystal. Further, it was not possible to determine the amount of adsorption dependence overall mass transfer resistance.

Diffusion coefficient obtained by the experiment was able to extrapolate the experimental diffusion coefficient obtained by chromatographic method.

It was possible to calculate the diffusion coefficient from the MD simulation. However, the value of the diffusion coefficient obtained from experiments and MD simulation results gravimetric method was 6-8 orders of magnitude different.Whereas the experimental value by gravimetric method to measure macro will perform a measurement in seconds, tracking is carried out in the MD simulation of the microscopic information from picosecond nanosecond. Therefore,it is considered a direct comparison with the self-diffusion coefficient and must be carefully considered.

I hope that the future, due to improved crystalline structure and parameters, calculation conditions, the calculation accuracy of the simulation is more developed.

REFERENCES

K. Chihara., K. Saito, Measurements and molecular simulation of adsorption of chlorinated organic compounds to the type and MFI-type zeolite FAU chromatographic method (2008).

(396a) Investigation of Sorbents for Warm CO2 Capture by Pressure Swing Adsorption

Zan Liu, David Couling and William H. Green, Department of Chemical Engineering, Massachusetts Institute of Technology, Cambridge, MA

Integrated gasification combined cycle with CO_2 capture and sequestration (IGCC-CCS) emerges as a promising method for achieving higher energy efficiency and better emission control compared with the traditional pulverized coal plant. But the current CCS technology, especially the commercially-available solvent-based CO_2 capture process is energy intensive which makes the IGCC process unaffordable. To achieve DOE's goal concerning IGCC-CCS, which requires a 90% CO_2capture with less than 10% increase in cost of electricity, some energy-efficient capture process needs to be developed.

Recent research conducted in our group indicates that the efficiency of IGCC can be improved if a warm CO_2 capture process based on solid sorbent can be employed. However, no commercial sorbents or sorption process exists for this purpose. To address this problem, in this paper, we present a sorbent material and an 11-step pressure swing adsorption (PSA) process which could be potentially used for warm CO_2 capture in IGCC plant. The metal oxide based sorbent material developed in our lab shows a stable working capacity, fast kinetics, and good cyclic stability in the temperature range of interest. Then, we propose an 11-step PSA process containing three pressure equalization steps and one high pressure purge step which can ensure a large H_2 recovery and lower compression work for CO_2 stream. The PSA process simulation is based on the measured performance specifications of the sorbent. In addition, the relationship between H_2 recovery and CO_2 capture is explored and improved by adjusting key operating parameters. Finally, the PSA model is integrated into an IGCC process simulation within Aspen Plus in order to determine optimal operating conditions and the power plant efficiency. A sensitivity analysis is conducted to suggest necessary material properties for improvement, which could potentially provide direction for next-step experimental work on sorbent development.

(396b) High-Throughput Isotherm Measurements On Functionalized Cr-MIL-101 Structures for Biogas Purification

Dirk Meister[1], Frieder Dreisbach[1], Angela Puls[1], Jens Rother[2], Tobias M. Fieback[2], Stephan Bernt[3] and Norbert Stock[3], (1)Rubotherm GmbH, Bochum, Germany, (2)Inst. of Thermo- and Fluiddynamics, Ruhr-University Bochum, Bochum, Germany, (3)Inst. for Inorganic Chemistry, Christian-Albrechts-University, Kiel, Germany

During the last decade the study of adsorption properties in a technical relevant pressure and temperature range came more and more in the focus of interest. The fast increase in the number of new porous materials, e.g. Hybrid materials as MOF's or MIL's as well as activated carbons or zeolites leads to a large variety of potentially candidates for technical adsorption applications.

For these applications a fast screening of these materials not only regarding to specific surface area and porosity but also at elevated pressures and temperatures with various technically relevant gases is necessary. Although the gravimetric method has some advantages to the volumetric method it is much easier to parallelize a volumetric measurement.

Many different adsorbent materials (AC, zeolites, MOF, IL) were screened with a new multi-sample volumetric instrument for their potential in adsorptive bio-gas cleaning and purification. The new instrument allows measuring the adsorption isotherms of up to six sample materials in one experiment. We have analyzed the adsorption of the typical biogas components (CH_4, CO_2, N_2, H_2, H_2S) in the pressure range up to 25 bar.

In this presentation we will focus on the results for a series of materials from the Cr-MIL-101 family[(1)]. Adsorption isotherms of the various technically relevant gases on the functionalized Cr-MIL-101-R (R = -OH, NH_2, NO_2, SO_3H, Br) will be presented. We will compare the results for the functionalized MOF materials with the results of commercially available MOF and classical adsorbents. Finally the results will be discussed in respect to the influence of the functionalization on the adsorption properties.

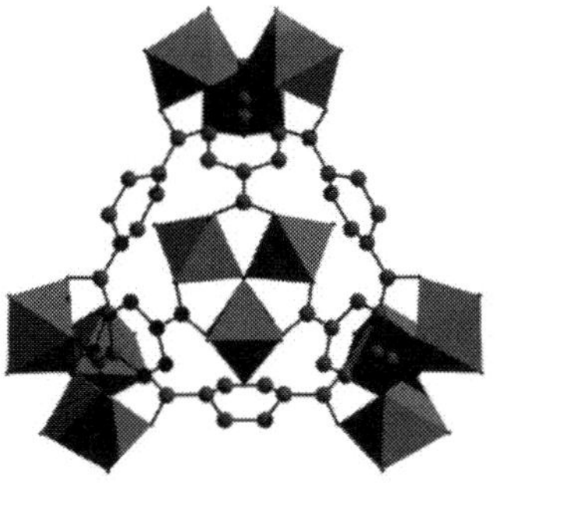

Graph 1: Supertetrahedron in the Cr-MIL-101 Structure.

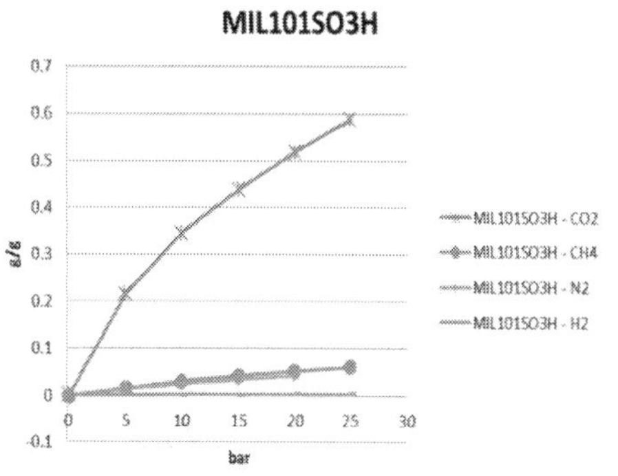

Graph 2: High pressure adsorption of various gases on Cr-MIL-101-SO_3H.

(1) G. Férey, C. Mellot-Draznieks, C. Serre, F. Millange, J. Dutour, S. Surblé, I. Margiolako, *Science* **2005**, *309*, 2040.

(396c) Adsorption Properties of Biochars From Swtichgrass, Hardwood and Softwood with and without Activation

Yanxue Han, U.S. Department of Agriculture, Crop Conversion Science & Engneering, Agricultural Research Service, Wyndmoor, PA, Akwasi A. Boateng, U.S. Department of Agriculture, Crop Conversion Science & Engineering, Agricultural Research Service, Wyndmoor, PA and Isabel M. Lima, U.S. Department of Agriculture, ARS, Agricultural Research Service, New Orleans, LA

Biochar can immobilize heavy metal ions and toxic organic contaminants from aqueous solution and soil. The adsorption properties of the chars typically depend on the porous structure and surface chemical properties which are, in turn, a function of the feedstock and pyrolysis conditions. We employed the biochars produced in a fluidized-bed fast pyrolysis (FP) reactor at 500 °C with a few seconds of residence time, and slow pyrolysis at 500 °C for 1h and 700 °C for 0.5h with and without activation to study zinc (Zn^{2+}) and copper (Cu^{2+}) adsorption in both aqueous solution and soil, as well as phenol adsorption in aqueous solution. Biochars were activated via steam at 800°C for 45 min at 3 ml/min water flow rate. The feedstocks from which these biochars were produced included switchgrass (SG), hardwood (HW) and softwood (SW). A quantitative determination of the adsorption properties was obtained using linear Langmuir isotherms, ATR-FTIR, Boehm Titration, BET surface area and SEM. These were used to explain the function of biochar surface functional groups in metal adsorption for biochars with and without activation.

(396d) Moisture Removal for Enhanced CO2 Separation From Post-Combustion Flue Gas

James C. Fisher II, National Energy Technology Laboratory, Department of Energy, Morgantown, WV and Ranjani V. Siriwardane, National Energy Technology Laboratory, U.S. Department of Energy, Morgantown, WV

Moisture removal for enhanced CO2 separation from post-combustion flue gas

James C. Fisher II[1,2], Ranjani V. Siriwardane*[1]

1. National Energy Technology Laboratory, United States Department of Energy, 3610 Collins Ferry Road, Morgantown, West Virginia 26507,

2. URS, 3604 Collins Ferry Road, Morgantown, West Virginia 26507

*ranjani.siriwardane@netl.doe.gov. Tel.: (304) 285-4513. Fax: (304) 285-4403 <>Abstract

Development of efficient CO_2 removal sorbents from flue gas is very critical for achieving successful processes for CO_2 emission control from existing coal combustion power plants. The moisture content of the power plant's flue gas has long been an issue for many types sorbents used for CO_2 separation because the sites which attract the CO_2, also attract H_2O. Materials such as zeolite 13X, have shown great promise for CO_2 capture with acceptable CO_2 capture capacities, low heat of reactions, and low sensible heat which contribute to low energy penalty for CO_2 removal. However, when moisture is present in the gas stream a high temperature is required to regenerate the strongly adsorbed H_2O molecules on the zeolite making the process more energy intensive. Many amine based sorbents also suffer from high moisture content via irreversible reactions resulting in unstable performance.

The purpose of this study is to find a cost effective method for removing the moisture from the flue gas stream to effectively use the sorbents for CO_2 separation. Molecular sieve 4A from Van Air Industries has shown great promise for moisture removal in this study with a capacity over 11 moles of H_2O per kilogram of sorbent (mol_{H2O}/kg). However, to achieve this high of capacity requires high regeneration temperatures reducing the overall efficiency of the power plant. In order to reduce the energy consumption a partial regeneration by heating the sorbent to 120 °C was used and it resulted in an acceptable CO_2 capture capacity of 4 mol_{H2O}/kg. Furthermore, removing the moisture from the gas stream allows an effective use of a sorbent like zeolite13X for CO_2 capture. . Preliminary energetic calculations indicate this process will be at least equal to, and potentially lower in energy consumption than current process modeled in NETL's baseline study.

(396e) Modeling Gas Adsorption-Induced Swelling of Coals

Pongtorn Chareonsuppanimit, Sayeed A. Mohammad, Robert L. Robinson Jr. and Khaled A. M. Gasem, School of Chemical Engineering, Oklahoma State University, Stillwater, OK

Gas adsorption-induced swelling of coals can impact the recovery of coalbed methane and design of CO_2sequestration processes. The swelling of coals can alter the cleat permeability and affect the flow behavior of gas in the coalbed reservoir. Studies have indicated that the cleat permeability is related to coal mechanical properties and changes to permeability can be estimated based on volumetric strain measurements on coals.

Pan and Connell (2007) presented a theoretical model to investigate adsorption-induced swelling of coals. In this study, we implemented the Pan and Connell swelling model through application of the simplified local-density (SLD) adsorption model. The theoretically rigorous SLD model, when combined with the Pan and Connell swelling model, was found capable of representing precisely the swelling behavior of coals due to adsorption of gases such as carbon dioxide, methane and nitrogen. The relation between strain and surface potential was found to be linear for all three gases mentioned above, confirming similar observations in the literature. An analysis of the literature data on swelling and its treatment with the SLD/Pan-Connell model will be presented. Further, the effect of coal swelling on changes to cleat permeability will also be presented.

Reference:

Pan, Z; Connell, L; 2007. A theoretical model for gas adsorption-induced coal swelling, International Journal of Coal Geology, 69, 4, 243-252.

(396f) Testing of Amidoxime-Based Adsorbent for the Recovery of Uranium From Seawater

Jungseung Kim[1], Yatsandra Oyola[2], Richard Mayes[3], Christopher Janke[2], Sheng Dai[3] and **Costas Tsouris**[4], (1)Energy and Transportation Science Division, Oak Ridge National Laboratory, Oak ridge, TN, (2)Oak Ridge National Laboratory, Oak ridge, TN, (3)Chemical Sciences Division, Oak Ridge National Laboratory, Oak Ridge, TN, (4)Energy and Transportation Science Division, Oak Ridge National Laboratory, Oak Ridge, TN

A polymeric adsorbent with amidoxime functional groups, synthesized at Oak Ridge National Laboratory (ORNL), has been employed in batch 5-gallon seawater experiments; flow-through 110-gallon seawater experiments; and field experiments at two marine sites. The effect of temperature on uranium adsorption kinetics and equilibrium in the range of 10 to 35°C has been investigated in batch experiments. From the experimental data, the adsorption kinetics and isotherm parameters will be determined. Kinetic models, coupled with mass transfer and equilibrium models, are employed to analyze the experimental data. A thermodynamic analysis will be also performed to provide changes in enthalpy, entropy, and free energy through the van't Hoff equation. Flow-through experiments with continuous recycle through 110-gallon seawater tanks have been conducted to test the effect of adsorbent packing in columns. Based on the results of laboratory experiments, marine testing has been conducted at two sites: (1) the Marine Sciences Laboratory of the Pacific Northwest National Laboratory at Sequim, WA, and (2) the Rosenstiel School of Marine & Atmospheric Science of the University of Miami, FL. Several weeks have been allowed for equilibrium to be reached in all types of the experiments. The parameters investigated in marine tests are temperature and flow-rate. Results are compared with experimental data obtained in parallel with Japanese adsorbents. The ORNL adsorbent has shown a capacity of 4 mg U per g of adsorbent, which is five times the capacity of the Japanese adsorbent. This study is expected to improve our current understanding of the influence of temperature on uranium adsorption by polymeric amidoxime-based adsorbents form seawater and provide guidelines for ocean deployment.

Experimental and Theoretical Comparative Study of Methane Adsorption on Activated Carbons and MOF-Basolite

Yuguo Wang, Cemal Ercan, Mohammed Hashim, Anwar Khawajah, Rashid Othman, Research & Development Center, Saudi Arabia Oil Company, Dhahran, 31311, Saudi Arabia

Abstract

In this paper, results of experimental and theoretical study of methane adsorption on activated carbons and MOF-basolite are presented. Five different activated carbons (named as AC1, AC2, AC3, AC4, AC5) and MOF-basolite were used in the study. The adsorbed amount of methane based on v/v, the critical factor in selecting materials for adsorptive storage, increases in the order of Basolite < AC4 < AC2 < AC1 < AC3 < AC5, which is not the same as the order of BET surface area – Basolite (867 m^2/g) < AC4 (999 m^2/g) < AC1 (1235 m^2/g) < AC3 (1426 m^2/g) < AC5 (1510 m^2/g) < AC2 (1589 m^2/g).. This reflects that fact that the synergistic effect of BET surface area, micropore volume, packing density and pore size distribution decides the adsorbed amount based on v/v.

One of the key design parameters for an adsorber, isosteric heat of adsorption is obtained at different fractional loading at 21°C. The initial isosteric heat of adsorption for methane-Basolite system is higher than that for methane-AC5 system, but it decreases faster as fractional loading increases. This indicates the higher degree of heterogeneity and could explain the close-to-rectangular isotherm shape of the methane-Basolite system. The degree of heterogeneity of an adsorbent-adsorbate system is expressed mathematically by using the slope of the isosteric heat of adsorption.

Introduction

Depending on the source and geographical location of production, natural gas can contain up to 95 mol% methane (CH4), the remaining being carbon dioxide (CO2), nitrogen, and small amounts of higher molecular weight hydrocarbons, such as ethane (C2H6), propane (C3H8) and butane (C4H10). It is well known that natural gas burns cleaner than gasoline, diesel and the other fuels. Because of this environmentally friendly behavior, it is commonly used for heating and its use continues to grow. It is also used in some vehicles by storing it as compressed natural gas (CNG) at pressures up to 270 atm. In current practice, natural gas is mainly stored as CNG for vehicle use or liquefied natural gas (LNG) for shipment. However, CNG requires expensive vessels and multi-stage compression and LNG also requires expensive cryogenic process. An attractive alternative to CNG and LNG might be adsorbed natural gas (ANG), where the gas is stored on porous material packed into a vessel at much lower pressure. ANG utilizes microporous adsorbents inside a vessel, which offers higher energy density and higher v/v storage capacity compared with CNG at typical natural gas pipeline conditions, pressure less than 50 bars and temperature less than 55°C1-9. So, at these conditions, ANG has high potential for exploitation not only in transportation but also in large-scale application such as ANG storage close to natural gas consumers.

In ANG operation, the most important item is the selection or development of a microporous material with high storage capacity and stability under cyclic operation. Potential microporous materials, which can meet these requirements, include activated carbon, metal-organic frameworks (MOF) and other organic or inorganic solids[10-20]. Although MOFs and the other organic solids display attractive sorption properties with a v/v adsorption capacity as high as 230 of absolute methane adsorption at 290 K and 35 bar, which also exceeds the DOE target of 180 v/v, methane is strongly entrapped in MOFs' structures and heating up to 100 °C is necessary for desorption[13]. The other factors, which may also affect the applicability of MOFs as natural gas adsorbent, include their stability and ability to tolerate impurities such as H_2S, black powder, mercaptans etc., which are common in industrial applications. Inorganic solids like zeolites have lower performance for methane storage than activated carbons. Moreover, zeolites present much more hydrophilic surfaces than carbons and tend to adsorb water preferentially. Therefore, activated carbon currently remains the only commercially viable adsorbent for natural gas storage in terms of adsorption capacity and stability.

In this paper, both experimental and theoretical investigations were carried out to determine the methane adsorption capacity of five different granular activated carbons and MOF-basolite in the temperature range of 10 to 56 °C. The practicality of using MOF-basolite as an adsorbent for methane storage is compared with activated carbons. Temperature dependent Toth equation was used to fit the adsorption data and to derive the isosteric heat of adsorption. The degree of heterogeneity is defined quantitatively.

Experimental

In this study, the volumetric method was used to measure the methane adsorption on adsorbents. The major components of the experimental set-up are shown in Figure 1. A vacuum pump, not shown in Figure 1, is also connected to the system for degassing. The adsorber has a thermal jacket connected to a heater/chiller so that the temperature of the adsorber can be set to desired temperature for measurement.

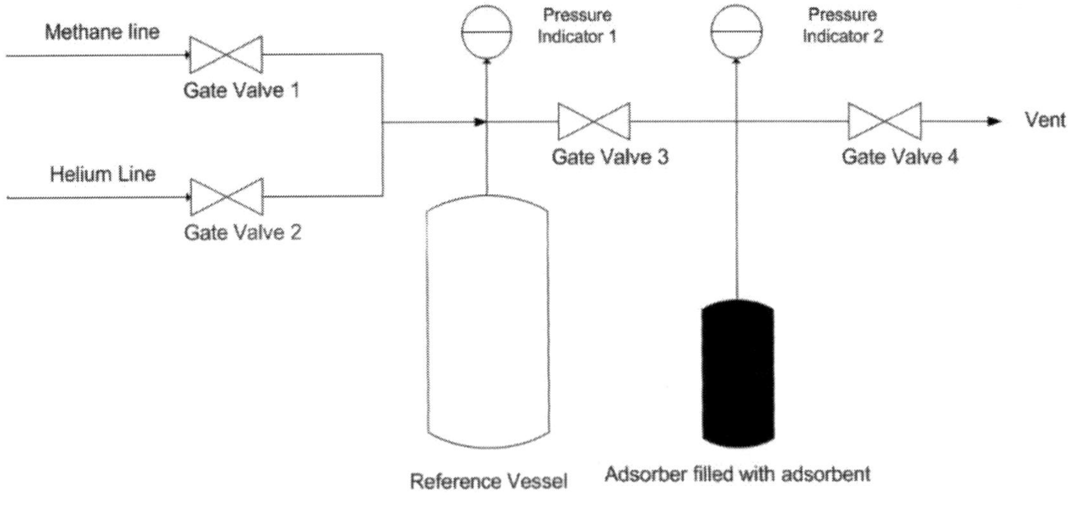

Figure 1. Schematic diagram of the experimental set-up.

To determine the adsorption isotherm, the adsorber was first filled up with 50 grams of granular activated carbon and was degassed at a pressure of 2.5×10^{-4} torr at 120 °C for 4 hours. Then, the adsorber was cooled down to and maintained at desired temperatures of 10, 21, 38 or 56 °C. The reference vessel has a volume of two liters and the adsorber has a volume of 120 cm^3. The reference vessel was first charged with methane to a certain pressure, then, valve 3 was opened and the pressure between two vessels was left to equalize. Readings of the two pressure indicators were taken after 20 minutes of reaching equilibrium.

Materials and Characterization

Five types of commercial granular activated carbons, labeled as AC1, AC2, AC3, AC4, AC5,with different physical characteristics and MOF basolite A 520 W in the pellet form (cylinder 3×3 mm) were used. The high purity methane (>99.9%) and ultra-high purity helium (99.99%) were used without further purification. Micromeritics ASAP surface area and porosity analyzer were used for N_2 adsorption/desorption at 77 K. The physical properties measured are in Table 1. Here, the total pore volume from nitrogen porosimeter is for the pores that have diameter range of 0 to 20 nm and the micropores are defined as the pores that have diameters range of 0 to 2 nm.

Table 1. Physical properties of the granular activated carbon samples

Sample	AC1	AC2	AC3	AC4	AC5	MOF-basolite
ASTM mesh size	8×16	30×70	2×60	12×40	6×60	3×3 mm cylindrical pellet
Bulk density (g/cm^3)	0.47	0.39	0.49	0.54	0.50	0.187
Skeletal density (g/cm^3)	2.299	2.363	2.402	2.059	2.286	1.817
Nitrogen Porosimetry (77 K)						
BET surface area (m^2/g)	1235	1589	1426	999	1510	867
Total Pore Volume (cm^3/g)	0.629	0.747	0.599	0.500	0.682	0.633
Micropore Volume (cm^3/g)	0.600	0.706	0.560	0.456	0.451	0.367
BJH average pore width (Å)	18.00	18.70	17.47	20.64	26.08	--

Determination of adsorption of methane

After the adsorber was filled with adsorbents, it was tapped gently until the level of the particles did not go down anymore. At this point, the particles was packed to its bulk density. The total volume of the adsorber cell ($V_{adsorber}$) consists of the adsorbent skeletal (V_{sk}) and void space (V_{void}) volumes. For precise determination of the total void volume, helium was used since it is essentially inert for adsorption at the experimental conditions of a temperature range of 0 to 56 °C and a pressure range of 0 to 50 bars. To calculate the amount of methane or nitrogen adsorbed per gram of activated carbon, non-ideal gas behavior is considered and Equation 1 is used. This volumetric measurement of adsorption is similar to the ones used by Gummar and Talu[21] and Salehi et al[22].

$$n_{adsorbed} = \frac{\left[\frac{P_1 V_r}{ZRT} - \frac{P_2 \times V_{void}}{ZRT}\right]}{M_{ac}} \quad (1)$$

where Z is compressibility factor, T is temperature, R is the ideal gas constant and M_{ac} is the mass of the adsorbent used. Z is determined from the well-known Soave-Redlich-Kwong (SRK) equation of state.

Results and Discussion

Figure 2.1 and 2.2 shows the plots of the amount of adsorbed methane versus pressure at 21 °C for the five different activated carbons. All the plots have the shape of Type-I isotherm, which indicates that these samples are microporous materials, typical of activated carbons.

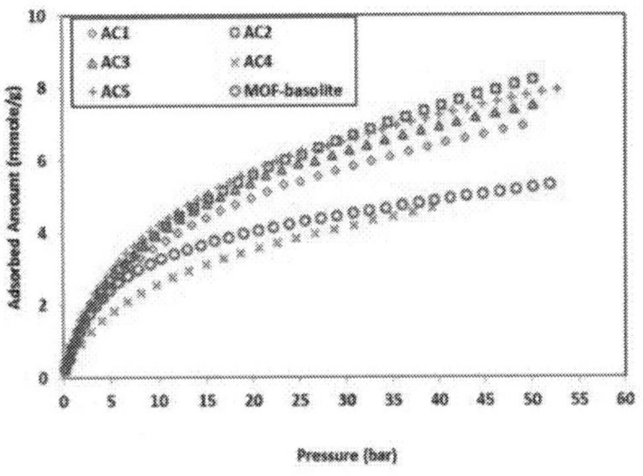

Figure 2.1. Comparison of the methane adsorption capacity among the five granular activated carbons based on mass to mass.

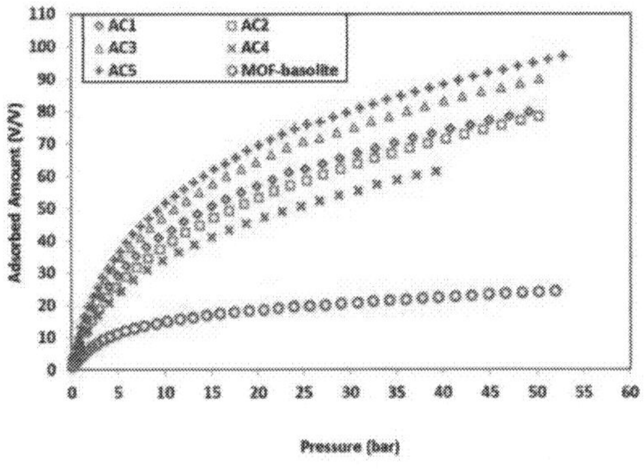

Figure 2.2. Comparison of the methane adsorption capacity among the five granular activated carbons based on volume to volume.

In Figure 2.1, Based on the moles of methane adsorbed per gram of activated carbon (mass to mass base, m/m), the capacity of methane adsorption on five activated carbons increases in the order of AC4 < AC1 < AC3 < AC5 < AC2. AC2 has the highest BET surface area of 1589 m^2/g while AC5 has the second highest BET surface area of 1510 m^2/g. Although the surface area of AC2 and AC5 are close to each other, this adsorbed amount difference may be due to micropore volume, different pore size distribution and AC2 has more slit pores that close to the optimal pore size for methane adsorption. Earlier theoretical studies[23-25] found that for methane adsorption on slit-pore activated carbon, the optimum pore width is 11.2 to 11.4 Å to create the maximum density for the adsorbed phase.

In Figure 2.2, based on the volume of methane adsorbed per volume of activated carbon in Figure 2.2, the capacity for methane adsorption increases in the order of AC4 < AC2 < AC1 < AC3 < AC5. This capacity for methane adsorption change can be attributed to the fact that the bulk density in volume to volume comparison becomes a more important factor. The bulk density for the five activated carbons are AC1(0.47 g/cm^3), AC2(0.39 g/cm^3), AC3(0.49 g/cm^3), AC4(0.54 g/cm^3), AC5(0.50 g/cm^3). Although AC2 has the highest BET surface area, it has the lowest bulk density. Then, for the same volume, AC2 will have the lowest mass packed. AC5 has the second highest BET surface area and bulk density, and AC4 has the highest bulk density but lowest BET surface area. Therefore, it is reasonable to say that the synergetic effect among BET surface area, micropore volume, micropore size distribution and bulk density makes AC5 have the highest volume to volume adsorption capacity.

To compare the adsorption capacity of MOF-basolite with activated carbons of similar BET surface area, AC4 is used. Although AC4 has lower m/m adsorption capacity than MOF-basolite as shown in Figure 2.1, its v/v adsorption capacity is much higher than MOF-basolite. Therefore, for fixed volume adsorber, the bulk density (packing density) of the adsorbent is also critical in deciding the adsorption capacity and storage amount.

The isotherms in Figure 2.1 and 2.2 are all Type-I adsorption isotherms. Many semi-empirical approaches have been proposed in describing Type-I adsorption equilibrium data[26]. These isotherm equations include Freundlich, Sips, Toth and Unilan equations. Among them, Toth equation, Equation 2–5, is the attractive one because of its correct behavior at low and high pressure.

$$C_\mu = C_{\mu s}\frac{bP}{[1+(bp)^t]^{1/t}} \quad (2)$$

$$b = b_0 exp\left[\frac{Q}{RT_0}\left(\frac{T_0}{T}-1\right)\right] \quad (3)$$

$$t = t_0 + \alpha\left(1-\frac{T_0}{T}\right) \quad (4)$$

$$C_{\mu s} = C_{\mu s,0} exp\left[\chi\left(1-\frac{T}{T_0}\right)\right] \quad (5)$$

Here, t is a parameter that characterizes the heterogeneity of the adsorbent-adsorbate system and its value is usually less than 1.0; $C_\mu, C_{\mu s}, C_{\mu s,0}$ are adsorbed amount, saturation adsorbed amount and saturation adsorption amount at the reference temperature respectively; b *and* b_0 are the adsorption affinity constant, adsorption affinity constant at reference temperature; R is ideal gas constant; P is adsorption equilibrium pressure; x is a parameter and Q is a measure of heat of adsorption. Q is also the isosteric heat of adsorption at zero fractional loading in Toth equation.

Average relative error (ARE) for the modeling is estimated from Equation 6.

$$Averagerelativeerror(ARE)\% = \frac{\sum\left|\frac{Y_i^{exp}-Y_i^{modeled}}{Y_i^{exp}}\right|}{N} \quad (6)$$

Here, N is the number of experimental data points, superscripts *exp* and *modeled* stand for the experimental and modeled values respectively, Y represents the amount of methane adsorbed.

Because AC5 has the highest adsorption capacity based on v/v, the data for AC5 and MOF-basolite is used in fitting the temperature dependent Toth equation. In Figure 3, isothermal adsorption data for AC5 and MOF-basolite is fitted with high accuracy of low average relative error, 1.4532% and 5.1070%, respectively for methane-AC5 and methane-Basolite adsorption systems.

The numerically optimized parameters from Toth equation were used in calculating the isosteric heat of adsorption at different fractional loading at 21°C using the equation 7, 8 and Toth equation 2-5[27]. The isosteric heat of adsorption versus fractional loading, Figure 4, shows that at less than 0.3 fractional loading, the initial isosteric heat of adsorption for methane-Basolite system is higher than that for methane-AC5 system, but it decreases faster as fractional loading increases. This indicates the higher degree of heterogeneity and could explain the close to rectangular isotherm shape of the methane-Basolite system in Figure 2.1 and 2.2.

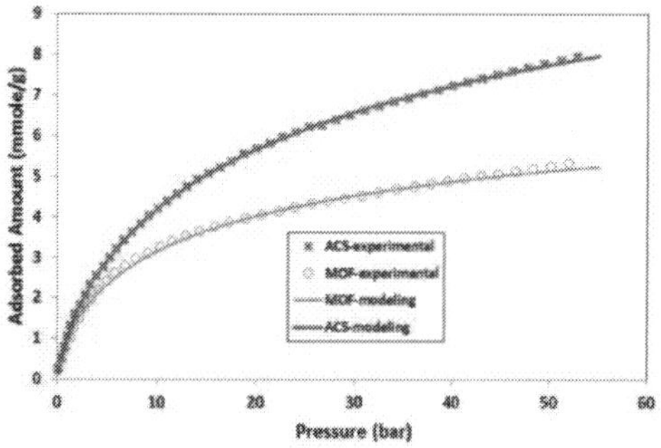

Figure 3. Adsorption of methane and its modeling at 21 °C on AC5 and MOF-basolite.

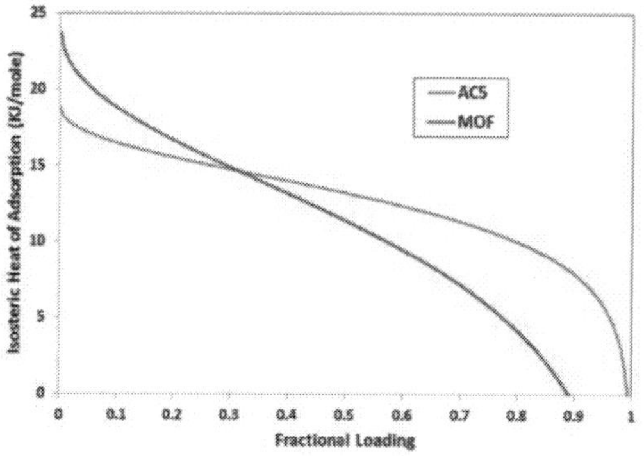

Figure 4. Isosteric heat of adsorption for methane on AC5 and MOF.

The degree of heterogeneity defines how the isosteric heat of adsorption changes as adsorption progresses to high coverage of the surface adsorption sites. As adsorption progresses, adsorbate molecule takes sites from high to low interaction energy sites, therefore, the amount of energy released per mole of adsorbate adsorbed decreases. This degree of heterogeneity can be defined by the absolute value of the rate of change of the isosteric heat of adsorption per mole of adsorbate adsorbed. Figure 5 is

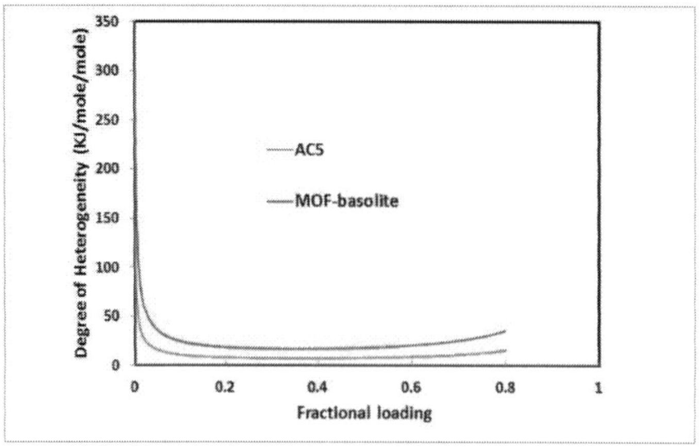

Figure 5. Degree of heterogeneity comparison between AC5-methane and MOF-basolite-methane systems.

the comparison of degree of heterogeneity between AC5 and MOF-basolite. MOF-basolite-methane adsorption system always has higher degree of heterogeneity than the AC5-methane adsorption system. This shows that different location of adsorption sites inside the pores of MOF-basolite has large difference of interaction energy with methane molecules. As adsorption progresses, from 10% of fractional loading up to 70%, degree of heterogeneity remains almost constant, which means that difference of interaction energy with the methane molecule changes at a constant degree. At higher degree of fractional loading over 70%, degree of heterogeneity increases slowly meaning the low interaction energy adsorption sites do not differ as largely as those at the beginning of the adsorption.

Conclusions

Five different activated carbons (named as AC1, AC2, AC3, AC4, AC5) and MOF basolite were used in the study. The adsorbed amount of methane based on v/v, the critical factor in selecting materials for adsorptive storage, increases in the order of Basolite < AC4 < AC2 < AC1 < AC3 < AC5, which is not the same as the order of BET surface area – Basolite (867 m^2/g) < AC4 (999 m^2/g) < AC1 (1235 m^2/g) < AC3 (1426 m^2/g) < AC5 (1510 m^2/g) < AC2 (1589 m^2/g).. This is due to the synergistic effect of BET surface area, micropore volume, packing density and pore size distribution decides the adsorbed amount based on v/v. Light bulk density is a disadvantage for MOF-basolite to be used as an adsorbent for methane storage. Activated carbons show high methane adsorption capacity based on v/v, and it the practical adsorbent for methane storage.

The initial isosteric heat of adsorption for methane-Basolite system is higher than that for methane-AC5 system, but it decreases faster as fractional loading increases. This indicates the higher degree of heterogeneity and could explain the close-to-rectangular isotherm shape of the methane-Basolite system. The degree of heterogeneity of an adsorbent-adsorbate system is expressed mathematically and different adsorbent-adsorbate systems can be compared directly.

References

1. Cook TL, Komodromos C, Quinn DF, Ragan S, Adsorbent storage for natural gas vehicles. In: Burchell TD. Carbon Materials for Advanced Technologies. Oxford: Elsevier Science Ltd., 1999:269-302.
2. Lu X-C, Li F-C, Ted, WA, Adsorption measurement in Devonian shales. Fuel. 1995; 74:599-603.
3. Chang KJ, Talu O. Behavior and performance of adsorptive natural gas storage cylinders during discharge. Applied Thermal Engineering. 1996;16:359-374.
4. Inomata K, Kanatawa K, Urabe Y, Hosono H, Araki T. Natural gas storage in activated carbon pellets without a binder. Carbon. 2002; 40:87-93.
5. Wegrzyn J, Gurevich M. Adsorbent storage of natural gas. Applied Energy. 1996;55:71-83.
6. Goetz V, Biloé S. Efficient dynamic charge and discharge of an adsorbed natural gas storage system. Chem. Eng. Comm. 2005;192:876-896.
7. Menon VC, Komarneni S, Porous adsorbents for vehicular natural gas storage: a review. Journal of Porous Materials. 1998;5:43-58.
8. Biloé S, Goetz V, Mauran S. Dynamic discharge and performance of a new adsorbent for natural gas storage. AIChE Journal. 2001;47:2819-2830.
9. Basumatary R, Dutta P, Prasad M, Srinivasan K. Thermal modeling of activated carbon based adsorptive natural gas storage system. Carbon. 2005;43:541-549.
10. Celzard A, Fierro V. Preparing a suitable material designed for methane storage: a comprehensive report. Energy & Fuels. 2005;19:573-583.

11. Pfeifer P, Ehrburger-Dolle F, Rieker TP, Gonzalez MT, Hoffman WP, Molina-Sabio M, Rodriguez-Reinoso F, Schumidt PW, Voss DJ. Nearly space-filling fractal networks of carbon nanopores. Phy. Rev. Letters2002;88:115502-1-115502-4.
12. Shao X, Wang W, Zhang X. Experimental measurements and computer simulation of methane adsorption on activated carbon fibers. Carbon. 2007;45:188-195.
13. Seki K. Design of an adsorbent with an ideal pore structure for methane adsorption using metal complexes. Chem. Commun.2001;16:1496-1497.
14. Eddaoudi M, Kim J, Rosi N, Vodak D, Wachter J, O'Keeffe M, Yaghi OM. Systematic Design of Pore Size and Functionality in Isoreticular MOFs and Their Application in Methane Storage. Science. 2002;295:469-472.
15. Dűren T, Sarkisov L, Yaghi OM, Snurr RQ. Design of new materials for methane storage. Langmuir. 2004;20:2683-2689.
16. Ma S, Sun D, Simmons JM, Collier CD, Yuan D, Zhou H-C. Metal-Organic framework from an anthracene derivative containing nanoscopic cages exhibiting high methane uptake. J. Am. Chem. Soc. 2008;130:1012-1016.
17. Atwood J L, Barbour LJ, Jerga A. Storage of methane and Freon by interstitial van der Waals confinement. Science. 2003;296:2367-2369.
18. Harlick PJE, Tezel FH. Adsorption of Carbon Dioxide, Methane and Nitrogen: Pure and Binary Mixture Adsorption by ZSM-5 with SiO_2/Al_2O_3 Ratio of 30. Separation Science and Technology. 2002;37:33-60.
19. Li P, Tezel FH. Adsorption Separation of N_2, O_2, CO_2, and CH_4 Gases by β-Zeolite. Microporous and Mesoporous Materials. 2007;98:94-101.
20. Li P, Tezel FH. Pure and Binary Adsorption Equilibria of Carbon Dioxide and Nitrogen on Silicalite.Journal of Chemical & Engineering Data. 2008;53:2479-2487.
21. Gumma S, Talu O. Gibbs Dividing Surface and Helium Adsorption.Adsorption. 2003;9:17-28.
22. Salehi E, Taghikhani V, Ghotbi C, Lay EN, Shojaei A. Theoretical and Experimental Study on the Adsorption and Desorption of Methane by Granular Activated Carbon at 25 °C. Journal of Natural Gas Chemistry. 2007;16:415-422.
23. Tan Z, Gubbins KE. Adsorption in carbon micropores at supercritical temperatures. J. Phy. Chem.1990;94:6061-6069.
24. Matranga, K. R.; Myers, A. L.; Glandt, E. D. Chem. Eng. Sci. 1992, 47, 1569-1579. Matranga KR, Myers AL, Glandt ED. Storage of natural gas by adsorption on activated carbon. Chem. Eng. Sci. 1992;47:1569-1579.
25. Chen X S, McEnaney B, Mays TJ, Alcaniz-Monge J, Cazorla-Amoros D, Linares-Solano A. Theoretical and experimental studies of methane adsorption on microporous carbons. Carbon. 1997;35:1251-1258.
26. Do DD. Adsorption Analysis: Equilibria and Kinetics, Series in Chem. Eng. Lodon:Imperial College Press, 1998.
27. Wang, Y, Ercan C, Khawajah A, Othman R. Experimental and Theoretical Study of Methane Adsorption on Granular Activated Carbons. AIChE J. 2012;58:782-788.

(396h) Structured Adsorbents with High CO2 Over N2 Selectivity and CO2 Uptake Capacity

Farid Akhtar, Department of Materials & Environmental Chemistry , Stockholm University, Stockholm, Sweden

Structured adsorbents with high CO_2 over N_2 selectivity and CO_2 uptake capacity

Farid Akhtar, Neda Keshavarzi, Arto Ojuva, Niklas Hedin, Lennart Bergström

Department of Materials and Environmental Chemistry, Stockholm University, Stockholm

SE-10691, Sweden

Berzelii Center EXSELENT on Porous Materials, Stockholm University, Stockholm 10691, Sweden

Abstract

Structured adsorbents with high CO_2 adsorption capacity, CO_2 over N_2 selectivity and rapid adsorption and desorption kinetics are ideal for CO_2 capture from power-plant flue gas at a low cost. We report here binder-free zeolite monoliths with high CO_2 over N_2 selectivity (> 500) and uptake capacity (4mmolg^{-1}) at 298 K and 101 kPa. Binder-free zeolite monoliths were consolidated by a rapid and facile processing approach called pulse current processing (PCP) from partially K exchanged NaA powders. The pore widow size of NaA was optimized by partially exchanging Na with K cations to achieve a high CO_2 over N_2 selectivity while maintaining a high CO_2 uptake capacity. The CO_2 uptake from binary mixtures of $0.10CO_2$-$0.90N_2$ was obtained from the single component CO_2 and N_2 adsorption isotherms by applying Ideal adsorption solution (IAS) and Fast IAS theories. The binder-free adsorbents with an optimum degree of ion exchange display extraordinarily high CO_2 over N_2 selectivity and high CO_2 uptake, together with a rapid CO_2 adsorption and desorption kinetics and high mechanical stability. The performance of the new adsorbent has been compared with other potential candidates for efficient swing adsorption processes by a figure of merit.

Key words: CO_2 Capture; Zeolite monoliths; Pulse current processing; Selectivity; Adsorption; Ideal adsorbed solution (IAS) theory.

(397a) Modeling of CO2 Absorption In and Stripping From Aqueous Amine, Amino Acid and Carbonate Blends

Jerry H. Meldon and Yi Gu, Tufts University, Medford, MA

An approximate analytical method based on steady-state Film Theory has been demonstrated to enable accurate, computationally efficient prediction of carbon dioxide absorption and desorption rates in packed columns. The method is based on a particular scheme for linearizing the coupled differential equations that govern steady-state diffusion with multiple reversible reactions. Interphase transport rates calculated based on Film Theory agree closely with those obtained via exact numerical methods. Notably, the same accuracy is observed whether the reactions are fast or slow; when their characteristic times differ by orders of magnitude; and irrespective of the magnitudes of transport and reaction driving forces. The method was recently applied [Meldon and Morales-Cabrera, *Chem. Eng. J.*, **2011**, *171*, 753] to the analysis of carbon dioxide absorption in and desorption from aqueous solutions of monoethanolamine, prior modeling of which had relied for accuracy on numerical methods of analysis. It is extended here to carbon dioxide capture with aqueous solutions of blends of amines, amino acids and alkali. The relative merits of candidate scrubbing solutions, for purposes of flue gas CO_2 capture at coal-fired power plants, are evaluated based on interphase transfer rates, rich and lean loadings, and stripper reboiler duties.

(397b) Optimization of Reactive Absorption Using an Evolutionary Algorithm

Chinmay Kale[1], Philip Lutze[2] and Andrzej Górak[2], (1)Laboratory of Fluid Separation, Department of Biochemical and Chemical Engineering, TU Dortmund University, Dortmund, Germany, (2)Department of Biochemical and Chemical Engineering, Laboratory of Fluid Separations, TU Dortmund University, D-44227 Dortmund, Germany

Optimization of reactive absorption using an evolutionary algorithm

Chinmay Kale, Philip Lutze, Andrzej Górak

TU Dortmund University, Department of Biochemical and Chemical Engineering, Laboratory of Fluid Separations, Emil-Figge-Strasse 70, D-44227 Dortmund, Germany; phone: +49 (0) 231 755-3034, e-mail: chinmay.kale@bci.tu-dortmund.de

Keywords: Column design, operating conditions, non-equilibrium stage model, operating cost

Suggested theme: Separations division: Distillation and absorption - Advances in Absorption (02a02)

Increasing amount of CO_2 is considered as the main reason for the global warming. Hence, carbon capture and storage has received a lot of attention over the past decade. Post combustion capture of CO_2 by reactive absorption using aqueous amines is an attractive solution to mitigate the CO_2 emissions from coal-fired power plants. This process has been extensively studied in terms of thermodynamics, reaction kinetics and the selection of amine(s). Nevertheless, for economic design of the absorption columns, rigorous analysis of the process based on non-equilibrium stage model and process optimization is necessary. In this contribution, results of detailed model validation of the reactive absorption together with the results of the sensitivity analysis and optimization are presented.

The process optimization is based on a generic non-equilibrium stage model which considers multicomponent mass and heat transfer, chemical reactions, non-ideal behaviour of the phases and hydrodynamics in the column. The model has been validated using the experimental results from the pilot scale absorption column and it predicts the experimental results with acceptable accuracy [1].

The optimization study was done by assuming 90% removal of CO_2 from feed gas as the design criterion for absorption column. The optimization problem is solved by applying an evolutionary algorithm [2] to minimize the annualized operating costs for absorption of CO_2 as objective function. The optimization variables are mainly column dimensions, flow rates and concentration of absorption solvent. The optimization method mentioned above allows the simultaneous determination of the dimensions of the column and operating conditions required to achieve the design criteria with minimum possible operating cost.

References

[1] C. Kale, I. Tönnies, H. Hasse, A. Górak, *Computer aided chemical engineering* **2011**, *29*.

[2] R. Angira, B. Babu, *Chemical Engineering Science* **2006**, *61 (14)*, 4707.

(397c) Derivation and Validation of a Short-Cut Model for the Absorption of CO2 in an OCM Mini-Plant

Erik Esche, David Müller, Steffen Stünkel and Günter Wozny, Chair of Process Dynamics and Operation, Berlin Institute of Technology, Berlin, Germany

Derivation and Validation of a Short-Cut Model for the Absorption of CO_2 in an OCM Mini-Plant

Erik Esche*, David Müller**, Steffen Stünkel***, Günter Wozny****

Chair of Process Dynamics and Operation, Berlin Institute of Technology, Sekr. KWT-9, Str. des 17. Juni 135, D-10623 Berlin, Germany

*erik.esche@tu-berlin.de, **david.mueller@tu-berlin.de, ***steffen.stuenkel@tu-berlin.de, ****guenter.wozny@tu-berlin.de

The Oxidative Coupling of Methane (OCM) presents a possibility of catalytically turning methane into longer hydrocarbons, especially ethylene and ethane. Hence, it could be an opportunity for replacing oil with methane from natural or biogas for producing base chemicals such as polymers, ethylene oxide etc.

As part of the Cluster of Excellence "Unifying Concepts in Catalysis", funded by the German Research Foundation, a mini-plant has been built at Berlin Institute of Technology (Technische Universität Berlin). The mini-plant features several types of reactors, e.g. fixed-bed, membrane, and fluidized-bed reactors, thus implementing the OCM process. For the subsequent product gas separation an absorption-desorption process for the extraction of carbon dioxide (CO_2), two gas separation membranes, and an adsorption-desorption unit are installed.

It is rather difficult to find the optimal combination and operation conditions of these reaction and separation units only through experimental investigations. On top of that, there are fluctuations in all mass and energy flows, which have to be taken into regard to ensure a safe and reliable operation.

Consequently, simulative optimization under uncertainty of the entire superstructure is proposed. Using rigorous models for all plant components would create a large-scale MINLP with complex non-linearities for example caused by reaction kinetics. Preliminary, deterministic optimization studies of separate reactors already showed convergence difficulties with the number of variables for a conventional packed-bed membrane reactor surpassing 130,000. To mitigate the size, complexity, and convergence time validated short-cut models should be introduced for all units in the entire mini-plant.

As a first step, the stand-alone optimization of the CO_2 absorption process is dealt with. Extensive experimental data for both monoethanolamine (MEA) and piperazine-activated methyldiethanolamine (aMDEA) for synthetic OCM product gas in the mini-plant described above exist and simulations have been set-up in Aspen Plus mimicking the plant behavior. The rate-based model in Aspen Plus also shows convergence difficulties, which highlights the necessity for simpler, more stable short-cut models.

The absorption desorption process is modeled in AMPL using an equilibrium-based approach for each theoretical plate of the packed columns. The number of components appearing in the reaction mechanism for the chemical equilibrium is reduced using dummy species representing the left out components. The heat of absorption is modeled similarly. Furthermore, basic equations for heat loss, pressure drop, and efficiencies for each theoretical plate are introduced. Preliminarily, AMPL and Aspen Plus simulations are compared to assess the behavior of the absorption in the short-cut model using the rigorous model as a reference case.

The derived model is compared to own experimental data for gases containing 15 to 25 vol% of CO_2, gas and liquid load factors of 0.3 to 0.7 $Pa^{0.5}$ and 20 to 40 m^3/m^2h respectively at pressures from 10 to 32 bar. The parameters for heat transfer, pressure drop, and efficiencies are fitted accordingly.

To prepare for the introduction of the model for the CO_2 absorption process into the whole superstructure, the stand-alone operation is optimized. The OCM product gas with a predefined composition is fed to the absorption column, which is specified to remove 90% of the CO_2 contained in the feed flow. The objective function for the stand-alone operation is the minimization of the energy required for the separation of CO_2, which is mainly the heat needed for regenerating the scrubbing fluid in the desorption column.

The optimal operating conditions are then applied to the mini-plant and validated by testing the plant behavior over several hours and observing the required adherence to the specifications. The optimality of the results is checked with the help of an experimental sensitivity analysis.

As a next step uncertainties observed in the experimental data will be added to feed streams etc. of the absorption process and the optimization problem will be reformulated for optimization under uncertainty.

Acknowledgements

This work is part of the Cluster of Excellence "Unifying Concepts in Catalysis" coordinated by the Berlin Institute of Technology (Technische Universität Berlin). Financial support by the Deutsche Forschungsgemeinschaft (DFG) within the framework of the German Initiative for Excellence is gratefully acknowledged.

(397d) Validation of a Process Model of CO2 Capture in an Aqueous Solvent, Using an Implicit Molecular Based Treatment of the Reactions

Charles V. Brand, Javier Rodriguez-Perez, Amparo Galindo, George Jackson and Claire S. Adjiman, Centre for Process Systems Engineering, Dept of Chemical Engineering, Imperial College London, London, United Kingdom

Validation of a process model of CO_2 capture in an aqueous solvent, using an implicit treatment of the reactions

C. V. Brand, J. Rodriguez-Perez, A. Galindo, G. Jackson, C .S. Adjiman

Departement of Chemical Engineering, Centre for Process Systems Engineering,

Imperial College London, London SW7 2AZ, United Kingdom

Carbon dioxide (CO_2) emissions are considered by the majority of the scientific community to play a major role in climate change and particularly in global warming. In this context, the development of carbon capture systems is a necessity that must be addressed in the short term. The most promising early stage technology, both in terms of performance and applicability, is currently thought to be post-combustion CO_2 absorption using amine solvents. There are, however, a number of concerns with the large scale deployment of this technology, including energy requirements, solvent degradation and the environmental and health impact resulting from loss of solvent and solvent degradation products. Modelling studies can play a useful role in addressing some of these issues and identifying the choice of solvent and operating conditions that yield the best performance. A key challenge is to develop models that can predict accurately the behaviour of the process, in the presence of limited experimental data.

We address this challenge by developing a complete CO_2 absorber-desorber model that incorporates state-of-the-art thermodynamics integrated into a rate-based process model. A characteristic of the model is that all reactions are treated within the thermodynamic model, based on the assumption that the reaction kinetics are not rate-limiting. This greatly reduces the amount of experimental data required to model the interactions between the solvent and CO_2 and therefore makes the approach ideally suited to study new solvents. Furthermore, no enhancement factor is used in our process model. Due to the transferability of the thermodynamic description, the same model is used for the absorber and the desorber. This approach is applied to CO_2 capture using an aqueous monoethanolamine (MEA) solution. The extent to which pilot plant data can be modelled is examined in detail, allowing the validity of the assumptions to be verified

The thermodynamic model used to determine the vapour-liquid equilibrium and the chemical equilibrium is based on the statistical fluid association theory for potential of variable range (SAFT-VR) [1]. In the SAFT-VR thermodynamic description, the reactions are treated implicitly within a physical perspective, with the products of the chemical reaction treated as associated aggregates of the reactant molecules, so that there is no need to incorporate explicit rates of reaction, temperature-dependent equilibrium constants, or mass balances on the reaction products. This greatly simplifies the model. It has been shown that this approach yields a good representation of the vapour-liquid and chemical equilibria of mixtures of CO_2, water and MEA under a wide range of conditions [2]. Different correlations for the rate-based equations governing the heat and mass transfer in the absorber are considered [3, 4].

The absorber-desorber process model is implemented in the gPROMS software and is validated using published pilot plant experimental data. The predictive capabilities of the mass transfer correlations used in this model are assessed through a sensitivity analysis and a scaling of the liquid-phase mass transfer coefficient is proposed. This scaling of the mass transfer is transferable to different operating conditions for both the absorber and the desorber and good predictions are obtained for the temperature and composition profiles in the gas and liquid phases.

[1] A. Gil-Villegas, A. Galindo, P. J. Whitehead, S. J. Mills, G. Jackson, and A. N. Burgess. *J. Chem. Phys.*, 106(10):4168–4186, 1997.

[2] N. Mac Dowell, F. Llovell, C. S. Adjiman, G. Jackson, and A. Galindo. *Ind. Eng. Chem. Res.*, 49, 1883-1899, 2003.

[3] K. Onda, E. Sada, and H. Takeuchi. *J. Chem. Eng. Japan*, 1(1):62-66, 1968.

[4] J. A. Rocha, J. L. Bravo and J. R Fair. *Ind. Eng. Chem. Res.,* 32(4):641-651, 1993.

(397e) CO2 Absorption in Aqueous Alkanolamine Systems : CO2 Concentration and Henry Constant Determination

Elise Provost, Vanessa Gonzalez and Walter Fürst, UCP, ENSTA ParisTech, Paris, France

For both energy and environmental purposes, the removal of acid gases such as carbon dioxide (CO_2) is involved in numerous processes such as natural gas treatment, as well as in petroleum refining, coal gasification, and hydrogen production. For such purification, reactive absorption with aqueous solutions of alkanolamines remains the most mature and widely implemented industrial technology.

Accurate understanding and modeling of the transfer occurring during the carbon dioxide absorption in alkanolamine solutions is of vital importance for optimizing the design of absorption units and reducing the absorber size, especially when high CO_2 loading are achieved. For the reason, knowing the concentration of molecular CO2 c_{CO2}, the transferring species, is crucial. Because CO2 reacts with the solvent to form HCO3-, CO32- and carbamate if a primary or secondary amine is used, c_{CO2} cannot be simply deduced from VLE measurements. A thermodynamic modeling is needed, and the Henry constant of molecular CO2 is deduced from the N2O analogy. An original experimental device has been developed allowing the direct measurement of c_{CO2} [1]. In this work, original measurement of c_{CO2} and HCO2 are presented for the CO2 loaded following systems (water + methyldiethanolamine MDEA), (water + MDEA + diethanolamine DEA), (water + DEA + 2-Amino-2-Méthyl-1-Propanol AMP). Differences are observed with the values deduced from the N2O analogy or predicted using electrolyte NRTL.

[1] Anas Archane, Walter Fürst and Elise Provost, Influence of Poly(ethylene oxide) 400 (PEG400) on the Absorption of CO_2 in Diethanolamine (DEA)/H_2O Systems, JCED, 2011 56(5) 1852-1856

Differential Solubility of Ethylene and Acetylene in Room Temperature Ionic Liquids by Computational Analysis

Xu Zhao, Huabin Xing, Qiwei Yang, Baogen Su, Zongbi Bao, Yiwen Yang and Qilong Ren**

Key Laboratory of Biomass Chemical Engineering of Ministry of Education, Department of Chemical and Biological Engineering, Zhejiang University, Hangzhou 310027, China

Ethylene and acetylene are very important chemical feedstocks for many products in petrochemical industries such as plastics, lubricants, intermediates, polymers and so on. The separation and purification for mixture of ethylene and acetylene has practical significance. Ethylene, obtained mainly from naphtha or natural gas cracking, usually contains less than 3 percent mole fraction of acetylene as an impurity due to their close boiling points. Acetylene production through coal cracking by plasma, a developing technology, also contains a small amount of ethylene (nearly 0.93 % mole fraction in a unit of acetylene). Solvent absorption of acetylene using an organic solvent, such as *N, N*-dimethylformamide (DMF) or *N*-methylpyrrolidinone (NMP), has the disadvantages of solvent loss and environmental pollution. Therefore, developing novel capture and separation methods for ethylene and acetylene is desired.

The room temperature ionic liquids (RTILs) have potential in realizing the ethylene (C_2H_4) and acetylene (C_2H_2) separation and avoiding solvent loss and environmental pollution compared with traditional solvents. The interaction mechanisms between gases and RTILs are significative for the exploration of new RTILs for gas separation, thus it was studied by quantum chemical calculation and molecular dynamics simulation in this work.

In the quantum chemical calculation, the optimized geometries were obtained for the complexes of C_2H_4/C_2H_2 with anions (Tf_2N^-, BF_4^- and OAc^-), cation ($bmim^+$) and their ion pairs, and the analysis for geometry, interaction energy, natural bond orbital (NBO) and atoms in molecules (AIM) was performed. The quantum chemical calculation results show that the hydrogen bonding interaction between gas molecule and anion is the dominant factor in determining the solubility of C_2H_2 in RTILs. However, the hydrogen bonding interaction, p-π interaction in C_2H_4-anion and the π-π interaction in C_2H_4-cation are weak and comparable, which all affect the solubility of C_2H_4 in RTILs with comparable contribution. The calculated results for the distance of $H_{gas}\cdots X$ (X= O or F in anions), BSSE-corrected interaction energy, electron density of $H_{gas}\cdots X$ at the bond critical point (ρ_{BCP}) and the relative second order perturbation stabilization energy ($E(2)$) are consistent with the experimental data that C_2H_2 is more soluble than C_2H_4 in the same RTILs and the solubility of C_2H_4 in RTILs has the following order: [bmim][Tf$_2$N] > [bmim][OAc] > [bmim][BF$_4$]. The calculated results also agree with the order of C_2H_2 solubility in different RTILs that [bmim][OAc] > [bmim][BF$_4$] > [bmim][Tf$_2$N]. Furthermore, the calculation results indicate that there is strong C_2H_2-RTIL interaction, which cannot be negligible compared to RTIL-RTIL interaction, thus the regular solution theory is probably not suitable to correlate

* To whom correspondence should be addressed. Telephone & Fax: +86 571 87952375.

E-mail address: xinghb@zju.edu.cn (Huabin Xing), renql@zju.edu.cn (Qilong Ren)

C_2H_2 solubility in RTILs.

In the molecular dynamics simulation, the diffusion coefficients of the ions have been found to increase after C_2H_4/C_2H_2 joins in. The interaction energy is consistent with the quantum chemical calculation results that C_2H_2 have a stronger electrostatic interaction with anion, and C_2H_4 have comparable electrostatic interaction of C_2H_4-anion and van der Waals interaction of C_2H_4-cation. The radial distribution functions show that C_2H_4 and C_2H_2 both have high distribution on the alkyl chain of imidazolium cation and the difference of C_2H_4 and C_2H_2 in ILs are attributed to the specific interaction with anions.

(414a) Technological and Economic Analysis of Adsorption Processes for Direct Capture of CO2 From Air

Ambarish R. Kulkarni, Stephanie A. Didas, Christopher W. Jones and David S. Sholl, School of Chemical & Biomolecular Engineering, Georgia Institute of Technology, Atlanta, GA

Direct capture of CO_2 from air is a concept that can be potentially used to account for CO_2 emissions from mobile and disperse sources that are inaccessible to flue gas capture. Initially, we study equilibrium process models to determine the viability of adsorption-based air capture technologies using an amino-modified silica adsorbent, TRI-PE-MCM-41, and a structured monolithic contactor unit. Using the purity of CO_2 and annual product throughput as metrics for comparing process performance; we identify some of the operational parameters, adsorbent characteristics and sources of parasitic energy losses that have a significant effect on the performance of the process. Specifically, we will discuss a process with a daily throughput of ~1.1 tCO_2 at 88.5 $\%CO_2$ purity using standard shipping container sized air capture modules. By accounting for the available sources of energy (low pressure steam and electricity) and geographic location, we obtain a net operating cost for air capture of CO_2 to be ~ \$100/tCO_2. Recent results from more detailed adsorption models and experiments will also be presented. Finally, potential strategies for reducing the energy and monetary cost of these processes will be discussed. Our analysis supports continued work to establish the technological and economic feasibility of adsorption-based air capture.

(414b) Dynamic Modeling and Control of a Solid-Sorbent CO_2 Capture Process with Two-Stage Bubbling Fluidized Bed Adsorber-Reactor

Srinivasarao Modekurti, Department of Chemical Engineering, West Virginia University, Morgantown, WV, Debangsu Bhattacharyya, Department of Chemical Engg., West Virginia University, Morgantown, WV and Stephen E. Zitney, AVESTAR Center, National Energy Technology Laboratory, Morgantown, WV

Solid-sorbent-based CO_2 capture processes have strong potential for reducing the overall energy penalty for post-combustion capture from the flue gas of a conventional pulverized coal power plant. However, the commercial success of this technology is contingent upon it operating over a wide range of capture rates, transient events, malfunctions, and disturbances, as well as under uncertainties. To study these operational aspects, a dynamic model of a solid-sorbent-based CO_2 capture process has been developed.

In this work, a one-dimensional (1D), non-isothermal, dynamic model of a two-stage bubbling fluidized bed (BFB) adsorber-reactor system with overflow-type weir configuration has been developed in Aspen Custom Modeler (ACM). The physical and chemical properties of the sorbent used in this study are based on a sorbent (32D) developed at National Energy Technology Laboratory (NETL). Each BFB is divided into bubble, emulsion, and cloud-wake regions with the assumptions that the bubble region is free of solids while both gas and solid phases coexist in the emulsion and cloud-wake regions. The BFB dynamic model includes 1D partial differential equations (PDEs) for mass and energy balances, along with comprehensive reaction kinetics. In addition to the two BFB models, the adsorber-reactor system includes 1D PDE-based dynamic models of the downcomer and outlet hopper, as well as models of distributors, control valves, and other pressure-drop devices. Consistent boundary and initial conditions are considered for simulating the dynamic model. Equipment items are sized and appropriate heat transfer options, wherever needed, are provided. Finally, a valid pressure-flow network is developed and a lower-level control system is designed.

Using ACM, the transient responses of various process variables such as flue gas and sorbent temperatures, overall CO_2 capture, level of solids in the downcomer and hopper have been studied by simulating typical disturbances such as change in the temperature, flowrate, and composition of the flue gas. To maintain the overall CO_2 capture at a desired level in face of the typical disturbances, two control strategies were considered–a proportional-integral-derivative (PID)-based feedback control strategy and a feedforward-augmented feedback control strategy. Dynamic simulation results show that both the strategies result in unacceptable overshoot/undershoot and a long settling time. To improve the control system performance, a linear model predictive controller (LMPC) is designed. In summary, the overall results illustrate how optimizing the operation and control of carbon capture systems can have a significant impact on the extent and the rate at which commercial-scale capture processes will be scaled-up, deployed, and used in the years to come.

(414c) CO2 Separation Using Mg(OH)2 Solutions – Dissolution Kinetics and Vapor-Liquid-Solid Equilibrium Studies of CO2- Mg(OH)2 System

Hari Krishna Bharadwaj[1], Joo-Youp Lee[1] and Tim.C Keener[2], (1)Chemical Engineering Program, University of Cincinnati, Cincinnati, OH, (2)Environmental Engineering Program, University of Cincinnati, Cincinnati, OH

Coal-fired power plants are responsible for a considerable amount of total anthropogenic CO_2 emissions throughout the world. CO_2 from flue gases can be absorbed by $Mg(OH)_2$ slurry at 52 °C in a column followed by a regeneration step in a stripper operated at 65 °C.

One of the most important steps in the CO_2 scrubbing process is the rate at which magnesium hydroxide dissolves from the solid part of the slurry and becomes available for reaction with absorbed CO_2. The dissolution rate kinetics of $Mg(OH)_2$ was studied using a pH stat and the dissolution rate was determined based on acid consumption required to maintain a preset pH value. The effects of pH, temperature, mixing and particle size on the dissolution rates of $Mg(OH)_2$ will be presented. A dissolution rate kinetic expression obtained under different operating conditions will also be presented.

Knowledge of Vapor liquid solid equilibrium (VLSE) of CO_2 in $Mg(OH)_2$ is essential for the proper design and operation of absorption and desorption equipment. The VLSE data for the $Mg(OH)_2$ – CO_2 system under various operating conditions of pressure, temperature, and concentration of $Mg(OH)_2$ will be presented. A suitable model for fitting the VLSE data will also presented.

(414d) CO_2 Capture Using Particulate Silica Aerogel Immobilized with Tetraethylenepentamine (TEPA)

Jerry YS Lin, Robert Pfeffer and **Nick Linneen**, Chemical Engineering, Arizona State University, Tempe, AZ

CO_2 capture using particulate silica aerogel immobilized with tetraethylenepentamine (TEPA)

Nick Linneen (presenter), Jerry YS Lin, Robert Pfeffer

School for Engineering of Matter, Transport, and Energy

Arizona State University

Tempe AZ 85289

Abstract:

Post-combustion capture of CO2 using solid amine-based sorbents is attracting significant attention due to their high selectively, reduced corrosive properties, and low energy regeneration requirements. Research has shown that sorbents that have large pore diameter and pore volume generally achieve the highest CO2 adsorption capacities. This is a result of the support's greater potential to retain large amounts of amine, while also allowing effective CO2 transport within the sorbent. We have developed a novel sorbent using particulate aerogel as a support. Aerogel, also known as "solid smoke," is a high porosity, high surface area, silica-based material that has very large pore volumes and pore diameters compared to other supports. Therefore aerogels can retain large amounts of amine while also maintaining some mesoporosity for effective CO2 transport. By impregnating tetraethylenepentamine (TEPA), which is an ultra-low molecular weight linear polyethylenimine (PEI), into commercially available 10 micron size aerogel (Cabot Corp. Nanogel), we have obtained an adsorption capacity as high as 6.1mmol/g which is very competitive relative to other CO2 sorbents. The aerogel-amine based sorbent was also found to achieve fast kinetics reaching 90% of its equilibrium capacity within the first 10min and also maintaining its adsorption capacity after 10 adsorption/desorption cycles.

(414e) Development of High Performance Carbon-Based Adsorbents for CO2 Adsorption

Placidus B. Amama, RXBT/UDRI, Air Force Research Laboratory, Wright-Patterson AFB, OH and Patrick J. Shamberger, RXBT, Air Force Research Laboratory, Wright-Patterson AFB, OH

Thermal management of high-power systems is a pressing challenge for many USAF platforms. These systems require rapid absorption and ultimately rejection of low-quality (low temperature) heat. Activated carbon is capable of adsorbing pressurized CO_2 gas with a high heat of adsorption, offering the possibility of rapidly absorbing and rejecting heat at low temperatures. The activated carbon adsorbent has to meet important performance criteria for high efficiency of the system. The performance criteria include high adsorption capacity, high adsorption kinetics, excellent stability and mechanical properties, and high thermal conductivity. For activated carbon adsorbents to meet all of these criteria, the CO_2 adsorption capacity and the adsorption-desorption kinetics must be improved. The adsorption of CO_2 on activated carbon can be significantly enhanced by the introduction of basic nitrogen functionalities on the surface. In this work, surface modification of a series of activated carbon adsorbents via gas phase amination under various conditions has been performed. The surface area, pore structure and pore volume were characterized using N_2 physisorption while the types of chemical functionalities present on the surface after amination were characterized by X-ray photoelectron spectroscopy (XPS), Fourier transform infrared spectroscopy (FTIR), and temperature programmed desorption (TPD). CO_2 adsorption on the modified activated carbon samples have been measured gravimetrically using a high-pressure TGA.

(432a) Vacuum-Enhanced Direct Contact Membrane Distillation: Modeling and Effect of Membrane Support Layer On Mass Transfer

Guiying Rao, Civil and Environmental Engineering, University of Nevada, Reno, Reno, NV, Andrea Achilli, Humboldt State University, Arcata, CA and Amy Childress, University of Nevada, Reno, NV

Vacuum-enhanced direct contact membrane distillation: modelling and effect of membrane support layer on mass transfer

G. Rao[1], A. Achilli [2], and A. E. Childress[1]

[1]Department of Civil and Environmental Engineering, University of Nevada Reno, Reno, NV 89557

[2] Environmental Resources Engineering Department, Humboldt State University, Arcata, CA 95521

Membrane distillation (MD) is a promising technology to produce drinking water because of its lower pressure and membrane mechanical strength requirements and almost complete rejection of salts and organics compared with pressure driven membrane processes such as reverse osmosis (RO). Practical application of MD is limited by its relatively low water flux. Vacuum-enhanced direct contact MD (VEDCMD) could be an attractive alternative because it shows 2-30 times higher water flux than direct-contact MD (DCMD) [1]. This study aims to develop a model for water flux prediction in VEDCMD, evaluate the value of vacuum enhancement, and investigate the effect of membrane support layer on mass and heat transfer.

Two mathematical models (Model 1 and Model 2) were developed for VEDCMD. Model 1 assumes that viscous flow and Knudsen diffusion are combined in parallel, and then they are combined with molecular diffusion in series. Model 2 assumes that Knudsen diffusion and molecular diffusion are combined in series, and then they are combined with viscous flow in parallel. Water fluxes predicted by both models using different temperatures and membranes were compared with experimental results. The experiments were performed using four different single-layer membranes under feed temperatures of 40 and 60 ^{o}C and distillate temperature of 20 ^{o}C in a bench-scale VEDCMD system.

The effects of vacuum-enhancement were also evaluated. In cases of low fouling solutions, vacuum-enhancement is desirable as it results in higher flux and does not significantly increase the rate of membrane fouling. The cost of vacuum enhancement was also compared with other methods to increase flux.

The effect of support layer on mass transfer was investigated as well by comparing experimental results of seven intact membranes (active layer plus support layer) with experimental results obtained with active layers only from each of these membranes. It was found that water fluxes of the membranes with active layers only were up to 60% higher than water fluxes of the intact membranes. Membrane characteristics were linked to water flux and results will be presented.

Reference:

[1] T.Y. Cath, V.D. Adams, Amy E. Childress, Experimental study of desalination using direct contact membrane distillation: a new approach to flux enhancement. Journal of Membrane Science, 2004. 228: 5 -16.

(432b) MD for Mineral Recovery From the Great Salt Lake

Tzahi Cath and Kerri Hickenbottom, Civil and Environmental Engineering, Colorado School of Mines, Golden, CO

As the demand for high-quality water increases, membrane technologies such as reverse osmosis are becoming widely used as treatment processes for tertiary wastewater treatment, water reclamation, and desalination. However, along with high-quality product water, a concentrated brine stream is produced as a byproduct of these treatment processes. Membrane distillation (MD) is an emerging membrane technology that has the capability to facilitate desalination and further concentrate brines. The advantages of MD are two fold: a high-quality water is produced for industrial and municipal uses and minerals can be extracted from the brine for agricultural and other beneficial uses (e.g., fertilizers, road salts). Additionally, MD has small footprint and ability to utilize low-grade heat, thus making it an attractive process over conventional processes for resource recovery.

Membrane distillation (MD) is a thermally driven membrane process in which the driving force is the difference in partial vapor pressures across a hydrophobic microporous membrane. One configuration of MD is direct contact MD (DCMD). In this process, brine solution and distillate streams are in direct contact on either sides of the MD membrane. The difference in partial vapor pressures of the two streams controls the mass transport of water vapors in MD: water evaporates from a heated feed stream of high salinity, diffuses through the pores of the membrane, and condenses into a cooler distillate stream on the opposite side of the membrane. MD has the potential to sustainably operate at or above saturation concentrations, thus possibly achieving close to zero liquid discharge. However, understanding of system performance under these conditions is limited. The current study aims at optimizing system performance through understanding the impact of membrane fouling/scaling (and their reversibility) on heat and mass transport through these membranes under supersaturated conditions.

DCMD experiments were performed with a feed solution of high salinity water from the Great Salt Lake (150 g/L, more than four times the salinity of seawater) and a distillate stream of deionized water. Optimal operating conditions were experimentally determined to be 50 and 30 °C for the feed and distillate streams, respectively, and the system was operated in a batch mode to closely monitor the performance of the membrane and vapor flux decline as the feed solution became more concentrated. DCMD was able to concentrate the feed solution to twice its original concentration, recovering 50% of the feed volume, and achieved 100% inorganic salt rejection! During the experiments, the vapor flux declined to 80% of its original value (11 to 2 $L\text{-}m^{-2}\text{-}hr^{-1}$), in which the concentrated feed neared supersaturation and scaling occurred at the membrane surface, indicating that increased concentration polarization limits heat and mass transfer through the membrane.

Scaling mitigation techniques were investigated to reverse the vapor flux to its original level and sustain the operations. Real-time microscopy was used to elucidate the onset of membrane scaling and revealed that membrane scaling contributes to the decline in vapor flux. Preventing scale formation on membrane surfaces sustained water flux and membrane integrity, and eliminated chemical consumption used for membrane cleaning. A system model was developed to evaluate the effect of operational and material parameters on system performance and to optimize operations, increasing water production and reducing energy demand

(432c) Treatment of Flowback/Produced Water From Gas Shale Plays with Membrane Distillation

Emily Wolff[1], Shiyun Zhu[1,2] and Radisav D. Vidic[1], (1)Department of Civil and Environmental Engineering, University of Pittsburgh, Pittsburgh, PA, (2)School of Environmental Science and Engineering, Shanghai Jiao Tong University, Shanghai, China

To reduce the amount of freshwater used in hydraulic fracturing and the overall cost of water management in a well field, shale gas development companies are striving to reuse most of the hydraulic fracturing flowback water. However, there are numerous instances when the flowback and produced water cannot be reused and must be disposed. Furthermore, flowback and produced water reuse is feasible only as long as there is sufficient number of new wells where this water can be utilized for hydraulic fracturing, which is not necessarily the case for mature fields. It has become increasingly evident that the future of the unconventional onshore gas industry is critically dependent on technical solutions that would enable efficient and cost effective management of flowback and produced waters. This study is designed to evaluate the promise of membrane distillation (MD) technology to meet this objective.

The total dissolved solids content of flowback and produced water can reach nearly 350,000 mg/L, ten times the average TDS concentration in seawater. Sodium and chloride are the primary ions present, followed by calcium, barium, strontium and magnesium. The water also contains some heavy metals at low concentrations. The extremely high TDS concentrations render the most prevalent desalination technologies employed today for seawater desalination ineffective. Even with favorable energy prices, reverse osmosis (RO) treatment of brine water becomes economically infeasible above 40,000 mg/L TDS (Cline et al., 2009). High salinity of the feed water decreases the driving force for mass transport in RO considerably but has minimal impact on MD because of the limited impact on vapor pressure of water, which is the driving force in MD, (Cath et al., 2004). Complete rejection of ions and dissolved non-volatile organics is achieved with MD as long as the membrane pores are not wetted (Nghiem et al., 2011). The other principal advantage of MD over other treatment technologies is a much lower expected energy demand because MD can operate at near atmospheric pressure and in the temperature range of 50-100 °C, which is significantly lower than typical temperatures used in conventional thermal treatment processes (e.g., distillation, mechanical vapor recompression). As a result, associated costs are predicted to be drastically lower than their thermal and membrane processes counter-parts, (Al-Obaidani et al., 2008).

In this present study, a flat sheet DCMD module with a membrane contact surface area of 0.004 m^2 was constructed to test the feasibility of MD technology for treating flowback and produced water from unconventional shale plays. Acrylic copolymer membrane with a mean pore size of 0.45μm was used in these experiments. Initial tests were conducted with this module using feed NaCl concentrations up to 30% at temperatures between 40-65°C and the results in terms of permeate flux and quality were similar to those found in the literature. Further testing was conducted using concentrated synthetic and actual flowback water from a well located in Westmoreland County, PA. The results are promising in that the permeate fluxes obtained are comparable to those obtained with NaCl only. The test with real flowback water, which had a TDS of 3.9%, yielded a permeate flux of 33.8 kg/(hr m^2) for an operating temperature gradient of just 31.5°C. In addition, no serious irreversible fouling or wetting of the membrane was observed during the 70-minute test with real flowback water and clean water permeability was restored after membrane rinsing with clean water. The results of ongoing studies to further demonstrate the feasibility of MD for flowback/produced water treatment at different feed compositions and temperature gradients will be presented at the conference.

References:

1. Al-Obaidani, S., Curcio, E., Macedonio, F., Profio, G.D., Al-Hinai, H., Drioli, E., (2008). "Potential of membrane distillation in seawater desalination: thermal efficiency, sensitivity study and cost estimation." Journal of Membrane Science **323**: 85–98.

2. Cath, T.Y., Adams, V.D., Childress, A.E., (2004). "Experimental study of desalination using direct contact membrane distillation: a new approach to flux enhancement." Journal of Membrane Science **228**(1): 5-16.
3. Cline JT, Kimball BJ, Klinko KA, Nolen CH. (2009). Advances in Water Treatment Technology and Potential Affect on Application of USDW. GWPC: 2009 Underground Injection Control Conference. San Antonio, TX January 26-29, 2009.
4. Nghiem, L.D., Cath, T.Y., (2011). "A scaling mitigation approach during direct contact membrane distillation." Separation and Purification Technology **80**(2): 315-322.

(432d) A Multiple-Effect Membrane Distillation Process for Concentration of Aqueous Hydrochloric Acid

Yingjie Qin[1], Rongling Liu[2], Xiaojun Li[2], Liqiang Liu[3] and Yanping Zhang[3], (1)Chembrane Research & Engineering,Inc., Newark, NJ, (2)Department of Biochemical Engineering, School of Chemical Engineering and Technology, Tianjin University, Tianjin, China, (3)Chembrane Engineering & Technology, Inc., Tianjin, China

It is well known that hydrochloric acid (HCl) is usually used as pickle liquor to remove surface oxide in a steel production industry. When pickling cannot be accomplished effectively and the quality of the treated metal surface deteriorates, the pickle liquor is discharged from the pickling tank, and the pickling tank is replenished with fresh acid solution. According to the data provided by the World Steel Association in 2011, the total world crude steel production was 1,490.1 million tonnes. And about 60 kg of spent pickle liquor was generated with the production of each ton of steel, and thus annual emission of spent pickle liquor was up to several million tons. As an example, 1000 ton of pickle liquor is being produced daily in a steel plant in north China, which contains 7% HCl and 15% $FeCl_2$. The spent pickle liquor usually contains 2-8 wt% hydrochloric acid and is considered a hazardous waste. The spent pickle liquor from steel processes is usually neutralized with lime and disposed in a landfill, which results into the following problem: high-salinity wastewater and sludge volume, the remaining salt residue processing difficulties, and most importantly, the acid unrecoverable. Therefore, there is an urgent need to recover and enrich hydrochloric acid to achieve economical and ecological benefits. Since the 1960s, hydrochloric spent pickling liquor is often treated in a hydrochloric acid regeneration system such as ion-retardation, diffusion dialysis and electro-dialysis, which recovers some of the hydrochloric acid and ferric oxide. Nevertheless, these regeneration processes produce lots of dilute hydrochloric acid solution. Thus, there still needs a novel and efficient technology to concentrate the recovered dilute HCl solution for further use. In the last few years, numerous studies have been performed to test the application of membrane distillation for concentrating dilute HCl solution. However, the high thermal energy consumption of the traditional MD process is one of the biggest barriers in its industrialization.

In the present study, multiple-effect membrane distillation (MEMD) based on AGMD module with function of internal heat recovery has been developed. This kind of MEMD process combines the advantages of MD process and conventional MSF process, avoids the disadvantages of MSF such as evacuation operation, and can provide a high PR value. The effects of feed-in concentration, cold feed-in temperature (T_c), hot feed-in temperature (T_h) and feed-in volumetric flow rate (F) on the performance of MEMD process were studied. The permeation flux (N) and energy efficiency, performance ratio (PR), and the average selectivity of water over HCl (β_{avg}) are the most important indicators for module performance evaluation. N indicates the productivity of this device; PR (performance ratio) is usually used to determine the thermal efficiency of evaporation-based process, which is defined as the amount of latent heat needed for evaporation of the produced water and the amount of heat provided to the system from an external energy source; β_{avg} is represents the measure of the preferential transport of water. The results showed that MEMD process could be used successfully for concentrating dilute HCl solution with the advantage of energy saving. The experimental data indicated that all N, PR and β_{avg} decreased with the increase of feed concentration. When the feed concentration was below 12 wt%, PR could achieve 6.0~9.6, and β_{avg} was about 10~190. As the concentration of HCl achieved 18 wt%, the values of PR and β_{avg} were still about 4.4 and 2.3, respectively. However, β_{avg} sharply decreased to a value around 1.0 when feed was further concentrated. It is also found that there exists trade-off phenomenon between N, PR and β_{avg} under experimental ranges, that is, the maximum N will be obtained with high temperature T_h, low temperature T_c and high flow rate F while the maximum PR is obtained with high temperatures T_h and T_c, as well as low flow rate F. the lowest β_{avg} will be obtained with low temperature T_h, low temperature T_c and low flow rate F. During an operational stability test lasting for 30 days, the performance of MEMD modules was kept in good condition.

(432e) Desalination by Airgap Membrane Distillation (AGMD) Using a Two Hollow-Fiber-Set Membrane Module

Dhananjay Singh and Kamalesh K. Sirkar, Chemical, Biological and Pharmaceutical Engineering, New Jersey Institute of Technology, Newark, NJ

Desalination by air gap membrane distillation (AGMD) combines in one device the processes of evaporation of water from hot brine through a porous hydrophobic membrane, condensation of the water vapor so generated on a cold condenser surface and recovery of heat in the condenser cooling liquid which may be the cold/cooled brine. Existing devices for AGMD are bulky due to the condenser arrangements adopted even though evaporation from brine may be carried out with porous hydrophobic hollow fibers. In this research we have developed a two-hollow fiber-set based compact device where all such functions are combined in a compact device. The first fiber set consists of porous hydrophobic hollow fibers of either polypropylene (PP) or polyvinylidene fluoride (PVDF). The second set of hollow fibers is of solid PP through the bore of each such hollow fiber the cooling liquid is passed and on the outside surface of which water vapor condensation takes place. A number of modules having different packing densities of hollow fibers have been studied for desalination of brine containing 1% NaCl. The performance of such modules have been investigated for a range of feed brine temperatures as well as the flow rates of the two liquid streams, hot brine and the cooling liquid. Enhanced water vapor productivity was achieved in small laboratory devices suggesting potential for scale up.

(432f) Selecting the Desired Solid Form by Membrane Crystallization: Crystals or Cocrystals

Gianluca Di Profio, Institute on Membrane Technology (ITM-CNR), National Research Council of Italy (CNR), Rende, Italy, Efrem Curcio, Department of Chemical Engineering and Materials, University of Calabria, Rende (CS), Italy and Enrico Drioli, Institute on Membrane Technology of the National Research Council - (ITM-CNR), Rende, Italy

In the last years, membrane crystallization technique has been recognized as a useful means for producing pharmaceutical crystals in controlled manner. Indeed, the interest in membrane technology arises from the possibility to improve control in the course of the crystallization process, particularly in the supersaturation stage, by using the right combination of membrane characteristics and operating parameters. The possibility to exploit membrane crystallization technology, operating in antisolvent configuration, to produce pharmaceutical cocrystals was recently verified for the Carbamazepine (CBZ) – Saccharin (SAC) system.

Cocrystallization of active pharmaceutical ingredients (APIs) with cocrystal formers (or coformers) is gaining increasing interest in the drug-development area since the resulting new solid forms are characterized by different physical-chemical properties compared to the original API. Indeed cocrystallization increases the diversity of solid state forms of an API and enhances pharmaceutical properties by modification of chemical stability' moisture uptake, mechanical behavior, solubility, dissolution rate, and bioavailability. Therefore, cocrystallization can directly impact scientific and legal aspects of drug development and life cycle management of the marketed products by providing alternative solid dosage forms and extended patent life.

Despite of its great potentialities, cocrystallization is by now mainly considered as an empirical technique based on "trial and error" strategies; this is because the basic mechanisms involved in cocrystals - and more general in crystal - formation are poorly understood to date; moreover cocrystallization process requires a special control in solution composition in order to avoid the overrunning in the phase diagram of the thermodynamic stability region of a specific specie, thus producing undesired or impure solid products. Such situation demands for the development of new robust and more efficient production technologies of solid dosage forms and, among them, particularly cocrystals .

In this paper, the application of membrane cocrystallization process was further investigated by performing a systematic study about the conditions promoting pure CBZ-SAC cocrystals or single components (CBZ/SAC) crystals from water/ethanol solvent mixtures. Particular attention has been devoted to verify the possibility to improve control during the supersaturation stage, operating in the proper zone of the solubility phase diagram of the chosen API/coformer/solvent system, in order to achieve full control in the crystallization of a specific solid form with the desired degree of purity.

Precipitates characterizations revealed that products from experimental trials where extremely sensitive to solution composition. When decreasing the initial CBZ/SAC molar ratio from 1.1 to 0.01 (while keeping all the other parameters constant) the composition of the precipitate drastically changes. CBZ 100% pure crystals are obtained for CBZ/SAC molar ratio higher than 0.3; further decreasing the molar ratio in the range 0.3-0.22 produces a mixture of cocrystals/CBZ; 100% pure CBZ-SAC cocrystals form I are obtained for solution composition consisting in 0.12-0.22 molar ratio, while a further decrease of molar ratio from 0.12 to 0.01 allowed to produce first cocrystals/SAC (up to 0.05) and then pure SAC crystals.

This observation confirms the great importance of the initial solution composition in addressing the final outcome of the cocrystallization process and it can be explained exploring the phase solubility diagram of the cocrystal system. However, in order to achieve the desired product with the right level of purity, it is necessary to operate in the proper conditions without overcoming its region of thermodynamic existence. To achieve this purpose, the role of the membrane was crucial to control the concentration mechanism of solution limiting the maximum level of supersaturation. The operating mechanism of the membrane-based strategy involves using a porous membrane as mean to control the solvent/antisolvent demixing in the crystallizing solution, by finely modulating solvent removal in vapor phase. Accordingly, the specific role of the membrane is to limit the maximum level of supersaturation reached in solution by controlling the solvent evaporation, so as providing specific routes in the phase diagram, thus avoiding the overrunning of the thermodynamic stability region of a specific specie, giving rise to the desired solid form.

Further Readings

G. Di Profio, E. Curcio and E. Drioli, *Ind. Eng. Chem. Res.*, 2010, **49,** 11878.

G. Di Profio, S. Tucci, E. Curcio and E. Drioli, *Cryst. Growth Des.*, 2007, **7,** 526.

G. Di Profio, C. Stabile, A. Caridi, E. Curcio and E. Drioli, *J. Pharm. Sci.,* 2009, **98,** 4902.

G. Di Profio, V. Grosso, A. Caridi, R. Caliandro, A. Guagliardi, G. Chita, E. Curcio and E. Drioli, *CrystEngComm*, 2011, **13,** 5670.

G. Di Profio, A. Caridi, R. Caliandro, A. Guagliardi, E. Curcio and E. Drioli, *Cryst. Growth Des.*, 2010, **10,** 449.

(432g) From Tailored Membranes to Crystal Design: How Physicochemical Properties of Polymeric Surfaces Affect the Nucleation of Organic Molecules

Gianluca Di Profio[1], Enrica Fontananova[2], **Efrem Curcio**[2,3] and Enrico Drioli[4], (1)Institute on Membrane Technology (ITM-CNR), National Research Council of Italy (CNR), Rende, Italy, (2)Department of Chemical Engineering and Materials, University of Calabria, Rende (CS), Italy, (3)c/o University of Calabria, Institute on Membrane Technology ITM-CNR, Rende (CS), Italy, (4)Institute on Membrane Technology of the National Research Council - (ITM-CNR), Rende, Italy

There is today an increasing interest in developing solid films acting as heteronucleants as a way to control the crystallization process of active pharmaceutical ingredients (APIs). While micro-fabrication methods have been developed so far to produce on lab-scale micropatterned supports with peculiar surface topography [1], here we describe routes for the production of macroscopic and handily films (membranes) from polymeric materials with a wide range of chemical functionalities. It is today well accepted that polymeric membranes used as heteronucleant surfaces can represent a promising alternative to metal, inorganic or biomaterials, since their structural and chemical properties are easily tunable by a variety of established preparation methods [2].

In this work, tailored membranes exhibiting different morphological characteristics and chemical functionalities were prepared by phase inversion technique from various polymers [poly(vinylidenefluoride-co-hexafluoropropylene), polyimide, sulphonated polyetheretherketone at different sulphonation degree, sulfonated polyetherethersulfone, polypropylene, polyethersulfone, cellulose acetate, polydimethylsiloxane]. These polymeric films, displaying specific surface patterning at the nanometric range (morphology characterized by SEM and AFM) and tailored physicochemical properties, were used as solid substrates for the heterogeneous nucleation of three representative APIs: acetaminophen, acetylsalicylic acid and glycine. The wide experimental investigation clarifies in which extent the chemical nature of the surface determines the possibility that a membrane acts as a nucleation-active substrate. It was found that, among various morphological parameters, roughness positively or negatively affects nucleation rate depending on the extent of interaction between the crystallizing solution and the surface (hydrophobic/hydrophilic behavior). In the case of hydrophilic surfaces, a significant fraction of interacting solute molecules is likely to accumulate at the solid-solution interface via adsorption, where molecular recognition events may induce partial realignment in the solute enriched layers (concentration polarization). In addition, molecules can further concentrate by physical entrapment in certain heterogeneous domains; in this case, increasing roughness would enhance the probability of molecular entrapment at the membrane surface, thereby leading to increased supersaturation level and accelerated nucleation rate. For hydrophobic supports, the Wenzel equation predicts an increased apparent contact angle at higher roughness; moreover, experimental results and theoretical analysis based on the Classical Nucleation Theory approach confirm a reduced tendency of organic molecules to nucleate [3].

[1] Y. Diao, T. Harada, A.S. Myerson, T.A. Hatton, B.L. Trout, *Nat. Mat.*, 2011, **10**, 867.

[2] G. Di Profio, E. Curcio, E. Drioli, *Ind. Eng. Chem. Res.*, 2010, **49**, 11878.

[3] E. Curcio, E. Fontananova, G. Di Profio, E. Drioli, *J. Phys. Chem.*, 2006, **B110**, 12438.

(433a) Environmentally-Responsive Membrane Adsorbers for Hydrophobic Interaction Chromatography

Heath Himstedt, Chemical and Biological Engineering, Colorado State Univeristy, Fort Collins, CO, Justin R. Weaver, Chemical and Biological Engineering, Colorado State University, Fort Collins, CO and S. Ranil Wickramasinghe, Department of Chemical Engineering, University of Arkansas, Fayetteville, AR

Hydrophobic interaction chromatography (HIC) is a technology capable of purifying high-value products such as proteins and pharmaceuticals with high selectivity. A feed stream containing the protein product is loaded at conditions which favor interactions between the hydrophobic proteins and the stationary phase (typically resin). The conditions are then changed, via introduction of a low conductivity solution, reducing these interactions. The protein which bound in the hydrophobic condition is then collected as a purified product. Unfortunately, HIC and other types of resin-based chromatography are very slow, yield only small volume of product, and are expensive to operate due to the large volume of chemicals used. Membrane based chromatography is an emerging protein purification process capable of quickly producing large volume of product. Specifically, membrane chromatography is appealing for purification of proteins in low concentrations (< 1 mg/mL).

Responsive membrane adsorbers for HIC were developed by grafting responsive polymer chains onto the surface of hydrophilic membranes, which were used to minimize nonspecific protein binding. The grafted polymer undergoes a conformational change (hydrophilic to hydrophobic) in the presence of ammonium sulfate salt. It is shown that the response of the grafted polymer chains yields improved HIC performance, specifically high binding capacities and recoveries. The effect of grafted polymer chain density on the stimulus response was investigated, and an optimal density determined. Membranes modified at this optimal density were tested at various loading buffer ionic strengths. Binding capacities and recoveries for these responsive HIC adsorbers compared favorably to, and in numerous instances surpassed, published values for industrial resin-based adsorbers.

(433b) The Effects of Intermolecular Interactions On Transmission of Plasmid DNA Through Ultrafiltration Membranes

Ehsan Espah Borujeni, Chemical Engineering, Penn State, University Park, PA and Andrew L. Zydney, Chemical Engineering, The Pennsylvania State University, University Park, PA

The effects of intermolecular interactions on transmission of plasmid DNA through ultrafiltration membranes.

There is considerable interest in developing new methods for the purification of plasmid DNA for use as gene therapy agents or DNA-based vaccines. Several previous studies have demonstrated that plasmid transmission through an ultrafiltration membrane occurs by elongation of the plasmid in the converging flow field into the membrane pore at large filtration flux. However, these data were obtained with very dilute DNA solutions, i.e., under conditions where there were no intermolecular interactions between the DNA molecules. The objective of this work was to examine the effect of DNA – DNA interactions during ultrafiltration processes.

Experimental data were obtained with plasmids from 3.0 and 16.9 kbp (kilo base pairs) in size using Omega (modified polyethersulfone) ultrafiltration membranes. In contrast to the behavior with dilute solutions, the transmission of the plasmid DNA showed a distinct maximum at intermediate values of the filtrate flux, with the sieving coefficient decreasing to near zero at very high values of the flux. The critical filtrate flux for plasmid transmission appears to remain independent of plasmid size, consistent with results for dilute plasmid solutions. However, the maximum plasmid transmission was a strong function of the plasmid size, with the extent of plasmid transmission decreasing with increasing plasmid size. The experimental results were analyzed using available models for concentrated polymer solutions to provide additional insights into the effects of intermolecular interactions on plasmid elongation during ultrafiltration. These results provide new insights into the factors governing the ultrafiltration of plasmid DNA solutions.

(433c) Transmission of Microgel Particles Through Porous Membranes

Si Pan, Mayra G Tzoc Torres, Todd R. Hoare and **Raja Ghosh**, Chemical Engineering, McMaster University, Hamilton, ON, Canada

Cross-linked poly(N-isopropylacrylamide) (PNIPAM) microgel particles could potentially be used as drug delivery systems. When used for such biomedical applications, these need to be in a highly pure form. Dialysis and centrifugation are the two most commonly used purification methods. Membrane filtration techniques such as ultrafiltration (UF) and microfiltration (MF) could also be used as purification methods. The microgel particles are stimuli-responsive and therefore their transmission behavior through porous membranes is significantly affected by operating and process variables. This study examines some of the factors affecting the transmission of 300~900 nm PNIPAM microgel through 0.2 micron microfiltration, and 3 and 50 kDa MWCO ultrafiltration membranes. The particles were labeled with a fluorophore to enhance their detectability and measurability. The observed effects are explained in terms of physicochemical properties of the microgel such as hydrodynamic radius and structural flexibility. The observations and inferences of this study are likely to be useful for designing and developing membrane based techniques for purifying microgels for biomedical application.

(433d) Affinity Membranes for Plasminogen Purification

Cristiana Boi, Claudia Castro and Giulio C. Sarti, University of Bologna, Bologna, Italy

Some eye conditions like diabetic rethinopaty, macular pukers, retinal detachment may benefit from vitreoctomy. Enzymatic vitrectomy with plasmin is envisaged to augment or even replace conventional vitrectomy by proposed means of less surgical risks, less surgeon time, lower costs. In addition, the use of autologous plasmin is beneficial since it avoids rejection problems and the search of compatible donors. Plasmin has properties to hydrolize a variety of glycoproteins, including laminin and fibronectin, by degrading the links between these components of the vitreoretinal interface and the inner limiting membrane, therapeutic posterior vitreous detachment has become possible.

Plasmin can be obtained by conversion of plasminogen using a variety of enzymes, including tissue plasminogen activator (tPA) and urokinase plasminogen activator (uPA). The purification of plasminogen from blood is normally performed with bead-based affinity chromatography. However, for ophthalmology applications, affinity membranes are ideally suited for the development of a disposable device to be used by the surgeon in the operating theatre, since they can be easily packed in a device to be used in a fast and economic process.

In this work we prepared affinity membranes for plasminogen purification using L-lysine as affinity ligand. To this aim regenerated cellulose membranes, Sartobind epoxy™, were used as the support for ligand immobilization. L-lysine was coupled by soaking the membranes in an aqueous solution of $Na_2B_4O_7$/NaOH and 1.4 dioxane at pH 10 and 80°C at different reaction times and L-lysine concentrations. Ligand density was measured using a colorimetric assay with Orange 7 as indicator.

The membranes have been characterized in batch and in dynamic experiments using both pure plasminogen and human serum. Pure plasminogen was obtained, in house, by purification from human serum with a commercial affinity chromatography resin, Pall Lysine HyperD, packed in 1 mL chromatography column and used to characterise the affinity membranes.

During experiments with human serum, fractions have been collected in all chromatographic steps, and analysed with both HPLC and SDS-PAGE electrophoresis. In particular, from the HPLC SEC analysis of the eluate a well defined plasminogen peak can be shown indicating that the lysine affinity membranes are suitable for the purification of plasminogen.

Aknowledgements

This work was financially supported by MIUR, Italian Ministry of Education, University and Research (PRIN 2008) and by the University of Bologna, Italy.

(434a) High-Flux Reverse Osmosis Membranes for Brackish Water Desalination

Lin Zhao, Department of Chemical & Biomolecular Engineering, The Ohio State University, Columbus, OH and W. S. Winston Ho, The Ohio State University, Columbus, OH

High-Flux Reverse Osmosis Membranes for Brackish Water Desalination

Lin Zhao, C.-Y. Chang, and W.S. Winston Ho, The Ohio State University

Novel high-flux reverse osmosis membranes with hydrophilic additives were synthesized and characterized under brackish water desalination conditions. Four selected hydrophilic additives, *o*-aminobenzoic acid-triethylamine salt, *m*-aminobenzoic acid-triethylamine salt, 2-(2-hydroxyethyl) pyridine, and 4-(2-hydroxyethyl) morpholine, were added into *m*-phenylenediamine (MPD) solution to react with trimesoyl chloride (TMC) during the interfacial polymerization between MPD and TMC. The effects of different concentrations for each hydrophilic additive were evaluated using brackish water desalination tests with 2000 ppm NaCl solution at 225 psi and 25°C. After the additive concentration was optimized, the synthesized membranes showed 36.7 gallons/ft^2/day water flux and 98.2% salt rejection or greater. After the post-treatment using aqueous solutions containing glycerol, sodium lauryl sulfate, and camphorsulfonic acid-triethylamine salt, the water flux of the membrane was further increased. The resulting membrane showed a flux of 52.6 gallons/ft^2/day and a salt rejection of about 98.8% or greater. This membrane outperformed the other membranes evaluated, and it also exhibited a good stability in terms of water flux (average 50 gallons/ft^2/day) and salt rejection (average 99%) during the entire period of a 30-day test with 2000 ppm NaCl solution at 225 psi and 25°C. The good stability was verified by comparing FTIR spectra of the same membrane sample before and after the 30-day test.

(434b) Zeolitic Imidazolate Framework-8 As a Reverse Osmosis Membrane for Water Desalination: Insight From Molecular Simulation

Zhongqiao Hu, Chemical and Biological Engineering, National University of Singapore, Singapore, Singapore and Jianwen Jiang, Chemical and Biomolecular Engineering, National University of Singapore, Singapore, Singapore

Over the past decade, metal-organic frameworks (MOFs) have emerged as a new class of porous materials and been considered as versatile materials for storage, separation, catalysis and other potential applications. In this work, a molecular simulation study is reported for water desalination in a MOF membrane composed of zeolitic imidazolate framework-8 (ZIF-8). The simulation demonstrates that water desalination occurs under external pressure, and Na^+ and Cl^- ions cannot transport across the membrane due to the sieving of small apertures in ZIF-8. The flux of water permeating the membrane scales linearly with the external pressure, and exhibits an Arrhenius-type relation with temperature (activation energy of 24.4 kJ/mol). Compared to bulk phase, water molecules in ZIF-8 membrane are less hydrogen-bonded and the life time of hydrogen-bonding is considerably longer, as attributed to the surface interactions and geometrical confinement. This simulation study suggests that ZIF-8 might be potentially used as a reverse osmosis membrane for water purification.

(434c) Novel Copper-Charged Anti-Microbial Reverse Osmosis Membranes

Sunitha Asapu, Chemical and Environmental Engineering, University of Toledo, Toledo, OH, Cyndee L. Gruden, Department of Civil Engineering, University of Toledo, Toledo, OH and Isabel C. Escobar, Chemical and Environmental Engineering, The University of Toledo, Toledo, OH

Water is essential for the survival of life on Earth, but the pollutants in water can cause dangerous diseases and fatalities. The provision of clean, drinkable water to people is a key factor. Over the years, natural sources of water have been falling short while, so the need for purified water has been increasing with increasing population. Reverse osmosis (RO) is a membrane-technology filtration technique that helps in the desalination and purification of water from lower quality sources, such as brackish water, seawater and wastewater. During the filtration of such sources, materials that are rejected by the membrane may accumulate on the surface of the membrane to foul it. Such materials include organic matter, colloids and microorganisms. The former two can be controlled via pretreatment; however, the accumulation of microorganisms is more problematic to membranes. Biofouling is the accumulation and growth of microorganisms on the surface of membranes and on feed spacers. After attachment, microorganisms excrete extra cellular polymeric substances (EPS), which form a matrix around the organism's outer surface as biofilm. These biofilms are detrimental and result in irreversible membrane fouling.

The purpose of this project is to investigate the charging of membrane surfaces with copper (Cu^{+2}) to control biofouling, and develop anti-biofouling membranes. The use of anti-biofouling membranes would reduce the cost associated with chemical additions as well as chemical storage. Copper and silver ions inactivate the bacterial cells and prevent the DNA replication in microbial cells. Previous studies using copper-charged feed spacers have shown the ability of copper to control biofouling without a significant amount of copper leached from the anti-biofouling polypropylene (PP) during crossflow filtration. Also, filtration using unmodified speed facers experienced almost 70% flux decline, while filtration using copper-charged feed spacers displayed only 25% flux decline. These intriguing results led to the hypothesis that the polymer chemistry can be extrapolated to produce membranes with increased biofouling resistance. To this end, glycidyl methacrylate (GMA), an epoxy, was used to attach a chelating agent, iminodiacetic acid (IDA), to membranes to facilitate the charging of copper to the surface. Cellulose acetate and polysulfone membranes were cast using the phase-inversion method. The membranes were then charged with copper ions to make them resistant to microbial growth. The pore size distribution analysis of cellulose acetate membranes, were conducted using various molecular weights of polyethylene glycol (PEG). The treated and original membranes were characterized for contact angle measurement, microbial resistance, chemically using Fourier Transform Infrared (FTIR), structural morphology using atomic force microscope (AFM) and copper dispersion on the membrane surface using scanning electron microscope (SEM).

The membranes were characterized with filtration of DI water and then subjected to protein rejection measurements. The permeation data over the five and a half hours of testing is shown in Figure 3. The permeation of the copper-charged membranes (treated) is initially lower than the non-treated membrane for the first two hours during the filtration of DI water. The membranes were then subjected to bovine serum albumin (BSA) and lipase filtration. The permeation of the BSA for both membranes was lower than the lipase because the particle size of lipase is significantly smaller than BSA. The non-treated membrane shows a decrease in permeation over time for both proteins.

(434e) Characterization of Aluminum-Neutralized Sulfonated Styrenic Pentablock Copolymer Films

Geoffrey M. Geise, Deptartment of Chemical Engineering & The Texas Materials Institute, The University of Texas at Austin, Austin, TX, Carl L. Willis, Kraton Performance Polymers, Inc., Houston, TX, Cara M. Doherty, CSIRO Manufacturing Science and Technology, Clayton, Australia, Anita J. Hill, CSIRO Manufacturing Science and Technology, Clayton, VIC, Australia, Timothy J. Bastow, CSIRO Manufacturing Science and Technology, Australia, Jamie Ford, University of Pennsylvania, Materials Science and Engineering, Karen I. Winey, Materials Science and Engineering, University of Pennsylvania, Philadelphia, PA, Benny D. Freeman, Department of Chemical Engineering, The University of Texas at Austin, Austin, TX and Donald R. Paul, Chemical Engineering, University of Texas at Austin, Austin, TX

Sulfonated polymers have been widely considered for membrane applications such as desalination, electrodialysis (ED), reverse electrodialysis (RED), and membrane capacitive deionization (CDI). Further optimization of these materials, however, is needed in order to extend and improve these technologies. Sulfonate groups are reactive sites that can be used as a platform for chemical modification of the polymer. In this work, an acid-form sulfonated styrenic pentablock copolymer was suspension-phase neutralized with triethylaluminum in order to prepare an aluminum crosslinked polymer, and a structure/property study was conducted. Aluminum neutralization was confirmed by ^{27}Al NMR. The morphology and free volume of the polymers were probed by small angle x-ray scattering (SAXS) and positron annihilation lifetime spectroscopy (PALS), respectively, and the stability of the aluminum-sulfonate group bond against ion exchange to the sodium form was probed by soaking the films in 1 mol/L NaCl for 30 days. The neutralized polymer sorbed considerably less water than the acid-form precursor due, primarily, to neutralization of the sulfonic acid groups. As a result, the hydrated neutralized polymer was mechanically stronger than the non-neutralized polymer, but the dry neutralized polymer was more brittle than the non-neutralized polymer. The material's permeability to water and salt decreased as a result of the neutralization process in a manner that is consistent with free volume theory. The material's water/salt sorption selectivity decreased due to neutralization of the polymer's fixed charge, but overall, the polymer's water/salt permeability selectivity increased upon neutralization due to an increase in the material's diffusivity selectivity, which stems predominantly from the decrease in water sorption and its effect on the polymer's free volume.

(434f) Membrane – Iron Oxide Nanoparticle System for Pollutant Degradation

Minghui Gui, Vasile Smuleac and Dibakar Bhattacharyya, Department of Chemical and Materials Engineering, University of Kentucky, Lexington, KY

The removal of chlorinated organic contaminants from the Superfund sites has always been a critical issue due to the toxicological and recalcitrant properties of those compounds. Some emerging ways employing the functionalized membranes and nanotechnology have attracted more and more attentions in the water remediation of hazardous waste sites. Here we developed the reactive membrane system with Fe/Pd and iron oxide nanoparticles (NPs) immobilized in polyacrylic acid (PAA) functionalized polyvinylidene fluoride (PVDF) microfiltration membranes for the degradation of chloro-organics such as polychlorinated biphenyls (PCBs) at the neutral pH.

PAA matrix inside the PVDF membranes is obtained by in-situ polymerization of acrylic acid (AA). The ion exchange capacity of the functionalized membranes allows the in-situ synthesis of iron/iron oxide core-shell and iron oxide NPs inside the pores. PCBs can be dechlorinated to biphenyls by using Pd doped zero-valent iron NPs made by borohydride reduction. However, biphenyl is still mildly toxic and can't be disposed to environment. On the other hand, toxic intermediates like chloroacetic acids can be generated during the direct PCB oxidation. By the development of controlled nanostructured membranes, we first dechlorinate 2,2'-dichlorobiphenyl (PCB4) to biphenyl. 2-chlorobiphenyl (PCB1) was detected as the intermediates. Depending on the Fe/Pd NP loading and composition, the complete dechlorination of PCB4 can be achieved. The further oxidative degradation of biphenyl by the heterogeneous Fenton reaction will lead to significant reduction in toxicity by forming aromatic ring breakdown products with no chlorine. The catalysis of iron oxide NPs (obtained by controlled air or H_2O_2 oxidation of iron NPs) can promote the decomposition of hydrogen peroxide (H_2O_2), generating the highly oxidative hydroxyl radicals (OH•) for the destruction of toxic organic compounds.

Convective flow experiments with membranes have also been conducted to study the intrinsic kinetics of PCB degradation, reducing the diffusion resistance of reactants through the membranes. The biphenyl obtained from the Fe/Pd loaded membranes was destructed in the iron oxide immobilized membranes by the addition of H_2O_2. Hydroxybiphenyl (dihydroxy-, trihydroxy-) and benzoic acid (hydroxy-, dihydroxy-) were detected by GC-MS and LC-MS/MS as the main intermediates during the biphenyl oxidation.

This research is supported by NIEHS-SRP grant and by the DOE-KRCEE programs.

(437a) Adsorption of Atomic and Molecular Fluids in a Porous Material Using Expanded Wang-Landau Simulations

Jerome Delhommelle and Caroline Desgranges, Department of Chemistry, University of North Dakota, Grand Forks, ND

We propose to apply Expanded Wang-Landau simulations to study the adsorption of atomic and molecular fluids in porous materials. This approach relies on a uniform sampling of the number of atoms and molecules adsorbed. The method consists in determining a high accuracy estimate of the grand-canonical partition function for the adsorbed fluids. Then, using the formalism of statistical mechanics, we calculate absolute and excess thermodynamic properties relevant to adsorption processes. In this paper, we examine the adsorption of Argon and carbon dioxide in the isoreticular metal-organic framework IRMOF-1. We assess the reliability of the method by showing that the predicted adsorption isotherms and isosteric heats are in excellent agreement with simulation results obtained from Grand Canonical Monte Carlo simulations. We also show that the proposed method is very efficient since a single Expanded Wang-Landau simulation run at a given temperature provides the whole adsorption isotherm. Moreover, this approach provides a direct access to a wide range of thermodynamic properties, such as e.g the excess Gibbs free energy and the excess entropy of adsorption.

(437b) Adsorption of Carboxylate On Calcite (101 4) Surface: Molecular Dynamics Simulation Approach

ByeongJae Chun, Seung Geol Lee, Giuseppe F. Brunello, Ji Il Choi and Seung Soon Jang, School of Materials Science and Engineering, Georgia Institute of Technology, Atlanta, GA

Adsorption of Carboxylate on Calcite ($10\bar{1}4$) Surface: Molecular Dynamics Simulation Approach

Byeong Jae Chun, Seung Geol Lee, Giuseppe F. Brunello, Ji Il Choi and Seung Soon Jang

Computational NanoBio Technology Laboratory, School of Materials Science and Engineering, Georgia Institute of Technology, 771 Ferst Drive NW, Atlanta, GA 30332-0245

To study the wettability alteration of calcite surface in oil reservoir, we investigate the molecular adsorption of carboxylate on the calcite $(10\bar{1}4)$ surface using full-atomistic molecular dynamics (MD) simulation as well as density functional theory (DFT). First, we investigate the most probable binding site on the calcite for the chemisorption of carboxylate using DFT calculations in DMol3. Once finding the binding site, we obtain the molecular interaction energy curve as a function of distance in which the dispersion interaction is calculated using DFT-D3. In order to take into account the molecular interaction between various molecules including oil, water and surfactant in our MD simulations, we determine a force field, especially off-diagonal van der Waals interactions: 1) Morse potential functions are fitted to reproduce the chemisorption binding energy obtained from the previous DFT calculation; 2) Morse potential function is also used to describe the non-chemisorption interactions between molecules and calcite surface; 3) Lennard-Jones potential functions are used to describe the interaction between the different types of molecules (e.g., oil-water pair); 4) Regarding van der Waals interactions between the same type of molecules such as calcite-calcite, oil-oil and water-water pairs, we employ the given original force fields. After developing the Force Field, we perform MD simulations to obtain the equilibrium structures and evaluate the thermodynamic stability of the chemisorbed carboxylate on calcite in comparison to that of the carboxylate in molecular assembly with surfactants. Through this study, we aim at characterizing the molecular mechanisms of the surfactant-driven carboxylate desorption.

(437c) Mechanistic Studies of Chiral Recognition of Solutes by Amylose Tris[(S)-á-Methylbenzylcarbamate]

Hung-Wei Tsui, Nien-Hwa Linda Wang and Elias I. Franses, School of Chemical Engineering, Purdue University, West Lafayette, IN

This amylose-based polymeric sorbent has the ability to separate a variety of chiral enantiomers (R and S), such as ethyl lactate, methyl mandelate, pantolactone, and benzoin.* These molecules contain OH and C=O functional groups, which may interact with the C=O and NH groups of the sorbent side chains via hydrogen (H-) bonding interactions. From the observed retention factors k_R and k_S and enantioselectivities ($S= k_R/ k_S$), as determined with High Performance Liquid Chromatography (HPLC), it was inferred that the molecular recognition of these molecules mechanisms were quite similar. Infrared spectroscopy (IR) data for polymer-solute and Density Functional Theory (DFT) simulations of the interactions of these solutes with the side chains of the polymer led to the following hypothesis for the chiral recognition mechanism. A strong H-bond forms between the solute OH groups and the sorbent C=O groups for both enantiomers as the primary interaction "anchor" point. Then a second H-bond forms between the solute C=O groups and the sorbent NH groups only for the R-enantiomer of the above four solutes. The S-enantiomer is prevented sterically from forming such a bond. Molecular Dynamics (MD) simulations were done for a 12-mer polymer model to determine the polymer structure, intrapolymer H-bonding, free functional groups, available for binding with the solute, and the geometry of the potentially enantioselective cavities. Monte Carlo (MC) and MD "docking" simulations were done to investigate the interactions of the benzoin enantiomers with the polymer. MD simulations were also done for the other solutes. The simulations are generally consistent with the above hypothesis, and they reveal the specific cavities which may allow enantioselective interactions. The results showed that S is lower as the solute molecular flexibility, as defined from the molecular geometric torsion angles distribution, increases. The results are useful in predicting how S varies with the solute molecular structure, and for selecting sorbents for specific chiral separations.

* H.-W. Tsui, J.N. Willing, R.B. Kasat, N.-H.L. Wang, E.I. Franses, J. Phys. Chem. B, 2011, 115, 12785-12800.

(437d) Molecular Screening of Alcohol and Polyol Adsorption Onto MFI-Type Zeolites

Ruichang Xiong[1], Stanley I. Sandler[2] and Dionisios G. Vlachos[1], (1)Chemical Engineering, University of Delaware, Newark, DE, (2)Department of Chemical Engineering, University of Delaware, Newark, DE

Configurational-bias grand canonical Monte Carlo (CB-GCMC) simulations and expanded ensemble (EE)-CB-GCMC simulations were performed to obtain adsorption isotherms of alcohols and polyols onto MFI-type zeolites from the gas phase and aqueous solution. In adsorption from both phases, Henry's constants and heats of adsorption at infinite dilution for straight-chain alcohols, diols and triols in silicalite-1 are found to increase, and the saturation loadings decrease with increasing carbon number. Adsorption of straight-chain alcohols is more favorable than that of branched-chain alcohols. Henry's constants increase with increasing number of hydroxyl groups for gas-phase adsorption, but decrease for adsorption from aqueous solution due to the strong hydrophilic solvent effect of water. The location of the hydroxyls does not affect significantly the adsorption from aqueous solution but does so in gas-phase adsorption. The saturation pressures for gas-phase adsorption decrease by orders of magnitude from the alcohols to the triols. Non-framework cations increase the adsorption of the small alcohols by an order magnitude at low concentrations (<1 mg/mL), but result in only a factor of 2 increase for larger alcohols like butanol at low concentrations (<0.03 mg/mL), and then decrease the adsorption at higher concentrations. Overall, the simulated results are in reasonable agreement with available experimental data.

(437e) Molecular Simulation Studies of CO2 Adsorption From Mixtures of N2, CH4, and H2O by Microporous Carbon

Yangyang Liu and Jennifer Wilcox, Department of Energy Resources Engineering, Stanford University, Stanford, CA

Carbon capture combined with CO_2 utilization and sequestration offers the possibility to reduce CO_2 emissions in the near future, while allowing for a continued use of fossil fuels. Adsorption of CO_2/CH_4, CO_2/N_2, and CO_2/H_2O mixtures has been studied in the current work. In the case of carbon capture, our studies show that functionalized carbon-based sorbents may be competitive with other sorbents in terms of the pure CO_2 adsorption capacity by modifying the surface functionality. The separation of CO_2 from CO_2/N_2 and CO_2/H_2O gas mixtures is involved in carbon capture for flue gas applications. In the case of carbon storage in coal and the organic matrix of gas shale, CO_2 injection into coalbeds and gas shale reservoirs may enhance subsequent methane (CH_4) recovery while simultaneously storing CO_2; therefore, the adsorption mixtures of CO_2/CH_4 is involved in CO_2 enhanced coalbed methane recovery (CO_2-ECBM) and shale gas productions. In this work, complex pore structures for general carbon-based porous materials and natural organic materials were modeled as a collection of independent, non-interconnected, functionalized graphitic slit pores with surface heterogeneities. Together with Bader charge analysis, electronic structure calculations coupled with *van der Waals*-inclusive corrections can provide the initial framework comprising both the geometry and corresponding charge information required to carry out statistical-based molecular simulations. Grand canonical Monte Carlo simulations are carried out to determine the adsorption isotherms and selectivity of CO_2 from CO_2/N_2 and CO_2/CH_4 gas mixtures at temperature/pressure conditions relevant to carbon capture and storage applications. The polarity induced by the surface functionalities and the interactions between the surface and the non-polar/polar molecules have been investigated. Molecular simulations of the adsorption of CO_2/CH_4 and CO_2/N_2 mixtures show that CO_2 preferentially adsorbs on oxygen-containing functionalized graphitic surfaces with an induced polarity due to the strong quadrupole moment of CO_2 compared with that of N_2 and the weak octupole moment of CH_4. For the mixtures where CO_2 has a stronger polarity, the difference in the quadrupole moment of the mixture component results in a different surface occupancy on the same surface, and the oxygen-containing functionalized surfaces not only increase the total adsorption of CO_2, but also enhance the separation selectivity of a given gas mixture. The investigation of different polar surfaces also shows that surface chemistry is tunable thereby influencing the selectivity and allowing for control of CO_2 separation based upon adsorption processes.

(437f) Pressure Enhancement in Nanopores: Effect of Pore Shape

Yun Long, Chemical and Biomolecular Engineering, North Carolina State University, Raleigh, NC, Keith E. Gubbins, Chemical Engineering, North Carolina State University, Raleigh, NC, Erich A. Muller, Department of Chemical Engineering, Imperial College London, London, United Kingdom, George Jackson, Centre for Process Systems Engineering, Dept of Chemical Engineering, Imperial College London, London, United Kingdom and Erik E. Santiso, Imperial College London, London, United Kingdom

Abundant experimental evidence suggests that adsorbates confined in nanoporous materials can exhibit high pressures. Examples include ice nanocrystals (ice VIII and ice IX, only existing at pressures of ~1 GPa for bulk water) in carbon nanotubes at 0.1 MPa bulk pressure[1], a bcc crystalline form of KI, found in the bulk phase only at pressures above 1.9 GPa, in carbon nanohorns at 0.1 MPa bulk pressure[2], chemical reactions that occur at low bulk pressures in nano-pores, but only at very high pressures in the bulk phase[3], and the deformation of pore walls confining the adsorbate in the course of adsorption[4].

We report molecular simulation studies of the pressure tensor for simple adsorbates (e.g. argon, water) in carbon nanopores of slit, cylindrical and spherical geometries. We calculate the pressure tensor by both the mechanical route and the thermodynamic route. Both routes give comparable results (especially for the slit geometry, the results from both routes agree perfectly). For modest bulk phase pressures of 1 bar or less, the pressures parallel to the pore walls (tangential pressure) were found to be positive and of the order 1 GPa, and because of the degree of confinement the tangential pressure in spherical pore is larger than that in the cylindrical pore, while the tangential pressure in the slit pore is the smallest. In slit pores, the pressure normal to the wall was constant throughout the pore and of the order of 0.1 GPa, and could be positive or negative depending on the pore size. For the cylindrical and spherical pores, the pressure normal to the wall is not constant over the radial direction, but the peak value is also ~ 0.1 GPa. Moreover, the tangential pressure is very sensitive to small changes in the bulk pressure, indicating a way to experimentally control the in-pore pressure.

[1] M. Sliwinska-Bartkowiak, M. Jazdzewska, L. Huang and K.E. Gubbins, *Physical Chemistry Chemical Physics*, (2008), **10**, 4909-4919; M. Jazdzewska, M. Sliwinska-Bartkowiak, A.I. Beskrovnyy, S.G. Vasilovsky, K.-Y. Chan, L.L. Huang and K.E. Gubbins, *Physical Chemistry Chemical Physics*, (2011), **13**, 9008-9013.

[2] K. Urita, Y. Shiga, T. Fujimori, T. Iiyama, Y. Hattori, H. Kanoh, et al, *J. Am. Chem. Soc*, (2011), DOI: dx.doi.org/10.1021/ja202565r

[3] e.g. K. Kaneko, N. Fukuzaki, K. Kakei, T. Suzuki and S. Ozeki, *Langmuir*, (1989), **5**, 960; O. Byl, P. Kondratyuk, J.T. Yates, *J. Phys. Chem. B*, (2003), **107**, 4277.

[4] e.g. Y. Fujiwara, K. Nishikawa, T. Iijima and K. Kaneko, J. Chem. Soc., Faraday Trans., (1991), **87**, 2763-2768; A.V. Neimark, F.-X. Coudert, A. Boutin, A.H. Fuchs, J. Phys. Chem. Lett., (2010), **1**, 445.

(437g) Exploring Solid-Fluid Phase Behavior of Adsorbed Water Nanoconfined Between Mica and Graphene Surfaces

Cody K. Addington, Department of Chemical and Biomolecular Engineering, North Carolina State University, Raleigh, NC, Yun Long, Chemical and Biomolecular Engineering, North Carolina State University, Raleigh, NC and Keith E. Gubbins, Chemical Engineering, North Carolina State University, Raleigh, NC

The behavior of an adsorbed water phase on the nanoscale is of particular importance to the field of nanofluidics and the lab-on-a-chip concept, where fluid structure and behavior at the nanoscale governs the systems under consideration. Experimental evidence supports the existence of a fluid water layer inside a graphene-mica slit pore with virtually no defects (Xu et al., *Science* **2010**, *329*, 1188), while defective surfaces cause liquid-solid transitions within the adlayer (Severin et al., *Nano Letters* **2012** *12*, 774-779). The structure of these "ices" is of interest due to the effects of confinement, which may cause exotic crystalline phases to form (Jazdzewska et al., ***Phys. Chem. Chem. Phys.*** **2011**, ***13***, 9008-9013)

Water adsorption in the graphene-mica slit pore was simulated by the grand canonical Monte Carlo method, mirroring experimental conditions. The mica surface was modeled using Steele's potential, while the graphene surface was modeled as either a smooth wall with Steele's potential or a flexible atomistic wall. The TIP4P/2005 model was used to model the interaction between water molecules. The final configurations of water were analyzed with specific order parameters tuned to detect the various crystalline ice phases. By varying the temperature of the simulation, freezing transitions can also be deduced.

Our simulation results support the experimental observations that defects can induce crystallization of adsorbed water. These simulations also provide an understanding of the freezing behavior of water confined within the graphene-mica slit pore system.

(442a) Sustainable Applications Through the Use of Reversible Ionic Liquids

Steven R. Saunders[1,2], Kyle M. Flack[1,3], Emily C. Nixon[1,3], Amy L. Rohan[1,2], Jackson R. Switzer[1,2], Elizabeth J. Biddinger[1,2], Pamela Pollet[1,3], Charles L. Liotta[1,2,3] and Charles A. Eckert[1,2,3], (1)Specialty Separations Center, Georgia Institute of Technology, Atlanta, GA, (2)School of Chemical and Biomolecular Engineering, Georgia Institute of Technology, Atlanta, GA, (3)School of Chemistry and Biochemistry, Georgia Institute of Technology, Atlanta, GA

Reversible ionic liquids (RevILs) are a novel class of switchable solvents that are being investigated for use in a wide variety of applications. Switchable solvents make use of an external physical stimulus to induce a chemical change which causes a drastic, abrupt change in the properties of the solvent. The RevILs discussed herein are formed when CO_2 is introduced to a silylamine (e.g., 3-(aminopropyl) triethylsilane) to form an ammonium-carbamate ion pair. These RevILs can be returned to their original molecular (i.e., non-ionic) form when heated or sparged with an inert gas. Through adjustments in the structure of the silylamine and the degree of conversion of the molecular form to the ionic form the properties (e.g., viscosity, density, polarity, etc.) of the solvent medium can be drastically tuned. This tuning ability makes these RevILs ideal solvent systems for a number of applications.

Two applications of particular interest where RevILs have made a significant impact in terms of reducing emissions and waste are carbon capture of CO_2-rich gas streams and nanoparticle synthesis. These RevILs provide several advantages over traditional CO_2 capture solvents (e.g., monoethanolamine) for recovery of CO_2 from, for example, coal fired power plant flue gas. The primary advantage involves a dual capture mechanism: (1) chemical absorption and (2) physical absorption which provides for the opportunity to reach similar recoveries as traditional solvents with lower energy requirements. Traditional synthesis routes for catalytically active nanoparticles can make use of harsh reactants and produce significant amounts of waste. Nanoparticles with relatively narrow size distribution and an average size less than 10 nm can be produced using these RevILs in a process akin to a reverse-micelle synthesis with the ability to recycle the RevIL These two applications offer a focus on waste reduction and the environmental sustainability of the processes.

(442b) Application of Ionic Liquid for the Efficient and Energy-Saving Synthesis of Ethylene Glycol

Jian Sun Sr., State Key Laboratory of Multiphase Complex System, Institute of Process Engineering, Chinese Academy of Sciences, Beijing, China, Weiguo Cheng, State Key Laboratory of Multiphase Complex System,, Institute of Process Engineering, Chinese Academy of Sciences, Beijing, China, Jinquan Wang Sr., Key Laboratory of Green Process and Engineering, Institute of Process Engineering, Chinese Academy of Sciences, Beijing, China, Junping Zhang, Institute of Process Engineering, Chinese Academy of Sciences, Beijing, China and Suojiang Zhang, 1.State Key Laboratory of Multiphase Complex System, 2.Key Lab of Green Process and Engineering, Institute of Process Engineering, Chinese Academy of Sciences, Beijing, China

Application of Ionic Liquid for the Efficient and Energy-Saving Synthesis of Ethylene Glycol

Jian Sun, Weiguo Cheng, Jinquan Wang, Junping Zhang, Suojiang Zhang*

Institute of Process Engineering, Chinese Academy of Sciences,100190, Beijing, PR China

**Corresponding author: sjzhang@home.ipe.ac.cn*

Ethylene glycol (EG) is chemical of importance for synthetic fibers, resins, and antifreeze. The most widely used commercial process to manufacture EG is the direct non-catalytic liquid-phase hydration of ethylene oxide (EO). A large excess of water (20-25 mol water/mol EO) must be used to obtain a high selectivity of EG in the hydration reaction mixture, which consumes a large steam input for products separations. Even so, an achieved typical product mixture is only approximately 90% EG. Therefore, the need to increase EG selectivity but reduce energy cost has spurred the development of alternative routes and the corresponding catalysts.

Recently, the manufacture of EG from EO via ethylene carbonate (EC) intermediate has been regarded as one of the most efficient routes to replace the current method. However, the catalysts still suffer from low activity, harsh reaction conditions and water/air sensitivity. In addition, the requirement of EC separation might be a response to an unsatisfactory compatibility of the two-step reaction catalysts.

Since ionic liquids (ILs) have shown an excellent performance for the synthesis of industrial chemicals substituting for traditional catalysts, herein, we designed and screened binary IL catalysts for the new process, between which one would be active towards the cycloaddition reaction and another could be effective for the hydration process. Effect of various parameters on the new process was studied in detail. It was found that in the presence of water, 1) EO conversion could be enhanced obviously, 2) the molar ratio of components in catalyst had a great effect on the selectivities of EC, EG and diethylene glycol (DEG), and 3) the catalyst was hydrothermally stable and could be reused over 10 times without loss of its activity and selectivity. Based on the results, an economic comparison of the direct hydration process and the new process was investigated by aspen-plus simulation. The total process avoids EC separation, and represents a simple, energy-saving, and cost-effective route to EG with high product quality, as well as easy catalyst recycling.

Keywords: ionic liquid, ethylene glycol, carbon dioxide, synthesis, energy-saving

Acknowledgments

This work was supported byNational Basic Research Program of China (2009CB219901), and National Science Fund of China (21006117 and 20936005).

(442c) A Comparative Study On Selective Alkylation of Naphthalene Catalyzed by Various Lewis Acidic Ionic Liquids

Ming Zhang, Zhiqiang He and Yun Fang, School of Chemical & Material Engineering, Jiangnan University, Wuxi, China

A series of double long-chain alkylnaphthalene were synthesized from naphthalene and long-chain α-alkenes catalyzed by various Lewis acidic ionic liquids in this paper. The Lewis acidic ionic liquids could be luckily used as catalyst instead of traditional acidic catalysts even zeolites in the process of alkylation with high activity because traditional acidic catalysts have several disadvantages, such as corrosion, low selectivity, and environmental pollution since difficultly recoved. More than that, the alkylated products and unreacted reactants could be isolated easily by decantation from the immiscible ionic liquid phase. In this paper, various Lewis acidic ionic liquids composed of $AlCl_3$-Et_3NHCl, $ZnCl_2$-Et_3NHCl or $FeCl_3$-Et_3NHCl were tested and the influences of type and dosage of catalyst, cationic-anionic proportion of ionic liquids, molar ratio of naphthalene to alkenes, reaction temperature, and reaction time were investigated. Excellent conversion of reactants and different selectivity of the alkylated product were achieved, and finally the catalyst could be recycled for several times without any obvious loss of its catalytic activity.

Use of Deep Eutectic Ionic Liquid for Biodiesel Preparation from Base-Catalyzed Transesterification of Rapeseed Oil

Wei Huang [a], *Shaokun Tang*, *[a] *Hua Zhao* [b]

[a] *Key Laboratory for Green Chemical Technology of Ministry of Education, School of Chemical Engineering & Technology, Tianjin University, Tianjin 300072, China*

[b] *Chemistry Program, Savannah State University, Savannah, GA 31404, USA*

**Corresponding author. Tel: +86-22-27408578; E-mail address: shktang@tju.edu.cn*

As a renewable and biodegradable fuel, biodiesel has been attracting research attention for the last decade in the face of diminishing oil supplies. To overcome hurdles of the conventional reaction and purification processes for biodiesel preparation, a number of new approaches have been actively pursued including new solid catalysts, novel reaction process, and co-solvents in recent years. However, these new methods are associated with new problems such as lower product yield, higher energy consumption, higher costs and so on.

In our work, a deep eutectic ionic liquid (DEIL), choline chloride and glycerol (1:2, molar ratio), was prepared and used in the transesterification reaction of rapeseed oil to biodiesel catalyzed by NaOH. Response surface methodology (RSM) and Box-Behnken Design (BBD) were employed to examine different reaction parameters such as methanol/oil molar ratio, catalyst concentration and DEIL concentration, and optimize the reaction conditions for achieving the maximum yield of fatty acid methyl ester (FAME).

Our data have suggested that the FAME yield of up to 88% could be obtained at the optimum conditions; this yield was higher than that of the reaction under the same condition parameters without the addition of DEIL (74%). A statistical analysis of variance (ANOVA) indicates that a second-order polynomial model appears to represent the actual relationship between the response and the significant variables, with a very low probability value (0.0001) and a high satisfying determination coefficient (R^2=0.9724).

The addition of DEIL to the transesterification reaction improved the FAME yield and enabled an easy and thorough separation of glycerol from the reaction mixture during and after reaction. This study exemplifies a cost-effective and environmental benign approach to biodiesel production by using the DEIL as a co-solvent for the transesterification reaction.

Keywords: deep eutectic ionic liquid, biodiesel, transesterification, base catalyst, response surface methodology

Acknowledgment

This project was financially supported by the National Natural Science Foundation of China (21176186).

(442e) Ionic Liquid Design and Techno-Economic Evaluation for Energy Efficient Extractive Recovery of Bio-Based Glycols

André B. de Haan, Boelo Schuur and Lesly Garcia-Chavez, Chemical Engineering, Eindhoven University of Technology, Eindhoven, Netherlands

Important bulk chemicals like glycols can be produced from renewable sources through fermentation processes. An important challenge is the energy efficient product recovery from diluted fermentation broths. Low molecular weight diols, such as Mono Ethylene Glycol, Propylene Glycols and Butane Diols are particularly difficult to separate due to their high water affinity. Their recovery by distillation or multiple effect evaporation is associated with high energy consumption. Liquid-Liquid extraction can be a promising alternative since it can be much more energy efficient. However, for conventional solvents the distribution coefficients are generally insufficient to achieve efficient extraction at low broth concentrations. For this reason advanced fluid extractants like ionic liquids are needed that improve the glycol distribution.

In this work COSMO-RS is used to support ionic liquid tailoring to optimize the glycol distribution coefficient (D) and selectivity (S) by employing the sigma electron profile. As a result the glycol distribution coefficients were increased up to >100 times that of conventional solvents. For the best performing ionic liquid tetraoctyl ammonium 2-methyl-1-napthoate liquid-liquid equilibrium data were determined and thermodynamic models constructed for various glycols. These thermodynamic models were used to develop conceptual process designs that demonstrated energy savings >50% compared to the current state-of-the-art three effect evaporation technology. Energy integration appeared crucial to achieve these energy savings in the extraction process. In spite of these significant energy savings the techno-economic evaluation demonstrated that the final economic feasibility of the extraction routes strongly depends on the capital expenditure for the respective glycols, which in turn is strongly dependent on the final glycol distribution coefficient achieved.

(442f) Dissolution of Activated Sludge Biomass in Imidazolium-Based Ionic Liquids

Patrisha J. Pham, Rafael Hernandez and Todd French, Dave C. Swalm School of Chemical Engineering, Mississippi State University, Mississippi State, MS

Activated sludge biomass (ASB), the solid produced during biological treatment of wastewater contains a network of polysaccharides and proteins collectively known as extracellular polymeric substance (EPS). The nature of ASB predicts that it may behave similarly with cellulose upon dissolution in ionic liquids (IL). Cellulose dissolution in ionic liquids is well-known and studied. In this work, ASB was dissolved in various ionic liquids such as 1-butyl-3-methylimidazolium chloride ([BMIM][Cl]) and 1-allyl-3-methylimidazolium chloride ([AMIM][Cl]), both of which are recognized to dissolve cellulose effectively. The biomass was recovered after precipitation with water. The regenerated ASB was characterized by Thermogravimetric Analysis (TGA). The bulk structure was observed through Scanning Electron Microscopy (SEM). Factors such as the nature of the ionic liquid, ASB homogeneity and temperature effects (microwave irradiation versus direct heating) on dissolution were investigated. Establishing successful dissolution parameters of ASB in imidazolium-based IL's will open doors for new ASB utilization i.e. sugar source for biofuels production.

(442g) Liquid-Liquid Extraction of Alcohols From Aqueous Solutions Using Ionic Liquids

Alexander Nann[1], Christoph Held[1], Martin Stoffers[2], Andrzej Górak[2] and Gabriele Sadowski[1], (1)Department of Biochemical and Chemical Engineering, Laboratory of Thermodynamics, TU Dortmund University, D-44227 Dortmund, Germany, (2)Department of Biochemical and Chemical Engineering, Laboratory of Fluid Separations, TU Dortmund University, D-44227 Dortmund, Germany

The production of alcohols and fuels via fermentation has been suggested as an alternative to the fossil-fuel based routes in recent years. Liquid-liquid extraction of organic compounds from dilute aqueous solutions using ionic liquids is a promising unit operation to replace energy- and cost-intensive separation techniques, e.g. distillation. The selection of appropriate ionic liquids (ILs) applied in chemical and separation processes requires reliable thermodynamic data. Therefore, systematic investigations of pure-component and mixture properties for systems containing ILs are essential. This work investigates the phase behavior of binary and ternary alcoholic solutions containing ILs as well as the extraction of 1-butanol from dilute aqueous solutions.

In a first step, physical and thermodynamic properties of pure IL and binary IL/alcohol mixtures of four ILs consisting of two different anions and two different cations were measured. The experimental investigations were carried out using vapor-pressure osmometry (VPO), gas-liquid chromatography (GLC), and a vibrating-tube densimeter at different temperatures and concentrations.

In a second step, pure-component IL model parameters for PC-SAFT were obtained from fitting to pure-IL densities and binary IL/alcohol data. The ILs were regarded as non-dissociated as well as non-aggregated. This approach allows for a quantitative modeling of densities and alcohol activity coefficients over a broad concentration and temperature range.

Subsequently liquid-liquid equilibrium (LLE) phase diagrams for ternary water/alcohol/IL systems were determined experimentally. It was found that all considered ionic liquids seem to be promising agents for liquid-liquid extraction of the considered alcohols from dilute aqueous solutions. Using the IL model parameters based on only pure and binary component data, PC-SAFT was able to predict the experimental LLEs of these systems precisely.

Finally, with knowledge of the liquid-liquid phase behaviour a multistage liquid-liquid extraction is investigated. Therefore an equilibrium model was implemented to Aspen Custom Modeler®. Important process parameters and the IL demand were identified by varying the number of theoretical stages, butanol fraction and phase ratio.

(443a) Evaporation Rate of Water in Hydrophobic Confinement

Pablo G. Debenedetti and Sumit Sharma, Chemical and Biological Engineering, Princeton University, Princeton, NJ

The drying of hydrophobic cavities is believed to play an important role in biophysical phenomena such as the folding of globular proteins, the opening and closing of ligand-gated ion channels, and ligand binding to hydrophobic pockets. We use forward flux sampling to compute the rate of capillary evaporation of water confined between two hydrophobic surfaces separated by nanoscopic gaps, as a function of gap, surface size and temperature. Over the range of conditions investigated (gaps between 9 and 14 Å and surface areas between 1 and 9 nm^2) the free energy barrier to evaporation scales linearly with the gap between hydrophobic surfaces, suggesting that line tension makes the predominant contribution to the free energy barrier. The exponential dependence of the evaporation rate on the gap between confining surfaces causes a ten order-of-magnitude decrease in the rate when the gap increases from 9 to 14 Å. The computed activation energies are of the order of $50kT$, and we present umbrella-sampling calculations of the free energy barriers. Evaporation rates per unit area are found to be two orders of magnitude faster in confinement by the larger (9 nm^2) than by the smaller (1nm^2) surfaces considered here, at otherwise identical conditions. We show that this is a direct consequence of the dependence of hydrophobic hydration on the size of solvated objects. For sufficiently large surfaces, the critical nucleus for the evaporation process is a gap-spanning cylindrical vapor tube.

(443b) Towards a Molecular Theory for Homogeneous Bubble Nucleation in Metastable Liquids

Korosh Torabi[1,2] and David S. Corti[1], (1)School of Chemical Engineering, Purdue University, West Lafayette, IN, (2)Department of Chemistry, Northwestern University, Evanston, IL

Homogeneous bubble nucleation refers to the process by which a first-order phase transition occurs within a bulk metastable liquid, in the absence of impurities or external perturbations. We report an equilibrium embryo definition for homogeneous bubble nucleation based on the following two order parameters: the number of particles n contained within a spherical volume v, which are in constrained equilibrium with the surrounding superheated liquid. The reversible work of formation, or the underlying free energy surface, of these (n,v) embryos within the superheated Lennard-Jones liquid is generated via several novel Monte Carlo simulation procedures. By doing so, we also emphasize several key aspects that any well-formulated equilibrium based theory of homogeneous nucleation must address. Namely, within the suggested embryo definition any configuration of the particles at a fixed location in space must be capable of being mapped onto one and only one point in the order parameter space. Moreover, the suggested equilibrium-based model must be able to identify the transitional configurations that, based on a committor probability analysis, are dynamically relevant to the nucleation process. Another novel characteristic of the developed theory is that we define a limit for the configuration space of the metastable phase. Hence, those configurations that have a zero committor probability are not included in the free energy surface. Finally, we develop a simulation-based algorithm that non-redundantly relates the number density of the critical nucleation sites to their free energies of formation (The critical nuclei are located close to the boundary of the metastable phase configuration space where the committor probability is about one half). In turn, this allows us to determine the forward rate coefficient of the critical nuclei based on their average life times calculated via MD simulation of a large number trajectories initiated from the transitional region of the configuration space. We compare our nucleation rate predictions to those obtained from other methods, and discuss the limitations of all of these approaches.

(443c) Crystal Nucleation and Growth in Molecular Systems

Jerome Delhommelle and Caroline Desgranges, Department of Chemistry, University of North Dakota, Grand Forks, ND

Controlling the crystallization of molecules from supersaturated solutions is a long-standing issue in solid state chemistry. Since polymorphs may have very different physical properties, it is crucial to control in which polymorph a molecule crystallizes in many applications, such as e.g. in the pharmaceutical industry. We discuss in this presentation our recent work on the molecular simulation of the crystallization of molecules from the liquid phase. We simulate the nucleation step using a combination of molecular dynamics simulation with the so-called umbrella sampling technique. This sampling method, when used together a reaction coordinate (or order parameter) characteristic of the crystalline order for the chosen system, is particularly well suited to study activated processes (or rare events) such as e.g. the nucleation event, which requires the system to overcome a large free energy barrier of nucleation. Throughout the nucleation and growth processes, we focus on examining how the selection of a specific polymorph takes place.

Loop Heat Pipes Exploiting Nano-porous Wicks
- Development and Application of Materials Sustaining Fluid under Tension

Among many different technologies for transferring heat, the heat pipe (HP) – closed-circuit systems in which a working fluid transfers its latent heat as it cycles between an evaporator and a condenser by capillary pumping force – is one of the most efficient systems known today and has offered an attractive alternative in an array of applications[1]. However, two main limitations have constrained the application of conventional HPs: 1) the maximum capillary head developed by the micro-porous wicks of ~1 bar that limits the maximum heat loads, the length, and their orientation with respect to gravity or accelerations.; and 2) the saturation state in the condenser region creates an unbounded liquid-vapor interface that causes undesired transients and additional resistance to heat transfer.

The transpiration process in vascular plants operates by a mechanism similar to the recirculation of liquid through the wick of HPs[2]. Unlike in HPs, though, the water pressure in plants is able to reach large negative values (down to -100 bars) to overcome gravity, viscous drag in the liquid path, and the reduced chemical potential of water in sub-saturated soils. The secret behind the amazing abilities of plants is a the structure of plants cell walls, which serve as nano-porous membranes, allowing the water pressure to be less than zero (under tension). Inspired by plants, our research is focused on the development of a nano-porous loop heat pipe (NPLHP) exploiting nano-porous wicks, including a theoretical model of the coupled transport and thermodynamic phenomena and an experimental platform with which to test design concepts for NPLHPs.

After presenting the design considerations of a NPLHP, we will focus on a central experimental challenge in its development: nano-porous membranes that allow for the pure liquids to come to metastable equilibrium at negative pressures with sub-saturated vapors. We will discuss several membrane structures developed in our laboratory and our experimental methods for characterizing them. We will then turn to an intriguing open question about water under tension: several experimental studies focusing on cavitation thresholds (the pressure at which the thermodynamically metastable liquid turns into vapor by nucleation) of water under tension have been reported, yet there is no consensus[3]. Many experiments using various methods have found the stability limit of water at room temperature is between -20 ~ -30MPa[4-6], while experiments in quartz inclusions appear to reach a stability limit of -140 MPa (with the use of an extrapolated equation of state). We will describe how measurements based our distinct experimental approach compare with those in the literature and discuss the implications for our understanding of the phase diagram of water and on the practical limitations on the use of water at negative pressure in technologies such as NPLHPs.

1 Faghri, A. *Heat Pipe Science and Technology*. (Taylor and Francis, 1995).

2 Nobel, P. S. *Physicochemical and Environmental Plant Physiology*. 2nd edn, (Academic Press, 1999).

3 Caupin, F. & Herbert, E. Cavitation in water: a review. *Comptes Rendus Physique* **7**, 1000-1017 (2006).

4 Briggs, L. J. Limiting Negative Pressure of Water. *Journal of Applied Physics* **21**, 721-722 (1950).

5 Wheeler, T. D. & Stroock, A. D. Stability Limit of Liquid Water in Metastable Equilibrium with Subsaturated Vapors. *Langmuir* **25**, 7609-7622, doi:10.1021/la9002725 (2009).

6 Davitt, K., Arvengas, A. & Caupin, F. Water at the cavitation limit: Density of the metastable liquid and size of the critical bubble. *EPL* **90**, 16002 (2010).

(443e) Energetic Origins of Metastable Zone, Induction Supersaturation, and Implications towards Nucleation

Venkateswarlu Bhamidi, Paul J.A. Kenis and Charles F. Zukoski, Department of Chemical & Biomolecular Engineering, University of Illinois at Urbana Champaign, Urbana, IL

For decades, the metastable zone in solution crystallization – a supersaturated region of the phase diagram in which no significant nucleation of crystals occurs – has been viewed as a kinetic phenomenon that originates from the strong dependency of the rate of nucleation on solute concentration. Here we show that, for several different compounds, when the system is slowly supersaturated, the observed metastability limit (MZL) becomes independent of both the rate of change of supersaturation and the time required to reach MZL, a result that suggests an MZL that in fact has energetic origins.

Our experiments are based on the crystallization of a variety of compounds from aqueous solutions. In these studies we employed an evaporation-based microfluidic crystallization platform, in which several small (5 μL) droplets of aqueous solutions were subjected to slow, regulated evaporation. By exploring the metastability limit exhibited by a large number of droplets, we show that the observed probability distribution of nucleation is well described by the classical descriptions of the work required to produce a critical cluster. From this probability distribution we extract an *induction supersaturation* that occurs when the energy barrier to nucleation has a magnitude of $1kT$, the product of Boltzmann's constant and the absolute temperature. This induction supersaturation is determined by the energetics of cluster formation and can be regarded as the limiting lower boundary of metastability. This surprising result raises important concerns regarding the current interpretation of the processes responsible for the nucleation of crystals from solution.

(443f) Ostwald-Like Ripening of the Anomalous Mesoscopic Clusters in Protein Solutions

Peter G. Vekilov[1], Ye Li[2] and Vassiliy Lubchenko[2], (1)Chemical Engineering, University of Houston, Houston, TX, (2)University of Houston, Houston, TX

Anomalous clusters of mesoscopic dimensions composed of protein-rich liquid exist in protein solutions, both in the homogenous region of the solution phase diagram and in the region supersaturated with respect to an ordered solid phase, such as crystals. They are of interest because in supersaturated solutions they are crucial nucleation sites for the ordered phase. We monitor over 71 hours the evolution of the anomalous clusters in solutions of the protein lysozyme employing dynamic and static light scattering and Brownian microscopy. For the sake of experimental clarity, we choose conditions, under which no condensed phases, liquid or solid, are stable or present as long-lived metastable domains. We show that cluster formation is reversible and the clusters are in near equilibrium with the solution, up to a capillary correction. We observe that en route to full equilibration, the average cluster size grows with time following a universal law $t^{0.26\pm0.03}$, independent of the cluster volume fraction. We show that the cluster size distribution is scale-invariant at all stages of the cluster evolution. These observations indicate that, similar to classical Ostwald ripening of new phases, cluster evolution is driven by minimization of the surface free energy and occurs by the exchange of single molecules. The observed cluster evolution may help to understand nucleation mysteries, such as variable nucleation rates under steady conditions, and others.

(450a) Molecular Design of Thin Film Composite (TFC) Hollow Fiber Membranes for Isopropanol Dehydration Via Pervaporation

Jian Zuo, Chemical and Biomolecular Engineering, National University of Singapore, Singapore, Singapore, Yan Wang, Department of Chemical & Biomolecular Engineering, National University of Singapore, Singapore, Singapore, Shi Peng Sun, National University of Singapore, Singapore, Singapore and Tai-Shung Chung, Department of Chemical and Biomolecular Engineering, National University of Singapore, Singapore, Singapore

High performance hollow fiber membranes consisting of a thin film composite (TFC) polyamide layer and a porous Torlon® polyamide-imide substrate have been fabricated via interfacial polymerization for pervaporation dehydration of isopropanol. The effects of substrate characteristics on the formation of the thin polyamide layer and the pervaporation performance are investigated. It is found that the resultant TFC membranes show superior separation performance if the substrates have a fine surface pore size and desirable sub-structural morphology. The effects of amine monomers, i.e., m-phenylenediamine (MPD) and hyperbranched polyethyleneimine (HPEI), for interfacial polymerization with trimesoyl chloride (TMC) on separation performance are also compared. The TFC membrane prepared from HPEI exhibits a higher pervaporation flux than that from MPD possibly due to the fact that the former has higher amine content than the latter. As a consequence, the polyamide selective layer of the former has a greater hydrophilicity than the latter. X-ray photoelectron spectroscopy (XPS) and contact angle measurements have confirmed the hypothesis. The molecular weight of HPEI also affects the pervaporation performance of the resultant TFC membranes. With the optimal interfacial polymerization conditions, the TFC membrane fabricated from HPEI with a molecular weight of 2 kg/mol shows the best separation factor of 624 with a flux of 1282 g/m^2-h and a permeate water concentration of 99.1 wt% using a feed composition of 85/15 wt% IPA/water at 50 °C. After post methanol treatment, the flux can be enhanced to 1920 g/m^2-h with a slightly decrease in permeate water concentration to 98.4 wt%. The mechanism of flux enhancement has been examined by slow beam positron annihilation spectroscopy (PAS). It is found that the removal of the excess amine from the TFC layer during the methanol treatment is the main cause.

(450b) Microfiltration of Lignocellulosic Hydrolyzates Using Dead End Filtration to Separate Lignin

Bhavin Bhayani, Empire State Paper Research Institute, Syracuse, NY and Ramarao Bandaru, Paper and Bioprcess Engineering, State University of New York, Syracuse, NY

Microfiltration of lignocellulosic hydrolyzates using dead end filtration to separate lignin and other inhibitors

B. Bhayani, B.V. Ramarao

Department of paper & Bioprocess engineering, SUNY ESF

In the production of biofuels from lignocellulosics, the biomass is first extracted using dilute mineral acids or hot water. This hemicellulose rich oligomers are hydrolyzed under treatments at higher temperature and pressure to yield sugar monomers which are fermented into biofuels such as ethanol, biobutanol or bioplastics. The initial treatment with either dilute acids or hot water also allows some of the lignin to pass in particulate, colloidal or dissolved forms into the sugar solution. Lignin and other phenolic compounds derived from it are strongly inhibitory to microbial fermentation processes. They also adhere to polymeric membranes, carbohydrates and other surfaces, fouling separation membranes. Their removal from the liquor stream not only facilitates downstream separations but improves the fermentation profiles of the resulting sugar hydrolyzates. Therefore, methods to separate lignin from liquors and hydrolyzates form valuable additional operations in the biorefineries.

A new approach has been developed which can separate inhibitory components from the extracts based on the fact that reduction of temperature of the biomass hydrolyzates can result in the native agglomeration and separation of lignin containing solids by simple settling and/or filtration which thus can result in large increases in separation efficiencies for hydrolyzates. By filtering the extracts at lower temperature (10°C) it is possible to obtain higher flux rate than at higher temperatures (20°C). As shown in Figure 1 and Figure 2, the cumulative flow for extracts at 10°C is higher than the extracts at 20°C at two different pressures. The turbidity of permeate for any given combination of pressure and temperature was less than 10 NTU. It has been hypothesized that the natural flocculation tendency of extracts increases with decreasing temperature.

In addition to the temperature, the effect of addition of a flocculent (Polyethylene oxide, PEO) has been investigated. Substantial enhancement in filtrate flux upon addition of 20 ppm of PEO as seen in Figure 3 is observed.

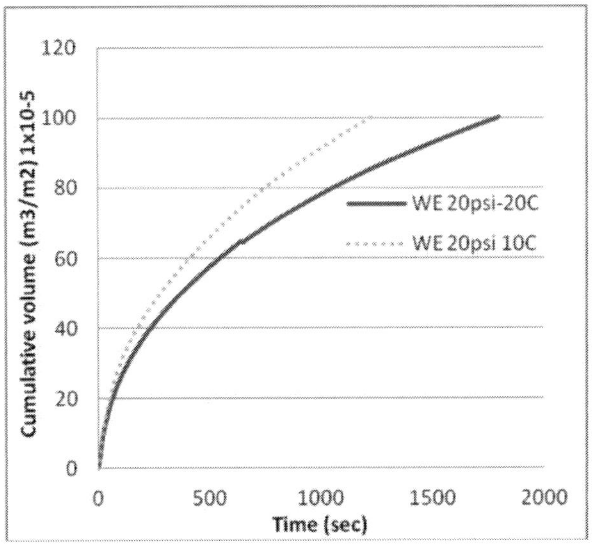

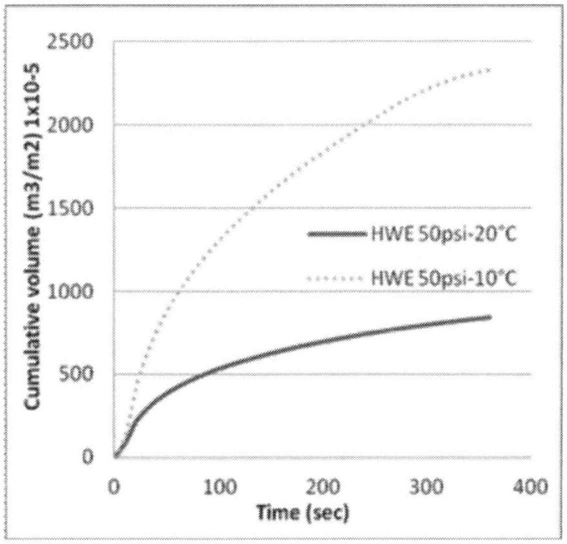

Figure SEQ Figure * ARABIC 1: Cumulative flow at different temperature at 20psi

Figure SEQ Figure * ARABIC 2: Cumulative flow at different temperature at 50psi

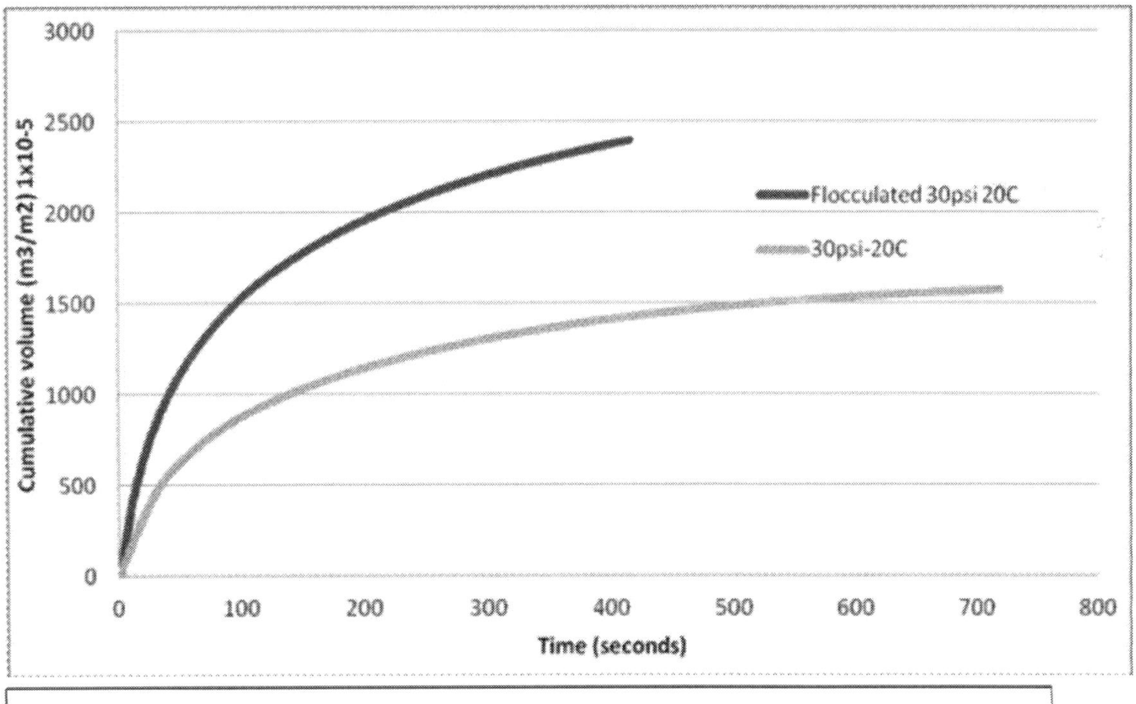

Figure SEQ Figure * ARABIC 3: Effect of addition of PEO

(450c) Mathematical Modeling of the Filtration of Lignocellulosic Hydrolyzates

Bhavin Bhayani, Empire State Paper Research Institute, Syracuse, NY and Ramarao Bandaru, Paper and Bioprcess Engineering, State University of New York, Syracuse, NY

Hot water extracts (HWE) of wood are a complex mixture of various compounds. Depending upon the time of extraction, the size of the colloidal particles decreases from 500 nm to 250 nm. Their surface potential also decreases with increased extraction levels (Duarte et al., 2010). Separation of lignin from lignocellulosic hydrolyzates can be achieved by membrane separation. Dead end filtration of the hydrolyzates has shown the formation of a cake layer consisting of the colloidal particulates in the suspensions. In this work, we present mathematical models of fouling behavior considering the combined actions of membrane pore constriction, pore blockage and cake formation under constant pressure conditions. Parameters representing each of these mechanisms were obtained by fitting to experimental filtrate flux data.

Experimental flux data were obtained for dead end filtrations over ranges of pressure, temperature, concentration, extraction time and molecular weight cutoff of filters. The models provide a clear picture of the transition from the pore blockage to cake filtration regimes. The results provide important insights into the underlying mechanisms of HWE fouling during microfiltration.

(450d) Sugar Concentration for Continuous Enzymatic Saccharification Using Nanofiltration and Reverse Osmosis Membranes

Mohammadmahdi Malmali, University of Arkansas, Fayetteville, AR, Jonathan Stickel, National Bioenergy Center, National Renewable Energy Laboratory, Golden, CO and S. Ranil Wickramasinghe, Department of Chemical Engineering, University of Arkansas

Cost effective conversion of lignocellulosic biomass into bioethanol requires maximizing ethanol yields. Enzymatic hydrolysis of polysaccharides, e.g. cellulose, starch, is usually conducted in a batch reactor. Disadvantages of the classical batch reactor are: product variation from batch to batch, higher overall investment costs due to larger reactor volumes, higher running costs due to frequent startup/shut down, one time use of enzymes as well as catalyst/enzyme separation costs. Development of a continuous saccharification process overcomes the limitations of batch operation. Additional potential advantages include: recovery and reuse of enzymes, improvement of product yield and kinetics, and reduction in inhibition of enzymes. During a continuous saccharification process it is desirable to concentrate the product sugar stream prior to fermentation.

Here we focus on sugar concentration using nanofiltration and reverse osmosis membranes. Nanofiltration, which originated in the 1970s, is one of the newest pressure driven membrane filtration processes. Low pressure reverse osmosis membranes came to be known as nanofiltration membranes with some of the earliest applications being described in the 1980s. Characteristics of nanofiltration membranes include greater than 99% rejection of multivalent ions, 0-70% rejection of monovalent ions and greater than 90% rejection of small organic compounds with molecular weights in the range 150-300.

We have investigated concentration of aqueous sugar streams using a number of different commercially available nanofiltration and low pressure reverse osmosis membranes. Experiments have been conducted using model feed streams as well as corn stover derived biomass hydrolysates. Model feed streams consisted of DI water containing D-Xylose and D-glucose at a ratio of 3:1. In addition compounds that negatively impact ethanol yields during fermentation such as acetic acid, furfural and 5-hydroxymethylfurfural were also added to the feed stream in the ratio 6:2:1. The effect of feed pH was determined. Membrane performance was characterized by determining the level of sugar (xylose and glucose concentration) as well as removal (passage through the membrane) of acetic acid, furfural and 5-hydroxymethylfurfural. The percentage decrease in permeate flux was also determined. Corn stover hydrolysates obtained from the National Renewable Energy Laboratory, Golden Colorado, were also tested.

Membrane surfaces have been characterized using X-ray photoelectron spectroscopy. Differences in membrane performance have been related to interactions between the various species present in the feed and chemical groups present on the membrane surface. The result presented here may be used to identify suitable membranes and operating conditions for sugar concentration and removal of species such as acetic acid, furfural and 5-hydroxymethylfurfural.

(450f) Algae Dewatering Using an Inclined Gravity Settler

Joanne Belovich, Cleveland State University, Cleveland, OH, Zhaowei Wang, Newland Biotech, Cleveland, OH and Jing Hou, Newland Biotech, Inc., Cleveland, OH

One of the main barriers to the cost-effective production of biofuel from microalgae culture is the difficulty of harvesting the algae from dilute cell suspensions. Because of light limitation, the microalgae concentration in large-scale ponds typically does not exceed 1-2 gdw/L. A downward flow inclined gravity settler was evaluated for its effectiveness in algae culture dewatering, using two stains of algae, *Scenedesmus dimorphus* and *Chlorella vulgaris*. Experimental results showed that *S. dimorphus* can be concentrated up to eight-fold using a single settler stage, with a biomass recovery of 80%. Separation efficiency was independent of biomass concentration between 1 and 6 gdw/L, suggesting that a two-stage sequential system of settlers may be effective to maximize biomass recovery and concentration. Efficiency of separation of *C. vulgaris* was slightly lower than that of *S. dimorphus*, due to the fact that *S. dimorphus* cells exist in aggregates of up to eight or more cells and thus settle more easily. The system demonstrated consistent results and steady state operation for up to three weeks of continuous operation, and is low in energy consumption. This separation system may be useful as a first stage for removing 95% of the water from large-scale algae cultivation systems.

(450g) Separation and Identificaiton of Microbial Lipids From Oleaginous Microorganism As Biodiesel Feedstock

Guochang Zhang[1], Todd French[2] and Rafael Hernandez[2], (1)Chemical Engineering, Mississippi State University, Mississippi State, MS, (2)Dave C. Swalm School of Chemical Engineering, Mississippi State University, Mississippi State, MS

Oleaginous yeasts are microorganisms that accumulate high lipids up to over 56% of dry cell mass, and can be used as a feedstock for biodiesel production. However, determination of the oil contents in the fatty cells by lipid extraction places a great role in the research on microbial lipids. The most commonly accepted method of lipid extraction in the laboratory uses chloroform, methanol and water as solvents. In the performance of the method, a lot of solvent is needed and the operation is difficult to handle, and the repetition is not high. The purpose of this experiment is to find a new method that is much easier and simpler for extracting lipid from oleaginous yeast grown on glucose or xylose as the sole carbon and energy source. In the experiments, several lipid extraction methods were investigated to determine the effectiveness of the method in terms of lipid recovery and composition. One method is the gravimetrical Bligh-Dyer method using water, methanol and chloroform as solvent. The second method evaluated used *tert*-butyl methyl ether or hexane as the solvent instead of chloroform. Since the terminal objective is to produce biodiesel, following extraction, the lipids have to be transesterified to fatty acid methyl ester (FAME). Therefore, in-situ transesterification of the fatty cells was also studied, where the lipids were extracted and transesterified simultaneously. The results showed that all the methods are effective to extract the lipids from the *R. glutinis* grown on glucose or xylose. In terms of the FAME yield based on cell mass, difference among the three methods is negligible. While using MTBE as solvent, the process is easier to operate.

(450h) Process Development of Fermentable Cellulosic Sugars from Biomass for Biofuels and Chemical Production

Sarad Parekh, Research and Development, Sweet Water Energy, Rochester, NY and Carl Felice, Sweetwater Energy

The recent volatility of petroleum prices is stimulating the biorefining industry to diversify its raw material feedstock, using a broad-based sugar platform technology. Corn silage and energy crops such as sweet sorghum are widely recognized as the promising high-volume, low-cost lignocellulosic feedstocks from which to derive C5- and C6-rich sugar streams that can be bioconverted into biofuels and chemicals.

This presentation will review Sweetwater Energy's overall approach and concept of integrating a decentralize model, whereby a modular, feedstock-flexible bioprocessing machine pretreats and processes ligocellulosic biomass, and generates adequate quantities of C5 and C6 sugar-rich streams at the biomass-generation site. The sugars are then transferred to a regional refining site for the production of biofuels. This talk will highlight: 1) the importance of innovative biomass harvesting and storage technologies to efficiently and economically transfer ensiled biomass from the field to the bioprocessing site, 2) characterization of the physical and chemical properties that enable optimum bioconversion process, 3) specific pretreatment conditions and various process strategies for deriving streams rich in C6 and C5 sugars for biofuels production, and 4) review recent advances made at Sweetwater Energy in enhancing and improving the ability to process and ferment a wide range of pretreated biomass into ethanol chemicals and advanced fuels, such as bio-jet fuel.

(460a) High-Throughput Evaluation of Nanoporous Adsorbents for Post-Combustion Carbon Dioxide Capture

Tae-Hyun Bae[1], Kenji Sumida[1], Jarad A. Mason[1], Justin J. Dutton[2], Ken J. Micklash[2], Steven S. Kaye[2] and Jeffrey R. Long[1], (1)Chemistry, University of California, Berkeley, CA, (2)Wildcat Discovery Technologies Inc.

There is significant interest in developing solid adsorbents that selectively adsorb a large amount of CO_2 at partial pressures applicable to CO_2 capture processes, because the significantly lower heat capacities of solids compared to aqueous amine solutions can reduce the amount of energy required for desorbing CO_2. The performance of adsorbents is generally evaluated by gas adsorption isotherms, which are phase-equilibrium analyses at many different partial pressures. Acquiring adsorption isotherms for multiple samples at many different conditions can be a tedious step that limits the rapid screening of materials. Here we report the development and use of a 28-channel high-throughput adsorption analyzer that greatly accelerates this screening process.

Initially, we evaluated a series of zeolite adsorbents for application in post-combustion CO_2 capture. Based on the CO_2 and N_2 adsorption isotherms obtained, the performance of zeolites at dry flue gas composition and temperature, such as CO_2 uptake and CO_2/N_2 selectivity, was estimated. Then the performance of materials in a fixed-bed adsorber was predicted through breakthourgh simulations. Adsorption kinetics, an important consideration in designing a fixed-bed adsorber, were also investigated. The high-throughput evaluation methodology developed in this work can be applied in the discovery of robust metal-organic frameworks, a new class of nanoporous adsorbents, for the efficient capture of CO_2 from flue gas.

(460b) Characterization of an Ion Exchange Resin for CO_2 Capture

Alexander Hallenbeck, Chemical Engineering, Carnegie Mellon University, Pittsburgh, PA and John R. Kitchin, Department of Chemical Engineering, Carnegie Mellon University, Pittsburgh, PA

We have investigated the CO_2 capture performance of a commercially available ion exchange resin based on a polystyrene functionalized primary amine for post-combustion CO_2 capture applications. The resin shows moderate capture capacities in the range of 1-2.5 mol/kg depending on the capture/regeneration conditions. Significantly, the resin showed low moisture capacity, and stability over multiple cycles in the presence of oxygen. The resin is fully regenerated in inert gas at temperatures in the range of 100-120 degC. Vibrational spectroscopy indicates that the primary capture mechanism is through carbamate formation, although there is evidence of a significant, secondary capture mechanism as well. The resin can capture CO_2 under conditions ranging from atmospheric CO_2 concentrations to 1 bar, and can be fully regenerated in 1 bar CO_2 at 200 degC. We will present our work in characterizing the CO_2 capture performance of the resin, and our assessment of its suitability for large scale utilization and other CO_2 capture applications.

(460c) Efficient CO2 Sorbents Based On Cost-Effective Silica Foams with Ultra-Large Mesopores

Genggeng Qi, Liling Fu, Brian Hyun Choi and Emmanuel P. Giannelis, Materials Science and Engineering, Cornell University, Ithaca, NY

Increases in atmospheric carbon dioxide have been implicated as one of the main causes of global climate change. Post-combustion capture shows the greatest near-term potential for reducing CO_2 emissions. The current benchmark process for post-combustion capture is absorption using amine solutions, which suffers from relatively low energy efficiency and issues associated with the use of liquid amine solvents such as equipment corrosion, solvent loss, and toxicity. Adsorption using supported amine sorbents provides a promising alternative to conventional amine scrubbing. However, CO_2 capture in practical applications has been impeded primarily by limited sorbent capacity and recyclability, and relatively high sorbent cost. Here we present a series of high-capacity amine sorbents supported on a cost-effective silica foam with ultra-large mesopores. The sorbents exhibit fast CO_2 capture kinetics, high adsorption capacity, as well as good stability over multiple adsorption-desorption cycles. A theoretical analysis is provided for the first time relating the support structure to sorbent performance, which may direct the structural optimization of sorbent supports.

(460d) Graphene As a Catalyst/Support for Hydrogenation/Hydrogen Storage

Nick Stuckert, Chemical Engineering, Univeristy of Michigan, Ann Arbor, MI and Ralph T. Yang, Department of Chemical Engineering, University of Michigan, Ann Arbor, MI

Recently our group found that graphene had a unique ability to dissociate hydrogen, and that this dissociation led to an increase in storage capacity (Wang, Stuckert and Yang, 2010). In this work, the same graphene is analyzed in order to characterize the hydrogen dissociation and recombination ability. A hydrogen isotope scrambling experiment was run for all different conditions of temperature and pressure. Isotherms were measured precisely across different temperatures (0, 30, 60 °C) up to 100 atm. The results encompass the complete characterization of the dissociation and recombination catalysis. The elementary steps are shown and solved. The rate limiting steps are isolated and the activation energies determined (through characterization from 0-80 °C). The effect of pressure (varied from 10-400 psi) is demonstrated as well. These results are intended to provide the information necessary to determine the potential applications of graphene in hydrogenation and dehydrogenation catalysis. Additional new results will be presented as well.

(460e) Denitrogenation of Diesel Fuel by Adsorption On MCM-41, Al MCM-41 and Cu-MCM-41: A Kinetic Study

Jesus Fabricio Guayaquil-Sosa, Chemical Engineering, Autonomous Metropolitan University-Iztapalapa, Iztapalapa, Mexico

Denitrogenation of diesel fuel by adsorption on MCM-41, Al MCM-41 and Cu-MCM-41: a kinetic study

J.F. Guayaquil-Sosa, C.O. Castillo-Araiza, F. López-Isunza and J.A. de los Reyes-Heredia
Autonomous Metropolitan University-Iztapalapa
Av. San Rafael Atlixco 186 Col. Vicentina. C.P. 09340. Iztapalapa, México City;
tel. +521 5558044600-1243,
e-mail: fabricio@xanum.uam.mx;

Summary

The remotion of nitrogen compounds from diesel fuel before hydrodesulfurization (HDS) has captured attention because these compounds strongly inhibit the HDS of some refractory sulfur molecules such as 4-methyldibenzothiophene (4-MDBT), 4,6-dimethyldibenzothiophene (4,6-DMDBT), etc.

In this work a set of adsorbents based on mesoporous molecular sieves; for instance, MCM-41, Al-MCM-41 and Cu-MCM-41 were synthesized, characterized and evaluated, adsorbing model molecules (quinoline and indole) for remotion of nitrogen compounds present in diesel fuel. These experiments were carried out in a laboratory batch stirred tank (15 ml) agitated at 300 rpm, room temperature, and a concentration of nitrogen-containing molecule of 300 ppmw dissolved in a mixture of hexadecane-toluene (1:1).

The performances of synthesized solids were compared with that for a commercial adsorbent (selexsorb CDX) based on γ-Al_2O_3-zeolite. Materials were characterized by using DRX, N_2 Physisorption, Temperature-Programmed Desorption of Pyridine (TPD-Py) and High-Resolution Transmission Electron Microscopy (HR-TEM).

Kinetic models based on Langmuirian mechanism were developed. The adsorption parameters were estimated by using the Levenberg-Marquardt method, and they suggested the quinoline was selectively more adsorbed than indole. Quinoline adsorption constants for Al-MCM-41 adsorbent were higher than the ones for the commercial adsorbent, suggesting that the synthesized material could be used as industrial adsorbent to remove quinoline. On the other hand, indole adsorption constants for Cu-MCM-41 material were higher than the commercial one.

Materials based on mesoporous molecular sieves such as Al-MCM-41 and Cu-MCM-41 appeared promising adsorbents for the removal of nitrogenated compounds, compared to the commercial CDX.

(460f) Selective Adsorption From Liquid Fuels by Silver Loaded Adsorbents

Xueni Sun, Chemical Engineering, Auburn University, Auburn, AL and Bruce Tatarchuk, Department of Chemical Engineering, Auburn University, Auburn, AL

The fuel cell systems such as PEMFC require "zero-sulfur" fuels with ultra-low sulfur contents. Deep desulfurization is applied in order to get ultra low sulfur level. Adsorption desulfurization (ADS) is one of the simplest technologies to remove the sulfur aromatics from the transportation liquid fuels under ambient temperature and pressure. Silver has shown the highest sulfur capacity and regenerability in adsorptive sulfur removal among those transition metals used for ADS. And the sulfur capacity and adsorption rate can be affected by the surface area and pore volume of the adsorbents according to our research. Activated carbon (AC) with high surface area and large pore volume has been demonstrated high sulfur adsorption rate and high sulfur capacity according to many literatures. In this work, high surface area activated carbon was applied as the support in order to get high silver dispersion which was measured by O2 chemisorption. The surface characteristics of this new silver loaded activated carbon were investigated by X-ray Diffraction (XRD), SEM and X-ray Photoelectron Spectroscopy (XPS). The performance of the high surface area activated carbon and silver loaded adsorbents were analyzed by using model fuels with different Ag loading weight. Then the performance of this activated carbon supported silver adsorbent was compared with Ag/TiO_2. By analyzing those samples, the activated carbon can provide significant high surface area support which can increase the dispersion of silver on the support. And the oxygen functional groups on the AC surface play an important role in ADS. However, the aromatic hydrocarbons will inhibit the desulfurization effect of AC by the electron interaction. The Ag loaded activated carbon adsorbents were designed to enhance sulfur adsorption and selectivity in terms of acid-base interaction. Also not so much research has been carried out on regeneration of activated carbon and this silver adsorbent. Thus, the thermal regeneration of this adsorbent under a relative low temperature needs further investigated.

(460g) Kinetics and Mechanism of Adsorption of Aromatic Sulfur Compounds On MOFs From Liquid Fuels

Alexander Samokhvalov, Chemistry Department, Rutgers University, Camden, NJ and Muslum Demir, Chemistry Department, Rutgers University, Camden, NJ

Metal-Organic Frameworks (MOFs) constitute the novel kind of well-ordered 3D "hybrid" organic-inorganic polymers that contain metal coordination centers surrounded by organic "linker" structural units. MOFs are of a strong current interest to the Chemistry and Chemical Engineering communities due to unique combination of 1) well-defined nano- and mesoporous structure; 2) extremely high specific surface area, up to 5,000 m^2/g, and 3) presence of metal cations that can form adsorption or catalytic active sites. MOFs feature moderate to high stability toward oxygen, temperature and non-water solvents, and they have been extensively tested for selective adsorption and separation of many gases of industrial importance such as hydrogen, carbon dioxide, acetylene etc. However, separation of organic compounds from non-water solutions was studied in much lesser extent.

Large-ring aromatic sulfur compounds e.g. benzothiophene (BT), dibenzothiophene (DBT), 4-methyldibenzothiophene (4-MDBT) and 4,6-dimethyldibenzothiophene (4,6-DMDBT) are present in raw petroleum, refinery napthas and commercial liquid fossil fuels. Aromatic sulfur compounds cause air pollution, degradation of catalytic converters of combustion engines, and deactivation of fuel-reforming catalysts of fuel cells. Adsorption capacity of some large-ring aromatic sulfur compounds from model liquid fossil fuels with certain MOFs was previously reported. One popular group of MOFs contains anions of 1,3,5-benzenetricarboxylic acid as organic "linker"; two well-known MOFs of this group are Cu-containing Basolite C300 *aka* HKUST-1 and Fe-containing Basolite F300 *aka* MIL-100(Fe).

Here, we report thermodynamic and kinetic studies of molecular mechanisms of temperature-dependent reversible adsorption/desorption of BT, DBT, 4-MDBT and 4,6-DMDBT on C300 and F300 MOFs from model liquid fuel in tetradecane n-$C_{14}H_{30}$ (model diesel fuel or refinery naphtha). Adsorption of BT, DBT, 4-MDBT and 4,6-DMDBT on Basolite C300 and F300 proceeds without formation of molecular products in liquid phase, as found by the RP-HPLC with the UV absorbance detection. Adsorption capacity follows the trend at 25 and 75 °C: BT > DBT > 4-MDBT > 4,6-DMDBT, consistently with molecular size and diameters of major kinds of nanopores present in C300 MOF. Kinetics of adsorption at 25 and 75 °C proceeds on timescale of hours, is more complex than either zeroths, first or second order rate law, and is due to adsorption on different sites in C300 and F300 MOFs. Two types of adsorption kinetics are observed, depending on initial molar concentration of aromatic sulfur compound: i) preferential adsorption of aromatic sulfur compound and ii) preferential adsorption of solvent n-$C_{14}H_{30}$, followed by preferential adsorption of aromatic sulfur compound. Experimentally determined excess adsorption isotherms of binary solutions "aromatic sulfur compound in n-$C_{14}H_{30}$" on MOF and calculated thermodynamic equilibrium constants support the assignments of adsorption sites.

(460h) Lithium-Modified Mesoporous Silica for Desulfurization and Denitrogenation From Raw Diesel Fuel and Natural Gas

Anton Koriakin[1], Yo-Han Kim[2] and **Chang-Ha Lee**[2], (1)Department of Chemical and Biomolecular Engineering, Yonsei University, Seoul, South Korea, (2)Chemical and Biomolecular Engineering, Yonsei University, Seoul, South Korea

Due to increasingly stringent environmental regulations, desulfurization and denitrogenation of fossil fuels have become very important research topics in the production of clean fuels with ultra-low sulfur levels. Selective adsorption appears to be a promising approach to ultra-deep desulfurization and denitrogenation for gas or liquid fuels because of the ambient operating condition without consuming pressurized hydrogen or oxygen. However, for selective adsorption to be viable, an adsorbent should selectively adsorb sulfur and nitrogen compounds and should regenerate efficiently.

In this study, two kinds of lithium-modified mesoporous silica adsorbents (YSP-Li and MCF-Li) were prepared. As a liquid fuel, residue hydrodesulfurization diesel (RHDS DSL: 271.3ppmw of N-compounds and 430.2ppmw of S-compounds) provided by a refinery factory was used. The solvent washing regeneration (MIBK) was applied. As a model municipal gas, a mixture of methane and methyl mercaptan (291 μmol/mol) was used. A hot purge gas (CH_4 or N_2) at lower than 100 °C was applied to the regeneration. Their adsorptive properties, such as adsorption capacity, adsorption rate, and regeneration ability, will be presented in liquid and gas fuels and will be compared to the performance of Si-Zr cogel.

(460i) In-Situ Spectroscopic Analysis of Adsorbents for Desulfurization of Industrial Reformate Gases for Application in Low Temperature PEM Fuel Cells

Achintya Sujan, Chemical Engineering, Auburn University, Auburn, AL and Bruce Tatarchuk, Department of Chemical Engineering, Auburn University, Auburn, AL

Chemical industries have supplied us with plastics, fertilizers, pharmaceuticals, detergents etc. These products are manufactured in large chemical plants & the production is monitored via temp, pressure & flow alone. However, chemical composition is mostly analyzed separately. This implies that sample analysis could be done at a later time and/or even away from the process. The time for analysis may cause a waste in raw materials & energy if an undesired product quality is obtained. With the advent of fiber optic spectroscopy, you can now look into the process. This would enable to adjust the process variables to prevent waste of material and energy. This technology is environmentally friendly too. Our focus is on the desulfurization process of reformate gases for applications in low temp PEM fuel cells as one of the clean energy sources in the future. Industrial hydrocarbon gases may contain between less than 1 PPM & close to 10,000 PPM of hydrogen sulfide (H_2S) depending on the geographical sources of these gases. H_2S is a poison for metallic catalysts in other fuel processing units & its concentration in feed stocks must be reduced to sub PPM levels. To achieve this, zinc oxide based sorbents have been employed for decades in the industry. At low temp (<100°C), nanostructured ZnO exhibits higher reactivity with H2S in comparison with classical sorbents made of micro-meter ZnO particles. Recently, this reactivity at low temp was improved by doping the ZnO based sorbents with transition metals like Mn, Fe, Co, Ni or Cu oxides allowing an increase in breakthrough capacity for H_2S capture. There is a color change with the adsorption of sulfur compounds providing a visual indication of the impending breakthrough. The dopant could be used as an efficient center for optical absorption in the UV-Vis region. Thus, optical Spectroscopy in this region is useful to study the process. Real-time synthesis, adsorption & desorption tests on these particles are discussed with the idea of operando spectroscopy emphasized for similar domains of work.

(480a) Recovery of Hydrocarbons From Froth Treatment Tailings of Oil Sand Bitumen Processing

Yuming Xu, CanmetENERGY, Natural Resources Canada, Devon, AB, Canada and Jianmin Kan, Total E&P Canada, Calgary, AB, Canada

Bitumen is separated from surface-mined oil sand ore in a warm water extraction process that produces bitumen froth typically containing 60 wt% bitumen, 30 wt% water, and 10 wt% mineral solids. The water and solids are then removed from the froth by solvent dilution followed by gravitational settling. If light alkanes are used as the solvent in the froth treatment process, some of the asphaltenes in the bitumen are precipitated and form aggregates with solids and water droplets. These aggregates quickly separate from the diluted bitumen phase leaving very clean diluted bitumen. After solvent recovery, the asphaltenes, water, and mineral solids are discharged as froth treatment tailings.

The asphaltenes contained in the tailings constitute valuable hydrocarbons that, if recovered, could be useable as coker feed or paving material. We conducted investigations of the feasibility of recovering asphaltenes from the tailings. Two methods were tested – *solvent extraction* and *agglomeration*. In *solvent extraction*, aromatic solvent is mixed with froth treatment tailings to dissolve the asphaltenes followed by centrifugation to remove the mineral solids and water. The experimental results demonstrated that more than 90% of the asphaltenes in the tailings can be recovered. The recovered asphaltenes contain only a small fraction of mineral solids and might be usable as coker feed. In *agglomeration*, the tailings are agitated at elevated temperature (80°C) causing the asphaltene particles to form large agglomerates that separate from the tailings to give an asphaltene-rich sediment phase and tailings that are almost asphaltene-free. The agglomerates still contain significant fractions of mineral solids and water and require further treatment.

(480c) Molecular Dynamics Simulations of Fly-Ash Based Geopolymers: Conformational Changes At High Temperatures

Fernando A. Soto[1], Ancy Kunjumon[1], Kunal Kupwade-Patil[1], Erez Allouche[1] and Daniela S. Mainardi[2], (1)Louisiana Tech University, Ruston, LA, (2)Chemical engineering, Louisiana Tech University, Ruston, LA

Molecular Dynamics Simulations of Fly-Ash based Geopolymers: Conformational Changes at High Temperatures

Fernando Soto, Ancy Kunjumon, Kunal Kupwade-Patil, Erez N. Allouche and Daniela S. Mainardi

Geopolymers are an emerging class of cementitious materials, which offer a sustainable, low energy consuming; low carbon footprint and 100% substitute to Portland cement as a cementitious binder in engineering applications. In addition, geopolymers offer an intriguing mixture of superior physical properties, extensive environmental benefits and strong prospects for commercial implementation in high performance refractory applications. The objective of this study is to gain an atomistic understanding of the structural evolution of geopolymeric gels. In particular, the ultimate goal is to develop insights as to how the molecular structure of geopolymers change with time at elevated temperatures. Density Functional Theory (DFT), Molecular Mechanics (MM) and Molecular Dynamics (MD) approaches are employed to model and simulate geopolymeric behaviors.

In this work, two models of hydrated sodium aluminosilicates (NASH) polymers were considered. NASH models with Si:Al ratio of 1:1 and 2:1 were simulated at selected temperatures ranging from 273 K to 4273 K. The study provided information on stability, and structural and transport properties of geopolymers. Furthermore, Gibbs Free-Energy barriers were calculated to gain an in-depth knowledge of reaction rates that are essential for understanding the geopolymerization mechanism.

(480d) Effects of Farm Wastes Co-Digestion On Microbial Community, Methane Production, and Fiber Quality As Biorefining Feedstock

Rui Chen[1], Fan Yang[2], Zhengbo Yue[3], James MacLellan[1], Terence L. Marsh[2], Yan Liu[1] and Wei Liao[1], (1)Biosystems and Agricultural Engineering, Michigan State University, East Lansing, MI, (2)Microbiology and Molecular Genetics, Michigan State University, East Lansing, MI, (3)Resources and Environmental Engineering, Hefei University of Technology, Hefei, China

The performance of anaerobic digestion is highly correlated to the compositions of its substrate and microbes. A lab-scale anaerobic digestion process of a mixture of dairy manure and corn stover in continuous stirred-tank reactor (CSTR) was performed to investigate the influence of additional organic carbon on methane biogas production, lignocellulose-degrading efficiency, microbial community, and fiber quality of solid digestate for ethanol production. Highest biogas production of 3,158 mL day^{-1} was observed at hydraulic retention time (HRT) of 40 days in the digester fed on manure and corn stover. Biogas productivity was closely correlated with the populations of polymer degrading bacteria *Bacteroidetes* (particularly *Unclassified Bacteroidetes*) and *Clostridia,* as well as methanogenic archaea such as *Methanosarcina* and *Methanosaeta*. The solid digestate (AD fiber) from co-digestion process showed better conversion efficiency in the enzymatic hydrolysis. Results also indicated that anaerobic co-digestion of corn stover and dairy manure significantly enhanced the energy balance of an integrated anaerobic digestion/bioethanol production process.

(480e) Continuous Hydrothermal Liquefaction of Livestock Waste Using a Plug-Flow Reactor

Mitchell James Minarick, Agricultural and Biological Engineering, University of Illinois at Urbana-Champaign, Urbana, IL

Direct liquefaction of biomass is a technology that has shown good potential for producing oil from biomass, especially for wet biomass sources like algae and livestock waste. Previous research at the University of Illinois at Urbana-Champaign proved that up to 70% of swine manure volatile solids could be converted into oil while reducing the manure's chemical oxygen demand by up to 75%. A continuous stirred-tank reactor design was implemented and produced similar conversion rates to the batch process. Algae conversion rates are similarly high for batch processes, and integrated systems that utilize a manure source to provide nutrients for enhanced algae growth could simultaneously clean the waste stream while producing algae for oil production.

However, the high pressures required for the direct liquefaction process preclude the scale-up of continuous stirred tank reactor systems. A plug-flow reactor was designed and tested to explore the feasibility of a scaled-up continuous reactor for the liquefaction of the biomass from this integrated bioenergy system.

(480f) Aeration Induced pH Elevation of Anaerobically Digested Wastewater for Enhanced Ammonia Stripping and Recovery

Quanbao Zhao, Department of Biological Systems Engineering, Washington State University, Pullman, WA and **Craig Frear**, Washington State University, Pullman, WA

During mesophilic or thermophilic anaerobic digestion, a considerable fraction of organic nitrogen is mineralized to ionic ammonia, producing an effluent with higher concentrations of ammonia nitrogen as well as elevated temperatures, both of which can be useful in dowstream processing of ammonia via ammonia stripping. Unfortunately, release of the ammonia also requires a suitable increase in pH well above that normally existing in digested wastewaters. Chemical dosing for pH elevations at or above 10 is possible but economically problematic. Our research has shown that limited, non-biological aeration at laboratory conditions of 0.5-1 L/min of the digested wastewater (dairy and poultry manure) at varying temperature regimes (35, 55, 70C), results in intriguing chemical shifts strongly related to the high levels of carbon dioxide, bicarbonates and carbonates present in digested effluent. These chemical shifts, ultimately release carbon dioxide and raise the pH of the effluent to levels near 10, which combined with elevated operating temperatures (from waste engine heat in commercial scenario), can lead to 70-90% shift from ionic to free, gaseous form of ammonia and subsequent recovery of the ammonia through acid contact. Length and extent of aeration time, percent recovery of ammonia, and final pH attained is influenced by wastewater source and in particular its levels of suspended solids, chemical composition, buffering capacity, and level of ammonia binding to struvite or other salt compounds. The resultiing stripped effluent shows an important enhancement in degree of suspended solids (TSS) and total phosphorus (TP) settling, with 50 and 80% reductions, respectively. This oral presentation is a summary of laboratory results with dairy manure wastewater, with particular emphasis on the chemical relationships and equilibrium shifts associated with the aeration process and its subsequent release of gases. Additional information will also summarize scale-up and demonstration of the process with industrial partners.

(480g) Aspen Simulation of Mobile Biomass to Liquid Facility – the Beauti-Fuel Concept

Xinying Liu[1], Bilal Patel[1], Diane Hildebrandt[2] and David Glasser[2], (1)Centre of Material and Process Synthesis, University of the Witwatersrand, Johannesburg, South Africa, (2)Centre of Material and Process Synthesis (COMPS), University of the Witwatersrand, Johannesburg, South Africa

The BeauTi-fueL facility is a mobile biomass to liquid facility, where the key equipment of the biomass to liquid process is placed into a commercial container and can be transported to a biomass site for liquid fuel production. This facility can be used to process the biomass such as agriculture and forestry waste as well as mulniciple solid waste, to convert them into energy products and can be useful for energy needed remote area. In the facility, biomass is fed into a plasma gasifier. The gas existing the gasifier is then cooled and cleaned before sent to the Fischer Tropsch synthesis unit. Liquid hydrocarbons formed in the Fischer Tropsch Synthesis is knocked out in the cooling and separation system and the tail gas with light hydrocarbons and unreacted CO and H2 is sent to a gas engine to generate electricity. The electricity generated is to be used for the process itself.
The reason why a plasma gasifier was selected was because of its compactness and low tar formation because of the high temperature achieved. The gasification chamber of the plasma gasifier with the needed capacity is small and can be fit into the container, where the size of the fixed bed gasifier and fluidized gasifier are all too large for this situation. The disadvantage of using a plasma gasifier is it utilizes a high quality energy form, which is electricity, to provide energy needed for the gasification process. This raises the question of the overall energy efficiency of the whole process.
An aspen simulation is conducted on this process concept at the claimed capacity of 1 barrel per day liquid hydrocarbon production. Process conditions such as air feeding rate, air enrichment, plasma gasification operation temperature and pressure, the Fischer Tropsch synthesis operation pressure, syngas conversion, hydrocarbon selectivity and the gas engine efficiency are studied. It is found that with the optimized process condition and the right biomass feed, only 1 dry ton per day biomass fed is needed for 1 barrel per day production.

(481a) Integration of Post-Combustion Capture with Compression and Liquefaction Process of CO2

Yeong Su Jeong[1], Ung Lee[1], Seeyub Yang[1], Youngsub Lim[1], Chi Seob Lee[2] and Chonghun Han[1], (1)School of Chemical and Biological Engineering, Seoul National University, Seoul, South Korea, (2)KEPCO E&C Company, Yongin, South Korea

Along with the worldwide efforts to reduce the greenhouse gases, Carbon Capture and Storage/Sequestration (CCS) has emerged as one of the most promising technologies in South Korea. Capture process is known to occupy 70 to 80 percent of the overall energy demands in CCS technology, and consequently, most of up-to-date research has been dedicated to this aspect by developing innovative CO_2 absorbents or modifying the process itself to reduce the regeneration energy of the absorbent. Each constituent of the technology can then be integrated with one another in order to mitigate the overall energy consumption. However, there is no possible candidate for CO_2 storage site in Korea at this time, so the capture process must be integrated with an intermediate compression process instead which is always required to transport $CO_2$from capture plant to the storage site.

In this study, a process simulation including both post-combustion capture of CO_2 using MEA absorbent and multi-stage compression was performed using Aspen Plus®. Different physical property methods were selected for each process to allow more stable and accurate simulation. [1] It was assumed that the compressed (liquefied) CO_2 will be transported via CO_2 carrier, which requires appropriate specifications of CO_2 for ship loading. [2] In addition, 4-stage compression system with intercooling was adopted where both cold and hot utilities are available from compression as well as isenthalpic expansion. A quantitative analysis of how these utilities can be used was carried out so that the energy required for capture as well as the overall process was reduced.

This research was supported by a grant from the LNG Plant R&D Center funded by the Ministry of Land, Transportation and Maritime Affairs (MLTM) of the Korean government.

References:

[1] Van Wagener, D.H., Rochelle, G.T., 2011. Stripper configurations for CO_2capture by aqueous monoethanolamine, Chem. Eng. Res. Des. 89, 1639-1646.

[2] Aspelund A., Jordal, K., 2007. Gas Conditioning–The interface between CO_2 capture and transport, Int. J. Greenhouse Gas Control, 1(3), 343-354.

(481b) A Hybrid Model-Based Framework for the HAZOP Analysis of Geological Storage of Carbon Dioxide

Greeshma Gadikota[1], Babji Srinivasan[1], Ah-Hyung Alissa Park[2] and Venkat Venkatasubramanian[1], (1)Chemical Engineering, Columbia University, New York, NY, (2)Earth and Environmental Engineering & Chemical Engineering, Columbia University, New York, NY

With ever increasing energy demand, carbon capture, utilization and storage (CCUS) has become one of the greatest challenges facing the global community since carbonaceous fuels are still the most abundant and inexpensive sources of energy. CCUS is a complex multistep process which involves (i) capture of CO_2 from flue gas generated in coal-fired power plants (ii) transportation of CO_2 using long pipelines to the storage site, and (iii) injection of CO_2 into the geological formations. Each of these steps pose a number of potential risks such as impurities associated with CO_2 capture, leakage in pipelines, corrosion in pipelines, leakage and/or migration of CO_2 in the geological formations due to faults and fractures, contamination of freshwater aquifers etc.Therefore, it is essential to perform a systematic HAZOP analysis for these processes before implementation of any of the CCUS technologies.

In this work, a hybrid model-based framework is proposed to perform HAZOP analysis for CCUS. First, detailed signed directed graph-based qualitative models are constructed for various components with specific emphasis on CO_2 storage. Further, dynamic simulations of the various tasks involved in the complex CCUS process are performed using quantitative models. These quantitative dynamic simulations with an emphasis on CO_2 storage, supplement the overall graph theoretical model built using process-knowledge. Together, these models are utilized to understand the behavior of the entire complex process under various conditions and more importantly, to evaluate the risks present in the overall system. Furthermore, these graph theoretic models with dynamic simulations are explored to perform fault detection, diagnosis and monitoring in the overall CCUS process with emphasis on CO_2 storage in geological media.

(481c) Impact of Alkalinity Sources On the Life-Cycle Energy Efficiency of Mineral Carbonation Technologies

Abby Kirchofer, Earth, Energy, and Environmental Sciences, Stanford University, Stanford, CA, Adam Brandt, Energy Resources Engineering, Stanford University, Stanford, CA, Sam Krevor, Petroleum Engineering, Imperial College, London, England, Valentina Prigiobbe, Petroleum and Geosystems Engineering, University of Texas at Austin, Austin, TX and Jennifer Wilcox, Department of Energy Resources Engineering, Stanford University, Stanford, CA

Impact of alkalinity sources on the life-cycle energy efficiency of mineral carbonation technologies

Abby Kirchofer[1], Adam Brandt[2], Sam Krevor[3], Valentina Prigiobbe[4], and Jennifer Wilcox[2]

[1] Earth, Energy, and Environmental Sciences, Stanford University

[2]Energy Resources Engineering, Stanford University

[3]Petroleum Engineering, Imperial College

[4]Petroleum and Geosystems Engineering, University of Texas at Austin

Mineral carbonation is a Carbon Capture and Storage (CSS) technology where gaseous CO_2 is reacted with alkaline materials (such as silicate minerals and alkaline industrial wastes) and converted into stable and environmentally benign carbonate minerals (Metz et al., 2005). Previous studies on mineral carbonation have principally focused on identifying process conditions optimal for enhancing the chemical rates of reactions. No study, however, has yet performed a scheme that optimizes these conditions with respect to engineering and economic consideration for the goal of producing a viable carbonation process.

Here, we present a holistic, transparent life cycle assessment model of aqueous mineral carbonation built using a hybrid process model and economic input-output life cycle assessment approach. We compared the energy efficiency and the net CO_2 storage potential of various mineral carbonation processes based on different feedstock material and process schemes on a consistent basis by determining the energy and material balance of each implementation (Kirchofer et al., 2011). In particular, we evaluated the net CO_2 storage potential of aqueous mineral carbonation for serpentine, olivine, cement kiln dust, fly ash, and steel slag across a range of reaction conditions and process parameters. A preliminary systematic investigation of the tradeoffs inherent in mineral carbonation processes was conducted and guidelines for the optimization of the life-cycle energy efficiency are provided.

The LCA model allows for the evaluation of the tradeoffs between different reaction enhancement processes while considering the larger lifecycle impacts on energy use and material consumption. All main process stages are included in the tool, and comprehensive system boundaries are applied throughout the model. The process-model core of the mineral carbonation LCA tool includes 8 process stages defined generically to be applicable to a variety of mineral carbonation technologies (see Figure 1). Because our tool aims to compare process schemes that vary significantly in input resource and process design, the model is built at a general, first-order level. The methodology accounts for the following three types of energy consumption: on-site energy consumption, energy of material and energy inputs consumed in the sector of interest (embodied direct energy), and energy of material and energy inputs consumed in all other sectors (embodied indirect energy). Including both on-site and embodied energy allows a full accounting of the total greenhouse gas (GHG) reduction benefits of each process scheme.

Figure 1. Life cycle process model schematic for aqueous mineral carbonation based on the Olivine – 155 ûC case; line thickness is scaled to the energy and mass fluxes.

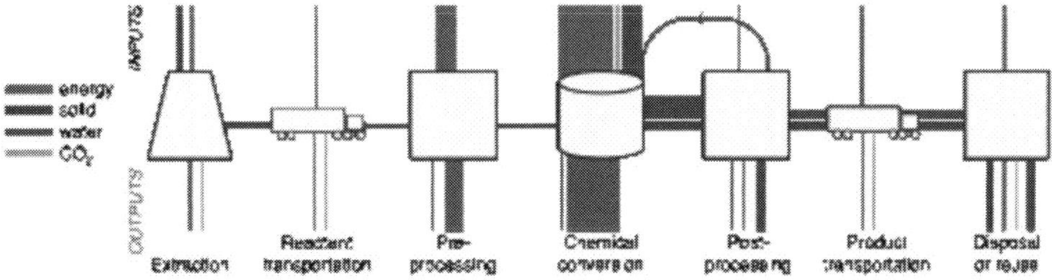

The life-cycle assessment of aqueous mineral carbonation suggests that a variety of alkalinity sources and process configurations are capable of net CO_2 reductions. The maximum carbonation efficiency, defined as mass percent of CO_2 mitigated per CO_2 input, was 83% for CKD at ambient temperature and pressure conditions. In order of decreasing efficiency, the maximum carbonation efficiencies for the other alkalinity sources investigated were: olivine, 66%; SS, 64%; FA, 36%; and serpentine, 13%. Within the range of parameters tested, for all cases maximizing the extent reacted of the alkalinity source unambiguously improves process efficiency. In general, maximizing the extent reacted is critical to optimizing the processes because it minimizes the material handling requirements which contribute negatively to the energy budget. In particular, we observed that:

á the process efficiency is maximized by increasing extent reacted through the most energetically favorable enhancement measure;

á mixing, heating, and grinding are the main energy drivers across all processes;

á reuse of carbonate as aggregate does not necessarily improve life-cycle energy efficiency;

á not all alkalinity sources benefit from high reaction temperatures;

á any steps to increase reaction rates would dramatically improve process efficiency.

Additionally, the CO_2 storage potential of mineral carbonation was estimated using the life-cycle assessment results and alkalinity source availability. The annual storage potential for a given alkalinity source was calculated by multiplying its availability (Mt/yr) by the CO_2 sequestration efficiency of mineral carbonation of that alkalinity source (t-CO_2/t-alkalinity source). For industrial alkalinity sources, availability was based on U.S. production rates (Kelly et al., 2011). For natural alkalinity sources, availability is estimated based on U.S. production rates of a) lime (18 Mt/yr) or b) sand and gravel (760 Mt/yr) (USGS, 2011). The low estimate assumes the maximum sequestration efficiency of the alkalinity source obtained in the current work and the high estimate assumes a sequestration efficiency of 85%. The total CO_2 storage potential for the alkalinity sources considered in the U.S. ranges from 1.3% to 23.7% of U.S. CO_2 emissions, depending on the assumed availability of natural alkalinity sources and efficiency of the mineral carbonation processes.

References

Metz, B., O. Davidson, H. C. de Coninck, M. Loos, and L. A. Meyer (eds.) (2005) IPCC Special Report on Carbon Dioxide Capture and Storage, Prepared by Working Group III of the Intergovernmental Panel on Climate Change. Cambridge University Press, Cambridge, United Kingdom and New York, NY, USA.

Kirchofer, A., A. R. Brandt, S. Krevor, V. Priogiobbe and J. Wilcox (2011) Impact of alkalinity sources on the life-cycle energy efficiency of mineral carbonation technologies. In preparation.

Kelly, K. E., G. D. Silcox, A. F. Sarofim and D. W. Pershing (2011) International Journal of Greenhouse Gas Control, **5**, 1587-1595.

USGS (2011) Mineral Commodity Summaries 2011, United States Geological Survey.

(481d) Uncertainty Quantification in CO2 Sequestration Using Surrogate Models From Polynomial Chaos Expansion

Yan Zhang and Nick Sahinidis, Chemical Engineering, Carnegie Mellon University, Pittsburgh, PA

In this paper, surrogate models are iteratively built using polynomial chaos expansion (PCE) and detailed numerical simulations of a carbon sequestration system. Output variables from the numerical simulator are approximated as polynomial functions of uncertain parameters. Once generated, PCE representations can be used in place of the numerical simulator and often decrease simulation times by several orders of magnitude. However, PCE models are expensive to derive unless the number of terms in the expansion is moderate, which requires a relatively small number of uncertain variables and a low degree of expansion.

To cope with this limitation, instead of using a classical full expansion at each step of an iterative PCE construction method [1], we introduce a mixed-integer programming (MIP) formulation to identify the best subset of basis terms in the expansion. Our approach makes it possible to keep the number of terms small in the expansion. We compare the proposed MIP-based subset selection method against the classical stepwise regression algorithm for building sparse PCE models [2]. The PCE models from our proposed method are simpler than the ones using stepwise regression.

Monte Carlo (MC) simulation is then performed by substituting the values of the uncertain parameters into the closed-form polynomial functions. Based on the results of MC simulation, the uncertainties of injecting CO2 underground are quantified for a saline aquifer. Moreover, based on the PCE model, we formulate an optimization problem to determine the optimal CO2 injection rate so as to maximize the gas saturation (residual trapping) during injection, and thereby minimize the chance of leakage.

References cited:

1. Oladyshkin, S.; Class, H.; Helmig, R.; Nowak, W. An integrative approach to robust design and probabilistic risk assessment for CO2 storage in geological formations. Computational Geosciences 2011, 15, 565–577.

2. Blatman, G.; Sudret, B. An adaptive algorithm to build up sparse polynomial chaos expansions for stochastic finite element analysis. Probabilistic Engineering Mechanics 2010, 25, 183-197.

(481e) Geo-Chemo-Mechanical Studies for Permanent CO2 Storage in Geologic Reservoirs

Greeshma Gadikota[1], Juerg Matter[2], Peter Kelemen[2] and Ah-Hyung Alissa Park[3], (1)Chemical Engineering, Columbia University, New York, NY, (2)Earth and Environmental Sciences, Columbia University, Palisades, NY, (3)Earth and Environmental Engineering & Chemical Engineering, Columbia University, New York, NY

Increasing concentration of CO_2 in the atmosphere is attributed to rising consumption of fossil fuels around the world. The development of solutions to reduce CO_2 emissions to the atmosphere is one of the most urgent needs of today's society. CO_2 injection into geological formations is one of the CO_2 storage options with a large capacity. If non-carbonate minerals such as basalt and olivine exist in the CO_2 injection site, the injected CO_2 can react with those minerals to form thermodynamically stable solid carbonates. In nature, mineral carbonation takes place via dissolution of the mineral to release Ca/Mg ions followed by precipitation of Ca/Mg in the presence of CO_2. The rate limiting step of this carbon mineralization process is considered to be the dissolution of the mineral.A question has been raised regarding the effect of in-situ carbon mineralization on the stability of the geologic storage of CO_2. Both the kinetics and extents of mineral dissolution and carbonation reactions within the geological reservoir are important factors to evaluate for the long term stability of the geologically injected CO_2.Therefore, this work focuses on understanding CO_2-mineral-water interactions at a wide range of reaction conditions (T_{max} = 200°C and $P_{CO2,\ max}$ = 200 bar). Both deionized water and synthetic brine were used as the aqueous reaction media. Minerals such as olivine (Mg_2SiO_4), labradorite ($(Ca,Na)(Al,Si)_4O_8$) and basalt (mixture of silicate minerals), were selected based on their abundance in nature and both dissolution and single-step carbonation experiments were performed to investigate fast and long-term kinetics of mineral weathering. Two different regimes (i.e., surface reaction vs. mass transfer limited regimes) of mineral weathering were identified based on the analyses of the activation energy and reaction kinetics.

(481f) Comparison of CO2 Storage Resource Methodologies

Angela Goodman, National Energy Technology Laboratory, Pittsburgh, PA and Grant Bromhal, National Energy Technology Laboratory, Morgantown, WV

Estimating storage resources for geologic CO_2 sequestration requires a high-level, first-cut assessment of potential reservoirs that can serve as viable storage sites. CO_2 storage resource estimates are based on physically accessible CO_2 storage volume in formations without consideration of regulatory or economic constraints. These high-level resource assessments may be used for broad energy-related government policy and business decisions. Subsurface storage volume estimates depend on geologic properties (area, thickness, and porosity of formations) and the efficiency of storage (the fraction of the accessible pore volume that is available to injected CO_2). Currently there is no standard method to estimate CO2 storage potential and reported CO2 estimate widely vary. In this work, publically available resource methodologies for predicting how much CO2 can be stored in deep saline formations are compared. Resource methodologies applied are from the United States Department of Energy, United States Geological Survey, Carbon Sequestration Leadership Forum, Massachusetts Institute of Technology, and Lawrence Berkeley National Laboratory. The methodologies are applied to formations throughout the United States.

(487a) Novel Hydrophilic Cast Nylon 6,6 Microfiltration (MF) Membrane Supported Thin Film Composite (TFC) Membrane for Engineered Osmosis (EO)

Liwei Huang[1], Ngoc Bui[1], Mark T. Meyering[2] and Jeffrey R. McCutcheon[1], (1)Chemical, Materials and Biomolecular Engineering, University of Connecticut, Storrs, CT, (2)3M Purification INC., Meriden, CT

Clean water and energy are essential for public health and economic prosperity. Recent studies show that Engineered Osmosis (EO) may be a viable technology for desalination, water and wastewater treatment, and power generation. However, the absence of a membrane designed for such processes is a significant obstacle hindering further advancements of this technology. Conventional thin film composite (TFC) reverse osmosis (RO) membranes yield very poor permeate water flux in EO due to their thick, dense and hydrophobic support layers resulting in internal concentration polarization (ICP). In this study, a novel TFC EO membrane has been designed using a conventional cast nylon 6,6 microfiltration (MF) membrane from 3M® as the support in order to minimize ICP effect. The ultra-thin polyamide (PA) selective layer has been built via *in situ* interfacial polymerization. Results show that **for the first time**, an intrinsically hydrophilic cast-film supported TFC membrane has been successfully fabricated with defect-free PA film. The resulting TFC membranes made on piperazine and m-phenylenediamine showed matched permselectivity performance with two commercial RO membranes (The Dow Water & Process Solutions NF 270 and SW-XLE), respectively. When this new TFC membrane was applied in EO, it showed higher osmotic flux and 10X lower salt flux than commercial forward osmosis (FO) membranes. The relatively high osmotic flux of this new TFC membrane can be attributed to its good wetting ability and open porous structure of the support. These qualities make it an excellent candidate for use in engineered osmosis processes including in seawater desalination and power generation.

(487b) Integrating Forward Osmosis Into Microbial Fuel Cells for Sustainable Wastewater Treatment

Zheng Ge, Fei Zhang and **Zhen He**, Civil Engineering, University of Wisconsin-Milwaukee, Milwaukee, WI

Osmotic microbial fuel cells (OsMFCs) are an emerging concept that integrates forward osmosis into an MFC for simultaneous wastewater treatment, bioenergy recovery, and water extraction. Previous studies of OsMFCs have demonstrated the proof of concept and examined different draw solutes as catholytes. This study investigated the performance of an OsMFC treating actual domestic wastewater. The OsMFC achieved water flux of 1.06-1.49 LMH and reduced wastewater effluent by 24.3-72.2% depending on hydraulic retention time. Increasing the recirculation rate of the catholyte from 100 to 3000 mL/min slightly improved water flux to 2.15 LMH. The low water flux was likely due to inefficient reactor configuration, membrane fouling, and concentration polarization. The OsMFC constantly produced bioelectricity and achieved a maximum power density of 28.2 W/m^3 from an acetate solution, or 4.5 W/m^3 from domestic wastewater. Preliminary energy analysis suggested the OsMFC has the potential to produce more energy than was consumed by the pumping system, thereby creating an energy-neutral wastewater treatment system. After operating the OsMFC for more than 100 days, the FO membrane was analyzed for fouling using electrochemical techniques, imaging, and water flux tests, and the results revealed microbial fouling and abiotic scaling on the surface and inside the FO membrane.

(487c) Removal of Tcep From Aqueous Solutions by Adsorption with Zeolites

Scott Grieco, O'Brien & Gere Engineers, Inc., Syracuse, NY and Bandaru V. Ramarao, Empire State Paper Research Institute, Syracuse, NY

Emerging contaminants (ECs) and micropollutants are becoming a global environmental problem as there is increasing information linking them to environmental and human health hazards. One micropollutant of interest is tris-2-chloroethyl phosphate (TCEP), an organo-phosphate ester used as a plasticizer and flame retardant. The European Union (EU) classifies TCEP as dangerous to the environment, toxic to aquatic organisms, because it may cause long-term adverse health effects in the aquatic environment. The EU report also suggests that TCEP is potentially carcinogenic and may cause impaired fertility in humans. Canada concluded that there may be a probability of harm to humans at any exposure level. Within the US, Minnesota considers TCEP a chemical of "emerging concern", California has included TCEP on the "Chemicals known to the State to Cause Cancer or Reproductive Toxicity", and New York has recently instated a law to prohibit the sale of any child care product containing TCEP.

Recent studies have focused on the presence of TCEP in environmental and drinking waters, and several other previous studies indicated that TCEP was the most frequently detected compound of the ECs investigated. Transport of TCEP from wastewater effluents to raw water sources is supported by research that identifies the presence of TCEP in surface waters at concentrations (0.1 mg/L to 0.6 mg/L) similar to those detected in wastewater effluents. This supports the premise that the TCEP is not being removed from the water cycle.

TCEP is one of the more recalcitrant compounds to remove with available wastewater technologies. Activated sludge processes studied in Europe achieved up to 20% removal of TCEP. Oxidation processes that include chlorination, ozonation, and UV radiation and UV/hydrogen peroxide each achieved less than 40% removal of TCEP. The published data for removal with activated carbon suggest less than 50% removal.

Therefore, TCEP is of special interest not only due to its potential toxicity and prevalence in the environment but it presents a difficulty in removal through conventional wastewater treatment operations. There are limited studies available to support removal of TCEP using adsorption. This paper presents a novel approach of using microporous zeolite adsorbants to remove TCEP from aqueous solutions. Equilibrium adsorption isotherms of TCEP on the selected zeolites were obtained experimentally. Surface hydrophbicity measurements and NMR analysis of the zeolite-TCEP system were conducted and provide valuable insight to the adsorption mechanism.

The results show excellent TCEP removal (>99%) as well as provide a foundation to extrapolate the zeolite adsorption application to other difficult to remove polar hydrophilic ECs from water and wastewater sources.

(487d) Enhanced Effectiveness of Dispersants in Oil Spill Remediation Through the Use of Modified Polysaccharides

Jingjian Tang[1], Pradeep Venkataraman[1], Etham Frenkel[1], Gary L. McPherson[2], Jibao He[3], Srinivasa R. Raghavan[4], Vladimir L. Kolesnichenko[5] and **Vijay T. John**[1], (1)Department of Chemical and Biomolecular Engineering, Tulane University, New Orleans, LA, (2)Tulane University, New Orleans, LA, (3)Central Instrumental Facilities, Tulane University, New Orleans, LA, (4)Department of Chemical and Biomolecular Engineering, University of Maryland, College Park, MD, (5)Department of Chemistry, Xavier University of Louisiana, New Orleans, LA

Dispersants are used in oil spill remediation to minimize harmful effects of floating oil on the environment. One of the key challenges in the treatment of oil spill with dispersant is to break oil slicks into stable small droplets which remain suspended in the water column and are biodegraded. COREXIT® 9500, widely used dispersant in marine oil spills, typically reduces the oil-water interfacial tension by 2-3 orders of magnitude and is effective at creating small oil droplets with minimal energy. However, it does not provide an effective barrier against coalescence of the droplets. The objective of this research is to enhance effectiveness of dispersant through the application of environmentally benign biopolymers that greatly improve the colloidal stability and minimize the usage of the dispersant. We specifically focus on hydrophobically modified polysaccharides where alkyl groups are attached to the polymer backbone. We hypothesize that the hydrophobic side chains preferentially anchor themselves in the oil phase resulting in the formation of a protective polymer layer around the droplet. Cryogenic Scanning Electron Microscopy images of the colloidal dispersion of oil in saline water created by the use of dispersant and the biopolymer prove the presence of a polymer layer around oil droplets. The turbidimetric measurements of the dispersions confirm the significantly enhanced colloidal stability being consequent on a combination of biopolymer and a lower usage of the chemical dispersant. Interfacial tension measurements show that addition of hydrophobically modified biopolymers does not impact the reduction of oil-water interfacial tension. Measurement of zeta potential of the emulsions shows a reversal of charge on the oil droplet. The enhanced stability of the emulsion can be attributed to increased electrostatic repulsions as well as presence of polymer layer as a barrier to coalescence.

The molecular weight of the hydrophobically modified chitosan affects the viscosity of the emulsion significantly. We show that for high volume ratios of crude oil to water, high molecular weight modified chitosan was able to form an oil-in-water gel without the aid of the dispersant. Ability of the modified biopolymer to tether the oil droplets in a polymer gel matrix, as confirmed by cryo-SEM images, has a great potential in near shore oil spill remediation and under conditions of high volume ratios of oil to water.

(487e) Electrooxidation of Organic Pollutants with Boron-Doped Diamond Anodes: Evaluation of Process Parameters in Pilot Scale

Dieter Woisetschläger[1], Michael Koncar[2] and Matthäus Siebenhofer[1], (1)Institute of Chemical Engineering and Environmental Technology, Graz University of Technology, Graz, Austria, (2)VTU Engineering, Grambach/Graz, Austria

Anodic oxidation of organic constituents with boron-doped diamond (BDD) electrodes has demonstrated its high potential for the degradation of persistent pollutants. Due to the electrochemical decomposition of water molecules, oxidizers are formed at the electrode surface. Especially hydroxyl radicals provide an efficient oxidizing environment for the purification of wastewater effluents.

In addition to the removal of organic substances, this project addressed the formation of chlorinated organic reaction product (AOX) and the reaction of ammonia in the wastewater. The electrochemical reactions were investigated with synthetic solutions of organic and inorganic substances at boron-doped diamond (BDD) electrodes. The process was then validated with wastewater from dye industry, pharmaceutical industry, chemical industry and landfill leachate. Experiments were performed in laboratory and pilot scale. Figure 1 exemplarily shows the results of landfill leachate degradation. The construction of the electrolysis cells in pilot scale was based on charge and mass transport considerations with a total electrode surface of 2.1 m^2 BDD electrode area. The analysis of experimental results and cost estimation of operation and capital cost confirms the high reliability and sustainability of electrochemical oxidation in wastewater treatment.

(487f) Hydrophilic Electrospun Nanofibrous Mat: An Effective Support for High Osmotic Water Flux Thin-Film-Composite Membrane

Ngoc Bui, Chemical, Materials & Biomolecular Engineering, University of Connecticut, Storrs, CT and Jeffrey R. McCutcheon, Department of Chemical, Materials and Biomolecular Engineering, University of Connecticut, Storrs, CT

***Hydrophilic electrospun nanofibrous mat*: An effective support for high osmotic water flux thin-film-composite membrane**

Nhu-Ngoc Bui, Jeffrey R. McCutcheon*

Department of Chemical, Materials & Biomolecular Engineering and Center for Environmental Sciences and Engineering; University of Connecticut; Storrs, Connecticut, USA

Abstract

Electrospun nanofibers supported polyamide composite membranes have been recently found to be an effective candidate for the next generation of engineered osmosis (EO) membranes [1]. The highly porous nanofibrous supports with interconnected pores structures play a critical role in minimizing the internal concentration polarization effect occurring inside the support layer during osmosis processes. It enables the capability of producing high osmotic water flux using EO technology. In this study, robust thin film composite membranes supported by hydrophilic nanofibers electrospun from blends of polyacrylonitrile (PAN) and cellulose acetate (CA) at different weight ratios were reproducibly fabricated and investigated. These membranes exhibited two to three times higher osmotic water flux than the standard commercial forward osmosis membrane made by Hydration Technology Innovation (HTI). Results shown that the nanofibrous mat is more hydrophilic when more PAN was blended. The contact angles of PAN and CA nanofibrous mats were 69.86 ± 15.83^{o} and 107.43 ± 5.60^{o}, respectively. The hydrophilicity of the electrospun nanofibrous supports was found to correlate well with the flux performance of TFC membranes. The more hydrophilic PAN-supported TFC membranes performs the highest average osmotic water flux of 55 L/m^2h in pressure-retarded osmosis mode and about 30 L/m^2h in forward osmosis mode when using 1.5M NaCl as the draw solution at $25^{o}C$. The corresponding reverse salt flux ranged from 0.5 to 2 g/m^2h. To the best of our knowledge, these membranes outperformed almost all forward osmosis flat sheet membranes have been reported in open literature. These results suggest that electrospun nanofibrous supported polyamide composite membranes may effectively enable applications like forward osmosis where internal concentration polarization is the performance-limiting factor.

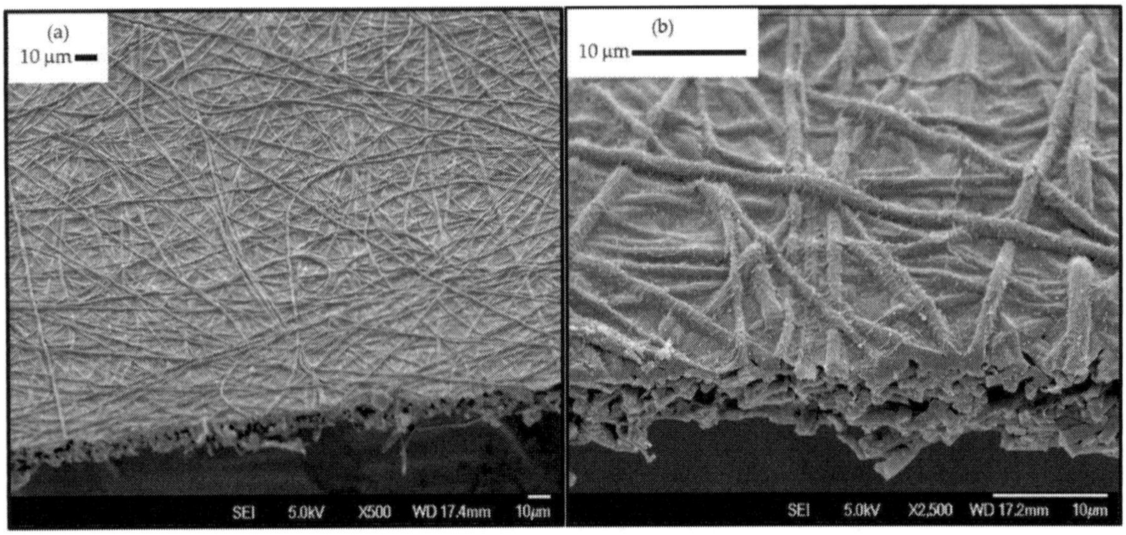

Figure 1. Cross-sectional SEM image of polyamide thin-film-composite (TFC) membrane supported on a polyacrylonitrile (PAN) nanofibrous mat.

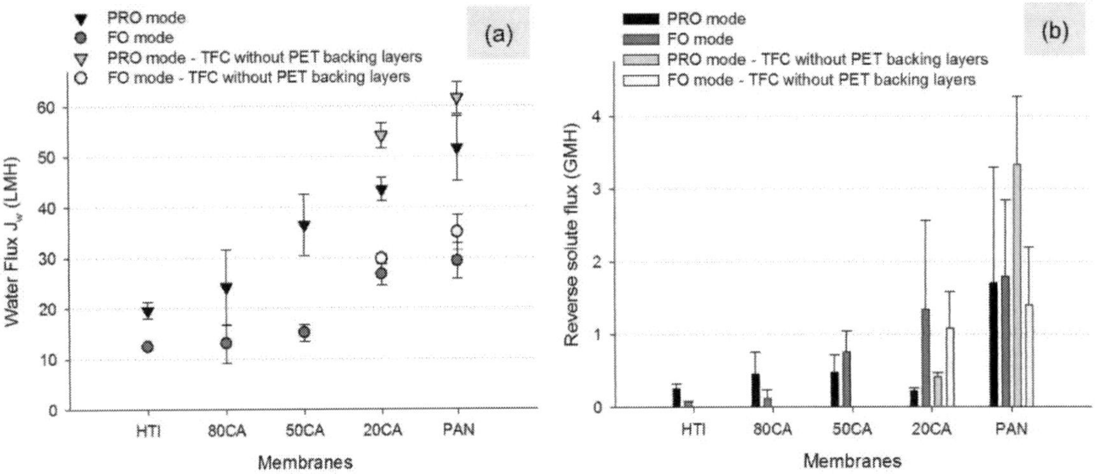

Figure 2. Osmotic water fluxes (a) and reverse salt fluxes (b) of TFC membranes supported on nanofibrous mats electrospun from blends of cellulose acetate and polyacrylonitrile at different weight ratios: 80/20, 50/50, 20/80 and pure PAN. *Experimental conditions:* 1.5M NaCl draw solution, deionized water feed solution, measured volumetric flow rate of the feed and draw solution were 0.5 lpm, temperature of both feed and draw solution was 25.0±0.5 °C.

References

[1] N.N. Bui, M.L. Lind, E.M.V. Hoek, J.R. McCutcheon, *Journal of Membrane Science*, 385-6, 10-19, (**2011**).

(487g) Hollow-Fiber Membrane Bioelectrochemical Reactor for Wastewater Treatment

Zheng Ge and **Zhen He**, Civil Engineering, University of Wisconsin-Milwaukee, Milwaukee, WI

Microbial fuel cells (MFCs) have been widely studied as a potential approach for simultaneous wastewater treatment and bioenergy production. The primary function of an MFC is wastewater treatment; however, the issues about the MFC effluent quality are not sufficiently addressed, possibly because most studies fed the MFCs using synthetic solutions that are "cleaner" than actual wastewater. A few studies that adopted actual wastewater have demonstrated an effective organic removal by MFCs, but whether the treated effluent meets the discharge standards needs further clarification. MFCs are designed to replace the secondary treatment, which usually requires a tertiary treatment such as filtration and/or disinfection for further purifying the treated effluent. To produce a high-quality effluent for either direct discharge or water reuse will make MFCs more competitive as a wastewater treatment technology.

The existing wastewater treatment accomplishes high-quality effluents using membrane bioreactors (MBR), which integrate membrane filtration using microfiltration (MF) or ultrafiltration (UF) membranes with suspended growth bioreactors. The MBR technology has been commercialized for municipal or industrial wastewater treatment. The filtration elements can be either submerged (internal) in a bioreactor such as an activated sludge tank, or installed as a sidestream (external). To recover energy from wastewater, anaerobic MBR is developed with the production of methane gas. More details about the MBR technology can refer to several in-depth review publications.

A possible strategy to involve filtration membranes in MFCs is to use MF/UF membranes as filtration media, installed either inside or outside the anode compartment similarly to that in MBRs. Ion exchange membranes are still required to separate the electrodes, unless a membrane-less MFC is employed. In this study, we developed a membrane bioelectrochemical reactor (MBER) by installing hollow-fiber membranes inside a tubular MFC. The MBER is expected to produce high-quality water while generate electric energy from organic oxidation. A recent study reported a bioelectrochemical membrane reactor using biofilm formed on stainless steel mesh as the filtration material in an MFC, which is very different from our MBER that uses actual MF membrane as the filtration material. We have investigated the performance of the developed MBER with both synthetic solution and actual wastewater for over 150 days. Electricity generation was recorded, trans-membrane pressure was monitored and the quality of the effluent was characterized.

(504a) How Much Water Can Be Recovered From Evaporated Waste Gaseous Streams?

Francesca Macedonio[1,2], Adele Brunetti[1,2], Giuseppe Barbieri[1] and Enrico Drioli[1,2], (1)Institute on Membrane Technology of the National Research Council - (ITM-CNR), Rende, Italy, (2)University of Calabria, Rende, Italy

How much water can be recovered from evaporated waste gaseous streams?

F. Macedonio[1, 2], A. Brunetti[1,2], G. Barbieri[1], E. Drioli[1, 2]

[1] National Research Council - Institute on Membrane Technology (ITM–CNR), Via Pietro BUCCI, c/o The University of Calabria, cubo 17C, 87036 Rende CS, Italy

[2] University of Calabria - Department of Chemical Engineering and Materials, cubo 44A, Via Pietro BUCCI, 87036 Rende CS, Italy

Keywords Dehydration, Membrane Condenser, Process Design

In most part of the globe, the available fresh water is disappearing at an exponential rate as population and technology grows. Industrial water withdrawals account for approximately 22% of global water consumption. Major industrial users include hydroelectric dams, power plants, ore and oil refineries, and manufacturing plants. In industrial processes, recycling and reusing of process water streams is useful to reduce fresh water requirements. More important is the possibility for the industry to close the water cycle by capturing evaporated water thus minimizing the request of fresh water and keeping more water available for other purposes.

In the present work, microporous hydrophobic membranes in an innovative membrane contactor configuration are proposed for the selective recovery of evaporated water exiting industrial plants in waste gaseous streams. In particular, hydrophobic membranes are employed in a *membrane condenser* (Figure 1). In the proposed system, the feed (super-satured industrial gas) is brought into contact with the retentate side of a hydrophobic microporous membrane. The hydrophobic nature of the membrane prevents the penetration of the liquid into the pores while the gases pass through the membrane. Therefore, the liquid is recovered on the retentate side of the membrane, whereas the dehydrated gases on the permeate side of the membrane.

Figure 1. Scheme of the membrane condenser process.

An experimental set up was built (Figure 2) and dehydration measurements were carried out for effectively evaluating the capability of the membrane to retain the liquid water.

Figure 2. Photo of the experimental set up.

In the built lab plant, the wet gaseous stream is fed directly to the membrane module placed into a furnace and containing PVDF commercial membranes. Before entering in the module the relative humidity (RH) of the feed stream is measured with an RH sensor positioned just at the inlet of the membrane module. RH and composition of the two (retentate and permeate) exiting streams are measured by using RH sensors and a gas chromatograph. The process is characterized by a very low pressure difference between the two membrane sides (0.1-0.3 bar).

The dehydration (water recovery) measurements were carried out at different values of temperature and feed pressure. The results achieved with the experimental tests proved the capability of the proposed system to capture the water present in the feed stream [1, 2].

A simulation study of the process was then carried out for comparing the data measured in the experiments with the results obtained through the simulation.

The comparison between the experimental tests and the model indicate a good agreement, with deviations less than 2.24% [1, 2]. This confirmed the validity of the simulation study done and its suitability for a preliminary screening of the potentialities offered by the membrane condenser in the dehydration of gaseous streams. In Figure 3 some of the results achieved through the modelling are shown.

Figure 3. Recovered water vs temperature reduction for the flue gas with RH_{feed}=100%, 50°C<T_{feed}<90°C.

Figure 3 highlights the amount of water that can be recovered from the flue gas, at different temperature of the inlet flue gas and constant RH. As it can be seen, for an inlet temperature of 90°C, a reduction of ca. 2°C is sufficient to recover the 20% (the amount to make the plant self sufficient) of water whereas this value growths up to 3.6°C only when the flue gas enters with a temperature of 60°C. The obtained results confirm the potentialities offered by the membrane condenser in the dehydration of gaseous streams.

It must be pointed out that the results here reported are focused on the analysis of the dehydration of flue gas stream; however the same approach can be suitable for the study of the dehydration of other gaseous streams, such as the one coming out from cooling towers, coal gasification, paper and mills, kilns factory, etc.

Acknowledgments

The EU-FP7 is gratefully acknowledged for co-funding this work through the project *"CapWa - Capture of evaporated water with novel membranes"* (GA 246074). We also wish to acknowledge Dr. Wolfgang Ansorge (Membrana GmbH) for supplying us samples of hollow fibres PVDF membranes.

Relevant bibliography

[1] E. Drioli, F. Macedonio, A. Brunetti, G. Barbieri, *Membrane Condenser for the recovery of evaporated "waste" water from industrial processes*. International Workshop on Membrane Distillation and Related Technologies, October 9 - 12, 2011, Auditorium Oscar Niemeyer - Ravello (SA).

[2] F. Macedonio, A. Brunetti, G. Barbieri, E. Drioli, Membrane Condenser as a new technology for water recovery from humidified "waste" gaseous streams. IECR, 2012. (Submitted)

(504b) A Novel Technology for Post-Combustion CO2 Capture

Gordana Obuskovic, Tripura Mulukutla and **Kamalesh K. Sirkar**, Otto York Department of Chemical, Biological and Pharmaceutical Engineering, New Jersey Institute of Technology, Newark, NJ

Conventional method for CO2 capture from post-combustion flue gas generally employs an absorption and stripping process. The CO2 is absorbed for example in an aqueous solution of amine e.g., monoethanolamine and then stripped out in a separate tower with steam at 120°C . Such a process is highly energy intensive. The low partial pressure of CO2 in the flue gas inhibits the application of CO2-selective membranes unless methods are employed to increase the CO2 partial pressure in the flue gas to be treated. Adsorption processes relying on adsorbent regeneration require among others the need to switch the flue gas between different columns. A number of other techniques are being studied to reduce the capital and operating costs for CO2 capture from post-combustion flue gas. We will describe a novel technique to potentially bypass the shortcomings of many existing approaches. We have employed a simulated humidified flue gas containing around 14% CO2 and demonstrated successful removal of bulk of the CO2 and its recovery in a CO2-concentrated stream.

(504c) Mathematical Modeling of Hollow Fiber Gas-Liquid Membrane Contactors for Acid Gas Removal

Ajay Makkuni[1], Shiguang Li[1], Dennis Rocha[1], S. James Zhou[1], Howard S. Meyer[1], Benjamin Bikson[2] and Yong Ding[2], (1)Energy Conversion, Gas Technology Institute, Des Plaines, IL, (2)PoroGen Corp., Woburn, MA

The removal of acid gases such as CO_2, SO_2 and H_2S from industrial and natural gases is an important separation process. Membrane gas absorption is a well known hybrid approach wherein the advantages of both absorption and membrane processes are combined for cost-effective separation and removal of acid gas components from industrial and natural gases. Gas Technology Institute (GTI) and PoroGen Corporation are jointly developing novel gas-liquid membrane contactors for CO_2 and H_2S removal. Extensive laboratory and bench-scale measurements are underway at GTI to bring this technology to commercialization. In these membrane contactors, the acid gas stream passes through the hollow fiber membrane tubes while a chemically reactive solvent flow on the shell side of the module. The acid gas components diffuse through the membrane walls and are finally removed in the selective solvent by reactive absorption. A variety of models with increasing complexity are compared for the simulation of the membrane gas absorption process. The tube side flow includes plug flow and parabolic velocity profiles. Flow patterns reflecting plug flow, Happel's free surface model and random distribution of fiber model are considered for the shell side of the membrane module. Solution methods include analytical and semi-analytical approaches, apart from numerical schemes. The coupled differential equations are solved using an Excel VBA platform combined with Aspen Plus process simulator.

(504d) Dense Membrane Contactors for Physical Solvents Regeneration: Proof of Concept Study and Modeling

Elodie Chabanon[1], Denis Roizard[1] and Eric Favre[2], (1)Laboratoire Réactions et Génie des Procédés LRGP (UPR 3349 CNRS), Nancy Université, Nancy, F-54001, France, (2)Laboratoire Réactions et Génie des Procédés LRGP (UPR 3349 CNRS), Nancy Université, Nancy F-54001, France

Dense membrane contactors for physical solvents regeneration: Proof of concept study and modeling

E. CHABANON, D. ROIZARD, E. FAVRE*

E-mail: eric.favre@ensic.inpl-nancy.fr

Membrane contactors have received increased attention for several years and are already used in different industrial applications, particularly for intensified gas liquid processes. Studies and applications are almost systematically based on microporous hydrophobic membranes. In that case, the achievement of non wetting conditions on long term range imposes the use of a liquid phase with a high surface tension, such as aqueous solutions. Consequently, low surface tension liquids, such as physical solvents, cannot be used. This study intends to explore the potentialities of dense asymmetric polymeric hollow fibers as membrane contactors for physical solvents applications. In that case, theoretical computations show that promising performances could be obtained, due to the possibility to simultaneously sustain a high transmembrane pressure (thanks to the dense skin layer which is not subject to wetting pressure limitations) and offer high mass transfer coefficients. A case study, dissolved carbon dioxide removal from a physical solvent (NMP), has been investigated at lab scale with a module based on PPO hollow fibers. Solvent regeneration by depressurization has been investigated and several operating parameters have been studied and modeled (pressure difference, solvent flowrate, CO_2 solubility). It will be shown that the concept offers attractive performances, especially with regard to the possibility to maintain the liquid under high pressure conditions, contrarily to flash drums which are classically used for physical solvents degassing operations. The improvement in terms of energy efficiency of the regeneration process based on this peculiarity (no liquid recompression step on the solvent loop) will be addressed. Several applications of the concept at industrial scale will be discussed.

(504e) Membrane Distillation with Carbon-Based Membranes

Ali Ashraf[1], Seyed A. Dastgheib[2], Glennys A. Mensing[1] and Mark A. Shannon[1], (1)Mechanical Science and Engineering, University of Illinois at Urbana-Champaign, Urbana, IL, (2)Illinois State Geological Survey, Prairie Research Institute, University of Illinois at Urbana-Champaign, Champaign, IL

Membrane distillation (MD) is an emerging technology for water desalination, industrial wastewater and radioactive waste treatment, removal of volatile organic compounds, concentration of agro-based and organic solutions, and other separation applications. In spite of having high potentials, MD is still an emerging separation technology mainly due to the short comings of available membrane materials and limitations of existing MD systems. Available membranes are relatively expensive and cannot be used at high temperatures. Carbon-based membranes (CBMs) can be alternative types of membrane materials for the MD process. Surface chemistry of CBMs can be tailored for different MD separation or simultaneous MD separation/catalytic applications even at high temperatures under harsh chemical conditions.

Different types of carbon membranes are developed by using different support materials including porous graphite sheet, carbon nanofiber mat, porous alumina disc, and quartz fiber tissue, and coated with chemical vapor deposition of pyrolytic carbon and/or deposition of amorphous carbon by pyrolysis of sucrose. CBM were heat treated in hydrogen or grafted with methyl groups by sonication in acetone to enhance their hydrophobicity. These membranes (e.g., Figure 1) had different porosities ranging from 100 nm to 1000 nm, demonstrated high degree of hydrophobicity, and maintained their properties at temperatures up to 400 °C under harsh conditions. Membranes were characterized by water contact angle measurement, nitrogen permeability, SEM, and XPS.

The CBMs were tested in a sweeping gas MD system for water desalination at a temperature range of 50-150 °C. Developed CBMs can be potentially used for water desalination and other separation applications at elevated temperatures where the existing polymer-based membranes cannot be properly used due to structural changes and thermal decomposition effects.

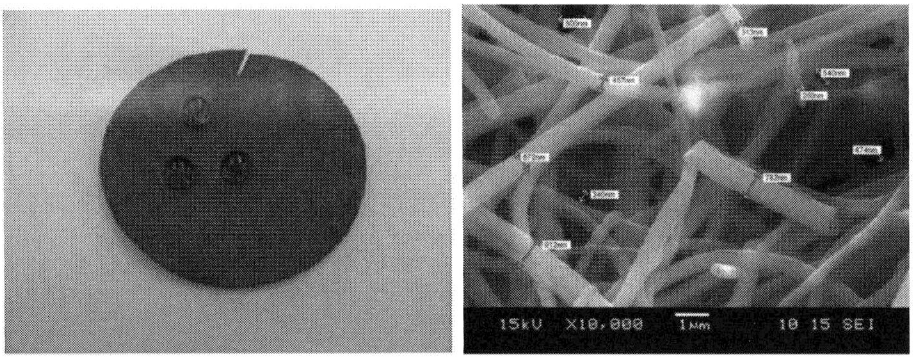

Figure 1: An example of developed CBM: Pyrolytic carbon-coated tissue quartz membrane.

(505a) Ionic Liquid Diblock Copolymers

Brian Adzima[1,2], Steven C. Taylor[1], Brian Kail[3], Hunaid Nulwala[1,2], Krzysztof Matyjaszewski[2] and David Luebke[4], (1)NETL, Pittsburgh, PA, (2)Department of Chemistry, Carnegie Mellon University, Pittsburgh, PA, (3)URS, South Park, PA, (4)US DOE/NETL, Pittsburgh, PA

Polymer membranes formed from diblock copolymers have been shown to give improved CO_2/N_2 and CH_4/N_2 separations. The observed increase in selectivity is attributed to the orthogonal orientation of the cylindrical morphology relative to the membrane surface.[1] While successful, these initial attempts have been based on polymers such as styrene-butadiene-styrene (SBS) which have low CO_2 solubility. We have developed techniques for preparing diblock copolymers specifically designed for CO_2 separation that contain ionic liquid blocks. This was accomplished using a combination of controlled radical polymerization techniques and click chemistry.[2] A detailed synthesis methodology will be presented. The core synthesis centers on the polymerization of c-vinyl triazole[3] monomers using free radical and reversible addition-fragmentation chain-transfer (RAFT)[4] polymerizations followed by quaternization of the triazole group and ion-exchange reactions. Approaches for producing block copolymers from these pILs will be discussed, as will strategies for synthesizing, and subsequently crosslinking block copolymers to form membranes. Alternative approaches such as the direct polymerizing of c-vinyl triazolium salts, and the incorporation of ionic liquid groups unto polymer backbones using the thiol–ene reaction will also be discussed.

[1] Buonomenna, M.G., Golemme, G., Tone C.M., Penelope De Santo, M., Ciuchi F., and Perrotta, E. *Adv. Funct. Mater.*, **2012**, 22, 1759–1767.

[2] Kolb, H.C., Finn, M.G., Sharpless, K.B., *Angew. Chemie. Int.-Ed.* **2004**, 40, 2004-2044.

[3] Nulwala, H., Takizawa, K., Odukale, A., Khan, A., Thibault, R.J., Taft, B.R., Lipshutz, B.H., Hawker, C.J., *Macromolecules*, **2009**, 42, 6068-6074.

[4] Moad, G., Rizzardo, E., Thang, S.F. *Aus. J. Chem.* **2005**, 58, 379-410.

(505b) Ionic Liquid Containing Mixed Matrix Membranes for O2 Separation

Shan Wickramanayake, URS Energy and Construction/NETL, Pittsburgh, PA, David Hopkinson, National Energy Technology Laboratory, Morgantown, WV, Christina Myers, National Energy Technology Laboratory, Pittsburgh, PA and David Luebke, US DOE/NETL, Pittsburgh, PA

Polymeric mixed matrix membranes containing ionic liquid were fabricated. Matrimid® and 1-hexyl-3-methylimidalzolium bis(trifluoromethyl sulfonyl) imide ([hmim][Tf_2N]) were selected as the polymeric and ionic liquid materials respectively. Inorganic zeolite sieves were selected as the sieving material. Membranes with four different morphologies, named *a*, *b*, *c* and *d* were prepared via a dry-wet nonsolvent-induced phase separation procedure. All membranes except type *a* membranes contain zeolite sieves, and the porosity of the membranes was tuned using additives. The membrane type *a* had a low separation performance for O_2/N_2; however, upon adding zeolite sieves and changing the fiber morphology, better permeance and selectivity were achieved. The morphology of the membranes was characterized by measuring the volume fraction of the polymer, macro-voids, and micro-voids for each type of membrane. The calculated porosities of both the bulk and the micro-porous region of the membrane were then related to the transport properties of membranes. Permeance and selectivity for O_2/N_2 separation are reported for ionic liquid containing mixed matrix membranes from 22 to 75° C.

(505c) Separation of Sugars From Biomass Hydrolysate Using Plasticized Liquid Membranes

Kelly Marbaugh[1], Sasidhar Varanasi[2] and Patricia Relue[1], (1)Department of Bioengineering, The University of Toledo, Toledo, OH, (2)Department of Chemical and Environmental Engineering, The University of Toledo, Toledo, OH

Pretreatment and saccharification of lignocellulosic biomass results hydrolysate that is rich in monomeric sugars such as glucose and xylose. However, hydrolysates often contains potential contaminants, such as phenolics, which can have deleterious effects on downstream process steps such as fermentation. Less than optimal utilization of biomass sugars due to these contaminants may lead to unfavorable process economics.

Plasticized liquid membranes (PLM) can be used to selectively remove sugars from the hydrolysate, thereby eliminating the problems associated with contaminants and producing a desirable sugar feedstock. PLMs are lipophilic in nature and are robust in a variety of aqueous environments. In this research, PLMs containing carrying agents such as quaternary ammonium salts and/or complexing agents such as aryl boronic acids are tested with model glucose/xylose solutions, the sugars most commonly found in biomass hydrolysate. Composition of the membranes was varied by changing concentrations of the carrying agents and plasticizing agents. The effects of the membrane porosity, thickness, and composition on sugar flux and membrane stability are reported.

(505d) Development of Flat-Sheet Membranes for C1-C4 Alcohols Dehydration Via Pervaporation From Sulfonated Polyphenylsulfone (sPPSU)

Yupan Tang[1], Natalia Widjojo[2], Gui Min Shi[3], Tai-Shung Chung[3], Martin Weber[4] and Christian Maletzko[4], (1)Chemical and Biomolecular Engineering, National University of Singapore, Singapore, Singapore, (2)Biomolecular and Chemical Engineering, National University of Singapore, Singapore, Singapore, (3)Department of Chemical & Biomolecular Engineering, National University of Singapore, Singapore, Singapore, (4)BASF SE Germany, Ludwigshafen, Germany

Abstract

In this work, pervaporation membranes made of sulfonated polyphenylsulfone (sPPSU) materials were developed and applied for C1-C4 alcohols dehydration. Compared to their non-sulfonated polyphenylsulfone (PPSU) counterpart, the membranes from sulfonated materials can achieve both a higher permeability and a higher selectivity. Fundamentals of PPSU membranes with different degrees of sulfonation are investigated by employing different analytic instruments. Polymer chains of sPPSU membranes get stiffer than those of PPSU membrane due to the favorable interaction between $-SO_3H$ groups. Moreover, sPPSU membranes maintain the similarly excellent mechanical properties and display slightly higher hydrophilicity compared with non-sulfonated PPSU membrane. The pervaporation characteristics and the temperature dependence of sPPSU membranes are examined with respect to the dehydration of an ethanol/water mixture. It is found that, for sPPSU membranes, permeation flux and permeability increase while separation factor and selectivity decrease with an increase in temperature. Activation energies based on flux (E_J) and permeability (E_P) are also found to rise with increasing degree of sulfonation, which reveal that dehydration performance of sPPSU membranes is more sensitive to operating temperature than that of PPSU membranes. Sorption analyses indicate that membrane separation performance is mainly attributed to the preferential diffusion of feed components through membranes. Dehydration of C1-C4 alcohols by pervaporation through these PPSU based membranes was subsequently studied. Interestingly, it can be observed that flux can be maintained above 30 g/m^2h while separation factor follows the order of IPA > n-BuOH > EtOH > MeOH. Especially, it can reach to a separation factor above 11 at 60 for methanol dehydration, which is better than most other polymeric membranes, indicating a great potential to be applied for methanol purification in the industry.

(505e) Development of High Performance Dual-Layer Hollow Fiber for Dehydration of Ethanol Through Novel Immiscibility Induced Phase Separation (I2PS) Process

Yee Kang Ong and Tai-Shung Chung, Department of Chemical & Biomolecular Engineering, National University of Singapore, Singapore, Singapore

This pioneer study discloses a novel hollow fiber spinning technique, namely immiscibility induced phase separation (I^2PS) process to fabricate high performance dual-layer hollow fibers for dehydration of ethanol via pervaporation. As opposed to the conventional composite membrane fabrication method, the I^2PS process takes the advantages of phase separation phenomena occurring in immiscible blend dopes made of incompatible polymers. The immiscible blend dopes were simultaneously extruded through a triple orifice spinneret and fabricated into dual-layer hollow fibers consisting of an outer protective layer and an inner selective layer. A defect-free selective layer is formed at the outer surface of the inner layer due to the incompatibility between the polymer solutions of both inner and outer-layers as well as the shielding effect of the outer protective layer that provides a sufficient time for the formation of defect-free selective layer during the phase inversion process. It is found that the phase inversion kinetics of the outer-layer dope solutions as well as the nature of outer-layer materials play great roles in determining the morphology of the outer protective layers and subsequently affects the permeance of the resultant hollow fibers. On the other hand, the selectivity of the as-spun fibers was found to be controlled by the outer skin of the inner-layer. The novel dual-layer hollow fiber possesses an ultra-high water permeance coupled with reasonable water/ethanol selectivity as compared to the pervaporation membranes available in the literatures. Heat treatment may enhance the selectivity of the aforementioned hollow fibers with some sacrifices of permeance. The present study may provide an innovative approach for hollow fiber fabrication in the future.

(505f) Bendable Ceramic Paper Membranes

Xinjie Zhang[1], Anthony Allegrezza[2] and Zhilong Wang[1], (1)Novarials Corporation, Woburn, MA, (2)Congress Street Consulting LLC, MILFORD, MA

Advanced porous membranes are of significant importance to dozens of industries including chemicals, oil and gas, automobiles, batteries, semiconductors, food and beverage, biotech and pharmaceutics, waste treatment, environmental protection and remediation by means of filtration or separation. Organic polymeric membranes currently dominate these markets due to their bendable feature and relatively low manufacture cost; however, these organic polymeric membranes suffer from several drawbacks, such as low stability at elevated temperature, low stability in acids and bases, swelling in organic solvents, and fast degradation in oxidative environments. Ceramic membranes, with the advantages of high chemical, thermal and mechanical stability, should be the attractive alternatives to polymeric varieties; however, ceramic membranes have not been widely adopted due to their high manufacture cost, rigid structure, low packing density, and low throughput. It has been a long time dream to make a membrane with both the advantages of polymeric membranes and ceramic membranes.

The advantages of nanoscience and nanotechnology make such a dream a reality. Using titania nanowires, Novarials has developed a Bendable Ceramic Paper Membrane platform technology. Titania nanowires are thin strands having diameters in the range of about 10nm to about 100nm, and lengths of from a few microns to tens and even hundreds of microns. Titania nanowires are prepared by cost effective hydrothermal process, and the paper-like ceramic membranes are fabricated by wet laid process. The membranes are porous, bendable/flexible and sufficiently strong, suitable for nanofiltration and ultrafiltration. Potential advantages of these membranes are low manufacturing cost, high porosity, superior chemical and thermal stability and for TiO2, the ability to react with UV light to remove organic material. The presentation will provide detailed information of Novarials Bendable Ceramic Paper Membranes.

(505g) Preparation and Characterization of Porous Poly(vinylidene fluoride) Membranes Using New Pore Forming Additives

Shada Salem[1], Rinku Thomas[2], Elena Guillen[2], Hassan Arafat[2] and Stephen W. Thiel[3], (1)School of Energy, Enviornmental, Biological, and Medical Engineering, University of Cincinnati, Cincinnati, OH, (2)Water and Environmental Engineering Program, Masdar Institute, Abu Dhabi, United Arab Emirates, (3)School of Energy, Environmental, Biological, and Medical Engineering, University of Cincinnati, Cincinnati, OH

Much attention has been paid to semicrystalline poly(vinylidene fluoride) (PVDF) as a membrane material due to its high resistance to harsh environments (including both both acidic and basic conditions), its high thermal and UV stabilities, and its affordable cost at large scale. [1-5]. PVDF membranes can be produced using wet phase inversion. Recently, organic and inorganic pore forming additives have been included in the membrane casting solution; these additives include poly (vinyl pyrrolidone), polyethylene glycol, $LiClO_4$, LiCl, and $ZnCl_2$. Changes in membranes morphology, performance, and mechanical strength due to the use of pore forming agents have been reported in the literature. [6-12].

This paper investigates the effect of new pore forming additives, trimeythelbenzene (TMB) and sodium chloride to the poly(vinylidene fluoride)-N,N-dimethylacetamide system. The use of these additives significantly influenced the final morphology of the membranes. The total porosity and the average pore diameter increased with the addition of TMB and NaCl. Scanning electron microscopy, water intrusion porosimetery, capillary flow porometry, and water contact angle were used to characterize the morphology, porosity, average pore size, liquid entry pressure and hydrophobicity of the membranes.

References

[1] P. Witte, P. J. Dijkstra, J. W. A. Berg, and J. Feijen, J. Membrane Sci.,117,1 (1996).

[2] L. Wu, J. Sun, Q. Wang, *J. Membr. Sci.* 2006, *285*, 290.

[3] D. J. Lin, C. L. Chang, T. C. Chenand, L. P. Cheng, *Desalination* 2002, *145*, 25.

[4] E. Fontananova, J. C. Jansen, A. Cristiano, E. Curcio, E. Drioli, *Desalination* 2006, *192*, 190.

[5] D. Wang, K. Li, W. K. Teo, *J. Membr. Sci.* 1999, *163*, 211.

[6] A. Bottino, G. C. Roda, G. Capanelli, S. Munari, *J. Membr. Sci.* 1991, *57*, 1.

[7] D.L.Wang, K. Li,W.K.Teo, *J. Membr. Sci.*163 (1999) 211.

[8] W.Y. Chuang, T.H. Young, W.Y. Chiu, C.Y. Lin, *Polymer*41 (2000) 5633.

[9] G. Arthanareeswaran, P. Thanikaivelan, K. Srinivasn, D. Mohan, M. Rajendran, *Eur. Polym. J.*40 (2004) 2153.

[10] M.L. Yeow, Y. Liu, K. Li, *J. Membr. Sci.*258 (2005) 16.

[11] D.J. Lin, C.L. Chang, F.M. Huang, L.P. Cheng, *Polymer*44 (2003) 413.

[12] J.H. Kim, K.H. Li, *J. Membr. Sci.* 138 (1998) 153.

Performance decline due to cake enhanced concentration polarization in cross flow membrane filtration: An transient electrokinetic model and experimental observation

Md. Abdullaha-Al-Mamun, Mohtada Sadrzadeh, Subir Bhattacharjee

Department of Mechanical Engineering, University of Alberta, Edmonton, AB, Canada T6G 2G8

September 10, 2012

Abstract

Fouling of nanofiltration membranes by multiple rejected components adversely increases the operating pressure, decreases the performance and requires cleaning. Considerable attention has been devoted toward understanding the fouling phenomena and developing models. However, the models are mostly empirical and fail to establish mechanistic link between the different phenomena. In this context, a transient model has been developed to predict the performance decline due to colloidal fouling of nanofiltration membranes. The model combines the transient growth of cake layer and cake enhanced concentration polarization (CECP) phenomenon to predict the permeate flux and observed rejection decline during cross flow membrane filtration. The cake layer has been considered as a swarm of non-interacting, incompressible, spherical, charged particles and represented using the Kuwabara cell model. The model also provides fundamental insight to the development of streaming potential and electroosmotic back flow, due to transport of ions around the charged colloidal particles of the cake layer, in light of classical Levine-Neale model of electrophoresis in concentrated dispersions. The CECP phenomena has been accounted by considering the hindered back diffusion of ions through the tortuous path of the cake layer. These phenomena are then combined with the film theory of cross flow filtration for predicating the permeate flux and observed rejection. In order to validate the developed model fouling of nanofiltration membrane by colloidal silica particles were carried out and the results were compared with the model predictions. Statistical analysis reveals the model is in good agreement with experimental results with only one adjustable parameter, cake porosity.

Keywords: Membrane Filtration, Nanofiltration, Cake Enhanced Concentration Polarization, Film Theory, Kuwabara Cell Model, Levine-Neale Electrophoretic Mobility, Electroosmotic Back Flow

1 Introduction

Over the past decades, several models have been postulated to elucidate the mechanism of permeate flux and observed rejection decline due to colloidal fouling of salt rejecting membrane processes. Hoek and Elimelech [1] successfully explained the performance decline mechanism for nanofiltration (NF) and reverse osmosis (RO) processes, and reported the flux decline due to cake enhanced osmotic pressure (CEOP) or cake enhanced concentration polarization (CECP). They explained the fouling mechanism as hindered back-diffusion of salt ions within colloidal cake layers resulting in an increase of concentration polarization (CP) as well as the transmembrane osmotic pressure (TMOP). Later, Wang and Tarabara [2] proposed another approach to determine the contribution of CP and cake resistance on performance decline based on salt permeability constant measurement. These approaches, quantify the relative contribution of CP and cake resistance during constant pressure filtration experiments. However, none of these models considered the coupled effect of transient growth of cake layer and CP phenomenon to establish a mechanistic model of performance decline independent of experimental results. Besides, these models used Kozeny-Carman equation to determine the cake resistance instead of cell model of swarm of particles which restricts the applicability for fixed cake structures only.

Earlier, CECP models also neglected the zeta potential effect of charged colloidal particles. Fluid flow and charge transport through the colloidal cake layer are interrelated due to the formation of electric double layer (EDL) around the particles. Transport of electrolyte through the charged packed bed develops electroviscous resistance [3, 4]. Recently, De and Bhattacharjee [5] proposed an electrokinetic model for flux decline during the filtration of charged colloids and electrolyte. The model assumed the cake layer as a bundle of charged cylindrical nano-pores and considered the electroviscous effect created by them. This model does not provide mechanistic links between the transient growth of cake layer and CP of salt rejecting membranes. Besides, the electroosmotic flow through a bed of stationary colloidal particles get affected by the surrounding particles because the presence of neighboring particles affects the fluid velocity field and ion distribution around each particle [6]. It is therefore important to know how the concentration of cake layer affects the electroosmotic flow of electrolyte. This problem can be solved by applying the cell model approach to the electroosmotic flow of electrolyte.

In this study, a mechanistic model is presented for transient growth of cake layer and CECP considering the cell model of cake structure and electroosmotic back flow (EBF) across the cake layer. To validate the model with experimental results, cross flow NF was conducted with silica particles and NaCl solution over a range of operating conditions.

2 Mathematical model

During filtration of electrolyte solution and charged particles above the critical flux of particle deposition, a stationary cake layer forms at the membrane surface. The cake layer resistance consists of the hydrodynamic resistance of the cake layer and electroviscous resistance due to the charged colloids. Also, the total resistance to permeate flow comprises membrane resistance (R_m) and total cake layer resistance (R_c). Since NF membrane also reject salt, the developed TMOP (π_m) also opposes the applied pressure. Therefore, the effective driving force for permeation is the difference between the operating pressure and TMOP [7].

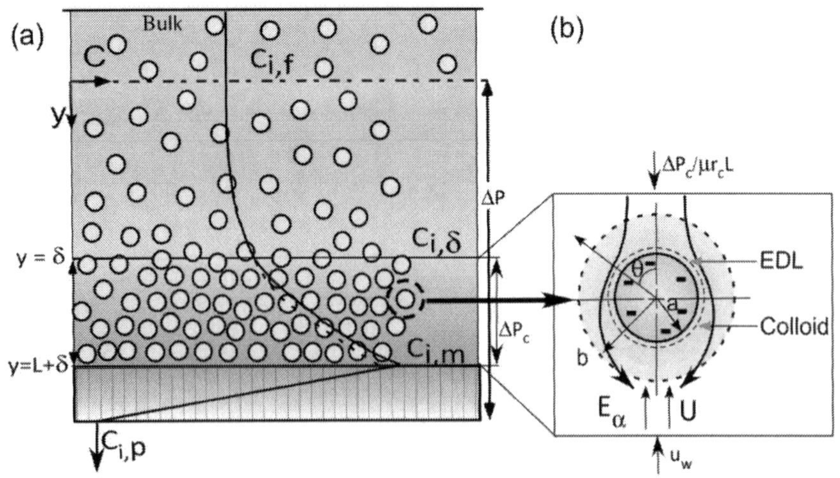

Figure 1 – a: Transport of electrolyte through CP Layer, cake layer and membrane b: Levine-Neale electrophoretic model based on Kuwabara Cell model

The CP layer and cake layer in cross flow membrane filtration is illustrated in Fig. 1a. The retained solutes at the membrane surface form cake and CP layers. To elucidate the transport of solvent and electrolyte through the interstices of cake layer, consider a single colloidal particle within the cake illustrated in Fig. 1b. According to the figure, a colloidal particle of hydrodynamic radius a (m) and zeta potential ψ_p (mV) is considered stationary in a spherical coordinate. The colloidal cake structure is represented using the Kuwabara cell model. The cell model defines a spherical fluid shell of radius b which is related to cake volume fraction (φ_c) as, $b = a/\varphi_c^{1/3}$.

The permeate flux due to the hydrodynamic Stokes drag of the cake layer can be written based on the resistance in series and Kuwabara cell models as follows:

$$v_w = \frac{\Delta P_c}{\mu r_c L} \tag{1}$$

here, ΔP_c is the pressure drop across the cake layer, L is the cake layer thickness, $r_c = \frac{9\varphi_c A_K}{2a^2}$ is the specific cake layer resistance, μ is fluid viscosity, and A_K is Kuwabara correction factor.

To calculate the electroosmotic flow of the salt solution pass through the stationary cake layer, we use the Levine-Neale electrophoretic model. According to the Levine and Neale [8] cell model, the electrophoretic mobility is given by:

$$\eta = \frac{U}{E_\infty} = \left(\frac{\epsilon\psi_p}{\mu}\right) g_2(\kappa a, \varphi_c) \tag{2}$$

here, ϵ is electrical permittivity, U is the mean relative velocity of the interstitial fluid with respect to the stationary colloidal particle at the cell boundary, E_∞ is the electric field induced by the streaming potential developed across the cake layer, $g_2(\kappa a, \varphi_c)$ accounts for the presence of neighboring particles, and κ is the inverse Debye length.

The expression of electric field (E_∞) developed by the induced streaming potential at zero current condition, is written based on the Onsager principle of reciprocity and Debye-Hückel approximation $(ze\psi_p/k_BT \leq 1)$ as follows [6]:

$$E_\infty = \frac{\Delta P_c}{L}\left(\frac{\epsilon\psi_p}{\mu\sigma^\infty}\right)\left(1 - \frac{2\, I_1(\kappa a)}{\kappa a\; I_0(\kappa a)}\right) \tag{3}$$

Here, $I_0(\kappa a)$ and $I_1(\kappa a)$ are zeroth and first order modified Bessel function of the first kind, respectively. The expression for bulk electroosmotic flux across the cake layer can be written from interstitial fluid velocity using the continuity assumption as follows:

$$u_w = U(1-\varphi_c) = \frac{\Delta P_c}{L\sigma^\infty}\left(\frac{\epsilon\psi_p}{\mu}\right)^2 (1-\varphi_c) g_2(\kappa a, \varphi_c)\left(1 - \frac{2\, I_1(\kappa a)}{\kappa a\; I_0(\kappa a)}\right) \tag{4}$$

Therefore, the total flux across the cake layer at no current condition due to effective cake pressure ΔP_c and reverse electroosmotic flow, is given as follows:

$$v_w = \frac{\Delta P_c}{\mu r_c L} - u_w = \frac{\Delta P_c g^*}{\mu r_c L} \tag{5}$$

The parameter g^* accounts for electroosmotic effect in swarm of charged colloidal particles and expressed as follows:

$$g^* = 1 - 9\varphi_c(1-\varphi_c) A_K \beta g_2(\kappa a, \varphi_c)\left(1 - \frac{2\, I_1(\kappa a)}{\kappa a\; I_0(\kappa a)}\right) \tag{6}$$

This parameter, g^*, is accounted as the electroviscous resistance, $R_c = r_c L/g^*$. The parameter β quantifies the effect of the colloid zeta potential (ψ_p) and solution conductivity (σ^∞) on EBF, and expressed as:

$$\beta = \frac{(\epsilon\psi_p)^2}{2a^2\mu\sigma^\infty} \tag{7}$$

For 1:1 electrolyte concentration, the TMOP is expressed by the van't Hoff equation [7],

$$\Delta\pi_m = 2\Re T(C_{i,m} - C_{i,p}) \tag{8}$$

The expression of $C_{i,m}$, derived by integrating the mass balance of electrolytes within the CP and cake layers with appropriate boundary conditions, is:

$$\frac{C_{i,m}}{C_{i,f}} = \frac{\exp\left[v_w\delta/D_i + v_w L/D_i^*\right]}{R_r + (1 - R_r)\exp\left[v_w\delta/D_i + v_w L/D_i^*\right]} \tag{9}$$

here, $R_r = 1 - C_{i,p}/C_{i,m}$ is the real rejection of the membrane. Hindered diffusivity of electrolyte is calculated from the tortuosity (ς) of the cake layer as $D_i^* = D_i\varepsilon\varsigma^{-1}$ where, $\varsigma = [1 - \ln(\varepsilon^2)]$ [9].

Transient growth of cake layer is obtained by integrating the mass balance equation of colloids in the cake layer with appropriate boundary conditions [10]:

$$\rho_p\frac{dL}{dt} = v_w\frac{C_{p,m} - C_{p,f}\ \exp(v_w/k_p)}{1 - \exp(v_w/k_p)} \tag{10}$$

Due to difficulty of determining the colloid mass transfer coefficient k_p at specific length of the slit channel and as the cake layer thickness is usually very small compared to the hydrodynamic diameter of the channel, constant length average k_p for laminar flow in channel has been considered using Leveque expression [11]. The thickness of the cake layer can be known by integrating the Eq. 10 over the time. The cake layer mass is obtained using $M_c = L\rho_p\varphi_c A_m$ where, A_m is the effective membrane area and ρ_p is particle density. The permeate flux for cake filtration can be written using the modified Darcy,s law [12]:

$$v_w = \frac{\Delta P - \Delta\pi_m}{\mu(R_m + R_c)} \tag{11}$$

where, ΔP is applied pressure and R_m is membrane resistance. The Eqs. 8, 10 and 11 are solved simultaneously to obtain the permeate flux and observed rejection. The observed rejection (R_o) can be written as follows:

$$R_{i,o} = 1 - \frac{C_{i,p}}{C_{i,f}} \tag{12}$$

Overall pressure drop during operation is the summation of trans-cake hydraulic pressure, cake-enhanced TMOP and transmembrane pressure. Therefore,

$$\Delta P = \Delta P_c + \Delta P_m + \Delta\pi_m \tag{13}$$

where, ΔP_m is the pressure drop across the membrane.

3 Results and discussion

The resulting set of differential-algebraic equations are solved using a coupled differential algebraic equation solver (DASSL) in Fortran. Colloidal fouling experiments in presence of NaCl solution were conducted for NF90 (Dow FilmTec) membrane to compare with the electrokinetic model. The transient permeate flux, observed rejection, and deposited mass were recorded as a function of time for 5 hrs. Simultaneously, the embedded parameters, namely, TMOP and cake porosity were calculated following the Hoek and Elimelech [1] procedure. The experiments were conducted at constant temperature (25°C) and pH (7.0), which allowed fixed zeta potential (-35 mV) for a given particle (100 nm Snowtex ZL, Nissan Chemical). In all plots, permeate flux has been presented as normalized flux (v_w/v_i) and the time of silica addition is considered as zero. The steady state salt water flux (v_i) and observed rejection (R_o) was obtained by solving Eqs. 8, 9, and 11 in absence of cake layer (i.e., $L = 0$ and $R_c = 0$). The pressure drops are

Table 1 – Experimental conditions for fouling experiment at 25 ± 1°C and pH=7.0

Experiment No.	Operating Parameters				
	ΔP (kPa)	$C_{i,f}$ (mM)	$C_{p,f}$ (ppm)	u_c (m/s)	$2a$ (nm)
1	965	10	300	0.1	100
2	689	10	300	0.1	100
3	1033.5	20	300	0.1	100
4	965	10	500	0.1	100
5	965	10	500	0.2	100

presented in normalized form namely normalized trans-membrane pressure ($\Delta P_m{}^* = \Delta P_m/\Delta P$), trans-cake hydrodynamic pressure ($\Delta P_c{}^* = \Delta P_c/\Delta P$) and cake-enhanced TMOP ($\Delta\pi_m{}^* = \Delta\pi_m/\Delta P$). The experimental results, raw data and calculated parameters, are shown as symbol in the graphs. The model predictions (lines) are superimposed on experimental results to compare with each other. The average porosity ($\varepsilon = 1 - \varphi_c$) obtained from each experiment was considered as the only adjustable parameter in the model. For the silica colloidal particles bulk concentration of 0.2 kg/m^3, the cake concentration was reported to be about 200 kg/m^3 [13] which was considered throughout the analysis.

3.1 Comparison of model and experiment

Five experiments were conducted varying the salt concentration, cross flow velocity, operating pressure, and silica concentration. Each parameter varied at two levels, and the results obtained from the experiments after 2 hrs and 4 hrs were used to demonstrate the goodness of fit of the model. Four responses namely flux decline, observed rejection, deposited mass, and normalized CEOP were investigated using the statistical parameters including mean squared error (MSE), root mean squared error (RMSE), correlation coefficient (R), and coefficient of determination (R^2). The statistical results for the responses are listed in Table 2. As can be seen, the model predictions are in good agreement with experimental results with RMSE value of less than 0.1 and correlation coefficient of greater than 0.9.

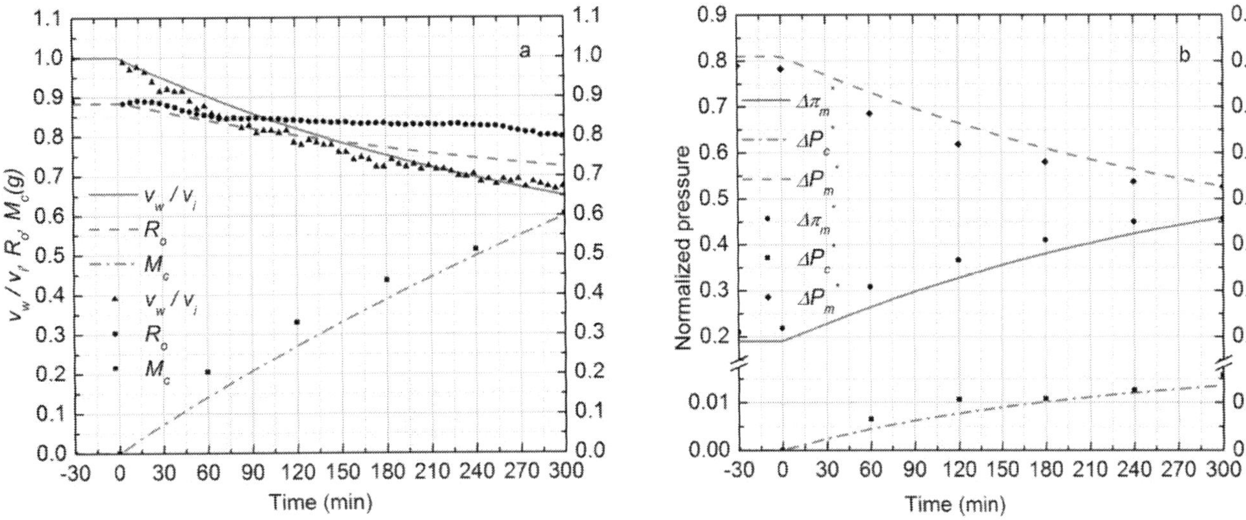

Figure 2 – a) Normalized permeate flux (v_w/v_i), Observed rejection (R_o) and Deposited cake mass (M_c) on NF90 membrane. Fouling experiment conducted at 965 kPa, cross flow velocity 0.1 m/s, 10 mM NaCl solution, 300 ppm of Silica, temperature of 25°C and pH 7.0, b) Comparison of $\Delta P_m{}^*$, $\Delta P_c{}^*$ and $\Delta\pi_m{}^*$.

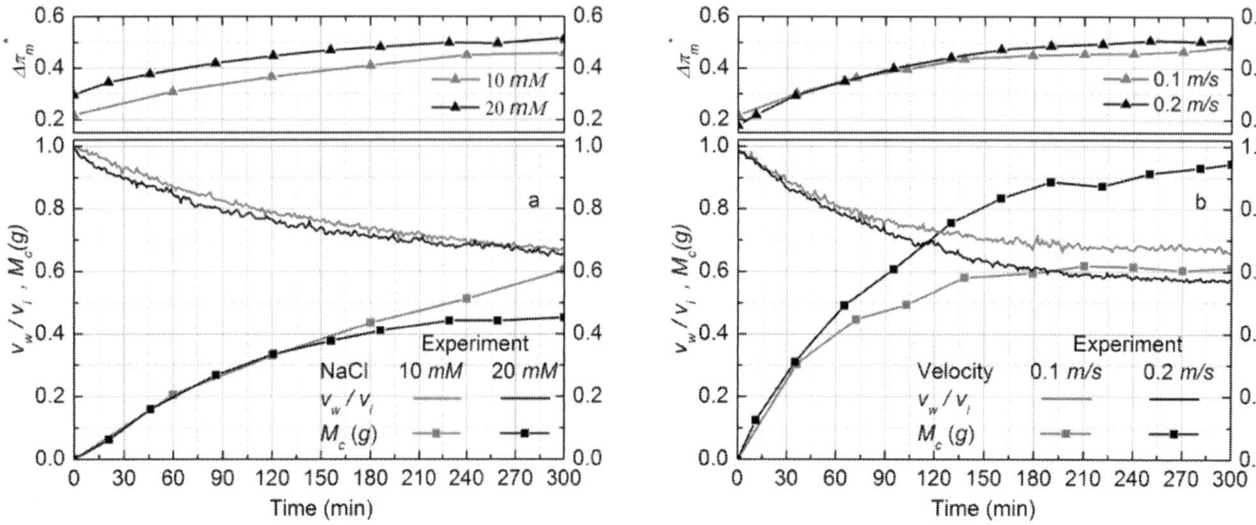

Figure 3 – Effect of a) salt concentration and b) cross flow velocity on water flux, mass deposition, and CECP.

Fig. 2 shows the comparison of model predictions and experimental results for experiment 1 (Table 1). The permeate flux and observed salt rejection declined over the filtration time with the increase of cake layer thickness. Contribution of cake layer resistance on pressure drop is insignificant compared to TMOP validating the primary mechanism of flux decline as CECP. The model captures the coupled effect of cake layer growth and CECP phenomena which allows accurate prediction of experimental results. The experimental average porosity, 0.48, was used for the model prediction as well.

Table 2 – Statistical criteria for validation of the developed model

Criterion	Response			
	Flux decline, (1-v_w/v_i)	Observed salt rejection, R_o	Deposited silica mass, M_c	Normalized CEOP, $\Delta\pi_m^*$
MSE[a]	1.11×10^{-3}	1.85×10^{-3}	9.34×10^{-3}	8.45×10^{-4}
RMSE[a]	0.033	0.043	0.096	0.029
R	0.959	0.952	0.901	0.965
R^2	0.92	0.906	0.813	0.931

[a] The MSE has the unit of squared of response unit while RMSE has the same unit as response

3.2 Effect of salt concentration and cross flow velocity

The effect of salt feed concentration and cross flow velocity on water flux, mass deposition, and CECP are shown in Fig. 3a and 3b, respectively. To study the effect of salt concentration on cake structure and CECP, two experiments (1 and 3 in Table 1) were conducted at same initial flux condition. The permeate flux decline is higher for 20 mM NaCl concentration even though the mass deposition is less. This is due to denser structure of cake layer and enhanced CP at higher salt concentration. Average experimental porosity for 20 mM was less, 0.45, as compared to 0.48 for 10 mM. Experimental results for two cross flow velocities (4 and 5 in Table 1) are shown in Fig. 3b. Flux decline is higher for increased cross velocity condition which seems to be counter intuitive. The result can be attributed to the fact of less initial TMOP and higher initial salt flux resulting in excessive mass deposition and increase in TMOP. Hence, the coupling between cake layer growth and CECP is significantly important for the development of a mechanistic model which has been accounted in the present study.

3.3 Electroosmotic back flow

The effect of EBF is captured in parameter g^* which quantifies the increase in cake pressure drop due to EBF in terms of pressure drop ratio $(\Delta P_c^*/\Delta P_{c,h}^*)$. This is the ratio of total pressure drop including EBF to the pressure drop due only to hydrodynamic resistance of cake layer. Fig. 4 shows the effect of zeta potential (ψ_p) and cake volume fraction (φ_c) on pressure drop ratio as a function of κa. The ratio exhibits a maximum value with respect to a smaller κa value for constant φ_c and it increases with the increase of ψ_p. However, the maximum value shifts to larger κa values and increases with the increase of φ_c for constant ψ_p. This indicates that at higher φ_c values the EBF becomes important at larger κa values. Therefore, the EBF is more important in presence of denser cake layer having higher zeta potential for 2< κa <15. In case of denser cake layer the outer cell envelop, b, becomes comparable with the EDL thickness. At this condition, fluid flow modifies the ion distribution around the particles of cake layer and creates the EBF.

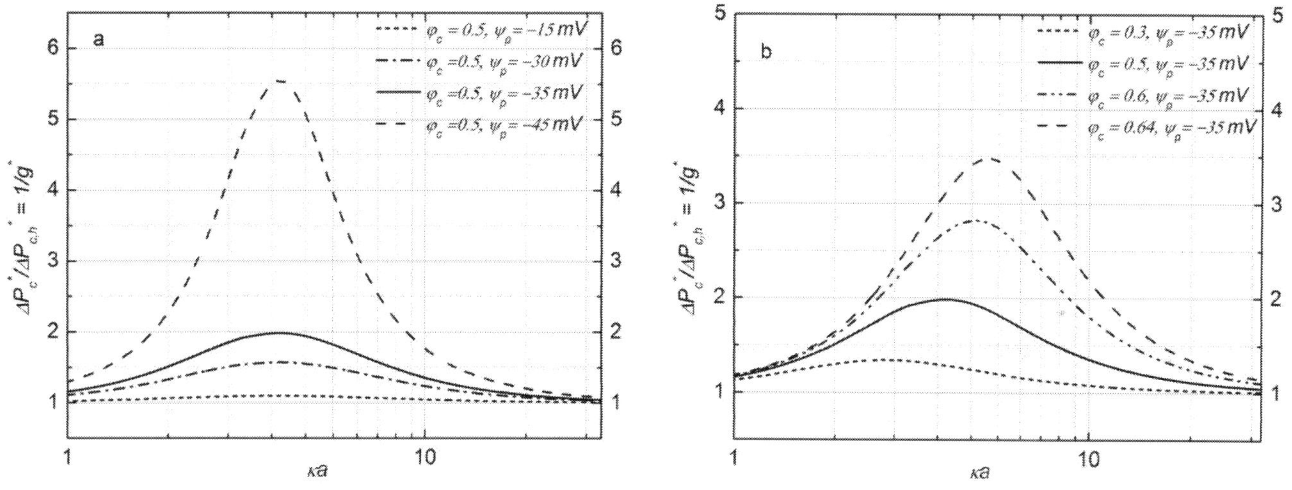

Figure 4 – Effect of a) zeta potential (ψ_p) and b) volume fraction (φ_c) on pressure drop ratio $(\Delta P_c^*/\Delta P_{c,h}^*)$ as a function of κa.

4 Conclusion

A mechanistic model has been developed for investigating the coupled effect of transient cake layer growth and CECP on performance decline of salt rejecting membranes. The model also captures the effect of cake volume fraction on EBF in light of Levine-Neale cell model of electrophoresis. In presence of denser and charged cake layer during low ionic strength operation the effect of EBF become important. Simultaneously, experimental study was conducted to investigate the effect of operating parameters and goodness of fit of the model. The model can predict the experimental results with reasonable accuracy with only one adjustable parameter, cake volume fraction. The RMSE values were lower than 0.1 and correlation coefficients were higher than 0.9.

References

[1] E. M. V. Hoek and M. Elimelech. Cake-Enhanced Concentration Polarization: A New Fouling Mechanism for Salt-Rejecting Membranes. *Environmental Science and Technology*, 37:5581–5588, 2003.

[2] F. Wang and V. V. Tarabara. Coupled effects of colloidal deposition and salt concentration polarization on reverse osmosis membrane performance. *Journal of Membrane Science*, 293:111–123, 2007.

[3] S. Levine, R. Marriott, G. H. Neale, and N. Epstein. Theory of Electrokinetic Flow in Fine Cylindrical Capillaries at High Zeta-Potential. *Journal of Colloid And Interface Science*, 52:136–149, 1975.

[4] W. R. Bowen and F. Jenner. Electroviscous effects in charged capillaries. *Journal of Colloid and Interface Science*, 173:388–395, 1995.

[5] S. De and S. Bhattacharjee. Flux decline during cross flow membrane filtration of electrolytic solution in presence of charged nano-colloids: A simple electrokinetic model. *Journal of Colloid and Interface Science*, 353:530–536, 2011.

[6] J. Masliyah and S. Bhattacharjee. *Electrokinetic and Colloid Transport Phenomena*. John Wiley and Sons, Inc., 2006.

[7] M. Elimelech and S. Bhattacharjee. A novel approach for modeling concentration polarization in crossflow membrane filtration based on the equivalence of osmotic pressure model and filtration theory. *Journal of Membrane Science*, 145:223–241, 1998.

[8] S. Levine and G. H. Neale. The prediction of electrokinetic phenomena within multiparticle systems. I. Electrophoresis and electroosmosis. *Journal of Colloid And Interface Science*, 47:520–529, 1974.

[9] B. P. Boudreau. The diffusive tortuosity of fine-grained unlithified sediments. *Geochimica et Cosmochimica Acta*, 60:3139–3142, 1996.

[10] S. De and P. K. Bhattacharya. Modeling of ultrafiltration process for a two-component aqueous solution of low and high (gel-forming) molecular weight solutes. *Journal of Membrane Science*, 136: 57 – 69, 1997.

[11] S. Kim and E. M. V. Hoek. Modeling concentration polarization in reverse osmosis processes. *Desalination*, 186:111–128, 2005.

[12] E. M. V. Hoek, A. S. Kim, and M. Elimelech. Influence of crossflow membrane filter geometry and shear rate on colloidal fouling in reverse osmosis and nanofiltration separations. *Environmental Engineering Science*, 19:357–372, 2002.

[13] S. K. Zaidi and A. Kumar. Experimental analysis of a gel layer in dead-end ultrafiltration of a silica suspension. *Desalination*, 172:107 – 117, 2005.

(506b) 2-Step Interfacial Polymerization of Trimesoly Chloride(TMC)/Metaphenylenediamine(MPD)-Derived Polyamide for Preparation of Organic/Inorganic Hybrid Reverse Osmosis Membranes

Toshinori Tsuru, Shoichi Sasaki, Masakoto Kanezashi and Tomohisa Yoshioka, Department of Chemical Engineering, Hiroshima University, Higashi-Hiroshima, Japan

Water purification technology has attracted increasing attention because of water shortages and the lack of access to safe drinking water. Recently, new classes of functional materials have been applied to develop reverse osmosis (RO) and nanofiltration (NF) membranes with increased water permeability, high salt rejection, less fouling and excellent stability. Inorganic-organic nanocomposite membranes are quite interesting because they showed increased flux by incorporating inorganic materials as such as zeolite [1] and TiO_2 [2] without significant decrease in rejection.

Thin film composite (TFC) PA membranes are typically prepared by interfacial polymerization (IP) of m-phenylene diamine (MPD) in aqueous phase and trimesoyl chloride (TMC) in organic phase on polysulfone ultrafiltration supports. Our strategy to enhance RO performance of TFC-PA membranes includes (a) addition of additives in organic phase [3, 6], (b) inorganic-organic nanocomposite membranes with zeolite nanoparticles [4], (c) inorganic-organic nanocomposite membranes with metal alkoxides [5], and (d) two-step interfacial polymerization.

In this presentation, as a novel technique, two-step interfacial polymerization via spray-coating will be discussed, and, in addition, the preparation of organic/inorganic hybrid membranes will be introduced.

References:

[1] B. H. Jeong, E. M. V. Hoek, Y. Yan, A. Subramani, X. Huang, G. Hurwitz, A. K. Ghosh, A. Jawor, Interfacial polymerization of thin film nanocomposites: A new concept for reverse osmosis membranes, Journal of Membrane Science, 294 (2007) 1-7.

[2] Lee et al., Polyamide thin-film nanofiltration membranes contaning TiO_2 nanoparticles, Desalination 219 (2008) 48.

[3] C. Kong, M. Kanezashi, T. Yamomoto, T. Tsuru, Controlled synthesis of high performance polyamide membrane with thin dense layer for water desalination, *Journal of Membrane Science*, 362 (2010) 76-80.

[4] C. Kong, T. Shintani, T. Tsuru, "Pre-seeding"-assisted synthesis of high performance polyamide-zeolite nanocomposie membrane for water purification, *New Journal of Chemistry*, 34(2010)2101-2104.

[5] C. Kong, A. Koushima, T. Kamada, T. Shintani, M. Kanezashi, T. Yoshioka, T. Tsuru, Enhanced performance of inorganic-polyamide nanocomposite membranes prepared by metal-alkoxide-assisted interfacial polymerization, Journal of Membrane Science, 366 (2011)382-388.

[6] C. Kong, T. Shintani, T. Kamada, V. Freger, T. Tsuru, Co-solvent-mediated synthesis of thin polyamide membranes, Journal of Membrane Science, 382(2011)10-16.

(506c) Multiple-Effect Membrane Distillation for Desalination of Seawater and Deep Concentration of the Brine Drained From Conventional Desalination Plants

Yingjie Qin[1], Ben Wang[2], Junbao Jin[2], Bin Wang[2], Liqiang Liu[3] and Dongsheng Cui[3], (1)Chembrane Research & Engineering,Inc., Newark, NJ, (2)School of Chemical Engineering & Technology, Tianjin University, Tianjin, China, (3)Chembrane Engineering & Technology, Inc., Tianjin, China

Many countries in the world suffer from a shortage of fresh water. Water shortage limits the socioeconomic sustainable development of the world. Desalination of seawater is an effect way to obtain fresh water. At present, three important commercially available desalination technologies are multi-stage flash (MSF), multiple-effect distillation (MED) and reverse osmosis(RO). However water recovery ratio of those technologies is usually 40-55%. The concentrated brine by-produced in the routine desalination plant by using RO, MSF, or MED is usually discharged into the sea. After pretreatment and desalination, the brine by-produced has a rise in salinity and temperature while contains many harmful chemicals such as scale inhibitor and biocide. The discharge of the brine by-produced was actually a salt, heat and chemical pollution to the sea which has negative impacts on marine environment.

Membrane distillation is a separation method in which volatile species diffuse through a porous hydrophobic membrane. The driving force for transport is the partial pressure difference across the membrane. Depending on the method employed to impose the vapor pressure across the membrane, membrane distillation system can be classified broadly into four categories: direct-contact membrane distillation (DCMD), vacuum membrane distillation (VMD), sweeping gas membrane distillation and air gap membrane distillation (AGMD). Potential advantages of membrane distillation over traditional evaporation processes include operation at low pressures, lower temperature as well as ease of process scale-up, avoid of corrosion as a result of inertness of hydrophobic polymer membrane material, and so on.

However, traditional membrane distillation process is with extremely low thermal efficiency. It is a reason that MD is not currently use in commercial scale. The performance ratio (PR) is usually used to determine the thermal efficiency of evaporation-based process, which is defined as the amount of latent heat needed for evaporation of the produced water and the amount of heat provided to the system from an external energy source. The PR of conventional MD process typically ranged from 0.2 to 1 while the PR of MSF and MED usually ranged from 7 to 15. Another problem with MD is wetting or fouling of microporous membrane surface, which leads to the decrease of permeate flux and leakage; leakage further leads to the contamination of the permeate product by the impurities in the feed. Wetting or fouling is a more grievous problem in a DCMD or VMD process, since the intimate contact of the permeate with the membrane in DCMD or the large trans-membrane pressure difference in VMD. On the other hand, the AGMD process with mild operation condition and thus with potentially longer operation stability was rarely reported in literature, it might be attributed the difficulty to fabricate large AGMD modules.

In the present study, multiple-effect membrane distillation (MEMD) based on AGMD module with function of internal heat recovery has been developed to desalinate seawater and to deep concentrate the brine drained from the conventional desalination plant. This kind of MEMD process combines the advantages of MD process and conventional MSF process, avoids the disadvantages of MSF such as evacuation operation, and can provide a high PR value. The water vapor flux (J), perform ratio (PR) and thermal efficiency (©¯) are the most important indicators for evaluation of MEMD modules. Experimental were conducted to investigate the performance of MEMD module by using fresh seawater after preliminary pretreatment such as ultrafiltration, using the brine from a MED desalination plant without further pretreatment until the salt concentration in the retentate went up to 15°Bé, and by using more concentrated brine of >15°Bé as a feed after deep de-calcium step. Under optimized operational conditions, when the concentration of salt varied between 3.5 - 27 °Bé, the value of J decreased from 6.5 to 2.3 kg/m^{2}h; the value of decreased from 0.91 to 0.45; and the value of PR decreased from 15 to 2.8. The value of PR is the highest value experimentally obtained for MD process. The long-term stability test was continuously performed for two than months.

In the late stage of the test, when a brine of 25°Bé was used as the feed, the conductivity of the permeate was still less than that of tap water. All these results demonstrated that MEMD process can be efficiently used to produce drinkable water and highly concentrated brine when fresh water or brine from conventional desalination plants was used as the feed to this MEMD process.

(506d) Membrane Fouling Behavior Evaluated by Constant Flux and Constant Transmembrane Pressure Filtration

Daniel J. Miller, Sirirat Kasemset, Donald R. Paul and Benny D. Freeman, Chemical Engineering, University of Texas at Austin, Austin, TX

Membrane systems can be operated in two ways: constant permeate flux with variable transmembrane pressure (TMP) or constant transmembrane pressure with variable permeate flux. Membrane fouling may be measured using either of these two methodologies. This presentation shows a comparison of constant-flux and constant-TMP operation of commercial flat-sheet membranes during oily water filtration. The critical flux—the flux below which minimal fouling occurs—was determined by a flux-stepping method using constant flux crossflow filtration. The membrane performance during fouling was evaluated using both operating modes at fluxes higher than and lower than the critical flux. Membrane resistance evolution during fouling is compared for constant-flux and constant-TMP operation. At fluxes lower than the critical flux, where fouling was relatively minor, the membrane resistances during constant-flux and constant-TMP fouling developed comparably. At fluxes higher than the critical flux, the membrane resistance evolution differed for the two operating modes. Reasons for similarities and differences in membrane behavior in each operational mode will be discussed. Surface modification is a popular means of attempting to alleviate membrane fouling. Polydopamine is currently of wide interest as a substrate-independent surface modification agent. Polydopamine-coated surfaces universally show high hydrophilicity and, in early studies, polydopamine-modified membranes have shown improved resistance to fouling when challenged with synthetic oily water. Surface-modified microfiltration and ultrafiltration membranes were tested in constant-flux and constant-TMP fouling experiments. Fouling of modified membranes will be compared to unmodified membranes and the ramifications of testing protocol on membrane fouling behavior will be discussed.

(506e) Effect of Mixed Monovalent and Divalent Ion Feeds On Sulfonated Polysulfone Desalination Membrane Performance

Joseph R. Cook[1], Geoffrey M. Geise[2], Habis Al-Zoubi[3], Benny D. Freeman[4], Kwansoo Lee[5], Benjamin Sundell[5] and James E. McGrath[6], (1)Chemical Engineering, The University of Texas at Austin, Austin, TX, (2)Deptartment of Chemical Engineering & The Texas Materials Institute, The University of Texas at Austin, Austin, TX, (3)Al-Hussein Bin Talal University , Ma'an, Jordan, (4)Department of Chemical Engineering, The University of Texas at Austin, Austin, TX, (5)Virginia Tech, (6)Macromolecules and Interfaces Institute, Virginia Tech, Blacksburg, VA

Modern water desalination membranes are largely based on crosslinked polyamides chemistry, which are susceptible to degradation by exposure to chlorine compounds, which are often necessary to act as a biocide to prevent fouling of the membranes. Sulfonated polysulfones show promising water/salt separation properties and significant resistance to degradation by chlorine compounds. These materials consist of randomly copolymerized charged and uncharged monomers, and these charge groups directly affect the salt and water transport and separation performance. One challenge of these materials is that they show reduced sodium chloride rejection when even relatively low amounts of divalent salts (such as calcium chloride or magnesium chloride) are added to the feed. This work investigates the effect of the chemistry and concentration of salts in the feed on different transport properties: water and salt permeability, water and salt partition coefficients, and water diffusivity. These properties are determined first in a feed of sodium chloride in water, with the competing ion added in increasing amounts to quantify the effect of competing ion concentration on performance. Tests are also performed with increasing amounts of sodium chloride, and different monovalent competing ions (such as potassium chloride) are investigated to demonstrate the prevalence of the phenomenon with divalent ions.

(509a) Using SAFT-FMT-DFT to Model the Adsorption of Light Gases and Hydrocarbons in Activated Carbon

Lucas A. Mitchell[1], Bryan J. Schindler[1], Carolina dos Ramos[1], Clare McCabe[1], Peter T. Cummings[1,2] and M. Douglas LeVan[1], (1)Chemical and Biomolecular Engineering, Vanderbilt University, Nashville, TN, (2)Center of Nanophase Materials Sciences, Oak Ridge National Laboratory, Oak Ridge, TN

The use of molecular simulation, the development of accurate models, and the comparison of theoretical predictions with experimental results help to advance our understanding of adsorption and its application. One prominent molecular model is density functional theory (DFT), which is commonly used to treat the adsorption of spherical molecules in slit-shaped carbon pores. In this paper, we couple DFT with an accurate molecular-based equation of state to calculate the fluid thermodynamic properties using fundamental measure theory (FMT), which is a rigorous theory for homogeneous and non-homogeneous hard-sphere chain fluids. An accurate model is obtained with adsorbing molecules treated as hard-sphere chains with square well interactions. The Mansoori-Carnahan-Starling-Leland and Carnahan-Starling-Boublik equations of state are used for the hard sphere interactions, and the statistical associating fluid theory for potentials of variable range (SAFT-VR) is used to describe the square-well fluid. For adsorption in pores, various potentials can be used for the interaction of the chain molecules with the wall. After comparing model predictions with results from Monte Carlo simulations, the model is used to obtain a pore size distribution for activated carbon from experimental data for adsorption of nitrogen, which is treated as a non-spherical molecule. The adsorption of n-pentane in slit-shaped carbon pores of various widths is then predicted from the model, and together with the pore size distribution, adsorption isotherms are obtained for comparison with experimental data.

(509b) Removal of Oxygenated Species From Liquid Toluene by Adsorption

Cristian Brunchi, Andrzej I. Stankiewicz, Herman J. M. Kramer and Thijs J. H. Vlugt, Process & Energy, Delft University of Technology, Delft, Netherlands

Removal of oxygenated species from liquid toluene by adsorption

Cristian C. Brunchi, Andrzej I. Stankiewicz, Herman J.M. Kramer and Thijs J.H. Vlugt

Process& Energy Laboratory, Delft University of Technology, NL-2628 CA Delft, the Netherlands

Corresponding author e-mail: c.c.brunchi@tudelft.nl

The goal of this study is to develop a general methodology that can allow one to select a good zeolite adsorbent for any specific case study that requires removal of traces of mixtures from bulk liquid systems. As a case study, we are investigating the removal of mixtures of oxygenated odor compounds from liquid toluene using adsorption as a separation technology. The typical odor compounds present in the toluene stream are: butanal, ethyl-hexenal, dimethylcyclohexanone, trimethylphenol and trimethylanisole. Pure component adsorption isotherms in Na-Y zeolite are computed using Monte Carlo simulations. Binary, ternary and 6-component mixture isotherms are generated using the Adsorbed Solution Theory (AST). The fugacity of the liquid bulk mixture is calculated with two approaches: using the Peng-Robinson EOS and by calculating the activity coefficients in the bulk liquid mixture (gama-phi approach). Finally the generated adsorption isotherms are compared with experimentally obtained adsorption isotherms. Preliminary results show that the system behaves strongly non-ideal in the adsorbed phase.

(509c) Multicomponent Adsorption of Alcohols Onto Silicalite-1 From Aqueous Solution: Assessment of Ideal Adsorbed Solution Theory

Peng Bai[1], Michael Tsapatsis[1] and J. Ilja Siepmann[1,2], (1)Chemical Engineering and Materials Science, University of Minnesota, Minneapolis, MN, (2)Chemistry, University of Minnesota, Minneapolis, MN

Configurational-bias Monte Carlo (CBMC) simulations in the isobaric-isothermal version of the Gibbs ensemble (GE) were carried out to probe the adsorption from aqueous solutions of methanol and/or ethanol on silicalite-1. This methodology requires neither specification of the chemical potential nor any reference to activity models based on experimental data. The CBMC-GE methodology can be applied to the complete range of mixture compositions from pure water to pure alcohols, and can also be used when multiple solute types are present at high concentrations. The simulations demonstrate high selectivities for the alcohols (alpha (ethanol) > alpha (methanol)) almost over the entire composition range. The ideal adsorbed solution theory is found to substantially under predict the amount of adsorption for water and leads to very large errors in selectivity at low alcohol concentrations in the solution phase. The simulations indicate that, at lower loadings, the adsorbed alcohol molecules can serve as seeds for water adsorption but, at higher loadings, alcohols displace water molecules from their preferred region.

(509d) Molecular Simulations for Adsorption, Diffusion and Separation of Noble Gas Mixtures In Metal Organic Frameworks

Yeliz Gurdal, Koc University, Istanbul, Turkey and Seda Keskin, Chemical and Biological Engineering, Koç University, Istanbul, Turkey

Since the number of different types of metal organic frameworks (MOFs) that has been synthesized is enormous, molecular modeling of these materials has played an important role in providing information about the performance of MOFs for specific applications. MOFs have garnered much attention for gas separation and gas storage due to their tunable pore sizes. A large number of molecular simulation studies investigated adsorption of CO_2, CH_4, N_2 and H_2 in addition to separation of binary mixtures of these gases. On the other hand, only very few publications addressed capture and separation of noble gases, although noble gas separations play significant role in industry. For example, Xe has several usage are including lighting, imaging, anesthesia, nuclear magnetic resonance, ion propulsion engines. The capture and separation of Xe from Kr is crucial to obtain high purity Xe. Identifying and developing materials which show higher capture and separation capacity for noble gas mixtures would contribute to noble gas separation industry by reducing energy consumption.

In this work, we used atomically detailed modeling methods to examine both single component and mixture adsorption of Xe, Kr and Ar in ten different MOFs. We used both GCMC simulations and Ideal Adsorbed Solution Theory to compute adsorption of mixtures in MOFs. In addition to calculating adsorption-based selectivities, we also computed self diffusivities of each component in Xe/Kr and Xe/Ar mixtures using molecular dynamics and predicted permeation selectivity of MOFs for these mixtures for the first time in the literature. The working capacity and gas permeability of each MOF were calculated using the data obtained from molecular simulations and the performances of MOFs in noble gas separations were compared with well known porous materials. Based on this comparison, we addressed which material properties are desirable to develop more selective MOF materials for Xe/Kr and Xe/Ar separations.

(509e) CO2 /CH4 Separation in Zeolite-Like Metal Organic Frameworks

Baris Demir, Department of Chemical Engineering , Istanbul Technical University, Istanbul, Turkey and M. Göktug Ahunbay, Department of Chemical Engineering, Istanbul Technical University, Istanbul, Turkey

Zeolite-like metal organic frameworks with sodalite-type framework (sod-ZMOFs) are promising candidates for natural gas (CO_2/CH_4) separation and purification applications. They have anionic frameworks and extra-framework ions in the pores of molecular dimensions increasing the interactions with guest molecules and improving their separation, storage, or ion-exchange capability. The indium-based sod-ZMOF, with imidazoledicarboxylic acid as the organic linker, has higher CO2 adsorption capacity than ZIF-8, which has an analogous structure based on sodalite-topology. The charge-compensating extra-framework ions inside the cavities of these MOFs can be ion-exchanged with alkali metals to improve CO_2/CH_4 separation performance. Furthermore they can be used in the preparation of polyimide-based mixed-matrix membranes due their good compatibility and dispersion properties.

In this study, adsorption and diffusion of CO_2 and CH_4 gases in the sod-ZMOFs were studied using Grand Canonical Monte Carlo and Molecular Dynamics simulations. The simulations were carried out with "as-synthesized" as well as Li^+, Na^+ and K^+ exchanged sod-ZMOFs. The selectivity CO_2 /CH_4 selectivity of the sod-ZMOFs were compared based the type of the cation and the ion-exchange ratio over a wide range of pressure. Adsorption and diffusion simulations showed that the selectivity of these MOFs increases two orders of magnitude when it is fully exchanged with Na^+ ions. Furthermore, the selectivity depends not only on the cation type and ion-exchange ratio but also on the location of the cations in the framework.

(509f) Carbon Dioxide Adsorption On Mesoporous Amorphous Silica Surfaces with Tethered Propylamine Groups

Aleksey Vishnyakov, Chemical and Biochemical Engineering, Rutgers, The State University of New Jersey, Piscataway, NJ, Aatto Laaksonen, Department of Materials and Environmental Chemistry, Stockholm Univerisity, Stockholm, Sweden and LiJun Chen, Chemistry, Stockholm University, Stockholm, Sweden

Using a combination of classical and ab initio modeling methods, we study adsorption of carbon dioxide and nitrogen on amorphous silica grafted with propylamine chains at moist conditions that may be found in post-combustion gases. We model physisorption and chemisorption aspect separately, using classical and ab initio tools, respectively, and consider different chemisorption mechanisms leading to formation either carbamates or bicarbonates. The carbamate mechanism requires two amine groups per each CO2 molecule chemisorbed, while bicarbonate formation requires a single amine. Thus, bicarbonate formation is desired from the practical point of view, but favorable only when reactants are surrounded by water molecules. Earlier experimental results indicated bicarbonate mechanisms; however, a few recent studies insist on the carbamate mechanism of CO2 chemisorption.

Ab initio modeling shows that activation barriers are rather low for both chemisorption mechanisms. Therefore, in these systems, physisorption mechanisms strongly influence chemisorption mechanisms, because physisorption precedes chemisorption. Since published experimental results on water sorption at amine-grafted silica vary dramatically, and we modeled CO2 physisorption at various levels of hydration using classical molecular dynamics and grand canonical Monte Carlo methods. Generally, we conclude that the physisorption scenarios are unfavorable to bicarbonate formation at all but highest hydration levels, which are not typically observed in experiments. In particular, water resides closer to the silica surface than CO2. CO2 molecule in the process of chemisorption reaction act is typically surrounded by alkylamines that are known to favor carbamate mechanism.

(509g) First-Principles Study of H2O Adsorption On the NaBH4(100) Surface

Ping Li, University of Pittsburgh, Pittsburgh, PA, J. Karl Johnson, Dept. of Chemical and Petroleum Engineering, University of Pittsburgh, Pittsburgh, PA and Michael A Matthews, Chemical Engineering, University of South Carolina, Columbia, SC

Sodium borohydride ($NaBH_4$) is a promising hydrogen storage material for niche applications due to its high gravimetric (10.7 wt%) and large volumetric (115 kg of H_2 m^{-3}) hydrogen content. The hydrolysis of $NaBH_4$ releases hydrogen with both fast kinetics and high extent of reaction under technical conditions by using steam deliquescence of $NaBH_4$. The $NaBH_4$ crystal is strongly hygroscopic, rapidly absorbing water vapor from the atmosphere and then dissolving into solution, even at fairly low relative humilities. The hydrolysis reaction takes place in solution at sufficiently high temperatures to produce H_2 and hydrated $NaBO_2$. To the best of our knowledge, there are no experimental data on the adsorption energy or structure of adsorbed water on $NaBH_4$ surfaces. It is difficult to characterize the initial adsorption processes on $NaBH_4$ surfaces from experiments because the adsorption and dissolution of the crystal occurs rapidly. We have therefore used first-principles density functional theory (DFT) to probe the details of the initial adsorption process at the atomic level. Standard DFT functionals do not include an adequate description of van der Waals interactions, which are expected to be important for the energetics and structure of H_2O adsorption on $NaBH_4$. We have therefore used two different approaches for including van der Waals forces within DFT calculations and have compared our calculations with standard generalized gradient approximation DFT calculations. We have computed the adsorption energies and geometries for various coverages of H_2O on the $NaBH_4$(100) surface. We characterize the H_2O-H_2O and H_2O-surface interaction energies as the number of H_2O molecules on the surface increases. Diffusion of water on the surface is also an important process for cluster growth. We have used the Climbing Image-Nudged Elastic Band method to compute the diffusion pathways for an isolated water molecule on $NaBH_4$(100).

(514a) Crystals Are Like People: Growth and Defects Are What Make Them Interesting

Michael F. Doherty, Chemical Engineering, UC Santa Barbara, Santa Barbara, CA

"Crystals Are Like People: Growth and Defects Are What Make Them Interesting"

There is a growing interest in predicting and controlling the size and shape of crystalline particles. Multidimensional population balances have been developed to accomplish this task but they suffer from the drawback of needing the absolute growth rate for every family of faces that may appear on the crystal surface. Such growth rates are known for only a handful of crystalline materials and prospects are bleak for extending the library of growth rate data. This raises the question of where the surface growth rates for all the families of faces will come from to drive multidimensional population balance engineering technology. One answer is "from first principles." In this lecture it will be shown how to reformulate multidimensional population balances in terms of relative growth rates and how to create first principles mechanistic models to calculate these quantities for real molecular crystals as a function of supersaturation.

(514b) Determination of Crystal Growth Rates for Solids Growing From Multiple-Conformers Solutions and Evidence of Growth Inhibition by the Wrong Conformer

Lotfi Derdour, Drug Product Science and Technology, Bristol-Myers Squibb co., New Brunswick, NJ and
Chandrasekar Sivakumar, Chemical Engineering, Carnegie Mellon University, Pittsburg, PA

The validity of a model derived for crystal growth rate for solids crystallized from solutions containing multiple conformers is evaluated. A simple and straightforward method for determining crystal growth rates using FBRM and FTIR/ATR is utilized to maximize growth rate data collected from a single experiment. Crystal growth rate data obtained experimentally were in good agreement with model predictions. In particular, the data indicated that when the approach of the right conformer (Derdour et al[[1]]) applies, crystal growth rate is inversely proportional to crystal size. Further, the results indicate that growth rates determined from supersaturation measurements can be over-estimated because of surface nucleation and/or agglomeration that can result from supersaturation spikes during anti-solvent addition. Lastly, crystal growth data obtained at different temperatures indicated that the slow growth rate observed at low temperatures may be caused in part by surface poisoning by the wrong conformer.

---

1 Derdour et al., Crystallization from solutions containing multiple conformers: A new modeling approach for solubility and supersaturation, Chem Eng Sci., 66(1), pp. 88-102, 2011.

(514c) Molecular Driving Forces of Crystallization: A Study of the Crystal Growth Rate From the Fluctuation-Dissipation Theorem

Jia Fu, Chemical and Environmental Engineering, University of California, Riverside, Riverside, CA and Jianzhong Wu, Chemical and Environmental Engineering, UC Riverside, Riverside, CA

A perennial challenge for crystal engineering is identification of thermodynamic conditions favoring crystal growth with controlled size and morphology. While experimental investigations are instrumental to understanding the effects of macroscopic variables on the kinetic processes, analysis and utilization of the experimental data toward a rational design require a fundamental knowledge linking molecular driving forces to thermodynamic variables dictating crystallization. In this work, the crystal growth rates are calculated from NVT molecular dynamics simulations based on the fluctuation-dissipation theorem. Special attention is given to the effect of intermolecular forces on the interfacial behavior and the kinetics of crystallization. By calculating of crystal growth rates at the (001) direction of model fluid-solid systems over a broad range of the parameter space, we find that crystal growth in protein solutions or in colloids can be qualitatively different from that in simple fluids and that the kinetics of colloidal crystallization can be achieved by an effective control of the potential of mean force in addition to thermodynamic conditions.

(514d) Simulation of the Crystal Shape Distribution in the Presence of Growth Modifiers

Aniruddha Majumder[1], Zoltan K. Nagy[1], Sean K. Bermingham[2] and Hassan Mumtaz[2], (1)Department of Chemical Engineering, Loughborough University, Loughborough, United Kingdom, (2)Process Systems Enterprise, London, United Kingdom

Crystal shape and its distribution are of great interest due to the crucial role it plays in determining the quality of the final products and the efficiency of downstream processes. Selected additives as growth modifiers (GMs) are often used in the industries to achieve the desired crystal shape distribution. Thus, prediction and control of crystal shape distribution are necessary for product design, quality control and economical operation point of view. Morphological population balance modelling (PBM) is widely used as a tool to describe the evolution of crystal shape distribution in a crystallizer [1,2,3]. Despite the fact that the effect of impurities (often called additives or growth/nucleation modifiers) on the crystal shape has been demonstrated in many experimental studies, most of the previous works related to the modelling of crystal shape distributions consider only the effect of operating conditions such as supersaturation, cooling profile and seed distribution on the crystal shape distribution. Recently the effect of GMs has been incorporated in single dimensional PBM model, to explain the effect of GMs on the overall growth of particles and on the crystal size distribution (CSD) [4]. However, to the best of our knowledge, no work has been presented in the literature, which incorporates the effect of growth modifiers on the evolution of crystal shape distribution.

In this work, we present a modelling framework and simulation results that can be used to predict the evolution of crystal shape distribution in a crystallizer in the presence of GM. The growth rate of the crystals depends not only on the concentration of the GM, but also the time during which the crystals are exposed to the GM in the mother liquor [5]. These two aspects have recently been incorporated in a mono-dimensional PBM proposed by Fevotte and Fevotte [4]. We further extend this model to a multi-dimensional PBM so that morphological PBM describing the evolution of crystal shape distribution can also be incorporated.

Following the model by Kubota and Mullin [6], the growth rate dependence on the GM concentration is described by surface coverage of the adsorbed species on the growing crystal and a proportionality constant denoting the effectiveness of the adsorbed species. On the other hand, in order to track the contact time of the crystals with GM, the density function is redefined with an additional property that takes into account the exposure time of the crystals to impurity. It is assumed that the transient coverage process can be described by a first order dynamics. The resulting PBMs are simulated using commercially available software gCRYSTAL (Process Systems Enterprise, UK) for 1D problems. On the other hand, well-established high resolution finite volume technique is used for simulation of all the 1D and multidimensional problems [7]. This study would be very useful in understanding the effect of GM as well as in achieving tailored crystal shape distribution.

References

[1] C. Borchert, N. Nere, D. Ramkrishna, A. Voigt and K. Sundmacher. On the prediction of crystal shape distributions in a steady-state continuous crystallizer. *Chem. Eng. Sci*., 64(4): 686-696, 2009.

[2] G. Yang, N. Kubota, Z. Sha, M. Louhi-Kultanen and J. Wang. Crystal shape control by manipulating supersaturation in batch cooling crystallization. *Cryst. Growth Des.*, 6(12):2799-2803, 2006.

[3] C. Y. Ma and X. Z. Wang. Model identification of crystal facet growth kinetics in morphological population balance modeling of l-glutamic acid crystallization and experimental validation. *Chem. Eng. Sci*., 70(0):22-30, 2012.

[4] F. Fevotte and G. Fevotte. A method of characteristics for solving population balance equations (PBE) describing the adsorption of impurities during crystallization processes. *Chem. Eng. Sci.*, 65(10):3191-3198, 2010.

[5] N. Kubota. Effect of impurities on growth kinetics. *Cryst. Res. Technol.*, 36(8-10):749-769, 2001.

[6] N. Kubota and J. W. Mullin. A kinetic model for crystal growth from aqueous solution in the presence of impurity. *J. Crys. Grow.*, 152(3):203-208, 1995.

[7] R. Gunawn, I. Fusman and R. D. Braatz. High resolution algorithm for multidimensional population balance equations. *AIChE J*, 50:2738-2749, 2006.

(514e) On Predicting Nuclei Shape Distribution

Meenesh R. Singh and Doraiswami Ramkrishna, School of Chemical Engineering, Purdue University, West Lafayette, IN

Modeling crystallization processes must account for the rate of nucleation as well as the shape distributions of nuclei. Most models that are based on the classical nucleation theory have limited considerations for non-uniform shapes and shape distributions of nuclei, anisotropic surface energies, and non-ideal solution behavior. We present a formalism to predict nuclei shape distribution accounting for the foregoing factors. The methodology includes formulation of Gibbs free energy function for all possible morphologies of a nucleus followed by multi-dimensional maximization to obtain critical shape and size of the nucleus at some supersaturation. Furthermore, the local fluctuations in the supersaturation due to Brownian motion of solute molecules can cause variations in the size and shape of nucleus. Such variations can be modeled using the Fokker-Planck equation provided their transition probabilities are known. The transition probabilities for supersaturation fluctuations are obtained from the framework of density fluctuations.[1, 2] The corresponding Ito's equation is solved following multidimensional optimization in morphology domains to obtain nuclei shape distribution of Potassium Acid Phthalate.

References:

1. Smoluchowski, M., Drei vortrage uber diffusion, brownsche bewegung und koagulation von kolloidteilchen. *Zeitschrift fur Physik* 1916, 17, 557-585.

2. Chandrasekhar, S., Stochastic problems in physics and astronomy. *Reviews of modern physics* 1943, 15, (1), 1-89.

(514f) Molecular Dynamics Simulations of Vapor Bubble Growth and Detachment On A Heating Surface

Jinyong Bao, Department of Chemical Engineering, City College of New York ,City University of New York, New York, NY, Joel Koplik, Department of Physics, Levich Institute, City College of New York and Graduate Center,City University of New York, New York, NY and David S. Rumschitzki, Department of Chemical Engineering, The City College of The City University of New York, New York, NY

Nucleate boiling is an attractive alternative for systems that require large heat transfer from a hot solid due to the large latent heats that are accessible at relatively low wall superheat above the fluid's saturation temperature. Based on our previous simulations of conductively-driven quasi-static vapor bubble growth in an axisymmetric, cylindrical cell comprised of solid and liquid phases of finite thicknesses with fluid motion and heat transfer obeying small Reynolds, Peclet, Capillary and Bond numbers, we found that the appearance and the motion of three-phase contact line (CL) is a critical element, central to physics needed to explain the large heat enhancement in nucleate boiling. However, our simulations also suggested that different ad hoc models for the poorly-understood motion of the CL as the vapor bubble grows do have a significant effect on vapor bubble deformation and detachment when the Bond number is no longer small. Therefore, we have used molecular dynamics (MD) to simulate a nanoscale version of our three-phase system. Molecular interaction physics is the only input, to this calculation and we then observe the resulting CL motion. Because MD not only includes heat transfer, but also fluid flow, this simulation removes many of the restrictions of our earlier continuum calculation for this nano-sized system. Since terrestrial gravity is utterly negligible at the nanoscale, we introduce a fictitious body force to mimic relevant Bond numbers. In our MD simulations, we study not only the heterogeneous birth of a vapor nucleus on the solid-fluid interface at constant pressure by virtue of heat transfer through the solid and its achievement of stability, but also its subsequent growth, deformation and detachment.

Our MD simulation domain of interest is a cuboid, which is composed of the fluid region sandwiched between two parallel solid walls. The fluid phase is made of argon atoms interacting via a 12-6 Lennard-Jones (LJ) potential. The atoms of two solid walls are tethered via springs to their lattice positions and also interact with both the fluid and each other via LJ potentials. By considering the effects such as the wettability of the solid surface, we find the appropriate interaction parameters between the solid and fluid atoms in LJ form, which are critical parameters for being able to observe nucleation and heat transfer. Periodic boundary conditions are applied in 4 vertical boundaries of the domain. After achieving thermal equilibrium of all the phases in the simulation domain (solid walls and liquid argon) at the uniform reduced temperature of T=0.75 (the fluid saturation temperature), we start to expend the top wall gradually at constant reduced temperature T=0.75 to some specified pressure at which no vapor bubble appears. Next, we instantaneously increase the temperature at the bottom layer of the bottom wall while keep the original lower temperature at the top wall, both maintained by thermostating. This maintains a temperature gradient through the fluid phase. Meanwhile, the pressure at the top wall is kept fixed by allowing the top surface to freely slide up and down. Our simulations show that after some vapor patches appear and disappear on the solid surface randomly in space and time, one of these patches successfully grows into a stable vapor bubble, whose growth and CL motion we then trace.

As noted, by artificially applying a uniform body force that creates a relevant Bond number, we have successfully nucleated and grown a vapor bubble to a moderate size and observed bubble necking and detachment. We found that, despite the existence of temperature slip between solid and liquid, the vapor bubble volume still grows at $t^{3/2}$, t being time, as in our earlier conduction-only continuum calculations. This scaling is also in agreement with numerous experimental observations. Furthermore, our temperature profiles show that vapor bubble surface and the three-phase CL appear to be at the liquid saturation temperature, which confirms the assumptions made in our continuum calculations' boundary condition at liquid-vapor interface. By further increasing this artificial uniform body force (i.e., Bond number) and keeping it constant, we find the body force indeed deforms the bubble and causes it to detach from the surface. During this process, we reset/raise the temperature at the top wall to avoid vapor condensation at the top of vapor bubble as the bubble grows, detaches and rise into regions of lower temperature, without changing the magnitude of the uniform body force. These simulations result in, among other things, the time evolution of the CL motion. The radius of CL initially expands with the growth of the bubble at low Bond number. Then, as the bubble grows to a size where the body force becomes significant, the CL contracts sharply but continuously as the bubble deforms until detachment. In addition, we also consider the effects of different parameters, such as those that influence the temperature slip between solid and liquid, the magnitude of the body force, the temperature of the bottom wall and the thickness of the bottom wall on vapor bubble growth and detachment.

(528a) Development of an Adsorption Model to Evaluate the Efficiency On Multiple Adsorption-Elution Cycles of Monolith-Supported IMAC Chromatography

Milena Vega[1], Eva M. Martín del Valle[1], Miguel Ángel Galán[1] and **Ramon Cerro**[2], (1)Department of Chemical Engineering, University of Salamanca, Salamanca, Spain, (2)Chemical Engineering, University of Alabama in Hunstville, Huntsville, AL

Ceramic monoliths are a natural choice of support for the IMAC technique, because their large surface/volume ratios, low pressure drop during flow, strong mechanical properties and simple methods used for preparation and handling of active chromatographic columns. In addition monolith columns are relatively inexpensive and can be easily regenerated.

In this work the possibility multiple adsorption-desorption cycles on the separation efficiency of ceramic monoliths covered with agarose type D-5 using IMAC is investigated, using a well-known protein.. An adsorption model has been developed, in order to describe the dynamics of adsorption of proteins. The model is used to validate the experimental results obtained at equilibrium concentration and predict equilibrium concentrations of future cycles. The theoretical/experimental model developed in this research will allow determination of kinetic parameters and provide a tool for scaling up separation processes in industrial IMAC for a safe and effective use of ceramic monoliths as active media supports.

(528b) Adsorptive Removal of HMF From Fructose/DMSO Mixtures On Porous Carbons

Nafiseh Rajabbeigi and Michael Tsapatsis, Department of Chemical Engineering & Materials Science, University of Minnesota, Minneapolis, MN

5-Hydroxymethyl furfural (HMF) is considered to be an important precursor in producing a range of biomass-derived chemicals. In aqueous media, the yield of HMF from fructose dehydration is low due to its rehydration to levulinic acid and formic acid and parallel reactions of fructose to products other than HMF. A well-known solvent that allows for highly selective fructose dehydration to HMF is dimethyl sulfoxide (DMSO). However, the corresponding disadvantage is concerned with separation of HMF from DMSO due to the high boiling point of DMSO which poses a challenge in the low cost separation of HMF as evaporation would be energy intensive.

Here, we investigate the room temperature separation of HMF from DMSO for a cost effective recovery of the product from the reaction mixture that contains unreacted fructose. Various carbon materials were tested as selective sorbents capable of separating HMF from DMSO. For all carbons tested, BP2000 and Norit1240 exhibit high capacity for HMF. Depending on relative concentrations, HMF/fructose separation factors up to 45 for BP2000 and 13 for Norit1240 were obtained. Under certain conditions, by increasing the concentration of fructose in the solution, HMF adsorption increased significantly. Adsorption experiments were performed in both batch and fixed-bed column systems. The granular activated carbons preserved their shape and were easily recovered.

(528c) The Effect of Silica Surface Functionalizations On Lysozyme Adsorption

Amina M. Darwish, Chemical and Materials Engineering, University of Cincinnati, Cincinnati, OH and Stephen W. Thiel, Chemical & Materials Engineering, University of Cincinnati, Cincinnati, OH

Chromatography is a standard protein separation method in the pharmaceutical industry. A deeper understanding of the surface interactions between the protein and the stationary phase has the potential to increase the efficiency of these chromatographic separations. Therefore, the effect of different surface functionalities on protein adsorption was studied using model adsorption systems. Silica of known pore size and geometry was functionalized with different surface groups to test the effects of the surface chemistry on lysozyme adsorption. Lysozyme adsorption experiments were conducted using 0.02M potassium phosphate buffer at pH 7 at room temperature; these conditions were chosen to minimize solution interaction. Adsorption on acid-washed silica was used as a control. For example, samples of acid-washed silica were functionalized with either carboxyl or amine groups. The carboxyl group changed the adsorption isotherm from Type II (observed for acid-washed silica) to Type I. The amine group prevented lysozyme adsorption. The amine-functionalized surface was also used to graft amino acids onto the surface using peptide synthesis methods, creating an amide bond with the surface amine group. Amino acids with similar R groups to the carboxyl and amine functionalities were selected. Lysozyme adsorption on glycine (R = H), aspartic acid (R = CH_2-COOH), and lysine (R = $(CH_2)_4$-NH_2) is compared to adsorption on the native acid-washed silica as well as the carboxyl- and amine-functionalized surfaces. The lysozyme was shown to follow different patterns of adsorption depending on the surface functionalities.

(528d) Influence of Ammonium Sulfate and Sodium Chloride On Adsorption of Pegylated Lysozyme On a Hydrophobic Resin

Albert Werner, Tim Blaschke and Hans Hasse, Laboratory of Engineering Thermodynamics, University of Kaiserslautern, Kaiserslautern, Germany

PEGylation of pharmaceutical proteins, i.e. the covalent binding of poly(ethylene glycol) (PEG) to the protein, can lead to improved biocompatibility, stability and residence time of the drug in the body. Usually, the protein is already purified before PEGylation. Therefore, the purification task after PEGylation is reduced to a separation of the PEGylated isoforms or a fraction with a certain PEGylation degree from the unPEGylated protein. The PEGylation changes the hydrophobicity of the proteins which leads to different interactions in hydrophobic chromatography systems. Therefore, hydrophobic interaction chromatography (HIC) is a promising technique for the separation of PEGylated protein isoforms. Lysozyme is a protein which is often used as a model protein due to its well-known structure and surface properties. Lysozyme can be PEGylated at three lysine residues, resulting in mono-, di- and tri-PEGylated isoforms.

In the present work, the adsorption of different forms of PEGylated lysozyme as well as native lysozyme and pure PEGs on a hydrophobic resin was studied by measuring both the adsorption equilibrium isotherms and the specific enthalpies of adsorption along the isotherms. This gives a complete thermodynamic picture of the adsorption as both Δg^{ads} and Δh^{ads} are determined experimentally so that $T\,\Delta s^{ads}$can be calculated. The measurements of the adsorption equilibrium isotherms were carried out using a fully automated robotic liquid handling station. For the calorimetric measurements, a high precision isothermal titration micro-calorimeter was used.

The different forms of PEGylated lysozyme studied in the present work are: mono-, di- and tri-PEGylated lysozyme for 5 and 10 kDa PEG each. The adsorption material is Toyopearl PPG-600M, a mildly hydrophobic resin. The studied salts are sodium chloride and ammonium sulfate; a sodium phosphate buffer was used to adjust the pH value. It was found, that adsorption mechanism changes completely upon variation of the salt. A molecular picture is discussed which describes the complex adsorption mechanisms.

(528e) Surfactant Imprinting of St ber Silica Particles for Application in Saccharide and Chiral Separations

Suvid Joshi[1], Alex Rao[2], Barbara L. Knutson[3] and Stephen E. Rankin[3], (1)Chemical and Materials engineering, University of Kentucky, Lexington, KY, (2)University of Massachusetts Amherst, Saugus, MA, (3)Chemical and Materials Engineering, University of Kentucky, Lexington, KY

Synthesis of monodisperse silica particles by precipitation from an ethanol/water ammonia solution was reported by St ber nearly four decades ago. The particles are easy to synthesize and readily tunable in the submicron particle size range, which has led to many application based on modifications of the synthesis. Here we report the synthesis and characterization of St ber particles imprinted at their surface with n-octyl-β-D-glucopyranoside (C8G1), with the aim of creating selective adsorption sites for separation of D-glucose. The strategy is to add the surfactant shortly after the precipitation of the silica nanoparticles so that the silicates are still in a liquid-like state but the surfactants are not incorporated inside of the particles to generate mesopores. Particles have been synthesized with various ratios of C8G1 and cetyltrimethylammonium bromide (CTAB) to find the most suitable surfactant combination for enhanced selective adsorption. Selectivity of the material has been demonstrated for separation of glucose and xylose. Chiral separation of D-glucose from L-glucose will also be discussed.

The synthesized particles are characterized using nitrogen adsorption, scanning electron microscopy and dynamic light scattering analysis to determine their size and the absence of mesopores. Attenuated total reflection Fourier transform infrared spectroscopy and high-performance liquid chromatography will be used to analyze adsorption of sugars through a series of depletion measurements. Adsorption isotherms will be presented for saccharides and the data will be modeled based on one- and two-site Langmuir adsorption models. Large increases in affinity and selectivity are demonstrated to be possible by this approach. For instance, imprinting St ber particles with a 1:1 mixture of C8G1 and CTAB leads to an enhancement in binding affinity towards D-glucose of 49 and a selectivity for D-glucose over D-xylose of over 70 based on the single-component adsorption parameters.

(528f) Metal Ion Adsorption by Chitosan Nanoparticles and Alginate Gel Microparticles for Water Purification Applications

Kun Yu and Nina C. Shapley, Chemical and Biochemical Engineering, Rutgers, the State University of New Jersey, Piscataway, NJ

Chitosan and alginate are both low-cost natural materials used for the removal of metal ions from aqueous solutions. In this research, three different concentrations of chitosan gel nanoparticles were synthesized by ionic cross-linking with sodium tripolyphosphate and characterized by zeta-sizing and TEM. The equilibrium adsorption capability and adsorption kinetics of Cu^{2+} from copper sulfate solution on to chitosan gel nanoparticles, calcium-alginate gel microbeads and alginate-chitosan combination particles at fixed pH have been explored in batch system.

Results show that the equilibrium adsorption properties of chitosan nanoparticles have similar trends to those of alginate microbeads, and the adsorption capacity varies with the synthesis concentration of chitosan nanoparticles. The Langmuir isotherm fits the low concentration region well for both chitosan nanoparticles and alginate microbeads and correlation coefficients were determined. Alginate-chitosan combination particles were found to have intermediate maximum adsorption capacity and Langmuir adsorption equilibrium constants similar to those of alginate microbeads in low concentration copper sulfate solutions. The adsorption capabilities of combination particles can be predicted by the weight ratio of calcium-alginate gel microbeads and chitosan gel nanoparticles.

In kinetics experiments, it was observed that as the adsorption process proceeded, a rapid uptake occurred first and was followed by a slow increase approaching equilibrium for both chitosan nanoparticles and calcium-alginate gel microbeads. Calcium-alginate gel microbeads need a longer time to reach equilibrium than chitosan nanoparticles, likely due to the significant difference in particle size. The adsorption kinetics also vary with the initial concentration of copper sulfate solution. The results will aid the design of fixed-bed columns for continuous adsorption, which are extensively used in industry.

Encapsulation Mechanism of DNA Nucleobases Onto Carbon Nanotubes: Free Energy Landscapes

Fernando J.A.L. Cruz[1,2], *José P. B. Mota*[1] *and Juan J. de Pablo*[2]
[1] Requimte/CQFB, Dept. Chemistry, Universidade Nova de Lisboa, 2829-516 Caparica, Portugal. [2] Dept. Chem. Bio. Eng., University of Wisconsin-Madison, Madison, WI.
fj.cruz@fct.unl.pt

I. Introduction

Ever since the unravelling of the double helix structure of DNA, the molecule has been considered as the building block of biological organisms. Because DNA is the agent that carries genetic information from one generation to the next, the knowledge of its physical and chemical properties is of paramount importance. Much of that information has to do with the way the nucleobases (NBs)—DNA's own building blocks—interact with their surrounding environment. Owing to their nanoscale dimensions, carbon nanotubes are able to selectively interact with DNA almost at the atomic level: homo- and heteropolymer ssDNA will readily adsorb onto the external walls of small diameter SWCNTs,[1] according to a π–stacking mechanism of the NBs on the walls. We have already characterized the SWCNTs energetic landscape regarding exo- and endoadsorption, using small probe molecules.[2,3] While exoadsorption on small diameter nanotubes (< 1 nm) is reasonably understood, confinement driven adsorption/binding is a phenomenon that remains quite unexplored. Okada[4] determined the encapsulation energy of (in vacuum) ssDNA as a function of nanotube diameter, and suggested an intricate relationship between both, particularly in the range $1.25 < D$ (nm) < 1.8. As far as we are aware, there are no previous studies of intratubularly confined aqueous nucleobases. We shall address this issue using the well-tempered metadynamics scheme proposed by Parrinello,[5] estimating the free-energy landscapes associated with the encapsulation process of Adenine (purine) and Thymine (pyrimidine) onto two different diameter (zig-zag) nanotubes, (16,0) with D = 1.25 nm and (23,0) with D = 1.8 nm. These two topologies correspond to the narrowest and largest nanopores obtained by electric-arc discharge.[3] Because nanotubes can be electrically charged,[6] the effect of charge density will be addressed considering purely hydrophobic solids and nanotubes with electric charges of $q = +0.05$ e^-/C. Physiological conditions (T=310K) are mimicked by explicitly including H_2O molecules and Cl^- ions in the calculations.

II. Results and Discussion

The encapsulation mechanism of both purine- and pyrimidine-types onto hydrophobic SWCNTs is energetically favorable and irreversible during the time windows spanned in the simulations (0.1 μs); the resulting NB/SWCNT hybrids exhibit lower free energies compared to the unconfined systems (Fig.1). The highly symmetrical free-energy landscapes obtained are essentially one dimensionally isotropic regarding the ξ_1 order parameter, in direct relationship to the nanotubes *quasi* 1-D symmetry. Once confined inside the nanotube, a nucleobase can explore a region whose boundaries are comprised between the entrances, $0.15 < \xi_1$ (nm) $< L_z$ (where L_z is the SWCNT length), and the inner surface of the wall, never returning to the bulk. It is very interesting to observe that confinement effects are remarkably stronger at the nanopore center, where free energy differences are largest, $[\xi_1, \xi_2]^{(16,0)}$ = (1.1, 0.08) nm and $[\xi_1, \xi_2]^{(23,0)}$ = (1.7, 0.08) nm. Independent umbrella sampling calculations identified a unique probability distribution maximum corresponding to a domain comprised between the solid walls and the nanopore center (Fig.2). Once confined, molecules are in direct contact with the walls at a minimum distance of closest approach of 0.26–0.28 nm. No layer of solvent could be observed between hydrophobic slabs. This is due to a robust pi-stacking mechanism of the NB onto the graphitic mesh, leading to strong dispersive energies of interaction with both the (16,0) and (23,0) topologies (Table I). These results have been compared with DFT and experimental data for graphene. Contrary to exoadsorption, intratubular binding affinity increases with nanotube curvature, as a result of enhanced interaction with the wall immediately opposite to the adsorptive one. We have estimated a diameter of D = 2.05 nm above which the SWCNT curvature mimics a flat graphene sheet and confinement effects tend to be negligible. Adenine and thymine evidence similar interaction energies with the hydrophobic nanopores, however, adenine confinement results in slightly favored energetics of *ca.* 0.9 – 1.4 kJ/mol (Table I). This previously unobserved finding indicate a binding affinity order: purines > pyrimidines.

Electrically charged nanotubes evidence a completely different behavior. The narrowest nanopore employed here, (16,0), inhibits encapsulation of either the purine or pyrimidine moiety. Instead, that nanotube favors adsorption at a region very close to the pore entrance as indicated by the probability distribution maxima at Ω = 0.97 nm and Ω = 1.04 nm (Fig.2). When the nucleobases are put in contact with the (23,0) solids, several encapsulation/exit events can occur (Fig.3); moreover, the confined systems exhibit lower free energies. The probability distribution curves indicate two maxima: *i)* close to the pore center, overlapping the one obtained for the corresponding hydrophobic solid, and *ii)* at the pore entrance. This second region of confinement is energetically less stable than the pore center (Table I), particularly in the case of adenine where deltaE=40 kJ/mol. Because the pyrimidine molecule has a larger dipole moment, its exposure to the bulk solution carries smaller electrostatic penalties, and therefore its location at the pore center is stabilized by only *ca.* 12 kJ/mol compared to the entrance.

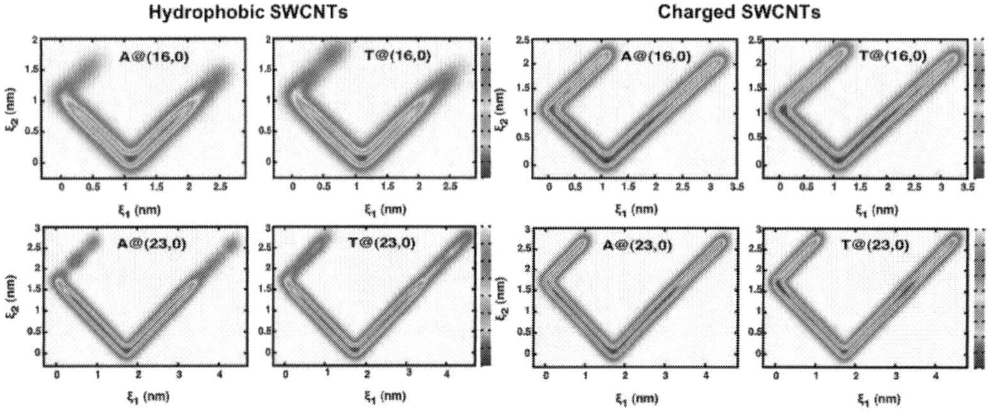

FIGURE 1. Free energy landscapes, $F(\xi_1, \xi_2)$, of the nucleobases in contact with (16,0) and (23,0) nanotubes. The two collective variables, ξ_1 and ξ_2, correspond to the distance between the nucleobase center of mass and the nanotube entrance or geometrical center, respectively, projected along the *z*-axis. For the hydrophobic solids, the absolute free energy minima at $[\xi_1, \xi_2]^{(16,0)}$ = (1.1, 0.08) nm and $[\xi_1, \xi_2]^{(23,0)}$ = (1.7, 0.08) nm coincide almost exactly with the SWCNT center. When the SWCNTs are charged, two free energy minima can be observed: at the nanotube center $[\xi_1, \xi_2]^{(16,0)}$ = (1.09, 0.08) nm and $[\xi_1, \xi_2]^{(23,0)}$ = (1.72, 0.08) nm, and at the entrance, $[\xi_1, \xi_2]^{(16,0)}$ = (0.08, 1.08) nm and $[\xi_1, \xi_2]^{(23,0)}$ = (0.08–0.33, 1.41–1.70) nm. Note that the nanopore entrances are symmetrical about the SWCNTs center of mass and thus the sets of collective variables are equivalent: $[\xi_1, \xi_2]^{(16,0)}$ = (0, 1.1) = $[\xi_1, \xi_2]^{(16,0)}$ = (2.2, 1.1) and $[\xi_1, \xi_2]^{(23,0)}$ = (0, 1.7) = $[\xi_1, \xi_2]^{(23,0)}$ = (3.4, 1.7).

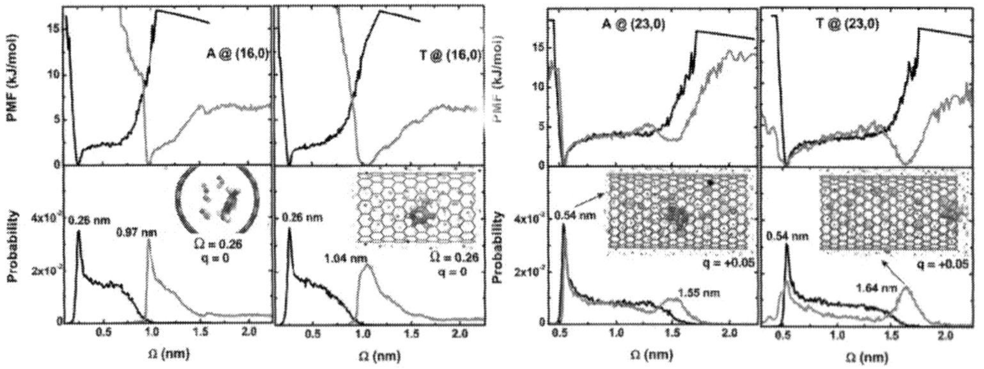

FIGURE 2. Potential of mean force (PMF) and probability distribution profiles: *black)* hydrophobic and *red)* electrically charged nanotubes. The order parameter, Ω, corresponds to the absolute distance between the center of mass of the nucleobase and that of the solid. Notice the parallel alignment of the nucleobases with the hydrophobic solid walls and the existence of two maxima in the probability curves for the electrically charged (23,0) nanotubes.

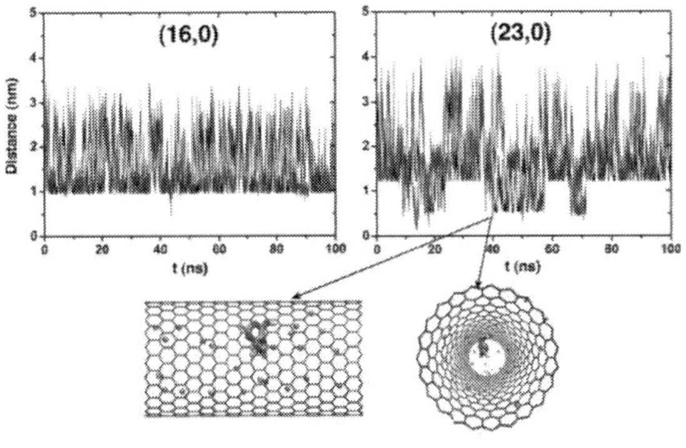

FIGURE 3. Distance between a nucleobase and electrically charged SWCNT center: *black)* adenine, *blue)* thymine. The distance plateaus at *ca.* 1nm (16,0) and 1.35nm (23,0) essentially correspond to the nanopore entrances.

Table I. Interaction energies between the nucleobases and the carbon nanotubes, $E^{int} = E^{vdw} + E^{Coul}$, where E^{vdw} is the dispersive energy and E^{Coul} corresponds to the electrostatic contribution.

		E^{vdw} (kJ/mol)		E^{Coul} (kJ/mol)		E^{int} (kJ/mol)	
		A	**T**	**A**	**T**	**A**	**T**
Unch.	(16,0)	-83.1 ± 0.1	-82.2 ± 0.1	0.0	0.0	-83.1	-82.2
	(23,0)	-68.6 ± 0.1	-67.2 ± 0.1	0.0	0.0	-68.6	-67.2
Charged	(16,0)[a]	-35.2 ± 0.2	-31.8 ± 0.3	-19.8 ± 0.3	-20.6 ± 0.3	-54.9	-52.3
	(23,0)[b]	-65.3 ± 0.1	-51.4 ± 0.3	-38.8 ± 0.2	-37.6 ± 0.3	-104.1	-89.0
	(23,0)[c]	-40.6 ± 0.2	-45.8 ± 0.3	-24.2 ± 0.2	-31.4 ± 0.3	-64.7	-77.2

[a]results for the (16,0) electrically charged nanotubes correspond to exoadsorbed nucleobases.
[b]confinement close to the nanopore center and [c]at the pore entrance

References
(1) Johnson, R. R.; Johnson, A. T. C.; Klein, M. L. *Nano Letters* **2008**, *8*, 69.
(2) Cruz, F. J. A. L.; Mota, J. P. B. *Phys. Rev. B* **2009**, *79*, 165426.
(3) Cruz, F. J. A. L.; Esteves, I. A. A. C.; Agnihotri, S.; Mota, J. P. B. *J. Phys. Chem. C* **2011**, *115*, 2622.
(4) Kamiya, K.; Okada, S. *Phys. Rev. B* **2011**, *83*, 155444.
(5) Barducci, A.; Bussi, G.; Parrinello, M. *Phys. Rev. Let.* **2008**, *100*, 020603.
(6) Zhao, X.; Johnson, J. K. *J. Am. Chem. Soc.* **2007**, *129*, 10438.

(539a) Metal-Organic Framework/Polymer Mixed-Matrix Membranes for H2/CO2 Separation: A Fully Atomistic Simulation Study

Liling Zhang, Soochow University, Suzhou, China, Zhongqiao Hu, Chemical and Biological Engineering, National University of Singapore, Singapore, Singapore and Jianwen Jiang, Chemical and Biomolecular Engineering, National University of Singapore, Singapore, Singapore

A fully atomistic simulation study is reported to investigate H_2/CO_2 separation in mixed-matrix membranes composed of polybenzimidazole (PBI) and zeolitic imidazolate framework-7 (ZIF-7). PBI is mimicked by the polymer consistent force field (PCFF) and ZIF-7 by a new force filed developed here. The density and glass transition temperature of PBI from simulation and experiment are in good agreement. The simulated lattice constants, bond lengths and angels of ZIF-7 match well with measured data. PBI/ZIF-7 membranes have stronger mechanical strength than PBI as evidenced by the enhanced bulk modulus upon adding ZIF-7. With increasing ZIF-7 loading, the fractional free volume and percentage of large voids (> 3 Å) become larger; the solubilities of H_2 and CO_2 also increase because of stronger affinity with ZIF-7. H_2 diffusion in PBI/ZIF-7 is not retarded by the addition of ZIF-7, in contrast to CO_2 diffusion. Compared to PBI, the permeabilities of H_2 and CO_2 in PBI/ZIF-7 are enhanced, as well as the permselectivity of H_2/CO_2. The performance of PBI/ZIF-7 is found to surpass the Robeson's upper bound. For the first time, this simulation study provides atomistic insight into the sorption, diffusion and permeation of gas molecules in MOF-filled membranes; and suggests high separation could be achieved in this type of novel membranes.

(539b) Molecular Simulation of CO2-Induced Plasticization in 6FDA-Based Polyimides in the Presence of Residual Solvent

Sadiye Velioglu, M. Göktug Ahunbay and S. Birgül Tantekin-Ersolmaz, Department of Chemical Engineering, Istanbul Technical University, Istanbul, Turkey

Polyimide membranes have a strong potential for CO2 capture in natural gas purification, coal gasification, and flue gas treatment processes. The principle issue in these applications is the plasticization of the polyimide membranes at high partial pressures of CO2, which can lead to reduced membrane selectivity and unpredictable membrane properties. Previous experimental studies on gas permeabilities of polyimide membranes showed that the presence of the residual solvent in the membrane has a significant effect on the membrane free volume size distribution, and it alters the gas sorption behavior and gas permeability. While a small amount of solvent in membrane acts as an anti-plasticizing agent, larger amounts exhibit an opposite effect. Molecular simulation methods were previously used to investigate the effect of the residual solvents on free volume changes of the polyimides.[1]

In this study, we investigated the change in the CO_2 sorption capacity of 6FDA-DAM and 6FDA-ODA as a function of residual solvent type and amount. A low-boiling-point solvent, tetrahydrofurane (THF) (b.p. 66ºC), and a high-boiling-point solvent, n-methyl-2-pyrrolidone (NMP) (b.p. 202ºC), were selected for this purpose. For each polyimide, two model matrices containing 4% either NMP or THF molecules were constructed. Following an equilibration process, the solvent molecules were removed gradually to obtain new matrices with 2% and 1% solvent content and finally with no solvent. The matrices were equilibrated again after each removal step.

The CO_2-induced plasticization effect was reproduced by applying sorption-relaxation cycles until the CO_2 concentration converges: At each cycle, the polymer matrix was loaded with CO_2 through Grand Canonical Monte Carlo simulations at the considered pressure, and then molecular dynamics simulation runs in the *NPT* ensemble were applied to obtain an equilibrated matrix. The cycle was repeated until CO_2 concentration converges. The change in the linkage flexibility due to the presence of solvent and its effect on CO_2 sorption capacity and plasticization were analyzed.

The structural properties of solvent-free 6FDA-DAM and 6FDA-ODA polyimides were estimated in good agreement with the experimental data. Estimated sorption coefficients of these gases in both polymers are within the same order of magnitude with the experimental data in the literature. Simulation revealed also that CO_2 solubility is solely a function of the FFV of the plasticized polyimides, and the increase in the fractional free volume and the flexibility of the diamine-dianhydride linkage are correlated.

Radial distribution function (RDF) analyses showed that THF affects the nitrogen in the imide ring of 6FDA-ODA and hinders chain packing; resulting in high fractional free volume (FFV) and sorption capacity. Between the four sorption sites (the oxygen and nitrogen in the imide, the fluorine in the $-CF_3$ group, and the carbon in $-CH_3$ group of the diamine) in 6FDA-DAM, the oxygen in the imide is the preferential one for both CO_2 and NMP indicating that NMP sorption at and around this site would lower the sorption capacity of CO_2. Consequently, residual solvents reduce CO_2 sorption capacity and alter CO_2/CH_4 selectivity of both polyimides. Relationship between backbone rigidity and plasticization resistance is altered in the presence of solvents.

Abbreviations: 6FDA: 4,4-hexafluoro isopropylidene diphthalicanhydride; ODA: 4,4 oxydianiline; DAM: 2,4,6-trimethyl-m-phenylene diamine.

[1]Chang, K. S.; Tung, C. C.; Lin, C. C.; Tung, K. L. Residual Solvent Effects on Free Volume and Performance of Fluorinated Polyimide Membranes: A Molecular Simulation Study *J. Phys. Chem. B.* 2009, 113, 10159.

(539c) Understanding and Controlling Fractional Free Volume in Ionic Liquid-Based Polymer Membranes for CO2 Capture

Jason E. Bara, Matthew S. Shannon, A. Christopher Irvin, Michelle S. Hindman and W. Jeffrey Horne, Chemical & Biological Engineering, University of Alabama, Tuscaloosa, AL

Research using ionic liquids (ILs) in gas separation membranes is now entering its second decade. The tunable structures of ILs have provided a number of unique materials being produced for membrane-based CO_2 capture. Approaches to IL-based membrane materials include supported liquid membranes, polymerized ILs or *poly(ILs)*, block co-polymers interfaced with ILs, poly(IL)-IL composites and gels. Given the advancements over the past several years in the state of IL materials development, we now see a need to re-visit some fundamental aspects of ILs in order to better outline a path forward for IL-based membranes.

While fractional free volume (FFV) is known to play a significant role in the properties and performances of polymer gas separation membranes, little focus has been given to developing relationships between FFV and IL-based materials. We recently calculated FFV values for 165 combinations of 1-*n*-alkyl-3-methylimidazolium ILs ([$C_n$mim][X]) using the COSMOtherm package. FFV was found to be a function of IL molar volume, and as such, earlier models for gas solubility based on Regular Solution Theory can now be reinterpreted in terms of free volume. In concert with experimental data, trends and relationships of gas solubility for various cation/anion pairs were directly linked to computational results. For gases such as CH_4 and N_2, it was determined that as FFV increases, solubility increases in a manner that is directly proportional to free volume. However, within a given IL family (e.g. same anion with various cations), solubility of CO_2 was observed to decrease with increasing FFV. Using experimental data, the influence of free volume on gas solubilities has been determined, and suggests that CO_2 selectivity is a function of the free volume to the (-3/2) power.

Furthermore, we are now exploring new ways to control IL FFV as a means of improving CO_2 solubility, selectivity and permeability in IL-based membranes, using COSMOtherm as a means of guiding materials design. Simulations suggest that maintaining a highly polar free volume around the IL is key to maximizing CO2 permeability and selectivity. Initial experimental and computational results obtained for these new generations of IL-based membranes will also be presented.

(539d) Recent Developments in Evapoporometry for Characterizing the Pore-Size Distribution of Membranes

William B. Krantz, Chemical and Biological Engineering, University of Colorado, Boulder, CO, Alan R. Greenberg, Department of Mechanical Engineering, University of Colorado Boulder, Boulder, CO, Elmira Kujundzic, Mechanical Engineering, University of Colorado, Boulder, CO, Adrian Yeo, Singapore Membrane Technology Center, Nanyang Technological University, Singapore, Singapore and Seyed S. Hosseini, Chemical Engineering Department, Tarbiat Modares University, Terhan, Iran

Recent Developments in Evapoporometry for Characterizing the Pore-Size Distribution of Membranes

William B. Krantz, Alan R. Greenberg, Elmira Kujundzic,

Adrian Yeo, and Seyed S. Hosseini

Evapoporometry (EP)is a novel means for characterizing the pore-size distribution in membranes. It is based on the principle that the vapor pressure is affected by the curvature of a volatile liquid contained in a porous material, which is described by the Kelvin equation. If a wetting liquid is used to saturate the pores, the vapor pressure will be depressed, whereas if a non-wetting liquid is used, the vapor pressure will be enhanced. EP involves first saturating the membrane with a volatile liquid that is not a solvent or swelling agent for the membrane material. The sample is then placed in a specially designed test cell that is placed on a microbalance that permits measuring sample mass as a function of time. The slope of the mass versus time curve provides the evaporation rate, which can then be related to the vapor pressure at the interface between the liquid in the porous material and the ambient gas phase. The vapor pressure in turn can be related to the pore diameter via the Kelvin equation. If the porous material is pre-saturated with a wetting volatile liquid, the evaporation rate will monotonically decrease as a function of time. This behavior results because the liquid will evaporate progressively from the largest to the smallest pores.

The accuracy and reproducibility of EP are confirmed by characterizing the pore-size distribution of 100 nm Anopore membranes. The latter are an alumina-based inorganic membrane that has very regular non-interconnected columnar pores. The results of EP characterization are compared with SEM measurements and with SEM and AFM analyses of other investigators. EP is also used to characterize 20 nm and 50 nm commercial PVDF membranes and compared with an independent characterization of the pore-size distribution via displacement porometry. This comparison reveals substantial compaction effects during the displacement porometry analyses that shift the pore-size distribution towards smaller pores. This observation is not surprising since pressures as high as 30 bar were required for the liquid displacement. However, it underscores one of the advantages of EP in that it permits characterizing the pore-size distribution of fragile and compressible membrane materials since it can be done at atmospheric pressure.

Other advantages as well as the limitations of EP will be discussed. In particular, a detailed error analysis is presented that shows that EP is a very accurate method for characterizing small pores, 2-100 nm, but incurs significant error for pores larger than approximately 200 nm. Although EP is applied to flat sheet membranes in the work presented here, the use of this technique for other applications will be outlined.

(539e) A Molecular Marker Approach for Real-Time Characterization of RO Membrane Integrity

Sirikarn Surawanvijit[1], John Thompson[2], Anditya Rahardianto[3], Val Frenkel[4] and Yoram Cohen[2], (1)Chemical&Biomolecular Engineering, University of California, Los Angeles, Los Angeles, CA, (2)Chemical and Biomolecular Engineering, University of California, Los Angeles, Los Angeles, CA, (3)Chemical & Biomolecular Engineering, University of California, Los Angeles, Los Angeles, CA, (4)ARCADIS U.S. Inc., Emeryville, CA

The use of reverse osmosis (RO) processes has grown significantly over the past few decades in water treatment and reuse applications to safeguard water supplies against harmful pathogens and impurities, as well as reducing water salinity to enable indirect potable water reuse. Maintaining membrane transport characteristics (e.g., flux and rejection) are paramount to ensuring process performance, particularly with respect to the passage of organics and nanosize enteric viruses that may pose a potential health hazard. Loss of membrane integrity (e.g., due to exposure to oxidants such as chlorine) is of particular concern in water reuse applications where passage of nano-size contaminants (e.g., enteric viruses) can occur due to membrane integrity breaches. Accordingly, given the need for online membrane performance monitoring, an online fluorescent marker-based approach was developed for membrane integrity detection and characterization. The approach involves dosing the feed stream with a molecular marker and monitoring its concentration in the permeate and concentrate streams. The change in marker concentration in response to a controlled marker dose to the feed is then monitored online. Application of the relevant membrane solute flux model, along with analysis of the marker's residence time distribution, then enables assessment of the detection of membrane integrity breaches, given previous characterization of the membrane with respect to the marker permeability. In the present study, detection and quantification of the effect of membrane breaches (induced by controlled exposure to sodium hypochlorite) on membrane transport characteristics was evaluated for commercial polyamide RO membranes. Membrane integrity breaches were first characterized by imaging via scanning electron microscope (SEM) and atomic force microscope (AFM) imaging. The characteristics of integrity breach were then correlated with the level and mode of chlorine exposure. Subsequently, experimental studies were carried out, with a specially-designed online fluorescent marker detection system, using both flat-sheet membranes (in a plate-and-frame RO cell) as well as in a spiral-wound membrane pilot. Results from the study demonstrated the technical feasibility for real-time online characterization of RO membrane integrity breaches as well as assessment of RO module performance.

(539f) Comparative Study of the Pervaporation of Methanol-Methyl Acetate Binary Mixtures with Hydrophilic and Organophilic Membranes

Thomas Winkler, Susanne Lux and Matthäus Siebenhofer, Institute of Chemical Engineering and Environmental Technology, Graz University of Technology, Graz, Austria

In recent years, there has been a significant rise in the number of pervaporation processes for industrial applications. The field of interest for the application of this unit operation has a span from dehydration processes to equilibrium limited reactions. Former applications conducted by pervaporation already represent the state of the art in the dehydration of organic solvents. The main benefits of pervaporation are based on low operation temperatures, partial evaporation and the nature of the driving force. The outlined advantages may result in energy savings compared to conventional unit operations when dealing with thermodynamic limitations.

In this research activity we focus on the separation of azeotropic mixtures. The methyl acetate synthesis serves as a model reaction, since it forms two low boiling azeotropes with methanol and with water. The reverse reaction, the hydrolysis of methyl acetate, as well as transesterification reactions are facing the same issues. The azeotropic mixture of methyl acetate and methanol represents the bottleneck in all these reactions. The major focus of our investigations is to overcome this limitation by applying a pervaporation process. We conducted a screening of various commercially available hydrophilic and organophilic membranes. Hydrophilic membranes enhance the permeate flux of polar components. The active layer of the investigated hydrophilic membranes consists of crosslinked PVA. An increase in the degree of crosslinking leads to a significant improvement in selectivity. The maximum permeate flux of the investigated PVA-membranes ranges from 0.9 to 9.4 $kg \cdot m^{-2} \cdot h^{-1}$. In the case of organophilic membranes a rise in the permeate flux up to values of 72 $kg \cdot m^{-2} \cdot h^{-1}$ was observed. Organophilic membranes with an active layer consisting of POMS show a better separation performance than PDMS based layers Although the difference in permeate flux is not that pronounced. A model for the permeate flux based on an adapted solution-diffusion model was proposed.

(539g) Perfluorinated Membranes for Pervaporation Separation of Solvent Mixtures in Pharmaceutical Processing

John Tang and Kamalesh K. Sirkar, Chemical, Biological and Pharmaceutical Engineering, New Jersey Institute of Technology, Newark, NJ

The pharmaceutical industry manufactures large quantities of drug products while employing a variety of organic solvents with a wide range of chemical properties. Mixtures of these solvents are often created as byproducts of the manufacturing process, which may include methanol, ethyl acetate, tetrahydrofuran (THF), toluene, dimethylformamide (DMF), dimethylsulfoxide (DMSO), dimethylacetamide (DMAc) and water. Pervaporation of binary mixtures was carried out using a perfluoro-2,2-dimethyl-1,3-dioxole-tetrafluoroethylene copolymer membrane, known as CMS-3, over a range of compositions and temperatures of 30, 50, and 60°C. This membrane has a significantly high fractional free volume, 0.342, while possessing reliable thermal, mechanical and chemical resistance. The membrane has been found to typically favor solvents with smaller molecular dimensions. Separation of organic-organic systems yields limited results. The membrane is selective for methanol in methanol-toluene systems, giving an overall selectivity value of 4.5 and producing methanol flux of 1.1 g/m^2-h. Consistent water-selective performance, comparable to that of zeolites for alcohol-water systems, was obtained when separating high boiling aprotic solvents from water. Using a 90 wt% DMF-10 wt% water feed, CMS-3 is able to achieve a separation factor of 10,766 at 60°C.

(542a) Development of a Novel Hot-Carbonate Process for Post-Combustion CO2 Capture: Role of Organic Promoters in Enhanced Absorption and Bicarbonate Crystallization

Manoranjan Sahu, Advanced Energy Technology Initiative, University of Illinois at Urbana-Champaign, Champaign, IL, Qing Ye, University of Illinois at Urbana-Champaign, Champaign, IL and Yongqi Lu, Prairie Research Institute, University of Illinois at Urbana-Champaign, Champaign, IL

Development of cost-effective new technologies are essential for separating and capturing CO_2 from power plant flue gases and retrofitting the CO_2 capture and sequestration (CCS) technology with minimum cost penalty. Currently available monoethanol amine (MEA)-based processes are highly energy-intensive and expensive. A novel hot-carbonate absorption process with bicarbonate crystallization-enabled high pressure stripping (Hot-CAP) is being developed to reduce the energy requirement for CO_2stripping and compression.

In the Hot-CAP, CO_2 is absorbed into a concentrated potassium carbonate solution (about 40wt%) at elevated temperatures (60-80°C). A carbonate–based system tends to have a slower CO_2 absorption rate compared with the MEA. A key technical issue is thus to enhance the absorption rate to a level that is comparable to MEA-based processes. In this study, five organic catalysts and three potassium amino acid salts were investigated as organic promoter to accelerate the absorption rate of CO_2 into the carbonate solution. The rate of CO_2 absorption was measured using a stirred tank reactor (STR) reactor. Results indicated that some selected organic catalysts enhanced the absorption rate significantly and the absorption rates were comparable with that of 5M MEA. The absorption rate depended on the promoter dose and CO_2 loading level in the solution. Additionally, the crystallization of the CO_2 rich solution without and with organic promoters was further investigated under various process conditions. Results from crystallization revealed that potassium bicarbonate prevailed in crystal solids formed.

This presentation will provide a summary of results from the studies of absorption and crystallization in the presence of promoters.

(542b) Process Development for Phase-Changing Aminosilicone-Based CO2-Capture Absorbents

Tiffany Westendorf[1], Benjamin R. Wood[1], Sarah Genovese[1], Irina Sultanova[1], Robert J. Perry[1], Rachel Farnum[1], Michael J. O'Brien[1], Malgorzata Rubinsztajn[1], Ravi-Kumar Vipperla[2] and Robert Enick[3], (1)Global Research, General Electric, Niskayuna, NY, (2)GE Energy, Greenville, SC, (3)University of Pittsburgh, Pittsburgh, PA

Liquid aminosilicone-based absorbents have been developed that have the potential to significantly reduce the cost of CO_2 capture from pulverized coal power plants. Upon exposure to CO_2, these novel absorbents solidify, thereby enabling innovation in CO_2 capture process design. Compared to MEA-based CO_2 capture absorbents, the phase-changing aminosilicones have the potential to reduce CO_2 capture cost through higher CO_2 absorption capacity, greater thermal stability, and reduced solvent losses due to volatility. Furthermore, the phase-changing behavior of these new absorbents and the absence of aqueous diluent enable the use of more energy-efficient separations process units. Finally, by enabling pressurized desorption of CO_2, the energy required to compress the clean CO_2 stream pre-sequestration is significantly reduced.

A process is being developed in which CO_2 is removed from simulated flue gas in a spray absorber, and the solid CO_2-sorbent complex is slurried in liquid absorbent for transport to a pressurized desorber. In the desorber, the solids are heated to release CO_2 and regenerate the liquid absorbent for recycle. Experiments at the lab scale focusing on spray absorption, slurry rheology, and pressurized desorption are in progress to generate data needed to scale up the process. These results and their effect on estimated CO_2 capture economics will be discussed.

(542c) A Techno-Economic Evaluation of Various Stripper Configurations in Amine Absorption Technology for Post-Combustion Carbon Capture

Sumedh S. Warudkar, Kenneth R. Cox, Michael S. Wong and George J. Hirasaki, Chemical and Biomolecular Engineering, Rice University, Houston, TX

As a part of this work, we are comparing the techno-economic performance of the amine absorption process under a variety of stripper operating conditions – ranging from vacuum strippers operating at 30 kPa to high pressure strippers operating at 300 kPa. We are modeling the amine absorption process using the process simulation software ProMax® with TSWEET®. The equipment sizing and capacity data thus generated are the feed parameters for Aspen Process Economic Analyzer which provides the cost estimates for the process equipment.

In our previous studies, we have reported the comparisons of energy consumption and absorber-stripper column sizing for monoethanolamine (MEA), diethanolamine (DEA) and diglycolamine (DGA) at varying concentrations and stripper operating conditions [1]. Of the three amines compared, we've identified 40 wt% DEA and 60 wt% DGA as the more suitable amines for carbon capture application than MEA. We have also shown that an increase in the stripper operating pressure results in a reduced reboiler energy duty and a smaller stripper column [2]. Use of very low pressure steam - waste heat, results in a reduced 'parasitic energy duty' at the power plant but results in larger stripper columns and increased load on the compression train.

It can be seen that there are two contrasting factors involved in deciding the optimal operational configuration – potential energy savings by reduction of parasitic duty and an increase in the capital expenditure with increasing the equipment sizes. Thus, in spite of the detailed technical analysis conducted previously; it hasn't been possible to predict the optimal configuration amongst those compared. The aim of the techno-economic analysis undertaken as a part of this study is to predict with some clarity the better performing absorbent and the corresponding system configuration.

References

1. S. S. Warudkar, K. R. Cox, M. S. Wong and G. J. Hirasaki, "Comparison of alkanolamines for post-combustion carbon capture from a pulverized coal fired power plant", AIChE Fall Meeting, Minneapolis, MN (2011).

2. S. S. Warudkar, K. R. Cox, M. S. Wong and G. J. Hirasaki, "Amine absorption with vacuum stripping for post-combustion carbon capture", AIChE Spring Meeting, Houston, TX (2012).

(542d) Novel Solvent Selection and Solvent Stripping for CO2 Capture From Power Plants

Juan M. Salazar[1], Urmila Diwekar[1], Kevin G. Joback[2] and Abhoyjit Bhown[3], (1)Vishwamitra Research Institute, Center for Uncertain Systems: Tools for Optimization and Management, Clarendon Hills, IL, (2)Molecular Knowledge Systems, Inc., Bedford, NH, (3)Electric Power Research Institute, Palo Alto, CA

Novel Solvent Selection and Solvent Stripping for CO_2 Capture from Power Plants

Juan Salazar and Urmila Diwekar

Center for Uncertain Systems: Tools for Optimization & Management

Vishwamitra Research Institute

Clarendon Hills, IL 60514

E-mail: urmila@vri-custom.org; Tel: 630-886-3047

And

Kevin Joback

Molecular Knowledge Systems, Inc.

Bedford, NH 03110-0755

And

Abhoyjit S. Bhown

Electric Power Research Institute, Inc.

Palo Alto, CA 94304

This paper presents a new approach to solvent selection for post-combustion CO_2 capture from fossil fuel fired power plants. Two main questions arise in a solvent capture process: how to select an effective separating agent and how to design and synthesize this separation process. In previous studies, few solvents were considered for CO_2 capture. Further, researchers arrived at these solvents using experimental methods with limited capacity to screen a large number possible candidates. In this work, we are using computer aided molecular design (CAMD) method to obtain a large number of possible solvent candidates. These candidates are then evaluated by integrating the solvent selection and CO_2 capture process to ensure improved energy, environmental, and economic performance. In this paper we present the first attempt at using group contribution based CAMD to derive new solvents for CO_2 capture.

CO_2 absorption in a solvent-based capture process is essentially based on the reversible, selective nature of the chemical reaction between the liquid solvent and CO_2 in the flue gas. The process flow diagram shown in Figure 1 has been widely used in literature. Flue gas from the boiler is brought into contact with the solvent in the absorber after the removal of impurities such as NOx, SOx and particulate matter using processes that also cool the flue gas. A blower is used to compensate the pressure drop experienced in the absorber. In the absorber, CO_2 selectively absorbs into the solvent, and flue gas leaving the absorber is relatively free of CO_2. The CO_2-rich solvent from the absorber is then pumped to the top of a stripper (or regeneration vessel), via a heat exchanger. Solvent is regenerated in the stripper at elevated temperatures and near atmospheric pressure. The desorption heat required for removing the absorbed CO_2 is provided to the reboiler section of the stripper. The CO_2-lean solvent, containing far less CO_2 is then pumped back to the absorber via a lean-rich heat exchanger to cool it to the operating temperature of the absorber. We used this process diagram to evaluate new solvents.

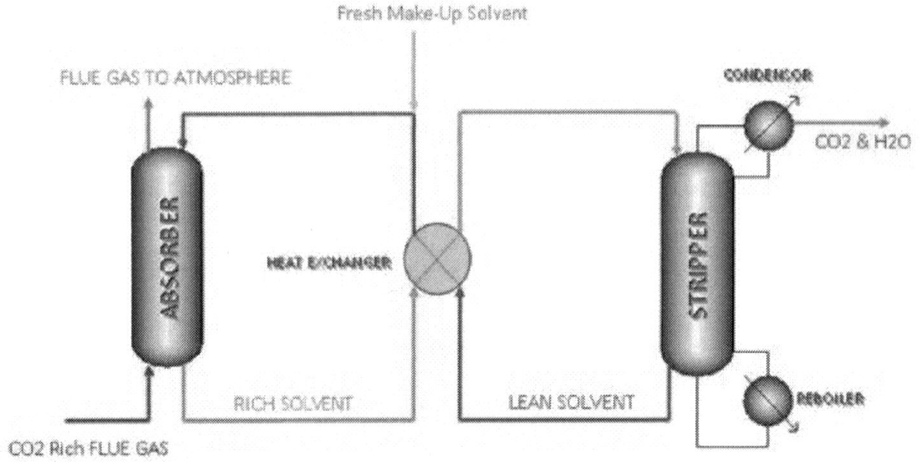

Figure SEQ Figure * ARABIC \s 1 1
Base Absorption Process Flowsheet for CO_2 Capture

In our CAMD approach, we concentrated on amines. In order to use group contribution methods, at first we analyzed the equilibrium performance for these new amines for absorption (CO_2 capture) and for regeneration (solvent stripping). The CAMD resulted in a list that included a group of more than 50 alkyl alcohol amines whose CO_2 solubility properties have not been experimentally determined as far as we know. We evaluated these solvents using ASPEN Plus models and group contribution methods like UNIFAC to evaluate the nonideality. Figure 2 08D0C9EA79F9BACE118C8200AA004BA90B02000000080000000E0000005F005200650066003300320033003000320037003400390030000000 shows the preliminary results of these solvents. The figure shows the energy requirement versus boiling point for 18 solvents simulated both as ideal an nonideal solvents. In general, we observe the nonideality related to these solvents reduces the energy consumption. We will be generating and evaluating additional solvents in order to determine optimum solvents.

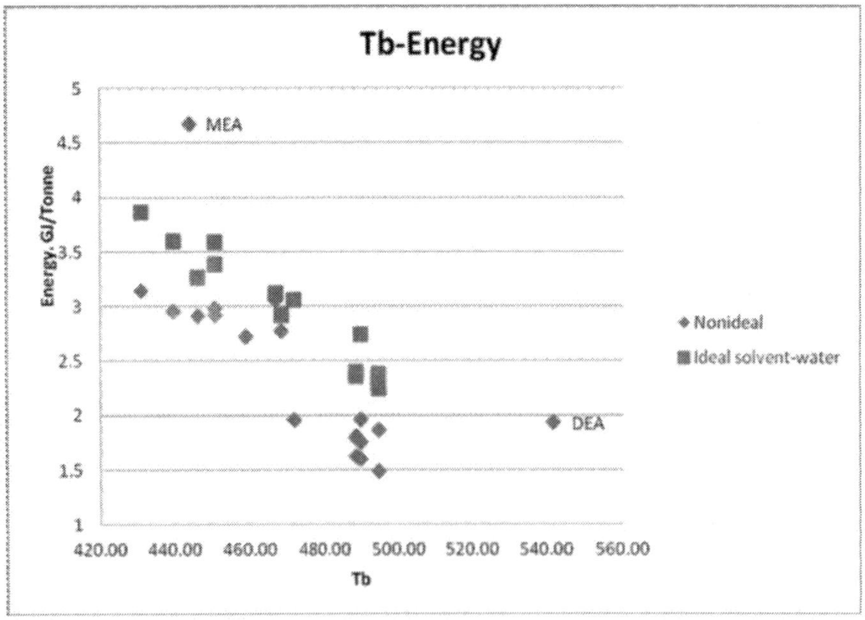

Figure SEQ Figure * ARABIC \s 1 2
Energy Requirements for the New Solvents

(542e) Low Energy Demanding CO2 Capture Process Using Aqueous Amine Solvent

Huiyong Kim, Sogang.ac.kr, Seoul, South Korea and Kwang Soon Lee, Sogang University, Seoul, South Korea

Amine based CO2 capture process has been studied and intensified using the heat integration concept. While many studies on aqueous absorbent-based carbon capture process focus on developing new solvent system, there is much less studies on process intensification to save required energy for CO2 separation.

The purpose of this research has been placed on development of a new process configuration that can achieve a considerable energy saving for CO2 capture. Sensitivity analysis for the absorber and desorber which are two main units of the process provided ideas and possible routes for process modification. With these possibilities, heat flows of the original process were studied for further heat integration to minimize unnecessary heat or coolant usage. With some modifications of the process configuration, heat requirements for solvent regeneration has been reduced considerably. Since minor modifications are proposed, results of this research are belived to be easily applicable to current amine-based carbon capture facilities. The new process has been suggested through a rate-based simulation study using AspenPlus.

(542f) Designing Reversible Ionic Liquids for CO2 Capture

Jackson R. Switzer[1], Amy L. Rohan[1], Kyle M. Flack[2], Emily C. Nixon[2], Amber C. Rumple[2], Elizabeth J. Biddinger[1], Pamela Pollet[2], Charles L. Liotta[2] and Charles A. Eckert[1], (1)School of Chemical and Biomolecular Engineering, Georgia Institute of Technology, Atlanta, GA, (2)School of Chemistry and Biochemistry, Georgia Institute of Technology, Atlanta, GA

Reversible ionic liquids (RevILs) have been developed as alternatives to traditionally cited CO_2 capture solvents such as monoethanolamine (MEA). RevILs capture CO_2 through chemical reaction of CO_2 with an amine functionality to form an ionic liquid. The resulting ionic liquid is then capable of absorbing additional CO_2 through physical absorption. Physical absorption is attractive because it has a relatively low energy penalty for reversal. RevILs also present additional energy advantages over MEA. Whereas MEA is used as a dilute aqueous solution, reversible ionic liquids are designed to be non-aqueous solvents, utilizing only the residual water present in the emission source. This strategy allows for significant energy savings by eliminating the need to heat large amounts of unused water during regeneration, improving the overall efficiency of the CO_2 capture process.

The design of reversible ionic liquids has been an important focal point in our research group. Using structure-property relationships we have shown an ability to tune the relevant CO_2 capture properties by altering the structure of the starting RevIL. Properties targeted include CO_2 capture capacity, ionic liquid viscosity, temperature of CO_2 release, temperature of RevIL evaporation, and the enthalpy of CO_2 absorption. In example, we have observed that by varying the degree and length of alkyl chain branching in the RevIL, we can favorably affect CO_2 capacity, ionic liquid viscosity, as well as the temperatures of CO_2 release and RevIL evaporation. We have also examined several additional modifications to the structure of our RevILs and investigated their effects on CO_2 capture properties. Our progress on developing these structure-property relationships to further the design of optimized solvents for CO_2 capture will be discussed.

(542g) CO2 Post Combustion Capture Using Aminosilicone Solvents

Irina Sultanova[1], Benjamin Wood[2], Sarah E. Genovese[1], Paul Wilson[3], Rachel Farnum[1], Paul Buckley[3] and Anthony Furman[3], (1)Global Research, General Electric, Niskayuna, NY, (2)General Electric, Niskayuna, NY, (3)GE Global Research Center, Niskayuna, NY

CO_2 post combustion capture using aminosilicone solvents

Irina Sultanova, Benjamin R. Wood, Sarah E. Genovese, Paul Wilson, Rachel Farnum, Paul Buckley, Anthony Furman (GE Global Research),

A post-combustion CO_2 capture process using aminosilicone solvents offers advantages over conventional MEA-based aqueous solvent methods. The aminosilicone-based process can potentially decrease the overall size of the system, decrease the amount of required energy for desorption, and therefore reduce the overall cost of CO_2 capture process. A bench-scale system is currently being developed to demonstrate the technical feasibility of continuous CO_2 absorption and desorption with liquid aminosilicones, including recovery and reuse of the solvent. Upon completion of the design and construction, the system will be used to generate the data needed for further scale-up of the process. Some physical properties of the aminosilicones including surface tension, viscosity, heat of CO_2 absorption, and desorption and absorption isotherms were measured and are currently used in design of unit operations. These results and their effect on the cost of CO_2 capture process will be discussed.

(548a) Molecular Simulation of Gas Transport and Entrance Effects in Micro and Mesoporous Carbons

Mahnaz Firouzi[1], Abby Kirchofer[2] and Jennifer Wilcox[1], (1)Department of Energy Resources Engineering, Stanford University, Stanford, CA, (2)Earth, Energy, and Environmental Sciences, Stanford University, Stanford, CA

Molecular dynamics simulations have been carried out to study gas slippage and the Klinkenberg effects of methane and carbon dioxide, in addition to their mixtures, confined in carbon slit micro- and mesopores. We used non-equilibrium molecular dynamics (NEMD) simulations with an external driving force imposed on the system. Simulations were conducted to determine the effect of pore size and exposure to an external potential on the velocity profile and slip-stick boundary conditions.

The results indicate that for small channels the particle-wall collisions may influence the velocity profile, which deviates significantly from the Navier-Stokes hydrodynamic prediction. In small pores, unlike in large pores where continuum flow occurs, the fluid velocity at the walls is nonzero. It is shown that the velocity profile is uniform for pore sizes less than 2 nm (micropores) where the transport is mainly due to molecular streaming and, to a lesser extent, molecular diffusion. As pore sizes increase to 10 nm parabolic velocity profiles are observed due to reduced interaction of gas molecules with carbon atoms of the pore walls. Also, the shape of the velocity profile is found to be independent of the applied pressure gradient in micropores. The simulation results are compared with available experimental Klinkenberg data, in order to understand both the merit and limitations of the simulation approach.

Extensive NEMD simulations have also been carried out to investigate entrance effects of gas diffusion in micropores. The dependence of pore filling and transport of carbon dioxide on the pore size, external pressure gradient, and temperature, have been investigated in carbon slit pores. The results indicate the significance of the pore structure and the fluids' state to their transport through a porous material. The goal of this work is to determine how physical characteristics (e.g., pore diameter) influence mass transfer, and to determine the dominant mass-transfer mechanisms in pores of varying diameter.

These fundamental studies indicate the significance of molecular-scale phenomena, which lead to an understanding of the mechanism associated with methane and carbon dioxide transport in the confined spaces for carbon capture and sequestration applications.

Transport properties of hierarchical porous materials

R. Valiullin, D. Mehlhorn, J. Kärger

Faculty of Physics and Earth Sciences, University of Leipzig, Leipzig, Germany

Ordered multiporous materials have drawn tremendous interest due to their tunable properties as compared to microporous or disorderdered mesoporous ones. Current or potential applications of hierarchical porous hosts include catalysis, drug delivery, protein separation, etc. In many processes involving porous solids, dynamic properties of the confined fluids play an important or even decisive role. As concerning dynamics, hierarchically-organized porous materials serve a few purposes. First of all, embedding mesoporous transport pathways into microporous frameworks, such as zeolites, may help to overcome transport limitations inherent to these materials by preserving, at the same time, their chemical functionality. Secondly, by the use of the confinement size-dependent properties, such as shifts of the phase transition points, an external control of the effective transport may be achieved. In materials with multiple mesoporosities, mass transfer of the confined fluids via molecular self-diffusion may, in most easy way, be tuned by changing the phase equilibria, gas-liquid or liquid-solid, depending on a particular application considered. Predicting transport in such complex systems is not trivial because of different phenomena and transport mechanisms involved. Using pulsed field gradient techniqie of NMR and by using two different hierachically-organized multiporous materials we experimentally demonstrate different regimes of transport and their relationships with the pore space organization and external conditions [1,2]. The results obtained are rationalized using a model considering trajectory time series of individual molecules in pore spaces [3].

[1] Mehlhorn, D., Valiullin, R., Kärger, J., Cho, K. & Ryoo, R. (2012) Intracrystalline Diffusion in Mesoporous Zeolites. *ChemPhysChem* 13, 1495.

[2] Mehlhorn, D., Valiullin, R., Kärger, J., Cho, K. & Ryoo, R. (2012) Exploring Mass Transfer in Mesoporous Zeolites by NMR Diffusometry. *Materials* 5, 699.

[3] Zeigermann, P. *et al.* (2012) Diffusion in Hierarchical Mesoporous Materials: Applicability and Generalization of the Fast-Exchange Diffusion Model. *Langmuir* 28, 3621.

(548c) Mass Transfer Rates of Argon, Oxygen, and Nitrogen in Carbon Molecular Sieve Measured by Pressure-Swing Frequency Response

Timothy J. Giesy and M. Douglas LeVan, Department of Chemical and Biomolecular Engineering, Vanderbilt University, Nashville, TN

The adsorptive separation of air into pure gases is complicated by similarities in isotherms for oxygen and argon on many adsorbents. Molecular sieves separate gases based on differences in effective molecular size, and thus have promise for separating oxygen and argon. In this work, mass transfer of the three main atmospheric gases (N_2, O_2, and Ar), has been studied in a carbon molecular sieve (CMS) material using pressure-swing frequency response. Pure gases were studied at pressures ranging from 0.125-1 bar over the frequency range from $2 \cdot 10^{-5}$ to 0.2 Hz. Measured frequency response spectra are compared with both the linear driving force model (corresponding with a surface barrier resistance) and a micropore diffusion model. Relevant model parameters are reported. The experimental results show that oxygen diffuses through the CMS much more quickly than the other two gases, with argon being the slowest of the three. These results suggest that this CMS material could be used as part of a process designed to generate a high-purity oxygen product from air.

(548d) Diffusion of CO2 in Ion-Exchanged Zeolites Rho Studied by the ZLC Technique

Enzo Mangano[1], Stefano Brandani[1], Magdalena M. Lozinska[2] and Paul A. Wright[2], (1)Institute for Materials and Processes, School of Engineering, University of Edinburgh, Edinburgh, United Kingdom, (2)School of Chemistry, University of St. Andrews, St. Andrews, United Kingdom

Here we present the use of the ZLC technique to investigate the diffusion of CO_2in different Ion-exchanged Zeolites Rho samples. The sample were synthesised at the University of St. Andrews and tested at the University of Edinburgh using the ZLC method as part of the Innovative Gas Separations for Carbon Capture (IGSCC) project.

The ZLC technique, first introduced for diffusion measurements in zeolites, has been continuously developed proving its validity in the investigation of the kinetic and equilibrium properties in solid adsorbents. The very small size of the column offers several key advantages compared to other experimental methods: foremost, the small required sample quantity (1-2 mg for kinetic measurements to 10-15 mg for equilibrium measurements) greatly reduces the synthesis costs of the materials and allows rapid testing of prototype materials. The small size of the column reduces the complexity of the analysis methods because heat and external mass transfer resistances are negligible and axial dispersion is maximised achieving a uniform external concentration. The dynamic response of the ZLC can be interpreted easily.

In this study we have extended the use of the ZLC to very slow diffusional time constants by reducing the gas flowrates to less than 3 cc/min. When very slow kinetics are measured it is very difficult to extract the diffusivities from the traditional long time asymptotic analysis. We will show how the combination of the full saturation and partial loading experiment can provide un-ambiguous diffusional time constants. The results for the diffusivity of CO_2in zeolite Rho samples will be shown to be strongly influenced by the framework structure as well as the nature and the position of the different cations in the framework.

We would like to acknowledge the financial support from the Engineering and Physical Sciences Research Council under grant EP/F034520/1 (Innovative Gas Separations for Carbon Capture).

(548e) Effective Medium Theory for the Transport of Fluid Mixtures Through Porous Media

Suresh K. Bhatia, School of Chemical Engineering, The University of Queensland, Brisbane, Australia and Mauricio Rincon Bonilla, The University of Queensland, Brisbane, Australia

The transport of fluids in the narrow confined pore spaces disordered materials with complex pore topologies is central to many emerging technologies for gas separation and storage, as well as in nanofluidics. Optimizing the performance of such technologies requires a deep understanding of how the flow of these mixtures is affected by the topology of the pore space, particularly its pore size distribution and pore connectivity. While there exist accurate techniques such as the hybrid effective medium-correlated random walk theory for the estimation of the effective diffusivity characterizing the transport of single fluids through porous materials, equivalent approaches for the case of fluid mixtures are yet to be developed. Existing alternatives for multi-component systems rely on computationally intensive solution of the pore network equations, or *ad hoc* adaptations of the single fluid theories which are useful for a limited variety of systems. We present here a hybrid Effective Medium-Correlated Random Walk Theory for the calculation of the matrix of effective transport coefficients of fluid mixtures diffusing through porous materials. The theory is suitable for those systems in which component fluxes at the single pore level can be related to the chemical potential gradients of the different species through linear flux laws, and corresponds to a generalization of the classical single fluid effective medium theory for the analysis of random resistor networks. Comparison with simulation of the diffusion of binary CO_2/H_2S and ternary $CO_2/H_2S/C_3H_8$ gas mixtures in membranes modeled as large networks of randomly oriented pores with both continuous and discrete pore size distributions demonstrates the power of the theory, which was tested using the well known generalized Maxwell-Stefan model for surface diffusion at the single pore level.

(548f) Adsorption and Diffusion of Water-Acetonitrile Mixtures At a Bare Silica Surface: A Molecular Dynamics Study

Sergey M. Melnikov[1], Alexandra Hoeltzel[2], Andreas Seidel-Morgenstern[1,3] and Ulrich Tallarek[2], (1)Physical and Chemical Foundations of Process Engineering, Max-Planck Institute for Dynamics of Complex Technical System, Magdeburg, Germany, (2)Dept. of Chemistry, Philipps-University, Marburg, Germany, (3) Physical and Chemical Foundations of Process Engineering, Max-Planck Institute for Dynamics of Complex Technical System, Magdeburg, Germany

A bare silica surface is the most popular stationary phase for separation of polar analytes in hydrophilic liquid chromatography (HILIC). Knowledge of adsorption behaviour and dynamic features of water-acetonitrile mixtures (the most common choice of the mobile phase in HILIC) at the surface is a key prerequisite for entire understanding of separation mechanism in HILIC. In practice adsorption is often characterized by excess adsorption isotherms. However, this information is essentially an integral over the whole chromatographic support, and it requires further interpretations to get an actual spatial distribution of the solvent mixture in the adsorpted phase. In this work we employ molecular dynamics (MD) simulation to make a link between experimentally obtainable (macro) characteristics of adsorption and a picture at the micro-level (which is hardly can be measured directly in experiment).

We have performed MD simulation of the water-acetonitrile mixture (for whole range of compositions) at the bare flat silica surface. There have been studied three types of idealized surfaces: a surface covered only by single (isolated) silanol groups, a surface covered only by geminal (partially bridged) silanol groups and the last one with only siloxane bridges. Iterative simulation runs have been preliminary performed to adjust necessary amount of water and acetonitrile molecules in that adsorption/desorption system. Our results demonstrate that at mixture compositions with the low water fraction water shows the positive surface excess for all three considered types of silica surface. Acetonitrile appears to have the positive surface excess mixtures with the high water content only at the surface with siloxane bridges. In our presentation we will also discuss how these three studied types of silica surfaces, and especially their combinations, affect diffusion of the solvent molecules at the surface.

(548g) Molecular Modeling of Protein Transport in Polymer-Grafted Ion Exchangers

Joseph Basconi, Giorgio Carta and Michael R. Shirts, Department of Chemical Engineering, University of Virginia, Charlottesville, VA

Ion exchange matrices containing charged grafted polymers have been shown to provide both increased protein binding capacity and faster protein adsorption kinetics as compared to traditional macroporous matrices. Experimental evidence of such improved characteristics has been obtained from macroscopic measurement of the rate of protein adsorption, and from intraparticle protein concentration profiles observed by confocal microscopy. Both of these measurements demonstrate that not only the rate, but also the mechanism of protein transport in these materials, is drastically different from ordinary pore-diffusion, exhibiting a different dependence on protein molecular size and concentration, salt concentration, and buffer type. However, the reasons behind these observations are not fully understood. Additional investigation is necessary to study the molecular mechanisms by which adsorption is facilitated in these systems.

In this work, we use molecular dynamics simulations to model a surface coated with a charged grafted polymer, as well as the interactions between this matrix and model proteins. We use coarse-grained representations of the polymer chains and proteins, along with an implicit solvent environment, as we are primarily interested in dynamical processes occurring over long length scales with respect to individual atoms. The polymer matrix is composed of chains representing dextran functionalized with sulfate groups to different extents (from 5% to 20% of the chain), and grafted to the surface at various densities (from 25 to 81 nm^2 per chain). We simulate the matrix in the absence of proteins, as well as with different concentrations of model lysozyme proteins interacting with the matrix.

Simulations of the grafted polymers alone, under a variety of experimental conditions, were used to validate the behavior of the model. The effective height of the grafted layer increases with increased graft density and the charge content of the dextran chains. We also explore the effects of different ionic strengths, represented by tuning the strength of electrostatics by the use of a reaction-field with separate short range and long range dielectric constants.

We estimate how the protein diffusivity varies with concentration, and investigate whether there is a critical concentration at which higher loadings significantly diminish the transport rate. We vary the effective ionic strength by tuning the dielectric constants, and vary the pH by tuning the surface charge on the protein, to determine how these factors affect protein diffusion through the matrix. Finally, we analyze the radial distribution functions to better understand the mechanism by which proteins interact with the chains. Molecular insights from this work are compared qualitatively with experimental measurements, and should aid in the optimization and design of future ion-exchange materials that utilize charged polymer grafts to facilitate protein adsorption and transport.

(548h) On the Influence of Pore Size, Pore Loading and Surface Charge Density On the Structure and Dynamics of the Ionic Liquid [EMIM+][TFMSI-] Inside a Slit Graphitic Nanopore

Nav Nidhi Rajput[1], Joshua D. Monk[2], Ramesh Singh[2] and Francisco R. Hung[2], (1)Cain Department of Chemical Enginering, Louisiana State University, Baton Rouge, LA, (2)Cain Department of Chemical Engineering, Louisiana State University, Baton Rouge, LA

Ionic liquids (ILs) are gaining relevance as alternative electrolytes in energy-related devices such as electrochemical double-layer capacitors (EDLCs) and dye-sensitized solar cells (DSSCs). A fundamental understanding of the structure and dynamics of ILs inside nanopores is essential to advance these technologies beyond their current state of the art. Structural properties of an IL confined inside a nanoporous electrode affect macroscopic properties such as the capacitance in an EDLC; and the dynamical properties of the confined IL is one of the factors that determines the macroscopic electrical resistance in electrochemical devices.

Molecular dynamics (MD) simulations were performed to study the IL 1-ethyl-3-methylimidazolium bis (trifluoromethylsulfonyl) imide [$EMIM^+$][$TFMSI^-$] confined inside a slit graphitic nanopore. Our goal is to understand the effect of pore size, pore loading and surface charge density on the structure and dynamics of the confined IL. As previous studies have largely concentrated on the structure and electric properties of ILs confined inside charged electrodes, we have especially focused on the dynamics of the system, which were studied in detail by measuring mean squared displacements (MSDs), van Hove self correlation functions (VHSCFs) and self-intermediate scattering function (SISFs).

Our results indicate that structural properties such as the local density profiles, the local orientation, and the radial distribution functions of the cations and anions vary strongly with pore size, pore loading and charge density in the pore walls. These variables also affect the dynamical properties of the confined IL. Our results suggest that the dynamics of the ions inside the slit pore are highly heterogeneous and depend strongly on their position with respect to the pore walls, as well as on the surface charge density of the walls. For example, our results show that the layer of cations that is closer to a negatively charged wall, as well as the layer of anions closer to a positively charged wall, exhibit extremely slow dynamics, long relaxation times and strong deviations from Gaussian dynamics. In contrast, the ions in the center of the pore have mobilities in the x-y plane that are comparable and sometimes even larger than those exhibited by the ions in a bulk IL.

(548i) Dynamics of Amine-Oxide Hybrid Adsorbents for Post-Combustion CO2 Capture

Praveen Bollini and Christopher W. Jones, School of Chemical & Biomolecular Engineering, Georgia Institute of Technology, Atlanta, GA

Amine-oxide adsorbents have emerged as promising candidates for post-combustion CO_2 capture. Although several amine adsorbents with exceptionally high CO_2 adsorption capacities and selectivities have been synthesized, dynamic properties of these materials are not well understood. The limited studies reported in the literature use isothermal models to interpret kinetic data. Given the high heats of CO_2-amine adsorption, this assumption needs to be tested rigorously. In this work, we carry out a detailed investigation into the heat and mass transfer dynamics of amine adsorbents in a packed bed operating in breakthrough mode to better understand the impact of heat evolution on CO_2 adsorption kinetics. Mass and heat transfer models are used to interpret concentration and temperature front measurements obtained from packed bed experiments. Both heat and mass transfer dynamics are modeled successfully. We found that amine loading has a significant effect on CO_2 diffusion properties, which in turn influences the breakthrough performance of these materials. The kinetic properties of these amine adsorbents provide insights that may help guide the design of optimal adsorbents for practical applications.

(555a) Self-Adaptive Operation of Ultrafiltration (UF)/Reverse Osmosis (RO) System for Desalination of Brackish Water and Coastal Seawater

Han Gu[1], Larry Gao[1], Anditya Rahardianto[1], Panagiotis D. Christofides[2] and Yoram Cohen[3], (1)Chemical & Biomolecular Engineering, University of California, Los Angeles, Los Angeles, CA, (2)Department of Chemical and Biomolecular Engineering, University of California, Los Angeles, Los Angeles, CA, (3)Chemical and Biomolecular Engineering, University of California, Los Angeles, Los Angeles, CA

The effectiveness of feed water pretreatment in reverse osmosis (RO) membrane desalination is essential to the operability of RO plants. Feed pre-filtration and chemical treatments (e.g., additives to suppress mineral scaling and reduce biofouling) are key methods to mitigate RO membrane fouling. In recent years, microfiltration (MF) and ultrafiltration (UF) systems have been increasingly explored as a method for effective RO feed pre-filtration. While the concept of MF/UF integration with RO has achieved a reasonable level of success, there are technical challenges in situations where feed water quality fluctuates (with respect to time), and where it is desired to reduce the use of chemical feed additives and membrane cleaning chemicals (for the purpose of both reducing cost and minimizing environmental impact). In meeting the above challenges, a novel concept of self-adaptive integrated UF-RO operation was developed and field tested. The self-adaptive UF operation integrates a number of novel model-based control and optimization methods and RO/UF design principles that were implemented in a unique "smart" (i.e., autonomous) MF-UF-RO system capable of processing up to about 45,000 gallons per day of raw feed water to a desalination module. This integrated UF/RO system is robust and capable of self-adapting (autonomously) in real-time to fast changing feed water quality (with respect to natural organic matter concentration, salinity, temperature and RO fouling propensity) by selecting the appropriate UF filtration and backwash strategies. Real-time individual UF module resistance data are utilized in the control algorithm to determine the appropriate inline coagulant dose, backwash frequency and intensity for the given feed water conditions. Another unique feature of the UF backwash strategy employed in the current study is the utilization of the RO concentrate for backwashing, which reduces substantially system foot print and energy usage. This system also allows real-time characterization of the fouling mitigation and contaminant removal performance of the UF membranes. Results from extensive field studies for treatment of cooling tower blow down (CTBD) water for reuse and coastal seawater successfully demonstrate the benefits of self-adaptive UF backwashing. The potential use of the present integrated MF-UF-RO approach to distributed deployment of smart water filtration and desalination systems will be discussed with specific focus on operability for ensuring fouling free operation.

(555c) Model Predictive Control Strategy for Optimizing Biological Nitrogen Removal (BNR) Processes Accounting for Greenhouse Gas Emissions

Babji Srinivasan, Department Chemical Engineering, Columbia University, New York, NY, Kartik Chandran, Earth and Environmental Engineering, Columbia University and Venkat Venkatasubramanian, Chemical Engineering, Columbia University, New York, NY

Biological Nitrogen Removal (BNR) is a widely practiced and increasingly mandated process for treating wastewater. BNR involves the sequential oxidation of ammonia to nitrate (termed nitrification) and subsequent reduction of nitrate to nitrogen gas (denitrification). Nitrification and denitrification are typically promoted under aerobic and anoxic conditions, respectively. However, recent studies have shown that subjecting nitrifying and denitrifying bacteria to alternating aerobic and anoxic conditions- although useful for BNR, can lead to the release of copious amounts of nitrous oxide (N_2O). N_2O is a greenhouse gas, 300 times as potent as CO_2 and is also the most powerful ozone depleting substance emitted during the 21stcentury. Therefore, optimization, balancing and control of BNR processes is essential to: (i) improve the effluent quality, (ii) reduce the energy consumption, (iii) increase the amount of wastewater processed, and (iv) reduce the greenhouse gas emissions from wastewater plants.

In this work, a model predictive control strategy is utilized to control and optimize the BNR process. A hybrid model of the BNR process is developed using the detailed studies conducted by Yu et al. on *Nitrosomonas europaea* (a widely studied nitrifying organism) under aerobic and anoxic conditions. The hybrid dynamic model is used to predict the effluent ammonia, nitrite, nitrous oxide and nitric oxide concentrations and fluxes produced in the BNR process for the given input conditions (namely, two substrates, influent ammonia and oxygen loadings). This model is used in a MPC framework to achieve the aforementioned objectives by manipulating the oxygen supply, rate of recycles in the system, and scheduling of aerobic and anoxic conditions in a modified BNR process.

(555d) Improved Emulsion Breaking for Treatment of Bilge Water by Induced-Air Flotation

Alexander Castillo[1], Ailenn D. Nieva[2] and **Alvin R. Caparanga**[2], (1)School of Graduate Studies, Mapua Institute of Technology, Manila, Philippines, (2)School of Chemical Engineering and Chemistry, Mapua Institute of Technology, Manila, Philippines

Disposal of bilge water from ship engine room has been one of the serious environmental issues the maritime industry today is concerned about. Emulsified oil-in-water poses problem in bilge water treatment because of the high stability of the emulsions causing difficulties in separation by conventional gravitational processes. Hence, an alternative technique to pre-treat emulsified bilge water from ship engine room was developed: induced-air flotation with poly(aluminium chloride) as destabilizing agent. The removal of oil and grease and turbidity from synthetic bilge water samples by this technique was carried out with test fluid prepared by mixing industrial diesel oil, heavy fuel oil, surfactant, iron oxide and distilled water. The pH of the test fluid was adjusted to 9, the typical pH of bilge water, to stabilize the emulsion. Experiments were carried out in an induced-air flotation system to evaluate the individual and combined effects of PAC dosage, air flow rate, and settling time in removal efficiencies for oil and grease and supernatant turbidity. It was observed that the PAC concentration, compared to air flow rate and settling time, has a greater influence on the removal efficiency. Although different treatment conditions were used, the results showed excellent removal efficiencies (close to 99%) attained at 0.3 ml/L PAC dosage, 9 L/min air flow rate and 40 min settling time. Using this set of conditions, the treated water had a residual oil and grease concentration that was less than 10 ppm, a value which satisfies current maritime regulations. The PAC could not be too high, otherwise, the water tend to become cloudy, resulting in poor removal efficiency. It is believed that this kind of improved emulsion breaking appears to have a great potential as pre-treatment for treating emulsified bilge water on board.

(555e) Bimetallic Nanoparticles for Water Purification

Lauren F. Greenlee, Materials Reliability Division, National Institute of Standards and Technology, Boulder, CO and Nikki Goldstein, Materials Reliability Division, National Institute of Standards & Technology, Boulder, CO

Bimetallic nanoparticles are of interest for water purification due to their versatility in contaminant adsorption and degradation. Studies have now demonstrated that metallic nanoparticles can be used to transform and remove a variety of contaminant types, including heavy metals, organic chemicals, pharmaceutical compounds, and pesticides. Many of these studies now use wet chemistry, either with an organic solvent or with water, to synthesize these nanoparticles, where an organic stabilizer is often used during synthesis to control nanoparticle size and prevent agglomeration. However, little has been done to understand why certain stabilizers work better than others with certain metals to produce specific particle sizes and controlled particle size distributions. Furthermore, there has been no systematic study of specific stabilizer properties that might affect nanoparticle synthesis and reactivity. These properties, such as stabilizer size, chelation ability, and the presence of specific functional groups, are likely to affect nanoparticle growth and agglomeration; the surface properties of the nanoparticles themselves (e.g., surface charge and hydrophilicity) will affect the stabilizer-metal association as the nanoparticles are synthesized. In this study, we begin to understand how specific properties of organic stabilizers affect metallic nanoparticle synthesis and reactivity; the properties of molecular weight and chelation strength will be discussed for several stabilizer molecules.

(555f) Integrated Electrokinetic and Microbial Fuel Cell Technologies for Enhanced Transport and Bioremediation of Hexavalent Chromium in Groundwater

Ryan Thacher, Lewis Hsu and Massoud Pirbazari, Sonny Astani Department of Civil and Environmental Engineering, University of Southern California, Los Angeles, CA

Groundwater contamination threatens potable water supplies around the world, and this fact has encouraged the progression of novel research for groundwater treatment. Hexavalent chromium (Cr^{VI}) is an EPA priority contaminant due to its high toxicity, and its prevalence in groundwater around the world as a result of improper disposal practices from a variety of industries. Cr^{VI} is highly mobile in water systems due to its solubility, making treatment by conventional methods difficult and costly.

Electrokinetic remediation is a technique that can be used for transporting ionic contaminants, including Cr^{VI} (as $HCrO_4^-$) by the application of an electric potential across a contaminated aquifer. Electrokinetic remediation is often limited by a pH change due to electrolysis reactions. These reactions create OH^- at the cathode and H^+ at the anode, which migrate towards the opposite electrodes by electrokinetic phenomena. At the point of intersection of these fronts a significant drop in electrical conductivity occurs, causing transport to slow or stop completely. Additionally, the dynamic pH change affects precipitation and dissolution reactions, which govern the solution chemistry and availability of the contaminant for transport. Many enhancement techniques have been investigated to mitigate this problem, however they can add significant economic and environmental cost.

This study investigates the potential to enhance electrokinetic transport of Cr^{VI} in groundwater while also promoting its reduction to Cr^{III} by integrating an electrokinetic system with microbial fuel cell (MFC) technology. Recent studies have shown MFCs to be effective in Cr^{VI} reduction, as Cr^{VI} can act as an electron acceptor when introduced to the MFC cathode compartment. This reaction is catalyzed by the presence of biofilm on the cathode, which facilitates a complex electron transport system from the cathode to Cr^{VI} in solution. In this context, *Shewanella putrefaciens* MR-1 bacteria have been documented to be effective biocatalysts for this application, and different carbon sources including lactate, acetate, formate, and pyruvate have demonstrated to be effective electron donors in the anode compartment.

In this study, the integration of these technologies is evaluated using a soil column under an applied electric potential, and an MFC with anode and cathode reservoirs with a 300 mL capacity. The electric potential applied across the soil column is varied in different experiments from 0.5 – 2.0 V/cm in an effort to optimize transport by counteracting the advective pull of Cr^{VI} towards the cathode by water flow. The soil column is operated with a continuous flow of simulated groundwater spiked with Cr^{VI} at a rate comparable to that of groundwater in sandy soils. The flow direction is from the anode towards the cathode, while Cr^{VI} is transported by electrokinetic phenomena in the opposite direction. During operation, the soil column effluent adjacent to the anode reservoir is pumped out into the cathode of the MFC. Recirculating solution to and from this location in the column will mitigate the dramatic drop in pH, which occurs in the absence of enhancement strategies from H^+ generation at the anode. Effluent from the cathode reservoir of the soil column is sampled and analyzed for Cr^{VI} by ion-exchange chromatography to determine the extent of treatment.

The optimization of the electrokinetic and MFC integration will be discussed in this presentation. Key points of discussion will be Cr^{VI} reduction rates and mechanisms at the biocathode, cathode effluent Cr^{VI} concentrations, effects of solution circulation to enhance Cr^{VI} transport, and energy production by the MFC.

(555g) Scaling Microbial Fuel Cells for the Treatment of Municipal Sludge

Orianna Bretschger and Greg Wanger, Microbial and Environmental Genomics, J. Craig Venter Institute, San Diego, CA

Wastewater treatment is an energy intensive process that requires technology improvements to reduce treatment costs, and ultimately, water rates. A 2005 report issued by the California Energy Commission (CEC) entitled "California's Water-Energy Relationship" elucidates the energy costs of wastewater treatment and emphasizes the need for more efficient, less energy intensive, treatment strategies. The State's wastewater processes in 2001 (including delivery, treatment, and disposal) resulted in the consumption of over 2000 GWh of electricity and 2.7 trillion BTUs of natural gas. As the state faces growing populations, changing environmental regulations, and limited water resources in Southern California, energy consumption for wastewater treatment is expected to rise along with energy costs.

Decreasing energy consumption and promoting energy conservation during wastewater treatment can be accomplished through several strategies including: 1) implementing energy efficient equipment and practices; 2) recovery of energy during treatment processes; and 3) optimizing treatment methods to minimize overall disposal costs of wastewater effluents and biosolids. The evaluated microbial fuel cell system offers a way to address all three strategies for achieving energy conservation with one technology. When optimized, microbial fuel cell systems can decrease overall energy consumption and wastewater treatment costs, and provide direct benefit to California rate-payers.

Here we present data from two different microbial fuel cell reactors scaled to either treat 38 or 54 liters of primary sludge per day.

Project Outcomes

1) Material costs were decreased from \$450,000/m^3 to \$40,000/m
designs and materials.

2) The evaluated microbial fuel cell system achieved a 70% removal efficiency of total suspended solids and volatile suspended solids, and an 88% removal efficiency of biological oxygen demand in a 7 day period. Degradation rates were found to increase when the microbial fuel cell was operated under a higher current draw (lower applied external resistance).

3) Odor reduction was observed over the course of treatment but not quantified. Additionally, methanogenesis was not quantified.

4) The microbial fuel cell achieved a maximum sludge reduction rate of 138 mg L^{-1} day ^{1} and an observed maximum power density of 1 W/m^3.

5) The results indicate that microbial fuel cells are not yet cost competitive with other alternative technologies.

6) Microbial fuel cell technology may ultimately provide a great benefit to California rate-payers. However, further technology development and evaluation is necessary to quantify what the cost-benefits will be.

Conclusions

The results of this project indicate that microbial fuel cell systems can accelerate degradation rates relative to conventional treatment methods, and potentially offer an energy neutral solution to sludge degradation. However, the technology must improve in terms of power output and capital cost reductions to replace conventional technologies in municipal wastewater treatment applications.

(559a) A New Method for Analyzing and Designing Filter Press Separations

Raymond M. Collins, Engineering and Process Sciences, Dow Chemical Co., Midland, MI

A New Method for Analyzing and Designing Filter Press Separations

Raymond Collins, Engineering and Process Sciences, The Dow Chemical Co., Midland, MI

A non-traditional model describing the separation of particulate solids from a liquid suspension by pressure-driven flow through a permeable barrier has been successfully applied to the design of several filter press applications. This flow model was derived by applying fundamental physical principles to the unsteady, mixed-phase system represented by cake filtration with a filter cloth. The analysis has proven suitable for operating modes employing both constant applied pressure and constant feed flow rate. The design procedure uses small-scale experiments to determine a governing permeability function, which has a reproducible, characteristic dependence on the collected volume of filtrate with parametric dependence on the properties of the solids, liquid and filter cloth. The experimental method and various lab-scale test filters will be described, and the utility of this method will be demonstrated for a few common, large-scale applications in chemical processing. These include the removal of polymer fines from wastewater, the deliquoring of biomass from a concentrated sediment of fermentation broth, and the isolation and washing of micron-sized solids used to manufacture energy materials.

(559b) Deposits Formed From the Evaporation of Sessile Droplets

Pei-Fang Sung, Chemical Engineering, Purdue Universtiy, West Lafayette, IN and **Michael T. Harris**, School of Chemical Engineering, Purdue University, West Lafayette, IN

The deposition of particles onto a surface from an evaporating sessile droplet was investigated using the Galerkin finite element method to solve the equations governing fluid motion and particle transport. In the computer code that was developed to simulate liquid, vapor and particle phases simultaneously, a convective-diffusive equation along the drop/substrate interface was successfully added into the simulation for more accurate particle deposition profile calculations. An edge deposition pattern was observed in an experiment with water droplets sitting on a glass substrate suspended with silica particles. The analysis of the interaction potential between silica particles and a glass substrate shows that the deposition rate constant is very small for such case. A solid phase is considered in this case when the particle concentration reaches the maximum-packing density and the movement of the solid/fluid interface was captured. Simulation results show that under this case, the simulated and experimental edge patterns are in good agreement.

(559c) Design and Optimization of Sea Salt Particles Filter for Naval Fuel Cell Using Microfibrous Materials (MFM)

Pengfei Zhao[1], Yanli Chen[1], Peng Cheng[1] and Bruce J. Tatarchuk[2], (1)Chemical Engineering, Auburn University, Auburn, AL, (2)CM3, Dept. of Ch. E. , Auburn University, Auburn, AL

Presence of sea salt particles (SSP) in the ocean atmosphere could cause severe degradation in fuel cell performance on Naval Ships. Therefore, the removal of these particles from the air stream fed to the cathode side of the fuel cell is essential. Current approach is using coalescors and inertial separators to remove the particles. But these devices are only effective for particles larger than 8 microns in size. To remove much smaller particles in a more efficient manner, we focused on the design and optimization of a new type of salt particles filter. The filter media in this case is a composite material known as Microfibrous Materials (MFM), which was manufactured by us in facility and has shown great applicability in removing particulate matters. However, MFM has substantially higher flow resistance than traditional filtration media due to decreased fiber diameter and tighter packing density. Thus, MFM is limited to be applied as media for air filters. Multi-Element Structured Arrays (MESA's) that developed by our group represent a new approach that can increase the filtration area and reduce the pressure drop across the filter by integrating several filters into a single unit. In this study, both single filter and MESA's were tested on an 8''x 8'' test rig for the filtration performance, including particle removal efficiency and pressure drop across the system. After comparing the results, MESA's could reduce the initial pressure drop by almost 70% and provide higher removal efficiency. The effects of varying fiber type, fiber diameter, media thickness, MESA's counts were also experimentally investigated. Additionally, we entrapped activated carbon particles (<200 microns) with different diameter sizes and quantities into Microfibrous Materials (MFM); then filtration performances were evaluated and compared. By modifying previous semi-empirical model and changing different parameters, optimization of filter design to get the highest particle removal efficiency with lowest energy consumption could be achieved.

(559d) An Immersed Method to Predict the Pressure Drop Over a Dynamic Bed of Particles

Jeremiah N. Mpagazehe and C. Fred Higgs III, Mechanical Engineering, Carnegie Mellon University, Pittsburgh, PA

The power input needed to perform particle-fluid separations is critical to many industries such as water treatment and energy production. This power input is dependent upon the pressure drop over the particle bed as the fluid moves through it. Traditionally, the Ergun equation, and variations of it, have been used to predict this pressure drop given a variety of parameters such as the particle shape, packing characteristics, and the fluid viscosity. Recently, efforts have been made to elucidate some of the important physics which dominate this process though numerical modeling using computational fluid dynamics (CFD). Though some success has been obtained with this method, past implementations of it did not allow for the easy modeling of a dynamic bed of particles as the filter cake is being formed. In this work, an immersed method is explored in which the particulates are dynamically coupled to the CFD mesh. The model is used to predict the pressure drop over the particle bed and the results are compared to experiments. Moreover, transient analysis of the flow characteristics is performed as the particles settle into a filter cake.

GEOMETRIC OPTIMIZATION OF CYCLONES FOR COMBINATION OF NONLINEAR MATHEMATICAL PROGRAMMING AND COMPUTATIONAL FLUID DYNAMICS TECHNIQUES (CFD)

Oscar L. Sgrott Jr.[*1]**, Karina K. da Costa**[1]**, Dirceu Noriler**[1]**,**
Waldir P. Martignoni[2]**, Henry F. Meier**[1]

[1] Regional University of Blumenau - Blumenau, Santa Catarina, Brazil
[2] PETROBRAS/AB-RE/TR/OT - Rio de Janeiro, Rio de Janeiro, Brazil
[*]e-mail: oscar.sgrott@gmail.com

Keywords: Cyclone, Gas-Solid Flow, CFD, Optimization.

ABSTRACT - The objective of this work is to propose a new methodology to optimize the geometrical characteristics of the cyclone separator to operate with high collection efficiency and low pressure drop. The proposed method basically consists in coupling the CFD code *CYCLO-HEXA*, "in-house" simulator of gas-solid flows in cyclones developed by Meier *et al.* (2011), in the optimization algorithm *COMPLEX* (BOX, 1965), with nonlinear objective functions and inequality constraints. From the results of CFD simulations, it is observed that the cyclones generated along the iterations converge to a common geometry, due to the restrictions set forth in the objective function, and this geometry gives the best results relative to maximize the collection efficiency with lower pressure drop.

1. INTRODUCTION

The cyclones are widely used on separating particulate from a gas stream due to its wide range of operation and simplicity in constructive form, which it makes a device with a low investment cost and maintenance. However, it should be noted that the collection efficiency of cyclones is directly related to operating conditions and geometrical characteristics. Despite this simplicity, the fluid dynamics of turbulent flow in cyclones is very complex, with various phenomena including recirculation zones, high-intensity turbulent, high conservation of vorticity, among others. As much as the principles that govern the operation of cyclones have not changed since its inception in the late nineteenth century, more than a century of interventions through research and industrial applications, resulted in projects of cyclones with significant improvements, as illustrated in Figure 1.

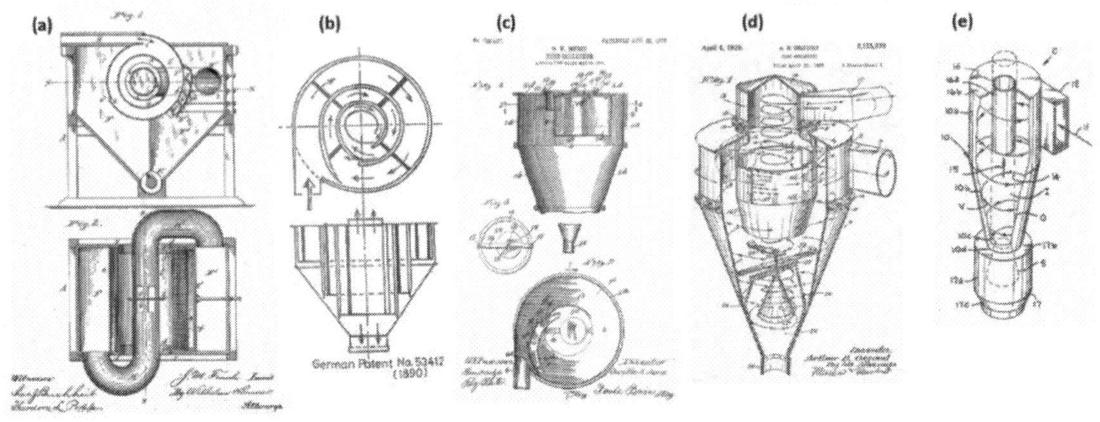

Figure 1. *Geometric evolution of cyclonic separators over more than one century. (a) Cyclonic separator patented by J. M. Finch in 1885; (b) German patent of 1890; (c) Cyclone patented by O. M. Morse in 1905; (d) Cyclone patented by A. B. Osgood in 1939; (e) Representation of a modern cyclone. (Source: Adapted from OGAWA, 1997 and HOFFMANN & STEIN, 2008).*

Before the 1960's, the optimization methods and equipment designs were empirical, based on experiments, intuition or semi-empirical calculations, which were used in the similarity laws and testing models. These empirical methods and semi-empirical, due to its simplicity, are still widely used in the design and evaluation of various equipment used in the industry, for example, cyclones. The objective of this work is develop a methodology for optimizing the geometrical characteristics of the cyclone separators, to operate with high collection efficiency and low pressure drop through CFD techniques coupled with optimization methods.

2. METHODOLOGY

The proposed methodology basically consists in coupling of the simulator of gas-solid flow in cyclones *CYCLO-HEXA* code, in the optimization algorithm *COMPLEX*. The *CYCLO-HEXA* code is the result of improvement of earlier versions developed by Meier (1998) in his Ph.D. work on the development of a general model for the simulation of cyclones. The good results obtained with the Eulerian-Eulerian model (MEIER and MORI, 1998; VEGINI *et al.*, 2008, MEIER *et al.*, 2011) were crucial for this approach was adopted as the standard in the *CYCLO-HEXA* code. The solid phase is treated like a hypothetical fluid and drag forces between phases are responsible for the gas-solid interaction. The *CYCLO-HEXA* code makes possible the use of up to five solid phases, each one with size of particle, density and specific volumetric fraction. The finite volume methods have been used to discretize the partial differential equations of the model using the SIMPLEC (SIMPLE Consistent) method for pressure-velocity coupling in a 3-D-space domain with a 3-D symmetric cyclone inlet. The modeling of turbulence is formulated by combining the k-ε model for radial and axial components of the Reynolds stress, and the mixture Prandtl theory for the tangential components.

The optimization algorithm *COMPLEX* used in this work, consists of maximizing an objective function subject to inequality constraints (Figure 2-a), that for this case, means the maximization of the collection efficiency establishing a maximum limit for the corresponding pressure drop. As variables, we chose five of the eight basic geometric characteristics of a cyclone (Figure 2-b), where each of the five variables has a restriction for upper and lower limits.

(a)

$$\begin{cases} \textit{Maximize} \quad F_{\text{Objective}} = \eta, \ \textit{if} \ \Delta P \le \Delta P_{\max} \\ \qquad\qquad\quad F_{\text{Objective}} = \dfrac{\eta}{\Delta P}, \ \textit{if} \ \Delta P > \Delta P_{\max} \\ \text{Subject to the restrictions,} \\ \qquad \begin{cases} (Ds)_{Lower} \le Ds \le (Ds)_{Upper} \\ (Dl)_{Lower} \le Dl \le (Dl)_{Upper} \\ (Ls)_{Lower} \le Ls \le (Ls)_{Upper} \\ (Lc)_{Lower} \le Lc \le (Lc)_{Upper} \\ (Lco)_{Lower} \le Lco \le (Lco)_{Upper} \end{cases} \end{cases}$$

(b)

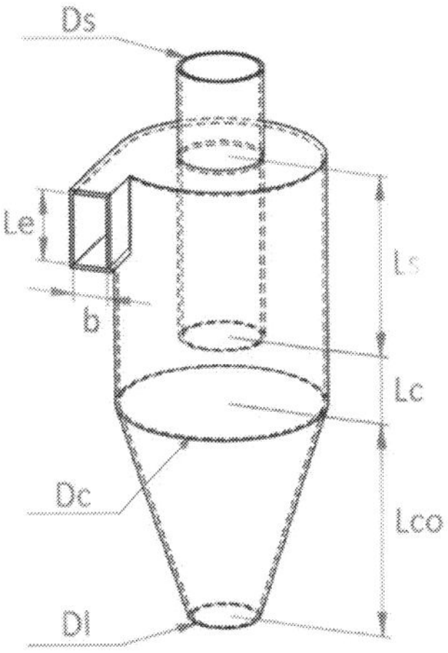

***Figure 2.** (a) Objective functions subject to inequality constraints, where η is the collection efficiency and ΔP is the pressure drop; (b) Geometrical characterization of a cyclone.*

The optimization method consists in construct a geometric figure with a number of dimensions at least equal to the number of manipulated variables more one. In this particular case, we have five variables, and from there was chosen a ten-dimensional optimization algorithm *COMPLEX* (twice the number of variables). The generation of these ten points is conducted randomly by a random number generator, where each point represents one different geometric configuration of a cyclone according to Figure 3.

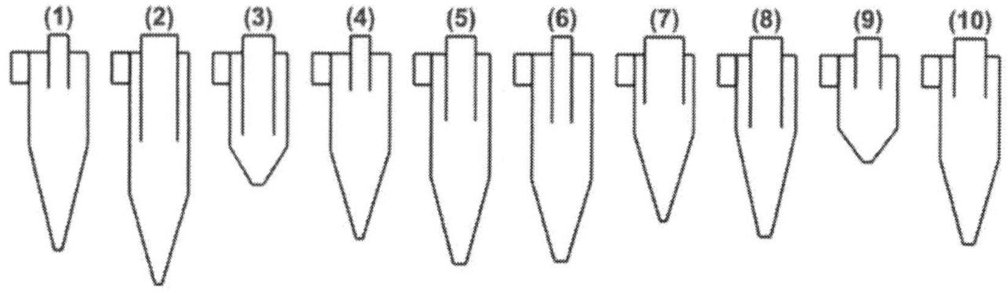

Figure 3. *First iteration of optimization algorithm COMPLEX, where each point represents one different geometric configuration of cyclone.*

The procedure starts with the generation of the initial condition, which corresponds to the ten initial geometry of the first iteration of the system. All of ten initial geometries are simulated through the *CYCLO-HEXA* code, and results of collection efficiency and pressure drop are subjected to evaluation of the objective function and constraints. The cyclones that present the worst performance in terms of the objective function are discarded and a new one is generated by the optimization algorithm *COMPLEX*. This procedure is repeated until the objective function is maximized, while the error or deviation of the iteration is minimized.

As a operational condition was used a gas flow of 220 m^3/h at 180°C with a loading ratio of 5000 mg of particles per m^3 of gas (diluted flow). The particulate system was analyzed using a Particle Size Analyzer by Laser Diffraction Mastersizer S, S-MAM 5005 Model of the Malvern brand. Through the particle size analysis, it was possible to obtain the diameter of the solid phases to the desired fractions, as shown in Table 1.

Table 1. *Size of the particulate phase.*

Solid phase	***Size***	***Fraction***
d_{s1}	24 µm	15.870%
d_{s2}	16 µm	17.065%
d_{s3}	12 µm	34.130%
d_{s4}	8 µm	17.065%
d_{s5}	5 µm	15,870%

Where d_{sn} is the average diameter to the solid phase n and the density (ρ_{sn}) adopted for all solid phases are 1400 kg/m³.

3. RESULTS AND DISCUSSIONS

As a first analysis, it is possible to verify the geometric evolution of cyclones along the iterations, as shown in Figure 4, where the cyclone in blue had the best performance of the interaction, while the reds color had the worst. The cyclones that presenting the worst results of collection efficiency or higher results of pressure drop will be dropped giving rise to a new geometry. Through the Figure 3, it is possible to observe that the geometries of cyclones become more similar along the iterations of the optimization process, confirming the search for an objective function maximized.

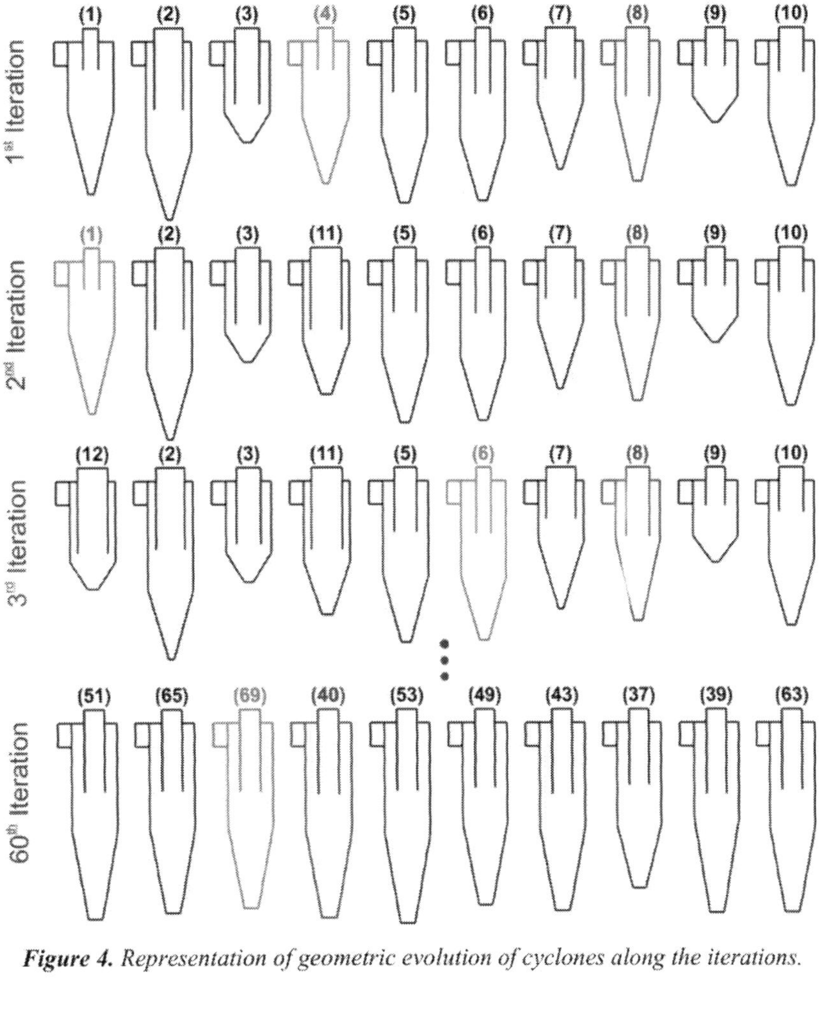

Figure 4. *Representation of geometric evolution of cyclones along the iterations.*

It is also possible to observe the maximization the objective function through graphics of collection efficiency and pressure drop of individual cyclones of the first and last iteration. In the first iteration (Figure 5), it can be seen a great variety of cyclones, some operating with a high collection efficiency, and consequently high pressure drop, while others have very low efficiency of collection, with pressure drops too low. This diversity was caused by the random generation of cyclones within the geometric constraints imposed. However, when analyzing the results obtained in the last iteration (Figure 6), we can see a great homogeneity of the results, besides maximizing the collection efficiency with pressure drop within the constraints imposed on the method.

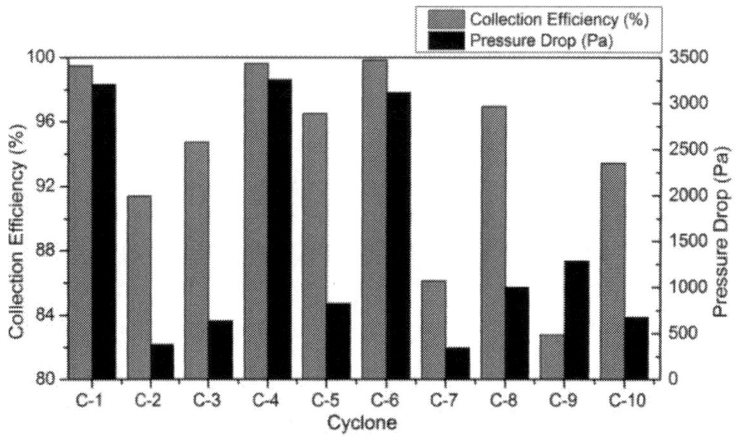

Figure 5. *Collection efficiency and pressure drop obtained at the first iteration.*

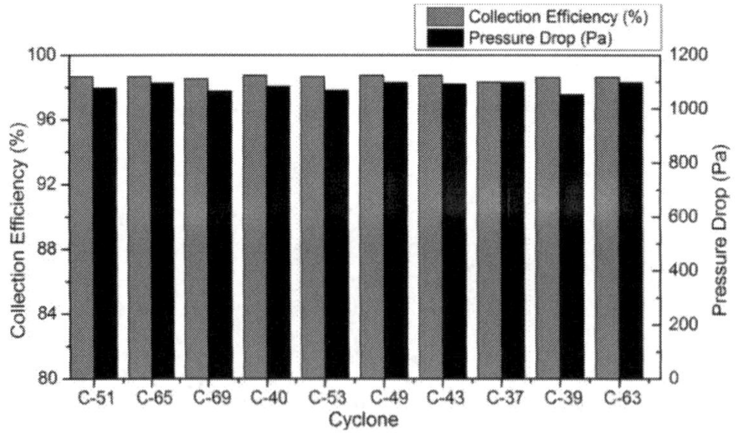

Figure 6. *Collection efficiency and pressure drop obtained at the last iteration.*

Finally, it is possible to observe the evolution of the average of the collection efficiency and pressure drop for each of the iterations of the iterative method. When it is analyzed the average of the collection efficiency of each iteration (Figure 7), it is possible to note that at the beginning of the method, there is a decrease in average efficiency, reaching lower values in the fourth iteration. This behavior can be explained due to the restrictions imposed on the Objective Function, since the method tends to eliminate the cyclones that are presenting pressure drop exceeding the maximum pre-established. After eliminating the cyclones with greater pressure drop, the method begins to search within the restrictions imposed, the greater Objective Function possible. In Figure 9, it is observed that during the iterative process, some cyclones with high pressure loss are still generated, but in the next iteration since they are replaced by corresponding one with lower pressure drop. This effect may be noted by the standard deviation peaks obtained at the iteration. Another factor that can be highlighted is that in recent iterations the increase of collection efficiency was in the order of 98.5%, according to the restrictions of maximum pressure drop established.

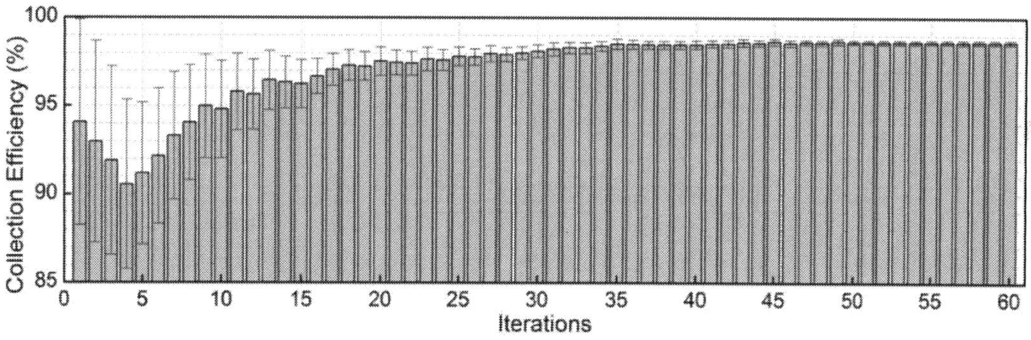

Figure 7. *Evolution of the collection efficiency average along the iterations.*

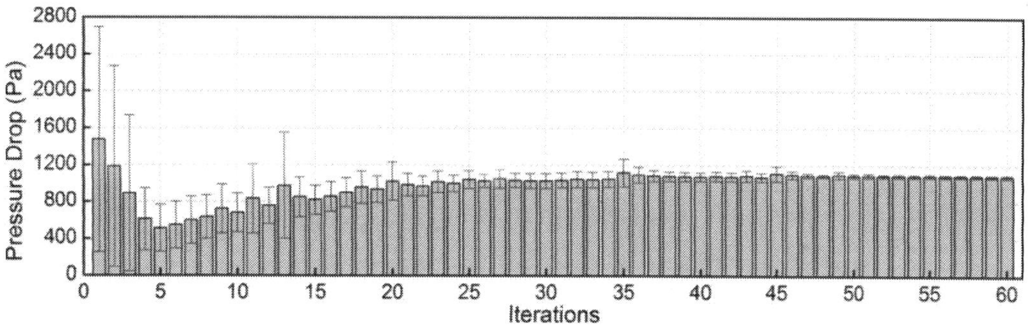

Figure 8. *Evolution of the pressure drop average along the iterations.*

Further evidence that proves the efficiency of the optimization method is the steady decline of the standard deviation along the iterations.

4. CONCLUSIONS

From the results presented in this work, it is possible to affirm that the proposed methodology for the geometrical optimization of cyclones showed satisfactory results within the operational limits established. It is also worth emphasizing the importance of using *CYCLO-HEXA* code for simulation of cyclones, because it presents reliable results at a computational cost much lower than required by current commercial software, being possible to resolve a large number of experiments in a relatively short time, since the complete simulation of each cyclone takes about 36 hours to complete in a personal computer.

REFERENCES

BOX, M. J., *A new method of constrained optimization and comparison with other methods*, The Computer Journal, 8, 42 - 52, 1965.

HOFFMANN, A. C. ; STEIN, L. E. *Gas Cyclones and Swirl Tubes: Principles, Design and Operation*. 2nd Edition. Springer. Berlin Heidelberg New York. 422 p. 2008.

MEIER, H. F. *Modelagem fenomenológica e simulação bidimensional de ciclones por técnicas da fluidodinâmica computacional*. Tese de Doutorado, UNICAMP, Campinas, São Paulo, 1998.

MEIER, H. F. ; MORI, M. *Gas-solid flow in cyclones: The Eulerian-Eulerian approach*. Computers & Chemical Engineering, v. 22, n. Suppl., p. S641-S644, 1998.

MEIER, H. F. ; VEGINI, A. A. ; MORI, M. *Four-Phase Eulerian-Eulerian Model for Prediction of Multiphase Flow in Cyclones*. The Journal of Computational Multiphase Flows, v. 3, p. 93-106, 2011.

OGAWA, A. *Mechanical Separation Process and flow patterns of Cyclone Dust Collectors*. Ind. Applied Mech. Ver, vol 50, no 03, pg. 97-130, 1997.

VEGINI, A. A. ; MEIER, H. F. ; IESS, J. J. ; MORI, M. *Computational Fluid Dynamics (CFD) Analysis of Cyclone Separators Connected in Series*. Industrial & Engineering Chemistry Research, v. 47, p. 192-200, 2008.

(559f) DNS-Based Analytical Model to Predict Particle Abatement Efficiency in Axial Separators

Marina Campolo and Francesco Zonta, Center for Fluid Dynamics and Hydraulics, University of Udine, Udine, Italy

Axial separators are widely used in the process industry to separate solid particles or droplets from the flow while it moves along a pipeline. Axial separators consist of fixed swirling vanes able to induce a rotational motion in the fluid and centrifugal forces on the heavy phase dispersed in the flow. The industrial design of these devices is based on proprietary criteria or uses semiempirical equations with an uncertain range of applicability. In this work, we develop a semi-analytical model which can be used to predict the separation efficiency based on the size of particles and characteristics of swirling flow (intensity and rate of decay in the axial direction). The model has been validated against accurate numerical data. These data include (i) DNS data of fluid flow at Re=5000 calculated for two different intensities of swirling motion (S 0.085 and S 0.17) and (ii) particle dispersion data calculated by Lagrangian tracking for three representative particle sizes (dp = 6.8, 21.4 and 67.8 μm). Results of the semi analytical model are in good qualitative agreement with deposition data calculated from the database, indicating that the essential physics of the turbulent dispersion process has been
correctly captured.

(565a) A Reverse Osmosis-Pressure Retarded Osmosis (RO-PRO) Hybrid Process for Low Energy Desalination

Chuyang Y. Tang, Singapore Membrane Technology Centre and School of Civil and Environmental Engineering, Nanyang Technological University, Singapore, Singapore

Reverse osmosis (RO) is a main stream technology for desalination. In seawater RO desalination, a pressure greater than the osmotic pressure of the seawater is used to drive water through the semi-permeable membrane. Current RO plants generate large quantities of concentrated brine, and its disposal is a major challenge due to the significant disposal cost and adverse environmental impact. Pressure retarded osmosis (PRO) has attracted great interest in recent years for its potential of energy generation from concentrated brines. PRO is the reverse process of RO, where the applied pressure on the concentrated brine side is less than the trans-membrane osmotic pressure difference such that water flows from a low osmotic pressure feed water to the high osmotic pressure brine. This osmotically driven water flux is against the pressure gradient, which means that the PRO process converts the chemical potential energy (the osmotic energy) into mechanical energy. In an RO-PRO hybrid process, the waste brine from the RO stage can be utilized for the subsequent PRO stage, and the osmotic energy harvested is utilized to offset the desalination energy demand. The potential of this technology in desalination and its challenges will be discussed in this presentation.

(565b) Seawater Desalination by Integrated Forward and Reverse Osmosis: Improved Product Water Quality for Potentially Less Energy

Devin Shaffer, Chem. and Env. Eng., Yale Univ., New Haven, CT, Ngai Yin Yip, Chemical & Environmental Engineering, Yale University, New Haven, CT, Jack Gilron, Zuckerberg Institute for Water Research, Ben Gurion University, Beer Sheva, Israel and Menachem Elimelech, Yale University, New Haven, CT

Seawater desalination for agricultural irrigation will be an important contributor to satisfying growing water demands in water scarce regions. Irrigated agriculture for food production drives global water demands, which are expected to increase while available supplies are further diminished. Implementation of reverse osmosis, the current leading technology for seawater desalination, has been limited in part because of high costs and energy consumption. Because of stringent boron and chloride standards for agricultural irrigation water, desalination for agriculture is more energy intensive than desalination for potable use, and additional post-treatment, such as a second pass reverse osmosis process, is required. In this perspective, we introduce the concept of an integrated forward osmosis and reverse osmosis process for seawater desalination. Process modeling results indicate that the integrated process can achieve boron and chloride water quality requirements for agricultural irrigation while consuming less energy than a conventional two-pass reverse osmosis process. In addition, an initial scoping out of the economics, shows that employing an FO-RO hybrid may lead to reduced overall costs relative to the two pass option presently employed.

(565c) Hybrid Water Treatment Processes Utilizing Forward/Reverse Pressure Driven Processes and Reverse Electro-Dialysis

Rahul Patil and **Glenn Lipscomb**, Department of Chemical and Environmental Engineering, University of Toledo, Toledo, OH

Reverse osmosis (RO) has become the preferred process for seawater desalination. The preference for reverse osmosis is due to the substantial energy savings that can be realized.

Forward osmosis (FO) recently has generated substantial interest as a lower cost alternative to reverse osmosis. In forward osmosis water is pulled across the membrane into a solution of higher osmotic pressure in contrast to reverse osmosis where water is forced across the membrane by pressurizing the higher osmotic pressure stream.

We examine the potential use of hybrid RO/FO for desalination. RO and FO can be used synergistically for desalination. However, FO also can be used to generate power by mixing the concentrated brine with seawater. The potential for integration of FO's separation and power generation into a desalination plant will be discussed. Process optimization relies on tracking osmotic pressure differences through the process and evaluating the effectiveness of the separation and power generation capabilities of the two unit operations. The trade-offs between use of FO and reverse electro-dialysis (RED) for power generation are discussed.

(565d) Membrane Engineering for Zero Liquid Discharge in Seawater Desalination

Enrico Drioli[1,2], Francesca Macedonio[1,2] and Aamer Ali[2], (1)Institute on Membrane Technology of the National Research Council - (ITM-CNR), Rende, Italy, (2)Department of Chemical Engineering and Materials, University of Calabria, Rende, Italy

Membrane engineering for zero liquid discharge in seawater desalination

E. Drioli[1, 2, 3], F. Macedonio[1, 2], A. Ali[2]

[1] National Research Council - Institute on Membrane Technology (ITM–CNR), Via Pietro BUCCI, c/o The University of Calabria, cubo 17C, 87036 Rende CS, Italy

[2] The University of Calabria - Department of Chemical Engineering and Materials, cubo 44A, Via Pietro BUCCI, 87036 Rende CS, Italy

[3] Hanyang University, WCU Energy Engineering Department, Room 917 9th Floor FTC Bldg., 17 Haengdang-dong, Seongdong-gu, Seoul 133-791 S. Korea

The reliability on seawater for industrial, drinking and household purposes has enormously increased in many parts of the world. Traditionally, pressure driven membrane based processes have been employed for seawater desalination. However, the pressure driven processes have some associated drawbacks that need to be addressed, operation at high pressure and the disposal of brine being major hurdles that adversely affect the process economy and cause environmental concerns. The use of new membrane based operation can't only resolve the problem of waste handling but also possesses the potential to boost the overall economy of the process. Membrane distillation (MD) is one of the advanced membrane based processes that can utilize waste grade energy including the solar and geothermal energy and can greatly enhance the recovery of the process. MD can be used to produce ultrapure water and can also be employed to recover the valuable crystals of different salts from the brine. However, MD is not fully mature process yet and requires further attention to understand the process and to improve its efficiency. Preparation of membranes specifically designed for MD applications, thermal polarization and coupling of flux with concentration at high feed concentration require special focus to utilize the potential of MD in more efficient way.

(565e) A Techno-Economical Analysis of Membranes, Amine Absorption and Membranes-Amine Hybrid Process for CO2 Removal From Flue Gases

Prodip Kundu, Amit Chakma and Xianshe Feng, Chemical Engineering, University of Waterloo, Waterloo, ON, Canada

In this study, a technical and economic analysis of CO_2 removal from flue gases with competing technology such as membranes, amine absorption and their hybrid process has been conducted. Polymer membranes have found applications in variety of gas separation processes. However, there exists considerable confusion and argument about the preferential use of membranes over alternative processes. Amine absorption is still considered a state of the art technology for acid gas removal. However, amine units use more energy when the CO_2 concentration increases. Membranes, on the other hand, use partial pressure as the driving force and are most effective at high concentrations of CO_2. Membranes can easily be combined with amine absorption to yield hybrid processes with cost/performance advantages that neither process could achieve individually. Membrane/absorption hybrid systems take the advantage of bulk CO_2 removal properties of membranes with a relatively small area while the amine is used for the final cleanup to achieve the required specification. Several cascade configurations of cellulose acetate based asymmetric hollow fiber membranes, and amine absorption with MEA and DEA as solvents were studied individually for a target recovery and purity, both for the product value of CO_2 (e.g., enhanced oil/coal bed methane recovery) and for environmental reasons (reducing the greenhouse effect). A mathematical model was first developed for multicomponent gas separation by high-flux asymmetric hollow fiber membranes which takes into account the permeate pressure build up inside the fiber bore, and also considers cross flow with respect to the membrane skin, irrespective of the flow direction of the bulk permeate stream outside of the porous layer. Amine unit was simulated in AspenPlus® and membrane model was later incorporated into AspenPlus® to study the hybrid process. The developed membrane model was incorporated into AspenPlus® as a user-model for an overall membrane process study and for the design and simulation of hybrid processes for CO_2 removal. The study was conducted to find the extent of optimal conditions for which the aforementioned technologies are best suited, and to evaluate the capital investment and processing cost.

(580a) Performance of a Pilot Scale Membrane Aerated Biofilm Reactor for the Treatment of Landfill Leachate

Eoin Syron, School of Chemical and Bioprocess Eng,, University College Dublin, Belfield, Dublin, Ireland and Eoin Casey, School of Chemical and BioProcess Engineering, University College Dublin, Dublin, Ireland

Performance of a pilot scale membrane aerated biofilm reactor for the treatment of landfill leachate

Eoin SYRON, Eoin CASEY,

School of Chemical and Bioprocess Engineering, University College Dublin, Belfield, Dublin 4, Ireland

Keywords: Oxygen Permeable Membranes, Bubbleless Aeration, Biofilm, Wastewater treatment

Aeration accounts for over 60% of the total energy spend in wastewater treatment plants, and even with recent technical improvements, bubble aeration can only achieve 50% oxygen transfer efficiency. Due to increasing energy costs and increasing regulatory pressure, there is a substantial growing interest in the application of diffusive gas transfer membranes for bubbleless aeration. These membranes avoid the mass transfer limitations associated with bubble-liquid interactions and increase the gas residence time within the reactor allowing more of the supplied oxygen to be transferred. Unlike conventional membranes in wastewater treatment, the concept of biofouling is not discouraged and the resultant biofilm on the membrane plays an active role in the degradation of COD and Nitrogenous constituents in wastewater.

Landfill leachate is water which has percolated through the waste bed in a landfill and leached organics and nutrients from the solid waste. This wastewater must be treated before being discharged to a water course to prevent pollution of the receiving water. Typically leachate has a high ammonia concentration due to the anaerobic degradation which takes place in the landfill. The Membrane Aerated Biofilm Reactor (MABR) is ideally suited for treatment of a high strength ammonia wastewater firstly as to the slow growing nitrifiers are attached to the membranes there is no need of an extra biomass retention operation, secondly the location of the nitrifying organisms on the membrane surface and their high oxygen demand exploit the oxygen transfer advantages of the MABR. While previous studies have reported on the MABR for ammonia nitrification, use of the MABR for the treatment of leachate has not previously been undertaken.

A 60 litre pilot plant (figure 1) was constructed to treat leachate from Arthurstown landfill, Co Kildare, Ireland. The pilot uses hollow fibre PDMS membranes supplied by SilPro (Broadstairs, UK) which have an internal diameter of 300mm and an external diameter of 500 mm giving a wall thickness of 100 mm and a total surface area of 4.2 m^2. PDMS membranes were chosen for this application due to their durability, ease of supply, and high oxygen permeability. In contrast to porous membranes, PDMS does not suffer from problems such as intra membrane wetting and pore clogging. Although the surface roughness of PDMS is quite low, there have been few reports of poor adhesion or biofilm formation on these membranes. The hollow fibres were potted in bunches of 60 fibres 0.4m long into 10mm PVC piping using polyurethane as the potting agent. Bunches were then arranged in cassettes of 11 bunches and 10 cassettes were placed in the reactor. The pH is controlled via a feedback loop and maintained between 7.5 and 8 with the addition of Sodium Hydroxide or Hydrochloric acid. Initial Oxygen Transfer Rates OTR experiments determined by re-oxygenation test gave results of 50gO_2 m^{-2} bar^{-2} day^{-1}.

Currently the reactor is removing 2.5g N-NH_4 m^{-2} day^{-1}which is comparable with previously reported systems Syron & Casey 2008, but it is foreseen that this rate of reaction will continue to increase with time and thicker biofilm growth, a theoretical maximum of 15 g N-NH_4 m^{-2} day^{-1} if all of the oxygen supplied is used for oxidation to nitrite. The current pilot is one of the few pilot scale studies of a MABR. Previous attempts have encountered problems with the control of excess biofilm resulting in clogging in the membrane module. In the present study the biofilm control strategy is allowing the pilot to continue with stable performance over a prolonged period.

References

Casey, E.; Glennon, B.; Hamer, G. Oxygen mass transfer characteristics in a membrane-aerated biofilm reactor. *Biotechnol.*

Bioeng. **1999**, *62* (2), 183–192.

Syron, E.; Casey, E. Membrane-aerated biofilms for high ratebiotreatment: Performance appraisal, engineering principles,scale-up, and development requirements. *Environ. Sci. Technol.***2008**, *42*, 1833–1844.

(580b) Symbiotic Hollow Fiber Membrane Photobioreactor for Microalgae Growth and Bacterial Wastewater Treatment

Linh Vu T. K., Department of Chemical and Biomolecular Engineering, National University of Singapore, Singapore 117576, Singapore and Kai-Chee Loh, Department of Chemical and Biomolecular Engineering, National University of Singapore, Singapore, Singapore

For decades, microalgae have gained much interest for production of high value compounds for food, feed, cosmetic and pharmaceutical products. With rapid depletion of fossil fuels and severe effect of greenhouse gas emissions on global climate change, many microalgae strains are potential sources for biodiesel production. Much attention is now focused on the development of a suitable photobioreactor that can assure mass production of high quality microalgal biomass. One of the crucial factors in designing a photobioreactor is the efficient gaseous transfer of CO_2 supply and O_2 removal.

Symbiotic microalgal-bacterial processes which incorporate microalgae and bacteria in one closed system could provide a potential solution for CO_2 supply and O_2 removal with low energy consumption. In this system, microalgae can produce sufficient O_2 for bacteria to biodegrade organic matter, while bacteria concomitantly release CO_2 for microalgae during photosynthesis. This process, so called photosynthetic oxygenation, is a promising approach to reduce the energy cost for intensive mechanical aeration and potentially eliminate CO_2 emission. For years, photosynthetic oxygenation has been applied not only for economical treatment of wastewater but also for degradation of hazardous contaminants. However, when growing microalgae and bacteria together, slow growth rate of microalgae and unexpected interactions between microalgae and bacteria may result in process failure. More importantly, microalgal biomass from the microalgal-bacteria consortia cannot be used flexibly for the production of food, feed or other high value compounds.

In this research program, we proposed to develop a hollow fiber membrane photobioreactor (HFMP) for symbiotic microalgal growth and bacterial wastewater treatment. In the proposed bioreactor, a gas exchange membrane will be used to physically separate microalgae and bacteria cultures as well as to solely facilitate transfer of CO_2 and O_2. The overall benefits of the proposed membrane bioreactor are to reduce the energy cost for mechanical aeration in an activated sludge process, reduce CO_2 emission to the atmosphere and most importantly, harvest microalgal biomass for extracting other value-added products.

Results obtained from the developed hollow fiber membrane photobioreactor (HFMP) demonstrate a proof-of-concept of the symbiotic relationship that existed between microalgae and activated sludge in wastewater treatment. In the absence of aeration in the synthetic waste water, BOD of synthetic wastewater was completely degraded while microalgae grew continuously in the absence of the external CO_2 supply. This concept hence can be applied for existing activated sludge treatment tanks in order to reduce the demand of aeration energy, and reduce CO_2 emission to the atmosphere. Most importantly, additional produced microalgal biomass can be harvested for other purposes such as for food, feed or biodiesel production.

(580c) Increased Operational Flexibility and Robustness of Anaerobic Wastewater Processes Via Bioreactor Coupling

Satish J. Parulekar, Chemical and Biological Engineering, Illinois Institute of Technology, Chicago, IL

Anaerobic wastewater purification processes have been increasingly used in the recent years. Anaerobic digestion is a biological process in which organic matter is transformed by a community of microorganisms into biogas (primarily methane and carbon dioxide) in the absence of oxygen. This process appears to be a promising method to solve some energy and ecological problems in agriculture and industry. Advantages this process has over the other common waste treatment process, the aerobic activated sludge process, include (i) much lower energy requirement; (ii) generation of less sludge, curtailing thereby the cost of energy-intensive post-treatment and disposal; and (iii) generation of biogas, which can be used as an energy source. In the past few years, a number of experimental studies aimed at improving the performance of anerobic digestion and kinetic models for the process have been reported. The process involves a consortium of acid generating bacteria (host population) and methane generating bacteria (commensal population). The initiation of this process requires hydolysis of complex insoluble organics by extracellular enzymes synthesized by the host population to produce simple soluble organics, which can be utilized by the host population. A comprehensive analysis of static and dynamic behavior of a mixed culture of involving acidogens (X) and methanogens (Y) in two coupled bioreactors is presented here. The two-bioreactor system is considered here to make the activated sludge process more robust with respect to wide fluctuations in feed conditions. A single continuous culture may operate at up to seven steady states, including up to four coexistence steady states, with only one coexistence steady state being locally stable. The one-way interaction between X and Y allows for compartmentalization of the system for a stand-alone bioreactor and two coupled bioreactors into two subsystems, which facilitates the analysis of steady state types and stability characteristics of these and classification of dynamic behavior. The bioreactors in the two-reactor system are identical only in terms of feed composition and bioreactor space time. The reactors need not have identical volumes and volumetric feed, effluent, and exchange rates. A two-reactor system may admit up to forty nine steady states, which are comprised of up to forty coexistence steady states, at least at very low interaction rates. The interaction rates between the two bioreactors, R_1 and R_2, are considered to be unequal. The steady states are comprised of symmetric steady states and asymmetric steady states. The symmetric steady states correspond to identical composition in the two bioreactors. The remaining steady states, the asymmetric steady states, correspond to different composition in the two bioreactors and are admissible over a range of interaction rates. The static and dynamic analysis of the two-reactor system is facilitated by appropriate grouping of large number of steady states arising for very low R into nine clusters. Numerical illustrations reveal the rich steady state structure of the bioprocess in coupled bioreactors. While a single bioreactor can operate at only one locally stable coexistence steady state, the coupled bioreactors can operate at up to five locally stable coexistence steady states over certain ranges of R_1 and R_2. The two-reactor system is operationally more flexible and more robust vis-a-vis single reactor as concerns maintenance of mixed culture. Emergence of four additional steady state clusters and additional coexistence and partial washout steady states at intermediate values of interaction rates reveals that the coupled bioreactors are an example of a complex system.

(580d) Two Phase Biodegradation of Phenol in a Hollow Fiber Supported Liquid Membrane Bioreactor

Prashant Praveen, Chemical and Biomolecular Engineering, National University of Singapore, Singapore, Singapore and Kai-Chee Loh, Department of Chemical and Biomolecular Engineering, National University of Singapore, Singapore, Singapore

During biodegradation of toxic aromatic compounds, microorganisms often experience substrate inhibition wherein the cell growth and metabolism are inhibited at higher concentration of the toxic substrate. Several bioreactors have been proposed for the alleviation of substrate inhibition. Most of these are based on the use of immobilized cells but with the advent of two phase partitioning bioreactor (TPPB), this is set to change. TPPBs are based on the equilibrium distribution of pollutants between cell culture medium and an immiscible, biocompatible and non-biodegradable organic solvent with high affinity for the substrate. Although TBBPs attain very high growth and biodegradation rates and would have made an ideal system; but various problems such as foaming and emulsification, arising from phase dispersion limit their applicability.

The operational challenges encountered in conventional TPPBs can be obviated by non-dispersive contacting between the two phases using semi-permeable membranes. In this research, a hollow fiber supported liquid membrane bioreactor (HFSLMB) was developed for simultaneous extraction and biodegradation of phenol from wastewater using *Pseudomonas putida*. In a semi-dispersive approach, extraction was carried out by dispersing the organic solvent into feed solution whereas the transport of substrate from solvent to the cells through the liquid membrane was non-dispersive. The dispersed solvent droplets facilitated liquid membrane renewal and resulted in a stable mass transfer flux while in the absence of any solvent, cell growth environment resembled that in a monophasic system. *P. putida* in suspension was able to biodegrade inhibitory phenol concentrations at 1000-4000 mg/L without experiencing severe substrate inhibition. For example, biodegradation of 4000 mg/L was achieved within 76 hours while the specific growth rate and biomass yield were 0.31 hr^{-1} and 0.26 g/g, respectively. Phenol biodegradation in the HFSLMB occurred in two stages: an exponential removal rate was observed in the beginning which subsequently assumed a linear profile under diffusion limitation. The specific growth rates during the exponential growth phase could be controlled by increasing the volume of the solvent while the removal rate under mass transfer limitation improved by changing the operating conditions and the interfacial area. In evaluating the long-term stability of the HFSLMB, the bioreactor was repeatedly operated in batch mode with 1000 mg/L phenol. The biodegradation performance improved during the first few runs due to the presence of attached cells on the membrane. But the performance deteriorated subsequently at higher biofilm thickness. The biofilms could be removed by alkaline washing lasting less than 5% of the operation time and resulting in the restoration of the biodegradation performance. With two washing cycles, HFSLMB operation could be sustained for 416 hours in which 20 batch runs were conducted.

The HFSLMB offered many advantages that included a better growth environment for the microorganisms, lower energy demand, high separation factor, high removal rates and elimination of secondary waste. The modular design is easier to scale-up and can also be used for continuous operation. Modeling the HFSLMB will provide further insights into substrate mass transfer and metabolism.

The novel anaerobic-circulating fluidized bed bioreactor (A-CFBBR) For High rate biological nutrient removal from high strength wastewater

MEHRAN ANDALIB , GEORGE NAKHLA *, JESSE ZHU

Introduction

High strength wastes produced by fertilizer production, explosive manufacturing, and recovery of nuclear fuels as well as landfill leachate contain in addition to organic matter, nitrogen in the forms of ammonia, nitric and nitrous acids (Chen and Lin, 1993). Similarly, many food processing wastewaters such as rendering wastes also contain high ammonia concentrations as a result of protein digestion. Nitrogen removal from such wastes, which require anaerobic processes to recover energy, can be challenging.

The Circulating Fluidized Bed Bioreactor (CFBBR), introduced and developed by Nakhla and his coworkers (Cui et al., 2004), was tested for biological nutrient removal (BNR) from municipal wastewaters in both lab and pilot scales at OLR of 5 kg COD/m^3·d and NLR of 0.5 kg N/m^3·d (Chowdhury et al., 2010). The CFBBR consists of an anoxic riser and an aerobic downer with fast and conventional fluidization regimes respectively. More than 90% organic, 70%-80% total nitrogen and 50%-70% phosphorous removal were reported, at hydraulic retention times (HRTs) of 2-3 h and an observed biomass yield of 0.12-0.16 g VSS/g COD. The CFBBR was reported to treat high strength wastewater such as landfill leachate and septage as well at the aforementioned OLR and NLR (Eldyasti et al., 2010). However, in order to accomplish high rate biological nutrient removal from high strength waste containing nitrogen, a newly developed bioreactor referred to henceforth as anaerobic-circulating fluidized bed bioreactor (A-CFBBR) was designed and tested to treat synthetic high strength wastewater with TCOD of 10600±270 mg/L, NH_3-N of 250±10 mg/L. The advantages of the A-CFBBR over conventional anaerobic treatment is simultaneous high rate nitrogen and carbon removal as well as energy recovery as opposed to only methanation.

Materials and Methods

System description and operating conditions

The A-CFBBR (**Figure 1**) is comprised of a conventional anaerobic fluidized bed (AF) with an overall height of 3.6 m followed by a CFBBR comprising anoxic and aerobic fluidized bed bioreactors (FBRs) with heights of 2 m and 3.6 m respectively. Zeolite particles (3 kg) with an average diameter (d_m) of 425-610 μm were used as carrier media in the anaerobic column and 3 kg zeolite particles with average diameter of 610-825 μm, were circulated between the aerobic and anoxic fluidized bed columns of CFBBR. The particle weight hold up in the aerobic column of the CFBBR was maintained at 2.4 kg continuously. In the riser, heterotrophic bacteria grow on the media and the biofilm becomes thicker. At a certain biofilm thickness, depending on the superficial liquid velocity, the biofilm-coated particles reach the height where they can be transferred to the downer through the inclined pipe. An intermediate graduated container was placed between the two columns, as shown in **Figure 1,** to monitor the particle transfer rate. After exposure to the high shear force in the gas-liquid-solid phase in the downer, the biofilm detaches and leaves the system along with the effluent. Particles from the bottom dense phase of the downer with a thin biofilm (< 100 μm) were transferred back to the riser manually through a gate valve that was occasionally opened at the time of particle transfer and to make up the particles in the riser. The hydrodynamic pressure difference between the riser and downer facilitated particle transfer.

The system was operated for about six months on synthetic wastewater (SWW). **Table 1** shows the operating conditions of the A-CFBBR system in Mode 3 in three different phases of VI-I to VI-III. Complementary information regarding the operating conditions in other Modes and Phases can be found Elsewhere (Andalib et al., 2012). At the maximum volumetric loading rate (Mode 3), the system was operating at an overall hydraulic retention time (HRT) of 12.8 h, empty bed contact time (EBCT) of 3.2 h and AF and CFBBR estimated sludge retention times (SRTs) of 32 d, and 28 d, respectively. The AF and CFBBR were operated at 37°C and 20°C, respectively.

Acclimatization and start-up

Initially, the clean media was fluidized in the anaerobic, anoxic, and aerobic columns at superficial liquid velocities (u_l) of 1.4 cm/s, 1.2 cm/s and 1.6 cm/s respectively. The anaerobic column was seeded with 16 L of secondary anaerobic digester sludge (ADS) from the St. Mary wastewater treatment plant (Ontario, Canada) with TSS and VSS concentrations of 15900 mg/L and 12180 mg/L respectively. The CFBBR was seeded with enriched nitrifiers, acclimatized in the lab using 10 L of returned activated sludge (RAS) from the Adelaide Pollution Control Plant, London, Canada, with TSS and VSS concentrations of approximately 7000 and 4800 mg/L respectively. Thereafter, the SWW, with the composition shown in **Table 2a**, was fed at a flow rate of 1.8 L/d and 0.1 L/d into the anaerobic and anoxic columns on day eight corresponding to OLR 0.28 kg $COD/m^3_{AF}\cdot d$ and NLR of 0.01 kg $N/m^3_{AF}\cdot d$.

Analytical methods

Samples from the feed tank, the anaerobic column, the anoxic column, and the final effluent were collected for analysis. Total suspended solids (TSS), volatile suspended solids, (VSS) were analyzed in accordance with Standard Methods 2540D, 2540E respectively (APHA 1998). HACH methods and testing kits (HACH Odyssey DR/2800) were used to analyze total and soluble chemical oxygen demand (TCOD and SCOD), total nitrogen (TN), NH_4-N, NO_2-N, NO_3-N, and PO_4. Alkalinity was measured by titration with 0.01 N H_2SO_4 in accordance with the Standard Method no 2320 (APHA 1998). DO and ORP were measured onsite using an Oakton DO 6 meter, and an Oakton ORPTestr 10 (Oakton, Singapore). Based on Standard Method no 2540G (APHA, 1998), the attached biomass on the carrier media was measured and expressed as mg VSS/g clean particles. The rate of biogas produced in the methanogenic column was measured by a wet tip gas meter connected to the top of anaerobic column. Methane, nitrogen, and hydrogen content of the biogas were determined using a GC.

Results and discussion

System biomass inventory

Figure 2a illustrates the buildup of attached biomass in the anaerobic, anoxic, and aerobic columns. The process of biomass buildup comprises two steps: the attachment of biomass during acclimatization and the growth of biomass during the process. The biomass attachment during acclimatization under the same hydrodynamic conditions should result in relatively the same biomass attachment. As shown in **Figure 2a**, the attachment of biomass in anaerobic, anoxic, and aerobic column, labeled as A, B and C, are 12.5 mg VSS/g media, 14 mg VSS/g 10.8 mg VSS/g media respectively. The aforementioned attached biomass densities correlated inversely with u_l of 1.4, 1.2, and 1.6 cm/s respectively. As shown in **Figure 2a**, the concentration of immobilized biomass in the anaerobic reactor increased slowly between days 1 and 120 as a result of the gradual start-up strategy adopted which limited the availability of substrate in the reactor, and the low growth rate associated with the methanogenic bacteria.

Nutrient removal

The anaerobic fluidized bed (AF) was operated as a mesophilic methanogenic reactor at a temperature of 35.7±0.8 °C and alkalinity of 3034±228 as mg $CaCO_3$/L. **Table 2b** illustrates the steady-state characteristics of the anaerobic, anoxic and aerobic effluents in Mode 3 when A-CFBBR was operational. The complementary information regarding the steady-state characteristics in Mode 1 when only the AF was operational and Mode 2 when the AF and only the aerobic fluidized bed were operational were depicted in **Table B** of the SD. As shown in **Table 2b,** although the feed pH was 4.6, pH inside the AF remained constant at 7 throughout the experiment due to the high rate of acetic acid degradation in the column. ORP in the AF was observed in the range of -158 mV to -193 mV.

Figure 2b illustrates the temporal biogas production in AF during phases I to VI. As shown in **Figure 2b**, the percent methane of the produced biogas fluctuated between 45% and 52% resulting in methane production rates of 103±6 L/d, 144±5 L/d, and 188±6 L/d in Phases IV, V and VI respectively. Based on the CH_4 equivalent of COD converted by aceticlastic methanogens at standard conditions (0°C and 1 atm) of 0.35 L CH_4/g COD, the theoretical methane production at 37°C was also calculated and depicted in **Figure 2b**. Discrepancies between the theoretical and measured methane of 3.8%-6.4% were observed. It should be noted that the ratio of biogas produced in the system to the volume of the reactor in Phase VI was 24 L_{gas}/L_{rea}.d which was handled without the need for a dedicated solid/liquid/gas separator at the top of the reactor.

Figure 2c depicts the COD removal rate in AF in different phases. As shown in **Table B** of the **SD**, the influent TCOD concentration was constant at 10600±270 mg/L. The effluent SCOD in phases I to IV, when the anaerobic attached biomass was not fully developed, was observed to be in the range of 218 mg/L to 260 mg/L. In phase IV, SCOD decreased to 124 mg/L at OLR of 26 kg COD/m^3·d. As seen in **Table 2b**, in Phase VI at the maximum OLR of 35 kg COD/m^3·d the system reached 98.3% TCOD removal, with effluent SCOD increasing marginally to 180 mg/L. TSS concentrations in the bulk liquid after seeding in phase I decreased gradually from 927 mg/L to the minimum level of 124 mg/L in Phase V at OLR of 26 kg COD/m^3_{AF}·d. However, increasing the OLR to 35 kg COD/ m^3_{AF}·d in Phase VI resulted in an increase in TSS to 188 mg/L. TCOD decreased from 10600 mg/L to an average of 50±20 mg/L in the A-CFBBR effluent throughout all phases, corresponding to the overall TCOD removal of 99.3%. The influent TCOD to the CFBBR comprised the AF effluent, which were 181 mg/L and 390 mg/L in phases V and VI respectively and 5% feed with a TCOD of 10600 mg/L. The TCOD influent to the CFBBR was 740±20 mg/L in phase VI corresponding to OLR of 3.3-4.8 kg COD/m^3_{CFBBR}·d based on the overall volume of the CFBBR in Phase VI (**Table 1**). The final effluent SCOD of the A-CFBBR was 21±10 mg/L and 35±10 mg/L in phase V and VI respectively. In Phase IV, the suspended VSS in the anaerobic column averaged 45 mg/L, which increased to 65 mg/L and 146 mg/L in Phases V and VI. The increase in the anaerobic effluent VSS as a result of higher VSS detachment rates in Phases V and VI (**Table 1** and **Table A** of the SD) coincided with higher OLRs in those Phases.

In Phase V, the anoxic column was not operational and nitrogen removal was solely based on biomass assimilation in the AF where NH_3-N decreased from 250 mg/L in the feed to 213±10 mg/L in the AF effluent, and nitrification in the aerobic column reduced ammonia from 213±10 mg N/L to < 1 mg NH_3-N/L in the system effluent. Assuming the overall anaerobic yield of 0.016 g VSS/g COD and 9±1.1% nitrogen content of biomass (**Figure 2d**), the nitrogen assimilated into biomass was calculated as 20±3 mg /L. The nitrate and nitrite concentrations in AF remained below 3.9 m Ng/L and 0.2 mg N/L in Phase V respectively. The NH_3-N remaining in the AF effluent was sequentially completely nitrified in the aerobic column to an average concentration of 0.9±0.5 mg N/L at NLR of 0.7 kg N/m^3_{CFBBR}·d. SNR batch tests in Phase V showed nitrification rate of 0.25-0.27 g N/g VSS·d, which confirmed the aerobic online nitrification rate within 10% discrepancy. The influent STN to the aerobic column in Phase V was 217±11 mg N/L in the form of NH_3-N and the effluent STN from the aerobic column was 205±9 mg/L in the of NO_x-N. Nitrogen assimilation in the aerobic biomass was estimated at 1.4 mg N/L in Phase V with OLR of 0.5±0.1 kg COD/m^3_{CFBBR}·d and biomass nitrogen content of 9%, thus, indicating that a partial denitrification, of 10 mg N/L may have occurred simultaneously in the aerobic column. The anoxic column was operated with three aerobic to anoxic recirculation flows to feed ratios of 2.3, 4, and 6.2 in Phases VI-I, VI-II and VI-III respectively (**Table 1**). The influent NH_3-N and TN to CFBBR in Phase VI were 215 mg N/L and 232 mg N/L corresponding to NLR of 1.1 kg N/m^3_{CFBBR}·d. As shown in **Table 2b**, the effluent NH_3-N throughout Phase VI remained 1-3 mg N/L, within 13% of the batch SNR range of 0.25-0.31 g N/g VSS·d. The recirculated nitrate to the anoxic column was denitrified to effluent concentrations of 76±22 mg NO_3-N/L, 39±2 mg NO_3-N/L and 88±5 mg NO_3-N/L in Phases VI-I, VI-II and VI-III respectively. In Phase VI-II when the effluent NO_3-N and NO_2-N were 39 mg N/L and 2 mg N/L, total nitrogen removal of 82.3% was achieved with effluent NH_3-N of 1.3 mg N/L. The batch SDNR test showed a range of 3.4-3.8 g NO_3-N/g VSS·d. In Phase VI-III, when the aerobic to anoxic recirculation flows to feed ratio increased to 6.2, aerobic DO dropped to 5.6 mg O_2/L from the initial value of 6.3 mg O_2/L and anoxic DO increased to 0.7 mg O_2/L in Phase VI-II, resulting in incomplete nitrification in the aerobic column as reflected by the average nitrite concentration of 20 mgN/L (**Table 2b**) as well as insufficient denitrification in the anoxic column. Therefore, effluent nitrite and nitrate increased to 20 mg N/L and 88 mg N/L, respectively (**Table 2b**). Phosphorus removal occurred as a result of biomass assimilation in the three columns. As shown in **Table 2b**, in Phase VI, 2.8±0.2 mg P/L and 3.0±0.3 mg P/L, were removed in AF and CFBBR respectively as compared with 3.3 mg P/L and 0.8 mg P/L estimated based on 1.8% biomass phosphorus content by weight of TSS.

At the optimum performance, Phase VI-II of Mode 3, the CFBBR in this work demonstrated 85% COD removal and 84% nitrogen removal at OLR of 4.8 kg COD/m^3_{CFBBR}·d, NLR of 1.1 kg N/m^3_{CFBBR}·d, EBCT of 1.6 h and HRT of 5 h. In comparison, Chowdhury et al. (2010) reported 90% COD and 80% nitrogen removal from municipal wastewater (MWW) using CFBBR at OLR of 4.2 kg COD/m^3_{CFBBR}·d, NLR of 0.26 kg N/m^3_{CFBBR}·d, EBCT of 1.5 h and HRT of 2.4 h. Elsewhere Eldyasti et al. (2010) observed 87% COD and nitrogen removal from landfill leachate using the same system at OLR of 2.6 kg COD/m^3_{CFBBR}·d, NLR of 0.8 kg N/m^3_{CFBBR}·d, EBCT of 11 h and HRT of 18 h. Other versions of the CFBBR developed by the same research group such as CFBBR with a bigger anoxic column called TCFBBR (Andalib et al., 2011b) and enhanced bioparticle circulation CFBBR (Li et al., 2012) also showed 88% and 95% COD as well as 84 and 88% nitrogen removal from MWW at OLR of 4.3 and 3.4 kg COD/m^3_{CFBBR}·d, NLR of 0.5 and 0.44 kg N/m^3_{CFBBR}·d, EBCT of 1.0 h and HRT of 2.2 h respectively. Thus, the A-CFBBR achieved comparable organics and nitrogen efficiencies, albeit at 2-4 times the NLR of MWW at roughly the same EBCT. Chen et al. (2011) also reported 87% COD and nitrogen removal from MWW using an A2/O system incorporating a biological aerated filter at OLR of 1.23 kg

COD/m^3·d, NLR of 0.23 kg N/m^3·d and HRT of 8 h. A longer HRT in the CFBBR in this study (of 5 h versus 2.2 h) and the study conducted by Eldyasti et al. (2010) seemed to be conducive to higher nitrogen loading rates of 1.1 and 0.8 kg N/m^3_{CFBBR}·d respectively. On the other hand, using a finer aeration system in the aerobic column of this work resulted in a lower biomass detachment rate of 0.035 1/d and a higher attached biomass of 20 mg VSS/ gmedia (**Table 1**) compared to the detachment rate of 0.093 1/d and attached biomass of 7 mg VSS/ gmedia in the work of Chowdhury et al. (2010), and the detachment rate of 0.18 1/d and attached biomass of 4.7 mg VSS/ gmedia in work conducted by Andalib et al. (2011b), which resulted in a higher nitrification rate.

Biomass yield

Figure 2d illustrates the linear regression of cumulative VSS produced, based on the sum of the effluent biomass, the net change in attached biomass and biomass wasted, versus cumulative COD removed in anaerobic fluidized bed, CFBBR and A-CFBBR. The highest observed yield for the AF utilizing acetic acid was 0.026 g VSS/g COD in phase II at OLR of 5.3±0.2 kg COD/m^3_{AF}·d. Along with increasing the OLR to 34.5 kg COD/m^3_{AF}·d in AF, the observed yield decreased to 0.016 g VSS/g COD. The very low observed yield of the anaerobic column was expected, due to the estimated long sludge retention time (SRT) in the column of 27-32 days (**Table 1**) and the true yield for methanogens utilizing acetate of 0.05 g VSS/g COD (Metcalf and Eddy 2003). Very low observed yields of 0.116 g VSS/g COD to 0.138 g VSS/g COD were measured in Phase VI-I to VI-III in the aerobic/anoxic CFBBR. Due to the fact that 95% of the overall COD was consumed in the anaerobic fluidized bed, the overall observed yield of A-CFBBR was calculated as 0.017 g VSS/g COD, close to the yield of AF.

Conclusions

- The newly developed A-CFBBR affected 99.7% COD, 84% nitrogen removal with a very low sludge yield of 0.017 g VSS/g COD while treating a high strength SWW containing 10700 mg COD/L and 250 mg NH_3-N/L. The system performance at a high OLR of 35 kg COD/m^3_{AF}·d and 1.1 kg N/m^3_{CFBBR}·d with an overall HRT of less than 12 h is remarkable.
- Microbial communities analysis using DGGE confirmed the presence of AOB and NOB in both the aerobic and anoxic columns and *Pseudomonas putida* responsible for SND in the aerobic column.

List of Tables and Figures

Table 1- Operating condition of the system in Mode 3 (A-CFBBR under operation)

Table 2- **(a)** Composition of the synthetic wastewater **(b)** Steady-state characteristics of anaerobic, anoxic and aerobic effluents in Mode 3 (A-CFBBR under operation)

Figure 1- Schematic of anaerobic-circulating fluidized bed bioreactor (A-CFBBR)

Figure 2 - **(a)** Biomass inventory **(b)** methane production in AF **(c)** COD removal in AF **(d)** biomass yields in AF, CFBBR and A-CFBBR

References

1. Andalib, M., Nakhla, G., Zhu, J., 2010 b. Dynamic testing of the twin circulating fluidized bed bioreactor (TCFBBR) for nutrient removal from municipal wastewater. Chem. Eng. J. 162, 616-625.
2. Andalib M, Nakhla G, , Zhu J. 2012. High-rate biological nutrient removal from high strength wastewater using novel Anaerobic Circulating Fluidized Bed Bioreactor (A-CFBBR). Bioresource Technology 118, 526-535.
3. APHA; AWWA; WEF. 1998. Standard methods for the examination of water and wastewater. 20th Edition, American Public Health Association, Washington D.C.
4. Chen, K.C., Lin, Y.F., 1993. The relationship between denitrifying bacteria and methanogenic bacteria in a mixed culture system of acclimated sludges. Wat. Res. 27(12), 1749-1759.
5. Chowdhury, N., Nakhla, G., Zhu, J., Islam, M., 2010. Pilot-scale experience with biological nutrient removal and biomass yield reduction in a liquid-solid circulating fluidized bed bioreactor. Wat. Environ. Res. Doi: WER 08-11-1541RR.
6. Cui, Y., Nakhla, G., Zhu, J., Patel, A., 2004. Simultaneous carbon and nitrogen removal in anoxic-aerobic circulating fluidized bed biological reactor (CFBBR). Environ. Technol. 25, 699-712.
7. Eldyasti, A., Chowdhury, N., Nakhla, G, Zhu, J., 2010. Biological nutrient removal from leachate using a pilot liquid–solid circulating fluidized bed bioreactor (LSCFB). J of Haz. Mat. 181, 289–297.
8. Metcalf and Eddy, 2003. Wastewater Engineering: Treatment and reuse, Fourth ed., McGraw-Hill.
9. Chen, Y., Peng, C., Wang, J., Ye, L., Zhang, L., Peng, Y., 2011, Effect of nitrate recycling ratio on simultaneous biological nutrient removal in a novel anaerobic/anoxic/oxic (A2/O)-biological aerated filter (BAF) system. Bioresource Technol. 102(10), 5722-5727.

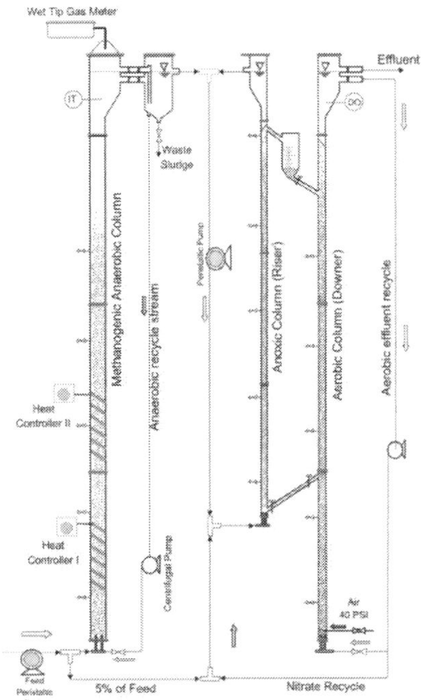

Figure 1

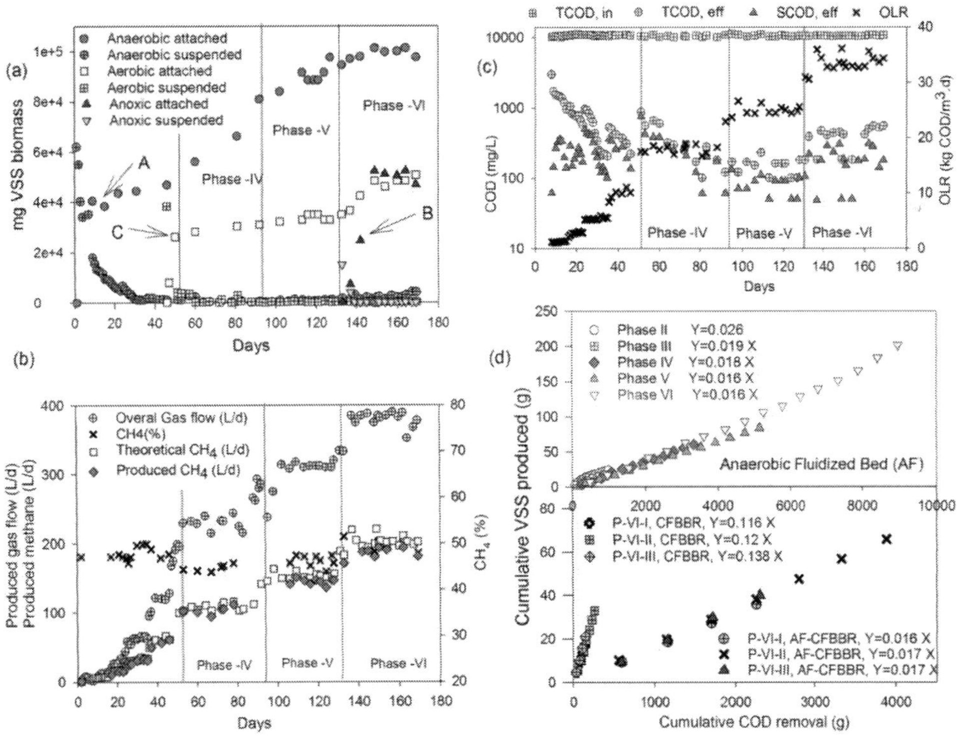

Figure 2

Table 1- System operating conditions in Mode 3 (A-CFBBR under operation)

	A-CFBBR under operation (**Mode 3**)		
	Phase-VI-I	**Phase-VI-II**	**Phase-VI-III**
Day	129-145	145-160	160-172
Feed flow rate (L/d)	47-55[a]	55[a]	55[a]
Anaerobic-Anaerobic circulation flow (Q/Q_{in})	109	107	105
Anoxic-Anoxic Circulation flow (Q/Q_{in})	2	2	2
Aerobic-Aerobic circulation flow (Q/Q_{in})	64	59	64
Aerobic-Anoxic Circulation flow (Q/Q_{in})	2.3	4	6.2
OLR based on anaerobic reactor (kg COD/m^3.d)	33.3±1.5	33.5±1.2	34.5±0.8
NLR based on anaerobic reactor (kg N/m^3.d)	0.8±0.1	0.8±0.1	0.85±0.0
OLR based on CFBBR reactor (kg COD/m^3.d)	3.8±0.7	3.3±0.1	4.8±0.2
NLR based on CFBBR reactor (kg N/m^3.d)	1.11±0.0	1.12±0.0	1.15±0.0
Anaerobic EBCT (h)=$V_{compact}/Q_{in}$	1.6	1.6	1.6
Aerobic EBCT (h)=$V_{compact}/Q_{in}$	1.3	1.3	1.3
Anoxic EBCT (h)=$V_{compact}/Q_{in}$	0.27	0.26	0.26
Anaerobic HRT(d)	0.32	0.32	0.32
Aerobic HRT(d)	0.17	0.17	0.17
Anoxic HRT (d)	0.04	0.04	0.04
Anaerobic attached biomass (mg VSS/g media)	32±0.5	33±0.2	33±0.4
Aerobic attached biomass (mg VSS/g media)	16.5±1.7	20.6±0.5	20.5±0.5
Anoxic attached biomass (mg VSS/g media)	1-41	85±1.5	82.2±2
Total anaerobic attached biomass (g VSS)	96.4	100.2	99.3
Total aerobic attached biomass (g VSS)	37.9	47.5	47.8
Total anoxic attached biomass (g VSS)	0.3-24.6	51.2	49.5
F/M in AF (g COD/g VSS.d)	5.5	5.4	5.5
F/M in CFBBR (g COD/g VSS.d)	0.66	0.35	0.55
Anaerobic col. detachment rate (1/d)		0.035	
Aerobic col. detachment rate (1/d)		0.035	
Anoxic col. detachment rate (1/d)		0.02	
Aerobic DO (mg/L)	6.2±0.1	6.3±0.1	5.6±0.0
Anoxic DO (mg/L)	0.35±0.1	0.4±0.1	0.7±0.2
AF estimated SRT (d)		32	
CFBBR estimated SRT (d)		28	
Superficial liquid velocity in anaerobic col. (cm/s)		1.42±0.04[b]	
Superficial liquid velocity in aerobic col. (cm/s)		1.7±0.1	
Superficial liquid velocity in anoxic col. (cm/s)		1.2±0.1	
Specific nitrification rate (**SNR**) (g NH_3-N/g VSS·d)-Downer		0.25-0.31	
Specific nitrification rate (**SNR**) (g NH_3-N/g VSS·d)-Riser		0.05±0.01	
Specific denitrification rate (**SDNR**) (g NO_3-N/g VSS·d)-Riser		3.4-3.8	
Specific denitrification rate (**SDNR**) (g NO_3-N/g VSS·d)-Downer		0.3±0.1	

[a] 5% of overall feed was directly fed to the bottom of the anoxic column.
[b] The terminal settling velocity of Zeolite with d_m=600 μm and 1000 μm biofilm thickness is 2.8 cm/s, still significantly lower than the operating u_L (Andalib et al., 2010 a).

Table 2a. Composition of the synthetic wastewater

Feed composition [a]	CH_3COOH (mL/L_{Feed})	NH_4Cl (g/L_{Feed})	K_2HPO_4 (g/L_{Feed})	$MgSO_4.7H_2O$ (g/L_{Feed})	$CaCl_2.2H_2O$ (g/L_{Feed})	Yeast extract (g/L_{Feed})	$NaHCO_3$ (g/L_{Feed})	Trace element (mL/L_{Feed})
Concentrations	9.5	1.7 and 0.93	0.1	0.03	0.03	0.03	5.7-6.7	1
Trace element Composition	$FeCl_2.4H_2O$ (mg/$L_{Trace\ E}$)	$MnCl_2.4H_2O$ (mg/$L_{Trace\ E}$)	H_3BO_3 (mg/$L_{Trace\ E}$)	$ZnCl_2$ (mg/$L_{Trace\ E}$)	$CuCl_2$ (mg/$L_{Trace\ E}$)	$AlCl_3$ (mg/$L_{Trace\ E}$)	$CoCl_2.6H_2O$ (mg/$L_{Trace\ E}$)	$NiCl_2$ (mg/$L_{Trace\ E}$)
Concentrations	2000	500	50	50	30	50	50	50

[a] All the above reagent were ACS Grade, 99.5% min.

Table 2b. Steady-state characteristics of anaerobic, anoxic, and aerobic effluents in Mode 3 (A-CFBBR operational)

		A-CFBBR operated (**Mode 3**)						
Parameter (mg/L)		**Phase-VI-I to VI-III**	**Phase-VI-I**		**Phase-VI-II**		**Phase-VI-III**	
Days (d)		*129-172*	*129-145*		*145-160*		*160-172*	
	Influent	Anaerobic Effluent	Anoxic	Effluent	Anoxic	Effluent	Anoxic	Effluent
Produced biogas(L/d)	-	333 to 405	-	-	-	-	-	-
TCOD	10600 ± 270	390± 140	61±14	49±8	80±5	58±7	79±24	67±1
SCOD		181± 86	30±11	31±5	50±16	40±18	46±16	34±12
TSS	-	188±85	28±8	22±10	42±15	32±10	29±8	30±10
VSS	-	146±56	21±5	16±8	27±12	19±7	23±6	23±8
NH_3-N	250	215±8	70±1	3±1	45±12	1.3±0.3	57±19	3.1±0.6
NO_3-N	2	1.8±1	4.7±4	76±22	2±0.5	39±2	16±11	88±5
NO_2-N	1	1.3±1	1.3±0.6	0.9±0.1	2±0.5	2±0.5	38±2	20±2
TN	255(430)[b]	231±12	-	78±21	-	45±10	-	113±10
PO_4-P	26.3	23.5±2.6	19±0.5	18.8±0.8	20.5±2	20±2	21±1	19.3±0.5
Alkalinity[a]	-	3065±115	2115±21	1875±35	2630±180	2017±280	2095±100	1885±120
ORP(mV)	-	-193±8	-	38±13	-	12±2	-	2±1
pH	4.6±0.1	7.09±0.0	8.2±0.1	8.4±0.1	8.2±0.2	8.4±0.1	8.1±0.1	8.3±0.0
Temp (°C)	20	35.7±0.8	22.4±0.1					

[a] as mg $CaCO_3$/L Nitrogen and phosphorus content of biomass were measured 9±1.2% and 2.1±0.4% respectively

Reduction of VOC Emissions in High Purity Oxygen Activated Sludge Wastewater Treatment Process: Toxchem™ Based Fate & Modeling Case Study

Malcolm Fabiyi, Richard A. Novak, Praxair, Inc., Burr Ridge, IL
Rajeev Goel, Spencer Snowling, Hydromantis, Inc., Hamilton, ON

ABSTRACT

In the last decade, stringent regulations have been imposed on the permissible level of Volatile Organic Compound (VOC) emissions from industrial facilities. The clean air act amendments of 1990 and related regulatory extensions (e.g., National Emissions Standards for Hazardous Air Pollutants, NESHAPs) have significantly reduced the acceptable level of Volatile Organic Compound (VOC) emissions from industrial facilities (Woodward & Curran, 2006). A total of 188 organic compounds have been designated Hazardous Air Pollutants (HAPs); facilities which generate or handle these compounds are subject to permitting, monitoring and reporting requirements. Although extensive emissions controls have been implemented in the production process of many industrial facilities, significant amounts of VOCs may appear in the wastewater. These VOCs can be stripped into the air during conveyance through collection systems or during biological treatment.

Fate and Transport models are used for a variety of tasks including design, emissions analysis and regulatory reporting. The development of fate & transport models in wastewater collection and treatment systems were spurred by regulatory drivers like the Clean Air Act (Melcer, 1994). A variety of computer-based fate & transport models such as BASTE, EPA Water Models (versions 7, 8 and 9), CORAL, PAVE, SIMS, TORONTO, INTERCEPTOR and Toxchem™ have been applied for addressing the fate of volatile contaminants in collection systems, drop structures, weirs, quiescent surfaces, and wastewater treatment processes (Quigley et al, 2006; Melcer, 1994).

EPA Water Models (7, 8 & 9) and Toxchem™ are the most commonly utilized fate and emissions transport modeling platforms in the wastewater industry. In both models, the default aeration options exclusively model the mass transfer behavior as diffused air processes or surface aeration style mechanical mixers. Attempts at modeling the performance of High Purity Oxygen (HPO) based devices have typically required finding the most appropriate approximation to the mass transfer and VOC stripping behavior of HPO aeration systems by modifying default parameters in the diffused air or surface aeration modules (NYSERDA, 2000; Levine et al, 2010; Rodieck et al, 2001). In contrast, an improved modeling approach is employed in the recently released Toxchem™ 4.1 (Hydromantis, 2012), which extends the simulation capabilities of the platform to include the modeling HPO and Sequencing Batch Reactor (SBR) process systems.

This paper discusses results for VOC emissions modeling obtained by applying Toxchem™ 4.1 to VOC reduction projects that utilize SBR and HPO systems. We provide a methodology for modifying the default parameters in diffuser and mechanical surface aeration modules in fate and transport models in order to enable VOC emissions characteristics associated with HPO systems to be modeled, and provide comparative analyses on the effect of the approach adopted on simulation results.

KEY WORDS

VOC emissions, mass transfer, mechanically agitated contactors, Toxchem, Fate & Transport models, High Purity Oxygen

BACKGROUND

The mass transfer models in fate and transport emissions models are based on ASCE standard model parameters (Chern & Yu, 1995). They model the three major mechanisms through which VOC removal from wastewater occurs: (i) biosorption of VOCs onto activated sludge floc and subsequent removal via waste sludge, (ii) volatilization and stripping of VOCs into the atmosphere during aeration, and (iii) biological degradation.

Mechanisms for VOC Stripping

There are two distinct mechanisms for VOC stripping in aerated activated sludge biological wastewater treatment processes: (i) *turbulence induced stripping*, in which turbulence generated by the aeration system induces surface renewal leading to the transfer of VOCs from the bulk liquid across the air-water interface; and (ii) *bubble aided stripping,* in which VOCs in the wastewater diffuse into un-dissolved gas bubbles and are stripped as the bubbles leave the water surface.

Aeration Systems and Their effect on VOC Stripping

Three general classes of aeration devices exist (Melcer, 1994; Mueller et al, 2002): (i) mechanical surface aerators (ii) diffused aeration systems, and (iii) subsurface mechanical aerators with gas sparging.

Mechanical Surface Aerators: Mechanical surface aerators spray water droplets into the headspace above the aerator; gas-liquid exchange occurs as gases diffuse from the surrounding atmosphere into the droplets, as well as from the droplets into the atmosphere. In these devices, VOC emissions occur in three ways: (i) from the droplets to the air, (ii) from the bulk liquid into air bubbles that have been entrained under the surface of the water, as the air bubbles rise and reach the liquid surface, and (iii) across the turbulent free surface of the basin.

Diffused Aeration Systems: Fine Bubble Diffuser systems are commonly utilized for aeration in wastewater treatment processes because they provide relatively high oxygen transfer efficiencies (OTE) (in the range of 20%-30% for most common applications) relative to alternatives like coarse bubble systems and surface aerators (Mueller et al, 2002). Considering the fact that air typically has about 21% oxygen by volume, OTE values for aeration systems in the range of 20%-30% implies that only about 4%-6% of the volumetric air flow supplied to a basin using a diffused aeration system is actually dissolved. The undissolved gases, which make up about 94%-96% of the gas supplied, provide a large volumetric flow of gas which strips VOCs from the basin.

Subsurface mechanical aerators: In subsurface mechanical aeration systems, process gases such as oxygen are injected below the water surface in combination with a mechanical mixer which serves to maintain efficient gas-liquid contacting. Oxygen diffuses into the wastewater, and VOCs transfer from the liquid phase to the gas bubbles. The bubbles eventually rise to the surface and VOC release occurs as the bubbles break the surface. Subsurface mechanical aerators generate some turbulence at the water surface, and also provide subsurface sparging of gases. The mass transfer and VOC stripping behavior of such systems therefore combines elements of both mechanical surface aeration (surface renewal aided VOC diffusion across the air-water interface due to turbulence) as well as diffuser aeration (off-gas VOC stripping.)

High Purity Oxygenation Systems for VOC Emissions Reduction

The VOC stripping potential of the large un-dissolved volume of gases (approximately 94%-96% of gas supplied) associated with the conventional aeration process has long been recognized. Process options that have high oxygen transfer efficiencies, such as high purity oxygenation systems, have been selected for several wastewater treatment applications where reduction in VOC emissions from the wastewater treatment process is a key project goal (e.g., Rodieck et al, 2001; Lubkowitz-Bailey et al, 2002; Levine et al, 2010; NYSERDA, 2000; Melcer, 1994). High purity oxygen (HPO) systems have been used in wastewater treatment applications for over four decades (Mueller et al, 2002). High efficiency oxygenation systems such as Praxair's In-Situ Oxygenation (I-SO™) system (Mueller, 2002) have been demonstrated to provide up to 97% reduction in VOC emissions relative to a comparative diffused air based aeration system (Rodieck et al, 2001).

Praxair's In-Situ Oxygenation (I-SO™) system is a subsurface, down-pumping mixing and oxygenation system used for providing oxygen in open basin HPO operations (Figure 1). The down-pumping action of the helical impeller generates a vortex which entrains pure oxygen gas supplied to a floating hood. The high shear in the draft tube region breaks up the entrained gas into fine bubbles. The intimate gas-liquid contact in the draft tube significantly enhances mass transfer. Dissolved oxygen is dispersed by the mixing action induced by the high velocity liquid jet ejected from the draft tube. Undissolved gas bubbles detach from the liquid flow and rise; some bubbles are recycled through the draft tube, allowing high oxygen transfer efficiencies to be attained. Standard aeration efficiencies of up to 10 lbs O_2/HP-hr and oxygen transfer efficiencies of up to 90%-95% can be obtained using the I-SO™ mixing system (Praxair, 2003; NYSERDA, 2000). HPO systems have high aeration efficiencies (8-10 lbs O_2/HP-hr), low mixing intensities (as low as <10 w/m^3) and high standard oxygen transfer efficiencies (+90%), and have been associated with significant reductions in VOC emissions when used in the wastewater treatment process (Mueller et al, 2002; Levine et al, 2010; Lubkowitz-Bailey et al, 2002; Rodieck et al, 2001; and NYSERDA, 2000).

METHODOLOGY

Given that commercially available modeling platforms offer only diffused aeration and mechanical surface aeration options as the default modules, the approach taken in this paper is to adapt aeration models (based on air) for use in modeling the behavior of high purity oxygen systems. Using Toxchem 4.1 as the fate & transport modeling platform, critical model parameters such as comparative gas flow rates, oxygen transfer efficiencies, O_2 mass fraction in gas source, ratio of the gas and liquid film mass transfer coefficients (K_g/K_l), aeration power, standard oxygen transfer rate (lbs O_2/HP-hr), and mass transfer correction factors (alpha) are varied for both diffused air and mechanical surface aeration systems to establish which variables to modify to adequately model the VOC emissions characteristics of pure oxygen systems.

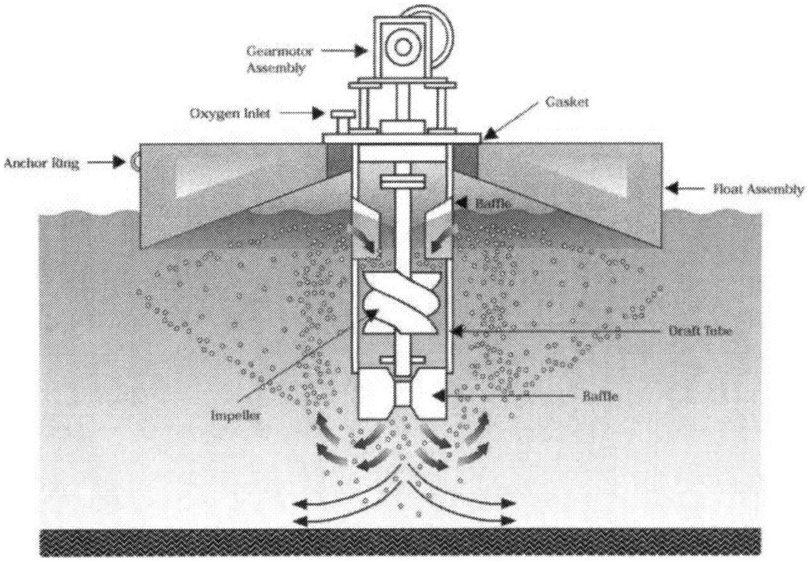

Figure 1: Praxair's In-Situ Oxygenation (I-SO™) system

Published data on comparative VOC emissions reductions from pilot and full scale applications using HPO vs. conventional air-based aeration systems are used as the data source for the analysis (Lubkowitz-Bailey et al, 2002; Rodieck et al, 2001).

In Lubkowitz-Bailey et al, 2002 and Rodieck et al, 2001, pilot and full scale data are presented for a SBR activated sludge process, used for the treatment of Active Pharmaceutical Ingredient (API) wastewater from a pharmaceutical plant. The data evaluated in this study is from the pilot system, which comprised a single SBR system operated in either diffused aeration or HPO mode (see Tables 1 & 2). The pilot scale SBR was operated in 6 hour cycles with 5 phases (Mix/Fill, React/Fill, React, Settle, and Decant). Wastewater flow of 250 gpd was supplied to the process. The reactor diameter was 10 ft, with a height of 15 ft, giving a total SBR volume of 6,000 gals. Fill volume during the tests was 4,650 gals, giving an effective liquid depth of about 11.6 ft. Gas flowrates supplied during the tests ranged from 0.32-0.67 cfm during HPO operations, and 28.5 to 32.5 cfm for diffused aeration operations. Under all aeration modes, enough oxygen was provided to ensure that COD removal targets were met. The VOCs contained in the API wastewater included Acetone, Toluene, MTBE, MEK, Methylene Chloride, 1,1-Dichloroethane, Ethanol, Isopropanol, Methanol, 1,1,1-Trichloroethane, Butyl Acetate and Heptane. During pilot testing, comparative VOC treatment performance was carried out in three baselines tests, and three spike tests (Rodieck et al, 2001). The spike tests were used as the basis for the modeling exercise in this study. Average MLSS was approximately 3,500 mg/l. Detailed monitoring of VOC emissions was undertaken by monitoring VOC in influent, effluent and off-gas samples.

Acetone and toluene were the major HAPs in the influent, with concentrations ranging from 7to 5,2200 and 49 – 320 mg/l respectively at pilot scale. Influent COD ranged from 4,069 mg/l (Rodieck et al, 2001) to 4,800 mg/l (Lubkowitz-Bailey et al, 2002).

The full scale system was designed for 240,000 gpd peak daily flow. Treatment was carried out in three (3) HPO SBR systems, with a design F/M basis of 0. 1 Ib BOD/lb MLSS-day. High Purity Oxygen supply to each SBR was provided using a single 40 HP Praxair I-SO™ system in each reactor, capable of transferring about 300 Ibs O_2/hr.

Model Configuration
VOC fate modeling was undertaken using the new Toxchem™ 4.1 modeling platform. This latest version of Toxchem™ (Hydromantis, 2012) includes High Purity Oxygen and Sequencing Batch Reactor Modules. Given that SBRs were utilized with pure oxygen and conventional aeration in the studies that were used as a basis for this paper (Rodieck et al, 2001; Lubkowitz-Bailey et al, 2002) the updated Toxchem™ platform enables the effective simulation of VOC and odor emission in treatment processes that involve the application of pure oxygen in subsurface mixing and dissolution systems. An earlier version of Toxchem™ which did not have custom SBR and HPO modules was used to support the modeling work in Rodieck et al (2001).

Since an integrated SBR and HPO module does not exist in Toxchem™ 4.1, the HPO SBR system was modeled as a SBR with modified diffused aeration parameters. To capture the dynamics of HPO operations, the effective oxygen transfer efficiencies and gas flowrates that apply to HPO operations were utilized in the diffuser models. Previous work has demonstrated that this approach to modeling HPO systems is effective (Fabiyi, 2012; Rodieck et al, 2001; NYSERDA, 2000). The three Spike tests reported in Rodieck et al (2001) were used as the basis for the simulations. The simulations were set up to evaluate HPO vs. diffused aeration system performance in parallel trains. Toxchem™ also allows visual displays that match emissions values around process units (emissions hotspots) to preset threshold values that are characterized by representative color codes (see Figure 2).

Model Calibration
Test data for the spike test runs were input into the representative model layouts and run using test values for process flowrates, oxygen mass transfer efficiencies and reactor conditions. A systematic analysis of the impact of various process parameters on the predictive accuracy of the model was undertaken using the Sensitivity Analysis tool in Toxchem™. This tool allows the evaluation of the impact of four broad classes of factors on model output, including: contaminant properties (e.g., Van't Hoff coefficients for estimation of Henry's Law Constant, adsorption characteristics, bio-kinetic properties of the compounds); site properties (elevation, wind speed, pH); and biological treatment system properties (e.g. liquid depth, oxygen transfer efficiencies, aeration power input, DO levels, Kg/Kl, aeration, etc.)

The sensitivity analysis indicated that the Van't Hoff estimates were the most important factor for ensuring that the model output was reliable. The default values for the Van't Hoff variables were reviewed and updated where necessary. Given that the reactor was operated at 80-100°F (Rodieck et al, 2001), default Henry's law constant data that were reported outside of this temperature range were dropped and replaced with the Van't Hoff estimates, which allow adjustment of Henry's Law constants for temperature.

A simplified form of the Van't Hoff relationship is given as

$$LnH_{cc,T} = A - \frac{B}{T}$$

Where $H_{cc,T}$ is the Henry's Law Constant, while A and B are the Van't Hoff coefficients. The value for A varies with the exact unit specifications used for Henry's Law Constant, but the numerical value estimated for B does not usually change for the commonly utilized unit expressions for $H_{cc,T}$.

Henry's Law Constant (HLC) values can be affected by a variety of factors such as temperature, pH, compound hydration, compound concentration, co-solutes, co-solvents, dissolved salts, suspended solids, dissolved natural organic materials, surfactants and water composition (Staudinger & Roberts, 1996). Starting Van't Hoff estimates were obtained using average values from a comprehensive compilation of published studies reporting experimental determinations of Henry's Law Constant temperature dependence relationships (Staudinger & Roberts, 2001). Table 3 summarizes the HLCs for the VOCs utilized in the model.

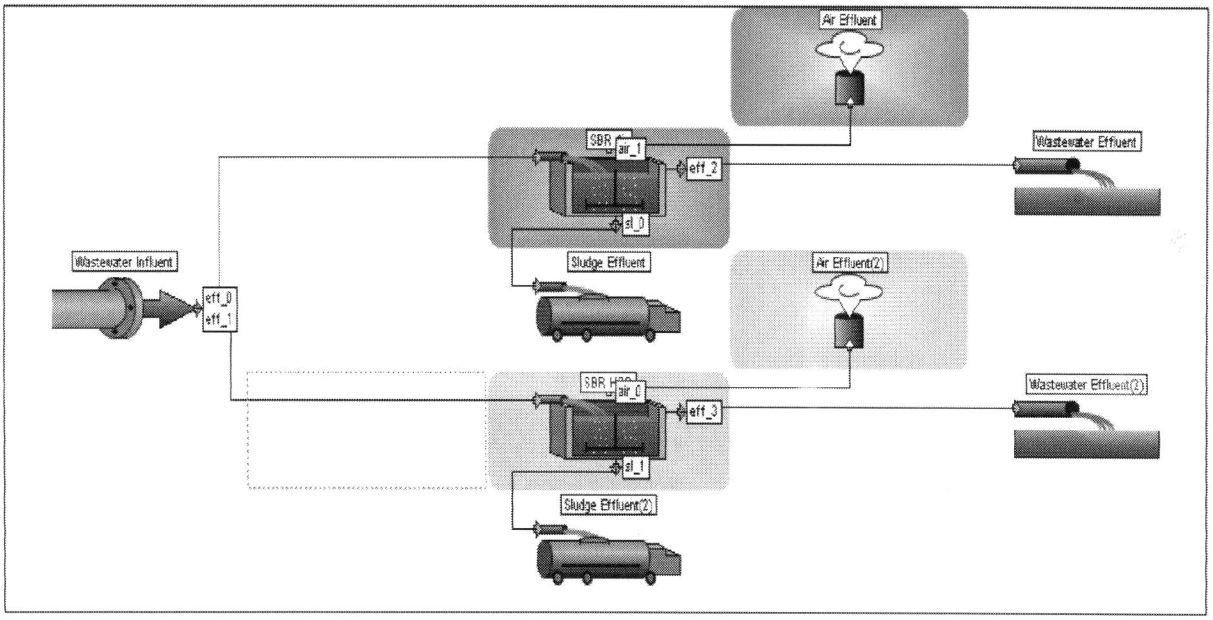

Figure 2: Layout depicting parallel trains used for evaluating VOC emissions behavior of HPO vs conventional diffused aeration SBR processes.

Table 1: Summary of Comparative VOC Emissions from pilot tests used in this study (Source: Rodieck et al, 2001).

FIELD MEASUREMENT RESULTS FOR TOLUENE

Testing Event	Percent of Influent Toluene Load Emitted to the Atmosphere			
	Measured		Model	
	HPO	DA	HPO	DA
HPO Baseline Test #1	0.1	---	0.3	---
HPO Baseline Test #2	0.5	---	0.4	---
DA Baseline Test	---	10.5	---	7.6
Spike Test #1	0.5	18.7	0.2	4.9
Spike Test #2	0.3	9.5	0.2	3.7
Spike Test #3	0.2	11.5	0.2	4.2
Average	0.3	12.6	0.2	5.1

Table 2: Model input variables (Source: Rodieck et al, 2001).

FATE AND TRANSPORT MODEL INPUT PARAMETERS

Category	Parameter	Units	HPO	DA
Ambient Conditions	Elevation	ft	329	
	Temperature	°F	80	
	Wind speed	mph	6.2	
Influent	VOC concentration	mg/L	variable	
	Suspended solids (SS)	mg/L	10	
	Volatile SS Ratio	%	10	
	Temperature	°F	80-100	
Reactor	Reactor Depth	ft	8.33	
	Reactor Surface Area	ft^2	30.6	
	Mixed liquor suspended solids (MLSS)	mg/L	3,000 – 4,400	
	MLSS volatile SS ratio	%	50	
	MLSS dissolved oxygen content	mg/L	2.0	
	Oxygen supply gas flow rate	cfm	0.33-0.80	15-35
	Oxygen transfer efficiency	%	95	10
	Mole fraction of oxygen source	---	1.0	0.2

Table 3: HLC = Henry's Law Constant. S&R (2001) refers to Staudinger, J & Roberts, P.V. A Critical Compilation of Henry's Law Constant Temperature Dependence Relations for Organic Compounds in Dilute Aqueous Solutions. Chemosphere 44 (2001).

		Default			S&R (2001)		Modified	
		HLC	Van't Hoff A	Van't Hoff B	Van't Hoff A	Van't Hoff B	Van't Hoff A	Van't Hoff B
Toluene			5.133	3024	5.271	1745	5.271	2800
	°C/St.Dev				(±1.11)	(±436.3)		
Methyl-Tertiary-Butyl Ether		0.022688			9.07	3178	6	3178
	°C/St.Dev	(@25°C)						
Methylene Chloride		0.122618			4.561	1644	4.561	2900
	°C/St.Dev	(@25°C)			(±0.50)	(±148)		
Butyl Acetate,n-		0.00671			6.4	2486	6.4	3846
	°C/St.Dev	(@25°C)						

RESULTS & OBSERVATIONS

a. Effect of HPO Operations on VOC Emissions

Rodieck et al (2001) carried out three spike tests using numerous VOCs that were intended to stress test the HPO system and evaluate its capacity for VOC reduction versus a comparative air based system. Those spike tests have been modeled in this paper, and good agreement was found between model predictions and test results. The VOCs of particular interest across all the spike tests that were carried out are Toluene, MTBE, $MeCl_2$ and n-Butyl Acetate. The model provided reliable estimations of the effect of aeration mode (conventional vs. HPO) on VOC stripping behavior (see Figures 3 – 6).

Spike Test 1 (Figure 3). 97% and 93% reductions in Toluene and MTBE emissions respectively were observed in Spike Test 1. The model also provides reliable estimates of VOC emissions reduction, with 98% and 97% reductions in Toluene and MTBE emissions respectively. . Comparative gas supply ratio V_{Air}/V_{HPO} (Air volume vs. HPO volume) is 91.2 (i.e., 77 ft3 HPO, 7024 ft3 air).

Spike Test 2 (Figure 4). 97% and 96% reductions in Toluene and Methyl Chloride emissions respectively were observed. Model estimates of VOC emissions reduction tracked test results at 97% for Toluene and 96% for Methyl Chloride. Comparative gas supply ratio V_{Air}/V_{HPO} (Air volume vs. HPO volume) is 54.6 (i.e., 125 ft3 HPO, 6830 ft3 air)

Spike Test 3 (Figure 5). 96% and 98% reductions in Toluene and Butyl Acetate emissions respectively were observed during the test. Model estimates of VOC emissions reduction was 97% for Toluene and 93% for Butyl Acetate. Comparative gas supply ratio V_{Air}/V_{HPO} (Air volume vs. HPO volume) is 48.4 (i.e., 161 ft3 HPO, 7800 ft3 air)

The data and simulation results across all the spike tests consistently demonstrate the significant reduction in VOC emissions obtained by using HPO instead of air as the source of oxygen in the process gas supply. Toluene was common to all the spike tests, and the average reduction in

Toluene related VOC emissions across all tests was 97% for HPO vs. conventional diffused aeration based systems (see Figure 6). Differences in the comparative volumes of gas introduced into the system during air versus HPO operations ranged from 48-92.

b. Comparison of HPO and SBR Modules

The predictive capacity of the new HPO module in Toxchem™ 4.1 was tested by comparing the results from simulations of the Batch 3 spike tests of Rodieck et al (2001) using the SBR and HPO modules. As previously indicated, the modeling of HPO operations in the SBR is done using diffused aeration principles. Although the I-SO™ system is a mechanically agitated mixing and gas dissolution system, it differs fundamentally from surface aerators because it is a subsurface mixer. The stripping behavior of subsurface mixing systems in the wastewater treatment process is modeled most accurately by the off-gas volumes. The mass transfer characteristics of the new HPO module in Toxchem™ is modeled on the hydrodynamics and mass transfer behavior of the I-SO™ system (Praxair, 2003).

Effect of Depth and Surface Area: The default HPO and Activated Sludge Diffusion Aeration modules in Toxchem™ are complete mix systems. In order to model an SBR system using these modules it could be necessary to modify the sequencing batch steps in a SBR as separate process units. We investigated the effect of depth and surface area on the stripping behavior in the HPO and Diffuse Aeration process layouts to determine whether the modification of either of these two parameters could be tolerated when trying to model the SBR using complete mix systems (see Figure 7). For the sensitivity analysis, depth ranged from 2 to 20 ft, and surface area from 30 to 200 sqft. Actual SBR depth and surface area were 11.6ft and 78.5 sqft respectively. The analysis indicated that while depth and surface area had no major effect on the stripping behavior in the HPO system, the diffused aeration system was significantly affected by these parameters. The sensitivity of the VOC stripping behavior in diffused aeration systems to surface area and depth, suggest that the modification of either of these parameters when trying to model the SBR using complete mix systems could lead to issues with the interpretation and accuracy of the model results.

Due to this observation, a simplified layout was selected, which comprised only of the HPO and Diffused Aeration module with no additional process units (Figure 8). The process units had the same depth and surface area values as the actual SBR. A key limitation to the method could stem from high VOC losses during the fill cycle of the SBR. Rodieck et al (2001) minimized VOC losses and volatilization during the fill cycle by feeding the liquid below the liquid surface (Geiling, 2012).

The results obtained for VOC reduction for HPO vs. conventional aeration processes using this approach compared well with results from the SBR modeling approach (see Figure 9).

c. *Effect of Gas Flow rate on VOC Emissions*

There are significant differences in gas flows between HPO and conventional aeration systems. As previously observed, the use of HPO systems can reduce gas flow as much as 50-100 times compared with air based process. Reduced gas supply and off gas rates in HPO systems is a key factor in the enhanced VOC reduction observed in HPO systems (Fabiyi, 2012, Rodieck et al, 2001; Levine et al, 2010).

A sensitivity analysis was run in Toxchem™ to determine the effect of gas flowrates on VOC stripping. A linear correlation was observed between VOC stripping and the volumes of airflow supplied to the wastewater treatment process within the range of flowrates analyzed (see Figure 10). The linear relationship observed between gas flowrates and VOC stripping provides strong support for the identification of bubble-aided stripping as a key mechanism for VOC emissions in diffused aeration and subsurface mechanically aerated systems (Levine et al, 2010).

CONCLUSIONS

VOC emissions are affected by a variety of factors including mixing intensity, gas flowrates, surface renewal rates and surface turbulence (Chern & Yu, 1995). This implies that VOC emissions can be reduced by using aeration devices that (i) have a relatively low power input, hence minimizing excessive turbulence and surface renewal, and (ii) provide high oxygen transfer efficiencies (OTE), thereby minimizing the amount of un-dissolved gas bubbles that will strip VOCs from the liquid.

We find that the VOC emission behavior of the HPO system is predicted accurately by modeling the mass transfer characteristics as that of a fine bubble diffuser system, with adjustments for gas flows (actual O_2 feed to the HPO unit is used), O_2 purity and oxygen transfer efficiency.

Attempts to utilize the default mechanical aeration module indicated that the mass transfer characteristics of surface aeration systems are not well suited to model HPO systems (Fabiyi, 2012). The ready adaptability of diffuse aeration models for capturing the dynamics of VOC emission from mechanical HPO oxygenation systems suggests that the dominant VOC emission mechanism in both diffused air and subsurface aeration systems is similar – VOC transfer to un-dissolved gas bubbles. In contrast, mechanical surface aeration systems introduce a third mechanism into VOC stripping dynamics. In addition to volatilization due to surface renewal from turbulence, and VOC emissions due to volatile transfer to un-dissolved gas bubbles, VOC losses also occur due to exchange between the air and volatiles in the airborne water droplets. Gross air and water conditions such as temperature and wind velocities therefore become more important in accurately estimating VOC emissions in mechanical surface aeration systems (Chern & Chou, 1999; Roberts & Dandliker, 1983).

Despite the potential for VOC emission reduction afforded by high purity oxygen systems for wastewater treatment, commonly available fate & transport models do not have default modules that enable their modeling (Levine et al, 2010; Lubkowitz-Bailey et al, 2002; Rodieck et al, 2001). Toxchem™ 4.1 is the first commercially available fate modeling tool that has a default HPO module. Our work has demonstrated that this new tool is robust and reliable, accurately modeling both pilot and full scale application of HPO aeration for VOC emission reduction.

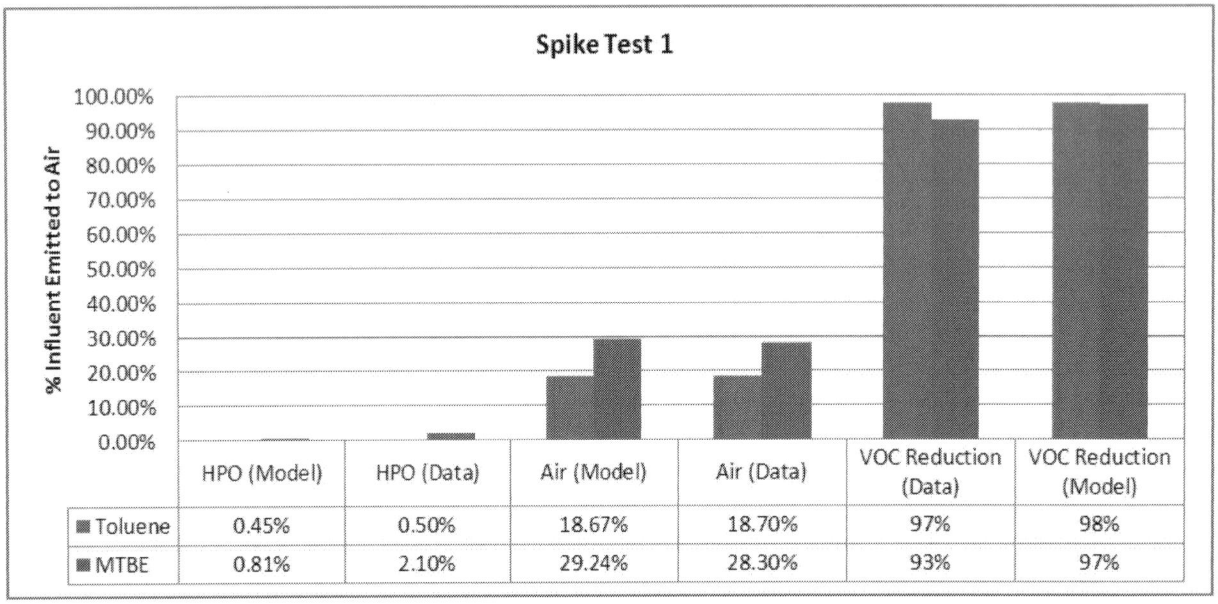

Figure 3: Effect of aeration mode on VOC Stripping in Spike Test 1

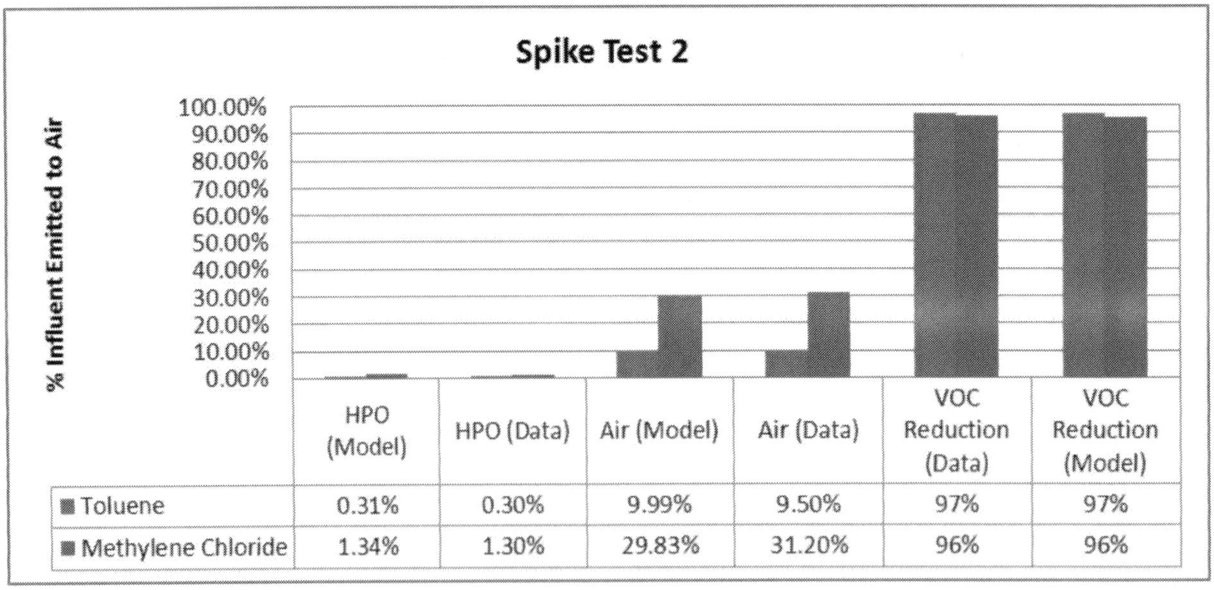

Figure 4: Effect of aeration mode on VOC Stripping in Spike Test 2

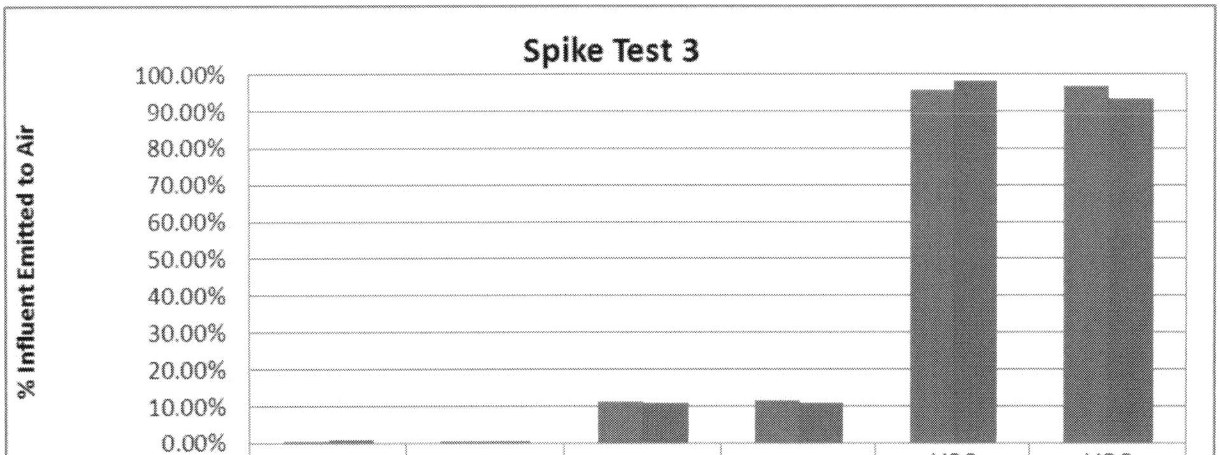

	HPO (Model)	HPO (Data)	Air (Model)	Air (Data)	VOC Reduction (Data)	VOC Reduction (Model)
■ Toluene	0.38%	0.50%	11.18%	11.50%	96%	97%
■ Butyl Acetate, -n	0.72%	0.20%	10.88%	10.80%	98%	93%

Figure 5: Effect of aeration mode on VOC Stripping in Spike Test 3

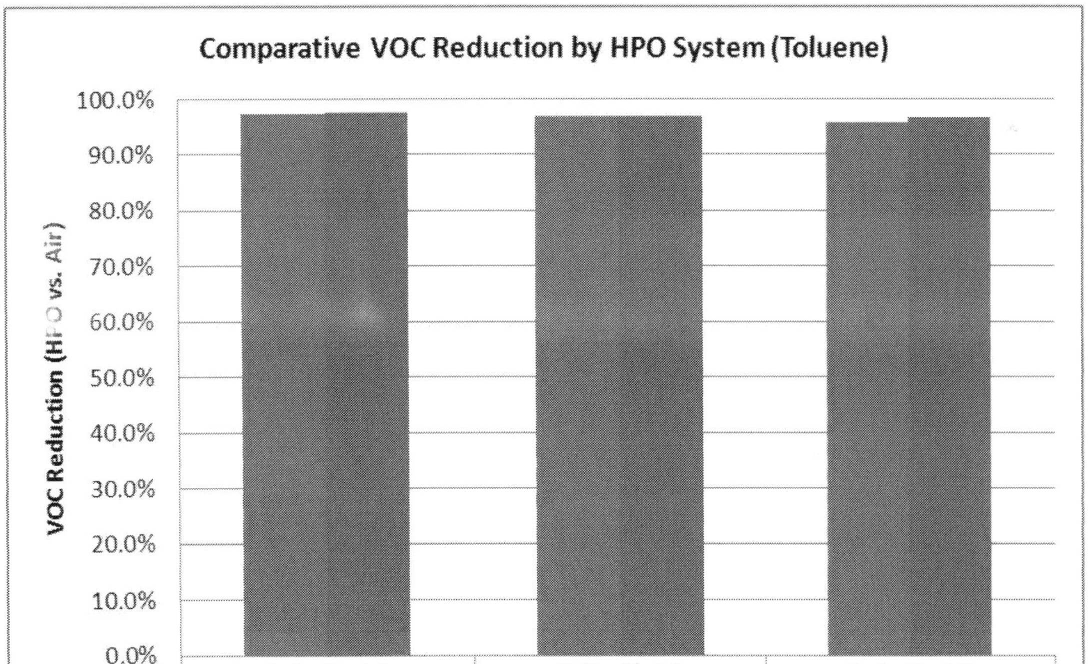

	Spike Test 1	Spike Test 2	Spike Test 3
■ Data	97.3%	96.8%	95.7%
■ Model	97.6%	96.8%	96.6%

Figure 6: Summary of Modeling and Empirical Values for VOC Emissions reduction obtained using I-SO™ HPO Systems

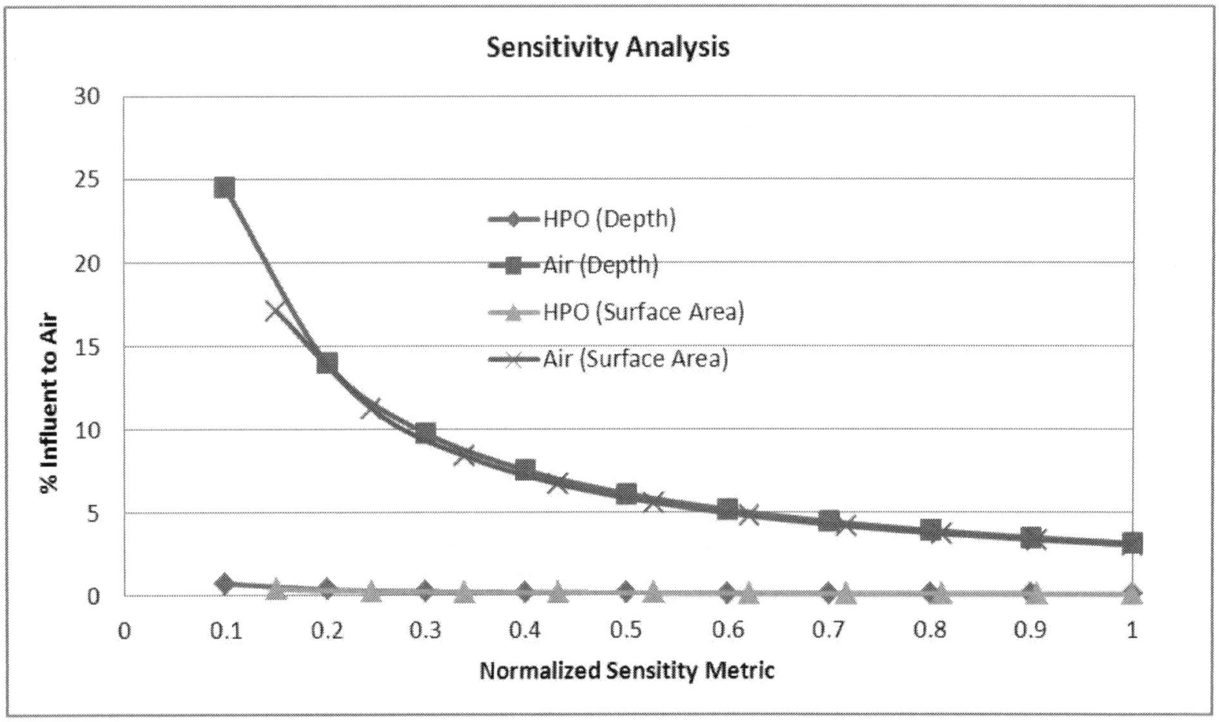

Figure 7: Effect of depth and surface area on VOC emissions for Toluene.

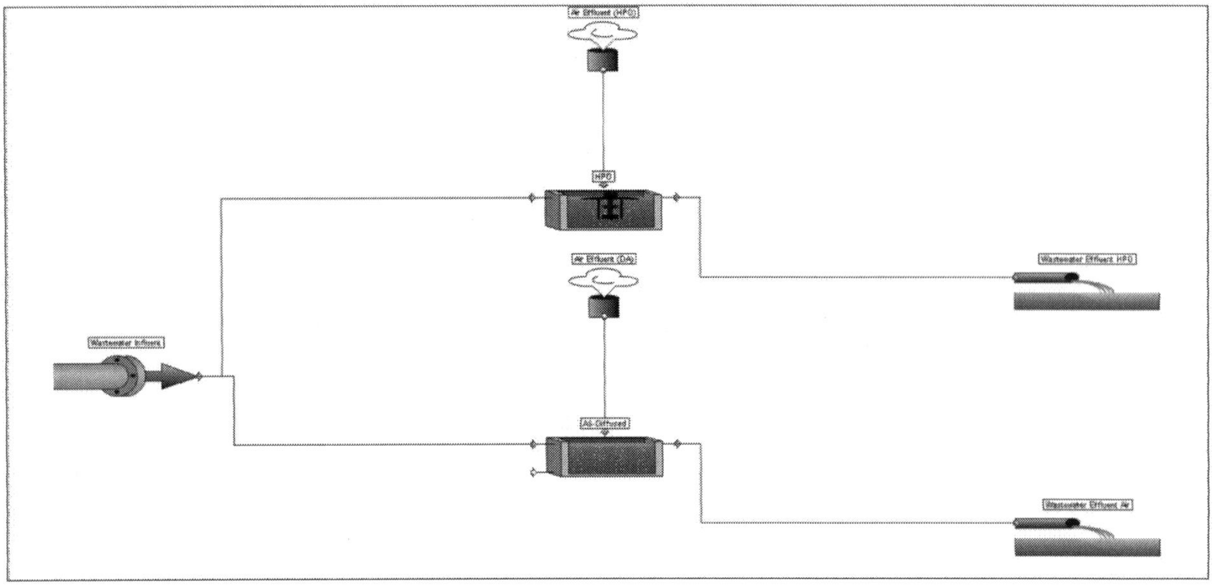

Figure 8: Simplified process layout with HPO and Diffused Aeration modules.

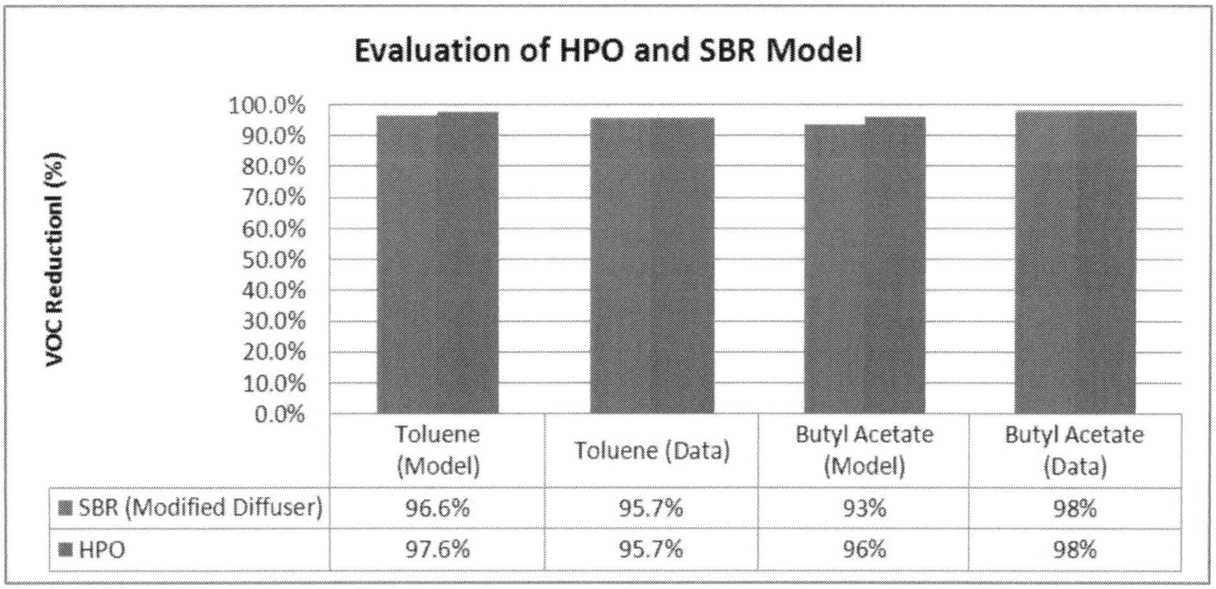

Figure 9: VOC reduction for HPO vs. Air based processes using the SBR model with modified diffused aeration parameters and the new HPO process layout.

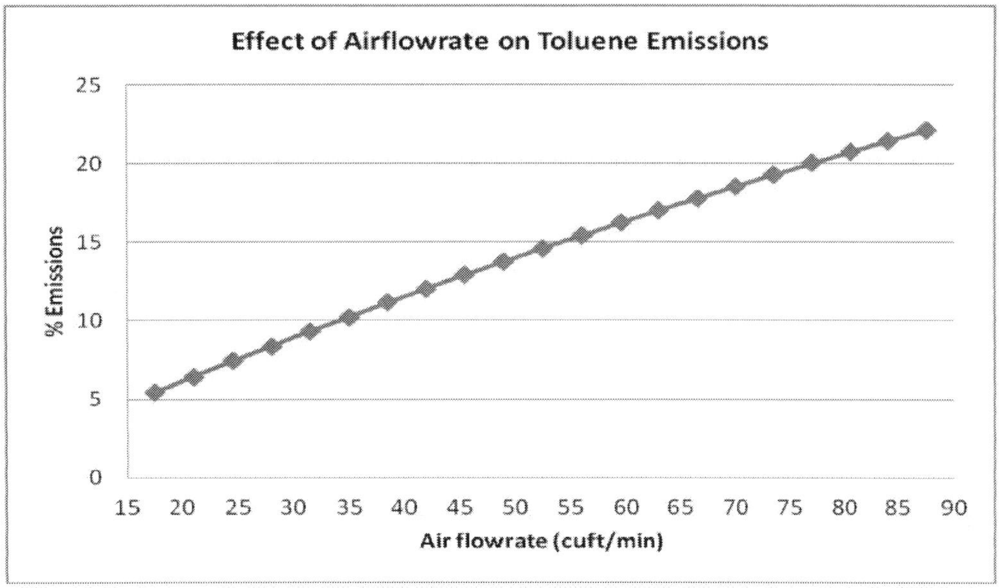

Figure 10: Effect of gas flowrates on Toluene emissions in API pharmaceutical wastewater

REFERENCES

1. Geiling, C. Personal Communications. May 2012.
2. Hydromantis, 2012. Toxchem Version 4.1 Release Notice. http://www.hydromantis.com/news_new.html
3. Jia-Ming Chern and Cheng-Fu Yu. Volatile Organic Compound Emission Rates from Diffused Aeration Systems. 1. Mass Transfer Modeling. Ind. Eng. Chem. Res. 1995, 34, 2634-2643.
4. Jia-Ming Chern and Shun-Ren Chou. Volatile Organic Compound Emission Rates from Mechanical Surface Aerators: Mass Transfer Modeling. Ind. Eng. Chem. Res. 1999, 38, 3176-3185.
5. Levine et al, 2010. Pure Oxygen Activated Sludge Unit Effectively Controls Volatile Organic Compound Emissions from a Mixed Petrochemical Wastewater, WEFTEC 2010
6. Lubkowitz-Bailey et al, 2002. Start-up of a High Purity Oxygen Sequencing Batch Reactor System for Treatment of an Active Pharmaceutical Ingredient Wastewater. WEFTEC 2002
7. Melcer, H. Monitoring and Modeling VOCs in Wastewater Facilities. Environmental Sci. Tech. Vol 28, No. 7, 1994.
8. Mueller et al, 2002. Aeration: Principles & Practice.
9. NYSERDA Report, 2000. In-Situ Oxygenation (I-SO™) for Volatile Compound Emission Control.
10. Praxair Report of Clean Water Tests at Philadelphia Mixing Solutions, 2003.
11. Quigley, C., Card, T., Monteith, H., Witherspoonm, J., Adams, G., Pettit, W., & Torres, E. A Comparison of Three Wastewater Collection System VOC Emissions Models. WEFTEC, 2006.
12. Roberts, P. V., & Dandliker, P. G. Mass Transfer of Volatile Organic Contaminants from Aqueous Solution to the Atmosphere during Surface Aeration. Environmental Sci. Tech. Vol. 17, No. 8, 1983.
13. Rodieck et al, 2001. Alternative Treatment Strategy for High Strength Pharmaceutical Wastewater. WEFTEC 2001.
14. Staudinger, J & Roberts, P.V. 1996. A Critical review of Henry's Constant for Environmental Applications. Critical Rev. Environ. Sci. Technol. 26 (3), 205-297
15. Staudinger, J & Roberts, P.V. A Critical Compilation of Henry's Law Constant Temperature Dependence Relations for Organic Compounds in Dilute Aqueous Solutions. Chemosphere 44 (2001) pp 561 – 576.
16. Woodward & Curran, 2006. Industrial Waste Treatment Handbook

(582a) The Growth of a Vapor Bubble On a Heated Surface Using the Lattice Boltzmann Calculation: Flow, Heat Transfer and Phase Change

Lin Zheng, Chemical Engineering, The City College of City University of New York, New York, NY, Taehun Lee, Mechanical Engineering, The City College of City University of New York, New York, NY and David S. Rumschitzki, Department of Chemical Engineering, The City College of The City University of New York, New York, NY

Nucleate boiling is a very attractive method for cooling down a hot surface because, due to typically large latent heats, evaporation of a small amount of liquid can absorb a large amount of heat from the solid. The danger, however, is that the solid surface can even temporarily dry out, and the consequent hiatus in cooling could be catastrophic. One would therefore like to understand the physics of nucleate boiling well enough to prevent dryout.

Previously we reported conduction-only calculations for this problem that, surprisingly, yielded results that agreed with a large amount of phenomenology and experimental observations for these systems. These calculations showed, however, that nearly all of the evaporation takes place at the three-phase contact line (CL). Moreover the poorly-understood motion of the CL is critical to the evolution of vapor bubbles as gravity deforms them and drives them toward detachment from the solid surface, processes that determine which boiling regimes the system is in and that are clearly central to system's deciding whether dryout will occur. Since CL motion with bubble growth is an input to a traditional continuum calculation, and since, even in the absence of phase change, is a subject of much disagreement in the literature, we seek to employ a method that obviates the need for ad hoc assumption about CL motion.

As such, we employ and extend the lattice Boltzmann equation (LBE) method, one of a class of methods known as diffuse interface methods, to include heat transfer, fluid flow and phase transition phenomena and implement it in two dimensions. The advantage of this method is that it considers the liquid and the vapor as a single phase of variable density. As such, the liquid-vapor interface is in the fluid domain and, in particular, the CL is part of its solid-fluid boundary, and their positions are outputs of the calculation, rather than input to it. The price that one pays for this approach is that the interface no longer remains nanoscopic, but rather is resolved over 3-5 gridpoints. The way that we employ this method is by using phase field theory to determine the transport properties of the multiphase flow such as pressure tensor and chemical potential from the Helmholtz free energy. This uses a known LBE formulation for the flow field. For the temperature field, we derive a LBE that, via Chapman-Enskog expansion techniques, can recover the correct energy equation (in an entropy formulation that allows a straightforward incorporation of the latent heat) from thermodynamic theory by ignoring the viscous dissipation term. We verify the accuracy of these calculations by comparison with problems having known analytic (e.g., similarity) solutions, such as Laplace's law, Stephan problem and Scriven's growth of a single bubble in an infinite fluid.

We look at the growth of a vapor bubble on an infinitely thin heated solid surface. Since we require the solid surface to be isothermal, it fixes the CL temperature as the surface temperature, unlike our earlier conduction-only calculation that included the solid and set the CL temperature equal to the saturation temperature. We calculate the temperature and flow fields and the motion of the CL at various time points up to and beyond bubble detachment and compare with (3D) data in the literature. We consider the size of the escaping bubble and of the vapor residue and its fate, including the descent and impingement of colder fluid that replaces the escaping detached bubble. Time permitting, we consider the case with a finite thickness solid as well and how issues of finite interface thickness play out in the solutions.

(582b) Rare Events Calculations of Methane Clathrate Hydrate Nucleation From Solution At Moderate Supersaturations

Brandon C. Knott[1], Valeria Molinero[2], Michael F. Doherty[1] and Baron Peters[1,3], (1)Chemical Engineering, UC Santa Barbara, Santa Barbara, CA, (2)Chemistry, University of Utah, Salt Lake City, UT, (3)Chemistry and Biochemistry, UC Santa Barbara, Santa Barbara, CA

Clathrate hydrates are crystalline ice-like inclusion compounds in which water molecules form cages around hydrophobic guest molecules. Methane clathrate hydrates form at high pressures and low temperatures and are found in abundance on the ocean floor and under permafrost. Additionally, their formation in oil and natural gas pipelines can severely affect pipeline economics and safety. Past computational studies aimed at elucidating the mechanism of hydrate nucleation used "brute force" molecular dynamics (MD) trajectories. These trajectories are run at extremely high driving force (i.e. supersaturation) compared to driving forces that exist in nature. We used a coarse-grained model of methane hydrates[1] to perform phase coexistence calculations and to compute supersaturations. Then we used rare events methods to examine methane hydrate nucleation at moderate supersaturation. These calculations enable us to estimate nucleation rates and draw conclusions regarding the reaction mechanism.

References

[1] L.C. Jacobson and V. Molinero, "A Methane-Water Model for Coarse-Grained Simulations of Solutions and Clathrate Hydrates", J. Phys. Chem. B, **2010**, 114, 7302-7311.

(582c) Quantifying Clathrate Hydrate Nucleation and Growth From Molecular Simulations

Amadeu K. Sum[1], David T. Wu[2], Somendra Nath Chakraborty[1] and Eric Grzelak[3], (1)Chemical & Biological Engineering, Colorado School of Mines, Golden, CO, (2)Chemical Engineering and Chemistry Departments, Colorado School Of Mines, Golden, CO, (3)Chemical & Biological Engineering Department, Colorado School of Mines, Golden, CO

Despite their industrial significance and ubiquity, the formation mechanism of hydrates remains poorly understood. The challenges associated with describing hydrate nucleation and growth mechanisms remain at the forefront of fields such as hydrogen storage, carbon dioxide sequestration, and hydrocarbon transportation and recovery. Various mechanisms of hydrate formation have been postulated, most involving the annealing of an amorphous or partially-ordered aggregation of cages into the thermodynamically-preferred crystal structures. In this talk, we will present an overview of our efforts in understanding clathrate hydrate nucleation and growth using molecular simulations, including defining the thermodynamics phase space and direct simulations capturing the different stages through the nucleation and growth processes. The molecular events during nucleation and growth are quantified by applying different order parameters to represent the structuring of water and gas molecules. One of the approaches is based on a Voronoi tessellation analysis, which provides information on the cooperative ordering of gas and water molecules. In applying the tessellation to the nucleating trajectories, we observe that several of the Voronoi properties (polyhedra volume, total surface area, asphericity parameter) show strong signatures of change when hydrates nucleate. The results suggest that the Voronoi tessellation analysis is a powerful and efficient tool that can be applied to obtain further insight into the mechanisms associated with the ordering of methane and water molecules in the nucleation and growth processes of hydrates.

(582d) Insights Into Hydrate Nucleation Using Molecular Simulations

Sapna Sarupria, Chemical and Biomolecular Engineering, Clemson University, Clemson , SC and Pablo G. Debenedetti, Chemical and Biological Engineering, Princeton University, Princeton, NJ

Gas hydrates represent an important class of crystalline solids that have applications in wide ranging fields such as petrochemical industry, environment and energy. While the thermodynamics of hydrate formation is well understood, we have very little insights into the kinetics of hydrate nucleation. Nucleation studies in experiments are challenging due to the difficulty in accessing the nanometer length and nanosecond timescales that are relevant to nucleation processes. Conventional molecular dynamics simulations, which sample nanometer length scales and nano-to-microsecond time scales are ideally suited to probe this phenomenon. However, the long waiting times involved in observing nucleation events in simulations make it computationally challenging to sample statistically relevant number of trajectories. Therefore, we approach this problem from two perspectives - first we generated several microsecond long trajectories of methane hydrate nucleation, and performed extensive analysis of the events leading to nucleation. This provides valuable insights into various factors that feature in hydrate nucleation. Secondly, based on the insights from the nucleation trajectories, we carried out detailed studies of these factors. Together, our studies provide insights into aspects governing hydrate nucleation, and will help design further studies focused on the kinetics of hydrate nucleation, and growth.

(582e) A Microscopic View of Nucleation in the Anatase-to-Rutile Transformation

Ya Zhou and Kristen Fichthorn, Dept. of Chemical Engineering, Pennsylvania State University, University Park, PA

We use molecular simulation techniques to investigate the anatase-to-rutile tansformation in titania nanocrystals. A thermodynamic analysis indicates that edge and corner atoms significantly influence the critical size at which rutile nanocrystals become energetically preferred over anatase. Using molecular dynamics simulations, we probe kinetics of the transformation in individual anatase nanocrystals and in nanocrystal aggregates. We follow structural evolution using simulated X-ray diffraction. Additionally, we develop a local order parameter to distinguish individual Ti ions as anatase, rutile, or anatase {112} twin-like. We apply our local order parameter to track the formation and growth of rutile nuclei. Anatase {112} twins form easily at surfaces and interfaces of nanocrystal aggregates and rutile nucleates among the twins. Stable rutile nuclei maintain {101} facets during growth, as a result of nucleation from layers of alternating anatase {112} twins, until the growth is inhibited at surfaces and grain boundaries. Our results are in agreement with experiment and indicate the central role of {112} twin-like anatase in the transformation.

(582f) Gel-Induced Selective Crystallization of Polymorphs

Ying Diao[1], Allan S. Myerson[2], T. Alan Hatton[1] and Bernhardt L. Trout[3], (1)Chemical Engineering, Massachusetts Institute of Technology, Cambridge, MA, (2)Department of Chemical Engineering, Massachusetts Institute of Technology, Cambridge, MA, (3)Massachusetts Institute of Technology, Cambridge, MA

Although nanoporous materials have been explored for controlling crystallization of polymorphs in recent years, polymorphism in confined environments is still poorly understood, particularly from a kinetic perspective, and the role of the local structure of the substrate has largely been neglected. Herein, we report the use of a novel material, polymer microgels with tunable microstructure, for controlling polymorph crystallization from solution and for investigating systematically the effects of nanoconfinement and interfacial interactions on polymorphic outcomes. We show that the polymer microgels can improve polymorph selectivity significantly. The polymorphic outcomes correlate strongly with the gel-induced nucleation kinetics and are very sensitive to both the polymer microstructure and the chemical composition. Further mechanistic investigations suggest that the nucleation-templating effect and the spatial confinement imposed by the polymer network may be central to achieving polymorph selectivity. We demonstrate polymer microgels as promising materials for controlling crystal polymorphism. Moreover, our results help advance the fundamental understanding of polymorph crystallization at complex interfaces, particularly in confined environments.

(582g) Model Based Modified Brust-Schiffrin Method for Synthesis of Monodispersed Gold Nanoparticles

Sivaramakrishna Perala, Department of Chemical Engineering, Indian Institute of Science, Bangalore, India and Sanjeev Kumar, Department of chemical engineering, Indian Institute of Science, Bangalore, India

The synthesis of thiol capped gold nanoparticles using Brust-Schiffrin Method[1] (BSM) produces sub 5 nm gold nanoparticles on gram scale. The high yield of extremely stable particles and varied options for surface functionalization have made BSM a widely used synthesis protocol in a number of fields. One of its key shortcomings, high polydispersity of particles, is overcome by size fractionation, a tedious procedure which also decreases the yield of particles in the desired size range. In this work, we have developed a comprehensive model for BSM of synthesis and used it to develop modified synthesis protocols to produce particles with improved monodispersity. The new protocols have been verified experimentally.

The two-phase BSM involves complex reduction reactions facilitated by phase transfer agents, nucleation, and simultaneous growth and capping of nanoparticles. The existing body of experimental findings, including the most recent one implicating micelles of phase transfer agent as the reaction sites[2], was analyzed and subjected to additional experimentation to develop a mechanistic picture of the synthesis process. Since populations of particles with different sizes and different free surface area available for growth need to be tracked, the framework of bi-variate population balance equations was used to capture the particle synthesis process. The model successfully captures the experimental findings available in the literature. It unravels competition between capping and continued nucleation as key to size and polydispersity control. The model is next used to simulate particle synthesis for alternative, non-obvious mixing protocols. Two improved strategies are proposed through insight gained from the model developed. Experiments conducted by using modified protocols exhibited better control over the monodispersity than the original BSM protocols.

References:

1.   Brust, M.; Walker, M.; Bethell, D.; Schiffrin, D. J.; Whyman, R., *J. Chem. Soc., Chem. Commun.*, 1994, **7**, 801-802. 2.   Li, Y.; Zaluzhna, O.; Tong, Y., *Langmuir*, 2011, **27**, 7366-7370.

(620a) Computational Characterization of MOF Pores and Determination of Molecular Selectivity

Eric L. First, Chrysanthos E. Gounaris, James Wei and Christodoulos A. Floudas, Department of Chemical and Biological Engineering, Princeton University, Princeton, NJ

We previously reported on a computational approach for the systematic characterization of the three-dimensional porous networks of zeolites [1]. Starting with only the crystallographic information for a structure, we automatically identify all void space that can potentially accommodate a guest molecule and classify it into a network of interconnected channels and cages. We have now extended our methodology to support the broader class of microporous materials known as metal-organic frameworks (MOFs).

Due to their large pore volumes and surface areas, MOFs are attractive materials for applications including carbon capture and hydrogen storage. The syntheses of hundreds of MOFs have been reported in the literature [2], and over a hundred thousand hypothetical MOFs have been generated [3]. While MOFs have the potential to be systematically synthesized, the vast number of possibilities calls for computational methods for database screening, so that only the most promising structures for a desired application need to be studied experimentally.

We present the first computational method capable of automatically identifying the windows, channels, and cages of a MOF and describing their geometry and connectivity. In addition, we perform a variety of quantitative calculations including window sizes, pore size distribution, accessible volume and surface area, pore limiting diameter and largest cavity diameter, and coordination number. We have applied our method to a large selection of MOFs, including zeolitic imidazolate frameworks (ZIFs) and hypothetical MOFs.

We have also developed a procedure to screen databases of microporous materials and identify those most suitable for a desired molecular separation. The approach is an extension of our previous work on zeolites [4-6], where we calculate the selectivity of a microporous material for a separation by comparing the activation energy required for two different molecules to pass through the adsorbent's windows. We demonstrate our method by identifying MOFs with the greatest potential to separate carbon dioxide from nitrogen and hydrogen from methane.

References:

[1] First, E. L., Gounaris, C. E., Wei, J., and Floudas, C. A. Computational characterization of zeolite porous networks: an automated approach. Physical Chemistry Chemical Physics, 13(38):17339-17358, 2011.

[2] Ockwig, N. W., Delgado-Friedrichs, O., O'Keeffe, M., and Yaghi, O. M. Reticular Chemistry: Occurrence and Taxonomy of nets and Grammar for the Design of Frameworks. Accounts of Chemical Research, 38(3):176-182, 2005.

[3] Wilmer, C. E., Leaf, M., Lee, C. Y., Farha, O. K., Hauser, B. G., Hupp, J. T., and Snurr, R. Q. Large-scale screening of hypothetical metal-organic frameworks. Nature Chemistry, 4(2):83-89, 2012.

[4] Gounaris, C. E., Floudas, C. A., and Wei, J. Rational design of shape selective separation and catalysis–I: Concepts and analysis. Chemical Engineering Science, 61(24):7933-7948, 2006.

[5] Gounaris, C. E., Wei, J., and Floudas, C. A. Rational design of shape selective separation and catalysis–II: Mathematical model and computational studies. Chemical Engineering Science, 61(24):7949-7962, 2006.

[6] Gounaris, C. E., Wei, J., Floudas, C. A., Ranjan, R., and Tsapatsis, M. Rational design of shape selective separations and catalysis: Lattice relaxation and effective aperture size. AIChE Journal, 56(3):611-632, 2009.

(620b) In Situ ATR-FTIR Analyses of CO2 Adsorption Processes in Porous Coordination Polymers

Jeffrey Culp[1,2], Angela Goodman[1], Danielle Chirdon[1], Kristi Kauffman[1], Lang Sui[1], David Luebke[1] and Christopher Matranga[1], (1)U.S. Department of Energy, National Energy Technology Laboratory, Pittsburgh, PA, (2)URS Corporation, South Park, PA

The adsorption/desorption behaviors for CO_2 on several different porous coordination polymers were studied using *in situ* attenuated total reflectance infrared spectroscopy in a high pressure cell. The significant IR adsorption cross section of CO_2 and the sensitivity of its IR adsorption bands to the adsorption pore environment make the molecule well-suited for *in situ* IR investigations. Give the potential benefits of the spectroscopic technique, literature reports of its use in the study of CO_2 adsorption processes in porous coordination polymers is surprising limited. The experimental technique is shown to be a powerful tool for investigating the CO_2 adsorption mechanisms in a wide range of porous materials and composite compounds including structurally dynamic materials, porous spin crossover compounds, and mixed matrix coordination polymer/organic polymer composites. The *in situ* measurements under variable pressures of CO_2 also allow transitions in the host structures to be correlated with the CO_2 adsorption process making the technique particularly valuable in deriving a detailed description of the adsorption process in structurally dynamic materials.

NMR cryoporometry as a complementary characterization technique

D. Kondrashova and R. Valiullin

Faculty of Physics and Earth Sciences, University of Leipzig, Leipzig, Germany

Gas sorption measurements are routinely used for characterizations of mesoporous solids. With better understanding of fluid behavior in confined spaced of materials with complex pore structures and with the development of novel DFT approaches, this method nowadays provides a considerable accuracy in structural analysis. In recent two decades, NMR cryoporometry has emerged as a novel experimental technique, which can provide better spatial resolution as compared to conventional nitrogen sorption measurements. Quantitative analysis of the data obtained using NMR cryoporometry, however, still remain quite approximate due to an incomplete understanding of underlying physical processes and the lack of microscopic approaches similar to those used for modeling gas-liquid equilibria. By means of experiments and theoretical modeling, we have obtained deeper insights into mechanisms controlling solid-liquid transitions in disordered mesoporous solids. Thus, using NMR cryoporometry as an experimental technique and Vycor porous glass as a model system, we have performed first detailed experimental studies of melting and freezing scanning behavior, namely melting/freezing patterns following incomplete freezing/melting of the confined fluid [1]. This type of experiments has provided some additional information on thermodynamics of the transition processes. The experimental results have been analyzed with the help of a microscopic model of crystal growth adopted for confined fluids. In this way, we have qualitatively reproduced most of experimental findings reported so far in the literature.

[1] Kondrashova, D., Reichenbach, C. & Valiullin, R. (2010) Probing pore connectivity in random porous materials by scanning freezing and melting experiments. *Langmuir* 26, 6380.

(620d) Characterization of Novel Ordered Mesoporous Carbons

Richard T. Cimino[1], Gennady Gor[2], Katie Cychosz[3], Matthias Thommes[3] and Alexander V. Neimark[1], (1)Department of Chemical and Biochemical Engineering, Rutgers, The State University of New Jersey, Piscataway, NJ, (2)Department of Civil and Environmental Engineering, Princeton University, Princeton, NJ, (3)Quantachrome Instruments, Boynton Beach, FL

Characterization of Novel Ordered Mesoporous Carbons

Richard Cimino[1], Gennady Gor[3], Katie A. Cychosz[2], Matthias Thommes[2] , and Alexander V. Neimark[1]

(1) Rutgers University, Department of Chemical and Biochemical Engineering,

98 Brett Road, Piscataway, NJ 08854, USA

(2) Quantachrome Instruments, 1900 Corporate Dr., Boynton Beach, FL 33426, USA

(3) Princeton University, Department of Civil and Environmental Engineering, Princeton, NJ USA

During last decade, the density functional theory (DFT) methods of pore size analysis were found to be superior to the traditional thermodynamic methods, like BJH method. DFT methods enable material characterization based on certain assumptions about the underlying geometry of the pore network. After the isotherm calculations are performed for individual pores of given shape within a large range of pore sizes, these theoretical isotherms are compiled into a 'kernel', from which the pore size distribution of a material can be determined from the experimental data. Until recently, most DFT methods relied on a single pore shape for kernel computation – either slit, cylindrical, or spherical. However, many materials possess hierarchical structures with micro- and mesopores, which form pore networks of various morphology. For example, the pores in newly discovered three-dimensional ordered mesoporous (3DOm) carbon materials form a regular cubic network of cage-like pores, the size of which can be tailored from 10 to 40 nm. In order to better characterize these and other templated carbons, hybrid kernels were developed with double and triple pore geometries (slit-cylindrical, spherical-cylindrical, slit-cylindrical-spherical). The hybrid methods can approximate the highly complex nature of possible pore structures. To improve our understanding of these materials, kernels were created for both nitrogen and argon adsorption and equilibrium desorption measurements. The validity of these calculations has been checked against traditional methods and previous density functional approaches, and show good agreement. Current research is expected to further optimize the ratio of pore geometries in these hybrid kernels and to improve our understanding of the complex nature of the pore structure of these materials.

(620e) A Novel Method for Pore Size and Porosity Analysis of Meso-and Macroporous Solids Using Electroacoustics and High Frequency Conductivity Measurements

Andrei Dukhin, Dispersion Technology Inc, Bedford Hills, NY and **Matthias Thommes**, Quantachrome Instruments, Boynton Beach, FL

We present a new method of determining porosity, pore size of meso-and macroporous materials by transmitting ultrasound pulses of a single or multiple frequencies through the porous material. Propagation of ultrasound through porous materials saturated/wetted with a liquid generates an electric response. This electroacoustic/electrokinetic effect is called seismoelectric current [1- 2]. The liquid and the porous body matrix are compressible at high frequencies of ultrasound at MHz range. Their compressibilities are different, which leads to the relative phase motion. This in turn causes displacement of the diffuse layer and the surface charge within the double layers the (wetting) liquid forms on the pores surfaces. This charge displacement shows up as oscillating electric current. It is possible to measure the seismoelectric current with existing electroacoustic devices initially designed for characterizing liquid dispersions, but we demonstrate that it is possible to utilize such a technique also for a porosity/pore size characterization of porous materials [2, 3]. Based on older theoretical predictions [1,2] it is expected that the magnitude of the seismoelectric signal is independent of the pore size if the liquid inside the pores only forms thin and isolated double layers on the pore walls. However, in case one uses a liquid with sufficiently low ionic strength these double layer are very thick so that they overlap and fill-out the complete pore space. For such a scenario we find that the seismoelectric current depends on the pore size [2,3]. In addition we utilize the correlation between seismoeclectric current magnitude and pore diameter due to hydrodynamic relaxation effects for the pore size analysis of wider macropores (pore width up to 8 μm. [1] We have performed experiments for various systems covering a wide range of pores sizes covering a wide pore size range from ca. nanometers to micrometers (e.g. porous glasses, porous titania, materials, geological cores etc., organic hydrogels)Further, we will demonstrate in our paper that information about the porosity of this materials can be obtained in a simple way by measuring the experimentally determined ratio of the electrical conductivity of the porous material to the equilibrium conductivity of the media (supernate); using this data the porosity can be calculated by applying the well-known Maxwell-Wagner theory [3].Summarizing, in our contribution we will discuss a novel, fast experimental technique (experiments can often completed in less than 10 minutes) for pore size/porosity characterization.

References

[1] A.S. Dukhin, P.J.Goetz, M. Thommes, Patent Pending, A1 20110283800 ,M. Thommes, Method for determining porosity, pore size and zeta potential, US Patent, Patents pending:

[2] A.S. Dukhin, P.J.Goetz, M. Thommes, Journal of Colloid and Interface Science 345, (2010) 547"Method for determining porosity, pore size and zeta potential of porous bodies"

[3] A.S. Dukhin, Phil Goetz. A1 20110012627 "Method for determining porosity with high frequency conductivity measurement"

(620f) Adsorptive Desulfurization of Hydrocarbon Fuels by Ag/TiO2-Al2O3 Adsorbents: Mechanism of Sulfur Adsorption At Ambient Conditions

A. H. M. Shahadat Hussain and Bruce J. Tatarchuk, Chemical Engineering, Auburn University, Auburn, AL

Sulfur removal by adsorption from liquid petroleum fuels at ambient conditions is a feasible approach for its low energy and zero hydrogen requirements. Silver supported on titania-alumina mixed oxide have demonstrated excellent regenerable sulfur adsorption capacity from hydrocarbon fuels at room temperature and pressure. In this work, the mechanism of sulfur adsorption onto mixed oxide supported silver adsorbent at ambient conditions was investigated. Breakthrough experiments, N_2 physisorption, NH_3 chemisorption, and infrared (IR) studies were conducted to understand the adsorption phenomenon. In the breakthrough tests, model fuels consisting of various sulfur and non-sulfur species were used as challenges to investigate sulfur selectivity. Ammonia chemisorption was employed to classify different acid sites on the adsorbent surface. Ammonia, 2, 6-Lutidine, and trimethyl chlorosilane (TMCS) were adsorbed on the adsorbent surfaces as probe molecules and the resulting surface complexes were observed via IR. Titania dispersed alumina support provided the highest concentrations of acid centers over individual metal oxide supports. Increase in surface hydroxyl groups was observed after silver impregnation on the support. IR spectra of thiophene (T), benzothiophene (BT), dibenzothiophene (DBT), and 4, 6-dimethyldibenzothiophene (4, 6-DMDBT) treated adsorbent samples revealed the presence of multiple sites for sulfur adsorption. There was less selectivity toward 4, 6-DMDBT because the attached methyl groups created steric hindrance. Both breakthrough and IR studies indicated that the sulfur compounds bind with the surface acid sites on silver adsorbent surface at ambient conditions.

(632a) Diffusion Behavior of Inert Gases in Activated Carbon Columns

John W. Stahl[1], Nathaniel Godfrey[2], Samuel Shouse[2] and Mario A. Oyanader[3], (1)Chemistry and Chemical Engineering, Geneva College, Beaver Falls, PA, (2)Chemical Engineering, Geneva College, Beaver Falls, PA, (3)Environmental Engineering, Geneva College, Beaver Falls, PA

Activated carbon beds to adsorb and retain inert gases have been widely used in the nuclear industry to allow sufficient decay time for radioactive isotopes such as xenon-131m, xenon-133, and krypton-88 before atmospheric discharge. Newer simplified designs of nuclear plants require much less frequent purging of radioactive waste gas from coolant water, such that adsorbed gases remain stationary in the adsorption bed for long periods of time, with the attendant concerns for diffusion of radioactive gases within the bed during this static period. This contribution describes experimental work with non-radioactive gases and a mass spectrometer detector to measure zone diffusion behavior of inert gases such as xenon, krypton, and argon. The inert gas is eluted through either an open tube or a packed column of activated carbon by a hydrogen or helium carrier gas. During elution, the flow is stopped for a fixed wait time and the resulting broadening of the inert gas zone is measured. Methodology and concepts from chromatography are found to be useful in this work. A theoretical model that includes the free gas diffusivity, the void fraction of the column, a tortuosity or pore size factor, and the adsorption equilibrium constant is found to provide a reasonably good match to the experimental data. Effective transport parameters are estimated using this model and compared with their equivalent obtained from experimental data.

(632b) Mercury Adsorption and Oxidation On Brominated Carbon Fibers in Combustion Flue Gas

Erik C. Rupp[1], Erdem Sasmaz[2] and Jennifer Wilcox[1], (1)Department of Energy Resources Engineering, Stanford University, Stanford, CA, (2)Chemical Engineering, University of South Carolina, Columbia, SC

Coal-fired power plants provided between 45 and 50% of the United States net energy generation in the years between 1999 and 2010. While alternate sources of energy, including natural gas and renewable sources, will continue to grow in importance, coal will remain as a necessary item in the energy portfolio of the US. Furthermore, the rapidly expanding economies of China and India are increasingly reliant on coal power, with upwards of 80% and 60%, respectively, of power consumption in the two countries provided by coal-fired power plants.

The use of coal has several drawbacks, including: the release of hazardous air pollutants (SO_2, NO_x, particulate matter); the release of gases, such as CO_2 with Global Warming Potentials; and the release of trace metals, with high levels of human toxicity, such as mercury. Hazardous air pollutants are largely regulated by the United States Environmental Protection Agency (US EPA) and climate change is currently a research area with intense monetary and manpower support. The release of mercury, Hg, from coal-fired power plants has only recently been regulated by the US EPA in the Mercury and Air Toxics Standards (MATS) ruling in December 2011 and has a direct impact on environmental health, especially in densely populated urban areas surrounded by coal-fired power plants, such as the Great Lakes region and the Eastern Seaboard of the US, as well as many Chinese and Indian urban centers.

In this study, bromine-impregnated carbon fiber (Br-AC) was tested for Hg capture and oxidation. An experimental system has been designed to simulate the combustion flue gas environment (CO_2, H_2O, NO_x, SO_2) where the sorbent is tested. Elemental and total (oxidized + elemental) Hg vapor is introduced from a mercury calibration system (PSA 10.536 Mercury Calibration System, PS Analytical, England) with the concentration of outlet flow measured using a commercial Hg analyzer (PSA 10.525 Sir Galahad System, PS Analytical, England). In this manner, the amount of oxidized Hg (Hg^{2+}) can be calculated. Breakthrough curves are obtained for the PBR experiments by measuring the mercury concentration at the reactor outlet. Outlet concentration is plotted as a function of time, and as the outlet concentration approaches that of the inlet, breakthrough is achieved. Results indicate that NO_x (NO and NO_2) promote the oxidation of Hg, while SO_2 prevents the adsorption or oxidation of Hg on the Br-AC. Currently, work is underway in an attempt to determine the oxidation state of Hg, which will help elucidate the mechanism of adsorption and oxidation, and create a better understanding of the Hg adsorption system.

(632c) Mercuric Chloride Vapor Adsorption On Cupric Chloride-Impregnated Activated Carbon Sorbents

Xin Li, Zhouyang Liu and **Joo-Youp Lee**, School of Energy, Environmental, Biological, and Medical Engineering, University of Cincinnati, Cincinnati, OH

Carbon sorbent injection is the most promising control technology in reducing mercury emissions from coal-fired power plants. $CuCl_2$-impregnated activated carbon sorbent has demonstrated excellent mercury oxidation and adsorption performance in our previous studies. Our previous study using XAFS shows that mercuric chloride ($HgCl_2$) is a major resultant oxidized mercury compound generated over the $CuCl_2$-impregnated activated carbon sorbent as a result of the reaction between elemental mercury and cupric chloride. This study focused on the kinetic adsorption behavior of mercuric chloride vapor onto the $CuCl_2$-impregnated activated carbon sorbents.

The experiments were performed in our lab-scale fixed-bed system at 140 °C by varying inlet $HgCl_2$ concentrations in the range of 5~20 ppbv. Three different sorbents were tested, i.e. raw commercial activated carbon (Norit's DARCO FGD), 4%(wt) $CuCl_2$-impregnated activated carbon, and 8%(wt) $CuCl_2$-impregnated activated carbon. The adsorption capacities of $HgCl_2$ onto raw, 4%, and 8% sorbents were 30 mg, 17 mg, 9 mg $HgCl_2$ per g sorbent, respectively, at 10 ppbv $HgCl_2$. The adsorption isotherms of $HgCl_2$ on DARCO-HG activated carbon and $CuCl_2$-impregnated activated carbon were found to be of the Langmuir type, and therefore the maximum of $HgCl_2$ adsorption concentration and the equilibrium constant were correlated based on the Langmuir equation. The Langmuir adsorption equilibrium constants of three sorbents were calculated to be 2.1×10^4 m^3/g for raw AC, 5.0×10^4 m^3/g for 4% $CuCl_2$-AC, 26.0×10^4 m^3/g for 10% $CuCl_2$-AC. The kinetic adsorption constants were estimated by fitting the model simulation with experimental data. The breakthrough data from experiments are in good agreement with the calculation results from the modified kinetic model. The simulation results indicate that pore diffusion resistance significantly increases with an increase in sorbent particle size. For large particles (> 30 μm), pore diffusion controls the total adsorption rate. For the sorbents used in this work (dp=15 μm), both pore diffusion and local surface adsorption was found to be a rate-limiting step.

(632d) Non-Conventional Technologies for Mercury Removal From Coal-Fired Electric Power Plants

Noah Meeks, Research & Environmental Affairs, Southern Company Services, Inc., Birmingham, AL

With the successful control of nitrogen oxides through catalytic reduction, sulfur dioxide through flue gas desulfurization units, and particulate matter through electrostatic precipitators and/or fabric filters, one of the last remaining pollutants from coal-fired power plants is mercury. Specifically, it has been shown that while particulate-bound and many forms of oxidized mercury are removed in existing pollution control equipment, elemental mercury requires a specialized approach. With the release of EPA Electric Utility MACT rule, and sometime more stringent state and local emissions rules, electric utilities such as Southern Company have heavily invested in research to achieve maximum emissions control and environmental protection.

This talk will include an overview of conventional mercury-specific technology options, such as activated carbon injection and oxidation. We will also discuss new approaches and results from our testing of non-carbon sorption, oxidation and precipitation, membrane-based sorption, multicontaminant removal, and final fate of mercury species (life cycle analysis). The removal of mercury is also affected by loading of sulfur trioxide and fine particulate matter (PM), so we will explain approaches which simultaneously address these issues as well.

(632e) Mercury Traces Removal From Natural Gas: Optimization of Guard Bed Adsorption Properties

Fabien Porcheron, Karin Barthelet, Jean-Marc Schweitzer, Antoine Daudin and Arnaud Baudot, IFP Energies nouvelles, Solaize, France

Mercury (Hg) is a natural contaminant found in the earth's crust where its concentration can range from 10 to 20 000 ppb. Mercury is thus released in the environment from a variety of natural sources, including volcanic, geothermal activities or wildfires; but also from anthropogenic activities as a total of about 2000 tons of mercury is estimated to be released each year from fossil fuel combustion or metal production [1]. For instance, natural gas production frequently generates hydrocarbons streams containing traces level of mercury [2], especially in the Southeast region of Asia where Hg concentration can reach up to 300 ppb. In these conditions, mercury is predominantly found as elemental mercury (Hg°) in the gas phase. Though these levels are rather low, the impact on the industrial equipment and human health can be serious. For instance, mercury has a strong ability to form amalgams with Al-based LNG (Liquefied Natural Gas) cryogenic exchanger leading to corrosion issues and potential industrial disaster like the one encountered in 1973 in Skikda, Algeria [3]. In addition, mercury is harmful for human health and numerous incentives have been issued to control and limit its emissions from anthropogenic sources.

Hence, mercury removal is an ongoing issue and natural gas streams are usually decontaminated using guard beds protecting downstream equipment. Traditionally, these guards beds are made of beads of activated carbon impregnated with elemental sulfur (S) [4]. Mercury will chemically bond with sulfur to form mineral cinnabar (HgS) within the porosity of the activated carbon. Mercury is thus immobilized in a non-hazardous form and guard beds are designed to decrease traces level of Hg down to 1 ppb for several years before being disposed and retreated.

However, these types of guard beds suffer from many drawbacks such as possible sulfur loss during mercury removal operation and are prone to capillary condensation for wet gas [4]. One way to avoid theses problems is to optimize the interaction between the carrier support and the active phase deposited within the solid. For instance, pore size distribution can be tuned to avoid capillary condensation issues and the nature of the active phase and mineral support can be adequately chosen to obtain strongly bound and finely dispersed mercury-reactive nanoparticles.

In this work, mercury adsorbents are prepared by dispersing copper sulfide (CuS) within the porosity of alumina carrier supports (Al_2O_3). In practice, the solids are prepared by wetness impregnation method with a solution of copper precursor, following by thermal treatment leading to supported copper oxide. The finely dispersed copper sulfide nanoparticles phase is finally obtained by a sulfidation procedure before mercury removal evaluation. We investigate the effect of copper oxide dispersion by varying the nature and/or the textural properties of the solid support. Scanning electron microscopy and X-ray diffraction experiments showed that the deposited copper oxide particles are more efficiently dispersed on the alumina surface for highly porous volumes and specific mineral phases. Mercury removal performances are then assessed through pilot plant experiments where a feed of elemental mercury is passed through a fixed bed of adsorbents. These experiments are finally interpreted using a dedicated mass transfer model and mercury removal mechanisms are studied through numerical simulations.

[1] UNEP Chemicals branch, **2008**. The global atmospheric mercury assessment: sources, emissions and transport. UNEP-Chemicals, Geneva.

[2] U.S. Environmental protection agency, **2001**. Mercury in petroleum and natural gas: estimation of emissions from production, processing and combustion. EPA-600/R-01-066.

[3] Kinney, G. T. Skikda LNG plant solving troubles. *Oil & Gas J. Sept*, 15, **1975**.

[4] McNamara, J.D. and Wagner, M.J. Process effects on activated carbon performance and analytical methods used for low levels mercury removal in natural gas applications. *Gas. Sep. Purif.*, **1996**, 10, pp. 137-140.

(635a) Stretched Perfluorosulfonic Acid Membranes for Hydrogen/Air and Direct Methanol Fuel Cells

Wenjing Zhang, Remington E. Fischer and **Peter N. Pintauro**, Department of Chemical and Biomolecular Engineering, Vanderbilt University, Nashville, TN

To develop a durable and high performance proton exchange membrane for hydrogen/air and direct liquid methanol fuel cell applications, we have systemically examined uniaxial stretching effects on perfluorosulfonic acid (PFSA) polymers. PFSAs represent a family of commercially available polymers with a perfluorinated backbone and side chains that terminate in hydrophilic sulfonic acid groups. Three PFSA materials were studied: DuPont's Nafion® (with a long and branched side chain of $-O-CF_2CFCF_3-O-(CF_2)_2-SO_3H$), 3M Company PFSA (a short-side-chain version of Nafion with $-O-(CF_2)_4-SO_3H$ pendant groups), and Aquivion® from Solvay Solexis (with a very short side chain of $-O-(CF_2)_2-SO_3H$). The effect of uniaxial stretching on the crystallinity and fuel cell properties of these polymers was probed, where solution-cast membranes were stretched at a temperature above the polymer's α-transition temperature, followed by a polymer annealing step.

Nafion 1100EW, 3M 1000EW and 825EW, Aquivion 830EW PFSA show an increase in crystallinity upon uniaxial stretching, with a decrease in methanol permeability and no change in proton conductivity. A direct methanol fuel cell employing a stretched 3M 825 EW PFSA membrane with a draw ratio (DR) of 4, for example, exhibits a 31% improvement in power density at 0.4V, as compared to a fuel cell with a Nafion 117 (1100 EW) membrane. The higher power output is associated with lower methanol crossover.

Interestingly, 3M's 733 EW PFSA polymer has no crystallinity before or after stretching. The nanostructure of ionic domains, however, varied with draw ratio. From wide-angle x-ray diffraction studies, the ionomer peaks of a stretched membrane shift to higher values of 2θ and a new ionomer peak appears when the draw ratio is greater than 4. For a stretched 733 EW membrane (DR=4), there is a 120% increase in the storage modulus with a proton conductivity equal to that of an unstretched film. This combination of properties makes stretched 3M 733 EW PFSA a promising PEM material for a hydrogen/air fuel cell. From an open circuit voltage humidity cycling fuel cell test at 80°C, for example, it was found that the lifetime of a stretched 3M 733 EW PFSA membrane (DR=4) was extended by a factor of 3.5, as compared to that of an unstretched membrane.

(635b) Influences of Confinement and Annealing On the Behavior of Polymer Electrolyte Materials

Brandon W. Rowe, Kirt Page and Christopher Soles, Polymers Division, National Institute of Standards and Technology, Gaithersburg, MD

Polymer electrolyte materials are of great interest in the continuing development of fuel cell and battery technologies, which offer cleaner, more efficient alternatives to power generated through tradition technologies. While substantial progress has been made in improving the performance of polymer electrolyte membrane (PEM) fuel cells, challenges in their development remain. Due to the complex structure present in these PEM materials, including ionic clusters and crystallites, their performance is intimately linked to the specific morphology of the system. As film thickness is decreased to improve performance and lower costs, the morphology of the system (*i.e.*, size and connectivity ionic domains and crystallites) will become more strongly influenced by interactions at the material interfaces, impacting the performance of the membrane layer and the polymer electrolyte in the catalyst layer. Nafion, the most prevalent PEM material, has been studied in detail in its bulk form; however, relatively little information has been published on the behavior of thin films of Nafion. The focus of this work is to quantify how the morphology and material properties of PEM materials change under confinement at interfaces, determine how these interfacial effects are influenced by processing, and to establish a correlation between these effects and fuel cell performance.

Using a variety of experimental techniques, including: quartz crystal microbalance (QCM), grazing incidence small angle x-ray scattering (GISAXS), ellipsometry, and x-ray reflectivity, we have investigated the influences of confinement and thermal annealing on the behavior of thin Nafion films. Consistent with previous studies, the sorption of water in thin (40 nm) Nafion films is lower than in bulk membranes. Additionally, the previous history of these films, i.e., thermal annealing, has a significant impact on water uptake behavior. The nanostructure within the Nafion films, as studied using GISAXS, was also influenced by confinement and annealing. In addition to changes in the ionic domain morphology in this material, which can be connected to the water uptake behavior, changes in the crystalline regions were also observed that relate to changes in the film strength and stability. To complement the studies on structure and transport properties, we have also investigated the mechanical properties and stress change upon hydration of these films using a cantilever bending technique. The observed deviations from bulk behavior in these thin Nafion films highlight the importance of understanding the influences of confinement in detailed modeling efforts, which typically use bulk-film parameters to describe transport in thin PEM layers and at catalyst interfaces.

(635c) Computation Study On Water and Charge Transport in the Cross-Linked and Sulfonated Poly (1,3-Cyclohexdiene) (xsPCHD) Based Proton Exchange Membranes

Qifei Wang, Chemical and Biomolecular Engineering, University of Tennessee Knoxville, Knoxville, TN and David J. Keffer, Chemical and Biomolecular Engineering, University of Tennessee, Knoxville, Knoxville, TN

Abstract

Using different computational approaches (Molecular Dynamics Simulation, Confined random walk (CRW) simulation and an analytical theory), this work studied the cross-linked and sulfonated Poly (1,3-Cyclohexdiene) (xsPCHD) hompolymer, xsPCHD-polyethylene glycol (PEG) block copolymer and xsPCHD/PEG blend proton exchange membrane (PEM) molecular level structures and water and charge transport in the membranes. The membrane structure was first studied through pair correlation functions (PCFs) from MD simulation. Interesting results are found for the copolymer and blend membranes. The PEG-water PCF shows PEG has greater affinity to water in the blend than that in the copolymer membrane, suggesting PEG is well mixed with water and serves as a cosolvent in the blend membrane. The PEG-H_3O^+ PCFs show that the end OH group of PEG has much stronger affinity to H_3O^+ while the backbone of PEG has weaker affinity to H_3O^+ in the copolymer membrane than that in the blend membrane. The features of the PCFs suggest very different proton transport mechanisms in the copolymer and blend membranes. The CRW simulation results show that bigger water clusters formed in the copolymer membrane. The existence of PEG in the blend membrane slows down the water diffusion but helps the structural diffusion of charge. Using a recently developed analytical theory, the proton conductivity is calculated and compared with that of experimental measurements. Good agreement is obtained for homopolymer membrane but the accuracy of the theory is affected when applied to copolymer and blend membrane due to the existence of PEG in these membranes.

(635d) Multi-Scale First-Principles Approach to Three-Phase Model of Polymer Electrolyte Membrane Fuel Cell

Giuseppe F. Brunello, Ji Il Choi and **Seung Soon Jang**, School of Materials Science and Engineering, Georgia Institute of Technology, Atlanta, GA

Multi-scale First-Principles Approach to Three-Phase Model of Polymer Electrolyte Membrane Fuel Cell

Giuseppe F. Brunello, Ji Il Choi and Seung Soon Jang

Computational NanoBio Technology Laboratory, School of Materials Science and Engineering, Georgia Institute of Technology, 771 Ferst Drive NW, Atlanta, GA 30332-0245

It is desirable to achieve fundamental understanding of the three-phase system consisting of carbon support, polymeric ionomer, and Pt catalyst nanoparticle, where the most essential electrochemical events for fuel cell operation take place in the presence of other molecular species such as water, proton, and oxygen. Although variety of cutting-edge in-situ and time-resolved probing techniques would be available to investigate nanometer-scale systems, the structure and dynamic behavior of the three-phase system of fuel cell has not been thoroughly analyzed at molecular level, which is mainly due to such complicated multi-component feature of the three-phase system in several tens of nanometers. In this context, we simulate the three-phase system using multi-scale first-principles modeling approach consisting of quantum mechanical density functional theory (DFT) and molecular dynamics (MD) simulations in order to investigate the nanophase-segregation of polymeric ionomers and water molecules, surrounding Pt nanoparticle on graphitic carbon support. For this, we develop a force field based on DFT computations and ran large-scale MD simulations. The nanophase-segregated structures and transport properties of water, proton, and oxygen are analyzed, and especially the spatial arrangement of molecular species around Pt particle is characterized in detail. All the simulation results are compared to those of bulk Nafion phase with the same water contents.

(635e) Hybrid Membranes for Proton Exchange Membrane Fuel Cell

Zhongyi Jiang, Hong Wu, Jingtao Wang, Lingli Nie, Weikang Yuan, Bin Zheng, Xiaohong Zheng, Yabo Wang, Weiqiang Hou, Lulu Xiao, Yuning Zhao, Tao Xu and Xiaohui Shen, Key Laboratory for Green Chemical Technology of Ministry of Education, Sch Chem Engn & Technol, Tianjin Univ, Tianjin, China

Hybrid membranes prepared by incorporating inorganic fillers into the polymer matrix are promising candidates for proton exchange membranes (PEMs) and have spurred increasing attention. The work in our group mainly focuses on the regulation of membrane microstructure (e.g. free volume) and the intensification of membrane performances (e.g. methanol barrier property, proton conductivity). Inorganic fillers can regulate the free volume of the resultant hybrid membrane by interfering the mobility of polymer chains. It is found that the polymer-inorganic interfacial interaction governed the membrane microstructure: i) attractive force would generate strong interfacial stress, which inhibits the mobility of polymer chains and leads to dense chain packing, in turn reducing free volume cavity size; and ii) repulsive force enhances the mobility of polymer chains, which allows for a loosened chain packing and increased free volume cavity size. The methanol transport of through PEM can be described by solution-diffusion mechanism. Inorganic fillers can change the hydrophilic/hydrophobic nature of membrane surface, which in turn tunes the methanol dissolution. Meanwhile, inorganic fillers can tune the methanol diffusion by adjusting the free volume and methanol affinity properties of membrane bulk. Both the filler structure (e.g. shape, particle size, pore size) and filler composition (e.g. hydrophilic/hydrophobic nature, acid/base property) can affect methanol dissolution and diffusion properties of hybrid membranes. Accordingly, the methanol barrier property of hybrid membrane can be intensified. The less-conducting inorganic fillers generally dilute the concentration of proton conducting group and reduce water uptake, resulting in reduced proton conductivity of hybrid membranes. Therefore, the filler surface is modified to introduce proton-conducting groups, which render additional pathways for proton transfer. Furthermore, the proton transfer barrier can be reduced by optimizing proton donor/acceptor ability of the conducting groups. In addition to polymer-inorganic hybrid membrane, inspired by the water storage mechanisms in plant cells, polymeric microcapsules are synthesized and incorporated into polymer matrix to prepare polymer-organic "hybrid membrane". The microcapsules serve as water reservoirs and enhance water retention properties in particular under low humidity, endowing hybrid membranes with high proton conductivity.

IMPROVEMENTS OF THREE METHODS OF THICKENER AREA CALCULATION

A. I. A. Salama
CanmetENERGY-Devon, Natural Resources Canada, Devon, AB, Canada

Merta and Ziolo (MZ) method

For a typical continuous thickener and following Merta and Ziolo (1985) developments, at zone z of a suspension and after some manipulations we can arrive at

$$Q_o = \frac{Q_z C_z}{\rho_o}\left(\frac{\rho_z}{C_z} - \frac{\rho_u}{C_u}\right) \tag{1}$$

where Q_o, Q_z, C_z, and C_u designate overflow volumetric flowrate (VFR-m^3/s), zone z suspension VFR, zone z solids mass concentration (SMC-kg/m^3), and underflow SMC, respectively. Assuming the overflow density ρ_o equals to 1 gm/cm^3. And let the upward water velocity be equal to solids settling velocity $\omega(C_z)$, or

$$\omega(C_z) = \frac{Q_o}{A} \tag{2}$$

where Q_o is the overflow VFR . Using solids and water mass balance and Eq. (2), the thickener cross sectional area, A, can be expressed as

$$A = \frac{Q_f C(0)}{\omega(C_z)\rho_l}\left(\frac{\rho_z}{C_z} - \frac{\rho_u}{C_u}\right) = \frac{Q_f C(0)}{\omega(C_z)}\left(\frac{1}{C_z} - \frac{1}{C_u}\right) \tag{3}$$

where $C(0)$ is the suspension initial SMC and assuming the liquid density ρ_l equals 1 gm/cm^3. Conducting solids settling (SS) test in graduated cylinder the mass balance conditions at time τ requires (Kynch, 1952 and Foust et al., 1960).

$$C(0)\,h_o = C_z(\tau)z(\tau) \;\&\; z(\tau) = h'(\tau) = h(\tau) + \omega(\tau)\tau \tag{4}$$

where $h(\tau)$ is the interface at time τ, and $\omega(\tau)$ is the rate of change in the interface level

$$\omega(\tau) = -\frac{\mathrm{d}h(\tau)}{\mathrm{d}\tau} \tag{5}$$

A typical settling curve (SC, $h(\tau)$~τ), $h(\tau)$, and $h'(\tau)$ are shown in Figure 1. Using Eq. (4) the SMC $C(\tau)$ is expressed as

$$C_z(\tau) = \frac{C(0)h_o}{h(\tau) + \omega(\tau)\tau} \tag{6}$$

Substituting Eq. (6) into Eq. (3) and at time τ we get

$$A = \frac{1}{\omega(\tau)}\left(\frac{Q_f}{h_o}h(\tau) - \frac{Q_f C(0)}{C_u}\right) + \frac{Q_f}{h_o}\tau = -\frac{1}{\frac{\mathrm{d}h(\tau)}{\mathrm{d}\tau}}\left(\frac{Q_f}{h_o}h(\tau) - \frac{Q_f C(0)}{C_u}\right) + \frac{Q_f}{h_o}\tau \tag{7}$$

Using Eq. (7) the derivative of A with respect to $h(\tau)$ is

$$\frac{\mathrm{d}A}{\mathrm{d}h(\tau)} = \frac{1}{\frac{\mathrm{d}h^2(\tau)}{\mathrm{d}^2\tau}}\left(\frac{Q_f C(0)}{C_u} - \frac{Q_f}{h_o}h(\tau)\right) - \frac{1}{\frac{\mathrm{d}h(\tau)}{\mathrm{d}\tau}}\frac{Q_f}{h_o} + \frac{1}{\frac{\mathrm{d}h(\tau)}{\mathrm{d}\tau}}\frac{Q_f}{h_o}, \text{ or} \tag{8}$$

$$\frac{\mathrm{d}A}{\mathrm{d}h(\tau)}=\frac{1}{\dfrac{\mathrm{d}h^2(\tau)}{\mathrm{d}^2\tau}}\left(\frac{Q_f C(0)}{C_u}-\frac{Q_f}{h_o}h(\tau)\right) \tag{9}$$

Assuming $\mathrm{d}h^2(\tau)/\mathrm{d}^2\tau \neq 0$, the local maximum of A is obtained by setting Eq. (9) equals zero, or

$$h^*=\frac{C(0)}{C_u}h_o \tag{10}$$

Using Eq. (10) and substituting into Eq. (7), the local maximum of A is expressed as

$$A^*=\frac{Q_f}{h_o}\tau(h^*)=\frac{Q_f}{h_o}\tau^* \tag{11}$$

where τ^* corresponds to h^* on the SC (see Figure 2). It can be demonstrated that A'' is negative at the conditions stated in Eqs. (10 and 11), or A^* is local maximum.

Talmadge and Fitch (TF) method

The details of TF method are found elsewhere (Talmadge and Fitch, 1955) and using Eq. (4). The thickener cross sectional area is

$$A=Q_f A_{unit}=\frac{Q_f}{h_o}\tau_u \tag{12}$$

where A_{unit} is unit area and see τ_u in Figure 3. Comparing Eqs. (11 and 12) and noting that for a given τ^*, examination of Figure 3 indicates that $\tau_u \geq \tau^*$ which implies that the TF method overestimate A^*. Therefore, the MZ method is the sound choice.

Solids sedimentation flux (SSF) Method

Rewriting Eq. (3) we get

$$A=\frac{Q_f C(0)}{\omega(C)}\left(\frac{1}{C}-\frac{1}{C_u}\right)=Q_f C(0)\left(\frac{1}{G(C)}-\frac{1}{G_u(C)}\right) \tag{13}$$

Where the solids sedimentation flux (SSF), $G(C)=C\omega(C)$, underflow SSF, $G_u(C)=C_u\omega(C)$, and C and C_u are the solids concentration in suspension and underflow. In the following analysis, the quantity $\omega(C)$ and $G(C)$ are designated ω and G, respectively. The derivative of A with respect to C is expressed as

$$\frac{\mathrm{d}A}{\mathrm{d}C}=\frac{Q_f C(0)}{C_u}\frac{\omega'}{C^2\omega^2}\left(C^2-(C_u)C-\left(\frac{\omega}{\omega'}\right)C_u\right) \tag{14}$$

Assuming that the term $((Q_f C(0)\omega'/(C_u C^2\omega^2))\neq 0$ and equating Eq. (14) to zero results in

$$C^2-(C_u)C-\left(\frac{\omega}{\omega'}\right)C_u=0 \tag{15}$$

Given the SC information, ω, ω', and C_u, Eq. (15) is quadratic equation in C and its solution is

$$C^*=\frac{C_u}{2}\pm\frac{\sqrt{C_u^2+4\left(\dfrac{\omega}{\omega'}\right)C_u}}{2} \tag{16}$$

It should be emphasized that in MZ method the local maximum A is determined by using Eqs. (10 and 11) (i.e., only the SC data is used) and the solution is a single point on the SC. However, in the SSF method the

SC and SSF curves are used and we end up with a quadratic equation (Eq. (15)), i.e., minor complication but not significant.

The local maximum A is obtained by substituting C^* into Eq. (13) and we get

$$A^* = \frac{Q_f C(0)}{\omega(C^*)}\left(\frac{1}{C^*} - \frac{1}{C_u}\right) = Q_f C(0)\left(\frac{1}{G^*} - \frac{1}{G_u^*}\right) \tag{17}$$

where $G_u^* = \omega^* C_u$.

Richardson and Zaki (RZ) correction factor

Richardson and Zaki (1954) proposed that the ω is expressed as a product of the particle free settling velocity, U_s, and a correction factor (CF) expressed in terms of the C

$$\omega = U_s(1-\gamma C)^m \tag{18}$$

where γ is a conversion factor from mass to volume solid concentrations and m is an empirical parameter.

In the SSF method series of cylinder settling tests are conducted at different initial C. The initial $\omega(C)$'s are recorded against the initial C's to form the input data. Using Eq. (15) and evaluating the term ω/ω' we get

$$C^2 - \left(\left(1+\frac{1}{m}\right)C_u\right)C - \left(\frac{1}{m\gamma}\right)C_u = 0 \tag{19}$$

where γ and its solution is given by

$$C^* = \frac{1}{2}\left(1+\frac{1}{m}\right)C_u \pm \frac{1}{2}\sqrt{\left(\left(1+\frac{1}{m}\right)C_u\right)^2 - \left(\frac{4}{m\gamma}\right)C_u} \tag{20}$$

Note that C_u and m are positive, $m >$ unity, and by inspection the +sign is selected in the ± square root term because: i) the 2nd term is less than the 1st term, ii) the 1st term is less than C_u, and iii) Eq. (20) implies $C^* < C_u$.

Christian correction Factor

Christian (1994) suggested to model ω as

$$\omega = 10^{mC+b} = \exp(kmC + kb) \tag{21}$$

where $k = \ln(10)$ and m and b are regression or empirical parameters. Again, using Eq. (15) and after manipulating the term ω/ω', we obtain

$$C^2 - (C_u)C - \left(\frac{1}{mk}\right)C_u = 0 \tag{22}$$

and its solution is

$$C^* = \frac{1}{2}C_u \pm \frac{1}{2}\sqrt{C_u^2 + \left(\frac{4}{mk}\right)C_u} \tag{23}$$

Note that C_u and mk are positive and the + sign is selected in the ± square root term because: i) m is negative, consequently, the 2nd term is less than the 1st term and ii) Eq. (23) implies $C^* < C_u$.

Conditions at local maximum thickener area

In MZ method, It has been shown that the local maximum A is determined using Eqs. (10 and 11). Let us explore similar condition for local maximum A in the SSF method. Rewriting Eq. (15) we obtain

$$G^2 - GG_u - \left(\frac{\omega^2}{\omega'} G_u\right) = 0 \quad (24)$$

Note that in Eq. (24), the flux G_u is defined as $C_u\omega$. Evaluating, Eq. (24) at different C's and it equals zero at A^*.

Demonstration

Single settling-test

Gravity settling in oil sands froth treatment is considered. A cascaded counter-current two-stage gravity-settlers flowsheet is used. The data set selected is focused on the streams around the 1st gravity settler. Data were collected during a testing campaign at CanmetENERGY-Devon using the froth treatment pilot plant. The feed to the PP may be grouped as light component (diluted bitumen, i.e., bitumen plus solvent) and heavy component (water plus clays). The two-stage raw data set is balanced to estimate the operating conditions (i.e., stream mass flowrates). The feed and intermediate data are summarized in Table 1. Separately, in the lab a simulated suspension, froth and solvent, was prepared to conduct settling test and the resulting SC is shown in Figure 4. Using the SC and MZ method, we obtain

$$h^* = \frac{C(0)h_o}{C_u} = \frac{0.157\text{x}102.0}{0.867} = 18.5\,\text{cm} \quad (25)$$

and $\tau^* \sim 12$ min. Substituting in Eq. (11), we get

$$A^* = \frac{Q_f}{h_o}\tau^* = \frac{4.165x1000}{102}12.0 \cong 490.0\,\text{cm}^2 \quad (26)$$

which corresponds to 10" diameter (size), the size of the settler used in the test.

Another single batch-settling test using limestone suspension (Foust, et al., 1962) is considered. The corresponding SC is presented in Table 2. The limestone is fed to a thickener at a rate of 50 tons of dry solids /h to produce a thickened sludge of 550 gm/litre. The test was made using 236.0gm/litre of suspension (or 236.0 kg/m^3). Assume the limestone density ρ_s is 2.61gm/cm3. The observed SC is fitted using the formula

$$h(\tau) = h(\tau_f) + \left(h(0) - h(\tau_f)\right)\exp(-\alpha\tau) \quad (27)$$

where h_o and $h(\tau_f)$ are the initial interface level of the suspension and the final sediment interface level at the end of the test. In Eq. (27), α is a fitting parameter, the optimal $\alpha^* = 0.745$ is obtained. The observed and estimated SCs are shown in Figure 5. The 50 tons/h=45400 kg/h or Q_f=45400/236=192.37m^3/h using the MZ method and substituting in Eq. (10) and replacing $h(0)$ by h_o=0.36m, we obtain

$$h^* = \frac{C(0)h_o}{\phi_u} = \frac{236x0.36}{550} = 0.1545\text{m} = 15.45\text{cm} \quad (28)$$

From the SC the corresponding τ^*=1.8 hr. Substituting in Eq. (11), we get

$$A^* = \frac{Q_f}{h_o}\tau^* = \frac{192.37}{0.36}1.8 \cong 962.0\text{m}^2 \quad (29)$$

which is close to the results (10,320 ft^2 = 958.8m^2) reported of by Foust et al. (1962).

The SSF method can be used in single SC cases. Assuming that the SC and ω~C can be modelled mathematically by using nonlinear regression. The estimation of ω may lead to large errors. Using the RZ model the results obtained are given in Table 3. The cross-sectional A^* is 1319.0m^2, which is higher than 962.0m^2 (MZ method). The reason for the overestimate in A^* is due to the limited number of data points. In case of Christian (1994) model reasonable results were obtained as shown in Table 4 which indicates that A^*=973.0m^2.

Multi settling-tests

The MS test data reported by Christian (1994) is used to demonstrate the SSF methods. The data set used is presented in the left sections of both Tables 5 and 6. The RZ model (see Eq. (18)) is used to design a thickener or settler and the results are presented in Table 5. Note that Eqs. (18, 20, and 20) are used to determine A^*=240.9m^2. The Christian (1994) model (see Eq. (21)) is used to design a thickener or settler and the results are presented in Table 6. Note that Eqs. (21, 22, and 23) are used to determine the A^* and equals 180.0m^2. Examination of Tables 5 and 6 indicates that, on one hand the RZ and Christian (1994) CFs produced C_{tang}=202.0 and 196.0kg/m^3, respectively, which are very close. On the other hand, the ω in the RZ model is 0.0024m/s compared with 0.004m/s in Christian model. Consequently, the RZ model produced overestimation in A^*.

Acknowledgement

The author would like to acknowledge support from the Government of Canada's interdepartmental Program of Energy Research and Development, PERD POL

References

1. Christian, J.B. (1994), "Improve clarifier and thickener design and operation," *Chem. Eng. Prog*, July: pp. 50-56.
2. Foust, A.S, Wenzel, L.A., Clump, C.W., Maus, L. and Andersen, L.B. (1962), "*Principals of Unit Operations*," John Wiley, NY.
3. Kynch, G.J. (1952), "A theory of sedimentation," *Trans. Faraday Society*, 48 pp. 166-176.
4. Merta, H and Ziolo, J (1985), "Calculation of thickener A and depth based on the data of batch-settling test," *Chemical Engineering Science* 40: 7: pp. 1301-1303.
5. Perry, J. (1963), "*Chemical Engineers' Handbook*," McGraw-Hill, New York.
6. Richardson, J.F. and Zaki, W.N. (1954), "Sedimentation and fluidization 1," *Trans. Inst. Chem. Eng.* (London) 32: pp. 35-53.
7. Talmadge, W.P. and Fitch, E.B. (1955), "Determining thickener unit areas," *Ind. Eng. Chem*. 47:1: pp. 38-41.

Table 1 – 1st stage data extracted from a two-stage settlers test data (CanmetENERGY)

Com / Str	B (%)	M (%)	W (%)	Solvent (%)	Sum (%)	Q (kg/min)
1st F	37.50	3.83	13.85	44.82	100.00	3.69
1st UF	18.92	16.91	61.50	2.67	100.00	0.83
ρ_b	~ 1.000	$\rho_{1st\ F}$	0.886			
ρ_m	2.650	$\rho_{1st\ UF}$	1.106			
ρ_w	1.000	$Q_{1st\ F}$ (litre/min)	4.165	$C_{1st\ F}$ (kg/litre)	0.157	
ρ_{solv}	0.746	$Q_{1st\ UF}$ (litre/min)	0.750	$C_{1st\ UF}$ (kg/litre)	0.867	
h_o (cm)	102.0	h^* (cm)	18.4	τ^* (min)	~ 12	
A = 490.0 (cm^2)						

Table 2 - A limestone SC data (Foust et al., 1962)

Fitting $h(\tau) \sim \tau$	
$h(0)$ / cm	36.00
$h(\tau_f)$ / cm	8.80
α	0.745

τ (hr)	$h_o(\tau)$ Observed (cm)	$h_e(\tau)$ Estimated (cm)	error^2
0.0	36.00	36.00	0.00
0.3	32.40	31.38	1.05
0.5	28.60	27.54	1.13
1.0	21.00	21.71	0.50
1.8	14.70	16.18	2.19
3.0	12.30	11.71	0.35
4.8	11.55	9.59	3.85
12.0	9.80	8.80	0.99
20.0	8.80	8.80	0.00
		Φ	10.06

Table 3 – SSF results using RZ CF – Table 2 data

DESIGN DATA				
$h(0)$ (m)	$h(\tau_f)$ (m)	C_o (kg/m^3)	C_u (kg/m^3)	Q_f (m^3)
0.360	0.087	236.000	550.000	192.370

Time τ (h)	o (m) / $h_o(\tau)$	e (m) / $h_e(\tau)$	$(o-e)_k^2$ / $(h_o(\tau)-h_e(\tau))^2$	ω_o (m/h) / $\omega(\tau)$ $-dh_e(\tau)/d\tau$	C (kg/m^3) / $\phi(\tau)=\phi_o h_o/(h(\tau)+\omega(t)*\tau)$	$(1-10^{-3}C/\rho_s)$ Fixed U_s	ln $(1-10^{-3}C/\rho_s)$ Fixed U_s	$G(\tau)=C(\tau)\omega(\tau)$	ln (ω_o)	ln (ω_e)	$(o-e)^2$	$h(\tau^*)$	τ	A	C_{max}	ω_{max}	G_{max}	A
0.00	0.3600	0.3600	0.0000	0.2018	236.00	0.91	-0.09	47.64	-1.60	-1.42	0.033	0.3600	0.00	544.11	465.92	0.01129	5.26	1319.0
0.25	0.3240	0.3139	0.0001	0.1678	238.74	0.91	-0.10	40.86	-1.79	-1.45	0.111	0.3240	0.25	673.52				
0.50	0.2860	0.2756	0.0001	0.1395	246.00	0.91	-0.10	34.31	-1.97	-1.54	0.181	0.2860	0.50	771.13				
1.00	0.2100	0.2173	0.0001	0.0964	270.83	0.90	-0.11	26.10	-2.34	-1.86	0.228	0.2100	1.00	842.27				
1.75	0.1470	0.1619	0.0002	0.0553	328.39	0.87	-0.13	18.18	-2.89	-2.61	0.080	0.1470	1.75	862.98				
3.00	0.1230	0.1167	0.0000	0.0220	465.28	0.82	-0.20	10.22	-3.82	-4.47	0.431	0.1230	3.00	837.41				
4.75	0.1155	0.0951	0.0004	0.0060	686.52	0.74	-0.31	4.13	-5.11	-7.75	6.967	0.1155	4.75	-919.53				
12.00	0.0980	0.0870	0.0001	0.0000	972.33	0.63	-0.47	0.03	-10.47	-12.59	4.500	0.0980	12.00					
20.00	0.0880	0.0870	0.0000	0.0000	976.53	0.63	-0.47	0.00	-16.39	-12.67	13.809	0.0880	20.00					
α			0.0011					IC	m	ln(U_s)	26.340	Φ	82.54	534.36				
1	IC								1.000	1.000			A	B				
0.73935									30.101	1.436								

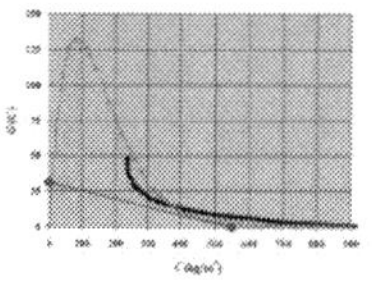

Table 4 – SSF results using Christian CF – Table 2 data

DESIGN DATA				
$h(0)$ (m)	$h(\tau_f)$ (m)	C_o (kg/m^3)	C_u (kg/m^3)	Q_f (m^3)
0.360	0.087	236.000	550.000	192.370

Time τ (h)	o (m) / $h_o(\tau)$	e (m) / $h_e(\tau)$	$(o-e)^2$ / $(h_o(\tau)-h_e(\tau))^2$	S velocity ω_o (m/h) / $\omega(\tau)$ $-dh_e(\tau)/d\tau$	C (kg/m^3) / $C(\tau)=C_o h_o/(h(\tau)+\omega(t)*\tau)$	flux kg/m^2/h / $G(\tau)=C(\tau)\omega(\tau)$	o log(ω_o)	$(o-m*C(\tau)-b)^2$	$h(\tau^*)$	C_{max}	$\omega(\tau)_{max}$	G_{max}	A
0.00	0.3600	0.3600	0.0000	0.2018	236.00	47.64	-0.6950	0.0006	0.3600	431.46	0.02	10.05	973.2
0.25	0.3240	0.3139	0.0001	0.1678	238.74	40.06	-0.7753	0.0018	0.3240				
0.50	0.2860	0.2756	0.0001	0.1395	246.00	34.31	-0.8555	0.0079	0.2860				
1.00	0.2100	0.2173	0.0001	0.0964	270.83	26.10	-1.0161	0.0179	0.2100				
1.75	0.1470	0.1619	0.0002	0.0553	328.39	18.18	-1.2569	0.0112	0.1470				
3.00	0.1230	0.1167	0.0000	0.0220	465.28	10.22	-1.6583	0.0175	0.1230				
4.75	0.1155	0.0951	0.0004	0.0060	686.52	4.13	-2.2202	0.3643	0.1155				
12.00	0.0980	0.0870	0.0001	0.0000	972.33	0.03	-4.5482	0.1517	0.0980				
20.00	0.0880	0.0870	0.0000	0.0000	976.53	0.00	-7.1169	8.6359	0.0880				
α		Φ	0.0011			m	b	0.5730	Φ				
1	IC				IC	-0.01	1						
0.73935						-0.00467	0.38242						

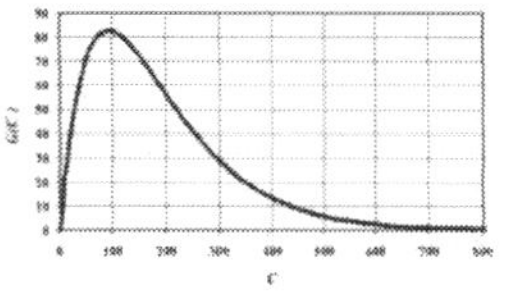

Table 5 – SSF results using RZ CF – Christian (1994) data

DESIGN DATA		
C_f (kg/m³)	C_u (kg/m³)	Q_f (m³)
5.0	232.0	177600.0

ω (m/s) / $\omega_s = -d(h_s(\tau))/d\tau$	C (kg/m³) / ϕ	$(10^3\text{-}C)$	$\ln(10^3\text{-}C)$	G (kg/m2/s) / $C*\omega_s$	a / $\ln(\omega_s)$	e / $\ln(\omega_s)=m*\ln(10^3\text{-}C)+b$	ε^2 / $(a-e)^2$
0.018	140.0	860.0	6.76	2.52	-4.02	-3.78	0.06
0.110	120.0	880.0	6.78	13.20	-2.21	-3.08	0.77
0.034	110.0	890.0	6.79	3.74	-3.38	-2.74	0.41
0.107	110.0	890.0	6.79	11.77	-2.23	-2.74	0.25
0.183	96.0	904.0	6.81	17.57	-1.70	-2.27	0.32
0.210	81.0	919.0	6.82	17.01	-1.56	-1.77	0.04
0.284	80.0	920.0	6.82	22.72	-1.26	-1.73	0.23
0.227	73.0	927.0	6.83	16.57	-1.48	-1.50	0.00
0.293	70.0	930.0	6.84	20.51	-1.23	-1.41	0.03
0.217	53.0	947.0	6.85	11.50	-1.53	-0.86	0.45
0.239	53.0	947.0	6.85	12.67	-1.43	-0.86	0.33
0.177	52.0	948.0	6.85	9.20	-1.73	-0.82	0.82
0.326	51.0	949.0	6.86	16.63	-1.12	-0.79	0.11
0.248	41.0	959.0	6.87	10.17	-1.39	-0.47	0.85
0.382	38.0	962.0	6.87	14.52	-0.96	-0.38	0.34
0.944	31.0	969.0	6.88	29.26	-0.06	-0.16	0.01
0.897	24.0	976.0	6.88	21.53	-0.11	0.06	0.03
1.100	19.0	981.0	6.89	20.90	0.10	0.21	0.01
4.534	5.3	994.7	6.90	24.03	1.51	0.63	0.77
2.560	3.3	996.7	6.90	8.45	0.94	0.70	0.06
6.045	3.1	996.9	6.90	18.74	1.80	0.70	1.21
					m	$\ln(C_r)$	7.10
				IC	0.01	0.5	Φ
					30.13421	-208.75	

$A=Q_f*C_f*(C_u-C_{tang})/G_{tang}/C_u/1000$			
C_{tang}	ω_{tang}	G_{tang}	A
201.74	0.00	0.45	240.93
Area calculation around the tangent			
210.00	0.00	0.37	230.70
205.00	0.00	0.43	239.52
200.00	0.00	0.51	240.58
195.00	0.00	0.60	236.17
190.00	0.00	0.71	228.82
185.00	0.00	0.83	217.44
Equate slope from C_u and G curve = G' local max A at tangent pt			

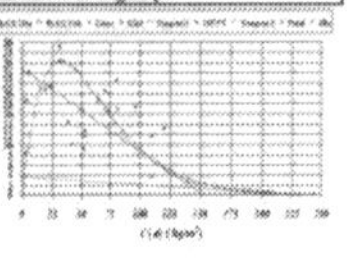

Table 6 – SSF results using Christian CF – Christian (1994) data

DESIGN DATA		
C_s (kg/m³)	C_u (kg/m³)	Q_f (m³)
5.0	232.0	177600.0

ω (m/s)	C (kg/m³)	G (kg/m²/s)	$LOG_{10}(\omega_s)$	$(LOG_{10}(\omega_s)-m*C-b)^2$
0.0180	140.00	2.52	-1.7447	0.0168
0.1100	120.00	13.20	-0.9586	0.1392
0.0340	110.00	3.74	-1.4685	0.0776
0.1070	110.00	11.77	-0.9706	0.0481
0.1830	96.00	17.57	-0.7375	0.0645
0.2100	81.00	17.01	-0.6778	0.0102
0.2840	80.00	22.72	-0.5467	0.0475
0.2270	73.00	16.57	-0.6440	0.0005
0.2930	70.00	20.51	-0.5331	0.0081
0.2170	53.00	11.50	-0.6635	0.0793
0.2390	53.00	12.67	-0.6216	0.0575
0.1770	52.00	9.20	-0.7520	0.1477
0.3260	51.00	16.63	-0.4868	0.0178
0.2480	41.00	10.17	-0.6055	0.1551
0.3820	38.00	14.52	-0.4179	0.0619
0.9440	31.00	29.26	-0.0250	0.0020
0.8970	24.00	21.53	-0.0472	0.0058
1.1000	19.00	20.90	0.0414	0.0035
4.5340	5.30	24.03	0.6565	0.1311
2.5600	3.30	8.45	0.4082	0.0073
6.0450	3.10	18.74	0.7814	0.2078
		m	b	1.2892
	IC	0.01	0.5	Φ
		0.0142	0.3696	
		0.0142	0.3700	
		see Christian paper		

C_{tang}	ω_{tang}	G_{tang}	A
195.68	0.00	0.77	180.35
Area calculation around the tangent			
195.00	0.00	0.79	180.32
195.68	0.00	0.77	180.35
196.00	0.00	0.76	180.34
197.00	0.00	0.74	180.23
198.00	0.00	0.72	179.97
199.00	0.00	0.70	179.57
Equate slope from C_u and G curve = G' Max A at tangent pt			

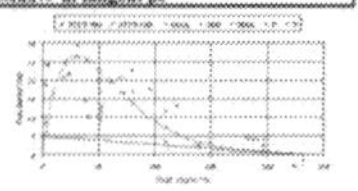

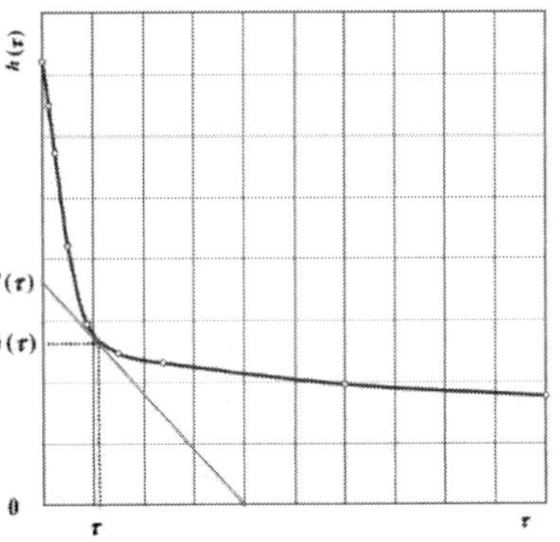

Figure 1 – SC $h(\tau)\sim\tau$

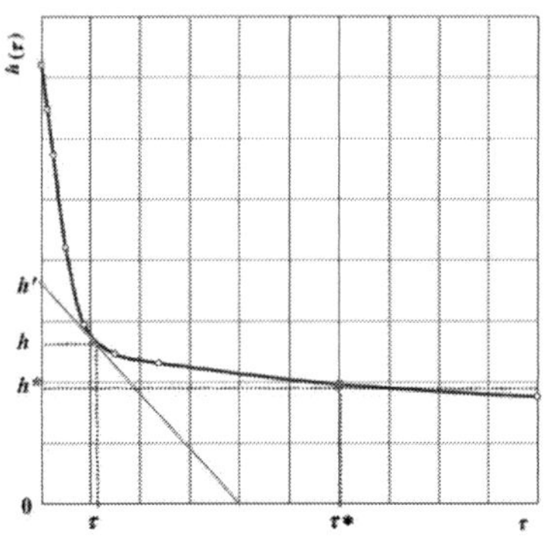

Figure 2 – Optimal h^* and τ^* on a SC

MZ method

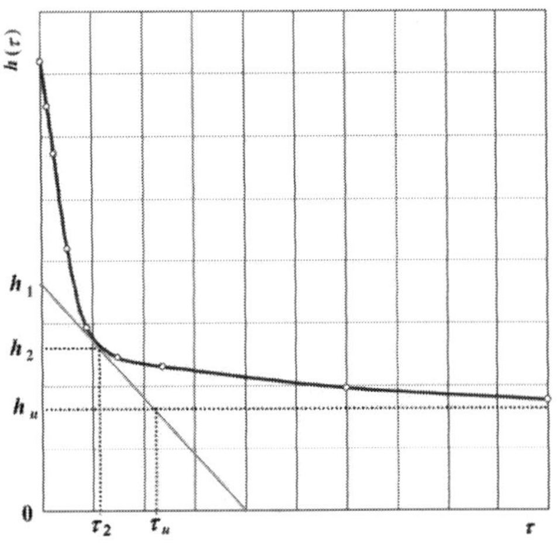

Figure 3 – h_1, h_2, h_u, τ_2, and τ_u of TF method

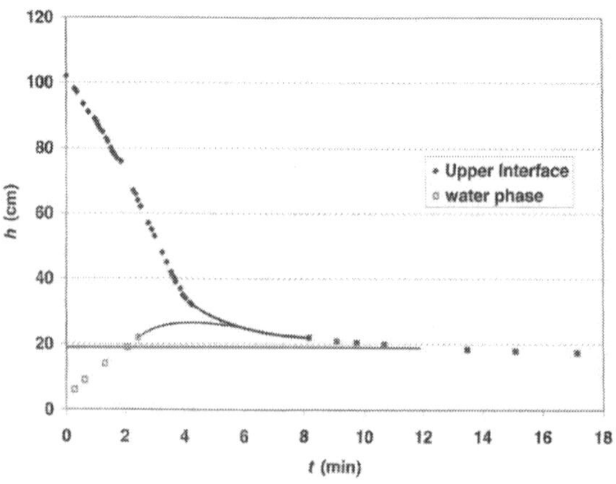

Fig 4 - Optimal h^* and τ^* on a SC – diluted bitumen gravity settler

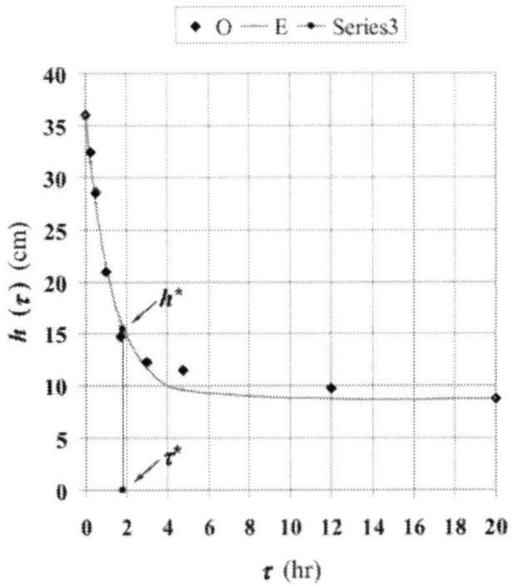

Fig 5 Observed SC – (Foust et al., 1962)

Color Removal and Biosolids Reduction with Ozone in Textile Plant Wastewater Treatment

Andre Rambor, Paulo Bon, White Martins South America (SAWM). Sao Paulo, SP, Brazil.
Malcolm Fabiyi, Richard A. Novak, and Karen Connery, Praxair, Inc., Burr Ridge, IL, USA

Abstract

In order to reduce the costs associated with tertiary physical-chemical color removal, an ozone decolorization process was developed at textile wastewater plant which employed high-purity oxygen activated sludge treatment. Prior to the adoption of the ozonation process, about 40% color removal was achieved through the biological process while the rest of the color removal was obtained through a tertiary, physical–chemical flocculation and coagulation process. Although the process was effective in removing color, large quantities of inert sludge were generated, making up as much as 60% of the total sludge from the plant. The cost for handling this sludge and for the purchase of the chemicals required for the flocculation & coagulation process was significant. The ozone based decolorization process has been effective in meeting effluent targets of 200 mg/l Pt-Co color and 50 NTU, at lower cost than the tertiary chemical treatment process. The residual gas stream from the ozonation process comprising about 3.5% ozone (w/w) and about 2.5 metric tons/day of recovered oxygen is recycled back to the biological basin. The reuse of the vent gas stream from the contactor provides about 70% of the oxygenation demand at the facility. This type of recycle scheme which leverages the O_2 content in the vent gas stream for supplementing the O_2 demand in the biological basin has been implemented at a number of SAWM operated wastewater facilities, and is an effective tool for ensuring cost effective operations when ozone is used as an oxidizing agent in biological wastewater treatment systems.

Keywords

Ozonation, decolorization, sludge reduction, vent gas reuse, High Purity Oxygen (HPO)

Introduction

Biological treatment of textile wastewater often requires non biological means for the decolorization of treated effluent in order to meet regulatory requirements. Manufacturers and consumers of industrial colorants, dyes and pigments have also come under increasing regulatory requirements (Reife and Freeman, 1996; Arslan-Alaton, 2004) not just for the removal of organic load, but also to meet specific effluent qualities such as color and turbidity. Ozone is effective in decolorizing many types of dyes, especially reactive and basic dyes (Dubrow et al, 1996). The decolorizing action of ozone occurs via the oxidative cleavage of conjugated systems of chromophoric molecules (Matsui, 1996; Davis, 2001)

Overview

The industrial textile wastewater plant reported in this paper uses a High Purity Oxygen (HPO) activated sludge biological treatment process. The color in the wastewater results from reactive dyes used in the textile manufacturing process. The plant treats 4,200 m^3/day of wastewater with an average influent COD value of 1,650 mg/l. Based on the average flow and a basin mixed liquor concentration of 2,500 mg/l, the corresponding loading rates (F/M) range from 0.5 – 0.6 kg COD/kg SS/day. Influent COD and color (reported as equivalent Pt-Co units in mg/l) values are summarized in Table 1. About 85% COD removal is achieved through the HPO biological process.

The biological basin is a single 4,590 m^3 reactor which has been separated by mixer placements into 3 cells of equal volume, in series. There is a DO meter in each cell, and O2 supply to each cell is controlled with a set-point of 2 mg/l. A total of four (4) Praxair In-Situ Oxygenation (I-SO™) systems (see Figure 1) are used to provide mixing and oxygenation, with two units placed in the first section. Each of the 4 X 25 HP I-SO™ units is operated at 270 rpm (Figure 1 and 2). Automatic pH adjustment is performed in an influent basin upstream of the biological treatment.

The activated sludge biological process was able to effectively remove as much as 45% of the influent color. Secondary clarifier effluent color was about 1,100 mg/l Pt-Co. In order to effectively meet effluent color targets, a physical chemical flocculation coagulation process was implemented using Ferric Sulfate and a combination of polymers. Although the physical chemical process was effective in eliminating color, a significant quantity of inert sludge was generated. The inert sludge from the physical chemical color removal process contributed about 6-7 tons per day of the total plant sludge of 10 - 11 dry tons per day. Because of the high costs associated with both purchased chemicals and sludge disposal, a cheaper and equally effective solution was needed.

Methodology

To achieve required color removal, 12 kg/hr of ozone at 10% w/w, produced using a Wedeco SMO 800S ozone generator, is applied to the secondary clarifier effluent, achieving an ozone dosage of 70 mg/l. Ozone is supplied through ceramic fine bubbles spargers to 3 x 7.5 m high cylindrical tanks that provide an average liquid retention time of about 30 minutes (see Figure 3). The three (3) reactors that make up the ozone contactor are operated in series. The residual gas from the ozonation process containing up to 3.5% ozone (w/w) and about 2.5 metric tons/day of recovered pure oxygen is recycled back to the biological basin. The DO level in the ozonated tertiary effluent is raised to an average of about 15 mg/l, providing adequate dissolved oxygen for discharge.

The energy required for generation of ozone at this facility is about 11 kWh/kg O_3. The gas feed to the generator is pure oxygen from a liquid oxygen supply source, supplied by S.A White Martins (SAWM). All water quality measurements were carried out using Standard methods (APHA, 1998). UV-VIS absorbance measurements were taken using a Hach 550 Spectrophotometer which was zeroed at 395 nm. Absorbance measurements for each sample were done across a wavelength spectrum from 395 to 905 nm.

Results & Observations

Vent Gas Recycle

The recycled O_2-O_3 stream is dissolved into the biological basin using a mechanically agitated contactor which is able to aspirate the vent gas stream (Figure 1). Praxair's In-Situ Oxygenation (I-SO™) system, which is used for oxygenation at this facility and the aspiration of the vent gas, is a subsurface, down-pumping mixing and oxygenation system used for providing oxygen in open basin HPO operations. The down-pumping action of the helical impeller generates a vortex which aspirates the vent gas, or entrains pure oxygen gas that is supplied to a floating hood. The high shear in the draft tube region breaks up the entrained gas into fine bubbles. The intimate gas-liquid contact in the draft tube significantly enhances mass transfer. Dissolved oxygen is dispersed by the mixing action induced by the high velocity liquid jet ejected from the draft tube. Undissolved gas bubbles detach from the liquid flow and are recycled through the draft tube, allowing high oxygen transfer efficiencies to be attained. Standard aeration efficiencies of up to 10 Ibs O_2/HP-hr and oxygen transfer efficiencies of up to 90%-95% can be obtained using the I-SO™ mixing system (Praxair, 2003; NYSERDA, 2000). The ability of the I-SO™ system to use low-pressure vent gas while providing high oxygen transfer efficiency makes it particularly well suited for recycle of vent gas from tertiary ozone treatment systems.

Effect of Ozonation on Excess Solids

The use of ozone has led to the complete elimination of the inert tertiary sludge. Prior to ozonation, the inert sludge from the physical chemical color removal process contributed about 6-7 tons per day of the total plant sludge of 10 - 11 wet tons per day. Waste activated sludge was also reduced by about 53% corresponding to a reduction in sludge yields from 0.163 to 0.076 kg SS/kg COD removed due to the lysis action of ozone in the biological basin (Figure 4).

Effect of Biological and Ozone Treatment on Color

Across both the ozone contactor and the biological basin, the specific color removal from ozone treatment is 13.3 kg Pt-Co/kg O_3. UV-VIS absorbance measurements indicate that the treatment steps are clearly effective for removing constituents that are strongly absorbing in the 400-750 nm range, which overlaps with the visible spectrum. The ozone recycle also enhanced the color removal in the activated sludge basin (Figure 5). Very little divergence in the absorbance measurements of the raw vs. treated samples is observed in the 750-905 nm range.

Effect of Ozonation on Treatment Costs

Ozonation had a significant impact on treatment costs. The use of ozone eliminated the inert sludge generated by the Flocculation-Coagulation process. Overall, the operating cost associated with the tertiary color removal process was reduced by 82% (Figure 6).

Effect of Ozonation on Biological Treatment Capacity

The addition of a strong oxidizing agent like ozone to an activated sludge process has previously raised concern for potential impact on the vitality of the microbial population (Jarvik et al, 2010). In the present study, however, the system's capacity for wastewater treatment has been retained. No deleterious effect of ozonation on carbonaceous removal has been observed.

Conclusion

Tertiary ozone treatment of an industrial textile wastewater, with re-use of ozone vent gas for biological treatment, was demonstrated. The process provided effective color removal at substantial cost savings over conventional physical chemical treatment, with further environmental benefits of reduced sludge production.

The application of ozone to the biological basin enabled both color removal and sludge reduction. The mechanisms by which these phenomena occur are significantly different indicating that separate mechanisms were at work. Decolorization is believed to proceed via the ozone aided oxidation of the conjugated double bonds of chromophores (Beltran, 2004). The ozone is also effective for achieving cell lysis. No further enhancement to COD reduction was observed, despite the fact that a significant improvement in decolorization was observed.

Although significant levels of sludge reduction are observed, the system's overall O_2 demand was not significantly changed. Prior to ozonation, the oxygen supply balance indicated that the oxygen consumption was about 0.6 kgs O_2/kg COD removed. Post ozonation, the oxygen supply is now about 0.55 – 0.6 kgs O_2/kg COD removed. These enhancements to oxygen utilization observed could potentially be due to surfactant induced enhancements to mass transfer as well as the possibility that the oxidizing action of ozone might lead to the oxidative quenching of ozone to oxygen.

This work demonstrates that full scale ozonation for color removal is a viable approach for industrial wastewater facilities. The use of ozone also provides the additional benefit of a vent gas stream that is available for use in the aerobic treatment process.

Table 1: Summary of loading rates and treatment outcome across the process

	Prior	Ozonation	
Flow (ave)	155	175	m3/hr
TSS Basin (ave)	2,500	2,500	mg/l
COD in (ave)	1,650	1,650	mg/l
BOD in (ave)	400	400	mg/l
COD out (ave)	248	248	mg/l
Color EQ (ave)	1,850	1,850	mg/l
Color 2o (ave)	1,100	650	mg/l
Color 3o (phy chem) ave	200		mg/l
Color 3o (O3) ave		185	mg/l

Note: EQ = Equalization Tank. 2° = secondary activated sludge basin; 3° = Tertiary flocculation – coagulation process (prior) and ozonation process. Color is reported as Platinum – Cobalt color equivalent, mg/l.

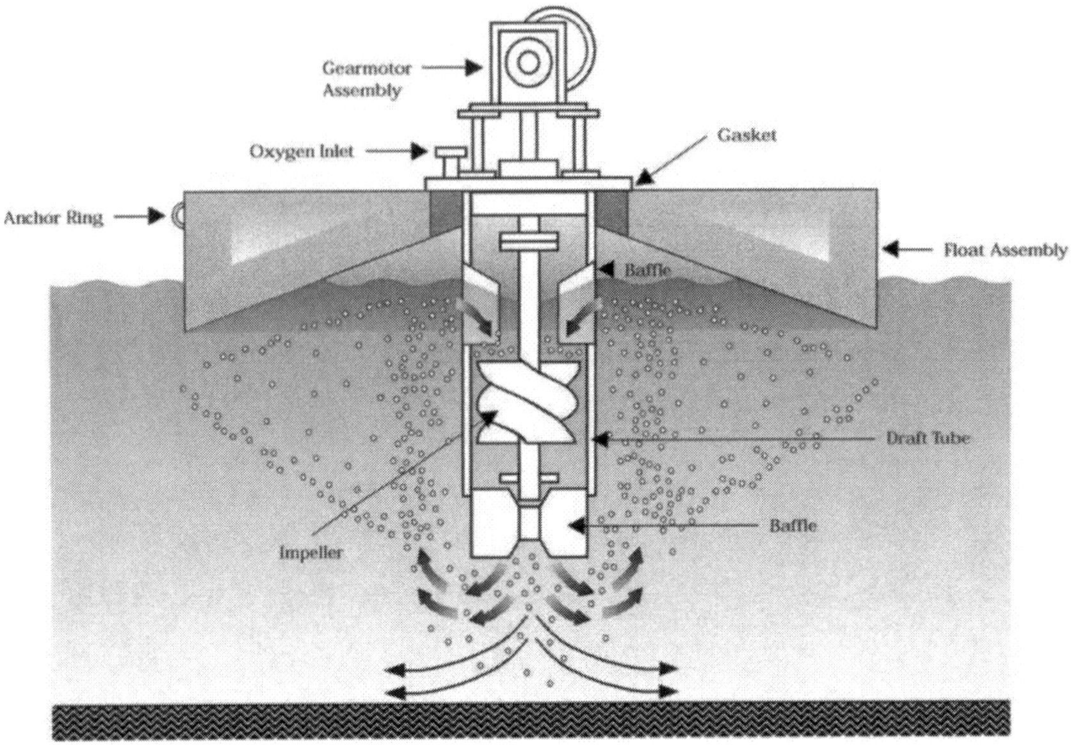

Figure 1: Praxair's In-Situ Oxygenation System (I-SO™)

Figure 2: Praxair's In-Situ Oxygenation System (I-SO™) in operation at the facility

Figure 3: Ozone contactor at Textile Wastewater Facility.

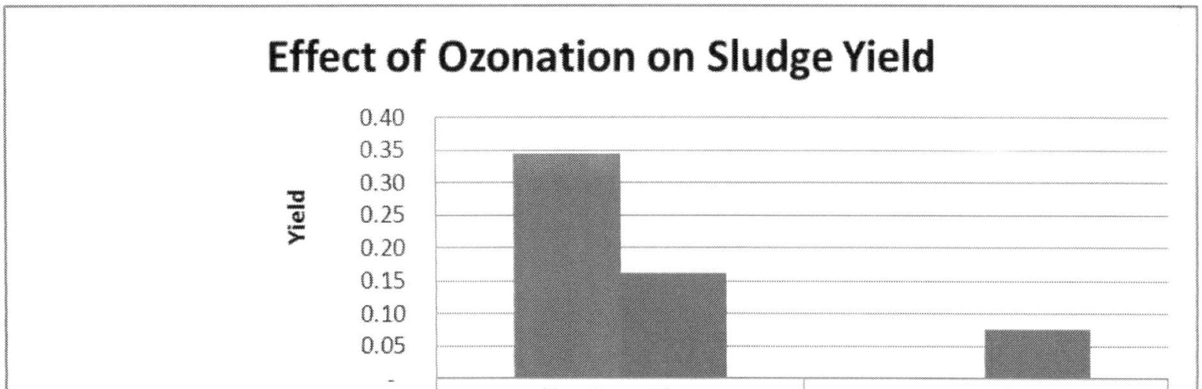

	Pre-Ozonation	Ozonation
Inert Yield (kg Inert/kg Color Removed in PtCo units)	0.34	-
Excess Biological Sludge (kg SS/kg COD removed)	0.163	0.076

Figure 4: Inert and Biosolids solids reduction as a result of ozonation

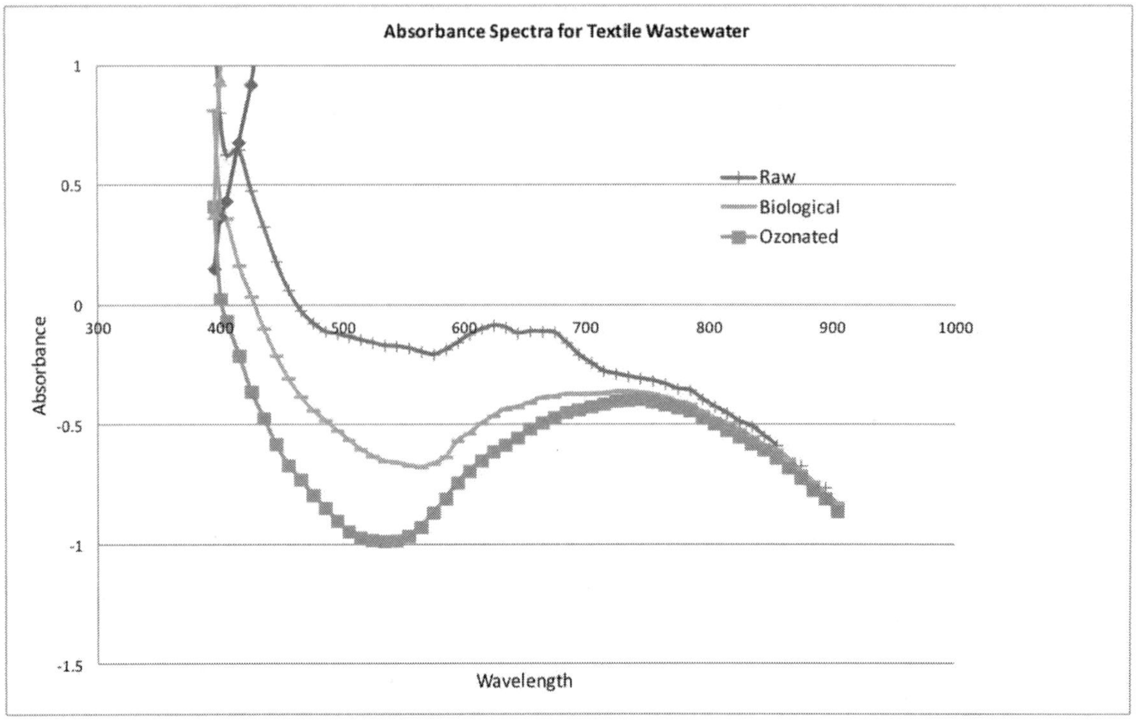

Figure 5: Effect of treatment on UV-VIS absorbance. The impact of treatment on absorbance and light transmittance (i.e., decolorization) in the 400-750nm range of the visible spectrum can be observed.

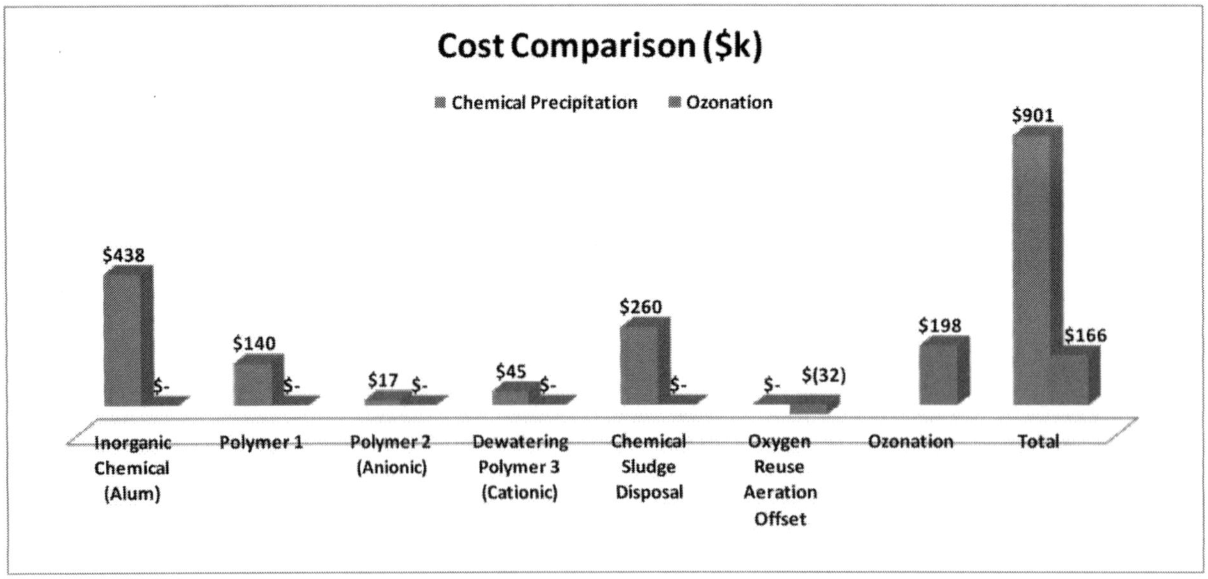

Figure 6: Summary of savings enabled by ozone decolorization

References

1. APHA, AWWA and WEF, 1998. Standard Methods for the Examination of Water and Wastewater, 20th ed. APHA, Washington, DC, USA.
2. Arslan-Alaton, I. Pretreatment of Textile Dye Carriers with Ozonation and Enhanced Ozonation. Environmental Engineering Science. Vol. 21, Number 4, 2004
3. Beltran, F. J. Ozone Reaction Kinetics for Water and Wastewater Systems, Lewis Publishers, Florida 2004.
4. Davis, B. The Chemistry of Color Removal: A Processing Perspective. Proc. Of the South African Sugar Technol. Assoc. Vol 75, 2001
5. Dubrow, S.F., Boardman, G.D., & Michelsen, D.L. Chemical Pretreatment and Aerobic-Anaerobic Degradation of Textile Dye Wastewater. Chapter 5. Environmental Chemistry of Dyes and Pigments. John Wiley & Sons, Inc, NY, 1996. Eds. Reife, A., & Freeman, H.S
6. Matsui, M. Ozonation. Chapter 3. Environmental Chemistry of Dyes and Pigments. John Wiley & Sons, Inc, NY, 1996. Eds. Reife, A., & Freeman, H.S
7. NYSERDA Report, 2000. In-Situ Oxygenation (I-SO™) for Volatile Compound Emission Control.
8. Praxair Report of Clean Water Tests at Philadelphia Mixing Solutions, 2003.
9. Reife, A., & Freeman, H.S. Environmental Chemistry of Dyes and Pigments. John Wiley & Sons, Inc, NY, 1996

(640c) Study On the Mechanism of Cr(VI) Biosorption by Flocculating Yeast SPSC01

Lijie Chen, School of Life Science and Technology, Dalian University of Technology, Dalian, China and Zhicun Wang, School of Life Science and Technology, Dalian University of Technology, Dalian, China

Biosorption of heavy metal have higher advantages than physical and chemical methods, particularly in the treatment of low concentrations of heavy metals in wastewater. The flocculating yeast SPSC01 in this article with good flocculation characteristics has been put into the industrial production of fuel ethanol. The mechanism of Cr(VI) biosorption by flocculating yeast SPSC01 has been studied. The results showed that, pH was the most significant parameter on the process of Cr(VI) reduction-adsorption. In the certain range, Cr(VI) was more favorable reduced to Cr (III) with the lower pH, and the process was reached equilibrium much earlier. Cr(III) was more favorable adsorbed on biosorbents with the higher pH, and the total chromium removal rate was higher. pH 2 was the optimal pH, considering all the factors. It was found that the Cr adsored to SPSC01 was Cr(III) by digesting SPSC01 after adsorption. SPSC01 was approached by chemical modification, and SPSC01 was compared with the different biosorbents on Cr(VI) adsorption.The results indicated that the major functional groups (amino, carboxyl and amide) may contribute to the absorption of Cr(VI). FTIR results also proved this result. Most of the Cr was absorbed on the cell wall, from the result of TEM and Cr composition on the components of SPSC01 after separation. Cell wall prevented the entry of lots of Cr ions, reducing the toxicity of Cr. When the Cr(VI) adsorption process reached equilibrium, SPSC01 still has biological activity, so could be adsorbed after desorption, and improved efficiency. The method of 1 mol/L HNO_3 at 35 ^{o}C was a good desorption method. SPSC01 could be recycled under this way, and the method was a favorable way for recovery of Cr.

(640d) Enhanced Degradation of Fats, Oils and Greases in Domestic Wastewater Sewer Networks and Grease Interception Systems Using Peat Humic Substances

Matthew G. Hunnemeder, Hoshizaki, Z. Otero Gephardt, Chemical Engineering, Rowan University, Glassboro, NJ and Gregory B. Hecht, Biological Sciences, Rowan University, Glassboro, NJ

The efficacy of peat humic substances in enhancing the degradation of fats, oils and greases (FOG) was investigated under controlled laboratory conditions using bench scale well-mixed bioreactors. An experimental design was used to evaluate the effects of temperature and peat humic substance (PHS) concentration on FOG degradation in domestic wastewater for a temperature range from 10°C to 30°C and a PHS concentration range from 0 to 20 ppm (v). Factors and interactions significantly affecting the rate of FOG degradation were identified, and models to predict FOG degradation rates as a function of PHS concentration and temperature were developed. The models were used to develop a PHS dosage calculation technique for field operations. Results indicate that PHS can enhance FOG degradation rates by up to a factor of 2, and microbial cell growth rates by up to a factor of 3. Atmospheric hydrogen sulfide generation increased with high PHS concentration at the higher temperature. The rate of FOG degradation using grease interceptor material was studied at 25°C and a PHS concentration of 500 ppm (v). In these systems, PHS was observed to increase the rate of FOG degradation by up to a factor of 2, and microbial colony growth rates by up to a factor of 5. This work indicates that PHS can enhance FOG degradation rates and increase microbial growth rates in wastewater treatment systems. These results have significant implications for wastewater treatment applications.

Improved Volatile Solids Reduction in Aerobic Digesters by Application of Sludge Ozonation using the Lyso™ Process

Malcolm Fabiyi, Richard A. Novak, Praxair, Inc., Burr Ridge, IL, USA
Alessio Colletti, Rivoira, S.p.A., Rome, Italy

Abstract

Ozonation of a recirculated stream from an aerobic digester has been employed to reduce volatile solids. The Praxair Lyso™ process, which utilizes plug flow principles for the ozone – sludge lysis reaction was used as the ozone contactor. Sludge drawn from the aerobic digester was circulated through the Lyso™ system where the sludge and ozone are contacted. Plug flow operations minimizes excessive ozone consumption by enhancing the selectivity of ozone for cell lysis rather than for random COD oxidation. Sludge ozonation enabled up to 80% sludge reduction with less than 10 days of HRT in the digester, indicating significant enhancement of Volatile Solids Reduction (VSR). The specific ozone utilization values obtained during the tests ranged from 0.01-0.03 kg O_3/kg SS removed. This specific ozone consumption value is about an order of magnitude lower than alternative sludge ozonation concepts.
We develop a conceptual model of sludge ozonation process using the GPS-X™ process modeling software platform. A simplified metric for reporting ozone effectiveness rates in sludge ozonation processes is proposed.

KEYWORDS

Sludge Ozonation, Lysis, Digesters, Lyso™

Introduction

The activated sludge process produces excess biological solids, incurring significant costs associated with the treatment, management and disposal of excess biosolids (Tchobanoglous et al, 2003). Strict regulatory guidelines for disposal options, such as EU directive 1999/31/EC and 86/278/EEC, impose stabilization requirements for the landfill or land application of sludge (Perez-Elvira et al, 2006).

Sludge minimization strategies have focused on two broad approaches. These include (i) Lysis-Cryptic growth processes and (ii) end of pipe solutions at the wastewater facility such as incineration and wet oxidative treatment of the excess sludge.

Lysis-Cryptic growth uses chemical, mechanical, electrical, thermal or biological means for minimizing excess sludge generation. The approach requires the release of the highly biodegradable cellular content of biomass by compromising cellular integrity. The released cellular material, called lysate, then serves as substrate to other viable microorganisms in the process. The process is termed Lysis-Cryptic growth to differentiate it from microbial growth associated with the original substrates (Chua et al, 2009).

Sludge Ozonation

Sludge ozonation is an effective means for achieving sludge minimization via Lysis-Cryptic in wastewater treatment facilities (Vergine et al, 2007). The process can be applied directly to the activated sludge basin, to Return Activated Sludge (RAS), or to aerobic or anaerobic digesters. The choice of where to apply sludge ozonation is influenced by a variety of factors, including (i) phosphorous management, (ii) quantities of sludge to be reduced, (iii) loading capacity of the activated sludge process, and (iv) VSR observed in digesters. In addition to excess sludge minimization, the use of sludge ozonation leads to a number of process benefits, such as enhanced dewatering and the mitigation of foaming and bulking problems (Fabiyi et al, 2007b).

Sludge Ozonation in Digesters

In the present study, sludge ozonation was applied to the aerobic digesters in a municipal plant in Italy. The plant treats up to 30,000 Person Equivalents (PE) of wastewater in two parallel trains. One train treats a third of the flow (10,000 PE) while the other handles two thirds (20,000 PE) of the influent capacity. The plant has a 155 kW compressor that provides the aeration requirements of the combined secondary activated sludge and aerobic digester systems. Airflow to the digester is controlled to maintain DO of 1 ppm, using a Fine Bubble Diffuser system.

Sludge digestion is a critical unit operation for achieving stabilization of sludge. The traditional digestion process relies on natural hydrolytic processes to break down the bacterial cell wall to facilitate the release of intracellular organic compounds. Sludge stabilization at the facility is handled in an aerobic digester of 900 m^3 volume. About 100 m^3/day of WAS is fed to the system, with TSS of approximately 7000 mg/l. VSS reduction in the digesters was minimal. To enhance VSS reduction, a sludge ozonation process using Praxair's Lyso™ process was adopted.

Sludge ozonation is an effective means of lysing cells to facilitate the rupture of the bacterial cell. This paper will present data from the Pilot and Full Scale implementations of sludge ozonation at the WWTP. The paper also discusses the effectiveness of sludge ozonation for enhancing VSS destruction in digesters and elaborates on the mechanisms that enable the very low level of ozone required for achieving VSS destruction using the Lyso™ sludge ozonation process (Table 1).

Methodology

Sludge drawn from the aerobic digester on the 20,000 PE line was circulated through a Lyso™ sludge ozonation system comprising a pump and a plug flow reactor for contacting ozone with the sludge. The Lyso™ system can handle up to 70 m^3/hr of sludge flow (Figure 1). Liquid oxygen supplied by Rivoira S.p.A. was provided to a Wedeco ozone generator. The ozone was aspirated into the contactor using a venturi, with ozone flow ranging from 0.06 to 0.67 kg/hr. Pilot testing was undertaken from February to August 2010.

Results

The Lyso™ process provided significant reduction in VSS. About 64%-85% reduction in sludge was achieved (see Figure 2, also D'Amato, 2011) with an estimated cost savings of €63,000/year. Significant enhancements to sludge settling and dewatering characteristics were also observed, in line with previous observations from sludge ozonation using the Lyso™ process (Vergine et al, 2007).

The specific ozone utilization values obtained during the tests were about 0.01-0.03 kg O_3/kg SST eliminated, while the effective ozone dosage rate was 0.002 gO_3/g SST (D'Amato et al, 2010). This study demonstrated improved ozone utilization rates (i.e., lower amounts of ozone required per unit of sludge reduced) when compared with alternative sludge ozonation approaches reported in the literature. The effectiveness of ozone utilization value is about an order of magnitude lower than alternative sludge ozonation concepts (see Table 1). The Lyso™ process uses a novel plug flow reactor approach for sludge ozonation that minimizes excessive ozone consumption by enhancing the selectivity of ozone for cell lysis rather than random COD oxidation (Fabiyi et al, 2007a, Fabiyi et al, 2007b).

By providing as much as 85% sludge reduction within less than 10 days of SRT in the digester, sludge ozonation offers the possibility for enabling the construction of smaller digesters. In contrast, conventional designs for aerobic digestion (without ozone treatment) would achieve only 30% maximum solids reduction for a 10 day SRT at 20°C (Metcalf & Eddy, 2003).

GPS-X™ Software Modeling of Sludge Ozonation in Aerobic Digesters

A conceptual model of the sludge ozonation system coupled with the aerobic digestion system was developed using the mantis2 model of the Hydromantis' GPS-X™ software platform (Goel et al, 2011). The default sludge pretreatment module in GPS-X™ allows for mechanical, electrical and chemical based Lysis-Cryptic growth systems to be modeled (see Figure 3).

Prior to ozonation very little digestion was obtained in the digester, consistent with a 9-10 SRT. Two scenarios were investigated. The first scenario replicated normal operating conditions in the digester, during which a flow of 100 m^3/day of Waste Activated Sludge (WAS) was charged to the aerobic digester, with an active volume of 900 m^3. There was no ozone conditioning of the sludge under this scenario. The second scenario considered the addition of a lysate agent to a sidestream contactor that draws liquor from the digester and discharges back to the digester following contacting (see Figure 3). Table 2 summarizes the parameter values that were utilized in the model. The specific chemical intensity (ozone dosage) utilized was the actual ozone dosage of 0.002 g/kg TSS used in the pilot study (D'Amato et al, 2010). Default values were used for all variables, except for the maximum growth rate of ammonia oxidizers (1/d) during sludge ozonation. Nitrifier activity plays an important role in mediating nitrification and pH conditions in the biological systems, and can be impacted by sludge conditioning (Vergine et al, 2007).

For the scenario with no ozone conditioning of the sludge, estimated Volatile Solids Reduction (VSR) and Total Solids Reduction (TSR) values were 26% and 34% respectively, in agreement with design estimates (Metcalf & Eddy, 2003). Model results for the sludge ozonation case gave VSR and TSR values of 54% and 82% respectively, which agree well with the results obtained during the test. All runs were implemented under steady state conditions (Figure 4).

Addressing the Confusion over Terminology for Ozone Usage Effectiveness

Many discussions of sludge ozonation confuse the ozone dosage and the specific ozone utilization rate. The ozone dosage rate refers to the amount of ozone brought into contact with a given quantity of solids in the ozone-solids contacting system. The dosage values are usually estimated as the ozone mass supply rate divided by the mass of solids in the system, with units of kg O_3/kg TSS. Most discussions of dosage do not specify how effective the amount of ozone injected has been in causing cell lysis. Without understanding the specific ozone utilization rate, which we define as the mass of ozone required per mass of solids eliminated, i.e., kg O_3/kg TSS eliminated, the true economics of the sludge ozonation process cannot be determined. Reviews of sludge ozonation processes routinely combine dosage values and specific ozone utilization rates without distinction (e.g., see Chu et al, 2009, where specific ozone utilization rates (kg O_3/kg TSS eliminated) reported by Vergine et al, 2007, are combined with ozone dosage rates from other studies).

It is less confusing to report the ozone dosage value as kg O_3/kg Solids, while reporting the specific ozone utilization rate in its inverse form, as kg TSS eliminated/kg O_3 utilized. We define this metric as the ozone effectiveness rate (OER). We believe the usage of this metric, relating directly to the efficiency of ozone use in sludge reduction, will be an effective means of ensuring that there is clarity in the interpretation of data from sludge ozonation studies (see Table 1). Because this metric increases as the effectiveness of ozone action increases, it is also more intuitive. In past practice, lower values for the specific ozone utilization rate corresponded to higher effectiveness, which can lead to confusion.

Conclusion

We have reported the successful application of the Praxair Lyso™ process for sludge minimization in aerobic digesters. The very high efficiencies of ozone utilization obtained in this study with the use of the plug flow Lyso™ system provides further validation of the importance of using a plug flow scheme to provide optimum selectivity for ozone lysis reactions. The commercial viability of the sludge ozonation process is also demonstrated.

A conceptual model of the process, developed using the sludge pretreatment module in GPS-X™ software modeling platform, accurately simulated the effect of sludge treatment with ozone on the VSR performance of aerobic digesters. As the adoption of sludge pretreatment and conditioning increases, modeling tools that enable the mechanistic dynamics of Lysis-Cryptic growth using various methods to be evaluated will become increasingly important.

We propose the adoption of a new metric, the ozone effectiveness rate (OER, kg TSS eliminated/kg O_3) as a means of helping to provide clarity in the reporting and interpretation of data from studies on sludge ozonation.

Figure 1: Lyso™ Sludge ozonation skid showing the plug flow contactor.

Table 1: Summary of specific ozone consumption and the corresponding sludge reduction. Adapted from Sievers et al, 2004.

Reference	Ozone dose kg O_3/kg $TSS_{treated}$	Ozone Utilization Rate kgs O_3/TSS $_{Eliminated}$	Ozone Effectiveness Rate kgs TSS $_{Eliminated}$/kg O_3
Yasui et al. (1996)	0.05	0.165	6.1
Sakai et al. (1997)	-	0.133-0.178	5.6-7.5
Kobayashi et al (2001)	-	0.25	4
Ried et al (2002)	0.05	0.395	2.5
Lyso™ (D'Amato et al, 2010)	0.002	0.01 - 0.03	33-100

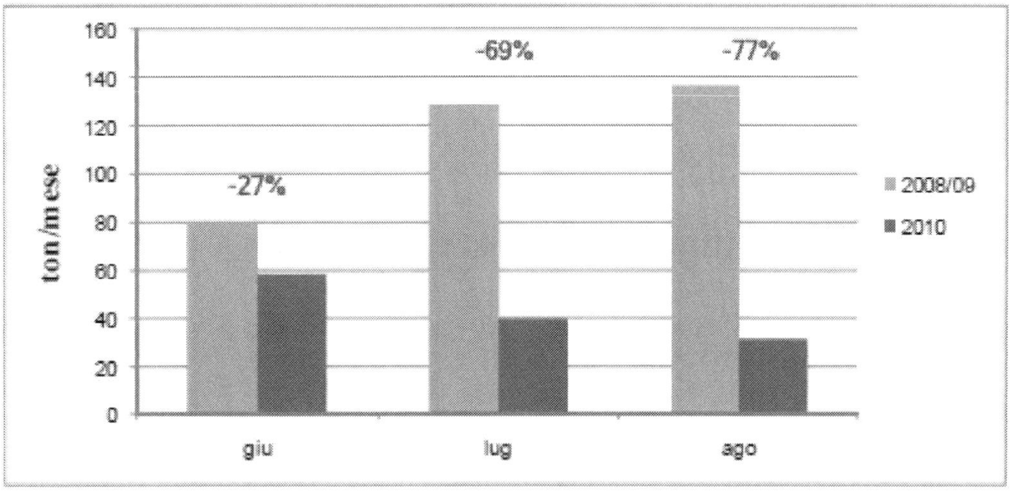

Figure 2: Sludge production during 2010 tests compared to same period sludge production in 2008/2009. Source: D'Amato et al, 2011.

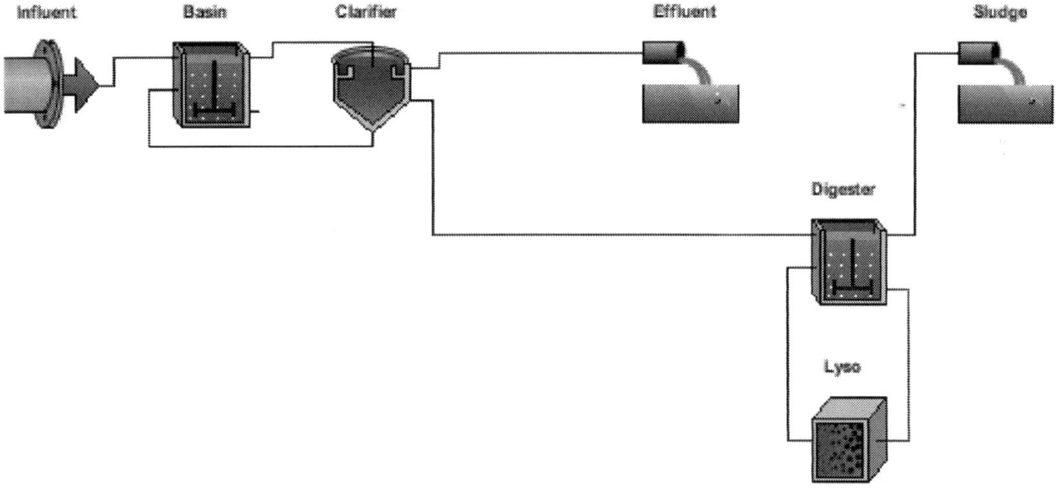

Figure 3: Layout for the conceptual GPS-X™ model of the Aerobic Digestion Process with Sludge Ozonation.

Table 2: GPS-X™ Model Inputs.

		Ozonated	No Treatment	Default
WAS Flow to Digester	m3/d	100	100	
Lyso™ System Feed	m3/h	70	0	
Specific Chemical Intensity	g/kg TSS	0.024	0	30
Half Saturation for Chemical effectiveness	g/kg TSS	5.00E-05	-	100
Threshold chemical dose for biological inactivation	g/kg TSS	50	-	50
Equivalent oxidative fracion in chemical	-	0.33	-	0.33
Fraction COD loss during treatment	-	0.01	-	0.01
Maximum solubilization factor	-	0.4	-	0.4
Maximum inert solid reduction factor	-	0.6	-	0.6
Factor for soluble biodegradable substrate	-	0.25	-	0.25
Maximum Growth Rate for Ammonia Oxidizers	1/d	0.285	0.9	0.9

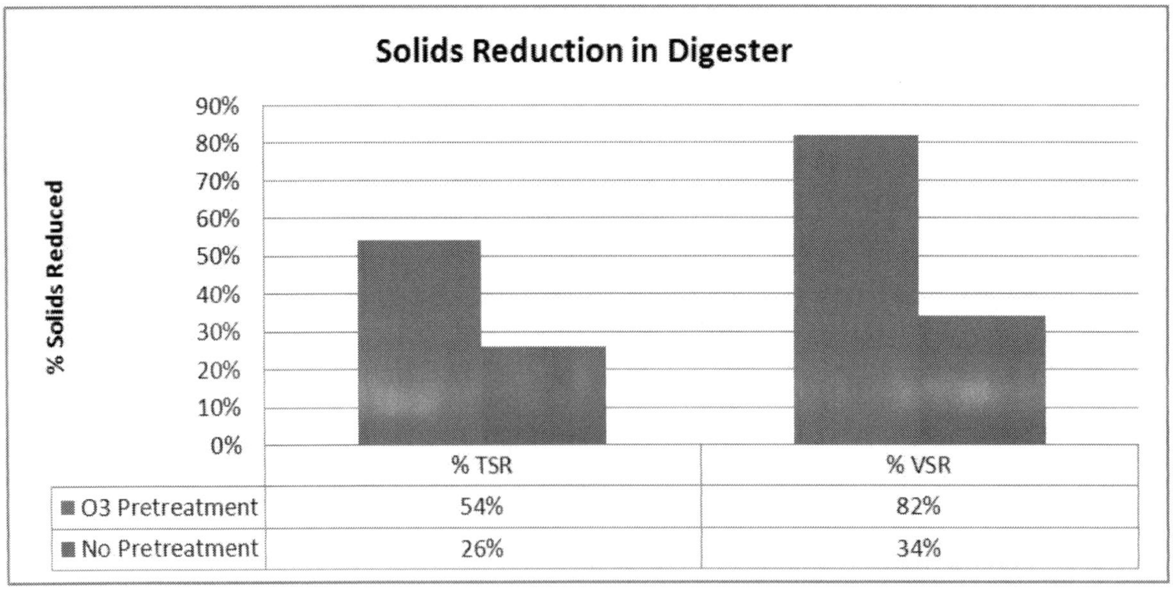

Figure 4: Results of GPS-X™ modeling of Sludge pretreatment process.

References

1. D'Amato, E., Sirini, P., Gori, R., Lubello, C., Chiavola, A., Mariani, F., Cima, E., & Liberati, L. Application of Ozone in the Digestion Tank of a Full Scale WWTP for Sludge Minimization. IWA Specialist Group on Sludge Management: Newsletter January 2011
2. D'Amato, E., Sirini, P., Gori, R., Lubello, C., Chiavola, A., Mariani, F., Cima, E., & Liberati, L. l'ozonolisi per la riduzione della produzione dei fanghi. esperienza presso l'impianto di depurazione di sabaudia. Perugia 2010
3. Fabiyi, M., & Novak, R. System and Method For Eliminating Sludge Ozonation. US Patent 7309432 B1. 2007a
4. Fabiyi, M., Novak, R., Ried, A., Wieland, A., Capra, R., and Sandon, A. Sludge Reduction Using Ozone Induced Lysis. Aug. 2007b World Congress, IOA-UV, Los Angeles, CA
5. Goel Rajeev, Schraa Oliver, Snowling Spencer, Andre Hank (2011) Mantis2 Model, Hydromantis GPS-X Technical Reference, 2011
6. Kobayashi et al (2001). Study on Sludge reduction and other Factors by Use of an Ozonation Process in Activated Sludge Treatment. Proc. 15th Ozone World Congress, London 2001. IOA 321.
7. Libing Chu, Sangtian Yan, Xin-Hui Xing,, Xulin Sun, Benjamin Jurcik. Progress and perspectives of sludge ozonation as a powerful pretreatment method for minimization of excess sludge production. Water Research 43 (2009) 1811 - 1822
8. Metcalf & Eddy, 2003. Wastewater Engineering: Treatment & Reuse. 4th ed. McGraw Hill Publishing.
9. Ried, A., Stapel, H., Koll, R., Schettlinger, M., Wemhöner, F., Hamann-Steinmeier, A., Miethe, M. and Brombach, A. (2002). Optimierungsmöglichkeiten beim Betrieb von biologischen Kläranlagen durchden Einsatz von Ozon. *Korrespondenz Abwasser*, **49**(5), 648–661
10. Sakai, et al. (1997). An Activated Sludge Process without Excess Sludge Production. Wat Sci Tech. 36 (11) 163 – 170
11. Sievers et al. (2004). Sludge Treatment by Ozonation – Evaluation of Full Scale Results. Water Science & Tech. 49 (4) 247 – 253
12. Tchobanoglous, G., Burton, F., & Stensel, H. D (2003). Metcalf & Eddy Wastewater Engineering: Treatment and Reuse. McGraw Hill, NY.
13. Vergine, P., Menin, G., Canziani, R., Ficara, E., Fabiyi, M., Novak, R., Sandon., A., Bianchi, A., Bergna, G. Partial Ozonation of Activated Sludge to Reduce Excess Sludge Production: Evaluation of Effects on Biomass Activity in a Full Scale Demonstration Test. IWA Conference, Moncton, Canada. 2007
14. Yasui et al. (1996). A Full Scale Operation of a Novel Activated Sludge Process Without Excess Sludge Production. Wat. Sci. tech. 34 (3-4) 395 – 404

(640f) Two-Phase Fluid Separation in Microscale-Based H-S Cell; Experiment and Modeling

Frederick Atadana, chemical engineering, oregon state university, Corvallis, OR and Goran Jovanovic, Oregon State University, Corvallis, OR

Two-Phase Fluid Separation in Microscale-Based H-S Cell;

Experiment and Modeling

Frederick Atadana[1], Ray Wheeler[2] and Goran Jovanovic[1]

[1]Oregon State University, Microproducts Breakthrough Institute, Corvallis, Oregon

[2]Umpqua Research Co.; Myrtle Creek, Oregon

In this work a novel design of the Hele-Shaw cell in the form of micropost lamina was used to achieve efficient and low cost separation of immiscible fluids. The device makes use of capillary pressure gradient, which is the source of driving force, for separation of two-phase fluid systems in micro scale structures. This novel approach is most applicable for separation of discrete bubbles/droplets from continuous fluid phase. The capillary pressure can be manipulated by the geometric design of device, changing the surface wetting properties of the device, and establishment of a temperature gradient. The device was tested on separation of water and oil, and water and air. Numerical simulation and preliminary experimental results of 50% aqueous and 50% air by volume, showed successful separation of the two phases. The application was extended to liquid-liquid extraction of shikimic using organic solvent from an aqueous solution. The results showed significant extraction of shikimic acid from aqueous solution.

(640g) The Oxidative Power of Corona Discharge

Matthäus Siebenhofer[1], Peter Letonja[2] and Theo Friedrich[2], (1)Institute of Chemical Engineering and Environmental Technology, Graz University of Technology, Graz, Austria, (2)Institute of Chemical Engineering and Environmental Technology, Graz University of Technology, Graz, Austria

The oxidative power of corona discharge

P. Letonja, T. Friedrich, M. Siebenhofer

Graz University of Technology, Institute of Chemical Engineering and Environmental Technology, Graz/Austria

Corona discharge, well known from electrostatic precipitation of particulate matter, has proven applicable too in wastewater treatment as well as off-gas purification.

Electrical discharge in gaseous medium depends on the basic gas composition, the temperature, the gas velocity and the type of discharge electrode. While precipitation of particulate matter needs plain electrodes to avoid formation of noxious substances, application of corona discharge in wastewater treatment is driven by needle discharge since oxidation power of corona discharge depends on the effect of very high field intensity. Brush electrodes are therefore the electrode type of choice. Because of the very different current/voltage characteristic in comparison with plain wire type electrodes brush electrodes provide significant UV-irradiation and ozone formation.

In the ongoing project oxidative precipitation of sulfur dioxide and nitrogen oxide was investigated in a wet tube type electrostatic precipitator, tube diameter was 76 mm, equipped with an 8 mm brush electrode of 0.7 m length. Corona start up voltage was 10 kV. Close to break through operation voltage of 21 kV the maximum specific corona current of 6 mA/m was achieved. The ozone formation potential, which depends on the gas flow rate, was about 300 ppm at 0.38 m/s gas velocity. The UV_{254} -irradiation intensity was equivalent 5% of the electrical power.

Nitrogen oxide is specified by extremely poor absorption properties. At low concentration it can hardly be oxidized. For oxidative absorption nitrogen oxide as well as sulphur dioxide need activation support.

Ambient air was spiked with either nitrogen oxide or sulfur dioxide and fed to the "corona discharge reactor", which was equipped with an absorption tower in series. The concentration of both constituents was varied between 100 ppm and 400 ppm. Figure 1 shows the result of a basic experiment for NO-treatment.

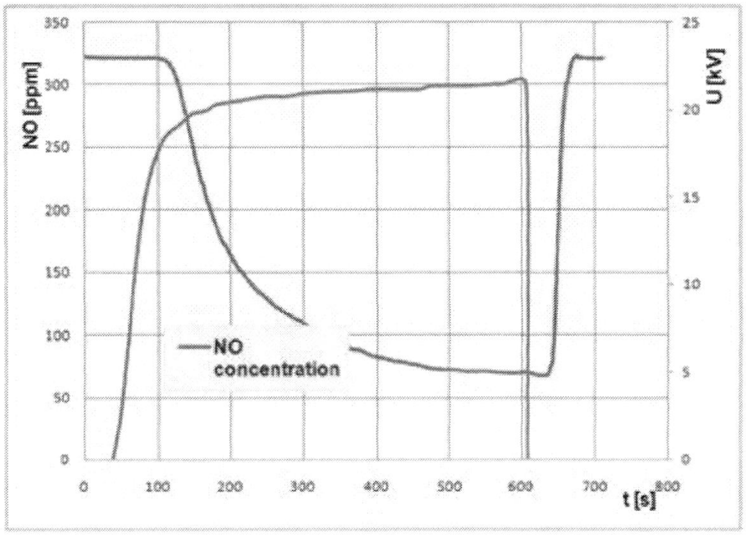

Figure 1: Oxidation and absorption of nitrogen monoxide. Operation conditions: ambient air

spiked wit NO, gas velocity 0.25 m/s

The final product of the constituents is either aqueous sulphuric acid or nitric acid.

Depending on the operation conditions the clean gas concentration of both substances could even be kept at < 10 ppm. Since oxidation is overall rate limiting modeling and scale up can refer to the design of plug flow reactors, considering space charge effects via the gas velocity in the reaction rate.

Based on the outcome of the project corona discharge seemingly offers a powerful tool in advanced pollution control of gaseous as well as liquid constituents.

(646a) Surface Modification of Membranes to Improve Fouling Resistance

Benny Freeman, Deptartment of Chemical Engineering , The University of Texas at Austin, Austin, TX

Across many platforms of membranes, fouling mitigation is a major challenge to be addressed to achieve the most energy-efficient, cost-effective membrane filtration processes. Previously, many surface modifications and functionalized polymers were reported to prevent fouling. However, most of these techniques and materials are practically difficult to implement in water purification membranes. We have discovered surface treatment methodologies that can be used to prepare high permeability polymeric membranes from all common water purification membrane classes.[1] These surface-modified membranes have persistent tolerance to fouling by proteins and emulsified oil, two ubiquitous contaminants in a variety of wastewaters. These membranes were prepared by depositing bio-inspired, self-polymerized hydrophilic polydopamine nanolayers on their surfaces. To demonstrate scalability of this modification, the nanolayers were also applied to entire membrane modules. Upon nanolayer deposition, the membranes could be further functionalized using fouling-resistant macromolecules, such as poly(ethylene glycol), to promote improved fouling resistance and, therefore, high membrane flux. More recent studies have focused on further characterizing the chemistry of formation of polydopamine,[2] exploring alterative molecules, such as hydroquinone and catechol,[3] which exhibit similar effects to dopamine when deposited on membrane surfaces, and characterizing the ability of these surface modifications to improve performance of forward osmosis membranes.[4]

[1] McCloskey, B.D., H.B. Park, H. Ju, B.W. Rowe, D.J. Miller, B.J. Chun, K. Kin, and B.D. Freeman, "Influence of Polydopamine Deposition Conditions on Pure Water Flux and Foulant Adhesion Resistance of Reverse Osmosis, Ultrafiltration, and Microfiltration Membranes," *Polymer*, **51**, 3472-3485 (2010).

[2] Dreyer, D.R., D.J. Miller, B.D. Freeman, D.R. Paul, and C.W. Bielawski, "Elucidating the Structure of Polydopamine," Langmuir, submitted.

[3]B.D. Freeman, D.J. Miller, and B.D. McCloskey, "Polymer Deposition and Modification of Membranes for Fouling Resistance," U.S. Patent Application Number US 2010/0059433 A1, March 11, 2010.

[4] Arena, J.T., B. McCloskey, B.D. Freeman, and J.R. McCutcheon, "Surface Modification of Thin Film Composite Membrane Support Layers with Polydopamine: Enabling Use of Reverse Osmosis Membranes in Pressure Retarded Osmosis," *Journal of Membrane Science*, **375**, 55-62 (2011).

(646b) Assessment of Hydrophilic and Antimicrobial Surface Modifications for Biofouling Control

Daniel J. Miller[1], Paula Araújo[2], Patricia Correia[2], Donald R. Paul[1], Benny D. Freeman[1] and J.S. Vrouwenvelder[3], (1)Chemical Engineering, University of Texas at Austin, Austin, TX, (2)WETSUS Centre of Excellence for Sustainable Water Technology, Leeuwarden, Netherlands, (3)Water Desalination and Reuse Center, KAUST, Thuwal, Saudi Arabia

Biofouling represents a major limitation in the use of membranes for water purification. Biofilms are notoriously difficult to prevent or to eliminate. Chlorination has been successfully used to control microbial growth, but chlorine's deleterious effects on some membranes forces costly de-chlorination before membrane treatment and re-chlorination before end use. The use of biocides and cleaning protocols for biofouling control may be reduced by membranes resistant to biofouling. This work evaluated the effect of several surface modifications on biofouling susceptibility of membranes and feed spacers. Polydopamine, a hydrophilic surface modification agent which can be deposited on nearly any substrate from aqueous dopamine solution, has recently been used to reduce organic fouling on many membranes. Polydopamine has also been used as a platform for conjugation of amine-terminated molecules, such as poly(ethylene glycol) monoamine. Poly(ethylene glycol) has a long history as an anti-fouling material in the membrane and biomedical fields. Copper coatings reportedly have antimicrobial effects and "anti-biofouling" feed spacers are currently available from some membrane manufacturers. Polydopamine and polydopamine-g-poly(ethylene glycol) coatings were applied to commercial polysulfone ultrafiltration membranes and polypropylene feed spacers. Copper coatings were applied to a polypropylene feed spacers and a commercial anti-biofouling feed spacer was also used.

Many biofouling studies reported in the literature consist of experiments which are very short term and/or are carried out under conditions not relevant to the operation of industrial membrane modules. Consequently, many anti-fouling strategies have been reported as effective in the academic literature while biofouling remains a significant hurdle in industrial applications. Modified membranes and feed spacers were tested for short-term, static bacterial adhesion and for longer-term biofilm development under realistic membrane module flow conditions. Membrane fouling simulators, containing membranes and feed spacers, have been developed which mimic the hydrodynamic conditions of full-size modules. This study showed that short-term, static adhesion tests are not predictive of ultimate biofouling potential and that the anti-fouling surface modifications employed here are insufficient for biofouling control/prevention. Reasons for the discrepancy between the short-term and long-term testing protocols will be discussed.

(646c) Design and Construction of Antifouling and Self-Cleaning Membranes Surfaces

Zhongyi Jiang, Yanlei Su, Wenjuan Chen, Jinming Peng and Xueting Zhao, Key Laboratory for Green Chemical Technology of Ministry of Education, Sch Chem Engn & Technol, Tianjin Univ, Tianjin, China

Design and Construction of Antifouling and Self-cleaning Membranes Surfaces

Zhongyi Jiang, Yanlei Su, Wenjuan Chen, Jinming Peng, Xueting Zhao*

[*] Prof. Z. Y. Jiang

Key Laboratory for Green Chemical Technology of Ministry of Education, School of Chemical Engineering and Technology, Tianjin University, Tianjin 300072, P.R.China

E-mail: zhyjiang@tju.edu.cn; suyanlei@tju.edu.cn; divinechwj@163.com; pjmlwy@126.com; zhaoxt365@gmail.com

Abstract

Membrane fouling constitutes a bottleneck limiting the wide application of membrane filtration technology in the environmental protection, biological pharmaceutical, food processing as well as water treatment industries.

Theoretically, constructing porous membrane surfaces faces more severe challenges compared with nonporous surfaces due to the additional requirements of ensuring three-dimensional functionalization and avoiding sacrifice of membrane separation performance. In recent years, the surface segregation method emerges as more popular method compared with surface grafting for fabricating more efficacious antifouling brush layer, achieving *in-situ* and three-dimensional surface modification (including membrane surface and internal pores) during the non-solvent induced phase separation membrane formation process.

Compared with the traditional hydrophilic antifouling membrane surfaces, amphiphilic membrane surfaces engineered by the "forced surface segregation" method were demonstrated to exhibit ultralow membrane fouling. The amphiphilic antifouling surfaces bearing a mixed brush architecture comprised both hydrophilic blocks and low surface energy blocks. Hydrophilic blocks generated fouling resistant hydration layers; whereas low surface energy blocks formed amounts of non-adhesive microdomians randomly distributing on the membrane surfaces, which played major role in repulsing the foulants away from the surfaces with the assistance of near-surface agitation, thus endowing the surfaces with outstanding fouling release potential. During the dynamic filtration process, the fouling resistant ability impeded the irreversible flux decline; the fouling release ability prevented the coalescence, migration and spreading of the holistic hydrophobic oil droplets, remarkably reducing or even eliminating the reversible flux decline. This review will present the design and construction of antifouling and self-cleaning membrane surfaces.

(646d) Biofouling and Cleaning Effectiveness of Surface Nanostructured, Reverse Osmosis Membranes

Kari J. Varin, Nancy Lin and Yoram Cohen, Chemical and Biomolecular Engineering, University of California, Los Angeles, Los Angeles, CA

Adequate suppression of membrane biofouling is necessary for the successful use of reverse osmosis (RO) membranes in brackish desalination and water reuse applications. In this study, the recently developed atmospheric pressure plasma-induced graft polymerization (APPIGP) method was utilized to generate a high surface density of active surface sites on a polyamide membrane for subsequent graft polymerization using suitable water soluble vinyl monomers. Surface structuring via graft polymerization was then employed to form a brush layer of polymeric chains that are terminally and covalently attached to the polyamide surface of an optimized (with respect to flux and salt rejection) of a thin-film composite (TFC) polyamide (PA) membrane. The chemical and physical features of the resulting surface nano-structured (SNS) TFC-PA membrane was first tuned with respect to wettability, by the selected vinyl monomer chemistry, as well as the reaction conditions, to achieve the desired surface architecture. Performance evaluation of the SNS-TFC-PA RO membranes was subsequently carried out by examining their biofouling propensities as assessed via flux decline studies using secondary treated municipal waste water effluent, permeability recovery by DI water and chemical (Na_2·EDTA) cleaning, and residual biofilm thickness. The results of the present study demonstrate that it is feasible to tailor-synthesize SNS-TFC-PA membranes of high permeability and salt rejection, with the added benefit of increased fouling resistance.

(646e) Antifouling Ultrafiltration Membranes Via Post-Fabrication Grafting of Biocidal Nanomaterials

Meagan Mauter[1], Yue Wang[2], Kaetochi Okemgbo[3], Chinedum Osuji[4], Emmanuel P. Giannelis[2] and Menachem Elimelech[5], (1)Harvard Kennedy School, Cambridge, MA, (2)Materials Science and Engineering, Cornell University, Ithaca, NY, (3)Yale University, (4)Chemical Engineering, Yale University, New Haven, CT, (5)Yale University, New Haven, CT

Ultrafiltration (UF) membranes perform critical pre-treatment functions in advanced water treatment processes. In operational systems, however, biofouling decreases membrane performance and increases the frequency and cost of chemical cleaning. The present work demonstrates a novel technique for covalently or ionically tethering antimicrobial nanoparticles to the surface of UF membranes. Silver nanoparticles (AgNPs) encapsulated in positively charged polyethyleneimine (PEI) were reacted with an oxygen plasma modified polysulfone UF membrane with and without 1-ethyl-3-(3-dimethylaminopropyl) carbodiimide hydrochloride (EDC) present. The nucleophilic primary amines of the PEI react with the electrophilic carboxyl groups on the UF membrane surface to form electrostatic and covalent bonds. The irreversible modification process imparts significant antimicrobial activity to the membrane surface as measured in two distinct biofouling tests. Post-synthesis functionalization methods, such as the one presented here, maximize the density of nanomaterials at the membrane surface and may provide a more e cient route for fabricating diverse array of reactive nanocomposite membranes when compared to similar membranes with nanoparticles dispersed throughout the membrane matrix.

(666a) Oxidation & Degradation Studies of Aminosilica Adsorbents

Stephanie A. Didas[1], Nicholas Brunelli[2], Praveen Bollini[1] and Christopher W. Jones[1], (1)School of Chemical & Biomolecular Engineering, Georgia Institute of Technology, Atlanta, GA, (2)School of Chemical & Biomolecular Engineeering, Georgia Institute of Technology, Atlanta, GA

The direct extraction of CO_2 from the ambient atmosphere, or "air capture," has the potential to reduce the global atmospheric CO_2 concentration and has been gaining significant attention in the academic and industrial world as of late. It has been demonstrated by our group and others[1-11] that amine oxide materials are a good candidate for air capture applications. From initial studies it has been determined that primary amines are the most promising amine type for air capture as it has the highest adsorption capacity and amine efficiency at ultra-low CO_2 partial pressures, as well as enhanced water uptake properties which can lead to cooperative adsorption and therefore even greater CO_2 uptake.[7] Additionally, previous work in our group has shown that class 2 primary aminosilicas possess oxidative stability under high temperature treatments.[12] Therefore, in this work, the effect of different primary amine structures towards oxidative stability as well as CO_2 adsorption capacity are being performed to further elucidate favorable structural properties for optimal air capture adsorbent design.

References

[1] S. Choi, J. H. Drese, P. Eisenberger, C. W. Jones, oral presentation, *AIChE Annual Meeting*, Nashville, TN, November (2009).

[2] S. Choi, J. H. Drese, P. M. Eisenberger, C. W. Jones, *Env. Sci. Technol.* (2011) **45**, 2420-2427.

[3] S. Choi, M. L. Gray, C. W. Jones, *ChemSusChem* (2011) **4**, 628-635.

[4] W. Chaikittislip, J. D. Lunn, D. F. Shantz, C. W. Jones, *Chem. Eur. J.* (2011) **17**, 10556-10661.

[5] W. Chaikittisilp, H.-J. Kim, C. W. Jones, *Energy Fuels* (2011) **25**, 5528-5537

[6] W. Chaikittisilp, R. Khunsupat, T. T. Chen, C. W. Jones, *Ind. Eng. Chem. Res*. (2011) **50**, 14203-14210.

[7] S. A. Didas, A. Kulkarni, D. S. Sholl, C. W. Jones, (2012) manuscript submitted.

[8] C. Gebald, J. A. Wurzbacher, P. Tingaut, T. Zimmermann, A. Steinfeld, *Env. Sci. Technol.* (2011) **45**, 9101-9108.

[9] C. Gebald, J. A. Wurzbacher, A. Steinfeld, *Energy Env. Sci.* (2011) **4**, 3584-3592.

[10] Y. Belmabkhout, R. Serna-Guerrero, A. Sayari, Chem. Eng. Sci. (2010) **65**, 3695-3698.

[11] N. R. Stuckert, R. T. Yang, *Env. Sci. Technol.* (2011) **45**, 10257-10264.

[12] P. Bollini, S. Choi, J. H. Drese, C. W. Jones, *Energy Fuels*. (2011) **25**, 2416-2425.

(666b) Adsorption in TB-Polymers of Intrinsic Microporosity

Alexandra Gonciaruk and **Flor R. Siperstein**, School of Chemical Engineering and Analytical Science, University of Manchester, Manchester, United Kingdom

Polymers of Intrinsic Microporosity (PIMs) display some features from traditional rigid adsorbents and polymeric materials. They are formed from monomers that contain contortion sites and prevents the polymer from packing efficiently, creating a microporous volume accessible for adsorbents. Due to the rigidity of the polymer chain, often formed by a series of fused aromatic rings, their swelling is significantly smaller than conventional polymeric materials. Some properties in PIMs can be tuned by functionalising the monomers to create specific interactions between the framework and adsorbed molecules. For example, the incorporation of amino groups is expected to increase the affinity of the polymers towards polar molecules such as carbon dioxide. In this work, we present the adsorption properties of a family of PIMs that contain Troger base (TB) units as part of their monomers. TB units are expected to provide basic adsorption sites where carbon dioxide can adsorb preferentially. We have analysed the effect of TB in a series of PIMs with different structures (linear chains, branched and networks) and have demonstrated that the presence of TB units favours adsorption of CO2 at low pressures, but reduces the capacity of the material at high pressures. The reduction in the capacity of the material at high pressures is related in some cases to a smaller initial microporous volume or to a smaller capacity of the polymer to swell, which depends on the structure of the polymer rather than on the details of its chemistry.

(666c) Novel Macroporous Adsorbents with Variable Functionalities: From Preparation to Application

Alexandros Lamprou, Miroslav Soos, Betrand de Neuville, Giuseppe Storti and Massimo Morbidelli, Institute for Chemical and Bioengineering, ETH, Zurich, Switzerland

Macroporous chromatographic supports are ideal for preparative applications, where good throughput, i.e. speed of separation, is required. In this regard, large throughpores provide high bed permeability, which allows operation at high flow rates. Meanwhile, if these large pores are able to accommodate a fraction of convective flow, so-called perfusive pores, then separation efficiency is independent of the flow rate.

To this end, by combining elements from colloidal and polymer reaction engineering, a novel macroporous packing material with advanced mass transfer properties was developed. Specifically, aggregation and breakage under flow of polymeric microclusters allows the control final cluster size and pore size distribution, while post-polymerization imparts high mechanical stability.

Owing to surface-bound ATRP initiators a versatility of functionalities can be realized, in order to employ different separation modes, relying on different types of substrate-solute interaction. So far, conventional cation and anion exchange chromatography have been addressed, as well as an alternative temperature-responsive modality, where instead of ionic strength or polarity of the mobile phase, temperature is utilized to control interactions. This is advantageous from the environmental and cost-reduction viewpoint, as well as for preserving bioactivity of the purified compounds.

As polymer brushes are grafted from the substrate surface, multiple active sites are available per unit area, offering increased binding capacities. Meanwhile, the flexibility of such chains allows for favorable steric distributions, so that large molecules like proteins, can be better accommodated. As a matter of fact, application examples demonstrate that our functionalized adsorbents may compete with existing commercial chromatographic supports.

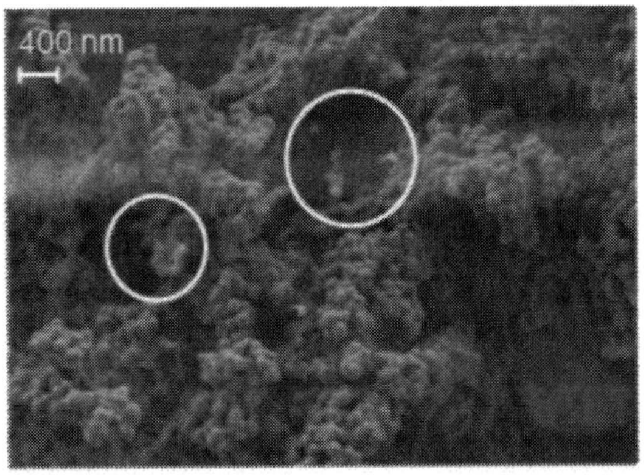

Figure 1. SEM image of supporting microcluster. White circles indicate large throughpores.

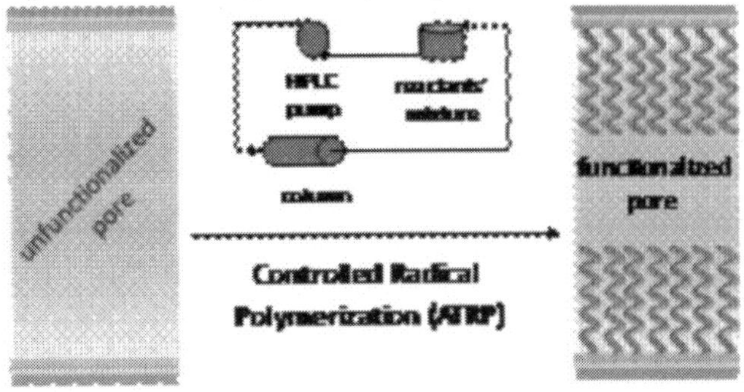

Figure 2. Pore functionalization scheme.

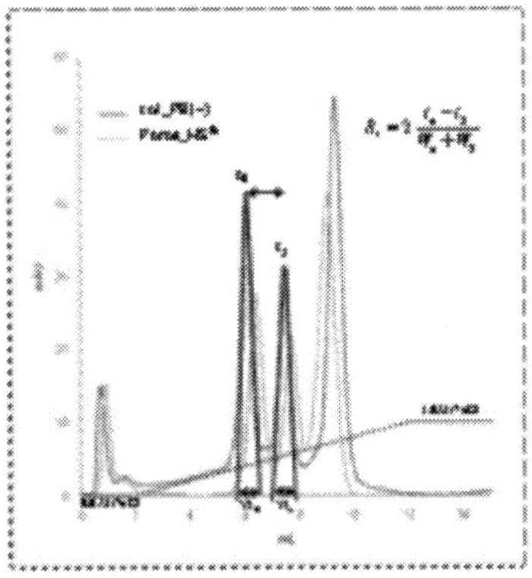

Figure 3. Illustrative separation of Lysozyme, α-Chymotrypsinogen and Cytochrome-c (elution order) by an anionic-PE functionalized column versus a commercial exchanger. Conditions: 30mM phosphate buffer pH7.0, linear gradient elution from 0.0M to 1.0M NaCl (dashed line) at 1ml/min. Peak resolution was calculated using the indicated formula.

(666d) Absolute Molecular Sieve Separation of Ethylene/Ethane Mixtures with a Silver Zeolite

David Farrusseng, Sonia Aguado and Cindy Aquino, IRCELYON, CNRS, Villeurbanne, France

Ethylene is an important petrochemical product used mostly as a raw material in the manufactures of polymers. In a typical ethylene plant, the separation of ethylene and ethane is a key stage in the whole production chain. Several methods are known for separation of ethylene from gaseous mixtures. Cryogenic distillation and liquid adsorption have disadvantages such as high capital cost and high operating expenses. Liquid absorption techniques suffer from solvent loss and need a complex solvent make-up and recovery system. Separation by adsorption is an energy-efficient alternative. Two separation concepts can be applied. The most studied adsorbents have been developed to promote selective uptake of the ethylene. Adsorbents for the separation of ethylene from C2 often include high surface area, microporous materials which have been treated with metal species capable of pi–complexation with olefins. The double bond of the olefin form pi–complexes and a different in adsorption affinity between olefin and paraffin can be achieved. Potential transition metals are copper and silver salts. The second concept relies on steric size exclusion. The later is more difficult to achieve given the close molecular diameter of C_2H_4 and C_2H_6 (4.163 and 4.443 Å, respectively). Kuznicki and coworkers has exploited also the use of crystalline titanium silicate molecular sieves for this separation. These materials have controlled pore size by ceasing thermal treatment during calcinations.[1] In 2010, they patented this separation with Na-ETS-10 zeolite modified with a mono-, di-, or tri-valent cation and mixtures thereof. Several of the modified ETS-10 materials show improved adsorption capacity, but the selectivity decreased compared with the raw material Na-ETS-10.

Here we report the absolute separation of ethylene from ethane with a silver exchanged zeolite A (Ag-Zeolite). The results are compared with silver Faujasite zeolite (AgX). The exceptionality of this behavior is the combination of preferential adsorption of olefin over paraffin with steric size exclusion.

Based on isotherm data, ethane is excluded from Ag-Zeolite whatever the pressure whereas it is adsorbed on AgX. In contrast, the shapes of the isotherms for the ethylene adsorption are very similar for both silver exchanged zeolites. The value of capacity at 100 kPa reaches 2 $mol{\cdot}kg^{-1}$ which is among the highest values found for zeolite based adsorbents. Other adsorbents with molecular sieving properties, like silver salts dispersed in ETS-10, reach about 1 $mol{\cdot}kg^{-1}$ at the same pressure range. Ethylene adsorption enthalpy can be estimated at DH_{303K}=-65kJ/mol which corresponds to values found in the literature for silver zeolites. The loading of ethylene at low pressures is attributed to strong interactions between ethylene and silver clusters inside the zeolite. As a consequence, the ethylene/ethane ideal selectivity is infinite for Ag-Zeolite and only 2,5 for AgX (based on capacity ratio at 10kPa).

The outstanding adsorption properties of AgA are confirmed by breakthrough experiments carried out on ethane/ethylene/N2 mixture (10:10:80) at 303 K and 1 atm. After the first seconds the mixture is fed, ethane breaks through the Ag-Zeolite column (1g). The exclusion of ethane in the Ag-Zeolite column is consistent with the absence of adsorption of ethane observed in volumetric measurements. During the first 15 min, the adsorbent retained all the ethylene and the outlet stream is composed of pure ethane. After 25 min, the concentration of the outlet gas reflected the feed gas composition, indicating the saturation of the column. By integration, we find the amount of ethylene adsorbed is 57 $cm^3{\cdot}g^{-1}$ which also corresponds to volumetric measurements.

Powder X-Ray Diffraction and Ar adsorption studies were carried out to elucidate the exclusion mechanism. Calculation of crystallographic parameters gives free diameter values of 0.436nm and 0.421 nm which are right in between kinetic diameters of ethylene (0.416 nm) and ethane (0.444nm).

[1] Shi, M.; Avila, A. M.; Yang, F.; Kuznicki, T. M.; Kuznicki, S. M. *Chem. Eng. Sci.* **2011**, *66*, 2817.

This paper reports work undertaken in the context of the project "OCMOL, Oxidative Coupling of Methane followed by Oligomerization to Liquids". OCMOL is a Large Scale Collaborative Project supported by the European Commission in the 7th Framework Programme (GA n°228953). For further information about OCMOL see: http://www.ocmol.eu

(666e) Selective Adsorption of Biofuels From Dilute Aqueous Solution with Grafted Calixarenes

Anthony Boone Thompson, Northwestern University, Evanston, IL

Fermentation-derived bioalcohols such as *n*-butanol and ethanol have potential as sustainable fuels, but their production is hindered by inefficiencies in separating from the dilute fermentation broth. The standard method of distillation is a major energy sink because of the low concentrations and unfavorable vapor-liquid equilibrium properties typical of A.B.E. fermentation broths. Thus, more efficient methods are needed in order to realize the potential of bioalcohols as renewable fuels. Adsorption is one such method, but conventional sorbents such as activated carbons have only limited ability to be fine-tuned at the molecular level, and as such, selectivities are typically low. Herein we investigate the use of calixarenes, intrinsically porous organic macrocycles, covalently immobilized to porous hydrophilic oxide supports, as tunable and uniform adsorption sites for separation of small organics from aqueous solutions. We previously showed that with nonpolar substituents at the upper rim of the calixarene, adsorption on these materials is entropy-driven, characterized by low net energy differences, and practically independent of temperature, indicating van der Waals interactions as the governing force. Presently we explore the effects of adding polar substituents to the calixarene structure to closer match the dipole moment of the intended guests, as well as the addition of a Lewis acid site at the bottom of the cavity, which is predicted to act as a coordination center for the guest's alcohol moiety. In addition to their potential uses in separations, these materials enable the study of calixarenes in water, where they are otherwise generally insoluble. Moreover, calixarenes are believed to be constrained to a cone-like conformation due to multiple lower-rim surface attachments as well as steric hindrance from the surface itself. These materials provide uniform and uniquely tunable adsorption sites on a bulk hydrophilic surface, and thus have great potential as selective adsorbents for biofuel separations as well as a broad range of other aqueous systems.

(666f) Biosorptive Dehydration of Alcohol/Water Azeotropes Using Compound Starch-Based Adsorbent

Peng Wu, Chunmei Gong, Shuai Zheng, **Jinsheng Sun** and Wenping Wang, School of Chemical Engineering and Technology, Tianjin University, Tianjin, China

Water removal represents a serious problem in most monohydric alcohol production because most monohydric alcohols form azeotropes with water at atmospheric pressure, except for methanol. Various techniques have been developed to break the azeotropic point, such as salting-out method, azeotropic distillation, extractive distillation, reactive distillation, adsorptive distillation and pervapration. Among the methods, the adsorptive option is most attractive, especially the biosorption, which has distinct advantages over conventional methods as it is environmentally friendly, highly selective, easily available, easy to operate, cost effective and reusable in repeated cycles in the treatment of aqueous organics. In this work, a specially formulated compound starch-based adsorbent (CSA), which consists of corn, sweet potato and foaming agent, was developed, and biosorptive dehydrations of ethanol, isopropanol and tert-butyl alcohol (TBA) using CSA were investigated, respectively.

In the first step, a systematic study of the biosorptive process was performed. The net retention time and separation factor of alcohol and water were obtained using inverse gas chromatography (IGC), the results indicated that the net retention time of water on CSA is significantly longer than that of alcohol which attested the feasibility of this adsorptive separation purpose and low temperature could be propitious to the process which was consistent with the traditional theory of adsorption.

Then, through a batch adsorption experiment, the optimum adsorption condition was determined under different bed depths, bed temperatures and kettle temperatures by single-factor method and orthogonal design method, which indicated that the optimization condition for ethanol was at Vapor superficial velocity of 0.14m/s, bed temperature of 81°æ and feed concentration of 92.5wt%, isopropanol at kettle temperature of 87°æ, bed temperature of 88°æ and bed depth of 26 cm and TBA, kettle temperature of 102°æ,bed temperature of 84°æ and bed height of 30cm. And take TBA for instance, the optimal regeneration condition was obtained under different air flow rates, air temperature and regeneration time by treating CSA with hot air by single-factor method and orthogonal design method and the results showed that the optimal regeneration condition was at air flow rate of 60L/h, air temperature of 120°æ and regeneration time of 40min. In order to characterize the service life of CSA, five cycles experiments of adsorption-regeneration were carried out at the optimum adsorption and regeneration condition, the results showed that there is practically no change in the regeneration efficiency of CSA after the cyclic process adsorption¨Cregeneration. It is due to the properly chosen regeneration conditions assuring complete desorption of the water adsorbed without changes in the CSA structure. This unchanged adsorption capacity after a number of cycles demonstrates reusability of the adsorbent.

Finally, field emission scanning electron microscopy (FESEM) and mercury porosimetry (MP) were applied to investigate the change of surface morphology and microstructure of CSA before and after the adsorption, as well as after regeneration, as shown in Fig.1 and Fig.2. The FESEM images show that the dry CSA is composed of wizened, irregular particles with many void spaces between them. After an adsorption experimental run, the particles become enlarged, the void spaces become smaller, and some particles adhered to each other. There is a significant change in the surface morphology of the CSA before and after adsorption, which indicated that the CSA was saturated with water. However, the particles of CSA are wizened as the nature adsorbent after a regeneration experimental run, which indicated that nine tenths of water was desorbed form CSA. The MP data indicated that the total pore volume, total pore area, average pore diameter and porosity of the CSA before adsorption are larger than those parameters after adsorption. In other words, a large numbers of water molecules were adsorbed on the surface of the pores in the adsorption process, which resulted in decreases of the above parameters after adsorption. However, all the above parameters of the CSA just slightly lower than the natural adsorbent after regeneration experimental run, which demonstrated that almost all the adsorbed water molecular was desorbed from the surface of the pores. The pore size distribution (PSD) curves show that below 5000 nm, the behavior of the two curves of the CSA before and after adsorption is very similar. Before adsorption, the curve exhibits two peaks at 6000 and 17000 nm. However, after adsorption, the peak at 6000 nm was no longer present, and the peak at 17000 nm became significantly smaller, which indicates that the CSA is mainly composed of macropores that play a significant role in water adsorption. Numerous macropores in the CSA were saturated with water molecules, which resulted in the disappearance of macropores of certain sizes and a reduction in total pore volume. After regeneration, the two peaks at 6000 and 17000 nm were reappeared, the PSD curve of the CSA was almost identical to the curve of before adsorption, which further manifested that the spent adsorbent can be regenerated completely and restored to the natural structure.

Thermodynamic discussions are subsequently done with parameters DH_s and DG_s for the adsorption determined respectively. The adsorption enthalpy of water for ethanol/water azeotrope was -31.6kJ/mol, for isopropanol/water azeotrope was -55.7kJ/mol and tert-butyl alcohol/water azeotrope, -38.3 kJ/mol, which indicated that the biosorption was physisorption in nature and exothermic. And the adsorption enthalpy of water significantly higher than the corresponding value of alcohol, which indicated that the CSA shows stronger preferential sorption of water over alcohol. The negative values of *DGs* indicate the spontaneous of the adsorption process, and the larger negative values for water at lower temperatures indicate that adsorption of water is favored at lower temperatures. Alternatively, the larger negative values of D*Gs* for water compared to those values for alcohol further confirm the preferential adsorption of water.

On the basis of the above conclusions, a more energy-efficient hybrid program that combines normal distillation with this crucial biosorptive pathway to break the azeotropic point was developed for the production of anhydrous alcohol, and the simulation results showed that the hybrid option was definitely reduced the energy and material requirements by a large margin compared to the prevailing azeotropic or extractive distillation schemes. These results indicated that the starch-based adsorbent can be considered as an effective and low-cost biosorbent to dehydrate alcohol.

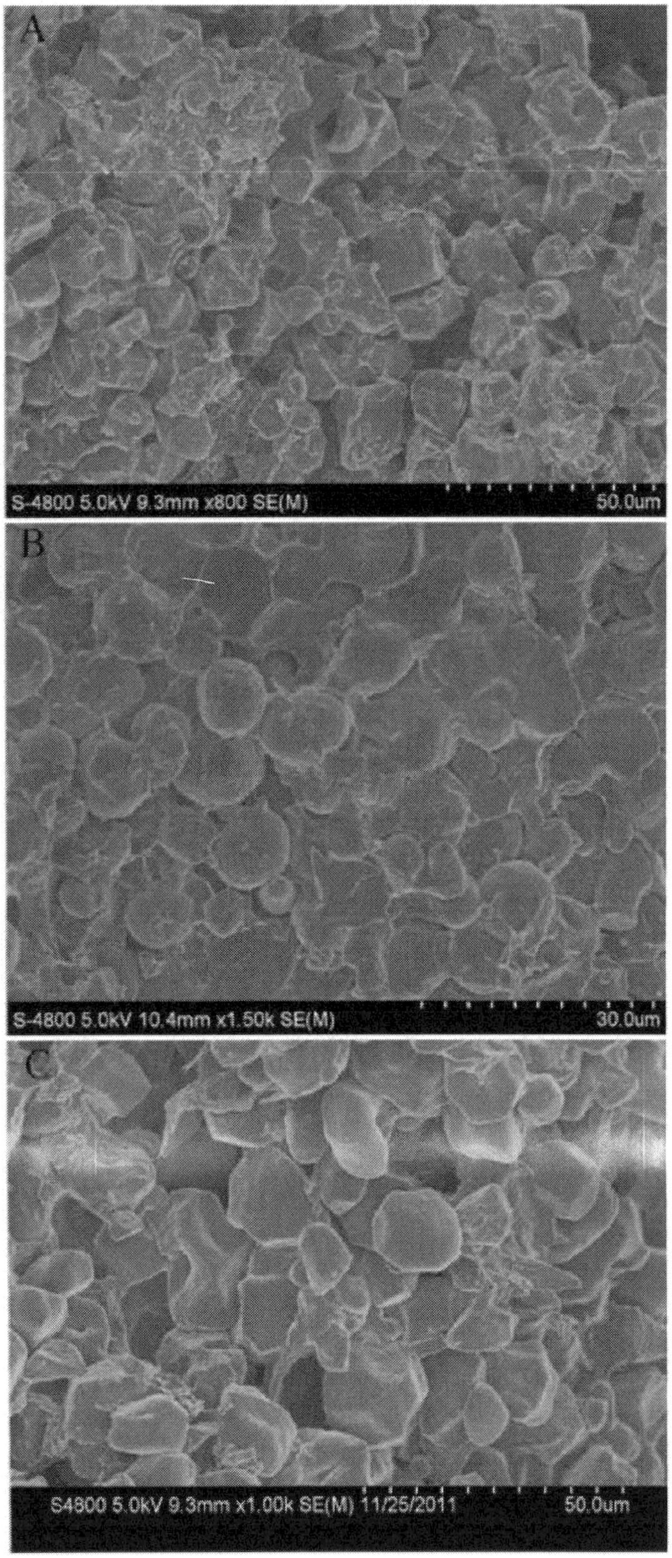

Fig.1 FESEM images of the CSA before (A) and after adsorption (B) as well as after regeneration (C).

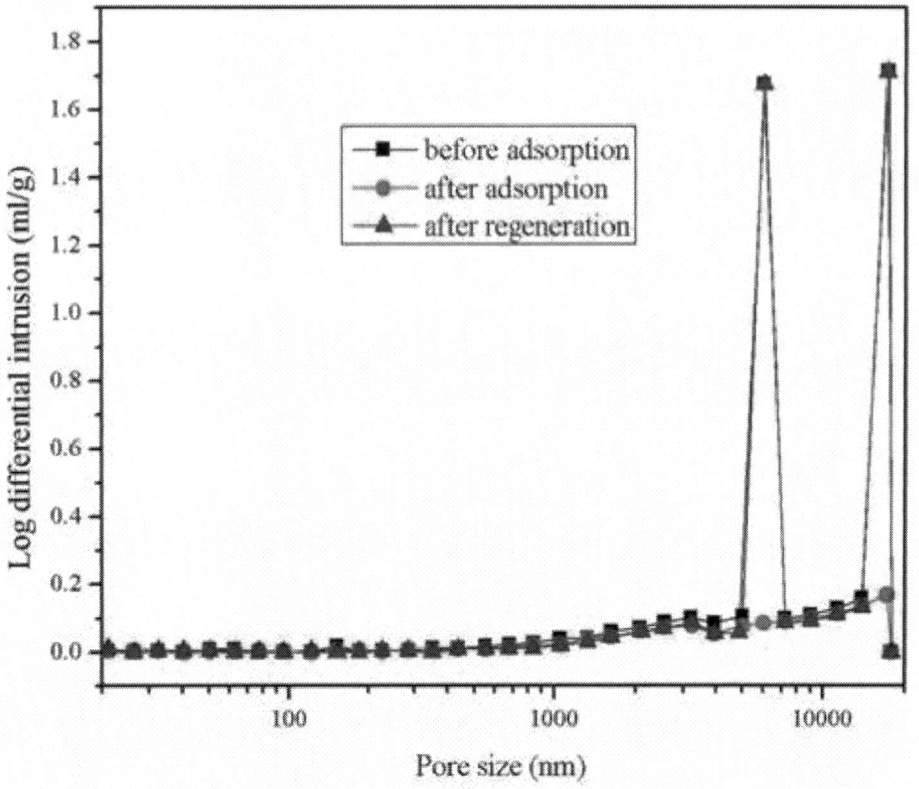

Fig.2. Pore size distributions (PSD) of the CSA before and after adsorption as well as after regeneration.

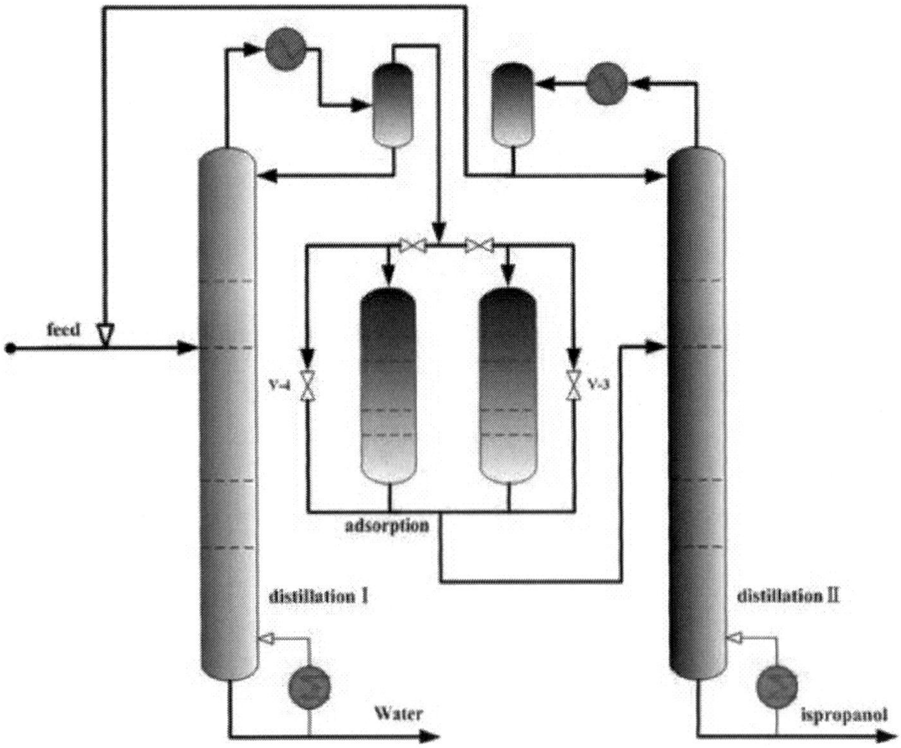

Fig.3. Diagram of the coupling process of biomass-based fixed-bed adsorption

Preparation of $H_8Nb_{22}O_{59}\cdot 8H_2O$ and Ion-exchange Properties for Na^+ and K^+

Zhang Xianglong[1], Zhai Yanlong[1], Wang Chang[2], Hao Qinglan[1]*

(1. College of Material Science & Chemical Engineering, Tianjin University of Science & Technology, Tianjin 300457, China; 2. College of Marine Science & Engineering, Tianjin University of Science & Technology, Tianjin 300457, China)

Abstract

The ion-exchanger $H_8Nb_{22}O_{59}\cdot 8H_2O$ with almost the same structure of $Rb_8Nb_{22}O_{59}$ was synthesized. The uptake amount for Na^+ and K^+ is very close to theoretical value of 2.55 $mmol\cdot g^{-1}$ calculated from the composition of $H_8Nb_{22}O_{59}\cdot 8H_2O$. The affinity of $H_8Nb_{22}O_{59}\cdot 8H_2O$ for Na^+ and K^+ is similar, while the results of distribution coefficient (K_d) showed the K^+ selectivity is higher than that of Na^+ at about pH 2.5.

Keywords: Ion-exchanger; $Rb_8Nb_{22}O_{59}$; distribution coefficient; protonated compound

1. Introduction

High-purity sodium chloride (99.99%) could be extensively applied on food industry, medicine, the medical field, etc. Very similar properties for Na^+ and K^+ results in difficulty for separation K^+ from sodium chloride solution. It is impossible to obtain a high-purity sodium chloride though conventional methods, such as precipitation, ion exchange, and extraction [1-3].

In order to improve the selectivity of inorganic ion-exchanger, more attention is focused on the ion-sieve compound with high selectivity for K^+ in recent decades. Ion-sieve compound is a metal oxide porous crystals (MOPCS)[4]. The ion-sieve compound show high selectivity for specific ion depending on the template ion. Because of the differences of occupying sites, $Na_2Ti_2O_3SiO_4 \cdot 2H_2O$ exchanger with a tunnel structure exhibited a high ion exchange capacity for small alkali metal ions and extremely high affinity for large alkali metal ions Cs^+ and Rb^+. It could be regarded as a promising treatment material for nuclear waste [5]. $MnO_2 \cdot 0.5H_2O$ showed a markedly high ion-exchange capacity (5.3 mmol/g) for lithium ions due to its ion-sieve property and is the most promising adsorbent for lithium in seawater [6]. $Rb_8Nb_{22}O_{59}$ was synthesized since 1960s, however, more attention was paid for the structure and physic-chemical properties of $Rb_8Nb_{22}O_{59}$[7,8].

The present paper describes the synthesis of $Rb_8Nb_{22}O_{59}$ and its ion-exchange properties for Na^+ and K^+. The structural characteristics are studied by X-ray diffraction (XRD), Thermogravimetric differential scanning calorimetry (TG-DSC), and Scanning electron microscope (SEM), respectively.

2. Experimental

2.1 Preparation and protonated $Rb_8Nb_{22}O_{59}$

Compound containing Rb_2CO_3 (99.9%, m.p.723 °C) and Nb_2O_5 (99.99%, m.p. 1520 °C) with Rb_2CO_3/ Nb_2O_5 molar ratio of 4/11 were completely mixed, followed by ground, and then calcined at different temperatures from 700 °C to 1200 °C for 8 h in air. The obtained $Rb_8Nb_{22}O_{59}$ samples were denoted as T_n (n=700, 800, 900, 1000, 1100, and 1200, respectively).

About 0.5 g $Rb_8Nb_{22}O_{59}$ was first treated with 50 mL of 10 M HNO_3 solution for 72 h and filtered, treated $Rb_8Nb_{22}O_{59}$ was then washed by deionized water, and finally dried at 70 °C for 12 h. The obtained $H_8Nb_{22}O_{59}\cdot 8H_2O$ were labeled as TH_n (n=700, 800, 900, 1000, 1100, 1200, and 1200, respectively).

2.2 physic-chemical analysis

XRD patterns of the different $Rb_8Nb_{22}O_{59}$ samples was carried out on a XD-3 X-ray powder diffractometer (Purkinje, China) equipped with Cu Kα radiation (λ=0.15406 nm) and scanning rate of 4 °/min.

TG-DSC analysis was performed using a TGA/DSC1/1100 instrument (Mettler Toledo Co. Ltd., Sweden). And the temperature was raised from room temperature to 1073 K at a heating rate of 10 °C/min.

The morphology of various samples was obtained using a SU-1510 (Hitachi, Japan) scanning electron microscope (SEM).

The amount of Na^+ and K^+ in the different solutions was measured by atom absorption spectrometry (AAS) using WFX-120 Elemental analyzer (China).

2.3. Ion-exchange capacity

The mixed solution was prepared using 0.1 M NaCl and 0.1 M NaOH (pH>12.40) or 0.1 M KCl and 0.1 M KOH (pH>12.40), respectively. The uptake of Na^+ and K^+ from above mixed solution (5 mL) was carried out by intermittent shaking 50 mg of the $H_8Nb_{22}O_{59}\cdot 8H_2O$ sample for 7 days at room temperature. The concentration of

Na^+ and K^+ in the supernatant was determined by titration with 0.1 M HCl. The ion-exchange properties were evaluated from the difference of Na^+ and K^+ concentration in the initial solution and in the supernatant solution, ion-exchange capacity was calculated by the following equation:

$$Q=(5\times c(OH^-) - 0.1\times V_{HCl})/m \qquad (1)$$

In which, Q is the amount of adsorbed Na^+ (K^+) at equilibrium (mmol/g); $c(OH^-)$ is the initial concentration of the mixed solution (mol/L); V_{HCl} is the volume of HCl (mL); m is the weight of adsorbent $H_8Nb_{22}O_{59}\cdot 8H_2O$ (g).

2.4. pH titration

A 50 mg $H_8Nb_{22}O_{59}\cdot 8H_2O$ sample was immersed in a mixed solution (5 mL) of 0.1 M NaCl and 0.1 M NaOH (or 0.1 M KCl and 0.1 M KOH) with different pH values. After intermittent shaking for 7 days at room temperature, the pH of the supernatant solution was measured with a pH meter (HANNA, Italy).

2.5. Distribution coefficient (K_d)

The mixed solution with different pH was prepared using NaCl and KCl ($Na^+/K^+=1$, mole ratio), and the pH values were adjusted though replacing NaCl and KCl with NaOH and KOH. A 50 mg $H_8Nb_{22}O_{59}\cdot 8H_2O$ sample was immersed in 5 mL of the above mixed solution and shaken for 48 h and then filtered. The K_d values were calculated according to the uptakes for Na^+ or K^+ according to the following formula:

$$K_d(mL/g) = \frac{mental\ ion\ uptake\ (\mathrm{mmol/g})}{mental\ ion\ concentration\ after\ adsoption\ (mmol/mL)} \qquad (2)$$

3. Results and Discussion

3.1. The characteristic of $Rb_8Nb_{22}O_{59}$ and $H_8Nb_{22}O_{59}\cdot 8H_2O$

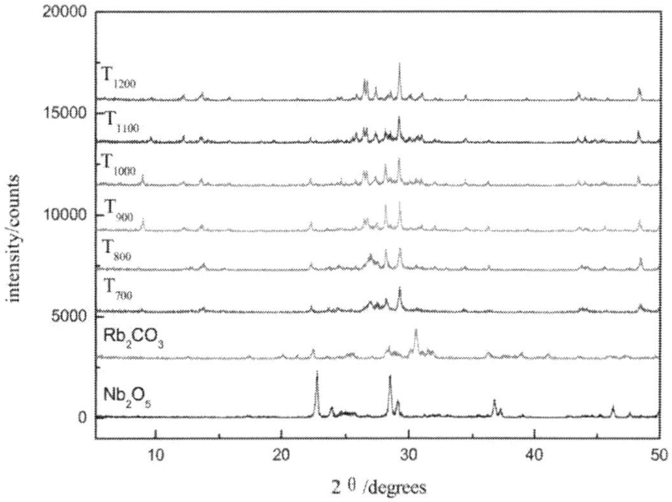

Figure 1. XRD patterns for Nb_2O_5, Rb_2CO_3, and $Rb_8Nb_{22}O_{59}$ at different calcination temperatures

XRD patterns for Nb_2O_5, Rb_2CO_3, and $Rb_8Nb_{22}O_{59}$ at different calcination temperatures for 8 h are shown in Figure 1. The obtained XRD patterns have been plotted for 2θ values ranging from 5° to 50° since the most intensive peak of Nb_2O_5, Rb_2CO_3, and $Rb_8Nb_{22}O_{59}$ fall within this range. The characteristic peaks of 2θ values for Rb_2CO_3 are 22.50°, 24.43°, 30.21°, 30.62°, 31.84°, 36.31°, and 41.16°, respectively. The characteristic peaks of 2θ values for Nb_2O_5 are 22.60°, 28.31°, 28.87°, 29.06°, 37.12°, and 46.06°, respectively. However, the characteristic peaks of Rb_2CO_3 and Nb_2O_5 completely disappeared when $Rb_8Nb_{22}O_{59}$ samples are calcined at different temperatures. Simultaneously, diffraction peaks of $Rb_8Nb_{22}O_{59}$ which have the characteristic peaks at 2θ values of 12.33°, 14.23°, 15.90°, 26.51°, 26.75°, 27.42°, 28.58°, 29.35°, 30.16°, 31.14°, and 34.47° are distinctively observed. The XRD results show that the characteristic peaks of $Rb_8Nb_{22}O_{59}$ calcined at 700 °C are weak, and the all intensities of the characteristic peaks of $Rb_8Nb_{22}O_{59}$ increase with the increase of the calcination temperature. These results indicate crystallinity increases with the increase of the calcination temperature.

3.2 Effect of calcination temperature

Table 1 Adsorptive capacity of $H_8Nb_{22}O_{59}\cdot 8H_2O$ samples for Na^+ and K^+

Samples	Uptake amount, Q/mmol·g^{-1}	
	Na^+	K^+
TH_{700}	2.32	2.24
TH_{800}	2.55	2.55
TH_{900}	2.19	2.03
TH_{1000}	2.50	2.49
TH_{1100}	2.45	2.43
TH_{1200}	2.60	2.55

Ion-exchange capacities of $H_8Nb_{22}O_{59}\cdot 8H_2O$ at different calcination temperatures for Na^+ and K^+ were determined by equation (1) from mixed solutions of 0.1 M NaCl and 0.1 M NaOH (pH>12.40) or 0.1 M KCl and 0.1 M KOH (pH>12.40), respectively. The capacity of Na^+ and K^+ of $H_8Nb_{22}O_{59}\cdot 8H_2O$ samples increases with the increase of the calcination temperatures. The results suggest that affinity of $H_8Nb_{22}O_{59}\cdot 8H_2O$ for Na^+ and K^+ with ion-exchange sites also enhances with the increase of the calcination temperatures. The uptake amount of $H_8Nb_{22}O_{59}\cdot 8H_2O$ for Na^+ is slightly higher than that of K^+, independent of calcination temperatures. The uptake amount of TH_{1200} for Na^+ and K^+ is 2.60 mmol·g^{-1} and 2.55 mmol·g^{-1}, respectively, which is very close to theoretical value of 2.55 mmol·g^{-1} calculated from the composition of $H_8Nb_{22}O_{59}\cdot 8H_2O$.

To avoid the steric interactions between the inserted metal ions, the mixed solution of NaOH and KCl ($Na^+/K^+=1$, mole ratio) was prepared with low concentration of Na^+

and K^+ (<12mmol/L) for K_d measurement. K_d reflects the intrinsic affinity between ion-exchange sites and guest ions, because of extremely low concentration of Na^+ and K^+. Also separation factor ($\alpha = \frac{K_d(K^+)}{K_d(Na^+)}$) is evaluate the difference of ion-exchange sites with Na^+ and K^+, however, the results of K_d and separation factor are further investigated in our research group.

3.3 Effect of pH values for adsorptive capacity of $H_8Nb_{22}O_{59}\cdot 8H_2O$

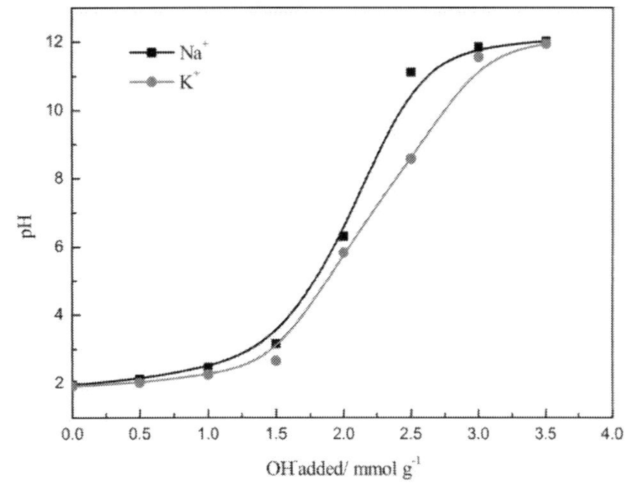

Figure 2 pH titration curves of $H_8Nb_{22}O_{59}\cdot 8H_2O$ calcined at 1200 °C for 16 h

The pH titration curves of $H_8Nb_{22}O_{59}\cdot 8H_2O$ sample calcined at 1200 °C for 16 h toward sodium and potassium ion are shown in Figure 2. The titration curves show that $H_8Nb_{22}O_{59}\cdot 8H_2O$ sample display the similar exchange capacities for Na^+ and K^+ over the pH range studied, indicating the similar affinity. The adsorptive capacity for Na^+ and K^+, evaluated from the amount of OH^- added, increases with pH values, which indicating the solutions with high pH values is helpful for the protonic dissociation from $H_8Nb_{22}O_{59}\cdot 8H_2O$. And this is the general feature of inorganic ion exchanger.

4. Conclusion

The synthesis of $Rb_8Nb_{22}O_{59}$ and its ion-exchange properties for Na^+ and K^+ was studied. The structural characteristics are studied by XRD, TG-DSC, and SEM, respectively. The all intensities of the characteristic peaks of $Rb_8Nb_{22}O_{59}$ increase with the increase of the calcination temperature. The calcination durations have little effect on the crystal structure when $Rb_8Nb_{22}O_{59}$ calcined at 1200 °C. The protonated sample $H_8Nb_{22}O_{59}\cdot 8H_2O$ shows markedly high selectivity for the adsorption of K^+ at pH 2.5. Ion-sieve exchanger $Rb_8Nb_{22}O_{59}$ is the promising adsorbent for K^+ for purification from sodium chloride solution. The detailed performances of $H_8Nb_{22}O_{59}\cdot 8H_2O$ need to be further investigated.

Acknowledgement

We thank Tanggu Municipal Science and Technology Commission of Tianjin Binhai New Area (China) for financial support (special funds).

References

1. Kielland J. Process for the recovery of potassium salts from solutions [P]. DE 691366, 1940-05-24
2. Yasuyuki Takeda, Aiko Yasui. Talanta, 2002, 3(56): 505-513
3. Kamatsu M. Potassium-selective adsorbent and its production [P]. JP 03-205315, 1991-09-06
4. Xiaojing Y, Yoji M, Junji H, Kohji S, Kenta O. Chem. Mater., 2005, 17: 5420 542
5. Clearfield A., Bortun L.N., Bortun A.I. Reactive & Functional Polymers, 2000,43: 85–95
6. Ramesh C., Hirofumi K., Yoshitaka M., Kenta O. Chem. Mater. 2000, 12: 3151-3157

7. Arnold R., Frederic H. J. Phys. Chem. 1960, 64; 748-753

8. John C. Dewan, A., Edwards J. Gordon R. J. J. Chem. Soc., Dalton Trans., 1978, 968-972

Experimental Results of Cryogenic CO_2 Removal from Natural Gas by Desublimation in a Staged Heat Exchanger

Mark Jensen[1], Chris Bence[2], Christopher Hoeger[2], & Kyler Stitt[2], David Bergeson[1], & Larry L. Baxter[1,2]

[1]Brigham Young University, Provo UT USA

[2]Sustainable Energy Solutions, Orem UT USA

Abstract

This paper describes and provides initial experimental results for a new purification method for raw natural gas (NG) streams. The method removes CO_2, NG liquids, H_2S, and most other NG impurities as desublimation or condensation products. The focus of this paper is on CO_2. Desublimation occurs in a patent-pending, desublimating heat exchanger that operates efficiently and continuously and uses a contacting liquid as an intermediate stream. The contacting liquid provides a non-fouling desublimation surface and direct-contact heat transfer, making it both robust and efficient in continuous operation. The solid CO_2 and contact liquid form a slurry until separated through filtration or other solid-liquid physical separation methods. The contact liquid cools in a conventional heat exchanger and recycles to the desublimating heat exchanger. The cold purified CO_2 and purified natural gas products pass through a regenerative heat exchanger to cool incoming raw natural gas, melting the purified solid CO_2 stream in the process. This process is capable of operating at various pressures and temperatures, both of which affect the final impurity levels. Compared to conventional CO_2 removal processes, this process does not require changes in process pressure and does not involve condensation or reboiling in a distillation column. It also combines what currently are several operating steps into a single process. These features decrease overall energy losses and both capital and operating expenses.

Introduction

Conventional CO_2 separation technologies for raw NG streams commonly involve amine-based solvents in absorption and stripping columns (Rochelle 2009). These processes consume about 1300-1600 MJ of work equivalent energy per ton of CO_2 separated, with most of the energy going into heating and dissociating the CO_2-laden solvent in the stripping tower (US Department of Energy 2007; US Department of Energy 2007; US Department of Energy 2008; Rochelle 2009). CO_2 frequently limits natural gas liquids recovery from processes using this technology as they necessarily must avoid freezing CO_2.

A recent alternative process involves a pressurized distillation column which forms a central region in which CO_2 freezes out of solution with the NG (Thomas and Denton 1987; Victory and Valencia 1987;

Haut, Denton et al. 1988; Harper 1989; Valencia, Northrop et al. 2008; Hoeger, Bence et al. 2010; Parker, Northrop et al. 2010; Kelley, Valencia et al. 2011). This cryogenic separation process removes carbon dioxide and hydrogen sulfide from raw natural gas in a process that controls CO_2 freezing within the distillation tower.

The process described in this paper represents application of the Cryogenic Carbon Capture™ process to natural gas treatment. This patent-pending technology from Brigham Young University and Sustainable Energy Solutions includes desublimating heat exchangers and heating/cooling recover heat exchangers that efficiently remove CO_2 from gaseous streams (Baxter 2007; Baxter 2008; Baxter 2008; Baxter 2008; Baxter 2010; Baxter 2010; Baxter 2011). The process economics and performance have been explored in detail both theoretically and experimentally for applications to flue gas treatment, where it exhibits substantial economic and energy advantages relative to alternatives. Theoretical analyses of the process have also been completed for producer gas and natural gas treatments. This paper reports the first experimental investigation of the process for natural gas treatment.

Process Description

The process described in this paper involves desublimating CO_2 from raw NG streams without pressurization and without distillation, absorption, or stripping columns. Pressurized, dried, raw natural gas cools to near its frost point (typically -80 °C or cooler, depending on CO_2 content) in a conventional heat exchanger, warming the products as it cools. This heat exchanger can also remove NG liquids as a separate stream. The cool NG stream then flows into the base of a multi-stage heat exchanging column with a contact liquid flowing countercurrent. Sieve plates create bubbles and high surface area between liquid and gas phases. As the gas cools in this column, CO_2 and other impurities (HCl, SO_2, Hg, etc.) desublimate and remain in the contact liquid for separation in subsequent units (Figure 1). The process occurs at constant (except for line losses) and arbitrary pressure, with condensation and desublimation points varying with pressure but process conceptual design remaining the same. If high pressures are necessary or preferred, such that the desublimation temperature of CO_2 drops to or below the condensation temperature of methane, subsequent heat transfer systems must accommodate methane evaporation in addition to solid CO_2 melting and or vaporization. This process decouples heat exchange from separation, increasing the range of operating conditions in the desublimating heat exchanger. Compared to conventional CO_2 removal processes, this process does not require changes in process pressure and does not involve condensation or reboiling in a distillation column. It also combines what currently are several operating steps into a single process. These features decrease overall energy losses and both capital and operating expenses.

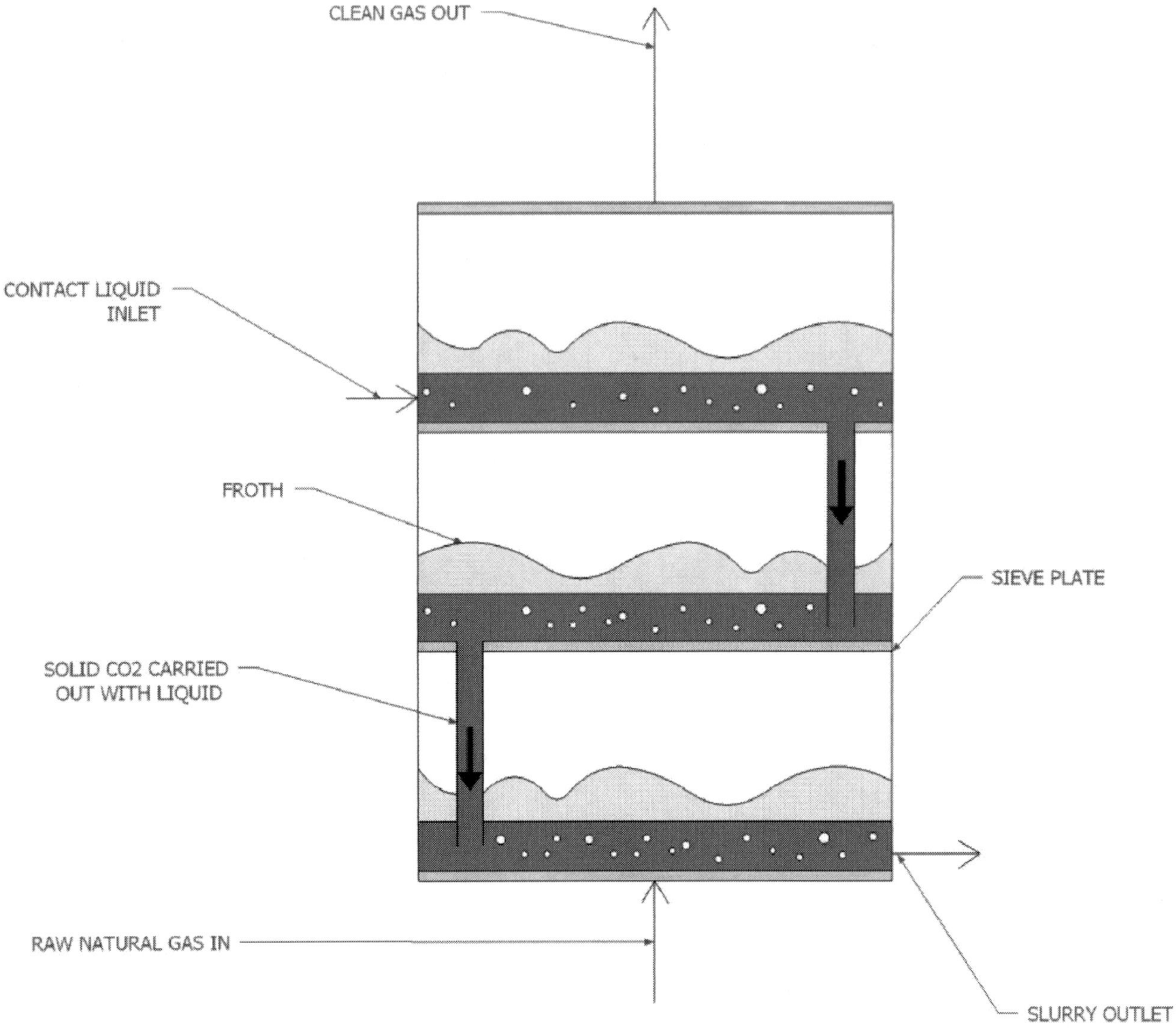

Figure 1 Multi-Stage Bubbling Heat Exchanger Flow Schematic

Experimental Apparatus

An optical-access, multi-stage heat exchanger provided most of the data in this paper. The optical access provides highly useful but qualitative information during its operation. The heat exchanger exterior surface comprises two vacuum-separated concentric acrylic tubes. Internal stages of perforated Teflon sheeting create bubbles in a manner similar to sieve plates in conventional distillation. Stainless steel provides structural rigidity, a weir, and a downcomer. O-rings seal the stages against the exterior and the steel frame provides stage separation and support (Figure 2). The acrylic construction and vacuum-sealed walls are for experimental convenience. A commercial system would use steel components and would not need vacuum sealing.

Mass flow controllers provide an arbitrary simulated raw NG composition, with the focus of this paper being a 50/50 mixture from cylinders containing 99% pure $CO_2$and methane. The gases flare in the local atmosphere, and in these experiments the process operates at local pressure (0.85 bar). An Enerac emissions analyzer quantified inlet and outlet CO_2 concentrations through gas sampling ports.

The contact liquid passes through a cryogenic cooling loop and enters the top of the cascade heat exchanger. Contact liquid flows counter current to the rising gas stream. Thermocouples monitor the top, middle, and bottom temperatures.

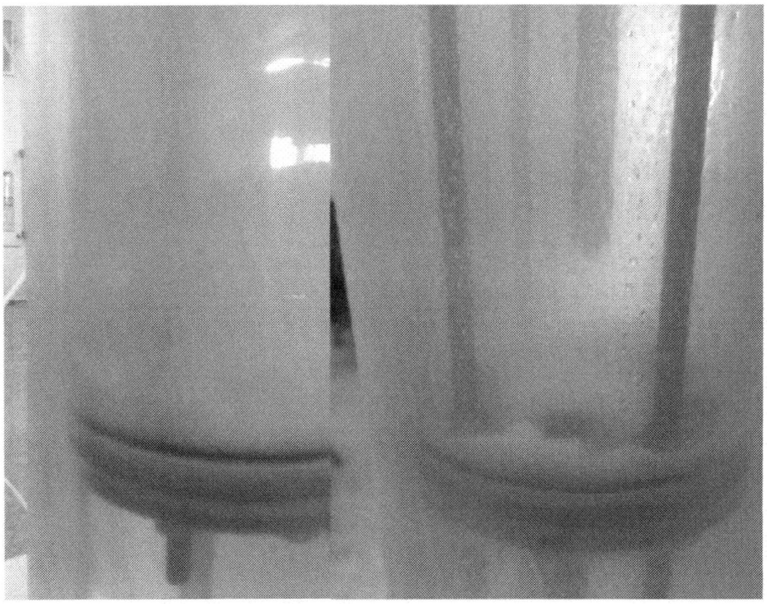

Figure 2 Single Stage of Heat Exchanger a) during operation with solid CO2 particles suspended in contact liquid and b) post-operation, drained, with solid CO2 remaining.

Experimental Results

Results from a two-hour run demonstrate CO_2 separation from 50% methane, 50% CO_2 gas mixture. The flow rate of mixed gas was 0.5 scfm. Solid CO_2 remained in the contacting liquid with increasing concentration over time, helping to establish flow behaviors over a range of solids loadings. The temperature varied between -128 and -133 °C. The gas sampling of the post treated gas showed that the stream contained between 0.3 to 0.4 % CO_2. Pure methane flowed into the system 12 minutes into the experiment shown in Figure 3, which produced an initial temperature spike in the heat exchanger fluid associated with sensible cooling of the methane flow. The cooler adjusted to this load and the temperatures stabilized over the following 8 minutes. The 50:50 methane/CO_2 mixture entered the system 32 minutes into the run, producing another temperature spike. In this case, both the sensible heat of the CO_2 stream and CO_2 desublimation increased the contact fluid temperature, the second being a much larger influence. The larger heat load in the fluid caused a correspondingly larger and especially longer change in contact fluid temperature (Figure 3). The temperature spike at 94 minutes occurred during inlet gas resampling to verify mixture concentrations. The temperature data are somewhat smoothed to remove thermocouple noise.

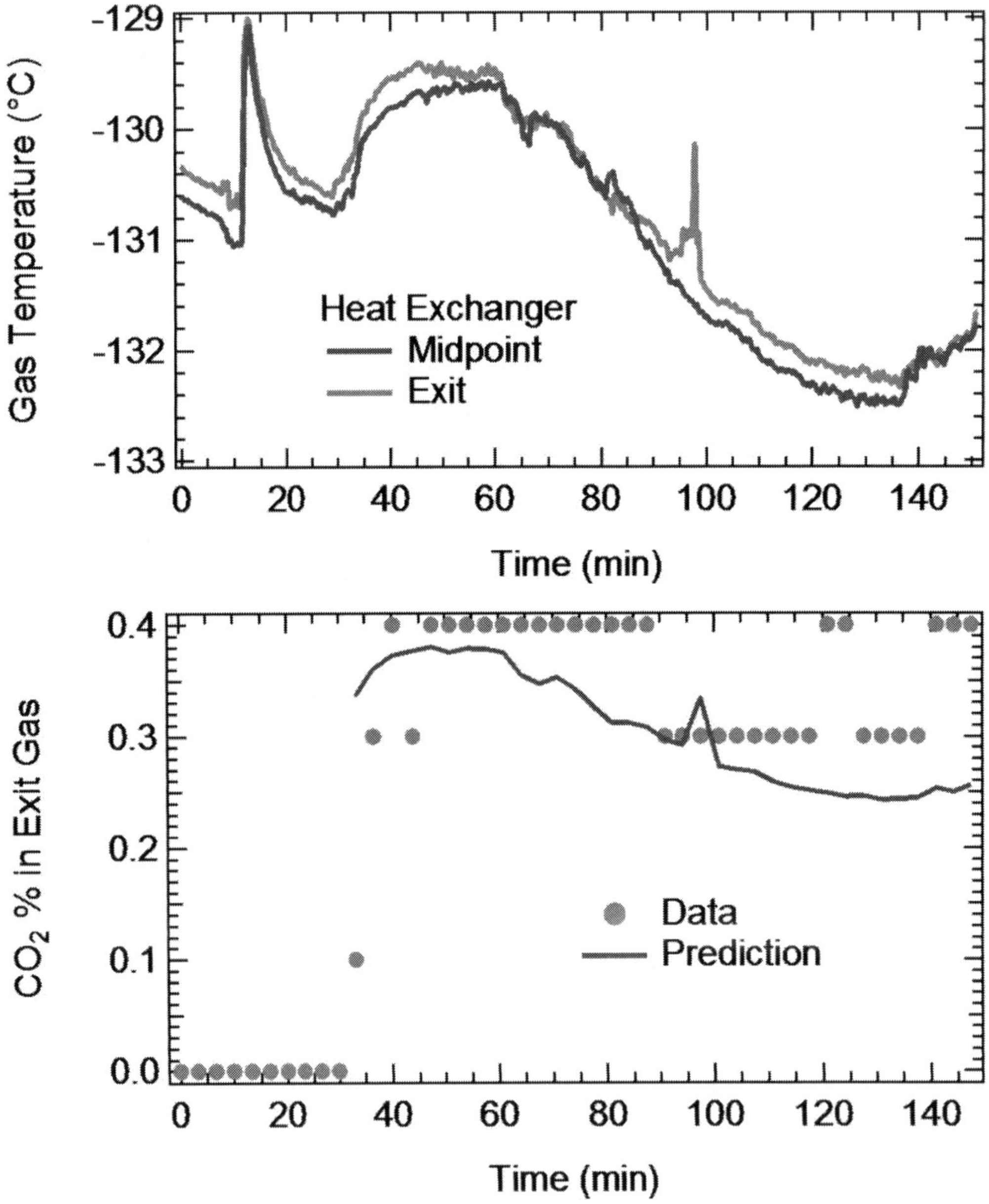

Figure 3 Temperature (°C) and CO_2 concentration (%) in exit stream vs. runtime.

The measured exit gas CO_2 concentrations (as percent) also appear in the graph. The illustrated data represent one measurement about every 20 seconds which is an undersampling of the actual measurements. The raw data collection rate (60 Hz) exceeds the process and gas analyzer response times. The data over the first 30 minutes indicate zero since only pure methane was flowing through the system at that time. Beginning at about 32 minutes, when the CO_2-methane mixture began to flow, the data show between 0.3 and 0.4% residual CO_2 in the exit gas. As evident in the data, the analyzer results are approaching its detection limit as it is reporting a single digit of precision.

Theoretical model results for CO_2 exit concentration in this system based on heat exchanger exit gas temperature and pressure also appear in the figure. This model depends on CO_2 vapor pressure, data and correlations which appear in Figure 4. There is no significant difference between the data and the correlations or between the differing correlations. The mole fraction of CO_2 in the gas phase is predicted as a function of temperature and system pressure as

$$y_{CO_2} \equiv \frac{p_{CO_2}}{p_{tot}} = \frac{\bar{f}_{CO_2}}{\bar{\phi}_{CO_2} P}$$

where $\bar{f}_i$, $\bar{\phi}_i$ and P represent partial fugacity, partial fugacity coefficient, and total pressure, respectively. The partial fugacities come from the vapor pressure correlations and are equal to partial pressures at low pressure. Equations of state correlations, such as Peng Robison, determine the fugacity coefficient, which is unity at low pressure.

The model predictions appear only over the range in which the process was capturing CO_2, that is, from about 30 minutes on. The model predictions lie within the range of data with sufficient precision to validate the experiment. Given how close these results are to the detection limit, the agreement between the experimental data and the predictions is better than expected.

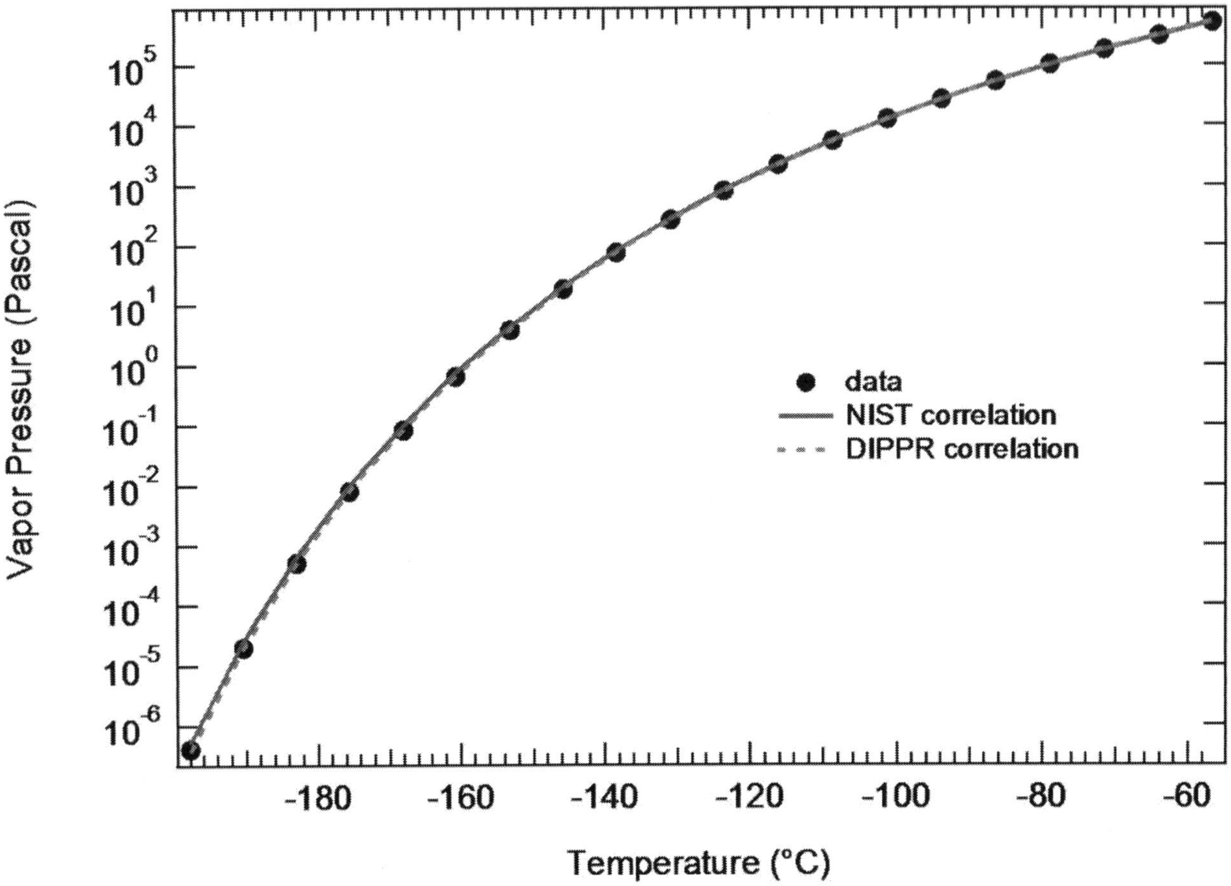

Figure 4 **CO_2 vapor pressure vs. temperature data and correlations. (Design Institute for Physical Properties (DIPPR), 2012), (National Institue of Standards and Technology (NIST), 2012)**

Conclusions

The Cryogenic Carbon Capture™ process removes CO_2 and other impurities from natural gas using processes and equipment similar to those for flue gas treatments. This process does not require pressurization or distillation of the gas stream, saving capital and operating costs and energy. It also does not involve stripping and absorption towers, though it does involve a staged heat exchanger. Finally, it combines several process steps into one, simplifying the overall treatment system.

The first data from the application of CCC™ to raw natural gas treatment indicate that the process behaves as predicted by a theoretical model based on process temperature, pressure, and gas composition and should provide a cost effective and energy efficient alternative to traditional gas treatments.

Industry Sponsorship

Research and experimentation was carried out with the assistance of Sustainable Energy Solutions, LLC in Orem, Utah. Sustainable Energy Solutions is an exclusive licensee of the Brigham Young University owned Cryogenic Carbon Capture (CCC) technology portfolio.

References

Baxter, L. L. (2007). Carbon dioxide capture from flue gas. US Patent Office. US, Brigham Young University. **PCT/US2008/85075**.

Baxter, L. L. (2008). Fossil and biomass power systems with stable CO2 sequestration. US Patent Office. USA, Brigham Young University. **2**.

Baxter, L. L. (2008). Simultaneous CO2 capture and energy storage. US Patent Office. USA, Brigham Young University. **2**.

Baxter, L. L. (2010). Carbon dioxide capture from flue gas. US Patent Office. US, Brigham Young University. **12/745,193**.

Baxter, L. L. (2010). Carbon dioxide capture from flue gas. Euorpean Patent Office. US, Brigham Young University. **08855143.7**.

Baxter, L. L. (2011). The cryogenic carbon capture process. **2009:740158(Patent)**.

Design Institute for Physical Properties (DIPPR). (2012). DIPPR Project 801. Provo, UT.

Harper, J. L. (1989). "Cost-effective technology: What's ahead offshore." Ocean industry **24**(1): 38-41.

Haut, R. C., R. D. Denton and E. R. Thomas (1988). Development and application of the controlled freeze zone process. Proceedings - Gas Technology Symposium: Gas Supply - A Critical Balance of Politics, the Market Place, and Technology, June 13, 1988 - June 15, 1988, Dallas, TX, USA, Publ by Soc of Petroleum Engineers of AIME

Hoeger, C., C. Bence, S. S. Burt, A. Baxter and L. Baxter (2010). Cryogenic CO2 capture for improved efficiency at reduced cost. 2010 AIChE Annual Meeting, 10AIChE, November 7, 2010 - November 12, 2010, Salt Lake City, UT, United states, American Institute of Chemical Engineers

Kelley, B. T., J. A. Valencia, P. S. Northrop and C. J. Mart (2011). Controlled freeze zone for developing sour gas reserves. 10th International Conference on Greenhouse Gas Control Technologies, September 19, 2010 - September 23, 2010, Amsterdam, Netherlands, Elsevier Ltd

National Institue of Standards and Technology (NIST). (2012). NIST Standard Reference Database. NIST.

Parker, M. E., S. Northrop, J. A. Valencia, R. E. Foglesong and W. T. Duncan (2010). CO2 management at exxonmobil's labarge field, wyoming, USA. SPE International Conference on Health, Safety and Environment in Oil and Gas Exploration and Production 2010, April 12, 2010 - April 14, 2010, Rio de Janeiro, Brazil, Society of Petroleum Engineers

Rochelle, G. T. (2009). "Amine scrubbing for CO2 capture." Science **325**(5948): 1652-1654.

Thomas, E. R. and R. D. Denton (1987). Conceptual studies using the controlled freeze zone (cfz) process. Papers Presented at the Summer National Meeting of the American Institute of Chemical Engineers., Minneapolis, MN, USA, AIChE

Thamibuthu, J. D. (2009). An overview of technologies and costs of carbon dioxide capture in power generation. *Journal of Power and Energy*, 201-212.

US Department of Energy, N. E. T. L. (2007). Cost and performance baseline for fossil energy plants volume 1: Bituminous coal and natural gas to electricity

US Department of Energy, N. E. T. L. (2007). Pulverized coal oxycombustion power plants volume 1: Bituminous coal to electricity

US Department of Energy, N. E. T. L. (2008). Pulverized coal oxycombustion power plants volume 1: Bituminous coal to electricity: Revision 2.

Valencia, J. A., P. S. Northrop and C. J. Mart (2008). Controlled freeze zonetmtechnology for enabling processing of high c02 and h2s gas reserves. International Petroleum Technology Conference, IPTC 2008, December 3, 2008 - December 5, 2008, Kuala Lumpur, Malaysia, Society of Petroleum Engineers

Victory, D. J. and J. A. Valencia (1987). "Methane-CO2 fractionation." Hydrocarbon Processing **66**(5): 44-46.

(668b) Optimize Solids-Liquid Separation with Inline Particle Size Measurement

Jack Shu, Des O'Grady, Benjamin Smith and Anjan P. Pandey, METTLER TOLEDO, Columbia, MD

Water recycle and tailing settling rates are optimized through solids flocculation. Yet as the incoming feed concentration and particle size distribution varies, so does the solids separation efficiency. Tracking changes to the real time particle size distribution and fine particle population enables fast and efficient water recycle. Offline measurement techniques require samples to be extracted and manipulated through dilution or dispersion which alter or destroy flocculated components, and offline measurements cannot be applied to make real-time process control decisions. *In situ* particle measurement technology, such as FBRM® and PVM®, measure changes to particle size, inline without the need for sampling. This technology is applied across the petroleum and mining industries to optimize tailings dewatering.

In this paper, flocculation performance, dosage costs, and solid-liquid separations were optimized using FBRM® and PVM® technology. These rugged probe based tools were critical to improve water recovery laboratory and on large scale gravity thickener tanks and pipe reactors. The ability to track changes on both the fine and coarse ends of the particle distribution enabled separation performance optimization based on flocculent type, slurry feed distribution, and shear rate.

(668c) Model of Shale Shaker Performance

Hesam H. Nejad[1], Vidya Raja[1] and **G. G. Chase**[2], (1)Chemical and Biomolecular Engineering, The University of Akron, Akron, OH, (2)Department of Chemical and Biomolecular Engineering, The University of Akron, Akron, OH

Shale shakers separate coarse solid particles from drilling muds in well bore-hole drilling operations. Drilling muds are engineered with rheological properties needed for efficient transport to the ground surface the coarse solid particles generated by the drill cutting into the soil and rock. Effective models are needed to improve the design and operation of shale shakers to optimize the separation of particles and to maximize the recycling of the mud.

The model designed in this work is based on a continuum model of cake filtration in which the coarse solids form a filter cake on the surface of the screen. Incorporated into the model are the mechanisms of the screen motion and the motion of the cake along the screen surface. The mud is modeled as a yield-stress fluid. The model calculations are compared with experimental measurements on a full scale shaker. This model is used to identify parameter sensitivity.

(668d) Morphology and Surface Properties of Polyethylene Oxide / Syndiotactic Polystyrene Aerogels

Xiao Wang, Polymer Engineering, The University of Akron, Akron, OH and Sadhan C. Jana, Department of Polymer Engineering, University of Akron, Akron, OH

Polyethylene oxide (PEO) was used to render hydrophobic syndiotactic polystyrene (sPS) aerogels hydrophilic. Such materials are useful in absorption of moisture and removal of particulate solids. Solutions of sPS and PEO in tetrahydrofuran were converted into gels by cooling at room temperature. The PEO/sPS aerogels were obtained by drying the gels under supercritical condition. The effect of PEO molecular weight and concentration on morphology and surface properties of PEO/sPS aerogels were investigated. It was seen that the introduction of PEO helped regulate the pore structure of PEO/sPS aerogels, the extent of which depended on the molecular weight and concentration of PEO. New micron pores were created when high molecular weight of sPS was used, and the size of micron pores increased with increasing PEO concentration; while the morphology of the aerogels did not change when low molecular weight PEO was incorporated. The presence of PEO increased surface hydrophilicity and maintained high surface area of the aerogels, which contributed to the high moisture absorption properties of PEO/sPS aerogels.

(672a) Quarternized Chitosan Electrospun Membranes for Virus Removal

Caryn L. Heldt and Bingyu Bai, Chemical Engineering, Michigan Technological University, Houghton, MI

For remote or personal water systems, small pore size membranes can be added to reject bacteria and viruses from water purely based on size; but, the small pore size required to remove these pathogens often increases membrane fouling and leads to large membrane areas, high transmembrane pressures, low water flux, and frequent changing of membranes. This increases the cost of water filtration. In order to make clean water accessible to everyone, there is an urgent need for improved water disinfection technology. Our goal is to create functionalized, large pore-sized membranes that have the ability to absorb virus and bacteria. Membranes with pores on the order of microns have the advantage of increased flow rate and reduced membrane fouling. As a first step in our goal to create water filtration membranes, we have created membranes made of biocompatible, inexpensive, and environmentally friendly chitosan, which has the ability to be functionalized to increase the membrane disinfection capacity. Nanofiber, chitosan membranes have been made by electrospinning. These membranes have an increased surface area as compared to microfibers and are on the same size scale as viruses, creating a curved surface that has the potential to improve virus absorption. Chitosan membranes were originally made, but the virus adsorption capability was low. To improve the virus absorption, we functionalized the membranes with a quaternized amine, forming a N-[(2-hydroxy-3-trimethylammonium) propyl] chitosan chloride (HTCC) blend. The HTCC cationic polymer has an increased positive charge, as compared to chitosan, and therefore, has increased virus absorption capability. Additives have been used to increase the ability to electrospin HTCC, by changing the viscosity and conductivity of the electrospinning solution. HTCC membranes have the ability to remove greater than 99% of virus present in solution, completing the first step to the creation of adsorption membranes that can remove pathogens from water.

MEMBRANE DEVELOPMENT FROM BANANA PEEL FIBERS FOR WASTE WATER TREATMENT AT LOW COST

Shounak Datta, Sudeb Karmoker and Dr. Md. Tanvir Sowgath,
Bangladesh University of Engineering & Technology (BUET)

Introduction

The growing problem of drinking water is a serious issue in urban areas of Bangladesh. The greater problem is the lack of water sources that can be used to produce potable water. Even in the Dhaka city, the capital of Bangladesh, people from many areas are complaining about the quality of water they get from WASA. The problem is much more severe in other coastal areas such as Chittagong, Barisal, and Khulna for presence of brackish water. According to Census 2001, tap water access in Bangladesh is 25.5%, where in coastal areas this figure decreases to 16.7%

Fresh water shortage increases the risk of conflict, public health problems, reduction in food production, inhibition of industrial product expansion, and these problems threaten the environment. The discharge of inadequately treated industrial waste water can therefore have far-reaching consequences; physical effects, toxic effects on aquatic life, eutrophication. Therefore, necessary preparation should be taken right now to face this challenge of getting fresh water for drinking, industrial and agricultural use.

During the last two decades high-performance membrane technologies such as nano-filtration (NF) and reverse osmosis (RO) has been used for water treatment worldwide. There are two basic problems for which this technology is still not much famous in third world countries; first, high production cost, and second, need of much more support for high pressure operation. Even the industries in coastal areas have to use reverse osmosis systems to protect their equipment from salinity. This causes a high cost for purchasing membranes and high pressure vessels.

Again, textile industries in Dhaka face high concentration of suspended solids (loose fiber), Biological Oxygen Demand (BOD), heavy metals and color in the final effluent. Aromatic and metal chlorides are present in the dye. These compounds are toxic to the aquatic environment. Allergic dermatitis and skin irritation are found in textile dye polluted area. The present study also investigates the suitability of banana peel fiber membrane as low cost media for post treatment of treated effluent. The waste water samples were taken from Nice Cotton Ltd., Zirani, Kaliakoir, Gajipur.

The aim of this project is to use local raw materials to produce cheap membranes for water treatment. For this purpose, *phase inversion technique* is chosen to be used. Banana peel fibers are to be extracted by *Kraft Process* and to be used as raw material for the membrane preparation. Membrane is then to be used for the waste water treatment.

Why Banana Peel Fibers?

The purpose of this project was to study the feasibility banana peel fiber membrane in order to remove dye from textile wastewater at low cost. The excellent ability and economic promise of the banana fiber to remove color from the textile waste water is shown in this work. The project is in initial phase.

The aim is to replace all expensive chemicals with locally available raw materials, if possible. The choice of banana peel fibers from immediate inner layer of banana peel was made for various reasons and water selectivity consideration.

The first reason was to select something cheap, or something that is considered as waste. The fact with banana peel is a type of food waste and it also has high fiber content (almost 40%). Besides, less chemical is needed to prepare the membrane for the water purification procedure. Since the membrane is produced from fewer chemicals they are less likely to introduce extra contaminants. The fibers are seen to have good water selectivity, and that is quite a good point for this selection.

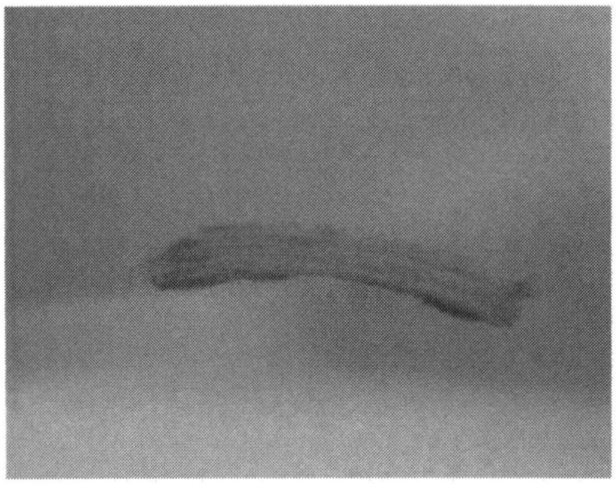

Fig. 1: Banana peel inner layer

Theory

Fiber Extraction

The fibers of banana peel are to be extracted and collected to be used in membrane preparation. There are three types of methods for fiber extraction from banana peel.

a) *Mechanical method* is the oldest method for fiber extraction. It has an advantage of requiring no chemicals but has a huge drawback as degumming is not properly done.

b) *Chemical method* uses chemical treatment for degumming. Two types of chemical processes are available, Kraft process and SLS process. Among them, SLS process is a bit more expensive where Kraft process can be done using only sodium hydroxide solution.

c) *Biological method* is the latest and most effective and sensitive method, as it reduces the chance of environmental pollution that is posed by chemical method. For this, use of micro-organism is required. Then the solids are allowed to be fermented. The next step is to extract using enzymes.

Phase Inversion Technique (PIT)

The main concept of phase inversion technique is to create porosity within a thin film. The mixture used for film preparation by phase inversion technique is given below

- Cellulose acetate(E- 398.3, acetyl content 39.8%)=22.2%
- Acetone = 66.7%
- Water = 10%
- Magnesium perchlorate = 1.1%

Among these items, purchase of cellulose acetate and magnesium perchlorate are the most expensive. [6]

The steps of phase inversion techniques are

- Partial evaporation of acetone for 3-4 minutes in glass plate at 0 to -3°C
- Cooling of film in ice bath for 1 hour, this is called gelation.
- Using permeation cell, simultaneous heating and compressing the film to get desired pore size and water selectivity
    - Heating temperature- 80°C
    - Compression Pressure- 20% more than operating pressure

Experimental Procedure

Apparatus

1. Water bath
2. 0.025 cm sided glass
3. Glass rod rolling from top
4. Cold Box
5. Refrigerator

Chemicals

1. Acetone
2. Magnesium Perchlorate
3. 10% NaOH solution

Raw Material

Banana peel fibers (replacing cellulose acetate)

Procedure

The initial approach is presented here.

Kraft process

- 10% NaOH solution was added with Banana peel and banana fruit
- Solution was cooked for 1 hour at 70°C
- The fibers were washed by distilled water
- The fibers were dried in sunlight/dryer

Phase Inversion Technique

- Fibers were mixed with appropriate portion of acetone, magnesium perchlorate and water.
- A glass plate was taken and a cast of uniform thickness was ensured by rolling glass rod over the mixture
- The plate kept in refrigerator for 5 minutes.
- The plate along with the attached membrane was immersed into ice bath and kept for one hour.
- After one hour, the membrane was detached from the plate. It was then collected and dried for use.

Why did PIT fail?

The reason for which the proposed Phase Inversion Technique (PIT) did not work was the difference of physical properties of cellulose acetate and banana peel fibers. Banana peel fibers definitely contain cellulose, but cellulose acetate is soluble in acetone where banana peel fibers are not. For this reason gelation could not be achieved from the mixture after allowing partial evaporation of acetone in cold box.

Modification of Phase Inversion Technique: Cold Binding Method

During the development of this project, Phase Inversion Technique was experimentally studied to develop the membrane. Finally, this process was modified to develop desired membrane characteristics. The final modified approach following which the achieved membrane was produced is given here.

- The extracted fibers were collected and washed by shaking in a beaker with distilled water.
- The water was soaked from fibers without heat drying.
- The remaining fibers were collected after 15 minutes.
- The fibers were weighed and specific amount of magnesium perchlorate and acetone were added in a beaker.
- The mixing was done by shaking, not stirring.
- The mixture was then filtered to remove extra acetone.
- The remaining filtrate was taken on a glass plate and a glass rod was rolled over it to ensure a cast of uniform membrane layer.
- The plate was kept on a large ice piece in the refrigerator to ensure cooling at 0^oC. The cooling was allowed for 10 minutes.
- The glass plate was brought out from the refrigerator and the membrane layer was collected carefully without damaging the shape.
- The remaining water was soaked from the membrane, and it was allowed to remain exposed to atmosphere for 8 hours.

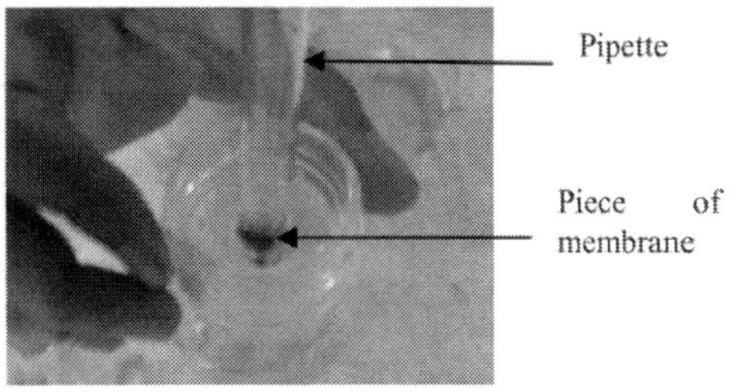

Fig. 3: Setup for testing effectiveness

a

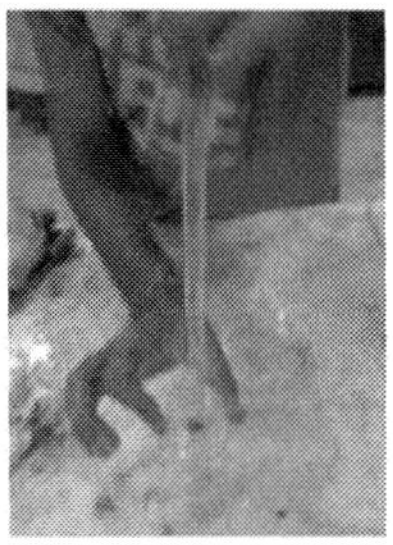

b

Fig. 4: Setup for testing effectiveness and collecting filtered sample (a, b)

Results and Discussions

Using the proposed steps, it was possible to produce a membrane which can be acceptable for waste water treatment problem. This membrane shows a good effectiveness in case of dye/color separation. The achieved membrane that we can use for waste water treatment is given in figure 5.

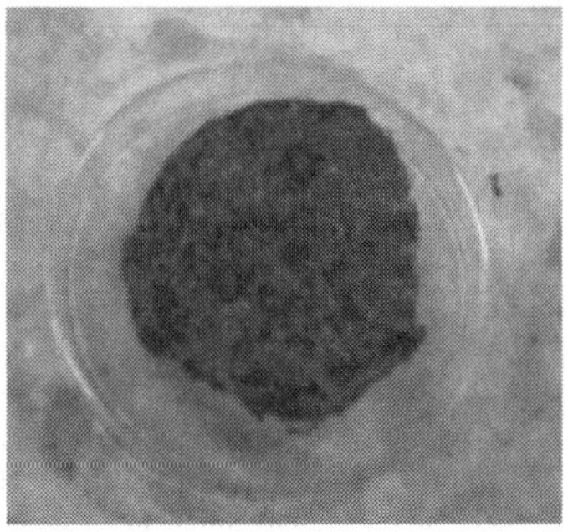

Fig. 5: The produced membrane film

To find the eligible source of fibers, both the fruit and the peel were tested by the Kraft process. The film formed by banana fruit and fibers extracted by Kraft process was analyzed by a microscope and presented here in figure 6 and 7. The brownish colors of both films are because of not using bleaching and additives. The experimented data and observations are recorded in Table 1. Among these samples, sample 5 was used to successfully prepare the film. Its microscopic view is presented in figure 10.

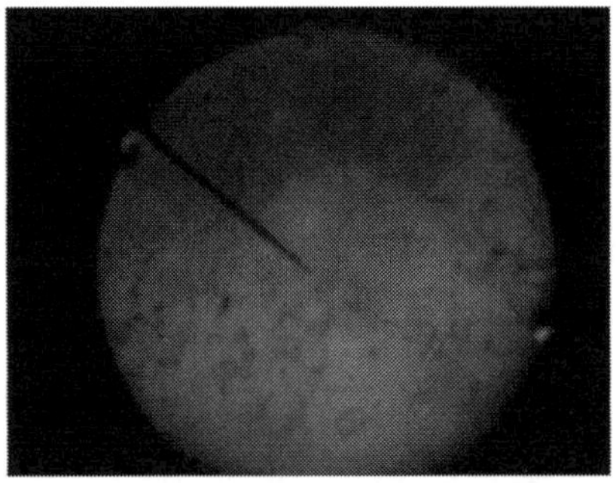

Fig. 6: Banana fruit film layer formation

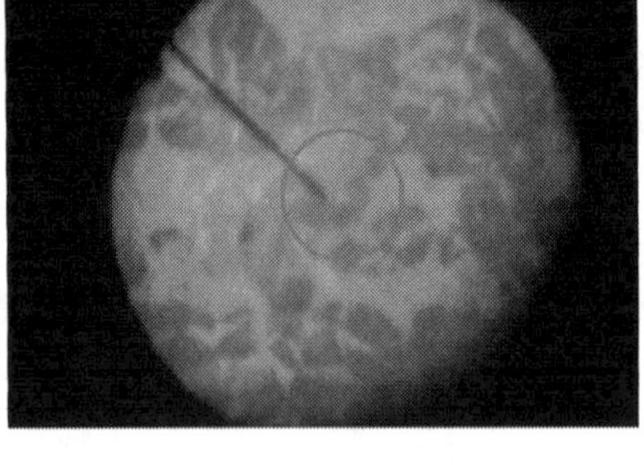

Fig. 7: Banana peel fiber film layer formation

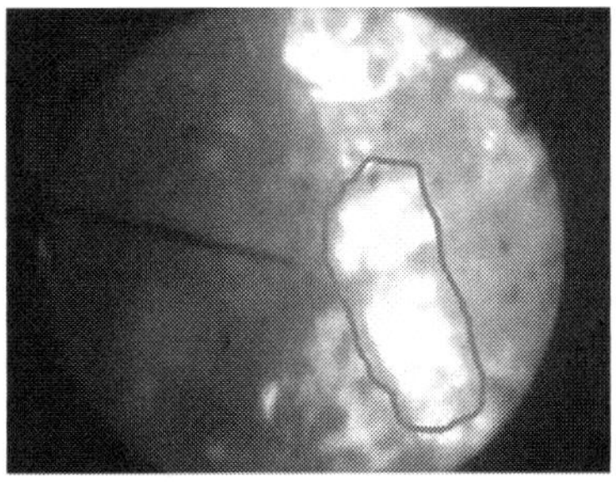

Fig. 8: Processed film layer formed by heat treating (*pores marked in red*)

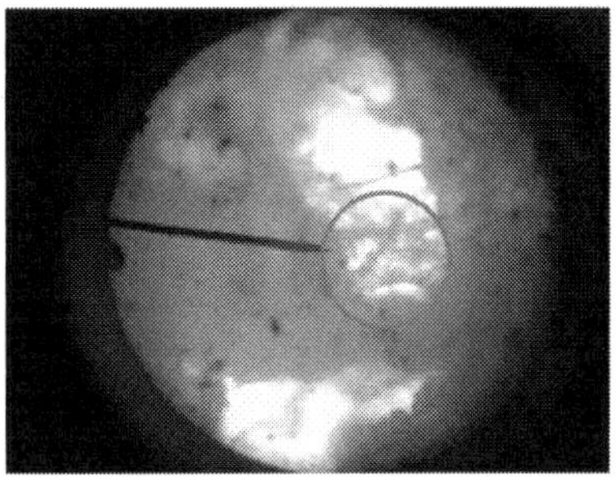

Fig. 9: Bonding between fibers into the film (*fiber interconnection marked in red circle*)

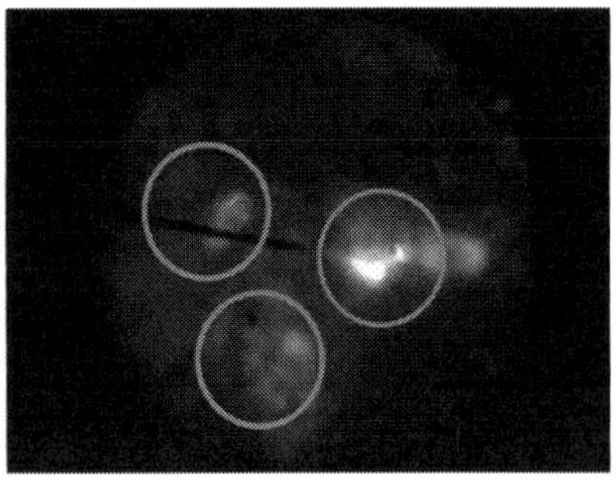

Fig. 10: Processed film layer formed by cooling in refrigerator and drying (*pores marked in red circle*)

These findings proposed whole new steps of procedures. The phase inversion technique had to be modified to prepare the produced membrane. This membrane is then used to test its effectiveness in filtration. The results are presented in table 2.

Characteristics:

- High pressure tolerance.
- Good tensile strength.
- Wetting does not affect tensile strength.
- Dye separation (please refer to *figure 11*).
- Temperature tolerance is 50-60°C
- No special preservation system is needed.

Limitations:

- Shrinkage of the prepared membrane was noticed. This shrinkage caused the membrane to be more pressure tolerant.
- The pores generated by partial evaporation of acetone of the mixture were not uniform.
- At 70°C temperature, the membrane film tends to be disintegrated.

(a) (b) (c)

Fig. 11: The effect of filtration on a sample from dying factory. (a) Sample from the factory, (b) after first pass through the membrane, (c) after second pass through the membrane.

Table 1. Various Mixture Specifications Used in the Experiment

Sample No.	**Banana Peel Layer Taken (g)**	**10% NaOH(aq) Taken (ml)**	**Fibers acquired (g)**	**Magnesium Perchlorate Added (g)**	**Acetone Added (ml)**	**Type of Treatment**	**Observation**
1	9.46	50	2.486	0.123	4.5	Heating	Layer burned
2	24.51	200	4.44	0.2	10	Freezing and keeping in ice bath	Bonding not formed
3	4.2	40	1.2	0.2	3.2	Soaking water and then heating	Layer formed, hard and brittle
4	15.2	100	4.4	0.5	8.5	Heating for 10 minutes at 50°C	Brittle layer formed
5	17.7	150	5.4	0.6	13.4	Keeping over ice piece in refrigerator for 10 minutes, after that remaining liquid is soaked	Soft layer formed, which shrunk and hardened after 7-8 hours

Table 2. Test Results Before and After Membrane Filtration (double pass)

Sample → / Properties ↓	Before Filtration	After Filtration (Double pass)
pH	8.1	7.35
Conductivity (mS/cm)	3.58	2.75
TDS (g/l)	1.79	1.31

The results provided here are the average results of five tests using the sample of the same source.

Conclusion

The success of this process depends on the appropriate and logical selection of raw material for the membrane preparation. The more the expensive chemicals can be replaced, the more the essentiality of this experiment can be established. The absence of appropriate apparatus is a huge disturbance in this case. Still, the formation of a porous layer, though not uniformly porous, can be said to be a partial success. The test results can be used to show that this membrane can be used for water treatment. The TDS data of table 2 shows that dissolved solids can be separated to an extent. The question still remains whether the shrinking of the membrane is beneficial or a problem that requires notice. The future plan is to prepare the proposed waste water treatment process and study it.

Acknowledgement

This is to express gratitude to our laboratory assistants for helping with their experience and brilliance. It is also to thank Bayzid Haider Tuhin for supplying the waste water sample to test the effectiveness of the membrane.

References

1. Ahmad , Mohiuddin (2005), "Living in the Coast Series 4-Urbanization", *Water Resource Planning Organization, Ministry of Water Resources*, Bangladesh.
2. Baker, Richard W. (2000), "Membrane Technology and Application, 2nd Edition", *McGraw-Hill*.
3. Hillis, P.(2000) ; "Membrane technology in water and wastewater treatment", *The Royal Society of Chemistry*, Cambridge, UK,
4. Li, Kun et al, (2010), "Analysis of chemical composition and Morphological structure of banana pseudo-stem". China
5. Prema, Parukuttyamma and Nicemol Jacob (2008), "Novel Process for the Simultaneous Extraction and Degumming of Banana Fibers under Solid State Cultivation", 2008.
6. Matsuura, Takeshi (1994); "Synthetic Membranes and Membrane Separation Processes", *CRC Press Inc.*, USA.

(672c) Engineering Biochars for Plant-Available Water Soil Applications

Catherine E. Brewer and Robert Brown, Center for Sustainable Environmental Technologies, Iowa State University, Ames, IA

Water availability for agriculture is a serious challenge in many areas of the United States and around the world. This challenge is expected to become worse as precipitation events intensify and dry periods lengthen with climate change. Several studies have shown increased water holding capacity in soils amended with biochars. In this research, biochar properties were varied by varying biomass feedstocks and reaction conditions in order to identify production conditions that create biochars with favorable pore structures and surface chemistries for soil water retention. The biochars studied included corn stover slow pyrolysis biochars made over a 200-800°C temperature range under an inert or a 5% oxygen reaction environment, as well as biochars from fast pyrolysis. The methods used to evaluate biochar properties included proximate analysis, elemental analysis, nitrogen gas adsorption for BET surface area and micropore volume, solid-state ^{13}C nuclear magnetic resonance spectroscopy (NMR), and soil water retention by both pressure plate and centrifuge.

Characterization and Assessment of Industrial Wastewaters in Egypt

Mohamed Mostafa* and Robert W. Peters, Ph.D., P.E.
Department of Civil, Construction and Environmental Engineering
University of Alabama at Birmingham
Birmingham, Alabama 35294-4440

Abstract

Poor water quality is one of the biggest problems faced by most countries especially, developing countries such as Egypt. Egypt is the most populous country in the Middle East with a population of more than 82 million people. Egypt's population has grown significantly since the beginning of the 20th century. Per capita water availability is declining in Egypt over time due to rapid population growth. The mismanagement of water resources is one of the main reasons for the depletion of water. The rapid population growth in Egypt also results in a significant increase in the size and number of industries due to increased demand for industrial products. The industrial areas in Egypt suffer from a number of environmental problems including management of sewage and disposal of pollutants from the surrounding factories into surface water pathways in the area. Results have shown that industrial and domestic wastewater samples often contain toxic levels of several heavy metals (Cd, Co, Pb, and Ni), which have adversely impacted surface waterways. Egypt's 329 major industries discharge as much as 2.5 x 10^{6} cubic meters/day of untreated effluent into Egyptian waters, with 50% to 64% of the industrial activity located in Cairo. It has been estimated that 80% of the country's industrial effluent is discharged untreated into the Nile River, canal, wells, municipal sewage systems, and ultimately the Mediterranean Sea. The analyses of Nile River samples (water and sediment) have shown the presence of high concentrations of Mn and Cr, and low concentrations of Cu and Pb. The Nile River also suffers from the effects of recurring oil spills.

Wastewater samples are being collected from various industrial sectors for analysis. In particular, we are focusing on industries that generate wastewater with high concentrations of contaminants. Among the various industrial sectors, the organic pollutants are mainly generated from food processing, pulp and paper, pharmaceuticals, textile, and tannery industries, while a substantial portion of effluents containing heavy metals are generated from steel, petroleum, textiles, tanneries, and chemical industries. In addition, wastewater generated from these industries typically contains high concentrations of BOD, COD, suspended solids, and nitrogen. The loads of these contaminants contained within sewage are as follows: BOD, COD, oil and grease, TDS, and heavy metals are 235, 423, 168, 296, and 1.65 tons/day, respectively. About 60 percent of the heavy metals discharge comes from the chemical industry and about 50 percent of the BOD load comes from food processing. The wastewater samples are being tested in the laboratory for different physical, chemical and bacteriological parameters. The parameters are pH, biochemical oxygen demand (BOD), Chemical Oxygen Demand (COD), oil and grease, total suspended solids (TSS), total dissolved solids (TDS), Zn, Cu, Phenol, Cr, and temperature. The most suitable treatment techniques will be identified based on the data collected.

Pollution Prevention (P2) is a positive and crucial step aimed at reducing, avoiding, or eliminating the generation of wastes or any hazardous substance, prior to recycling, disposal or treatment. A corporate policy statement is a very important factor to consider when starting a pollution prevention program. A successful P2 program requires support from all members of a company; especially, the management levels. Many economic and environmental benefits result from the implementation of a pollution prevention project. These benefits include reducing operating costs, reducing regulatory compliance issues, improving employee morale and participation, and reducing liability.

This research aims to improve water quality and the *environment* in Egypt by exploring ways to reduce, reuse, and recycle industrial wastes; and finding appropriate techniques of treating the Nile River oil spill. Specific objectives of this research are to: (1) develop a comprehensive literature data base of pollution prevention case histories in Egypt; (2) develop preliminary pollution prevention opportunities for the facility based on historical records, process flow diagrams, wastewater analyses, and mass balances; and (3) conduct on-site plant pollution prevention assessments to identify specific opportunities by which waste can be reduced at a specific facility. Results will be presented on the waste characterization and identification of appropriate treatment techniques to minimize disposal of industrial wastes in Egypt.

Introduction and Background

Egypt is a country located on the continent of Africa, divided into 27 governorates. Egypt's total area is about 1,001,450 km^2 (386,662 mi^2) and its population is more than 82 million people [CAPMAS, 2012]. Cairo has approximately 18 million people and is the largest city in the Middle East region [Higdon, 2012]. Alexandria, Egypt's second largest city has a population of more than 4 million people [CIA, 2011]. Typical of giant cities, they have continuous, if not exponential, population growth and spatial expansion. Per capita water availability is declining in Egypt over time due to rapid population growth. The rapid population growth in Egypt also results in a significant increase in the size of industries due to increased demand for industrial products [African Development Bank, 2009].

Surface water pathways in Egypt are exposed to pollution due to industrial activities in this region resulting in harm to human and aquatic health, in addition to a negative impact on the environment. It has been estimated that 80% of the country's industrial effluent is discharged untreated into surface water bodies. This leads to a significant increase in the proportion of pollutants such as BOD, Cr, oil and grease, Hg, *e. coli*, and heavy metals.

Egypt's 329 major industries discharge as much as 2.5 x 10^6 m^3/day of untreated effluent into Egyptian waters [Yousef and Sakr, 2007]. Cairo has approximately 18 million people and is the largest city in the Middle East region [Higdon, 2012]. Cairo's surrounding regions are burdened with increasing potable water demand and heavy wastewater discharges. The city's water resources are also affected by discharges from surrounding areas [Myllylä, 1995].

A large number of industries discharge their wastes on agricultural drains which are connected with the Nile River branches (Damietta and Rosetta), or directly connected with the Nile River [Abdel-Shafy and Aly, 2002]. This leads to a significant increase in the percentage of contaminants such as suspended and dissolved solids, nutrients, oil, grease, organic matters, and pesticides. The analysis of water samples in the Rosetta branch showed that the concentrations of Cu, Zn, and Cd are higher than the required standards, where the Egyptian standards for Cu and Zn in the receiving water must not exceed 1,000 μg/L [Ramirez *et al.*, 2004].

Studies indicate that only 65% of Egypt's population can access clean drinking water and 24% can access sewerage services [Abdel-Wahaab, 1995]. In rural areas, about 50% of the population is linked to drinking water supplies and only 5% to sanitation services [Abdel-Shafy and Aly, 2002]. Individual means of wastewater disposal (e.g., septic tanks and latrines) are used by people that don't have access to sewerage services.

There have been several studies conducted addressing water pollution in Egypt, including Saad *et al.* [1982, 2003], Moursy [1983], Lasheen and Ashmawy [1983], El-Sebae [1986], Badawy and Aly [1986], El-Gendy *et al.* [1991], Abd-Allah [1994], Ali and Soltan [1996], Yamashita *et al.* [1997], Al-Makkawy and Madbouly [1999], Abd-Allah *et al.* [1999], Ali *et al.* [2001], El-Iskandarani *et al.* [2004], Gerges and Smith [2004], Geriesh *et al.* [2008], Hamed and El-Moselhy [2009], Badawy and Emababy [2010], Ibrahim M. D.[2010], El-Otify *et al.* [2011], and El-Ghani *et al.* [2011].

Soltan and Awadallah (1995) collected water samples from different locations along the Nile River for analysis. The results showed elevated levels of TDS, TSS, TH, Na^+, Cl^-, Ca^{2+}, Li^+, K^+, HCO^{3-}, SO_4^{2-}, and Mg^{2+} in water samples taken from Damietta, due to the receipt of effluent from some industries, such as cheese making factories, fertilizers, textile, and pickled herring making. Statistical analysis of the database were established and a significant correlation coefficient values were obtained (r=0.61–1.0). Soltan (1995) collected wastewater samples from Kima drain (Aswan, Egypt) and also collected water samples from the Nile River at Aswan. The collected water and wastewater samples were analyzed to study the effect of discharging domestic and industrial wastewaters on water quality at the Nile River. The results showed that the quality of Nile water did not affected by domestic and industrial wastewaters discharged from Kima drain. Abdel-Moati and El-Samak [1997] collected forty two sediment samples from four lakes (Manzalah, Edku, Mariut, and Burullus) to assess the man-made impact on geochemistry of these lakes. The results showed a significant increase in the metals concentrations especially Pb and Cd, due to discharge untreated sewage, and industrial wastewater directly into these lakes. Abou-Elela and Zaher [1998] noted that there are a wide range of pollution prevention opportunities which could be implemented with significant financial advantages for the industry as well as reducing environmental pollution. Abdel-Sabour *et al.* [2001] collected industrial and domestic wastewater samples in the greater Cairo area. Industrial effluent samples collected from a fabric company had the highest amounts of total Zn, Cu, Cd, Co, and Pb, while a steel company discharged the largest amount of total Fe and Mo. A plastic company discharged the highest amount of total Ni. Efficient wastewater treatment and disposals are needed to minimize the risk of adverse effects of heavy metals on the environment.

Melegy *et al.* [2002] collected sediments and water samples from the Nile River in the southern area of greater Cairo for analysis. The results showed elevated levels of heavy metals in sediments and water samples taken near the pollution source. The concentration of heavy metals and nutrient salts increased in surface water in Wadi Hof, due to the receipt of effluent from Electrical Power Station. Wahaab and Badawy [2004] conducted a research to evaluate and assess the characteristics of the Nile River system, and identify major sources of Nile pollution. Soltan (2006) collected fifteen sediment samples from the Nile River at two stations (Mansoura city in the northern part of the Nile and Aswan city in the southern part of the Nile) to study the effect of Ethylenediaminetetraacetic acid (EDTA) within industrial discharge on the remobilization of heavy metal ions on bed sediments of the Nile River. Results showed that the remobilization of metal ions increased with increasing EDTA percentage on the Nile River.

In an effort to upgrade industrial pollution control in Egypt, preliminary engineering investigations were conducted at 13 plants to evaluate alternative control technologies to reduce wastewater pollution loadings, as well as to control air pollution, to improve solid waste management, and to address industrial hygiene problems [Tomalavage *et al.*, 1984]. Their study had as its focus to determine the most appropriate application of demonstrated technology to specific industrial problems in Egypt. The map in Figure 8 shows factories that discharge their effluents directly into the Nile River. Their geographical distribution is as follows: 18 units in Upper Egypt; 11 units in Greater Cairo; and 2 units in the Delta [Nile Basin Initiative, 2005].

Results and Discussion

This research will be carried out in the Nile River and the main purpose of this research is to assess the extent of pollution by the industrial activities. This research also focuses on removing contaminants from industrial waste water prior to discharge to surface water through in plant process modification. A thorough and comprehensive technical literature search is underway to develop a pollution prevention assessment case history project library, targeted to specific industrial applications.

According to the last survey of factories located near the Nile River conducted by Egyptian Environmental Affairs Agency (EEAA), some factories still discharging their wastes on the Nile River and the situation of the 31 factories located near the Nile River is as follows [Nile Basin Initiative, 2005]: (1) seven factories are recently discharging their wastes to the sewerage system instead of discharging directly to the Nile River; (2) four factories are recently discharging their wastes on lakes to be used for trees irrigation; (3) thirteen factories have units to reuse water; and (4) eight factories still discharging their wastes on the Nile River (Salt and Soda Factory at Kafr El-Zyaat, El-Nil Factory for Oil and Soap at Bany Korrah, Quina for paper at Couss, Kima Factory for fertilizers at Aswan, Edfu Factory for paper pulp and writing paper, Coca-Cola Factory at Bani-Sweif, and El-Kahera Factories for oil and Soap at El-Ayaat and El-Badrashein). The map in Figure 1 shows the location of the eight factories discharging their effluents directly into the Nile River and their flow.

The Google Earth will be used to examine the Nile River and identify major industries near the Nile River. The data will be collected from industries that still discharging their wastes on the Nile River. This data may include wastewater flow and heavy metals concentration in wastewater sample. This information is essential to create a statistical modeling of pollutant load in the Nile River. A multiple regression models will be created by using Microsoft Office Excel 2010 to identify those industries that are major polluters of the Nile River. A water samples will be collected from the Nile River upriver and downstream of industrial discharges (major industries) along the length of the river to help creating the regression models, as shown in the figure below. The collected water samples will be analyzed for different parameters during summer and winter season. The collected data will also be subjected to statistical analysis and the regression equations will be established between these parameters to predict the level of pollution in the Nile River.

A mass balances (mixing zones) will be involved to simulate/estimate the concentrations of the various pollutants in the river (seasonal basis). Pollution deposition onto river sediments may also be addressed to further refine the mass balance determinations. As a simple example, below are the two mass-balance equations for two industries discharge their waste directly into the Nile River.

Mass balances:

Equation1:

$$q_1 * C_1 + Q_{river,upstream} * C_{river,upstream} = Q_{river,downstream} * C_{river,\ downstream}$$

Equation 2:

$$q_2 * C_2 + Q_{river,downstream} * C_{river,\ downstream} = Q_3 * C_3$$

Overall balance:

$$Q_{river,upstream} + q_1 = Q_{river,downstream}$$

$$Q_3 = Q_{river,downstream} + q_2$$

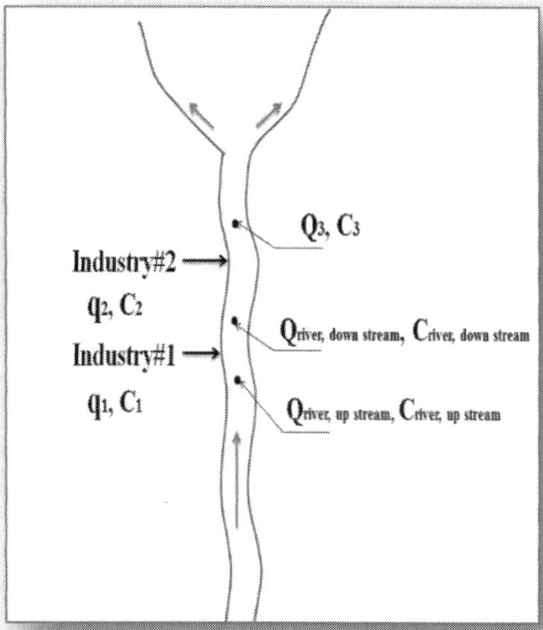

The physicochemical properties of industrial point sources between Aswan and El-kanater for the year 1999 are presented in the table below. According to the last survey of the Nile system pollution sources, most of these point sources are not complying with environmental standards, where the major sources of pollution are Coca-Cola and oil factories in Souhag, and sugar factories in Ques, Deshna, Kom-ombo, Aramant and El-Hawamdia [Nile Basin Initiative, 2005].

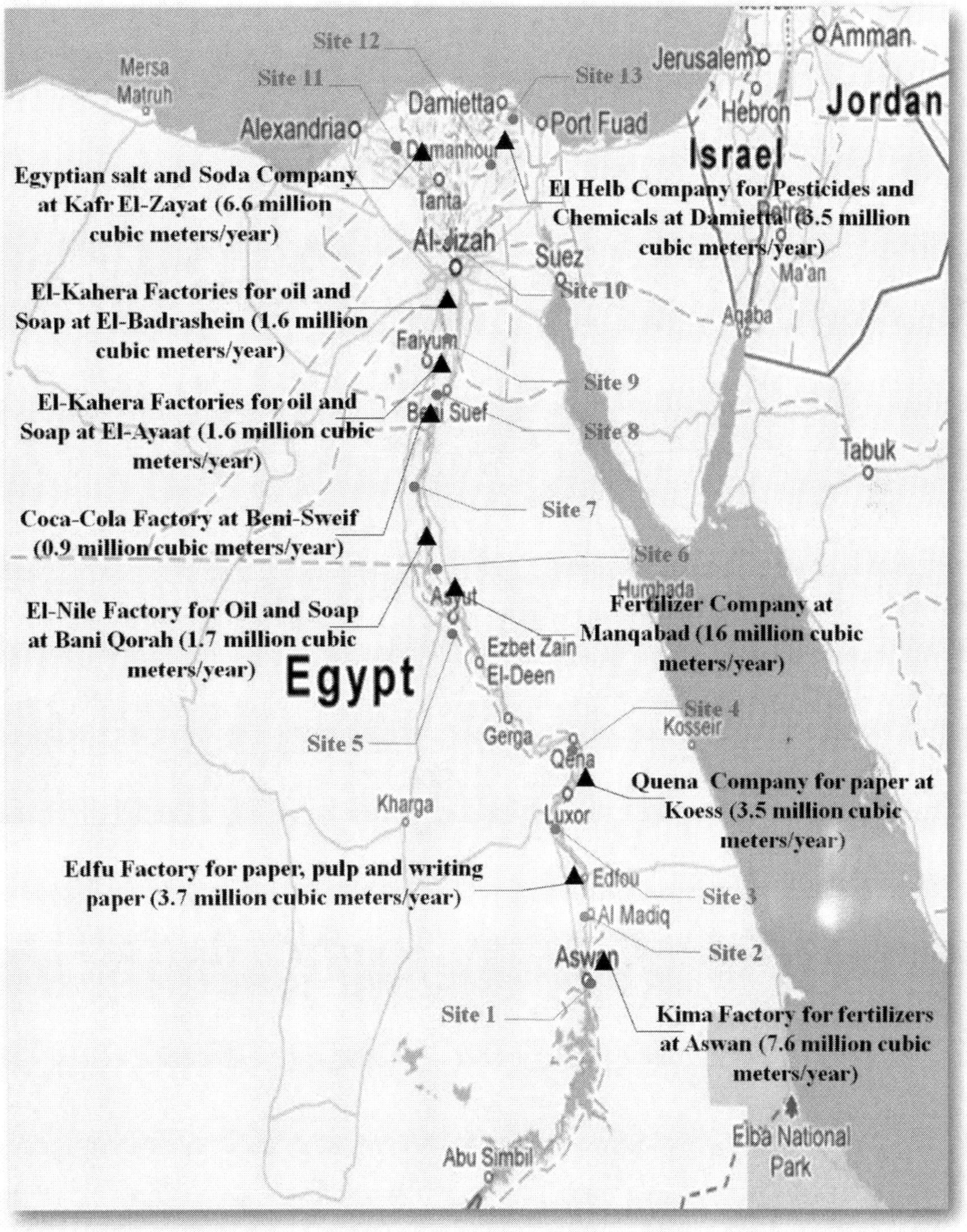

Figure 1. Factories that discharge their effluents directly into the Nile River (Source: Google Maps/Google Earth, 2012; Nile Basin Initiative, 2005)

Table 1. Physicochemical properties of industrial point sources discharging into the Nile River (From Aswan to El-kanater) [Nile Basin Initiative, 2005]

Point Source	Discharge, (Mm³/year)	BOD, mg/l	COD, mg/l	TSS, mg/l	Oil and Grease, mg/l
Consent standard, mg/l		30	40	30	5
River Nile main branch:					
Komombo Sugar Industry	22.6	760	1,500	46	9.3
Edfu Paper and Pulp	8	78	622	158	11.1
Edfu Sugar Industry	16.5	260	370	35	7.4
Armant Sugar Industry	17.4	70	161	15	9.6
Ques Sugar Industry	22.6	33	59	20	3.36
Deshna Sugar Industry	12.1	67	800	20	8.32
Aluminium Industry	5	760	N/A	N/A	N/A
Naga Hamadi Sugar Industry	22.6	54	117	23	4.04
Onion Industry	<1	296	572	N/A	N/A
Souhag Oil Industry	<1	75	260	61	5.87
Coca Cola Industry	<1	75	260	61	5.87
Seflak Industry	N/A	5.4	9	41	0.7
Mankabad Pipe	11.2	1.5	22	28	2.04
Hawamdia Chemical	13.3	N/A	N/A	N/A	N/A
Chemical Industry	5	420	5,600	79	48.4
Hawamdia Sugar Moulas	<1	445	6,000	166	50.2
Hawamdia Sugar Pipe	18	440	3,850	285	17.6
Kotsica Starch and Glucose	1.5	1,603	2,120	N/A	N/A
El Nasr Glass Tube	<1	2	27	36	4.39
Rosetta branch:					
El Malaya Company	4.6	15	1,886	N/A	N/A
Salt and Soda Company	6.6	1,580	4,402	N/A	N/A
Damietta branch:					
Talkha Fertilizers	18.8	N/A	1,200	N/A	N/A

Summary and Conclusions

This research should contribute significantly to improve water quality in Egypt by reducing waste generation. However, the success of this study requires support from factories' officials, government, and other stakeholders. The regression equations established between two parameters, can be used to predict the value of unknown parameter if the value of the other parameter is known. This study will help identifying the extent of pollution in the Nile River and the value of physicochemical parameters theoretically, consequently saving time and money.

Some new ideas will be addressed in this research, such as reusing wastewater generated by paints and textile manufacturing industries in the manufacture of cement bricks and tiles. Different samples will be prepared and tested to evaluate the effect of using wastewater on the properties of the bricks and tiles. These testes include compressive strength, bending strength, abrasion strength, and water absorption. TCLP leaching test was also conducted to ensure that the concentrations of heavy metals in the solutions below U.S. EPA TCLP limits.

New technologies may be identified for application to specific industrial sectors, such as using the BioAmp system for the food industry, where this system is very effective in reducing BOD levels. Low-cost techniques to reduce or reuse waste will be highlighted so that the industrial firms can realize short payback periods for implementation in the plant.

LIST OF REFERENCES

Abd-Allah, A.M.A., 1994. "Residual Levels of Organochlorine Pollutants in Fish from Abu-Quir Bay and Idku Lake, Alexandria, Egypt", ***Toxicol. Environ. Chem.***, ***44*** (1–2): 65–71.

Abd-Allah, A.M.A., M.A. Hamed, and M.M. Hassanain, 1999. "Quality Criteria of the River Nile Water, Damietta Branch between Mansoura City and Damietta Province", ***Modelling, Measurement and Control C, 60*** (1): 1.39–1.51, ISSN-07612524.

Abdel-Moati, M.A.R., and A.A. Sammak, 1997. "Man-made Impact on the Geochemistry of the Nile Delta Lakes. A Study of Metals Concentrations in Sediments", ***Water, Air, and Soil Pollution, 97***: 413–429, ISSN-00496979.

Abdel-Sabour, M.F., F.H. Rabie, T. Mostafa, and S.A. Hassan, 2001. "Impact of Industrial Wastewater Disposal on Surface Water Bodies in Mostord Area, North Greater Cairo", ***J. Environ. Sci., 13*** (4): 485–490.

Abdel-Shafy, H.I., and R.O. Aly, 2002. "Water Issues in Egypt Resources, Pollution and Protection Endeavors", ***Central European J. Occupational and Environmental Medicine, 8*** (1): 3–21.

Abdel-Wahaab, R., 1995. "Wastewater Treatment and Reuse; Environmental Health and Safety Consideration", ***Inter. J. Environ. Health Research, 5***: 2–13.

Abou-Elela, S.J., and F. Zaher, 1998. "Pollution Prevention in the Oil and Soap Industry: A Case Study", ***Water Sci. Technol., 38*** (4–5): 139–144.

African Development Bank, 2009. ***Egypt Private Sector Country Profile***. Available on-line at: http://www.afdb.org/fileadmin/uploads/afdb/Documents/Project-and-Operations/Brochure%20Egypt%20Anglais.pdf

Ali, G.H., G.E. El-Taweel, M.M. Ghazy, and M.A. Ali, 2001. "Microbiological and Chemical Study of the Nile River Water Quality", ***Int. J. Environ. Stud., 58*** (1): 47–69.

Ali, M.M., and M.E. Soltan, 1996. "The Impact of Three Industrial Effluents on Submerged Aquatic Plants in the River Nile, Egypt", ***Hydrobiologia***, ***340***(1–3): 77–83, DOI: 10.1007/BF00012737.

Al-Makkawy, H.K., and M.D. Madbouly, 1999."Persistence and Accumulation of Some Organic Insecticides in Nile Water and Fish", ***Resour. Conserv. Recycl., 27*** (1–2): 105–115.

Badawy, M.I., and O.A. Aly, 1986. "PCB Levels in Surface and Waste Waters in Egypt", ***Environment International, 12*** (5): 577–580, ISSN: 01604120.

Badawy, M.I., and M.A. Emababy, 2010. "Distribution of Polycyclic Aromatic Hydrocarbons in Drinking Water in Egypt", ***Desalination, 251*** (1–3): 34–40.

Central Agency for Public Mobilization and Statistics (CAPMAS), 2012. Cairo, Egypt. Available on-line at: http://www.capmas.gov.eg/pages_ar.aspx?pageid=893.

Central Intelligence Agency (CIA), 2011. "The World Factbook", Available on-line at: https://www.cia.gov/library/publications/the-world-factbook/geos/eg.html

El-Gendy, K.S., A.A. Abdalla, H.A. Aly, G. Tantawy, and A.H. El-Sebae, 1991. "Residue Levels of Chlorinated Hydrocarbon Compounds in Water and Sediment Samples from Nile Branches in the Delta, Egypt", ***J. Environ. Sci. Health Part B Pestic. Food Contamin. Agric. Wastes, 26*** (1): 15–36.

El-Otify, A.M., M.M. El-Sheekh, and H. Saber, 2011. "Detrimental Effects of Industrial Effluents on the Growth and Metabolic Products of the Nile Planktonic Cyanobacteria", ***Fresenius Environmental Bulletin***, ***20*** (11): 2911–2919, ISSN-1018-4619.

El-Ghani, M.A., A.M. El-Fiky, A. Soliman, and A. Khattab, 2011. "Environmental Relationships of Aquatic Vegetation in the Fresh Water Ecosystem of the Nile Delta, Egypt", ***African Journal of Ecology***, ***49*** (1): 103–118.

El-Iskandarani, N.S., M. Okbah, and A. Jensen, 2004. "Principal Component Analysis for Quality Assessment of the Mediterranean Coastal Water of Egypt", ***Modell. Meas. Control C., 65*** (1–2): 69–83.

El-Sebae, A.H., 1986. "Egyptian Mediterranean Coastal and Estuarine Pollution by Pesticides and Heavy Metals", ***Water Sci. Technol.***, ***18*** (4–5): 311.

Geriesh, M. H., K. Balke, and A.E. El-Rayes, 2008. "Problems of Drinking Water Treatment Along Ismailia Canal Province, Egypt", ***Journal of Zhejiang University Science B***, ***9***: 232–242, ISSN-1673-1581.

Gerges, A.A.M., and E. Smith, 2004. "Cost/Benefit Analysis of Oily Water Discharge Regulations in Egypt", ***Proc. World Water Environ. Resour. Congr. Crit. Transitions Water Environ. Resour. Manage.***, 3947 – 3956, Salt Lake City, UT, June 27 – July 1).

Hamed, M.A., and K.M. El-Moselhy, 2009. "Levels of Some Metals in the Coastal Water and Sediments of the Red Sea, Egypt", ***Modell Meas. Control C., 61*** (1–2): 43–57.

Higdon, H., 2012. "The Land of the Pharaohs", ***The Beacher Weekly Newspaper***, ***28*** (23), Available on-line at: http://www.thebeacher.com/pdf/2012/BeacherJun14.pdf.

Ibrahim, M.D., T. Puestow, A.A. Khan, and L.M. Lye, 2010. "Satellite Water Quality Monitoring Validation: Case Study of Lake Manzalah, Egypt", ***Annual Conference of the Canadian Society for Civil Engineering 2010, 1***: 138–147, ISBN-13: 9781617389061.

Lasheen, M.R., and A.A. Ashmawy, 1983. "Selected Heavy Metals in Fishes from the Nile River", ***Internat. Conf. – Heavy Metals in the Environment***, 1174–1177.

Melegy, A.A., M.A. Ahmed, and G.A.E. Mostafa, 2002. "Environmental Impacts on Water and Sediments of the River Nile in Greater Cairo (near Helwan City)", Modelling, ***Measurement and Control C***, ***63*** (3–4): 41–53, ISSN-07612524.

Moursy, A.S., 1983. "Oil Pollution Studies in the Nile River. I. Survey of Oil and Grease in Nile River", ***Environ. Int.***, ***9*** (2): 107–111.

Myllylä, S., 1995. "Cairo - A Mega-City and Its Water Resources", Paper presented at the third Nordic conference on Middle Eastern Studies: Ethnic encounter and culture change, Joensuu. Available on-line at: http://www.hf.uib.no/smi/paj/myllyla.html.

Nile Basin Initiative, 2005. "Nile Basin National Water Quality Monitoring Baseline Study Report for Egypt", Nile Trans boundary Environmental Action Project. http://nile.riverawarenesskit.org/English/NRAK/Resources/Document_centre/WQ_Baseline_report_Egypt.pdf.

Ramirez, C.E., N.P. Mijaylova, and T.E. Diaz, 2004. "Treatment Techniques for the Recycling of Bottle Washing Water in the Soft Drink Industry", ***Wat. Sci. Tech, 50, 2***:107–112.

Saad, M.A.H., A.I. Beitagy, and W.M. Mahmoud, 2003. "Total Dissolved and Particulate Lead in the Western Harbor of Alexandria, a Mediterranean Basin under Stress", ***Mar. Pollut. Bull.***, *47*(1–6): 52–58.

Saad, M.A.H., M.M. Ebu Elamayam, A.H. El-Sabae, and I.F. Sharaf, 1982. "Occurrence and Distribution of Chemical Pollutants in Lake Mariut, Egypt. I. Residues of Organochlorine Pesticides", ***Water Air Soil Pollut.***, *17*(3): 246–252.

Soltan, M.E., and R.M. Awadallah, 1995. "Chemical Survey on the River Nile Water from Aswan into the Outlet", ***Journal of Environmental Science and Health. Part A: Environmental Science and Engineering and Toxicology***, ***30*** (8): 1647–1658, ISSN-1077-1204.

Soltan, M.E., 1995. "Effect of Kima Drain Wastewaters on Nile River Waters", ***Environment International***, ***21*** (4): 459–464, ISSN-0160-4120.

Soltan, M.E., 2006. "Remobilization of Selected Metal Ions from Nile Sediment (Egypt) According to Sequential Extraction and Metal–EDTA Complex", ***Chemistry and Ecology***, ***22*** (5): 359–378, ISSN-0275-7540.

Tomalavage, A.M., W.D. Sitman, and M. Ramanathan, 1984. "Industrial Waste Treatment in Egypt: Case Histories", ***Proc. Mid-Atlantic Ind. Waste Conf.***, ***16***: 147–157.

Wahaab, R.A., and M.I. Badawy, 2004. "Water Quality Assessment of the River Nile System: An Overview", ***Biomedical and Environmental Sciences***, *17*: 87–100, ISSN-0895-3988.

Yamashita, N., S. Masunaga. M.S. Rizk, and Y. Urushigawa, 1997. "Vertical Problems of Organochlorine Pesticides in Sediment Cores from Nile River and Manzala Lake, Egypt", ***Toxicol. Environ. Chem.***, ***58*** (1–4): 151–161.

Yousef, R.M., and M.L. Sakr, 2007. "Desalination Technology Roadmap 2030", (January), http://www.idsc.gov.eg/Upload/Documents/28/EN/Desalination_technology_Roadmap%5B1%5D.pdf.

(672e) Moringa Oleifera Tree for Producing Drinkable Water: Quantifying Biochemical Changes within the Seeds

Darrell Velegol[1], Stephanie B. Velegol[2], Danny Hoover[1], Georgia Karmee[1], Benjamin Kutz[2] and Kamil Siniakowicz[1], (1)Chemical Engineering, Penn State University, University Park, PA, (2)Civil Engineering, Penn State University, University Park, PA

The seeds of the Moringa oleifera tree have been used for centuries to produce drinkable water in the developing world. Simply crushing the seed into water releases a cationic, antimicrobial peptide (AMP) that flocculate turbidity and penetrate and disrupt the membranes of many pathogens. Previously, we have been able to demonstrate that when the AMP from the seed is adsorbed onto sand to create functionalized sand ("f-sand"), that the AMP retains its effectiveness both for turbidity removal and killing of certain pathogens. However, anecdotal evidence also indicates that these seeds work only when they are dried on the tree, pointing to a biochemical change within the seed itself as it dries. We quantify this change by testing the ability of the seeds at various stages of growth to remove turbidity and inactivate pathogens. We also compare turbidity and disinfection results of the active peptide dissolved in bulk solution with those from the active peptide adsorbed onto carrier surfaces.

(694a) Use of an Autosorb System for the Measurement of Diffusion of CO2 in an Ion-Exchanged Zeolite Rho

Enzo Mangano[1], Stefano Brandani[1] and Matthias Thommes[2], (1)Institute for Materials and Processes, School of Engineering, University of Edinburgh, Edinburgh, United Kingdom, (2)Quantachrome Instruments, Boynton Beach, FL

The Quantachrome Autosorb-iQ™ is a state-of-the-art volumetric system for the measurement of BET surface areas, micropore volumes and adsorption isotherms of gases on solid adsorbents. It comes with a library of software tools which will interpret automatically the experimental results and yield the required equilibrium properties.

As part of the Adsorption Research Industrial Consortium at the University of Edinburgh, we are interested in expanding the use of commercial equipment to functionalities for which the systems can be used in a non-conventional way. We present the use of the Quantachrome Autosorb-iQ™ for the measurement of diffusivities in a Zeolite Rho sample.

To interpret the dynamic response of the system we developed an analytical solution to a model which is adapted from the theoretical model developed by Brandani in 1998 for the analysis of the piezometric method. Due to the particular configuration of the instrument, modifications to this simplified model were applied in order to take account of the real experimental conditions. The results on carbon dioxide on Zeolite Rho, for which we have independent values measurements carried out using the ZLC technique, confirm the validity of the methodology in the analysis of slow diffusion processes.

We would like to acknowledge the financial support from the Engineering and Physical Sciences Research Council under grant EP/F034520/1 (Innovative Gas Separations for Carbon Capture) and M.M. Lozinska and P.A. Wright of the University of St. Andrews for providing the zeolite sample used in this study.

(694b) Experimental Characterization of Adsorbents Via Volumetric Frequency Response Analysis

Mohammad I. Hossain[1], Armin D. Ebner[2] and James A. Ritter[2], (1)University of South Carolina, Columbia, SC, (2)Department of Chemical Engineering, University of South Carolina, Columbia, SC

Frequency response methods have been developed as an alternative technique for the determination of mass transfer mechanisms controlling the transport of gases into and out of adsorbent materials. Recent results in the literature are showing that this technique seems to be the best for discriminating the controlling mechanism. It is also possible to reveal two or more different controlling mechanisms that may dominate in different frequency ranges. The present talk will focus on the experimental methodology for determining in and out of phase responses using the volumetric frequency response approach, which is most typically carried out with the use of a bellows connected to a reciprocating moving shaft in constant volume system. The newly constructed and commissioned experimental apparatus will be described in detail. The analysis and results will be discussed in light of experimental data obtained with CMS 172K with O_2, Ar, CH_4, N_2 and CO_2 at 1 atm and four different temperatures.

(694c) Retention Models and Interaction Mechanisms of Acetone and Related Molecules with Amylose Tris[(S)-α-Methylbenzylcarbamate] Sorbent

Hung-Wei Tsui, Margaret Hwang, Yiwen Chen, Lei Ling, **Elias I. Franses** and Nien-Hwa Linda Wang, School of Chemical Engineering, Purdue University, West Lafayette, IN

This polymeric sorbent is capable of separating chiral enantiomer solutes, such as benzoin.* HPLC results show a substantial effect of the solvent or mobile phase composition on the retention factors k_R and k_S and the enantioselectivity S= k_R/ k_S. A typical solvent is n-hexane containing isopropanol (IPA) with IPA volume fractions =0.01 to =0.1. The retention factor dependence on follows the empirical relationship ln (k) =A-B ln (1), where the intercept A and the slope B are dimensionless constants. Values of B range from 0.4 to 0.8 for benzoin and related chiral solutes. These slopes are related to solute-sorbent, IPA-sorbent, and IPA-solute interactions. Because such interactions are quite complex for chiral solutes, several simple non-chiral solutes, such as acetone and phenylacetaldehyde, each containing one C=O functional group which can bind with the NH group of the AS sorbent, were examined. The retention factors for these solutes also follow Eq. (1), with slopes B ranging from 0.25 to about 0.4. Thermodynamic models reveal that the above three types of interactions alone cannot lead to slopes B below 1.0. Chromatography simulations, which take into account competitive adsorption of solute and IPA and IPA-solute binding, support this conclusion. Our hypothesis for explaining values of the slopes B below 1 involves accounting also for substantial IPA aggregation in solution. This phenomenon leads to a reduction in the IPA thermodynamic activity, which affects IPA adsorption and IPA-solute interactions. Chromatography simulations with added IPA aggregation support this hypothesis, which is confirmed further by Infrared Spectroscopy (IR) data and Density Functional Theory (DFT) simulations of the IPA-IPA and IPA-solute interactions. The results apply also to the data of the chiral solutes.

* H.-W. Tsui, J.N. Willing, R.B. Kasat, N.-H.L. Wang, E.I. Franses, J. Phys. Chem. B, 2011, 115, 12785-12800.

(694d) Flux Response Technology (FRT) Applied in Zero Length Column Diffusivity Measurements

Ayodeji Sasegbon, Imperial College London, London, United Kingdom and **Klaus Hellgardt**, Chemical Engineering, Imperial College London, London, United Kingdom

Intracrystalline (or intraparticle) diffusion within porous materials is widely studied for its useful and important applications in molecular separation, heterogeneous catalysis, membrane technology, fuel cells, soil mechanics and petroleum engineering to name but a few. The well known automotive catalytic converter makes use of a catalytic monolith reactor in a 'honeycomb' configuration to oxidise carbon monoxide and unburned hydrocarbons while reducing oxides of nitrogen. Understanding the intracrystalline diffusion that occurs within the washcoat and substrate material of monolith supported catalysts is vital to the modeling and scale-up of reactors and consequently affects both performance and cost effectiveness of such systems.

The purpose of this paper is to report on the measurement of the intracrystalline diffusivity in the washcoat of monolith samples using Flux Response Technology (FRT). In previous studies, we have shown the versatility of the FRT method in measuring *in situ* adsorption, reaction and desorption processes in flow reactors. The FRT apparatus is an extremely sensitive tool which allows the measurement of flowrate changes of the order of 10^{-2} μl/min. By applying the Zero Length Column (ZLC) method to the analysis of concentration perturbation induced ad- and desorption transients recorded by the FRT technique, accurate rates of diffusion within washcoats can be arrived at.

Shown in Figure 1a is an example of a flux response profile obtained for a cordierite sample coated with a washcoat of alumina/CeZrOx when subjected to repeated concentration perturbations of propane in helium. For the investigation of diffusivities within the washcoat of the cordierite samples, it is necessary to monitor the release of the adsorbed species (C_3H_8 in this instance) into the effluent stream in order to apply the ZLC analysis. The signal of the DPT pressure responses in the flux response profile have been shown to be directly proportional to the amount of probe molecules ad/desorbed onto or from the surface of the porous adsorbent.

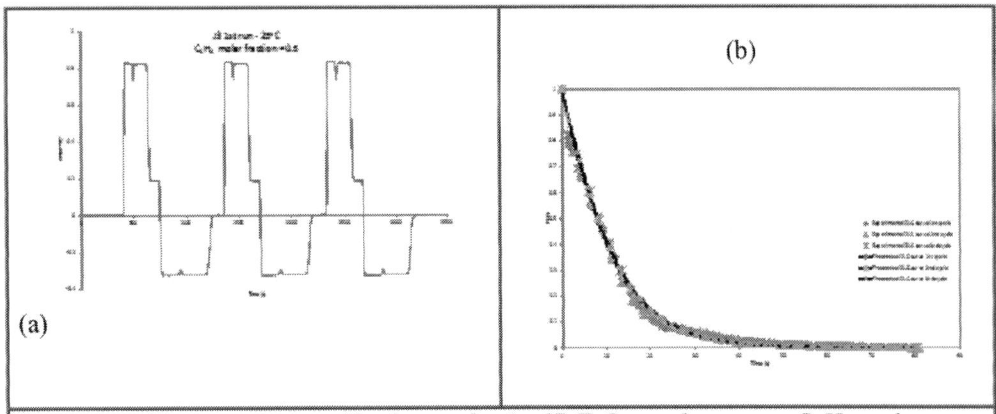

***Figure 1. a.** Flux response profiles for an alumina/CeZrOx washcoat at a C_3H_8 mole fraction of 0.5 showing three sorption cycles.* **b.** *Experimental and fitted desorption curves of C_3H_8 (mole fraction of 0.5) from alumina/CeZrOx washcoat sample at 25 °C*

ZLC response curves (nonlinear regression) were fitted to experimental, dispersion corrected FRT data (shown in Figure 1b) and allow the extraction of effective diffusivities of propane within the washcoat. At 25^oC an effective diffusivity of $7.04 \times 10^{-10}\ m^2/s$ was determined for the above sample which agrees well with published data measured by other macroscopic techniques. In particular, it is reassuring to observe that the calculated activation energy for propane diffusion in alumina/CeZrOx compares well to that predicted by Granato *et al.* based on theoretical molecular dynamics simulations of propane diffusion in 13X, and pulse field gradient nuclear magnetic resonance (PFG-NMR) data by Schwartz *et al.* The results also compare well to the values of Sun *et al.* who employed a Wicke-Kallenbach type method using a zeolite

o -10

membrane and report diffusivities of propane in silicalite in the temperature range between 30 to 70 *C* as 4-6 10 m^2/s (Table 1).

It has been observed that propane diffusivity in 13X is much smaller when measured with the ZLC approach as compared to PFG-NMR and it has been suggested that this may be due to the PFG-NMR technique returning intra-crystalline diffusivities whereas ZLC experiments result in the observation of combined diffusion and desorption processes.

We will show that with FRT these two processes can be compared and contrasted in one experiment by analysing the dispersion corrected adsorption and desorption transients in the FRT profile.

Table 1. *Comparison of Diffusivities and Activation Energies*

Reference	Method	Temperature (K)	D (m^2s^{-1})	E_a (kJ/mol)
Granato *et al.* (2010)	MD simulations	373	2.9×10^{-9}	7.9
Schwartz *et al.* (1995)	PFG-NMR	307	7.0×10^{-10}	7.9
Sun *et al.* (1996)	Wicke-Kallenbach	298-334	5.0×10^{-10}	8.0
This work	FRT-ZLC	373	1.4×10^{-9}	6 1

(694e) Measurements and Modeling of in-Situ Gas Densities and Adsorption Isotherms for CO2 On Wet Coals

Sayeed A. Mohammad, Pongtorn Chareonsuppanimit, Agelia Abudour, Robert L. Robinson Jr. and Khaled A. M. Gasem, School of Chemical Engineering, Oklahoma State University, Stillwater, OK

High-Pressure gas adsorption isotherms on wet coals are typically measured without the aid of an *in situ* gas density meter. As a result, the densities of bulk gas-phase are inferred either volumetrically or gravimetrically with the assumption that the gas-phase free from water vapor (or dry). However, the effect of the presence of water vapor in the gas-phase for adsorption isotherms of a supercritical gas such as CO_2 may be an over-simplification and may lead to significant errors in the data reduction of the adsorption isotherms. To test this hypothesis, a vibrating tube density meter was installed within a high-pressure adsorption apparatus to measure the *in situ* gas densities for CO_2 isotherms on wet coals. Further, a new data reduction method was implemented that utilizes the accurate measurements of wet gas densities rather than the densities of pure CO_2 (or dry gas density). In this manner, the adsorption isotherm measurements were aided by the inclusion of a gas-phase density meter.

Two data reduction approaches can be formulated and applied for adsorption isotherms based on these new measurements. The two approaches differ in the manner in which the effect of moisture on gas adsorption is taken into account. The two approaches and their relative merits will be presented. Further, the adsorption isotherm data obtained from the two approaches were modeled with the simplified local density (SLD) adsorption model. An analysis and interpretation of the SLD model results based on these two approaches will also be presented for these systems.

(694f) Pure and Binary Adsorption of 1,2,4 Tri Methyl Benzene and 1,3 Di Isopropyl Benzene On Alumina

Kevin F. Loughlin and Salam Taji, Chemical Engineering, American University of Sharjah, Sharjah, United Arab Emirates

ABSTRACT

The adsorption of large molecules on adsorbents is very limited in the adsorption literature. In this study we report on the adsorption of pure compounds 1,2,4 tri methyl benzene and 1,3 di isopropyl benzene on Alumina, followed by the adsorption of different combinations of mixtures of the components.

A new apparatus and experimental technique has been designed for this study. An adsorption column 12 inches in length and 1 inch diameter is filled with AA-300 2.5 X 4 mesh Alcan Activated Alumina and placed in an oven. Two helium lines are connected to the column. One helium line is passed through a 0-50 ml flowmeter, thence through a bubbler of either liquid the exit of which passes through a gas coalescer in preparation for entry into the column. The bypass helium line is fitted with a microvalve controller for flow adjustment, combined with the first helium line and then both streams enter the column as one. The outcome of the column is monitored with a PROLAB mass spectrometer.

The Alumina column is activated at 270 °C for 6 hours by passing helium only using the bypass line. After activation the column is isolated and the oven temperature adjusted to the isotherm temperature. For isotherm measurement, the bubbler flow is minimized and monitored and the bypass helium line flow is maximized. Fifteen minutes are used for coalescer wetting during which the outlet is not fed to the column. Then the system is opened to the alumina column and mass spectrometer and the starting time noted. The flow is fed to the column until steady state is achieved with the outlet pressure being recorded. The bypass flow is reduced by 2 units on the microvalve, and additional loading enters the column until a new equilibrium pressure is established. This procedure is continued for 5 to 7 steps until the bypass flow is turned off. All the time the flow through the bubbler is maintained constant and continuously monitored. The final outlet recorded pressure is the equilibrium pressure from the bubbler. From mass balances at each step, the equilibrium loading at the various pressures can be established after allowance for a dead volume calibration. The system is regenerated in preparation for the next isotherm measurement. For the measurement of binary isotherms a second bubbler is employed and a similar procedure employed..

The system will be modeled using the isotherms employed by Brandani et al. (2003).

Reference

Brandani, F., D. M. Ruthven, and C. G. Coe. "Measurement of adsorption equilibrium by the zero length column (ZLC) technique. Part 1: Single component systems." *Ind. Eng. Chem. Res.*, 2003: 1451-1461.

KEYWORDS:, pure isotherms, binary isotherms, 1,2,4 tri methyl benzene, 1,3 di isopropyl benzene Alumina,

(694g) Nonlinear Direct Inverse Method: A Way to Combine Nonlinear Calibration with Isotherm Determination

Monzure-Khoda Kazi[1], Bijan Medi[2] and Mohammad Amanullah[2], (1)School of Chemical and Biomedical Engineering, Nanyang Technological University , Singapore, Singapore, (2)School of Chemical and Biomedical Engineering, Nanyang Technological University, Singapore, Singapore

AIChE Annual Meeting 2012- Pittsburg

Separations Division

Adsorption and Ion Exchange(02E)

Nonlinear direct inverse method: a way to combine nonlinear calibration with isotherm determination

Monzure-Khoda Kazi , Bijan Medi, Mohammad Amanullah*

School of Chemical and Biomedical Engineering
Nanyang Technological University

62 Nanyang Drive, Singapore 637459

Singapore

(aman@ntu.edu.sg)

This work addresses a way to combine nonlinear calibration with isotherm determination. In this method, like classical inverse method, experimental elution profiles are compared with the results of a detailed model that accounts for nonlinearity in equilibrium, axial dispersion, and mass transfer kinetics. However, unlike classical inverse method, the calibration of detector is carried out simultaneously with isotherm determination thereby reducing cost and saving time. In this study no limitation is imposed on the linearity of the detector signal or on the overlapping of elution profiles in case of enantiomers. The method has been experimentally verified for the separation of a mixture of guaifenesin enantiomers over a wide range of operating parameters.

(697a) Hydroxide Conducting Polymerized Ionic Liquid Block Copolymers for Alkaline Fuel Cells

Yuesheng Ye, Department of Chemical and Biological Engineering, Drexel University, Philadelphia, PA, Tsen-Shan Wang, University of Pennsylvania, Philadelphia, PA, Karen I. Winey, Materials Science and Engineering, University of Pennsylvania, Philadelphia, PA and **Yossef A. Elabd**, Chemical and Biological Engineering, Drexel University, Philadelphia, PA

In previous work in our laboratory, we reported on the chemical stability of an imidazolium-based hydroxide conducting polymerized ionic liquid (PIL), poly(1-[(2-methacryloyloxy)ethyl]-3-butylimidazolium hydroxide) (poly(MEBIm-OH), over a broad range of humidities, temperatures, and alkaline concentrations using the combined techniques of electrochemical impedance spectroscopy and nuclear magnetic resonance spectroscopy [1]. This polymer provides a good building block for future investigations since chemical stability of hydroxide conducting polymers is a significant issue as it pertains to long-term alkaline fuel cell (AFC) operation. In this talk, we will present the hydroxide conductivity and morphology of a series of PIL diblock copolymers, poly(MMA-*b*-MEBIm-OH), composed of the IL monomer, MEBIm-OH, and the non-ionic monomer, methyl methacrylate (MMA), at various PIL compositions. PIL block copolymers conjoin the advantages of block copolymers and PILs: self-assembled nanostructured morphologies and unique electrochemical properties. To date, there are few reports of PIL block copolymers in the literature. Therefore, a fundamental understanding of morphology and ion transport in these materials will be highly beneficial, specifically for the optimization of AFC performance.

1. Ye, Y.; Elabd, Y.A. Relative Chemical Stability of Alkaline Exchange Polymerized Ionic Liquids. *Macromolecules* **2011**, *44* (21), 8494-8503.

(697b) Hydroxyl Functionalized Graphene: A Promising Anhydrous Proton Exchange Membrane

Pabitra Choudhury and J. Karl Johnson, Chemical and Petroleum Engineering, University of Pittsburgh, Pittsburgh, PA

We predict that graphene functionalized with OH groups can be used as an efficient proton exchange membrane capable of functioning in the complete absence of water. Fast anhydrous proton transport takes place on the surface of functionalized graphene as a result of the nanostructured spacing of the functional groups. We have considered two model membranes to test our hypothesis: (1) a membrane consisting of graphane (fully hydrogenated graphene) where a one-dimensional (1-D) chain of the hydrogen atoms has been replaced by OH groups (hydroxyl graphane), as shown in Figure 1, and (2) a membrane consisting of graphane completely saturated with OH groups on one side (i.e., every other C atom) of the graphane. This membrane forms a two dimensional (2-D) hydrogen bonded network, as shown in Figure 2. The fully hydroxylated structure has the advantage of providing redundant pathways for proton diffusion. However, the saturated OH coverage on the 2-D surface could reduce the proton diffusivity. We have used first principles density functional theory to study the structural, energetic, and transport properties of model membranes. Our calculations predict that the proton mobility on the surface of a 1-D hydrogen bonded graphane membrane will be close to the diffusivity of H^+ in bulk water and about 10 times faster than through Nafion at room temperature. In contrast, the proton diffusivity along the surface of a 2-D hydrogen bonded graphane membrane is about two orders of magnitude slower than the estimated diffusivity of proton across the 1-D hydrogen bonded chain of graphane membrane. We predict that functionalized graphene proton exchange membranes could operate over a wide temperature range, making them more efficient than membranes requiring hydration.

Acknowledgement: We gratefully acknowledge the support by DTRA under Contract No. HDTRA1-09-1-0008. We also gratefully acknowledge NSF XSEDE (TeraGrid) resources under allocation numbers TG-DMR100097, TG-DMR110091 and TG-SEE090006. We thank the Center for Simulation and Modeling at the University of Pittsburgh for providing computational support.

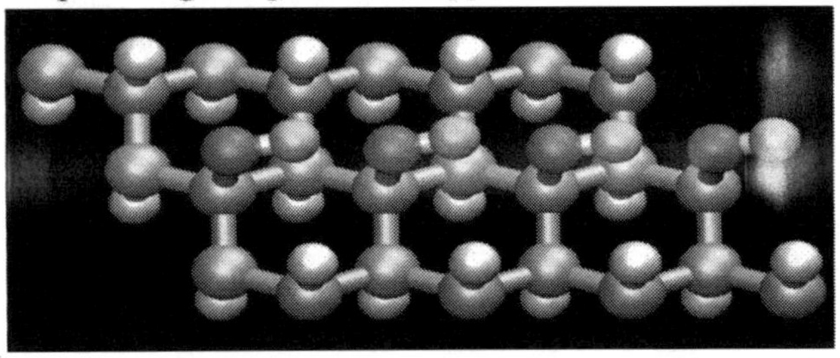

Figure 1. Functionalized 1-D hydrogen bonded hydroxyl graphene (sp^3 like structure). The cyan color represent C, white color represent hydrogen atom, and red-green combination represent O and H in OH groups, respectively.

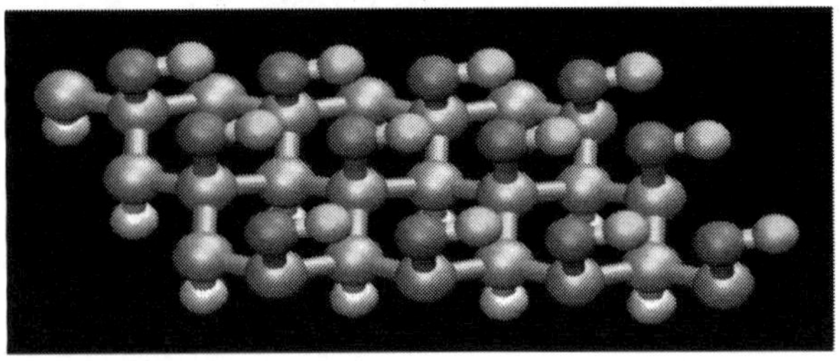

Figure 2. OH saturated 2-D hydrogen bonded graphane membrane. The cyan color represent C, white color represent hydrogen atom, and red-green combination represent O and H in OH groups, respectively.

(697c) Gas Permeation in Perflurosulphonated Membranes: Influence of Temperature and Relative Humidity

Marco Giacinti Baschetti[1], Matteo Minelli[2,3], Jacopo Catalano[4] and Giulio Sarti[1], (1)Chemical Engineering, Mining and Environmental Technology, University of Bologna, Bologna, Italy, (2)University of Bologna, Bologna, Italy, (3)Advanced Applications in Mechanical Engineering and Materials Technology Interdepartmental Center for Industrial Research, CIRI-MAM, Bologna, Italy, (4)Worcester Polytechnic Institute

Gas permeation in Perflurosulfonated membranes: influence of temperature and relative humidity

Marco Giacinti Baschetti, Matteo Minelli, Jacopo Catalano[1], Giulio C. Sarti

Dipartimento di Ing. Chimica, Mineraria e delle Tecnologie Ambientali (DICMA)

Alma Mater Studiorum - Università di Bologna, via Terracini 28, 40131, Bologna, Italy.

Perfluorosulfonated ionomer (PFSI) are of high interest in energy application due to their ability to be used as polyelectrolyte membranes (PEM) inside hydrogen or methanol based fuel cells (FC).[1,2] In fuel cells (FC) applications, the polyelectrolyte membranes (PEM) are exposed to different gases such as hydrogen, oxygen, nitrogen and carbon dioxide, depending on the type of FC considered, at varying operative conditions, both in terms of temperature and relative humidity. For these reasons, the knowledge of gas permeability of different gases in such materials as a function of these parameters is of interest for a proper description and modeling of the system. Different studies appeared in the literature addressing these issues [3-10], however systematic investigation of these phenomena are relative scarce and a thorough characterization is still missing for many of the PSFI now available.

In the present study, gas permeability in two different PFSI, namely Nafion® and Aquivion®, have been investigated as a function of temperature and relative humidity. In particular, permeation of gases such as O_2, N_2 and CO_2 and also Helium, used to mimic H_2, have been studied from 25 to 50°C and at relative humidity up to 80%.

The results showed that both materials posses very similar permeability values and a common behavior with respect to temperature and relative humidity. The permeability of the different gases in the two ionomers increases as the temperature or the humidity increase, the latter parameter having a really dramatic effect on the gas flow rate across the membrane. The measured permeability indeed showed increases of up to two orders of magnitude when humidified gas were flown in the system.

The effect of water on the membrane transport properties is rather complex since the permeability has a sharp increase already at very low humidity, and then shows a more gentle, exponential rise with RH that seems to lead towards values comparable with those of gas permeability in pure water.

The comparison between the two materials does not show a clear trend; in general, however, Aquivion seems to have higher permeability than Nafion at low relative humidity and at the lower temperatures inspected, while Nafion shows higher permeation rates at larger R.H. and high temperatures.

These findings can be related to the fact that at a given activity Nafion in general sorbs less water than Aquivion, due to the higher equivalent weight. Such difference seems to be significant at low activity, where the water content is still low. It becomes then less important at higher activity where interconnected hydrophilic domains are formed in both membranes.

The temperature dependence follows an Arrhenius type of behavior for all gases in the two polymers, as expected for permeation processes substantially related to a solution diffusion mechanism. The activation energies were calculated for both materials as a function of R.H, and showed higher values in Nafion than in Aquivion, in agreement with the permeability behavior described above.

In all cases, for both materials the activation energy ranges between the value of the dry polymer and that obtained for the gas permeation in pure polymer. In the case of Nafion, a well defined trend of permeability with activation energy can be found with water activity, at least in the case of nitrogen and oxygen. In the case of Aquivion, on the other hand, no clear trend was detected.

All the experimental evidences suggest that gas permeation in humid PFSI is driven by gas sorption and diffusion in the hydrophilic domains of the materials and ultimately in the water channels which are usually considered to form in the hydrated matrix. This hydrated phase indeed has a largely higher gas permeability than that of the hydrophobic phase and controls the overall permeation process.

References:

1. Fuel Cell Handbook (7th Edition), EG&G Technical Services, Inc. (2004) Contract No. DE-AM26-99FT40575

2. Marcinkoski, J., Kopasz, J., & Benjamin, T. Progress in the US DOE fuel cell subprogram efforts in polymer electrolyte fuel cells. Int. J. Hydrogen Energy, 2008;33: 3894.

3. Sakai T, Takenako H, Wakabayashi N, Kawami Y, Torikai E. Gas permeation properties of solid polymer electrolyte (SPE) membranes. J Electrochem Soc 1985;132:1328

4. Ogumi Z, Takehara Z, Yoshizawa SI. Oxygen permeation through Nafion and NEOSEPTA. J Electrochem Soc 1984;131:769.

5. Pellegrino J, Kang KS. CO2/CH4 transport in polyperfluorosulfonate ionomers: effects of polar solvents on permeation and solubility. J Membr Sci 1995;99:163.

6. Buchi FN, Wakizoe M, Srinivasan S. Microlectrode investigation of Oxygen permeation in perfluorinated proton exchange membranes with different equivalent weights. J Electrochem Soc 1996;143:927

7. Ma S, Odgaard M, Skou E. Carbon dioxide permeability of proton exchange membranes for fuel cells. Solid State Ionics 2005;176:2923.

8. Mohamed HFM, Ito K, Kobayashi Y, Takimoto N, Takeoka Y, Ohira A. Free volume and permeabilities of O2 and H2 in Nafion membranes for polymer electrolyte fuel cells. Polymer 2008;49:3091.

9. James Cw A, RoyMcGrath JE, Marand E. Determination of the effect of temperature and humidity on the O2 sorption in sulfonated poly(arylene ether sulfone) membranes. J Membr. Sci. 2008;309:141.

10. Catalano J., Myezwa T., De Angelis M.G., Giacinti Baschetti M., Sarti G.C. The effect of relative humidity on the gas permeability and swelling in PFSI membranes Int. J. Hydrogen Energy 2012;37:6308

---

[1] Present address: Center for Inorganic Membrane Studies, Department of Chemical Engineering, Worcester Polytechnic Institute, 100 Institute Rd, Worcester, MA 01609, USA.

(697d) Pore Filling Anion Exchange Membranes and Membrane Electrode Assemblies for Solid-State Alkaline Fuel Cells

Takeo Yamaguchi[1], Han Zhang[2], Hyangmi Jung[1], Hidenori Ohashi[2] and Takanori Tamaki[2], (1)Chemical Resources Laboratory, Tokyo Institute of Technology, Kanagawa, Japan, (2)Chemical Resources Laboratory, Tokyo Institute of Technology, Yokohama, Japan

In solid-state alkaline fuel cells (SAFCs), stability improvement of the membranes and water management of the MEAs are essential to achieve good performances. To improve the stability, suppressing water mobility by porous substrate matrix leads to improve the membrane durability. The pore-filling membranes exhibited higher stability compared with normal cast films made of the same polymer. Next, water management is important for SAFCs because water reacts at the cathode and is generated at the anode, in contrast to proton-exchange membrane fuel cells (PEMFCs). Thus, optimization is required to establish design methodology for SAFCs with pore-filling anion exchange membranes. Water movement in a SAFC system has been observed accurately using a water collection method. We confirm for the first time the movement of water from the anode to the cathode during an actual fuel cell performance test. Compared with a PEMFC, water movement occurs in the opposite direction leading to difficult water management in the SAFC. With the knowledge obtained, controlling membrane performances and careful control of humidity have achieved water management in the SAFCs.

(697e) Hybrid Proton-Conducting Membranes Doping with Various Functionalized Titania for Direct Methanol Fuel Cells

Hong Wu, Zhongyi Jiang, Tao Xu, Weiqiang Hou and Xiaohui Shen, Key Laboratory for Green Chemical Technology of Ministry of Education, Sch Chem Engn & Technol, Tianjin Univ, Tianjin, China

Organic-inorganic hybrid proton-conducting membranes have gained tremendous attention in the field of direct methanol fuel cells (DMFCs). The hybrid membranes combine the intrinsic physical and chemical properties of both inorganic and organic components. A variety of polymers have been doped with various functionalized inorganic particles or inorganic phase to achieve high proton conductivity, low methanol crossover, good mechanical and chemical properties. The tuning of the fuctionalized groups on the inorganic fillers, which could provide additional pathways for proton transfer, is a key issue in fabricating hybrid membranes. Chelation grafting of functional groups onto titania allows facile and fine modifications of the inorganic surface. Titania submicrospheres grafted with phosphonic acid groups (-PO_3H_2), sulfonic acid groups (-SO_3H), carboxylic acid groups (-CO_2H), amino groups (-NH_2) and amino acid groups (-CO_2H, -NH_2) were respectively incorporated into chitosan (CS) or sulfonated poly(ether ether ketone) (SPEEK) membrane matrix to fabricate hybrid membranes with improved proton conductivity. The conductivities at room temperature and 100% RH of the chitosan/phosphonic acid-modified titania membrane, the SPEEK/sulfonic acid-modified titania membrane, the SPEEK/carboxylic acid-modified titania membrane and the SPEEK/amino acid-modified titania membrane were 0.011, 0.053, 0.071 and 0.066 S cm^{-1}, respectively, with 15 wt% of fillers. Besides, due to the different proton acceptor/donor capability of functional groups, the hybrid membrane embedded with amino-functionalized titania exhibited higher values at elevated temperature of 60 ^oC compared with carboxylic-functionalized titania. And the hybrid membrane embedded with amino acid-functionalized titania displayed a superior proton-conducting ability with the proton conductivity of 0.258 S cm^{-1} at 60 ^oC. In addition, SPEEK/carboxylic acid-functionalized titania and SPEEK/amino acid-functionalized titania membrane hybrid membranes were also fabricated via in situ sol-gel method and their proton conductivities were 0.063 and 0.047 S cm^{-1}, respectively, nearly three times higher than the control membranes. In conclusion, all the hybrid membranes exhibited enhanced proton conductivity, low methanol crossover, better mechanical and chemical properties, and are promising candidates for DMFCs.

(706a) Responsive Ultrafiltration Membranes Composed of N-Isopropylacrylamide (NIPAAM), Superparamagnetic Iron Oxide (SPIO) Nanoparticles and Cellulose Acetate

Sneha Chede[1], Geoffrey D. Bothun[2] and Isabel C. Escobar[1], (1)Chemical and Environmental Engineering, The University of Toledo, Toledo, OH, (2)Department of Chemical Engineering, University of Rhode Island, Kingston, RI

Membrane fouling occurs when there is reversible or irreversible attachment of macrosolutes present in the water to the membrane surface. Membrane replacements due to the fouling can add significantly to the operating costs of the membrane system. Reversible fouling can be minimized by crossflow operation, backflushing during filtration and/or by chemical pretreatment, but it may take time and energy and as well as have to meet safety guidelines with respect to chemicals used. On the other hand, irreversible fouling cannot be corrected by backflushing or crossflow filtration, and it results into permanent flux decline. Among irreversible foulants, natural organic matter (NOM) is considered to be a major contributor. NOM is composed of wide range of hydrophilic and hydrophobic components; hence, any static hydrophobic or hydrophilic membrane can get fouled. Therefore, a dynamic membrane able to alternate between being hydrophobic and hydrophilic is hypothesized to decrease. Stimuli-sensitive polymers have been gaining attention in recent years due to their unique property that the polymer changes its conformation from a coiled and hydrophobic state to a uncoiled (or globular) and hydrophilic state in the presence of a stimulus. The stimulus can be pH, temperature, ionic strength or electric or magnetic fields. By alternating between the phases membrane surface can be made dynamic.

The purpose of this study is to develop stimuli responsive membranes to control fouling using membrane casting dopes made of cellulose acetate with N-isopropylacrylamide (NIPAAM). NIPAAM is a stimuli-responsive polymer, which offers the potential to collapse or expand the membrane with changes in temperature. As a respond to a temperature decrease, this polymer expands into a hydrophilic state, while a temperature increase causes it collapse into a hydrophobic state. The phase change arises from the existence of a lower critical solution temperature (LCST) such that the polymer precipitates from aqueous solution as the temperature is increased. By continuously activating the film, it is hypothesized that a dynamic stimuli-responsive surface will prevent foulants from attaching to the membrane surface.

However, alternating between hot and cold temperatures to stimulate the polymer is energy intensive, and during heating and cooling, the response time within the membrane is hindered by the heat transfer resistances. These problems are hypothesized to be eliminated if the heating is localized and intensified within the membrane matrix by embedding small superparamagnetic iron oxide (SPIO) nanoparticles within temperature sensitive membranes. SPIO nanoparticle heating is caused by Neel and Brownian relaxations. A temperature increase due to Neel relaxations occurs by the reorientation of magnetic moment of the particles after being rotated by an external AC field. Brownian relaxation contributes due to the SPIO nanoparticles in a liquid. Thus, temperature activation can be triggered through alternating electromagnetic heating of the embedded SPIO nanoparticles.

(706b) Hydrodynamic Permeability of Nanoparticle Modified, Thin Film Composite Membranes

Luis Escobar-Ferrand and Christopher J. Durning, Chemical Engineering, Columbia University, New York, NY

Hydrodynamic Permeability of Nanoparticle Modified, Thin Film Composite Membranes

Luis Escobar-Ferrand (1, 2); Christian Lubombo (1); Amrut Biswal (1); Ji Seung Kim (1) and Christopher J. Durning (1)

(1) Department of Chemical Engineering, Columbia University, New York, NY 10027

(2) Pall Corporation, Port Washington, NY 11050

Layer by Layer (LbL) deposition of primarily inorganic silica nanoparticles enabled surface modification of polymeric micro and ultrafiltration (MF/UF) membranes to produce thin film composite (TFC) membranes with potential nanofiltration (NF) capabilities. The hydrodynamic permeability of these TFC's is measured to evaluate their performance under typical operating conditions using dead-end permeation experiments. Several combinations of substrates and nanoparticles both spherical (cationic/anionic) and elongated (anionic) silica nanoparticles were tried and compared with realistic hydrodynamic models obtaining positive results. Surface modification results in permeability reduction of approximately two orders of magnitude with respect to the bare substrates to values comparable to those for typical commercial NF membranes. Comparison and good agreement with theoretical hydrodynamic models with no adjustable parameters was also established providing indirect confirmation that the surface layers are largely defect (crack) free. Imaging of our TFC membranes after permeation tests confirmed no significant mechanical damage, that is integrity and robustness of the TFC produced in typical applications.

The work supports that nanoparticle based LbL surface modification of MF/UF membranes can produce filtration quality media, achieving goals for particular and specific membrane water applications, such as nanofiltration (NF) softening processes, natural organic matter (NOM) elimination and possibly reverse osmosis (RO) desalination.

(706c) Optimization of Nano-Composite Membrane Applied in Electrodialysis

Sreedhar Sambamoorthi[1], Varghese Kurian[1], Jogi Ganesh Dattatreya Tadimeti[2] and Sujay Chattopadhyay[1], (1)Department of Paper Technology, Indian Institute of Technology Roorkee, Saharanpur, India, (2)Department of Paper Technology, Indian Institute of Technology Roorkee, Saharanpur , India

Electrodialysis is electro-membrane process to separate ions/electrolytes from their solution. This has been applied since last three decades but due to more popular methods (due to process simplicity) like reverse osmosis or ultra filtration, this has seen lesser attention in development of technology. Recently it has again become popular due to energy crisis and it is thought to be less energy intense than that of the already popular techniques. The process efficiency depends on membrane resistance, concentration of salts, type of ions and voltage applied. The synthesis technique, ion selectivity and cost involved in membrane manufacturing limits its application in commercial in large scale.

Recently a focus is towards minimization of the electrical resistance of the membrane being used in ED apparatus. Few have reported that incorporation of nano particles into the polymer matrix would lead to reduction of resistance and would help water uptake. In this direction, we have tried to find an optimum concentration of meso-porous silica nano particles (as cross linked with PVA and free blended mode) incorporation that may be utilized to obtain maximum ionic flux in separation of NaCl, maximum ionic conductance, and highest transport number. It is needless to mention that mechanical property of membrane was not compromised in this optimization process. Silica particle added physically to the polymer solution did not give desired mechanical stability and ignored for optimization. The cost analysis indicate that cross linked silica while blended with polymer matrix reduces the cost drastically if the quality of water is high value.

Keywords – Optimization, Electrodialysis membrane, silica content, Transport number, Conductivity, Ionic Flux.

(706d) Magnetically-Activated Membranes for Wastewater Treatment

Heath Himstedt[1], Troy Bauder[2], Greg Butters[2], Xianghong Qian[3] and S. Ranil Wickramasinghe[4], (1)Chemical and Biological Engineering, Colorado State Univeristy, Fort Collins, CO, (2)Colorado State University, (3)Ralph E Martin Department of Chemical Engineering, University of Arkansas, Fayetteville, AR, (4)Department of Chemical Engineering, University of Arkansas, Fayetteville, AR

Filtration membranes, whose filtration properties are controllable via application of an external magnetic field, have been developed. Poly(2-hydroxy ethyl methacrylate) (polyHEMA) chains are grafted onto the surface of commercially available nanofiltration membranes. Superparamagnetic nanoparticles are then covalently bound to the end of the polyHEMA chains. When a magnetic field is applied, the nanoparticles will experience a force as they align with the direction of the magnetic field. When place in an alternating magnetic field, the chains quickly align back-and-forth over the surface of the membrane. This motion creates small areas of micro-mixing directly above the membrane surface, decreasing the severity of concentration polarization. It is argued that this effect should reduce membrane fouling due to colloidal foulants.

Oily wastewaters were treated using these responsive membranes. The wastewaters used were oily produced water, a byproduct of oil and gas operations, from coal bed methane wells in Colorado and Wyoming. The severity of fouling, shown by a reduction in flux decline and improved rejection, is reduced by the micromixing in an alternating field. Additionally, the membranes are more easily cleaned, and the modified membranes recovered a greater fraction of their original flux after cleaning.

It is believed that the filtered produced water permeate could be suitable for irrigation use without the addition of chemicals. Produced water permeate were applied to soil columns to test the effect of this hypothetical irrigation water on soil properties. The permeate was then used to water a variety of seed currently grown in Colorado and Wyoming. These tests indicate that the filtered produced water permeate could be suitable as an irrigation water if the proper soil and crop were chosen.

(706e) High Performance Gas Separation Membrane From a Polymer of Intrinsic Microporosity by Photochemical Surface Modification

Qilei Song[1], Sanna K. Nataraj[1], Paul Zavala-Rivera[1], Shaheen A. Al-Muhtaseb[2] and Easan Sivaniah[1], (1)Cavendish Laboratory, Physics, University of Cambridge, Cambridge, United Kingdom, (2)Chemical Engineering, Qatar University, Doha, Qatar

The concept of design in next-generation chemical separation membrane is to develop porous materials by tuning the pore size approximate to the kinetic diameter of molecules (sub-nanometer scale) to achieve highly selective separation with high rates of permeation. Among many microporous organic polymers (MOPs) materials, solution-processable polymers of intrinsic microporosity (PIMs) have attracted significant interest in recent years, showing high gas permeability and reasonable selectivity as gas separation membrane.

In this study, we demonstrate a method of surface modification of a microporous polymer (PIM-1) membrane by combination of ultraviolet light irradiation and ozone (UVO) treatment to enhance the gas separation performance. The ultraviolet irradiation in the presence of atmospheric oxygen produced ozone and atomic oxygen which transformed the polymer chains and generated a nanoporous asymmetric thin layer over the pristine membrane. The chemical changes, structure and morphologies of the membranes exposure to UVO treatment were examined by various techniques. We found significant sensitivity of gas permeation properties with respect to the operation conditions, such as the atmosphere of UV irradiation and exposure time.

The pure PIM-1 membrane presents a CO_2 permeability of ~5600 Barrer and CO_2/N_2 selectivity of 20, and CO_2/CH_4 selectivity at ~13, which are in agreement with the literature data of PIM-1 membrane. Control experiments of UV irradiation in pure N_2 showed that the permeability and selectivity were approximate to that of neat polymer. Upon exposure to UV irradiation in air or controlled O_2/N_2 atmosphere, the permeability of CO_2 showed slight increase (up to 7000 Barrer) after short exposure for 5-10 min, but then decreased with extended exposure for 20-60 min, while the selectivity evidently increased. Particularly, the selectivity of H_2 over N_2 and CH_4 increased evidently (up to ~60) with moderate permeability of H_2 (~2000 Barrer). The CO_2 permeability reached ~2000 Barrer and both selectivity of CO_2/N_2 and CO_2/CH_4 increased to approximately 30 after exposure to UV/ozone for 30 min. Most of these data surpassed the Robeson's 2008 upper bound.

The mixed gases permeation properties of PIM-1 membranes with UV/ozone treatment were also investigated using certified gas mixtures of CO_2/N_2 and CO_2/CH_4 with feed pressure up to 30 bar. The surface modification by UV/ozone treatment could enhance the selectivities of mixed gases, which are higher and more stable than that of unmodified membrane, particularly for CO_2/CH_4 separation, at slight expense of CO_2 permeability.

In sum, the photochemical surface modification *via* the UVO treatment could enhance the selectivity of the composite-layered membrane while the permeability was still maintained at reasonable high level, showing high potential for CO_2 capture, natural gas separation and hydrogen purification. The concept of this process could offer a direction on improving the separation performance of various microporous polymer membrane materials.

Acknowledgements

We acknowledge the NPRP grant from the Qatar National Research Fund (QNRF), the Engineering and Physical Sciences Research Council (EPSRC, UK), and the China Scholarship Council.

Notes

Corresponding author: es10009@cam.ac.uk (Dr. Easan Sivaniah)

References

1. P. M. Budd, E. S. Elabas, B. S. Ghanem, S. Makhseed, N. B. McKeown, K. J. Msayib, C. E. Tattershall and D. Wang, *Advanced Materials*, 2004, 16, 456-459.

2. P. Zavala-Rivera, K. Channon, V. Nguyen, E. Sivaniah, D. Kabra, R. H. Friend, S. K. Nataraj, S. A. Al-Muhtaseb, A. Hexemer, M. E. Calvo and H. Miguez, *Nature Materials*, 2012, 11, 53-57.

(727a) Enhanced Stability of Metal-Organic Frameworks Via Plasma-Enhanced Chemical Vapor Deposition

Gregory W. Peterson[1], Jared DeCoste[2], Colin Willis[3], Martin Smith[3] and Corinne Stone[3], (1)Research & Technology Directorate, U.S. Army Research, Development and Engineering Command, Edgewood Chemical Biological Center, Aberdeen Proving Ground, MD, (2)ECBC, APG, MD, (3)Dstl, Porton Down, United Kingdom

Metal organic frameworks (MOFs) have become a leading class of porous materials for applications such as gas storage, catalysis, and toxic gas removal. Yet, many interesting MOFs, especially those with coordinatively unsaturated metal centers and/or carboxylate linkers, are prone to degradation from moisture, making them less useful for ambient operations. Through treatment of Cu_3BTC_2 MOF with plasma-enhanced chemical vapor deposition of perfluorohexane, we have sufficiently modified the structure such that it is more-resistant to moisture degradation. Characterization with ^{13}F NMR shows the presence of perfluoro groups within the pore structure, reducing water-clustering and increasing the metal-oxo bond stability. Ammonia breakthrough studies indicate that, not only does the material retain its high ammonia capacity post-treatment, but the stability actually increases retention and capacity under low and high humidity conditions. Further work is currently being explored on a variety of plasma/MOF systems to study the effects of perfluoro compound and metal-linker behavior.

(727b) Impact of Alkyl-Functionalized BTC On Properties of Copper-Based Metal-Organic Frameworks

Yang Cai[1], Yadong Zhang[2], Yougui Huang[1], Seth Marder[2] and Krista S. Walton[3], (1)Chemical and Biomolecular Engineering, Georgia Institute of Techonlogy, Atlanta, GA, (2)Chemistry, Georgia Institute of Techonlogy, Atlanta, GA, (3)Chemical and Biomolecular Engineering, Georgia Institute of Technology, Atlanta, GA

Highly porous metal-organic frameworks $[Cu_3(MBTC)_2(H_2O)_3]_n$ (where MBTC is methyl-1,3,5-benzenetricarboxylate) and $[Cu_3(EBTC)_2(H_2O)_3]_n$ (where EBTC is ethyl-1,3,5-benzenetricarboxylate) have been solvothermally synthesized. The MOFs are isostructural to each other but do not form the same framework as HKUST-1 (CuBTC) due to the steric impact of the alkyl-functionalized ligands. The new MOFs are formed by two different $[Cu_2(O_2CR)_4]$ units (where R is an aromatic ring), which create the three-dimensional framework with open metal sites and high surface areas. The pore size and adsorption properties are altered by introducing the organic groups methyl ($-CH_3$) or ethyl ($-C_2H_5$) groups. Compared with HKUST-1, these MOFs adsorb comparable amounts of carbon dioxide and methane at pressures up to 5 bar. However, water adsorption is significantly lower compared to HKUST-1 due to the hydrophobic functional groups. This work indicates that the strong adsorption preference of water on open metal sites can be negated with proper choice of hydrophobic functional group, without losing the strong interaction with other gases.

(727c) Polymorphism in Metal-Organic Frameworks: Effect of Functionalization

Ambarish R. Kulkarni, Yang Cai, Krista S. Walton and David S. Sholl, School of Chemical & Biomolecular Engineering, Georgia Institute of Technology, Atlanta, GA

Recently, considerable research has been focused towards development of novel Metal-Organic Frameworks (MOFs) that are targeted for specific applications. MOFs are 3-d nanoporous materials consisting of metal centers bonded to organic linkers, which offers ample opportunities for tuning fundamental properties such as diffusion and adsorption of small molecules. One commonly used approach involves the functionalization of the organic linker. This strategy is typically expected to yield a new iso-structural MOF with more favorable properties. Recently, alkyl substitution of BTC (BTC = 1,3,5-benzenetricarboxylate) was shown to form a topology that was completely different from the well-studied HKUST-1. Such substitution induced polymorphism has been previously shown to exist for Zeolitic Imidazole Frameworks (ZIFs).

In this talk, we report the results of Density Functional Theory (DFT) calculations that investigate the relative stability the two experimentally reported topologies of CuBTC. We extend these calculations to explore the hypothetical topologies for BTC MOFs. Our results show that the relative energies of the two experimental topologies for functionalized BTC MOFs are very similar, and that the difference is comparable to that found for the various ZIF polymorphs. In addition to methyl and ethyl, the effect of other substituents will also be discussed. The relative stabilities of BTC polymorphs is also studied using a classical flexible force field which is then compared to the DFT results.

(727d) Adsorption of Toxic Industrial Chemicals in MOFs Under Humid Conditions

Pritha Ghosh and Randall Snurr, Chemical and Biological Engineering, Northwestern University, Evanston, IL

Adsorption of Toxic Industrial Chemicals in MOFs under Humid Conditions

Pritha Ghosh, Randall Q. Snurr

Northwestern University, Department of Chemical & Biological Engineering

Activated carbons impregnated with metal salts are the current standard for adsorption and capture of toxic industrial chemicals (TICs). Metal-Organic Frameworks (MOFs) are an attractive alternative due to their chemically tunable nature and relative ease of synthesis. MOFs have been reported to outperform activated carbons in capturing TICs, especially MOFs containing open metal sites (Britt et al., *PNAS* **2008**, *105*, 11623). However, in the presence of ambient humidity, water molecules coordinate to the open metal sites, leading to a drastic decrease in the MOF's capacity to capture TICs (Grant Glover et al., *Chem. Eng. Sci.* **2011**, *66*, 163). Since TICs may need to be captured under humid conditions (such as in a gas mask), it is important to assess adsorption of TICs in MOFs with water vapor present.

In this work, we model the adsorption of TICs under humid conditions in a set of hydrophobic MOFs. Grand canonical Monte Carlo (GCMC) simulations are used to predict the pure and mixture isotherms of a toxic gas and ambient water. After validating the simulated water isotherms with experimental results, we investigate the adsorption of sulfur dioxide and ammonia with a fixed partial pressure of water vapor, corresponding to a relative humidity of interest. Hydrophobic MOFs do not experience competitive adsorption between water and toxic gas, as was observed in MOFs containing open metal sites. In the case of ammonia, our preliminary results suggest that there is cooperative adsorption in some MOFs. We are also investigating the mechanism by which the pores of hydrophobic MOFs achieve high selectivity for small, polar toxic gases over water to facilitate the design of MOFs capable of capturing TICs effectively even under humid conditions.

(727e) Metal Organic Framework Adsorbents for Control of Toxic Industrial Chemicals

Catherine Thibaud-Erkey[1], Krista S. Walton[2], Joshua A. Sheffel[1] and Paul M. Schoenecker[2], (1)Physical Sciences, United Technologies Research Center, East Hartford, CT, (2)Chemical & Biomolecular Engineering, Georgia Institute of Technology, Atlanta, GA

There is an increasing concern about the potential threat of chemical attacks in buildings. To protect occupants from the effect of hazardous contaminants, it is necessary to implement an efficient continuous filtering system for the air brought into the building and recirculated inside the building. Since most buildings are equipped with mechanical heating, ventilating, and air conditioning (HVAC) systems, the filters used in these systems can be replaced or supplemented by filters containing chemical adsorbents that can remove the targeted hazardous contaminants from the air, in particular Toxic Industrial Chemicals (TICs).

Treated and non-treated activated carbons are traditionally used to remove hazardous compounds from the air. ASZM-TEDA carbon, an impregnated activated carbon, has been used for over 15 years to remove TICs. ASZM-TEDA is a coal-based activated carbon, impregnated with copper, zinc, silver, and molybdenum compounds, as well as triethylenediamine. One drawback is that this type of carbon does not perform efficiently for low-molecular-weight compounds and is significantly affected by humidity. Therefore, many industrial chemicals and toxic gases are not well adsorbed by currently available filters.

The presentation will summarize the testing and development of novel adsorbents for TICs control. The selected materials are known as metal organic frameworks (MOFs) and consist of metal centers coordinated with organic ligands to form highly porous materials with well-defined nanometer pore sizes. Six MOF materials and three activated carbon materials were compared to ASZM-TEDA (baseline material) for the adsorption of eight TICs including ammonia, sulfur dioxide and chlorine. It was shown that using a metal organic framework could greatly improve the building protection against ammonia and methylamine. The effect of temperature and humidity on the adsorption performance was also investigated for promising adsorbent/adsorbate pairs.

The project is funded by the Technical Working Support Group (TSWG), which operates as a program element under the Combating Terrorism Technical Support Office (CTTSO).

(727f) Capture and Separation of CO2 From Binary Gas Mixtures Using Flexible Porous Coordination Polymers

Jeffrey Culp[1,2], Fan Shi[1,2] and Christopher Matranga[1], (1)U.S. Department of Energy, National Energy Technology Laboratory, Pittsburgh, PA, (2)URS Corporation, South Park, PA

Structurally dynamic coordination polymers or metal organic frameworks (MOFs) are a unique class of CO_2 sorbents that undergo structural phase transitions from low porosity "closed pore" structures to high porosity "open pore" structures once a critical threshold pressure of CO_2 is reached. The adsorption isotherms for these compounds show a pronounced step at the threshold pressure followed by a large increase in CO_2 uptake. Subsequent desorption of CO_2 from the saturated state is highly hysteretic. The threshold pressure and the width of the hysteresis loop in these compounds are not only functions of temperature and material structure, but also highly dependent on the identity of the adsorbate. As such, they are of intense interest for their potential as gas selective sorbents. One important fundamental question surrounding these structurally dynamic materials is whether the selectivity suggested by the pure gas isotherms is preserved in the presence of gas mixtures, or does the adsorption of one gas open the structure and allow other gases access to the newly created pore structure. Understanding the adsorption mechanisms in these materials under mixed gas conditions is therefore critical for evaluating their separation abilities and limitations. To address this question, a series of structurally dynamic porous coordination polymers have been evaluated for CO_2 capture from mixed gas streams using a variety of experimental techniques including *in situ* infrared spectroscopy, small angle neutron scattering, gas chromatography and high pressure breakthrough curves.

(727g) High-Throughput Screening of MOFs for CO2 Separation

Sangil Han, Yougui Huang, Taku Watanabe, Ying Dai, Krista S. Walton, Sankar Nair, David S. Sholl and **J. Carson Meredith**, School of Chemical & Biomolecular Engineering, Georgia Tech, Atlanta, GA

The capture of CO_2 from fossil fuel combustion flue gas is the focus of significant research efforts. Existing CO_2 capture technologies such as amine absorption are extremely energy intensive. Metal-organic frameworks (MOFs) are potential size-selective, high-capacity materials for adsorptive and membrane-based capture of CO_2 and other gases. MOFs exhibit nanoporous crystalline structure in which organic linker molecules self-assemble with metal centers to form crystals with well-defined pore size, high surface area, and low framework density. MOFs have attracted attention as possible components of CO_2 capture technologies based on either adsorption or membranes.

In principle, the enormous number of MOFs with various combinations of organic linker and metal cation creates an opportunity to tailor pore structure, size, and functionality for CO_2 capture from flue gas. Significant problems, however, frustrate attempts to discover or design such materials. In addition to the large, under-explored design space, some MOFs are unstable in water vapor, and the principles for designing water-stable MOFs are not yet well known. The adsorption properties of a MOF are critical to its value for CO_2 capture. Measurements for determining adsorption isotherms are tedious and time consuming. Almost all available data for gas adsorption in MOFs reports adsorption of dry single-component gases. For considering practical CO_2 capture applications, the performance of MOFs that have been exposed to water vapor and acid gases like SO_2 and NO_2 is critical, and almost no data addressing this issue are available. There are few published reports of MOF stability after exposure to water and none, to our knowledge, that examine stability as a function of exposure to acid gases.

To address the above challenges, we report here the efficient screening of MOF CO_2/N_2 adsorption and diffusion selectivity, as well as their sensitivity to water vapor and acid-gases, *via* the a novel parallel high-throughput (HT) sorption screening system. Our HT approach collects adsorption information on the gases directly relevant to CO_2 capture (CO_2 and N_2) in the pressure regime relevant for flue gas applications. A key element of our HT experiments is to collect adsorption data at only one relevant value of pressure, rather than measuring a complete adsorption isotherm. The HT-sorption instrument has an array of 36 chambers, each connected to a pressure sensor and a valve (Supporting Information). By monitoring pressure decay versus time, CO_2 and N_2 adsorption were measured at a single state point at 30 °C. We illustrate this approach by simultaneously investigating 8 candidate MOF materials, of which the best material was found to have a CO_2/N_2 membrane selectivity of 152 and a CO_2 permeability of 60 barrer for Co-NIC. This approach provides a significant increase in efficiency of obtaining the separation properties of MOFs. While we describe here the identification of novel materials for CO_2 capture, the methodology enables exploration of the performance and stability of novel porous materials for a wide range of applications. To our knowledge, this is the first high-throughput screening system which enables measurement of adsorption capacities and selectivities on MOFs, in addition to exposing the MOFs to water vapor and acid gases. The method leads to rapid selection of MOFs with high selectivity and stability to water vapor and acid gases. As a result, this talk reports data for acid gas stability previously unknown for these MOFs. The HT screening system can handle a variety of gas adsorbates and hence can be applied widely to screen adsorption and diffusion properties of other materials including inorganic compounds, polymers, and low-volatility liquids.

(727h) Reverse Shape Selectivity in Liquid Phase Adsorption of Xylene Isomers in the Zirconium Terephthalate MOF Uio-66

Mariana Moreira[1], **Alexandre Ferreira**[1], João Carlos Santos[1], José Miguel Loureiro[1], Florence Ragon[2], Patricia Horcajada[3], Kyu.-E. Shim[4], Young Kyu Hwang[5], U.- Hwang Lee[4], Jong-San Chang[6], Christian Serre[7] and Alirio E. Rodrigues[8], (1)Departamento de Engenharia Química, Faculdade de Engenharia, Universidade do Porto, LSRE - Laboratory of Separation and Reaction Engineering - Associate Laboratory LSRE/LCM, Porto, Portugal, (2)Université de Versailles Saint-Quentin-en-Yvelines, Institut Lavoisier , Versailles, France, (3)Université de Versailles Saint-Quentin-en-Yvelines, Institut Lavoisier, Versailles, France, (4)Korea Research Institute of Chemical Technology, Catalysis Center for Molecular Engineering, Taejon, South Korea, (5)Catalysis Center for Molecular Engineering, Korea Research Institute of Chemical Technology, Daejon, South Korea, (6)Korea Research Institute of Chemical Technology, Daejeon, South Korea, (7)Institut Lavoisier (UMR 8180), Université de Versailles Saint Quentin-en-Yvelines, Versailles Cedex, France, (8)Laboratory of Separation and Reaction Engineering, University of Porto - Faculty of Engineering, Porto, Portugal

Reverse shape selectivity in liquid phase adsorption of xylene isomers in the zirconium terephthalate MOF UiO-66

Mariana A. Moreira[†], João C. Santos[†], Alexandre F. P. Ferreira[†], José M. Loureiro[†], Florence Ragon[‡], Patricia Horcajada[‡], Kyu.-E. Shim[§], Young- K. Hwang[§], U.- Hwang Lee[§], Jong.-S. Chang[§], Christian Serre[‡], Alírio E. Rodrigues[†]*

[†] LSRE - Laboratory of Separation and Reaction Engineering - Associate Laboratory LSRE/LCM, Faculdade de Engenharia, Universidade do Porto, Rua Dr. Roberto Frias, 4200-465 Porto, Portugal; [‡] Institut Lavoisier (UMR CNRS 8180), Université de Versailles Saint-Quentin-en-Yvelines, 45 avenue des Etats-Unis, 78035 Versailles cedex, France; [§] Catalysis Center for Molecular Engineering, Korea Research Institute of Chemical Technology (KRICT), P.O. Box 107, Yusung, Taejon 305–600, Korea.

Abstract

Powder, agglomerates and tablets of the microporous zirconium (IV) terephthalate Metal-organic framework UiO-66 were evaluated for selective adsorption and separation of xylene isomers and ethylbenzene in liquid phase using *n*-heptane as eluent. Pulse experiments, performed at 313 K in presence of *n*-heptane, revealed the *o*-xylene preference of this material, further confirmed by binary and multi-component breakthrough experiments in the presence of *m*-, *p*-xylene, and ethylbenzene. The shaping of the material as tablets did not significantly change its selectivity towards the *o*-xylene isomer despite a loss in capacity. The binary and multicomponent equilibrium data of the xylene isomers and ethylbenzene was well modeled by a Dual Langmuir model. The simulations of the breakthrough and reverse breakthrough experiments were performed in the three materials, presenting good agreement between the simulation and experimental results.

(748a) Reverse Electrodialysis Revisited

Frank Leitz, Bureau of Reclamation, Denver, CO

More than three decades ago, John Weinstein and the author published a paper in Science magazine on Electric Power from Differences in Salinity. This paper provided design equations, experimental data and cost estimates for a process that constitutes a reversal of the process of electrodialysis: using differences in salinity to produce electric power. In the present climate of concern about CO_2 generation, this process deserves to be reconsidered. Using apparatus designed for water desalination, a cost of 0.97 cents per kilowatt-hour was developed. With modern manufacturing technology and apparatus designed specifically for this application, this could be improved upon. A number of lessons can be drawn from this experience, not the least of which is that virtually every "new" idea actually has a long history.

(748b) Selectivity Studies On Pilot-Scale Electrodialysis Reversal

Leila Karimi, New Mexico State University,Institute for Energy and the Environment/WERC, Las Cruces, NM, Ali Sharbat, Institute for Energy and the Environment, New Mexico State University, Las Cruces, NM and Abbas Ghassemi, Institute for Energy and the Environment, Chemical Engineering Dept, New Mexico State University, Las Cruces, NM

Although Electrodialysis Reversal (EDR) technology has been commercially used since the early 1960s, the fundamental understanding of this technology is less developed than that of other technologies such as distillation and reverse osmosis. Selectively removal of some ions of interest compared to other ions which present in water has not been studies in details specially in pilot-scale. Groundwater resources, which are very important sources of drinking water in many parts of the world as well as southwest region of the United States, have various water chemistries. Depending on several parameters, various amounts of anions such as chloride, sulfate, and alkalinity species that contribute to high total dissolved solids (TDS) and/or salinity can be found in groundwater samples. Understanding of selective removal of ions and implication of this understanding would help to reduce the cost of desalination by reducing the number of moles of ions that is removed to get appropriate product water. The objective of this study was to evaluate: a) effect of salt removal par electric stage, b) ionic strength and composition of feed water, c) operating conditions including temperature, and d) anionic exchange membrane fouling on selectively removal of some ions of interest compared to other ions.

Full pilot-scale studies were performed on a GE electrodialyser with the capacity of 10 gpm influent flow rate at the Brackish Groundwater National Desalination Research Facility (BGNDRF) in Alamogordo, New Mexico. Brackish feed-water was provided from aquifers in the Tularosa Basin with conductivity of 1700-6600 μS/cm, which are within the spectrum for water quality in the southwestern region of the United States. The experiment operated with single/double electrical/hydraulic stage(s) using CR67HMR or CR64LMR cation-exchange membranes, AR204 or AR908 anion-exchange membranes, and Spacer-Mark IV with an effective membrane surface area of 3,000 cm^2. Selectivity of EDR stack for different monovalent and divalent ions was evaluated by water analysis of feed, concentrate, and product water samples under various salt removals par stage, various ionic strengths and compositions of feed water, various operating conditions and new/aged anionic exchange membranes.

Preliminary data analysis indicated that all the above-mentioned parameters have direct effect on selectively removal of some species compared to other species. The results showed that higher temperature increases the selectivity for divalent ions compared to monovalent ions. Interestingly, the results indicated that generally aged anionic membranes have less selectivity compared to new anionic membranes. The results of this study could be used to help EDR designers/operators to selectively target the ions of interest and it would result in reducing the cost of desalination by reducing the number of moles of ions that is removed to reach the targeted product water.

(748c) Overcoming Electro Dialysis Reversal Commercialization Challenges in Tough-to-Treat Waters

Jamie Vinsant, GE Water, Boston, MA

As global water scarcity impacts customer throughout the world, customers are using Electro dialysis Reversal (EDR) technology to treat non-traditional waters, particularly in applications that are considered "Tough-To-Treat" in comparison to cleaner, more traditional sources of fresh water. Comparison of EDR in side-by-side pilots with other available technologies has shown potential advantages in treatment of wastewater due to the membranes resistance to residual chlorine and tolerance of higher suspended solids. Both qualities greatly improve the robustness of the treatment system while lowering the overall cost to treat due to lower energy consumption and less pretreatment.

This paper will summarize the data and operational experience from several EDR applications that have moved from technology development to field pilot and to a commercial system. We will connect both successes and ongoing challenges as a way to highlight areas that need ongoing technology investment and piloting. These case studies will provide operating examples where EDR can be more robust than other treatment options and provide a platform to achieve higher water recoveries on various types of wastewater at lower cost.

(748d) Reducing the Cost of Inland Desalination: Pushing the Conventional Design Limits in EDR and EDR / RO / NF Hybridization Efforts

Ali Sharbat[1], Neil E. Moe[2], Jim Loya[3], John Barber[2] and Abbas Ghassemi[4], (1)Institute for Energy and the Environment, New Mexico State University, Las Cruces, NM, (2)GE Water & Process Technogies, Las Cruces, NM, (3)New Mexico State University,Institute for Energy and the Environment/WERC, Las Cruces, (4)Institute for Energy and the Environment, Chemical Engineering Dept, New Mexico State University, Las Cruces, NM

The shortage of drinking water is an emerging crisis in many parts of the world. After surface water resources, groundwater supplies are a very important source of drinking water in many parts of the world as well as in the Southwest region of the United States. Reverse Osmosis (RO), Nanofiltration (NF), and Electrodialysis Reversal (EDR) are the dominant technologies for inland desalination of brackish groundwater, and each technology has its own unique characteristics, strengths, and weaknesses. Electrodialysis Reversal is usually capable of achieving higher recovery under fouling (COD) and scaling (hardness, CaSO4) conditions, and is resistant to silica scaling and chlorine residual. In EDR the salt removal level is flexible and depends on design and operating conditions. In RO and NF, CapEx is lower and does not strongly depend on feed chemistry and/or operating conditions. Energy consumption is relatively insensitive to feed water chemistry; very high salt removal is common in RO systems while NF strongly rejects divalent species. Hybrid RO/EDR or NF/EDR systems potentially provide benefits beyond what the individual technologies can ensure. The most important benefit would be higher combined recovery rate which contributes to lower concentrate disposal cost. Lower energy cost per unit of desalinated water is also another benefit.

Extensive studies were conducted using full sized pilot-scale EDR equipment by Institute for Energy and the environment (IEE), New Mexico State University (NMSU). The EDR pilot was designed and built by GE Water & Process Technologies (WPT), where the electrodes, spacers, and membranes were sized identical to commercial systems with capacity of 10 gpm, while the number of cell pairs was typically 40 per stage, compared to 600 at most commercial sites. The experiments operated with single/double electrical/hydraulic stage(s) using CR67 cation-exchange membranes, AR204 / AR908 anion-exchange membranes, and MkIV-2 spacers with an effective membrane surface area of 3,000 cm^2. A 2-stage RO/NF pilot system was also designed and built by GEWPT with capacity of 18 gpm using 4" elements. Different types of RO and NF membranes were characterized. The experiments were performed at the Brackish Groundwater National Desalination Research Facility (BGNDRF) in Alamogordo, New Mexico, using the various sources of brackish water with conductivity of 1700-6600 μS/cm available on site. Various hybridization scenarios were evaluated using EDR and RO simulation tools, and the most promising ones were demonstrated using the pilot system. In addition, the effects of operating EDR above its conventional design limits with respect to salt removal per stage and allowed supersaturation levels were explored. Water chemistry, temperature, recovery, velocity, volts/amps, pressure, saturation level of different salts, and recovery rates were monitored as experimental variables. Several water samples were collected for complete water analysis.

Data analysis demonstrated that hybrid systems in certain scenarios can reduce the cost of treated water by reaching higher recovery rates. In the case of inland brackish water desalination, higher combined recovery rates result in minimizing the concentrate disposal issue. Operation above the conventional design limits of EDR was also demonstrated on a practical level, contributing to both lower capital cost and higher recovery.

(748e) Maximizing Inland Desalination Concentrate Recovery Using Electrodialysis Reversal

Charlie He, Carollo Engineers, Phoenix, AZ

Bench and pilot testing confirmed the potential for the proposed process (ozonation, followed by biological filtration, followed by EDR with and without ion exchange softening) to reduce RO concentrate from reclaimed water RO concentrate by 79 to 91%. This reduction is achieved by removing organic compounds and reducing scaling potential. Organic compounds are removed by a two-step process where biodegradable organic compound is increased from 0 to ~20 mg/L by contact with ozone and then removed in a biological filter. Scaling potential is reduced by ion exchange softening. The proposed process reduces the evaporation pond size by up to 91% and improves the water reuse potential and salinity management for inland utilities.

Collaborative Development of Zero Discharge Desalination Technology

Tom Davis[1], Brad Biagini[2], Malynda A. Cappelle[1], Bernie Mack[2], and Emily Gilbert[2]

[1]Center for Inland Desalination Systems, University of Texas at El Paso
[2]Veolia Water Solutions & Technologies

ZDD (Zero Discharge Desalination) is a technology that was developed with research funding from the Bureau of Reclamation and Sandia National Laboratories and is being commercialized by Veolia with the support of the University of Texas at El Paso. The ZDD technology is directed to reduction in the volume of concentrate produced in the desalination of groundwater that contains sparingly soluble solutes, particularly calcium sulfate. Initial research to demonstrate feasibility was funded by Sandia National Laboratories. Subsequently Reclamation funded pilot studies, first for the treatment of irrigation drainage containing calcium sulfate and selenium and second for treatment of groundwater at the Brackish Groundwater National Desalination Research Facility (BGNDRF) in Alamogordo, NM.

ZDD technology utilizes electrodialysis metathesis (EDM) to treat the concentrate produced by another desalination process, usually RO or NF. EDM devices utilize the same types of anion exchange and cation-exchange membranes as normal electrodialysis, but there are four membranes and solution compartments in the repeating unit of an EDM stack. The RO or NF is operated under conditions in which the $CaSO_4$ content is at a safe level. The RO/NF concentrate is then sent to the EDM stack as one of the two feed streams. The salt-depleted diluate stream from the EDM is returned to the RO/NF for recovery of more water. The other feed stream to the EDM contains NaCl. The salts in the two feed streams change partners in the EDM stack to produce two highly concentrated streams, one rich in $CaCl_2$ and the other rich in Na_2SO_4. Because these two salts are highly soluble, they can be concentrated to high levels. Typically the volume of each concentrate stream is about 1% of the volume of feed water to the RO/NF. These two concentrate streams from the EDM can be processed separately to recover useful compounds, or they can be combined to precipitate $CaSO_4$, which has potential commercial value for soil augmentation or for production of gypsum board. There are two major benefits to combining the concentrate streams. The first benefit is that the troublesome ions in the feed water are removed as a solid byproduct instead of remaining in solution. Thus, the disposal of liquid salt solution is reduced substantially. The second benefit is that the NaCl reconstituted in the supernatant of the precipitation chamber is at a fairly high concentration. The NaCl can be recovered from the supernatant and returned to the EDM stack.

The pilot study on treatment of irrigation drainage was sponsored by the Bureau of Reclamation and took place on a grape farm in the Panoche Water District near Firebaugh, CA. The irrigation drainage water contains about 450 µg/L of selenium, nine times the MCL of 50 µg/L. RO reduced selenium below detection limits in the permeate, but the high level of $CaSO_4$ in the feed water limits RO recovery to 50%. Hence, RO alone is not an adequate solution to the treatment of the irrigation drainage, because disposal of the selenium-laden RO concentrate is a real problem. Treatment of the RO concentrate with EDM shown in Figure 1 produced a Na_2SO_4-rich concentrate with 4700 μg/L of selenium. Thus the selenium was contained in a volume less than 10% of the original feed.

Figure 1. ZDD Pilot Plant at Farm in Panoche Water District, California.

Because of natural gypsum deposits in the vicinity, the groundwater under Alamogordo, New Mexico, is nearly saturated in $CaSO_4$. In some wells, notably those that would be the water source for a desalination plant for the City of Alamogordo, $CaSO_4$ is the dominant solute. ZDD technology is applicable for concentrate management of this water that is difficult to treat by RO alone. The Bureau of Reclamation sponsored a project to demonstrate the effectiveness of ZDD technology in the treatment of this difficult water. In preliminary experiments in a 1-gpm pilot plant it was determined that EDM effectively removed $CaSO_4$, but silica became the solute that limited the yield of RO permeate. It was determined that certain nanofiltration (NF) membranes were permeable to silica, so the RO membranes were replaced with NF membranes in the demonstration plant. The demonstration plant consisted of two containers. A 40-foot container housed an NF system that was designed to treat the groundwater at a rate of 20 gpm of permeate. The EDM system was housed in a 20-foot container. Figure 2 shows the two containers at the Brackish Groundwater National Desalination Research Facility in Alamogordo. Figures 3 and 4 show the NF system and the EDM system in the containers.

Figure 2. ZDD Containers at Brackish Groundwater Nation Desalination Research Facility.

Figure 3. NF System in 40-Foot Container.

Figure 4. EDM Stack in 20-Foot Container.

A yield of 98% was obtained in pilot studies of the ZDD process treating the Alamogordo groundwater compared to 70-75% yield when this groundwater is treated with RO alone.

(748g) Electrosorption of Ions by Mesoporous Carbon Materials

Ketki Sharma[1], Sotira Yiacoumi[1], Jorge Gabitto[2], Richard Mayes[3], Jim Kiggans[4], David W. DePaoli[5], Sheng Dai[3] and **Costas Tsouris**[5], (1)Civil and Environmental Engineering, Georgia Institute of Technology, Atlanta, GA, (2)PVAMU, Prairie View, TX, (3)Chemical Sciences Division, Oak Ridge National Laboratory, Oak Ridge, TN, (4)Materials Science and Technology Division, Oak Ridge National Laboratory, Oak Ridge, TN, (5)Energy and Transportation Science Division, Oak Ridge National Laboratory, Oak Ridge, TN

The electrical double layer (EDL) formation near a charged surface has significance in electrochemical processes, such as electrosorption of ions and energy storage in supercapacitors. Electrosorption of ions from aqueous solutions is utilized in the capacitive deionization (CDI) process for water desalination. The main objective of the present work is to study the transport of ions in mesoporous carbon materials during the CDI process and synthesize efficient carbon electrode materials for application in CDI. The electrodes used in this study are made of mesoporous carbon materials of 8-20 nm pore size, which are synthesized at Oak Ridge National Laboratory. Neutron imaging is employed to quantify ion transport in charged mesoporous carbon electrodes. Specifically, neutron imaging is used as a tool to visualize neutron-absorbing ions in carbon electrodes during the CDI process, and image analysis is used to extract the transient concentration profiles of ions. From the time-dependent concentration profiles, the effective diffusion coefficient for neutron-absorbing ions is obtained at different values of the applied potential. It was found that the effective diffusivity increased seven times from 0 to 1.2 V applied potential. Neutron imaging also reveals interesting ion-transport phenomena, which can aid in the determination of optimal conditions for the operation of the CDI process.

(749a) Transient Multicomponent Models for Simulating Water and Nutrient Flow Through Soils Amended with Biochar

Hao Sun[1], Caroline A. Masiello[2] and **Kyriacos Zygourakis**[1], (1)Chemical and Biomolecular Engineering, Rice University, Houston, TX, (2)Earth Science, Rice University, Houston, TX

Biochar is charcoal generated for intentional soil amendment by pyrolyzing sustainable biomass feedstocks. In addition to providing an easily deployable method for carbon sequestration, properly "engineered" charcoals can increase the water holding and cation exchange capacities of soils, improving the ability of plants to survive under drought conditions and reducing fertilizer runoff into watersheds. Fertilizer runoff has become a serious problem, because as much as 70% of fertilizer applied to crop fields is leached into the groundwater or lost to streams and rivers, eventually leading to large hypoxic "dead" zones in the world's oceans (including the Gulf of Mexico).

We present here the development and testing of mathematical models that allow us to evaluate the environmental performance of biochars used for soil amendment. Our earlier work (1-2) has established that both the pore structure and the surface chemistry of biochars can vary widely depending on the composition of the biomass feedstocks and on the pyrolysis conditions employed during biochar production. The work presented here will demonstrate how these biochar properties affect the ability of soil/biochar mixtures to adsorb, retain and release water and nutrients. Our ultimate objective is to develop a comprehensive theoretical framework that can guide us to engineer biochars with optimal properties for every field application.

The first model describes the dynamic adsorption/elution of a model fertilizer (ammonium nitrate) in columns packed with mixtures of biochar and soil and perfused with aqueous solutions of the fertilizer. This model can be used to simulate a fertilization event that is followed by irrigation/rain events that elute the fertilizer adsorbed in the solid column. The dynamic model accounts for all the important processes occurring in this system: convection and dispersion of the aqueous solution through the bed, external mass transfer between the fluid phase and the soil and biochar particles, intraparticle diffusion and adsorption of the solute on the pore surface area of the sorbents. To our knowledge, this is the first model that accounts explicitly for the presence of two solid phases with widely different pore structures, surface chemistries and, therefore, nutrient holding capacities. It also accounts for the complicated pore structure of biochars that consists of interconnected networks of large macropores, mesopores and micropores. The transient mass balances lead to a system of partial differential equations that are discretized using orthogonal collocation on finite elements. The resulting differential equations are approximated by backward differentiation formulae and the nonlinear systems at each time-step are solved by Newton's method.

We will first discuss the approach used to determine the optimal mode of applying a given amount of fertilizer (e.g. 100 kg N per hectare). The solution of this optimization problem determines the flow rate and inlet solute concentration that maximizes the loading of solute in our column, or, equivalently, minimizes fertilizer runoff. The solute adsorbed on sorbent particles will then be released slowly during the subsequent irrigation or rain events. We then carry out a systematic parametric study to determine how the dynamic loading and elution processes are affected by the amount of biochar added to the mixture, the size, porosity and pore size distributions of biochar particles, and the adsorption equilibrium constants for the solute or the kinetic constants of solute adsorption. As expected, simulations predict that higher biochar amendment levels improve the ability of the soil/biochar mixture to retain nutrients and release them over a longer period of time. Simulation results also reveal that smaller biochar particles accelerate both the adsorption and the desorption of nutrients, while lower biochar porosities both increase the amount of nutrient adsorbed and prolong its subsequent release. The adsorption kinetic constants also play a prominent role in determining both the nutrient loading and the dynamics of the adsorption/elution processes.

Fertilization, irrigation and rain events are intermittent processes. Therefore and in order to complete our theoretical framework, we have also developed a model that describes the dynamics of water saturation and drainage for a column packed with soil/biochar mixtures. This model is important to describe the transients that precede and follow each nutrient adsorption or elution event described by the previous model. A systematic parametric study is also carried out to evaluate the importance of the hydraulic conductivity and capillary pressure of the biochar particles on the duration of the saturation and drainage transients. Simulation results show that these transients are short and do not affect significantly the dynamics of the nutrient adsorption/elution processes.

1) Hao Sun, W.C. Hockaday, C.A. Masiello and Kyriacos Zygourakis*, "Multiple Controls on the Chemical and Physical Structure of Biochars," *Industrial and Engineering Chemistry Research*, **51**, 3587−3597 (2012) dx.doi.org/10.1021/ie201309r.

2) T.J. Kinney, C.A. Masiello, B. Dugan, W.C. Hockaday, M.R. Dean, K. Zygourakis and R.T. Barnes, "Hydrologic properties of biochars produced at different temperatures," *Biomass and Bioenergy* (2012), doi:10.1016/j.biombioe.2012.01.033

(749b) Anchoring of Iron Hydro(oxide) Particles On Activated Carbon for Removing Hexavalent Chromium From Aqueous Solutions

Refugio Bernardo García Reyes, Alcione Garcia González, Jonathan Valentin Reyes, Arturo Muñiz Ovalle and Eliut Aguilar Sandoval, Facultad de Ciencias Químicas, Universidad Autónoma de Nuevo León, San Nicolás de los Garza, NL, Mexico

Adsorption processes are considered as a viable option for the removal of heavy metals from aqueous solutions. Metal oxide particles can be used for the adsorption of negatively-charged species such as hexavalent chromium but these particles cannot be packed in columns because of the high pressure and the possibility of losing these small particles in continuous operation. To overcome these shortcomings, in this research ferrous hydro(oxide) nanoparticles were anchored by forced hydrolysis at different conditions on commercial activated carbon for removing hexavalent chromium from aqueous solutions. Findings demonstrate that temperature, salt concentration as well as ageing time significantly affected the quantity of iron hydro(oxide) particles anchored to the porous activated carbon although surface area did not change notably after the forced hydrolysis process. Moreover iron-loaded activated carbon was capable of removing almost twice the quantity of hexavalent chromium per gram of adsorbent in comparison with the non-treated activated carbon. Finally, a high percentage of hexavalent chromium can be desorbed from iron-loaded activated carbon with alkaline solutions.

In conclusion, iron hydro(oxide) particles can be anchored by forced hydrolysis treatment in activated carbons increasing the removal efficiency of hexavalent chromium from aqueous solutions.

(749c) Preparation of Novel Zirconium Immobilized Carbon for Adsorption of Arsenate From Water

Narahari Mahanta and **Chen J. Paul**, Department of Civil & Environmental Engineering, National University of Singapore, Singapore, Singapore

Abstract

Arsenic is the naturally occurring carcinogenic metalloid and present in environment in several oxidation states. Arsenic contamination in natural water is a serious threat and challenging issue worldwide due to its acute and chronic toxicity, especially through ingestion through drinking water. In the present study, we explored the application of Zirconium immobilized carbon particle for the adsorption of arsenate from aqueous environment. For this, carboneous materials was prepared from old leaf by acid treatment. Raman spectra revealed that the prepared carbons are amorphous in nature. To increase the arsenic removal efficiency, Zr (IV) was chemically immobilized to these carbon based materials. For the immobilization process, the derived carbon materials were mixed in Zr (IV) salt solution and subsequently treated with NH_4OH. *Designed nano-adsorbent was screened under scanning e*lectron microscope (SEM) analysis to explore the morphology and the presence of Zirconium on the carbon was confirmed by Energy dispersive spectroscopy and elemental analysis. These designed nano-adsorbent was used for the removal of the arsenate from water. To investigate the adsorption efficiency, designed amount of nano-adsorbent was mixed with arsenate solutions and kept for constant shaking. The effect of pH on the adsorption behavior in the batch adsorption study was carried out at a wide range of pH ranging from 2 to 11. It was observed that the pH has a strong influence on the adsorption behavior of arsenate. The maximum adsorption was achieved in the pH value at 2.5 and the trend decreases with increasing in pH. The applicability of the Langmuir adsorption isotherms well described the adsorption process and at optimal pH the adsorption efficiency was about 117.6 mg/gm. From the adsorption study, it was observed that around 70 % of final efficiency was adsorbed at initial 30 minutes time and after that it slows down. This indicates that the rate of adsorption was quite faster. Fourier transform infrared spectroscopy of unmodified, Zirconium immobilized and arsenic adsorbed carbon was investigated in detail. Hence from the above results it can be concluded that, these novel zirconium immobilized carbon sorbents are efficient in removing arsenate from water and can be an attractive as low cost alternative for the treatment of arsenic contaminated waste water.

ION-IMPRINTED CHITOSAN HYDROGEL FOR RECOVERING Ag(I) FROM AQUEOUS SOLUTION

Dr. Vikas Mittal[*], *and Dr. Kean Wang*
The Petroleum Institute, Abu Dhabi, UAE

Abstract

Chitosan (CS) hydrogels were prepared and imprinted by Ag(I) with different mass ratios. Compared with the native chitosan hydrogel sample, the optimally imprinted sample (Agim-75) presents a significantly enhanced Ag(I) adsorption capacity as well as a high selectivity. This is attributed to the specific cavities formed by the template Ag(I) ions in the polymer matrix. The effect of adsorption conditions and desorption process were also studied.

Experimental

Preparation of Imprinted CS Hydrogels. The preparation process consist of 3 steps. In Step (I), different amount of silver salt ($AgNO_3$) was firstly dissolved in 1 % (volumetric) acetic acid. Then 2 g of CS powder (90 % DD) was added to the acid solutions and stirred overnight. CS hydrogels were prepared in Step (II) using the phase inversion method, in which the mixture solution was added drop-wisely into 1 M NaOH solution via a syringe. Most of the silver ions were converted to silver oxides which are then embedded in the CS matrix in this step. The obtained hydrogel beads (with the diameter of ~ 2 mm) were then dispersed into 30 ml of 1 M NaOH solution at 50 °C for 5 hours for crosslinking reaction to take place. 1 ml of epichlorihydrin (ECH) was added to NaOH solution as the crosslinker. In Step (III), the crosslinked hydrogels were washed with excess 5 M HNO_3 solution to remove the embedded/remaining silver oxides and Ag(I). This process was repeated until the leaching out solution contains virtually no Ag(I) and the color of the hydrogels becomes fully transparent. Finally, the hydrogels were regenerated with 0.2 M NaOH for 8 hours and washed with DI water to attain a pH ~ 6.

Six Ag(I) imprinted CS hydrogels were fabricated and named *Agim3, Agim10, Agim25, Agim50, Agim75* and *Agim100*, respectively, according to the mass ratio of $AgNO_3$ to CS (e.g. Agim3 = 3 mg $AgNO_3$ imprinted in 1 g CS, and so on). *Nonim* is the native CS hydrogels which is cross-linked but not imprinted. The hydrogels were blot-dried and then freeze dried overnight to obtain the weight percentage of pure CS in the hydrogels.

Ag(I) Adsorption/Desorption. The adsorption were conducted in batch adsorbers under the room temperature. The blot-dried hydrogels with the known weight of pure CS were dispersed in 25 ml of Ag(I) aqueous solution with the natural pH of 5.6. After 12 hours of adsorption under stirring, 5 ml sample solution was withdrawn from the batch and filtered via a PTFE syringe filter (filter pore size:

[*] The Presenter

0.45 μm). The sample solution was kept in 2 % HNO_3 for the subsequent ICP-AES analysis (Model: Perkin Elmer ICP Optima 2000 DV).

Results & Discussion

Figure 1 compares the adsorption capacities on the 7 hydrogel samples at the same experimental conditions (Dosage: 0.6 g/L; C_0 = 10 mg/L Ag(I)). We see that, compared with the non-imprinted CS hydrogel (Nonim), the Ag(I) adsorption capacity was increased by more than 40 % on the imprinted samples. However, there is no obvious difference between the capacities of the 6 imprinted samples.

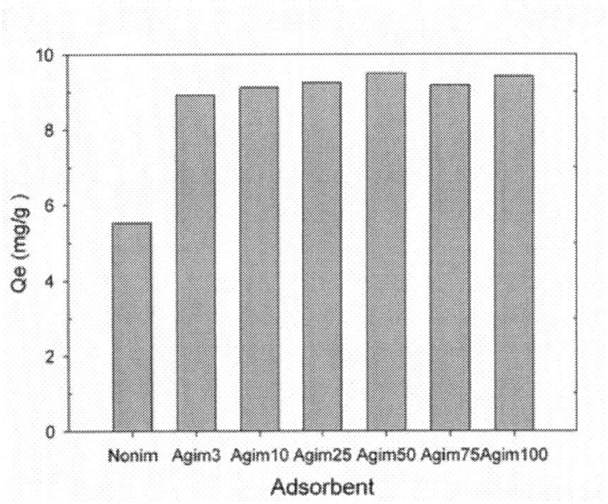

Figure 1 Adsorption capacities of Ag(I) on seven hydrogel samples (Dosage: 0.6 g/L; C_0 = 10 mg/L). The error bars represent the dual experiments.

Selectivity. The adsorption of the mixture solutions of Ag(I)-Cu(II) was then measured on each hydrogel sample and the adsorption selectivity was used as the benchmark to further scrutinize the performance of each sample. It is interesting to see in Figure 2 that the imprint ratio can drastically change the selectivity. The non-imprinted sample (Nonim) shows a higher selectivity of Cu(II) to Ag(I) (in term of mass) while the 3 imprinted samples (Agim3, Agim10 and Agim25) show little selectivity for the two ions. When the imprint ratio further increases, however, the imprinted hydrogels (Agim50, Agim75, and Agim100) demonstrated a higher adsorption selectivity for Ag(I). This observation can be generally explained as the follows: 1) Copper has two charges, which present stronger bond (affinity) with the protonated amino groups of CS and therefore is preferentially adsorb on pure CS hydrogels; 2) Upon imprinting, however, cavities were formed in CS matrix by the template Ag(I) and Ag_2O molecules. Such cavities (particularly the one formed by Ag(I), as will be further discussed later) would present stronger affinity to Ag(I), due to the favorable geometry.

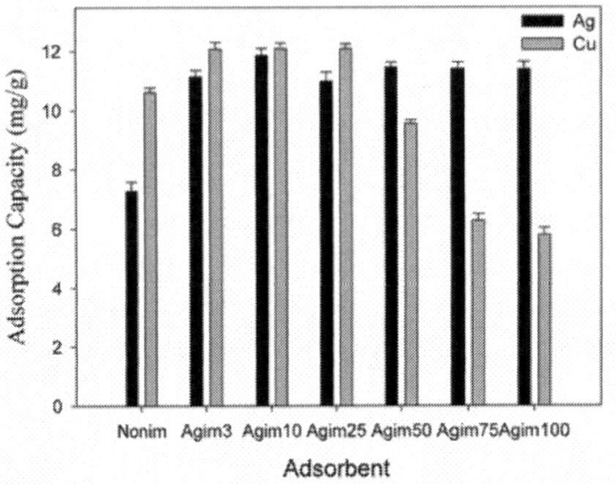

Figure 2 Competitive adsorption of Ag(I) and Cu(II) on 7 hydrogel samples. (Black bar: Ag(I), Grey bar: Cu(II)). The error bars represent the dual experiments.

Table 1 lists the distribution and selectivity coefficients calculated. It is noted that K_d and k of Agim75 and Agim100 are about 7 and 8 times those of Nonim. The difference between Agim75 and Agim100 is subtle as the K_d of Agim75 is slightly higher while the k of Agim100 is a bit better. Therefore, Agim75 was selected as the optimal adsorbents because of the less usage of the silver salt (imprint ratio: 75 mg-$AgNO_3$/g-CS) during the synthesis

Table 1 Competitive adsorption of Ag(I) and Cu(II) on 3 representative hydrogel samples

Hydrogels	Ions	Q_e (mg/g)	K_d	k
NonAgim	Ag(I)	8.66	2038	
	Cu(II)	10.61	3753	0.54
Agim75	Ag(I)	11.39	3955	
	Cu(II)	6.24	1056	3.75
Agim100	Ag(I)	11.43	3943	
	Cu(II)	5.80	935	4.21

Desorption. To examine the desorption efficiency on the imprinted hydrogels, 0.5 M and 1 M HNO_3 were used, respectively, to elute/desorb the Agim75 hydrogels pre-saturated with 25 ml of 2 mg/L Ag(I) solutions. We found that that more than 99 % of the pre-adsorbed Ag(I) can be recovered by either solution within the initial 5 minutes. It was also found that the re-generated adsorbents maintained the similar capacity after consecutive five adsorption-desorption cycles.

Conclusions

CS hydrogels imprinted with a high mass ratio of Ag(I), i.e.75 mg-$AgNO_3$/g-dryCS, present a much improved adsorption capacity as well as a good selectivity for Ag(I) in aqueous solution. The enhanced performance is attributed to the specific cavities created in CS matrix by the template Ag(I) ions. The optimal pH value for Ag(I) adsorption is ~ 5.0 and the adsorbed precious metal can be conveniently recovered by the acid wash.

References

1. Jiang, W.; Su, H. J.; Huo, H. Y.; Tan, T. W., Synthesis and Properties of Surface Molecular Imprinting Adsorbent for Removal of Pb(2+). *Applied Biochemistry and Biotechnology* 2010, 160, (2), 467-476.
2. X. Song, C. Li, R. Xu, and K. Wang, 'Molecular-Ion-Imprinted Chitosan Hydrogels for the Selective Adsorption of Silver(I) in Aqueous Solution', *IECR*, 51(34), 11261–11265, 2012.

Importance of mesopore of activated carbon in removing chromium(VI) from aqueous solution

Mingxia Fan[a,b], Shitang Tong[a*], Charles Q. Jia[c]

a. Department of Chemical Engineering, Wuhan University of Science and Technology, 947 Heping Avenue, Wuhan, Hubei, P. R. China, 430081

b. College of Chemical and Environmental Engineering, Hubei University of Technology, Wuhan, Hubei, P.R. China, 430068

c. Department of Chemical Engineering and Applied Chemistry, University of Toronto, 200 College Street, M5S 3E5, Toronto, Ontario, Canada

* Corresponding author, Tel: +86-27-86568662; Fax: +86-27-86568662; Email address: tongshitang@wust.edu.cn

ABSTRACT

In this study, a set of lab-made mesoporous activated carbons (MACs) were prepared from upgraded coal tar pitch with the nano-sized imprinting method for testing the ability to remove Cr(VI) from aqueous solutions. Adsorption of Cr(VI) was found to be highly pH dependent and the low pH was particularly favorable for Cr(VI) adsorption. The electrostatic interaction played a major role in adsorption of Cr(VI) ions from acidic solution on MAC. Three adsorption isotherm models namely, Langmuir, Freundlich and Dubinin–Radushkevich were used to analyze the equilibrium data. The adsorption kinetic data of Cr(VI) was analyzed and was found fitting well in a pseudo-second-order equation. Removal of Cr(VI) from aqueous solution increased with the increase of mesoporosity of activated carbon which highlighted the importance of mesoporosity in enhancing adsorption efficiency by avoiding steric hindrance..

Keywords: Chromium(VI), mesoporous activated carbon, adsorption isotherm, adsorption kinetics, mesoporosity

1. Introduciton

Chromium hexavalent ion (Cr(VI)) in wastewater is highly hazardous to the ecosystem[1]. Therefore, the treatment of Cr(VI) from wastewater is one of the most importantly environmental issues.

The methods employed for the removal of Cr(VI) from aqueous solutions include chemical oxidation–reduction, ion exchange, electrodialysis, electrochemical precipitation, evaporation, solvent extraction, reverse osmosis, chemical precipitation and adsorption [2–5]. Activated carbon adsorption is considered to be the simplest technique due to low cost and perfect removal. Many studies have been done on development of activated carbon adsorbents for Cr(VI) removal in the literature [6,7].

Compared to normal activated carbon (AC), mesoporous activated carbon (MAC) is regarded to be more favorable for removal of the large sized species from wastewater by adsorption due to easy access of the large sized species to mesopores. The objectives of the present work are to study the adsorption performance of Cr(VI) on MAC from aqueous solution to verify importance of mesopores on adsorption of Cr(VI).

2. Experimental

2.1 Preparation and characterization of mesoporous activated carbons (MACs)

The MAC samples tested were prepared from coal tar pitch by using the nano-sized imprinting method described in details by Tong et al [8].

The pore structural properties of the six MAC samples were examined using N_2 adsorption at 77K as seen in Table 1. The honeycomb-like porous structure was seen in the SEM image of the MAC (Figure 1) and the scale provided evidence of existing large pores in MAC.

Table 1 Pore structural properties of MACs.

Serial No.	S_{BET} m^2/g	S_{mes} m^2/g	V_{mic} mL/g	V_t mL/g	V_{mes}%
AC01	918.01	297.83	0.28	0.51	45.10
AC02	781.16	254.38	0.22	0.50	56.00
AC03	1197.39	582.62	0.27	0.78	65.38
AC04	1266.96	721.92	0.23	0.80	71.25
AC05	963.66	655.74	0.13	0.72	81.94
AC06	879.17	717.75	0.06	0.74	91.89

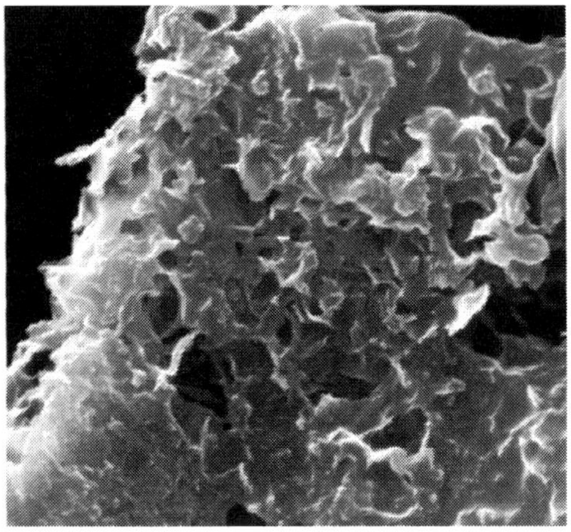

Figure 1. SEM image of MAC(AC04)

2.2 Adsorption of Cr(VI) on MACs

The working solution was prepared by dissolving potassium dichromate in double-distilled water. pH of the solution was adjusted by using either 0.1N NaOH or 0.1N H_2SO_4. Adsorption experiments were carried out in a thermostatic bath oscillator. 50 mL Cr(VI) solution of known concentration (C_0) and initial pH was taken in a 150mL erlenmeyer flask with a fixed dosage (2g/L) of MAC. The mixture was agitated at a speed of 200 rpm in the thermostatic shaker bath for a certain period of time. Except for the runs for investigating the influence of mesopore, AC02 was used in all experiments. Equilibrium adsorption isotherms were measured in 48 hours to ensure attainment of equilibrium. The concentrations of Cr(VI) in the solution were determined by using a UV-visible spectrometer according to the standard procedure (GB7467-87).

3. Results and Discussion

3.1 Effect of pH

Percent removal of Cr(VI) on MAC was examined for different pH values of the solution as shown in Figure 2. It is seen that the adsorption displayed strong pH dependence. A large percent removal of Cr(VI) at low pH can be explained by the strongly electrostatic interaction near the interfacial region between the charged metal ions in solution and the charged surface of MAC. In acidic solution, the charged Cr(VI) ion is predominantly in forms of $Cr_2O_7^{2-}$, $HCrO_4^-$, CrO_4^{2-}, $Cr_3O_{10}^{2-}$ and $Cr_4O_{13}^{2-}$ which are all negatively charged species. According to pH of the zero point of charge (pHzpc) measured, all MAC samples have the value of pHzpc around 7.35, indicating MACs display a positive charged surface. As a result, the electrostatic interaction played a major role in adsorption of Cr(VI) ions from acidic solution on MAC. With an increase of pH, the percent removal of Cr(VI) is rapidly decreased, ascribed to the competent adsorption of the negatively charged chromate ions and OH^- on the surface of MAC[9].

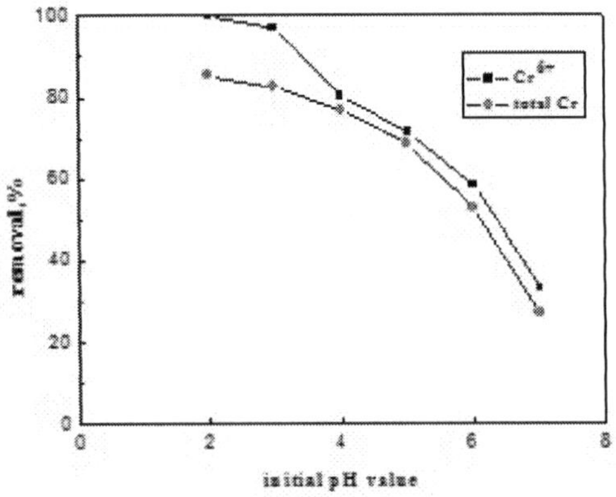

Figure 2. Effect of pH value on adsorption of Cr(VI) on MAC (initial concentration of Cr(VI), 50 mg/L; temperature, 303 K; contact time, 48 hrs; dosage, 2g/L)

On the other hand, chromium existing in solution is mainly in two states which are Cr(VI) and Cr(III). The stability of these forms in solution highly depends on pH of the solution. In knowledge of chromium chemistry, the negative charged Cr(VI) ions are easily reduced to Cr(III) at low pH in the presence of MAC as shown in following equations (1) and (2).

$$2\ Cr_2O_7^{2-} + 3C + 16H^+ = 4Cr^{3+} + 3CO_2 + 8H_2O \qquad (1)$$

$$4\ HCrO_4^- + 3C + 16H^+ = 4Cr^{3+} + 3CO_2 + 10H_2O \qquad (2)$$

It is noted from Figure 2 that total removal of chromium is always lower than that of Cr(VI). Particularly the difference increases with a decrease in pH. In fact, the removal of Cr(VI) at low pH is partially due to reduction of Cr(VI) into Cr(III) which are known to exist in various hydrolyzed forms of $[Cr(H_2O)_6]^{3+}$, $[Cr(H_2O)_5OH]^{2+}$and $[Cr(H_2O)_4(OH)_2]^+$depending on pH of the solution. Clearly, the positive charged surface of MAC does not favor to adsorption of the positive charged Cr(III) species due to electrostatic repulsion.

3.2 Adsorption isotherms

Adsorption equilibrium isotherms of Cr(VI) on MAC were measured as shown in Figure 3. The positive temperature dependence suggests the chemical nature of Cr(VI) adsorption on the MAC. Langmuir, Freundlich and Dubinin–Radushkevich(D–R) isotherms were used to interpret equilibrium adsorption data. The values of model parameters and correlation coefficients R^2 were presented in Table 2. The correlation coefficient for three models are all >0.95, likely Langmuir model is the best interpretation of the experimental data. The predicted maximum adsorption of Cr(VI) on MAC ranges from 42.52 to 53.79 mg/g when raising temperature from 303 to 323K.

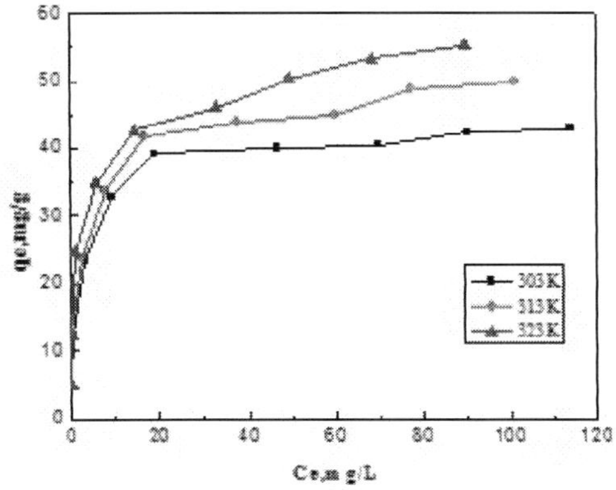

Figure 3. Effect of temperature on Cr(VI) equilibrium adsorption amount (initial concentration of Cr(VI), 10-200 mg/L; pH, 3; contact time, 48 hrs; dosage, 2g/L)

Table 2 Model parameters for adsorption Cr(VI) onto MAC at different temperatures

Tempe -rature K	Langmuir[d]			Freundlich[e]			Dubinin-Radushkevich[f]			
	K_L	q_m	R^2	K_F	1/n	R^2	B	q_s	E	R^2
303	16.74	42.52	0.9963	21.65	0.153	0.9483	0.00104	54.85	21.93	0.9658
313	21.54	47.44	0.9860	21.98	0.186	0.9662	0.00135	71.23	19.25	0.9712
323	16.98	53.79	0.9721	26.47	0.164	0.9977	0.00144	83.66	18.63	0.9934

d. $q_e=K_Lq_mCe/(1+K_LCe)$; q_m: the maximum adsorption capacity (mg/g); K_L: Langmuir constant (L/mg);

e. $q_e=K_FCe^{1/n}$; K_F: a constant indicative of the relative adsorption capacity of the adsorbent (mg/g); 1/n: the intensity of the adsorption ;

f. $q_e=q_De^{-B\varepsilon 2}$; $\varepsilon=RTexp(1+1/Ce)$; $E= (-2B)^{-1/2}$; q_D: the adsorption capacity(mg/g); ε: the Polanyi potential; B: the mean free energy E (kJ/mol) of adsorption per molecule of the adsorbate, (mol^2/kJ^2) .

3.3 Adsorption kinetics

Figure 4 shows the time dependence of removal of Cr(VI) on MAC. It is seen that the MAC displays a rapidly initial rate at three temperature levels. The fast initial rate is ascribed to the large mesopores of MAC and the gradually lowering rate to approaching equilibrium indicates the formation of steric hindrance in pores with progressing of adsorption of Cr(VI), that seems to lower the diffusive pace of the adsorbed Cr(VI) into deep pores of MAC.

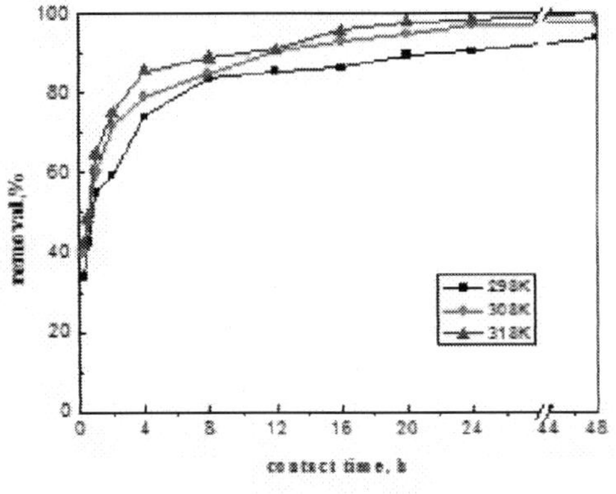

Figure 4. Effect of time on adsorption of Cr(VI) on AC (initial concentration of Cr(VI), 50mg/L; pH, 3; dosage, 2g/L)

The measured kinetic data were fitted into three kinetic models, the pseudo-first-order, the pseudo-second-order and the intra-particle diffusion. The kinetic model parameters for three models were shown in Table 3. The pseudo-second-order kinetics is the best suitable for representing the measured kinetic data of adsorption of Cr(VI) on MAC. Several research groups reported that the pseudo-second order kinetic model was well suited in describing adsorption kinetic of Cr(VI) onto activated carbons [10–12].

Table 3 Kinetic parameters for the adsorption of Cr(VI) onto MAC

Temperature K	qe	pseudo-first-order[g]			pseudo-second-order[h]			intra-particle diffusion[i]	
		K_1	qe	R^2	K_2	qe	R^2	K_P	R^2
298	22.65	2.97	11.109	0.9769	0.906	23.132	0.9996	0.3858	0.9346
308	24.26	2.6	11.112	0.9865	0.89	24.606	0.9994	0.3819	0.9344
318	24.59	3.5	11.820	0.9630	1.03	25.013	0.9995	0.3714	0.9116

g. $\log(q_e-q_t)=\log q_e-k_1t/2.303$; q_e: the amount of solute adsorbed at equilibrium per unit mass of adsorbent (mg/g); k_1: the rate constant (1/min).

h. $t/q_t=1/k_2q_e^2+t/q_e$; k_2: the equilibrium rate constant (g/mg min).

i. $q_t=k_pt^{1/2}+c$; kp: the diffusion reaction rate constant (mg/g $min^{1/2}$).

3.4. Steric hindrance: importance of mesoporosity.

Figure 5 shows an importance of mesopores for removal of Cr(VI). The percent removal increased with an increase of mesopore porosity for different length of contact time. With short contact time, the removal was more significantly influenced by the mesopore porosity than with long contact time. It means that the intra-particle diffusive resistance of the adsorbed Cr(VI) visually occurs due to formation of steric hindrance in pores. The larger mesopore prosity, the higher removal approached, but the difference of removal with different contact times was decreased. With an increase of mesopore porosity up to near 92%, the difference of removal completely disappeared regardless of the difference of the contact time, indicating that mesopores dominate the adsorption kinetics during removal of Cr(VI) due to accessibility of mesopores to the large sized adsorbate. In addition, surface oxygen groups in MAC may be one of important factors affecting the adsorption performance of Cr(VI), however, the experimental result will be discussed elsewhere.

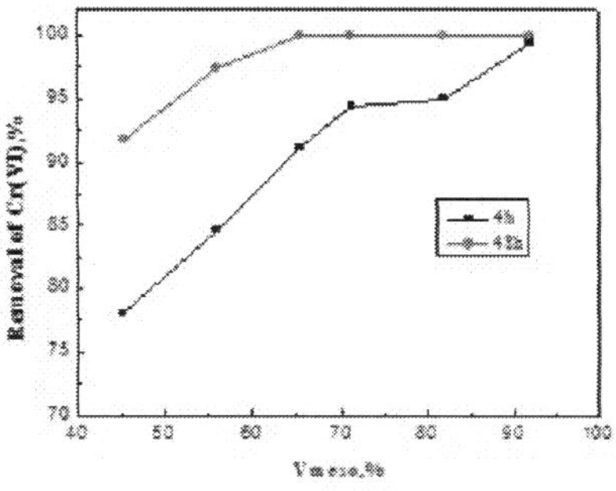

Figure 5. Importance of mesopore in adsorptive removal of Cr(VI) with MACs (initial concentration of Cr(VI), 50 mg/L; temperature, 303 K; pH,3; dosage, 2g/L)

4. Conclusions

A set of mesoporous activated carbons (MACs) were prepared from upgraded coal tar pitch with the nano-sized imprinting method. The MACs were found very effective adsorbents for the removal of Cr(VI) from aqueous solutions. The solution's pH has an important role in Cr(VI) adsorption and low pH are favorable for Cr(VI) adsorption on MAC. All MAC samples have the value of pHzpc around 7.35, indicating MACs display a positive charged surface under acidic conditions . As a result, the electrostatic interaction played a major role in adsorption of Cr(VI) ions. It is evident that high temperature resulted in an increase in equilibrium adsorption capacity, which suggests a chemosorption mechanism. The experimental data fitted well to the Pseudo-second-order kinetic models. Removal of Cr(VI) from aqueous solution increased with the increase of mesopore porosity of MACs, demonstrating the importance of mesoporsity in avoiding steric hindrance .

References

[1] Q. Diana, L. Oqliveira, Removal of As and Cr from aqueous solutions using solid waste from leather industry, J. Hazard. Mater. 151 (2008) 280–284.

[2] T. Wang, Z. Li, Removal of chromium from water and wastewater, Environ. Sci.Technol. 26 (2003) 85 in Chinese.

[3] B. Hua, B. Deng, Influences of water vapor on Cr(VI) reduction by gaseous hydrogen sulfide, Environ. Sci. Technol. 37 (2003) 4771.

[4] M. Pettine, L. Campanella, F.J. Millero, Reduction of hexavalent chromium by H2O2 in acidic solutions, Environ. Sci. Technol. 36 (2002) 901–907.

[5] C. Kim, Q. Zhou, B. Deng, E.C. Thornton, H. Xu, Chromium(VI) reduction by hydrogen sulfide in aqueousmedia: stoichiometry and kinetics, Environ. Sci. Technol. 35 (2001) 2219–2225.

[6] E. Demirbas, M. Kobya, A.E.S. Konukman, Error analysis of equilibrium studies for the almond shell activated carbon adsorption of Cr(VI) from aqueous solutions, Journal of Hazardous Materials 154 (2008) 787–794.

[7] M. Kobya, Removal of Cr(VI) from aqueous solutions by adsorption onto hazelnut shell activated carbon: kinetic and equilibrium studies, Bioresource Technology. 91 (2004) 317–321.

[8] Tong Shitang; Mao Lei; Zhang Xiaohua, et al. Synthesis of Mesoporous Carbons from Bituminous Coal Tar Pitch Using Combined Nanosilica Template and KOH Activation [J]. Industrial & Engineering Chemistry Research. 50(2011):13825-13830.

[9] K. Mohanty, M. Jha, B.C. Meikap, M.N. Biswas, Biosorption of Cr(VI) from aqueous solutions by Eichhornia crassipes, Chem. Eng. J. 117 (2006) 71–77.

[10] L. Khesami, R. Capart, Removal of chromium(VI) from aqueous solution by activated carbons: kinetic and equilibrium studies, J. Hazard. Mater. 123 (2005) 223–231.

[11] N. Zhao, N. Wei, J. Li, Z. Qiao, J. Cui, F. He, Surface properties of chemically modified activated carbons for adsorption rate of Cr(VI), Chem. Eng. J. 115 (2005) 133–138.

[12] C.A. Basar, A. Karagunduz, B. Keskinler, A. Cakici, Effect of presence of ions on surface characteristics of surfactant modified powdered activated carbon, Appl. Surf. Sci. 218 (2003) 170–175.

Acknowledgment

Financial support from National Science Foundation of China through project NSFC 51174150 and Research Center of Green Manufacturing and Energy-saving & Emission Reduction Technology (Wuhan University of Science and Technology) through project A1004 are gratefully acknowledged.

Adsorption of Aniline, Phenol, 4-chlorophenol and 4-nitrophenol by Granular Activated Carbon: Relationship with Solvatochromic Parameters modeling study

Dr. S. Suresh, Department of Chemical Engineering, Maulana Azad National Institute of Technology Bhopal, Bhopal-462 051, M.P., India.*
Dr. V. C. Srivastava, Dr. I. M. Mishra, Department of Chemical Engineering, Indian Institute of Technology Roorkee, Roorkee- 247667, India.
**Corresponding Author: Tel.: +91-8989005393; E-mail: sureshpecchem@gmail.com*

Abstract

In the present paper, equilibrium adsorption isotherm data generated in the temperature range of 288-318 K are reported for the adsorption of Aniline (AN), Phenol (P), 4-chlorophenol (CP) and 4-nitrophenol (NP) from aqueous solution onto granular activated carbon (GAC). Adsorption of these adsorbates onto GAC was found to be favourably influenced by an increase in temperature. The dispersion force between the π-electrons of GAC and the π-electrons in AN, P, CP and NP molecules is the main adsorption mechanism. The adsorption behavior of AN, P, CP and NP strongly depends upon the intrinsic molar volume (V_I), thus, steric effect has strong influence on observed adsorption coefficients values. This study also shows that hydrogen bonding between the adsorbates and the GAC plays an important role in the adsorption behavior.
Key words: Adsorption; Granular Activated Carbon; Aniline; Phenol; 4-chlorophenol; 4-nitrophenol; Solvatochromic Parameters.

Introduction

Phenol (P) and its derivatives and associated compounds are very important organic intermediates, used in the manufacture of many products in such units as drugs, rubber, pesticides, varnishes and also, dyestuffs, chemicals, petrochemicals, paper, wood, metallurgy and coking plants (1-10). P and its derivatives can also originate from diffuse emissions, e.g. from the tar coatings of the roads and pipes and from the use of pesticides including their transformation products. Phenols are extremely toxic to living species. Exposures to phenol can occur in a workplace, from environmental media, from contaminated drinking water or foodstuffs or from use of consumer products containing phenol (7). Phenol is absorbed through all routes readily: ingestion, inhalation, and dermal. Aniline is the common substance found in the effluents of pharmaceutical, dyestuff, photographic, and agrochemical industries. Phenols and aniline are mutagenic and also carcinogenic. Aniline reacts easily in the blood to convert hemoglobin into methaemoglobin, thereby preventing oxygen uptake (7). Total phenol concentration in the wastewater of a typical Indian refinery processing 5.0 million tons of crude per year. The concentration of phenolic compounds in the wastewater from resin plants is typically in the range of 12–300 mg l^{-1}. Ministry of Environment and Forests, Government of India; and United States Environmental Protection Agency has prescribed that the concentration of phenols (measured as phenol) should not exceed 1.0 mg l^{-1} for their discharge into the surface waters and 5.0 mg l^{-1} for discharge into the public sewers, on land for irrigation and marine coastal areas.

Various treatment methods such as adsorption (1-9) photo-degradation, coagulation and flocculation, chemical oxidation, biological process (9-10), etc. have been studied for the removal of phenols from wastewater. Conventional methods for the removal or destruction of AN from wastewaters include solvent extraction, biodegradation (11), catalytic oxidation (12), membrane separation, ultrasonic degradation, supercritical water oxidation, and electrochemical oxidation.

Adsorptive removal of Aniline (7) and phenolics (1-9) from wastewaters has been reported. The adsorption of P and substituted phenols from aqueous solution on activated carbons has been intensively investigated for several decades. Of all the methods proposed for the removal of phenols and associated compounds, the adsorption appears to offer the best prospects for overall treatment, especially for the effluents with moderate and low concentrations. Because of a good capacity for adsorption of organic matter, granular and powder activated carbons are the most widely used adsorbents for the adsorptive removal (1-9).

The present paper reports the solvatochromic parameters like intrinsic molar volume (V_I), the polarity/polarizability parameter (π), the hydrogen-bonding donor parameter (α_s) and hydrogen-boding acceptor parameter (β_m) were used to predict adsorption coefficient (K_D) values.

Experimental

Materials: All the chemicals used in the study were of analytical reagent grade. AN was procured from Qualigens Fine Chemicals, Mumbai. P was procured from Ranbaxy Fine Chemicals, New Delhi. CP, NP, NaOH, and HCl were obtained from S.D. Fine Chemicals, Mumbai. Stock solutions of the AN, P, CP and NP were made by dissolving an exact amount of the respective compound in double distilled water. The GAC supplied by S.D. Fine-Chemicals limited, (Mumbai, India) was manufactured from coconut shells.

Methods: The physico-chemical characterization of GAC was performed using standard procedures and the results are given elsewhere (1-8). The BET surface area of GAC was 977.05 m^2g^{-1}, whereas BJH adsorption/desorption surface area of pores was 45.03/47.63 m^2g^{-1}.

Analysis of AN, P, CP and NP: The initial and residual concentration of AN, P, CP and NP were determined by finding out the absorbance of the respective solutions at 230, 269, 279 and 317 nm, respectively, using UV/vis spectrophotometer (Lambada 35; PerkinElmer, MA 02451, USA). Detailed analyses are reported by Suresh et al. (8).

Batch adsorption: For each experimental run, 100 ml aqueous solution of known concentration of the respective compounds was taken in a 250 ml conical flask containing 1 g of GAC. These flasks were agitated at a constant shaking rate of 150 rpm in a temperature- controlled orbital shaker (Metrex Scientific Instruments, New Delhi) maintained at 288, 303 and 318 K as the case may be. In all the experiments, the pH of the solution was maintained constant using a buffer of potassium dihydrogen orthophosphate (KH_2PO_4) and sodium hydroxide (NaOH). The samples were withdrawn from the flasks after 24 h contact time which was more than the equilibrium time of 15 h determined in another study. The samples were centrifuged using a centrifuge (Remi Instruments, Mumbai) at 10000 rpm for 5 min. The removal of the given compound from the solution and the adsorptive uptake in solid phase (q_t) was calculated by using reported equation (1-9).

Theory

Adsorption equilibrium study: Equilibrium adsorption studies are required for the design of an adsorption system. For use of these studies in the design and optimization of the process, it is necessary to correlate the equilibrium sorption data. These are several equilibrium isotherm equations (1, 3-4, 6) have discussed the theory associated with the most commonly used isotherm models. Sum of square of errors (SSE) was used as a criterion for finding out the most appropriate model to represent the experimental data. The SSE is given as:

$$SSE = \sum_{i=1}^{n} (q_{e,\exp} - q_{e,cal})_i^2 \qquad (1)$$

Where, $q_{e,\exp}$ and $q_{e,cal}$ are the experimental and calculated equilibrium adsorbate uptake, respectively; and n is the number of data points.

2.5.2. Adsorption thermodynamics: ΔG_0, ΔH_0, ΔS_0 and temperature. The equilibrium adsorption constant (K_D) can be related to as follows (8). Apparent isosteric heat of adsorption ($\Delta H_{st,a}$) at constant surface coverage (q_e) is calculated using the Clausius-Clapeyron equation (13).

Results and discussion

The results for the equilibrium adsorption of AN, P, CP and NP onto GAC are presented (not shown here). In order to show the low concentration region more clearly, only results for

$20 < C_0 < 1000$ mg l^{-1} are shown in the figure. It is found that the adsorption of all the solutes increases with an increase in temperature. The adsorptive uptake, q_e rises sharply at a low C_e value and then gradually with an increase in C_e. An increase in temperature increases the chemical potential of the organic molecules to penetrate through the surface pores of GAC. Also, the mobility of the adsorbate increases with an increase in temperature. These together results in the enhancement in the adsorptive capacity of the GAC at higher temperatures. An increase in the phenol adsorption capacity of the carbonaceous adsorbents with an increase in temperature has also been reported by other investigators (1-9). Detailed results are reported by Suresh et al. (8). K_D represents the partition coefficient of adsorbate between the solution and the adsorbed phases at infinite dilution. A close look at the K_D values shows that it follows the trend as NP>CP>P>AN at all temperatures (values not shown here). Variation of K_D with σ and dipole moment (D) for P, CP and NP is shown in Fig. 1. A high correlation between K_D and σ and D is observed. It may, thus, be inferred that the substitution of more electron-withdrawing group onto phenol increases its K_D values. Also, the adsorabte having higher D value gets easily adsorbed onto GAC.

Correlation of adsorption coefficients with solvatochromic parameters

Solvatochromic parameters like intrinsic molar volume (V_I), the polarity/polarizability parameter (π), the hydrogen-bonding donor parameter (α_s) and hydrogen-boding acceptor parameter (β_m) can be used to predict K_D values. The linear solvation energy relationships (LSER) models based on solvatochromic parameters are generally used to predict the K_D values of adsorption of various compounds onto GAC.

The values of solvatochromic parameters of AN, P, CP and NP are listed in Table 1, whereas the K_D values of adsorption of these compounds onto GAC at 288, 303 and 318 K are listed in Table 2. Significant linear relationship was observed between K_D and β_m **or** α_s **or** V_I (Table 2 and Fig. 2). However, no relationship was observed between K_D values and π parameter (Fig. 2 and Table 2). Among the four solvatochromic parameters, β_m (R^2=0.760-0.876) and V_I (R^2=0.758-.873) showed high linear correlation with K_D at all temperatures. A significant relationship (R^2=0.573-0.620) was also observed between K_D and α_s. Considering the fact that R^2 values greater than 0.876 were not obtained using LSER with respect to individual parameters, quadratic solvation energy relationships (QSER) were also fitted to the data. However, even QSER relationship did not improve the correlation. Therefore, the LSER model only has been used.

Regression equations of the form of Eq. (2) given by Kamlet et al. (14-15) could be used to predict the behavior of adsorption of various adsorbates onto GAC.

$$K_D = a_0 + a_1 V_I + a_2 \beta + a_3 \alpha + a_4 \pi \qquad (2)$$

Where, a_0, a_1, a_2, a_3 and a_4 are the regression coefficients. These coefficients evaluate the contributions of the corresponding solvent effects on K_D. The following regression relationships were deduced after the fitting of the data at 288, 303 and 318 K, respectively.

288 K: $$K_D = -54.08 + 0.117V_I - 0.7\beta - 23.58\alpha \qquad R^2=1 \qquad (3)$$

303 K: $$K_D = 342.37 - 0.759V_I + 13.77\beta + 173.20\alpha \qquad R^2=0.999 \qquad (4)$$

318 K: $$K_D = 223.7 - 0.45V_I + 10.37\beta + 117.99\alpha \qquad R^2=1 \qquad (5)$$

Assuming the a_0 in Eq. (2) to be a function of temperature (of Arrhenius equation form), the following modified equation was used to predict the adsorption behavior at all temperatures:

$$K_D = K_{D0}\exp(-A/T) + a_1 V_I + a_2 \beta + a_3 \alpha + a_4 \pi \qquad (6)$$

Where, K_{D0} and A is a constant and T is the absolute temperature (K). The statistical confidence of regression relationships at a given significance value was determined by F-test: $F = R^2(N-k-l)/(k(1-R^2))$ where, N and k are the number of experimental K_D values and the solvatochromic indexes, respectively and R is the correlation coefficient and R^2 is the coefficient of determination (16).

The fitting of the Eq. (6) as shown in Fig. 3 was first carried out with all the parameters and later on by neglecting the π parameter. R^2 values showed insignificant difference when π parameter was also considered. Thus, it was be concluded that π parameter has negligible effect in the adsorption behavior and could be neglected in the correlation.

The final regression equation which considered the effect of temperature but neglected the effect of π is given as:

$$K_D = 76.82\exp(-211/T) - 0.3633V_I - 7.624\beta + 85.65\alpha;$$

$$R^2=0.787; \quad F(12,5)=10.8 \qquad (7)$$

Eq. (7) is statistically significant at 95% confidence interval because $F(12,5)=10.8 > F_{0.05}(12,5)=4.68$.

This study shows that adsorption behavior of AN, P, CP and NP strongly depended upon the V_I, thus, steric effect has strong influence on observed K_D values. This study also shows that the hydrogen bonding between the adsorbates and the GAC plays an important role in the adsorption behavior. It seems that the adsorbate and GAC act as both hydrogen bonding acceptor or donor, however, higher correlations for β_m show that adsorbates act more as hydrogen bonding acceptor and GAC surface acts as hydrogen bonding donor.

The Van't Hoff's plot for AN, P, CP and NP adsorption onto GAC from which ΔH_0 and ΔS_0 values have been estimated (figure not shown here). ΔG_0 values were negative indicating that the adsorption process led to a decrease in ΔG_0 and that the adsorption process is feasible and spontaneous (1, 3-4, 6). The positive values of ΔH_0 indicate the endothermic nature of the adsorption process. The adsorption for AN, P, CP and NP onto GAC, is thus both by physical adsorption and chemical transition (17). The values obtained in the present study are in the same range. The positive value of ΔS_0 suggests increased randomness at the solid/solution interface with some structural changes in the adsorbates and the GAC. Detailed results and discussions are reported by Suresh et al. (8). The variation in isosteric heat of adsorption with surface loading can also be attributed to the possibility of having lateral interactions between adsorbed AN, P, CP and NP. Detailed results and discussions are reported by Suresh et al. (8).

Conclusion

In the presented study, many isotherm models were tested for modeling the adsorption behavior of AN, P, CP and NP onto GAC. Solvatochromic parameters like intrinsic molar volume (V_I), the polarity/polarizability parameter (π), the hydrogen-bonding donor parameter (α_s) and hydrogen-boding acceptor parameter (β_m) were used to predict adsorption coefficient (K_D) values. The linear solvation energy relationships models based on solvatochromic parameters were used to predict the adsorption coefficient (K_D) values. This study shows that the V_I and hydrogen bonding plays an important roles in the adsorption behavior of AN, P, CP and NP onto GAC.

References

1. Suresh S, Srivastava V.C. and Mishra I.M (2011a), "Isotherm, Thermodynamics, Desorption, and Disposal Study for the Adsorption of Catechol and Resorcinol onto Granular Activated Carbon," *J. Chem. Eng. Data (ACS).* 2011, 56 (4), 811–818.
2. Suresh S, Srivastava V.C. and Mishra I.M (2011b), "Adsorptive removal of phenol from binary aqueous solution with aniline and 4-nitrophenol by granular activated carbon," *Chem. Eng. J.* 2011, 171 (3), 997-1003.
3. Suresh S, Srivastava V.C. and Mishra I.M (2011c), "Adsorption of Hydroquinone in Aqueous Solution by Granulated Activated Carbon," *J. Environ. Eng. (ASCE).*137(12), 1145-1157.
4. Suresh S, Srivastava V.C. and Mishra I.M (2011d), "Isotherm, Thermodynamics, Desorption, and Disposal Study for the Adsorption of Catechol and Resorcinol onto Granular Activated Carbon," *Sep. Sci. Technol.* 46(11), 1750 - 1766, 2011.
5. Suresh S, Srivastava V.C. and Mishra I.M (2012a), "Adsorptive removal of aniline by granular activated carbon from aqueous solutions with catechol and resorcinol," *Environ. Technol.* 33 (7), 773-781.
6. Suresh S., Vijayalakshmi G., Rajmohan B. and Subbaramaiah V (2012b), "Adsorption of Benzene Vapor onto Activated Biomass from Cashew Nut Shell: Batch and Column Study," Recent *Patents Chem. Eng.* (Accepted).
7. Suresh S, Srivastava V.C. and Mishra I.M (2012c), "Studies Of Adsorption Kinetics And Regeneration Of Aniline, Phenol, 4-Chlorophenol And 4-Nitrophenol By Activated Carbon," *Chem Ind. Chem. Eng. Q.* (Accepted).
8. Suresh S, Srivastava V.C. and Mishra I.M (2012d), "Adsorption of Aniline, Phenol, 4-chlorophenol and 4-nitrophenol onto Granular Activated Carbon: Isotherm, Thermodynamics study," *Brazilian J. Chem. Eng.* (under review).
9. Suresh, S. and Keshav A. (2012), "Textbook of Separation Processes", *Studium Press (India) Pvt. Ltd* (ISBN: 978-93-80012-32-2), pp.1-459.
10. Suresh S. (2011), "Biodegradation of Hydroquinone Using Sequential Batch Reactor: A Preliminary Study of Industrial Effluent," Res. *J. Chem. Environ.*15 (2).
11. O'Neill, F.J., Bromley-Challenor, K.C.A., Greenwood, R.J. and Knapp, J.S (2000), Bacterial growth on aniline: Implications for the biotreatment of industrial wastewater," *Water Res.*, 34, 4397.
12. Barbier J.J., Oliviero, L., Renard, B. and Duprez, D (2002), "Catalytic wet air oxidation of ammonia over M/CeO_2 catalysts in the treatment of nitrogen containing pollutants," *Catal. Today*, 75, 29.
13. Young, D.M. and Crowell, A.D (1962), " Physical adsorption of gases, London: Butterworths, 426.
14. Kamlet M.J.; Solomonovici, M.; Taft, R. W (1979), *Am. Chem. Soc.*, 101, 3734.
15. Kamlet, M.J., Abboud, L., Abraham, M. H. and Taft, R. W (1983), "Linear solvation energy relationships. 23. A comprehensive collection of the solvatochromic parameters, π^*, α, and β, and some methods for simplifying the generalized solvatochromic equation. *J. Org. Chem.*, 48, 2877.
16. Georgiev, G.S. and Dakova, I.G (1994), "Solvent effect on the methacrylic acid-methyl methacrylate radical copolymerization. Analytical estimation by linear and nonlinear solvation energy relationships," *Macromol. Chem. Phys.*, 195, 1695.
17. Jianguo, C., Li, A., Hongyana, S., Zhenghao, F., Chao, L. and Quanxing, Z (2005), "Adsorption characteristics of aniline and 4-methylaniline onto bifunctional polymeric adsorbent modified by sulfonic groups," *J. Hazard. Mater.*, 124, 173.
18. International Program on Chemical Safety (IPCS) (2005) Phenol. Environmental Health Criteria.161.
19. Yang, K., Wu, W., Jing, Q. and Zhu, L (2008), " Aqueous Adsorption of Aniline, Phenol, and their Substitutes by Multi-Walled Carbon Nanotubes," *Environ. Sci. Technol.*, 42, 7931.
20. Laszlo, K., Tombacz, E. and Novak, C(2007), "pH-dependent adsorption and desorption of phenol and aniline on basic activated carbon," *Colloids Surf. A: Physicochem. Eng. Aspects*, 306, 95.
21. Yogesh, K., Popat, M., Ganguly, B., Brahmbhatt, H. and Bhattacharya (2008), "A Studies on the separation performances of chlorophenol compounds from water by thin film composite membranes," *Macromol. Res.* 16, 590.

Table 1. Physic-chemical properties and solvatochromic parameters.

Parameters	AN	P	CP	NP	References
Molecular weight (MW) (g mol^{-1})	93.13	94.11	128.56	139.11	(18)
Water solubility (g l^{-1}) at 25 ^{0}C (C_s)	34	83	27	12.4	(18)
pK_a	4.63	9.95	9.38	7.15	(19)
Octanol/ water partition coefficient at 25 ^{0}C ($\log K_{ow}$)	0.94	1.46	2.39	1.91	(19)
Projected area (nm^2) (A_m)	0.47	0.437	0.482	0.519	(20)
Dipole moment (Debye) (D)	1.17	1.233	1.477	5.44	(21)
Polarity /Polarizability parameter (π)	0.73	0.72	0.72	1.01	(19)
Intrinsic molar volume (cm^3/mol) (V_I)	0.567	0.536	0.625	0.685	(19)
hydrogen-bonding donor parameter (α_s)	0.16	0.61	0.67	0.93	(19)
hydrogen-bonding acceptor parameter (β_m)	0.5	0.33	0.72	1.01	(19)

Table 2. Correlation of adsorption coefficients with solvatochromic parameters.

Temp. (K)	Linear equation	R^2	Quadratic equation	R^2
Hydrogen-bonding acceptor parameter (β_m)				
288	$K_D=1.664\beta_m-0.057$	0.593	$K_D=-1.8943\beta_m^2+4.613\beta_m-1.1255$	0.786
303	$K_D=10.54\beta_m-2.978$	0.573	$K_D=36.08\beta_m^2-35.478\beta_m+8.8603$	0.994
318	$K_D=11.08\beta_m-1.561$	0.620	$K_D=27.035\beta_m^2-22.198\beta_m+6.388$	1.000
Hydrogen-bonding donor parameter (α_s)				
288	$K_D=2.052\alpha_s-0.385$	0.761	$K_D=1.2119\alpha_s^2+0.3938\alpha_s+0.1764$	0.613
303	$K_D=13.3\alpha_s-5.242$	0.769	$K_D=36.852\alpha_s^2-28.09\alpha_s+4.1445$	0.999
318	$K_D=14.35\alpha_s-4.179$	0.876	$K_D=34.57\alpha_s^2-25.158\alpha_s+5.1203$	0.986
Intrinsic molar volume (cm^3/mol) (V_I)				
288	$K_D=63.98V_I-33.29$	0.873	$K_D=-0.4863\times10^{-4}V_I^2+0.069V_I-22.764$	0.842
303	$K_D=58.95V_I-32.29$	0.758	$K_D=8.0117\times10^{-4}V_I^2-0.92V_I+264.28$	0.991
318	$K_D=9.43V_I-4.76$	0.805	$K_D=5.9503\times10^{-4}V_I^2-0.66V_I+186.67$	0.999
Polarity /Polarizability parameter (π)				
288	$K_D=0.915\pi-0.338$	0.012	$K_D=-141.2\pi^2+179.02\pi-55.057$	0.949
303	$K_D=15.56\pi-6.574$	0.085	$K_D=-418.33\pi^2+543.24\pi-170.87$	0.283
318	$K_D=16.25\pi-5.461$	0.091	$K_D=-549.21\pi^2+709.01\pi-220.93$	0.425

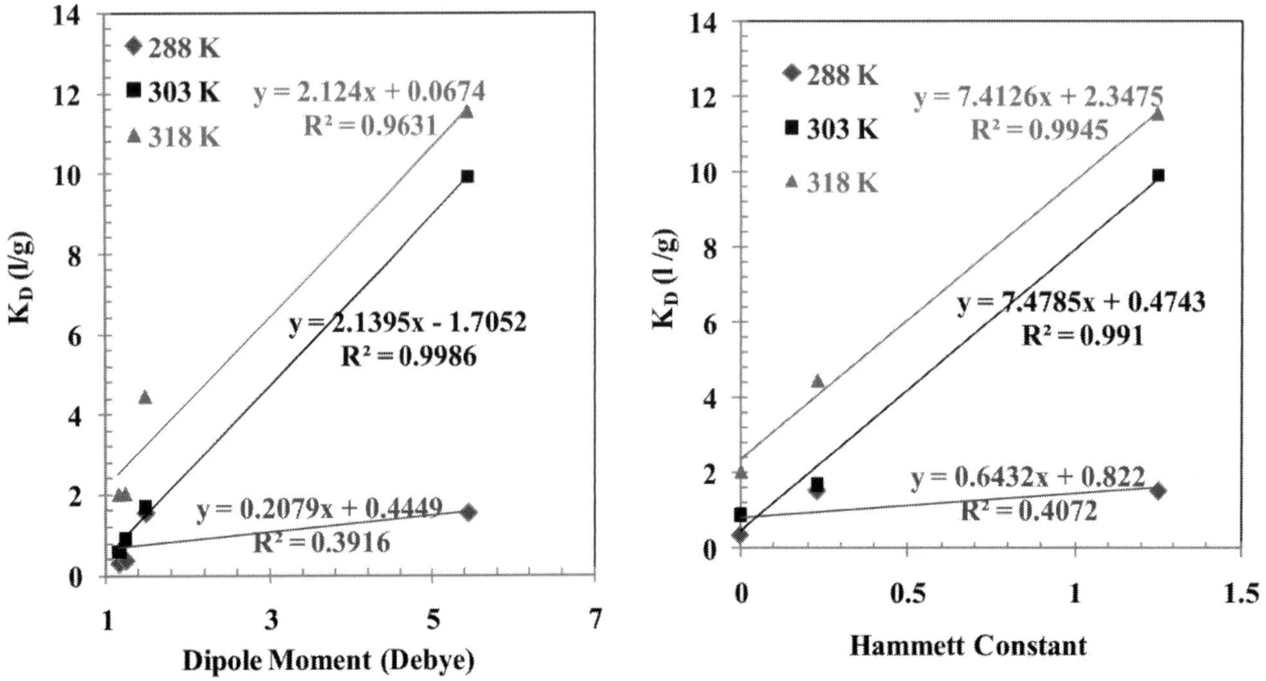

(a) Dipole Moment **(b) Hammett Constant**

Figure 1. Variation of adsorption coefficients with dipole moment and Hammett constant.

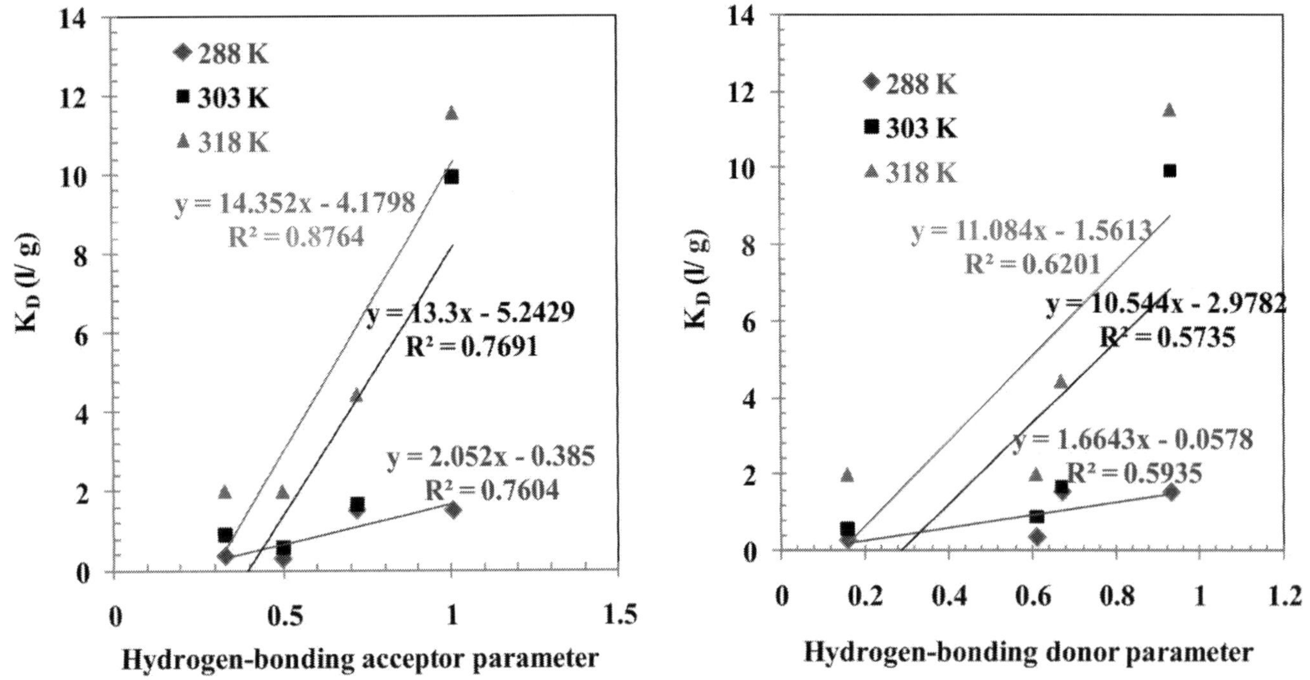

(a) Hydrogen-bonding acceptor parameter **(b) Hydrogen-bonding donor parameter**

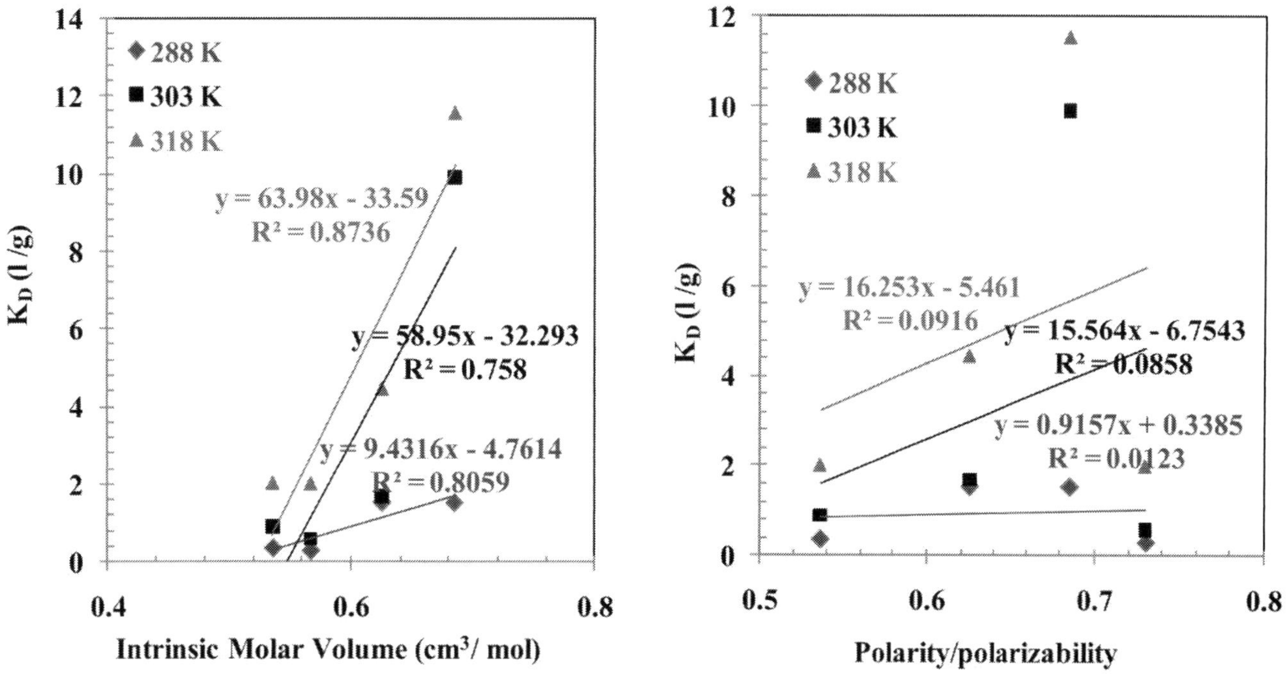

(c) Intrinsic Molar Volume (cm^3 mol^{-1}) **(d) Polarity/polarizability**

Figure 2. Effect of solvatochromic parameters on adsorption coefficients at different temperatures.

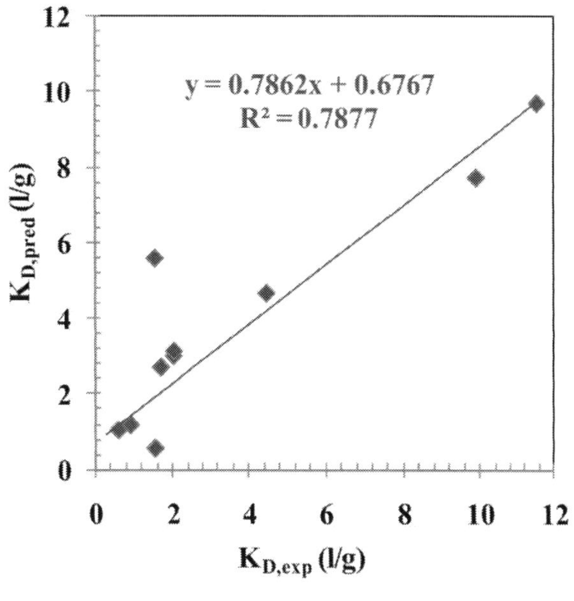

Figure 3. Correlation between experimental and predicted K_D values given by Eq. (11).

(761a) Design of a Simulated Moving Bed Plant for the Purification of Flavors From Complex Mixtures

Kiara Kochendoerfer, Mirjana Minceva and Wolfgang Arlt, Chair of Separation Science and Technology, University of Erlangen-Nuremberg, Erlangen, Germany

As a highly selective separation technology chromatography is almost an inevitable step in downstream processing schemes of mixtures containing components with similar physical properties. For a separation of a mixture in two fractions the continuous counter-current chromatography, i.e. Simulated Moving Bed (SMB), outperforms the conventional batch chromatography in terms of yield, productivity and eluent consumption. Currently, the SMB production scale applications involve mainly binary mixtures separations i.e. separations of enantiomers.

The objective of this work is the development of an approach for designing SMB purifications of multi-component mixtures. This is a particularly difficult task when dealing with feed stocks of plants, since the feed often contains hundreds of components, its composition is not fully characterized, and the physical and chemical properties of a high number of components present in the mixture are unknown. One of the particularities of these separation tasks is that high purities of a single feed component is often not of ultimate importance, more critical is ensuring the absence of certain critical components in the final product.

The lecture informs about the model-based approach supported by a restricted number of systematically selected experiments, which are combined with simulations of the SMB unit operation. The first step in the design of the SMB purification process was a simplification of the separation task, done by the assumption that the starting mixture can be presented as a model mixture containing the target component and the most critical impurities.

Furthermore, the chromatographic column hydrodynamics parameters, adsorption equilibrium and mass transfer parameters of the model mixture were determined experimentally. These parameters were used for the first selection of the operating parameters of a laboratory scale SMB unit for the separation of the model mixture. After an experimental validation of the simulated SMB unit performances, the SMB purifications of the complex mixture were performed. The results of these separation experiments were used to extend the SMB model and tune its parameters. The applicability of the extended model was confirmed experimentally.

(761b) Modeling and Simulation of an Industrial Scale SMB Unit for the Separation of p-Xylene

Marta S. P. Silva[1], José P. B. Mota[2] and Alírio E. Rodrigues[1], (1)Departamento de Engenharia Química, LSRE - Laboratory of Separation and Reaction Engineering, Faculdade de Engenharia, Universidade do Porto, Porto, Portugal, (2)Requimte/CQFB, Departamento de Química, Universidade Nova de Lisboa, Caparica, Portugal

The Simulated Moving Bed (SMB) process, developed by Universal Oil Products (UOP) in the 1960s, has been widely used for the separation of *p*-xylene with high purity (99.7 %) in industrial scale (Parex). Prior to the application of optimization techniques to an industrial scale unit, an accurate model has to be developed.

A detailed analysis of the construction of an industrial scale plant shows the existence of three types of dead volumes which interfere in the performance of the unit, and therefore should not be neglected. The Parex process has 24 adsorption columns divided into two adsorbent chambers, which are connected by two circulation lines. The internal piping and the support grids create dead volumes between adsorption columns. And finally, each bed is connected to the rotary valve by means of a bed line.

A mathematical model of the Parex unit accounting for the three types of dead volumes was developed. The mathematical model has proven to be capable of satisfactorily predicting the performance of the Parex unit in terms of purity and recovery.

The simulations were used to analyze the cyclic steady state of the simulated moving bed unit with 7 zones, and it was made a comparison with the 4 zones operation mode. For a chosen case, the effect of the dead volumes in the internal concentration profiles of the unit and in the extract history was studied. The analysis of the performance of the unit shows that the size of the each type of dead volume influences the performance of the Parex unit and consequently they have to be taken into account in the design and optimization of an industrial scale plant.

(761c) Making Decisive Decisions On Simulating Moving Bed Designs

Paul Bussmann, Floor Boon and Ronald Vroon, Dutch Institute for Applied Science (TNO), Zeist, Netherlands

The working of an simulating moving bed (SMB) is still poorly understood. Although the SMB is used in food, petrochemical, and pharmaceutical industries, the operation is suboptimal and more widely applicable. Determining the design (CAPEX) and operating parameters (OPEX) of SMB processes remains a challenge. The optimization problem is complicated by the relative large number of decision variables (such as flow rates, particle size, column length and configuration) and various conflicting objectives (such as maximization of the adsorbent productivity and minimization of the water use for a required product recovery and purity).

A multi-objective optimization is adopted opposed to a single-objective optimization. Using a single-objective optimization, several objectives are combined into a single scalar objective function using arbitrary weight factors. Drawbacks are (i) the sensitivity to the values of the weighting factors used (which are difficult to assign on an *a-priori* basis) and (ii) the risk of losing some optimal solutions. The solution of a multi-objective optimization problem is given by an optimum Pareto set, which is a set of operating conditions such that no operating condition can be found that would lead to objective values which are all better that those in the Pareto set.

The method consists of the following steps:

1. Formulation of product and process requirements (in terms of e.g. recovery, purity and dilution);
2. Adsorbent selection;
3. Determination of the SMB configuration (required number of columns (or transfer units) per section, VARICOL, PowerFeed);
4. Determining the theoretical working point using the triangle theory [Mazzotti *et al.*, 1997];
5. Sizing the SMB in the theoretical working point;
6. Multi-objective optimization, resulting in a Pareto plot. An optimization tool within gProms® was developed, enabling single and multi-objective optimization of an SMB.

The approach was successfully developed and validated in the separation of an industrial hydrolysate obtained from agricultural side streams in high value carbohydrate fractions.

(761d) Prediction-Correction Method for Optimization of Nonlinear SMB Separations

Jason Bentley and Yoshiaki Kawajiri, School of Chemical & Biomolecular Engineering, Georgia Institute of Technology, Atlanta, GA

Simulated moving bed (SMB) chromatography is a powerful continuous separation technique that is being used for more and more applications in the food, petrochemical, and pharmaceutical industries. In practice the SMB operation is difficult to optimize, especially in cases where the adsorbing components of a mixture have nonlinear isotherms. Due to the complex dynamics SMB operators commonly use heuristic approaches and manual tuning to find operating conditions that satisfy high purity constraints with some safety margins. These operating conditions may be conservative and achieve suboptimal productivity or desorbent consumption.

We propose the prediction-correction method which uses model-based optimization, SMB experimentation, and parameter estimation to optimize the SMB operation in a systematic manner. The prediction-correction (PC) algorithm (Figure 1) is designed to improve the process development stage of SMB operation by rapidly finding the optimal operating conditions that achieve design objectives. We use an iterative scheme where the optimal operating conditions are predicted by solving a model-based optimization problem and these are implemented in the SMB. Then the model parameters are refined by fitting our model to the SMB experimental results. The refined parameter set is used to again predict the optimal operating conditions and these are successively implemented until the termination criteria are satisfied. Thus, the PC method reduces the task of SMB process development into solving a sequence of computational problems.

We show that the PC method was successfully used to optimize the separations of linear and nonlinear isotherm systems. For the linear isotherm system we used uridine and guanosine, and for the nonlinear system we used cyclopentanone and cyclohexanone. For both systems the PC method was successfully used to find optimal operating conditions where greater than 96% purity is achieved in each product stream while maximizing productivity. The SMB experiments can be terminated before reaching the cyclic steady state, and the operating conditions can be updated online. In Figure 2 we show an example from the linear system results of how the optimal operating conditions are achieved, and the PC algorithm converges in only three iterations.

The PC algorithm can also be used to identify a sufficient adsorption isotherm model for SMB optimization. We show in a case study with cycloketones that a single-component Langmuir isotherm was updated to a competitive Langmuir isotherm, which led to better fitting. This isotherm model update allows the PC algorithm to find the optimal operating point quickly with greatly increased productivity, while it may result in suboptimal productivity without the isotherm update.

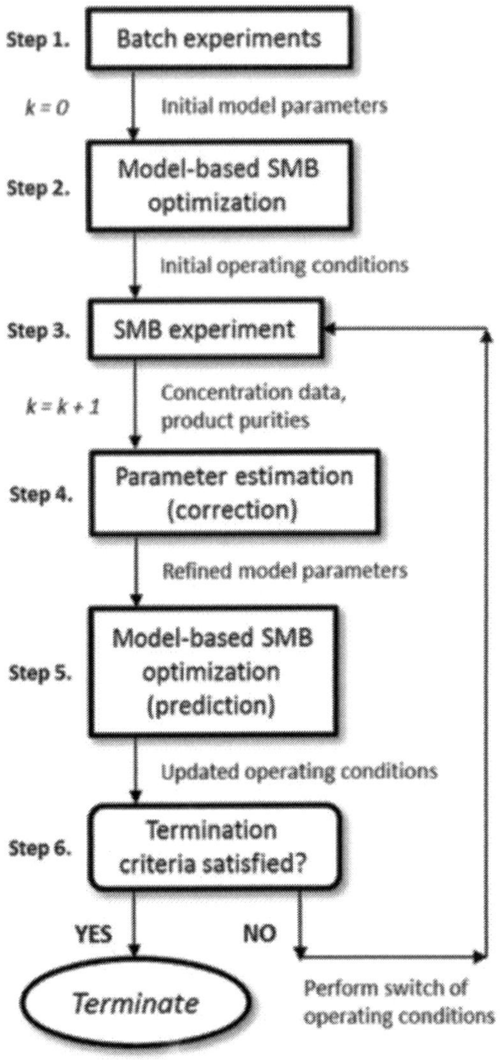

Figure 1. Prediction-correction algorithm

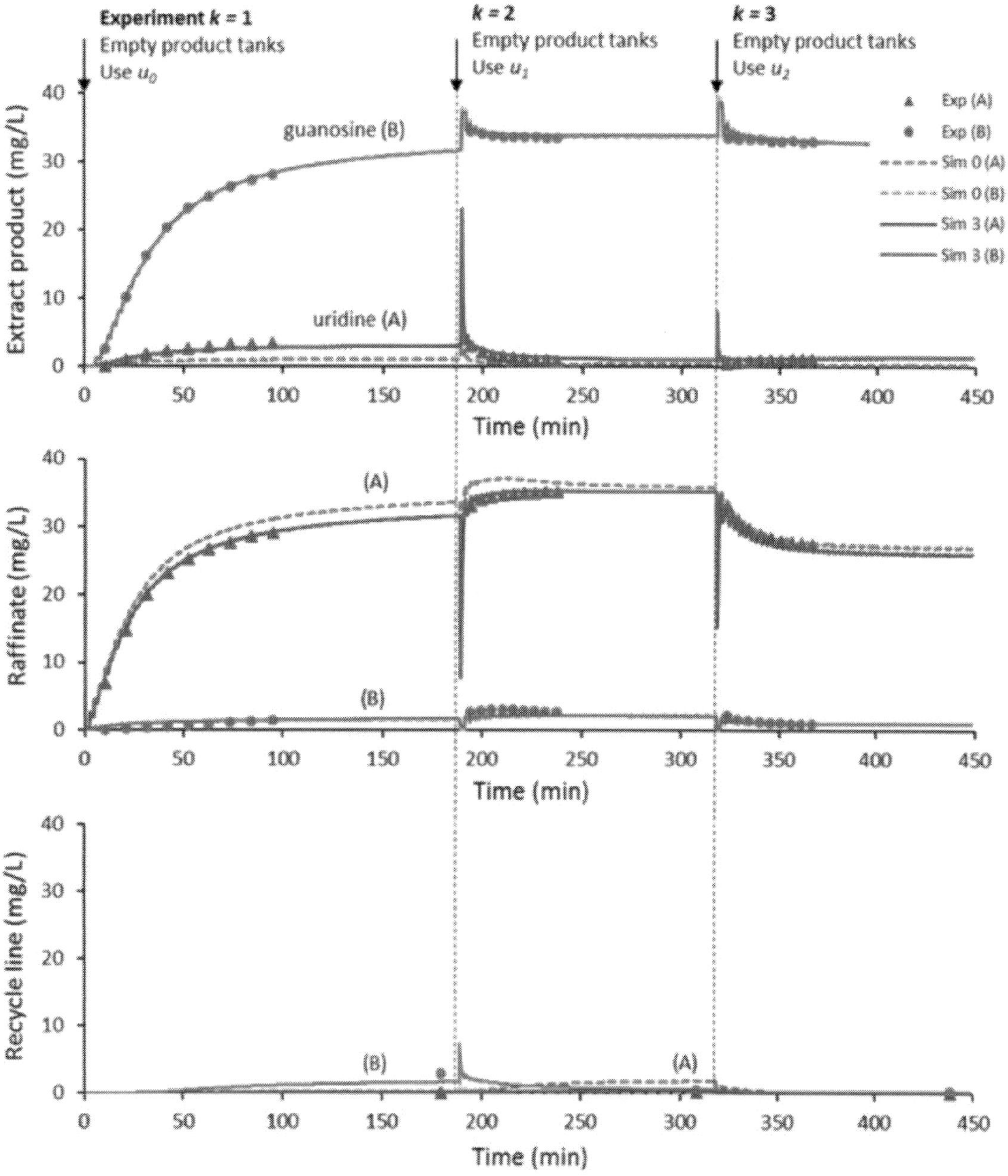

Figure 2. SMB experimental concentration data and simulated concentration profiles using our model. The product purities are improved by implementing optimal operating conditions, u_i once the model parameters are refined by fitting the previous experimental data set in each iteration k.

(761e) Continuous Centrifugal Partition Chromatography: Bridging the Gap Between Liquid-Liquid Extraction and Liquid-Solid Chromatography

Mirjana Minceva, Chair of Separation Science and Technology, University of Erlangen-Nuremberg, Erlangen, Germany

Centrifugal partition chromatography (also known to as continuous countercurrent chromatography) is a support-free liquid-liquid chromatographic techniques, which combines the principles of extraction and chromatography. The separation of a mixture is accomplished as a result of the different partitioning of the solutes between two liquid phases, whereas one of the phases is kept stationary by means of centrifugal forces in a specially designed "column".

Sequential centrifugal partition chromatography (sCPC) is a novel continuous cyclic chromatographic technology, which is used for the separation of a feed mixture in two products. One of the specifics of sCPC is that the feed stream is introduced continuously during one cycle and the products are collected sequentially in each of the two steps of the cycle.

In this work, the theoretical background and principle of sCPC are presented. A mathematical model is developed and employed in a systematic study of the influence of the operating parameters on the process separation performances. The results of this study are demonstrated experimentally using a model binary feed mixture. The advantages and limits of the sCPC technology are underlined, and used to define potential fields of sCPC application. The technology is compared with the continuous countercurrent liquid-liquid extraction and continuous countercurrent liquid-solid chromatography i.e. Simulated Moving Bed technology (SMB).

(761f) Method Development in the Separation of an Enantiomer with Supercritical Fluid Chromatography

Abhijit Tarafder, Fahimeh Kamarei and Georges Guiochon, Department of Chemistry, The University of Tennessee, Knoxville, TN

Supercritical fluid chromatography (SFC), which uses sub/supercritical carbon-dioxide as the principal solvent in chromatographic operations, is steadily replacing HPLC in various industrial separations, especially in the separation of enantiomers. The main advantage of SFC compared to HPLC is the increment in productivity by several times. Additionally, the usage of carbon-dioxide, a carbon neutral *green* solvent, increases the sustainability of the chromatographic processes significantly. The main problems in a wider application of SFC are its operational complexity and lack of physical understanding in designing more competitive industrial processes. This hinders the industrial practitioners from developing or transferring optimized operational methods for a particular separation.

The separation performance in SFC strongly depends on the solvent density, and to a lesser extent, on the temperature. This indicates that the effects of the operating conditions on chromatographic performances can be understood more clearly by following the changes in both the density and the temperature, rather than analyzing the operating variables i.e. the pressure and the temperature [1,2]. In other words, SFC operations can be more clearly analyzed based on the iso-density or isopycnic plots, drawn on the pressure-temperature plane, compared to an analysis based on the isobars and the isotherms.

The talk will demonstrate that guided by the solvent isopycnic plots on the pressure-temperature plane, one can develop effective separation methods quickly, without the help of detailed simulation studies. The demonstration will be based on the separation problem of the enantiomers of Naproxen, a nonsteroidal anti-inflammatory drug (NSAID). First the variation of the separation factors of the two enantiomers will be investigated as a function of the density and the temperature of the mobile phase. The optimum separation method in terms of the best combination of high yield and high productivity will then be selected based on this data.

1. A. Tarafder, G. Guiochon: Use of isopycnic plots in designing operations of supercritical fluid chromatography: I. The critical role of density in determining the characteristics of the mobile phase in supercritical fluid chromatography, Journal of Chromatography A, 1218 (2011) 4569– 4575.
2. A. Tarafder, G. Guiochon: Use of isopycnic plots in designing operations of supercritical fluid chromatography: II. The isopycnic plots and the selection of the operating pressure–temperature zone in supercritical fluid chromatography, Journal of Chromatography A, 1218 (2011) 4576– 4585.

(763a) Removal of Barium From Ground Water Using Clinoptilolite

Liie Lumiste[1], **Sundaravadivelnathan Ponnusamy**[2], Rein Munter[1] and Shuguang Deng[2], (1)Department of Chemical Engineering, Tallinn University of Technology, Tallinn, Estonia, (2)Chemical Engineering, New Mexico State University, Las Cruces, NM

Adsorption of barium (Ba^{2+}
was investigated. The adsorption equilibrium and kinetic studies have been performed. The adsorption equilibrium data can be described by the Freundlich isotherm, and the adsorption kinetics of barium follows the pseudo second order reaction model. The effects of pH, temperature, hardness and ammonia concentration were studied on the barium removal capacity of clinoptilolite. At higher concentrations of barium (100 ppm), hardness had a greater effect and ammonia had less or no effect on the adsorption of barium. At lower concentrations of barium (10 ppm), hardness had a less effect and ammonia had a more effect on the adsorption. Adsorption of barium onto clinoptilolite was independent of pH of the water. A maximum removal efficiency of 93% was obtained at an initial concentration of 10 ppm. All the experiments were performed at a ground water temperature at 8°C. ICP-MS was used to measure the concentration of barium in the water. SEM-EDS were used to characterize the adsorbent material.

(763b) Magnetic Polyaniline Nanocomposites for Environmental Remediation

Hongbo Gu[1], Sowjanya B. Rapole[1], Yudong Huang[2], Suying Wei[3] and Zhanhu Guo[4], (1)Integrated Composites Laboratory (ICL), Dan F Smith Department of Chemical Engineering, Lamar University, Beaumont, TX, (2)School of Chemical Engineering and Technology, State Key Laboratory of Urban Water Resource and Environment, Harbin Institute of Technology, Harbin, China, (3)Chemistry and Biochemistry, Lamar University, Beaumont, TX, (4)Dan F. Smith Department of Chemical Engineering, Lamar University, Beaumont, TX

Magnetic Polyaniline Nanocomposites for Environmental Remediation

Hongbo Gu, Sowjanya B. Rapole, Yudong Huang, Suying Wei and Zhanhu Guo

In this talk, the toxic Cr(VI) removal from polluted water by the magnetic conductive polyaniline (PANI) polymer nanocomposites (PNCs) was investigated. The PNCs were synthesized using a facile surface initiated polymerization (SIP) method. The effects of initial Cr(VI) concentration, PNC concentration and pH value on the Cr(VI) removal have been studied. The synthesized PNCs showed unique capability to remove Cr(VI) in solutions in a wide pH range. The saturation magnetization of these PNCs was observed to have no obvious decrease after treatment with Cr(VI) solutions and they can be easily recycled by a permanent magnet. The Cr(VI) removal kinetics in the different pH solutions was determined. The Cr(VI) removal mechanism was investigated by Fourier transform infrared spectroscopy (FT-IR), X-ray photoelectron spectroscopy (XPS), energy-filter transmission electron microscopy (EFTEM) and temperature dependent resistivity as well.

(763c) Alternative WATER Treatment Using Cactus Mucilage From Opuntia SPP.: Application in Bacteria Removal

Daniela M. L. Stebbins, Department of Chemical & Biomedical Engineering, University of South Florida, Tampa, FL

ALTERNATIVE WATER TREATMENT USING CACTUS MUCILAGE FROM OPUNTIA SPP.: APPLICATION IN BACTERIA REMOVAL

Daniela Stebbins, Dawn Fox, Audrey Buttice, Deni-Maire Smith Norma Alcantar

Department of Chemical & Biomedical Engineering, University of South Florida

Cholera is a rapidly dehydrating, diarrheal causing disease that can kill within hours if left untreated and affects millions of people every year (WHO 2011). The cholera outbreak in Haiti that followed a powerful earthquake in 2010 motivated this study. Our purpose is to investigate an alternative treatment to remove bacteria from water using a natural biomaterial that can be used by communities in need. The mucilage is produced by several cacti species. The specie that we are using is the *Opuntia ficus-indica*, commonly known as Nopal or Prickly pear cactus. The Nopal cactus pads are readily available, inexpensive and have been used to clean water by ancient communities in Mexico. Extensive research in our group has shown that the mucilage is efficient at removing turbidity, bacteria (*E. coli* and *Bacillus sp.*) and arsenic from contaminated water. Now we are investigating the conditions under which cactus mucilage will be able to reduce the levels of a surrogate for *Vibrio Cholerae*, named *Vibrio furnissii*. Lyophilized pellets of the surrogate bacteria were purchased from a testing laboratory specialized in environmental and public health threats. The pellets were suspended in Marine Broth following specification from the vendors, transferred to Marine Agar and kept at 4 °C until the experiments were performed. A single colony, incubated at 35 °C for 24 hours, was transferred to 5 mL of Marine Broth and kept in a shaker incubator for 24 hours at 35 °C. A 1 mL aliquot was diluted in 4 mL of Marine broth and kept in the same conditions for more 2 hours. The bacteria suspension was washed once in PBS to remove extracellular metabolites. The pellet was re-suspended in 50 mL of PBS and used as spike solution to set up water columns where the mucilage was tested. The final bacteria concentration in the testing water columns (10 mL) was ~ 10^6 cells. We have tested 3 concentration of the mucilage (0, 1 and 10 mg/L), with and without 10 mg/L of calcium carbonate in three repetitions. The experiments were conducted at room temperature and columns were left undisturbed for 2 hours. Aliquots of 1 mL were taken from the top of the experimental water columns and homogenized with a vortex. Four serial dilutions were placed and sub-aliquots of 100 μL were plated and incubated at 35 °C for 24 hours to plate count and to establish the mucilage removal abilities. Although the higher concentration of the mucilage (10 mg/L) did not decrease the bacteria concentration, the lower concentration of the mucilage (1 mg/L) removed 25% of the original concentration of the bacteria on the top of the water column. The treatment using 1 mg/L of mucilage plus calcium carbonate was determined to remove 40% of the bacteria. These preliminary studies suggest that mucilage from the *Opuntia ficus-indica* is a viable flocculation method for Cholera. Its low cost, accessibility, and current use as a food source in low income communities give mucilage great potential as a water treatment method for areas such as Haiti that have had devastating bacterial contaminations.

REMOVE OF TEXTILE DYES USING AGRICULTURE WASTES

Erika Montes[1], Sebastian Velasco[1] and Gabriel Camargo[2], (1)Environmental engineering, Universidad Libre, Bogota D.C., Colombia, (2)Ingenieria Mecanica, Universidad Libre, Bogota, Colombia.

Abstract

This research is searching for a low cost material to treat the wastewaters from the textile industries. The principal pollutants present in these waters are the dyes. Currently the most common proceedings to treat the wastewaters are very expensive. A solution is, use the agricultural wastes to treat them and give an adequate disposition. Here in the Andina's region one of the most popular economic activities is the production and trade of flowers, due to this activity lots of residues have been generated.

The adsorbents were characterized with equilibrium isotherms and kinetic adsorptions using sunflowers stalks without the skin. The stalks were kept inside of a room with constant conditions; temperature (20ºC) and relative humidity (RH) of 72% by 168 hours. It were analyze two different dyes (Congo red and methylene blue) The different times of the adsorptions were 40 minutes with constant temperature (20°C) .The solution concentration was measure by UV/VIS spectrophotometer. The superficial surface was calculated by Langmuir isotherms and probe using nitrogen adsorption. A pilot plan was tested for scaling purposes.

The dyes recovery has been evaluated with several solvents and it is being analyzed the reuse as a dye in other process.

Introduction

Many industries like the textile industry used dyes to color their products and thus produce wastewater containing organics with a strong color, where in the dyeing processes the percentage of dye lost wastewater is 50% of the dye because of the low levels of dye-fiber fixation [1]. Due to the environmental problematic about the wastewaters coming from the textile industries, many methods or techniques have been created to remove them, like adsorption, flocculation combined with flotation, membrane filtration, electro-kinetic, coagulation, ozonation, oxidation, precipitation and ion-exchange[2]. These methods are not very successfully due to suffering from many restrictions [3]. Some of these treatments are shown in table 1.

Table 1
Dye-remove treatments (examples).

Methods	**Advantages**	**Disadvantages**
Chemical treatments		
Photochemical	No sludge is produced and foul odors are greatly reduced	Formation of by-products
Electrochemical destruction	No consumption of chemicals and no sludge build up	Relatively high flow rates cause a direct decrease in dye removal
Anaerobic textile- dye bioremediation systems	Allows azo and other water- soluble dyes to be decolorized	Anaerobic breakdown yields methane and hydrogen sulfide
Physical Treatments		
Ion exchange	Regeneration: no adsorbent loss	Not effective for all dyes
Electrokinetic coagulation	Economically feasible	High sludge production

A small quantity of dye in water can be toxic to human health and some aquatic species [4], and affect the visible standards [5]; our government legislation requires textile wastewater to be treated but it is not specific in which kind of dyes or how much must be remove. Some studies tested the agricultural wastes as adsorbents for these pollutants [6], because of their economical obtaining for removal of different pollutants has proved to be an efficient at removing many types of pollutants such as heavy metals [7].

The principal raw materials in the productions of oils and grace are the oil palm, cotton, soy, sunflowers seeds, sesame and others, generating lots of agricultural wastes [8]. The sunflower stalks were used on this research to search for an economic and easy obtaining adsorbent to remove dyes present in textile wastewaters.

Materials and methods

Adsorbent:

The sunflowers (*Helianthus annuus*) stalks were obtained from a local market, then it were kept inside of a room with constant conditions; temperature (20°C) and relative humidity (RH) of 72% by 168 hours, cut into a small pieces and crushed, with a particle size of 850 mµ.

Reagents and solutions:

The methylene blue (CAS number: 97130-83-1) dye was obtained from MERCK, the chemical structure of the methylene blue is shown in Fig. 1; congo red (CAS number: 573-58-0) dye was obtained from MERCK. The structure of the congo red is shown in Fig. 2. The solutions were dissolved in distilled water. The pH was not adjusted.

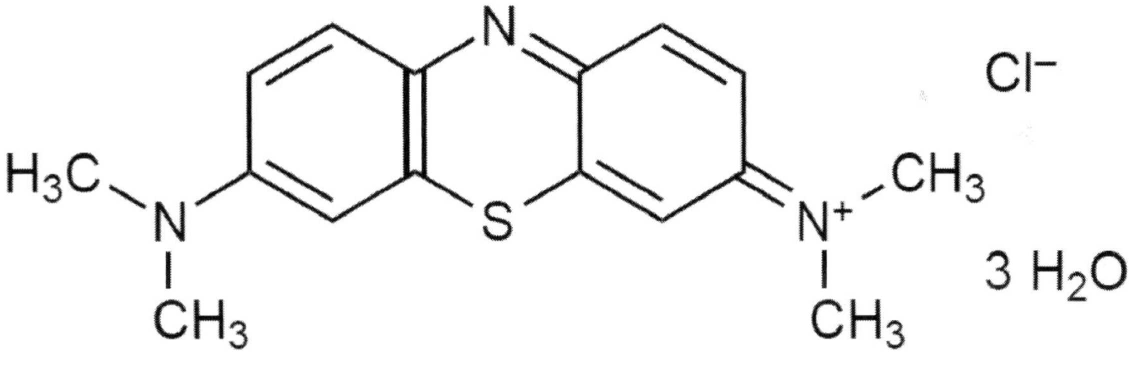

Fig. 1 Chemical structure of the methylene blue.

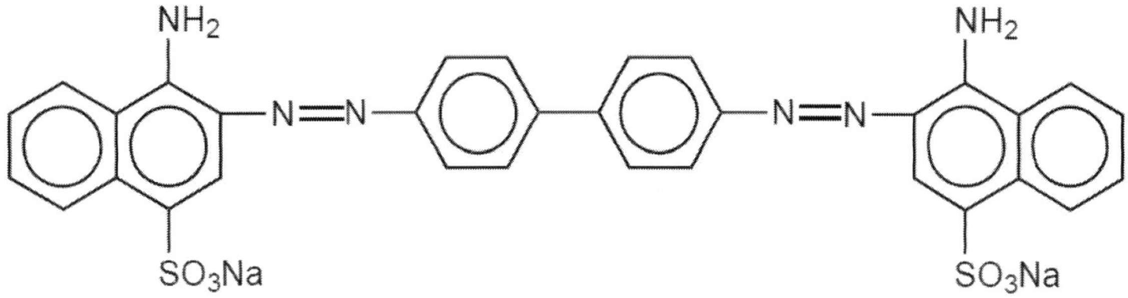

Fig. 2 Chemical structure of the congo red.

The wave lengths were determined by Thermo Scientific Genesys 5 UV-VIS Spectrophotometer and calibration curves which are 664nm and 496nm for methylene blue and congo red respectively whom adjust to the literature [9].

Determination of saturation curves (rates procedure):

The adsorption process were conducted at room temperature (20°C) and relative humidity (RH) of 72%, by mixing 0.250gr of sunflower stalks with an aquous solution of 25mL of the different dyes in 50mL-Erlenmeyer flask. The measures were 20, 40, 60, 80, 100, 120, 140 and 160 minutes for each dye at 257rpm in a Scilogex ms-n-s10 stirrer at a concentration of 100ppm with BRAND stir bars (25mm). To determinate the absorption of the resulting solutions after the mixing, they were separated with a nylon filter; the solutions were centrifuged at 1500rpm for 10 minutes and measured in spectroquant cells of 10mm with a Thermo Scientific Genesys 5 UV-VIS Spectrophotometer.

Adsorption procedure:

The adsorption process were conducted at room temperature (20°C) and relative humidity (RH) of 72%, with different concentrations: 20, 40, 60, 80, 100, 250, 500, 1000 and 1500 ppm for each dye at 257rpm in a Scilogex ms-n-s10 stirrer with BRAND stir bars (25mm) by 40 minutes. To determinate the absorption of the resulting solutions after the mixing, they were separated with a nylon filter; the solutions were centrifuged at 1500rpm for 10 minutes and measured in spectroquant cells of 10mm with a Thermo Scientific Genesys 5 UV-VIS Spectrophotometer.

TGA (Thermo Gravimetric Analysis) Examination:

The TGA was measured with a Differential Scanning Calorimeter Netzsch STA 409s with 0.0102gr sample of sunflower stalk at 10 °C per minute by one hour until it reaches 600°C.

Results and discussion

Ph:

The pH was measure at 20, 40, 60, 80 and 100 ppm for each dye after the adsorption process as shown in Fig. 4 and Fig. 5 for each dye.

ppm	pH
20	6,21
40	6,21
60	6,07
80	6,24
100	6,46

Fig. 4 pH for Methylene blue

ppm	pH
20	6,30
40	6,23
60	6,20
80	6,37
100	6,67

Fig. 5 pH for Congo red

Effect of the adsorbent on dye removal:

As shown in Fig. 5 and Fig. 6, the best adsorption time was at 40 minutes for each dye with 92.8% and 98.1% remotion of methylene blue and congo red respectively.

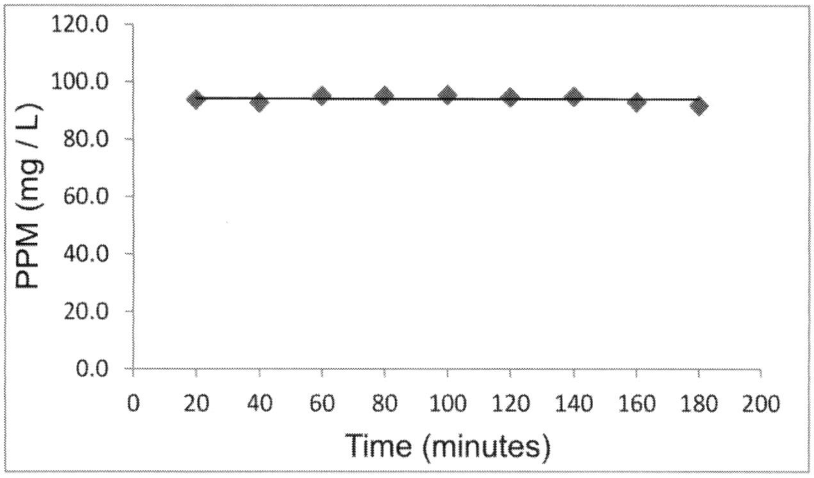

Fig. 5 Adsorption rate for Methylene blue

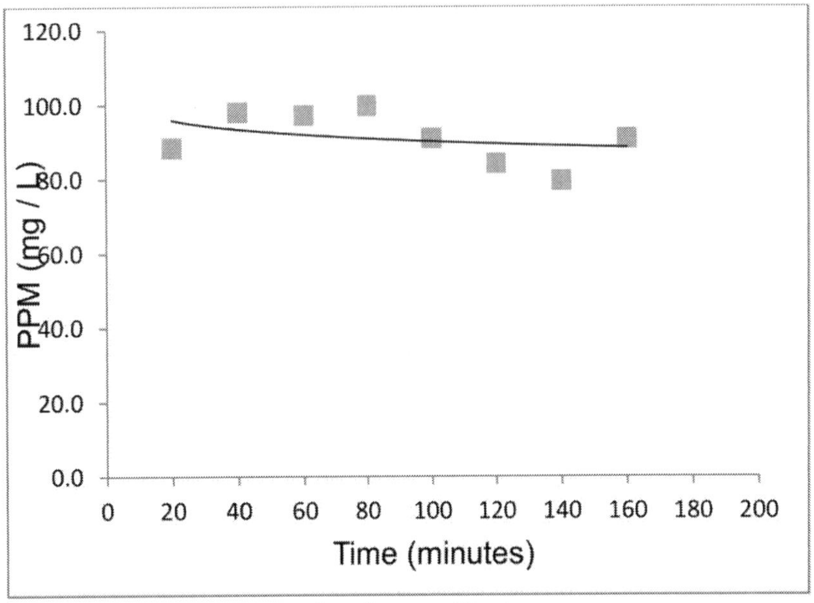

Fig. 6 Adsorption rate for Congo Red

Effect of the adsorbent on dye removal at different concentrations:

As shown in Fig. 7 and Fig. 8, the rate of adsorption was 40 minutes for each dye concentration. The Langmuir isotherm can not be apply as shown in Fig. 9 and Fig. 10, which shows a behavior of a freundlich isotherm. The coefficients of relation for the freundlich isotherms are (R^2) 1 for each colorant, so the adsorption is a physical adsorption.

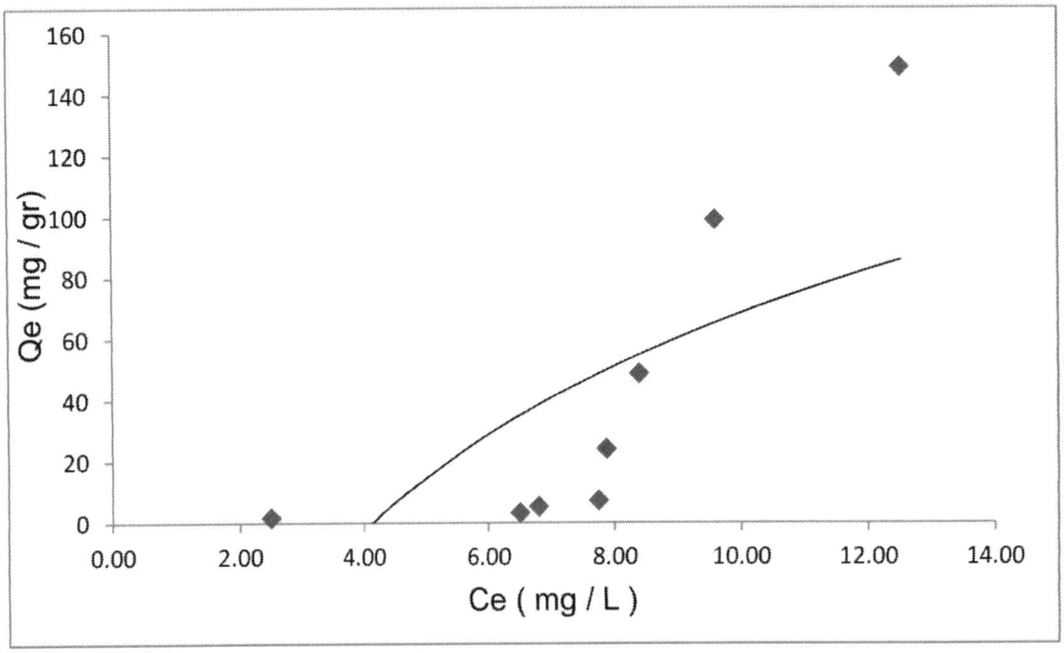

Fig. 7 isotherm adsorption for Methylene blue at 40 minutes

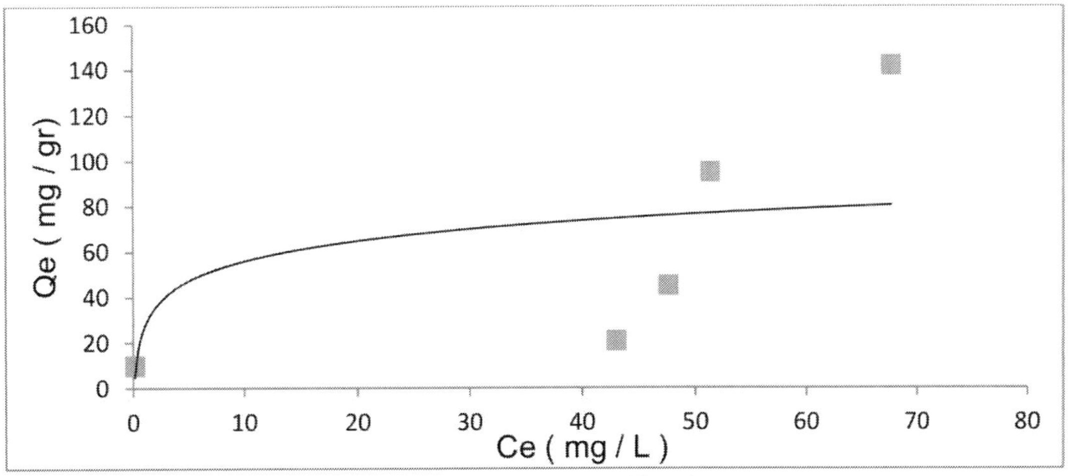

Fig. 8 isotherm adsorption for Congo Red at 40 minutes.

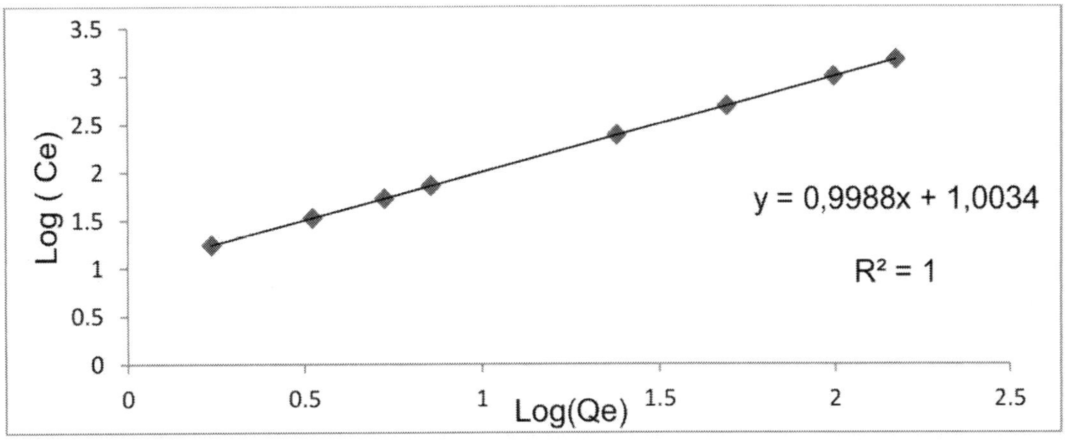

Fig. 9 Freundlich isotherm for Methylene blue

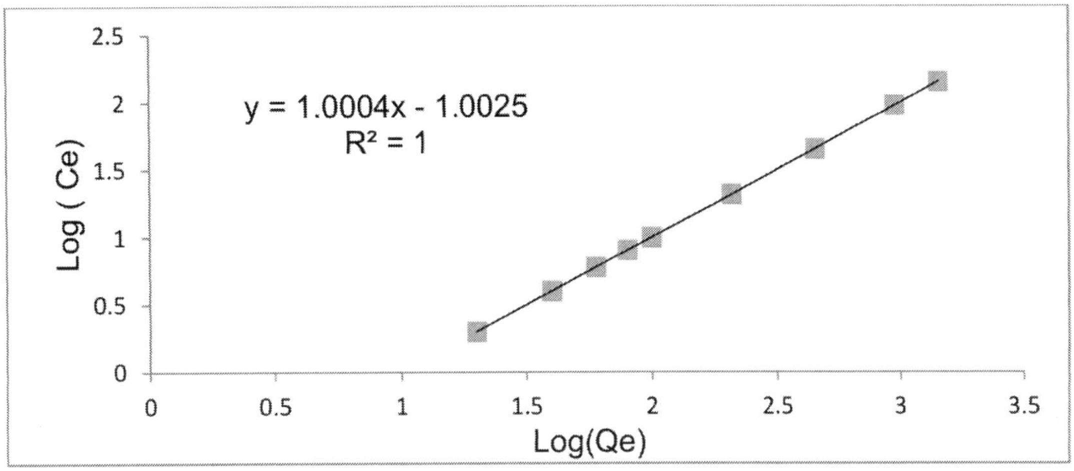

Fig. 10 Freundlich isotherm for Red Congo

TGA (Thermo Gravimetric Analysis):

According to the Fig. 9, the mass change it is considerable until it reaches 200°C where all the water present in the sunflower stalk evaporates; the 30% remainder represents a high content of carbon.

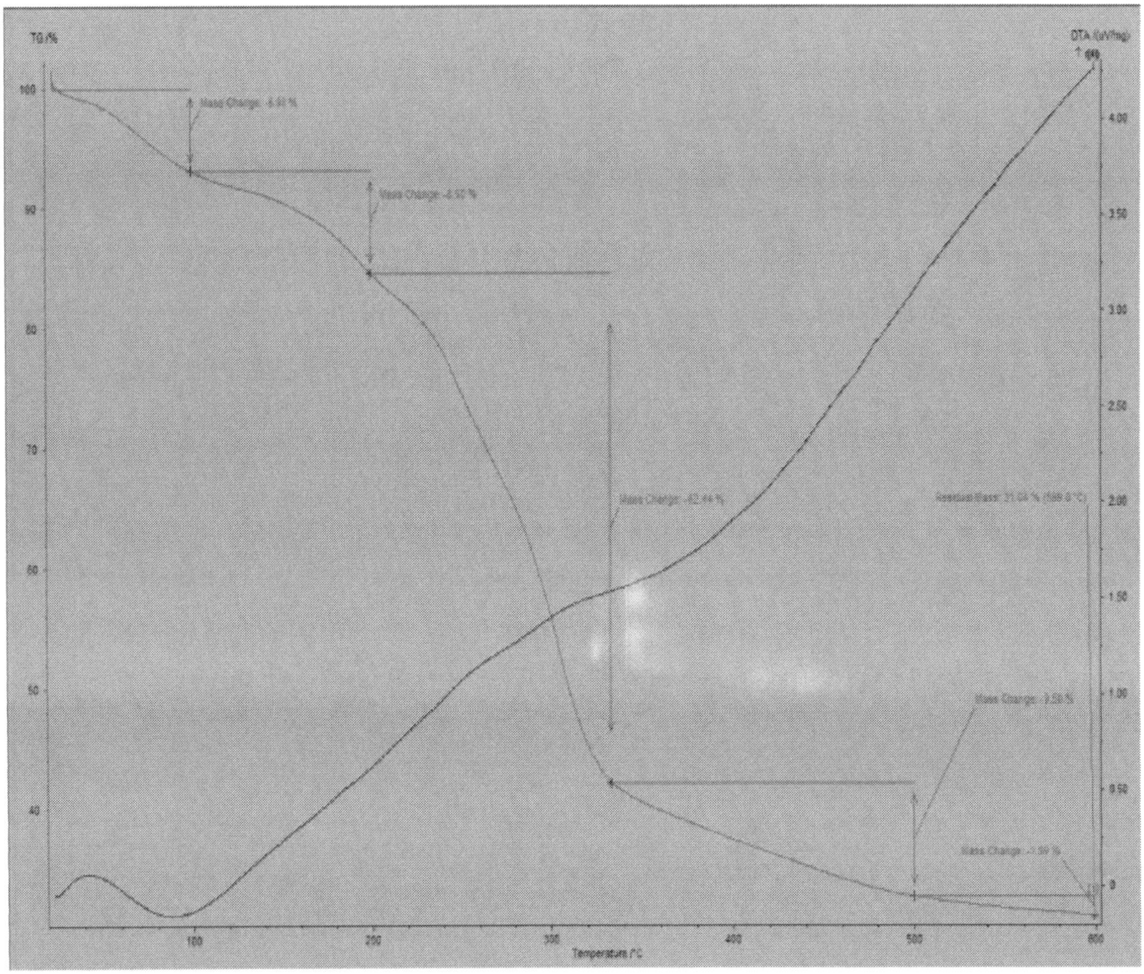

Fig. 9 TGA for sunflower stalk.

Conclusions:

The sunflowers stalks are a low cost adsorbent, it does not need any kind of induced water remotion to be prepared as an adsorbent and its obtaining is very easy and cheap. It is a very efficient adsorbent because of his high percents of remotion and low rates of adsorption. According to the results, the adsorption is a physical adsorption because of its freundlich behavior. The pH it is a factor that affect the measures and the behavior of the adsorptions.

Currently it is being analyzed a textural characterization with BET surface areas, TGA's of the sunflower stalks with the colorants and SEM testing.

References

- [1] N. Mohan, N. Balasubramanian, C.A. Basha, Electrochemical oxidation of textile wastewater and its reuse, J. Hazard. Mater. 147 (2007) 644–651
- [2] K. Xie, W. Zhao, X. He, Adsorption properties of nano-cellulose hybrid containing polyhedral oligomeric silses-quioxane and removal of reactive dyes from aqueous solution, Carbohyd. Polym. 83 (2011) 1516–1520.
- [3] T. Robinson, G. McMullan, R. Marchant, P. Nigam, Remediation of dyes in textile effluent: a critical review on current treatment technologies with a proposed alternative, J Biores. Technol. 77 (2001) 247–255.
- [4] J. Avom, J.K. Mbadcam, C. Noubactep, P. Germain, Adsorption of methylene blue from an aqueous solution onto activated carbon from palm-tree cobs, Carbon 35 (1997) 365–369.
- [5] M.S. Chiou, P.-Y. Ho, H.-Y. Li, Adsorption of anionic dyes in acid solutions using chemically cross-linked chitosan beads, J. Dyes Pig. 60 (2004) 69–84.
- [6]A. Ahmad, B. Hameed, Reduction of COD and color of dyeing effluent from a cotton textile mill by adsorption onto bamboo-based activated carbon, J. Hazard. Mater. 172 (2009) 1538–1543.
- [7]Sun G, Shi W. Sunflower stalks as adsorbents for the removal of metals from wastewater. Ind Eng Chem Res 1998;37:1324–8.

- [8] Carlos Federico Espinal G., Héctor J. Martínez Covaleda, Marcela Salazar Soler Camilo A. Barrios Urrutia, March of 2005, LA CADENA DE LAS OLEAGINOSAS EN COLOMBIA UNA MIRADA GLOBAL DE SU ESTRUCTURA Y DINAMICA 1991-2005, recover august 25 of 2012.

- [9] Sun, G. and Xu, X. 1997. Sunflower stalk as adsorbents for color removal from textile wastewater. Ind. Eng. Chem. Res., 36: 808.

(763e) Arsenic Adsorption On Xerogel, Aerogel, and Nanogel Activated Alumina

Lucy Mar Camacho[1], Sundaravadivelnathan Ponnusamy[2], Isaac Campos[1], Tom Davis[3] and Shuguang Deng[2], (1)Center for Inland Desalination Systems, University of Texas at El Paso, El Paso, TX, (2)Chemical Engineering, New Mexico State University, Las Cruces, NM, (3)Center for Inland Desalination Systems (CIDS), University of Texas at El Paso, El Paso, TX

Various forms of sol-gel activated alumina (AA) have been synthetized and tested as potential arsenic adsorbent materials. Some of them included xerogel, aerogel and Nanogel AA. Synthesis of the sol-gel was made following the Yoldas method which included hydrolysis, peptization, gelation and calcination. In the present study, modifications to the Yoldas method to produce aerogel and Nanogel AA included using supercritical-CO_2 extraction and addition of nanoparticles, respectively. Equilibrium experiments were conducted to determine optimum pH conditions for adsorption and maximum adsorption capacity. Results were compared with two commercial activated alumina adsorbent materials, namely H-156 and Axens. It has been observed that Xerogel removed arsenic at any pH within the range of 2.0 and 10.0. It has also been observed that the tendency to increase arsenic uptake for the synthesized adsorbents increased in the order or Xerogel AA > Axens > Aerogel AA > H-156 AA > Nanogel AA. The adsorption capacity of Xerogel can be modeled following the Freundlich adsoption isotherm. The pH independency and superior adsorption capacity for Xerogel AA makes this material a feasible and economic alternative for arsenic removal from drinking water without the need of adding buffer chemicals for pH adjustment.

(763f) Production of Activated Carbon From Fast Pyrolysis Biochar

Bernardo G. del Campo, CSET, Iowa State University, Ames, IA, Robert C. Brown, Mechanical Engineering, Iowa State University, Ames, IA, David Laird, Iowa State University, Ames, IA and Laura R. Jarboe, Department of Chemical and Biological Engineering, Iowa State University, Ames, IA

Fast pyrolysis, a thermochemical conversion of biomass, is rapidly developing. This technology focuses on the production of bio-oil, biochar and non-condensable gases for transportation fuels, chemicals, and energy. Fast pyrolysis yields approximately 10-20% biochar by mass, which is currently being explored as an agricultural soil amendment. However, one approach to developing value-added biochar products with potentially great economic and environmental benefits is the upgrading of biochar to activated carbon. The goal of this project is to produce activated carbon from fast pyrolysis biochar and use it for cleaning syngas from fast pyrolysis, as well as to detoxify the water-soluble fraction of bio-oil for the microbial production of biofuels. Different chemical (strong acid, bases and alkali carbonates) and physical (steam, carbon dioxide and ozone) treatments will be further explored to increase biochar surface area, porosity and adsorption capabilities. Preliminary results for detoxifying aqueous bio-oil extracts with biochar and activated carbon to produce rich bacterial growth media showed that activated carbon is a potent detoxifier, while biochar (without activation) is only moderately effective. A techno-economic analysis will also be performed to understand the impact of the carbon activation process on the internal rate of return of the fast pyrolysis platform.

(771a) What Chemicals Will We Need to Capture CO2?

Jason E. Bara, Chemical & Biological Engineering, University of Alabama, Tuscaloosa, AL

A number of solvents are currently being studied in the application of CO_2 capture from post-combustion flue gas streams. Of these, aqueous amine solutions are the most thoroughly developed and furthest along in process scaling, stemming a long history of experience within the natural gas industry. Monoethanolamine (MEA) is perhaps the most well-known, thoroughly studied, and readily available amine for post-combustion CO_2 capture. However, concentrated aqueous MEA solutions likely cannot meet the stringent energy and cost limitations required to economically achieve 90% CO_2 capture at coal-fired power plants. Also in development are piperazine (PZ) and 'advanced' amines, which show great potential to improve energy efficiency and stability, but are not yet produced on a scale approaching that of MEA. Among the non-conventional or "designer" non-aqueous solvents, ionic liquids (ILs) have been the most prominent. However, few non-aqueous solvents have been taken beyond bench-scale activities. Aqueous and non-aqueous solvents will also be important for pre-combustion CO_2 capture applications.

In the coming decades, as CO_2 capture processes are broadly deployed worldwide, the solvent(s) with the most favorable economics and performance will need to be manufactured in quantities of at least 1 megaton per year, creating at new chemical(s) of global importance. With the ultimate goal of deep reductions in greenhouse gas (GHG) emissions in mind, the processes by which the proposed chemicals for CO_2 capture are manufactured should also be considered in terms of their energy requirements, efficiencies, waste products, and CO_2 emissions. This presentation wil begin to address some of these technical and economic questions by examining the manufacture of amines, ILs and other physical solvents that have been proposed for post-combustion and pre-combustion CO_2 capture, as well as the fundamental building block of all amines, ammonia.

(771b) Membrane Processes Applied to Carbon Capture in Coal-Fired Power Plants: Multi-Stage Design and Economic Analysis

Davide Bocciardo, **Maria-Chiara Ferrari** and Stefano Brandani, Institute for Materials and Processes, School of Engineering, University of Edinburgh, Edinburgh, United Kingdom

In order to reduce the greenhouse gases emissions from coal-fired power plants, membrane separation seems a promising alternative to conventional carbon capture technologies. The aim of this work, in collaboration with ETP – Energy Technology Partnership – and Scottish Power, is to investigate the process design of the capture system, ranging from the analysis of the single membrane modules to a complete multi-stage scheme. The DOE case 9 – subcritical coal-fired power plant– has been assumed as reference case.

The modelling process involved the implementation of a membrane module simulation tool that we have developed to be able to predict the performance of countercurrent-sweep and cross-flow modules, which are the typical large scale industrial applications of membranes. The module simulator includes also a new 2D cross-flow model which accounts for both feed and permeate flow directions on both sides.

The module simulator has been implemented with a software interface that allows us to study the complete integration of the membrane modules with a flowsheet for a coal-fired power plant in UniSim Design®, the Honeywell process simulator. In order to find the optimal design in terms of CO_2 purity, energy consumption and membrane area, single and multi-stage simulations have been carried out. Various configurations have been studied to achieve the desired targets in terms of CO_2 recovery (0.9) and purity (above 0.95).

As a result of the multi-stage analysis and initial optimisation, 2 options have been identified: a retrofit and a possible modification including the recycle of part of the CO_2 to the boiler. In both the configurations, after the membrane system, the CO_2-rich stream has then been compressed up to 150 bar and a final refrigeration stage has been included in order to increase the CO_2 purity.

In order to give an overall idea of the application of this technology to carbon capture, particular attention has been focused on the economic analysis: the capture costs have been calculated and a complete evaluation of the Levelised Cost of Electricity (LCOE) has been carried out. Starting from the calculation of the base-case scenario, represented by the DOE case 9, the contribution of the capture scheme has been included and its effect on both operative and capital cost has been identified. The implementation is based on European (*ZEP - Zero Emission Platform*) and UK (*DECC - Department of Energy and Climate Change*) references and it also includes the possible contribution of the EU Emission Trading Scheme. It represents a key feature in order to have a complete comparison between membrane gas separation and amine adsorption, the actual carbon capture commercial technology.

(771c) Post-Combustion Carbon Dioxide Capture Using Membrane Processes: A Sensitivity Analysis

Bouchra Belaissaoui, UPR 3349 CNRS, LRGP, Nancy, France, David Willson, Stanbridge Capital, New York and Eric Favre, Laboratoire Réactions et Génie des Procédés LRGP (UPR 3349 CNRS), Nancy Université, Nancy F-54001, France

Energy saving is the main challenge for post-combustion CO_2 capture technology. Membrane process has attracting growing interest for CO_2 capture applications. This work focus on the potential of membrane processes for CO_2 capture for a wide range of process parameters in order to identify the right place and role of membrane processes in CCS technology.

A broad range of CO_2 inlet concentration (5%-70%) corresponding to various emission sources has been investigated. In order to explore the influence of membrane material selectivity CO_2 over N_2 (a) on the process performances, three different membrane selectivities are simulated: a=50 corresponding to commercially available membranes and a=100 and 200 corresponding to prospective membranes. A large range of process performances in term of CO_2 purity and recovery ratio are investigated and the corresponding energy requirement is compared to the reference technology, MEA absorption. Furthermore, two compression strategies: feed compression with energy recovery system on the retentate and vacuum pumping are investigated. Their performances in term of energy requirement and membrane surface area are evaluated and compared. Hereafter are some important conclusions :

- For concentrated CO_2 inlet flue gas (x_{in}>0.2), a standalone membrane process can fulfill the target of CO_2 purity and recovery ratio with a lower energy requirement than a standard MEA absorption process.
- For more concentrated flue gas (x_{in}>0.3), it is shown that there is no significant interest to increase the membrane selectivity up to 100 because much higher membrane surface area will be required with almost the same energy requirement.
- For very concentrated CO_2 (x_{in}>0.5), a membrane process could play a role of polishing step with very low energy requirement (0.2-0.5 GJ/ton CO_2).
- Energy requirement decreases significantly for moderate CO_2 purity in the permeate suggesting that membrane process could play a role of a preconcentration process in a hybrid process.
- For more diluted CO_2 flue gas (x_{in}<0.15), membrane process cannot fulfill the targets of CO_2 purity and recovery ratio (R=0.9 and y=0.9) whatever the membrane selectivity. In this case, a multistage membrane process is needed to achieve the capture target.
- Feed compression with ERS strategy leads to a much lower membrane surface area (a factor of ×6 could be attained) while energy requirement remains almost the same with the permeate vacuum energy being slightly lower.
- A trade-off exists between membrane and surface area regarding the CO_2 recovery ratio, the membrane selectivity and the compression strategy. A technico-economic analysis is needed in order to determine the optimum operating parameters and configuration, taking into account both the CAPEX and OPEX of the process.

(771d) Metal-Organic Framework-Containing Membranes for Carbon Dioxide Capture

Tae-Hyun Bae and Jeffrey R. Long, Chemistry, University of California, Berkeley, CA

Metal-organic frameworks (MOFs) are a new class of porous crystalline materials, composed of metal ions or clusters connected by organic linkers through strong chemical bonds. Owing to their large surface areas and tunable functionality, MOFs are of potential utility for gas separations relevant to clean energy technologies, such as carbon dioxide capture and hydrogen purification. So far, the majority of gas separation studies on MOFs have focused on adsorptive separations. However, MOF membranes can also have excellent separation performance due to the selective adsorption/transport of one component in the feed mixture through the pore channels of the material.

In particular, MOFs containing open metal sites are good candidates for the fabrication of CO_2 selective membranes, since those sites are known to selectively adsorb/transport CO_2 molecules over other components. Furthermore, CO_2 molecules adsorbed within MOF pores can inhibit the transport of other components in the feed mixture, such that this type membrane can potentially have high selectivity. In this work, thin films of MOFs were grown on porous alumina support and their CO_2 separation properties were investigated.

Meanwhile, it is often very hard to prepare defect-free thin MOF continuous films with large membrane surface area. One of technologically-viable options to achieve high separation performance is incorporating MOF crystals into a suitable polymer matrix. Such mixed-matrix membranes containing MOF nanocrystals were also fabricated and investigated for CO_2 capture.

(771f) Atomistic Simulation to Design MOF-Supported Ionic Liquid Membranes for CO2 Capture

Krishna Mohan Gupta, Chemical and Biomolecular Engineering , NUS , Singapore, Singapore, Chen Yifei, Department of Chemical and Biomolecular Engineering, National University of Singapore, Singapore, Singapore, Zhongqiao Hu, Chemical and Biological Engineering, National University of Singapore, Singapore, Singapore and Jianwen Jiang, Chemical and Biomolecular Engineering, National University of Singapore, Singapore, Singapore

Carbon capture and sequestration (CCS) is now not only a scientific interest, but a societal issue for environmental protection. There has been increasing interest to use neat as well as supported ionic liquids (ILs) for CO_2 capture. Nevertheless, most studies to date have used polymeric and inorganic supports to prepare supported ILs. In the last decade, metal-organic frameworks (MOFs) have emerged as a new class of porous materials. The enormous choices of organic linkers and metal oxides allow the structures and functions of MOFs to be tuned in a rational manner, thus we envision that MOFs are versatile supports to fabricate supported IL for CO_2 separation from flue gas.

In the current work, atomistic simulations have been performed to investigate CO_2 separation in IRMOF-1 supported IL membranes. In addition, the microscopic properties of ILs in IRMOF-1 are also investigated. The ILs consist of identical cation 1-*n*-butyl-3-methylimidazolium $[BMIM]^+$, but four different anions, namely hexafluorophosphate $[PF_6]^-$, tetrafluoroborate $[BF_4]^-$, bis(trifluoromethylsulfonyl)imide $[Tf_2N]^-$, and thiocyanate $[SCN]^-$. The cations and anions in IRMOF-1 are more packed compared with bulk phase due to the confinement effect. The simulation results also suggest that anion has a stronger interaction with IRMOF-1 than cation. The small anions $[PF_6]^-$, $[BF_4]^-$, and $[SCN]^-$ prefer to locate near metal-cluster, particularly the quasi-spherical $[PF_6]^-$ and $[BF_4]^-$. In contrast, the bulky and chain-like $[BMIM]^+$ and $[Tf_2N]^-$ reside near phenyl ring. Among the four anions, $[Tf_2N]^-$ has the weakest interaction with IRMOF-1 and thus the strongest interaction with $[BMIM]^+$.

Different IL loadings into IRMOF-1 are also considered to elucidate the performance of membranes towards CO_2 capture. With increasing the weight ratio of IL to IRMOF-1 ($W_{IL/IRMOF-1}$), the selectivity of CO_2/N_2 at infinite dilution is enhanced. At a given $W_{IL/IRMOF-1}$, the selectivity increases as $[Tf_2N]^- < [PF_6]^- < [BF_4]^- < [SCN]^-$. This is in acco dance with the prediction from COSMO-RS method and largely similar to the order of binding energy between CO_2 and anion. In [BMIM][SCN]/IRMOF-1 membrane with $W_{IL/IRMOF-1} = 1$, $[SCN]^-$ is identified to be the most favorable site for CO_2 adsorption. [BMIM][SCN]/IRMOF-1 outperforms polymer-supported ILs in CO_2 permeability and its performance surpasses the Robeson's upper bound. This simulation study reveals that anion has strong effects on the microscopic properties of ILs and suggests that MOF-supported ILs are potentially intriguing for CO_2 capture.

(771g) Mechanically Robust, Ionic Liquid Containing Hollow Fiber Membranes for CO2 Separation Applications

Shan Wickramanayake, URS Energy and Construction/NETL, Pittsburgh, PA, David Hopkinson, National Energy Technology Laboratory, Morgantown, WV, Christina Myers, National Energy Technology Laboratory, Pittsburgh, PA, Steven C. Taylor, Oak Ridge Institute for Science and Education, Pittsburgh, PA; NETL, Pittsburgh, PA and David Luebke, US DOE/NETL, Pittsburgh, PA

The inherent properties of ionic liquids (ILs), such as insignificant vapor pressure, thermal stability above 200°C, functional diversity, and the high selectivity and permeability for CO_2, make them ideal candidates for separation of CO_2 from H_2, N_2, O_2, and CH_4. Several research groups have successfully demonstrated the application of ionic liquids in the form of supported ionic liquid membranes (SILMs) in which an IL is suspended within the pores of a membrane support and serves as the active gas transport medium of the membrane. The efficiency of SILMs can be further enhanced by spinning into hollow fiber geometry. However, because of the small outer diameter of hollow fibers and its highly porous substructure, even relatively robust polymers such as Matrimid can have a low overall tensile strength. In particular, the addition of ionic liquid and increasing porosity both cause a decrease in the mechanical integrity of the fibers. The scope of this investigation is to fabricate mechanically robust IL containing hollow fiber membranes that can be used in industrial applications. Preliminary work was carried out using a well known and characterized IL, 1-hexyl-3-methyl-imidazolium bis(trifluoromethyl-sulfonyl)imide ([hmim][Tf_2N]) with Matrimid®, and Torlon as the polymeric support material. The mechanical properties were characterized using a Dynamic Mechanical Analyzer (DMA). The CO_2 transport properties were recorded for the IL containing membranes with enhanced morphological features from 37 to 200 °C.

(778a) Immobilization of Vesicles Incorporated with Aquaporins Onto Polymeric Membranes for Low Energy Water Purification

Wenyuan Xie and Yen Wah Tong, Department of Chemical and Biomolecular Engineering, National University of Singapore, Singapore, Singapore

AQPs are transmembrane water channel proteins. Their unique selectivity, high water transport capability, and low activation energy have garnered strong scientific interests with many studies concentrating on the fabrication of biomimetic membranes based on the reconstitution of water channel proteins into self-assembled amphiphilic lipid or polymer bilayers. It has been reported that the osmotic permeability of a single aquaporin is in the range from 6×10^{-14} to 24×10^{-14} m^3 s^{-1}, and the permeability of an AQPs reconstituted biomimetic membrane is 167 um/s/bar, which is by two orders of magnitude greater than the commercial polymeric membranes. It is clear that aquaporin-incorporated biomimetic membranes possess an extremely high potential for water purification and desalination. However, the chief limitation of the biomimetic membrane is that the ultrathin biomimetic bilayers are too fragile to withstand high hydraulic or osmotic pressure in the water purification application. Here we show an innovative yet simple and easy-to-implement method to overcome the membrane fragility by immobilizing water-channel proteins incorporated vesicles into pre-fabricated membranes to develop advanced membrane materials for purifying water at low pressures and low energy. We show that this novel membrane exhibits high mechanical strength and stability during nanofiltration process and AQPs made contribution to the water performance.

(778b) Membrane Bioreactor Process and Fouling Control for Water Reclamation and Groundwater Recharge Applications

Woonhoe Kim[1], Varadarajan Ravindran[2], Ryan Thacher[1] and Massoud Pirbazari[1], (1)Sonny Astani Department of Civil and Environmental Engineering, University of Southern California, Los Angeles, CA, (2)University of Southern California, Los Angeles, CA

The use of membrane bioreactor systems (MBR) for water treatment, wastewater treatment and water reclamation and reuse has drawn great interest from researchers and engineers. Despite its many advantages such as smaller footprint and better product quality, their wider application require better control of membrane fouling and permeate flux decline as foulant removal depends on chemical or physical cleaning. In spite of their excellent retention characteristics, there are still problems that slow down use of membranes in these appliactions. Over time, membrane fouling and subsequent cleaning cause deterioration of membrane materials, resulting in compromised permeate quality and reduced membrane life spans. Chemical and biological fouling are major problems in membrane filtration due to reduced permeate flux, increased energy costs, and system downtime for maintenance. Biofouling in MBR systems is caused by the buildup of organic chemicals, microorganisms, and microbial communities at the membrane surface. Biofilms attached to a surface begin with cell adhesion, and progress to thick layers of extracellular polymeric substances (EPS), other organic chemicals, and a complex community of microbial cells that are often difficult to remove. Organic fouling due to the presence of natural organic matter (NOM) often leads to surface and internal pore fouling. More importantly, organic fouling progressively leads to biological fouling in so far as providing the organic nutrients for biofilm growth and sustenance.

The emerging use of advanced oxidation processes such as ozone, hydrogen peroxide, and UV radiation as well as catalysts (such as titanium dioxide) in water treatment offers new opportunities, because the AOPs are able to decompose a number of membrane foulants very efficiently, particularly the NOM. The effect of these processes on membrane fouling is difficult to predict due to the complex nature of NOM, and the major influence of the conformation of NOM and the decomposition products on organic and biological fouling.

In the context of water reclamation and reuse, the MBR process with ultrafiltration can be made more energy efficient, economical and versatile than reverse osmosis or nanofiltration by using exhibiting better transport properties, higher permeate flux, and greater fouling resistance. The MBR process is specially designed for the control of membrane fouling and permeate flux decline by a combination of powder activated carbon (PAC) sorption and fluid management. The present research is intended to overcome organic and biological fouling problems, and to improve aqueous transport properties. This aspect of the study required the investigation of fouling mechanisms in the operation of MBR systems.

The flat-sheet membrane evaluation tests in batch-mode represent an economic means for evaluating membrane materials for fouling potential and rejection characteristics. Preliminary flat-sheet membrane tests were performed in plate-and-frame cells using polyether sulfone (PES) membranes to obtain *a priori* information on the nature of organic and biological fouling. These tests also included subjecting the feed solution to AOPs including ozonation, and combinations of ozone and hydrogen peroxide. The breakdown of NOM constituents into organics of lower molecular weights and different structural characteristics shall influence the extent of surface fouling and internal pore fouling. Another aspect was the evaluation of microbial fouling using Escherichia coli (E. coli) as model bacteria to obtain prior estimates of microbial fouling and permeate flux decline at different bacterial concentrations in the feed. In these tests, the permeate flux and the effluent total organic carbon (TOC) and UV absorbance at 254 nm were determined to evaluate membrane rejection and flux decline patterns. The reclaimed water used in these tests was obtained from a water reclamation facility in Southern California.

The MBR continuous flow tests that followed the flat-sheet membrane tests employed polyether sulfone hollow-fiber membranes fabricated from the same material as the flat-sheet membranes. These experiments simulated the conditions in an MBR system using PAC and acclimated microorganisms. In these studies, the permeate flux and the effluent TOC and UV absorbance (at 254 nm) were determined to evaluate flux decline patterns in the hollow-fiber membranes. The information has been useful from several standpoints including the nature and mechanisms of membrane fouling, and its impact on permeate flux decline, which will eventually lead to the design of superior hollow-fiber membranes for different MBR applications. Furthermore, the MBR tests using reclaimed water subjected to AOPs would provide a good evaluation on the impact of NOM oxidation on membrane fouling and rejection. It is anticipated that the insights gained from these studies will be useful in the development of novel membranes with superior fouling resistant and aqueous transport characteristics through material modification techniques.

(778c) Optimal Design and Operation of Real-Time Response Systems for Water Security

Angelica Mann[1], Gabriel Hackebeil[1], Jose Rodriguez[1], Katherine A. Klise[2], Carl Laird[1] and Terranna Haxton[3], (1)Chemical Engineering, Texas A&M University, College Station, TX, (2)Sandia National Laboratories, Albuquerque, NM, (3)National Homeland Security Research Center, U.S. Environmental Protection Agency, Cincinnati, OH

Drinking water distribution systems are vulnerable to accidental or intentional contamination. Contaminants can ingress into the system through pipe breakages, water sources, or outlet (e.g. fire hydrant, household faucets) backflows. One proposed approach is the installation of an contaminant warning detection system – a sparse layout of sensors to detect the presence of contaminant. Effective control and cleanup of the system is also essential to reduce population exposure to the contaminant and to return the system to an operational state. Modern mixed-integer programming strategies play a role in the design of protection systems and optimal execution of a response strategy. In conjunction with researchers at Sandia National Laboratories and the Environmental Protection Agency, we have developed several advanced problem formulations to protect the public against potential contamination events. These tools have been integrated into the EPA's Water Security Toolkit, an open-source software package being developed for use by water utilities. In this presentation, we will show three components of the Water Security Toolkit.

First, we developed a novel water quality simulation package called Merlion that increases the simulation speed of TEVASIM by about an order of magnitude for large numbers of contamination. This package is also used to efficiently generate an explicit mathematical model of the contaminant spread for use in mixed-integer programming formulations.

Second, we present a pair of mixed-integer programming (MILP) formulations for source inversion and optimal sampling. The source inversion problem involves locating the contamination source given a small set of measurements. Advancing existing techniques, we have developed an MILP formulation that makes use of discrete (yes/no) measurements from both fixed sensors and manual grab samples to successfully invert for the potential contamination sources. Using an integrated strategy, we then solve another MILP formulation to find the optimal manual grab sample locations for the next sampling cycle. This integrated strategy is effective in finding the contamination source using a small number of sampling cycles.

While identifying the contamination source, population exposure is continuing. One measure to mitigate this exposure is to increase chlorination of the system from a set of disinfectant booster stations. Finally, we propose a MILP formulation for optimally placing booster stations to help neutralize possible contamination events. Due to the large uncertainty in the contaminant injection location and time, this placement problem is a stochastic problem that would normally be solved using decomposition techniques. However, we propose a formulation and solution approach that is both fast and memory efficient. We use Merlion to rapidly solve the large number of possible contamination scenarios and formulate the stochastic programming problem. Using a set of reduction techniques that combine scenarios, but yield an exact mathematical transformation of the original stochastic problem, we can dramatically reduce the size of the complete stochastic programming problem. This approach has been used to place booster stations in a network with more than 3000 nodes and a stochastic programming formulation with more than 400,000 scenarios.

Sandia National Laboratories is a multi-program laboratory managed and operated by Sandia Corporation, a wholly owned subsidiary of Lockheed Martin Corporation, for the U.S. Department of Energy's National Nuclear Security Administration under contract DE-AC04-94AL85000.

(778d) Effects of the Different Stages of Superstructure Development On the Efficiencies and Designs of Heat-Integrated Process-Water Networks

Elvis Ahmetović[1,2] and Zdravko Kravanja[2], (1)Faculty of Technology, University of Tuzla, Tuzla, Bosnia, (2)Faculty of Chemistry and Chemical Engineering, University of Maribor, Maribor, Slovenia

In a recent paper by Ahmetović and Kravanja (2012), a novel superstructure and mixed-integer non-linear programming (MINLP) model was proposed for the simultaneous synthesis of process-water and heat exchanger networks. This superstructure combines the water network (WN) given by Ahmetović and Grossmann (2011) and the heat exchanger network (HEN) introduced by Yee et al. (1990). It includes both direct and indirect heat exchanges (isothermal and non-isothermal heat transfers) between hot and cold streams, as well as some additional opportunities for the splitting and mixing of freshwater and wastewater streams. This model's objective is to minimize the total annual costs of the overall network.

This contribution describes and discusses the usage of the simultaneous optimization model proposed by Ahmetović and Kravanja (2012) for studying its effects, during different stages of superstructure development, on the efficiencies and designs of heat-integrated process-water networks. We first present a base case-study of a water network synthesis problem and reports its freshwater and utilities' consumption. Then, different designs of heat-integrated process-water networks are synthesized and compared from the viewpoint of freshwater and utility consumption, total amount of indirect heat exchanged, and the total eat exchanger area, as well as the total annual cost. In addition, the complexities of the resulting different network designs are also discussed. Using the above-mentioned synthesis methodology, some novel water-network designs have been reproduced for certain literature case-studies with minimum total annual cost significantly better than that reported in the literature. We used different GAMS (2012) MINLP solvers (i.e. BARON, SBB, DICOPT) for model-solving during various stages of the superstructure's development. It is worth pointing out that only small-size MINLP problems regarding heat-integrated water-networks could be solved with global optimality. In other cases we used local solvers SBB and DICOPT to solve problems over reasonable computational times, and compared the obtained results. The results clearly indicated that, after allowing for additional degrees of freedom for heat transfer, and water-splitting and mixing within the network, better network designs with reduced total annual cost resulted, in comparison with those reported results in the literature.

Keywords: water and heat-exchanger networks, MINLP model, simultaneous superstructure optimization.

Acknowledgement: The authors are grateful to the Scholarship scheme for academic exchange between the EU and Western Balkan countries for providing a JoinEU-SEE postdoctoral fellowship. Financial support is also gratefully acknowledged from the Slovenian Research Agency (Program No. P2-0032), and Federal Ministry of Education and Science Bosnia and Herzegovina.

References:

Ahmetović, E. & Grossmann, I. E. (2011). Global superstructure optimization for the design of integrated process water networks. AIChE Journal 57(2): 434-457.

Ahmetović, E. & Kravanja, Z. (2012). Simultaneous synthesis of process water and heat exchanger networks. Manuscript in the preparation.

GAMS (2012). GAMS solvers. http://www.gams.com/solvers/index.htm.

Yee, T. F., Grossmann, I. E. & Kravanja, Z. (1990). Simultaneous optimization models for heat integration—III. Process and heat exchanger network optimization. Computers and Chemical Engineering 14(11): 1185-1200.

(778e) Global Optimization of Water Network Synthesis Under Uncertainty with Risk Management

Cheng Seong Khor[1,2] and Nilay Shah[2], (1)Chemical Engineering Department, Universiti Teknologi PETRONAS, Tronoh, Malaysia, (2)Centre for Process Systems Engineering, Department of Chemical Engineering, Imperial College London, London, United Kingdom

This work addresses the problem of integrated water network synthesis under uncertainty with risk management. We develop a superstructure composed of water sources and sinks with membrane and non-membrane separation-based water regenerators, which leads to a mixed-integer nonlinear program (MINLP). The MINLP involves continuous variables on water flowrates and contaminant concentrations, and 0–1 variables on existence of interconnections of the network elements. Detailed nonlinear models are adopted for membrane regenerators such as reverse osmosis. To address uncertainty in the problem, we employ the classical recourse-based two-stage stochastic programming formulation by using discrete scenarios to approximate the underlying probability distribution of the uncertain parameters. The representation gives rise to a multiscenario two-stage stochastic MINLP, or more specifically, a mixed-integer quadratically-constrained quadratic program (MIQCQP). The framework comprises network design in the first stage and its operation in the second stage acting as recourse. In addition, we extend the model to address risk management using the Conditional Value-at-Risk (CVaR) metric. However, a large number of scenarios are often required to capture the underlying uncertainty of the problem, causing the model to suffer from the curse of dimensionality. To circumvent this issue, we propose a stepwise solution strategy mainly driven by solving the risk-neutral model using a small number of scenarios to reduce the computational load in the scenario-ordering step of simulating the Value-at Risk (VaR) parameter, as well as to provide initial values for solving the risk-averse mean–CvaR model. The presence of nonconvex bilinear terms necessitates the implementation of global optimization techniques to ensure reliable solutions. In this regard, we conduct computational experiments on both the risk-neutral and risk-averse formulations to study the performance of a new global MIQCQP solver, GAMS 23.8.1/GloMIQO 1.0.0, in comparison with the latest releases of two state-of-the-art commercial global solvers, namely BARON (version 10.1.3) and LindoGLOBAL (version 6.1.1.488). By varying the relative optimality tolerance, the solver performances are assessed mainly in terms of CPU time and solution quality of the resulting water network generated. Where possible, we analyze the performances by inferring on the strengths of the algorithmic components in the solvers.

(778f) Cooling Tower Wastewater Treatment and Reuse

Anditya Rahardianto, Han Gu, Larry Gao, John Thompson, Panagiotis D. Christofides and Yoram Cohen, Chemical & Biomolecular Engineering, University of California, Los Angeles, Los Angeles, CA

Cooling towers are ubiquitous across many industries, from large-scale power generation to commercial buildings. Cooling towers consume large volumes of water due to evaporative and windage/drift losses. To limit the accumulation of dissolved and suspended solids (and thus to mitigate mineral scaling, corrosion, and biological growth) in cooling water systems, periodic discharge of cooling water (i.e., blowdown) is needed, contributing to additional loss of cooling water that must also be replenished. Applications of advanced membrane treatment technologies (ultrafiltration (UF) and reverse osmosis (RO)) for cooling tower wastewater (CTW, i.e. blowdown water) treatment and reuse, while environmentally and economically desirable, remain limited. CTW treatment and reuse reduce blowdown water loss, while at the same maintain cooling water quality at levels necessary for effective mitigation of cooling water system corrosion/scaling. In the present work, a fundamental computational framework was developed to quantify and assess the operational costs and benefits of CTW treatment via integrated UF-RO processes. Potentially effective UF-RO operational strategies for CTW treatment were assessed and their optimal process conditions were identified. The potential benefits of advanced adaptive UF-RO operation that optimally handles fluctuations in cooling tower system conditions and UF-RO operational costs (e.g., cooling water quality, peak/non-peak electricity costs) were examined, including both steady-state and intermittent operational modes. Examination of UF-RO process operational requirements for CTW treatment will be presented focusing on the results of a recent field study using a novel UF-RO pilot system operation in a co-generation plant at the University of California, Los Angeles. Results from the field study demonstrated both a significant level of water savings as well as monetary savings for the facility. The overall analysis approach and field study results demonstrate that there is a range of scenario for beneficial CTW treatment and reuse, which would be best served by a self-adaptive treatment process.

Improving the quality of Pulp & Paper Mill waste water employing Electroflocculation Technology

Rita Tandon, Satya Dev Negi, R.M.Mathur

Abstract

Pulp and Paper Industry is one of the highest polluting industries. The wastewater from this industry has very high proportion of organic pollutants and significant color loads. Although there is no stringent legislation to the tolerance limits for discharged color in effluents but some state pollution control authorities have imposed tolerance limits for the color in discharged effluents. The present practice of color removal involves precipitation and coagulation of colored substances using aluminium salts. Although such chemical treatment methods are expensive but the industry has perforce adapted these methods to contain the tolerance limits for color discharge.

Lot of research has been carried out worldwide on various EOP options for removal of color from pulp & paper mill effluent. One such option is electroflocculation (EF) technology which has been extensively studied by Central Pulp & Paper Research Institute (CPPRI). The present paper highlights the findings of pilot plant trials conducted by CPPRI on electroflocculation technology at a mill site to assess the techno-economic viability of the technology for treating different colored streams for improving the quality of pulp & paper mill waste water.

Introduction

Public demand for cleaner environment is forcing the industry to take adequate steps to reduce pollution loads in discharge effluents. In the recent times, the presence of color in pulp & paper mill effluents has emerged as an major environmental issues due to public perception and this has prompted the Regulatory Bodies to include color as a pollutant in the charter on **Corporate Responsibility for Environmental Protection (CREP)** for compliance by Indian Pulp & Paper Industry.

The Indian paper industry uses on an average 100 to 200 m^3 of fresh water per ton of paper and nearly 75% of water used is discharged as effluent. In addition to high proportion of inorganic and organic pollutant the effluent from paper industry contains significant color loads which varies from mill to mill depending upon the raw material used, process employed, type of end products and extent of system closure in the mill.

Broadly the color reduction strategies include at source control through process modification or through end-of-pipe (EOP) treatment methods. Since the process modification demands proper planning and timely execution with substantial investments, the focus is more on EOP options as an immediate measure to control effluent color. Lot of research was done worldwide on various EOP options, which included physico-chemical treatment method, physical separation method, UV irradiation method, biological method, etc., however, among various options tried, chemical precipitation using coagulants and flocculants is the most accepted option for color removal as it is easy to implement the process in existing ETP's without requiring any major investment, but the process has limitation particularly in sludge handling.

In an endeavor to identify a suitable technology for color removal from pulp & paper mill effluents, CPPRI focused on EF technology and extensively studied the technology on lab and pilot scale to establish the techno-economic viability of the process for removal of color from pulp & paper mill effluents.

The present paper highlights the findings of up scaling of the electro-flocculation technology to pilot scale and the viability of the technology for treating pulp & paper mill effluents for color removal.

Results & Discussions

Sources and Magnitude of Color in Paper Industry

Pulp and paper mill effluents have a characteristic brownish black color, which is mainly due to lignins, tannins and other extractives bearing, chromophoric groups. The magnitude of color depends on the raw material used process employed and type of end products produced. Lignin and extractives are highly polymerized substances and are difficult to biodegrade.. The color in small mills is mainly due to lignin going through spent liquors. It is estimated that the quantity of lignin going through spent liquors varies from 300-400 kg/ton of pulp, generating color load of about 1400-1500 kg PCU / ton of pulp. It is estimated that 90% of the color is due to lignin.

In alkali extraction stage effluent only about 50-60 kg of lignin / ton of paper is going into effluents and the combine effluents has color load of about 1500 PCU. In newsprint mills where eucalyptus constitutes the raw material for production of mechanical pulp component, very high color load are noticed in the effluents. Eucalyptus contains about 3-6 % extractives, mostly tannins, which are leached out during pre steaming and refining stages. The washing of CMP pulp is highly colored and the color intensity is several times higher when compared to color due to lignin compounds. **Table-1** summarizes the magnitude of color load in different effluent streams of pulp & paper industry. **Table-2** summarizes the molecular size of chromophores present in the effluent which indicates that except for black liquor, the molecular size of color bodies varies between 5,000-13,000 Dalton units.

Table- 1
Magnitude of Color Load and respective Chromophores in different Effluent Streams of Indian Pulp & Paper Industry

Streams	Total Solids (g/l)	Color (PCU)	Chromophores
Agro-based Mills (Chemical Pulping)			
Black liquor (Bleached Grade)	**40**	**82000**	**Lignin & its degraded products**
Black liquor (Unbleached Grade)	**13**	**6333**	**Lignin & its degraded products**
Combined Mill Effluent (Bleached Grade)	**3.3**	**3100**	**Lignin & its degraded products**
Combined Mill Effluent (Unbleached Grade)	**3.5**	**3030**	**Lignin & its degraded products**
Wood Based Mills (Chemical Pulping)			
Ep Stage Effluent (Eucalyptus, Poplar)	**3.8**	**1600**	**Chlorolignins**
Combined Mill Effluent (Eucalyptus, Poplar)	**3.2**	**2000**	**Degraded lignin products**
Rayon Grade Washing filtrate (Mixed hardwood)	**0.9**	**1200**	**Degraded lignin products**
Wood Based Mills (Mechanical Pulping)			
Pulping Effluent (Eucalyptus)	**1.3**	**8816**	**Extractives (Poly phenols & tannins)**

Source- CPPRI

Table- 2

Molecular Size of Chromophores in different Effluent Streams of Indian Pulp & Paper Industry

Streams	Total Solids (g/l)	Color (PCU)	Molecular Weight* (Dalton)
Agro-based Mills (Chemical Pulping)			
Black liquor Bleached	**40**	**82000**	**5,000-50,000**
E/Ep Stage Effluent	**2.6**	**3800**	**12,000-13,000**
Combined Mill Effluent	**3.3**	**3100**	**7,000**
Wood Based Mills (Chemical Pulping)			
Ep Stage Effluent	**3.8**	**1600**	**7,000-13,000**
Combined Mill Effluent	**3.2**	**2000**	**5,000-7,500**

Source- CPPRI

Application of Electroflocculation (EF) Technology for Color Removal from Pulp & Paper Mill Effluents

Electro-coagulation and electro-flocculation are two techniques involving the electrolytic addition of coagulating metal ions directly from sacrificial electrodes. These ions coagulate with pollutants in water, in a similar manner to that of adding coagulating chemicals such as alum and ferric chloride and allow the easier removal of the pollutants. The system produces about half to one third volume of the sludge.

The Electroflocculation (EF) process involves the application of an electric current to sacrificed metal electrodes having coagulating properties, like Al, Fe, or Mg, generating metal ions and gas bubbles simultaneously. The metal ion released combines with pollutant and gas bubbles generated capture the coagulant agglomerates resulting in most of the pollutant being floated to surface.

The total process is accomplished in three steps as shown below.

1st Step

At Anode $$M \xrightarrow{\text{Electrolytic Dissociation}} M^{2+} + 2e^-$$

At Cathode $$2e^- + 2H_2O \xrightarrow{\text{Electrolytic Dissociation}} 2OH^- + H_2\uparrow$$

2nd Step

$$M2+ \quad + OH^- \longrightarrow M(OH)_2 . MOH^+$$

Insoluble metal hydroxide

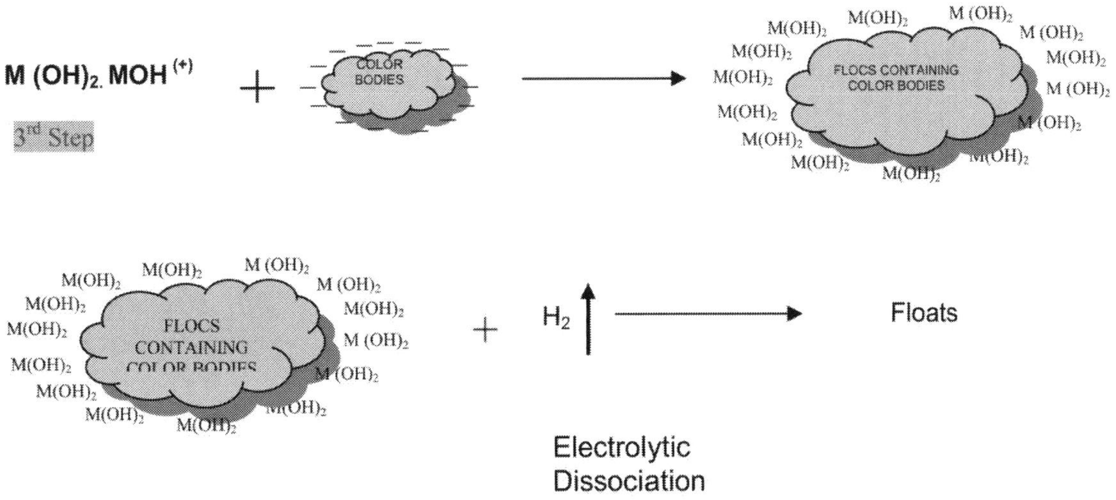

Optimization of Process Variables

To optimize the process conditions extensive research was conducted at lab scale and following variables were studied extensively-

- Type of electrode
- pH
- Temperature
- Total Solids Concentration
- Reaction time
- Optimum Voltage
- Optimum Current
- Electrode Spacing
- Current Density
- Conductivity

The effect of pH, temperature, retention time is shown in **fig. 1, 2 & 3**.

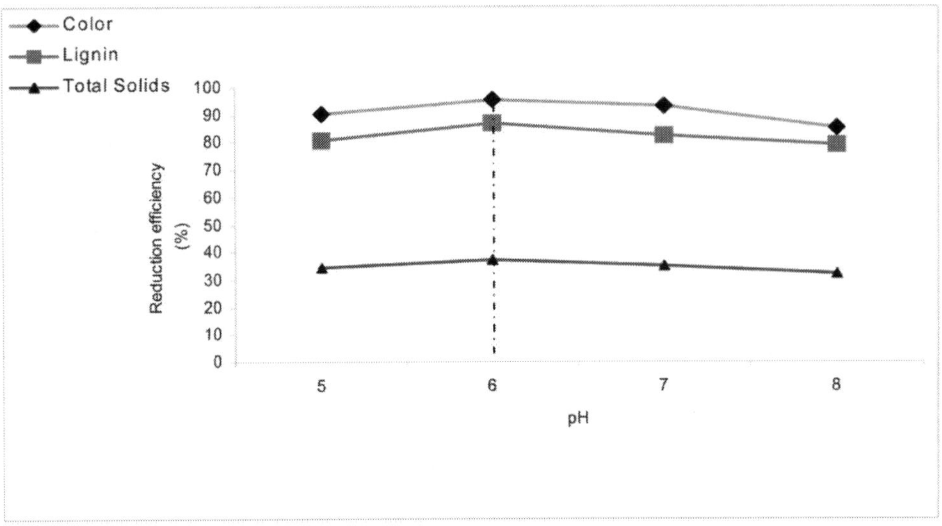

Fig-1 pH VS Process Efficiency

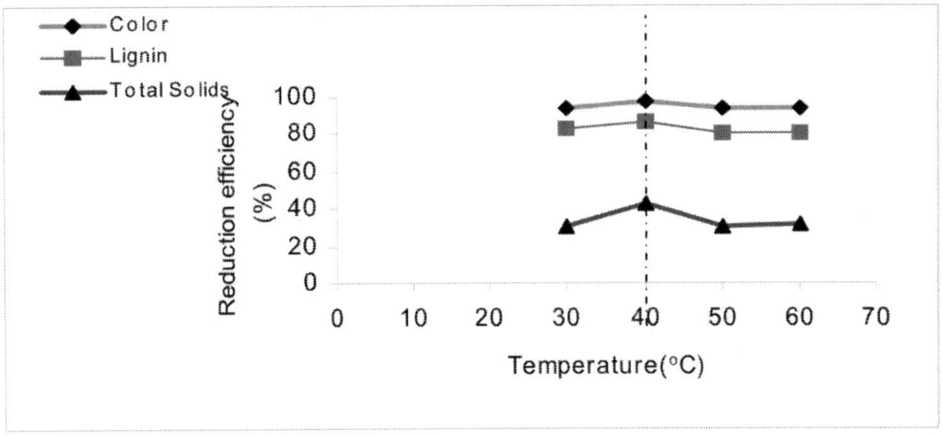

Fig-2 Temperature Vs Process Efficiency

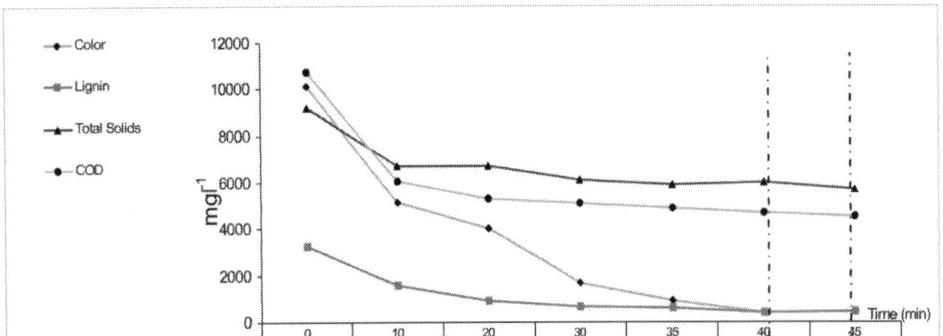

	0	10	20	30	35	40	45
Color	10124	5137	4000	1686	932	405	436
Lignin	3240	1600	890	640	630	410	480
Total Solids	9200	6700	6700	6100	5900	6000	5700
COD	10712	6034	5299	5106	4874	4680	4526

Fig-3 Reaction Time Vs Process Efficiency

Optimization of electrode spacing

Experiments were conducted using different electrode spacing viz 0.5 cm, 1cm and 2 cm. It is observed that the efficiency with respect to removal of color and lignin has increased from 59%, 71% at 0.5 cm spacing to 95%, 84% to 1 cm spacing and then leveled off. As shown in **fig. 4** the optimum electrode spacing was found to be 1 cm.

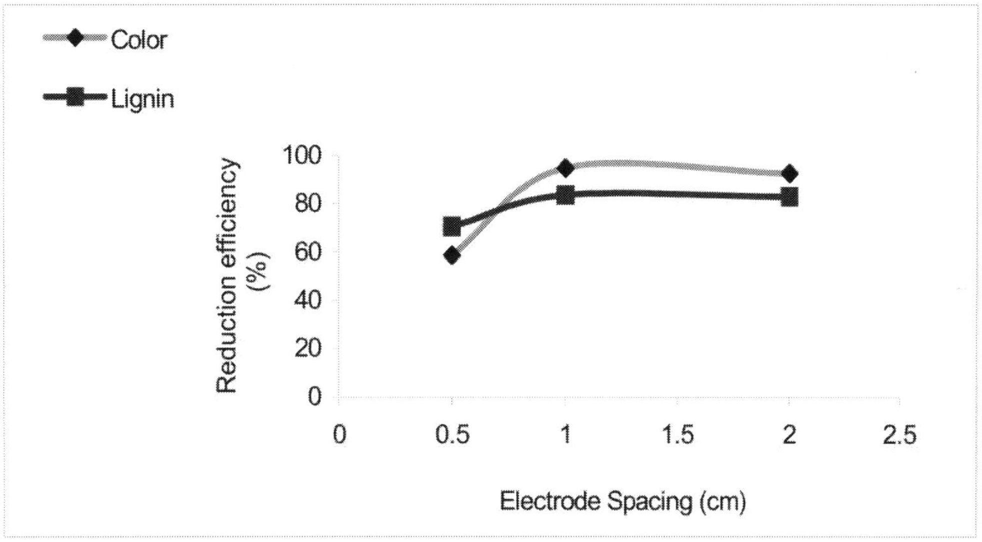

Fig-4 Electrode Spacing Vs Process Efficiency

Pilot Plant Trials on Electroflocculation Technology Conducted in wood based Paper Mill, India

Having established the technical feasibility of the process in lab scale, it was felt necessary to assess the economic viability of the technology on pilot scale. A pilot plant of $1m^3$/hr capacity was fabricated, installed and commissioned at Wood based paper mill, India and pilot plant trials were conducted on different colored effluent streams.

A schematic of the pilot plant is depicted in **fig.5**

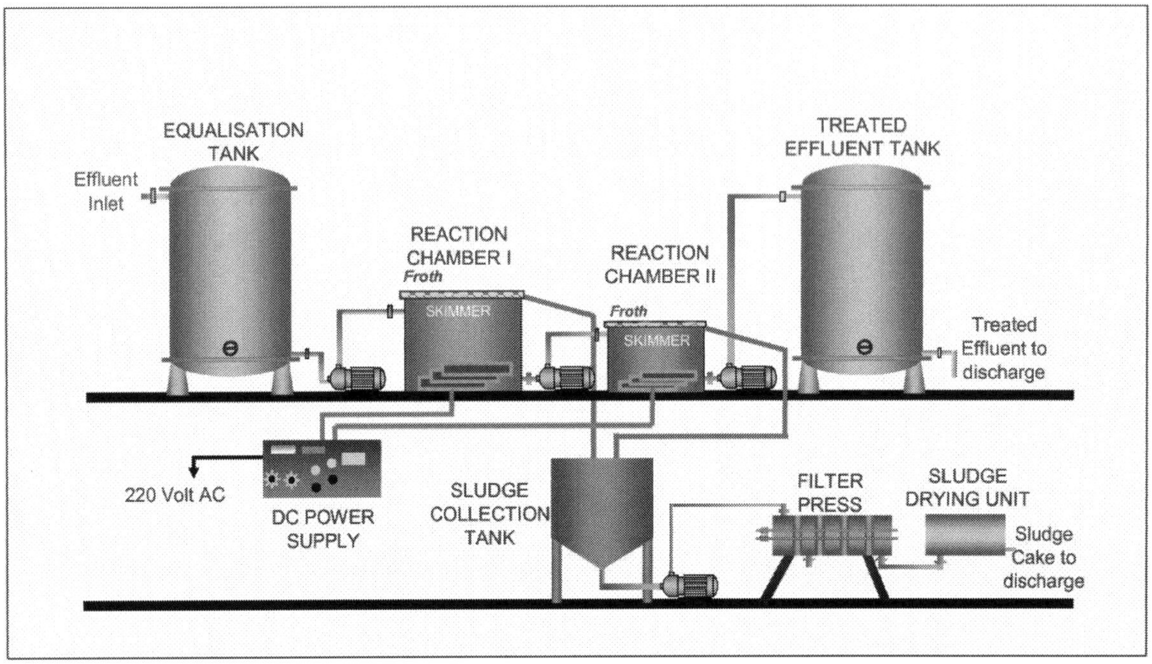

Fig-5 Schematic Of Pilot Electroflocculation Plant of 1m3/hr Capacity Installed in a Wood based Paper Mill, India

Pilot Trials with Bleach Plant Effluent

Trials conducted with bleach plant effluent having initial color load of 518 PCU had resulted into reduction on color by 86%, lignin by 86%, COD by 77% and AOX by 73%. The final achievable color load was 70 PCU. Finding of the trials are depicted in **table- 3**.

Table-3

Trials on Bleach Plant Effluent

	Untreated	Treated	% Redn.
Color, PCU	518	70	86
Lignin, mg/l	137	19	86
COD, mg/l	768	176	77
AOX, mg/l	15	4	73
Flow rate- $1m^3/hr$, Voltage- 6.0 volts, Current Density- 11-44mA/cm^2, Power consumed- 1.4 kWh/m^3			

Pilot Trials with RGP Washing Effluent

Trials conducted with RGP washing effluent having initial color load of 1143 PCU had resulted into reduction on color by 88%, lignin by 67% and COD by 38%. The final achievable color load was 140 PCU. Finding of the trials are depicted in **table- 4**.

Table-4

Trials on RGP Washing Effluent

	Untreated	Treated	% Redn.
Color, PCU	1143	140	88
Lignin, mg/l	153	51	67
COD, mg/l	600	372	38
Flow rate- 1m^3/hr, Voltage- 6.0 volts, Current Density- 11-44mA/cm^2, Power consumed- 1.16 kWh/m^3			

Pilot Trials with Bagasse Black Liquor

Trials conducted with bagasse black liquor having initial color load of 1178 PCU had resulted into reduction on color by 89%, lignin by 71% and COD by 47%. The final achievable color load was 135 PCU. Finding of the trials are depicted in **table- 5**.

Table-5

Trials on Bagasse Black Liquor

	Untreated	Treated	% Redn.
Color, PCU	1178	135	89
Lignin, mg/l	350	100	71
COD, mg/l	1151	611	47
Flow rate- 1m^3/hr, Voltage- 6.0 volts, Current Density- 11-44mA/cm^2, Power consumed- 1.55 kWh/m^3			

Techno-Economic Viability of EF Technology

Based on the findings of the pilot plant trials the techno-economic viability of the process was assessed for treatment of different colored effluents. The results are summarized in **table -6**.

It is estimated that-

- **The cost of treatment for treating bleach plant effluent with input color of 518 PCU and discharge color of 70 PCU works out to be Rs.22/ m^3.**

- **The treatment cost for treating combined mill effluent with input color of 1800 PCU and discharge color of 250 PCU works out to be Rs. 18/m^3.**

- **The treatment cost for treating agro-based black liquor with input color of 3000 PCU and discharge color of 450 PCU works out to be Rs. 53/m^3.**

Table-6
Economic viability of EF Technology for treating different colored effluents

Effluent type	Bleach Plant Effluent	Combined Mill Effluent	Agro-based black liquor
Initial Color, PCU	518	1800	3000
Final Achievable Color, PCU	70	250	450
Color Redn. Efficiency, %	85	85	85
Total Treatment Cost Rs./m^3	22	18	53

Conclusion

1. The electro-flocculation process is technically feasible to remove color from all types of colored effluents generated in Pulp & Paper Industry. Color removal efficiency achieved is in the tune of 85-90%. The voltage requirement is 6.0 volts.
2. Besides color, the process also removes lignin to the tune of 65-80%, COD 45-65%, and Toxicity in terms of AOX 70-75%.
3. Based on pilot scale studies, economics for treatment of pulp & paper mill effluent has been worked out to be between Rs. 18 to Rs.53 per m^3 to achieve a discharge color in the range of 70-450 PCU, depending upon the input color load.
4. The technology needs to be demonstrated on commercial scale, however, it is recommended that individual pulp & paper mills shall make their own evaluation and conduct pre-feasibility studies before adoption in each case.
5. EF technology may also be tried for other industries and small scale industries as well, which generates colored effluents viz. textile, dyes manufacturing units, tanneries, ink manufacturing units, etc.

References

1. Subrahmanyam, P. V. R., Parekh, R.C. and Mohan Rao, j. G., Ippta, Vol. 9, No. 1, 1972, p. 14.
2. Sharma, G. D., Singh, M.N. and Agarwal, N.R., Ippta , Convention: 124(78).
3. 'Pollution Engineering practice Handbook' Ann Arbor Science Publishers. p. 419-426 (1975).
4. Hillis, W. E., Wood extractives and their significance to pulp and paper industry. p.59, Academic Press, New York-1962.
5. Kulkarni A.G., Mathur R. M., Gupta Abha and Pant, Rajesh. Ippta, Vol. 24, No. 3, !987.

6. Metijevic, E. and Janauer, G. E., J. Colloidal Interface Science, 21:68 (1966).

7. Matijevic, E. and Stryker, L. J., J. Colloidal Interface Science, 22:68 (1966).

8. Fernandes, J. D., Thampi, K. P., Prahaladhan, V. M. Singh, Sarju, Ippta, Vol. 22, No. 2, June 1985.

9. Kulkarni, A. G., Mathur, R. M., Tandon Rita, Naithani, S. and Pant, Rajesh, Ippta, Vol.25, No. 1, 1988.

10. Kulkarni, A. G., Mathur, R. M., Abha Gupta, Naithani, S. and Pant, Rajesh, Ippta, Seminar, 1989.

(778h) A Combined Cost Model for Analysis of Degraded Water Utilization in Thermoelectric Power Plants

Iman Safari[1], Michael E. Walker[1], Javad Abbasian[2], Hamid Arastoopour[3], Ming-Kai Hsieh[4], Ranjani B. Theregowda[4], David A. Dzombak[4], Liu Wenshi[5], Radisav Vidic[6] and David C. Miller[7], (1)Chemical and Biological Engineering, Illinois Institute of Technology, Chicago, IL, (2)Chemical and Biological Engineering Department, Illinois Institute of Technology, Chicago, IL, (3)IIT, Chicago, IL, (4)Department of Civil and Environmental Engineering, Carnegie Mellon University, Pittsburgh, PA, (5)Civil and Environmental Engineering, Carnegie Mellon University, Pittsburgh, PA, (6)Department of Civil and Environmental Engineering, University of Pittsburgh, Pittsburgh, PA, (7)U.S. Department Of Energy, National Energy Technology Laboratory, Morgantown, WV

The thermoelectric power industry in the U.S. uses a large amount of freshwater. The large water demand is increasingly a problem, especially for new power plant development, as availability of freshwater for new uses diminishes in the United States. Reusing degraded water sources, such as treated municipal wastewater, provides one option to mitigate freshwater usage in the thermoelectric power industry. The amount of freshwater withdrawal that can be displaced with degraded water sources at a particular location requires evaluation of the water management and treatment requirements, considering the quality and abundance of the non-traditional water sources.

This paper presents a modular framework for evaluation of the economic impacts of utilization of degraded waters for cooling in thermoelectric power systems. Various models are presented for estimation of water quality in a recirculating cooling system, fouling rate in the condenser, and the costs associated with fouling given the make-up water quality, the treatment strategy, and the operating conditions. The results of the overall economic analysis performed on several case studies are discussed.

(784a) Ammonia-Based Cox Free Hydrogen Production by Membrane Reactors

Gang Li, Tomohisa Yoshioka, Masakoto Kanezashi and Toshinori Tsuru, Department of Chemical Engineering, Hiroshima University, Higashi-Hiroshima, Japan

Ammonia decomposition in a bimodal catalytic membrane reactor (BCMR) consisting of a Ru/γ Al_2O_3/α Al_2O_3 bimodal catalytic support and a hydrogen selective silica membrane in a single unit was proposed for CO_x free hydrogen production in the present study. The bimodal catalytic membrane showed a H_2 permeance of 6.2×10^{-7} mol/(m^2 s Pa) at 500 °C, with H_2/NH_3 and H_2/N_2 permeance ratios of 200 and 720, respectively. Ammonia conversion was surprisingly enhanced form 45 to 95% at 450 °C in the BCMR after selective H_2 extraction. The BCMR showed excellent stability with respect to both gas permeation properties and catalytic activities. Moreover, based on quantitative discussion on the effect of bimodal structure on the gas permeation, pore structure and catalytic activity of the bimodal catalytic membrane reactors, the optimal preparation conditions for the membrane reactor was then obtained. An interesting phenomenon that NH_3 conversion increased as the feed pressure was increased in the membrane reactor was observed, although high pressure was thermodynamically not beneficial to the NH_3 decomposition.

A mathematical model was formulated for the simulated study of NH_3 decomposition in a co-current configuration membrane reactor, and the effect of Damköhler number (Da), permeation number (θ), and membrane selectivity on membrane reactor performance was discussed theoretically. The required membrane performance for autothermal decomposition of NH_3 in a bimodal catalytic membrane reactor was calculated based on theoretic simulation. Moreover, the effect of various operating conditions on the membrane reactor performance was studied experimentally, and the results showed good agreement with those calculated by the proposed simulation model.

(1) Gang Li, Masakoto Kanezashi, Toshinori Tsuru. Highly enhanced ammonia decomposition in a bimodal catalytic membrane reactor for CO_x-free hydrogen production. *Catalysis Communications*, 15 (2011) 60–63.

(2) Gang Li, Masakoto Kanezashi, Tomohisa Yoshioka, Toshinori Tsuru. Ammonia decomposition in catalytic membrane reactors: Simulation and experimental studies. *AIChE Journal*, doi: 10.1002/aic.13794, in press.

(3) Gang Li, Masakoto Kanezashi, Hye Ryeon Lee, Tomohisa Yoshioka, Toshinori Tsuru. Preparation of a novel bimodal catalytic membrane reactor and its application to ammonia decomposition for CO_x-free hydrogen production. *International Journal of Hydrogen Energy*, under review.

(784b) Membrane Reactors for Carbon Capture and Hydrogen Production in IGCC Plants: Simulation, Cost Analysis and Design Optimization

Fernando V. Lima, Department of Chemical Engineering and Materials Science, University of Minnesota, Minneapolis, MN, Prodromos Daoutidis, Chemical Engineering and Materials Science, University of Minnesota, Minneapolis, MN and Michael Tsapatsis, Department of Chemical Engineering & Materials Science, University of Minnesota, Minneapolis, MN

In recent years, incentives for the sustainable use of fossil fuels (e.g., coal) have risen in the United States and worldwide. These incentives are motivated by the changes in global climate that have been associated with the emissions of greenhouse gases, such as carbon dioxide (CO_2), to the atmosphere during fossil fuel combustion. Particularly, for coal-based power generation, the challenge is to mitigate such emissions efficiently and economically. Integrated Gasification Combined Cycle (IGCC) plants are an emerging technology for coal-based electricity production that allow CO_2 capture with higher efficiency and lower penalty on cost of electricity when compared to conventional pulverized-coal plants. This presentation focuses on the alternative of using water gas shift (WGS) membrane reactors for pre-combustion carbon capture and hydrogen (H_2) production in IGCC plants. Specifically, a stand-alone membrane reactor model, which will be ultimately integrated into the IGCC flowsheet, is introduced for simulation, cost analysis and optimization studies.

Regarding the modeling task, a one-dimensional and nonisothermal WGS membrane reactor model is developed. This membrane reactor is equipped with a H_2-selective zeolite membrane that holds potential to be used in the WGS environment of coal-based gasification plants. Specifically, a shell and tube reactor design is considered, in which the catalyst is packed in the tube side, a thin membrane layer is placed on the surface of the tube wall and the sweep gas flows in the shell side to lower the H_2 partial pressure of this side. Also, an inner tube, where a coolant flows, is used to provide the necessary cooling that ensures the reactor temperature remains within certain bounds set by the high-temperature WGS catalyst (Fe-Cr-based) operation. The model assumes the reactor operates at steady state and considers constant pressure, adiabatic shell with respect to the environment, plug-flow operation and neglects axial dispersion. Also, the flux through the zeolite membrane is described by a combination of activated and Knudsen diffusion components.

The developed model is simulated using operating conditions consistent with the IGCC plant, including the syngas feed from the gasifier after steam injection, WGS reactor temperature and pressure. The simulation results showed that configurations with co/counter-current flow for the shell (sweep) and co-current flow for the jacket (coolant), relative to the tube (feed), produced a CO_2 rich stream (retentate) and a H_2 rich stream (permeate) that can be further processed for storage and gas turbine power generation, respectively. These product streams satisfied the constraints reported by the U.S. DOE, such as the R&D goal of 90% CO_2 capture and the desired H_2 recovery value of 95% [1], for the retentate and permeate, respectively. These configurations also maintained the reactor temperature under the operating limit associated with the selected WGS catalyst. Using the successful case studies as benchmark, novel R&D targets for the zeolite membrane characteristics (permeance, selectivity) and cost to meet these constraints are established and compared to the requirements for other materials reported by the DOE [1]. Finally, the optimization of the membrane reactor design, considering the obtained membrane characteristics from the cost analysis, is performed. This optimization has the objectives of maximizing the reactor performance and minimizing the membrane use (and thus its cost) by placing it in the optimal location. The previously introduced optimization problem for the isothermal case [2,3] is extended to incorporate the nonisothermal nature of the problem, including the addition of the constraint on the upper limit of the reactor temperature.

References

[1] John J. Marano. Integration of H_2 separation membranes with CO_2 capture & compression. Report to DOE, Contract No. DE-AC26-05NT41816, 2010.

[2] Fernando V. Lima, Prodromos Daoutidis, Michael Tsapatsis, and John J. Marano. Modeling and optimization of membrane reactors for carbon capture in Integrated Gasification Combined Cycle units. Ind. Eng. Chem. Res., 51(15):5480-5489, 2012.

[3] Fernando V. Lima, Prodromos Daoutidis, Michael Tsapatsis, and John J. Marano. Modeling and optimization of membrane reactors for carbon capture in IGCC units. In AIChE Annual Meeting, Minneapolis, MN, October 2011.

(784c) A MULTI-Functional Membrane Reactor System for the Destruction of Chemical Warfare Agents

Yousef Motamedhashemi[1], Majid Monji[1], Doug Parsley[2], Nitin Nair[1], Paul K. T. Liu[3], Fokion Egolfopoulos[4] and **Theodore Tsotsis**[5], (1)Mork Family Department of Chemical Engineering and Material Science, University of Southern California, Los Angeles, CA, (2)Media and Process Technology, Inc., Pittsburgh, PA, (3)Media and Process Technology, Pittsburgh, PA, (4)Department of Aerospace and Mechanical Engineering, University of Southern California, Los Angeles, CA, (5)Mork Family Department of Chemical Engineering and Materials Science, University of Southern California, Los Angeles, CA

The possibility of use of chemical weapons has increased in recent years, both as a result of potential terrorist attacks as well as of ongoing international conflicts. The successful application of a flow-through catalytic membrane reactor (FTCMR) as an individual protection (IP) system for the destruction of dimethyl methylphosphonate (DMMP), which is known as a chemical precursor (and used to simulate its characteristics) for Sarin (GB), a toxic chemical warfare agent (CWA) was previously investigated and reported by this group[1].

In this current effort, a novel hybrid system is being developed that combines the FTCMR with a surface-flow membrane system (SFMS). The main advantage of this hybrid system, combining the SFMS (capable of the physical removal of a large fraction of the CWA from contaminated air streams) and the FTCMR (which completely oxidizes the remaining amount) is the continuous CWA destruction for extended time periods which are appropriate for both IP and collective protection (CP) applications.

As part of the study, the effect of the DMMP concentration in the feed on its complete combustion in the FTCMR was investigated. The studies indicate that the protection (complete conversion) time is a function of the DMMP concentration, with longer times observed for the lower concentrations, and shorter protection times associated with the higher concentrations. These observations validate the role for the FTCMR as a second stage in a hybrid system, following a bulk-toxin removal SFMS first stage. We also investigate the performance of the SFMS. Experiments indicate that a high removal rate can be achieved with a proper choice of (feed rate/membrane surface area) and the sweep ratio. The pressure ratio (P/P_s) limit for the DMMP removal is << 1, supporting the proposed SFMS concept for the effective removal of condensable toxins, i.e., steady-state adsorption/desorption *via* nanoporous carbon membrane films, as opposed to the traditional batch-wise adsorption with subsequent regeneration. The low (P/P_s) limits indicate that the selected carbon membrane provides a favorable isotherm relationship; with an optimal selection of (feed rate/membrane surface area) and sweep ratio, a high degree of removal can be achieved, so that only trace residual amounts may require further catalytic degradation in the FTCMR. The performance of the combined system is currently under investigation, and experimental results and the modeling of the system performance will be presented at the meeting.

References:

1. Motamedhashemi, M.M.Y., Egolfopoulos, F.N., and Tsotsis, T.T., "Application of a Flow-Through Catalytic Membrane Reactor (FTCMR) for the Destruction of a Chemical Warfare Simulant," *J. Membrane Sci.*, 376, 119, 2011.

(784d) Non Isothermal Model for Theoretical Investigation of Hydrogen Purification by Water Gas Shift Reaction From Reforming Mixtures Using Catalytic Membrane Reactor

Bhanu Vardhan Reddy Kuncharam, Artie McFerrin Department of Chemical Engineering, Texas A&M University, College Station, TX and Benjamin Wilhite, Artie McFerrin Dept. of Chemical Engineering, Texas A&M University, College Station, TX

Conventional membrane reactors for hydrogen separation by water gas shift reaction from reformate mixtures employ hydrogen permselective Polymeric or Pd based materials in tandem with catalysts. Polymeric membranes provide promising hydrogen permselectivities, but they lack thermal stability at temperatures above 200^0C which limits their application for low temperature WGS reaction. Dense Palladium membrane has infinite inherent hydrogen permselectivities but they are expensive, prone to poisoning by carbonaceous materials and suffer embrittlement over multiple thermal cycles. In order to address these problems our research group is exploring the use of catalytic membranes to achieve hydrogen separation without relying upon costly and delicate permselective materials.

A detailed 2D Non-Isothermal model was developed to explore the use of externally applied axial and/or radial thermal profiles to enhance H_2 recovery rates while maintaining competitive (>100:1) permselectivities. 2D Isothermal simulation of a catalytic membrane reactor receiving a typical reformate mixture (9% CO, 3% CO_2, 28% H_2, 15% H_2O) with pure steam as sweep co-reactant gas predicted permselectivities of 90:1 to >200:1 corresponding to hydrogen recoveries of 20% to 40% [1]. This talk will detail model development and results of both isothermal and non-isothermal membrane reactor simulations.

[1] B.Kuncharam, B.A.Wilhite, Cost-Effective Catalytic Membranes for H2 Purification. 67b. Presented at AIChE Spring Meeting 2012.

(784e) Biocatalytic Membranes for Carbon Dioxide Capture From Flue Gas

Shada Salem, School of Energy, Enviornmental, Biological, and Medical Engineering, University of Cincinnati, Cincinnati, OH and Stephen W. Thiel, School of Energy, Environmental, Biological, and Medical Engineering, University of Cincinnati, Cincinnati, OH

Absorptive capture of CO_2 from flue gas includes a number of steps: dissolution in the aqueous phase, hydration, dissociation to form bicarbonate, and finally formation of carbonate. The rate-controlling step in this mechanism is usually the hydration of CO_2; if a viable means to accelerate the hydration of CO_2 could be found, it should be feasible to fix rapidly large quantities of CO_2 into solution. Fortunately, a catalyst for the rapid and reversible hydration of CO_2 exists in biological systems: carbonic anhydrase (CA) [1-4]. Recent studies have confirmed that the hydration of CO_2can be improved using CA [5-7].

This paper presents preliminary results investigating the possibility of using a biocatalytic membrane incorporating CA to capture CO_2. Poly(vinylidene fluoride) (PVDF) was used as a membrane material due to its high resistance to harsh environments, both acidic and basic conditions, its high thermal and UV stabilities, and its reasonable cost [8-13]. PVDF membranes were cast using wet phase inversion [14]; casting parameters were changed to modify the hydrophobicity of the membrane to permit use in biocatalytic CO_2 capture. Bovine carbonic anhydrase (BCA) was physisorbed on the PVDF membranes and the effects of pH, temperature, substrate concentration, and enzyme loading on enzyme activity were studied. The stability of the enzyme was also studied. Measurement of enzyme activity using CO_2 as the substrate is difficult due to the high reaction rate, so CA activity was measured using hydrolysis of nitrophenal as a model reaction. Immobilization improved the reactivity and stability of the immobilized enzyme. PVDF is a promising candidate for carbonic anhydrase immobilization for carbon dioxide capture using a biocatalytic membrane system.

References

[1] Brown, R.S., in Aresta, M. and J.V. Schloss, editors, *Enzymatic and Model Carboxylation and Reduction Reactions for Carbon Dioxide Utilization, NATO ASI Series, Series C: Mathematical and Physical Sciences, vol. 314,* Kluwer Academic Publishers, Dordrecht (1990), 145-180.

[2] Dodgson, S.J., R.E. Tashian, G. Gros and N.D. Carter, editors, *The Carbonic Anhydrases: Cellular Physiology and Molecular Genetics,* Plenum Press, New York (1991).

[3] Pocker, Y., in Aresta, M. and J.V. Schloss, editors, *Enzymatic and Model Carboxylation and Reduction Reactions for Carbon Dioxide Utilization, NATO ASI Series, Series C: Mathematical and Physical Sciences, vol. 314,* Kluwer Academic Publishers, Dordrecht (1990), 129-143.

[4] Khalifah R.G., *The Journal of Biological Chemistry,*246(8), (1971) 2561-2573.

[5] Bond G.M., Stringer J., Brandvold D.K., Simsek F.A., Medina .G., and Egeland G.. *Energy & Fuels*, 15, (2001) 309-316. 7

[6] Cheng Li-H., Zhang L., Chen Huan-L., and Cong-Jie Gao Cong-J. *J. Membr. Sci.* 324 (2008) 33–43.

[7] Trachtenberg, M.C., Lihong, B., CO2 Capture: Enzyme vs. Amine, Proc. The 4th Annual Conference on Carbon Capture and Sequestration DOE/NETL (2005).

[8] P. Witte, P. J. Dijkstra, J. W. A. Berg, and J. Feijen, *J. Membr. Sci.*, 117 (1996) 1.

[9] L. Wu, J. Sun, Q. Wang, *J. Membr. Sci.* 285, (2006) 290.

[10] D. J. Lin, C. L. Chang, T. C. Chenand, L. P. Cheng, *Desalination* 145, (2002) 25.

(784f) Layer-by-Layer Assemblies for Membrane-Based Enzymatic Catalysis

Andrew Tomaino[1], Thomas D. Dziubla[2] and D.B. Bhattacharyya[2], (1)Chemical Engineering, University of Kentucky, Lexington, KY, (2)Chemical and Materials Engineering, University of Kentucky, Lexington, KY

Abstract: Biomolecule immobilization within the porous domain of microfiltration membrane media is a classic, well-studied approach for applications in such areas as enzymatic catalysis and separations. While considerable progress has been made in regards to understanding the optimal conditions for the preservation of the activity of these immobilized biomolecules, much more work is required in order to fully quantify these precise conditions for each type of biomolecule. An effective approach to biomolecule immobilization is layer-by-layer (LbL) functionalization, which holds promise due to the electrostatic nature of these assemblies. Since, in nature, enzymes and proteins exist in an electrostatic environment, the layer-by-layer approach offers a relatively high magnitude of immobilized biomolecular activity, as opposed to other methods of immobilization such as covalent attachment.

The goal of this study is to define and quantify the optimal electrostatic environments for enzymatic immobilization in order to maximize enzyme activity and overall membrane reaction effectiveness. Glucose oxidase (GOx) was used as a model enzyme within a poly(acrylic acid) (PAA)-poly(allylamine hydrochloride) (PAH)-functionalized poly(vinylidene fluoride) (PVDF) microfiltration membrane pores. The local activities of enzymes need to be quantified as a function of the properties of positively and negatively charged functional species within the porous domain, surface charge, enzyme immobilization configuration, and bulk membrane properties. The study is mainly focused on the analyses of these data, particularly the comparison of functionalized surface chemistry with local enzymatic configuration and activity. The understanding of the effect of these surfaces and enzyme locations will be important for advancing the LbL methods. Enzyme activities are quantified using batch experiments under convective modes through pressure modulations.

(788a) Adsorption Based Process for CO2 Capture and Biogas Purification

Ravi Jain and Donald Satchell, InnoSepra LLC, Middlesex, NJ

Technologies based on amines such as MEA and KS-1 have been extensively evaluated and tested for CO_2 capture from power plants. They have also been commercialzed for biogas purification. Numerous studies have indicated that the amine-based CO_2 capture and removal remains fairly expensive with a CO_2 capture cost of over $80/ton for a coal-fired power plant, and an increase in LCOE (levelized cost of electricity) of over 80% over the based plant without CO_2 capture. The current development trajectory indicates that the solvent-based CO_2 capture is not expected to achieve the DOE target of <35% increase in LCOE with CO_2 capture in the near future. Another commercial target, a CO_2 capture cost of <$40/ton to enable widespread CO_2 use in EOR applications in the absence of any climate legislation, is also likely to remain elusive for the solvent-based processes in the foreseeable future. In this paper preliminary experimental results, and techno-economic analysis of an adsorbent-based CO_2 capture process will be presented. The process is based on physical sorption and has the potential to reduce the cost of CO_2 capture by more than 40% over MEA resulting in a CO_2 capture cost of less than $50/ton. The process also has the potential to limit the increase in LCOE with CO_2 capture to less than 50%. With additional process optimization and integration with the power plant the process has the potential to meet the commerical CO_2 EOR target of <$40/ton, and the DOE CO_2 capture target of <35% increase in LCOE with CO_2 capture in a timeframe which is signficantly shorter than the solvent based processes. The technology is also applicable to CO_2 removal from raw biogas produced by anaerobic digestion and results in more than 50% reduction in the natural gas purification cost compared to amines in this appliaction. The technology could be potentially commercialized in this application in less than four years with learnings utilized to reduce the CO_2 capture costs in power plant applications.

(788b) Preliminary Studies of CO2 Removal From Simulated Pre-Combustion Syngas Via a Pressure Swing Membrane Absorption Process

Xingming Jie, John Chau, Gordana Obuskovic and **Kamalesh K. Sirkar**, Otto York Department of Chemical, Biological and Pharmaceutical Engineering, New Jersey Institute of Technology, Newark, NJ

In a conventional integrated gasification combined cycle (IGCC) for coal as a solid fuel, pre-combustion CO_2 capture from post-shift reactor synthesis gas is of significant interest since CO_2 (~40% of the dry gas) is present at a much higher partial pressure. The low temperature water gas shift reactor product stream is likely to be available at ~20 atm and around 150-200^{0}C. In a conventional IGCC plant the post-shift reactor synthesis gas is cooled, humidified and expanded to recover some electrical energy and then sent to Claus and Selexol units for H_2S and CO_2 separation and a purified H_2 stream. The object of this research is to carry out CO_2 absorption at a high temperature and pressure to purify the hydrogen and obtain simultaneously a purified CO_2 stream. Helium was used here as an inert surrogate for H_2.

Laboratory-based preliminary experiments were carried out via an advanced pressure swing absorption-based device and a cyclic process to produce purified helium at a high pressure and simultaneously obtain a purified CO_2 stream useful for subsequent sequestration steps. The device employed was a hydrophobized porous membrane absorber containing a CO_2-selective high-temperature stable absorbent on one side and a simulated post-shift reactor synthesis gas introduced periodically on the other side. We will illustrate the nature of the cyclic process and the usefulness of various steps in producing the desired level of purification.

(788c) Hybrid Membrane Cryogenic Process for Post-Combustion CO2 Capture

Bouchra Belaissaoui, UPR 3349 CNRS, LRGP, Nancy, France, Yann Le Moullec, EDF R&D , Chatou, France, David Willson, Stanbridge Capital, New York and Eric Favre, Laboratoire Réactions et Génie des Procédés LRGP (UPR 3349 CNRS), Nancy Université, Nancy F-54001, France

In Post-Combustion Carbon Capture and Storage (CCS) technology, the main challenge is the reduction of the energy requirement for the capture step.

Different capture processes have been investigated in the literature, such as absorption, adsorption, cryogenic and membrane separation. Among these processes, cryogenic separation is known to be energy efficient once a concentrated CO_2 flue gas is treated (approximately > 40%). Moreover, membrane separation has been shown to offer attractive performances when a moderate purity in the permeate is desired. This suggests that a synergy between the two processes could occur leading to a minimal overall energy requirement.

The aim of this work is to evaluate, through a simulation study, the potential for a hybrid process combining membrane and cryogenic separation to achieve efficient Carbon capture. The hybrid process combines a first step CO_2 pre-concentration with a membrane unit and a second step CO_2 cryogenic concentration.

Because large range of CO_2 content in the flue gas can be found in post-combustion capture, the impact of CO_2 feed content (in the range 5-30%) has been analysed in this study. Moreover, in order to explore the potential interest of improved membrane materials on the energy requirement of the hybrid process, a hypothetical membrane showing a CO_2/N_2 selectivity of 100 and available membrane selectivity of 50 has been simulated for this purpose. Furthermore, two compression strategies: (i) Vacuum Pumping, and (ii) Feed Compression with Energy Recovery System (ERS) have been investigated and the corresponding process performance in term of energy requirement and membrane surface area are evaluated and compared.

It is shown that for CO_2 feed concentration ranging between 15 and 30% (corresponding to large proportion of emission sources), the hybrid process shows a reduced energy requirement compared to the reference technology, namely, chemical absorption in MEA. Compared to standalone cryogenic approach, the hybrid process effectively improves the energy efficiency when diluted (i.e. < 30%) CO_2 inlet concentrations have to be treated. The percentage energy decrease increases significantly for lower inlet CO_2 mole fractions.

From the membrane material point of view, an improved selectivity (100 in place of 50) does not offer significant benefit in term of energy requirement while the membrane surface area increases. Thus, in this hybrid process, available membranes of selectivity of 50 seem to be appropriate. Moreover, the use of feed compression with Energy Recovery System (ERS) in the membrane module shows close energy requirement comparing to vacuum pumping option, while the membrane surface area is significantly decreased.

(788d) Synthesis of Zeolitic Imidazolate Framework ZIF-7 and Application for CO2 and CH4 Separation

Xiaofei Wu[1], Mehdi Niknam[2], Bin Yuan[3] and Shuguang Deng[3], (1)Chemical Engineering, New Mexico State University, Las cruces, NM, (2)New Mexico State University, Las Cruces, NM, (3)Chemical Engineering, New Mexico State University, Las Cruces, NM

Synthesis of Zeolitic Imidazolate Framework ZIF-7 and Application for CO_2 and CH_4 Separation

Xiaofei Wu, Mehdi Niknam, Bin Yuan, Shuguang Deng

New Mexico State University

Chemical Engineering Department

Las Cruces, NM 88003

Zeolitic imidazolate framework ZIF-7 is synthesized by both conventional hydrothermal method and microwave-assisted methods. The cubic structures of these two ZIF-7 samples obtained in this work are confirmed by X-ray powder diffraction results. The scanning electron microscopy (SEM) images of these two samples reveal that they have different particle sizes. The pore textural properties obtained by carbon dioxide adsorption at 0°C also indicate that they have different pore structures. Adsorption equilibrium and kinetics of CO_2 and CH_4 on these two samples are determined at various temperatures and gas pressures up to 1 bar, which match the simulated results. The combined equilibrium and kinetic selectivity is calculated from the adsorption isotherms and the adsorption kinetic plots, the results show that both of these two ZIF-7 samples have a good selectivity of separating CO_2 and CH_4.

Keywords: Adsorption, CO_2/CH_4 separation, microwave-assisted synthesis, ZIF-7

(788e) Silicalite Adsorbent Membranes for Separation of CO2 and N2 Gases

Muhammad Tawalbeh, F. Handan Tezel and Boguslaw Kruczek, Department of Chemical and Biological Engineering, University of Ottawa, Ottawa, ON, Canada

Silicalite-1/ceramic composite membranes were prepared on the inner surface of zirconium oxide and/or titanium oxide tubular supports by pore plugging hydrothermal synthesis. The effect of support pore size on the membrane's morphology and permeation performance was investigated. Five supports with different pore sizes in the range of 0.14 – 1.4 mm were studied. The synthesized membranes were characterized by X-ray diffraction (XRD), scanning electron microscope (SEM), electron diffraction spectrometer (EDS), and permeation experiments with pure and mixed CO_2 and N_2 gases.

The SEM of the membranes revealed a dense and continuous surface morphology with a highest crystal size of around 3.0 mm on the surface of all supports. SEM micrographs also showed a continuous layer grown over four supports out of five, with different pore sizes, that were investigated, with no layer observed on the support with 1.4 mm pore size. The uniformity of the zeolite layer was increasing with decreasing pore size of the support, except for the membrane synthesized using the support that has a pore size of 0.14 mm. SEM micrographs also showed that the surface coverage with zeolite crystals decreased by increased pore size of the support except for the membrane with 0.45 mm support which showed a complete surface coverage. The EDS analysis confirmed the formation of pure silicalite-1 layer at the top as well as inside the pores of all supports. The silicalite-1 crystals penetration increased with increased pore size of the support.

Pure gas permeation experiments with CO_2 and N_2 gases at 293 K revealed that the permeance values for N_2 were constant with pressure difference across all the membranes while the CO_2 values were decreasing with the increasing pressure difference across all the membranes prepared with supports with different pore sizes. The highest ideal selectivity was around 2.7 for the membrane with the support that had a pore size of 0.45 mm. The mixed gas permeation tests with CO_2 and N_2 (50/50 mixture) performed at different feed total pressures showed that increasing the feed pressure decreases the permselectivity for all supports. The highest permselectivities of around 24 were observed for the membranes prepared using the supports with 0.2 and 1.4 mm pore sizes. The highest CO_2 permeances of around 6.4×10^{-6} (mol/m^2 Pa s) were obtained for the membranes with the 0.45 and 0.8 mm pore size supports and the lowest CO_2 permeances of around 1.1×10^{-6} (mol/m^2 Pa s) were obtained for the membrane with the 0.2 mm pore size support. The very high permeances did not allow the testing of the membranes for pressures higher than 3 atm. Despite these very high permeances, the membranes prepared provided reasonably high selectivity. The membranes with the 0.45 pore size supports showed the best reproducibility among all membranes studied.

(788f) Preferential CO2 Separation Over H2 with Poly(amidoamine) Dendrimer Immobilized in a Poly(ethylene glycol) Network

Ikuo Taniguchi, Teruhiko Kai, Shuhong Duan and Shingo Kazama, Chemical Research Group, Research Institute of Innovative Technology for the Earth, Kizugawa, Japan

Preferential CO_2 Separation over H_2 with Poly(amidoamine) Dendrimer Immobilized in A Poly(ethylene glycol) Network

Ikuo Taniguchi,* Teruhiko Kai, Shuhong Duan, Shingo Kazama,

Chemical Research Group, Research Institute of Innovative Technology for the Earth,

9-2 Kidzugawadai, Kizugawa, Kyoto 619-0292, Japan

Tel: +81-774-75-2305, Fax: +81-774-75-2318

1. Introduction

Dendrimers are diverse families of highly branched macromolecules with well-defined molecular architecture and have been developed with various chemistries.[1] The dendritic molecules have gained particular interest in the area of nanotechnology, macromolecular science and engineering.[2,3] Among a wide variety of dendrimers, poly(amidoamine) (PAMAM) dendrimers are the first successfully developed and one of the most investigated dendritic structures.[4] Potential applications of the dendrimers have been investigated, including biomedical and pharmaceutical uses. Besides the versatile potentials, another fascinating feature of PAMAM dendrimer was found by Sirker et al. which was that the dendrimer exhibited preferential permeation of CO_2 over a gas mixture with a liquid immobilized membrane system.[5] The sorption of CO_2 was facilitated by the specific interaction between CO_2 and the primary amines of the branching end of PAMAM dendrimer. PAMAM dendrimer would be an essential agent in the selective CO_2 separation.

In this research group, further extension of CO_2 separation with PAMAM dendrimer has been investigated. For example, PAMAM dendrimers with lower generations flow at ambient condition, and thus the liquid dendrimer should be stably immobilized in a matrix especially for use under a pressurized condition.[6] Recently, an effective incorporation of PAMAM dendrimer in a cross-linked poly(vinyl alcohol) (PVA) was developed.[7] PVA network was formed with a cross-linker in the presence of PAMAM dendrimer in aqueous media, so that the dendrimer content in the membrane was readily controlled. The polymeric membrane showed a preferential CO_2 permeation over even smaller H_2 with preferable mechanical properties, and the CO_2/H_2 selectivity exceeded 30 at 0.56 MPa of CO_2 partial pressure (total pressure: 0.70 MPa). The dendrimer containing polymeric membranes hold potential for CO_2 separation, for example in an integrated gasification combined cycle (IGCC) plant with CO_2 capture and storage (CCS). In the coal-fired plant, the syngas consisted of H_2 (60 vol%) and CO_2 (40 vol%) after the water-gas shift reaction with 2.5 MPa in total pressure. Selective separation of CO_2 has been examined in the IGCC plant, and a solvent absorption with Selexol has gained current acceptance. Based on our simulation, a CO_2 separation with a membrane could be competitive or even more preferable and cost effective in comparison to the solvent absorption method.[8] However, in the membrane formulation of PAMAM dendrimer in a cross-linked PVA network, the time-consuming drying process required further improvement and optimization in terms of large-scale production and processability.

As for the selective CO_2 separation over H_2, an alternate immobilization of PAMAM dendrimer has been exploited, which is photocross-linking of poly(ethylene glycol) dimethacrylate (PEGDMA) in the presence of the dendrimer.[9,10] The cross-linking reaction takes place very rapidly under mild conditions, and the polymeric membrane is obtained in a couple of minutes. Thus unfavorable side reactions would be suppressed, such as Michael addition reaction between methacrylate and primary amine. The obtained polymeric membrane demonstrated an excellent CO_2 selectivity over H_2.[9] The CO_2 separation properties are readily tunable by altering the dendrimer concentration and generation.

2. Experimental

Fabrication of polymeric membranes

PAMAM dendrimer (2.0 g, 3.9 mmol), and PEGDMA (2.0 g, 2.7 mmol) were dissolved in ethanol (2.0 g) containing HCPK (9.1 mg, 44 mmol) under shading. The reaction mixture (2.5 g) was poured into a Teflon dish (ID: 60 mm) and exposed to the UV light for 90 s. The thickness of the obtained membrane was ca. 500 mm after dried.

CO_2 separation experiment

CO_2 separation properties over H_2 of the polymeric membrane were measured at 298 K under atmospheric pressure. A CO_2/H_2 gas with 5 vol% of CO_2 was humidified at 298 K and fed at a flow rate of 100 mL/min. Helium was supplied to the permeate side of the cell as a sweep gas at a flow rate of 10 mL/min. Compositions of the permeate and the feed sides were analyzed on a gas chromatograph.

3. Results and Discussion

For selective CO_2 separation, facile fabrication of polymeric membranes, in which PAMAM dendrimer is stably entrapped, has been investigated in this research group. Photopolymerization of PEGDMA in the presence of the dendrimer is one of the successful examples, which is motivated by the several considerations: (1) the reaction takes place rapidly and efficiently under mild conditions, minimizing unfavorable side reactions and providing macroscopically homogenous membranes. (2) PAMAM dendrimer was immobilized with a desired mixture ratio without remarkable leakage of the dendrimer. Thus, (3) the CO_2 separation properties can be readily controlled. (4) The thickness of the membrane is also tunable, which is associated with permeance of gases. (5) The compatible solvent can be eliminated after membrane formulation.

With PAMAM dendrimer immobilized PEG network, following results are obtained.

1. The CO_2 separation properties are increased with increase in PAMAM dendrimer content in the polymeri membrane.
2. 0th generation PAMAM dendrimer exhibits the highest CO_2 separation properties when the dendrimer content is kept to 50 wt%.
3. The polymeric membrane is opaque, which indicates macrophase separation between the dendrimer and cross-linked PEG.
4. The phase separation takes place in a couple of microns scale and the phase-separated structure is related to CO_2 separation properties.
5. PEG length, corresponded to the distance between cross-linking points, determines the CO_2 separation properties.
6. Humidity is also a key to characterize the CO_2 separation properties, and higher separation properties results

from operations under higher humidity.

For example, the separation factor increases with PAMAM dendrimer concentration and reaches about 500 when the dendrimer content is 50 wt% at 298 K and 5 kPa of CO_2 partial pressure under 80 % of relative humidity. The resulting polymeric mixture would be promising as a novel CO_2 separation material.

In this presentation, structural analysis of the polymeric membrane and interplay with CO_2 separation performance will be discussed.

4. Acknowledgements

This work was financially supported by the Japanese Ministry of Economy, Trade and Industry. The authors are also thankful for the assistance of Ms. Hiromi Urai and Rie Sugimoto throughout the membrane fabrication processes and gas chromatography measurements.

5. References

(1) Newkome, G. R.; Moorefield, C. N.; Všgtle, F. *Dendritic Molecules: Concepts, Syntheses, Perspectives*; Wiley-VCH; Weinheim, 1996.

(2) Rosi, N. L.; Mirkin, C. A. *Chem. Rev.* **2005**, *105*, 1547-1562.

(3) Lee, C. C.; MacKay, J. A.; FrŽchet, J. M. J.; Szoka, F. C. *Nature Biotech.* **2005**, *23*, 1517-1526.

(4) Tomalia, D. A.; Baker, H.; Dewald, J.; Hall, M.; Kallos, G.; Martin, S.; Roeck, J.; Ryder, J.; Smith, P. *Polym. J.* **1985**, *17*, 117-132.

(5) Kowali, A. S.; Chen H.; Sirkar, K. K. *J. Am. Chem. Soc.* **2000**, *122*, 7594-7595.

(6) Kouketsu, T.; Duan, S.; Kai, T.; Kazama S.; Yamada, K. *J. Membr. Sci.* **2007**, *287*, 51-59.

(7) Duan, S.; Kai, T.; Taniguchi, I; Kazama S., Y. *J. Membr. Sci.* submitted to *J. Membr. Sci.*

(8) Nagumo, R.; Kazama, S.; Fujioka, Y. *Energy Procedia*, **2009**, *1*, 4089-4093.

(9) Taniguchi, I.; Duan, S.; Kazama, S.; Fujioka, Y. *J. Membr. Sci.* **2008**, *322*, 227-280.

(10) Taniguchi, I.; Kazama, S.; Jinnai, H. *J. Polym. Sci. Part B, Polym. Phys.* in press.

(788g) Pvam-PIP/PS Composite Membrane with High Performance for CO2/CH4 Separation

Zhi Wang, Zhihua Qiao, Chenxin Zhang, Shuangjie Yuan, Jixiao Wang and Shichang Wang, Chemical Engineering Research Center, School of Chemical Engineering and Technology, Tianjin University, Tianjin, China

Polyvinylamine(PVAm)-piperazine(PIP)/polysulfone(PS) composite membrane was prepared in our previous work [1]. The membrane exhibits high CO2/N2 separation performance and could well meet the requirement of flue gas separation [1]. In this work, CO2/CH4 separation performance of the PVAm-PIP/PS composite membrane was investigated. The membrane was developed by coating PVAm-PIP mixed solutions with different mass ratios of PIP/PVAm (mPIP/mPVAm) on the PS ultrafiltration membrane. The effects of mPIP/mPVAm in the coating solutions and wet coating thickness on the CO2/CH4 performance of the PVAm-PIP/PS composite membrane were investigated. The PVAm-PIP/PS composite membrane prepared in this work showed higher CO2/CH4 performance than other membranes reported in the literature due to the large increase of the introducing carrier concentration and low crystallinity. For CO2/CH4 gas mixture (10/90 by volume), the PVAm-PIP/PS composite membrane with the mPIP/mPVAm of 1.430 and wet coating thickness of 30μm showed CO2 permeance of 4510 and 697 GPU [1 GPU = 10-6 cm3 (STP)/ (cm2 s cmHg) = 3.35×10-10 mol/(m2 s Pa)] at feed pressure of 0.11 and 3.0MPa, respectively, and the corresponding CO2/CH4 selectivity of 121 and 29, respectively. Moreover, the separation performance stability of the PVAm-PIP/PS composite membrane was investigated and no deterioration in the membrane perm-selectivity was observed. Finally, the economic evaluation of the membrane with the highest CO2/CH4 performance prepared in this work was carried out. PVAm-PIP/PS composite membrane could well meet the requirement of CO2/CH4 separation.
[1]Qizo ZH, Wang Z, Zhang CX, Yuan SJ, Zhu YQ, Wang JX, Wang SC. AIChE Journal.2012; doi: 10.1002/aic.13781.

Keywords: fixed carrier membrane, piperazine, cross-linking, polyvinylamine, CO_2 permeance, CO_2/CH_4 selectivity

AUTHOR INDEX

AUTHOR INDEX

AUTHOR INDEX

AUTHOR INDEX

AUTHOR INDEX

AUTHOR INDEX

AUTHOR INDEX

AUTHOR INDEX

AUTHOR INDEX

AUTHOR INDEX

AUTHOR INDEX

AUTHOR INDEX

AUTHOR INDEX

AUTHOR INDEX

AUTHOR INDEX

AUTHOR INDEX

AUTHOR INDEX

American Institute of Chemical Engineers
3 Park Avenue
New York, New York 10016-5991

CURRAN ASSOCIATES INC
proceedings.com